Thin Films and Nanostructures

Cu(In$_{1-x}$Ga$_x$)Se$_2$ Based Thin Film Solar Cells

Volume 35

Serial Editors

VLADIMIR AGRANOVICH
Institute of Spectroscopy
Russian Academy of Sciences
Moscow, Russia

DEBORAH J. TAYLOR
Freescale Semiconductors
Austin, Texas

Honorary Editors

MAURICE H. FRANCOMBE
Department of Physics
and Astronomy
Georgia State University
Atlanta, Georgia

STEPHEN M. ROSSNAGEL
IBM Corporation,
T. J. Watson Research Center
Yorktown Heights, New York

ABRAHAM ULMAN
Alstadt-Lord-Mark Professor
Department of Chemistry
Polymer Research Institute
Polytechnic University
Brooklyn, New York

Editorial Board

David L. Allara
Pennsylvania State University

Allen J. Bard
University of Texas, Austin

Franco Bassani
Scuola Normale Superiore, Pisa

Masamichi Fujihira
Tokyo Institute of Technology

George Gains
Rensselaer Polytechnic Institute

Phillip Hodge
University of Manchester

Jacob N. Israelachivili
University of california Santa Barbara

Michael L. Klein
University of Pennsylvania

Hans Kuhn
MPI Gottingen

Jerome B. Lando
Case Western Reserve University

Helmut Mohwald
University of Mainz

Nicolai Plate
Russian Academy of Sciences

Helmut Ringsdorf
University of Mainz

Giacinto Scoles
Princeton University

Jerome D. Swalen
International Business Machines Corporation

Matthew V. Tirrell
University of Minnesota, Minneapolis

Claude Weisbuch
Ecole Politechnique, Paris

George M. Whitesides
Harvard University

Anvar Zakhidov
University of Texas at Dallas

Recent volumes in this serial appear at the end of this volume

Thin Films and Nanostructures

Cu(In$_{1-x}$Ga$_x$)Se$_2$ Based Thin Film Solar Cells

Subba Ramaiah Kodigala
*Solar Cell & LED Technology,
Thousand Oaks, CA, 91360 and
Department of Physics & Astronomy
California State University
Northridge, CA, 91330*

Volume 35

AMSTERDAM • BOSTON • HEIDELBERG • LONDON
NEW YORK • OXFORD • PARIS • SAN DIEGO
SAN FRANCISCO • SINGAPORE • SYDNEY • TOKYO
Academic Press is an imprint of Elsevier

Academic Press is an Imprint of Elsevier
30 Corporate Drive, Suite 400, Burlington, MA 01803, USA
525 B Street, Suite 1900, San Diego, CA 92101-4495, USA
The Boulevard, Langford Lane, Kidlington, Oxford, OX5 1GB, UK
Radarweg 29, PO Box 211, 1000 AE Amsterdam, The Netherlands

Copyright © 2010 Elsevier Inc. All rights reserved.

No part of this publication may be reproduced, stored in a retrieval system or transmitted in any form or by any means electronic, mechanical, photocopying, recording or otherwise without the prior written permission of the publisher.

Permissions may be sought directly from Elsevier's Science & Technology Rights Department in Oxford, UK: Phone (+44) (0) 1865 843830; Fax (+44) (0) 1865 853333; Email: permissions@elsevier.com. Alternatively you can submit your request online by visiting the Elsevier website at http://elsevier.com/locate/permissions, and selecting *Obtaining permission to use Elsevier material*.

Notice
No responsibility is assumed by the publisher for any injury and/or damage to persons or property as a matter of products liability, negligence or otherwise, or from any use or operation of any methods, products, instructions or ideas contained in the material herein. Because of rapid advances in the medical sciences, in particular, independent verification of diagnoses and drug dosages should be made.

ISBN: 978-0-12-373697-0
ISSN: 1543-5016

For information on all Academic Press publications
visit our website at books.elsevierdirect.com

Transferred to Digital Printing in 2010

Working together to grow
libraries in developing countries

www.elsevier.com | www.bookaid.org | www.sabre.org

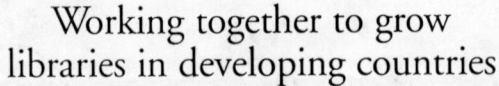

In memory of My beloved grandmother **Narasamma Kodigala**

Contents

Preface		xi
Acknowledgements		xiii

1 Introduction   1

    1.1 Role of Solar Energy   1
    1.2 Spectral Distribution of Solar Radiation   3
    1.3 State of the Art of Thin Film Solar Cells   4
    1.4 Photovoltaic Conversion   9
    1.5 Homojunction   12
    1.6 Schottky Barrier   12
    1.7 Criteria for the Choice of Heterojunction Pair in the Thin Film Solar Cells   13
    1.8 Window Materials   14
    1.9 Absorber Materials   15
    1.10 Copper Based I–III–VI$_2$ Semiconductors   16
    References   17

2 Growth Process of I-III-VI$_2$ Thin Films   21

    2.1 Deposition of CuInSe$_2$ 21
    2.2 Deposition of CuGaSe$_2$ Thin Films   28
    2.3 Growth Process of Cu(InGa)Se$_2$ Thin Films   28
    2.4 Deposition of CuInS$_2$ Thin Films   44
    References   48

3 Surface Analyses of I–III–VI$_2$ Compounds   55

    3.1 Atomic Force Microscopy Analysis   55
    3.2 Scanning Electron Microscopy Analysis   63
    3.3 Electron Probe Microanalysis   78
    3.4 Secondary Ion Mass Spectroscopy Analysis   81
    3.5 Auger Electron Spectroscopy   84
    3.6 X-ray Photoelectron Spectroscopy   90
    References   109

4	**Structural Properties of I–III–VI$_2$ Absorbers**	**115**
	4.1 Crystal Structure of I–III–VI$_2$ Compounds	115
	4.2 CuInSe$_2$	117
	4.3 CuGaSe$_2$	143
	4.4 Cu(InGa)Se$_2$	147
	4.5 CuInS$_2$	169
	4.6 CuIn(Se$_{1-x}$S$_x$)$_2$	184
	References	188

5	**Optical Properties of I–III–VI$_2$ Compounds**	**195**
	5.1 Band Structure of I–III–VI$_2$ Compounds	195
	5.2 Photoluminescence	212
	5.3 Raman Spectroscopy	276
	References	307

6	**Electrical Properties of I–III–VI$_2$ Compounds**	**319**
	6.1 Conductivity of CuInSe$_2$	319
	6.2 Conductivity of CuGaSe$_2$	339
	6.3 Conductivity of CuInGaSe$_2$	345
	6.4 Conductivity of CuInS$_2$	350
	6.5 Conductivity of CuIn(Se$_{1-x}$S$_x$)$_2$	358
	6.6 Hopping Conduction	360
	6.7 Thermoelectric Power	367
	6.8 DLTS of I–III–VI$_2$ Heterostructures, Homojunctions, and Schottky Junctions	370
	6.9 Admittance Spectroscopy	386
	References	387

7	**Fabrication and Properties of Window Layers For Thin Film Solar Cells**	**393**
	7.1 CdS Window Layer	393
	7.2 InS or In$_2$S$_3$ Thin Films	430
	7.3 ZnO	432
	7.4 ITO	488
	7.5 FTO	490
	References	495

8	**$Cu(In_{1-x}Ga_x)Se_2$ and $CuIn(Se_{1-x}S_x)_2$ Thin Film Solar Cells**	**505**
	8.1 Basic Theory of $CuInSe_2$ Based Solar Cells	505
	8.2 $CuGaSe_2$ Solar Cells	539
	8.3 $Cu(In_{1-x}Ga_x)Se_2$ Thin Film Solar Cells	547
	8.4 $CuInS_2$ Thin Film Solar Cells	643
	8.5 $CuIn(Se_{1-x}S_x)_2$ Based Thin Film Solar Cells	660
	References	664

Subject Index — 681

Recent Volumes In This Series — 685

Preface

Recent developments around the world have encouraged the already initiated research and development programs on renewable energy, for example solar, thermal, and wind energy, which are friendly to the environment. The Governmental budgets and private investments for the generation of electricity from solar cell modules or grids are steadily increasing with the vision to mitigate future energy demand. Several photovoltaic plants are mushrooming around the world aiming for the generation of large quantities of electricity compared with the past. Silicon wafers are much in demand for the production of integrated chips and in the solar cell industry, therefore, their cost has increased, as witnessed by the market. The developments of $Cu(In_{1-x}Ga_x)Se_2$ Based Thin Film Solar Cells may be an alternative to Si.

The book covers the fabrication and characterization of solar cell materials such as $CuInSe_2$, $CuGaSe_2$, $CuInS_2$, $Cu(In_{1-x}Ga_x)Se_2$, and $CuIn(Se_{1-x}Se_x)_2$ (I–III–VI$_2$) and their single-crystal or thin-film solar cells. Chapter 1 describes the energy demands and the invention of solar cells. The state-of-the-art of thin-film solar cells such as CIGS, multi-junction, Si, plastic, and dye-sensitized thin-film solar cells is briefly presented for understanding the principle behind the physics concerning the p–n junction. The growth of I–III–VI$_2$ thin-film absorbers by various vacuum and nonvacuum techniques for solar cell applications is described in Chapter 2.

The surface and cross section studies of absorber layers by AFM and SEM support to make abrupt p–n junction solar cells. The composition, chemical states, grading, and depth profiles of layers by EPMA, SIMS, AES, and XPS techniques are covered in Chapter 3, which are the main tools to assess their quality. Chapter 4 demonstrates the structural properties of I–III–VI$_2$ compounds with the aid of XRD and TEM techniques, including the role of secondary phases. The effect of deposition recipes on the compounds is mentioned.

The optical studies of Cu-based absorber layers by means of absorption, reflectance, and transmission are given in Chapter 5. The experimental details and working principles of photoluminescence and Raman spectroscopies are illustrated. In addition, the low-temperature PL and Raman spectroscopy analyses of absorbers provide defect levels, phases, structure, etc. Both the techniques contribute to determining defect levels, which degrade the efficiency of CIGS thin-film solar cells. Chapter 6 illustrates the electrical properties of I–III–VI$_2$ absorber materials such as low-temperature conductivity, mobility, hopping conduction, and thermoelectric power. Finally the DLTS of hetero-, homo- and Schottky-junctions made by Cu-based absorbers shows the nature of deep defect levels.

Chapter 7 presents the preparation and characterization of window materials such as CdS, ZnS, InS, and ZnSe as well as TCO compounds such as ZnO, ITO, and FTO for solar cell applications. In particular, the properties of CdS and ZnO layers are highly concentrated. Chapter 8 deals with fabrication and testing of I–III–VI$_2$ based single-crystal and thin-film solar cells. The variation of solar cell efficiency with the influence of composition, substrates, buffers etching, doping, etc., on the thin-film solar cells is described in detail. The I–V and C–V analyses of solar cells are extensively corroborated with and without buffer layers. In addition, the band structure of the CIGS heterojunction is explained with examples. The transformation of technology from the laboratory CIGS cell to monolithic modules is reported in some detail.

Acknowledgments

I am highly indebted to my beloved teacher Professor V. Sundara Raja who introduced Solar Energy Physics to me. Without him I would have not been standing to write a book on the important topic. I am highly grateful to Prof. T. Sudharsan whose help is unforgettable as well as to Prof. Khan at University of South Carolina, Columbia, SC, USA. I am thankful to Professors H. Markoc, D. Johnston, and M. Reshchikov, Virginia Commonwealth University, Richmond, VA , USA, for their help during my stay at VCU. I sincerely thank Professors I. Bhat, T. P. Chow, M. Schubert, and C. Wetzel, Rensselaer Polytechnic Institute, Troy, New York, USA, for enlightening discussions. I thank Dr. Shalini for fruitful discussions and her constant help to shape things as a book. A number of advices by Professors Y. K. Su, S. J. Chang, and F. S. Chuang, National Cheng Kung University, Tainan, Taiwan are invaluable. I am highly thankful to Professors G.V. Subba Rao, U. V. Raju, A. K. Bhatnagar, C. Sunandana, and M. Sharon for their help in the subject matters. I am extremely grateful to Professors R. D. Tomlinson, A. E. Hill, and R. D. Pilkington for their wholehearted help and discussions about new topics during my stay at University of Salford, UK. I sincerely thank Drs. Kazamerski, Contreras, David, Ramesh, Johnston, Tuttle, Deb, Postma, Venkataramana, Madhu, Raghu, Das, Ginley, Yang, Mark, Noufi, Ramanathan, Thomas, Narayana Rao for their help from my bottom of heart. I am highly grateful to Drs. Schock, Birkmire, Shafarman, Burgelman, Kapur, Niki, Site, Siebentritt, Tiwari, Neelakanth, Lux-Steiner, Lincot, Powalla, Nakada, Guillemoles, Olsen, Kundu, Kushiya, Scheer, Lewerenz, Ennaoui, Igalson, Chichibu, Nair, Shirakata, Anderson, Stanbery, Dharmadasa, Zang.

None the less I thank my wife Mitra Vinda, lovely children Ashok and Sri hari for their extraordinary help and patience during my work on this book without spending much time with them. I thank my mother Mrs. Sampoornamma, brother Mr. Chandraiah, brother-in-law Mr. Prasad, other family members, friends, and colleagues for their help directly or indirectly.

1 Introduction

1.1 Role of the Solar Energy

The global energy demand is in apex due to fast growth of industrialization and population while the conventional fossil fuels like coal, oil, natural gas, *etc.*, are fast depleting. It is estimated that fossil fuels, especially oil, will be exhausted in another 200–300 years. The oil reserves at different places around the world are geographically shown in Figure 1.1 [1]. The highest percentage of oil reserves is located in the Gulf of the Middle East. The impact of the oil shortage was felt early in 1973 as well as recently in the wake of the Gulf war. Awareness about the energy crisis has led to the search for alternative energy sources, particularly those are renewable and friendly to environment. According to the annual energy review of 2000, the generation of electricity in the USA is 52, 20, 19, 7, and 2% from coal, nuclear, natural gas-petroleum, hydroelectric and renewable energy, respectively. The data indicate that the percentage of renewable energy utilized is low. On the other hand, the emissions of CO_2 from fossil fuels, biomass, *etc.*, need to be reduced by adopting somewhat alternative sources, such as renewable energy. The reduction of CO_2 in the atmosphere is essential, as global warms due to emission by industries that has an impact on the ozone layer. It is a well known fact that the ozone layer blocks UV radiation, while the radiation travels from the Sun to the Earth. Therefore the solar energy is an attractive alternative source, which is inexhaustible and pollution free. The Earth receives an annual energy of 1018 kWh from the Sun, which is about 20,000 times more than the present annual energy consumption of the world. Even if a fraction of this energy is effectively utilized, the energy problems can be mitigated to some extent. However, it is a dilute and intermittent source. As such, solar energy systems generally occupy large areas and need proper energy storage systems. Secondly the Earth receives an average radiation of 5-1/2 h per day or less from the Sun in southern hemisphere of the Earth, whereas it is very less (0–1.5 h/day) in the northern hemisphere.

Methods of solar energy utilization can be broadly divided into two categories: (i) photothermal and (ii) photovoltaic. The photothermal systems convert solar radiation into thermal energy, which may be used directly or converted into electricity. The photovoltaic systems directly convert sunlight into electricity. The photovoltaic effect was first discovered by Becquerel [2] in 1839 who observed a photovoltage when light was directed onto one of the electrodes in an electrolyte solution. Adams and Day [3] were the first to observe the effect in solid selenium. About 40 years later Lange [4], Grondhal [5], and Schottky [6] did pioneering work on selenium and cuprous oxide photovoltaic cells. In 1954, Chapin *et al.*, [7] reported a single

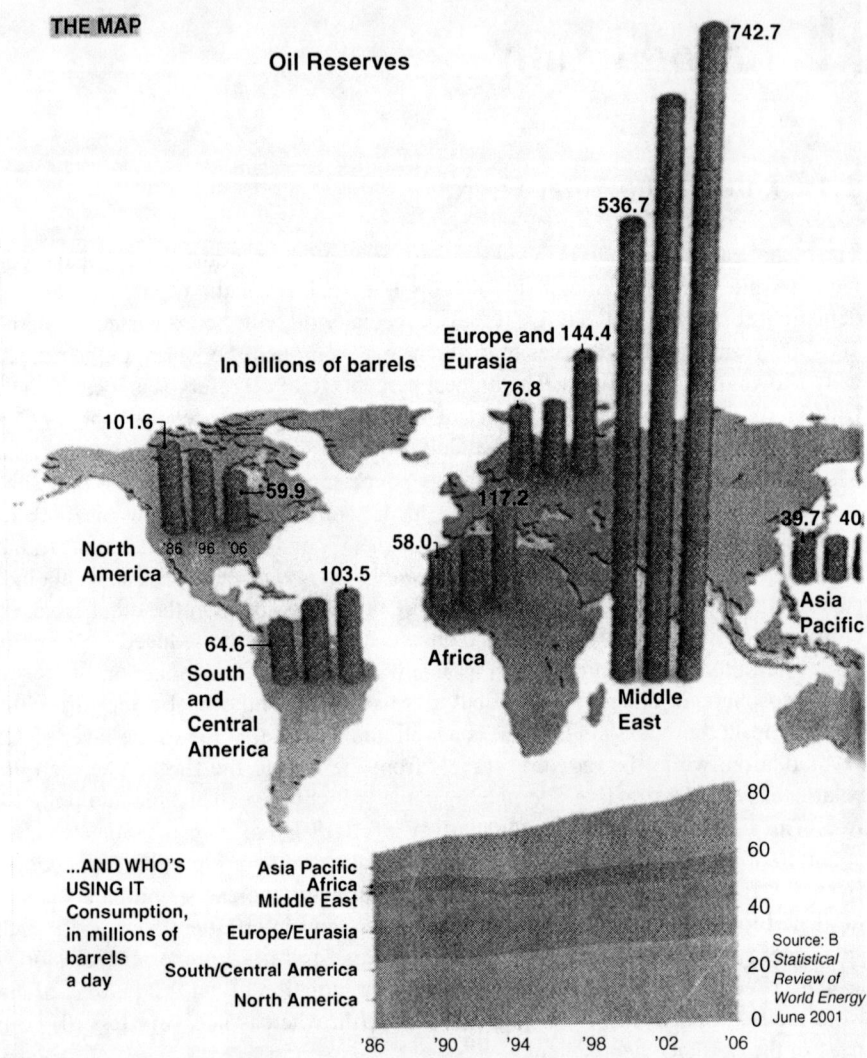

Figure 1.1 Oil reserves in the world.

crystal silicon photovoltaic cell with an efficiency of 6% and in the same year, Reynolds et al., [8] reported Cu$_x$S/CdS heterojunction with about 6% efficiency. The GaAs solar cells having efficiencies in excess of 6% were first reported in 1956 by Jenny et al., [9]. In 1958 Si based solar cell panels were deployed in Vanguard I satellite for space program. These are only a few milestones to cite the early developmental history of photovoltaics. Today photovoltaic modules are incredibly the prime source of power for satellites. However, the space quality solar cells are eventually expensive. The photovoltaic utilization for terrestrial applications demands a substantial reduction in the cost. The thin film solar cells essentially require only small amount of expensive semiconductor materials. Therefore they can be preferable as option for energy. The current emphasis in photovoltaics is directed toward the development of high performance, inexpensive, stable thin film solar cells that can serve in the long run as a viable alternative to single crystal silicon technology. The thin films are not only necessary for the photoelectrically active layers but also for optical windows, transparent conducting coatings, antireflection coatings, contacts, *etc* [10,11]. Most of the thin-film deposition techniques are amenable for mass production and play a vital role in the development of low-cost solar cells.

1.2 Spectral Distribution of Solar Radiation

The extraterrestrial spectral distribution of solar radiation is shown in Figure 1.2 [12,13]. It can be approximated to the spectrum of a black body at a temperature 5973 K. The solar radiation gets attenuated as it passes through the atmosphere due to scattering by air molecules, dust particles and absorption by ozone, CO_2 and H_2O. The extent of attenuation of the beam radiation depends on the path length of the ray through the atmosphere and is generally expressed in terms of air mass "m" [14]. The air mass is defined as $m = 1/\cos\upsilon$, where υ is the angle between a line

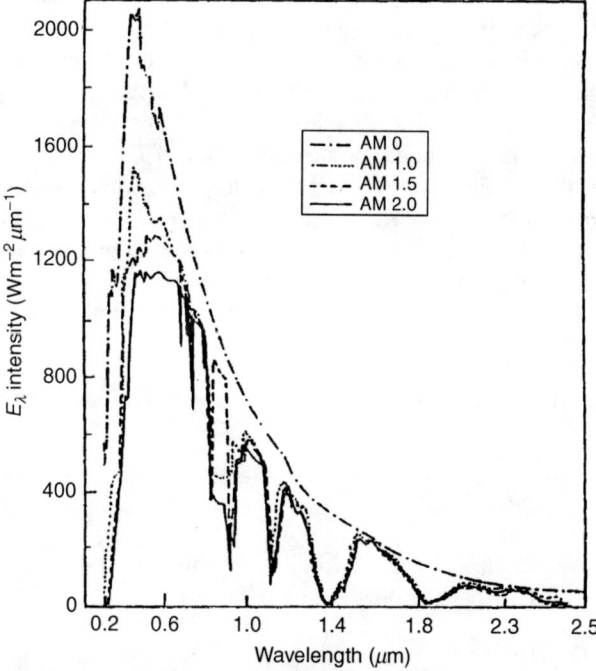

Figure 1.2 Spectral distribution of solar radiation with different air mases.

vertical to the observer and a line from the Sun to the observer. When $m=0$ means AM0 space condition. (i) the Sun is at zenith, that is, $m=1$ for $\upsilon=0$; (ii) $m=1.5$ for $\upsilon=48.2°$, and (iii) $m=2$ for $\upsilon=60°$. The standard solar spectrum commonly used to specify the efficiency of terrestrial cells is either AM1 (100 mW/cm^2) or AM1.5 (85 mW/cm^2) global spectrum. The AM0 (135 mW/cm^2) or space (extraterrestrial) solar spectrum is rich in ultraviolet than that in terrestrial solar spectrum. Suppose a typical solar cell exhibits efficiency of 28% in the space, whereas the same shows $\sim$32% in terrestrial field because of poor electrical conversion of ultra violet light [15]. On the other hand, harsh inherent electron or other radiations and wide variation of temperature from -180 to 40 °C in the space may be caused.

1.3 State of the Art of Thin Film Solar Cells

1.3.1 CIGS Thin Film Solar Cells

Recently CuInGaSe$_2$ (CIGS) thin film solar cell with an active area of 0.5 cm^2 made by ZSW company exhibits the highest efficiency of 20.3% (Table 1.1) [16]. A several companies such as First Solar, Nanosolar, Globalsolar, Miasolé, Solopower, Honda, Sharp, Avancis *etc.*, have been immensely involving to developing and producing CIGS based thin film solar cell mani-modules to target production of several GW/Yr range around the world. For example, the Solyndra company initiated a new approach to develop 1.8 m long cylindrical tube CIGS modules, as fluorescent tubes and panel width of 1 m to reach 110 MW [23]. Unlike other companies, Ascent Solar Inc. company developed CIGS monolithically interconnected thin film solar cells on flexible plastic substrates with module aperture efficiency of 11.9% and module efficiency of 10.5% while Solopower company made CIGS thin film solar cell panel on the metal flexible substrates, which exhibits aperture efficiency of 11%.

The Cu$_x$S/CdS thin film solar cells may be origin for the development of CIGS thin film solar cells in step by step process in which Cu$_x$S is unstable compound with time [8]. In order to improve stability of compound, In is added to it to make strong covalent bonding that turns into CuInS$_2$. The CuInSe$_2$ CIS, CuGaSe$_2$ CGS, and CuAlSe$_2$ are systerical compounds to CuInS$_2$. The theoretical models reveal that

Table 1.1 Highest reported efficiencies of solar cells with prominent materials

S.No	Solar cells	η (%)	Ref.
1	CIGS	20.3	[16]
2	CdTe	15.8	[17]
3	CZTS	10	[18]
4	GaAs multijunction	41.1	[19]
5	Si	24.5	[20]
6	Polymer	7	[21]
7	DSSC	11	[22]

1.55 eV band gap absorber materials are optimal to capture maximum solar spectrum. In this juncture, $CuInS_2$ is suitable material but its performance is inferior to $CuInSe_2$. The band gap of $CuInSe_2$ is 1.1 eV therefore band gap needs to be increased by adding either Ga or Al in it. Thus the CIGS thin film solar cells now occupies main stream of market as next generation of photovoltaics. The fabrication and characterization of CIGS thin film solar cells will be discussed in detail in the next chapters.

1.3.2 Multiple Junction GaAs Based Thin Film Solar Cells

The leading laboratories such as Fraunhofer Institute for Solar Energy Systems, National Renewable Energy Laboratory, Spectrolab, Emcore, *etc.*, have been developing III–V group based monolithic $Ga_{0.35}In_{0.65}P/Ga_{0.83}In_{0.17}As/Ge$ multijunction thin film solar cells. The electricity costs 3$/watt for multijunction solar cells, whereas it is 8$/watt for Si solar cells but much more complexity is involved for designing and fabrication of former [15]. The multijunction thin film solar cells are promising candidates for terrestrial and extraterrestrial applications owing to the highest conversion efficiency of 41.1% over the cell area of 0.5 cm^2 under 400–500 suns in the concentrated systems, whereas the same cells show efficiency of less than 30% under 1 sun (1.5 AM) [19]. Similarly the Spectrolab developed metamorphic $Ga_{0.44}In_{0.56}P/Ga_{0.92}In_{0.08}As/Ge$ multijunction thin film solar cells, which show efficiency of 40.7% under 240 suns, whereas lattice matched $Ga_{0.44}In_{0.56}P/Ga_{0.92}In_{0.08}As/Ge$ multijunction cells exhibit more or less same efficiency of 40.1% [24]. As we know the absorber with band gap of 1.55 eV in the single junction solar cell is optimal to absorb high intensities photons from the solar spectrum. However, the multijunction solar cells utilize all the wavelength photons of solar spectrum by graded band gap absorber cells that enhance efficiency of cells. Over all aspects of lattice mismatch and band gap grading, III–V compounds are best suit for multijunction solar cells comparing with other materials. The fundamental principle behind the Physics of multijunction solar cells is that the absorbers of top InGaP, second AlGa(In)As, third Ga(In)As and bottom Ge sub cells in the multijunction solar cells contain band gaps of (E_{g1}) 1.9, (E_{g2}) 1.6, (E_{g3}) 1.4, and (E_{g4}) 0.65 eV, respectively (Figure 1.3A and B) [24–26]. The absorber of top cell absorbs ultraviolet and partial visible region photons with energy greater than or equal to Eg_1. The near infrared region photons with less than Eg_1 energy hit second cell by passing through top cell where the photons greater than or equal to Eg_2 are absorbed by second cell AlGa(In)As absorber. The same phenomena is applicable to third cell Ga(In)As absorber. The bottom Ge cell absorbs rest of low energy or infrared region photons. The voltage of multijunction cell is combination of voltage of each cell, whereas the lowest current of the sub cell dictates current of device since the cells are connected in series. The efficiency of cell can be increased by properly equaling current of each cell in the multijunction. In order to improve cell current, the thickness of top cell is increased hence absorption by top absorber increases thus current increases in the cell. The current is high in Ge sub cell since most of the spectrum covers by Ge. The efficiency of the cell can also be increased by decreasing thickness of Ge absorber. The diffusion length of minority carrier is higher than

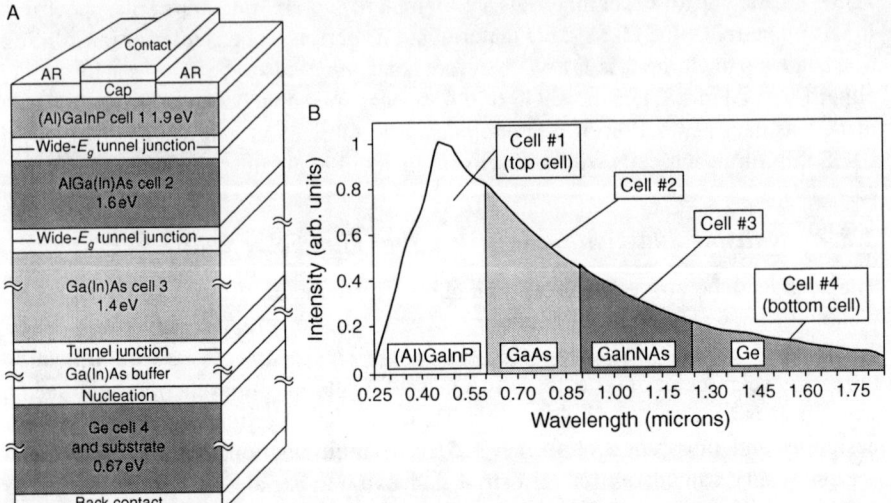

Figure 1.3 (A) Multijucntion thin film solar cells in which part of the spectrum observed by each sub cell noted in color and (B) participation of different band gap absorbers.

the thickness of Ge that causes to have higher current. In general the absorber absorbs photon energy that equal to band gap of absorber. If it is greater than band gap the remaining energy dissipates heat. The p-type InGaAs step graded buffer layer is introduced in the place of n-type InGaAs buffer as metamorphic configuration in third cell, which creates more defect levels to relax other layers. In the past, researchers thought that properly lattice matched cells are prospectable for high efficiency. However lattice mismatched buffer configuration improves efficiency of cell [27]. The cell with higher number of multiple junctions such as five or six junctions may also enhance efficiency.

1.3.3 Plastic Solar Cells

Recently tremendous research work on the plastic or polymer based light weight thin film solar cells has been gained. A number of major companies such as Konarka, Solarmer, Fiber Inc, Plextronics, *etc.*, have sincerely devoted their effort to increase efficiency of the cells. The efficiency of polymer based solar cells steadily reaches from 3 to 10% as a milestone [22,28]. The Konarka company first introduced its polymer based thin film solar cell prototype modules to the market even though the efficiency of module is low around ~3% that would be also profitable because of low cost production. The low efficiency flexible prototype modules have been using in calculators, laptop computers, cell phones, iPods, carrying bags, umbrellas, *etc.* So far the highest reported efficiency on polymer thin film solar cells is 6.8% [21].

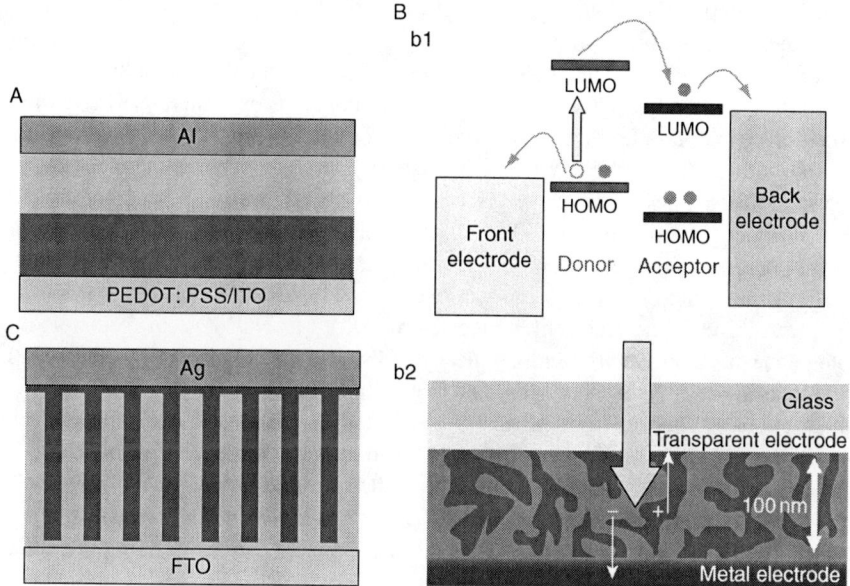

Figure 1.4 (A) Planar heterojunction, (B) bulk heterojunction and (C) ordered heterojunction.

There are different kinds of heterojucntion plastic solar cells (Figure 1.4); (i) one is planar heterojunction that the electron donor (copper phthalocyanine) and the electron acceptor (perylene tetracarboxylic derivative) are sand witched between reflecting metal (Al, Ag, or Au) and poly(3,4-ethylenedioxyl-enethiophene):polystyrene sulfonic acid (PEDOT:PSS) coated TCO (ITO or FTO) [29,31], (ii) bulk heterojunction consists of a typical structure of glass/ITO/40 nm PEDOT:PSS/ P3HT(1 wt.%)-PCBM(0.8 wt.%)TiO$_x$/100 nm Al thin film, which exhibits efficiency of 5% in which poly(3,4-ethylenedioxyl-enethiophene):polystyrene sulfonic acid (PEDOT:PSS) is used as a conducting layer for hole transport and poly (3-hexylthiophene), that is, P3HT with band gap of 1.6–2 eV acts as electron donor, whereas phenyl C$_{61}$-butyric acid methylester (PCBM) contributes acceptor. The TiO$_x$ with band gap of 3.7 eV grown by sol–gel technique is virtually used as an optical spacer [32]. The interface between n and p is randomly distributed all over the place in the bulk heterojunction, since both donor and acceptor contributors are mixed unlike conventional heterojunction solar cells. The schematic diagram of bulk heterojunction solar cell is shown in Figure 1.4B [29]. The generated electron–hole pair (exciton) has diffusion length of 10 nm or less. Hence, it has to find donor–acceptor interface within diffusion length of 10 nm otherwise it undergoes recombination without getting separation. In this case, the morphology of active layer plays a dominant role for separation of electron–hole pair to drive electrons and holes to their respective electrodes by avoiding suppression of excitons. In such a way the domain size of active layer should be kept double size of exciton diffusion length. The Coulombic attraction force between hole and electron can be defined as $e^2/4\pi\varepsilon_0\varepsilon_r$, which is inversely proportional to the dielectric constant (ε_r) of semiconductor. The dielectric constant of polymer semiconductor is ~ 3, whereas it is nearly thrice in inorganic semiconductiors. Therefore Coulombic attraction is negligible in the organic semiconductors but in order to separate exciton, more than 0.4 eV energy is essential in the polymer semiconductors. Nearly 300 nm thick active layer is enough to absorb most of the photons [33]. The inorganic semiconductor had high dielectric constant and low exciton binding energy. Therefore room temperature thermal energy of 0.028 eV is enough to dissociate exciton in the inorganic semiconductors. When the photon hits organic solar cell, the electron excited from LUMO to HOMO and forms electron hole pair, which relax with binding energy of 0.1–1.4 eV (Figure 1.4B). The exciton or electron–hole pair is separated by the donor–acceptor pair interface which contains chemical potential. The separated electrons and holes are collected by respective electrodes where diffusion is dominant [30,34]. Unfortunately the thickness of polymer is confined to less than 100 nm due to having low hole mobility in polymer semiconductors, which is for example in the typical range of 4×10^{-4} cm^2/Vs for poly[4,8-*bis*-substituted-benzo[1,2-b:4,5-b']dithiophene-2,6-diyl-alt-4-substituted-thieno[3,4-b]thiophene-2,6-diyl] (PBDTTT) [21] and electron mobility of 1 cm^2/Vs for PC$_{60}$BM. After annealing the plastic solar cell, the efficiency increases from ~ 3 to 6% due to formation of nanodomains [35]. (iii) The order heterojunction solar cells are made with nanostructured TiO$_2$ templates/P3HT polymer that the polymer is inserted into TiO$_2$, as shown in Figure 1.4C. The radius of straight and continuous pores should be slightly less than diffusion length of exciton of polymer. The ordered heterojunction solar cells exhibit efficiency of 5%. Recently Heliatek company claims efficiency of 7.7% for tandem organic photovoltaic cell over the area of 1.1 cm^2 [36].

The dye sensitized solar cells (DSC) first developed by Gratzel in 1991 is also known as Gratzel cells. The typical configuration of laboratory DSC consists of TCO (SnO$_2$:F) coated glass substrate/TiO$_2$/Ruthenium polypyridyl complex (N3 (RuL2(NCS)$_2$, L=2,2',-bipyridyl-4,4'-(COOH)$_2$)) dye/electrolyte/Pt/TCO glass, which shows efficiency of 10.8%. In the cell Pt is counter electrode (Figure 1.5). The interface is between organic donor and inorganic nonporous TiO$_2$ acceptor. When the light hits cell, the ruthenium dye absorbs it and electron injection takes place into TiO$_2$ conduction band. An I^-/I_3^- redox couple regenerates photooxidized dye molecules in organic solvent. The positive charge transportation takes place by electrolyte to the metal electrode, where I_3^- involves to transfer an electron from the Pt counter electrode [22,37,38]. The demerit is that the fast degradation in efficiency is observed in the polymer based thin film solar cells and the life time of cells may be around 3 years. If the span of life time of cells is high around a decade that would be highly beneficial to the market. The low efficiency cells cost is high for example electricity costs 11–12 US\$/W$_p$ that needs large volume of production of cells but the expecting cost would be less than 1\$/W$_p$ [39]. Recently the stability and damp heating tests prove that the performance of organic cells increases.

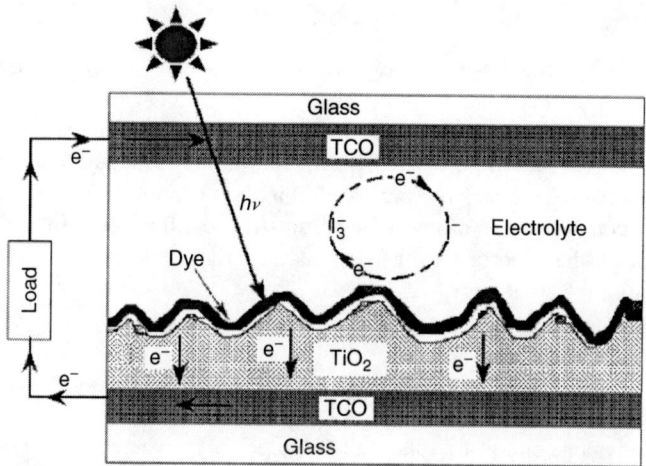

Figure 1.5 Dye sensitized solar cells.

1.3.4 Si Solar Cells

The Si based solar cells with highest efficiency dominate the market, despite Si owing indirect band gap because of its good stability and suitable electrical, chemical properties. On the other hand Si is technologically well developed material. A 300-μm thick layer is essential for thin film solar cells due to low absorption coefficient. There are three varieties of Si based solar cells such as amorphous, polycrystalline and single crystal in which single crystal solar cells show the highest efficiency of 24.5% compared to other cells because of minimization of recombination losses. The boron doped single crystal Si ingot is grown by Czochralski technique. The sliced p-type crystalline or polycrystalline wafers with ~ 25 cm diameter are doped by phosphorous using diffusion process up to certain depth of wafer and then rapid annealing in order to make abrupt p–n junction. To complete cell configuration, the metal contacts on both n and p sides are made. The microcrystalline or amorphous Si based thin film solar cells are also developed but the cells show low efficiency of 9–10% [15]. On the other hand, efficiency degrades with time.

1.4 Photovoltaic Conversion

Solar cell or photovoltaic cell is nothing but p–n junction when the photons with energy greater than or equal to the band gap of p-absorber impenge on solar cells, the electrons excite from valence band to conduction band in the absorber. Thus electron-hole pairs are created. The electrons cross barrier from p- to n- region and holes from n- to p- region in order to release their energy before taking place recombination. In other words the electron–hole pairs, which are within one diffusion length from the built-in electric field of the junction are separated giving rise to photovoltage and photocurrent. The interface is usually homojunction or heterojunction or Schottky barrier.

1.4.1 Heterojunction

It is an interface formed between two semiconductors of different energy gaps, a small band gap semiconductor known as "absorber," in which optical absorption takes place and a large band gap semiconductor, known as "window," that is highly transparent to solar radiation. In the heterojunction, the front surface recombination loss is eliminated. However, interface recombination cannot be ignored. Two different kinds of configurations are possible: front-wall configuration in which photons first incident on the absorber layer, whereas they first incident on window layer in the back-wall configuration.

The output current of circuit can be related to

$$J = J_L - J_D - J_{SH} \tag{1.1}$$

from the equivalent circuit of solar cell (Figure 1.6A), where J_L is photogenerated current, J_D is diode current, and J_{SH} is shunt current [34]

$$\left(J_{SH} = V_j/R_{SH}\right). \tag{1.2}$$

The expression for output voltage can also be written as

$$V = JR_S - V_j, \tag{1.3}$$

where R_S is series resistance and V_j is voltage across diode and resistor R_{SH}. A solar cell behaves like a p–n junction in dark and the current–voltage relationship of the junction in dark is represented by the standard Shockley diode equation

$$J_D = J_0\{\exp[qV_j/Ak_BT] - 1\}, \tag{1.4}$$

where J_0 is reverse saturation current, A is diode quality factor, q is electron charge, k_B is Boltzmann constant, T is absolute temperature. By substituting Equation (1.2),

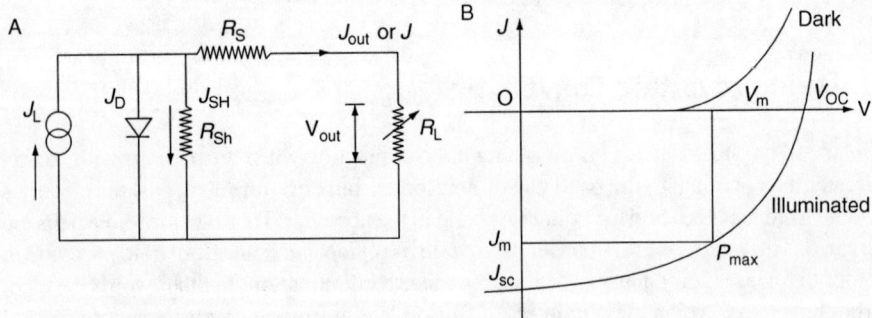

Figure 1.6 (A) Equivalent circuit of solar cell, and (B) J–V characteristics of solar cell under dark and light illumination.

Equation (1.3), and Equation (1.4) in Equation (1.1), the final derivation for output current results in combination of diode forward current and light generated current for solar cell under light illumination [40]

$$J = J_0 \left[\exp\frac{q}{Ak_BT}(V - JR_s) - 1 \right] + \left(\frac{V - R_sJ}{R_{sh}}\right) - J_L. \quad (1.5)$$

1.4.2 Open-Circuit Voltage

When $J = 0$, the corresponding voltage known as open-circuit voltage is denoted by V_{oc}. It is seen that the reverse saturation current increases, the open-circuit voltage decreases in the solar cell. The experiments on several CIS, CIGS, and CGS cells reveal that the V_{oc} increases linearly with increasing E_g up to certain point then it becomes nonlinear, as shown in Figure 1.7. In linear portion the expression follows as [41]

$$V_{oc} = E_g/q - 0.5 \text{ V}. \quad (1.6)$$

The V_{oc} equals to the length between quasi Fermi levels of window and absorber in the inorganic semiconductor solar cells. In the case of plastic solar cells, the V_{oc} is determined by the distance between the highest occupied molecular orbital (HOMO) level of the electron donors and the lowest unoccupied molecular orbital (LUMO) level of the electron acceptors, for which the equation is derived as [33]

$$V_{OC} = 1/q \left(\left| E_{HOMO}^{Donor} \right| - \left| E_{LUMO}^{Acceptor} \right| \right) - 0.3 \text{V} \quad (1.7)$$

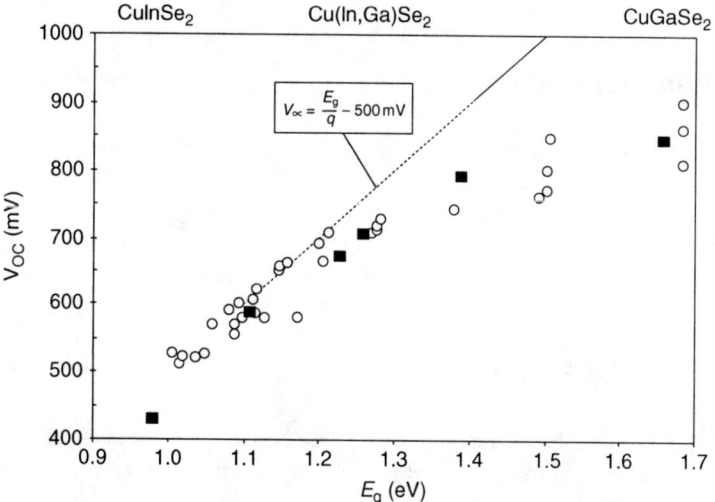

Figure 1.7 A plot of V_{oc} versus band gap (E_g) for different Ga composition of CuInGaSe$_2$ solar cells (O and ■ represent for two different results).

1.4.3 Short-circuit Current

When $V=0$, the corresponding current is known as short-circuit current and is denoted by J_{sc}. It depends upon the incident photon flux, absorption coefficient and collection efficiency.

1.4.4 Fill Factor (FF)

FF, defined as the ratio of maximum power delivered by the cell to the product of J_{sc} and V_{oc}, is given by

$$FF = \frac{J_m V_m}{J_{SC} V_{OC}}, \tag{1.8}$$

where J_m and V_m are the current and voltage corresponding to the maximum power point respectively (Figure 1.6B). The FF is one of the parameters to caliber stability of solar cell. Poor FF is always observed for low efficiency cells. The high series resistance and low shunt resistance result in low FF in the solar cells. The highest reported FF is 88% for multijunction solar cells.

1.4.5 Efficiency

The conversion efficiency (η) of the solar cell is given by

$$\eta = J_m V_m / P_{in}, \tag{1.9}$$

where P_{in} is the total input power to the cell.

1.5 Homojunction

It is an interface formed between the p- and n-regions of a single semiconductor. The best known examples are Si and GaAs solar cells. In general, homojunction exhibits high efficiency, particularly in direct band gap materials with high absorption coefficient but there are appreciable losses due to front surface recombination.

1.6 Schottky Barrier

It is a metal–semiconductor junction in which a blocking contact is formed. The performance is usually limited by large thermionic emission currents that reduce the open-circuit voltage. A thin insulating layer, usually, oxide is employed in metal–insulator–semiconductor (MIS) and semiconductor–insulator–semiconductor (SIS) configurations to reduce forward currents and improve the open circuit voltage [42].

1.7 Criteria for the Choice of Heterojunction Pair in the Thin Film Solar Cells

1.7.1 Band Gap of Absorber

The band gap is the first order parameter in deciding the semiconductor pair for the heterojunction. Based on theoretical considerations, Loferski [14] proposed a plot of theoretical conversion efficiency versus band gap of the semiconductor (Figure 1.8). It is seen that a semiconductor with a band gap of about 1.55 eV is ideal for absorber to achieve maximum conversion efficiency. In addition, it should be a direct band gap material with high optical absorption coefficient. It should preferably be a p-type material because of longer electron diffusion length.

1.7.2 Band Gap of Window Material

A wide band gap n-type semiconductor with a low series resistance is very essential to function solar cells well.

1.7.3 Electron Affinities

Electron affinity of the absorber and window materials should be compatible so that no potential spike occurs at the junction for the minority photoexcited carriers.

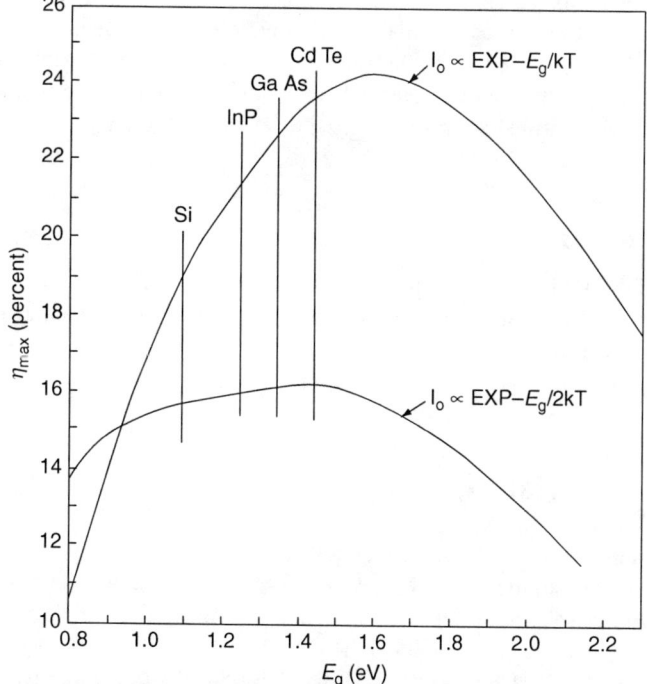

Figure 1.8 Theoretical efficiency versus semiconductor band gap.

1.7.4 Lattice and Thermal Mismatch

The lattice and thermal mismatch between the absorber and window material should be as small as possible to reduce interface state density and recombination losses through such states.

1.7.5 Electrical Contacts

It should be possible to form low resistance electrical contacts to both n- and p-type materials. So far the suitable metal electrical contact is not probably found for p-CdTe absorber.

1.7.6 Cell Stability and Life Time

The cell must have stable performance and long operating life at least more than a decade. The cell must pass damp heating or thermocycling test in order to compile market standards to introduce into public.

1.7.7 Deposition Methods

Suitable deposition methods for thin film formation with reproducibility and control are highly essential. The cost effective techniques such as ink based printings, screen printings, doctor blade, spray, *etc.*, are encouraging techniques while costly vacuum techniques such as thermal, sputtering, MOCVD, MBE, *etc.*, are part of the process for cutting edge technologies whereby cost is negligible. In the case of fabrication of multijunction solar cells, the MOCVD technique is more suitable to maintain or tailoring composition of the layers particularly phosphorus based materials.

1.7.8 Materials

Materials should be abundant, nontoxic with no environmental degradation. For example, Sn, Fe, Al, Si, *etc.*, based compound semiconductors are amenable for thin film solar cells. If toxic compounds are used, proper care must be taken suppose in the case of H_2Se, H_2S gases, Hydrozene, *etc.*

1.8 Window Materials

The wide band gap semiconductor materials usually n-type or doped with suitable n dopants to obtain required low resistivity are used as window materials. Table 1.2 gives some physical properties of wide band gap materials, which can be used in heterojunction solar cells. The CdS is the most widely studied window material with absorbers like Cu_2S, CdTe, InP, $CuInS_2$, $CuInSe_2$ $CuIn_{1-x}Ga_xSe_2$, *etc.*, others include CdZnS, In_2O_3:Sn, ZnO, SnO_2:F, *etc*. In the case of CdS, CdZnS, and ZnO

Table 1.2 Physical properties of some window materials

Material	E_g (eV)	χ (eV)	Lattice constants (Å) a	Lattice constants (Å) c	Crystal structure	Type	Ref.
CdS	2.42	4.5	4.14	6.72	Wurtzite	n	[43–45]
ZnS	3.70–3.58	4.5–3.9	3.82, 5.41	6.26	Wurtzite, Sphalerite	n, p	[46]
ZnSe	2.67	4.09	4.00	6.54	Wurtzite	n, p	
SnO$_2$	3.7	4.8–4.9	4.74	3.19	Tetragonal	n	[47,48]
In$_2$O$_3$	3.65	4.1–4.6	10.12	–	Cubic	n	
ZnO	3.2	4.2	3.25	5.21	Hexagonal	n	[49]
CdO	2.3–2.7	4.5	4.69	–	Cubic	n	
CdIn$_2$O$_4$	2.67–3.24	4.5	9.166–9.202	–	Cubic Spinel	n	[50,51]
Cd$_2$SnO$_4$	2.35–3.00	4.5	5.57 (b=9.89), 9.206	3.90	Orthorhombic, Cubic Spinel	n	

suitable dopants such as In or Al are used to obtain high conductivity with reasonably high transmission. Higher V_{oc} and J_{sc} are reported with CdZnS as a window instead of CdS in the heterojunction solar cells.

1.9 Absorber Materials

The performance of heterojunction thin film solar cell is basically dependent on the choice of the optimum absorber material. As discussed in earlier, the material should be a direct band gap semiconductor with a high absorption coefficient. A plot of absorption coefficient of different semiconductors as a function of photon energy (hν) is shown in Figure 1.9. There are a number of possible materials with band gap neighborhood of 1.5 eV. However other factors such as technology and stability are also important. The solar cells based on absorber materials like Si, a-Si, GaAs, CdTe, InP, CuInSe$_2$, CIGS, Zn$_3$P$_2$, CZTS, WSe$_2$, *etc.*, have received much attention [11,52]. The search for innovative materials has led to investigations on efficiencies in excess of 10%. The studies on ternary [53], quaternary [54–59], and pentenary [59–61] semiconductor materials with possibility of tailoring the band gap and lattice constants to match the window materials are essential. The family of ternary chalcopyrites offers numerous possibilities for obtaining absorber materials because of their favorable optical and electrical properties [53]. The materials are of current technological interest since they also find application in other solid state devices like visible and infrared light emitting diodes, infrared detectors, parametric oscillators, up converters and far infrared generators. Among the ternary chalcopyrite semiconductors, copper based I–III–VI$_2$ semiconductors have attracted many investigators. The basic criteria for thin film solar cells are that low lattice mismatch between absorber and window minimizes interfacial states. The lattice mismatch

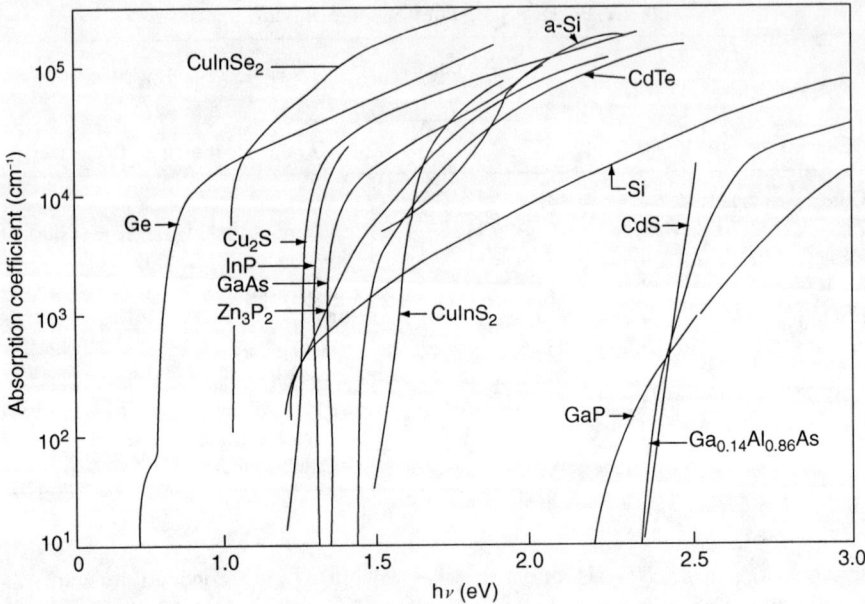

Figure 1.9 Absorption spectra of various semiconductors.

Table 1.3 Physical parameters of some absorbers

Compound	E_g (eV)	Lattice parameters (Å)		μ_n (cm²/Vs)	μ_p	Lattice mismatch with CdS (%)	Electron affinity (χ; eV)	Work function (ϕ; eV)	Ref.
		a	c						
CuInSe$_2$	1.01–1.04	5.782	11.62	320	10	1.16	4.7	5.2	[62]
CuGaSe$_2$	1.72	5.77	11.55	–	–	1.37	–	–	–
CuInS$_2$	1.55	5.523	11.12	200	15	5.59	–	–	–
CdS	2.42	4.136	6.716	250	–	–	3.8	4.7	[63,64]
ZnSe	2.46	–	–	–	–	–	–	–	–

between variety of chalcopyrite absorbers and CdS is given in Table 1.3. The direct band gap may enhance larger minority carrier diffusion length [65].

1.10 Copper Based I–III–VI$_2$ Semiconductors

The compound semiconductors, particularly CuInX$_2$ ($X=$ S, Se) or Cu(In$_{1-x}$Ga$_x$)Se$_2$ have several desirable features as absorbers in the thin film solar cells. (i) They are direct band gap semiconductors with a high absorption coefficient. (ii) They can be

Table 1.4 Physical parameters of $CuIn(Se_{1-x}S_x)_2$ system

Material	E_g (eV)	χ (eV)	Lattice constants (Å) a	c	Conductivity type	Ref.
$CuInS_2$	1.55–1.438	4.04	5.51	11.00	n, p	[70–73]
$CuIn(Se_{0.25}S_{0.75})_2$	–	–	5.58	11.15	n, p	
$CuIn(Se_{0.32}S_{0.68})_2$	1.390	–	–	–	p	[57]
$CuIn(Se_{0.5}S_{0.5})_2$	–	–	5.65	11.28	n, p	
$CuIn(Se_{0.54}S_{0.46})_2$	1.223	–	–	–	p	
$CuIn(Se_{0.75}S_{0.25})_2$	–	–	5.71	11.43	n, p	
$CuInSe_2$	0.96–1.04	4.58	5.78	11.62	n, p	[74]

deposited either in n- or p-type. However, the $CuGaSe_2$ always exhibits to be p-type. A wide variation in the conductivity can be achieved by controlling the atomic ratios of cations or by doping with different dopants. (iii) The electron affinities of these semiconductors appear compatible with CdS, CdZnS, indium tin oxide (ITO) such that deleterious conduction band spikes are unlikely upon heterojunction formation [66–68]. (iv) The lattice match is good with CdS and CdZnS minimizing interfacial state density [69]. (v) The materials can be deposited in thin film form employing suitable techniques for large scale production. (vi) The materials should be more stable than Cu_xS. In the ternaries, trivalent indium seems to bind copper tightly in the chalcopyrite lattice suppressing the undesirable copper migration into CdS, which is observed in the Cu_xS/CdS solar cells. The basic properties of $CuInS_2$, $CuInSe_2$, and $CuIn(Se_{1-x}S_x)_2$ are listed in Table 1.4.

References

[1] Statistical Review of World Energy, 06/2001, British Petroleum.
[2] E. Becquerel, Compt. Rend. 9 (1839) 561.
[3] W.G. Adams, R.E. Day, Proc. R. Soc. Lond., B, Biol. Sci. A25 (1877) 113.
[4] B. Lange, Zeit. Phys. 31 (1930) 139.
[5] L.O. Grondhal, Rev. Mod. Phys. 5 (1933) 141.
[6] W. Schottky, Zeit. Phys. 31 (1930) 913.
[7] D.M. Chapin, C.S. Fuller, G.L. Pearson, J. Appl. Phys. 25 (1954) 676.
[8] D.C. Reynolds, G. Leies, L.L. Antes, R.E. Marburger, Phys. Rev. 96 (1954) 533.
[9] D.A. Jenny, J.J. Loferski, P. Rappaport, Phys. Rev. B101 (1956) 1208.
[10] K.L. Chopra, Thin Film Phenomena, McGraw-Hill, New York, 1969.
[11] K.L. Chopra, S.R. Das, Thin Film Solar Cells, Plenum Press, New York, 1983.
[12] C.E. Backus, Thin Film Solar Cells, IEEE Press, New York, 1976.
[13] M.P. Thekaekara, Data on incident solar energy, in: The Energy Crisis and Energy From the Sun, Institute of Environmental Sciences, 1974.
[14] J.J. Loferski, J. Appl. Phys. 27 (1956) 777.

[15] J. Merrill, D.C. Senft, J. Mater. 59 (2007) 26.
[16] M. Powella, Zentrum fur Sonnenenergie- und Wasserstoff-Forschung Baden-Wurttemberg, Germany. www.pv-tech.org, August 29th 2010.
[17] J. Britt, C. Ferekides, Appl. Phys. Lett. 62 (1993) 2851.
[18] IBM. Private communication, 2010.
[19] F. Dimorth, et al., Fraunhofer Institute for Solar Energy Systems. www.nextbigfuture.com, May 19th 2010.
[20] J. Zhao, A. Wang, P. Altermatt, M.A. Green, Appl. Phys. Lett. 66 (1995) 3636.
[21] H.Y. Chen, J. Hou, S. Zhang, Y. Liang, G. Yang, Y. Yang, et al., Nature 3 (2009) 649.
[22] S. Ito, T.N. Murakami, P. Comte, P. Liska, C. Gratzel, M.K. Nazeeruddin, et al., Thin Solid Films 516 (2008) 4613.
[23] T. Cheyney, Photovoltaics International, second ed., (2008), www.pv-tech.org, (p.76).
[24] R.R. King, D.C. Law, K.M. Edmondson, C.M. Fetzer, G.S. Kinsey, H. Yoon, et al., Adv. OptoElectron. (2007), (ID 29523).
[25] N.H. Karam, R.R. King, M. Haddad, J.H. Ermer, H. Yoon, H.L. Cotal, et al., Solar Energy Mater. Solar Cells 66 (2001) 453.
[26] M. Yamaguchi, T. Takamoto, K. Araki, Solar Energy Mater. Solar Cells 90 (2006) 3068.
[27] S.H. Lee, K.L. Chen, S. Gallo, III-V Multijunction Solar Cells, Iowa State University, 2008. (Private communication).
[28] M.R. Reyes, K. Kim, D.L. Carroll, Appl. Phys. Lett. 87 (2005) 083506.
[29] R. Janssen, 'Introduction to polymer solar cells (3y280), 2010.' (Private communication)
[30] A.C. Mayer, S.R. Scully, B.E. Hardin, M.W. Rowell, M.D. McGehee, Mater. Today 10 (2007) 28.
[31] C.W. Tang, Appl. Phys. Lett. 48 (1986) 183.
[32] J.Y. Kim, S.H. Kim, H.-H. Lee, K. Lee, W. Ma, X. Gong, et al., Adv. Mater. 18 (2006) 572.
[33] W. Cai, X. Gong, Y. Cao, Solar Energy Mater. Solar Cells 94 (2010) 114.
[34] Wikipedia, Solar Cell.
[35] K. Kim, J. Liu, M.A.G. Namboothiry, D.L. Carroll, Appl. Phys. Lett. 90 (2007) 163511.
[36] Karl Leo. www.heliatek.com, (accessed June 2010).
[37] S. Dai, J. Weng, Y. Sui, C. Shi, Y. Huang, S. Chen, et al., Solar Energy Mater. Solar Cells 84 (2004) 125.
[38] M. Gratzel, Nature 414 (2001) 338.
[39] T.D. Nielsen, C. Cruickshank, S. Foged, J. Thorsen, F.C. Krebs, Solar Energy Mater. Solar. Cells 94 (2010) 1553.
[40] S.R. Kodigala, Ph.D. Thesis, Sri Venkateswara University, Tirupati, 1992.
[41] J. Malmstrom, J. Wemmerberg, M. Bodegard and L. Stolt 17th European PV conference 2001, 1265; H.W. Schock, U. Rau, T. Dullaber, G. Hamma, M. Balboul, T.M-. Friedelmeier, A. Jasemeck, I. Kotschau, H. Kerber and H. Wiesner, in proc. 16th Europ. Photovoltaic. Sol. Em. Conf. 2000, 301–308.
[42] R.H. Bube, Heterojunctions for thin film solar cells, in: Solar Materials Science, Academic Press, New York, NY, 1980.
[43] H.F. Wolf, Semiconductors, Wiley Interscience, New York, NY, 1971.
[44] M. Savelli, J. Bougnot, B.O. in: Seraphin (Ed.), Solar Energy-Solid State Physics Aspects, Topics in Applied Physics, Vol. 31, Springer-Verlag, Berlin, 1979.
[45] D.L. Feucht, J. Vac. Sci. Technol. 14 (1977) 57.
[46] C.J. Nuese, J. Educ. Mod. Mat. Sci. Eng. 2 (1980) 113.
[47] R. Singh, K. Rajkanan, D.E. Brodie, J.H. Morgan, IEEE Trans. Electron Devices ED-27 (1980) 656.
[48] E. Bucher, Appl. Phys. 17 (1978) 1.

[49] O. Lupan, T. Pauporte, L. Chow, B. Viana, F. Pelle, L.K. Ono, et al., Appl. Surf. Sci. 256 (2010) 1895.
[50] G. Haake, Solar Energy Mater. Solar. Cells 14 (1986) 233.
[51] T. Pisarkiewicz, K. Zakrzewska, E. Leja, Thin Solid Films 153 (1987) 479.
[52] R.H. Bube, A.L. Fahrenbruch, Fundamentals of Solar Cells, Academic Press, New York, 1983.
[53] J.L. Shay, J.H. Wernick, Ternary Chalcopyrite Semiconductors: Growth, Electronic Properties and Applications, Pergamon Press, New York, 1975.
[54] S.K. Chang, H.L. Park, H.K. Kim, J.S. Hwang, C.H. Chung, W.T. Kim, Phys Stat. Solidi (B) 158 (1990) K115.
[55] L. Roa, C. Rincon, J. Gonzalez, M. Quintero, J. Phys. Chem. Solids 51 (1990) 551.
[56] M. Robbins, V.G. Lambrecht Jr., Master. Res. Bull. 8 (1973) 703.
[57] H. Neff, P. Lange, M.L. Fearheiley, K.J. Backmann, Appl. Phys. Lett. 47 (1985) 1089.
[58] I.V. Bodnar, B.V. Korzun, A.I. Lukomski, Phys. Stat. Solidi (B) 105 (1981) K143.
[59] B.R. Pamplin, R.S. Feigelson, Thin Solid Films 60 (1979) 141.
[60] G.H. Chapman, J. Shewchun, B.K. Garside, J.J. Loferski, R. Beaulieu, Solar Energy Mater. 1 (1979) 451.
[61] J. Shewchun, J.J. Loferski, R. Beaulieu, G.H, Chapman and B.K. Garside, J. Appl. Phys. 50 (1979) 6978.
[62] T. Loher, W. Jaegermann, C. Pettenkofer, J. Appl. Phys. 77 (1995) 731.
[63] M. Gloeckler, MS Thesis, Colorado State University, Fort Collins, CO 80523, 2003.
[64] G. Liu, T. Schulmeyer, J. Brotz, A. Klein, W. Jaegermann, Thin Solid Films 431 (2003) 477.
[65] L.L. Kazmerski, F.R. White, M.S. Ayyagari, Y.J. Juang, R.P. Patterson, J. Vac. Sci. Technol. 14 (1977) 65.
[66] K.J. Bachmane, E. Buchler, J.L. Shay, S. Wagner, Appl. Phys. Lett. 29 (1976) 121.
[67] L.L. Kazmerski, P.J. Ireland, F.R. White, R.B. Cooper, in: Proc. 13th IEEE Photovoltaics Specialists Conference, 1978, p. 620.
[68] L.L. Kazmerski, P. Shelden, in: Proc. 13th IEEE Photovoltaics Specialists Conference, 1978, 541.
[69] L.L. Kazmerski, in: G.D. Holah (Ed.), Ternary Compounds 1977, Vol. 35, 1977, p. 217 (London; Inst. Phys.).
[70] M. Robbins, V.G. Lambrecht Jr., Mater. Res. Bull. 8 (1973) 703.
[71] I.V. Bodnar, B.V. Korzun, A.I. Lukomski, Phys. Stat. Solidi (B) 105 (1981) K143.
[72] S.J. Fonash, Solar Cell device Physics, Academic Press, New York, 1981.
[73] H.L. Hwang, C.Y. Sun, C.Y. Leu, C.C. Cheng, C.C. Tu, Rev. Phys. Appl. 13 (1978) 745.
[74] H. Hahn, G. Frank, W. Klinger, A.D. Meyer, G. Strorger, Z. Anorg. Aug. Chem. 271 (1953) 153.

2 Growth Process of I-III-VI$_2$ Thin Films

The famous growth techniques for the deposition of Cu(In$_{1-x}$Ga$_x$)Se$_2$ and CuIn(Se$_{1-x}$S$_x$)$_2$ absorbers have been developed such as thermal vacuum evaporation and low-cost nonvacuum process methods, which are discussed in this chapter giving priority to former absorber.

2.1 Deposition of CuInSe$_2$

There are several diversified techniques to grow semiconducting thin films, some of them are costly and others being cost effective. The plausible techniques such as vacuum evaporation, flash evaporation, sputtering, spray pyrolysis, chemical method, *etc.*, to grow CIS thin films are exploited.

2.1.1 Vacuum Evaporation

Michelsen and Chen [1] first reported that the deposition of CuInSe$_2$ thin films directly from three elemental sources offers the greatest degree of control of composition over the stoichiometry to produce high efficiency CuInSe$_2$ thin film solar cells. In order to deposit large area deposition, three elemental effusion cells each with two independent heaters are applied [2,3]. The Cu, In, and Se are co-evaporated from their independent carbon crucibles heating by tantalum filament. The temperature ranges of 1100–1250, 930–1100, and 240–320 °C for Cu, In, and Se are applied. The quartz crystal monitor controls thickness of the samples. In order to obtain chamber base pressure of 10^{-4}–10^{-5} Pa, oil-diffusion pump with liquid nitrogen trap or turbo molecular or cryogenic pump is used. The typical distance between source and substrate is kept to be 20 cm. Tungsten lamps with parabolic reflectors or resistive heaters are used for heating substrates to obtain temperature in the range of 175–400 °C [4,5]. In a dual source evaporation to compensate loss of chalcogen (Se or S) during evaporation two resistive heated crucibles are used, one for single phase CuInSe$_2$ charge and the other for the chalcogen. By controlling partial pressure of chalcogen either *p*- or *n*-type films can be obtained. This method is successfully employed to produce device quality films [6,7].

Unlike the Cu and In layers are successively evaporated onto either glass or Si (100) substrates from tungsten boats by thermal evaporation at RT and annealed at 150 °C then Se evaporated onto Cu–In alloy from graphite effusion source is annealed under argon flow at 450 °C for 15 min to have $CuInSe_2$ thin films [8]. The source temperatures of 1290–1380 and 990–1070 °C with corresponding fluxes of 0.62–1.76×10^{15} Cu/cm^2-s and 0.36–1.72×10^{15} In/cm^2-s for Cu and In, respectively, which result in total metal flux of 1.8×10^{15} at./cm^2-s, are employed [9].

The single phase CIS thin films with (112) preferred orientation are deposited onto glass substrates at substrate temperature of 150–300 °C by using CIS powder as a source. In order to make sintered CIS, high purity Cu, In, and 5% excess Se elements are sealed in quartz tube under pressure of 10^{-2} mbar and heated to 1127 °C at a rate of 3 °C/min then held at 1127 °C for 3 h. The ampoule is cool down to RT at the rate of 5 °C/min to make fine CIS powder [10–12]. The $CuInSe_2$ ingots are also prepared by similar fashion by keeping Cu–In–Se in an evacuated and sealed quartz tube at 1100 °C for 10 h and cooling down to RT [13].

2.1.2 Flash Evaporation

The flash evaporation technique can be used instead of evaporation by effusion cells for large area coatings. However proper designed evaporation boats should be used otherwise spitting takes place [14]. The $CuInSe_2$ sintered powder has been continuously dropped through either mechanically or electromagnetically vibrating glass tube with a typical frequency of 50 Hz onto molybdenum boat, which is heated at 1173–1500 °C to deposit CIS thin films onto glass substrates [11,15,16]. In order to compensate loss of Se in the films, an additional Se source containing graphite boat is added to the system, as shown in Figure 2.1 [17]. The flash evaporated layers show

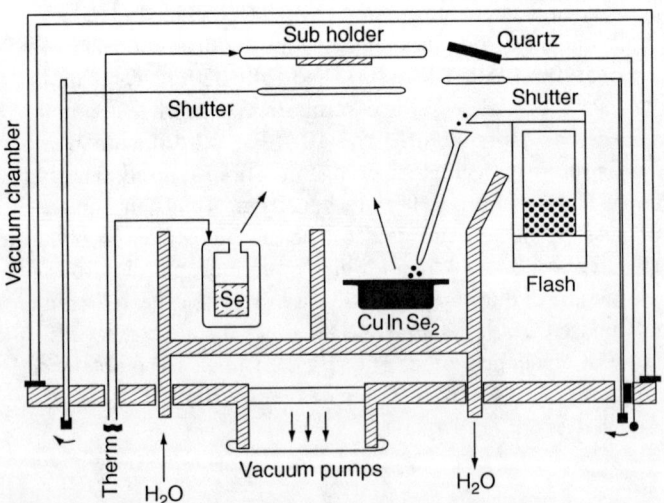

Figure 2.1 Schematic diagram of flash evaporation technique.

Cu poor composition because the low source temperature is used for the chalcopyrite CGS, CIS, and CIGS powder or sinter compound. The left over compound in the boat shows Cu rich composition. The suitable boat temperature is 1400 °C and the deposition rate is 1 nm/s. The Se composition is always less than 50% in the grown layers [18].

2.1.3 Sputtering

The advantage of sputtering is that the chamber gets heat up relatively low comparing with other techniques and the deposition can easily be controllable by plasma current. The loss is minimum in the film that means the composition of the film is more or less as same as target. In order to grow Cu/In/Se stack the schematic diagram of RF sputtering system is shown in Figure 2.2. After evacuation of chamber to 1×10^{-6} torr, pure Ar is introduced into chamber to sputter target and maintain pressure of 1×10^{-2} torr. The high purity Ar with flow rate of 6–15 sccm is applied. Prior to deposition of thin films, presputtering is done for 10 min to clean contamination under working pressure of 3×10^{-2} Pa. Pure Cu, In, and Se targets with diameter of 4 in. are used at 60, 75 mA DC current and RF power of 100 W, respectively. 0.27 μm Cu, 0.6 μm In, and 1.46 μm thick Se layers are subsequently deposited onto either Mo coated glass or glass substrates at RT then Cu/In/Se stack is selenized at 250–400 °C in an encapsulated carbon block to obtain CuInSe$_2$ thin films. It is found that good crystalline CuInSe$_2$ thin film results for above the annealing temperature of 320 °C [19]. In the sputtering, Cu$_x$Se and In$_2$Se$_3$ are found as contaminations on Se target that can be avoided by changing the configuration and or positions of the targets as well as optimizing the power levels of targets. The flux of Se is maintained more than three times of total metal fluxes resulting in growth rate of 4–5 Å/s. The typical distance between substrate and target is maintained to be 5 cm. The polycrystalline Se target is used rather than amorphous Se to avoid abnormal discharge of power due to lower thermal conductivity of latter [20]. Unlike the Cu, In, and Se individual targets with dimensions of $110 \times 110 \times 3$ mm^3 square shape are arranged like roller coaster style, as shown in

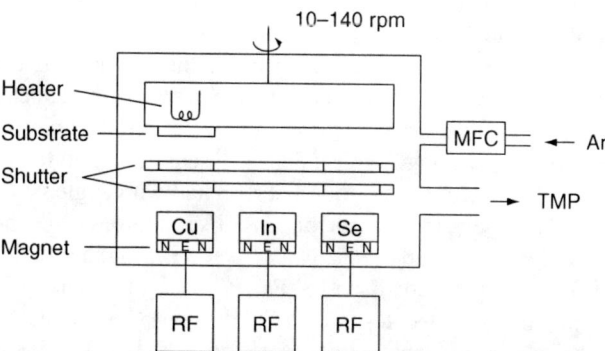

Figure 2.2 Schematic diagram of RF sputtering system for the growth of CuInSe$_2$ thin films.

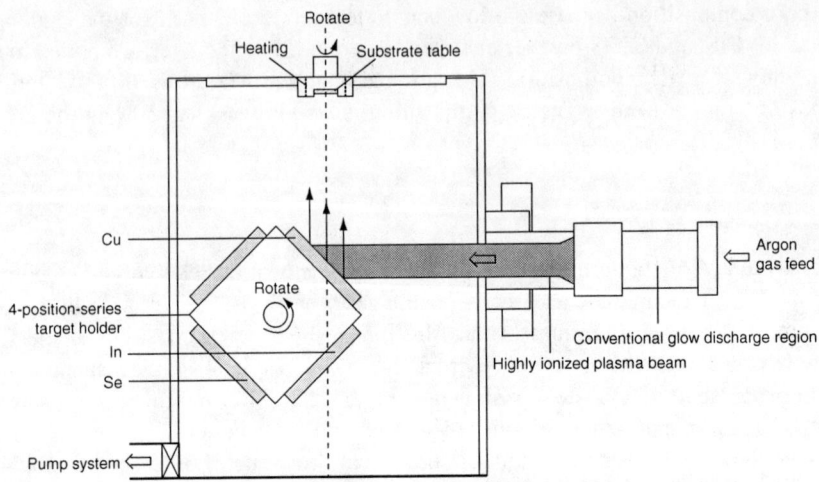

Figure 2.3 Schematic diagram of ion beam sputtering.

Figure 2.3. The composition of CIS layers is controlled by varying deposition time of each sputtering. The Cu, In, and Se are subsequently deposited and annealed under vacuum at 400 °C for 1 h. The highly ionized plasma beam generated by conventional glow discharge is directed towards target for sputtering. The anode voltage (V)/current (A) of 80/025, 80/0.25, and 75/0.2 as well as the cathode V/A of 10/12, 10/12, and 9/12 are applied for Cu, In, and Se depositions, respectively. The deposition times of 330 s, 360 s, and 30–180 min are used to deposit Cu, In, and Se, respectively [21].

There are several methods to grow CIS thin films in the hybrid process. The Cu and In are sequentially deposited onto Mo coated glass substrates by sputtering technique to have 1 μm thick layers using 6″ diameter target on which 1 μm thick Se is evaporated by thermal evaporation. The Cu/In/Se stack sample converts into $CuInSe_2$ after annealing in graphite box under vacuum of 1.333 Pa in the temperature range of 100–600 °C. The box is located into quartz tube [22]. On 10×10 cm^2 glass substrates, Mo, Cu, and In by DC sputtering or Cu, In, and Se by evaporation with Cu/In = 0.85–0.95 and 40% excess of Se are deposited to form $CuInSe_2$ thin films and annealed up to 550 °C in short time of few seconds or higher, using novel rapid-thermal processing (RTP) technique because the precursor is combination of $CuIn_2$, Cu, and $CuInSe_2$ phases. Single phase $CuInSe_2$ exists after annealing precursor by RTP [23]. In the reactive sputtering, the $CuInSe_2$ thin films are grown by planar magnetron coreactive sputtering. Typical $5″ \times 12″$ size Cu and In targets in adjacent are located perpendicular to the substrate but deviated with angle of 22–1/2° at distance of 22.8 cm. The H_2Se gas flow rate of 75 Pa-l/s is allowed while the partial pressures of Ar and H_2Se are 0.2 and 0.5 Pa, respectively. During the sputtering, a partial pressure of H_2Se is maintained to be 0.4 Pa. Because of plasma creation, by the involvement of number of molecules such as Cu, In, Se, HSe, *etc.*, finally $CuInSe_2$ thin film forms on the substrate at temperature of 450 °C. The base Cu-rich $CuInSe_2$ layer on Mo coated 7059 corning glass substrate is deposited with compositions of Cu:In:

Se = 27.2:23:48.9 at substrate temperature of 350 °C, which is p-type, $\rho = 1$ Ω-cm, $\mu = 2$ cm^2/Vs, $p = 10^{17}$–10^{20} cm^{-3}, $E_g = 1.0$ eV, and rough surface morphology. The top layer with In-rich CuInSe$_2$ is deposited at substrate temperature of 450 °C, which had compositions of Cu:In:Se = 24.3:25.2:50.2, $n = 10^{13}$ cm^{-3}, $\rho = 10^4$ ohm-cm, $\mu = 15$ cm^2/Vs, $E_g = 1.04$ eV, and smooth surface morphology. Obviously, the base layer is combination of CuInSe$_2$ and Cu$_2$Se but deposition of second layer consumes Cu$_2$Se phase to form CuInSe$_2$ layer [24,25]. In another occasion 800 nm thick Cu–In alloy is grown onto Mo coated glass substrates by sputtering individual Cu and In targets with working pressure of 0.2 Pa. The RF power of Cu is varied from 20 to 60 W while keeping RF power of 60 W for In at constant. Under Se pressure of 1.3 Pa, the Cu–In alloy layers are annealed in closed graphite crucible fitted in quartz tube, as shown in Figure 2.4 [26–29]. The Cu and In films sequentially grown onto glass substrates to have desired Cu/In ratio are annealed under 5–10% H$_2$Se and Ar ambient at 400 °C for 1 h in order to obtain CuInSe$_2$ thin films [30]. Alike the CuIn thin films deposited onto Mo coated glass substrates selenized in horizontal quartz tube at atmospheric pressure of H$_2$Se diluted in Ar at 400 °C for 10–80 min. The selinized samples are sulphurized at 550 °C for 30 min in H$_2$S + Ar to have CISS thin films [31]. Unlike the Cu/In stack layers formed at room temperature by sputtering of Cu and evaporation of In on Mo coated glass substrates are annealed under Ar at 380–425 °C to intermix as Cu–In alloy. The Cu–In alloy is converted into CuInSe$_2$ by annealing under Se atmosphere at the same temperature for which solid Se is heated to produce the flux rate of 70 nm/min under pressure of 5×10^{-7} torr [32]. In another case the Cu by sputtering, In and Se by vacuum evaporation are sequentially deposited onto Mo covered glass substrates at room temperature. The Cu/In/Se stack is annealed under N$_2$ atmosphere at 200–400 °C in carbon block to protect loss of Se [33].

Piekoszewski et al., [34,35] first prepared thin films by sputtering that the fine and coarse CIS targets (5 cm dia and 0.2 cm thickness) are made from fine powder and ungrounded small CuInSe$_2$ pieces, respectively. The CIS films are deposited onto 7059 corning glass under Ar pressure of 2×10^{-2}–8×10^{-2} torr, RF power of 100 W and substrate temperature of 25–506 °C. The distance between substrate and target is 3–6 cm. The CIGS single target is made by adding Ga to the CIS compound and almost similar growth conditions are used to deposit CIGS thin

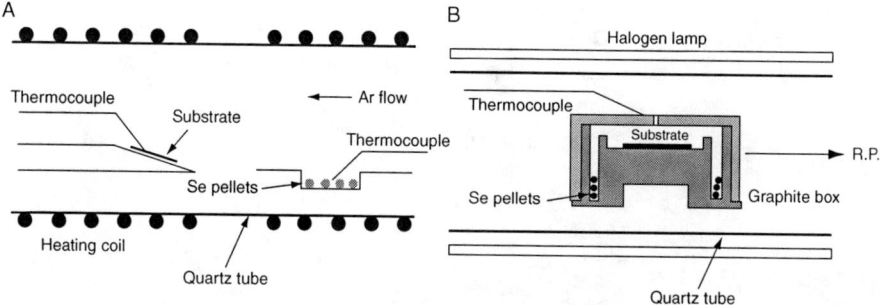

Figure 2.4 (A) Sketch of atmospheric selenization and (B) Vacuum system.

films [36]. In order to dope Na, the Na_2Se mixed with CIS or CIGS (Na/CIS = 5%) is cold pressed at 4.9×10^7 Pa, which is used to deposit CIS or CIGS thin films by sputtering with RF power of 100 W (3 W/cm^2) and Ar pressure of 1.33 Pa [37]. The poor structural performance is observed for the selenization of the layers without RTP therefore the RTP is used to ramp up temperature from 150 to 400 °C [38].

2.1.4 Pulsed Laser Deposition

The CIS films are deposited onto corning 7059 glass substrates at substrate temperature of 550 °C by PLD ArF excimer laser with a typical wavelength of 193 nm and pulse width of 25 ns, which hits CIS pellet through the quartz window, as shown in Figure 2.5. The chamber pressure of $1-3 \times 10^{-3}$ torr is maintained during deposition of films. The CIS pellet is made by pressing CIS sintered powder at 8.6×10^7 Pa. The target rotation speed of 15 rpm is maintained to avoid heating on it and for uniform film growth. The distance between target and substrate is 3 cm and substrate temperature being 500 °C [39,40]. Similarly XeCl laser ($\lambda = 308$ nm, 30–50 mJ/pulse, 4–5 J/cm^2, frequency 25 Hz, and $f = 100$ mm lens) beam is scanned on polycrystalline CIS target, which is kept at 45° angle with respect to base. The distance from substrate to target, chamber base pressure and substrate temperature are maintained to be 5 cm, 10^{-5} mbar and 50–550 °C, respectively [41].

2.1.5 Spray Pyrolysis

Spray pyrolysis is one of the low-cost techniques and has been employed by different groups for the preparation of $CuInSe_2$ films. Pamplin and Feigelson [42,43] successfully deposited sphalerite form of $CuInSe_2$ semiconductors and their solid solutions onto glass substrates at substrate temperature of 350 °C by spray using chemical solutions of 0.02 M cuprous, indium chlorides and N,N-dimethyl thiourea

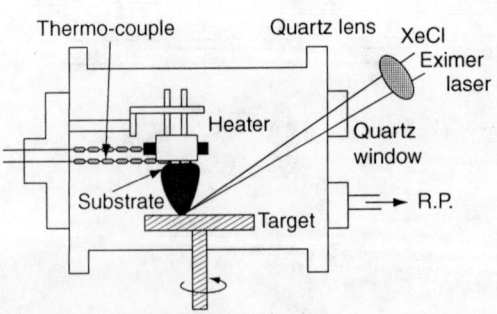

Figure 2.5 Schematic diagram of pulsed laser deposition system.

(NDTU) for the first time. Aqueous solution containing analar grade $CuCl_2·2H_2O$, $InCl_3$, N,N-dimethyl selenourea (NDSU) is used for spray deposition of $CuInSe_2$ thin films. The concentration of Cu, In, and Se in the solution is kept to be 1:1:3.75. The pH of the solution is adjusted to be 3.0 by adding a few drops of HCl. The solution is prepared just prior to the commencement of the spray. The films are deposited onto hot glass slides using compressed air as the carrier gas and the spray rate is kept to be 6.5 ml/min [44].

The chalcopyrite structured p-$CuInSe_2$ films along with some In_2O_3 impurity for the temperature range of 175–210 °C are grown by spray techniques [45]. The effect of NDSU concentration in the starting solution (keeping Cu:In ratio 1:1) on the properties of $CuInSe_2$ films is investigated [46]. Films coated onto glass substrates held at $T_S = 150$–200 °C with 2.2 parts of NDSU are p-type with a resistivity of 10–100 Ω-cm. The effect of pH of the starting solution on the production of chalcopyrite structure is investigated [47–49]. Both CuCl and $CuCl_2$ are used as source of copper. Films deposited onto glass substrates held at 300 °C exhibit chalcopyrite structure when solution pH 3.0 with 90% acid neutralized solution containing CuCl and pH 4.0 with $CuCl_2$ as a source of copper. Abernathy et al., [50] studied the kinetic effects in the film formation and observed that the solution with initial Cu:In ratio less than unity exhibited a second phase, $Cu_{2-x}Se$, contradictory to the thermodynamic predictions [51]. The thermodynamic calculations of the deposits under equilibrium conditions show that it is possible to spray deposit degenerate and chalcopyrite $CuInSe_2$ films [52]. Based on an equilibrium quaternary phase diagram of Cu–In–Se–O, Haba and Bates [53] concluded that an increase in oxygen incorporation, an increase in pH or a decrease in copper to indium ratio in the original spray solution results in the formation of copper selenide phases, instead of In_2O_3.

2.1.6 Electrodeposition

In order to deposit CIS layer either conducting or metal substrates are needed. The first attempt to electrodeposit $CuInSe_2$ films has been made by Bhattacharya [54]. Subsequently, several attempts are made to deposit $CuInSe_2$ films from aqueous solutions containing precursors of copper, indium, and selenium without [55–58] and with [59–62] complexion agents. A second approach is selenization of the electrodeposited Cu–In alloy film either by solution methods or by annealing under H_2Se [63,64]. The films prepared by this method are found to peel off resulting in the failure of the device [63,65]. In addition, H_2Se being highly toxic, creates serious safety concerns. Obtaining films with good morphology and stoichiometry are the major concerns. The effect of annealing on the structural properties of electrodeposited $CuInSe_2$ films are investigated [66]. The films are deposited onto FTO by electrodeposition using aqueous solutions of 5 mM $CuCl_2$, 5 mM $InCl_3$, 10 mM SeO_2, pH 1.5, and potentials at from −0.5 to −0.9 V [67].

2.1.7 Solvothermal Technique

First, Se powder is dissolved in anhydrous ethylenediamine and stirred for 2 h then CuCl and $InCl_3$ are added to it. The mixture is stirred for 2 h and heated in teflon lined stainless steel autoclave at 200 °C for 24 h. The chalcopyrite CIS nanoparticles from the solution are collected and cleaned with distilled water and ethylene [68].

2.2 Deposition of $CuGaSe_2$ Thin Films

The $CuGaSe_2$ (CGS) thin films are evaporated by vacuum evaporation using CGS ingot. In order to prepare CGS ingot, the Cu, Ga, and Se are sealed under vacuum at 10^{-5} torr in quartz tube, which is heated to 1100 °C and cooled down to 700 °C in slow pace whereby it is left for 3 days to have homogeneity then cooled down to RT at the rate of 20 °C/h [69]. Temperatures of Cu, Ga, and Se Knudsen effusion cells and substrate are set at 1070, 840–940, 180, and 480 °C in molecular beam epitaxy (MBE) system, respectively [70]. In some cases Se from tungsten boat is evaporated under pressure of 5×10^{-6} mbar. By employing bilayer process, first Cu-rich CGS layer is deposited onto SLG and in the second stage only Ga and Se are evaporated to obtain Ga-rich CGS sample at substrate temperature of 350 °C. After completion of deposition of CGS layer, the evaporation of Se has been continued until to cool down to substrate temperature of 250 °C and growth rate of film being 7 Å/s. The cells made with this absorber show efficiency of 7.9% [71], whereas the reported highest efficiency of CGS cells is 9.3% [72]. The Cu-poor CGS thin films are made by three-stage process that the Ga is added to the Cu-rich film in the presence of Se at the end of second stage [73].

2.3 Growth Process of $Cu(InGa)Se_2$ Thin Films

The $CuInGaSe_2$ (CIGS) thin films are prepared by various vacuum and nonvacuum techniques, which are illustrated below. In general the CIGS layers are grown onto the metalized glass substrates, hence the seed layer on the substrate determine growth of CIGS layer.

2.3.1 Growth of Mo Layer on Glass Substrates

The Mo layer is one of the influential candidates for the growth of CIGS layers, which acts as a bottom contact layer for thin film solar cell. On the other hand the Mo layer blocks migration of impurities from either glass or flexible substrates to absorber at some extent. The Mo layer is coated onto glass substrates by DC magnetron sputtering adopting bilayer process in which 0.1 μm thick layer is first layer at a chamber pressure of 10 m torr having resistivity of 60 μΩ-cm with good adhesion to the glass. The second layer with thickness of 0.9 μm grown at 1 m torr shows resistivity of 10 μΩ-cm, which is not good adherent to the glass. However, the final combined layer is in good adherent to the glass as well as having lower resistivity [74]. The Mo layers deposited at 2 m torr by Ar have small grain sizes and closed grain boundaries, whereas the layers grown at 8 m torr shows large grain sizes like elongate aligned and also open grain boundaries with lower density. The CIGS layers deposited on smaller grain size Mo layer coated at 2 m torr show smaller grain sizes of 0.5–1 µm, whereas layers grown on larger size Mo coated at 8 m torr show 3 µm larger grain sizes [75].

2.3.2 Physical Vapor Deposition of CIGS Thin Films

At the initial stage of development of CIS thin film solar cells, either stoichiometry or nonstoichiometry $CuInSe_2$ thin films grown by either vacuum evaporation or some other process have been applied to build thin film solar cells. The latest developments comprise graded composition layers such as top layer consist of Cu poor with respect to bulk or bottom thin film to have smooth surface. On the other hand, the surface layer contains lower carrier concentration with respect to bulk that make abrupt junction with window layer. This kind of bilayer phenomenon is inevitably borrowed from Si technology. The higher percentage of Ga lies at back of the film with respect to bulk film that create back surface field to avoid recombination of charge carriers at back. The physical vapor deposition or vacuum evaporation of CIGS thin films provide pure thin film without having scope of impurities as compared to that of nonvacuum process.

2.3.2.1 CIGS by Three-Stage Process

The bilayer process has been using to grow either CIGS or CIS absorber layers for high efficiency thin film solar cells. The layers are deposited onto Mo coated glass substrates with sizes of 10×10 cm^2 by three-stage process in line by vacuum coevaporation using four independent elemental sources whereby the substrates are moving with a speed of 1 cm/min over top of the elemental sources. The schematic diagram of in line vacuum evaporation process system to grow CIGS thin films is shown in Figure 2.6 [76]. The Knudsen effusion cells are typically employed at temperatures and fluxes of 1240 °C, 1×10^{-4} Pa-Cu, 845–870, 7×10^{-5}–1.3×10^{-4} Pa-In, 970, 5×10^{-5} Pa-Ga, and 210 °C, 8×10^{-3} Pa-Se for the deposition of CIGS thin films, respectively [77]. In brief, In–Ga–Se precursor layers are first deposited onto Mo coated glass substrates at substrate temperature of 350 °C, followed by evaporation of Cu and Se at substrate temperature of 500 °C. In the third stage, In, Ga, and Se are added to the precursors to form desired CIGS thin films [76]. Without using precursor layer, 2.3 µm thick Cu-rich CIGS layers are also first deposited onto Mo coated alumina substrates at little higher substrate temperature of 450 °C and 1.2 µm thick high Cu-poor CIGS layers are then deposited at substrate temperature of 550 °C. In the case of deposition of CIS layers, the lower substrate temperatures of 350 and 450 °C are employed for the deposition of first and second layers, respectively. The global composition of CIGS layers is Cu:In:Ga:Se=23.8:19.4:7.4:49.4 [78]. In MBE technique under chamber pressure of 5×10^{-8} torr the CIGS films on 1 µm thick

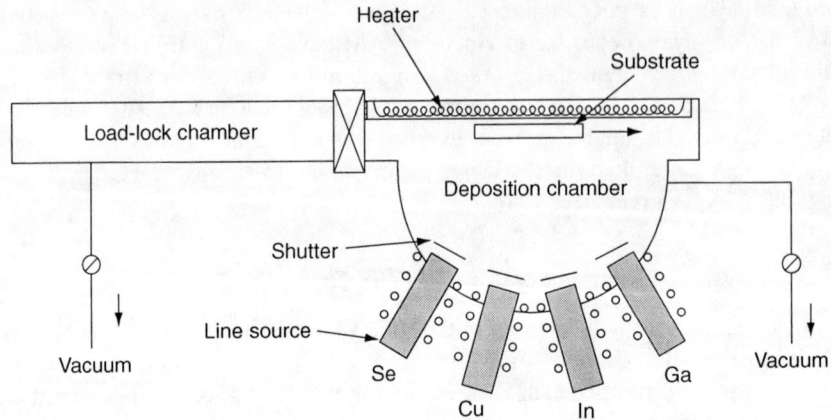

Figure 2.6 Schematic drawing of in line evaporation process system for 10×10 cm^2 area CIGS thin films.

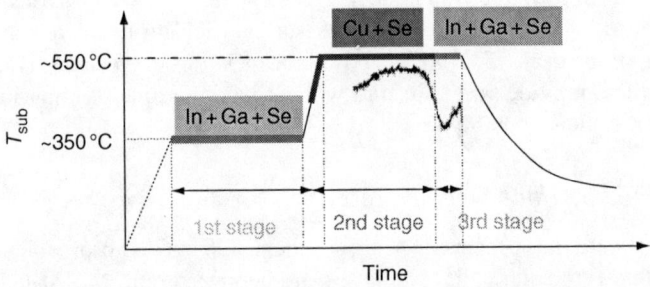

Figure 2.7 Temperature profile for the growth of CIGS layers with variation of time.

Mo covered 2 mm thick soda-lime glass substrates are prepared by three-stage process using Cu, In, Ga, and Se as source materials. In the first stage the In, Ga, and Se are normally evaporated at substrate temperature (T_S) of 350 °C that turns into (InGa)$_2$Se$_3$ precursor layer on which Cu and Se are evaporated at $T_S > 550$ °C, as a second stage. At the end of second stage slightly Cu-rich layers $1.2 < \text{Cu}/(\text{In}+\text{Ga}) < 1.10$ are formed. In the third stage, In, Ga, and Se are deposited to convert the films as InGa rich or Cu-poor films $0.80 < \text{Cu}/(\text{In}+\text{Ga}) < 0.85$. The grown layers are cooled down to 350 °C under Se flux as a final treatment and their typical compositions are Cu/III = 0.8 and Ga/(In + Ga) = 0.28, as shown in Figure 2.7. The beam equivalent pressure of Se is generally kept at constant of 1×10^5 torr. In the first stage, the Se/(In + Ga) flux ratio of 6.7–11.9 is maintained, which is critical point to decide the orientation of the layers and the Se/Cu ratio is maintained to be 60 in the second stage. The XRD analysis reveals that the (112) preferred orientation is observed in the final CIGS layers for Se/(In + Ga) ratio of < 7.1 kept in the first stage, whereas (220 or 204) is the preferred orientation for > 7.6. The In–Ga–Se layers prepared at 350 °C in the first stage is close to the composition ratio of $(In+Ga):Se \sim 2:3$, that is, $(InGa)_2Se_3$ phase. On the other hand the $(In_{1-x}Ga_x)_2Se_3$ phase is the defect wurtzite structure, which presents (006) reflection at $2\theta \sim 27°$ for the flux ratio of $Se/(In+Ga)=6.7$ that leads to (112) orientation in the final CIGS layer. In addition, the (300) peak at 46° appears for the flux ratio of $Se/(In+Ga)=9.4$. The intensity ratio of 1.35 for (300/006) supports formation of moderate a-orientated wurtzite crystallites those are enough to form (220) oriented layers indicating that (300) and (006) reflections are strongly related to (220) and (112) orientations, respectively. The orientation can be changed from (300) to (006) by increasing the substrate temperature but the grown layers are amorphous at below the substrate temperature of 200 °C. In the second stage, higher flux ratio of Se/Cu intends to give (220) as the preferred orientation. If the substrate temperature is lowered to below 450 °C, the surface of the layers becomes rough and had small grains. The lower Cu flux influences to have larger and smaller grains at the surface and back side of the CIGS layer, respectively, whereas higher Cu flux promotes larger grains throughout the CIGS layers. During the deposition of Cu and Se in the second stage, $(InGa)_2Se_3$ phase converts into CIGS layers, there is a formation of $Cu_{2-x}Se$ phase, which slightly reduces the intensity of (220) orientation. 2 μm thick CIGS layers had typical composition ratio of $Cu/(In+Ga)=0.82$ and $Ga/(In+Ga)=0.28$ showing smooth surfaces irrespective of (112) and (220) orientations. The recent experimental studies reveal that the cells made with (220) oriented CIGS absorbers exhibit higher efficiency, V_{oc} and fill factors (FFs) comparing with the (112) oriented cells [79,80].

Unlike previous three-stage processes, the present process slightly deviates from growth temperatures that a fabrication of 2.7 μm thick $Cu(InGa)Se_2$ thin films on Mo coated glass substrates by coevaporation of pure constituent elements of Cu, In, Ga, and Se, involving three-stage process lasts for about 48 min. As shown in Figure 2.8, first stage starts with deposition of $In+Ga+Se$ at 260 °C for about 20 min. In this stage, 80 and 90% of In and Ga, respectively are incorporated and the rest of the percentages will mix into the layers at the end of the second stage process. In the presence of Se flux, the temperature is ramped to 560 °C within 5 min duration. The $(In_{1-x}Ga_x)_2Se_3$ compound is formed confirming multiple reflections such as (101), (102), (110), (113), (201), etc., by XRD analysis. At 560 °C, the Cu and Se are evaporated to make nearly Cu-rich layers for 20 min. The composition in the layers is $0.97 \leq Cu/(In+Ga) \geq 1.08$. Presumably, the second stage process is enough to grow $Cu(InGa)Se_2$ thin films but it gives rough surface that may create poor performance

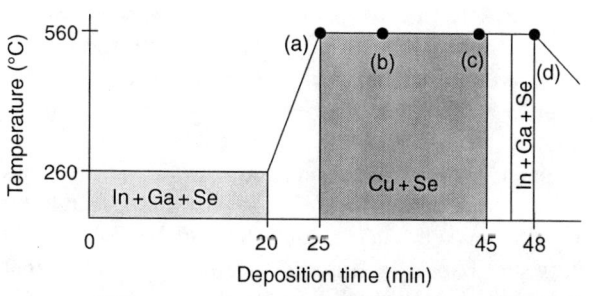

Figure 2.8 The variation of temperature with time for the growth of CIGS layers.

of the junction therefore third stage process is adopted to have smooth surface. After 1-1/2 min break, in the third stage, In, Ga, and Se are evaporated for 1-1/2 min to have Cu-poor layers. Finally, the temperature is ramped down to 350 °C under Se flux. The grain sizes of the films are smaller and larger at the end of the second and third stages, respectively. The composition of Cu(InGa)Se$_2$ film is Cu/(In+Ga) = 0.92 and the AES depth profile shows that the Ga concentration is high at the back surface of the films, which is gradually graded to lower concentration from back to front. The grading is difficult, if the Cu is rich. If the composition of the layers falls in between Cu$_2$Se and (InGa)$_2$Se$_3$ tie line that results in stable composition. According to tie line logic, which will be discussed in Chapter 4, either one In or Ga atom diffuses out for incorporation of every three Cu atoms, which reacts with other atoms to enhance thickness of the layers. During second stage, Cu$_x$Se phase forms for Cu-rich growth in which Cu(InGa)Se may be soluble to enhance grain sizes in the third stage that results in Cu(InGa)Se$_2$ thin film. In typical case, as Ga concentration, Ga/(In+Ga) is decreased from 0.4 to 0.22 in the third stage process, the efficiency of typical Cu(InGa)Se$_2$ solar cells gradually decreases from 15.1 to 14.2%. On the other hand the open circuit voltage decreases from 638 to 630 mV and FF also decreases from 0.767 to 0.745 mV but the short circuit current remains at more or less the same level of 30–31 mA/cm^2 [74,81,82]. The higher growth temperatures of 400 and 570 °C are employed for first and second + third stages, respectively [75].

The presence of Cu$_{2-x}$Se phase can be observed by means of variation of temperature induces in the CIGS layers due to its emissivity during the growth of CIGS by three-stage process. In fact the second stage of CIGS thin film growth leads to deposition of Cu and Se. After few minutes of growth the grown three samples are removed from the chamber with little time differences, the samples show Cu/(In+Ga) composition ratios of 0.83, 0.89, and 0.98. It is well known fact that the composition of Cu increases with time and at certain point the film composition is eventually stoichiometric, that is, Cu:(In+Ga):Se = 1:1:2. The film composition of Cu/(In+Ga) increases to further as 1.08 and 1.6 but the thing is that the temperature of film decreases from 550 to 540 °C during growth of CIGS that is unusual behavior. The In, Ga, and Se are deposited in the third stage process to nullify the excess Cu where the temperature is 550 °C and the film composition is 0.88. The substrate temperature is 550 °C for Cu/(In+Ga) < 1 and stoichiometric ratios in the film, whereas the substrate temperature decreases for Cu/(In+Ga) > 1. It can be concluded that in general the Cu-rich films had a Cu$_{2-x}$Se phase that causes to high emissivity, hence the film or substrate temperature decreases, whereas the Cu$_{2-x}$Se phase probably does not exist at low Cu composition therefore the film does not show emissivity under constant heating power. The films with Cu$_{2-x}$Se, which has semimetallic nature, had higher carrier concentration of 10^{19}–10^{20} cm^{-3} [83].

The incorporation of Na either by self diffusion from the soda-lime glass substrates or intentional doping reduces density of defect levels, grain boundaries and enhances grain sizes. The first stage consists of deposition of Ga, Ga$_x$Se$_y$, and CGS layers on Mo coated glass substrates at substrate temperature of 350 °C that encourages improving the adhesion between CIGS and Mo as well as makes Ga grading in the CIGS layer that results in band gap grading. As mentioned earlier, second stage leads to deposition of Cu-rich CIGS layers at substrate temperature of 550 °C in which $Cu_{2-x}Se$ liquid phase and CIGS form. The formation of $Cu_{2-x}Se$ enhances growth process. The Na_2Se is simultaneously evaporated to have 6 at.% Na in the total thickness of CIGS layer. An early stage of entry of Na into CIGS causes poor adhesion to the substrate. In order to mitigate poor adhesion of CIGS to the Mo layer, the Na_2Se is introduced in the second stage. In the third stage, slightly Cu-poor CIGS layers are made by evaporating In, Ga, and Se only on the previous Cu-rich CIGS layers at the same temperature of 550 °C. The Se flux has usually been continued until to cool down to substrate temperature of 350 °C. The Na concentration of about 7 at.% is detected up to 100 Å depth from the surface in 2.5 μm thick Cu(InGa)Se$_2$ thin films by XPS. The Ga grading is also observed in the CIGS layers and its concentration is high at backside of the layer. According to phase diagram, the $CuIn_3Se_5$ phase prevails for Cu/In ratio between 0.28 and 0.51 but the $CuIn_3Se_5$ and $CuInSe_2$ phases coexist for ratio of 0.5–0.75. The Cu(InGa)$_3$Se$_5$ phase is formed for the Cu/(In+Ga) ratio of 0.51, which continues to 0.71, evidencing by existence of (110), (202), (114) and (310) reflections. The carrier concentration of CIGS on SiO_x decreases with decreasing Cu/(In+Ga) ratio for constant ratio of Ga/(In+Ga) = 0.45, which is 1×10^{16} cm^{-3} for the ratio of 0.8. The carrier concentration increases to 10^{16}–10^{17} cm^{-3} for Cu/(In+Ga) range of 0.4–0.8, if Na_2Se is incorporated. An increased hole concentration is observed in the air exposure layers for several days due to chemisorptions of oxygen by Na that compensate Se vacancies in the layers. The band gap grading in the solar cells improves short circuit current because the back surface field reflects the carriers diffusing from the Mo. The quantum efficiency increases at long wavelength range. In the second stage, the incorporation of Na creates Cu vacancy that will act as acceptor. If Na occupies In site, it will create double acceptors [84].

Both the absorber and buffer layers are grown in continuous process in the same chamber without breaking process. First the stoichiometric CGS layers are grown at about 450 °C then the three-stage process continues to grow CIGS layers, unlike other process. Near the surface of CIGS layer, the Ga content of 30 at.% is maintained. On top of the CIGS layers, either In_xSe_y or $ZnIn_xSe_y$ buffer layers are coevaporated at substrate temperature of 300 °C [85]. The temperature profile with deposition time for the layers is shown in Figure 2.9.

Unlike conventional three source evaporation, the CIGS layers are deposited on 1 μm thick Mo coated glass substrates by electron beam evaporation using In_2Se_3, Cu, Ga, and Mo precursor pockets. Cu/(In+Ga) = 0.85, Cu/In = 1.16, and Ga/(In+Ga) = 0.26 are maintained in the precursor layer, followed by annealing under Se atmosphere at 500 °C for ½ h on which, second stage is started that 500 nm In_2Se_3, 80 nm Cu, and 50 nm Ga are deposited to make Cu gradient as Cu/In = 0.93 in the layers and the stacked layer is annealed at 400 °C for 1 h then second selenization is done at 500 °C. In third stage process, 30 nm Cu and 50 nm Ga are coated to maintain Cu/In = 1, followed by selenization [86]. The binary compounds instead of individual elements are used for three-stage process in order to develop CIGS layer. The In_2Se_3, Ga_2Se_3, and Se elements are evaporated

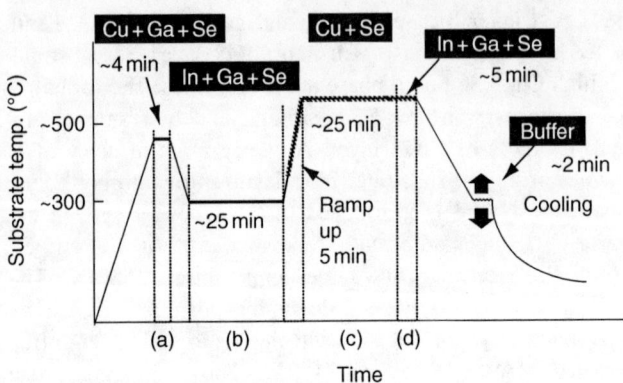

Figure 2.9 Temperature profile with function of time for the growth of CIGS layers.

at substrate temperatures of 150 and 325 °C to form $(InGa)_2Se_3$ layer. In the second stage Cu_2Se is only evaporated at 500 °C, followed by annealing under Se atmosphere for 10 min. In the third stage, a small quantity of In_2Se_3, Ga_2Se_3, and Se elements are evaporated in order to suppress $Cu_{2-x}Se$ phase [87]. In another version, the stoichiometric mixture of Cu, In and Se, and Ga_2Se_3 are evaporated under Se vapor at substrate temperature of 250 °C. The CIGS precursor layer is annealed under H_2Se/Ar at 250, 350, and 500 °C. The precursor annealed at 250 °C show multiple phases of InSe, $Cu_{2-x}Se$, Ga_2Se_3, CIS, Cu_9Ga_4, and GaSe, whereas the same precursor annealed at 350 °C shows CIS as majority phase along with GaSe, $CuGa_2$, Cu_9Ga_4, and GaSe phases indicating that Cu based phases are consumed, whereas single phase chalcopyrite structured CIGS layers with composition of Cu:In:Ga:Se=22.5:22.5:3:52 form after annealing precursor at 500 °C [88]. By DC sputtering the Cu–Ga–In metal precursors are grown onto Mo coated glass or glass substrates at RT or substrate temperature of 150 °C using Cu–Ga(23 at.%) alloy and In targets. 0.8 µm thick metal precursor layer is essential to obtain 2.3 µm thick CIGS layer. The Cu–In–Ga alloy precursor layers, which kept in graphite crucible is inserted in vacuum tight quartz tube and selenized at substrate temperature of 550–650 °C by heating Se pellets. The sequential multilayers on glass such as glass/In/Cu–Ga/In/Cu–Ga/In and glass/In/Cu–Ga selenized under Se vapor result in single phase chalcopyrite structure, whereas selenization of glass/Cu–Ga/In bilayer structure consists of secondary phase of In_2Se_3 along with CIGS. (a)glass/In/Cu–Ga/In/Cu–Ga/In (Cu:In:Ga:Se=27.8:2.84:25.03:44.44), (b)In/Cu–Ga/glass (26.07:6.96:21.29:45.68), and (c)Cu–Ga/In/glass (26.59:8.52:19.56:45.33) methods are also tried [89].

2.3.2.2 CIGS Thin Films by Two Stage Process

First metal layers or stack layers formed, followed by either selenization or sulfurization process is so called two stage process. The Cu, In, and Se or with Ga are evaporated onto Mo coated glass substrates at substrate temperature of 150–200 °C for 30 min from Knudson cells under vacuum. The temperature is ramped up to 400–500 °C within 5 min duration whereby the precursors are selenized for 60 min, as shown in Figure 2.10. The precursor preparation temperature is very critical otherwise the layers easily peel off from the Mo, unless they are prepared above the temperature of 150 °C in which Se flux plays an important role. The formation of Cu_xSe phase and In loss have to be minimized in the beginning because there is no additional In/Ga supply is done unlike three-stage process. The phase separation is found for the Se content of 15 at.%, the fine grains participate for 20 at.% and uniform growth occurs for 30 to 40 at.%. The precursor selenized at 350 °C for 60 min shows CIS along with Cu–In binary alloy as a secondary phase, whereas the CIS films with good crystalline and chalcopyrite structure form for above selenization temperature of 400 °C. At constant substrate temperature of 450 °C for 60 min, the Se beam flux of less than 1×10^{-5} torr leads to form Cu-rich CIS by losing In. On the other hand Se/(In+Cu) is less than unity and Cu–In islands also form. Therefore the Se flux more than 3×10^{-5} torr is required. The CIGS precursors prepared at 150 °C, with Se less than 50% and composition of Cu:In:Ga:Se=19.1:15.2:16.2:49.5 prevail CuSe, InSe, and GaSe phases. With Se greater than 50% and composition of 19:13.1:15.1:52.8, the precursors had extra phases of In_6Se_7, Se, and Ga_2Se_3. In the case of Cu–In–Se precursors prepared at 150 °C for 30 at.% Se, it demonstrates $Cu_{11}In_9$, InSe, and In phases and the same precursors at 500 °C form as

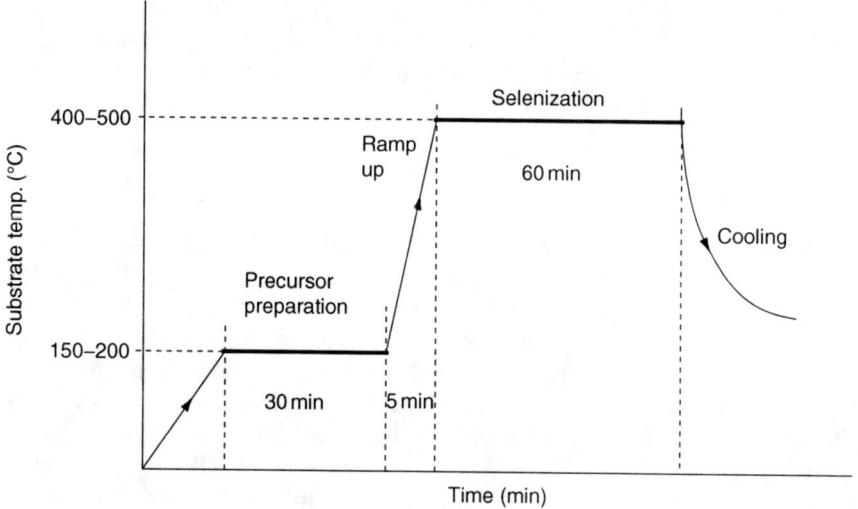

Figure 2.10 Variation of growth temperature with time for the deposition of CIS or CIGS layers by two stage process.

$$Cu_{11}In_9 + 2InSe + 20Se(g) \rightarrow 11CIS$$

other reactions are

$$2Cu_{11}In_9 + 29Se \rightarrow 11Cu_2Se + 18InSe, Cu_2Se + Se$$
$$\rightarrow 2CuSe, \text{ and } CuSe + InSe \rightarrow CIS.$$

Similar reaction path follows for CIGS

$$CuSe + InSe + Ga_2Se_3 + Se \rightarrow CIGS,$$

whereas in the case of low temperature of 450 °C, the In loss occurs by means of reevaporation of In_2Se as $2In + Se \rightarrow In_2Se$ for Se deficient and $2InSe + Se \rightarrow In_2Se + 2Se$; $2In + Se \rightarrow In_2Se$ for Se excess [90]. It is predicted that the thin film formation from multisource evaporation follows the reaction path as [91]

$$2Cu + Se \rightarrow Cu_2Se; \ 2In + Se \rightarrow In_2Se; \ In_2Se + Se$$
$$\rightarrow 2InSe; \ 2InSe + Cu_2Se + Se \rightarrow 2CuInSe_2$$

At substrate temperature of 200 °C, the Cu, In, and Se are sequentially evaporated onto Mo coated SLG to form stack as Mo/(Cu–Se)/(In–Se). In the case of Ga incorporation, the formed stack lies Mo/(Cu–Se)/(In–Se)/(Ga–Se) or Mo/(Cu–Se)/(In–Ga–Se) with Ga grading. In the stacked layers, Se to metal composition ratio of 0.4 is optimal. If the ratio is higher or lower than 0.4, the layers easily peel off from the substrate and the Ga diffusion takes place after the selenization, respectively. The selenization is done under Se vapor at 550 °C for 30 min [92]. The (In,Ga)Se precursor layers are deposited onto Mo coated SLG at 25–350 °C, which are exposed to Cu and Se fluxes at the substrate temperature of over 200 °C. Finally the In–Ga–Se/Cu–Se precursor is postannealed under Se flux at 500 °C to made CIGS layer [93].

2.3.3 Cu(InGa)Se$_2$ Thin Films by H$_2$Se Selenization Process

$CuInSe_2 + Ga_2Se_3$ grounded to make small crystallites filled in cylindrical shape graphite crucible are evaporated by applying source temperature of 1450 °C to form CIGS thin films on Mo coated glass substrates at substrate temperature of 250 °C. On the other hand, Se vapor is generated from Se containing stainless steel effusion cells. The grown precursor layers are treated with H$_2$Se under Ar at atmospheric pressure using heat treatment by halogen lamps at different temperatures of 250, 350, and 500 °C. The precursors treated at 250 or 350 °C accomplish $CuInSe_2$, $CuGa_2$, GaSe, and Cu_9Ga_4 phases, whereas single phase CIGS forms by treating at 500 °C. In order to increase Ga composition from 2 to 5–6 at.%, the substrate temperature has to be increased to over 500 °C or heating time has to be prolonged. The composition ratio of Cu/(In + Ga) varies from top to bottom of the layers as

0.95–0.5 [94]. The $Cu_9(In_{0.54}Ga_{0.36})_4$ intermetallic compound forms on SLG after annealing Cu–In–Ga sputtered layers under $Ar/Ar+O_2$ at 450 °C. 3490 Å thick $Cu_{0.8}Ga_{0.2}$, followed by 4890 Å thick In from the respective targets are deposited by sputtering to have $Cu/(In+Ga)=0.9$ and $Ga/(In+Ga)=0.2$. The sputtered precursor had combinations of Cu, Cu_3Ga, CuIn, and In. During the selenization and sulfurization of the precursors using H_2Se and H_2S gases, Cu_9Ga and $Cu_{15}In_9$ phases are found. The precursor selenized at 450 °C for 15 min, followed by sulfurization at 550 °C for 15 min result in uniform Ga composition in the layers, whereas the Ga grading is observed in the layers for 30 min selenization that means higher Ga accumulation is observed in the backside of the layers. The InSe and $Cu_9(In_{0.2}Ga_{0.8})_4$ form for 10 min selenization indicating that fast selenization of In takes place as compared to Ga. 10 or 30 min sulfurization gives Cu_xIn_y and $CuInGaS_2$. 15 min selenization and 15 min sulfurization contribute 2–5 at.% of sulfur in the layers, whereas 20 min selenization provides <1 at.% with CIGS single phase. H_2Se 15 min/H_2S 30 min process shows higher sulfur concentration on the surface of CIGSS [95]. The similar Cu–In and Cu–In–Ga intermetallic layers evaporated by e-beam evaporation are processed at 450 °C under 5% $H_2Se+5\%N_2$ to obtain $CuInSe_2$ and CIGS. The Cu/In ratio varies from 1.1 to 0.8 by varying the thickness of the layers. The sulfur diffuses into the absorbers by processing layers under $5\%H_2S+90\%N_2$ at 575 °C for 20 min. The Cu-rich films (Cu/In = 1.2) contain Cu_2Se in which the sulfur can be included because the CuSe is liquid at high temperature of 575 °C [96].

2.3.4 Cu(InGa)(SeS)$_2$ Layers by Selenization and Sulfurization or Hybrid Process

In-line pilot process, the alkaline barrier SiN, Mo layer and Na compounds are successively deposited onto 3 mm thick float glass. The multiple (Cu85%+Ga15%) and In layers by DC magnetron sputtering technique and Se layers by thermal evaporation are subsequently deposited onto glass/SiN/Mo/Na compounds. The stack Cu–Ga/In/Se layers are selenized and sulfurised using RTP reactor with ramp rate of 10 °C/s. The high Se vapor and partial S vapor are introduced by heating solid Se and S into the RTP chamber at ∼550 °C to obtain require Se to S ratio in the $CuInGaSeS_2$ (CIGSS) layers. During the selenization and sulfurization process the samples are heated at both sides of front and back by halogen lamps in order to avoid bending of the substrates. An average $Se/(Se+S)$ ratio of $6.1\pm0.6\%$ is determined in the CIGSS by XRFA over the large area of 60×90 cm^2. In situ X-ray diffraction studies on selenization of CuInGa layers are studied. The $Cu_{11}(InGa)_9$ is the starting compound for selenization to form $Cu(InGa)Se_2$ when Cu, In, and Ga are sputtered by DC magnetron sputtering technique. While selenizing, a part of the $Cu_{11}(InGa)_9$ compound dissociates into In_4Se_5, $Cu_{2-x}Se$ and, CuSe. The rest of $Cu_{11}(InGa)_9$ compound turns into $Cu_{16}(InGa)_9$ and at the same time the In_4Se_3 and CuSe decompose into InSe and $Cu_{2-x}Se$. Surprisingly, Ga based selenized compounds are not observed. The Ga and Se concentrations are high towards the back side of the layers. Finally, Ga-rich $Cu(InGa)Se_2$ phase is formed underneath of the Ga-poor Cu(InGa)Se$_2$ or CuInSe$_2$ [97,98]. Prior to prepare CIGSS layers by two stage process method, 80 Å thick NaF layer is deposited onto Mo coated soda-lime glass substrates by thermal evaporation on which 0.7 μm thick Cu–Ga and In metallic precursor layers are deposited by using Cu$_{0.78}$Ga$_{0.22}$ and In metallic targets at chamber pressures of 1.5×10^{-3} and 7×10^{-4} torr and DC magnetron sputtering power of 350 and 230 W, respectively. The grown metallic precursor layer had composition ratio of Cu:In:Ga = 46.6:40.96:12.44. The Se layer with twice composition of metals is evaporated by thermal process onto Cu–In–Ga metallic precursor. The grown CuInGaSe precursor layer is sulfurized by using RTP at process temperature of 550 °C under H$_2$S atmosphere with ramp rate of 4 °C/s to obtain final 2 μm thick CuInGa(SeS)$_2$. The grown CIGSS layer contains composition ratio of Cu:In:Ga:Se:S = 23.11:26.32:44.8:5.77, Cu/In = 0.88 and S/(S + Se) = 0.11 [99]. Alike 0.6 μm thick Cu–In$_{0.7}$Ga$_{0.3}$ alloy to have Cu/(In + Ga) = 0.9 and Ga/(Ga + In) = 0.25 is deposited onto 1500 cm^2 area glass substrates by DC magnetron sputtering in a vertical in-line AV 400 V7 sputter unit using Cu$_{0.75}$Ga$_{0.25}$ and In targets at 2.5×10^{-3} mbar and RT. The CuIn$_{0.7}$Ga$_{0.3}$ alloy is annealed in a horizantol tube at 450 °C for 15–45 min while flowing H$_2$Se:Ar gas to convert into binary compounds and continued annealing process under H$_2$S at 500 °C < 10 min. The Cu–(In,Ga) precursor is annealed under 1 mol%H$_2$Se + Ar at 450 °C for 20 min and annealed under H$_2$S/Ar at 500 °C for 10 min result in CuIn(Se$_{0.9}$S$_{0.1}$)$_2$ and CuGa(Se$_{0.8}$S$_{0.2}$)$_2$ quaternary systems. The quaternary systems are converted into Cu(In$_{0.75}$Ga$_{0.25}$)(Se$_{0.85}$S$_{0.15}$)$_2$ multinary system by further annealing under H$_2$S/Ar at 550 °C for 10 min. Similar way Cu(In$_{0.75}$Ga$_{0.25}$)(Se$_{0.6}$S$_{0.4}$)$_2$ can be prepared that the Cu–(InGa) precursor annealed under 1% H$_2$Se/Ar at 450 °C for 10 min to form binary compounds that converts into CuIn(Se$_{0.4}$S$_{0.6}$)$_2$ and CuGa(Se$_{0.8}$S$_{0.2}$)$_2$ quaternary phases by annealing under H$_2$S/Ar at 500 °C for 10 min. The quarternary phases convert into Cu(In$_{0.75}$Ga$_{0.25}$)(Se$_{0.6}$S$_{0.4}$)$_2$ after annealing them under H$_2$S/Ar at 550 °C for 10 min. The volume fraction of sulfoselenides dictates x and y positions in the quaternary alloys CuIn(Se$_{1-x}$S$_x$)$_2$ and CuGa(Se$_{1-y}$S$_y$)$_2$ [100]. Gossla and Shafarmann [101] prepared double and single layers that Cu, In, and Ga are evaporated using boron nitrade crucibles at source temperatures of 1100–1400 °C using PVD system, as shown in Figure 2.11. The S and Se are evaporated at source temperatures of 100 and 350 °C, respectively. Two different kinds of processed CIGS samples denoted as first and second folds are grown. In first fold, Cu–Ga–Se layers are deposited at 350 °C with combinations of Ga/(In + Ga) = 0.29 and Se/(In + Ga) = 3 then on which Cu–S film is deposited at 550 °C with S/Cu = 18. In second fold, In–Ga–S films are deposited at 250 °C with combinations of Ga/(In + Ga) = 0.28 and S/(In + Ga) = 19 then Cu–Se is deposited at 550 °C with Se/Cu = 2. The deposition times for group III layers and for Cu chalcogen layers are 60 and 90 min, respectively. The RBS analysis reveals three distinct phase layers of 470 nm thick Se free Cu$_{0.3}$In$_{0.23}$S$_{0.47}$ surface layer, 230 nm thick Cu$_{0.27}$In$_{0.08}$Ga$_{0.24}$Se$_{0.04}$S$_{0.37}$ interface layer and 2170 nm thick Cu$_{0.29}$In$_{0.16}$Ga$_{0.10}$Se$_{0.4}$S$_{0.05}$ bottom layer for first fold samples. However, the XRD analysis reveals that first phase layer contains Ga/(In + Ga) = 0.35 and S/(Se + S) = 0 for bulk thin film and second phase had Ga/(In + Ga) = 0.12 and S/(Se + S) = 1 but interface layer is not found. In the case of second

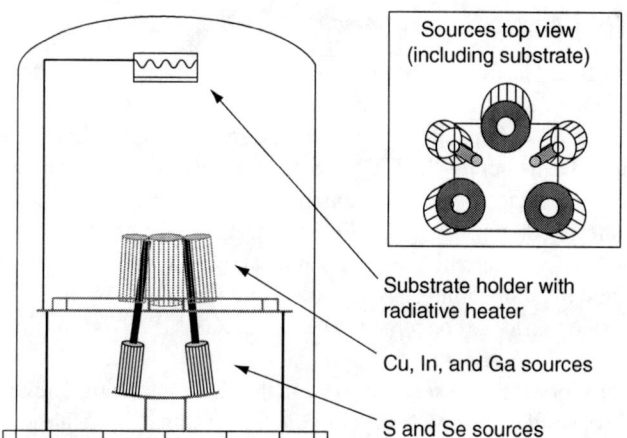

Figure 2.11 Schematic diagram of physical evaporation deposition system with five sources for CIGSS thin films.

fold process two phase layers (bilayer) such as the bottom one with 1660 nm thick $Cu_{0.22}In_{0.16}Ga_{0.09}S_{0.55}$ layer contains $Ga/(In+Ga)=0.34$, $S/(Se+S)=1$, $a=5.47$ and $c=10.91$ Å. 670 nm thick $Cu_{0.21}In_{0.19}Ga_{0.06}S_{0.45}Se_{0.09}$ top layer with $Ga/(In+Ga)=0.2$ and $S/(Se+S)=0.7$ contains $a=5.57$ and $c=11.17$ Å. The cells made with first fold processed CIGSS layer shows inferior short circuit current than second fold processed layer due to attaining multiple phase layers. By modifying the deposition conditions that means keeping all the deposition rates of Cu, In, Ga, S, and Se at constant and Cu poor with $Cu/(In+Ga) \sim 0.8-1.0$, CIGSS single layer is formed at 550 °C. In single step process, high vapor pressure of sulfur is needed. The cells made with single layer exhibits efficiency of 9.4% for $Ga/In+Ga=0.73$ without KCN treatment. The open circuit voltage of cell increases from 808 to 854 mV but efficiency decreases to 7.4% for increasing $S/(S+Se)=0.26$ indicating that the efficiency of cell decreases with increasing sulfur content [95,102].

2.3.5 Nonvacuum Process

The nonvacuum process is one of the advantageous methods to reduce fabrication cost of solar cell as compared to that of cells made with vacuum processed CIGS absorber layers. The Cu, In, Ga, and Se materials can be fully used for the growth of CIGS absorber layer, whereas 20–50% materials get waste in the vacuum process. The controlling of composition of absorber layer is easier in nonvacuum than vacuum process. However, so far, the maximum efficiencies of 13.95 [103], 13.6 [104], and 11.7% [105] are reported for the cells made with CIGS absorber by nonvacuum process, which are lower than the world record efficiency of 20.3% by vacuum processed cells [106].

2.3.5.1 CIGSS Absorber by Hydrazine

The Cu_2S, In_2Se_3, Ga, S, and Se are individually dissolved in hydrazine at room temperature, which are stirred and kept for several days. Appropriate quantities of solutions are mixed to deposit CIGSS layer by spin coating at 290 °C for 5 min on glass/Mo. In order to obtain thick CIGSS layers several depositions are cycled for example about 12 cycles have to be done to obtain ~ 1.3 μm thick layer, for which the growth rate is 100 nm per cycle. Finally the CIGSS precursor layer is annealed at 490 °C for 30 min. The coating experiment dealt with hydrazine is very highly toxic, the proper fume hood or scrubber is necessary and precautionary measurements have to be taken as far as safety concerned. The $CuIn_{0.7}Ga_{0.3}(SeS)_2$ thin films prepared at 450 °C shows chalcopyrite structure ($a = 5.699$ and $c = 11.421$ Å) along with (100) and (110) peaks for $MoSe_2$ hexagonal phase in the XRD spectrum. The SEM of CIGSS thin film shows similar columnar structure of conventional CIGS absorber [107].

2.3.5.2 CIGS Thin Films from Nanoparticles by Ink Based Spray or Spin Coating

The Cu, In, and Ga metal pieces are dissolved in acid digestion followed by adding NaOH to adequately control pH, *etc*. The fine mixed metal oxide powder is extracted, after cleaning the hydroxide precipitate. The metal oxide is dispersed in water by milling technique in order to make ink for printing in which the organic solvents may be added to make rheology by properly balancing viscosity of metal oxide ink. The oxide precursor ink is deposited onto metalized glass substrates or flexible substrates presumably by printing technique. The technique which used is not as simple as assume or may be complex. The Cu–In–Ga alloy is formed by heating Cu–In–Ga oxides under N_2 and H_2 gases at substrate temperature of 500–550 °C. The alloy is selenized by using H_2Se and N_2 gases at substrate temperature of 420–450 °C to attain CIGS thin film [108,109].

The NREL group processed CIGS absorber by different nonvacuum techniques that the Cu–In–Ga–Se nanoparticle colloids are prepared by mixing CuI, InI_3, and GaI_3 in pyridine and Na_2Se in methanol at low temperature under inert atmosphere. They are then mixed together. The reaction takes place as

$$1.1CuI + 0.68InI_3 + 0.23GaI_3 + 1.91Na_2Se \rightarrow Cu-In-Ga-Se$$
$$\rightarrow Cu_{1.1}In_{0.68}Ga_{0.23}Se_{1.91}.$$

The NaI by product is separated from Cu–In–Ga–Se, methanol/pyridine colloids, which is dissolvable in methanol. The methanol/pyridine is evaporated from nanoparticles by streaming N_2. The driven CIGS nanoparticles with support of organic solvent are sprayed onto heated metalized glass substrates at 144 °C to form Cu–In–Ga–Se thin film, which is annealed under Se flux of 15 Å/s at 550 °C for 10 min. In addition, Ga (flux ~ 1 Å/s) and In(flux ~ 3 Å/s) are evaporated at 550 °C to make Cu deficiency or stoichiometric CIGS absorber. The cell made with modified CIGS absorber layer delivers efficiency of 4.6% [110]. The nanoparticles

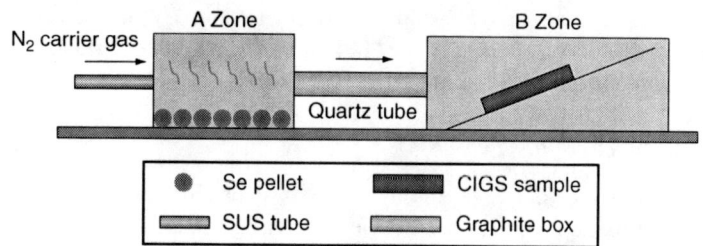

Figure 2.12 Schematic diagram of selenization process.

extracted by the similar kind of technique are mixed with organic solutions in order to obtain rheology of mixture. The nanoparticles mixed with methanol are sprayed onto heated metalized glass substrates at 160 °C under nitrogen ambient. The deposited mixture by doctor blade method or spray is heated at 70 °C for 5 min on hot plate under air to evaporate solvents and continued to heat at 330 °C for 5 min. The resultant porous CIGS layer on glass/Mo is selenized in two-zone quartz tube furnace by rapid-thermal process. As shown in Figure 2.12, similar to Figure 2.4A, the Se vapor is created at zone A at temperature of 400 °C and transformed to zone B by N_2 carrier gas with flow rate of 20 sccm whereby the sample is kept at 550 °C for 30 min that converts into $Cu_{0.9}In_{0.64}Ga_{0.23}Se_2$ thin film. The higher flow rates and higher temperatures of Se increase grain sizes of CIGS layers. The cells made with $Cu_{0.9}In_{0.64}Ga_{0.23}Se_2$ absorber shows efficiency of $<1\%$. The poor efficiency is due to small grain sizes in the range of few 100 nm [111–113]. Similarly, the $CuCl_2 \cdot 2H_2O$ and $In(NO_3)_3 \cdot 4H_2O$ are dissolved in deionized water. The citric acid (CT) with metal to CT ratio of 0.5 is added to the chemical solution while heating and stirring it. To obtain gel the solution is heated at 90 °C and further heated to 400 °C to made combustion. After cleaning fine powder of metal oxide is obtained. The ethyl cellulose as thickening agent and polyethylene glycol as dispersant are added to the metal oxide powder to form slurry with reasonable viscosity. The metal oxide thin film is coated by spin coating using precursor slurry, which is heated to remove and burn organic solvents. The samples are selenized in two-zone furnace at 500 °C for 1 h under Se vapor, which is developed at 330 °C using N_2 carrier gas. The Cu_2Sc and In_2O_3 secondary phases present in the $CuInSe_2$ film [114]. By choosing a different chemical solvent approach, the Cu nitrate hemipentahydrate, In chloride and Ga nitrate, hydrate are dissolved in methanol. The ethylcellulose is separately dissolved in 1-pentanol with ratio of 0.1. They are mixed with ratio of 0.5 to make paste to coat on metalized glass substrates by doctor blade technique, which is equally dispersed on the substrate. The sample is heated to 250 °C by putting on hot plate, followed by selenization at 560 °C under Se vapor for 10 min. In the layer, higher concentration of carbon or thick carbon layer is observed on the rear side of the sample and Ga grading is also found while Se concentration is higher on the surface. The Cl impurity determined in the sample could be due to undissolved CuCl. The thin film solar cell made with CIGS absorber shows efficiency of 6.7% [115]. Simply the CIGS nanoparticles are dispersed in tetrachloroethylene and sprayed onto heated glass substrates to form CIGS thin film. The cell made with this absorber exhibits efficiency of 1% [116]. The precautionary measurements have to be taken while handling Se, S and organic solvents.

2.3.5.3 Chemical Spray Pyrolysis

The spray pyrolysis technique is employed to deposit absorber, window, and transparent conducting oxide layers. A brief review covering the principle, deposition variables, advantages and limitations of the technique is presented here. The spray pyrolysis was used as early as 1910 to obtain transparent oxide films as reported by Foex [117]. In 1964, Chamberlin and Hill [118–121] extended this technique to grow sulfide and selenide films. Bube *et al.* [122–127], Savelli *et al.* [128–130], and Chopra *et al.*, [131,132] continued to work on spray pyrolysis experiments to develop semiconducting thin films. The excellent reviews have been published by several authors on spray deposited films [133–135]. The deposition of wide band gap oxide films for transparent conductor applications is also reported by Haacke [136].

In principle, spray pyrolysis is a simple technique where a solution, usually aqueous, containing soluble salts of the constituent elements of the desired compound is sprayed onto hot substrates. The sprayed droplets, on reaching the hot substrate surface undergo pyrolytic (endothermic) decomposition and form the desired film. The other volatile by-products and excess solvent escape in vapor phase. The substrate provides the necessary thermal energy for decomposition and subsequent recombination of the constituent species giving rise to a coherent film. It should be noted that various intermediate reactions and products may be formed and reaction kinetics is rather complex. A variety of soluble salts of the desired cations are available as their chlorides, acetates, nitrates, formats, and propionates. Likewise, soluble anion compounds such as thiourea, selenourea, NDTU, N,N-dithiocyanate are available. The atomization of the chemical solution into the fine droplets is effected by a spray nozzle with help of filtered carrier gas which may or may not be involved in the pyrolytic reaction.

The technique has the following advantages:

i) It is a simple and inexpensive technique. The process can be extended to large area coatings. It is a nonvacuum process that contributes certain thin films with quality comparable to other conventional techniques.
ii) The starting materials are analar grade inorganic salts and organic compounds which are inexpensive compared to the spec-pure starting materials employed in other techniques.
iii) The properties of films can be suitably modified by varying the atomic/ionic concentrations of the constituent species in the starting solution. However, stoichiometry is one of the serious problems in some cases. In order to compensate for the loss of more volatile constituents during pyrolysis they are excessively taken in the starting solutions.

The limitations are, in general, poor grain size of the film compared to that obtained by other vacuum processes. The properties of spray deposited films are influenced by spray dynamics, substrate temperature, chemical nature and composition of the starting solution and spray rate. The degree of atomization and droplet size are governed by the spray nozzle geometry, carrier gas, its pressure, air flow rate and solution flow rate. Lampkin [137] studied the aerodynamics of atomization and droplet impact process and correlated the dynamic features of spray process with the kinetics of film growth and surface topography. When both the size and momentum of the spray droplets are uniform, optically good and smooth films can be obtained in the case of CdS films. The commercial spray nozzles as well as nozzles made out of glass are generally used. Recently, there has been much interest in the use of ultrasonic transducer for the atomization of chemical solution [138,139]. The spray nozzle, substrate or both are oscillated to obtain uniform thick samples. The substrate temperature is perhaps the most critical parameter for successful deposition of thin film. The structural, optical properties, electrical, *etc.*, of spray deposited films are strongly dependent on the nature of substrate and its temperature. The composition of the starting solution, anion to cation ratio, dopants, type of solvent and pH of the solution control the properties of deposited films. Spray rate, type of carrier gas used, ambient atmosphere, postdeposition heat treatment are the other process parameters, which have significant impact on the properties of deposited films. The technique is mostly used for the deposition of sulfide, selenide, and oxide films. The details of chemical spray deposition technique and the process parameters are given below.

2.3.5.3a Description of the Chemical Spray Deposition Experimental Set up

A typical chemical spray deposition technique is shown in Figure 2.13. The spray system essentially consists of spray nozzle, substrate heater, temperature controller, air compressor, pressure regulator/air filter, air flow meter and solution reservoir. A commercial spray nozzle (model ¼ JAUSS, M/A spraying systems Inc., USA) made of stainless steel is used in the set up. Basically the nozzle has three ports viz., air port, liquid port and cylinder port. Air port is connected to an air compressor through a pressure regulator/filter and air flow meter, liquid port to a solution reservoir and cylinder port to the compressor through a solenoid valve, when the valve is activated, compressed air lifts a pneumatically operated air cylinder inside the spray nozzle allowing the solution to be atomized. This arrangement prevents the solution flow out of the nozzle before atomizing air is admitted. The spray solution can be fed to the liquid port either by suction or gravity. The former arrangement is most likely preferable. The nozzle is suspended from a rigid support which could be raised or lowered. The nozzle to substrate distance is 35 cm for optimum coverage. To obtain films with uniformity, the nozzle is moved to and fro through an arc of about 20° by a DC motor and eccentric drive arrangement. The nozzle and heater assembly are kept inside a fume cup-board with an exhaust duct and a fan. During the experiment, the spray process could be observed through the glass door of the cup-board.

2.3.5.3b Deposition of $Cu(In_{1-x}Ga_x)Se_2$ and $CuIn(Se_{1-x}S_x)_2$ Thin Films by Spray Pyrolysis

In order to deposit CISS thin films, NDTU is added to the starting solution of CIS chemical solution, as mentioned earlier. Excess NDTU and NDSU over stoichiometric

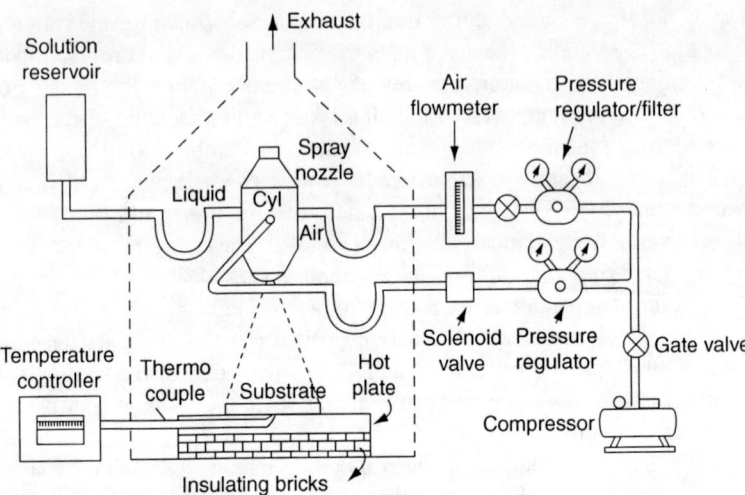

Figure 2.13 Schematic diagram of chemical spray deposition experiment.

amounts are taken to compensate for the loss of chalcogenides during pyrolysis. While the total concentration of NDTU and NDSU in the solution is kept constant, the relative concentration of either NDTU or NDSU in the solution is varied to achieve the desired value of x. The substrate temperature to obtain single phase $CuInS_2$ thin films is optimized to be 277 °C by a systematic investigation. Therefore, the substrate temperature in between 277 and 350 °C is varied for the deposition of $CuIn(Se_{1-x}S_x)$ thin films and after a few initial trials, the temperature is found to be 330 °C to obtain single phase CISS thin films. All the spray operations are done in a dark room to prevent dissociation of NDSU into elemental selenium. About 650 ml solution is required to deposit about 1 μm thick layer [44]. The $CuIn_{1-x}Ga_xSe_2$ thin films are grown by adding $GaCl_3$ chemical solution to the process of $CuInSe_2$, a little higher substrate temperature of 300–350 °C is applied and the rest of the things are almost the same [140].

2.4 Deposition of $CuInS_2$ Thin Films

2.4.1 Vacuum Evaporation

The $CuInS_2$ thin films are deposited by vacuum evaporation using single phase $CuInS_2$ powder (4N+) as a source material. In order to compensate loss of sulfur during growth, an additional sulfur source is added to the system for which Ta boat is used. If the temperature of sulfur source is exceeded beyond 600 °C, the sulfur reacts with Ta boat, whereas alumina boat is used for $CuInS_2$ powder. The growth rate is 20–50 Å/s and anneal of the as-deposited films under H_2S/Ar atmosphere (at 400 °C, 20 min) results in the sharpening of absorption edge indicating compensation of sulfur loss [141–145]. The CuInS$_2$ thin films are also prepared onto GaAs substrates by flash evaporation technique [146]. The single phase n-and p-CuInS$_2$ films with chalcopyrite structure are deposited at substrate temperature of 450 °C by three source evaporation technique for which Cu, In, and S effusion cells are used. The source temperatures of 1330–1350, 1020–1170, and 160–220 °C are applied for Cu, In, and S source bottles, respectively. The flux of materials can be controlled by varying temperatures. The fluxes of Cu, In, and S are 8.81×10^{14}, 8.75×10^{14} and 8.88×10^{14} molecules/s for Cu:In:S = 1:2.2:4.4, respectively. In order to avoid sulfur splattering, a small diameter orifice and several baffles containing number of holes with 0.1 mm diameter are used inside the bottle, as shown in Figure 2.14. Two independent heaters are used for each Cu and In effusion cells to properly monitor effusion, that is, one is at top another at bottom [147].

2.4.2 Sputtering

The single phase p-CuInS$_2$ films are obtained by RF sputtering using a 1-1/2 in. CuInS$_2$ target [148–150]. In some cases extra sulfur vapor is added while sputter depositing CuInS$_2$ thin films. A target is made that the stoichiometric Cu, In, and S elements are sealed in quartz tube and slowly heated to 600 °C for 1 day and held at 1000 °C for 2 days then slowly cooled down to RT in order to avoid sample cracking. The CuInS$_2$ powder is pressed under pressure of 60000 lbf/inch to make target of 1-1/2 in. diameter. The pressure of sputtering argon is 20–50 m torr. The preferred growth direction changes from (112) to (220) as the sulfur pressure is increased during sputtering. An overall sulfur deficiency in the films is observed, which is a function of DC bias voltage that induces on the target surface. As-grown films exhibit very high resistivity, however, after annealing under Ar (10^{-2} Torr) for 10–30 min at 410 °C reduces resistivity and enhances intensity of (112) peak along with the appearance of (103) peak. The single phase CuInS$_2$ films by ion beam sputtering deposited at substrate temperatures of 45–100, 100–360, and 420 °C are in amorphous, polycrystalline, and epitaxial in nature, respectively [151].

2.4.3 Sulfurization

The CuInS$_2$ thin films are prepared by sulfurization under H$_2$S or sulfur of Cu–In alloy grown by sputtering [152] or MBE [153] or electrodeposition [154] or electroless deposition [155]. The sputtered Cu–In films are sulfurized at 300 °C under H$_2$S ambient for 2 h. Alloy films with Cu/In ratio of 1.38 and 1.18 convert into single phase CuInS$_2$ on sulfurization while those with Cu/In = 0.94 yield multiphase films [152]. The single phase, sphalerite CuInS$_2$ films are obtained by sulfurization of Cu$_{1.0}$In$_{1.0}$ layers either under H$_2$S flow for T > 300 °C or in the presence of liquid sulfur for T > 375 °C. Likewise the Cu–In alloy films obtained by electroless deposition are sulfurized under H$_2$S gas at 550 °C to obtain n-and p-CuInS$_2$ films [155,156]. The CuInS$_2$ thin films deposited from chemical solutions containing Cu/In ratio in the range from 1:0.8 to 1:1.5 by sulfurization yields single phase, sphalerite films with (111) preferred orientation. By annealing Cu–In alloy film

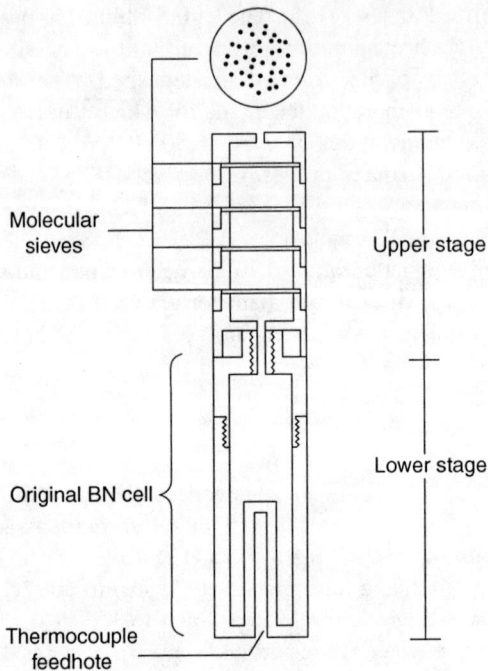

Figure 2.14 Schematic diagram of sulfur source bottle.

under H_2S at 400 °C for 60 min, single phase and chalcopyrite $CuInS_2$ films can be obtained [157].

During the preparation of $CuInS_2$ thin films, the growth process is observed at different stages by the influence of the substrate temperature with and without Na. The Cu and In sputtered by DC magnetron with Cu/In ratio of 0.8 onto SLG/ Mo/ 20 nm NaF and the sulfur evaporated by a Knudsen source had $Na/(Cu+In) = 0.02$. As the sulfurization is started, $CuIn_2$ and $Cu(111)$ phases appear. When the substrate temperature is reached to 110 °C, all the phases disappear and $Cu_{11}In_9$ starts to appear. At 140 °C, the $Cu_{16}In_9$ phase appears, whereas the $CuInS_2$ is found at 170 °C. At 200 °C, the $CuIn_5S_8$ and InS phases appear but the $Cu_{16}In_9$ phase disappears at 250 °C. The intensity of $CuInS_2$ is stronger, as temperature is reached to 260 °C. At 470 °C, the InS phase disappears and the intensity of $CuInS_2$ is found to increase. The presence of $CuIn_5S_8$ phase is continued up to the end of the process. With sodium doping, at 110 °C the $CuIn_2$ and Cu (111) disappear and $Cu_{11}In_9$ appears. The $CuInS_2$ is found to appear at 180 °C and $Cu_{16}In_9$ presents at 200 °C. At the same temperature of 200 °C, the intensity of $Cu_{11}In_9$ decreases and $CuInS_2$ increases until to disappearance of $Cu_{11}In_9$ phase. At 250 °C, all metallic phases are found to disappear and the intensity of $CuInS_2$ further increases. The spinal-phase $CuIn_5S_8$ is observed at 380 °C. At 550 °C, the chalcopyrite phase is constant. Finally the $CuInS_2$ and $CuIn_5S_8$ phases are constant. Cu–In alloy with $Cu/In = 9/11$, thickness of 0.25 in. and diameter of 4 in. target is employed to deposit $CuInS_2$ by introducing H_2S gas into the RF reactive chamber. The RF power can be varied from 200 to 300 W and chamber pressure from 2.13×10^{-1} to 4.53×10^{-1} Pa. The Cu–In alloy exists as $Cu_{11}In_9$ and CuIn for H_2S flow rate of 20 sccm and substrate temperature of 400 °C [158]. The $CuIn_2$ (211) becomes $Cu_{11}In_9$ (112) at substrate temperature of 300 °C then an increase of temperature to 500 °C, the $Cu_{11}In_9(313)/Cu_{16}In_9$ (811) phase forms. The Knudsen cell supplies sulfur with beam pressure of 1×10^{-2} Pa. The $CuInS_2$ layer starts to form at temperature of ~300 °C. On the other hand Cu_2S phase starts to form at substrate temperature of 450 °C while cooling down to room temperature, which converts into CuS at 260 °C. The entire absorber formation takes place about 50 min [160]. The process of thin film solar cell begins with cleaning of 1 cm width Cu tape, which is cleaned with different chemical solutions. The In layer with thickness of 0.7 μm is electrochemically deposited on the clean Cu-tape. Cu–In is reacted with sulfur vapor at high temperature to form $CuInS_2$ thin film on Cu-tape, which moves with speed of 6.5 m/min. The $Cu_{2-x}S$ phase on the surface of $CuInS_2$ is etched off by KCN treatment, followed by annealing under vacuum at moderate temperature to have fine $CuInS_2$ thin film on Cu foil [159].

2.4.4 Spray Pyrolysis

Pamplin and Feigelson [42] have successfully spray deposited $CuInS_2$ thin films with sphalerite structure for the first time. The $CuInS_2$ thin films are prepared starting with an aqueous solution containing analar grade $CuCl_2 \cdot 2H_2O$, $InCl_3$, and NDTU. The concentration of copper and indium chlorides in the solution is 0.00125 M. A few drops of HCl are added to improve the solubility. The pH value of the solution is maintained to be 3.5. The atomic ratio of Cu:In:S in the solution is kept to be 1:1:3. Excess NDTU is taken to compensate for the loss of sulfur during pyrolysis. The solution is prepared just prior to commencement of the spray. It is sprayed onto hot glass substrates using compressed air as the carrier gas. The spray rate is kept to be 6.5 ml/min. The substrate temperature is varied from 227–327 °C in steps of 25 °C. In order to investigate the effect of pH value of the starting solution on the growth of $CuInS_2$ thin films, the substrate temperature is kept constant at 277 °C and spray deposition is carried out with starting solution having pH 1.5, 2.4, 3.5 and 4.5. In all these processes, the deposition conditions as well as the starting solution concentration remain the same [161,162]. Sphalerite form of p-$CuInS_2$ films form onto alumina or glass substrates held at $T_S = 210$–230 °C [163]. Annealing under vacuum at 450 °C does not result in chalcopyrite peaks. On the other hand, In_2O_3 peaks are observed. An excess thiourea is used in the starting solution to compensate for the loss of chalcogen during pyrolysis [164]. Starting solution with 20–30% excess thiourea yields films with chalcopyrite structure when deposited at $T_S = 350$ °C. However, secondary phase is observed in the films [165]. Near stoichiometric films are obtained for substrates held at $T_S = 300$–400 °C from starting solutions containing Cu:In:S in the range 1:1:2.8 to 1:1:3.2 [166]. The effect of excess copper in the starting solution on the properties of $CuInS_2$ films is also studied [165]. The $CuInS_2$ thin films with 0–30% excess copper deposited at $T_S = 147$–

447 °C exhibits sphalerite structure with no impurity phases. Annealing samples at 397 °C under N_2 ambient transforms the structure from sphalerite to chalcopyrite.

2.4.5 Electrodeposition

The $CuInS_2$ films are prepared by electrodeposition from an aqueous solution of Cu^+, In^{3+}, and thiourea and subsequent annealing under H_2S/Ar at about 500 °C for 15 to 60 min [154]. For the deposition of films on Ti substrates, the $CuSO_4$, $In_2(SO_4)_3$ and $Na_2S_2O_3$ chemical solutions and potential of −0.9 V versus Ag/AgCl at pH 1.5 are also employed [168]. In some cases, Cu–In–S layers deposited by electrodeposition show chalcopyrite phase, In_2O_3, and $CuIn_5S_8$ [167]. The photovoltaic characterization carried out on a photoelectrochemical cell using polysulphide electrolyte show n-type photoresponse of $CuInS_2$ thin films. The polysulphide-p-$CuInS_2$ system does not give J–V characteristics. A strong variation in the morphology of these films is observed even on the same sample.

References

[1] R.A. Mickelsen, W.S. Chen, In: Proc. 15th IEEE Photovoltaic Specialists Conference, 1981, p. 800.
[2] R.W. Birkmire, R.B. Hall, J.E. Phillips, In: Proc. 17th IEEE Photovoltaic Specialists Conference, 1984, p. 882.
[3] N.G. Dhere, M.C. Lourenco, R.G. Dhere, L.L. Kazmerski, Solar Cells 16 (1986) 369.
[4] M. Varela, E. Bertran, M. Manchon, J. Esteve, J.L. Morenza, J. Phys. D Appl. Phys. 19 (1986) 127.
[5] S. Isomura, S. Shirakata, T. Abe, Solar Energy Mater. 22 (1991) 223.
[6] L.L. Kazmerski, G.D. Holah (Ed.), Ternary Compounds 1977, Vol. 35, 1977, p. 217 (London; Inst. Phys.).
[7] L.L. Kazmerski, Copper-Ternary Thin Film Solar Cells, NSF/RANN Rep. Washington, D.C, 1978.
[8] O. Aissaoui, S. Mehdaoui, L. Bechiri, M. Benabdeslem, N. Benslim, A. Amara, *et al.*, J. Phys. D Appl. Phys. 40 (2007) 5663.
[9] R.E. Rocheleau, J.D. Meakin, R.W. Birkmire, In: 19th IEEE Photovoltaic Specialist Conference, 1987, p. 972.
[10] N.M. Shah, J.R. Ray, K.J. Patel, V.A. Kheraj, M.S. Desai, C.J. Panchal, *et al.*, Thin Solid Films 517 (2009) 3639.
[11] H. Sakata, H. Ogawa, Solar Energy Mater. Solar Cells 63 (2000) 259.
[12] A.F. Fray, P. Lloyd, Thin Solid Films 58 (1979) 29.
[13] M.S. Sadigov, M. Ozkan, E. Bacaksiz, M. Altunbas, A.I. Kopya, J. Mater. Sci. 34 (1999) 4579.
[14] J.M. Merino, M. Leon, F. Rueda, R. Diaz, Thin Solid Films 361–362 (2000) 22.
[15] C.M. Joseph, C.S. Menon, J. Phys. D Appl. Phys. 34 (2001) 1143.
[16] R.D.L. Kristensen, S.N. Sahu, D. Haneman, Solar Energy Mater. 17 (1988) 329.
[17] N. Romeo, V. Canevari, G. Sberveglieri, A. Bosio, Solar Cells 16 (1986) 155.
[18] M. Klenk, O. Schenker, V. Alberts, E. Bucher, Thin Solid Films 387 (2001) 47.

[19] S. Yamanaka, M. Tanda, K. Horino, K. Ito, A. Yamada, M. Konagai, et al., In: 21st IEEE Photovoltaic Specialists Conference, 1990, p. 758.
[20] T. Nakada, K. Migita, A. Kunioka, In: 23rd IEEE Photovoltaic Specialist Conference, 1993, p. 560.
[21] P. Fan, G.-X. Liang, A.-H. Zheng, X.M. Cai, D.P. Zhang, J. Mater. Sci. Mater. Electron (Nov. 2009), DOI 10.1007/s10854-009-0013.2 (online).
[22] D. Bhattacharyya, I. Forbes, F.O. Adurodija, M.J. Carter, J. Mater. Sci. 32 (1997) 1889.
[23] F. Karg, V. Probst, H. Harms, J. Rimmasch, W. Riedl, J. Kotschy, et al., In: 23rd IEEE Photovoltaic Specialist Conference, 1993, p. 441.
[24] T.C. Lammasson, H. Talieh, J.D. Meakin, J.A. Thornton, In: 19th IEEE Photovoltaic Specialist Conference, 1987, p. 1285.
[25] A. Rockett, T.C. Lommasson, P. Campos, L.C. Yang, H. Talieh, Thin Solid Films 171 (1989) 109.
[26] F.O. Adurodija, J. Song, K.H. Yoon, S.K. Kim, S.D. Kim, S.H. Kwon, et al., J. Mater. Sci. Mater. Electron. 9 (1998) 361.
[27] S.D. Kim, H.J. Kim, K.H. Yoon, J. Song, Solar Energy Mater. Solar Cells 62 (2000) 357.
[28] A.E. Delahoy, J. Britt, F. Faras, F. Ziobro, A. Sizemore, G. Butler, et al., AIP Conf. Proc. 306 (1994) 370–381.
[29] A.E. Delahoy, F. Faras, A. Sizemore, F. Ziobro, Z. Kiss, AIP Conf. Proc. 268 (1992) 170–176.
[30] O.F. Yuksel, H. Safak, M. Sahn, B.M. Basol, Phys. Scr. 71 (2005) 221.
[31] C.J. Sheppard, V. Alberts, J. Phys. D Appl. Phys. 39 (2006) 3760.
[32] M. Tanda, S. Manaka, J.R.E. Marin, K. Kushiya, H. Sano, A. Yamada, et al., Jpn. J. Appl. Phys. 31 (1992) L753, (22nd IEEE Photovoltaic Specialist Conference, 1991, p. 1169).
[33] M. Tanda, S. Manaka, A. Yamada, M. Konagai, K. Takahashi, Jpn. J. Appl. Phys. 32 (1993) 1913.
[34] J. Muller, J. Nowoczin, H. Schmitt, Thin Solid Films 496 (2006) 364.
[35] J. Piekoszewski, J.J. Loferski, R. Beaulieu, J. Beall, B. Roessler, J. Shewchun, Solar Energy Mater. 2 (1980) 363.
[36] T. Tanaka, N. Tanahashi, T. Yamaguchi, A. Yoshida, Solar Energy Mater. Solar Cells 50 (1998) 13.
[37] T. Tanaka, T. Yamaguchi, A. Wakahara, A. Yoshida, Thin Solid Films 343 (1999) 320.
[38] V. Alberts, R. Swanepoel, M.J. Witcomb, J. Mater. Sci. 33 (1998) 2919.
[39] S. Kuranouchi, A. Yoshida, Thin Solid Films 343 (1999) 123.
[40] A. Yoshida, N. Tanahashi, T. Tanaka, Y. Demizu, Y. Yamamoto, T. Yamaguchi, Solar Energy Mater. Solar Cells 50 (1998) 7.
[41] J. Levoska, S. Leppavuori, F. Wang, O. Kusmartseva, A.E. Hill, E. Ahmed, et al., Phys. Scr. T54 (1994) 244.
[42] B.R. Pamplin, R.S. Feigelson, Thin Solid Films 60 (1979) 141.
[43] B.R. Pamplin, R.S. Feigelson, Mater. Res. Bull. 14 (1979) 1.
[44] S.R. Kodigala, V.S. Raja, Thin Solid Films 208 (1992) 247.
[45] M. Gorska, R. Beaulieu, J.J. Loferski, B. Roessler, J. Beall, Solar Energy Mater. 2 (1980) 343.
[46] P.M. Sarro, R.R. Arya, R. Beaulieu, T. Warminski, J.J. Loferski, In: Proc. 5th E.C. Conference On Photovoltaic Solar Energy, Athens, 1983, p. 901.
[47] C.W. Bates Jr., K.F. Nelson, J.B. Mooney, J.M. Recktenwald, L. Macintosh, R. Lamoreaux, Thin Solid Films 88 (1982) 279.

[48] C.W. Bates Jr., M. Uekita, K.F. Nelson, C.R. Abernathy, J.B. Mooney, Appl. Phys. Lett. 43 (1983) 851.
[49] C.R. Abernathy, C.W. Bates Jr., A.A. Anani, B. Haba, G. Smestad, Appl. Phys. Lett. 45 (1984) 890.
[50] C.R. Abernathy, C.W. Bates Jr., A.A. Anani, B. Haba, Thin Solid Films 115 (1984) L41.
[51] H. Lamoreaux, Spray Pyrolysis of CdS-CuInSe$_2$ Solar Cells, (1983), (SERI Technical Status Rep.).
[52] J.B. Mooney, R.H. Lamoreaux, Solar Cells 16 (1986) 211.
[53] B. Haba, C.W. Bates Jr., Thin Solid Films 158 (1988) 133.
[54] R.N. Bhattacharya, J. Electrochem. Soc. 130 (1983) 2040.
[55] R.P. Singh, N. Khare, S.L. Singh, Nat. Acad. Sci. Lett. 7 (1984) 347.
[56] Y. Ueno, H. Kowai, T. Sugiura, H. Minoura, Thin Solid Films 157 (1988) 159.
[57] S.N. Sahu, R.D.L. Kristensen, D. Haneman, Solar Energy Mater. 18 (1989) 1385.
[58] N. Khare, C. Razzini, L.P. Bicelli, Thin Solid Films 186 (1990) 113.
[59] G. Hodes, T. Engelhard, D. Cahen, L.L. Kazmerski, C.R. Herrington, Thin Solid Films 128 (1985) 93.
[60] R.N. Bhattacharya, K. Rajeswar, Solar Cells 16 (1986) 237.
[61] C.I. Qiu, I. Shih, Can. J. Phys. 65 (1987) 1011.
[62] F.J. Pern, J. Gorai, R.J. Matson, T.A. Gessert, R. Noufi, Solar Cells 24 (1988) 81.
[63] G. Hodes, D. Cahen, Solar Cells 21 (1986) 245.
[64] C.D. Lokhande, G. Hodes, Solar Cells 21 (1987) 215.
[65] V.K. Kapur, B.M. Basol, E.S. Tseng, Solar Cells 21 (1987) 65.
[66] F.J. Pern, R. Noufi, A. Mason, A. Franz, Thin Solid Films 202 (1991) 299.
[67] K.T.L. De Silva, W.A.A. Priyantha, J.K.D.S. Jajanetti, B.D. Chithrani, W. Siripala, K. Blake, et al., Thin Solid Films 382 (2001) 158.
[68] H. Chen, S.-M. Yu, D.-W. Shin, J.-B. Yoo, Nanoscale Res. Lett. 5 (2010) 217.
[69] S.A.A. El-Hady, B.A. Mansour, M.A. El-Hagary, Thin Solid Films 248 (1994) 224.
[70] A. Yamada, Y. Makita, S. Niki, A. Obara, P. Fons, H. Shibata, Microelectron. J. 27 (1996) 53.
[71] V. Nadenau, D. Hariskos, H.W. Schock, In: Proc. 14th European Photovoltaic Solar Energy Conference Bedford, UK, 1997, p. 1250.
[72] S. Schuler, S. Nishiwaki, M. Dziedzina, R. Klenk, S. Siebentritt, M.Ch. Lux-Steiner, Mater. Res. Soc. Symp. Proc. 668 (2001) H5.14.1.
[73] J.A. Abushama, S.W. Johnston, D.L. Young, R. Noufi, Mater. Res. Soc. Symp. Proc. 865 (2005) F11.2.1.
[74] A.M. Gabor, J.R. Tuttle, A. Schwartzlander, A.L. Tennant, M.A. Contreras, R. Noufi, In: 1st world Conference on Photovoltaic Conversion Energy, 1994, p. 83.
[75] F.S. Hasoon, Y. Yan, H. Althani, K.M. Jones, H.R. Moutinho, J. Alleman, et al., Thin Solid Films 387 (2001) 1.
[76] T. Negami, T. Satoh, Y. Hashimoto, S. Nishiwaki, S.-I. Shimakawa, S. Hayashi, Sol. Energy Mater. Sol. Cells 67 (2001) 1.
[77] S. Ishizuka, K. Sakurai, A. Yamada, K. Matsubara, P. Fons, K. Iwata, et al., In: 19th European Photovoltaic Solar Energy Conference, Paris, 2004, p. 1729.
[78] W.E. Devaney, W.S. Chen, J.M. Stewart, R.A. Mickelsen, IEEE Trans. Electron Devices 37 (1990) 428.
[79] S. Chaisitsak, A. Yamada, M. Konagai, Jpn. J. Appl. Phys. 41 (2002) 507.
[80] S. Marsillac, S. Don, R. Rocheleau, E. Miller, Solar Energy Mater. Solar Cells 82 (2004) 45.

[81] M.A. Contreras, J.H. Tuttle, A. Gabor, A. Tennant, K. Ramanathan, S. Asher, et al., In: 1st world Conference on Photovoltaic Conversion Energy, 1994, p. 68.
[82] A.M. Gabor, J.R. Tuttle, D.S. Albin, M.A. Contreras, R. Noufi, A.M. Hermann, Appl. Phys. Lett. 65 (1994) 198.
[83] N. Kohara, T. Negami, M. Nishitani, T. Wada, Jpn. J. Appl. Phys. 34 (1995) L1141.
[84] T. Nakada, D. Iga, H. Ohbo, A. Kunioko, Jpn. J. Appl. Phys. 36 (1997) 732.
[85] M. Konagai, Y. Ohtake, T. Okamoto, Mater. Res. Soc. Symp. Proc. 426 (1996) 153.
[86] N. Romeo, A. Bosio, V. Canevari, R. Tedeschi, S. Sivelli, A. Romeo, et al., In: 19th European Photovoltaic Solar Energy Conference, Paris, 2004, p. 1796.
[87] D.Y. Lee, B.T. Ahn, K.H. Yoon, J.S. Song, Solar Energy Mater. Solar Cells 75 (2003) 73.
[88] V. Alberts, M.L. Chenene, Semiconductor Sci. Technol. 18 (2003) 870.
[89] H.K. Song, S.G. Kim, H.J. Kim, S.K. Kim, K.W. Kang, J.C. Lee, et al., Sol. Energy Mater. Sol. Cells 75 (2003) 145.
[90] K. Kushiya, A. Shimizu, A. Yamada, M. Konagai, Jpn. J. Appl. Phys. 34 (1995) 54.
[91] W.N. Shafarman, R.W. Birkmire, S. Marsillac, M. Marudachalam, N. Orbey, T.W.F. Russell, In: 26th IEEE PVSC, Anaheim, 1997, p. 331.
[92] T. Nakada, R. Onishi, A. Kunioko, Solar Energy Mater. Solar Cells 35 (1994) 209.
[93] S. Nishiwaki, T. Satoh, S. Hayashi, Y. Hashimoto, S. Shimakawa, T. Negami, et al., Sol. Energy Mater. Sol. Cells 67 (2001) 217.
[94] M.L. Chenene, V. Alberts, B.C. Cuamba, In: 3rd World Conference on Photovoltaic Energy Conversion, 2003, p. 2PA815.
[95] G.M. Hanket, W.N. Shafarman, R.W. Birkmire, In: 4th World Conference on Photovoltaic Energy Conversion, 2006, p. 560.
[96] B.M. Basol, A. Halani, C. Leidholm, G. Norsworthy, V.K. Kapur, A. Swartzlander, et al., Prog. Photovolt. Res. Appl. 8 (2000) 227.
[97] V. Probst, W. Stetter, J. Palm, R. Toelle, S. Visbeck, H. Calwer, et al., In: 3rd world Conference on Photovoltaic Conversion Energy, 2003, p. 329.
[98] V. Probst, W. Stetter, J. Plm, S. Zweigart, M. Wendl, H. Vogt, et al., In: 17th European Photovoltaic Solar Energy Conference, Munich, 2001, p. 1005.
[99] S.S. Kulkarni, G.T. Koishiyev, H. Moutinho, N.G. Dhere, Thin Solid Films 517 (2009) 2121.
[100] V. Alberts, Thin Solid Films 517 (2009) 2115.
[101] M. Gossla, W.N. Shafarman, Thin Solid Films 480–481 (2005) 33.
[102] B. Sang, L. Chen, M. Akhtar, R. Govindarajan, A.E. Delahoy, J. Pankov, In: 31st IEEE Photovoltaic Specialist Conference, 2005, p. 215.
[103] J.K.J. van Duren, C. Leidholm, A. Pudov, M.R. Robinson, Y. Roussillon, Mater. Res. Soc. Symp. Proc. 1012 (2007), (1012-Y05-03).
[104] V.K. Kapur, R. Kemmerle, A. Bansal, J. Haber, J. Schmitzberger, P. Le, et al., In: 33rd IEEE Photovoltaic Specialist Conference, 2008, (627_080519130516).
[105] C. Eberspacher, K. Pauls, J. Serra, In: 29th IEEE Photovoltaic Specialist Conference, New Orleans, 2002, p. 684.
[106] M. Powalla, PV-tech.org, 26th August 2010.
[107] D.B. Mitzi, M. Yuan, W. Liu, A.J. Kellock, S.J. Chey, L. Gignac, et al., Thin Solid Films 517 (2009) 2158.
[108] V.K. Kapur, A. Bansal, P. Le, O.I. Asensio, Thin Solid Films 431–432 (2003) 53.
[109] V.K. Kapur, M. Fisher, R. Roe, Mater. Res. Soc. Symp. Proc. 668 (2001) H2.6.1.
[110] D.L. Schulz, C.J. Curtis, R.A. Flitton, H. Wiesner, J. Keane, R.J. Matson, et al., J. Electron. Mater. 27 (1998) 433.
[111] S. Ahn, K. Kim, K.H. Yoon, Current Appl. Phys. 8 (2008) 766.

[112] S.J. Ahn, K.H. Kim, J.H. Yun, K.H. Yoon, J. Appl. Phys. 105 (2009) 113533.
[113] M.E. Calixto, P.J. Sebastian, J. Mater. Sci. 33 (1998) 339.
[114] P. Luo, R. Zuo, L. Chen, Sol. Energy Mater. Sol. Cells 94 (2010) 1146.
[115] M. Kaelin, D. Rudmann, F. Kurdesau, H. Zogg, T. Meyer, A.N. Tiwari, Thin Solid Films 97 (2005) 486.
[116] B. Goodfellow, V. Akhavan, M. Panthani, Private communication.
[117] M. Foex, Bull. Soc. Chim. 11 (1944) 6.
[118] J. E. Hill, R. R. Chamberlin, U. S. Patent, 3, 148084, 1964.
[119] R.R. Chamberlin, J.S. Skarman, J. Electrochem. Soc. 113 (1966) 86.
[120] J.S. Skarman, Solid State Electron. 8 (1965) 17.
[121] R.R. Chamberlin, Am. Ceram. Soc. Bull. 15 (1966) 698.
[122] R.H. Bube, F. Buch, A.L. Fahrenbruch, Y.Y. Ma, K.W. Mitchell, IEEE Trans. Electron Devices 24 (1977) 487.
[123] C.S. Wu, R.S. Feigelson, R.H. Bube, J. Appl. Phys. 43 (1972) 756.
[124] C.S. Wu, R.H. Bube, J. Appl. Phys. 45 (1974) 648.
[125] Y.Y. Ma, R.H. Bube, J. Electrochem. Soc. 124 (1977) 1430.
[126] R.S. Feigelson, A.N. Diaye, S. Yin, R.H. Bube, J. Appl. Phys. 48 (1977) 3162.
[127] S.Y. Yin, A.L. Fahrenbruch, R.H. Bube, J. Appl. Phys. 49 (1978) 1294.
[128] J. Bougnot, M. Perotin, J. Marucchi, M. Sirkis, M. Savelli, In: Proc. 12th IEEE Photovoltaic Specialist Conference, 1976, p. 519.
[129] M. Savelli, In: Proc. Workshop II-VI Solar cells, Montpellier, PI-1, 1979.
[130] J. Boughnot, M. Savelli, J. Marucchi, M. Perotin, M. Marjin, O. Maris, et al., In: Proc. Workshop II-VI Solar Cells, Montpellier, P-II-1, 1979.
[131] K.L. Chopra, R.C. Kainthla, D.K. Pandya, A.P. Thakoor, G. Hass, R.E. Thun (Eds.), Physics of Thin Films, Vol. 12, Academic Press, New York, 1982.
[132] K.L. Chopra, S. Major, D.K. Pandya, Thin Solid Films 102 (1983) 1.
[133] M.S. Tomar, F.J. Garcia, Prog. Cryst. Growth Charact. 4 (1981) 221.
[134] J.B. Mooney, S.B. Radding, Ann. Rev. Mater. Sci. 12 (1982) 81.
[135] D.S. Albin, S.H. Risbud, Adv. Ceram. Mater. 2 (1987) 243.
[136] G. Haacke, Ann. Rev. Mater. Sci. 7 (1977) 73.
[137] C.M. Lampkin, Prog. Cryst. Growth Charact. 1 (1979) 405.
[138] J.C. Viguie, J. Spitz, J. Electrochem. Soc. 122 (1975) 585.
[139] G. Blandenet, M. Court, Y. Lagarde, Thin Solid Films 77 (1981) 81.
[140] K.T.R. Reddy, R.B.V. Chalapathy, In: 11th International conference on multinary and ternary compounds, 1997, p. 317.
[141] L.L. Kazmerski, M.S. Ayyagari, G.A. Sanborn, J. Appl. Phys. 46 (1975) 4865.
[142] L.L. Kazmerski, M.S. Ayyagari, F.R. White, G.A. Sanborn, J. Vac. Sci. Technol. 13 (1976) 139.
[143] L.Y. Sun, L.L. Kazmerski, A.H. Clark, P.J. Ireland, D.W. Morton, J. Vac. Sci. Technol. 15 (1978) 265.
[144] L.L. Kazmerski, D.L. Sprague, R.B. Cooper, J. Vac. Sci. Technol. 15 (1978) 249.
[145] L.L. Kazmerski, M.S. Ayyagari, G.A. Sanborn, F.R. White, A.J. Merrill, Thin Solid Films 37 (1976) 323.
[146] H. Neumann, B. Schumann, D. Peters, A. Tempel, G. Kuhn, Krist. Tech. 14 (1979) 379.
[147] Y.L. Wu, H.Y. Lin, C.Y. Sun, M.H. Yang, H.L. Hwang, Thin Solid Films 168 (1989) 113.
[148] H.L. Hwang, C.Y. Leu, C.C. Cheng, C.C. Tu, Rev. Phys. Appl. 13 (1978) 745.
[149] H.L. Hwang, C.L. Cheng, L.M. Liu, Y.C. Liu, C.Y. Sun, Thin Solid Films 67 (1980) 83.
[150] A.N.Y. Samaan, S.M. Wasim, A.E. Hill, D.G. Armour, R.D. Tomlinson, Phys, Phys. Status Solidi A 96 (1986) 317.

[151] J.R. Chen, C.C. Nee, H.L. Hwang, L. Lu, Y.C. Liu, Appl. Surf. Sci. 11/12 (1982) 544.
[152] S.P. Grindle, C.W. Smith, S.D. Mittlwman, Appl. Phys. Lett. 35 (1979) 24.
[153] J.J.M. Binsma, H.A. Van der Linden, Thin Solid Films 97 (1982) 237.
[154] G. Hodes, T. Engelhard, D. Cahen, L.L. Kazmerski, C.R. Herrington, Thin Solid Films 128 (1985) 93.
[155] A. Gupta, A.N. Tiwari, A.S.N. Murthy, Solar Energy Mater. 18 (1988) 1.
[156] J.J.M. Binsma, H.A. Van der Linden, Thin Solid Films 97 (1982) 237.
[157] J. Herrero, J. Ortega, Solar Energy Mater. 20 (1990) 53.
[158] Y.B. He, T. Kramer, A. Polity, M. Hardt, B.K. Meyer, Thin Solid Films 431–432 (2003) 126.
[159] M. Winkler, J. Griesche, I. Konovalvo, I. Penndorf, I. Wienke, O. Tober, Sol. Energy 77 (2004) 705.
[160] R. Scheer, R. Klenk, J. Klaer, I. Luck, Sol. Energy 77 (2004) 777.
[161] S.R. Kodigala, V. Sundara Raja, Mater. Lett. 12 (1991) 67.
[162] S.R. Kodigala, V. Sundara Raja, J. Mater. Sci. Mater. Electron. 10 (1999) 145.
[163] M. Gorska, R. Beaulieu, J.J. Loferski, B. Roessler, Solar Energy Mater. 1 (1979) 313.
[164] P. Rajaram, R. Thangaraj, A.K. Sharma, A. Raza, O.P. Agnihotri, Thin Solid Films 100 (1983) 111.
[165] A.N. Tiwari, D.K. Pandya, K.L. Chopra, Thin Solid Films 130 (1985) 217.
[166] H. Onnagawa, K. Miyashita, Jpn. J. Appl. Phys. 23 (1984) 965.
[167] R.N. Bhattacharya, D. Cahen, G. Hodes, Solar Energy Mater. 10 (1984) 41.
[168] T. Yukawa, K. Kuwabara, K. Koumoto, Thin Solid Films 286 (1996) 153.

3 Surface Analyses of I–III–VI$_2$ Compounds

The basic studies such as surface, compositional, depth, and chemical states are essential to know about the absorbers prior to insert in the thin film solar cells. In this chapter, the results on the I–III–VI$_2$ compounds such as surface and cross-sections of the absorbers recorded by AFM and SEM are illustrated. In addition, the composition, chemical states, grading, and depth profile studies by EPMA, SIMS, AES, and XPS are explored by citing several examples.

3.1 Atomic Force Microscopy Analysis

The atomic force microscope (AFM) is one of the most powerful techniques to investigate the status of surface conditions of either thin film or single crystals, which are being to employ for solar cell applications. In general, the tips are made by Si$_3$N$_4$ containing typical 4 side pyramidal, radius of 20–50 mm, apex angle of 45°, height of 2.8 μm and base of 4×4 μm^2 for the atomic force microscope. As shown in Figure 3.1, the tip is fixed underneath of the cantilever [1]. The cantilever has very low spring constant of 0.1 N/m, which bends or deflects for the forces between 10^{-8} and 10^{-6} N while scanning the surface of sample by the tip. The deflection of the cantilever is recorded by the detector. In this process the red laser beam is focused on the cantilever, which is coated with Au. The reflected laser beam detected by detector is mapped by computer as a surface topography in three dimensional that is based on the cantilever deflections. The x, y, and z piezo tubes are connected underneath of the sample table. The distance between surface of the sample and tip is constantly maintained by the function of z piezo tube. There are three kinds of modes such as contact, noncontact and tapping modes, as illustrated in Figure 3.2 [1]. The nature of the sample and research applications decide the selection of modes. In the "contact" mode, the tip lightly touches surface of the sample that causes to create repulsive force from the surface. The repulsive force between the sample and the tip is used to maintain the feedback that falls into repulsive regime. In the "noncontact" mode, the tip is very close to the sample but does not touch. The interaction force occurs between the tip and the surface that lies into the attractive regime containing quite small force in the order of pN (10^{-12} N). The constant frequency is generated by oscillating the stiff cantilever, which is used to control the feedback. The detecting signal is mainly depends on the changes of

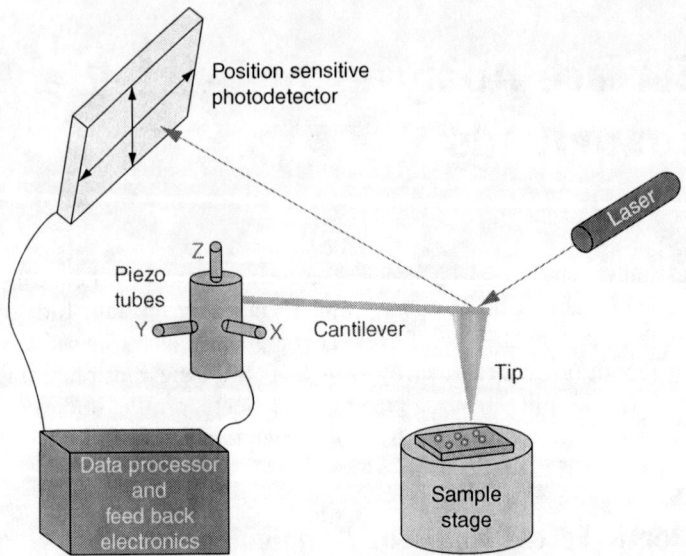

Figure 3.1 A typical schematic diagram of atomic force microscope (AFM) system.

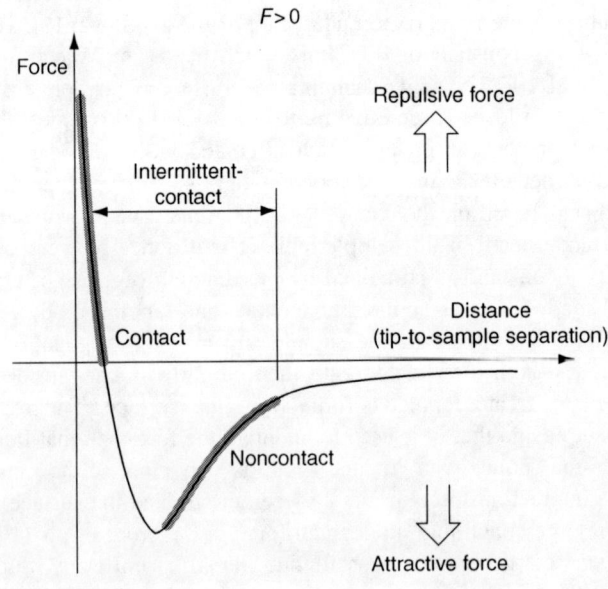

Figure 3.2 The Schematic digaram of force versus distance between tip and the sample. Courtesy of Bruker Corporation.

the resonant frequency of the cantilever. The mode lies in between contact and noncontact mode is so called tapping mode or intermittent-contact mode. In this mode, the stiff cantilever is oscillated closer to the sample than in noncontact mode. The tip intermittently touches or taps the surface that means a part of the oscillations rolls into repulsive regime. In order to avoid sticking of the tip into the soft material or liquid, the strong stiff cantilevers are normally employed for the applications of taping mode.

The surface roughness of absorber plays a vital role to form abrupt p–n junction between absorber and window layer. If the absorber is thin with rough surface causes to shunt path in the devices. The very smooth surface is also not applicable because it causes reflection losses. Therefore it is essential to study the surface roughness of Cu-based chalcopyrites or persistence of any other secondary phases on the surface. One can see number of dark spots randomly distributed on the surface of as-grown $CuInSe_2$ (CIS) layers, which could be due to $Cu_{2-x}Se$ grains. After KCN etching CIS, they disappear by leaving 200–300 nm deep holes in their native places, as shown in Figure 3.3 [1]. The CIS thin films grown onto conducting glass substrates by electrodeposition at potential of -2.03, -2.46 and -2.7 V have p^+, n, and n^+ conductivities, respectively. The grown layers with nonstoichiometric composition show coalescence shapes for the growth potentials of -2.03 and -2.46 V, whereas the CIS layers grown at potential of -2.7 V show round shape individual grains, as shown in Figure 3.4 [2]. The surface roughness of CIS layers grown onto SLG increases from 8 to 14 nm with increasing substrate temperature from 200 to 250 °C due to an increased grain sizes, as depicted in Figure 3.5. The reason is that the layers grown at low temperature shows densely packed and irregular grains due to in sufficient kinetic energy while layers grown at high temperature show improved crystalline structure and grain boundaries. The coalescence and large grains are observed for the growth temperatures of 200 and 250 °C, respectively [3,4]. The mean lateral grain area increases from 0.3 to 1.8 µm^2 with increasing substrate temperature from 400 to 550 °C for Cu-rich CIGS growth. The mean grain size is independent of growth process for low growth temperatures of 400 and 480 °C (Figure 3.6) [5]. The CIGS precursor layer made at substrate temperature of 350 °C, followed by annealing under Se vapor is so called two-stage process, as described in Chapter 2. The CIGS precursor and post-annealed two stage

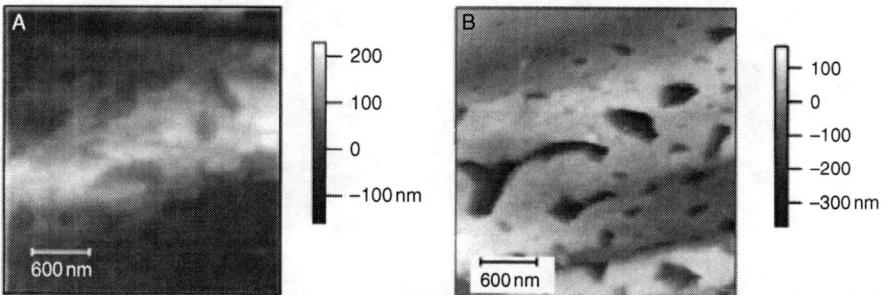

Figure 3.3 AFM images of $CuInSe_2$ thin film; (A) as-grown and (B) etched by KCN.

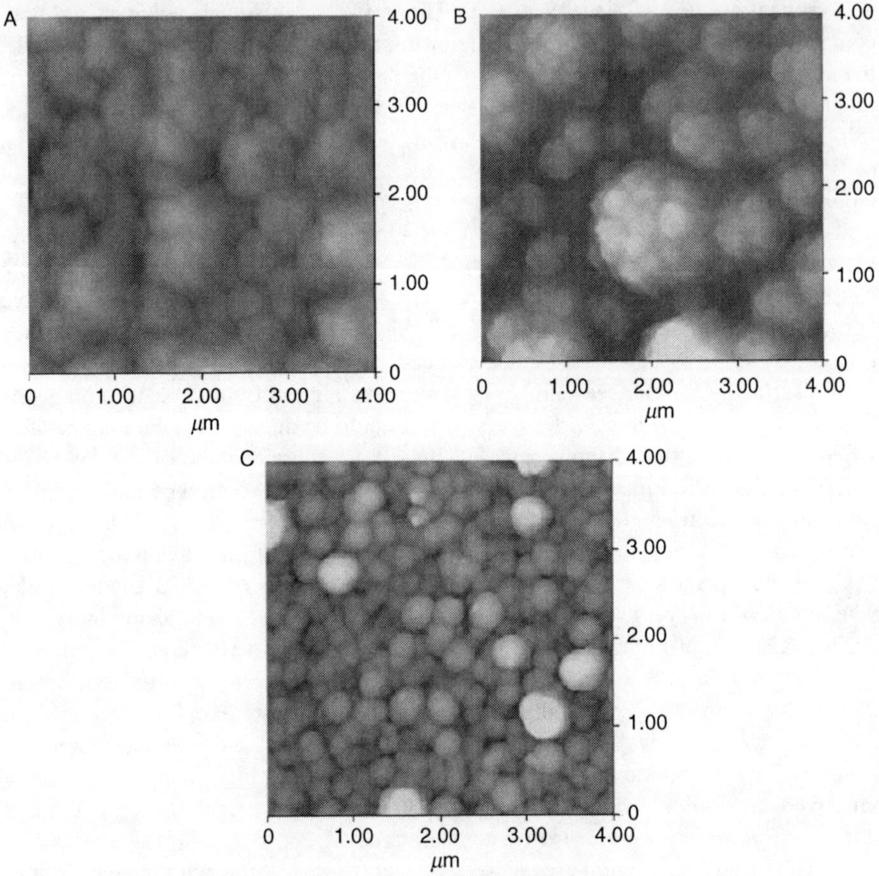

Figure 3.4 AFM of CuInSe$_2$ thin films grown onto SnO$_2$:F coated glass substrates by electrodeposition at (A) −2.03, (B) −2.46, and (C) −2.7 V.

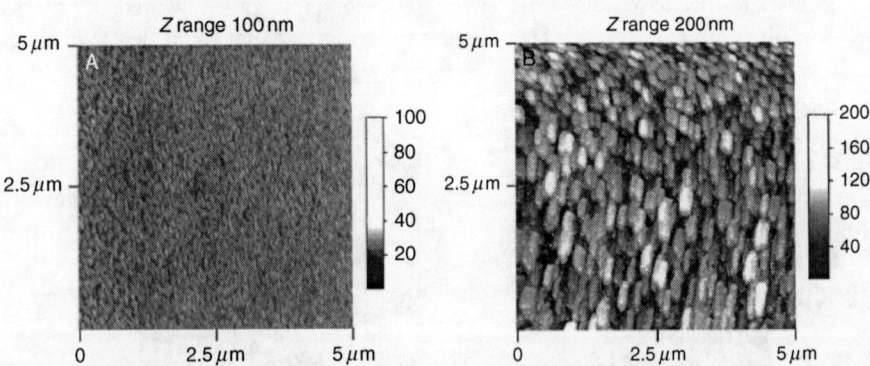

Figure 3.5 AFM of CuInSe$_2$ thin films deposited at substrate temperatures of (A) 200 and 250 °C.

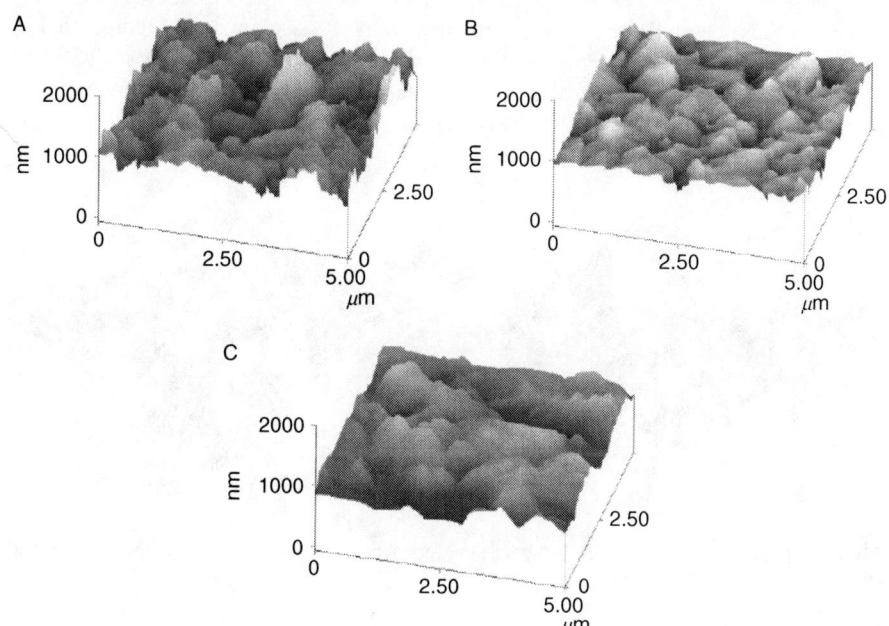

Figure 3.6 AFM images of CIGS thin films grown at different temperatures (A) 400, (B) 480, and (C) 550 °C.

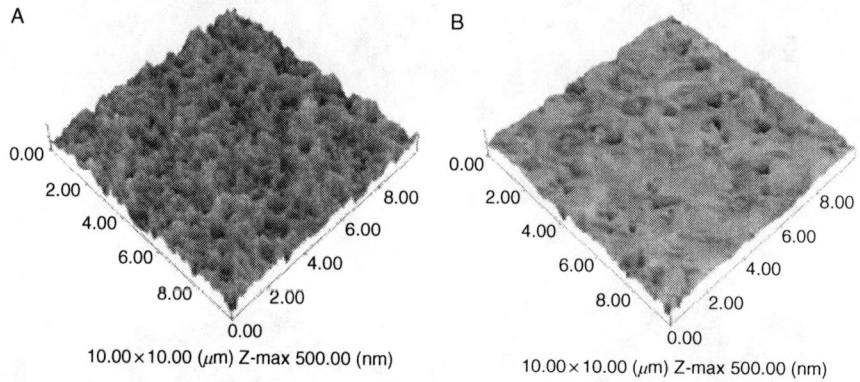

Figure 3.7 AFM images of as-grown CIGS precursor layer at 350 °C and (B) annealed under Se flux at 530 °C.

processed CIGS layers show rough surface with small grains on the surface and smooth or flat surface without any grain culture, respectively, as shown in Figure 3.7. The surface of CIGS layer developed by two-stage process is as same as device quality CIGS layer grown by three-stage process [6]. The same surface features are observed in the CIGS layers developed by standard three-stage process and in-line process. Both of them are in synonym, as shown in Figure 3.8 [7]. The similar kind of surface morphology is observed on both the SS/Mo/CIGS and SS/SiO$_2$/Mo/CIGS samples grown by three-stage process. In the latter the effect of SiO$_2$ buffer is nil (Figure 3.9) [8]. The surface of CIGS layers grown by three-stage

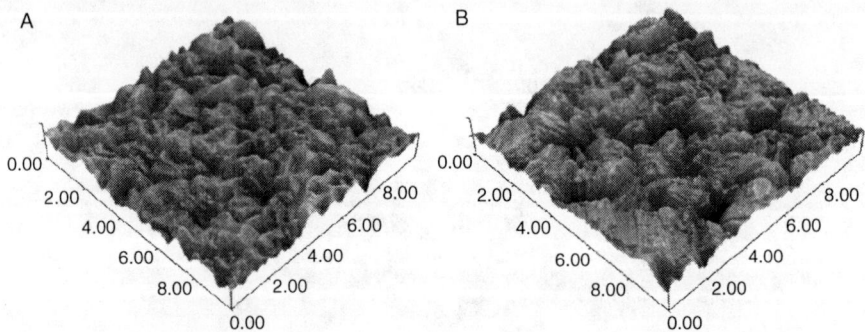

Figure 3.8 AFM images of CIGS layers developed by (A) standard three-stage process and (B) in-line evaporation process.

Figure 3.9 AFM images of (A) SS/Mo/CIGS and (B) SS/SiO$_2$/Mo/CIGS.

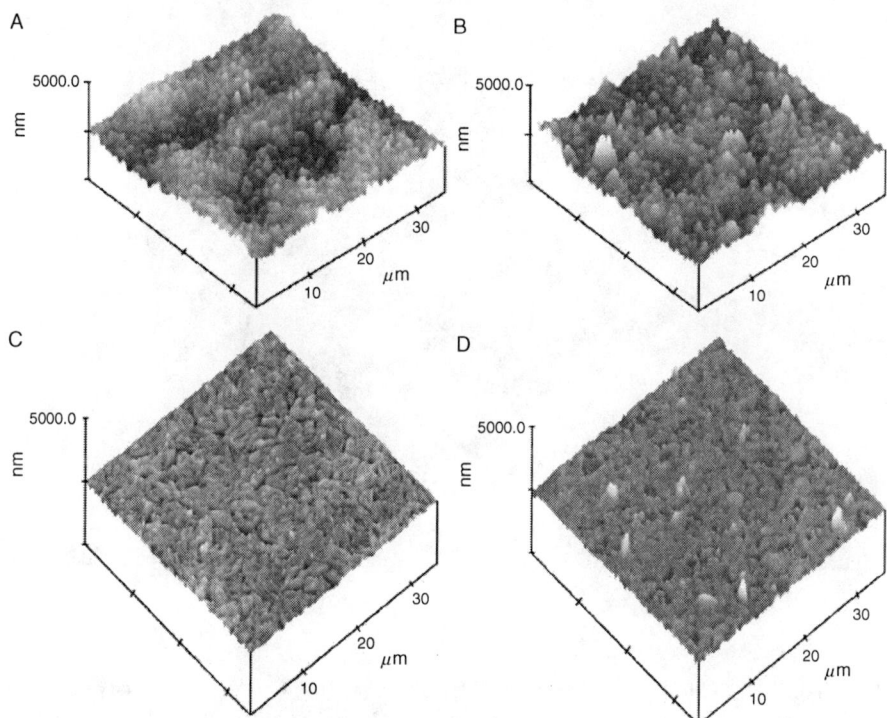

Figure 3.10 AFM scans of CIGS; (A) SP_1, (B) SP_2, (C) PV_1, and (D) PV_2 samples.

evaporation (PV_1 and PV_2) is smoother than that of sputtered CIGS layers (SP_1 and SP_2), as shown in Figure 3.10. The RMS values of typical SP_1, SP_2, PV_1, and PV_2 samples are 100.3, 114.1, 32.46, and 41.22 nm, respectively. The extra features on the surface of the layers in SP_2 and PV_2 can be seen, they are probably impurities [9]. A dense pyramidal grain with size of 0.5–2 μm, is observed in the CIGS layers grown with (112) orientation onto glass/Mo by electrodeposition, as shown in Figure 3.11 [10]. The AFM of CIGSS films (Shell Solar Inc) grown onto Mo coated glass substrates shows RMS values of 21.5 and 126.6 nm for low and high magnifications. Two and three dimensional graphs are given in Figure 3.12 [11].

The AFM analysis discerns that large size $Cu_{2-x}S$ grains are scattered on the surface of $CuInS_2$ layers grown onto Cu tape substrates by roll to roll process. After KCN etching, they disappear by leaving holes in their origins, as pointed out earlier (Figure 3.13) [12]. The (112) oriented grain of $CuInS_2$ thin film prepared by reactive RF sputtering using Cu–In (Cu:In = 9:11) alloy disc targets at power of 200 W under H_2S gas shows normal structure, as shown in Figure 3.14A and B. Of course one can see large grains about 0.3 μm on the $CuInS_2$ matrix those are absolutely Cu_xS phase (c and d) [13]. The $CuInS_2$ thin films are grown onto soda-lime glass substrates by vacuum evaporation at substrate temperature of 350 °C in the presence of sulfur evaporation with different configurations such as; (a) 1st In/2nd Cu (285 nm thick

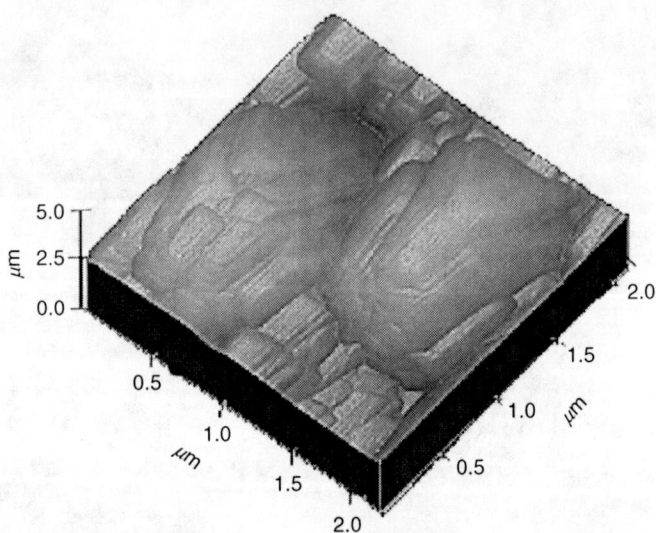

Figure 3.11 AFM image of CIGS layer grown by electrodeposition showing pyramidal shape grain.

$CuInS_2$), (b) 1st Cu/2nd In (230 nm thick $CuInS_2$), (c) In+Cu simultaneously (200 nm thick $CuInS_2$), (d) 500 nm thick $CuInS_2$, and (e) Cu-poor $CuInS_2$ thin films. As shown in Figure 3.15, all the $CuInS_2$ thin film samples contain CuS on the surface of the sample except sample-e. The CuS appears on the surface of the layer as large grains, which can also be seen by SEM images. The Raman analysis confirms that the CuS phase arises at 474 cm^{-1}. However, the CuS phase is invisible in thinner $CuInS_2$ thin films [14]. The grain sizes of $CuInS_2$ thin films grown onto glass substrates at substrate temperature of 250 °C increase with increasing thickness of the layers. The roughness also increases from 5 to 7 nm with increasing grain sizes, as shown in Figure 3.16 [15]. Alike the $CuInS_2$ thin films deposited onto ITO glass substrates by RF sputtering show that the grain sizes and RMS values are 25–100 and 4 nm, respectively, as shown in Figure 3.17 [16]. Similar grain sizes of 55 nm are observed in the $CuInS_2$ thin films deposited onto ZnO:Al coated glass substrates by sputtering using Cu–In alloy target under H_2S flow and pressure of 1.6–2.13 × 10^{-1} Pa at substrate temperature of 200 °C [17]. The CGS single crystals with Cu/Ga ratio of 1.05 grown onto (001) GaAs substrates by MOCVD technique are etched in KCN solution in order to remove $CuSe_x$. After etching the sample, the Cu/Ga ratio changes to stoichiometric composition of 1.0. The as-grown sample is sputter etched by Ar$^+$ for 3 h and annealed at 300 °C for 3 h then heated to 450 °C in steps of 50 °C is so called first cycle. The repetition of first cycle is considered as a second cycle. The as-grown and first cycle treated sample contains 1–4 nm height secondary profiles, in addition to 10–20 nm height grooves. After second cycle, those disappear, the smooth and 10–20 nm height grooves in the [110] direction appear, as shown in Figure 3.18 [18].

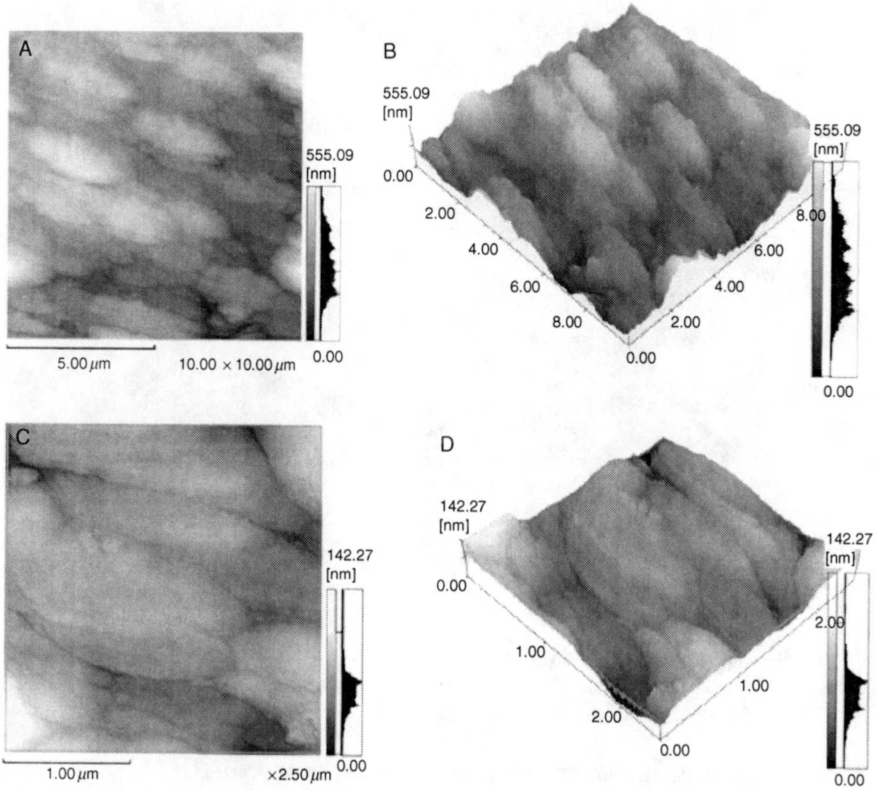

Surface Analyses of I–III–VI$_2$ Compounds

Figure 3.12 AFM images of Shell CIGSS layers with different magnifications.

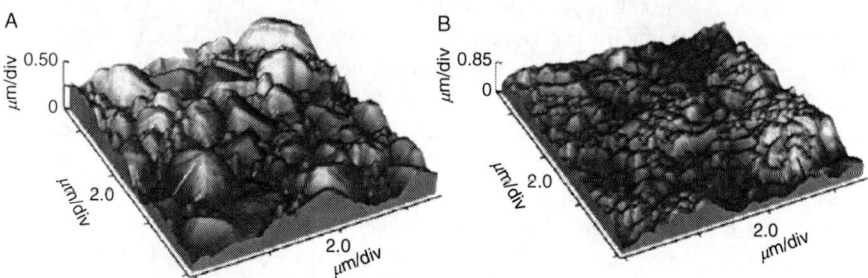

Figure 3.13 (A) Large size Cu$_{2-x}$S grains on CuInS$_2$ layers and (B) after KCN etching. They disappear by leaving holes.

3.2 Scanning Electron Microscopy Analysis

In the scanning electron microscope (SEM), the electron beam is used to scan the surface of the layer by applying a typical accelerating voltage of 10–25 kV. The SEM studies reveal that the CIS thin films deposited onto Au coated Al$_2$O$_3$

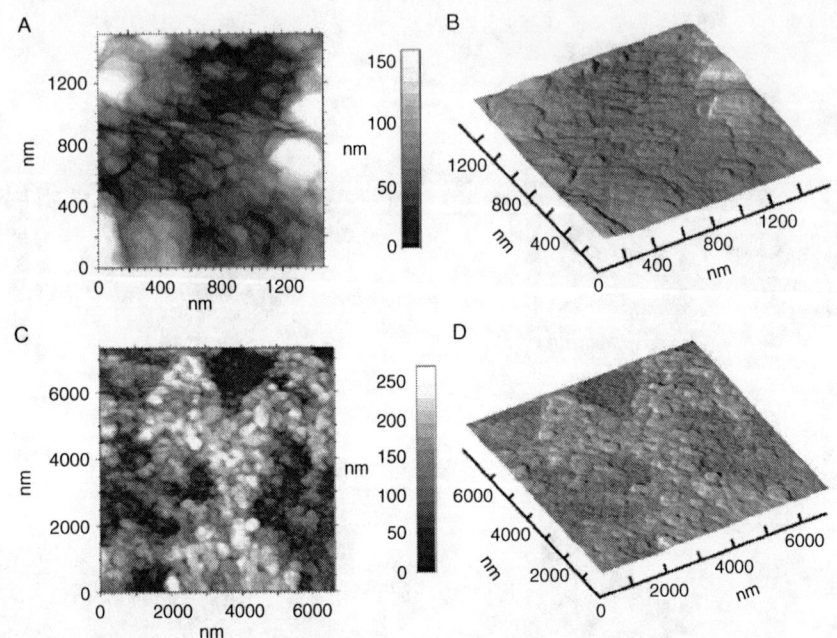

Figure 3.14 AFM of CuInS$_2$ thin films grown by RF reactive sputtering (A) (112) oriented grain surface two dimensional (B) the same in three dimensional, (C) Cu$_x$S phase on CuInS$_2$ matrix and (D) the same in three dimensional.

substrates by sputtering technique at substrate temperature of 330 °C show cauliflower like structures, whereas CIS layers deposited at substrate temperature of 505 °C exhibit triangular shapes (Figure 3.19). The same layers grown onto glass instead of Al$_2$O$_3$ at substrate temperatures of 480 and 505 °C show undeveloped and needle type grains, respectively [19]. As shown in Figure 3.20, the surface of CIS layer grown onto soda-lime glass substrates accomplishes a number of Cu$_{2-x}$Se pits. The CIS layers deposited at 300 °C for longer time of 3 h by RF sputtering causes to form secondary phases of Cu$_{2-x}$Se on the surface of CIS due to reevaporation of Se. The Cu/In ratio of 3 is observed on the islands or secondary phase regions [20]. If the CIS or CIGS layers not grown by two-stage or three-stage process but normally felicitate secondary phase on the surface of them. The InSe/Cu/InSe stacks grown onto Mo coated glass substrates at 200 °C by thermal evaporation, followed by annealing under Se vapor at 500 °C for 1 h result in CIS with Cu/In = 0.95 and large facetted grains. A different kind of InSe/Cu/GaSe stack is also grown onto glass/Mo at 300 °C by the same method, followed by annealing under Se vapor at 500 °C for 1 h to obtain CGS layers, as a two-stage process (Figure 3.21). The Ga/(In + Ga) composition ratio is adjusted by manipulating thickness of InSe and GaSe. The grown CuIn$_{0.7}$Ga$_{0.3}$Se$_2$ thin films with Cu/(In = Ga) = 0.9 also show similar facet grain structure [21].

In the MBE system, four Knudsen cell evaporation sources are fixed for Cu, In, Ga, and Se to grow CIGS layers onto Mo coated soda-lime glass substrates

Surface Analyses of I–III–VI$_2$ Compounds

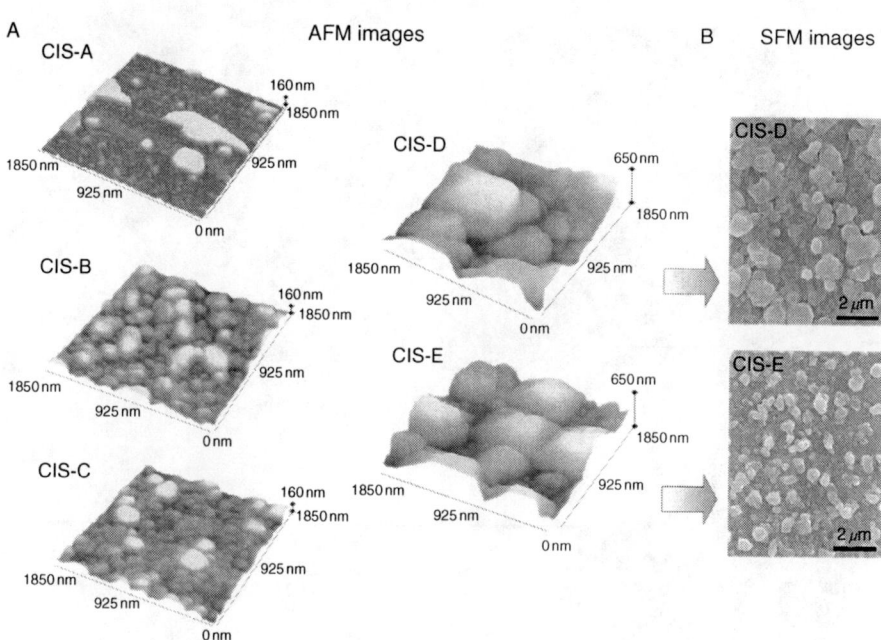

Figure 3.15 AFM and SEM of CuInS$_2$ thin films grown by vacuum evaporation; all of them have CuS phase except sample-*e*.

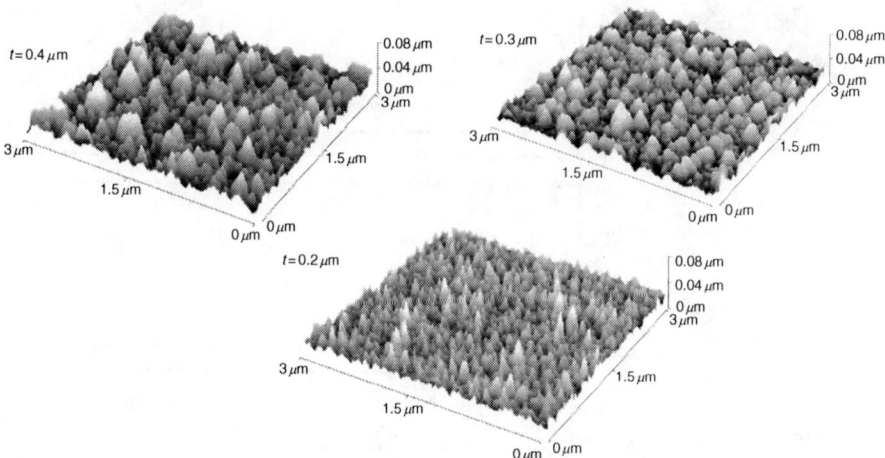

Figure 3.16 AFM of CuInS$_2$ thin films grown onto glass substrates by vacuum evaporation.

employing three-stage process. In order to reduce large quantity of Se consumption, RF-based Se radical source is developed, which uses 10 times less Se for the growth process than conventional evaporation sources. To generate plasma the RF power of

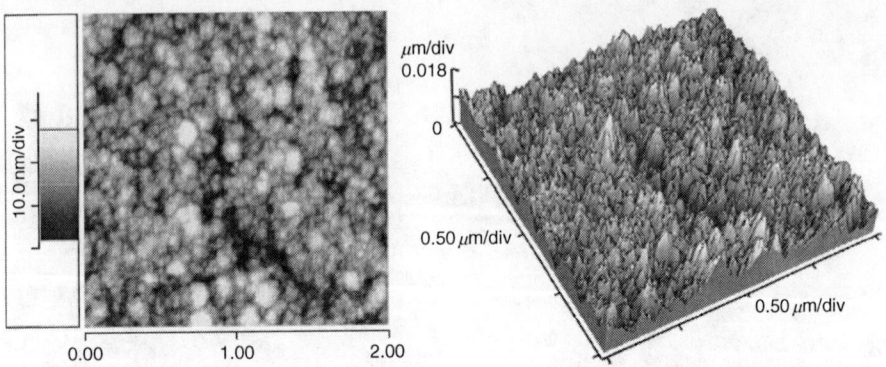

Figure 3.17 AFM of CuInS$_2$ thin films grown onto ITO glass substrates by RF sputtering.

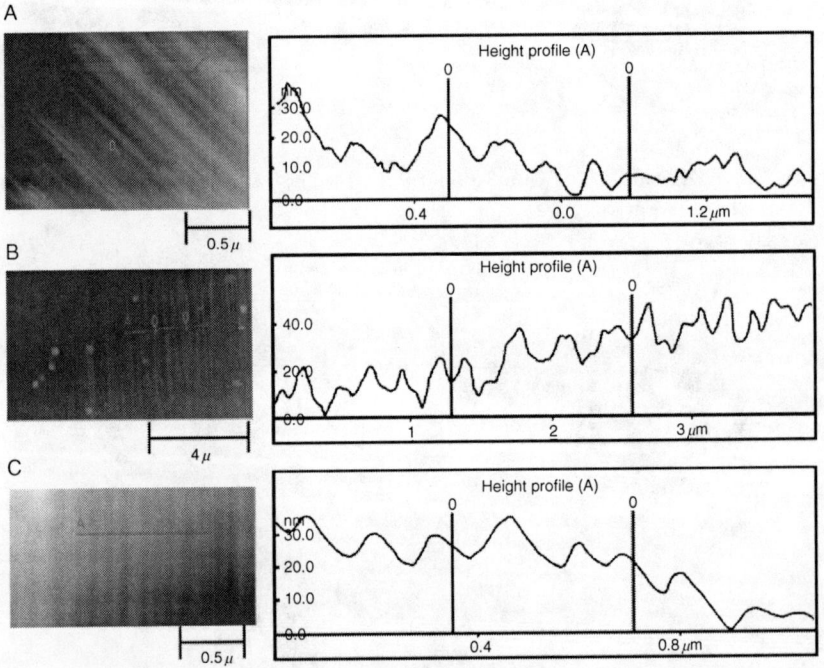

Figure 3.18 AFM image of CuGaSe$_2$ single crystal; (A) as-grown, (B) after first cycle, and (C) second cycle surface cleaning.

100 W and Ar as a working gas are used. While using RF-based Se radical source, the Se evaporation source is also continued. The CIS and CGS films are grown using three-stage process but at constant substrate temperature of 400 °C. In the case of CIGS growth, as the substrate temperature of 350 °C is applied for the first stage while the substrate temperature of 550 °C is employed for second and third stages.

Surface Analyses of I–III–VI$_2$ Compounds

Figure 3.19 SEM of CuInSe$_2$ thin films deposited at two different substrate temperatures of 330 and 505 °C.

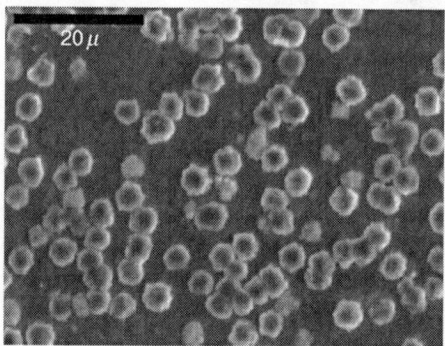

Figure 3.20 SEM image of CuInSe$_2$ thin films grown by RF sputtering technique; pits like structure on the surface is likely Cu$_{2-x}$Se phase.

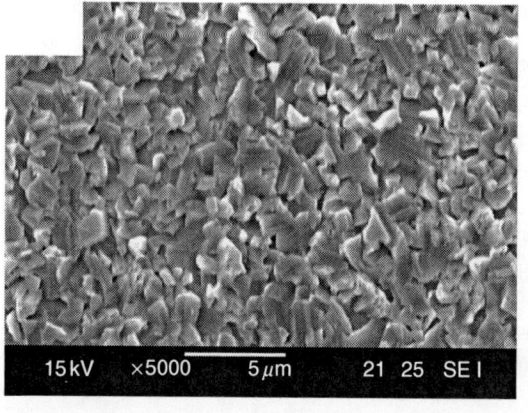

Figure 3.21 SEM of CIS layer grown by two-stage process.

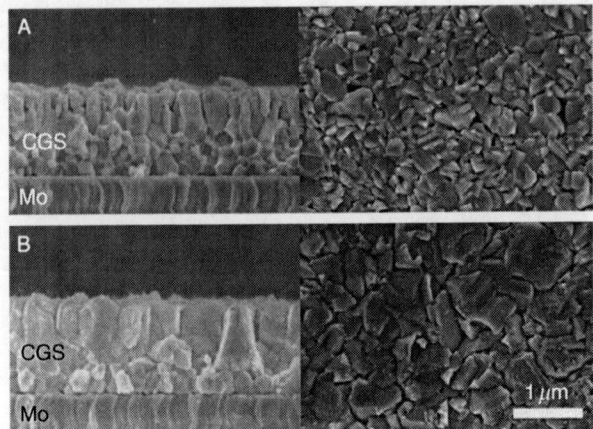

Figure 3.22 SEM cross-section and top view of CGS thin films grown onto Mo by; (A) Se radical source and (B) Se evaporation source.

The CGS films grown by three-stage process with Se radical source show better quality layers than that of conventional source methods. The films consist of densely packed large grains and smooth surfaces, as shown in Figure 3.22. The similar results are found in the CIS and CIGS layers [22]. The CGS thin films with composition $0.86 < Cu/Ga < 0.95$ are grown onto Mo coated corning 7059 glass substrates at elevated temperatures of 600, 630, and 660 °C. The selenium over pressure Se/(Cu + Ga) = 4 is employed during the growth of CGS layers in order to compensate loss of Se at high temperature. The grain sizes increase with increasing substrate temperature and surface morphology also changes, as shown in Figure 3.23 [23].

In the three-stage process, the substrate temperatures of 350 and 550 °C are constantly maintained to grow CIGS layers. The layers grown at 550 °C shows the low In concentration (Cu:In:Ga:Se = 23.5:18.3:7.8:50.4) as compared to that of layers (Cu:In:Ga:Se = 20.6:20.6:8.8:50) grown at 350 °C. The layers grown at low substrate temperature of 350 °C had low sticking coefficient. Over all, the large and columnar grain sizes are observed for the growth temperature of 550 °C, whereas the smaller grain sizes and less columnar structure are observed for 350 °C. The grain sizes are larger in the CIGS on the Ti/Mo than that on SL/Mo (Figure 3.24D and E). The quality of the films depends on the stress. Suppose, if the thermal expansion coefficient of the substrate is lower than that of the films, the tensile stress occurs while cooling the samples after the deposition. The compressive stress takes place in the films if the thermal expansion coefficient of the substrates is higher than that of films. The thermal expansion coefficient of $9 \times 10^{-6}\,K^{-1}$ for CIGS is close to that of SLG, whereas it is $5 \times 10^{-6}\,K^{-1}$ for Ti [24].

The $(InGa)_2Se_3$ precursor is a nucleation plat-form for the deposition of CIGS layers either two-stage or three-stage process. In the two-stage process, Cu–Se is deposited onto the precursor at deposition temperature of 550 °C. The $(InGa)_2Se_3$ precursor is first deposited onto Mo covered SLG substrates at various deposition temperatures of 50, 300, 400, and 500 °C correspondingly, the Se/(In + Ga) ratio

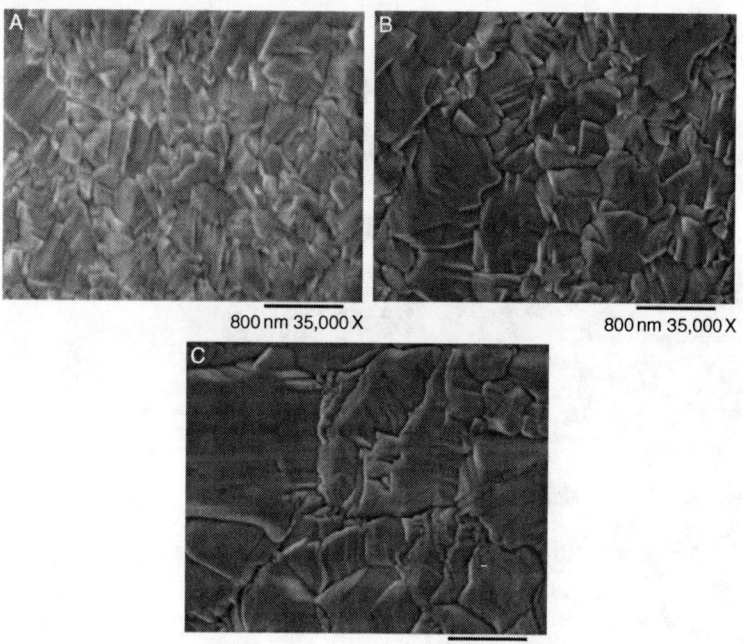

Figure 3.23 SEM of CGS thin films grown at different substrate temperatures of (A) 600, (B) 630, and (C) 660 °C.

decreases from 1.58, 1.44, 1.35 to 1.34 due to reevaporation of Se but Ga/(In+Ga) ratio of 0.27 is almost constant. A smooth surface of $(InGa)_2Se_3$ precursor consists of small grain sizes of 30–50 nm for the deposition temperature of 50 °C, whereas the densely packed columnar structure is observed for the deposition temperature of 300–400 °C. The precursor shows large grains with faceted planes for the deposition of temperature of 500 °C, as shown in Figure 3.25. The AFM analysis reveals the RMS values of 20, 65, and 140 nm for the $(In,Ga)_2Se_3$ precursor layer growth temperatures of 50, 300, 400 and 500 °C, respectively. After the deposition of Cu and Se on top of the precursor to obtain final CIGS film, which does not show rough surface due to consumption of Cu–Se by it and rearrangement of the species at high temperature. In addition, the Raman spectra show that the intensity of mode 152 cm^{-1} for γ-$(InGa)_2Se_3$ phase decreases with increasing growth temperature in the range of 300–500 °C. The mode at 145 and broad mode at 240 cm^{-1} are observed for the deposition temperature of 200 °C, which are concerned to alpha or beta meta-stable multiphases. The XRD analysis reveals that $(InGa)_2Se_3$ precursor layer deposited at 50 °C shows amorphous, whereas layers deposited at 200 °C show γ-phase with (006) orientation. Films deposited at 300 °C, the (110) and (300) lines are observed. The (110) line becomes stronger for further increasing deposition temperature. The CIGS layer shows (112) orientation for the precursor growth temperature of below 200 °C, whereas layers show (220, 204) orientation for the precursor deposition temperature of 300–400 °C [25].

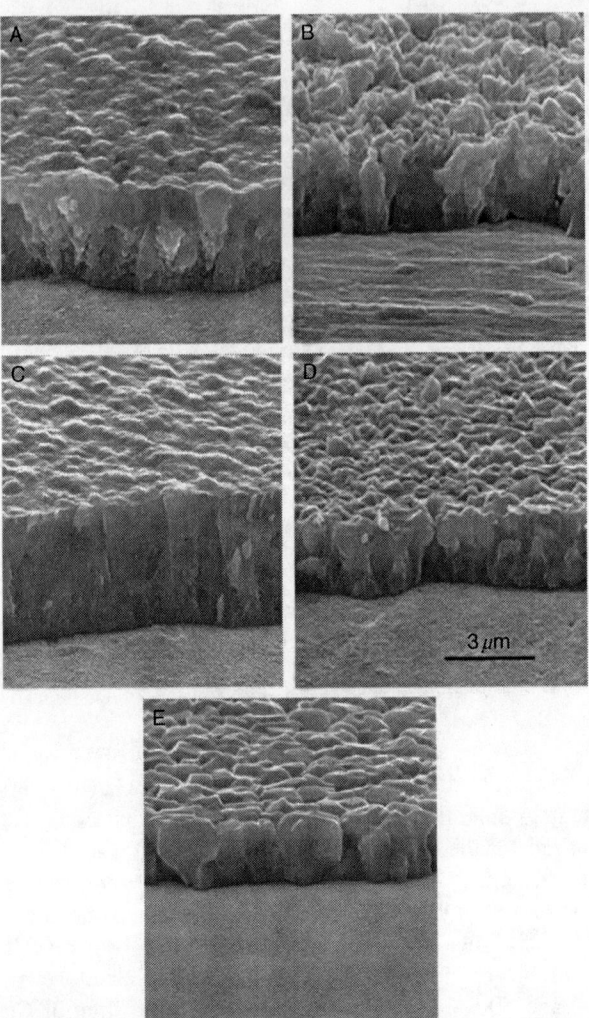

Figure 3.24 SEM of CIGS layers deposited onto Ti/Mo at (A) 350 °C, (B) 550 °C, (C) 350 °C with Na, (D) 550 °C with Na, and (E) SL/Mo with Na at 550 °C.

The CIGS layers deposited at substrate temperature of 520 °C by three state processes demonstrate that the grain sizes and grain boundaries of CIGS layers decrease with increasing Ga content [26], while the layers processed with single stage at substrate temperature of 450 °C show very narrow V-shaped grains. The upper half of the CIGS layer processed by three-stage process had larger grain sizes and the lower half shows smaller grain sizes [27]. The layers on SLG with Al_2O_3 barrier show larger and triangular shape grains with less preferred orientation of (204, 220) [28]. The layers with $Cu/(In+Ga)=0.9$ and $Ga/(In+Ga)=0.3$ are grown by three-stage processes at different substrate temperatures of 350, 450, 500, and 550 °C. With increasing growth temperature the grain sizes increase. The grain sizes of the layers grown at substrate temperature of 550 °C are larger and high dense due to liquid phase assistance of Cu_xSe phase that takes place

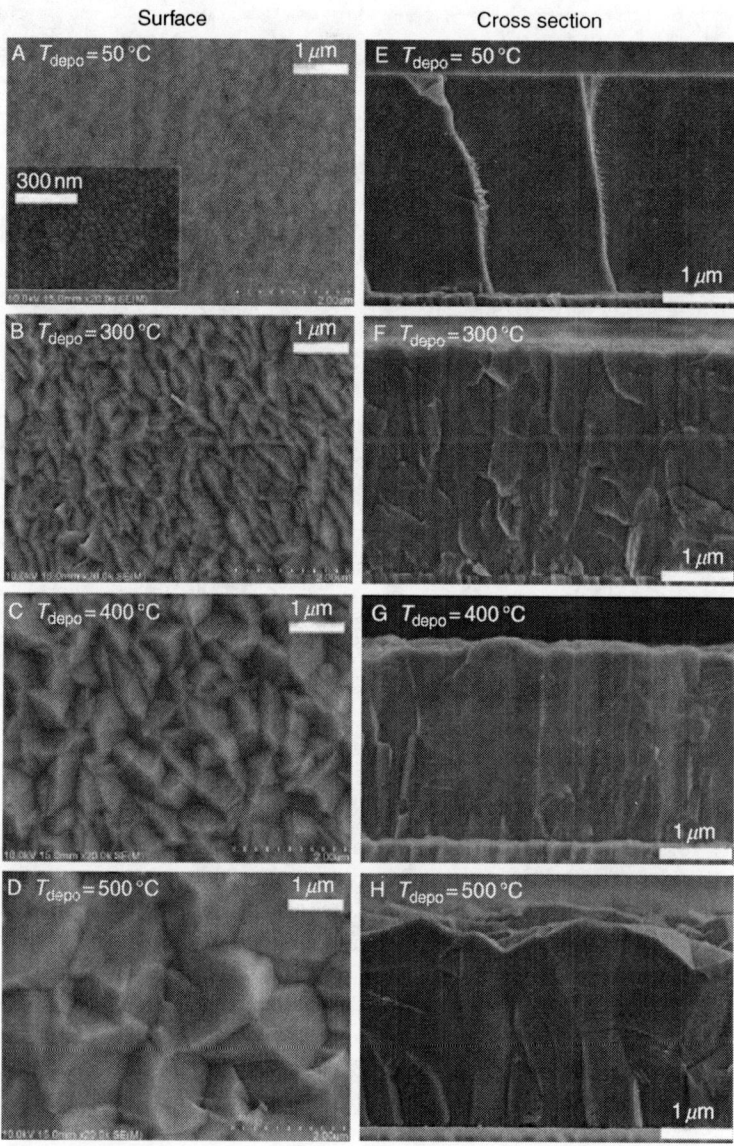

Figure 3.25 SEM of (InGa)$_2$Se$_3$ precursor grown at different temperatures.

above the substrate temperature of 523 °C [29]. SEM studies reveal that the polycrystalline Mo substrate shows oval shape polycrystalline structure on which the CIGS layers grown reveal bright dots for the growth thickness of 3 nm look like islands growth. The polycrystalline nature is observed in the CIGS layers with increasing thickness from 3 to 30 nm (Figure 3.26) [30]. The layers annealed at 280 °C have binary structure, whereas the CuSe phase dissolves in the CIGS for annealing temperature of 380 °C that make voids and rough surface. In order to obtain

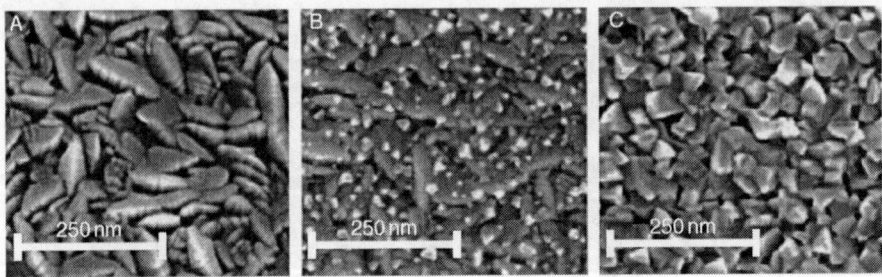

Figure 3.26 SEM of (A) Mo layer (B) 3 nm thick CIGS on Mo, and (C) 30 nm thick CIGS on Mo layer.

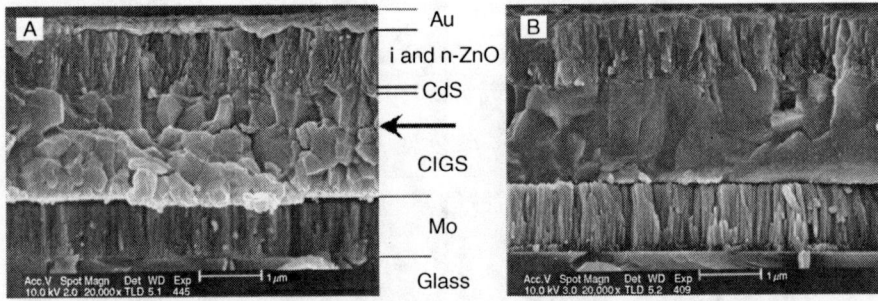

Figure 3.27 SEM cross-section of CIGS cell for which first-stage precursor layer grown at substrate temperature of (A) 150 and (B) 325 °C.

smoother surface, the films are annealed at 250 °C without Se vapor, followed by annealing under selenium at 575 °C. The absence of selenium prevents the formation of Cu_xSe phase. The fact is that the Cu-rich layers always perform large grain structure but which cause shunts in the device. The Cu-poor layers had small grains and smooth surface and no shunts in the device [31]. The quality of the CIGS layers quietly depends on the $(InGa)_2Se_3$ layer grown in the first-stage process. A $(InGa)_2Se_3$ layer at 150 °C and another one at 325 °C grown, followed by second and third processes are normally completed to obtain final CIGS layers on glass/Mo, as described in Chapter 2. The cross-section of CIGS cell is shown in Figure 3.27. The first layer which is grown at 150 °C shows self assembled bilayer structure with small grain sizes, whereas the same layer grown at 325 °C shows large grain sizes without bilayer structure. The cell with the self assembled bilayer structure degrades the performance of the cell [32].

The SEM surface, cross-section images of CIGS without and with Na are shown in Figure 3.28A–C. The grain sizes are larger in the Na doped films than in undoped films grown onto polyimide at substrate temperature of 450 °C [33]. A striking difference between 4 and 16 nm thick NaF layers inserted in the CIGS layers can be clearly seen that the grain sizes are larger in former than that in latter due to slow process of interdiffusion of In and Ga in the CIGS layers for higher Na

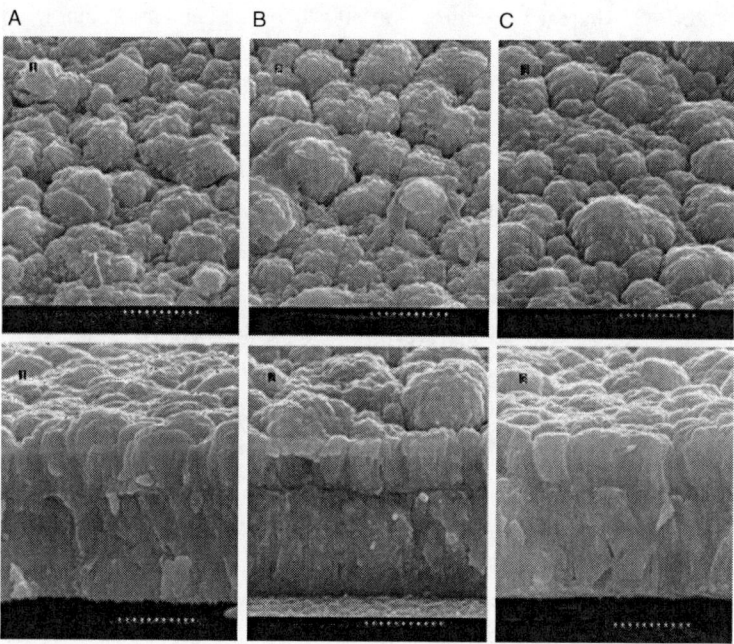

Figure 3.28 SEM of CIGS films grown onto polyimide substrates; (A) without Na (B) without Na and (C) with Na.

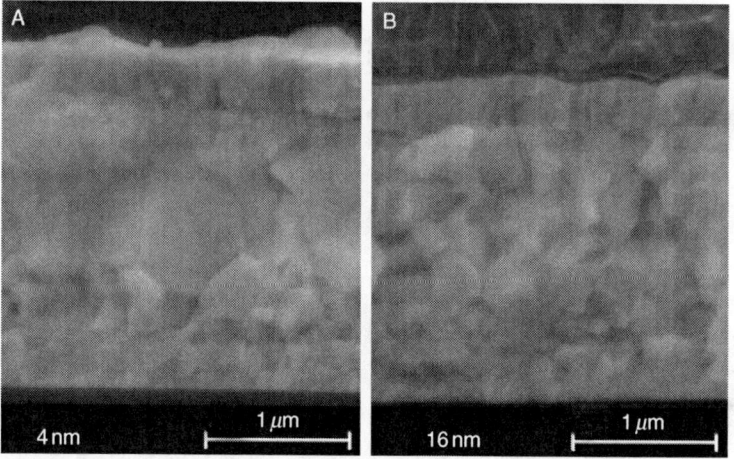

Figure 3.29 SEM cross-section of CIGS cell (A) with 4 nm NaF and (B) with 16 nm NaF.

content (Figure 3.29). In fact, higher percentage of Ga causes smaller grain sizes and grain boundaries in the samples. The higher percentage of Na reduces the rate of interdiffusion of In and Ga [34]. In the case of binary compounds, the Mo/1500 nm InSe/200 nm Cu/200 nm GaSe precursor layers are used in which InSe and Cu layers grown at 200 °C, whereas GaSe grown at 300 °C by sequential thermal evaporation show well packed round shape grains. After selenizing the precursor under Se elemental vapor at 550 °C for 1 h the CIGS layer shows well faceted columnar structure, as shown in Figure 3.30 [35].

Keeping first-stage temperature of 400 °C at constant, the CIGS layers for cell structures are grown onto Al foil at two different substrate temperatures of 400 and 550 °C in second and third stages. The cell with CIGS layers grown at 400 °C shows no photovoltaic activity owing to undeveloped structure in the sense the interdiffusion of In and Ga is not fully taken due to lack of temperature. The structure accomplishes three distinct regions as well as cracks. In the case of CIGS prepared at 550 °C, small grains are observed at the bottom of the CIGS layer closer to Mo layer while at top of the layer the large grains are observed. On top of the layers, some of the small cracks are arisen due to stress relaxation, as shown in Figure 3.31 [36]. The plane view of high efficiency cells shows faceted grain morphology. However, a remarkable difference can be attributed between two high efficiency thin film solar cells that the grain sizes are smaller in the CIGS layers with high Ga concentration of $Ga/(In+Ga)=0.31$ ($\eta=18.4\%$) compared to

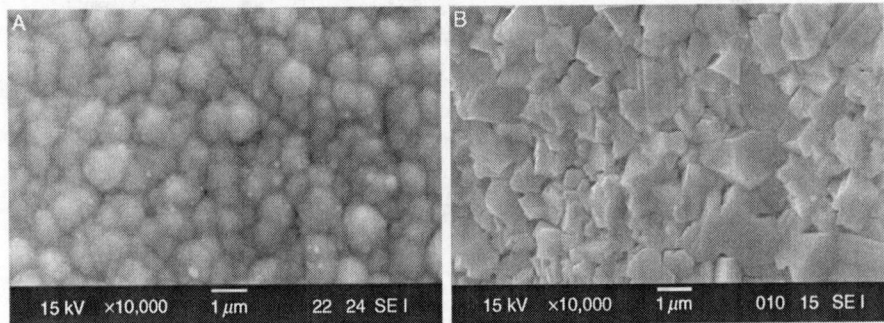

Figure 3.30 SEM of (A) CIGS precursor layers and (B) CIGS layers by selenization.

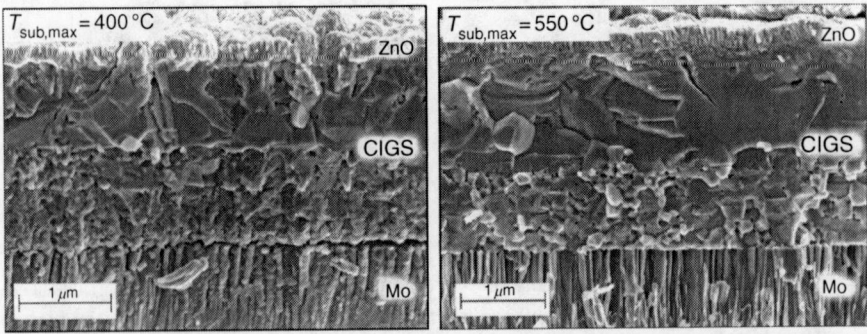

Figure 3.31 Cross-section of CIGS cells for which the CIGS layers grown at (A) 400 and (B) 550 °C.

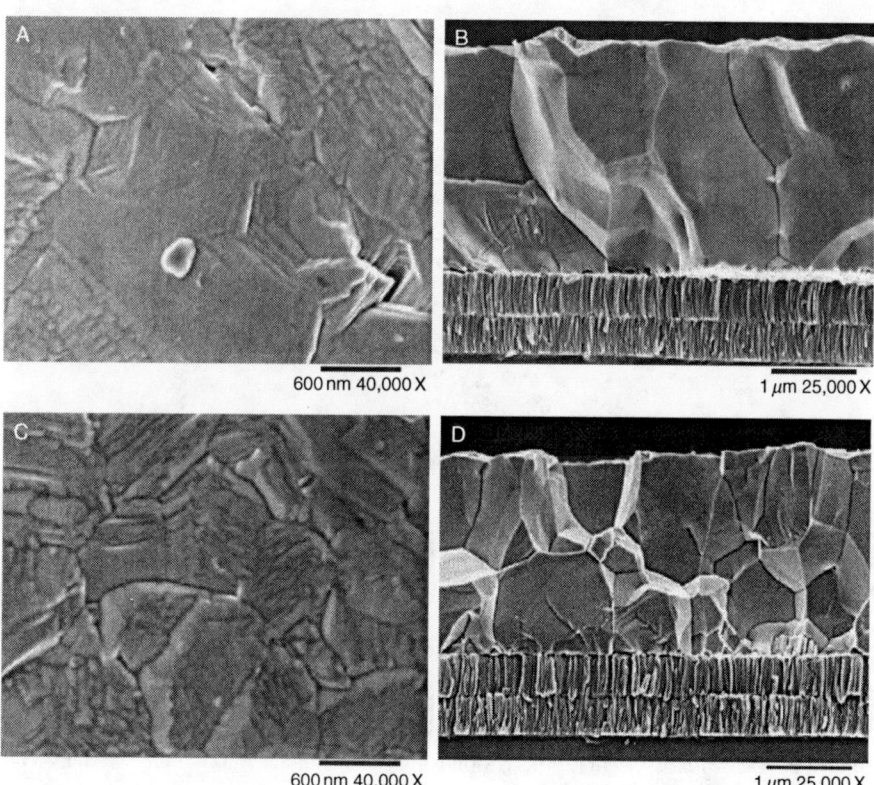

Figure 3.32 SEM of CIGS; (A) surface scan for Ga/(In+Ga)=0.26 and (B) cross-section of cell for 0.26, (C) surface scan for 0.31, and (D) cross-section of cell for 0.31.

that in Ga/(In+Ga)=0.26 layers (η=19.3%) at closer to the Mo layer, as shown in Figure 3.32 [37]. 1.5–2 μm thick CIGS layers with Cu/(In+Ga)=0.8–0.9 and Ga/(In+Ga)=0.3 are grown onto Mo coated glass substrates by elemental source evaporation using probably three-stage process. The CIGS layers grown at substrate temperature of 400 (sample-*a*) and 550 °C (sample-*b*) show smaller and larger grain sizes, respectively. The cells made with these absorbers exhibit efficiencies of 9.3 and 15.9%, respectively. The CIGS layers grown at 400 °C, post-annealed in the same chamber by ramping up to 550 °C in 10 min and kept at the same temperature for 1 minute (sample-*c*) and another sample being held for 60 min (sample-*d*). No remarkable change in the grain sizes is observed in sample-*c*. In sample-*d*, the grain sizes are as same as sample-*b*. The cells made with sample-*c* and *d* exhibit efficiencies of 15.3 and 15.7%, respectively. The CIGS layers grown at low temperature of 400 °C show (204,220) preferred orientation, whereas layers grown at 550 °C show (112) preferred orientation. The post-anneal samples show much more random orientation [38]. The CIGSS layers grown onto Mo coated glass substrates by two-stage method also show 1 μm size well faceted

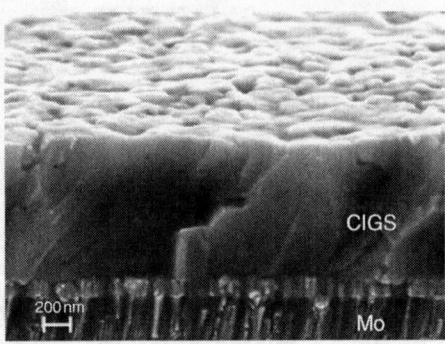

Figure 3.33 SEM cross-section of CIGSS cell.

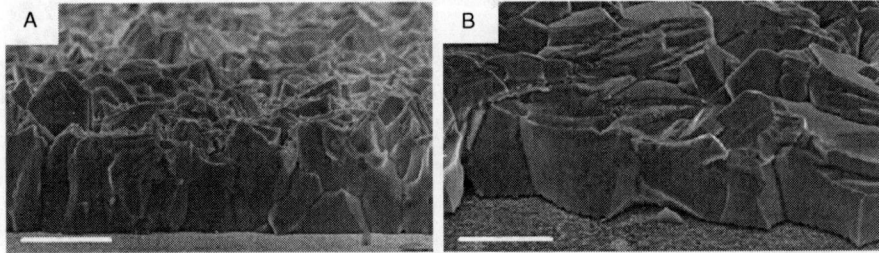

Figure 3.34 SEM cross-section of CIGS thin films grown by (A) in-line process and (B) multi stage in-line process.

grains [39]. The cross-section of CIGSS cells made with CIGSS layers grown by hydrazine method is as same as conventional cells but some voids between Mo and CIGSS are observed, as shown in Figure 3.33 [40]. In the in-line process, extra evaporation sources of Cu, In, Ga, and Se are added in order to deposit CIGS layers in one single run by controlling evaporation rates and temperatures while moving the substrates unlike conventional evaporation. The quality of CIGS layers grown onto substrates by multi stage inline process is more reliable than the one grown by inline, as shown in Figure 3.34 [41].

It is well known phenomenon that Cu-poor and stoichiometric $CuInS_2$ films show smaller grain sizes, whereas Cu-rich and annealed $CuInS_2$ layers exhibit well developed grains up to 5 μm [42]. The Cu-rich $CuInS_2$ thin films grown onto Si substrates by MBE technique show quasi layered growth and triangle pits, whereas In-rich $CuInS_2$ thin films contain round shaped droplets like structure on the surface [43]. Porous structure is also observed in the $CuInS_2$ thin films grown onto Mo coated glass substrates under $H_2S + Ar$ by reactive magnetron sputtering technique at substrate temperature of 350 °C. The surface morphology changes from porous to columnar with increasing substrate temperature from 350 to 500 °C for Cu/In ratio of 1.1. The oval shape pits can also be seen. The surface morphology of the grown films with Cu/In ratio of 2.3 at substrate temperature of 500 °C is even porous. After etching in KCN solution, the penetration of large holes can be seen up to depth of Mo layer. The films exhibit more dense and small crystallites with decreasing Cu/In ratio from 2.3, 1.1 to 0.9, as shown in Figure 3.35 [44]. The $CuInS_2$ thin films show large grains with size of 2 μm, densely packed and surrounded by 0.5–1 μm small size grains for Cu/In ratio of 1.8. The similar grain sizes are observed for the ratio of 1.4

Figure 3.35 SEM of $CuInS_2$ thin films grown by reactive magnetron sputtering at different temperatures and Cu/In ratios; (A) 350, (B) 420, (C) 500 °C for Cu/In = 1.1; (D) Cu/In = 2.3, (E) 1.0, and (F) 0.9 for the deposition temperature of 500 °C.

but three distinct layers can be discerned; bottom, middle and top layers contain small, columnar structure and larger grains, respectively. The grain sizes of 1 μm and porosity are observed for Cu/In ratio of 1.2. In addition, three distinct layers also exist [45].

3.3 Electron Probe Microanalysis

The electron probe microanalysis (EPMA), which works on the principle of wavelength dispersive analysis of X-rays, is one of the reliable analytical methods to determine the chemical composition of either thin film or bulk weighing as little as 10^{-14}–10^{-11} g and having a volume as small as about 1 μm^3. When a primary electron beam of sufficient energy impinges on the sample in which the constituents generate emission of characteristic X-rays. The diffracted photons are then detected by a photomultiplier tube or equivalent. The number of photons emitted is directly related to the number of atoms from which they are emitted. Different wavelengths corresponding to different elements are characterized by the angle of the crystal. The geometrical conditions are set to satisfy the Bragg relation $\lambda = 2d\sin\theta$, where d and θ are known and λ is the wavelength emitted by the unknown element.

The ratio of the count rates for an element in the sample to that of pure element is called k-ratio. This quantity, to a first approximation, gives the concentration of the element in the sample. However, there is a necessity to correct this since the behavior in an unknown material is expected to be different from that in a standard. The atomic number (Z) correction is required to account for (i) changes caused as a result of backscattering of electrons from the sample, (ii) electron energy loss due to inelastic scattering as the beam penetrates the material. Some of the X-ray photons released from the sample may be absorbed en-route to the detector or scattered out of the line to the detector. The corresponding correction factor is known as absorption (A) correction. The X-rays emitted by the sample on passing through it may cause secondary ionizations giving rise to further X-ray emissions. In such situation, the detected intensity is greater than the actual energy generated by the electron beam and the corresponding correction factor is known as fluorescence (F) correction. Models are now available for incorporating the average ZAF corrections.

The typical Cameca mbx electron microprobe equipment is used to determine the elemental composition of the films. The thickness of the film is more than 1 μm. An accelerating voltage of 15 kV with a regulated current of 60 nA is used for analysis. The electron beam is rastered over an area of 50 μm × 50 μm for a period of 10 s to obtain each data point. About 20 data points or more are carried out. The standards are used for the determination of corrections. The sulfur concentration is obtained from the difference. The relative accuracy of the microprobe data is 0.5 at.% on the basis of X-ray counting statistics. However, the observed standard deviation for each concentration is about 1 at.%.

The composition of CIGS sample is determined at various accelerating voltages showing different compositions. The reason is that at low accelerating voltage the electron beam penetration depth is low therefore it shows only the composition of surface layer, whereas the electron beam penetration depth is high for high

Surface Analyses of I–III–VI$_2$ Compounds

Table 3.1 Compositions in percentage of CIGS layers determined at different operating voltages

Sample	A.V (kV)	Cu	In	Ga	Se	Ga/(In+Ga)	Cu/(In+Ga)	Ref.
CIGS-1	7	23.09	22.27	2.83	51.81	0.11	0.92	[46]
	10	23.60	21.49	4.72	50.19	0.18	0.90	
CIGS-2	15	20.77	21.37	4.97	52.89	0.19	0.79	
	20	21.40	23.23	5.50	49.87	0.19	0.74	
CIGS-3	10	23.25	18.83	7.73	50.20	0.29	0.87	[47]
	20	22.82	19.67	6.96	50.56	0.26	0.86	

A.V = Accelerating Voltage

accelerating voltages therefore the obtained composition is scaled to total thick layer. As expected, the Cu composition decreases and Ga composition increases with increasing accelerating voltage. The CIGS layers are prepared in such way to have grading composition for thin film solar cell applications that means the Cu concentration is higher at front surface of the film, whereas the Ga concentration is higher at rear side of the film, as shown in Table 3.1 [46]. The similar impact is found on the different compositions of CIGS layer with variation of acceleration voltage, as shown in Table 3.1 [47].

In some cases to interpret the results well average compositions of the film are taken. For example, adding the deficient elements by physical vapor deposition to four CIGSS samples prepared by electrodeposition result in different compositions. The compositions of the samples are measured by EPMA and an average is taken at operating voltages of 10 and 20 kV (Table 3.2) [48].

The CuIn(Se$_{1-x}$S$_x$)$_2$ thin films prepared onto glass substrates by spray pyrolysis are examined by both EPMA at 15 kV and AES, which are given in Table 3.3. The data in the second row represent for each x value from AES analysis. The deposition temperature for growth of CuIn(Se$_{1-x}$S$_x$)$_2$ thin films is not one and the same in such a way the substrate temperature is tuned to have required compositions [49]. The chemical compositions of CuInS$_2$ thin films deposited by spray pyrolysis at 377 °C change after annealing under sulfur atmosphere for 1 h, as shown in Table 3.3. The compositions of the layers are determined at accelerating voltage of 10 kV, and current of 10 mA [50].

The Cu–In alloy with Cu:In = 50.4:49.6 is selenized under H$_2$Se + Ar flow at different temperatures in order to produce CIS thin films. As the selenization temperature is increased the In content decreases due to loss of In at high temperature. On the other hand the Cu content increases in the samples. The EPMA analyses of the samples (type-*a*) are given in Table 3.4 [51]. The CIS thin films deposited by pulsed laser technique experience similar loss of Indium, after annealing them in the same deposition chamber [52]. On contrast to EPMA, the XRF analysis provides little deviation in composition for the same sample because it scans mostly surface composition. The EPMA of CIS thin films (type-*b*) grown onto Mo substrates

Table 3.2 Average compositions of CIGS layers determined at two different accelerating voltages

Sample	Cu	In	Ga	Se	S	E_g (eV)
1	21.32	2.25	29.51	36.92	10	1.61
2	22.22	2.72	23.43	39.72	11.90	1.76
3	8.01	1.41	37.75	39.09	13.74	1.96
4	5.3	0.14	42.37	37.26	14.93	2.0

Table 3.3 Compositions of $CuIn(Se_{1-x}S_x)_2$ thin films determined by EPMA and AES

x	Cu	In	S	Se	O	Cl	Na	Cu/In	Ref.
0.0	26.7	24.5	–	48.8	–	–	–	0.9	[49]
	25.8	24.5	–	47.7	2.0	–	–	1.05	
0.3	25.5	24.6	14	35.9	–	–	–	1.04	
	25.5	24.3	13.3	35.5	1.4	–	–	1.05	
0.5	26.6	21.6	26.2	25.6	–	–	–	1.23	
	25.3	24.1	26.1	23.4	1.1	–	–	1.05	
1.0	24.6	26.2	49.2	–	–	–	–	0.94	
	24.9	25	48.6	–	1.5	–	–	1.00	
$CuInS_2$	26	22	41	–	7	2	2	–	[50]
Annealed	25	21	46	–	4	0	4	–	

Table 3.4 Compositions of $CuInSe_2$ thin films determined by EPMA at 15 kV and 10 nA

Sample	Annealing Temp. (°C)	Cu	In	Se	Cu	In	Se	Ref
					After heat treatment			
a_1	300	30.2	29.5	40.3	–	–	–	[51]
a_2	400	31.1	24.2	44.7	–	–	–	
a_3	500	31.3	23.3	45.4	–	–	–	
a_4	600	30.1	23.7	46.2	–	–	–	
b_1	350	20.6	25.2	54.3	22.6	26.3	51	[53]
b_2	400	20.3	26.4	53.4	22.5	26.6	50.9	
b_3	450	19.1	25.6	55.4	23.2	26.9	49.9	

by electrodeposition shows Se-rich. After heat treating the samples under vacuum at different temperatures for a short time of 20 min, the Se loss occurs but still in stoichiometric condition, as depicted in Table 3.4 [53].

The CIS thin films prepared by closed space chemical vapor transport (CSCVT) at different substrate temperatures show a variety of compositions due to effect of substrate temperature, as shown in Table 3.5. The samples exhibit *n*-type for extreme high growth temperature. The chemical compositions of the samples are determined by

Table 3.5 Compositions of CuInSe$_2$ thick layers grown by CSCVT

T_S (°C)	Cu	In	Se	I	Conductivity
300	27.48	19.54	47.18	5.8	p
350	26.3	21	51.7	1.0	p
370	24.32	23.08	52.3	0	p
400	22.5	25.	52.3	0	p
450	19.11	25.14	55.75	0	n or intrinsic
500	15.78	27.34	56.88	0	Intrinsic

energy dispersive X-ray spectroscopy (EDS). Surprisingly, the Cu composition decreases with increasing substrate temperature [54]. Obviously, the samples lack Indium or Se composition due to volatile nature of In and Se, if samples are grown at high temperature.

3.4 Secondary Ion Mass Spectroscopy Analysis

The secondary ion mass spectroscope (SIMS) is one of the versatile techniques to assess the distribution of the elements in the thin film solar cell structure or absorber. The precise Ga grading helps to create back surface field (BSF) in the thin film solar cells to reduce reverse saturation current and enhance the efficiency of the cell. Therefore either in such a way the Ga grading should be applied among the Cu, In, S, and Se elements either in the CIGS or CIGSS absorber layer. The SIMS depth profiles of CuInGaSe$_2$ thin films are studied for which the (InGa)$_2$Se$_3$ precursor deposited at different temperatures of 350, 200, 100, and 25 °C in order to investigate the distribution of Cu, In, Ga, and Se elements in the film. In the three-stage process, the development of (InGa)$_2$Se$_3$ precursor is first stage and the rest of second and third-stage processes follows the standard procedure. The separation between Cu–Se and InGaSe is observed for the precursor growth temperature of 25 °C. The interdiffusion between Cu–Se and InGaSe is in small quantity for the deposition temperature of 100 °C (Figure 3.36A). The interdiffusion between Cu–Se and In–Ga–Se is occurred and on the other hand Cu–Se/In–Ga–Se structure is also observed for the deposition temperature of 200 °C. In the case of deposition temperature of 350 °C for (InGa)$_2$Se$_3$ precursor, the In and Ga are distributed throughout the film. The higher concentration of In and lower concentration of Ga are determined at the surface of the film, as expected (Figure 3.36B). The In and Ga compositions decrease with decreasing deposition temperature of precursor. The Cu composition decreases and Se composition increases with decreasing temperature from 200 to 25 °C, evidenced by EDAX analysis [55]. The SIMS depth profile of glass/Mo/Cu(InGa)Se$_2$ thin films reveals that Ga counts increase and at the same time In counts decrease during the first sputtering time of 2–30 min, as shown in Figure 3.37 [56]. The similar reports on Mo/CIGS are explored by investigators, as shown in Figure 3.38. The distributions of In and Ga are in quite

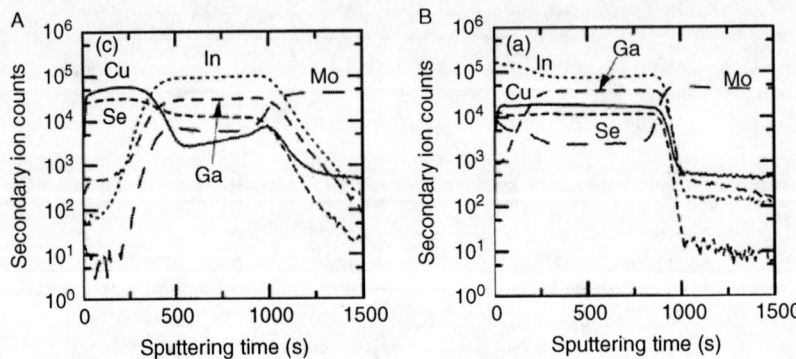

Figure 3.36 SIMS depth profiles with variation of sputtering time for Mo/CIGS layer for precursor growth temperatures A) 100 °C and B) 350 °C.

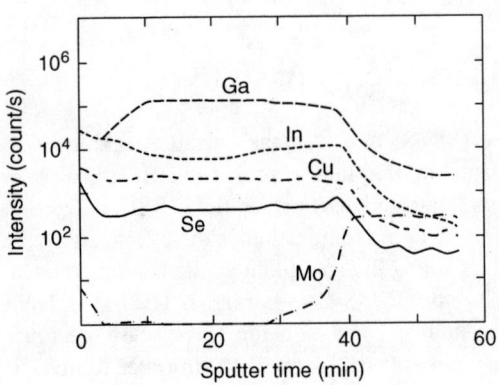

Figure 3.37 SIMS depth profiles of soda lime glass/Mo/CIGS layers with variation of sputtering time.

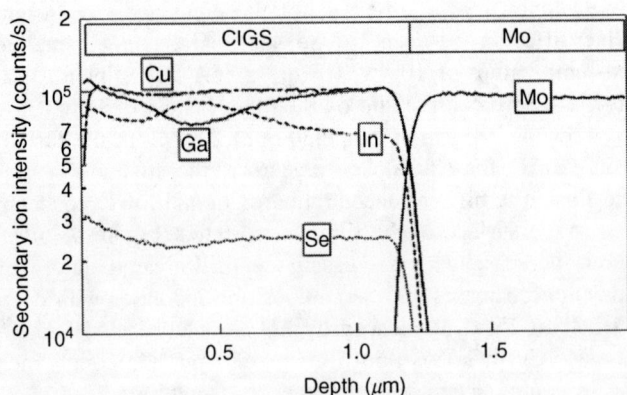

Figure 3.38 SIMS depth profile of distribution of different elements for CIGS on Mo with variation of depth.

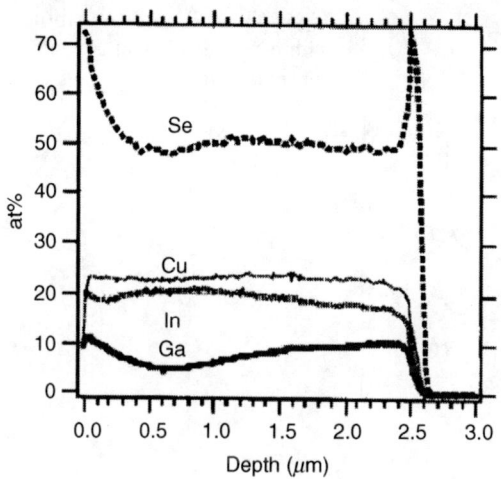

Figure 3.39 SIMS depth profile of double graded CIGS absorber.

opposite, as expected. The CIGS cells with Ga graded CIGS absorber show efficiency of more than 16% [57]. There are two ways to control Ga grading in the CIGS absorber layer by controlling substrate temperature or flux of Ga evaporation. The Ga dip is observed at the surface of CIGS layer (Figure 3.38) that slowly disappears, if the substrate temperature of CIGS is increased from 500 to 540 °C evidencing equal distribution of Ga. The diffusion of Ga is slow at lower substrate temperature [58]. On contrast to normal grading in the composition, the double grading contributes higher efficiency in the thin film solar cells. The notch appears at the near surface of the CIGS layers and the rest of the things are as usual, as shown in Figure 3.39. The normal grading will be discussed in the AES analysis [59]. The importance of band gap grading due to Ga will be discussed in Chapter 8.

The Na doping is done in the CIGS absorber layer to obtain quality absorber layers in the development of PI/Mo/CIGS/CdS/i-ZnO/ZnO:Al cell. Prior to deposition of CIGS layer by three-stage process onto Mo covered PI substrate, the NaF layer is deposited with various thicknesses of 0, 4, 8, and 16 nm. At 500 nm depth the Ga concentration is lower for 16 nm thick NaF layer as compared to that of other Na dopings indicating that Na does not encourage much interdiffusion between In and Ga, as shown in Figure 3.40. The distribution of elements in the CIGS for 4 nm thick NaF is as same as un-doped samples. The concentrations of Na in the layers are found to be 3, 56, 90, and 210 ppm for NaF thickness of 0, 4, 8, and 16 nm, respectively [35]. The Ga dip is well developed with increasing substrate temperature from 450, 500 to 550 °C due to proper Ga grading. The Cu dip is too found for low substrate temperature due to lack of temperature to mix elements properly, as shown Figure 3.41 [37].

In general, Mo is the back contact layer for CIGS thin film solar cells. The Se peak occurs as a spike at the interface of CIGS and Mo in the SIMS depth profile as compared to that in the bulk film, as shown in Figure 3.42 [60]. This observation supports

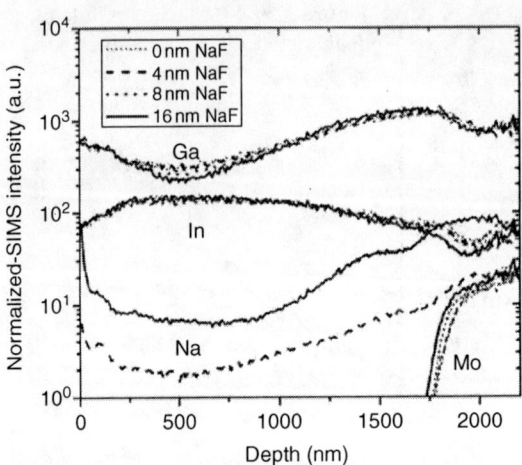

Figure 3.40 SIMS depth profiles with effect of different Na doping concentrations.

the generation of secondary phases of $MoSe_2$ or Mo_3Se_4 at the interface, as detected by XRD analysis. The CIGS cells fabricated on Fe/Ni alloy substrate with Cr barrier layer underneath of Mo show efficiency of 9.1%. The cell without barrier shows poor performance. The SIMS analysis reveals that the 1.7% Ni, 0.18% Fe, and 0.20% Co are found in the CIGS absorber layer diffusing through Mo layer from Ni–Fe alloy substrates. In order to prevent diffusion of these elements to some extent, the Cr layer is sand witched between substrate and Mo layer. Still lower percentages of Ni, Fe, and, Co are diffused into CIGS absorber, that is, one order of magnitude is lower. In addition, Cr is also diffused into CIGS absorber, as shown in Figure 3.43. The concentrations of diffused elements are higher in the CIGS than in Mo [61]. Both the Cu and Na diffuse into In_2S_3 buffer layer from absorber and substrate in the CIGS/In_2S_3 cell configuration, respectively. Their concentrations are high, if the deposition temperature of In_2S_3 is increased [62]. The distribution of Cu concentration is lower near the surface of CIS layers grown by using 3 mm diameter crucible nozzle for Cu evaporation source than that in the CIS layers deposited by evaporation employing 2 mm diameter crucible nozzle. The reason is that the reaction temperature for Cu is not enough to mobile Cu or Cu-based secondary phases to consume by CIS in the case of 3 mm diameter crucible nozzle [63]. The concentration of Ga is lower in the CIGSS absorber layer as compared to that of CIGS layer because of participation of S in the CIGSS layers that can be seen in SIMS analysis [40].

3.5 Auger Electron Spectroscopy

In order to attain the chemical compositions of the layers/compounds by quantitatively, different kinds of methods have been employed in Auger electron spectroscopy (AES) such as (i) utilization of published relative sensitivity factors, (ii) calculation of sensitivity factor including matrix effects, (iii) calibration with an

Surface Analyses of I–III–VI$_2$ Compounds

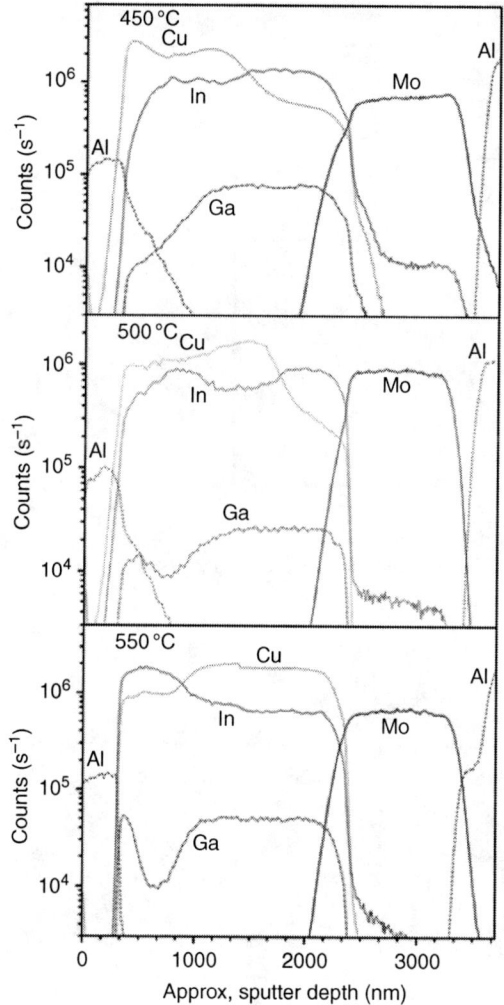

Figure 3.41 SIMS depth profile of CIGS layers grown at different substrate temperatures.

Ag standard, and (iv) calibration with identical ternary compounds. In the case of published sensitivity factors, the atomic concentration of any element "i" in a material containing j elements can be arrived from a formula

$$C_i = (I_i/S_i) / \left(\sum_j I_j/S_j \right), \tag{3.1}$$

where S is the relative sensitivity factor of a pertinent element and I is the normalized peak to peak differential Auger sensitivity [64]. The relative sensitivity factor can be written as

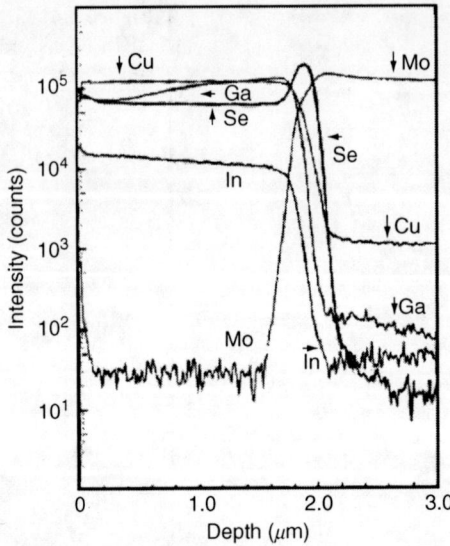

Figure 3.42 SIMS depth profile of CIGS film grown onto Mo coated glass substrates.

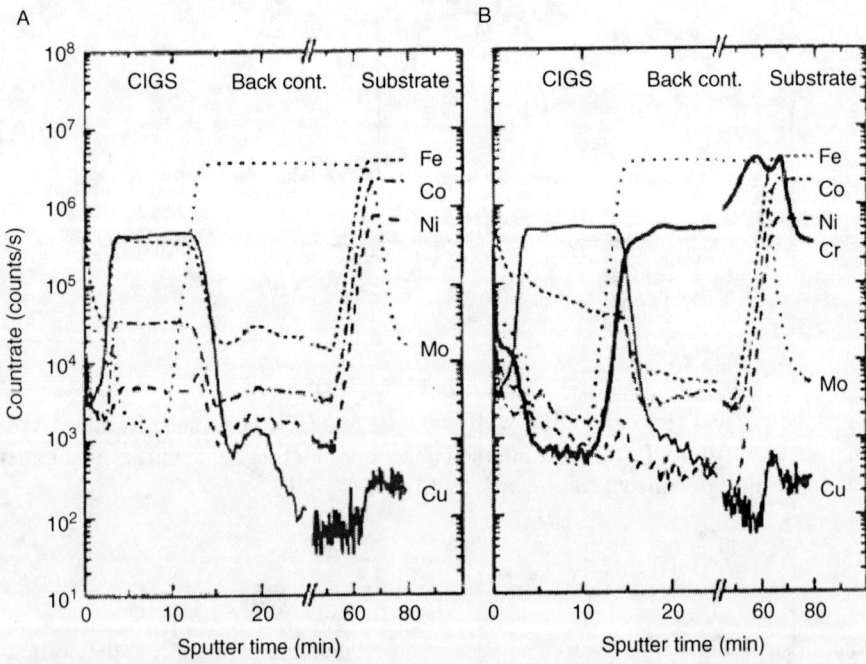

Figure 3.43 SIMS depth profile of CIGS film grown (A) with Cr barrier and (B) without Cr.

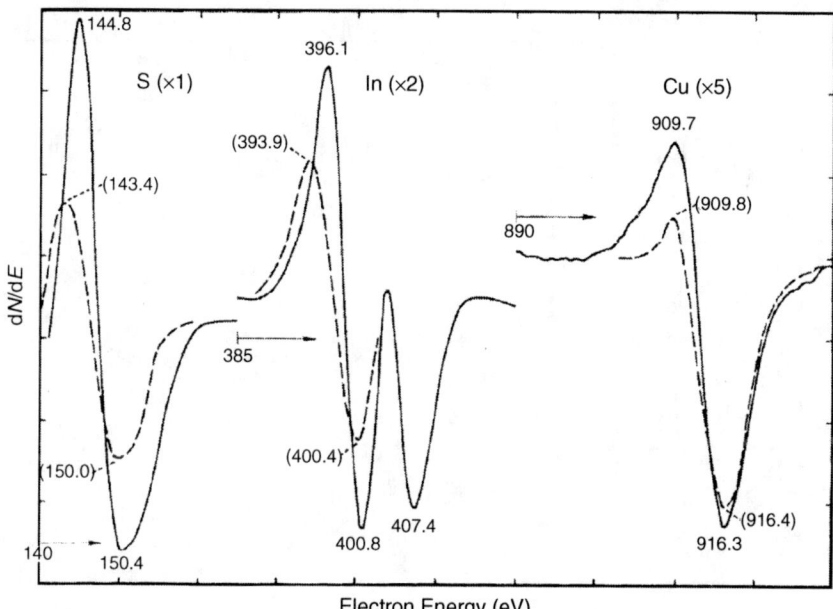

Figure 3.44 AES spectra of CuInS$_2$ in expanded version (solid line-single crystal, dotted line-thin film) $E_p = 3$ KV and $I_p = 6.5$ μA.

$$(P_n/P_m) = (E_m/E_n)^{1/2}(\delta_m/\delta_n)^2(1+r_n)/(1+r_m). \tag{3.2}$$

The last term is negligible, from the graph $\delta_S = 6.6$, $\delta_{In} = 6.5$, and $\delta_{Cu} = 6.6$ can be found for polycrystalline CuInS$_2$, where E, δ, and r are the energy of the pertinent peak, the separation between the maximum and minimum of the differential peak and the backscattering factor, respectively. Peak to peak height is 23.9 mm and 71.4 mm for I_{Cu} and I_{In}, respectively. For example, the copper concentration (C_{Cu}) can be obtained, $C_{Cu} = I_{Cu}/\{I_{Cu} + I_{In}(P_{Cu}/P_{In}) + I_S(P_{Cu}/P_{In})\} = 0.249$. In the similar way In and S concentrations can be obtained, as shown in Figure 3.44. The Auger electron spectra of the typical as-grown and etched CuIn(Se$_{0.5}$S$_{0.5}$)$_2$ thin films (Figure 3.45A and B) reveal that the carbon, chlorine, and oxygen are impurities in the former, whereas no traces of impurities except 1.1 at.% oxygen is determined in the latter. The compositions of etched CuIn(Se$_{0.5}$S$_{0.5}$)$_2$ thin film are Cu-25.3, In-24.1, S-26.1, and Se-23.4% [49]. The similar kind of AES spectrum for CuInS$_2$ thin films grown by three source evaporation shows Cu, In, and S with compositions of Cu:In:S = 1:1.21:2.13 and without any additional features such as oxygen and chlorine, *etc.*, as shown in Figure 3.46 [65]. The AES depth profile is also one of the tools to assess the composition distribution in the CIGS layers on par with SIMS analysis. The In and Ga lines cross over each other and a dip in the Ga line is quite opposite to In line at the surface of CIGS absorber layer in the AES depth profile, which is used in 15.1% efficiency cells, as shown in

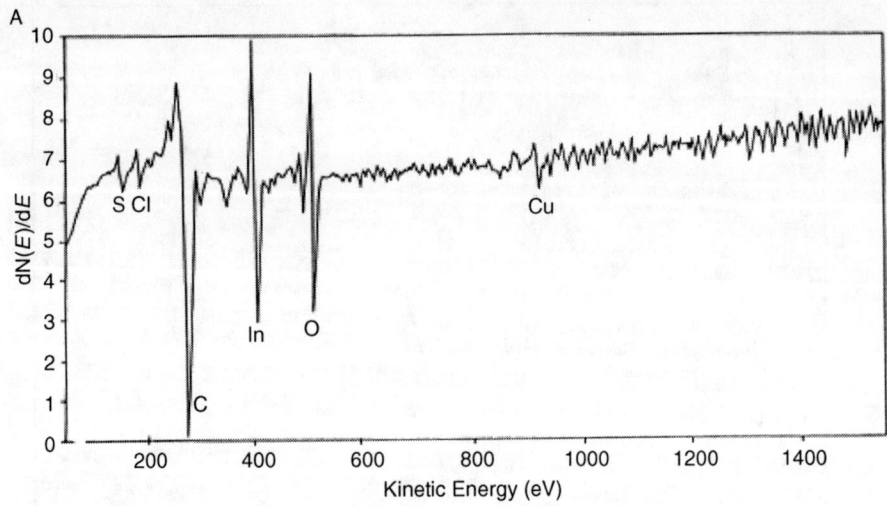
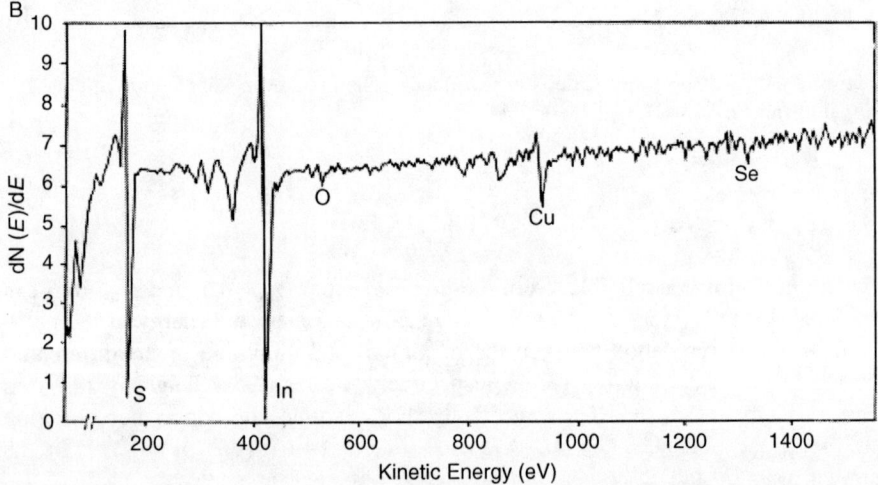

Figure 3.45 (A) Auger electron spectra of as-deposited $CuIn(Se_{0.5}S_{0.5})_2$ thin film and (B) sputter etched $CuIn(Se_{0.5}S_{0.5})_2$ thin film.

Figure 3.47 [59]. The Ga grading is observed from top to bottom in the CIGS layers for which the $(InGa)_2Se_3$ layer is grown at 325 °C, whereas more or less uniform Ga grading is observed for the substrate temperature of 150 °C, as shown in Figure 3.48 [32]. Four different CIGSS samples made by electrodeposition are enriched by PVD. As shown in Figure 3.49, in sample-1, two different regimes are seen for Cu; one is at surface of the sample and another one at constant throughout the film.

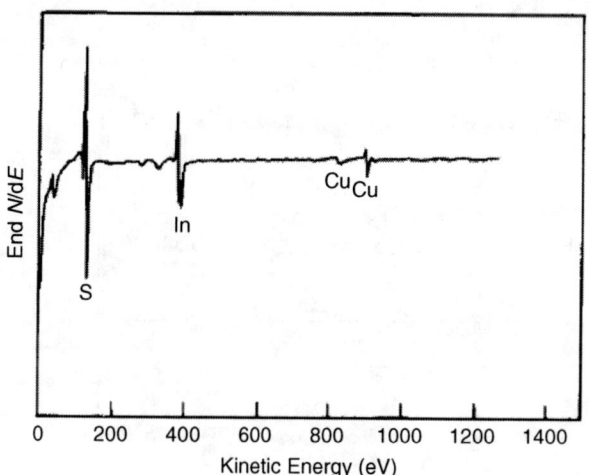

Figure 3.46 AES spectrum of CuInS$_2$ thin films grown by three source evaporation.

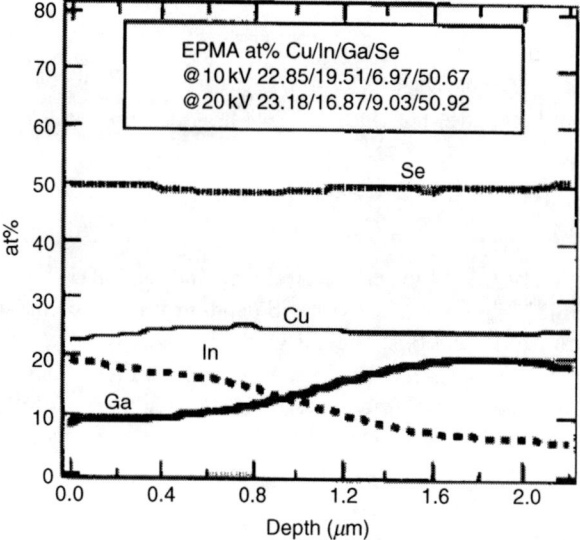

Figure 3.47 AES depth profile of CIGS absorber with normal grading used in 15.1% efficiency cells.

In sample-2, the S concentration is higher at the surface and the Se concentration varies inversely to S indicating that the S substitutes Se. The concentration of S is constant throughout the bulk film. At surface indium oxide signature is appeared. Sample-3 shows two selenium core levels at the surface. One type of Ga pattern at the surface and a different type of pattern in the bulk are exercised but the Cu concentration is low at the surface. In sample-4, low In in the sample with respect

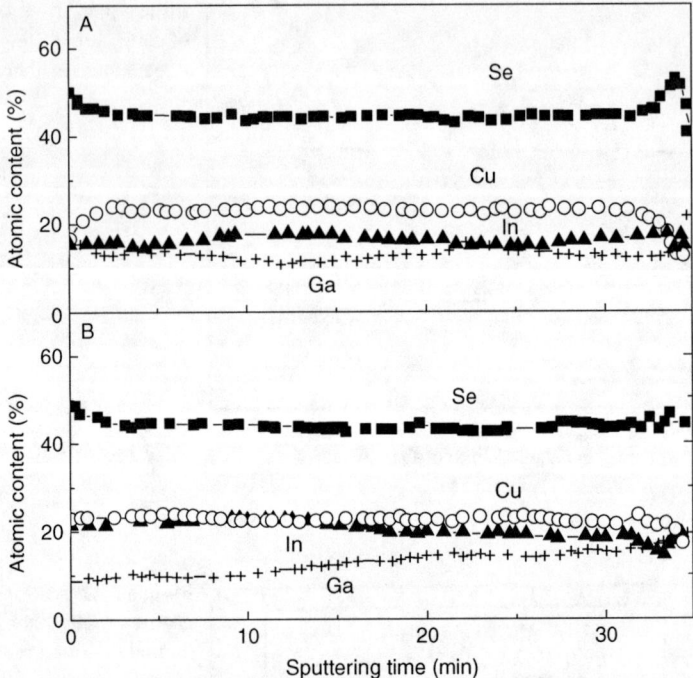

Figure 3.48 AES depth profile of CIGS layer for which precursor layer deposited at (A) 150 and (B) 350 °C.

to surface and low concentration of Cu at the surface are found and the XRD shows Ga(SSe) phase in the sample [48]. Figure 3.50 shows AES depth profiles of $CuInS_2$:Na thin films with Cu/In ratio of 1.1 and 0.8. The Cu grading gradually increases from surface to bulk but quite opposite trend in Na can be seen in Figure 3.50B indicating that the Na doping takes place more easily in the Cu-poor $CuInS_2$ (Cu/In = 0.8) thin films than in the Cu-rich $CuInS_2$ thin films. In fact, Na occupies Cu site in $CuInS_2$ or CIS thin films [66].

3.6 X-ray Photoelectron Spectroscopy

The X-ray photoelectron spectroscopy (XPS) or electron stimulated chemical analysis (ESCA) plays a vital role to estimate the composition, thickness, and in particular chemical state of the compound. The binding energy (BE) values of the compound and asymmetry of the XPS peaks decide whether the secondary phase is in the sample or not. If the secondary phase concentration is high, the BE of the single phase eventually shifts from its original place. Secondly, the concentration of the secondary phase is low, the shape of the peak is asymmetry but the BE value retains its original place

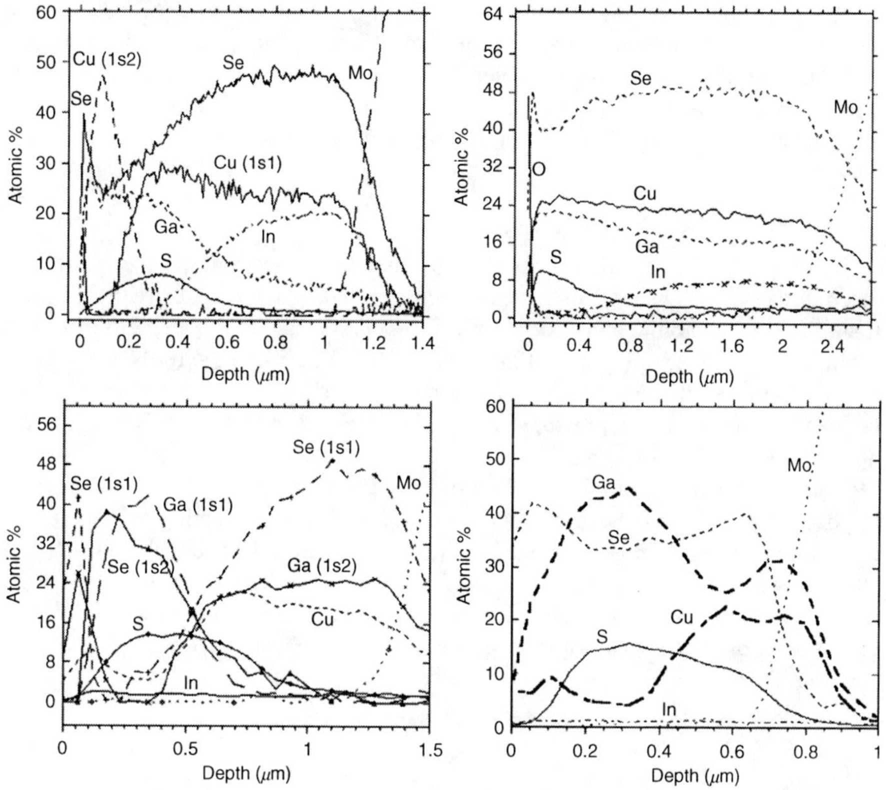

Surface Analyses of I–III–VI$_2$ Compounds

Figure 3.49 AES depth profiles of CIGS thin films with different compositions Cu:(In+Ga):(Se+S) (A) 1:1.49:2.2, (B)1:1.18:2.32, (C) 1:4.89:6.59, and (D) 1:8.02:9.84.

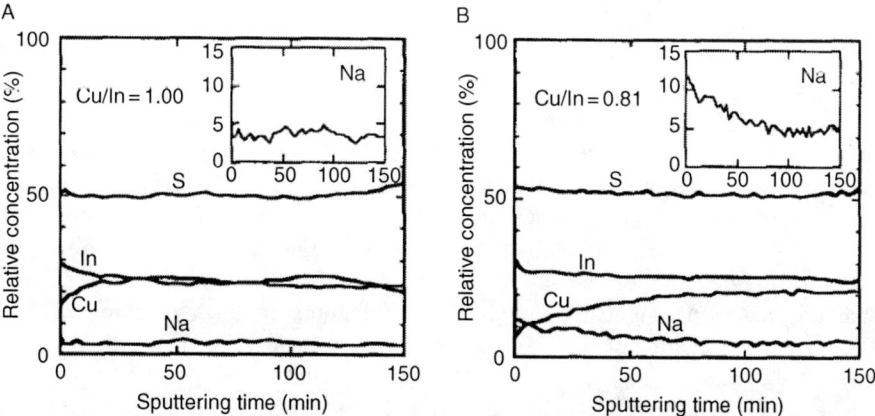

Figure 3.50 AES depth profiles of CuInS$_2$ thin films; (A) Cu/In=1 and (B) 0.81.

[67]. For example, the BE of Cu-2p$_{3/2}$ or Cu-2p$_{1/2}$ is not one and the same in the Cu$_{2-x}$Se and CIS compound whereby the former is secondary phase in the latter.

When a soft X-ray photon with energy of hv is impinged on atom, the photoelectron ejects with kinetic energy of E_{KE} from the atomic orbital where core electron contains binding energy (BE) of E_b. The relation can be written as

$$E_{KE} = hv - E_b. \tag{3.3}$$

By measuring the kinetic energy (KE) of the ejected electron, the BE of the core electron can be evaluated. Since the core electron BE is the characteristic of a particular atom, the atomic constituents of a compound can readily be determined. For example, using typical MgK$_\alpha$ = 1253.6 eV radiation in ESCA 3-MK 11 spectrometer, the X-ray photoelectron spectra can be recorded by a cylindrical mirror analyzer. The chamber Ar pressure of 1×10^{-7} torr, MgK$_\alpha$ as an X-ray source, applied current 20 mA, accelerating voltage 15 kV and band pass energy of 50 eV are typical parameters. The etching of Se starts to occur at 0.2 kV and Cu etching occurs at 1 kV and above for longer time of Ar ion beam etching over the raster area of 10×10 mm^2. In general, the binding energies are calibrated taking C-1S as a standard with a measured typical value of 248.6 eV. The XPS analysis is one of the marvelous techniques to find chemical state of the elements either in the compound or thin film but somewhat notorious to determine the chemical composition of thin film due to its sensitivity.

3.6.1 CuInSe$_2$

The XPS spectrum of (112) oriented p-type CIS single crystal shows core electron BE levels Cu2p$_{1/2}$-952.4, Cu2p$_{3/2}$-932.6, In-3d$_{3/2}$-452.5, In-3d$_{5/2}$-444.9, and Se-3d-54.3 eV. The atomic composition ratio of the elements can be obtained from XPS analysis using simple relation

$$N_1/N_2 = (I_1/S_1)/(I_2/S_2), \tag{3.4}$$

where N is the composition, I is the area of the peak, and S is the sensitive factor of the corresponding element. The composition ratios $N_{Cu}/N_{In}=0.75$ and $N_{In}/N_{Se}=0.49$ are determined from XPS while the EPMA gives an amount of 0.92 and 0.66 for N_{Cu}/N_{In} and N_{In}/N_{Se} in the typical CIS, respectively. A difference in atomic ratios between EPMA and XPS is due to difference in sensitivity factors of the machines or cleaning of surface of the sample by bromine in methanol solution for XPS. There may be another reason that Cu depletes or electro migrates into bulk or In segregation on the surface of the sample. The depletion of Cu is evident that after annealing Au coated sample, the Cu/In atomic ratio changes from 0.74 to 0.94 due to segregation or out diffusion of In from surface to Au that form In–Au to create Cu vacancies [68]. The similar core electron binding energies Cu2p$_{1/2}$-951.9, Cu2p$_{3/2}$-932, In3d$_{3/2}$-452.3, In3d$_{5/2}$-444.7, and Se3d-54.5 eV are observed for CIS thin films [69], whereas etched polycrystalline CIS bulk shows little lower core

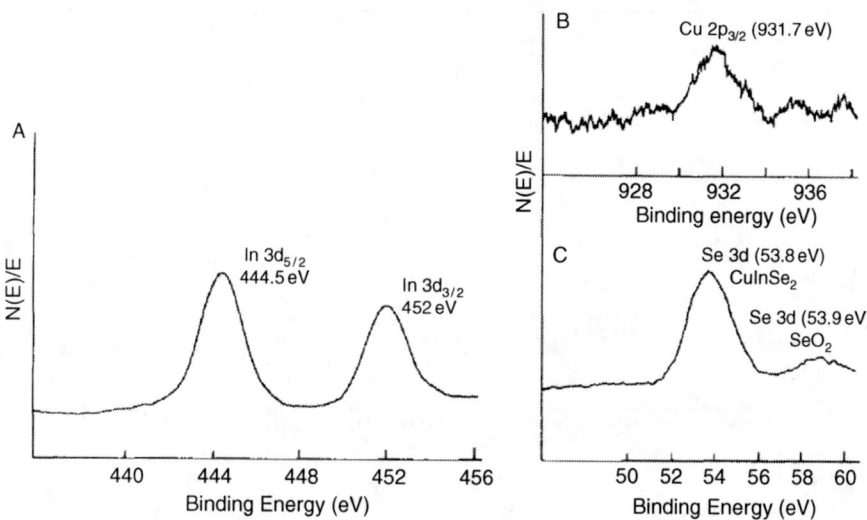

Figure 3.51 XPS spectra of (A) In, (B) Cu, and (C) Se for CuInSe$_2$ thin film grown by spray pyrolysis.

electron BE levels of Cu2p$_{3/2}$-931.6, In3d$_{5/2}$-443.9, and Se-3p$_{3/2}$-160.25 eV [70]. The BE of Cu2p$_{3/2}$-931.7, In3d$_{5/2}$-444.5, In3d$_{3/2}$-452, and Se-3d-53.8 eV for CIS thin films grown by spray pyrolysis technique are observed indicating that no signature of Cu$_2$Se (Cu2p$_{3/2}$-952.16 or CuL$_3$M$_{45}$M$_{45}$-918.2 eV) present. Regarding Se-3d, secondary phases such as In$_2$Se$_3$ (54.6 eV) and SeO$_2$ (Se-3d-53.9 eV) are observed but no traces of Cu$_x$Se phase (54.5 eV), which are virtually different from either elemental Se (55.3 eV) or CIS. Over all, the Cu2p$_{3/2}$ peak is not well pronounced either in the CIS, CuInS$_2$, or its solid solution thin films grown by spray pyrolysis technique due to depletion of Cu on the surface or segregation of In on the surface, as shown in Figure 3.51 [49,71]. The Se layers are deposited onto glass substrates by solution method, followed by vacuum evaporation of In and Cu layers as a stack. The XPS spectra of as-deposited Se/In/Cu stack show binding energies of 932.8, 444.7, and 55.5 eV for Cu2p$_{3/2}$, In3d$_{5/2}$, and Se3d$_{5/2}$, respectively, which are as same as elemental core levels amid Cu, In, and Se are elements in the stack. After annealing the stack layers under vacuum 10^{-5} mbar for 1 h at different temperatures of 150, 200, 250, 300, and 400 °C, the same core levels appear for Cu and In but the Se core level decreases from 55.5 to 54.5 eV indicating formation of CIS [72]. The BEs of In and Se increase to higher energy levels but BE levels of Cu remain reside at the same level with increasing thickness of CIS thin film on Mo coated soda-lime glass substrates. The binding energies of 932.6 and 932.7 eV for Cu-2p$_{3/2}$ peaks are more or less in equal after deposition times of 10 and 310 s for CIS, respectively, but In-3d$_{5/2}$-444.8 9 (10 s), In-3d$_{5/2}$-445.3 (310 s), and Se-3d$_{5/2}$-54.2 (10 s) and Se-3d$_{5/2}$-54.6 eV (310 s) are in different with the deposition time or thickness due to segregation. The In/Cu atomic ratios are 1.04 and 3.72 for 10 and 310 s, respectively [73].

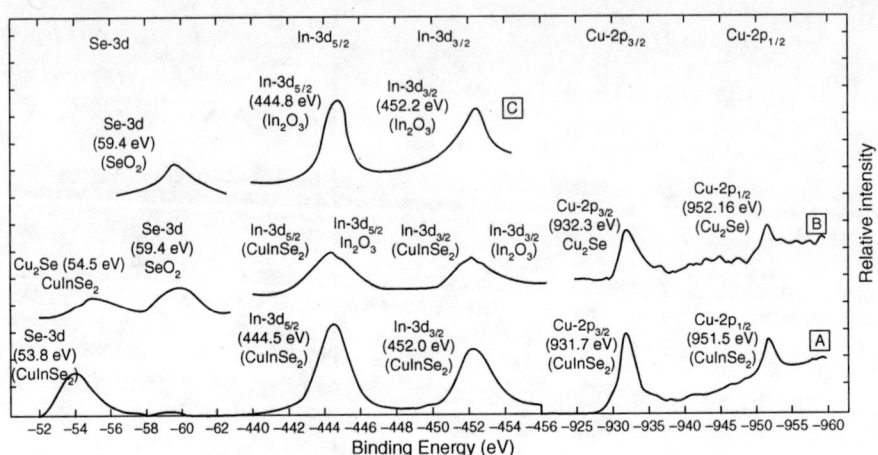

Figure 3.52 XPS spectra of CuInSe$_2$ pellets (A) as-grown, (B) annealed under air at 300 °C for 15 min, and (C) annealed at 350 °C.

The stoichiometric CIS pellets show Cu-2p$_{3/2}$-931.7, Cu-2p$_{1/2}$-951.55, In-3d$_{5/2}$-444.5, In-3d$_{3/2}$-451.9, and Se-3d-53.8 eV peaks in the XPS spectra. After annealing the sample under air at 300 °C for 15 min, the peak positions change to Cu-2p$_{3/2}$-932.3, Cu-2p$_{1/2}$-952.16, and Se-3d-54.5 eV from their original positions indicating formation of Cu$_2$Se phase. Beyond the doubtful reasons, a shoulder peak at 59.4 eV is due to SeO$_2$ phase. A change in In peaks evidences In$_2$O$_3$ phase. The Se and In peaks relate to SeO$_2$ and In$_2$O$_3$ phases become stronger and Cu peaks absent with further increasing annealing temperature to 350 °C, as shown in Figure 3.52 [74]. Alike the CIS sample annealed under dry oxygen at 180 °C for different timings of 0, 4, 10, and 30 min indicate that the Se-3d peak at 53.8 eV gradually disappears and a peak at 59.4 eV generates at the same time. This observation reveals transformation of phase from Se to SeO$_2$. The similar pattern is observed in the In that the transformation occurs from 451.9 and 444.5 eV to 452.2 eV and 444.8 for In-3d$_{3/2}$ and In-3d$_{5/2}$ at the same conditions, respectively. The oxidation layer formed on the CIS thin films with $p = 10^{17}/\text{cm}^3$ grown by three source evaporation is less uniform than that on single crystal due to rougher surface of thin films [67]. The Cu, In, and Se oxides such as CuO, In$_2$O$_3$, and Se$_2$O$_3$ are also observed in the CIS samples grown by electrodeposition. After annealing the samples under air at 350 °C for 30 min, changes in intensities and binding energies are observed indicating that the sharp symmetrical peaks of In and Se oxide phases such as In$_2$O$_3$ and Se$_2$O$_3$ are found to increase while oxide level of Cu decreases revealing fast oxidation of In and Se [75]. The anodic oxide grown on CIS has Cu-2p$_{3/2}$-932.2, Cu-2p$_{3/2}$-931.8, and Cu-2p$_{3/2}$-935.5 eV for CIS, CuO, and Cu$_2$O phases, respectively. They are similar to bulk copper oxides such as CuO and Cu$_2$O whereas thermal oxides, native oxides, and air grown oxides prefer to show In$_2$O$_3$ and SeO$_2$ phases only [76].

The (112) oriented p-type CIS single crystals with carrier concentration (p) of $2-5 \times 10^{17}/\text{cm}^3$ polished and cleaned with a 3% bromine in methonal are oxidized at 180 °C for 2 h under dry oxygen flow that forms indium oxide but no Cu oxide. The estimated thickness of oxidation layer is 700 Å. The depth profile of the layers reveal that the layers are sputtered for 0–24 min, in which either by AES or XPS, there are three regions can be divided as α region for 0–8 min sputter time, β region 8–16 min, and δ region 16–24 min depending on the nature of sputtering curves for Cu, In, and Se signal versus sputtering time. The AES survey scan in the 0–1500 eV range shows that the Cu, In, and Se signals are found at δ region, whereas In, O, and Se are observed but no Cu signal at α region. In the oxidized CIS, In positions of 397 and 404 eV are different from 401 and 407.5 eV in the virgin CIS. The similar kind of oxidation results is observed on InP. The XPS individual scans of In and Se relate to In-$3d_{3/2}$-452.2, In-$3d_{5/2}$-444.8, and Se-3d-59.4 eV indicating In_2O_3 and SeO_2 states. In the β region, the Cu-$2p_{3/2}$-932.3 and Se3d-54.5 eV are related to Cu_xSe phase. Regarding Indium, prime and shoulder peaks are related to In and In_2O_3 phases. In the δ region, Cu-$2p_{3/2}$-931.7, In-$3d_{5/2}$-444.5, In-$3d_{3/2}$-451.9, and Se-3d-53.8 eV are all corresponded to CIS, as shown in Figure 3.53 [67]. The XPS analyses are carried out on hydrogen ion implanted with doses of 120×10^{16} ions/cm^2 at substrate temperature of 300 °C into p-type CIS single crystal grown by Bridgman method. In $3d_{5/2}$-444.7 eV is observed in the as-grown sample, upon hydrogen ion implantation, the In peak position shifts to higher energy level with increasing implantation time from 10, 1330, 4330 to 7930 s. Beyond the doses of 120×10^6 ions/cm^2 hydrogen

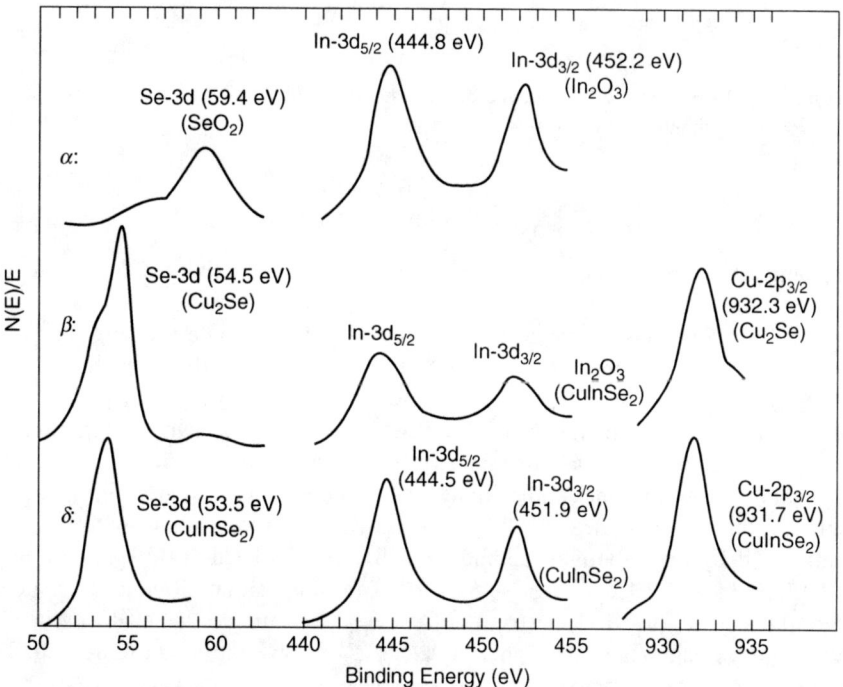

Figure 3.53 XPS spectra of air annealed $CuInSe_2$ with (α) 0–8 min, (β) 8–16 min, and (δ) 16–24 min sputter time.

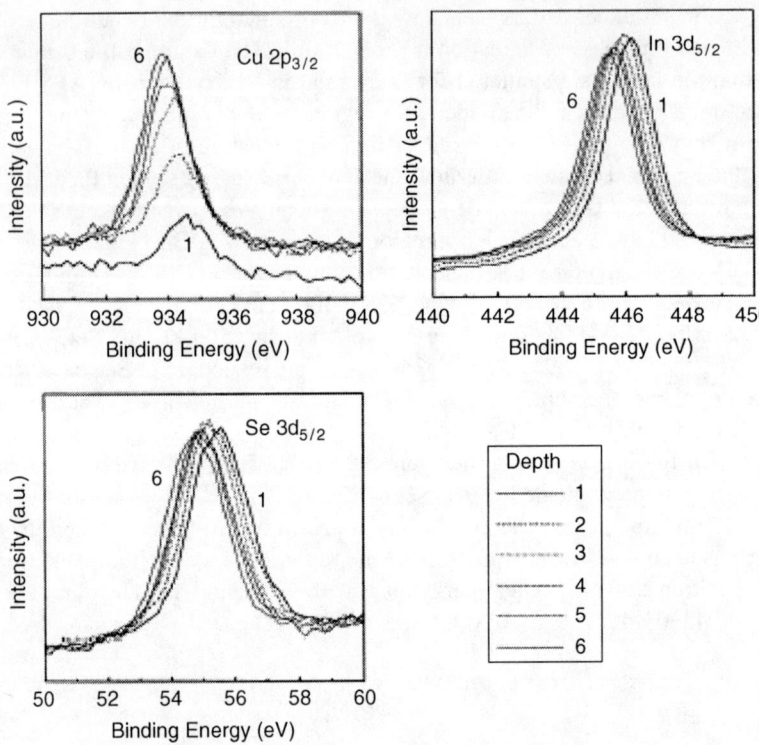

Figure 3.54 XPS spectra of Cu, In, and Se with sputter depth profiling for $CuIn_3Se_5/CuInSe_2$ layer (different depth regions of 1–6).

implantation, the metallic Cu and In are present and the conductivity of the sample also converts into n-type [77].

The $CuIn_3Se_5$ phase segregates on the surface of CIS while growing the layers for thin film solar cell applications. The core electron binding energies of Cu-$2p_{3/2}$, In-$3d_{3/2}$, and Se-$3d_{5/2}$ shift toward lower energy with sputter depth evidencing a phase change from $CuIn_3Se_5$ to CIS, as shown in Figure 3.54. Secondly, a gradual increase of intensity of Cu with depth profiling may be also suggesting a change in composition. The binding energies are higher in the $CuIn_3Se_5$ than those in the CIS [78]. The $CuIn_2Se_{3.5}$ nano-crystallites without Cu_2Se synthesized by hydrothermal method have BE levels of Cu-$2p_{3/2}$-931.5, In-$3d_{5/2}$-444.15, In-$3d_{3/2}$-452.1, and Se$3d_{5/2}$-53.9 eV. The composition of the layers matches with $CuIn_2Se_{3.5}$. The XRD spectrum of $CuIn_2Se_{3.5}$ is similar to the chalcopyrite CIS but the lattice parameters are higher in values as $a = 5.64$ and $c = 11.61$ Å [79]. The p-CIS layers grown onto Mo substrates heat treated for 5 min from room temperature to 240 °C under argon atmosphere show that the Mo forms MoO_3 phase giving peak position at 232.6 eV and another $MoSe_2$ phase is also observed at 229 eV. The Se3d position at 54.5 eV confirms the formation of $MoSe_2$ phase and the Mo $3d_{5/2}$ and $3d_{3/2}$ states are 227.8 and 231 eV, respectively [80].

3.6.2 $CuIn_{1-x}Ga_xSe_2$

The as-grown $CuIn_{1-x}Ga_xSe_2$ thin flim with composition of Cu:In:Ga:Se=22.94: 20.31:7.53:49.23 contains Cu_2Se and CuSe phases, which show Cu-LMM line positions at ∼917.75 and 918.25 eV in the XPS spectra, respectively. After etching sample in KCN, it's composition changes to 21:22:19:7.66:49.9 and the Cu-LMM position of CIGS is observed at 917.75 eV and other XPS peaks concerned to secondary phases disappear. After air exposing polycrystalline CVD $CuGaSe_2$ thin films for 4 months, the BE of $Cu2p_{3/2}$ changes from 932.2 to 932.6 eV due to oxidation. Surprisingly, a change in BE of 932.6 eV is due to contribution of $Cu_{2-x}Se$ phase. The Cu $(OH)_2$ phase forms for further oxidation by evidencing BE of 934.8 eV, which coincides with the reported standard value of 934.4 eV. After KCN etching, the sample retains its original BE of 932.2 eV, as shown in Figure 3.55 [81]. The binding energies of Se-3d, In-$3d_{5/2}$, Cu-$2p_{3/2}$, and Ga-$2p_{3/2}$ in the $CuIn_{1-x}Ga_xSe_2$ increase with increasing x and their bowing parameters being 0.668, 0.238, 0.6, and 0.821 eV, as depicted in Figures 3.56 and 3.57, respectively. The bowing parameter of Ga-$2p_{3/2}$ is high as compared to that of other elements due to larger electronegativity of Ga-2p [82]. BE levels of Cu-$2p_{3/2}$-931.53, In-$3d_{5/2}$-444.19, Ga-$2p_{3/2}$-1117.1, and Se-3d-53.44 eV for the CIGS layers grown by co-evaporation technique slightly shift to lower energy levels, as the thickness of ZnSe is increased on the CIGS layers due to chemical interaction of the elements. The BE levels are Zn-$2p_{3/2}$-1020.75 and Se-3d-52.86 eV for ZnSe [83].

The $CuInGaSe_2$ thin films grown onto conducting glass substrates by electrodeposition show Cu-$2p_{3/2}$-932, In-$3d_{5/2}$-444.8, In-$3d_{3/2}$-453, and Se-$3d_{5/2}$-54.9, Se-$3d_{3/2}$-57 eV but no Ga signal is observed. After etching the samples by chemical

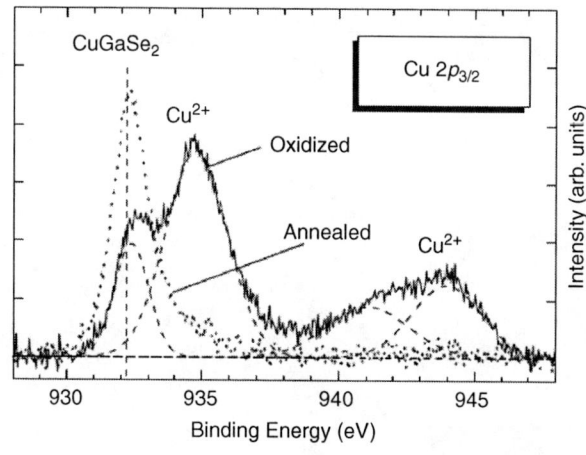

Figure 3.55 XPS spectra of CGS thin film.

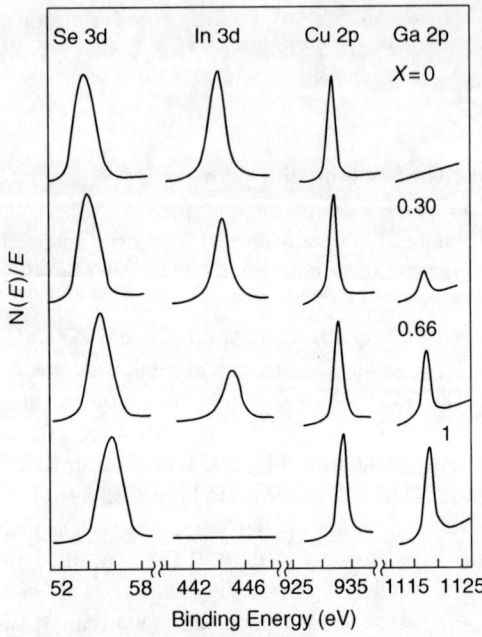

Figure 3.56 XPS spectra of CuIn$_{1-x}$Ga$_x$Se$_2$ thin films.

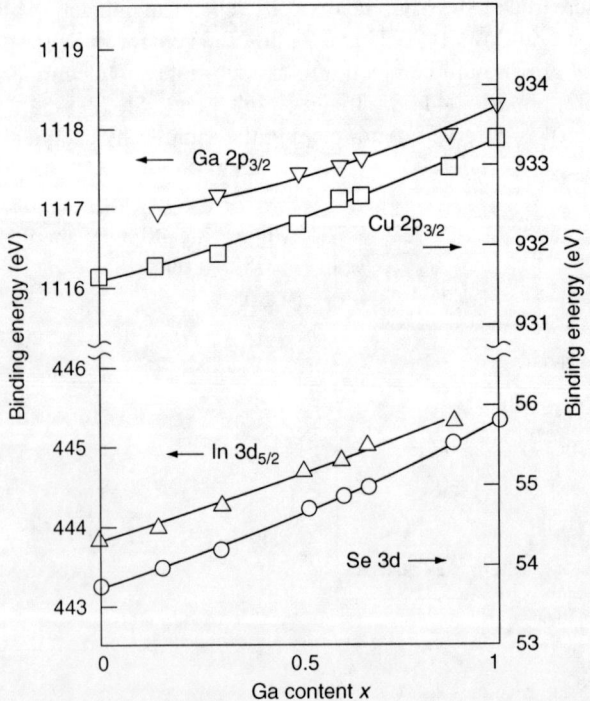

Figure 3.57 Variation in Binding Energies of Cu, In, Ga, and Se.

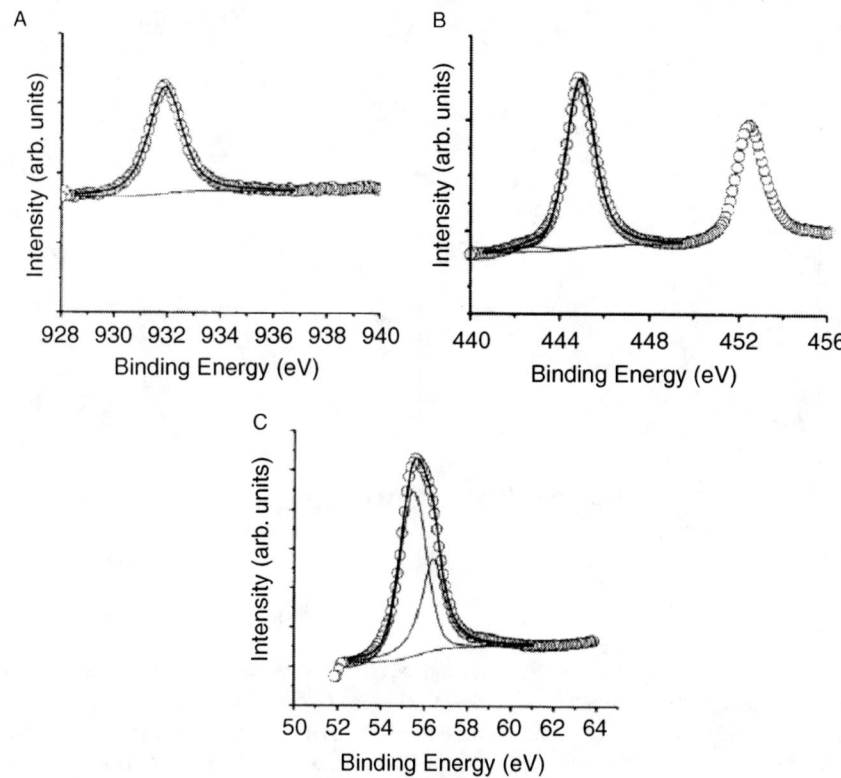

Figure 3.58 XPS spectra of CIGS thin films grown by electrodeposition.

process of various solutions (1)10 ml of HCl(37%) + 10 ml of HNO_3(70%) + 80 ml H_2O for 1 min, (2)1 g of NaOH in 1 ml of H_2O_2(30%) + 40 ml H_2O for 1 min, (3) NH_4OH (35%) for 5 min, (4)Etch-1 etched by NaOH and $Na_2S_2O_4$, and (5)1 g of $K_2Cr_2O_7$ in 10 ml of H_2SO_4 (95–98%) + 490 ml H_2O for 20 s, reveal that the Cu-$2p_{3/2}$ peak position shifts to higher value of around 933.6 eV, which is close to the Cu-$2p_{3/2}$-942 eV of CuO [84]. The similar observation is reported for In and Se. Figure 3.58 shows XPS spectra of as-grown CIGS thin film by electrodeposition. After paraoxide etching, a new peak generates at 58.7 eV in the XPS spectra confirming that the Se forms SeO_2 on the CIGS layers due to oxidation of Se. After etching, the conductivity of the samples turns into n-type [85]. The quality CIGS layers can be obtained by selenization of In/Ga/Cu/In precursor under Se + Ar vapor at 500 °C rather than under Se vapor or vacuum. The binding energies of In are observed to be ~444.4 and 444.2 eV for the CIGS layers formed in the former and latter processes, respectively. This observation indicates that the CIGS layers formed under vacuum contains oxidation of In. The oxidation of Se, that is, SeO_2 (Se3d-58.75 eV) is also observed in vacuum processed CIGS layers, as shown in Figure 3.59. The origin of oxygen could be from atmosphere [86]. The CIGS thin

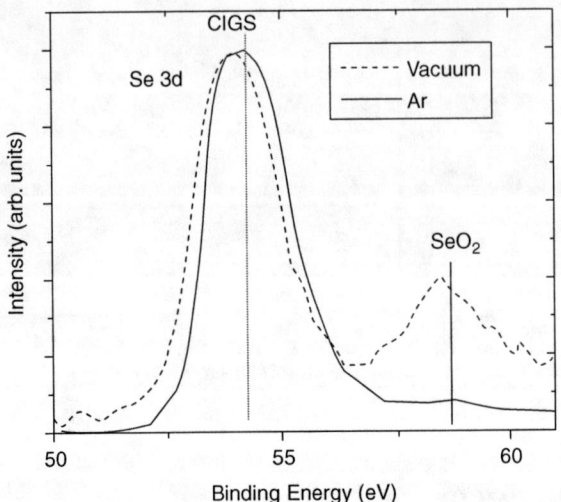

Figure 3.59 XPS spectra of Se for CIGS layers processed under Ar and vacuum.

films are grown by three and two-stage processes in which the three-stage process samples contain In-rich surface (sample-I), whereas two-stage sample had Cu-rich surface (sample-J). The In-rich CIGS sample may be having OVC on its surface. The XPS wide scan survey of CIGS thin film (sample-I) is shown in Figure 3.60A. As usual the oxygen and carbon are found as surface contaminations. The individual scans run for Se, In, Ga, and Cu for both the samples (Figure 3.60B and C) at different etching times such as 0, 20, 40, and 90 min corresponding etching depths of 0, 400, 800, and 1800 Å at the rate of 20 Å/min reveal that the positions of peaks shift with depth. The shifts in peak positions with depth are obviously evidence of change in chemical states or compositions that is nothing but composition grading in the samples. The Ga_2Se_3 phase is observed in both the samples irrespective of Cu-rich or poor [87].

The CIGS thin films left over for several days in outdoor atmosphere enhance the intensities of Se-3d relating to Se(O_x) at 59.5 eV and In3d at 405 eV corresponding to In(O_x) indicating that the In and Se are more affinity to oxygen than other elements do. In the as-grown samples, the source for oxygen is probably Na_2O from glass substrates [88]. After annealing the CIGS layers grown by three source evaporation under oxygen flow at 200 °C for 4 h, the line positions of Cu-$2p_{3/2}$, In-$3d_{5/2}$, and Ga-$2p_{3/2}$ in the XPS spectra shift to lower values from their original positions [89]. The CIGS layers are treated with chemical solution containing 1.4 mM CdAc and with different 0.5, 1, 1.5 M NH_3 solution concentrations at 60 °C for 15 min. With increasing NH_3 concentration, the Cu-$2p_{3/2}$ position shifts from ~932.5 to 932.8 eV and Se position shifts from ~54.5 to 54.9 eV but no change in In MNN peak at 406 eV is observed and the O1s position at 530.35 eV relates to OH^-. It could be speculated that the $Cd(OH)_2$ may be formed on the CIGS surface but more

Surface Analyses of I–III–VI$_2$ Compounds

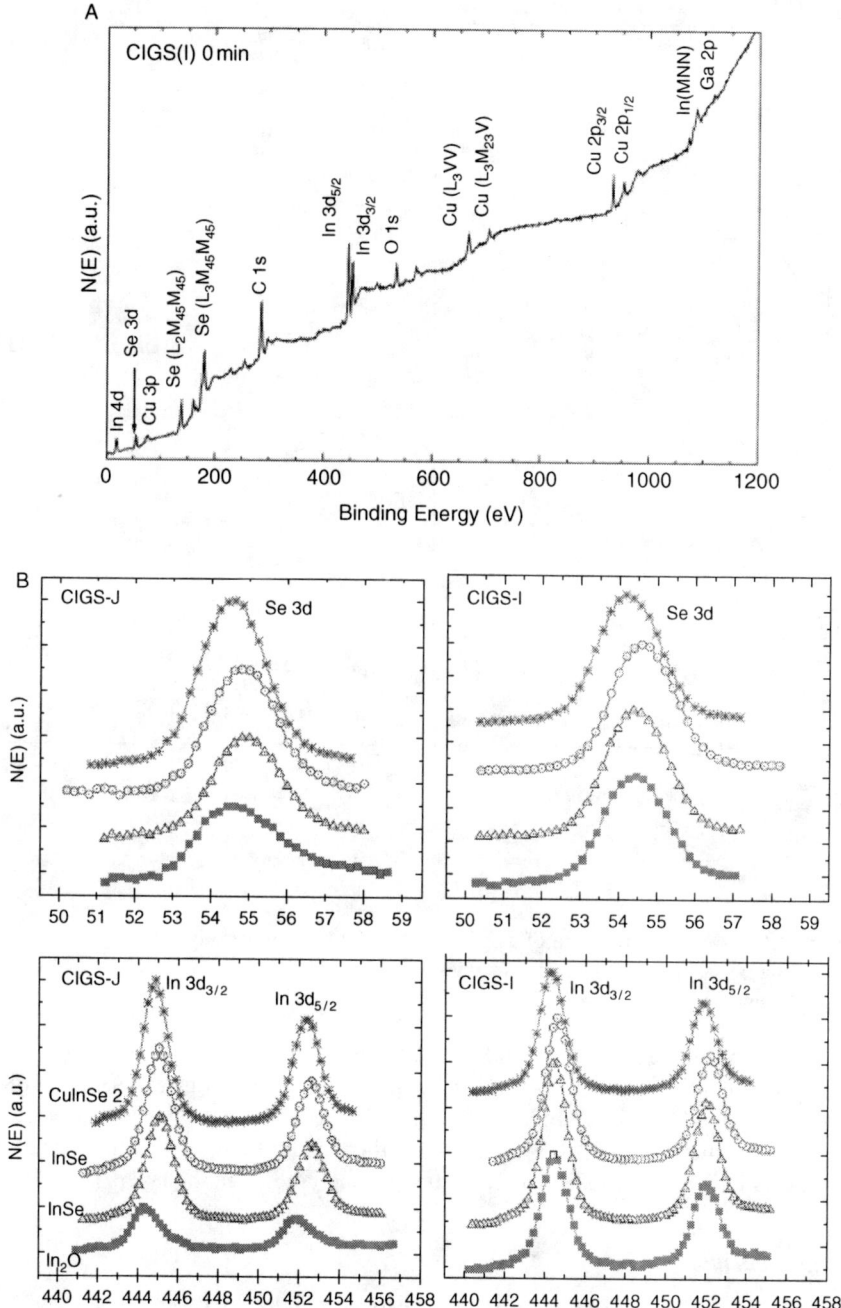

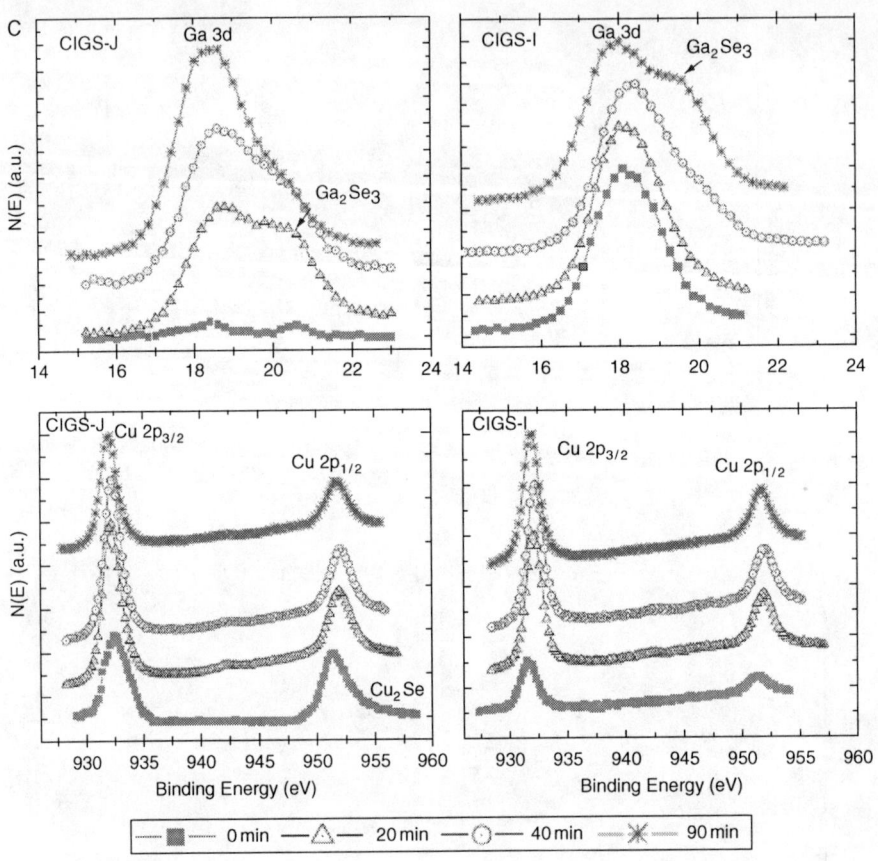

Figure 3.60 (A) XPS wide scan of as-grown CIGS by three-stage processes (sample-I), (B) Individual scans of Se and In for samples I and J, and (C) Narrow scans of Ga and Cu for samples I and J.

investigations are needed in this direction [90]. After etching CIGS/In$_2$S$_3$ for 6 min, the binding energies of S2p$_{1/2}$-163.1, S2p$_{3/2}$-161.9 eV for sulfur and Se3p$_{1/2}$-168.3 and Se3p$_{3/2}$-160.3 eV for selenium are observed. The Se BE peak positions gradually increase with increasing etching time. On the other hand sulfur peak position gradually disappears indicating interdiffusion of S and Se and grading in the layers [91].

3.6.3 CuInS$_2$

Figure 3.61 shows core electron BE values Cu-2p$_{3/2}$-932.5, Cu-2p$_{1/2}$-952.5, In-3d$_{5/2}$-444.8, In-3d$_{3/2}$-451.8, and S-2p has doublets as 161.2 and 162.4 eV for CuInS$_2$ nanorods. The sulfur core levels may probably be considered as Cu–S and In–S bonding, respectively. No other phases are found in the 400–450 nm

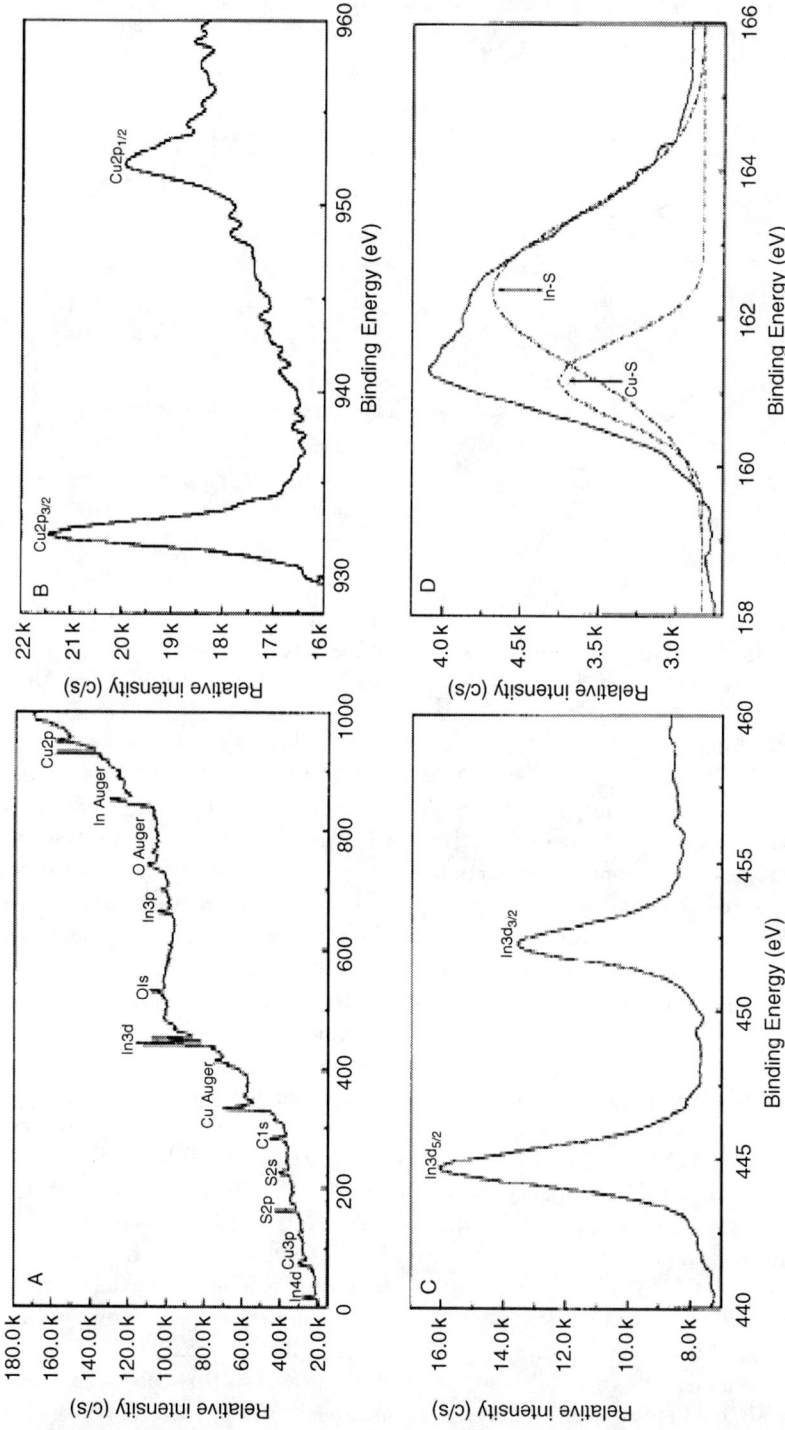

Figure 3.61 XPS spectra of $CuInS_2$ nanorods; (A) full scan, (B) Cu, (C) In, and (D) S.

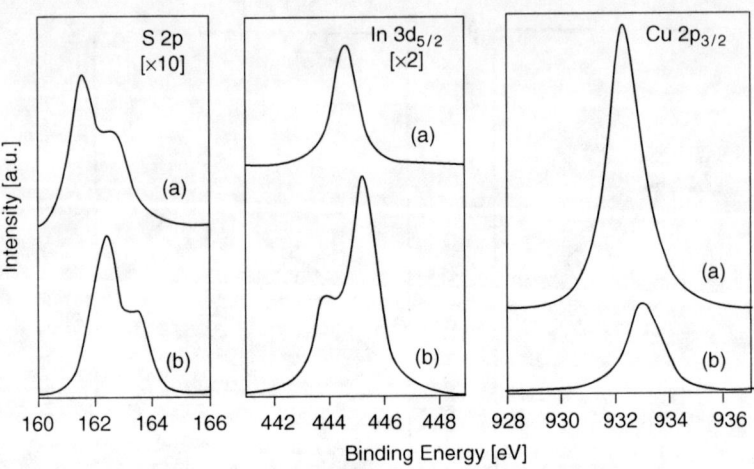

Figure 3.62 XPS spectra of CuInS$_2$ thin films grown by MBE with different compositions; (A) Cu/In = 2.8 ± 0.6, S/(Cu + In) = 0.8 ± 0.2, Cu-rich (p-type) (B) Cu/In = 0.4 ± 0.1 and S/(Cu + In) = 1.0 ± 0.2, In-rich (n-type).

length and 20–25 nm diameter CuInS$_2$ nanorods prepared by hydrothermal process at 180 °C for 15 h [92]. The similarity is found in the case of sulfur [93]. The Cu2p$_{3/2}$-932.6, In-3d$_{5/2}$-444.3, and In3d$_{3/2}$-451.9 eV are observed in the CuInS$_2$ thin films grown by spray pyrolysis technique matching with the standards [71]. The Cu-rich CuInS$_2$ thin films grown by MBE technique show Cu2p$_{3/2}$-932.35, In3d$_{5/2}$-444.63, S2p$_{3/2}$-161.53 eV, whereas they are 932.95, 445.3/443.89, and 162.34 eV in the In-rich CuInS$_2$ samples, respectively (Figure 3.62) [43]. The In-3d peaks shift to higher energy level with increasing In content in the CuInS$_2$ films grown onto Si (111) substrates but Cu and S peaks retain at the same positions [94]. The CuInS$_2$ films are grown by sputtering using Cu$_2$S and In$_2$S$_3$ cold pressed target with Cu$_2$S/In$_2$S$_3$ ratio of 2. The Cu2p$_{3/2}$ of CuInS$_2$ and Cu$_2$S as a bump are observed in the XPS spectrum. The BE value of Cu2p$_{3/2}$ is lower in the Cu$_2$S than that in the CuInS$_2$. After KCN etching, the bump disappears indicating that the Cu$_2$S phase is etched off, as shown in Figure 3.63 [95]. After etching the CuInS$_2$ thin films grown onto SnO$_2$:F coated glass substrates by spray pyrolysis the oxygen position at 530 eV still exists in the spectra due to oxidation of In or Cu metals but intensity of another oxidation peak at 531.8 eV decreases due to surface contamination. The peak positions of Cu-2p$_{1/2}$ (952.5 eV), Cu-2p$_{3/2}$ (932.6 eV), In3d$_{3/2}$ (452.5 eV), In3d$_{5/2}$ (445 eV), and S2p (162 eV) also shift to slightly higher energy levels and their intensities also increase as shown in Figure 3.64. The chlorine peak at 200 eV still exists but intensity decreases giving impression that the bulk layer is still contaminated with Cl [96].

After etching CuInS$_2$ thin films in K$_2$SO$_4$, followed by In acidic media solution treatment, there are slightly changes in Cu and In binding energies. However, no change is found in S (Table 3.6) but FWHM of sulfur line decreases [97]. The difference between Cu-2p$_{3/2}$ and In3d$_{5/2}$ is ~487.4 eV as well as between Cu2p$_{3/2}$ and Sp$_{3/2}$ is ~770 eV in the CuInS$_2$ thin films irrespective of whether Cu-rich or In-rich

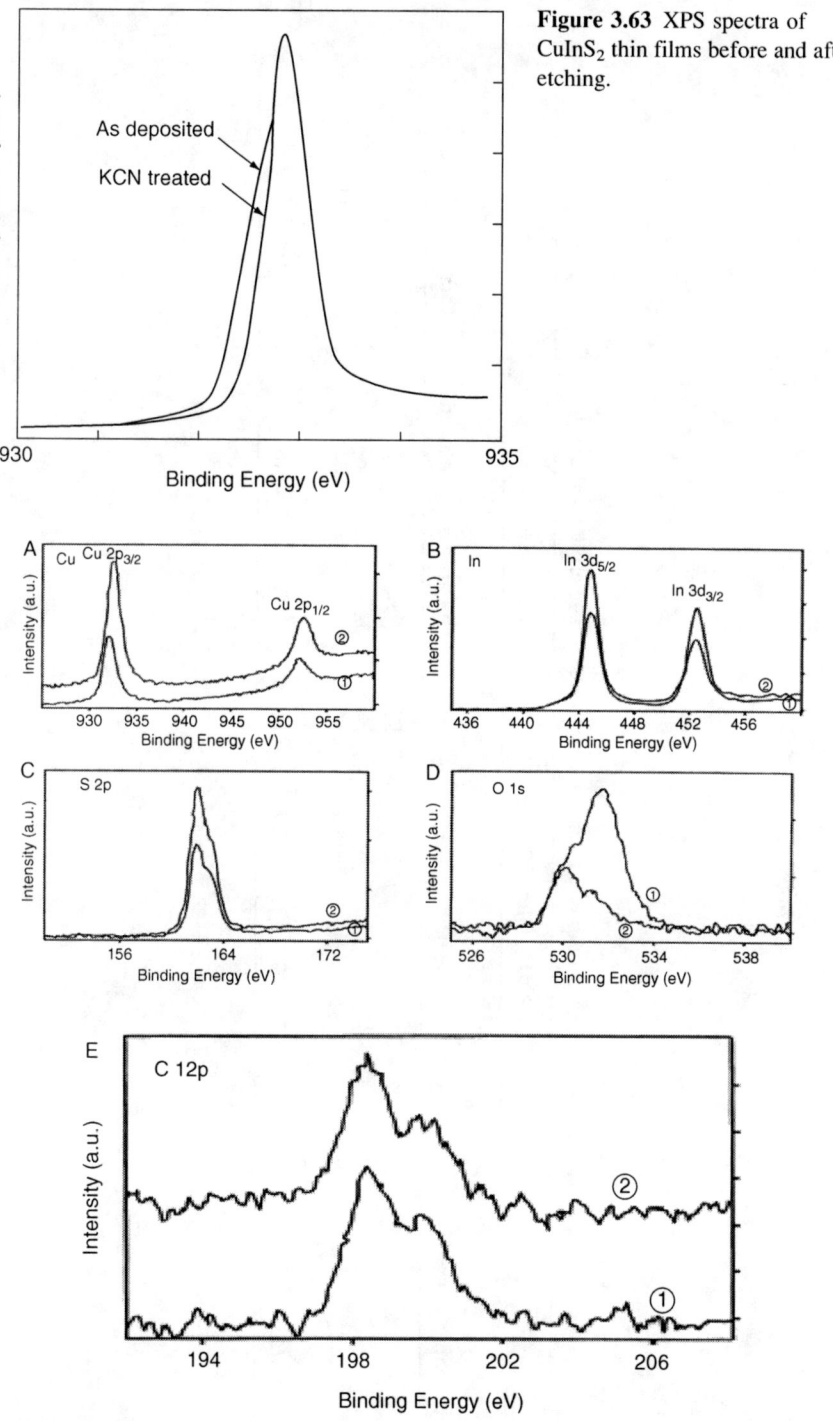

Figure 3.63 XPS spectra of CuInS$_2$ thin films before and after etching.

Figure 3.64 XPS spectra of CuInS$_2$ (1) as-grown and (2) after etching; (A) Cu, (B) In, (C) S, (D) O, and (E) Cl.

Table 3.6 Binding energies of Cu- and In-rich CuInS$_2$ single crystal and thin film

Line	Single crystal	CuInS$_2$ thin film [98–100]				CuInS$_2$ thin film [97]			
		Cu-rich Cu/In = 1.25	Cu-rich	In-rich		As-grown	Etched	CuS	Cu or In foil
Cu2p$_{3/2}$	932.2	932.6	932.1	932.8	932.5	932.1	932.4	932.0	932.5
Cu2p$_{1/2}$	–	–	–	–	952.5	–	–	–	–
In3d$_{5/2}$	444.7	445	444.7	445.4	444.6	444.4	444.8	–	443.7
In3d$_{3/2}$	–	–	–	–	452.5	–	–	–	–
S2p	162	161.7	162.1	162.7	162.1	161.4	161.3	162.2	–

Table 3.7 Binding energies of $CuIn(Se_{1-x}S_x)_2$, its binary and oxide compounds

S.No	Compound	Cu ($2p_{3/2}$)	In $3d_{3/2}$	In $3d_{5/2}$	S ($2S_{1/2}$)	Se($3d_{5/2}$)	Ref.
1	$CuIn(Se_{1-x}S_x)_2$, $x=0.0$	931.8	451.9	444.5	–	53.8	[49]
2	0.3	932.2	452	444.5	225.9	53.8	
3	0.5	932.4	452.1	444.6	226	53.9	
4	1.0	932.8	452.2	444.7	NA	–	
5	CuS	931.6	–	–	225.6	–	[101,102]
6	Cu_2S	931.8	–	–	225.3	–	
7	Cu_2Se	932.3	–	–	–	54.5	[103]
8	CuSe	932.2				54.5	[74]
9	Cu_2O	935.5					
10	Cu_2O	932.7					[104,105]
11	CuO	931.8					[74]
12	CuO	933.6					[101–103]
13	In_2S_3 (InS)		452.4	444.4			[106]
14	In_2S_3		445.1	444.3			[107]
15	In_2Se_3		452.6	444.9		54.6	[108]
16	In_2O_3		452.15	444.8			
17	SeO_2					58.9	
18	SeO_2					59.4	103, 104

$CuInS_2$. The FWHM of $Cu2p_{3/2}$ for Cu-rich and In-rich $CuInS_2$ thin films are 1.67 and 1.42 eV, respectively. After KCN etching Cu-rich $CuInS_2$ thin films, the FWHM value of 1.67 eV comes down to 1.4 eV evidencing the presence of CuS phase in the Cu-rich films. The XPS valence band spectra reveal that the $E_f - E_v = 1.3$ eV and 0.5 eV are in In-rich and the etched Cu-rich $CuInS_2$ films or single crystals, respectively. The valence band peak position at 3 eV disappears in etched Cu-rich $CuInS_2$ thin films due to removal of CuS phase [98]. The $Cu2p_{3/2}$-934 and S2p-162 eV are observed in spray deposited $CuInS_2$ thin films. The O1s state at 530 eV for Indium oxide and 531.8 eV for oxygen from surface contaminations are observed in the $CuInS_2$ thin films. After annealing the samples under H_2S, the In–O phase (530 eV) disappears but oxygen state remains exist. In fact the spray deposited films contain 6–8% of oxygen [50].

Na doped $CuInS_2$ layers are deposited onto Mo or Ti coated soda-lime glass substrates with Ga doping Ga/(In + Ga) = 0.18, Cu/(In + Ga) = 0.87 and without by hybrid process using H_2S gas. The XPS spectra of these films show no difference in $Cu2p_{3/2}$, $Cu2p_{5/2}$, $In3d_{5/2}$, $In3d_{3/2}$, and S2p peak positions or intensities by adding Ga or not. However, semicore level spectra of $CuInS_2$:Na show overlapping of Ga3d and In4d at 19 eV in the BE scan range of 15–23 eV. The intensity of valence band spectrum also decreases in the 4.5–11.5-eV range but no shift in band position. These results suggest that the VBM remains constant by adding Ga into $CuInS_2$:Na but CBM shifts to upper level by

0.03 eV. The CIS/CdS solar cells made with CuInS$_2$:Na:Ga absorber demonstrate higher open circuit voltage of 0.8 V otherwise it would be 0.76 V without Ga doping. The efficiency of cells increases from 10.6 to 11.2% with Na and Ga doping [93].

3.6.4 CuIn(Se$_{1-x}$S$_x$)$_2$

Since growth temperature of quaternary CuIn(Se$_{1-x}$S$_x$)$_2$ films grown by spray pyrolysis technique is same except $x=1$ there may be possibility to have secondary phases in the quaternary system amid lack of growth temperature with varying x value. Therefore BE values of all secondary phases are taken to rule out the phases, which are feasible to form in the system. Table 3.7 shows that the BE values of Cu-2p$_{3/2}$ peak for CuIn(Se$_{1-x}$S$_x$)$_2$ films increase from 931.8 to 932.8 eV with increasing x value from 0 to 1. The BE values of 931.8 and 931.6 eV for Cu$_2$S and CuS, respectively do not fall in between 931.8 and 932.8 therefore Cu$_2$S and CuS presence may be ruled out. The BE values of In-3d$_{3/2}$-452.6 and In-3d$_{5/2}$-444.9 eV for In$_2$Se$_3$ and In$_2$S$_3$ phases do not fall between 452.2 and 451.9 of CuIn(Se$_{1-x}$S$_x$)$_2$ system therefore the presence of In$_2$Se$_3$ and In$_2$S$_3$ can also easily be ruled out. However, there

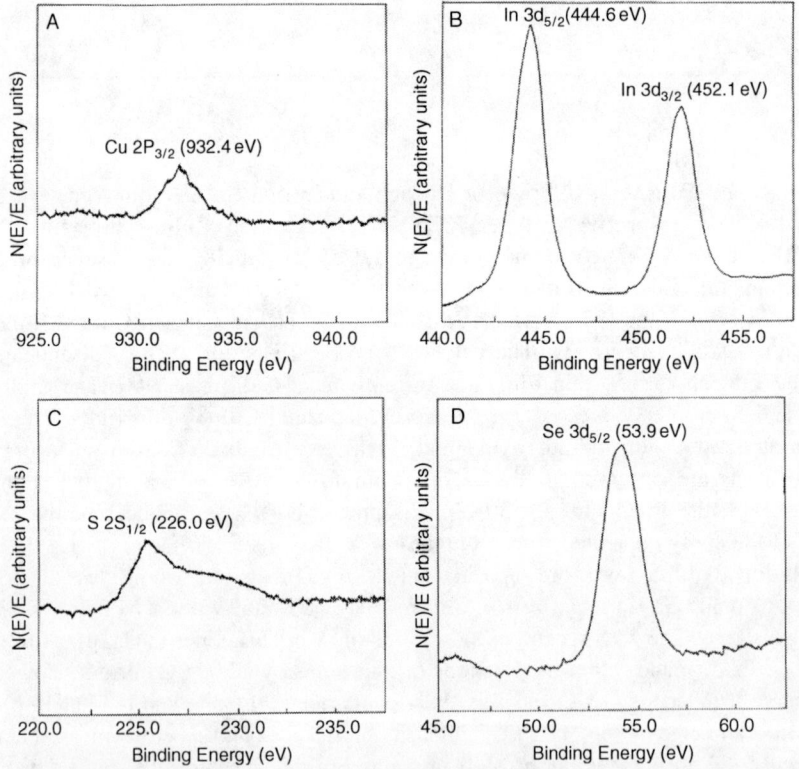

Figure 3.65 XPS spectra of CuIn(Se$_{1-x}$S$_x$)$_2$ thin film ($x=0.5$); (A) Cu, (B) In, (C) S, and (D) Se.

is a small asymmetry in the Cu and S peaks of $CuIn(Se_{0.5}S_{0.5})_2$ thin films that could be assigned to the secondary phases of Cu_2S or CuS. The other binary phases such as Cu_2O, In_2O_3, Cu_2Se, and $CuSe$ could be ruled out even though the BE values of those phases fall in the quaternary system, for example, no asymmetry is found in oxygen peak for Cu_2O and In_2O_3, whereas the Se peak does not exhibit asymmetry for Cu_2Se or $CuSe$. As well as In peak does not show asymmetry to assigning for In_2S_3 and In_2Se_3 phases. No type of secondary phases is detected from the optical absorption spectra and XRD analysis. The quantity of segregation of secondary phase might be low for XRD experimental detection limits. Only in $x=0.5$ samples, asymmetry is found, which might be due to the effect of growth temperature. The $CuInS_2$ and CIS films are deposited at the substrate temperature of 277 and 350 °C, respectively. In comparison, the deposition temperature for CuIn $(Se_{0.5}S_{0.5})_2$ is chosen higher than that of $CuInS_2$ therefore there could be possibility of segregation of secondary phases such as Cu_2S or CuS. Suppose, if some other secondary phases are in the samples, the shape of the peaks to the corresponding phases should have been asymmetry (Figure 3.65). The extraction of composition from XPS analysis is somewhat not amenable, since the technique is very sensitive but results from surface etched samples may be reliable [49].

References

[1] L. Zang, Private Communication (2010); R. Howland, L. Benatar, Private Communication (2010); S. Niki, P.J. Fons, A. Yamada, Y. Lacroix, H. Shibata, H. Oyanagi, et al., Appl. Phys. Lett. 174 (1999) 1630.
[2] I.M. Dharmadasa, R.P. Burton, M. Simmonds, Solar Energy Mater. Solar Cells 90 (2006) 2191.
[3] N.M. Shah, J.R. Ray, K.J. Patel, V.A. Kheraj, M.S. Desai, C.J. Panchal, et al., Thin Solid Films 517 (2009) 3639.
[4] N.M. Shah, C.J. Panchal, V.A. Kheraj, J.R. Ray, M.S. Desai, Sol. Energy 83 (2009) 753.
[5] W.N. Shafarman, J. Zhu, Mater. Res. Soc. Symp. Proc. 668 (2001) H2.3.1.
[6] T. Wada, S. Nishiwaki, Y. Hashimoto, T. Negami, Mater. Res. Soc. Symp. Proc. 668 (2001) H2.1.1.
[7] T. Negami, T. Satoh, Y. Hashimoto, S. Nishiwaki, S.-I. Shimakawa, S. Hayashi, Sol. Energy Mater. Sol. Cells 67 (2001) 1.
[8] T. Satoh, Y. Hashimoto, S.I. Shimakawa, S. Hayashi, T. Negami, Solar Energy Mater. Solar Cells 75 (2003) 65.
[9] A.M. Hermann, C. Gonzalez, P.A. Ramakrishnan, D. Balzar, N. Popa, P. Rice, et al., Sol. Energy Mater. Sol. Cells 70 (2001) 345.
[10] R.C. Valderrama, P.J. Sebastian, J.P. Enriquez, S.A. Gamboa, Solar Energy Mater. Solar Cells 88 (2005) 145.
[11] T. Delasol, A.P. Samantilleke, N.B. Chaure, P.H. Gardiner, M. Simmonds, I.M. Dharmadasa, Solar Energy Mater. Sol. Cells 82 (2004) 587.
[12] M. Winkler, J. Griesche, I. Konovalvo, I. Penndorf, I. Wienke, O. Tober, Sol. Energy 77 (2004) 705.

[13] Y.B. He, T. Kramer, A. Polity, R. Gregor, W. Kriegseis, I. Osterreicher, et al., Thin Solid Films 431 (2003) 231.

[14] A. Bollero, M. Grossberg, T. Raadik, J.F. Trigo, J. Herrero, M.T. Gutierrez, Mater. Res. Soc. Symp. Proc. 1165 (2009) M02–M06.

[15] C. Guillen, J. Herrero, M.T. Gutierrez, F. Briones, Thin Solid Films 480–481 (2005) 19.

[16] R. Cayzac, F. Boulch, M. Bendahan, P. Lauque, P. Knauth, Mater. Sci. Eng. B 157 (2009) 66.

[17] Y.B. He, W. Kriegseis, T. Kramer, A. Polity, M. Hardt, B. Szyszka, et al., J. Phys. Chem. Solids 64 (2003) 2075.

[18] Th. Deniozou, N. Esser, S. Siebentritt, P. Vogt, R. Hunger, Thin Solid Films 480–481 (2005) 382.

[19] J. Piekoszewski, J.J. Loferski, R. Beaulieu, J. Beall, B. Roessler, J. Shewchun, Solar Energy Mater. 2 (1980) 363.

[20] J. Muller, J. Nowoczin, H. Schmitt, Thin Solid Films 496 (2006) 364.

[21] F.B. Dejene, Solar Energy Mater. Solar Cells 93 (2009) 577.

[22] S. Ishizuka, A. Yamada, H. Shibata, P. Fons, K. Sakurai, K. Matsubara, et al., Solar Energy Mater. Solar Cells 93 (2009) 792.

[23] M.A. Contreras, M. Romero, D. Young, in: 3rd World Confernce on Photovoltaic Energy Conversion, 2003, S4OB124-2864.

[24] S. Marsillac, S. Dorn, R. Rocheleau, E. Miller, Solar Energy Mater. Solar Cells 82 (2004) 45.

[25] T. Mise, T. Nakada, Solar Energy Mater. Solar Cells 93 (2009) 1000.

[26] T. Tokado, T. Nakada, in: 3rd World Conference on Photovoltaic Energy Conversion, 2003 (2PD366-p539).

[27] D. Rudmann, D. Bremaud, M. Kaelin, H. Zogg, A.N. Tiwari, in: 25th European Photovoltaic Solar Energy Conference, 2005 (4BO.6.2).

[28] D. Rudmann, F.-J. Haug, M. Kaelin, H. Zogg, A.N. Tiwari, G. Bilger, Mater. Res. Soc. Symp. Proc. 668 (2001) H3.8.1.

[29] L. Zhang, Q. He, W.L. Jiang, F.F. Liu, C.J. Li, Y. Sun, Solar Energy Mater. Solar Cells 93 (2009) 114.

[30] T. Schlenker, V. Laptev, H.W. Schock, J.H. Werner, Thin Solid Films 480–481 (2005) 29.

[31] C.J. Hibberd, M. Ganchev, M. Kaelin, K. Emits, A.N. Tiwari, in: 33rd IEEE Photovoltaic Specialist Conference, 2005, (339-08050850533).

[32] D.Y. Lee, B.T. Ahn, K.H. Yoon, J.S. Song, Solar Energy Mater. Solar Cells 75 (2003) 73.

[33] H. Zachmann, S. Heinker, A. Braun, A.V. Mudryi, V.F. Gremenok, A.V. Ivaniukovich, et al., Thin Solid Films 517 (2009) 2209.

[34] R. Caballero, C.A. Kaufmann, T. Eisenbarth, M. Cancela, R. Hesse, T. Unold, et al., Thin Solid Films 517 (2009) 2187.

[35] F.B. Dejene, V. Alberts, J. Phys. D Appl. Phys. 38 (2005) 22.

[36] D. Bremaud, D. Rudmann, M. Kaelin, K. Ernits, G. Bilger, M. Dobeli et al., Thin Solid Films 515(2007) 5857.

[37] K. Ramanathan, G. Teeter, J.C. Keane, R. Noufi, Thin Solid Films 480–481 (2005) 499.

[38] J.D. Wilson, R.W. Birkmire, W.N. Shafarman, in: 33rd IEEE Photovoltaic Specialist Conference, 2008, (629_08051080553).

[39] S.S. Kulkarni, G.T. Koishiyev, H. Moutinho, N.G. Dhere, Thin Solid Films 517 (2009) 2121.
[40] D.B. Mitzi, M. Yuan, W. Liu, A.J. Kellock, S.J. Chey, L. Gignac, et al., Thin Solid Films 517 (2009) 2158.
[41] M. Powalla, G. Voorwinden, D. Hariskos, P. Jackson, R. Kniese, Thin Solid Films 517 (2009) 2111.
[42] S. Nakamura, A. Yamamoto, Solar Energy Mater. Solar Cells 75 (2003) 81.
[43] W. Calvet, C. Lehmann, T. Plake, C. Pettenkofer, Thin Solid Films 480–481 (2005) 347.
[44] T. Unold, T. Enzenhofer, K. Ellmer, Mater. Res. Soc. Symp. Proc. 865 (2005) F16.5.1.
[45] E. Rudigier, J.A. Garcia, I. Luck, J. Klaer, R. Scheer, J. Phys. Chem. Solids 64 (2003) 1977.
[46] C. Amory, J.C. Bernede, E. Halgand, S. Marsillac, Thin Solid Films 431–432 (2003) 22.
[47] M.A. Contreras, J. Tuttle, A. Gabor, A. Tennant, K. Ramanathan, S. Asher, et al., in: First World Conference on Photovoltaic Energy Conversion, Hawaii, 1994, p. 68.
[48] J.E. Leisch, R.N. Bhattacharya, G. Teeter, J.A. Turner, Solar Energy Mater. Solar Cells 81 (2004) 249.
[49] S.R. Kodigala, V. Sundara Raja, Scr. Mater. 44 (2001) 771.
[50] S. Marsillac, M.C. Zouaghi, J.C. Bernede, T.B. Nasrallah, S. Belgacem, Solar Energy Mater. Solar Cells 76 (2003) 125.
[51] V. Alberts, M. Klenk, E. Bucher, Jpn. J. Appl. Phys. 39 (2000) 5776.
[52] S. Kuranouchi, A. Yoshida, Thin Solid Films 343–344 (1999) 123.
[53] C.X. Qiu, I. Shih, Solar Energy Mater. 15 (1987) 219.
[54] A. Zouaoui, M. Lachab, M.L. Hidalgo, A. Chaffa, C. Llinares, N. Kesri, Thin Solid Films 339 (1999) 10.
[55] S. Nishiwaki, T. Satoh, S. Hayashi, Y. Hashimoto, S. Shimakawa, T. Negami, et al., Sol. Energy Mater. Sol. Cells 67 (2001) 217.
[56] T. Yamaguchi, Y. Yamamoto, A. Yoshida, Solar Energy Mater. Solar Cells 67 (2001) 77.
[57] S. Ishizuka, K. Sakurai, A. Yamada, K. Matsubara, P. Fons, K. Iwata, et al., Solar Energy Mater. Solar Cells 87 (2005) 541.
[58] T. Negami, T. Satoh, Y. Hashimoto, S. Shimakawa, S. Hayashi, M. Muro, et al., Thin Solid Films 403–404 (2002) 197.
[59] M.A. Contreras, J. Tuttle, A. Gabor, A. Tennant, K. Ramanathan, S. Asher, et al., Solar Energy Mater. Solar Cells 41–42 (1996) 231.
[60] T. Wada, N. Kohara, S. Nishiwaki, T. Negami, Thin Solid Films 387 (2001) 118.
[61] M. Hartmann, M. Schmidt, A. Jasenek, H.W. Schock, F. Kessler, K. Hertz, et al., in: 28th IEEE Photovoltaic Specialists Conference, 2000, p. 638.
[62] S. Spiering, A. Eicke, D. Hariskos, M. Powalla, N. Naghavi, D. Lincot, Thin Solid Films 451 (2004) 562.
[63] K. Kondo, H. Sano, K. Sato, Thin Solid Films 326 (1998) 83.
[64] L.L. Kazmerski, D.L. Sprague, R.B. Cooper, J. Vac. Sci. Technol. A 15 (1978) 249.
[65] Y.L. Wu, H.Y. Lin, C.Y. Sun, M.H. Yang, H.L. Hwang, Thin Solid Films 168 (1989) 113.
[66] T. Watanabe, H. Nakazawa, M. Matsui, H. Ohbo, T. Nakada, Solar Energy Mater. Solar Cells 49 (1997) 357.
[67] L.L. Kazmerski, O. Jamjoum, P.J. Ireland, S.K. Deb, R.A. Mickelsen, W. Chen, J. Vac. Sci. Technol. A 19 (1981) 467.

[68] H. Neumann, M.V. Yakushev, R.D. Tomlinson, Cryst. Res. Technol. 38 (2003) 676.
[69] S.T. Lakshmikumar, A.C. Rastogi, J. Appl. Phys. 79 (1996) 3585.
[70] D. Cahen, P.J. Ireland, L.L. Kazmerski, F.A. Thiel, J. Appl. Phys. 57 (1985) 4761.
[71] P. Rajaram, R. Thangaraj, A.K. Sharma, O.P. Agnihotri, Solar Cells 14 (1985) 123.
[72] K. Bindu, C.S. Kartha, K.P. Vijayakumar, T. Abe, Y. Kashiwaba, Solar Energy Mater. Solar Cells 79 (2003) 67.
[73] D. Schmid, M. Ruck, H.W. Schock, Sol. Energy Mater. Sol. Cells 41/42 (1996) 281.
[74] G.K. Padam, G.L. Malhotra, S.K. Gupta, Solar Energy Mater. 22 (1991) 303.
[75] K.T.L. De Silva, W.A.A. Priyantha, J.K.D.S. Jajanetti, B.D. Chithrani, W. Siripala, K. Blake, et al., Thin Solid Films 382 (2001) 158.
[76] L.L. Kazmerski, O. Jamjoum, J.F. Wagner, P.J. Ireland, K.J. Bachmann, J. Vac. Sci. Technol. A 1 (1983) 668.
[77] K. Otte, G. Lippold, D. Hirsch, R.K. Gebhardt, T. Chasse, Appl. Surf. Sci. 179 (2001) 203.
[78] S.H. Kwon, B.T. Ahn, S.K. Kim, K.H. Yoon, J. Song, Thin Solid Films 323 (1998) 265.
[79] Y. Jin, C. An, K. Tang, L. Huang, G. Shen, Mater. Lett. 57 (2003) 4267.
[80] R.J. Matson, O. Jamjoum, A.D. Buonaquisti, P.E. Russell, L.L. Kazmerski, P. Sheldon, et al., Solar Cells 11 (1984) 301.
[81] R. Caballero. C. Guillen, M.T. Gutierrez, C.A. Kaufmann, Prog. Photovoltaics Reg. Appl. 14 (2006) 145; R. Wurz, A. Meeder, D.F. Marron, Th. Schedel-NIedrig, K. Lips, Mater. Res. Soc. Symp Proc. 865 (2005) (F5.36.1;14099).
[82] Y. Yamaguchi, J. Matsufusa, A. Yoshida, J. Appl. Phys. 72 (1992) 5657.
[83] T. Schulmeyer, R. Hunger, W. Jaegermann, A. Klein, R. Kniese, M. Powalla, in: 3rd World conference on the Photovoltaic Energy Conversion, 2003, p. 364.
[84] K.H. Yoon, W.J. Choi, D.H. King, Thin Solid Films 372 (2000) 250.
[85] T. Delsol, M.C. Simmonds, I.M. Dharmadasa, Sol. Energy Mater. Sol. Cells 77 (2003) 331.
[86] R. Caballero, C. Maffiotte, C. Guillen, Thin Solid Films 474 (2005) 70.
[87] C. Calderon, P. Bartolo-Perez, O. Rodriguez, G. Gordillo, Microelectron. J. 39 (2008) 1324.
[88] M. Ruck, D. Schmid, M. Kaiser, R. Schaffler, T. Walter, H.W. Schock, Solar Energy Mater. Solar Cells 41–42 (1996) 335.
[89] U. Rau, D. Braunger, R. Herberholz, H.W. Schock, J.-F. Guillemoles, J. Appl. Phys. 86 (1999) 497.
[90] Q. Nguyen, G. Bilger, U. Rau, H.W. Schock, Mater. Res. Soc. Symp. Proc. 763 (2003) B8.17.1.
[91] S. Gall, N. Barreau, S. Harel, J.C. Bernede, J. Kessler, Thin Solid Films 480–481 (2005) 138.
[92] J. Xiao, Y. Xie, R. Tang, Y. Qian, J. Solid State Chem. 161 (2001) 179.
[93] T. Abe, S. Kohiki, K. Fukuzaki, M. Oku, T. Watanabe, Appl. Surf. Sci. 174 (2001) 40.
[94] W. Calvet, H.J. Lewerenz, C. Pettenkofer, Thin Solid Films 431–432 (2003) 317.
[95] Y. Yamamoto, T. Yamaguchi, T. Tanaka, N. Tanahashi, A. Yoshida, Solar Energy Mater. Solar Cells 49 (1997) 399.
[96] M.C. Zouaghi, T.B. Nasrallah, S. Marsillac, J.C. Bernede, S. Belgacem, Thin Solid Films 382 (2001) 39.
[97] M. Aggour, U. Storkel, C. Murrell, S.A. Campbell, H. Jungblut, P. Hoffmann, et al., Thin Solid Films 403–404 (2002) 57.
[98] R. Scheer, H.J. Lewerenz, J. Vac. Sci. Technol. A 12 (1994) 56.

[99] Y. Shi, Z. Jin, C. Li, H. An, J. Qiu, Appl. Surf. Sci. 252 (2006) 3737.
[100] S. Bini, K. Bindu, M. Lakshmi, C.S. Kartha, K.P. Vijajakumar, Y. Kashiwaba, *et al.*, Renewable Energy 20 (2000) 405.
[101] V.G. Bhide, S. Salkalachen, A.C. Rastogi, C.N. Rao, M.S. Hegde, J. Phys. D 14 (1981) 1647.
[102] S. Seal, L. Bracho, S. Shukla, J. Vac. Sci. Technol. A 17 (1999) 2950.
[103] K. Sieghbahn, C. Vordling, A. Fanlman, R. Nordberg, S.E. Karlson, I. Lindgren, *et al.*, ESCA-Atomic, Molecular and Solid State Structure Studies by Means of Electron Spectroscopy, Almquist and Wiksells, Uppsala, Sweden, 1967.
[104] P.J. Pinhero, J.W. Anderegg, D.J. Sordelet, T.A. Lograsso, D.W. Delaney, P.A. Thiel, J. Mater. Res. 14 (1999) 3185.
[105] S. Badrinarayana, A.B. Mandale, S.R. Sarkar, J. Mater. Res. 11 (1996) 1605.
[106] L.L. Kazmerski, P.J. Ireland, P. Sheldon, T.L. Chu, S.S. Chu, C.C. Lin, J. Vac. Sci. Technol. A 17 (1980) 1061.
[107] G.D. Nichols, D.A. Zatko, Inorg. Nucl. Chem. Lett. 15 (1979) 401.
[108] P.W. Palmberg, G.E. Raich, R.E. Weber, N.C. Mac Donald, Hand book of Auger Electron Spectroscopy, Physical Electronics Industries, Eden Prairie, MN, 1975.

4 Structural Properties of I–III–VI$_2$ Absorbers

The I–III–VI$_2$ compounds such as CuInSe$_2$, CuGaSe$_2$, CuInS$_2$, Cu(In$_{1-x}$Ga$_x$)Se$_2$, and CuIn(Se$_{1-x}$S$_x$)$_2$ solid solutions virtually exhibit either tetragonal with space group of $I\bar{4}2d$, that is, chalcopyrite or cubic (zinc blende) with space group of $I\bar{4}2m$, that is, sphalerite structure. In addition, CuAu (CA), CuPt, and order vacancy compound (OVC) structures exist in the compounds. In the chalcopyrite structure (tetragonal), (112), (204, 220), and (116, 312) are the prime intensity peaks in which (112) is the preferred orientation or the highest intensity peak in most of the cases, (204, 220) being the next intensity peak and (116, 312) the least one. Its mainly depend on growth recipes such as substrate temperature, composition of Cu, In, and Se, *etc*. The CuInSe$_2$ thin films with Cu/In ratio of 1.18 and 0.83 grown onto 7059 corning glass by MBE show (112) and (204/220) as intensity peaks, respectively [1]. The odd and low intensity reflections such as (101), (103), (211), (213), (105), (301), *etc.*, are the characteristic chalcopyrite peaks, which enable us to confirm the structure to be chalcopyrite. However, in some cases it is difficult to distinguish these peaks from noisy signals of X-ray diffraction (XRD) pattern because of their low intensities particularly in the thin films. On the other hand, splits in (204, 220) and (116, 312) reflections to (204), (220); (116), (312); and (316), (332) support the chalcopyrite structure. The Miller indices follow as (hkl) and ($h,k,l/2$) for chalcopyrite and sphalerite structures, respectively.

4.1 Crystal Structure of I–III–VI$_2$ Compounds

CuInSe$_2$, CuInS$_2$, and other I–III–VI$_2$ (ABC$_2$) compounds crystallize in the chalcopyrite structure. The name "chalcopyrite" (CH) originally came from the mineral CuFeS$_2$ compound. The lattice is a body centered tetragonal (space group $I\bar{4}2d$) with 4 atoms per unit cell. Each A and B-atom is tetrahedrally coordinated to four C-atoms while each C-atom is tetrahedrally coordinated to two A-atoms and two B-atoms in an ordered manner. The chalcopyrite structure can be regarded as a superstructure of zinc blende (sphalerite) structure arising from an ordered substitution of zinc blende metal atoms of valence Z (Figure 4.1). This ordered substitution leads to the doubling of the unit cell size in the z-direction. If A- and B-atoms are randomly distributed, the sphalerite structure would result. The precise location of C-atoms in the chalcopyrite structure depends on the strengths of A–C and B–C

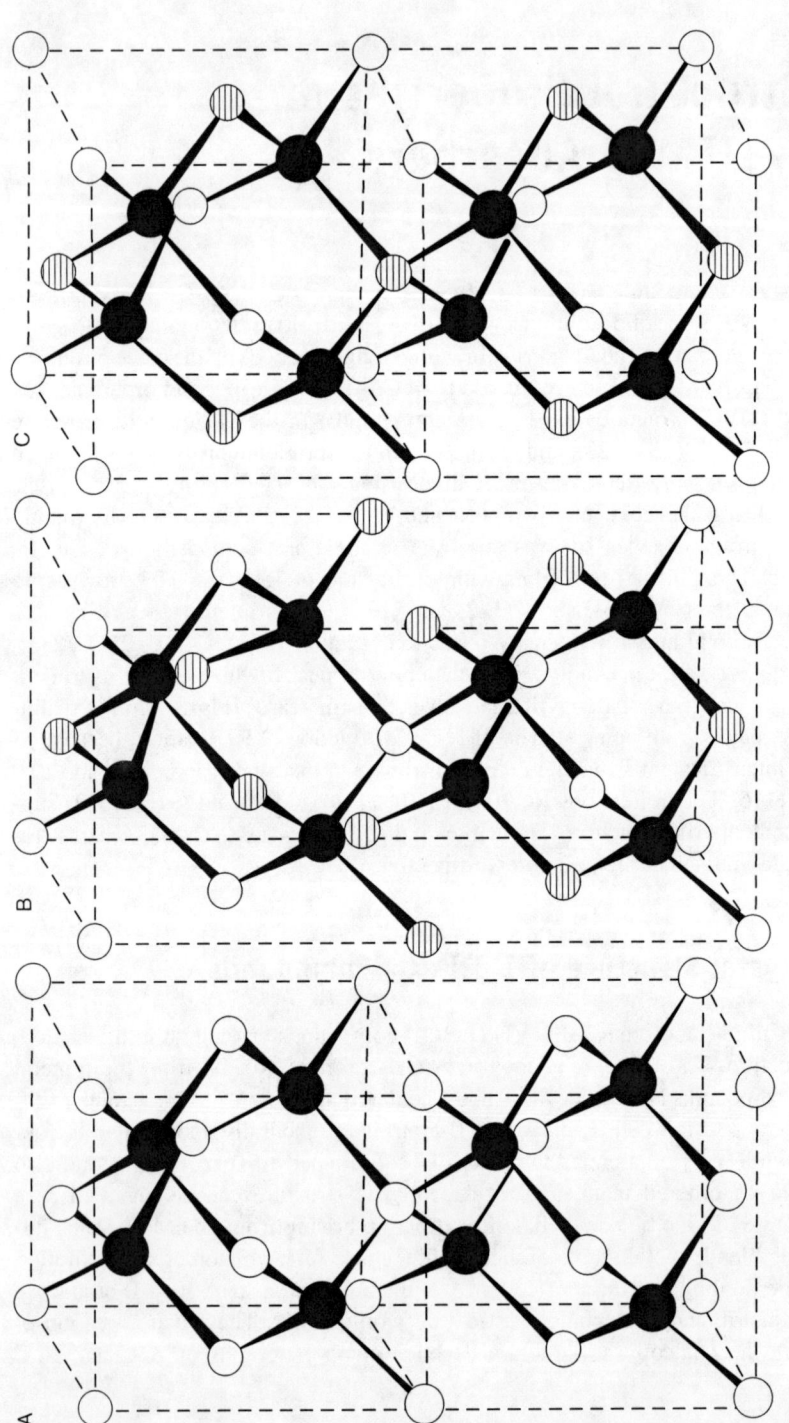

Figure 4.1 Crystal structure of CuInSe$_2$ (A) Sphalerite or Zinc blende (○ Zn atom (Cu or In) and ●Se atom), (B) Chalcopyrite (○Cu, ◎ In, and ●Se), and (C) CuAu (CA) (○Cu, ◎ In, and ●Se).

interactions. Robbins et al. [2] suggested that there is a stronger bonding of C-atoms with its two nearest A-atoms presumably because of A-atom d-electron contribution to the bonding. The result is that the C-atoms are shifted from their ideal positions by a small amount of λ, the free parameter of chalcopyrite structure [3]. The lattice constant ratio c/a in the chalcopyrite structure is generally different from the ideal value of 2. The quantity $[2-(c/a)]$ is a measure of the tetragonal distortion. Accordingly, the atomic coordinates of ABC_2 in a chalcopyrite structure are given by

A-atom (0,0,0), (1/2,0,1/4), (1/2,1/2,1/2), (0,1/2,3/4)
B-atom (1/2,1/2,0), (0,1/2,1/4), (0,0,1/2), (1/2,0,3/4)
C-atom (3/4,1/4+λ,1/8), (1/4,3/4-λ,1/8),
(1/4-λ,1/4,3/8),(3/4+λ,3/4,3/8),
(3/4,1/4-λ,5/8),(1/4,3/4+λ,5/8),
(1/4+λ,1/4,7/8),(3/4-λ,3/4,7/8).

The CH can be converted into CA by transforming alternating (100) A cation plane by $a/\sqrt{2}$ in a <100> direction, as shown in Figure 4.1C [4]. In the OVC, ($2V_{Cu}^- + In_{Cu}^{2+}$) defect pair plays a major role to construct the crystal structure. In the $(Cu_2Se)(In_2Se_3)_{1+x}$ formula, the x value decides the phase of OVC. For example, the compound is $CuInSe_2$ for $x=0$, $CuIn_2Se_{3.5}$ $x=1$, $CuIn_3Se_5$ $x=2$, $CuIn_4Se_{6.5}$ $x=3$, $CuIn_5Se_8$ $x=4$, etc.

The differential thermal analysis (DTA), XRD, and EPMA/EDAX sarcastically contribute to build phase diagram of $CuInSe_2$ [5]. The phase diagram runs in the Cu_2Se–In_2Se_3 tie line versus temperature from room temperature to 1100 °C. The single phase chalcopyrite $CuInSe_2$ exists in the 50–55 mol.% of In_2Se_3 correspondingly the Cu/In ratio lies between 1 and 0.82 as well as in the temperature range from room temperature to 800 °C. The $CuInSe_2$ phase along with Cu_2Se exists for the Cu/In > 1.0, whereas the $Cu_2In_4Se_7$ and $CuIn_3Se_5$ phases form for the Cu/In < 0.82, as illustrated in Figure 4.2A [6]. Recently, the quasi pseudo phase diagram has been constructed in the interest range of Cu/In ratio from 0.82 to 1. The α-$CuInSe_2$ chalcopyrite exists for the Cu at.% from > 22 to 24 in the temperature range of RT to 800 °C that is viable range for solar cell applications. There are α-$CuInSe_2$ and β-$CuIn_3Se_5$ phases at room temperature. The α-$CuInSe_2$ exists for the Cu atomic composition ratio between 24 and 24.5% at room temperature. However, no single phase $CuInSe_2$ is observed for the Cu 25 at.%, that is, stoichiometric composition, as depicted in Figure 4.2B. By adding Ga and Na to the $CuInSe_2$, the α-phase formation likely widens Cu at.% range [7]. There are also some other reports on the phase diagram [8].

4.2 CuInSe$_2$

The transmission electron microscopy (TEM) and XRD are important tools to investigate structure of $CuInSe_2$ compound. The TEM analysis essentially requires a very thin layer to conduct experiments otherwise the analysis would give spurious results. A thick sample can be made into thin sample by mechanical polishing and sputter etching. As much as thick samples are necessary for XRD analysis that would give fruitful results.

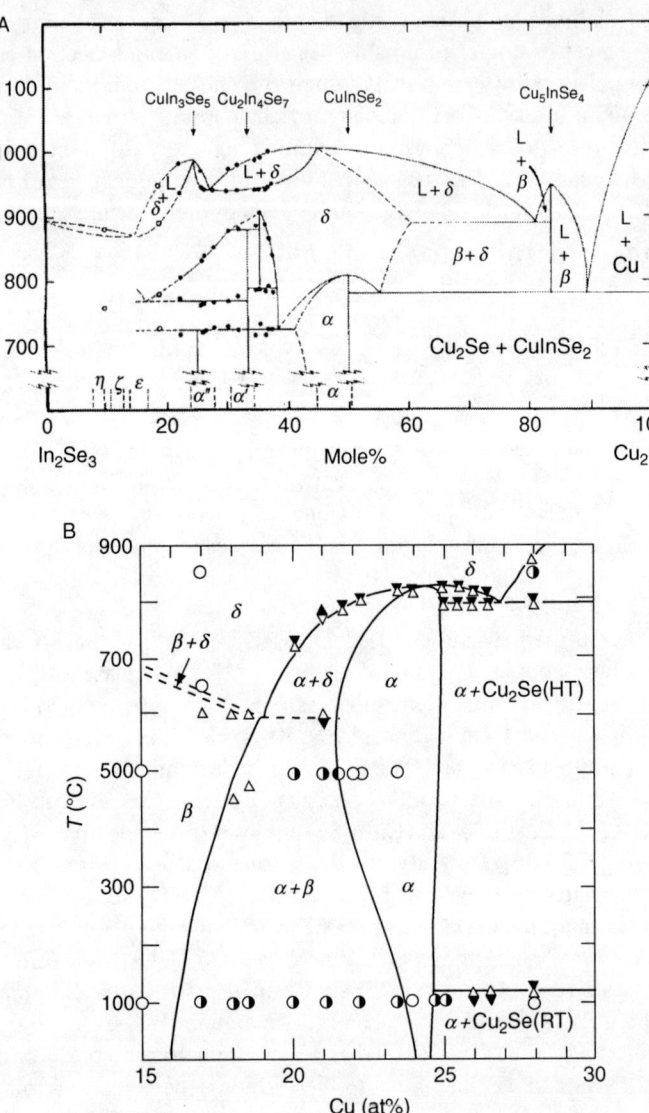

Figure 4.2 (A) The Cu_2Se–In_2Se_3 pseudobinary phase diagram. (B) The Cu_2Se–In_2Se_3 pseudobinary phase diagram; α-chalcopyrite, β-chalcopyrite, δ high temperature (ZnS type structure), α'- $Cu_2In_4Se_7$; α''- $CuIn_3Se_5$, η, ε and ζ - cubic and hexagonal; β - defect ○ single phase region, ● two phase region, ▲ DTA heating, ▼ DTA cooling.

4.2.1 TEM analysis of CuInSe$_2$

There are two modes in the TEM analysis; one is image mode and another one is diffraction mode. The image of the specimen is formed in the image plane then magnified. The diffraction pattern created by the electron beam in the back focal

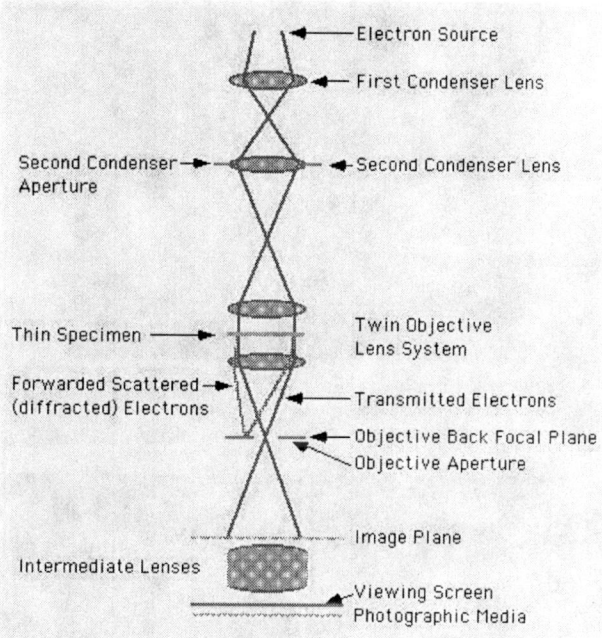

Figure 4.3 Transmission electron microscopy imaging method.

plane is transformed to the screen in the diffraction mode, as shown in Figure 4.3. The electron wavelength (λ_e) in Å can be derived at operating voltage (V) as $\lambda_e = (150/V)^{1/2}$, the interplanar distance d is given by

$$d = L\lambda_e/R, \tag{4.1}$$

where L is the camera constant and R is the radius of the diffraction ring. In order to determine L, selected area electron diffraction (SAED) of Au is used as a standard. There are several methods to extract the film from the substrate to carry out TEM analysis. (i) by sputter etching backside of the substrate for example if carbon is used as a substrate, (ii) KCl substrate easily dissolves in the water then an interested experimental thin film from KCl floats on the water, and (iii) an interested thin film is extracted from the glass substrate by peeling off in diluted HF solution, followed by cleaning in distilled water.

The CuInSe$_2$ (CIS) single crystal grown by traveling heater method treated in InCl$_3$ and CH$_3$CSNH$_2$ solution forms a thin CuInS$_2$ layer on it. The SAED pattern of CIS, which had composition of Cu:In:Se = 25:25:50 is shown in Figure 4.4 revealing chalcopyrite structure. The SAED pattern of CuInS$_2$ thin film formed on CIS with composition of Cu:In:S = 23.4:26:50.6 also shows chalcopyrite structure [9]. The Cu, In, and Se are evaporated by migration-enhanced epitaxial

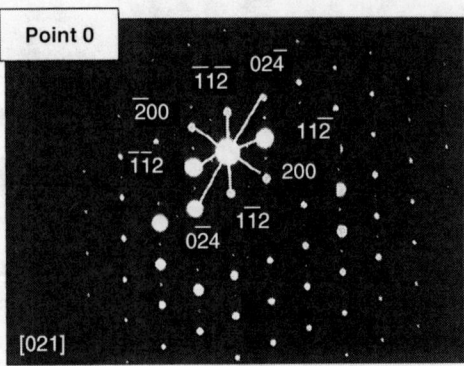

Figure 4.4 SAED pattern of CuInSe$_2$ single crystal.

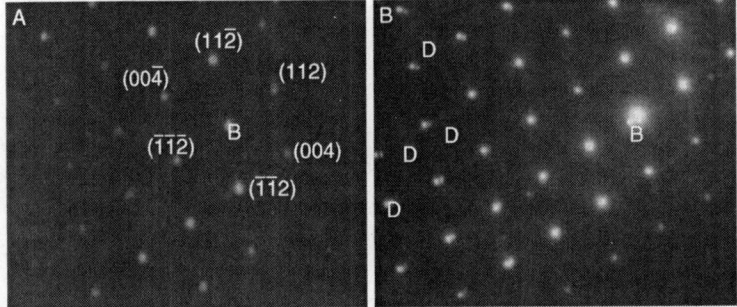

Figure 4.5 SAED pattern of CuInSe$_2$. (A) Matrix region and (B) Island region (zone axis B = $[\bar{1}\bar{1}0]$), D denotes double spotty pattern.

(MEE) reactor in a MBE chamber, whereby the substrates are rotating. The CuInSe$_2$ films with 8 nm thickness are first deposited onto GaAs (100) substrates at substrate temperature of 360 °C, followed by 40 nm thick same layer at 412 °C and again back to the normal growth temperature of 360 °C. This kind of deposition process encourages to sustain defect free layers to some extent. The Cu-rich CIS layers with Cu/In = 1.09 are deposited using above recipe onto GaAs containing 2° misoriented toward [011] direction with terrace width of ~80 nm that causes to reduce the defect density in the epitaxial layer for TEM analysis. The SAED patterns are recorded on matrix and island regions along the $[\bar{1}\bar{1}0]$ zone axis on the CIS thin film, which show spotty pattern indicating single crystal nature. However, double spotty is observed on island region on contrast to matrix regions indicating slightly misorientation, as shown in Figure 4.5 [10]. The SAED spotty pattern of CuInSe$_2$ thin film deposited onto glass substrates by spray pyrolysis technique indicates single crystal nature, as shown in Figure 4.6A. The periodicity follows in x-axis as (h0h), y-axis ($\bar{h}0h$) and z-axis (002h), as shown in schematic Figure 4.6B. The chalcopyrite characteristic peaks such as (101), (103), (301), and (501) are found but the

Structural Properties of I–III–VI$_2$ Absorbers

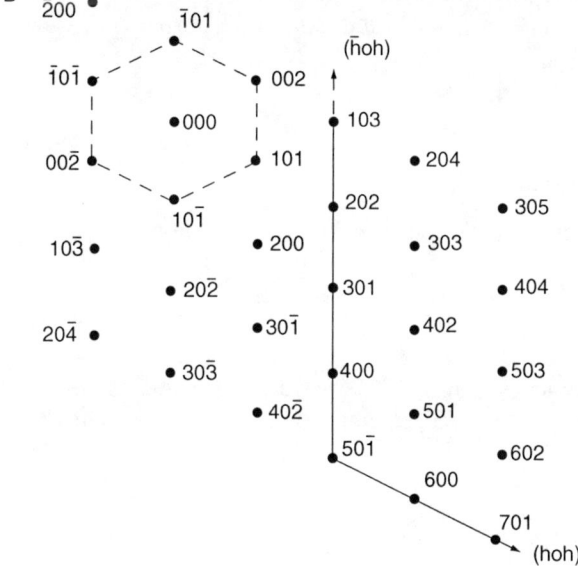

Figure 4.6 Selected area electorn diffraction of CuInSe$_2$ thin film grown onto glass substrates by spray pyrolysis technique; (A) experimental and (B) graphical.

most intensity peaks such as (112), (204, 220), and (116,312) absent in the SAED pattern unlike XRD pattern. Some of the diffraction indices match with the CuIn$_{2.5}$Se$_{3.5}$ phase. Therefore it can be concluded that the sample contains mixed phases of CuInSe$_2$ and CuIn$_{2.5}$Se$_{3.5}$. In the SAED pattern, the presence of some of the low intensity extra spots might be due to combination of microtwins, grain boundaries or secondary phases [11]. The typical SAED ring pattern of CuInSe$_2$ thin film grown by spray pyrolysis technique reveals polycrystalline nature, as shown in Figure 4.7, which is not bright and some of the rings are diffused due to over load of thickness for the electron diffraction [12]. The clear SAED ring pattern of the electrodeposited polycrystalline CIS thin films is also shown in Figure 4.7B [13].

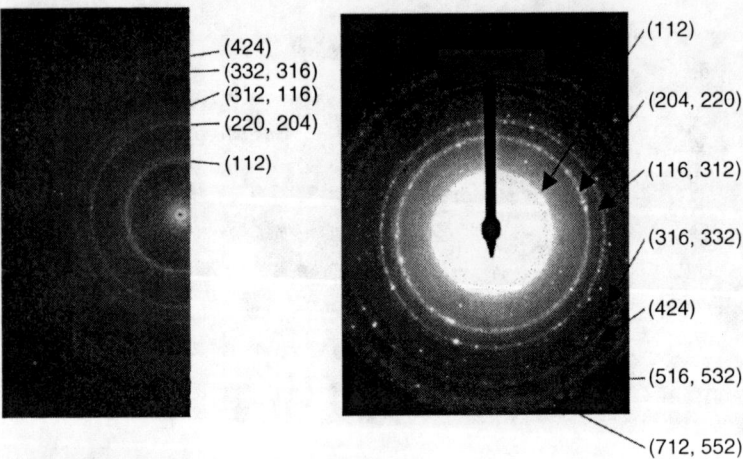

Figure 4.7 Selected area electron diffraction pattern of CuInSe$_2$ thin film grown by (A) spray pyrolysis and (B) electrodeposition.

4.2.2 XRD analysis of CuInSe$_2$

The interplanar distance d can be calculated from known diffraction angle θ using the Bragg's relation

$$2d\sin\theta = n_o\lambda_e, \tag{4.2}$$

where n_o is the order of diffraction and λ_e is the wavelength of radiation. The d spacing can be related to (hkl) indices for tetragonal and cubic structure to obtain lattice parameters of the crystal.

$$\frac{1}{d^2} = \frac{h^2+k^2}{a^2} + \frac{l^2}{c^2} \quad \text{(tetragonal)} \tag{4.3}$$

$$\frac{1}{d^2} = \frac{h^2+k^2+l^2}{a^2} \quad \text{(cubic)}. \tag{4.4}$$

In order to construct theoretical XRD pattern for chalcopyrite CuInSe$_2$, the reflected X-ray intensity of each (hkl) plane is calculated using the well known expression, as given below [14]

$$I = \left|F_{hkl}\right|^2 P\left(\frac{1+\cos^2 2\theta}{\sin^2\theta\cos\theta}\right)A(\theta)e^{-2M}, \tag{4.5}$$

where the structure factor $F_{(hkl)}$ can be obtained using the relation

$$F_{(hkl)} = f_{Cu}\sum e^{2\pi i(hx+ky+lz)} + f_{In}\sum e^{2\pi i(hx+ky+lz)} + f_{Se}\sum e^{2\pi i(hx+ky+lz)} \qquad (4.6)$$

by substituting the atomic positions of Cu, In, and Se for each (*hkl*) plane. The theoretically constructed XRD pattern of CuInSe$_2$ is given in Figure 4.8 [11], which is slightly deviated from the other reported intensities that could be due to selection of λ value. In the present case, the correction factor (λ) is considered to only *z* coordinator at C atomic position. There are several models employed by choosing correction factors to *x* and *y* coordinators at C atomic positions [15]. The entire tedious construction process of theoretical XRD pattern is carried out using computer program. The typical theoretical XRD results are also given in Table 4.1 [16]. The CuInSe$_2$ polycrystalline crystals with composition of Cu:In:Se = 24.7 ± 0.2:25.7 ± 0.2:49.7 ± 0.5 grown by hydrothermal solution show chalcopyrite structure exhibiting chalcopyrite characteristic reflections in the XRD, as given in Table 4.1 [17]. The CIS thin film grown by combination of electron beam evaporation of Cu and In, followed by selenization under H$_2$Se at 400 °C also show chalcopyrite structure, as depicted in Table 4.1 [18]. The experimental XRD pattern of CuInSe$_2$ single crystal with plethora of peaks is also shown in Figure 4.9 [19]. The CIS crystals with composition ratio of Cu:In:Se = 24.8:25.4:48.8 and 1% of iodine grown by CVT at 730 °C and cooled down to RT at the rate of 10 °C/s exhibit chalcopyrite structure with multiple reflections [20].

There are several reports on the growth of CuInSe$_2$ single crystal and existence of secondary phases. The pure Cu and In elements are alloyed as CuIn under vacuum

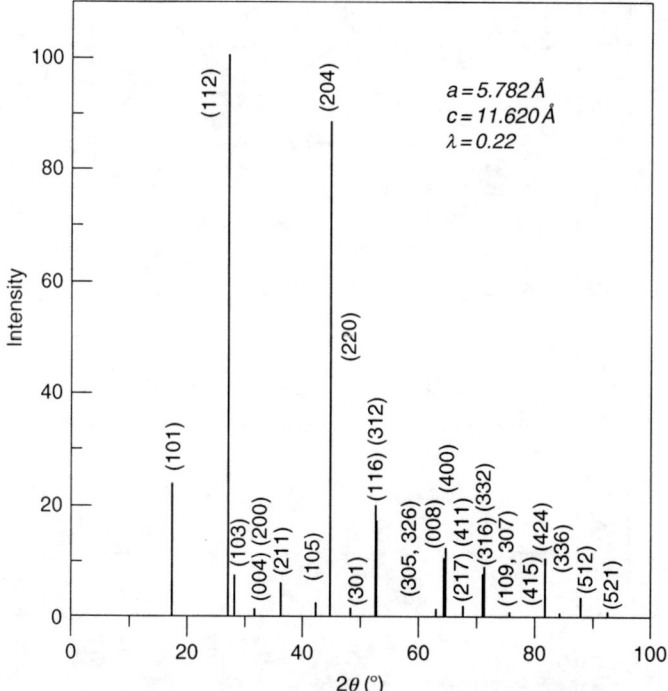

Figure 4.8 Theoretical XRD pattern of CuInSe$_2$.

Table 4.1 X-Ray diffraction data of chalcopyrite structure CuInSe$_2$

(hkl)	Calculated [16]			JCPDS 23-209			Hydrothermal [17]			Thin film [18]
	d (Å)	2θ	I/I$_0$	d (Å)	2θ	I/I$_0$	d (Å)	2θ	I/I$_0$	I/I$_0$
101	5.2	17.05	4	5.2	17.05	6	5.10	17.4	10	2
112	3.34	26.69	100	3.34	26.69	70	3.33	26.8	100	100
103	3.22	27.71	3	3.2	27.88	6	3.20	27.9	3	3
211	2.52	35.63	4	2.52	35.63	15	2.52	35.7	2	
105,213	2.16	41.92	2	2.15	42.02	6	NA	NA		4
204,220	2.04	44.42	70	2.04	44.41	100	2.04	44.3	20	1
301	1.90	–	5	1.900	47.88	6	NA	NA		55
116,312	1.743	52.50	40	1.743	52.50	85	1.74	52.5	9	1
305,323	1.48	–	5	1.480	62.78	6	NA	NA		20
400	1.445	64.49	11	1.446	64.44	25	1.44	64.5	2	
217,411	1.39		4	1.393	67.20	4	1.39	67.3	4	2
316,332	1.327	71.04	16	1.327	71.03	35	1.33	71.0	8	
109,307	NA			1.256	75.73	2	NA	NA		5
415	1.2		4	1.200	79.95	2	NA	NA		
424	1.181	81.48	14	1.181	81.50	60	1.18	81.5	4	
501,431	1.15		4	1.151	84.10	6	NA	NA		5
336,512	1.11		8	1.114	87.58	25	1.11	87.6	5	
417,521	NA			1.070	92.19	8	1.07	92.4	3	3

NA, not available.

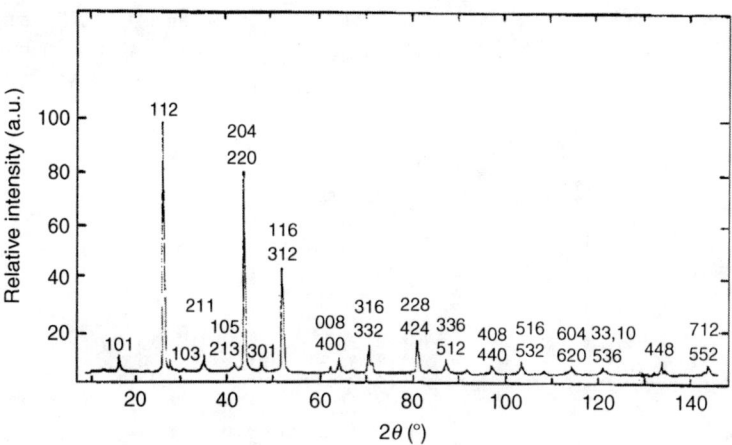

Figure 4.9 XRD pattern of a typical CuInSe$_2$ single crystal.

in sealed ampoule at 800 °C. The CuIn alloy and Se are sealed off in the carbon coated quartz tube under pressure of 3×10^{-6} torr. The charge is brought to 1100 °C at the ramp of 100 °C/hr and kept for 10–24 h. The ampoule moves with a speed of 4–8 mm/day for 10 days in order to quench the ampoule to room temperature. The concentration (mol.%) x' is defined as $x' = \text{CuInSe}_2/(\text{CuInSe}_2 + \text{CuSe})100$. The rapid quenching is stopped, when the tip and end parts of the ingots reached to 500 and 600 °C for $x' = 7\%$, respectively. The lower part of the ingot is single phase CuInSe$_2$ single crystal, middle part is combinations of α-Cu$_{2-x}$Se and Cu$_3$Se$_2$ and the end part being of α-CuSe and Se (hexagonal), as shown in the XRD spectrum (Figure 4.10). The CuInSe$_2$ and α-Cu$_{2-x}$Se for $2 \leq x' \leq 5$ mol.% and the chalcopyrite CuInSe$_2$ phase for $x' \geq 5$ mol.% in the tip portion of the crystal exist [21]. The CuInSe$_2$ single crystals grown by solute diffusion method show that the CuInSe$_2$ appears along with Cu$_2$Se phase for the Se temperature of 320 °C. Only the chalcopyrite CuInSe$_2$ is observed for the Se temperature of 400–470 °C. On the other hand, In$_2$Se$_3$ phase also sticks on the inner walls of the quartz tube [22].

The CuInSe$_2$ films formed by sputtering technique from fine targets show Se deficient, In-rich, and number of different phases. The films deposited from coarse target at substrate temperatures of RT, 50–330, 450–550 °C show amorphous, sphalerite and chalcopyrite structure, respectively, as shown in Figure 4.11. It is clear that the characteristic peaks such as (101), (211), (103), *etc.*, present for chalcopyrite structure, whereas they absent for sphalerite structure in the XRD pattern. The CIS targets used in this investigation are fabricated that high purity Cu, In, and Se elements are sealed in evacuated quartz tube and heated to 350 °C for 24 h then slowly brought to 1050 °C. After cooling down to RT, the ingot is grounded and again heated to 600 °C for 100 h to obtain homogeneity. The well grounded powder and ungrounded with broken small pieces are cold pressed to form fine and coarse targets, respectively [23]. The CuInSe$_2$ powder with 2% excess In exhibits polycrystalline chalcopyrite structure by evidencing presence of (211)

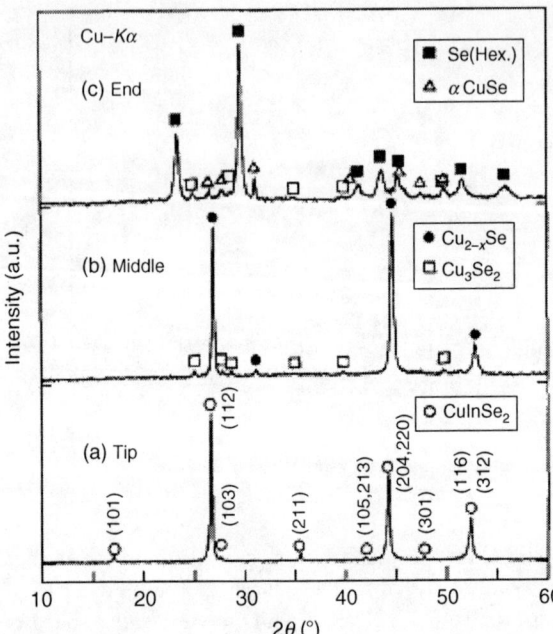

Figure 4.10 XRD pattern measured at different places on CuInSe$_2$ crystals.

reflection. The films deposited by evaporation of CIS charge at lower substrate temperature of 250 °C shows sphalerite structure, whereas films deposited at higher temperature of 350 °C show chalcopyrite structure by witnessing (211) peak. The chalcopyrite CuInSe$_2$ films deposited at substrate temperature of 350 °C for shorter distance of 3 cm between substrate and source shows no preferred orientation, however the films exhibit preferred orientation of (112) for longer distance of 5–15 cm [24]. It could be due to variation of composition flux.

The CuInSe$_2$ thin films grown onto Mo coated glass substrates by RF reactive sputtering technique under Ar atmosphere at substrate temperature of 450 °C using CuInSe$_2$ target with 5 wt.% excess Se show sphalerite structure with reflections of (111), (202/220), and (113/311). The high Se content inhibits the development of chalcopyrite structure, hence a small quantity of H$_2$ is introduced into the chamber to form volatile compounds such as H$_2$Se and HSe. The films grown at substrate temperature of either 350 or 400 °C in the presence of small quantity of H$_2$ shows chalcopyrite structure with characteristic peaks of (101), (103), (211), and (105/213) without any binary phases of Cu$_{2-x}$Se and In$_2$Se$_3$, whereas films deposited at lower substrate temperature of 214 °C shows sphalerite structure [25]. The Cu/In/Se stack grown by sequential evaporation at room temperature shows Cu and In peaks only. After annealing the stack at different temperatures of 200, 300, and 400 °C binary phases, sphalerite, and chalcopyrite CuInSe$_2$ exist, respectively, as shown in Figure 4.12. In the case of CuInSe$_2$ thin films prepared by spray pyrolysis method with Cu:In:Se composition ratio of 1:1:2–2.5 at substrate temperature

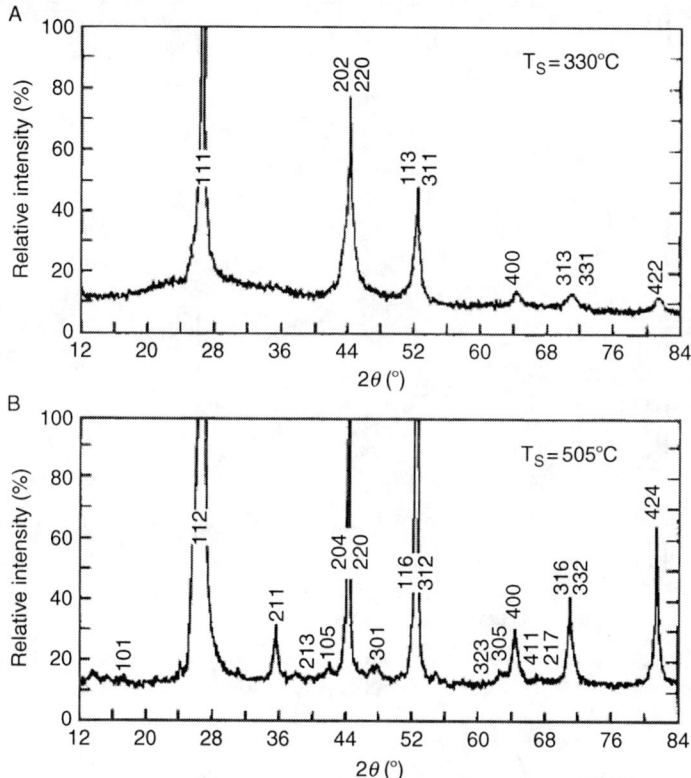

Figure 4.11 XRD pattern of CuInSe$_2$ thin films deposited by sputtering technique. (A) sphalerite and (B) chalcopyrite structure.

of 300 °C show chalcopyrite structure, whereas films grown at lower temperature of 290 °C present only CuInSe$_2$ single phase sphalerite structure. Keeping Cu:In:Se = 1:1:3 and T_S = 290 °C, the intensity ratio of (112)/(220) peaks increases with increasing Cu/In ratio from 0.8 to 1.2. The intensity ratio also increases with increasing substrate temperature for Cu:In:Se = 1:1:3. The films deposited at substrate temperatures of 162 and 220 °C corroborate secondary phases of CuSe$_2$, Se, and CuInSe$_2$ for Cu:In:Se = 1.1:1:4. The CIS films grown with Cu:In:Se = 1.1:1:4, T_S = 300 °C and neutralization of 75–90% spray solution exhibit chalcopyrite structure, whereas the films show sphalerite structure for lower deposition temperature of 300 °C and zero neutralization [26,27,28].

Chalcopyrite structure is observed in the CuInSe$_2$ thin films with composition ratio of Cu:In:Se = 1:1:3.75 in the solution grown by spray pyrolysis at substrate temperature of 320 °C. After annealing the samples under vacuum at 350 °C for 1 h the intensities of prime peaks increase because of an increased crystallinity, whereas the intensities decrease in γ-irradiated samples due to loss of selenium and distortion of crystal lattice [11]. The polycrystallinity increases with increasing thickness of the layers witnessing by several XRD peaks [29]. The characteristic chalcopyrite peaks

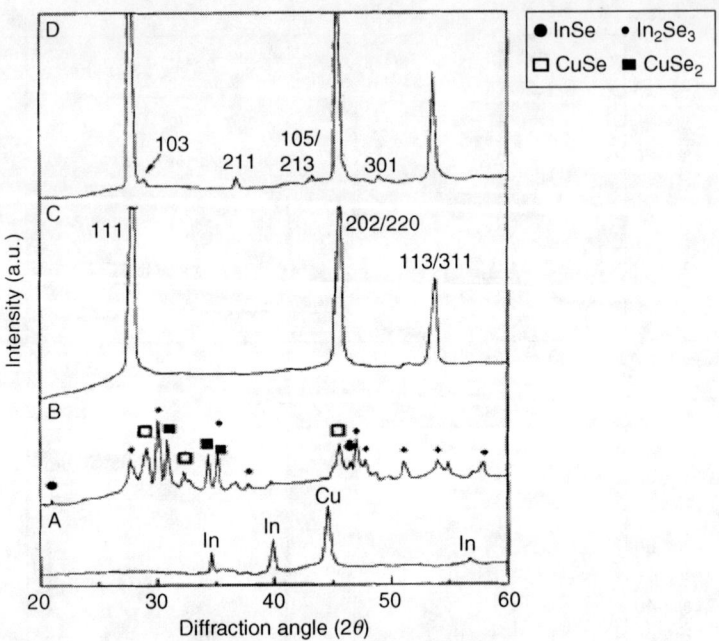

Figure 4.12 XRD pattern of (A) as-deposited Cu/In/Se stack and annealed to form CuInSe$_2$ at different temperatures of (B) 200, (C) 300, and (D) 400 °C.

such as (101), (103), (211), and (301) are observed in the Cu-rich CIS films, whereas they absent in the sphalerite structure or In-rich film. The chalcopyrite peak (211) disappears slowly with increasing potential from –0.55 to –0.8 V in steps of –0.05 in the electrodeposition correspondingly the composition varies from Cu-rich to In-rich. The calculated diffraction angles of (112) and (116) shift toward higher angle by 0.045° and 0.23° with moving structure from chalcopyrite (tetragonal) to sphalerite (cubic) respectively. The c/a ratio also changes from 2.01 to 2.0. After Annealing Cu-rich and In-rich CIS thin films with composition ratio of Cu:In:Se = 26.28:23.17:50.56 and 23.95:24.81:51.24 at 440 °C for 20 min, chalcopyrite and sphalerite structures present, respectively, as shown in Figure 4.13 [30]. The excess Cu in the CIS obviously favors chalcopyrite structure. The structure changes from sphalerite to chalcopyrite by varying deposition parameters and annealing conditions for the CuInSe$_2$ thin films deposited onto Mo coated glass substrates by electrodeposition (ED). The Cu/In ratio in the CuInSe$_2$ films can be increased by varying potential in the range from –0.55 to –0.75 V at pH 1.7. Excess Se is suppressed by annealing the samples under argon flow at 400 °C for 30 min. The XRD spectrum shows a broad and weak peak in the as-grown ED CIS samples. The (112) peak width decreases and intensity increases when the films are annealed at 230 °C for 25 min. This effect is high for further increasing the annealing temperature from 230 to 300 °C. Small intensity peaks such as (101) and (211) at 17.2 and 35.7° segregate for the annealing temperature of ≥ 350 °C in the spectra, respectively. The orientation factor $f(112)$ increases well, as compared to other

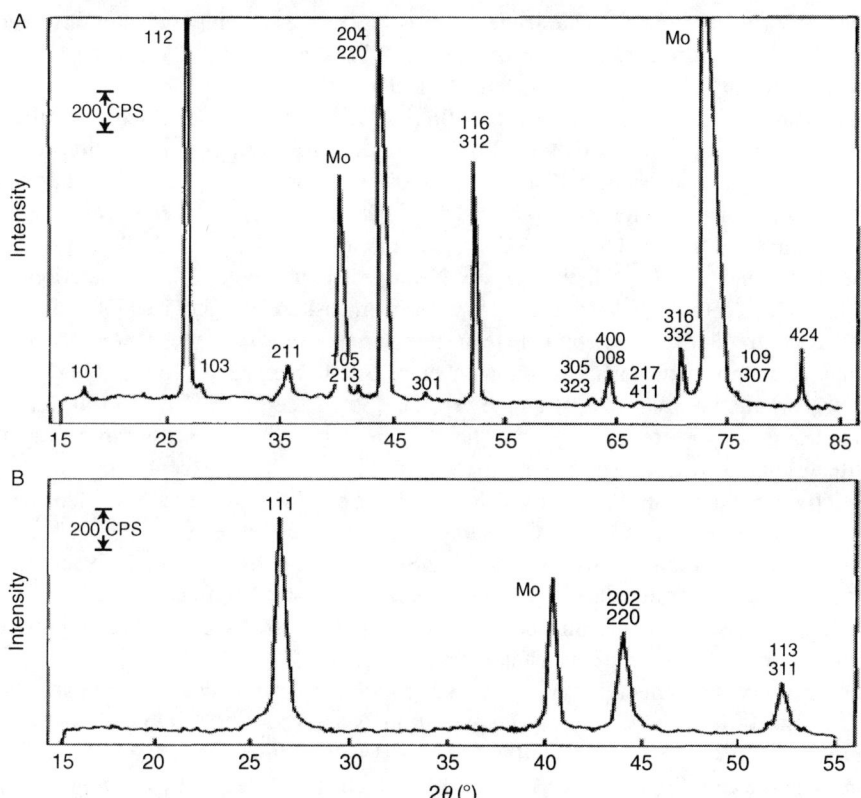

Figure 4.13 XRD pattern of (A) Cu-rich (chalcopyrite) and (B) In-rich (sphalerite) $CuInSe_2$ thin films.

factors $f(204/220)$ and $f(116/312)$. After annealing glass/Mo/Cu/In_2Se_3/Cu_{2-x}Se multilayers deposited by ED at 550 °C for 30 min under Ar flow, the stack converts into $CuInSe_2$ thin films, which show chalcopyrite structure with intensity ratio of (112)/(204,220) as unity. Along with (101), (112), (103), (112), (105,213), (204,220), (301), (312), and (400) reflections of CIS, the peaks due to Mo–Se present at 32 and 55° in the XRD spectrum [31].

The (112) preferred orientation peak slightly shifts toward higher diffraction angle from 26.63 to 26.93° with increasing In/(In+Cu) ratio from 0.54 to 0.78 for the films deposited at 360 °C by spray pyrolysis technique. The chalcopyrite characteristic peaks such as (101), (211), (105/213), (301), (417/217), and (501) present for In/(In+Cu)=0.54, whereas the diffraction peaks (301), (417/217), and (501) absent for In/(In+Cu)=0.6. The (110) peak presents for the composition $0.67 \leq$ In/(In+Cu) ≤ 0.78 that could be due to either In_2Se_3 phase or $CuIn_3Se_5$. The FWHM of (112) is larger for the films deposited at 400 °C than that for the films grown at 360 °C. The lattice constants $a=5.758$ and $c=11.536$ Å of $CuInSe_2$ thin films linearly decrease with decreasing In/(In+Cu) ratio from 0.5 to 0.8 [32].

Similarly the chalcopyrite characteristic peaks (101), (211), and (301) are observed along with primary peaks such as (112), (204,220), and (116,312) in the CuInSe$_2$ mono-grain layer for Cu/In = 1.1 and 0.91. The chalcopyrite peaks disappear and new phase Cu$_2$In$_4$Se$_7$ generates for Cu/In = 0.77 and 0.5 in the In-rich thin films. As Cu/In ratio is decreased, the (204,220) peak shifts to higher diffraction angle [33]. The electrodeposited CIS does not show any secondary phases for optimum conditions. In fact the electrodeposited CIGS precursor with composition of Cu:In:Ga:Se = 28.77:21.05:1.84:48.34 shows phases of CIGS and Cu$_2$Se (311). On the other hand, CIS films have intensity XRD peaks but the crystallinity deteriorates if Ga is added to it evidencing by decrease of intensities of XRD peaks [34].

On contradictory to previous results, the CuInSe$_2$ thin films deposited by either sputtering or vacuum evaporation with extra Se source show stronger (112) preferred orientation than that of single crystals [35]. It could be due to influence of some other deposition parameters. The stoichiometric and In-excess samples contribute chalcopyrite structure but In$_2$O$_3$ phase is seen in In excess samples [36]. Strictly speaking, slightly excess In containing CIS samples are chalcopyrite because after etching Cu-rich CIS samples to remove Cu$_x$Se phase, the left over composition is slightly In rich. The CuInSe$_2$ layers deposited onto 7059 corning or quartz substrates at substrate temperature of 550 °C by XeCl laser ablation maintaining the distance of 30 mm between target and substrates gives (112) preferred orientation and characteristic chalcopyrite peaks of (101), (103), and (211). The intensities of peaks decrease with increasing the distance between substrate and target due to lack of growth rates. The growth rates are 100–120 Å/min and 30–50 Å/min for the distances of 30 and 40 mm, respectively. The target is prepared by cold pressing the CuInSe$_2$ charge at 4.9×10^7 Pa, which is melt grown by reacting Cu, In, and Se in quartz tube under vacuum at high temperature [37].

The CA and OVC phases can be developed that CuInSe$_2$ thin films deposited onto (100) and (001) oriented with 2° off substrates by MBE technique show CH+CA and CA structures, respectively. The (002), (004), (006), and (008) reflections are observed for CA structured CuInSe$_2$ thin films, whereas the same reflections with high intensity also present due to formation of cation planes parallel to the surface in the CH samples and the Cu and In atoms are equally distributed. The In-rich CuInSe$_2$ thin films with ratio of Cu/In = 0.98 have chalcopyrite structure with (112) preferred orientation. The chalcopyrite CIS and CA CuInSe$_2$ are observed for the ratio of 1.02. No CuSe phase is detected in the moderate Cu-rich films with ratio of 1.09. It may not be in the detectable limits unlike other reports, as shown in Figure 4.14. However, combination of CH and β-Cu$_{2-δ}$Se phases form as a ripple along {110} direction in the Cu-rich CuInSe$_2$ (Cu/In = 1.06) thin films, whereas round mounds with random distribution are noticed by SEM analysis in the In-rich CuInSe$_2$ thin films [4,10].

The XRD spectrum of CuIn$_3$Se$_5$ (OVC) single crystals grown with starting compositions of Cu:In:Se = 1:3:5 by horizontal Bridgman technique is similar to that of CuInSe$_2$, however extra reflections such as (002), (110), (200,004), (202), and (114) appear, as shown in Figure 4.15. On the other hand, the peak position of (112) preferred orientation in 2θ is lower than that of CuInSe$_2$ that means the tetragonal

Structural Properties of I–III–VI$_2$ Absorbers

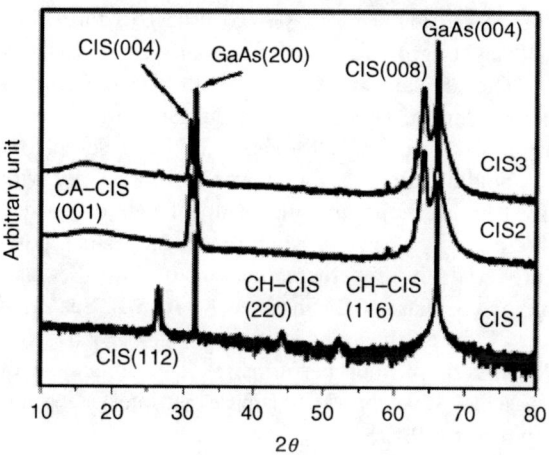

Figure 4.14 XRD pattern of CuInSe$_2$ thin films grown on GaAs substrates with 2° misoriented toward [110] direction; Cu/In = 0.98 (CIS1), Cu/In = 1.02 (CIS2), and Cu/In = 1.09 (CIS3).

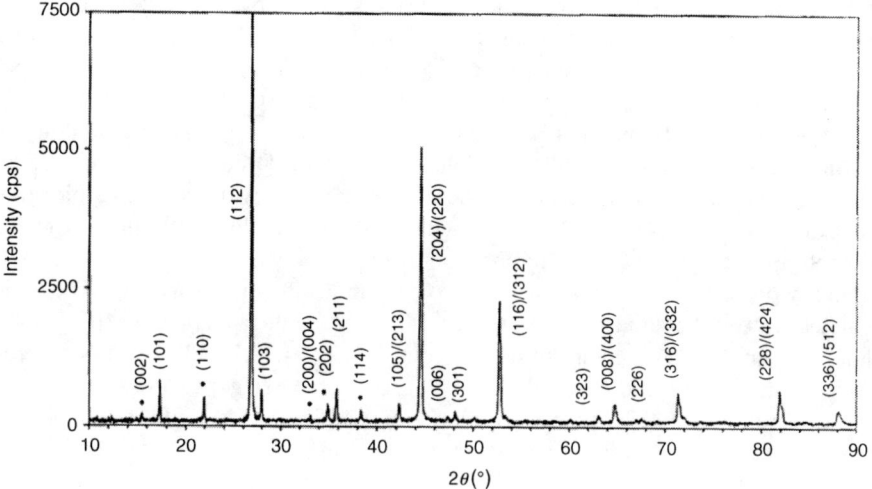

Figure 4.15 XRD pattern of CuIn$_3$Se$_5$ single crystal grown by horizontal Bridgman technique.

structure lattice constants $a = 5.759$ and $c = 11.524$ Å should be higher [38]. The CuIn$_3$Se$_3$ thin film can be prepared that first In and Se are coevaporated onto Mo coated glass substrates at 350 °C, followed by codeposition of Cu and Se at 550 °C with slightly Cu-rich then In and Se being deposited at 550 °C by three source evaporation to develop In-rich surface layer. The thick top layer is considered as CuIn$_3$Se$_5$ phase having (202) and (114) reflections. In addition, the secondary γ-In$_2$Se$_3$ phase is also found. The CuInSe$_2$ thin films experience with Cu$_{2-x}$Se phase as well as low intensity (202) and (114) peaks (OVC). The CIS thin films along with $CuIn_3Se_5$ phase or CIS layers grown by three stage process consists of composition of Cu:In:Se = 1:4:5. The 2θ value of $CuInSe_2$ with mixed phase of $CuIn_3Se_5$ is found to be higher than that of stoichiometric $CuInSe_2$ thin films in the XRD spectrum. The surface analysis confirms formation of $CuIn_3Se_5$ phase. The $Cu2p_{3/2}$, $In3d_{5/2}$, and $Se3d_{5/2}$ binding energies shift toward lower energy with increasing sputter-etching time for film indicating that the chemical state of the surface layer is different from the bottom one. The AES depth profile shows composition ratio as Cu:In:Se = 1:4:5, this could be either combination of $CuIn_3Se_5$ and In_2Se_3 phases or some other unknown phase. 0.2 cm^2 area Mo/CIS/In$_x$Se$_y$/ZnO/ITO/Ni–Al cell with very thin OVC layer shows efficiency of 8.46%. However, the cells with thick $CuIn_3Se_5$ layers show poor performance The advantage of OVC layer in the thin film solar cell is that the OVC suppresses interface states between the absorber and window layer [39,40].

The $CuInSe_2$ thin films are prepared not only by evaporating Cu, In, and Se but also by sequentially evaporating Cu_2Se and In_2Se_3 binary compounds. The deposited Cu_2Se/In_xSe double layers are annealed under Se vapor at 550 °C for 1 h to form $CuInSe_2$ layer on which $CuIn_3Se_5$ layer is grown by evaporation of In_2Se_3 and Se compounds at the same temperature for 10 min. The $CuInSe_2$ thin film along with γ-In_2Se_3 phase is formed for <0.35 µm thick Cu_2Se layer, whereas single phase $CuInSe_2$ thin film is appeared for >0.4 µm thick Cu_2Se layer. In the XRD spectrum, the (202) and (114) peaks at $\sim$34.8 and $\sim$38.3° corresponding to $CuIn_3Se_5$ are observed around the neighborhood of (211) peak located at $\sim$35.6° for $CuInSe_2$ phase. The intensities of these peaks increase with increasing thickness of $CuIn_3Se_5$ phase [41]. Unlike the Cu–In–Se precursor films with three different compositions grown at 225 °C by three source thermal evaporation onto Mo coated alumina polycrystalline substrates (Al_2O_3) are rapid thermal processed. The stoichiometric $CuInSe_2$ thin films processed at 700 °C with ramp rate of 10 °C/s for hold time of 30 s with composition of Cu:In:Se = 24.94:24.74:50.17 shows chalcopyrite structure with additional Cu_2Se phase. After processing the Cu-poor precursor at temperature of 500 °C with ramp rate of 5 °C/s for 30 s, the film accomplishes OVC as a main phase and secondary phases of Cu_2Se and In_2Se_3, which contains chemical composition of 24.94:24.74:50.17. The Cu-rich precursor also processed at 500 °C with ramp up rate of 150 °C/s for 30 s contains chalcopyrite phase along with secondary phases of $Cu_{2-x}Se$ and Cu_2Se, which had compositions of 28.81:22.79:48.4, as shown in Figure 4.16 [42].

The Cu_xSe ($05 \leq x \leq 2.0$) and In_ySe ($0.66 \leq y \leq 2.0$) phases are found at grain boundaries, free surface locations and in isolated grains. The films on the glass are smoother than those on Mo/Al_2O_3. The roughness and grain sizes of films increase by one order of magnitude with moving from Cu-poor to Cu-rich compositions in the $CuInSe_2$ thin films; for example, in the typical sample the grain sizes increase from 0.1 to 1.0 µm with increasing from Cu/In < 0 to Cu/In > 0. The binary phases $Cu_{2-x}Se$ and In_2Se_3 are individually prepared onto glass substrates at substrate temperature of 350–500 °C with excess Se of 2–3 times, in order to cross check phases in the CIS. The conductivity of $Cu_{2-\delta}Se$ is semimetallic behavior with

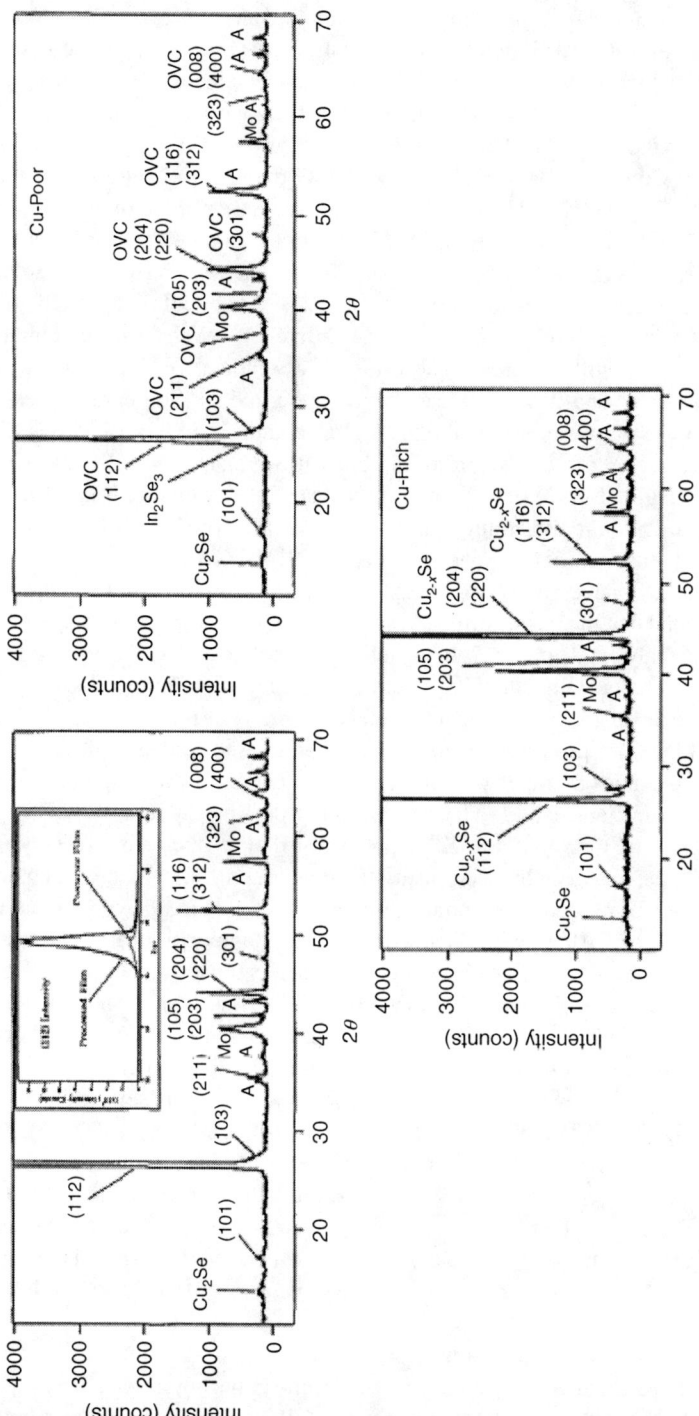

Figure 4.16 XRD pattern of stoichiometric, Cu-poor and Cu-rich CuInSe$_2$ thin films grown by three source thermal evaporation.

$\rho = 10^{-3}$ Ω-cm and $\alpha = 10^4$ cm^{-1} in the wavelength range of 500–2000 nm. The Cu$_{1.85}$Se films deposited at 500 °C with orientation of (111) have lattice constant of $a = 5.739$ Å, which is close to $a = 5.718$ Å of CuInSe$_2$, therefore it is little difficult to distinguish Cu$_{1-\delta}$Se ($\delta = 0.15$) phase fairly from CuInSe$_2$ phase by the XRD analysis, if it is in the films. The Cu-rich and slightly Cu-poor CuInSe$_2$ thin films with compositions of Cu:In:Se = 31.7:21:47.3, 26:24.3:49.6 and 23.6:25.4:51deposited at 500 °C generate Cu$_2$Se phase. The Cu$_{2-\delta}$Se appears as a minor phase on the shoulders of (112) and (204,220) peaks in the XRD, which can be removed off by KCN treatment. However, Cu$_2$Se phase cannot be etched off because, which is located at intragranular position. The films deposited with < 18% Cu has CuIn$_2$Se$_{3.5}$ as a major phase and In$_2$Se$_3$ as a minor phase. The OVC CuIn$_2$Se$_{3.5}$ consists of Se sublattice, Cu, and In sublattice with an order. The Cu$_x$Se is a minor phase for Cu content in the range between 19 and 25%, whereas the Cu$_{2-x}$Se and Cu$_x$Se phases appear for > 25 at.%. The CIS films deposited at 350 °C, In$_y$Se and Cu$_x$Se binary phases form. The diffraction angles of several orthorhombic Cu$_2$Se, cubic Cu$_{2-x}$Se and CuSe$_2$ phases fall into angle of (112) that means overlap each other therefore it is difficult to assign particular compound unambiguously. However, in very Cu-rich film, the peak appears at 30.93° that can be assigned to either (002,004) diffraction line of CuInSe$_2$ or Cu$_{2-x}$Se phase, which does not appear in Cu < 22 at.% CuInSe$_2$ thin films therefore it can be concluded that the phase formed in very Cu-rich CuInSe$_2$ thin films is possibly Cu$_{2-x}$Se phase. Only single phase CuInSe$_2$ felicitates for Cu < 22 at.%, whereas single phase CuInSe$_2$ presents along with Cu$_{2-x}$Se for 22 < Cu < 25 at.% range [43,44].

The presence of Cu$_{2-x}$Se phase can be detected from the asymmetry of (112) diffraction peak [45]. The peak represents by Cu$_{2-x}$Se phase resolves well and asymmetry in (112) becomes stronger with increasing Cu content from 3.6, 8.3 to 39.4% excess, as shown in Figure 4.17. The intense (112) peak is asymmetry if Cu$_{2-x}$Se presents along with CuInSe$_2$ in the XRD pattern [46]. The nucleation of Cu$_{2-x}$Se/Cu$_2$Se strongly depends on the composition, temperature, and type of substrate. The CuInSe$_2$ thin films deposited onto glass substrates by spray pyrolysis at substrate temperature of 270 °C with chemical Cu:In:Se composition of 0.8:1:4 show chalcopyrite CuInSe$_2$ + Cu$_{2-x}$Se phase, whereas layers deposited onto CdS even with higher Cu content chemical composition of 1.1:1:4 do not show Cu$_{2-x}$Se phase. The Cu-rich CuInSe$_2$ and Cu$_{2-x}$Se layers grown onto GaAs (001) substrates at substrate temperature of 450 °C by MBE are studied by low dimensional reciprocal space mapping that reveals CuInSe$_2$, Cu$_{2-x}$Se and GaAs phases, as shown in Figure 4.18 [47]. The Cu$_{2-x}$Se as a second phase with reflection of (115) gives an in-plane lattice constant of 0.5770 nm and an out-plane lattice constant of 0.5722 nm, which are comparable with the in-plane lattice constant of 0.5788 nm for CuInSe$_2$. The results demonstrate that Cu$_{2-x}$Se is strained with CuInSe$_2$. The similar mapping is done on only Cu$_{2-x}$Se layer, which shows an in-plane lattice constant of 0.5764 nm and an out-plane lattice constants of 0.5751 nm indicating as relaxed one.

A bilayer process is the potential method to grow CuInSe$_2$ for thin film solar cell applications. The Cu and In sputtered by magnetron sputtering and Se by evaporation are deposited at substrate temperature of $(T_S) \leq 550$ °C. The Cu-rich as a bottom

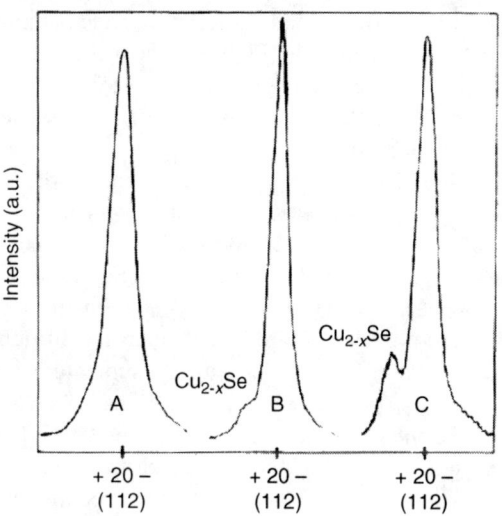

Figure 4.17 XRD pattern of CuInSe$_2$ for (112) peak along with Cu$_{2-x}$Se phase: (A) 3.6, (B) 8.3, and (C) 39.4% excess Cu.

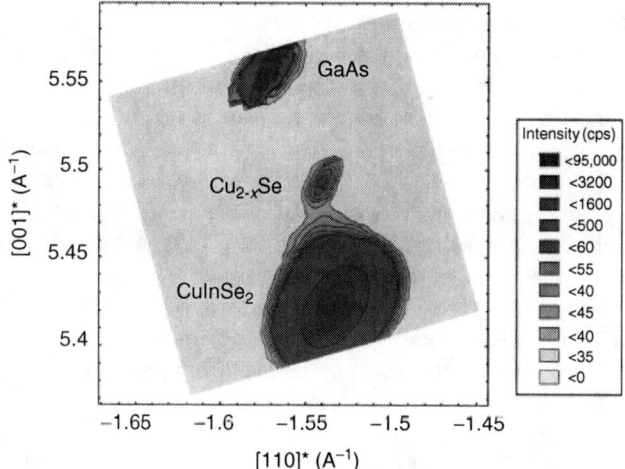

Figure 4.18 Reciprocal space mapping of CuInSe$_2$ and Cu$_{2-x}$Se phases in two dimensional.

layer and In-rich as a top layer are formed as a bilayer; 2 μm thick Cu-rich layers with composition of Cu:In:Se = 30:23:47 (Cu/In = 1.9) are first deposited onto Mo coated glass substrates at 400 °C, followed by In-rich layers at $350 < T_S < 500$ °C. The TEM of Cu-rich CIS bottom layer deposited at 400 °C shows that the Cu-rich CIS surrounds the Cu$_2$Se grain. The XRD shows (101) and (103) reflections indicating chalcopyrite structure. The (200) peak occurs at 30.9° due to Cu$_2$Se phase with lattice spacing of 0.288 nm for Cu/In > 1.2. As mentioned earlier, the lattice parameter of 0.5739 nm for Cu_2Se is close to 0.5781 nm of CIS from which the lattice mismatch is determined to be 0.73%. The moderate Cu-excess $CuInSe_2$ ($1.1 > x > 1$) thin films have partial cover of Cu_2Se layer on its surface. The Cu_2Se precipitates at larger quantity for Cu-rich conditions. The point defects segregate at higher deposition temperature of $(T_S) > 450$ °C, which help to form secondary phase of Cu_2Se, whereas the phase separation causes to increase the point defects dissolution at lower deposition temperature of $(T_S) < 400$ °C. The top In-rich layer deposited onto Cu-rich CIS layer at ≤ 400 °C shows a small quantity of Cu_2Se phase by XRD, whereas the In-rich surface layer is observed but Cu_2Se is buried for $T_S = 400$ °C. The chalcopyrite peaks (101) and (103) are observed in the In-rich $CuInSe_2$ thin films without any Cu_2Se phase for the deposition temperature of $T_S \geq 450$ °C. The intensity ratio of $I_{(101)}/I_{(112)}$ is 0.03 for In rich films with Cu/In = 0.8 deposited at 450 °C, whereas the intensity ratio of $I_{(101)}/I_{(112)}$ is 0.012 for 350 °C. The intensity ratio of $I_{(101)}/I_{(112)}$ is 0.01 and 0.003 for Cu-rich films deposited at 450 and 350 °C, respectively. The $CuIn_3Se_5$ phase contains reflections of (110), (202) and (114), and $CuInSe_2$ consists of chalcopyrite reflections of (101) and (103) for Cu/In ≤ 0.9. Keeping Cu/In = 0.5 the intensity ratio of $I_{(110)}/I_{(112)}$ is 0.009 and 0.0025 for $T_S = 450$ and 350 °C, respectively. The OVC is uniform in thickness and composition for the films deposited at 500 °C. The final $CuInSe_2$ bilayer is stoichiometric or In-rich for the flux ratio of Se to In ($F_{Se}/F_{In} = 5$), which is applied for the top layer (In-rich layer), whereas the films are Cu-rich and Cu_2Se for the ratio of Se to In (F_{Se}/F_{In}) = 2.5. The Se reevaporates at 400–500 °C, the In is left over as liquid droplets which do not react with Cu-rich $CuInSe_2$ and Cu_2Se due to lack of Se [48]. Akin the $CuInSe_2$ layers with Cu/In > 1.2 show chalcopyrite structure with secondary phases of Cu_xSe. The films with Cu/In < 0.69 also show chalcopyrite $CuInSe_2$ along with In_2Se_3 phase. The single phase $CuInSe_2$ films are obtained for the Cu/In composition ratio of 0.83–1.04. A number of small grains are observed for Cu/In = 0.69 while tetrahedral facet grains are observed for Cu/In = 1.29, which are related to Cu_xSe binary phase from SEM analysis. For this analyses 0.4–0.8 μm thick Cu–In layers are sequentially deposited onto Mo coated glass substrates by ion plasma sputtering or thermal evaporation using Mo boat. The layers grown by latter technique produce Cu_9In_{11} phase, whereas former technique films have mixed phases of Cu_xIn, Cu, and In in the layers. The $CuInSe_2$ thin films are developed by selenization of CuIn alloy at 280–300 °C for 20–50 min, followed by annealing at 500 °C for 20–90 min [49].

The Cu/In/Se and Cu/Se/In stack layers sequentially evaporated onto Mo coated 7059 corning glass but Se deposited from effusion cell at room temperature, followed by annealing at 400 °C for 15–30 min forms chalcopyrite CIS. However, the layers annealed at lower temperature of 200 °C show multiple binary compounds such as CuSe, $CuSe_2$ InSe and In_2Se_3 for Cu/In/Se stack and α-CuSe, CuSe, CuSe, $CuSe_2$, Cu_2Se, and α-In_2Se_3 for Cu/Se/In stack [26]. Similarly the Cu/In/Cu/In, In/Cu/In/Cu, and In/Cu/In metallic layers sequentially grown onto glass substrates by e-beam evaporation, followed by selenization at 500 °C show chalcopyrite structure and (112) preferred orientation, irrespective of the sequence, however at low temperature selenization, the existence and domination of different phases such as CuSe, $Cu_{2-x}Se$, $CuSe_2$ InSe, In_2Se_3, *etc.*, depend on the sequence [50]. In the case of electrodeposited $CuInSe_2$ thin films, which are annealed at different temperatures from room temperature to 500 °C for 1/2 h. The FWHM of (112) preferred orientation increases with increasing annealing temperature up to 400 °C indicating that an increased crystallinity of the layers for the deposition potentials of –0.5 V is observed, whereas the layers grown at higher potentials need higher annealing temperature of 500 °C. An increase in grain size with annealing time for the higher Cu/In ratio layers is higher than that for stoichiometric samples. The In_2O_3 phase is detected in the layers annealed at high temperature [51].

280 nm thick Cu, 130 nm In and three times thick Se of total Cu and In are deposited onto glass or Mo coated glass as a sequence of In/Cu/In/Se/In/Cu/In/Se by thermal vacuum evaporation at room temperature. The stack samples kept in graphite box and installed in quartz tube are annealed in the presence of with or without Se shots under vacuum at pressure of 10^{-2} Torr and 500 °C for 20 min. The Cu/In and Se/(Cu+In) ratios of 0.92–1.31 and 0.84–0.98 are found for annealing the stack without Se in the $CuInSe_2$ thin film, respectively, whereas they are 0.85–1.04 and 0.89–1.3 for annealing with Se, respectively. The Cu-rich $CuInSe_2$ films exhibit $Cu_{2-x}Se$ as a secondary phase and near stoichiometric In-rich films have splits in (204/220) and (116/312), as shown in Figure 4.19 [52]. The Cu/In layers sputtered and Se evaporated at room temperature onto glass substrates show Se, In, CuSe, Cu_2In, and Cu_2Se phases. The Cu/In/Se stack precursor layers annealed at 220 °C consist of CuSe, InSe, Cu_2Se, and Cu_2In, whereas the $CuInSe_2$ forms for the annealing temperatures between 220 and 350 °C. The $CuIn_2Se_{3.5}$ phase disassociates into $CuInSe_2$ and In_xSe_y for annealing temperature of 350 °C. The latter is unstable compound hence it gets evaporate. The chalcopyrite phase is formed above the annealing temperature of 350 °C and the grain sizes of the layers increase [53].

As shown in Figure 4.20, the grown Cu–In layers are selenized under vacuum and at 1 atm pressure under argon flow by heating Se pellets in graphite block at

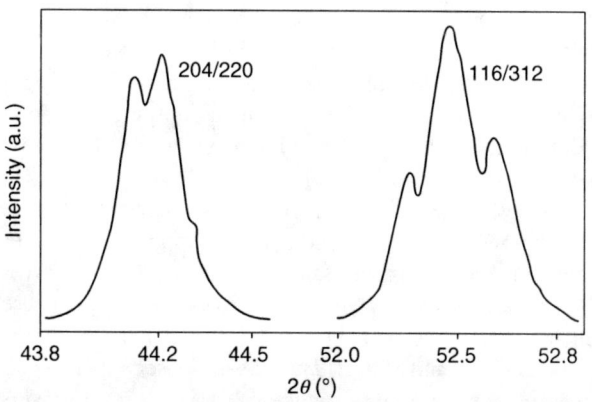

Figure 4.19 XRD pattern of resolved peaks (204), (220) and (116), (312) for near stoichiometric In-rich CIS thin films.

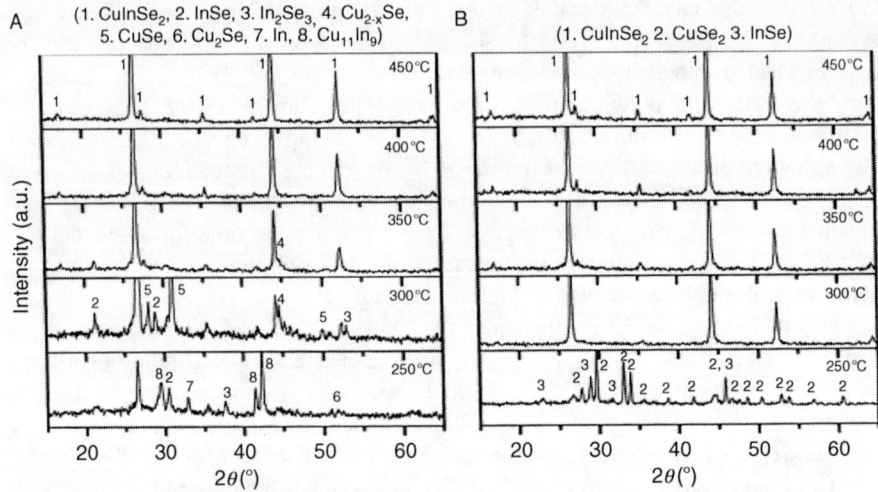

Figure 4.20 XRD pattern of Cu–In–Se stack layers selenized at different temperatures (A) under 1 atm and (B) under vacuum.

different temperatures to observe formation of phases. 800 nm thick Cu–In layers deposited onto glass substrates by D.C magnetron sputtering using Cu and In individual targets. $Cu_{11}In_9$ and $CuIn_2$ phases form at low Cu/In ratios. As Cu/In ratio is increased, $Cu_{11}In_9$ remains constant but $CuIn_2$ disappears. The Cu–In layers selenized under 1 atm pressure (a) at different temperatures; (i) between 450 and 400 °C result in $CuInSe_2$ films, (ii) 350 °C, $Cu_{2-x}Se$, (iii) 300 °C, InSe, In_2Se_3, $Cu_{2-x}Se$ and CuSe phases, and (iv) 250 °C, InSe, In_2Se_3, Cu_2Se, In, and $Cu_{11}In_9$ phases indicating that single phase $CuInSe_2$ forms only at higher temperature whereby sufficient kinetic energy is available. At low temperature, multiple phases arise that means lack of thermal energy to form $CuInSe_2$. In the case of vacuum selenization (b), between 300 and 450 °C, $CuInSe_2$ forms, whereas $CuSe_2$ and InSe phases appear at low temperature of 250 °C. The Raman studies confirm that the $CuInSe_2$ layers formed above the selenization temperature of 350 °C shows A_1 chalcopyrite characteristic peak. In addition to this, B_1 mode at 213 and B_2 or E mode at 230 cm^{-1} are observed. The secondary phases such as Cu–Se (185 and 258 cm^{-1}) and In–Se phase (115 cm^{-1}) are observed below the temperature of 350 °C [54]. The Cu–In films are selenized under Se vapor by heating Se pellets, the α-$CuInSe_2$ starts to develop at temperature ranges of 250–300 °C but $MoSe_2$ starts to form at above 440 °C. There is no evidence of formation of $MoSe_2$ provided that less Se vapor is employed [55].

A 50–200 Å thin amorphous Se layer is deposited as interface layer between substrate and metallic (Cu/In or In/Cu) layers, which are selenized under Se vapor at 500 °C for about 30 min until to obtain Se/metal ratio of 1.25. The selenization of Cu/In layers gives $CuInSe_2$ and other phase of In_2Se_3, whereas selenization of In/Cu exhibits single phase chalcopyrite structure evidencing (101), (103), (211), and (105/213) peaks. Apart from these reflections, splits in (220/204) and (116/312) peaks are observed. The CIS layers formed with Cu/In sequence peels off easily from the substrates, whereas layers with In/Cu sequence have good adherence. It is understood that first In deposition onto the substrates, followed by Cu as In/Cu structure gives good quality CuInSe$_2$ films even with thin Se bed layer because of good mixing of Cu with In, which contains high energy while depositing Cu by e-beam evaporation. The CIS layers used for analysis are prepared that the In and Se by thermal evaporation from independent crucibles at source temperatures of 1050 and 280 °C, respectively, and Cu by e-beam evaporation are done to form CuInSe$_2$ layers onto SiO$_2$ coated soda lime glass substrates or corning 7059 glass under chamber pressure of $1–5 \times 10^{-6}$ torr. The Cu and In layers are sequentially deposited with thickness of 0.2 and 0.44 μm, respectively [56]. The as-grown CIS precursor by ED at RT using -0.5 V had composition of Cu:In:Se = 39.18:17.46:43.36 and Cu$_2$Se phase. After selenizing precursor in tubular furnace at 550 °C for 50 min, the single phase CIS occurs by swallowing Cu$_2$Se secondary phase and the composition changes to Cu:In:Se = 39.5:16.64:43.86 but the percentage of selenium does not improve in the CIS layer [57].

Cu–In–Se precursor films grown onto Au coated glass substrates by ED demonstrate combination of CuInSe$_2$, InSe, In$_2$Se$_3$, and Cu$_{2-x}$Se phases. After annealing the precursor layers at 150 °C under N$_2$ atmosphere for 1 or 2 h, the intensities of standard (112), (204,220), and (116,312) peaks increase and the binary phases disappear that means, which are consumed by CIS phase, as shown in Figure 4.21. The composition of annealed film is Cu:In:Se = 25.57:25.01:49.42. The intensities of

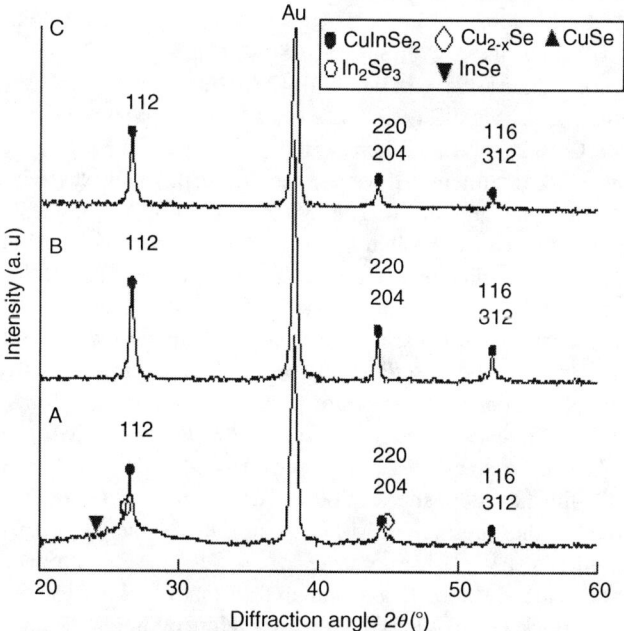

Figure 4.21 XRD pattern of (A) as-deposited CuInSe$_2$ thin films, (B) annealed at 150 °C under N$_2$ for 1 h, and (C) 2 h.

standard peaks decrease when the films are annealed for longer time of 2 h revealing losses of selenium that causes to reduce the quality of the layers. The segregation of In_2Se_3 or Se can be avoided by making Cu rich CIS thin films [58]. The intensities of (112) and (204/220) peaks steadily increase with increasing CIS annealing temperature from 200 to 400 °C. After the heat treatment, the Se concentration decreases due to reevaporation of Se and color of the film changes from dark gray to bluish. No change in Cu to In ratio is observed for the heat treatment of the samples at 300 °C, however crystallinity strongly improves [59]. The XRD analysis reveals that the $CuInSe_2$ thin films deposited by one step ED could be either in tetragonal $CuInSe_2$ or tetragonal $CuIn_2Se_{3.5}$ for the typical deposition concentrations of 10, 73, and 17 mM for Cu^{2+}, In^{3+}, and SeO_3^{2-}, respectively in the solution that results in composition of Cu:In:Se = 24.33:20.6:55.06 in the film. However, an increase in Cu concentration in the solution from 10 to 12 mM-Cu, 71 mM-In, and 17 mM-Se results in $CuInSe_2$ thin film composition of Cu:In:Se = 28.09:21.65:49.94, which show mixed phases of $CuInSe_2$ and α-Cu_2Se. This observation indicates that the Cu_2Se phase segregates with increasing Cu content in the deposition solution. The layer by layer growth is also done by ED, first Cu is electroplated onto Mo coated glass substrates using 0.005 M $Cu(SO_4)_2$ solution and In–Se using 0.025 M $InCl_3$ and 0.025 M H_2SeO_3, on which similarly Cu–Se is coated using 0.025 M $Cu(SO_4)_2$ and 0.025 M H_2SeO_3. The final glass/Mo/Cu(50 nm)/InSe/CuSe stack annealed at 450 °C under Ar leads to tetragonal $CuInSe_2$ or $CuIn_2Se_{3.5}$ with composition ratio of Cu:In:Se = 23.54:23.01:52.45. The InSe film adhesion to the substrate is poor if 50 nm thick Cu is not first coated as a bed layer. The Cu–In alloy deposited by electroless method using 0.025 M $CuCl_2$, 0.025 M $InCl_3$, TEA, and NH_4OH at pH 1.5, followed by selenization at 550 °C results in $CuInSe_2$/$CuIn_2Se_{3.5}$, Cu_2Se, and Cu phases in the film with composition of Cu:In:Se = 36.98:20.37:42.65 [60].

The as-grown films with Cu_xSe, In_ySe, and $CuInSe_2$ phases etched off in KCN solution shows $CuInSe_2$ and In_ySe phase indicating that Cu_xSe phase can be easily etched off by KCN treatment. However, the In_xSe phase mixes with $CuInSe_2$ in the films when the films are annealed at 230–250 °C [30]. After annealing the amorphous Cu–In–Se precursor deposited by electroless method under Se atmosphere at 530 °C for 2 h becomes phases of $CuInSe_2$ and Cu_2Se [61]. The Cu/In stack layers prepared by electron beam evaporation at 25 °C shows inhomogeneous structure evidencing In peak in the XRD spectrum, whereas layers grown at 200 °C, $Cu_{11}In_9$ forms. The $Cu_{11}In_9$ is one of the most important phases to make quality CIS layers. The intensities of chalcopyrite peaks are high for $CuInSe_2$ thin films prepared by selenization of Cu–In stack layers deposited at 200 °C rather than grown at 25 °C under H_2Se. The grain sizes of 2 and 1 µm are observed for $CuInSe_2$ thin films grown using Cu–In stack deposited at 200 and 25 °C, respectively. The voids or hollows are formed at the interface of Mo and $CuInSe_2$ for the Cu–In stack formed at 200 °C rather than at 25 °C. The intensity of $MoSe_2$ phase is stronger in the XRD for the stack deposited at 200 °C as compared to that of 25 °C [62].

The Cu/In/Se stack layers deposited at room temperature are selenized at 400 °C with elemental Se vapor exhibit single phase chalcopyrite $CuInSe_2$ thin films. At room temperature, the evaporation of Cu and In shows CuIn and In phases, whereas the stack layers deposited at below the temperature of 130 °C show phases of Cu_9In_4, CuIn, and In [63]. The Cu–In alloy precursors deposited onto Mo coated glass substrates at substrate temperature of 200 °C selenized by H_2Se under Ar at or above temperature of 400 °C present chalcopyrite $CuInSe_2$ and $Cu_{2-x}Se$ phase [64]. The CIS layers deposited by one step electro-deposition and annealed under Ar at 450 °C using chemical vapor transport gas system shows CIS, $CuIn_2Se_{3.5}$, and α-Cu_2Se phases. The films formed by layer by layer process such as Mo/Cu/In–Se and Mo/Cu/In–Se/Cu–Se annealed under Ar at 450 °C show In_2Se_3 and $CuInS_2/CuIn_2Se_{3.5}$, respectively. The Cu–In alloy layers formed by electroless deposition selenized under Se vapor at 550 °C show multiple phases of $CIS/CuIn_2Se_{3.5}$, Cu, and α-Cu_2Se phases [60]. The annealing chamber pressure also influence to form CIS or secondary phases. 2200 Å In/1000 Å Cu layers are thermally evaporated onto glass substrates at room temperature, followed by selenization in a separate chamber at temperature of 260 or 400 °C and different pressures between 0.2 and 10 mbar whereby N_2 is used as a transport agent for Se vapor that is created by heating Se pellets. The chalcopyrite structure films form for selenization temperature of 400 °C and pressure of 8 mbar for 10 min. The CuSe and α- and β-In_2Se_3 phases occur for the recipe of 260/340 °C, 8 mbar for 10 min, whereas CuSe and $CuInSe_2$ phases form for lower pressure of 0.7 mbar and only $CuInSe_2$ exist for further decreasing pressure to 0.3 mbar, keeping the same temperature and processing times. The reason for existing single phase $CuInSe_2$ at low pressure is due to decrease of collisions between N_2 and Se molecules [65].

The In_ySe layers behave insulating and $\alpha = 10^4 cm^{-1}$ in the 500–2000 nm range. The In_ySe layers deposited onto glass show two dominant peaks at $d = 3.248$ Å and 5.232 Å not coinciding with the reported JCPDS but closely matching with $CuInSe_2$ reflections of $d = 3.217$ Å (103) and 5.18 Å (101), which may be related to In_2Se_3 tetragonal structure. The peak at 25.4° represents by In_ySe phase is a shoulder to (112) peak, which slowly appears for annealing temperature of 350 °C or above the temperature [30]. Alike the layers grown onto glass substrates by sputtering method using $CuInSe_2$ single target at room temperature have a small peak before (112) peak relating to secondary phase of In_2Se_3 [66]. The films with similar secondary phase of In_2Se_3 along with $CuInSe_2$ onto Mo substrates exist for the deposition potential of -0.6 V versus SCE. The In_2Se_3 phase can be found on the both sides of (112) in as-grown electrodeposited CIS layers. After annealing the multiphase films under vacuum at 550 °C for 8 min, single phase chalcopyrite $CuInSe_2$ thin films with stoichiometric composition present, as evidenced by existence of (211) characteristic chalcopyrite peak, as shown in Figure 4.22 [67,68]. 2.2 µm thick $CuInSe_2$ thin films grown by ED, followed by selenization are annealed at different temperatures of 500, 550, 580, 600, and 620 °C for 10 min using RTP furnace under N_2 flow with ramp rate of 40 °C/s. XRD analysis reveals that the In_2Se_3 phase as a shoulder to (112) peak is gradually swallowed by CIS with increasing annealing temperature, as shown in Figure 4.23. The anneal temperature of 580 °C or above is high enough to suppress the In_2Se_3 phase [69]. The $CuInSe_2$ films grown by one step ED contain multiple phases of Cu_2Se, In_2Se_3, and $CuInSe_2$ recrystallized under Se vapor of 10^{-2} atm shows that the In-rich CIS films had intensity ratio of 10 for (112)/(220), whereas Cu-rich CIS films experience 2 [70].

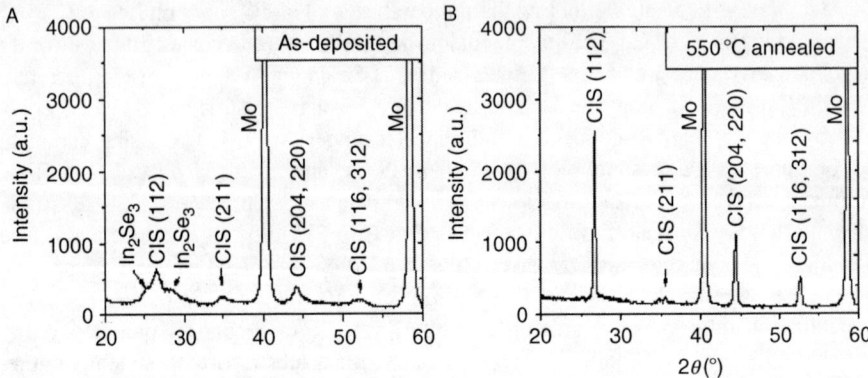

Figure 4.22 XRD pattern of electrodeposited CuInSe$_2$ thin films; (A) as-deposited and (B) annealed under vacuum at 550 °C for 8 min.

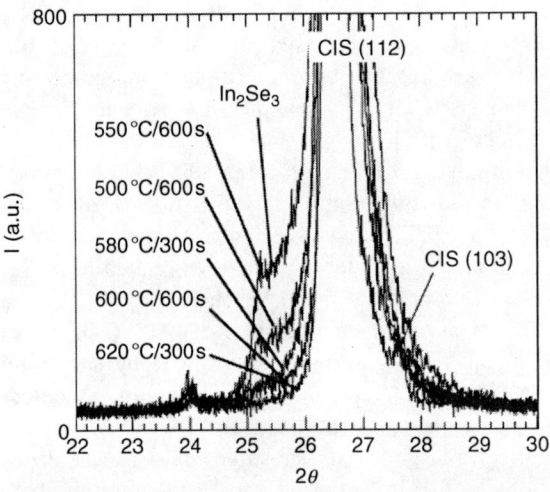

Figure 4.23 Narrow scan XRD pattern of electrodeposited CIS layer around neighborhood of (112) diffraction peak.

The exothermic reaction between In$_2$Se$_3$ and liquid Se generates heat causing poor crystallization and delamination [71]. In fact, other than Cu$_{2-x}$Se, In$_2$Se$_3$, *etc.*, phases, the In$_2$O$_3$ phase is found in the electrodeposited CIS thin films. The CuInSe$_2$ thin films deposited onto FTO by ED using aqueous solutions of 5 mM CuCl$_2$, 5 mM InCl$_3$, 10 mMSeO$_2$, pH 1.5 and potentials at from -0.5 to -0.9 V show (112), (204/220) and (116/312) reflections. After annealing the films under air at different temperatures of 200, 350, and 500 °C for 30 min show SeO$_2$ phase, only CuInSe$_2$ phase and other phases of In$_2$O$_3$, CuO, and Cu$_2$O$_3$H$_2$O, respectively [72]. The CuInSe$_2$ films annealed under air at 300 °C or above temperature indicate that the oxygen content in the films increases with increasing annealing temperature, resulting in a strong peak at 30.5° for In$_2$O$_3$ and several other peaks [30]. The as-grown and annealed CuInSe$_2$ thin films by spray pyrolysis technique exhibit chalcopyrite structure, however in both cases the

In$_2$O$_3$ is a secondary phase [73]. The CuInSe$_2$ thin films deposited onto FTO glasses by ED shows stoichiometric for the working electrode potential of –0.75 V with reference to Ag/AgCl electrode, Cu-rich CuInSe$_2$ thin films with a secondary phase of CuSe for –0.6 V and In-rich CuInSe$_2$ thin films with secondary phases of In$_2$Se$_3$ and In$_2$O$_3$ for –1.0 V [74]. After annealing glass/SnO$_2$/ED CIS layers grown at potentials of from –0.9 to –1.0 V under Ar atmosphere at different temperatures of 350, 450, and 550 °C for 120 min, the peaks appear at 26.6, 44.3, and 52.4° for CuInSe$_2$. Since no chalcopyrite characteristic peaks exist, the CIS layers may be considered as sphalerite. However, the layers annealed at higher temperature of 550 °C shows a peak at 30.6° corresponding to In$_2$O$_3$ phase [75]. The Cu-rich and stoichiometry CuInSe$_2$ thin films deposited onto Mo coated glass substrates by electrochemical method using chemical solutions of CuCl$_2$, InCl$_3$, and SeO$_2$ with potential of from –0.5 to –3.0 V versus SCE at room temperature show no secondary phases, whereas In-rich CuInSe$_2$ thin films show elemental indium (Figure 4.24). Broad XRD peaks evidence a poor crystallinity of the films; hence, the layers have to be annealed to improve the crystallinity [76]. The polycrystalline nature of CuInSe$_2$ thin films prepared by vacuum evaporation at 320 K increases with increasing deposition rate from 0.1, 3.5 to 10 nm/s, which are prepared from chalcopyrite polycrystalline CIS powder [77]. The typical lattice constants of CIS are given in (Table 4.2).

4.3 CuGaSe$_2$

A TEM of CuGaSe$_2$ thin films with Cu/Ga = 1.4 grown onto glass/Mo by close spaced chemical vapor transport technique shows single phase in which grain boundaries can be easily observed. The inset shows SAED pattern containing prime (112),

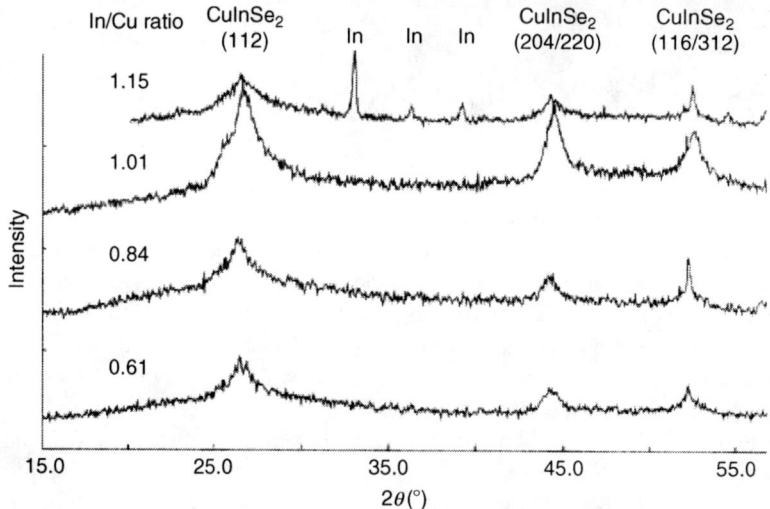

Figure 4.24 XRD pattern of CuInSe$_2$ thin films deposited at different In/Cu ratios by electrodeposition.

Table 4.2 Lattice constants of CuInSe$_2$ and CuIn$_3$Se$_5$ thin film

Sample	a (Å)	c (Å)	Ref.
CIS Single crystal	5.784	11.58	[78]
CIS Single crystal[a]	5.7783	11.5716	[79]
CIS thin film	5.789(6)	11.604(4)	[38]
CIS thin film	5.78	11.62	[41]
CuIn$_3$Se$_5$	5.75	11.52	
CIS	5.754	11.518	[42]
CuIn$_3$Se$_5$	5.789	11.62	

[a]Cu:In:Se = 24.4:23.7:51.9.

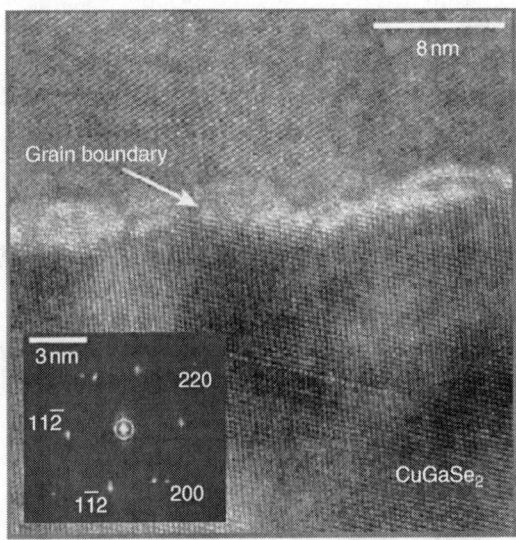

Figure 4.25 TEM of CuGaSe$_2$ thin film grown by closed space chemical vapor transport. SAED pattern is given in inset.

(220) reflections (Figure 4.25) [80]. (112) is preferred orientation in the CuGaSe$_2$ thin films grown by flash evaporation at source temperature of 1250 °C and substrate temperature of 150–450 °C while no preferred orientation is observed in the typical CuGaSe$_2$ single crystals grown by either Bridgman or chemical vapor transport techniques but both (112) and (204,220) peaks are at the same levels in intensity [81]. The similar (112) preferred orientation is found in the CuGaSe$_2$ films with Cu/Ga composition ratio of 1.34 grown onto 7059 corning glass by MBE. The (112) and (204/220), etc., peaks are at the same level in intensity for Cu/Ga = 0.72, in addition β-Ga$_2$Se$_3$ phase exists [1]. The stoichiometric CuGaSe$_2$ thin films grown onto ZnO intrinsic substrates by MOCVD technique show chalcopyrite structure with splitting of (204), (220) and (116), (312) peaks. A resolve between (204) and (220) or between (116) and (312) increases with increasing Cu/Ga ratio from 0.92 to 1.03. On the other hand, the (112) peak position also shifts toward higher angle and its intensity increases

with increasing Cu/Ga ratio. The secondary phase Cu_2Se (111) presents along with $CuGaSe_2$ phase for Cu-rich growth, that is, Cu/Ga = 1.8, whereas the $CuGa_3Se_5$ phase is detected for Cu/Ga = 0.34. No secondary phases are observed for stoichiometry $CuGaSe_2$ thin films with Cu/Ga = 1.03, as shown in Figure 4.26 [82]. The similar kind of observation is found for $CuGaSe_2$ thin films deposited onto (0002) oriented ZnO layers covered soda-lime glass substrates by evaporation of Cu, Ga, and Se from effusion cells. A peak angle 2θ at 27.1° relates to Cu_xSe phase in the Cu-rich $CuGaSe_2$ thin films is observed. The $CuGaSe_2$ thin films deposited on (0002) oriented ZnO layers show preferred orientation of (112). No doublet is observed in the Ga-rich $CuGa_3Se_5$ phase layers indicating inferior structural quality, as compared to CGS. The (112) diffraction angles of 27.8 and 28.2° for Ga-rich and Cu-rich CGS films are observed, respectively [83]. The FWHMs of 0.49 and 0.19° for (112) oriented $CuGaSe_2$ thin films grown by three source evaporation at two different temperatures of 530 and 630 °C are observed in the XRD spectrum, respectively. The films are continuously kept under Se vapor while cooling down the samples to 300 °C. The layers grown at higher temperature have good crystallinity evidencing by lower FWHM [84]. The c/a ratio decreases with decreasing Cu/Ga from 1.35 to 1.0 or lower. A change in c/a ratio is rapid for Cu/Ga ratio between 1.15 and 1.075 due to decrease of c with increasing Ga content in the sample. The unit cell of $CuGaSe_2$ contracts that lead to increase of point defect concentration of V_{Cu} with decreasing Cu content. Even though the Cu content is decreased in the sample, the chalcopyrite structure still continues but does not change into sphalerite structure evidencing by splitting of (204)/(220) and (116)/(312) reflections [85]. However the $CuGaSe_2$ thin

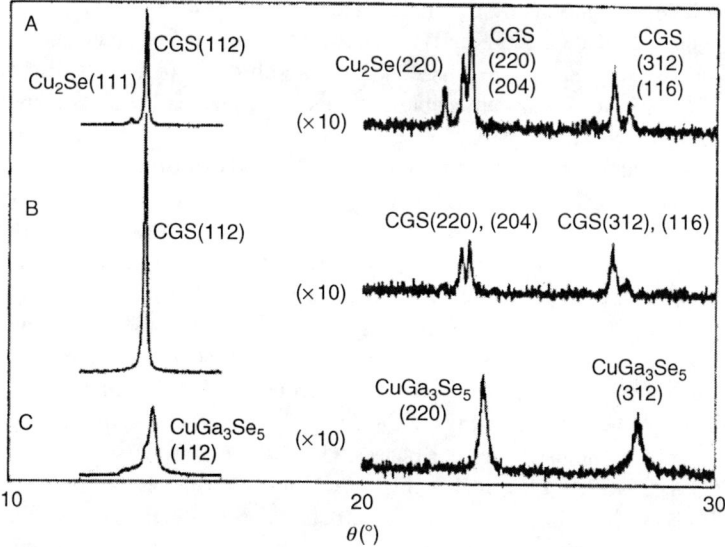

Figure 4.26 XRD pattern of $CuGaSe_2$ thin films deposited with different composition ratios. (A) Cu/Ga = 1.8, (B) 1.03, and (C) 0.34.

films deposited at substrate temperature of 450 °C by thermal evaporation on Al_2O_3 and glass substrates show chalcopyrite structure with (112) orientation for Cu-rich layers (Cu:Ga:Se = 24.11:24.26:51.63), whereas cubic structure (111) occurs for the Cu-poor layers (Cu:Ga:Se = 21.96:25.43:52.61). On the other hand, the (112) peak position changes from higher to lower angle for changing sample composition from Cu-rich to Cu-poor, that is, from 27.686 to 27.519 at.%. The peak position lies at lower angles for Cu-rich films due to stress in the films. In the Cu-rich films, secondary Cu_7Se_4 phase is observed. After etching this phase by NaCN the composition ratio of Cu/(Cu + Ga) is 0.5, that is, stoichiometry [86]. The H_2 flow of 0.56 l/min into cyclopendadienylcopper triethylphosphine (CpCuTEP), 0.9 µmol/min triethylgallium (TEGa) and ditertiarybutylselenide (DTBSe) of 36 µmol/min are processed on GaAs (001) oriented at substrate temperature of 570 °C to grow $CuGaSe_2$ thin films. Partial pressure of Cu/Ga is 2.0–2.7 in the gas phases that results in 1–1.1 in the film. The XRD spectrum shows (004), (008) $CuGaSe_2$ peaks along with (002) and (004) GaAs substrate reflections indicating that the c-axis of $CuGaSe_2$ is normal to the GaAs (001) substrate plane for 260 nm thick CGS layer. The number of groves of $CuGaSe_2$ runs in the (001) direction on the GaAs (001) substrate [87].

The $CuGaSe_2$ layers are deposited onto Mo coated SGS by CVD technique using Cu_2Se and Ga_2Se_3 sources correspondingly HI and HCl used as transport agents. The Cu-rich CGS layers by single stage run at 500 °C for 3 h are grown. In the two stage process, first nearly stoichiometric layers are grown at 500 °C for 4 h and in the second stage; Ga-rich process is done at 530 °C for 20 min followed by annealing under H_2 gas at the same temperature for 5 min at the reactor pressure of 100 mbar. In the second stage Ga dissolves easily into Cu-rich compound. The grown sample had composition of Cu:Ga: Se = 23.8:26.2:49.9. The XRD analysis reveals that $Cu_{2-x}Se$ and Cu_2Se phases appear in the $CuGaSe_2$ samples grown by single stage process. 1.5 µm thick films are in single phase $CuGaSe_2$ for Cu:Ga ratio of 1:15, whereas $Cu_{1.8}Se$, Cu_2Se_x, and $CuGaSe_2$ phases for 1:5 and Cu_2Se_x and $Cu_{1.8}Se$ phases for 1:3 are observed in the XRD spectra. The slightly Ga-rich samples are amenable for good efficiency cells, as is convinced by slightly In-rich CIS cells [88,89]. The Cu, In, and Ga deposited by electron beam evaporation and thermal evaporation, respectively are selenized to form $CuInSe_2$ and $CuGaSe_2$ thin films. In this process, the sequence of layers follows as Cu/Ga/Cu/Ga, Cu:Ga = 1:1, Cu/In/Cu/In and Cu:In = 1:1 and are annealed at 120 °C. The grown metallic layers are annealed in graphite box, which is kept in quartz tube to anneal at 500 °C under Se vapor using elemental Se as a source. The as grown Cu/In/Cu/In layers form $CuIn_{2-x}$ ($0 \leq x \leq 1$) as a leading phase and metallic Cu as a minor phase, after annealing at 120 °C, the $CuIn_{2-x}$ decomposes into $Cu_{11}In_9$ and In metallic. The In/Cu/In/Cu and In/Cu/In are the best sequences in which the intermetallic $Cu_{11}In_9$ quantity is higher in former than that in latter. Annealing the Cu/Ga/Cu/Ga at 120 °C becomes $CuGa_2$ as leading phase and Cu_3Ga and Ga minor phases. Cu/Ga/Cu/Ga layers selenized at 250 °C show $CuGa_2$, Cu_9Ga_4, $Cu_{2-x}Se$, Ga_2Se_3 and Ga phases, whereas the layers selenized at 500 °C show $CuGaSe_2$ phase as dominant and Ga_2Se_3 minor phase. In the case of Cu/In/Cu/In, no Cu–Se phase is detected (Figure 4.27). The faster selenization is occurred in Cu/In system as compared to that in Cu/Ga system at low temperatures of 120 and 250 °C [90].

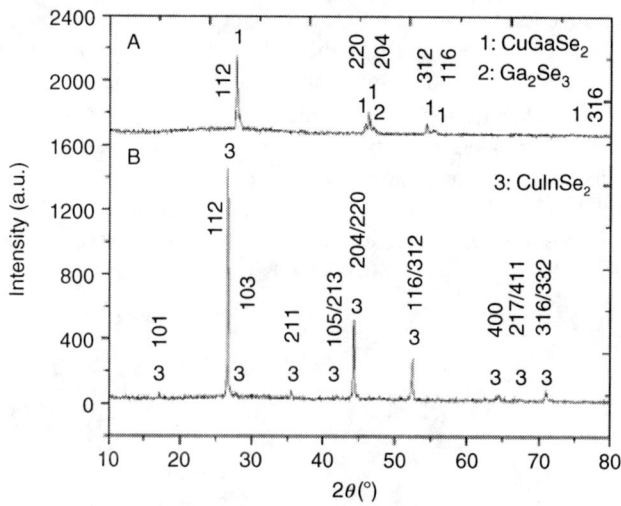

Figure 4.27 XRD pattern of CuGaSe$_2$, Ga$_2$Se$_3$ phase, and CuInSe$_2$ thin films.

Broad peaks consisting of amorphous structure, CGS peaks with δ-GaSe and single phase chalcopyrite CGS are observed in the CuGaSe$_2$ thin films deposited by sputtering technique employing CGS target with 2% excess Se made from ingot at different substrate temperatures of 170, 244, and 330 °C, respectively [91]. The CGS layers onto corning 7059 glass substrates are prepared by three source evaporation whereby the graphite crucibles are used for Cu and In evaporations, whereas the specially designed stainless steel container is employed for Se. The source temperatures of 1383, 1563, 663 °C for Ga, Cu, and Se are employed respectively to deposit CGS layers at substrate temperature of 748 K. The CGS and Cu$_x$Se phases are observed for the conditions of Cu-rich and high Se pressures, whereas the same phases are repeated along with extra Cu for low Se pressures. At the stoichiometric vapor composition of Cu and Ga, the CGS single phase is observed for higher Se pressure, whereas the extra GaSe and Cu phases are found for low Se vapor pressure. In the case of Ga-rich vapor pressure, (CGS)$_x$ + (Ga$_2$Se$_3$)$_{1-x}$ phases are appeared. The OVC (CGS)$_x$(Ga$_2$Se$_3$)$_{1-x}$ + GaSe phases are found for low Se pressure, as shown in Figure 4.28 [92]. The CuGaSe$_2$ layers deposited onto GaAs (100) substrates at different temperatures of 450, 500, 550, and 570 °C had a small amount of Cu$_x$Se but no additional phases are found. The Cu$_x$Se with reflections of (111) and (220) is appeared in the samples as a secondary phase. At or above 550 °C, the MoSe$_2$ phase with reflections of (100), (110), and (200) is seen in the XRD spectrum. The hexagonal structured MoSe$_2$ layer with c-axis formed parallel to the Mo layer [93] (Table 4.3).

4.4 Cu(InGa)Se$_2$

The Cu-poor CuInGaSe$_2$ (CIGS) layers are indeed suitable for the fabrication of CIGS based thin film solar cell but their structure may not be as same as stoichiometric CIGS thin film. The SAED pattern of Cu poor, stoichiometric CIGS thin film and large precession angle SAED of Cu-poor CIGS thin films are shown in Figure 4.29. The stoichiometric CIGS thin films show chalcopyrite structure with Miller indices of (004) and (008) concerning $I\bar{4}2d$ space group. The Cu-poor CIGS thin films accomplish additional (002) and (006) indices relating to $I\bar{4}2m$ space group, which are forbidden reflections in the chalcopyrite structure. In order to avoid ambiguity, a large precession angle SAED is carried out that confirms the presence of (002) and (006) reflections. If the reflections are forbidden they should not be in the precession angle measurement but in this case they present therefore the structure of Cu-poor CIGS thin films belongs to $I\bar{4}2m$ space group not to

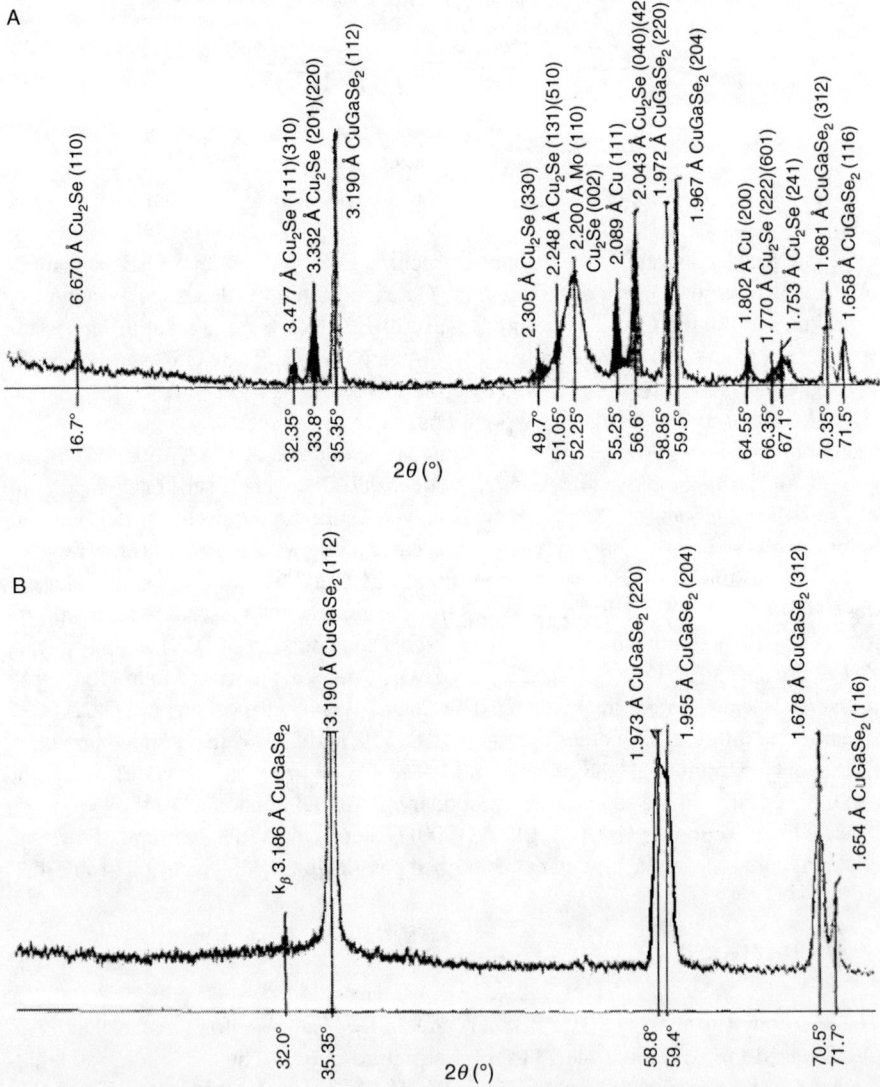

Figure 4.28 (Continued)

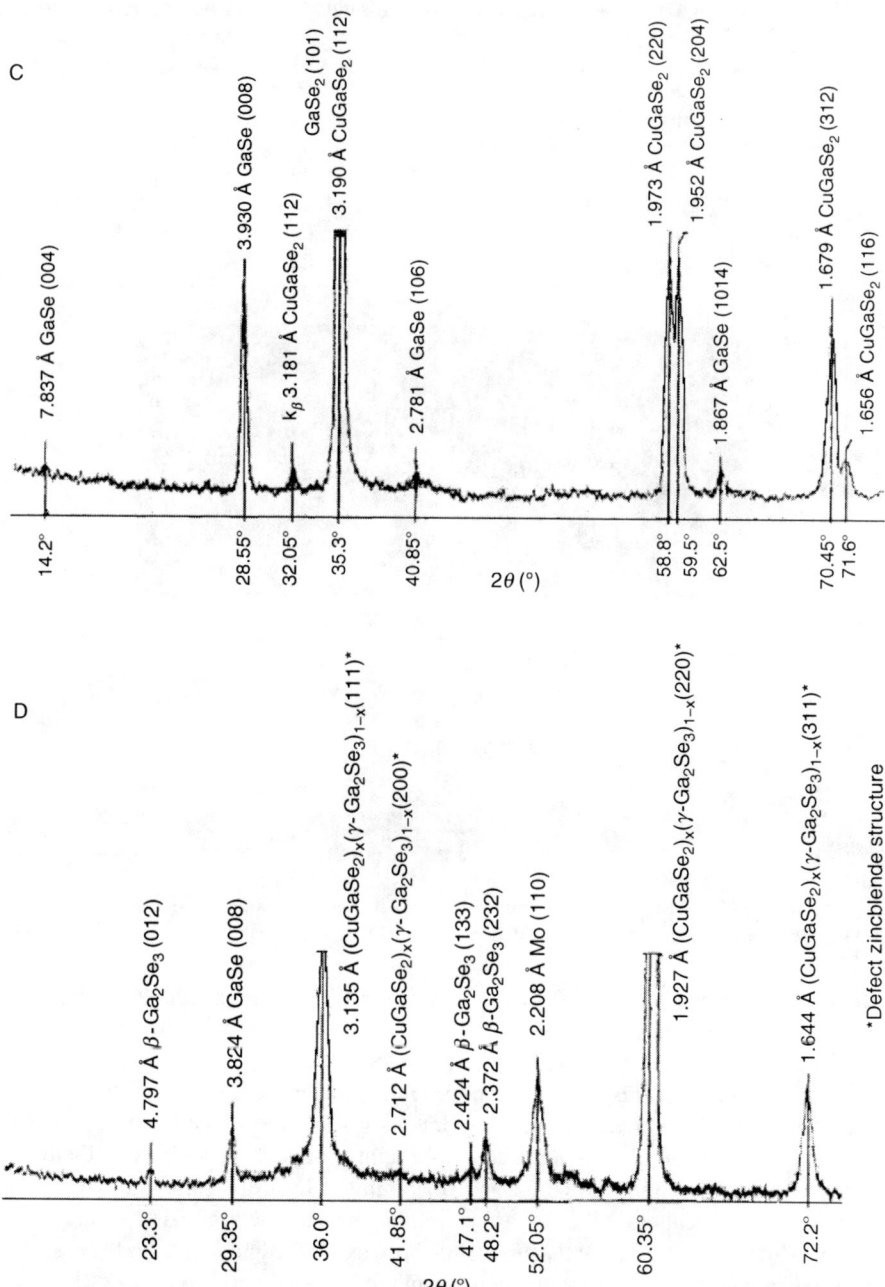

Figure 4.28 XRD pattern of CuGaSe$_2$ thin films grown at different conditions. (A) Cu rich and Se poor, (B) stoichiometric, (C) slightly Ga rich and Se poor, and (D) Ga rich and Se poor.

Table 4.3 Lattice parameters of CuGaSe$_2$ and CuGa$_3$Se$_5$ thin films

Sample	a (Å)	c (Å)	Ref.
CGS thin film	5.6	10.99	[81]
CGS thin film	5.61	11.03	[82]
CuGa$_3$Se$_5$	5.49	10.89	
CGS thin film	5.49	10.89	[83]
CGS thin film	5.61	11.01	[91]
CuGa$_3$Se$_5$	5.61	11.03	

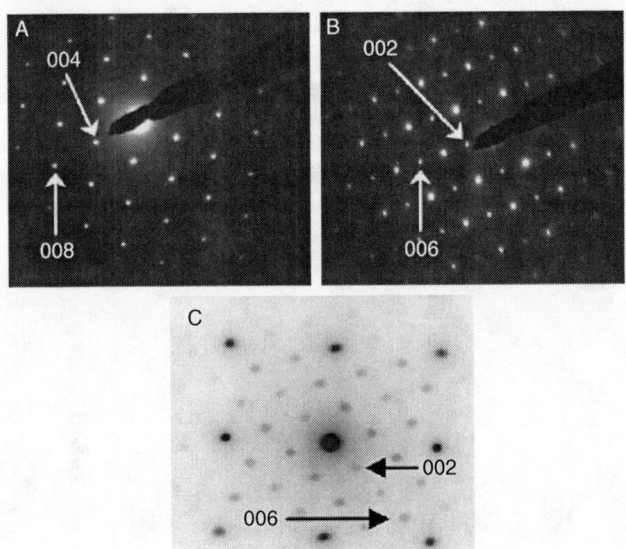

Figure 4.29 SAED of (A) stoichiometric CIGS (I$\bar{4}$2d), (B) Cu-poor CIGS (I$\bar{4}$2m), and (C) SAED with large precession angle of Cu-poor CIGS (I$\bar{4}$2m).

chalcopyrite structure [94]. The Cu$_{2-x}$Se is one of the notorious secondary phases in the CIGS that degrades quality of the thin film solar cell. Figure 4.30A and B show SAED pattern of CIGS grain and Cu$_x$Se particles on CIGS grains or at grain boundaries in the thin film, respectively. The bright spotty pattern is related to CIGS in both Figure 4.30A and B, whereas low intensity spots are related to Cu$_x$Se phase in Figure 4.30B. Both the low intensity and bright spot patterns are unique therefore it can be concluded that the CIGS and Cu$_x$Se might have the same structure [95]. The literature reveals fact that both the compounds show tetragonal structure.

The CuIn$_{1-x}$Ga$_x$Se$_2$ crystals grown by horizontal Bridgman method are single phase chalcopyrite structure in the entire range of $x = 0$–1 and a typical XRD spectrum of CuIn$_{0.56}$Ga$_{0.44}$Se$_2$ single crystal is shown in Figure 4.31. A linear variation in lattice parameters of a and c with varying x, so called Vegard's law that is observed in the grown CuIn$_{1-x}$Ga$_x$Se$_2$ single crystal solid solutions, as shown in Figure 4.32. A variation in lattice parameters with x follows as $a = (5.781 - 0.173x)$ and $c =$

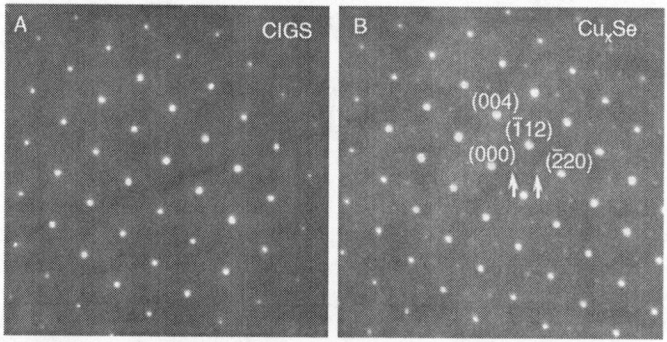

Figure 4.30 SAED pattern of (A) CIGS and (B) Cu_xSe particles on CIGS.

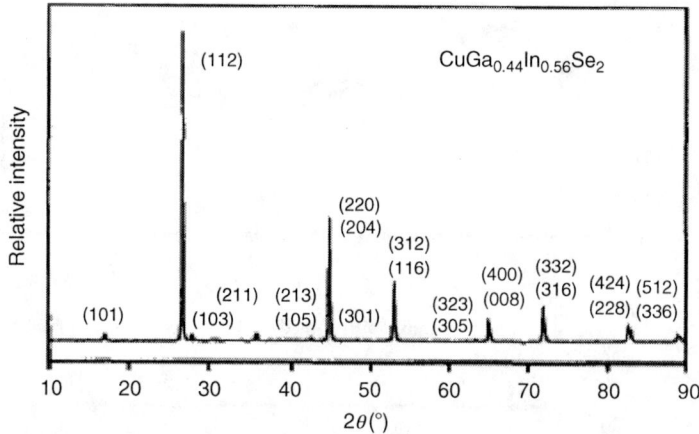

Figure 4.31 XRD pattern of $Cu(In_{0.56}Ga_{0.44})Se_2$ single crystal grown by horizontal Bridgman method.

$(11.614-0.594x)$ Å [96]. The $CuIn_{1-x}Ga_xSe_2$ crystals grown by normal freezing method also follow Vegard's law for a and c lattice constants in the entire $x = 0-1$, as shown in Figure 4.33 and all the samples show chalcopyrite structure [97]. Table 4.4 shows variation of lattice contants with composition.

The c/a ratio varies linearly from 2 to 1.96 with increasing x in the composition range $0.6 \leq x \leq 1.0$, whereas the c/a ratio is 2 in the range of $x \leq 0.4$. A variation in lattice constants with x follows as $a = 5.777-0.173x$ and $c = 11.59-0.581x$ Å in the $CuIn_{1-x}Ga_xSe_2$ crystals grown by chemical vapor transport method using iodine as transport agent with the amount of 6–8 g/cm^3 in the 20 mm dia and 140 nm long quartz tube in the two zone temperatures 800 and $800 + \Delta T$ °C (60–120 °C) [100]. The relations $a = 5.784-0.175x$ and $c = 11.608-0.572x$ Å are also observed for the $CuIn_{1-x}Ga_xSe_2$ thin films grown by RF sputtering using powder target. The c/a ratio varies linearly from 1.966 to 2.007 with increasing x from 0 to 1. The $CuIn_{1-x}Ga_xSe_2$ thin films show clearly the doublet peaks

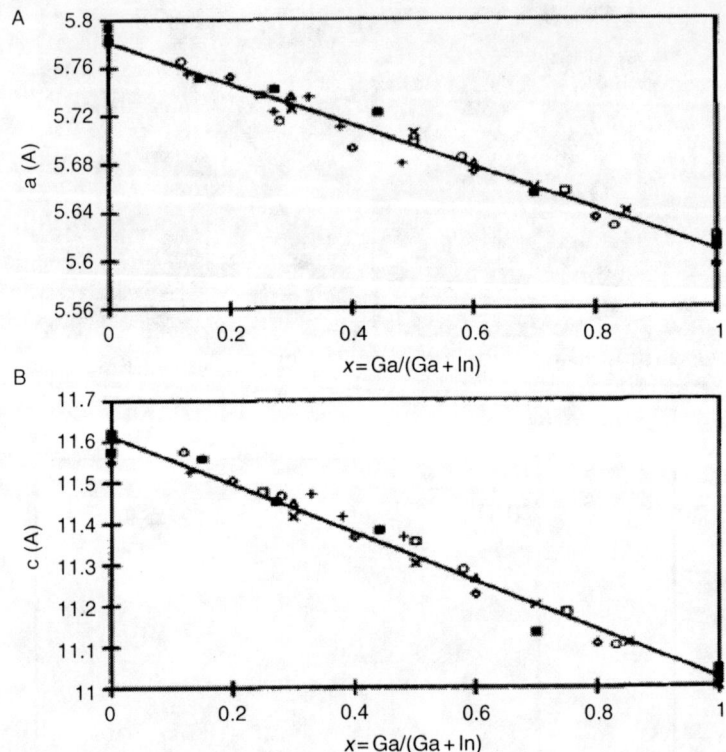

Figure 4.32 Variation of lattice constants with x in $CuIn_{1-x}Ga_xSe_2$ single crystal.

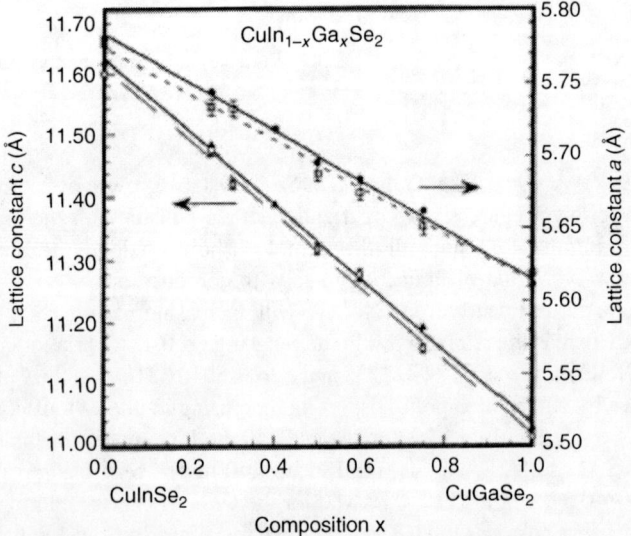

Figure 4.33 Change in lattice constants of $CuIn_{1-x}Ga_xSe_2$ with composition.

Table 4.4 Lattice constants (Å) of $CuIn_{1-x}Ga_xSe_2$ single crystals and polycrystalline ingots

x	0	0.2	0.3	0.39	0.48	0.6	0.62	0.68	0.74	1	Ref.
a	5.77	5.73	–	5.7	5.69	–	5.67	5.64	5.64	5.61	[98]
c	11.55	11.44	–	11.35	11.34	–	11.19	11.16	11.12	11.02	
a	5.783	–	5.736	–	–	5.680	–	–	–	5.612	[99]
c	11.618	–	11.448	–	–	11.264	–	–	–	11.032	
c/a	2.009	–	1.996	–	–	1.983	–	–	–	1.966	
d (g cm^{-3})	5.74	–	5.69	–	–	5.65	–	–	–	5.57	

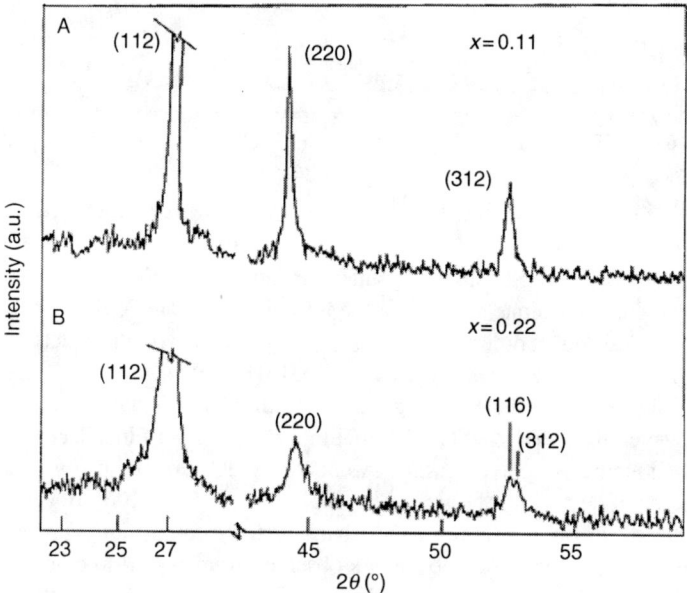

Figure 4.34 XRD pattern of $CuIn_{1-x}Ga_xSe_2$ thin films grown by coevaporation technique.

such as (116/312) and (400/008) for $x > 0.51$ and the doublets (220/204) and (332/316) for $x > 0.59$ [101]. The $CuIn_{1-x}Ga_xSe_2$ thin films grown onto glass substrates by coevaporation show a relation for lattice constants as $a = 5.812 - 0.322x$ and $c = 11.65 - 0.438x$ Å. The typical XRD patterns of CIGS for $x = 0.11$ and 0.22 are shown in Figure 4.34 [102]. The same kind of relations are also observed for CIGS by researchers as $a = 5.765 - 0.164x$ and $c = 11.557 - 0.555x$ Å [98]. The peak splitting or doublets in (204, 220) and (312, 116) present for $0.6 \leq x \leq 1.0$, whereas they disappear for $x \leq 0.4$ [100]. Alike the (116, 312) peak of $CuIn_{1-x}Ga_xSe_2$ thin films at diffraction angle of 52.25° gradually resolves well into (116) and (312) at 54.25 and 55.1°, respectively with increasing x value from 0 to 1 [103]. The sputter deposited CIGS thin films with Ga/(Ga + In) > 0.25 and 1.1 < Cu/(In + Ga) < 0.95 show $a = 5.703(7)$ and $c = 11.385(3)$ Å [104]. The p-type

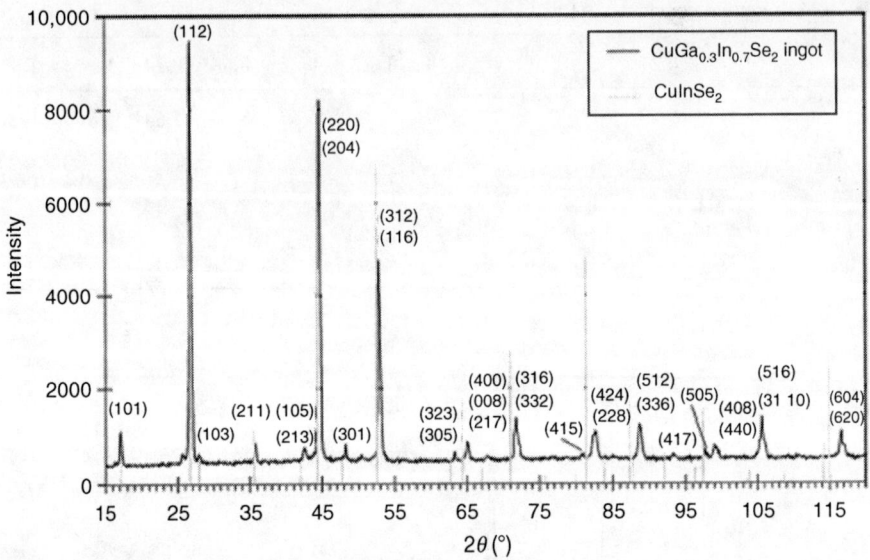

Figure 4.35 XRD pattern of $CuIn_{0.7}Ga_{0.3}Se_2$ single crystal grown by vertical Bridgman method.

chalcopyrite $CuIn_{0.7}Ga_{0.3}Se_2$ single crystals with lattice constant $a = 5.721$ Å grown by vertical Bridgman technique shows that the intensity (I) of (112), (204,220), (116,312) peaks is in the descending order $I_{(112)} > I_{(204,220)} > I_{(116,312)}$ and multiple peaks also present with low intensity, as shown in Figure 4.35 [105]. The FWHM of (112) peak increases with increasing x due to decrease of grain size with increasing Ga content in the films [101]. However, the FWHM of (112) for $CuIn_{1-x}Ga_xSe_2$ thin films decreases for up to $x = 0.25$ then increases with increasing x value due to decrease of quality of crystallinity or an increase of defects for the samples at higher level of Ga [106]. In general, the Cu is deficient in the grown crystals even though the starting composition is taken in stoichiometric ratio. A variation of composition is observed from crystal to crystal at different places in the same grown sample. An additional $CuIn_3Se_5$ phase is detected in the chalcopyrite structured $Cu(In_{0.5}Ga_{0.5})_3Se_5$ crystals grown by horizantal Bridgman method, which is detected by Laue method [107].

1.6 μm thick $CuIn_{0.75}Ga_{0.25}Se_2$ layers with composition of Cu:In:Ga: Se = 26.86:20.71:5.93:46.50, Cu/(In + Ga) = 1.01 and (Cu + In + Ga)/Se = 1.15 by flash evaporation from CIGS charge and 1.4 μm thick CIS films with composition of Cu:In:Se = 24.11:24.95:50.94, Cu/In = 1.05, (Cu + In)/Se = 0.96, and lattice constant (a) of 5.785 Å show (112) preferred orientation with additional odd reflections such as (101), (103), and (211) indicating chalcopyrite structure, as shown in Figure 4.36. Table 4.5 shows substrate temperature, composition, *etc.*, of CIGS layers grown with growth rate of 20–49 Å/min [108].

Unlike coevaporation of individual elements, the γ-In_2Se_3 film is formed by coevaporation of In and Se at substrate temperature of 330 °C on which Cu and Ga layers are deposited and intermixed at 425 °C then finally selenized to form CIGS layer. The XRD analysis of CIGS layers with composition of Cu:In:Ga:Se = 21.5:23:5.5:50 shows

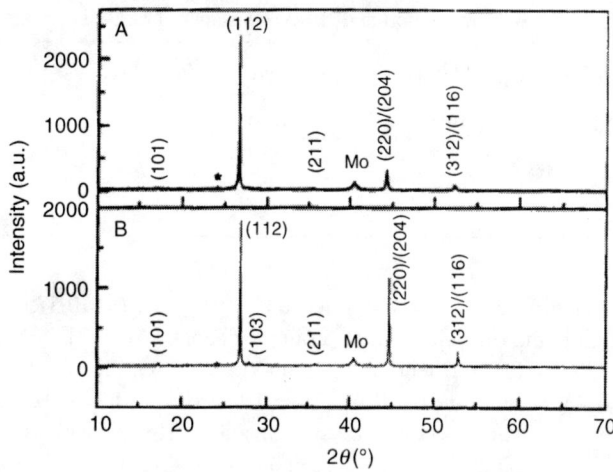

Figure 4.36 XRD pattern of CuIn$_{1-x}$Ga$_x$Se$_2$ thin films with different compositions; (A) CuInSe$_2$ and (B) CuIn$_{0.75}$Ga$_{0.25}$Se$_2$ thin film.

Table 4.5 Deposition conditions and physical parameters of CIGS thin films

T_C (°C)	T_S (°C)	Thickness (μm)	Cu	In	Ga	Se	a (Å)	E_g (eV)
1100	60	3.1	14.14	26.67	–	59.19	5.769	1.26
1250	200	2.1	24	24.46	–	51.55	5.78	0.86
1250	200	1.65	22.36	24.66	–	52.98	5.773	0.97
1250	200	2.4	23.33	25.46	–	51.21	5.771	0.98
1250[A]	340	1.4	24.11	24.95	–	50.94	5.785	–
1200[B]	350	1.6	26.86	20.71	5.93	46.5	–	–

T_C and T_S are source and substrate temperatures, CIS (A), CuIn$_{0.75}$Ga$_{0.25}$Se$_2$ (B).

(112) preferred orientation and chalcopyrite characteristic peaks (101), (103), and (211) [109]. The (112) preferred orientation is found in the CIGS films deposited at 550 °C by evaporation of Cu, In, Ga, and Se, whereas films deposited at 350 °C show random orientation [110]. After annealing flash evaporated CuIn$_{0.75}$Ga$_{0.25}$Se$_2$ thin films under Se vapor at 300 °C, followed by annealing under N$_2$:H$_2$ = 9:1 at the same temperature, the intensity of (112) preferred orientation increases due to an increased crystallinity of the samples [111]. The CIGS layers deposited by sputtering technique using low cost elemental targets on square foot area are post annealed under Se vapor at 560 °C for 1 h (P$_1$) and another set of samples are also post annealed under Se vapor at 580 °C for 4 h duration (P$_2$). The Cu deficiency is observed in P$_1$ samples that causes to have lower lattice parameters, as compared to that of P$_2$ samples. Unidentified impurity is found at 31.8° in P$_2$ sample, which may cause to deteriorate efficiency of the cells. The broad peaks are indication of small grain sizes in P$_1$ sample. The CIGS (P$_3$) and (P$_4$) samples grown by three

Table 4.6 Composition and lattice constants of CIGS samples

Sample	Cu/(In+Ga)	Ga/(In+Ga)	Se/(Cu+In+Ga)	a (Å)	c (Å)
P_1	0.74	0.49	0.92	0.57220	1.1396
P_2	0.87	0.4	0.88	0.57277	1.1457
P_3	0.53	0.46	0.85	0.57362	1.14492
P_4	1.52	0.5	0.87	0.57363	1.14568

stage process with different compositions had a $Cu_{1.8}Se$ phase ($a = 0.575$ Å) and its concentration is high in the Cu-rich CIGS (P_4) sample (Table 4.6) [112].

The (204, 220) preferred orientation is found in the CIGS layers with composition ratio of Ga/(Ga+In) = 0.30 and Cu/(In+Ga) = 0.81 grown by thermal evaporation onto Mo (110) and Mo (100) or (111) single crystal substrates. The intensities (I) of (204,220), (112), (116,312) reflections are in the descending $I_{(204,220)} > I_{(112)}, > I_{(116,312)}$ sequence [113]. After annealing ED CIGS or CIS layers grown onto Mo polished substrates/Mo coated glass substrates under vacuum at 450 °C for 20 min, the well developed diffraction reflections such as (112), (204,220), and (116, 312) occur in the XRD spectra [114]. The $\theta/2\theta$ scans of Cu(InGa)Se$_2$ thin films on glass/Mo and quartz/Mo reveal that the (112) preferred orientation peak splits into doublet indicating Ga grading in the layers (Figure 4.37A). The grazing XRD with an incident angle of 1.0° indicates low profile split, whereas normal $\theta/2\theta$ bulk scan shows well resolved split evidencing that Ga concentration is low at the surface and high in the bulk, as shown in Figure 4.37B. For this analysis the Cu(InGa)Se$_2$ thin films grown onto both Mo coated glass and quartz substrates at room temperature by evaporation of CIGS bulk with Ga/(Ga+In) = 0.6, followed by annealing under Se vapor at 550 °C for 1 h show Cu:Ga:In:Se composition ratio of 22.9:12.9:12.3:51.9 on the surface. The bulk is synthesized by reacting pure Cu, In, Ga, and Se elements in a sealed quartz ampoule under vacuum at high temperature [115].

The Cu(InGa)Se$_2$ thin films with composition of Cu:In:Ga:Se = 24.64:8.65:16.75:49.96 as a precursor grown by thermal evaporation at RT from Cu(InGa)Se$_2$ bulk, which is synthesized by combining pure elements with ratio of Cu:In:Ga:Se = 1:0.4:0.6:2 in sealed quart tube under vacuum and melt at high temperature then cooled down to low temperature. Two kinds of samples consist of different compositions in the films. The precursor sample is annealed under saturated Se vapor by heating Se shots so called sample-P_5. Another precursor thin film layer is annealed under vacuum without Se vapor at temperature of 500 °C for 1 h denoted as sample-P_6. The sample-P_5 and -P_6 exhibit chalcopyrite structure evidencing by presence of low intensity chalcopyrite characteristic peaks of (101), (103) and (211). Secondly, (400,008), (424,228), and (116,312) diffraction peaks split into doublets (Figure 4.38). The FWHM of 0.31 and 0.59° are observed for sample-P_5 and -P_6, respectively. A slight shift in diffraction angle toward lower angle in sample-P_6 occurs with respect to sample-P_5 due to differences in compositions. The physical parameters of both the samples are given in Table 4.7. It can be concluded that the band gap is higher in sample-P_5 owing to well crystallization [116]. Secondly, the saturated Se vapor is provided for sample-P_5 during annealing that prevents

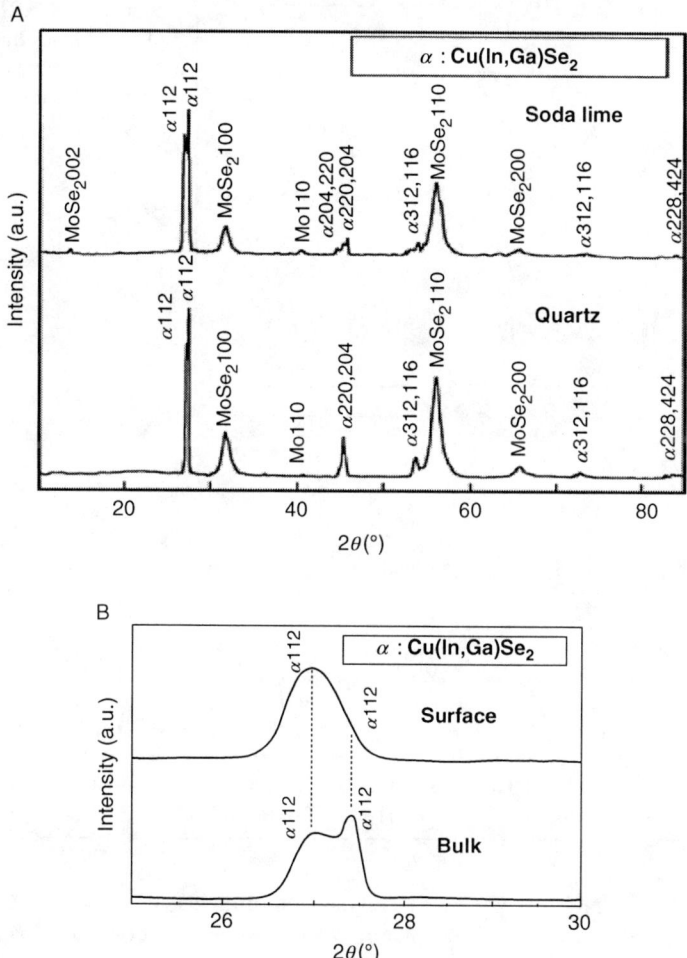

Figure 4.37 (A) XRD pattern of CuInGaSe$_2$ thin films deposited on soda lime glass and quartz substrates and (B) XRD pattern of surface and bulk CuInGaSe$_2$ thin films.

the Se loss. Therefore the composition differs in the samples may also cause to have different band gaps.

Eventually, the substrate temperature controls the percentage of (112) preferred orientation in the CIGS thin films deposited by coevaporation of Cu, In, Ga, and Se. As substrate temperature is increased from 475 to 550 °C in steps of 25 °C, the intensity ratio of (112)/(204,220) peak gradually increases from 1.85 to 5.88, as shown in Figure 4.39. The CIGS layers are deposited by CUPRO process as changing Cu flux from poor to rich and then to poor without changing fluxes of other elements at different substrate temperatures of 475, 500, 525, and 550 °C. The intensity ratio of (112)/(204,220) peak increases with increasing growth temperature of CIGS layers. They are 1.85, 1.53, 2.66, and 5.88 for the layers grown at

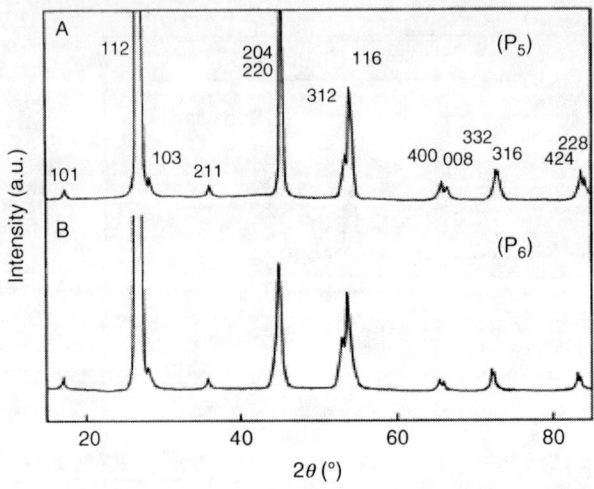

Figure 4.38 XRD pattern of CuInGaSe$_2$ thin films selenized under Se vapor (P$_5$) and vacuum (P$_6$).

Table 4.7 Physical parameters of CIGS layers

CIGS sample	Composition Cu:In:Ga:Se	Lattice constants (Å)		$p \times 10^{16}$ (cm^{-3})	μ (cm^2/Vs)	ρ (Ω-cm)	E_g (eV)	Grain size (μm)
		a	c					
P$_5$	25.45:8.51:15.43:50.61	5.671	11.237	4.94	38.2	3.32	1.402	0.5
P$_6$	25.58:12.96:14.1:47.36	5.694	11.312	23.1	8.46	3.19	1.304	1

475, 500, 525, and 550 °C, respectively. In addition the peak positions of (112), (204,220), and (116,312) shift toward lower angle, the doublet (204,220) and (116,312) peaks resolve well. The grain sizes also increase with increasing substrate temperature [117]. (112) is preferred orientation in the CuIn$_{0.75}$Ga$_{0.25}$Se$_2$ thin films grown at RT, 100 and 200 °C by flash evaporation technique. However, as temperature is increased, the (112) preferred orientation becomes stronger. On contrast to low temperature deposition, (204, 220) and (116, 312) reflections appear well for the layers deposited at substrate temperature of 200 °C [118]. Similarly, the Raman spectroscopy contributes the stronger chalcopyrite nature with increasing substrate temperature. The CIGS layers grown by sequential sputtering of Cu, In, and Ga but Se by thermal evaporation are annealed at 580 °C for 4 h and another sample at 560 °C for 1 h under Se atmosphere. For longer time annealing the CIGS sample at high temperature secondary amorphous phase prevails in the sample due to reevaporation of Se, *etc.*, that degrades the quality of the layers [119]. The CIGS films grown onto different oriented GaAs substrates at substrate temperature of 540 °C by hybrid sputtering using In and Cu$_{0.72}$Ga$_{0.28}$ or Cu$_{0.8}$Ga$_{0.2}$ targets and Se by thermal evaporation from effusion cell. The substrate temperatures of 540, 640, and 700

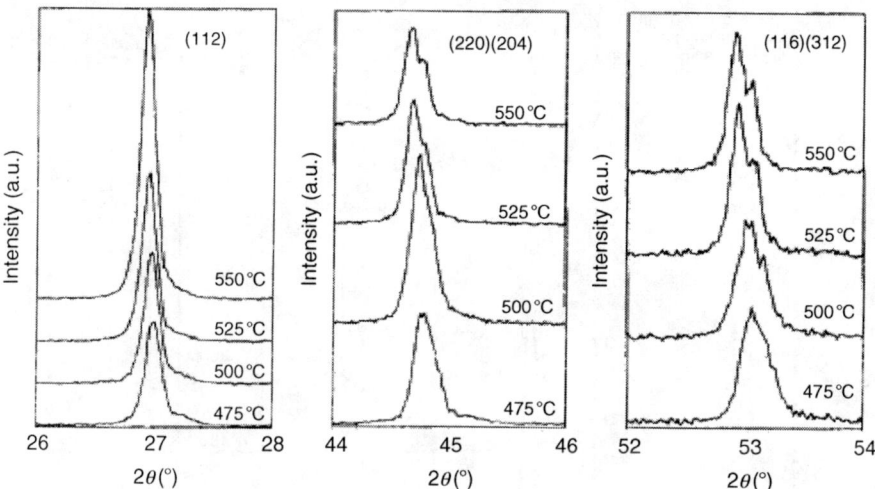

Figure 4.39 XRD pattern of CuInGaSe$_2$ thin films prepared at different substrate temperatures.

°C are found optimal for the growth of CIGS layers onto (110), (100), and (111) oriented GaAs substrates, respectively indicating that the orientation determines growth temperature. In the XRD spectrum, the broad peak is combination of (204) and (220) observed at 44.452 and 44.508° with FWHM of 0.088 and 0.064° for CIGS grown onto (110) GaAs at 540 °C, respectively [120]. The (112) preferred orientation depends on the growth rate of CIGS layers. The deposition time is reduced by increasing appropriate growth rate to maintain constant thickness of 2 μm, the intensity ratio of (112)/(204,220) peak for 2 μm thick sample is 42, 15, 4.2, and 4.0 for growth timings of 30, 15, 7.5, and 3.75 min, respectively. As growth time is decreased the preferred orientation of (112) decreases indicating occurrence of random orientation. It may be probably lack of volume of atoms to rearrange in the crystal sites or insufficient time/kinetic energy for the atoms to form (112) preferred orientation [121].

In the XRD spectrum, the Ga graded CIGS layers with Ga/(In+Ga)=0.2 grown by coevaporation technique show a broad peak at 27.6° along with (112) sharp peak. On contrast, the CIGS samples without Ga grading exhibit no additional peak. This study indicates that the CuGaSe$_2$ is formed at the back side of CIGS layer not exactly CuGaSe$_2$ phase but with lower concentration of In in the Ga graded samples [122]. The positions of (112), (204,220), and (116,312) peaks for the CIGS layers grown by sputtering of Cu and In and evaporation of Ga and Se at room temperature gradually shift to higher angle with increasing annealing temperature from 420, 500 to 550 °C indicating that the diffusion of Ga increases from backside of the precursor into film with increasing temperature. The distribution of Ga takes place equally in the CIGS film at higher annealing temperature. However, no change in the Mo peak is observed, as shown in Figure 4.40 [123]. The surface of CIGS samples show (112) preferred orientation for grazing incidence GIXRD angle of 0, 0.2°, whereas

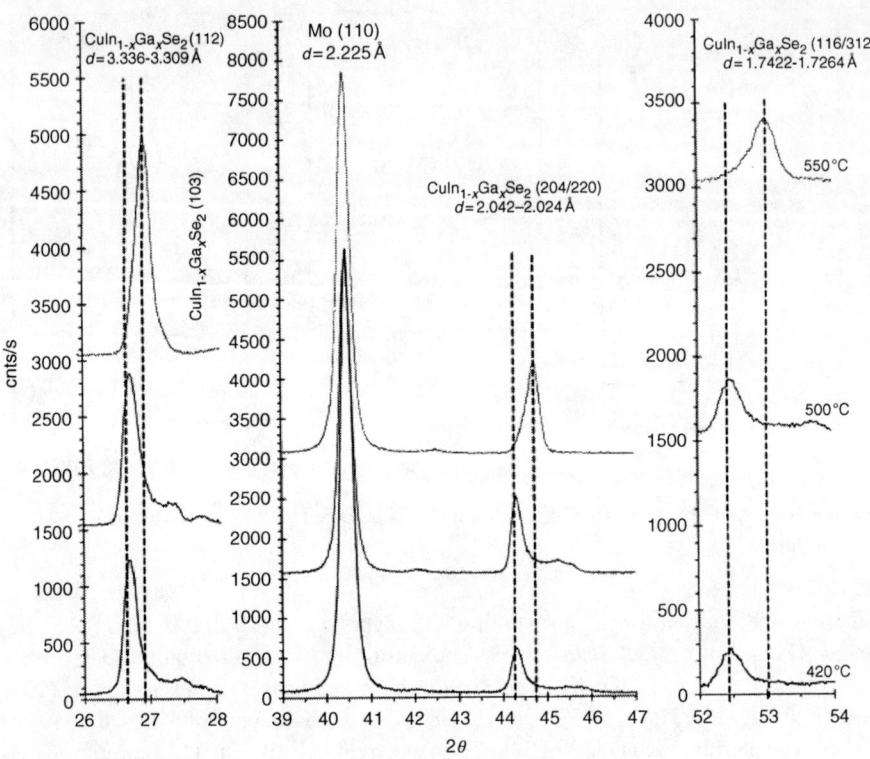

Figure 4.40 XRD pattern of CIGS layers recrystallized at different temperatures.

(204,220) is intensity peak for higher GIXRD. As the grazing incidence angle of (204,220) diffraction peak is decreased from 5, 2, 1, 0.5 to 0.2, the peak position shifts by small quantity toward higher angle in the XRD spectrum indicating that the Cu is depleted with lower concentration at the surface that means Ga grading takes place [124]. No difference in (112) preferred orientation is observed for the CIGS layers grown onto glass and stainless steel substrates [125]. The intensity of (112) peak increases and FWHM decreases with increasing Cu content in the layers due to an increased crystallinity. Beyond Cu/(In+Ga) ratio of 1.2, there is no improvement in intensity for the x values in the $CuIn_{1-x}Ga_xSe_2$ layers. For this analysis the Cu and In by electron beam evaporation and Ga by thermal evaporation are deposited in the sequence of In/Ga/Cu/In. In the second stage, the layers are selenized in the partially opened graphite box using quartz tube furnace by increasing the selenium temperature to 500 °C at the rate of 9 °C/min for 30 min [126]. The precursor is selenized either under Se vapor+Ar flow with flux of 0.5 l/m or under vacuum. The former process provides quality crystalline $CuIn_{0.7}Ga_{0.3}Se_2$ layers comparing with latter. In the sense, the lower FWHM and little higher diffraction angle are present for (112) intensity peak, as shown in Figure 4.41 [127]. The intensity of (112) preferred orientation is higher in the CIGS than in the CIS layers

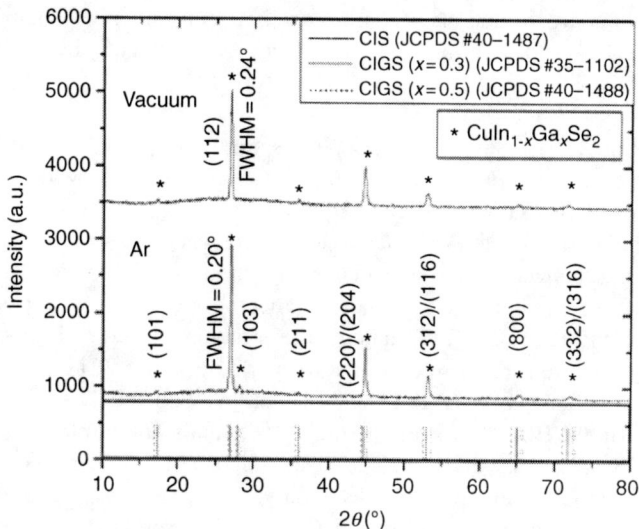

Figure 4.41 XRD pattern of CIGS layers grown under Se+Ar or Se only.

(Cu:In:Se = 19.1:23.96:56.95) for which the CIGS layers with composition of Cu:In:Ga:Se = 21.62:18.96:9.33:50.08 are prepared by two step ED; first Cu-Ga on Mo at −1.95 V, followed by CuInSe$_2$ at −1.25 V and annealed under Ar flow at 600 °C for 1 h. [128].

The thin film solar cells prepared with CIGS sample with intensity ratio of 15.4 for (204,220)/(112) preferred orientation shows the highest conversion efficiency of 19.5%. The (204,220) orientation is occurred for over pressure of Se vapor during CIGS growth. The cathode luminescence of (204,220) orientation sample shows emission line at 1.12 eV which shifts toward higher wavelength with increasing excitation power from 20 pA to 1 nA as a blue shift, whereas there is no shift with increasing excitation beam intensity for (112) preferred orientation sample, whereby the emission peak needs more excitation power or intensity to generate at 1.27 eV, which seems to be combination of two emission lines [129]. The CIGS layers show better results for introduction of H$_2$O vapor into the chamber at partial pressure of 9×10^{-3} Pa while growing CIGS layers by three stage process in MBE chamber. All the XRD patterns of CGS, CIGS, and CIS thin films show narrow FWHM as compared to those grown without introduction of H$_2$O vapor. The FWHM of (112) peak is 1493 and 1608 arc–sec for introduction of H$_2$O vapor and without introduction of H$_2$O vapor, respectively, whereas (204,220) peak shows 2172 and 2378 arc–sec for with and without H$_2$O vapor, respectively. The crystal quality of the layers increases for introduction of H$_2$O vapor. The SIMS analysis reveals that the concentrations of O and Na increase in the layers for introduction of vapor. It is likely possible that the oxygen reduces donor (In$_{Cu}$) concentrations by reacting with In to form In–O. None the less the acceptor concentration either V$_{Cu}$ or Cu$_{In}$ proportionally increases with decreasing donor concentration (In$_{Cu}$) that causes more p-type conductivity. On the other hand Na concentration also enhances the hole concentration. The optimized Se fluxes are 6×10^{-3}, 8×10^{-3} and 3×10^{-3} Pa for the growth of CIS, CIGS, and CGS layers while keeping Cu, In, and Ga beam fluxes of 1×10^{-4}, 1×10^{-4} and 5×10^{-5} Pa, respectively [130].

The additional $CuIn_2Se_{3.5}$, $CuIn_3Se_5$ phases are found in the $CuIn_{1-x}Ga_xSe_2$ thin films grown by selenization of magnetron sputtered precursors, which had composition ratio of Cu:In:Ga:Se = 18.9:28.12.5:50.5 from XRD analysis [131]. The Cu, In, and Ga deposited onto Mo coated soda lime glass substrates by ion-plasma sputtering are selenized using elemental selenium at 500 °C to form CIGS layers, which exhibit (112) preferred orientation. The other low intensity peaks such as (101), (103), and (211) confirm chalcopyrite structure, whereas In_2Se_3 segregates as a secondary phase for low selenization temperature < 300 °C [132]. The 2θ value of (112) peak shifts from 26.65 to 27.2° when cathodic potentials are increased from 0.6 to 1.1 eV in the ED. The lower and higher potentials favor to form CIS and In or Ga-rich CIGS layers, respectively. The layers deposited at lower potential of -0.6 V show Cu-rich CIS and CuSe as a secondary phase, whereas the layers grown at higher potentials, the In_2Se_3 (110) and Ga (101) phases at diffraction angles of 25.12 and 37.8°, respectively are observed as secondary phases in the CIGS layers. On the other hand, the n-type conductivity is observed. The layers grown at -0.8 V show no secondary phases [133]. The CIS and $CuInGaSe_2$ thin films deposited onto SnO_2 by ED technique show amorphous nature. After annealing under vacuum at 400 °C, the layers show polycrystalline nature with multiple reflections such as (112), (103), (211), (204,220), and (116,312). The intensities of XRD peaks increase with increasing deposition solution concentration of $Ga_2(SO_4)_3$ from 0.5 to 1.12 mM and 1.4 mM. The ED technique can dope Ga content upto 6% in the CIGS films [134]. The CIGS with $x=0.2$ shows (112), (103), (211), (204,220), (116,312), (400,008), (316,312), and (228,424) reflections. As x is increased from 0 to 1 in the CIGS, the peak (116,312) position constantly shifts from 52.3 to 54.25/55.1° and (116,312) peak is well resolved as doublets (312) and (116). In the entire range of $x=0$–1, no Cu_2Se or $(InGa)_2Se_3$ phases are observed. The sequential stack of 670 nm thick CuGa/CuIn/Cu layers are selenized under normal atmosphere using non-toxic diethyl selenium $(C_2H_5)_2Se$ and N_2 flow rates of 3 μmol/min and 2 l/min to make CIGS layers at substrate temperature of 515 °C for 90 min, respectively [135].

The amorphous CIGS films are grown at room temperature using the same CIGS compound with composition ratio of Cu:In:Ga:Se = 1:0.4:0.6:2 by thermal evaporation. The EPMA analysis reveals that only Cu and Se are detected for below selenization temperature of 300 °C due to formation of $Cu_{2-x}Se$, whereas Cu:(In+Ga):Se = 25:25:50 compositional CIGS is observed for above selenization temperature of 400 °C. The Ga content increases in the film for higher temperature range of 400–600 °C. The $CuSe_2$ and Cu_7Se_4 phases are observed for the selenization temperature of < 300 °C, whereas the chalcopyrite $Cu(InGa)Se_2$ phase is observed for > 400 °C in the XRD spectrum. The grain sizes of 0.4 μm remarkably increase for 400–600 °C processes [136]. In principle, Cu-rich and stoichiometric CIS films are in rough and Cu-poor or In-rich films are shiny. The Cu-rich and poor films had large and smaller grain sizes with different shapes, respectively. The Cu_2Se phase precipitates on the surface of the films for Cu/In > 1. In the same film, the lateral composition is observed that at one point the Cu/In ratio is 1 and another point being 1.06, which follow rough and smooth surfaces, respectively, and the similarity is observed in the CIGS layers. The $CuGaSe_2$ layer with Ga-rich shows (204,220) orientation. First $CuInSe_2$ layer with (112) orientation as a bottom layer deposited onto glass substrate at substrate temperature of 500 °C, followed by deposition of $CuGaSe_2$ layer, the ultimate final layer shows the same (112) orientation. The (204,220) peak intensity increases with increasing thickness of top layer to 3000 nm. In the final layer, no Cu_xSe phase is observed. The multilayers such as CIS/CGS/CIS with 3 μm thick show (112) preferred orientation [137]. After third stage deposition of $Cu_{0.9}(In_{0.7}Ga_{0.3})Se_{2.1}$ thin films, the influence of Se flux is studied. The intensity ratio of (112)/(204,220) is 5 for Se flux rate of 35 Å/s, whereas it is 14 for 15 Å/s indicating that the (112) orientation loses its ground with increasing Se flux. The V_{oc} and FF are low in the CIGS thin film solar cells for higher Se flux rate of 35 Å/s because of generation of pores on the surface. The poor diode quality factor of 2.01, series resistance of 0.51 Ω-cm^2 and shunt resistance of 470 Ω-cm^2 indicate that the intensity of (204,220) plane on par with (112) plane shows not much encouraging results in terms of PV parameters. The CIGS absorber with Se flux of 15 Å/s shows a strong (112) orientation. The cells made with CIGS absorber grown at Se flux of 15 Å/s exhibit good PV parameters as $V_{oc} = 0.655$ V, $J_{sc} = 36.48$ mA/cm^2, $FF = 73.52\%$, $A = 1.62$, $R_s = 0.26$ Ω-cm^2, and $\eta = 17.57\%$ over the area of 0.421 cm^2 [138]. The $CuIn_{0.75}Ga_{0.25}Se_2$ layers from sputtering with 1.1 < Cu/(In+Ga) < 0.95 has also shown chalcopyrite structure with lattice parameters of $a = 5.703(7)$ and $c = 11.385(3)$ Å [31].

Cu and In sequentially evaporated by e-beam evaporation and Ga by thermal evaporation as a sequence of In/Ga/Cu/In onto Mo coated SLG. The stack is selenized in graphite container under selenium vapor supplied by heating elemental selenium through the Ar flow of 0.5 l/min in three zone heating furnace. Cu_2Se and CuSe are found as secondary phases in the $CuInGaSe_2$ thin films, no matter whether films are Cu-rich or poor. The secondary phases are clearly seen in the Cu-rich films by XRD, as compared to that in the Cu-poor films [139]. It is well known fact that $Cu_{2-\delta}Se$ is a secondary phase and the (204,220) reflection splits in the Cu-rich CIGS films grown by evaporation. No split is observed in the stoichiometry and Cu-deficient films due to smaller lateral resolution and defect structure with a smaller lattice constants, respectively [140].

Additional In and Ga are added to the electrodeposited Cu-rich CIGS precursors by PVD at substrate temperature of 550 °C, followed by selenization under Se vapor. The XRD patterns of CIGS precursor layer grown by ED along with chemical composition adjusted CIGS with $x = 0.16, 0.26, 0.39$ are given in Figure 4.42. The precursor composes of CIGS phase and Cu_2Se, whereas all the composition adjusted CIGS layers show chalcopyrite structure [141]. The metal precursor Cu/Ga/In deposited onto Mo by DC magnetron sputtering at room temperature is selenized under $H_2Se/Ar/O_2$ at 450 °C for 90 min and post heat treated under Ar for 60–90 min at either 500 or 600 °C then again selenized

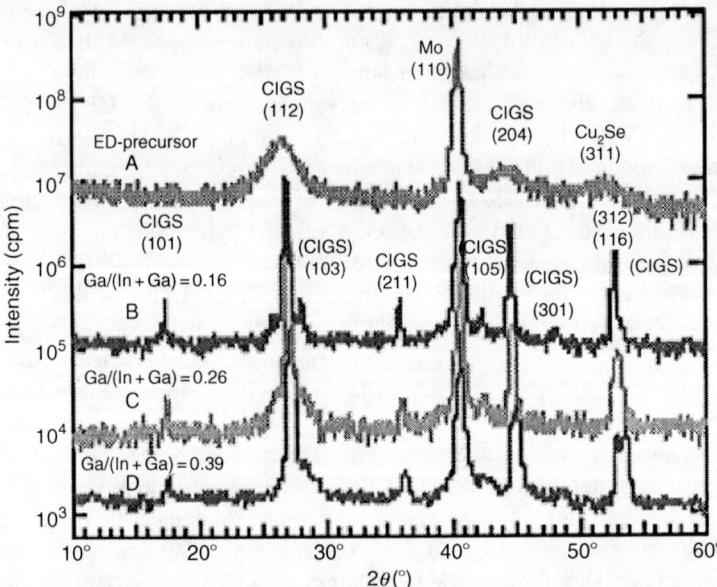

Figure 4.42 XRD pattern of CIGS precursor grown by electrodeposition and CIGS with different compositions.

to regain lost selenium in the samples. The precursors with $Ga/(In+Ga) = 0.25$ or 0.5 post heat treated at 500 °C, two distinct phases of $CuInSe_2$ and $CuGaSe_2$ appear at 44.4° for (204,220) and 45.6, 45.9° for (204), (220), respectively. If the annealing temperature is 600 °C, only the CIGS phase appears. On the other hand, after annealing the precursor with $Ga/(In+Ga) = 0.75$ at 500 °C, the only CIGS phase appears at 45 and 45.2° for (204) and (220) doublets, respectively. On contrast to higher annealing temperature of 600 °C, a higher percentage of Ga takes place solubility in the CIGS at lower temperature [142]. Slightly In-rich $CuInGaSe_2$ thin films with $Cu/(In+Ga) = 0.9$ and $Ga/(In+Ga) = 28\%$ grown onto Mo coated soda-lime glass substrates by simultaneous evaporation exhibit chalcopyrite structure, as shown in Figure 4.43. Air anneal CIGS layers indicate that the (112) peak position does not change for the annealing temperature of 300 °C, whereas it moves to lower angle by losing Ga 5 and 8% for the annealing temperature of 400 and 500 °C, respectively. A small peak presents at 24.1°, which may be related to In_2O_3 phase. The SIMS analysis reveals that the concentrations of Cu, In, Ga, and Se increase in the Mo region that means diffusion is taken place [143]. The CuInGaSe precursor prepared by sputtering and evaporation is annealed by RTP at 550 °C had $Cu/(In+Ga) = 0.92$. After air annealing at 400 °C under flow of Ar, N_2, and O_2 for 20 min, the In_2O_3 phase is observed in the CIGS, however no In_2O_3 phase is detected for less than 20 min annealing time [144].

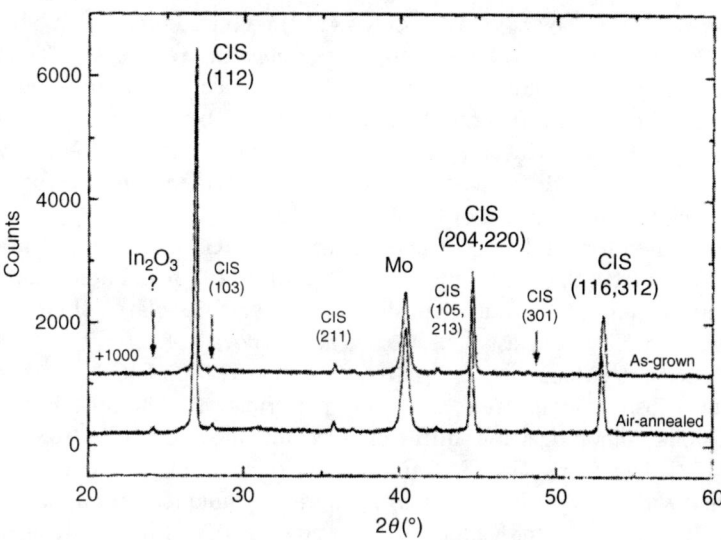

Figure 4.43 XRD pattern of CIGS thin films; (A) as-grown and (B) air-annealed.

4.4.1 Effect of Na on Cu(InGa)Se$_2$ thin films

The intensity ratio of (112)/(204,220) diffraction peaks is 3 for the total thick CIGS layers on glass/Mo/SiO$_x$ without Na. First 1/3 thick CIGS layer shows ratio of 30, the remaining 2/3 of final layer consists of ratio of 20 with silica barrier layers indicating that the Na does not diffuse into the layer through the barrier. The Na influences to increase the (112) orientation. The well resolved (204) and (220) peaks are good sign of chalcopyrite structure. The preferred orientation can also be controlled by Na impurities. Prior to deposition of CIGS, the NaF powder is deposited onto the substrates for Na doping into CIGS. The CIGS layers deposited at 450–510 °C by coevaporation and NaF with different thickness reveal that 30 Å thick NaF is good enough to obtain (112) preferred orientation. The intensity ratio of (112)/(204,220) is 40 for the CIGS layers grown onto Mo coated soda-lime glass substrates, whereas the ratio is one order of magnitude higher for Na doping in the form of 30 Å NaF layer. The similar trend is also observed for the substrates with alumina barrier. The ratio is 3, 490, 81, and 25 for without NaF coating, 30, 200, and 400 Å NaF thick layers, respectively. The ratio of 2.5 is even small comparing with the JCPDS 35-1102 powder data [145]. The intensity ratio of (112)/(204,220) is 4.76 for the CIGS layers on precursor layer of SLG/Mo/NaF(40 nm), whereas the ratio is less than unity for low content of Na [146]. Prior to deposition of Mo onto glass substrates, Al$_2$O$_3$ layer is deposited in order to protect diffusion of Na from the substrate. The CIGS layers are deposited onto different kinds of processed SLG/Al$_2$O$_3$/Mo, SLG/Mo(e-beam), SLG/Mo(sputtered) and SLG/Mo/NaF(100 Å) substrates. The (112) is a preferred orientation for the last one, whereas the (204,220) is preferred one for the first indicating Na diffusion favors for the growth of (112) oriented layers [147].

The In-rich CIGS layers with Na shows (112) preferred orientation but without Na still the layers show (112) preferred orientation with low intensity. The bilayer, that is, $CuInSe_2$ on $CuGaSe_2$ films are studied with effect of Na. In the presence of Na, the In and Ga inter diffusion is slow. The distance between (112) peaks represented by $CuInSe_2$ and $CuGaSe_2$ phases is small indicating that In-diffusion into $CuGaSe_2$ and Ga out diffusion from $CuGaSe_2$ is low. The inter diffusion of In and Ga is high in the absence of Na. Both the $CuInSe_2$ and $CuGaSe_2$ layers show (112) orientation for Na doping. The orientation of CIGS layers in the presence of Na is different from the one without Na [148]. The (112) diffraction peak is a preferred one for the CIGS films deposited onto soda lime glass (SLG) substrates rather than on Na free glass substrates. The Na does not diffuse into films for below the substrate temperature of 450 °C. The devices with CIGS absorber deposited at low substrate temperature show poor performance of the cells due to lack of Na diffusion or inhomogeneous diffusion of Na into the CIGS films from SLG substrates [149]. The $CuIn_{0.6}Ga_{0.4}Se_2$ thin films on 7059 corning and quartz substrates are studied with effect of Na doping by sputtering technique. The target is made by cold pressing of Na_2Se and $CuIn_{0.6}Ga_{0.4}Se_2$ powders. The films made using CIGS target, which had $Na/CuIn_{0.6}Ga_{0.4}Se_2$ ratio zero, show (In+Ga)/Cu and Se/Cu ratios of 1 and 2, respectively. As the Na molar concentration is increased to 5 or 10%, the (In+Ga)/Cu and Se/Cu ratios change to 2 and 3, respectively which is equal to the $Cu(InGa)_2Se_{3.5}$ phase. The lattice constants a and c decrease from 5.715 to 5.70 Å and from 11.385 to 11.35 Å with increasing Na concentration from 0 to 10%, respectively. On the other hand band gap also increases from 1.25 to 1.36 eV. The results confirm the existing of $Cu(InGa)_2Se_{3.5}$ phase [150]. The Na doped $Cu(In_{1-x}Ga_x)Se_2$ and $Cu(In_{1-x}Ga_x)_2Se_{3.5}$ thin film layers are grown by RF sputtering technique using cold pressed (2.7×10^{-5} Pa) $Cu(In_{1-x}Ga_x)Se_2$ and Na_2Se targets at Ar pressure of 1.33 Pa and power of 100 W (3 W/cm^2). The $Cu(In_{1-x}Ga_x)Se_2$ target is prepared using 5N pure Cu, In, Ga, and Se compounds, which are sealed in evacuated quartz tube, followed by melting at high temperature to form CIGS charge. The $Cu(In_{1-x}Ga_x)_2Se_{3.5}$ thin films are grown by controlling Na composition in the target. It is confirmed by the XRD analysis that the (110), (202), and (114) are the evident reflections for the OVC $Cu(In_{1-x}Ga_x)_2Se_{3.5}$. The characteristic peak (110) for the OVC disappears with increasing Ga content in the samples that may be due to an increased polycrystalline nature but the variation in E_g and lattice parameters follow as normal. The lattice parameters a and c are slightly lower in the $Cu(In_{1-x}Ga_x)_2Se_{3.5}$ than that in the $Cu(In_{1-x}Ga_x)Se_2$ and varies linearly with varying x from 0 to 1. The band gaps are slightly higher in the $Cu(In_{1-x}Ga_x)_2Se_{3.5}$ than that in the $Cu(In_{1-x}Ga_x)Se_2$ thin films and varies linearly with x [151].

4.4.2 Cu(InGa)(SSe)$_2$ thin films

The selenization and sulfurization of (CIGSS) absorbers are important criteria for thin film solar cells and modules. The CIGS thin films deposited onto 1 μm thick Mo coated glass substrates at substrate temperature of 550 °C by Cu, In, Ga, and

Se sources using four Knudsen-type cells maintaining Cu/(In+Ga)~0.8–0.9 and Ga/(In+Ga)~0.3. The more sulfur content can be added into CuInGaSe$_2$ by incorporating oxygen into the annealing reaction chamber while sulfurizing the layers using 2% H$_2$S in Ar. The ratio H$_2$S/(H$_2$S+H$_2$Se) of 0.98 or 0.97 and annealing temperature of 525 °C are maintained. The slower sulfur reaction with CIGS occurs when admitting H$_2$Se into the reaction chamber. The diffraction angle of (112) peak for as-grown CIGS sample is 27.04°. After sulfurization/annealing under H$_2$S or H$_2$S+O$_2$, the second peak generates at 28.25° relating to Cu(InGa)S$_2$ (CIGS$_2$) and third peak at 29° corresponds to (102) of CuInS$_2$. As the grazing incidence angle increases from 0.2, 0.5 to 1.0°, the intensity ratios of CIGS/CIGS$_2$ increases indicating the content of sulfur is higher on the surface. The In and Ga are more in affinity with selenium and sulfur, respectively than the other partners do during annealing [152]. The XRD determines (112) as a preferred orientation along with low intensity chalcopyrite (101) and (211) peaks in the Cu(InGa)(SeS)$_2$ layers grown onto Mo coated soda lime glass substrates by two stage process such as metal layers formed by DC-magnetron sputtering, followed by selenization under H$_2$Se at 450 °C and sulfurization under H$_2$S at 500 °C [153]. The CuInGaSe precursor layer onto Mo coated glass substrates prepared by one step ED, followed by physical vapor deposition of In$_2$S$_3$ and Ga under Se vapor at substrate temperature of 600 °C for 1 h allows developing CuInGaSe$_2$ thin film with different compositions. The incorporation of Ga by ED at required level is very difficult because it needs positive potential. The (112), (204,220), and (116,312) reflections are present in all four CIGSS samples. The CIGSS sample-P$_7$ with composition of Cu:In:Ga:Se:S = 21.32:2.25:29.51:36.92:10 exhibits only (112), (204,220), and (116,312) reflections, whereas extra GaSe phases in sample-P$_8$ (Cu:In:Ga:Se = 22.22:2.72:23.43:39.72:11.9:1.76) are observed, as shown in Figure 4.44. The Ga(Se,S) phases are segregated in sample-P$_9$, which had compositions of 8.01:1.41:37.75:39.09:13.74:1.93. The sample-P$_{10}$ with compositions of 5.3:0.14:42.37:37.26:14.93:2.0 exhibits multiple (hkl) reflections of Ga(SeS) phase in the XRD spectrum, in comparison with the results of sample-P$_9$ [154].

1 μm thick Cu–In–Ga layers grown onto corning glass substrates by electron beam evaporation are selenized and sulfurized in elemental selenium and sulfur under N$_2$ atmosphere at 260 °C for 20–30 min then the temperature is raised to 400–500 °C with ramp rate of 9 °C/min and held for 20 min, in order to grow CIGSS layers. The CIGSS layers show (101), (112), (211), (204/220), (116/312), and (316/332) reflections. The 2θ angles of 26.82 and 26.97° are observed for the partially resolved most intensity peak (112). The intensity of latter close to the 2θ of 27.62° for (112) CuGaSe$_2$ decreases with increasing growth temperature indicating that inter diffusion of Ga increases in the CIGSS lattice [155]. The sulfurization of glass/Mo/Cu-Ga/In is done that the temperature has been ramped up and keeping up at constant for some time and again ramped up to higher temperature. Suppose the sulfurization time is 90 min, if sulfurization time is kept for longer time then etching takes place on the CIGSS layer. It is essential to increase the temperature rather than increasing the sulfurization time to incorporate sulfur into the layers. In the absorber

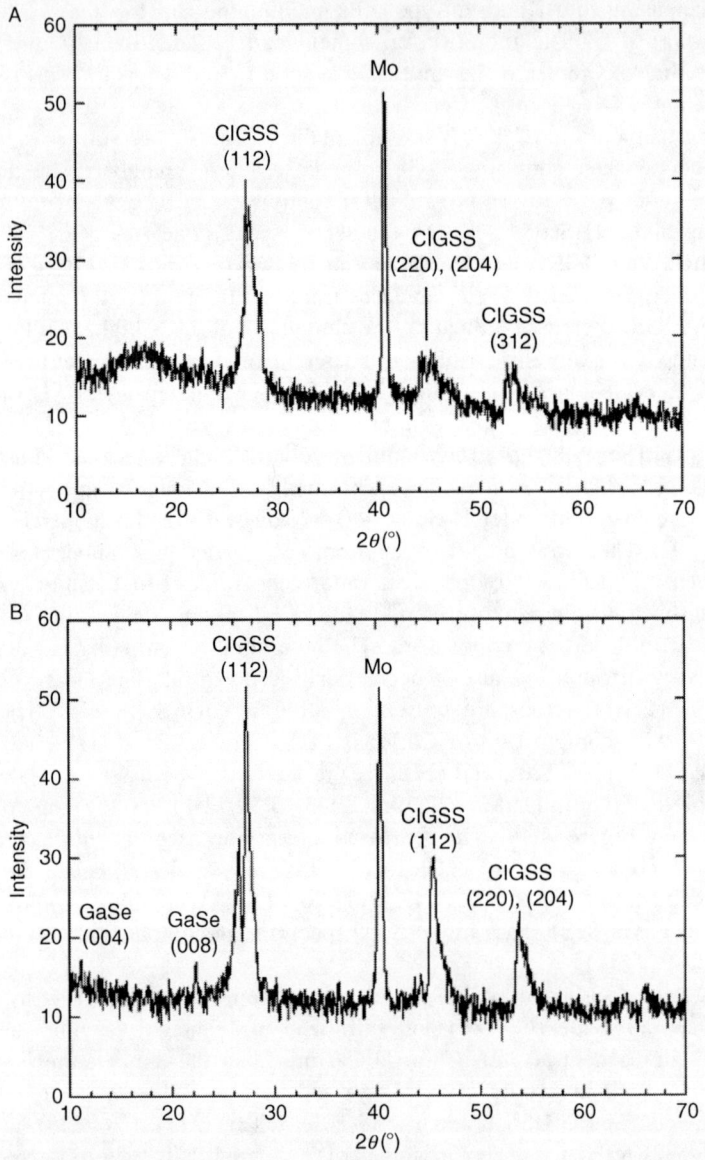

Figure 4.44 (A) XRD pattern of single phase CIGSS layer (sample-P$_7$), and (B) XRD pattern of CIGSS along with GaSe as a secondary phase (sample-P$_8$).

layers, the ratio of S/(S+Se) increases from 0, 16, 20 to 23 correspondingly the XRD peak moves from 26.5, 26.8, 26.9 to 27.2° and finally becomes broader [156]. Different solid solutions and their lattice parameters of CIGSS are given in Table 4.8. The (112) is the most intensity peak in the CIGSS layers. Intermediate quaternary phases of CuIn(Se$_{0.4}$S$_{0.6}$)$_2$ and CuGa(Se$_{0.8}$S$_{0.2}$)$_2$ are observed, after annealing glass/

Table 4.8 The CIGSS compound with different compositions and lattice constants

Compound	a (Å)	c (Å)
$Cu(In_{0.75}Ga_{0.25})(Se_{0.85}S_{0.15})_2$	5.7049	11.4032
$Cu(In_{0.75}Ga_{0.25})(Se_{0.75}S_{0.25})_2$	5.6693	11.3649
$Cu(In_{0.75}Ga_{0.25})(Se_{0.66}S_{0.35})_2$	5.6405	11.2984
$Cu(In_{0.75}Ga_{0.25})(Se_{0.55}S_{0.45})_2$	5.6221	11.2524

Mo/Cu-Ga/In under 1%H_2Se/Ar at 450 °C for 10 min, followed by annealing under H_2S/Ar at 550 °C for 10 min. The XRD pattern of intermediate phases and single phase CIGSS are shown in Figure 4.45A and B, respectively [157].

4.4.3 Role of Mo

The effect of Mo back contact on CIGS is also studied in the glass/Mo/CIGS structure. The XRD analysis reveals that $MoSe_2$ phase is formed between Mo and CIGS layers confirming by (100) and (110) peaks, whereas for the structure glass/SiO_2/Mo/CIGS, the SiO_2 layer acts as a barrier for diffusion of Na through the Mo layer. The Mo–Se compound forms two kinds of structures Mo_3Se_4 and $MoSe_2$. The dark field cross sectional analysis reveals the Mo, $MoSe_2$, and CIGS interfaces. The $MoSe_2$ layer formed in the structure had lattice spacing $d = 1.3$ nm. The c-axis of the grains parallel to the Mo layer and the $MoSe_2$ layers are perpendicular to the Mo layer otherwise it peels off easily. The Schottky behavior is confirmed on the glass/Mo/SiO_2/CIGS structure by I-V measurements; the I-V curve is depressed with decreasing the temperature from RT to 98 K. The standard cell with glass/Mo/CIGS gives p-n junction behavior and same slope with varying temperature from room to 120 K. The band gap of $MoSe_2$ is found to be 1.41 eV from the absorption measurements [158]. In addition to Mo (110), the hexagonal $MoSe_2$ phase with reflections such as (002), (100), (110), and (200) is also present. The EPMA analysis shows the composition of 68.7 and 30.4% for Se and Mo at the interface between Mo and $Cu(InGa)Se_2$, respectively. The analysis confirms formation of $MoSe_2$ phase at the interface [115].

4.5 $CuInS_2$

The selected area electron diffraction (SAED) of $CuInS_2$ thin film on Si substrate and the same layer at the interface of Si in the [001] direction reveal patterns of (001), (002), (003), etc., and (002), (004), (006), etc., respectively. The missing odd reflections or presenting only even reflections at the interface of $CuInS_2$ and Si in the electron diffraction are due to the epitaxial growth from the sulfur termination [001] direction. Both the patterns with either odd or even reflections lead to CA

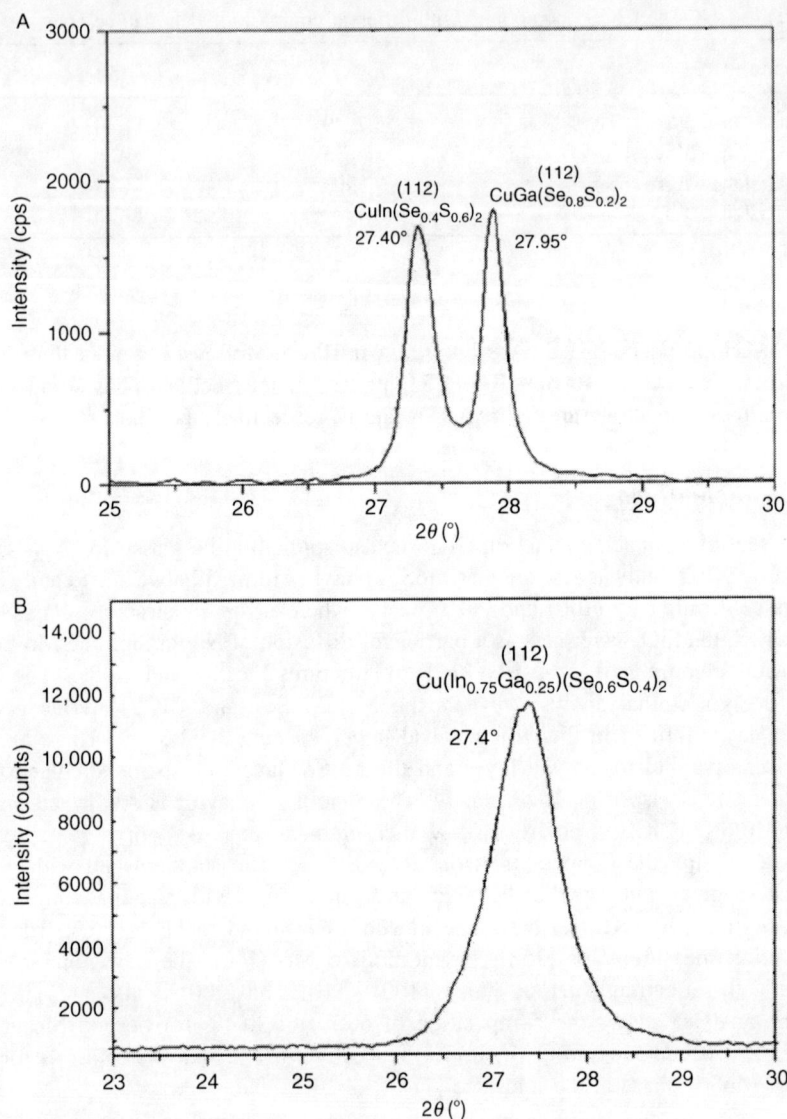

Figure 4.45 (A) XRD pattern of intermediate quaternary $CuIn(Se_{0.4}S_{0.6})_2$ and $CuGa(Se_{0.8}S_{0.2})_2$ phases, and (B) XRD pattern of single phase CIGSS thin film.

structure, as shown in Figure 4.46. The growth of CA phase is favorable to [001] direction, that is, alternate Cu and In atoms in the c direction. The CA phase is higher in the Cu-rich than in the Cu-poor $CuInS_2$. The XRD spectrum shows (001), (002), (003), (004), and (005) reflections for CA $CuInS_2$ thin films (Figure 4.46C).

Structural Properties of I–III–VI$_2$ Absorbers

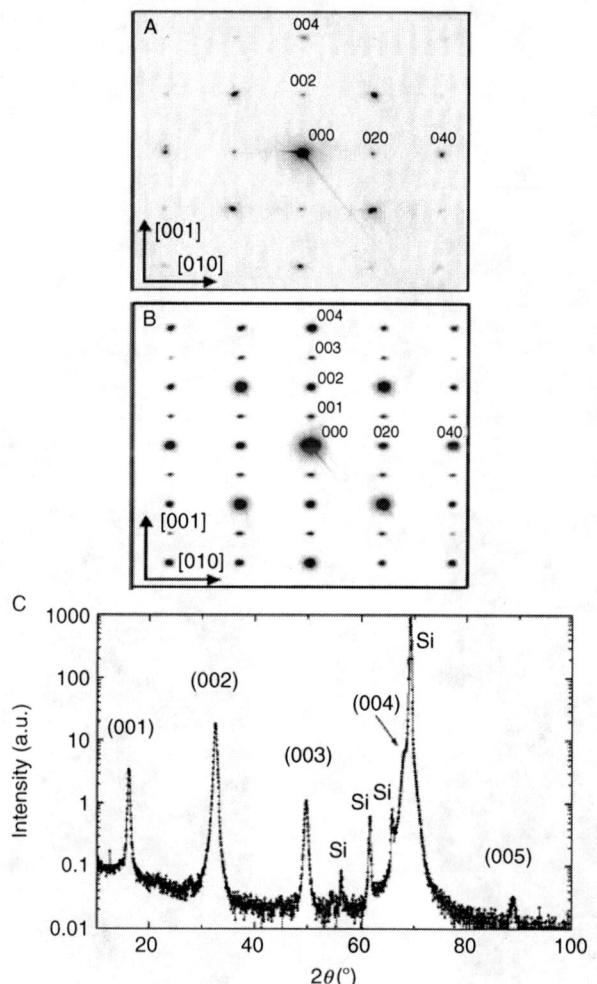

Figure 4.46 SAED of CuInS$_2$ epitaxial layer grown onto (1 × 1) sulfur terminated Si(001) substrates; (A) close to the interface and (B) thin film, and (C) XRD pattern of CuInS$_2$ epitaxial layer grown on (1 × 1) sulfur terminated Si(001) substrates.

The CuInS$_2$ thin films used for structural analysis are grown onto sulfur-terminated (001) Si substrates by MBE at substrate temperature of 350 °C for 45 min to obtain 0.1 μm thick layer [159]. Figure 4.47 shows both CH (213), (303) and CA (022), (110) structures in the CuInS$_2$ thin films grown onto Si(111) substrates for [111]$_{CH}$ zone axis [160]. The (112) preferred orientation at $2\theta = 27.94°$ and next low intensity (204, 220) peak at 46.9° in the XRD spectrum are observed for the CuInS$_2$ thin films prepared onto glass by modulated flux atomic layer deposition that means alternative deposition of Cu, In, and S under vacuum at substrate temperature of 250 °C with deposition rate of 5 Å/s. In addition, CA phases are also observed in the layers by

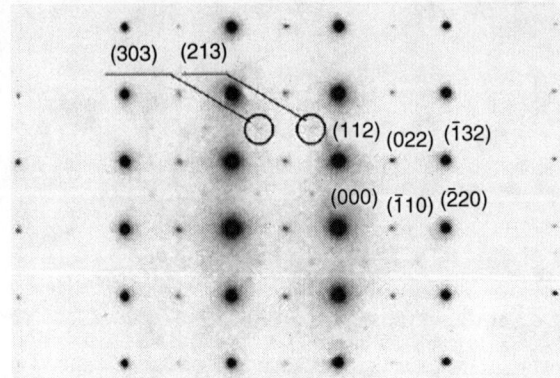

Figure 4.47 SAED of CuInS$_2$ thin films grown on Si (111) substrates indicates CH and CA structures.

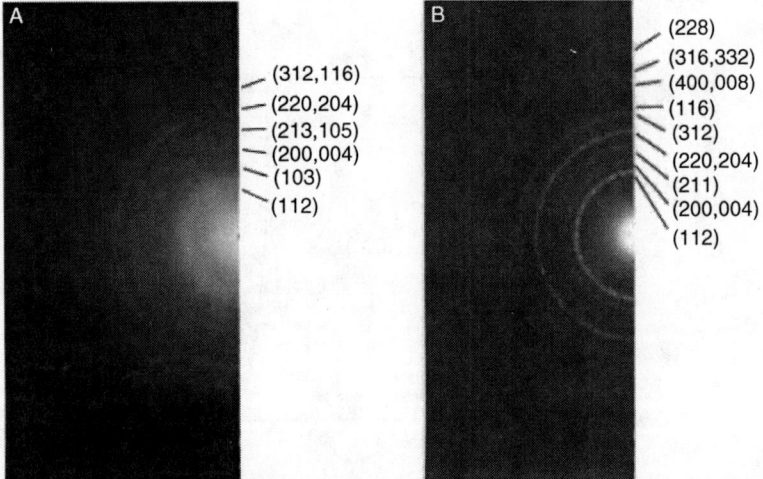

Figure 4.48 SAED pattern of CuInS$_2$ thin films grown onto glass by spray pyrolysis technique at T$_S$ = 277 °C; (A) Chemical solution pH = 3.5 and (B) pH = 4.5.

evidencing reflections (001), (002), (003), and (004) at 14.4, 28.98, 44.06, and 60.10°, respectively. However, there is a possibility of existence of CuIn$_5$S$_8$ phase in the CuInS$_2$ layers because of its reflections match with CA [161]. The SAED pattern of CuInS$_2$ thin films deposited at pH 3.5 and 4.5 show chalcopyrite structure, as shown in Figure 4.48, whereas Cu$_2$S films form for pH 2.5. The SAED pattern of CuInS$_2$ thin films is diffused for pH 3.5. It could be due to either fine grain structure or over loaded thickness of CuInS$_2$ for SAED analysis. In order to prepare CuInS$_2$ thin films, the chemical solutions of cupric chloride, Indium trichloride and thiourea and the composition ratio of Cu:In:S = 1:1:3 are employed [162,163]. As the substrate temperature of $CuInS_2$ thin films with Cu/In ratio of 1.1 deposited by spray pyrolysis is increased from 317 to 377 °C in steps of 20 °C, the intensities of prime peaks such as (112), (204,220), and (116,312) increase evidencing an increased crystallinity of the samples [164]. The intensities of (112), (204,220), (116,312), and (002) peaks in the $CuInS_2$ thin films also increase with increasing annealing temperature from 250, 350, 450 to 550 °C [165]. The $CuInS_2$ thin films are grown onto 7059 corning glass substrates at different substrate temperatures of RT, 400, 600 °C by MBE using three elemental sources, whereby a specially designed three stages Knudson type effusion cell is used for sulfur source. The films deposited at RT show a broad peak indicating amorphous nature. After annealing the samples in the temperature range from 200 to 600 °C, the (112), (204,220) and (116,312) peaks exist. The composition of annealed sample is likely to be $Cu_{1.0}In_{1.17}S_{2.24}$. The films deposited at substrate temperature of 400 °C, the prime intensity peaks such as (112), (204,220), (116,312), and (002,400) present. As the annealing temperature is further increased to 600 °C, the stoichiometric $CuInS_2$ films form and the intensities of all three peaks increase. The FWHM of (112) peak decreases from 1.5, 0.76, 0.58 to 0.55° with increasing annealing temperature from 200 to 600 °C. Less the FWHM means more the crystallinity [166].

After KCN etching $CuInS_2$ layers with Cu/In = 1.8 shows chalcopyrite structure confirming by presence of chalcopyrite characteristic add reflections such as (103), (211), (213), and (301) in the XRD spectrum. Resolves in (004,200), (204,220), and (116,312) peaks reveal ordered sublattice of Cu and In atoms of the roquesite phase, as shown in Figure 4.49. A tetragonal splitting is a striking evidence for chalcopyrite structure. The $CuInS_2$ thin films used for XRD analysis is deposited that the 550 nm thick Cu and 650 nm thick In are grown onto Mo coated glass substrates by DC sputtering, followed by sulfurization at 600 °C for 2 min using elemental sulfur [167]. Unlike the $CuInS_2$ thin films are grown onto Mo coated glass substrates by

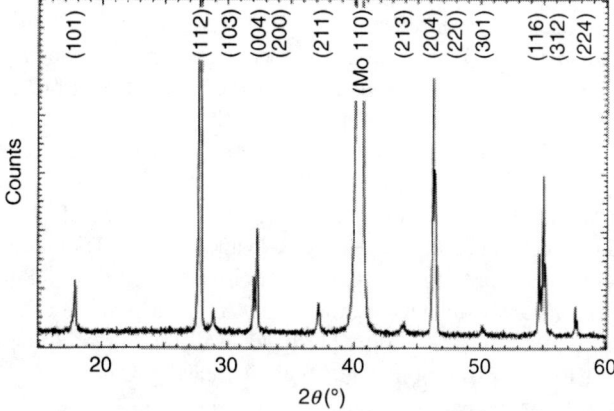

Figure 4.49 XRD pattern of KCN treated $CuInS_2$ thin films grown by hybrid process of sputtering and sulfurization.

E-beam evaporation technique using binary Cu_2S and In_2S_3 phase sources; in addition RF plasma is provided close to the substrates with power of 400 W. The single phase chalcopyrite $CuInS_2$ thin films form for Cu_2S/In_2S_3 flux ratio of 4. In order to improve the crystallinity of $CuInS_2$ thin films, the substrate is biased either positive or negative voltage. The films grown by floating potential of -14 V show good crystallinity evidencing by split in (116), (312) peaks. The similar resolve in (204)/(220) and (116)/(312) peaks for the $CuInS_2$ nanorods grown by hydrothermal process is observed [168]. The presence of Cu_xS_y phase enhances the crystallinity of $CuInS_2$ thin films (Figure 4.50). After etching the sample in 10% KCN solution for 3 min, some of the chalcopyrite peaks appear, as shown in Figure 4.50B [169]. Since Cu_xS_y phase forms on the surface of as-grown $CuInS_2$ layer as a blanket, which acts as a mask for low intensity peaks therefore the low intensity peaks could not be in the detectable range by XRD.

The $CuInS_2$ layers, which are blue in color, deposited onto Mo coated glass substrates by evaporation of 290 nm thick Cu and In and sulfur from Knudsen cells but In thickness is adjusted in such a way to have slightly Cu-rich films, which are etched in 10% KCN solution for 3 min at room temperature to remove CuS. A resolve in (204, 220) peak indicates roquesite $CuInS_2$, as shown in Figure 4.51 [170]. The $CuInS_2$ thin films with bilayer configuration are deposited onto Mo coated glass substrates by thermal vacuum evaporation at substrate temperature of 400–500 °C. A bilayer is a combination of a bottom layer with Cu/In=0.92 and a top layer with Cu/In=0.88. A single layer with Cu/In=0.92 shows chalcopyrite structure with multiple chalcopyrite characteristic reflections such as (101), (103), (211) and (213), along with (112) preferred orientation and a split in (116,312) is evidence for tetragonal distortion, whereas the $CuInS_2$ single layer with Cu/In=0.88 shows (204,220) preferred orientation. The $CuInS_2$ bilayer shows single phase chalcopyrite structure with resolving (204), (220) peaks [171]. The XRD analysis reveals that the front surface of $CuInS_2$ layers prepared onto Mo coated glass substrates by CVD method using single precursor of $(PPh_3)_2CuIn(Set)_4$ at substrate temperature of 390 °C exhibits secondary phase, whereas no secondary phase is observed at the back surface, as shown in Figure 4.52. The secondary phase may be $CuIn_5S_8$ spinal phase. It is evident that secondary phase always segregates on the surface. The preferred orientations of (220) and (112) at the front and back faces of the sample are observed, respectively [172]. (112) preferred orientation is observed in the $CuInS_2$ thin films deposited onto glass substrates by RF sputtering using $CuInS_2$ target without S vapor, whereas (204,220) orientation is found for the layers grown in the presence of S vapor. The intensities of (112), (204,220), and (116,312) peaks decrease for excess S vapor [173]. The intensity ratio of (112)/(220) increases from 2 to 6, as Cu/In ratio is increased from 0.8 to 1 thereafter the intensity is constant with further increasing Cu/In ratio in the chemical solution for spray deposited $CuInS_2$ thin films [174]. Overall the (112) peak position shifts from higher to lower diffraction angle with increasing Cu/In ratio indicating an increase of quality of the crystal (Table 4.9). As shown in Figure 4.53, the Cu-rich $CuInS_2$ thin film with composition of Cu:In:S=36.6:17.1:46.3 (sample-c) shows slightly preferred orientation ($I_{112}/I_{220}=6$) and chalcopyrite characteristic peak

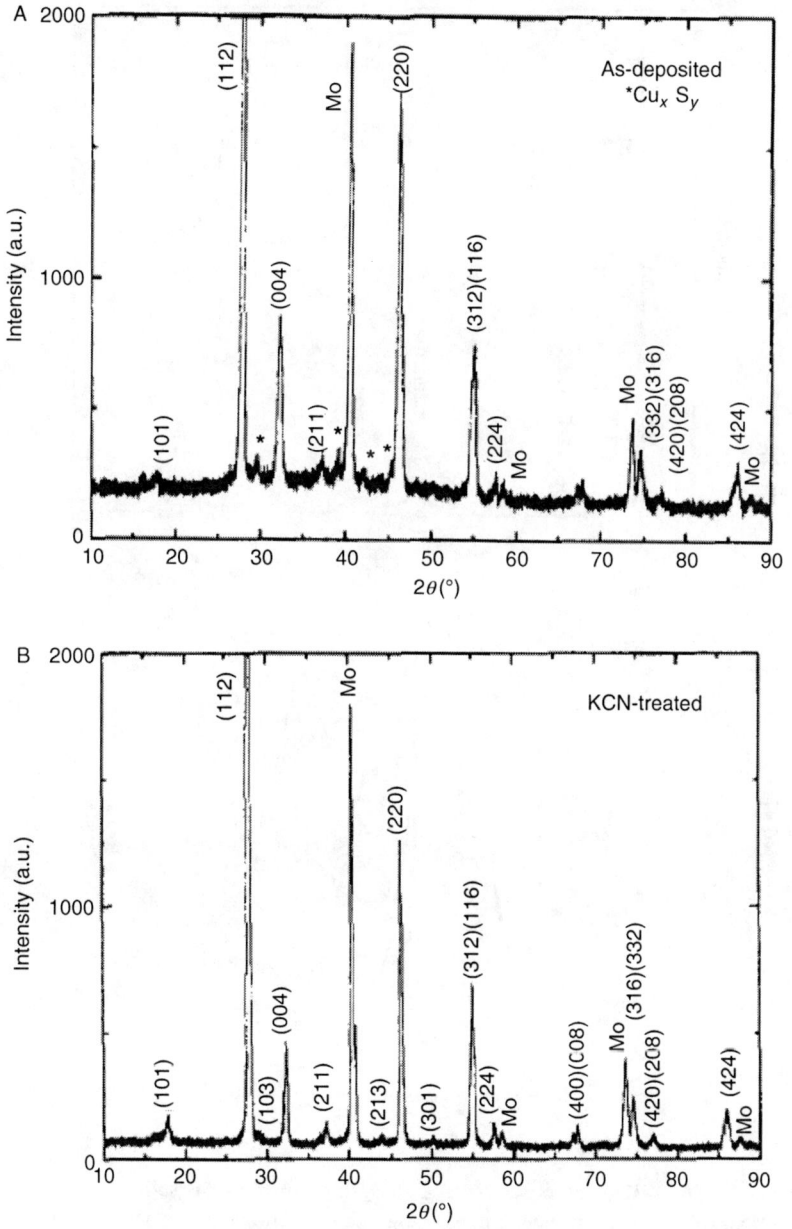

Figure 4.50 (A) XRD pattern of CuInS$_2$ thin films grown by E-beam using RF plasma and Cu$_2$S and In$_2$S$_3$ sources and (B) XRD pattern of CuInS$_2$ thin films treated by KCN to remove Cu$_x$S$_y$ phase.

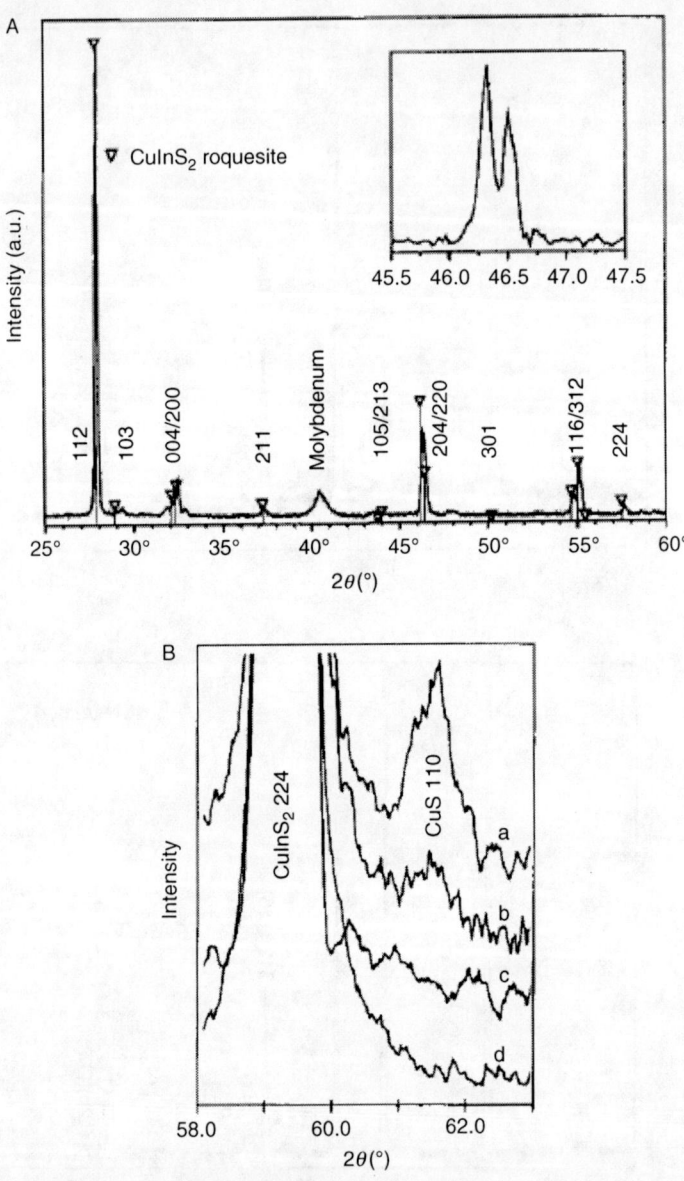

Figure 4.51 (A) XRD pattern of Cu-rich CuInS$_2$ thin films; inset is splitting of (204), (220); the symbol (▼) indicates chalcopyrite signature either splitting or characteristic peaks. (B) XRD pattern of CuInS$_2$ thin films: (a) Cu-rich single layer, (b) single layer with rough surface, (c) bilayer, and (d) KCN etched Cu-rich single layer.

(101) and 103) at 17.9 and 28.8°, respectively. On the other hand a split in (004,200) peak is observed that supports the signature of chalcopyrite structure [175].

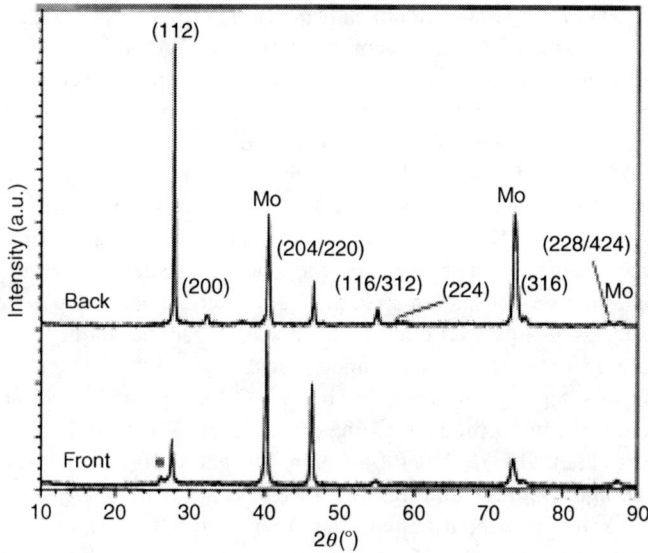

Figure 4.52 XRD pattern of CuInS$_2$ thin films deposited onto Mo coated glass substrates by CVD method.

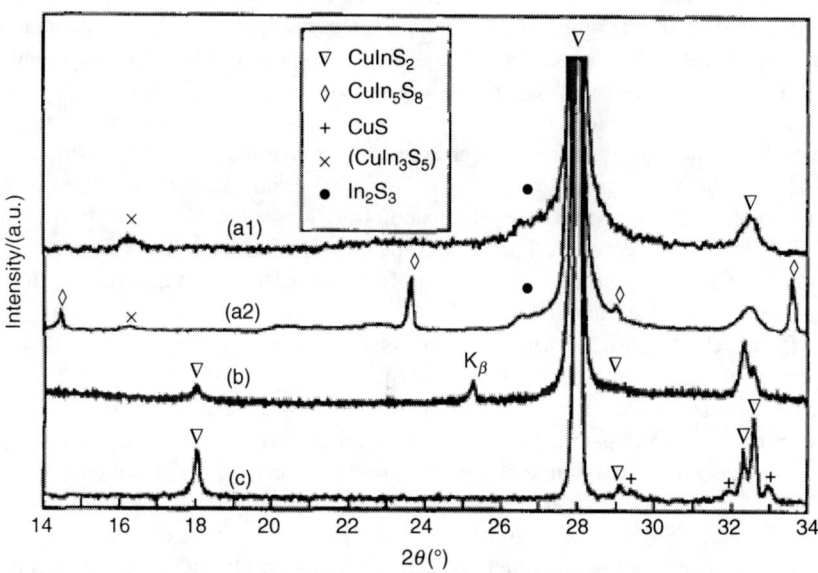

Figure 4.53 XRD patterns of CuInS$_2$ thin films with different compositions.

The CuInS$_2$ thin films grown onto Al$_2$O$_3$ by three source evaporation show chalcopyrite structure with (112) preferred orientation without any secondary phases [176]. The Cu and In are sputtered by DC magnetron with Cu/In ratio of 0.8 and Na/

(Cu + In) = 0.02 onto SLG/ Mo/20 nm NaF and the sulfur is evaporated by a Knudsen source. H$_2$S flow of 25–30 sccm, chamber pressure of 2.13–4.53 × 10^{-1} Pa, sputter power of 200 W and substrate temperature of 500 °C are optimum conditions to grow CuInS$_2$ films. The sulfur peaks appear in the CuInS$_2$ thin films for above H$_2$S flow rate of 30 sccm. The FWHM of (112) peak reduces from 0.239 to 0.162° that means the grain sizes increase from 59 to 132 nm with increasing substrate temperature from 400 to 500 °C. The layers deposited at 400 °C exhibit additional reflections such as (101), (316,332), and (228,424) at 17.95, 75, and 86.3°, respectively. The high (112) prefered orientation is observed for the layers deposited at 400 °C [177].

Twenty alternative Cu and In metallic layers sequentially deposited onto Mo coated glass substrates by sputtering under 2 × 10^{-5} torr partial argon pressure are annealed from 160 to 400 °C for 2 h under vacuum at 10^{-3} torr pressure, followed by sulfurization at temperature from 160 to 400 °C for 20 h under the same pressure using elemental sulfur to produce CuInS$_2$ thin films, which show well developed peaks such as (112), (103), (200/400), (204/220), and (116/312) [134]. (112) is an intensity peak in the CuInS$_2$ thin films deposited by spray pyrolysis with excess sulfur of 30% in solution comparing with the one grown with 20% excess sulfur. The films annealed under air at 400 °C for 1 h shows better results. No difference in the XRD pattern for Cu/In ratio in the range from 0.96 to 1.2 is observed but beyond 1.2 the structure of the samples deteriorates [178]. The (112) preferred orientation occurs in the CuInS$_2$ thin films for composition ratio Cu:In:S = 1:1:3.9 in the solution grown onto quartz and sital substrates by spray technique [179]. In general, the In-rich deposition impedes growth of CuInS$_2$ while the Cu-rich deposition enhances crystal structure and growth but secondary growth participation takes place [180].

The sphalerite structure dominates in the CuInS$_2$ films grown onto glass substrates by spray pyrolysis method at substrate temperature of 350 °C for Cu:In:S = 1:1:3, which changes to chalcopyrite with increasing sulfur ratio from 3, 4, 5 to 6 evidencing by appearance of chalcopyrite characteristic (103) peak [181]. The FWHM of (112) preferred orientation peak for CuInS$_2$ thin films decreases with increasing Cu/In ratio from 1, 1.1 to 2 but for 4 increases. Note that the Cu/In ratio is counted in the chemical solution [182].

After annealing Cu-rich CuInS$_2$ films under vacuum at high temperature Cu$_x$S phase generates well on the surface of the layer [166]. The intentionally grown Cu$_x$S layer onto CuInS$_2$ films is etched off in KCN solution, followed by annealing at 500 °C revealing disappearance of Cu$_x$S phase. The reason to anneal the sample is that if there is any leftover small quantity of Cu$_x$S in it segregates violently again [182]. An additional covellite CuS phase evidencing several reflections of d = 1.508 Å (110) at 61.4° (Figure 4.53B), 3.05 (102), 2.82 (103), 2.73 (006) and 2.32 (105) occurs in the Cu-rich CuInS$_2$ layers grown by single layer process, which is quasi-semimetallic. One of the diffraction angles of CuS peak is close to that of (224) peak. After KCN treatment the covellite phase on the surface of CuInS$_2$ is etched off, as evidenced by the XRD analysis. The doping of sulfur is easier in the Cu-rich CuInSe$_2$ than in the Cu-poor CuInSe$_2$ [171]. The Cu-rich CuInS$_2$ thin films show smooth surface owing to Cu$_x$S$_y$ secondary phase. After etching in KCN solution, the sample becomes rough because of removal of secondary phase [168]. The Cu$_2$S, CuS, and CuInS$_2$ thin films are individually prepared by sulfurization of either

Cu or Cu–In alloys in order to cross check their diffractions with $CuInS_2$. The diffraction angles of (101) and (202) for CuS phase closely match well with $CuInS_2$ (112) and (224), respectively. On the other hand, the Cu_2S phase (004) also closely matches well with $CuInS_2$ (116). Therefore it is somewhat difficult to differentiate them, if the secondary phases present in the $CuInS_2$ thin films [183]. The peaks at 29.3 and 31.8° relating to CuS (covellite phase) secondary phase are observed in sample-*c*. The single phase $CuInS_2$ thin films exist for nearly stoichiometric composition of Cu:In:S = 24.6:24.9:50.5 (sample-*b*). In the case of In-rich (sample-*a1*, Cu:In:S = 22:26:52 and sample-*a2*, 21.1:27.2:51.6) samples, the intensity of (112) peak decreases. The $CuIn_3S_5$ and In_2S_3 phases are common in both the samples *a1* and *a2* but $CuIn_5S_8$ spinal phase additionally appears in sample-*a2*. The $CuIn_3S_5$ phase appears at 16° and the $CuIn_5S_8$ phase presents at 14.4, 16, 23.6, 26.6, and 33.7°. The transition angle between (112) and (111) is 0.089° when the structure of $CuInS_2$ changes from chalcopyrite (112) to sphalerite (111) [175] (Figure 4.53). The $CuInS_2$ thin films by spray pyrolysis show reflections of (112), (204/220), and (116/312) along with secondary phase of Cu_xS. After annealing as-grown layers under H_2S at 450 °C followed by annealing under H_2 at the same temperature, (112) becomes preferred orientation. The H_2 anneal removes O_2 and decreases Cl content in the films. The Cl is one of the impurities in the spray deposited $CuInS_2$ thin films [184].

Cu and In layers with different composition ratios of Cu/In = 1.8, 0.9, and 0.8 sequentially deposited onto Mo coated float glass substrates by DC magnetron sputtering sulfurized at 500 °C show that the roquesite phase $CuInS_2$ (JCPDS 27-159) is observed for Cu/In = 1.8 and 0.9. The $CuIn_5S_8$ is secondary phase, peak at 26.7° may be due to In_2S_3 (27.4°) or Cu_7S_4 (26.6°) for Cu/In = 0.8. The sulfurization of Cu–In with Cu/In = 1.2 at 400 °C for 60 min show secondary phase of $CuIn_5S_8$ and additional $Cu_{11}In_9$ phase lines at 42.1 and 42.3° indicating that sulfurization is not completed. The as grown sample for 1.8 shows CuS phase blue in color that is not probably found in other cases [185]. Keeping the pH 4 the single phase $CuInS_2$ thin films are obtained for Cu/In = 1, 1.1, and 1.25 corresponding growth temperatures of 380, 350, and 290 °C indicating that the lower growth temperature is good enough to grow single phase $CuInS_2$ films, if high Cu/In ratio is used or vice-versa [186]. The chalcopyrite structure is dominant in the $CuInS_2$ films prepared with Cu/In = 1.2 and S/Cu = 3.9 at higher substrate temperature of 350–400 °C but sulfur deficiency is found. The high intensity (112) peak is observed for pH 3.9, Cu/In = 1.2 and T_S = 390 °C [187]. The $CuInS_2$ thin films deposited at substrate temperature of 320 °C with Cu:In:S = 1:1:3 composition in solution show low intensity (112) peak along with unknown secondary phase. After annealing the layers under H_2 at 400 °C for 30 min, the (112) peak intensity increases and still the secondary phase persists. The as-grown layers annealed under H_2 at 400 °C for 3 h or at 500 °C for 30 min show dominant peaks (112), (004/200), (204/220), and (116/312) while the layers annealed under vacuum at 500 °C for 1 h show similar peaks but secondary phase In_2O_3 exists. The layers grown at higher temperature of 380 °C with Cu:In:S = 1:1:6 composition in solution annealed under H_2 at 500 °C for 1 h shows dominant $CuInS_2$ structure peaks, whereas the low intensity peaks and secondary phase exist for growth temperature of below 500 °C. The secondary phase is commonly observed in the CuInS$_2$ thin films for Cu:In:S = 0.8:1:3 whether the layers are annealed or not [188].

The CuInS$_2$ thin films deposited by spray pyrolysis with low pH and high substrate temperature (T_S) lead to chalcopyrite structure. The low pH and low T_S lead to sphalerite structure. In- and Cu-rich films exhibit sphalerite structure CuInS$_2$ + In$_2$S$_3$ phase and chalcopyrite structure, respectively. The sphalerite CuInS$_2$ phase along with secondary In$_2$S$_3$ phase exist for the conditions of $T_S = 360$ °C, pH 3.5 and $0.7 \leq$ Cu/In ≤ 1.0. The In$_2$S$_3$ phase is a prime secondary phase in the In-rich CuInS$_2$ films [189]. The Cu$_7$S$_4$, CuS$_2$ secondary phases and chalcopyrite structure CuInS$_2$ phase with strong (112) peak occur for Cu/In > 1.0. The (112) peak becomes stronger with further increasing Cu/In ratio [190]. Alike the multiphases such as CuIn$_5$S$_8$, CuIn$_{11}$S$_{17}$, Cu$_7$S$_4$, In$_6$S$_7$, and In$_2$S$_3$ could be observed for $T_S = 320$ °C, pH 4, and Cu/In ≤ 1. The chalcopyrite CuInS$_2$ thin films are observed for $T_S = 380$ °C and Cu/In = 1–1.25 [191]. The CuInS$_2$ thin films grown at substrate temperature of 300 °C show sphalerite structure for 0–30% excess Cu, whereas chalcopyrite structure exists for more than 30% excess Cu but Cu$_x$S secondary phase presents. After annealing CuInS$_2$ thin films with more than 30% excess Cu under nitrogen flow at 397 °C for 2 h, the structure changes from sphalerite to chalcopyrite but Cu$_x$S phase pronounces well. After annealing single phase CuInS$_2$ thin flims grown with 20% excess Cu the structure changes from sphalerite to chalcopyrite, as shown in Figure 4.54 [192].

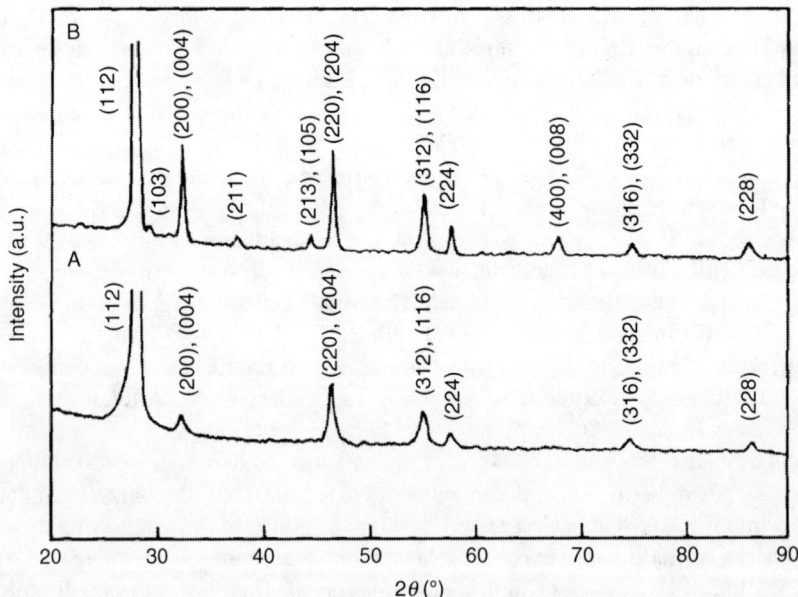

Figure 4.54 XRD pattern of (A) CuInS$_2$ thin films with 15% excess Cu in solution deposited by spray pyrolysis technique and (B) annealed under N$_2$ atmosphere at 397 °C for 2 h.

By sulfurizing Cu–In layers (Cu/In = 0.99–1.2) under H_2S atmosphere at temperature of 350 °C for the duration of 3 h single phase exhibit, whereas the sulfurization of Cu–In at temperature of 250 °C for 2 h consist of In_6S_7, InS, CuS, and $CuInS_2$ phases and sulfurization of Cu–In layers at higher temperature of 400 °C for 2 h contributes Cu_2S and $CuInS_2$ phases [193]. Alike the sulfurization of Cu–In layers by sputtering with Cu–In target under $H_2S + Ar$ atmosphere at the temperature of 300–400 °C for more than 2 h results in (112) oriented $CuInS_2$ thin film along with In_2S_3 phase [194]. The $CuInS_2$ films are sputter deposited at substrate temperature of 400 °C, RF power of 100–300 W (1.23–3.70 W/cm^2), H_2S flow rate of 5–10 sccm, chamber pressure of 4.4–8.2 × 10^{-4} torr and Cu–In alloy target with Cu:In ratio of 9:11. The as-deposited films have $CuInS_2$ and $Cu_{11}In_9$ phases. After annealing at 400 °C for 90 min, the $Cu_{11}In_9$ phase disappears but Cu_2S phase generates. As further enhancing annealing time of 180 min, the Cu_2S phase becomes little stronger. The single phase $CuInS_2$ films can be deposited for only optimum conditions [195]. Single phase chalcopyrite structure presents in the $CuInS_2$ thin films prepared by sputtering using cold pressed target of $Cu_2S + In_2S_3$, evidencing by resolve in (204), (220) peaks for $Cu_2S/In_2S_3 = 1.5$, whereas $CuInS_2$ along with In_2S_3 presents for $Cu_2S/In_2S_3 = 1$ [196]. Similarly the In_2S_3 phase with diffraction angle of 26.3° for (104,110) reflection is observed along with $CuInS_2$ for the Cu_2S/In_2S_3 flux ratio of 2. The intensity of secondary phase decreases with increasing flux ratio from 2 to 3. The In_2S_3 phase disappears for further increasing the flux ratio to 4 [168]. The Cu and In layers sequentially electrodeposited onto Ti substrates are annealed under air at 130 °C for 30 min, followed by sulfurization under saturated H_2S atmosphere at 550 °C for 30 min then etched in 0.1 M KCN solution. This experimental process produces single phase $CuInS_2$ thin film for Cu/In = 0.68, $CuIn_{11}S_{17}$ phase for 0.4 and both phases present for 0.51 [197]. The typical lattice constants of $CuInS_2$ are given in Table 4.10.

The $CuInS_2$ polycrystalline pellets can be formed by high pressure method at temperature of 700 °C and pressure of 22.5 MPa employing Cu_2S and In_2S_3 powders for the growth time of 1 h, however $CuIn_{11}S_{17}$ phase pellets are found for the temperature of 400 °C at the same pressure [198]. The $Cu_{1.8}S$ is found as a secondary phase in the Cu-rich $CuInS_2$ samples by evidencing diffraction angle at 51.56°, whereas Cu_2In secondary phase is detected by existing reflections such as

Table 4.9 Composition, diffraction angle of (112) peak and its FWHM for $CuInS_2$ thin films

Composition Cu:In:S	2θ (°)	FWHM (°)	Ref.
29.2:22.4:48.4	27.904	0.0905	[175]
28.4:23.4:48.3	27.907	0.0948	
21.2:25.7:53.2	27.908	0.2432	
19.7:26.7:53.6	27.899	0.3814	

(101), (002), (102), and (103) in the In-rich CuInS$_2$ samples. The Cu and In are deposited onto n-Si (111) substrates by evaporation from effusion cells at source temperatures of 1200 and 800 °C, respectively for which sulfur is supplied from di-*tert*-butyl disulfide C$_8$H$_{18}$S$_2$ (TBDS) organic source in the MBE system to develop CuInS$_2$ [199]. Amorphous nature with small crystallites is found in the CuInS$_2$ films deposited onto Ti substrates by ED using CuSO$_4$, In$_2$(SO$_4$)$_3$ and Na$_2$S$_2$O$_3$ at potential of −0.9 V versus Ag/AgCl and pH 1.5. After annealing the layers under vacuum at 400 °C, CuIn$_{11}$S$_{17}$ single phase with reflections (112), (024), and (132) present for Cu:In = 6:14 in the solution whereas In$_2$O$_3$ phase is observed as secondary phase for Cu:In = 3:2 and 1:1 [200]. In order to suppress the binary phases in the CuInS$_2$ absorber layer, the metal oxide layers such as Cu–In–O precursor layers sputtered from Cu$_2$In$_2$O$_7$ target are sulfurized under H$_2$S (56 torr) + Ar atmosphere at 550 °C for 1 h. The grown layers show phases of CuInS$_2$ and In$_2$O$_3$. The secondary In$_2$O$_3$ phase can be avoided by sulfurizing the precursor layers under H$_2$S + H$_2$ (150 torr) atmosphere instead of H$_2$ + Ar, as shown in Figure 4.55 [201]. No NaInS$_2$ phase is observed in the CuInS$_2$ thin films

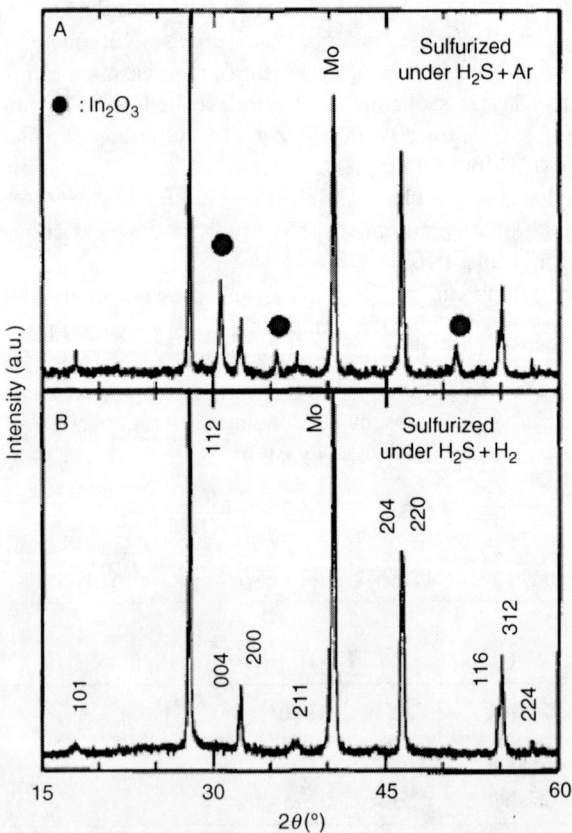

Figure 4.55 XRD pattern of CuInS$_2$ thin films formed by sulfurization of Cu–In–O; (A) under H$_2$S + Ar and (B) H$_2$S + H$_2$.

Table 4.10 Lattice constants of CuInS$_2$ and CuAu (CA) type CuInS$_2$

Compound	a (Å)	c (Å)	Ref.
CuInS$_2$	5.523	11.141	JCPDS 27-159
	5.514	11.141	[165]
	5.52	10.94	[175]
	5.50	11.05	[190]
	5.51	11.05	[192]
	5.52	11.17	[194]
	5.523	11.134	[195]
CA	5.508	–	[159]

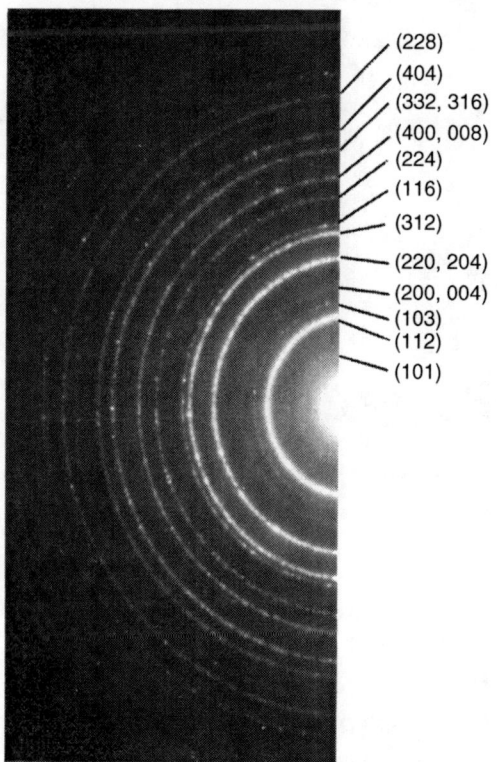

Figure 4.56 SAED pattern of CuIn(Se$_{0.7}$S$_{0.3}$)$_2$ thin film grown by spray pyrolysis technique.

for Cu/In = 0.73, if Na is doped, whereas NaInS$_2$ phase presents for the ratios of 0.77, 0.8, and 0.96 in the XRD spectrum [202]. The density of the samples increases from 4.67 to 4.79 g/cm^3 with increasing Cu/In ratio from 0.6 to 1.5 in the CuInS$_2$ single crystals [203].

4.6 CuIn(Se$_{1-x}$S$_x$)$_2$

Figure 4.56 shows the typical very bright SAED pattern of polycrystalline CuIn(Se$_{1-x}$S$_x$)$_2$ thin films for $x=0.3$ deposited by spray pyrolysis technique in which the characteristic chalcopyrite reflections such as (101), (103) are observed [12]. The XRD diffraction angle of (112) preferred orientation shifts toward higher angles with increasing x in the CuIn(Se$_{1-x}$S$_x$)$_2$ thin films, as shown in Figure 4.57. The diffraction angle of CuInS$_2$ is 27.87° for (112) peak. The FWHM of (112) diffraction peak decreases from 0.354 to 0.295° with decreasing S/(S+Se) ratio from 0.515 to 0.264 [204]. The (112) peak diffraction angle shifts from 26.70 to 27.98° with increasing sulfur content x from 0 to 1 in the p-CuIn(Se$_{1-x}$S$_x$)$_2$ single crystals grown by horizontal Bridgman method [205]. The intensity of (112) preferred orientation gradually decreases with increasing x from 0 to 1 in the CuIn(Se$_{1-x}$S$_x$)$_2$ thin films

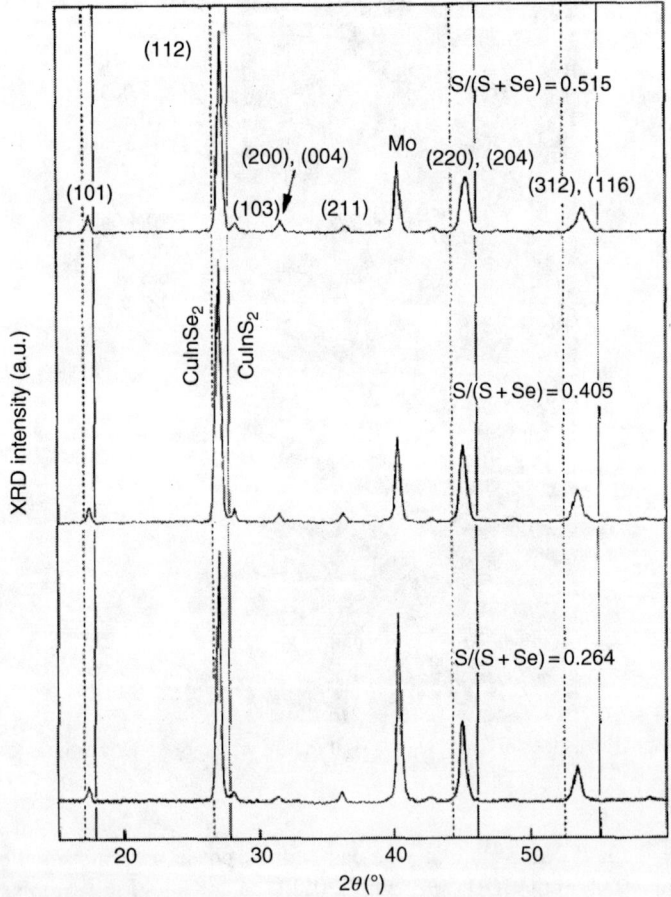

Figure 4.57 XRD pattern of CuIn(Se$_{1-x}$S$_x$)$_2$ thin films grown by sulfurization and selenization of CuIn alloy with different (S/S+Se) compositions.

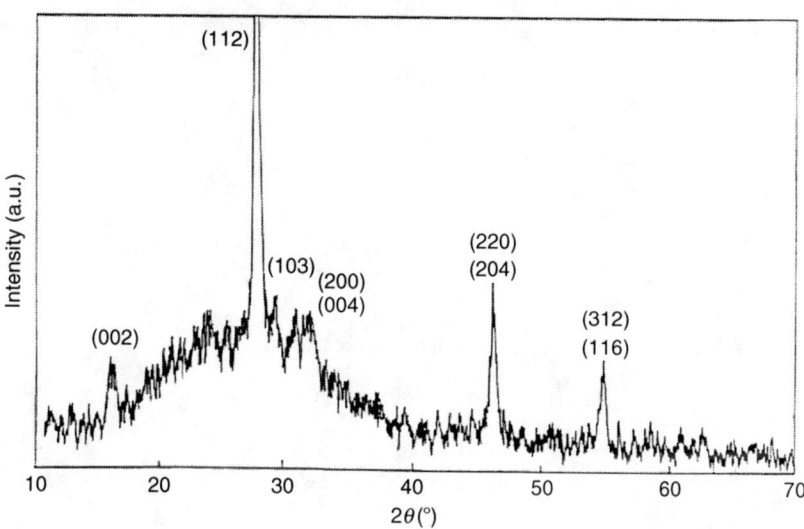

Figure 4.58 XRD pattern of $CuIn(Se_{1-x}S_x)_2$ thin for $x=0.7$ grown by spray pyrolysis technique.

prepared by spray pyrolysis at the substrate temperature of 400 °C. On the other hand, the intensities of (204,220) and (116,312) peaks gradually increase. It is evidenced that the growth temperature has to be kept high for $CuInS_2$ thin films, in order to obtain quality layers because that cannot be the same for both the $CuInSe_2$ and $CuInS_2$ compounds [206]. The XRD pattern of a typical single phase chalcopyrite $CuIn(Se_{1-x}S_x)_2$ thin film, $x=0.3$ is shown in Figure 4.58. Both the lattice constants follow Vegard's law with increasing x value in the spray deposited $CuIn(Se_{1-x}S_x)$ thin films from 0 to 1, as shown in Figure 4.59 [12]. Similarly the lattice constants a and c decrease linearly with increasing x (Figure 4.60) and follow the relation $a=5.769-0.253x$ and $c=11.726-0.669x$ in the single crystals. The tetragonal distortion ($\Delta=2-c/a$) increases linearly with increasing x as $\Delta=-0.0109+0.0012x$ [205]. Glass/Mo/0.25 μm thick In layer/1 μm thick $CuInSe_2$ layer samples are treated at 500 °C for 1 h by heating with S+Se shots. In the films, S/(S+Se) ratio varies from 0 to 1 and the composition of Cu:In:(S+Se)=27:27:46 is maintained in crystalline thin films. The $CuIn_5S_8$ (110) phase is observed for above S/(S+Se)=0.2. The (112) peak splits due to grading of the sulfur in the samples. The grain sizes decrease with increasing S/(S+Se) ratio in the samples. The (112) peak diffraction angle shifts to higher angle with increasing S/(S+Se) ratio and the lattice constants decrease with increasing S/(S+Se) ratio and follow Vegard's law as $a=5.801-0.291x$ and $c=11.61-0.470x$ [207,208]. The similar variation is observed in the $CuIn(Se_{1-x}S_x)_2$ alloy system as $a=5.7946-0.2747x$ and $c=11.5685-0.0473x$ [209]. The reason is that the ionic radius of sulfur (1.09 Å) is smaller than that of Se (1.39 Å) therefore the tetragonal unit cell sinks with increasing sulfur content in the $CuIn(Se_{1-x}S_x)_2$ solid solutions. The In/Cu/In/Cu layers with Cu=180 nm and In=420 nm thickness are deposited by e-beam

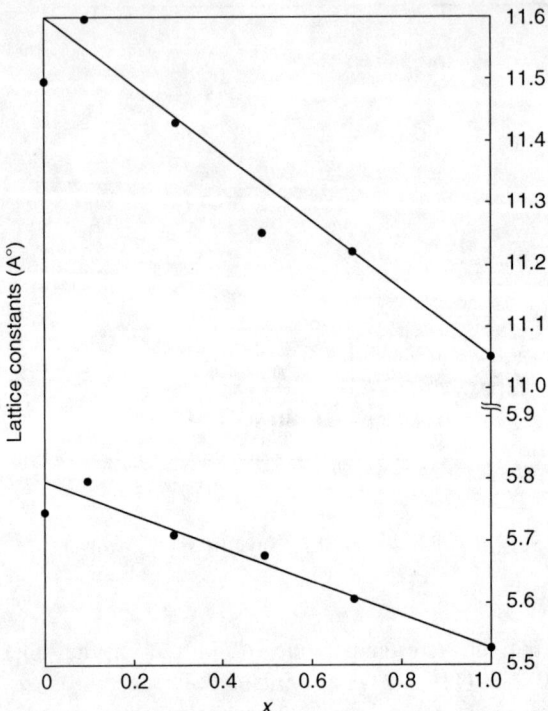

Figure 4.59 Variation of lattice parameters with x in $CuIn(Se_{1-x}S_x)_2$ thin films grown by spray pyrolysis.

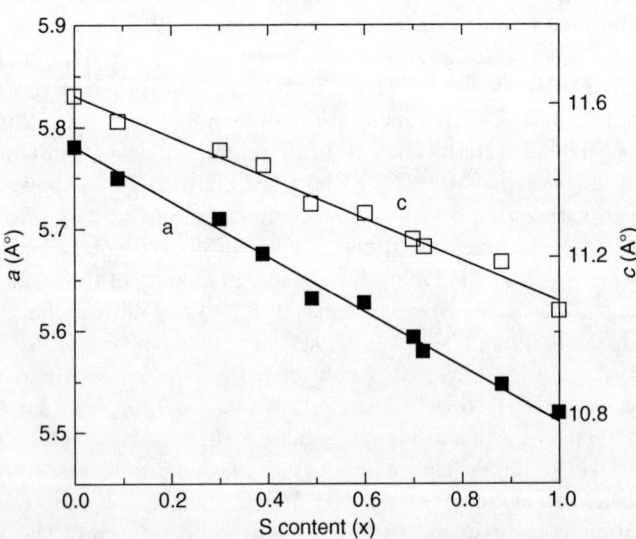

Figure 4.60 Variation of lattice constants with sulfur (x) in $CuIn(Se_{1-x}S_x)_2$ single crystals.

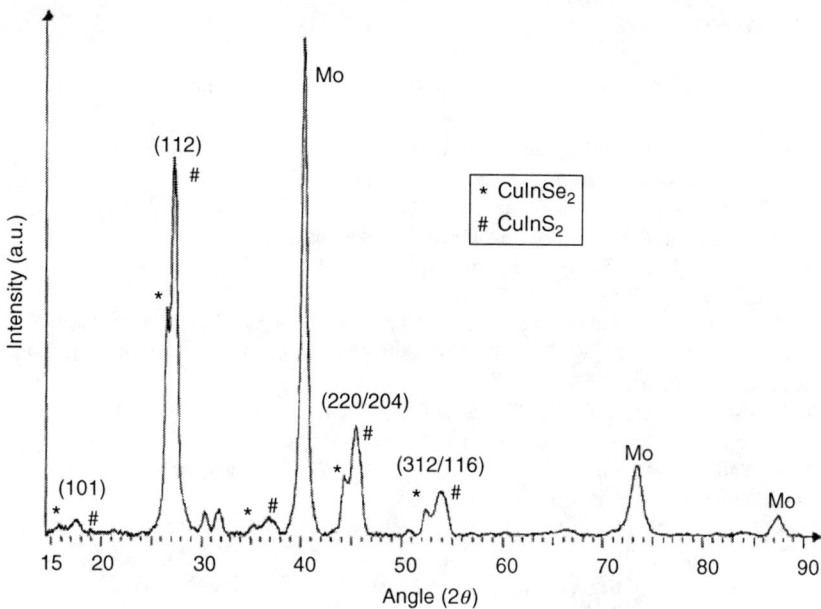

Figure 4.61 XRD pattern of combined $CuInS_2$–$CuInSe_2$ phase thin film.

evaporation onto Mo coated SLG, followed by annealing under vacuum at 120 °C for 1 h to intermix. The layers are annealed by rapid thermal annealing process under Ar, H_2Se, and H_2S gases of 200, 2.5, and 3 sccm, respectively at 450 °C for 5 min. Two kinds of processes are adopted to prepare $CuIn(S,Se)_2$ samples; samples under 1.25 vol% of H_2Se in Ar are rapid thermally annealed at 450 °C for 40 min then H_2Se is terminated and Ar is purged for 5 min. After that the samples are annealed under 1.5 vol% of H_2S in Ar for 5 min. The samples prepared by this method show $S/(S+Se)$ ratio between 0.5 and 0.8 and two phases as $CuInS_2$ and $CuInSe_2$, as shown in Figure 4.61. The Se is totally depleted for beyond 5 min sulfurization. In the second process, the metallic layers are rapid thermally annealed at 450 °C under H_2Se/Ar for 40 min then under $H_2Se/H_2S/Ar$ for 10 min. The CuIn$(SSe)_2$ layers grown by this method have single phase for $Cu/In = 0.9$ and $S/(S+Se) = 0.1$. The individual $CuInS_2$ or $CuInSe_2$ layers can be prepared by annealing Cu/In stack under H_2S/Ar or H_2Se/Ar at 450 °C for 40 min [210]. No other phases are observed in the Cu-rich films process. The near stoichiometric layers, which can be prepared from $Cu:In:Se:S = 1:5:2:20$ in the solution at wide range of potentials, show indium (101) as secondary phase. In addition, extra In peaks are observed in the In-rich layers. For this investigation the $CuIn(SSe)_2$ thin films are prepared by ED employing $CuSO_4$, $In_2(SO_4)_3$, SeO_2, and H_2NSNH_2. After adding thiourea to the rest of the chemical solution, the red precipitation occurs, which has to be filtered to use for the deposition. At pH 1.25, the potential from −1.0 to −3.0 versus SCE is employed to deposit the layers [211].

References

[1] S. Chichibu, T. Mizutani, K. Murakami, T. Shioda, T. Kurafuji, H. Nakanishi, et al., J. Appl. Phys. 83 (1998) 3678.
[2] M. Robbins, J.C. Phillips, V.G. Lambrecht Jr., J. Phys. Chem. Solids 34 (1973) 405.
[3] A. Tempel, B. Schumann, Krist. Tech. 13 (1978) 389.
[4] B.J. Stanbery, S. Kincal, S. Kim, T.J. Anderson, O.D. Crisalle, S.P. Ahrenkiel, et al., 28th IEEE Photovoltaic Specialists Conf, Anchorage, AK, 2000 p 440.
[5] L.S. Palatnik, E.I. Rogacheva, Sov. Phys. Dokl. 12 (1967) 503.
[6] M.L. Fearheiley, Solar Cells 16 (1986) 91.
[7] T. Haalboon, T. Godecke, F. Ernst, M. Ruhle, R. Herberholz, H.W. Schock, et al., Inst. Phys. Conf. Ser. 152 (1998) 249. (11th International Conference on Ternary and Multinary compounds 1997).
[8] B.J. Stanbery, C.-H. Chang, T.J. Anderson, Inst. Phys. Conf. Ser. 152 (1998) 915. (11th International Conference on Ternary and Multinary compounds 1997).
[9] T. Wada, Y. Hashimoto, S. Nishiwaki, T. Satoh, S. Hayashi, T. Negami, et al., Solar Energy Mater. Solar Cells 67 (2001) 305.
[10] S. Yoon, S. Kim, V. Craciun, W.K. Kim, R. Kaczynski, R. Acher, et al., J. Cryst. Growth 281 (2005) 209.
[11] S.R. Kodigala, V.S. Raja, A.K. Bhatnagar, F. Juang, S.J. Chang, Y.K. Su, Mater. Lett. 45 (2000) 251.
[12] S.R. Kodigala, V.S. Raja, Thin Solid Films 208 (1992) 247.
[13] F. Chraibi, M. Fahoume, A. Ennaoui, J.L. Delplancke, J.L. Delplancke, M. J. Condensed Matter 5 (2004) 88.
[14] B.D. Cullity (Ed.), Elements of X-ray Diffraction, Addision-Wesley Publishing Company Inc, London, 1978, p. 137.
[15] J.M. Delgado, G.D. Delgado, R. Guevara, Inst. Phys. Conf. Ser. 152 (1998) 139. (11th International Conference on Ternary and Multinary compounds 1997).
[16] M.L. Fearheiley, K.J. Bachmann, Y.H. Shing, S.A. Vasques, C.R. Herrington, J. Electron. Mater. 14 (1985) 677.
[17] J.L. Hurd, T.F. Ciszek, J. Cryst. Growth 70 (1984) 415.
[18] B.M. Basol, V.K. Kapur, R.C. Kullberg, Solar cells 27 (1989) 299.
[19] I. Shih, C.H. Champness, A.V. Shahidi, Solar Cells 16 (1986) 27.
[20] T.F. Ciszek, J. Cryst. Growth 70 (1984) 405.
[21] H. Miyake, H. Ohtake, K. Sugiyama, J. Cryst. Growth 156 (1995) 404.
[22] H. Matsushita, S.-I. Ai, A. Katsui, J. Cryst. Growth 224 (2001) 95.
[23] J. Piekoszewski, J.J. Loferski, R. Beaulieu, J. Beall, B. Roessler, J. Shewchun, Solar Energy Mater. 2 (1980) 363.
[24] A. Kumar, A.L. Dawar, P.K. Shishodia, G. Chauhan, P.C. Mathur, J. Mater. Sci. 28 (1993) 35.
[25] I. Martil, J. Santamaria, E. Iborra, G. Gonzalez-Diaz, F. Sanchez-Quesada, J. Appl. Phys. 62 (1987) 4163.
[26] A.G. Chowles, J.A.A. Engelbrecht, J.H. Neethling, C.C. Theron, Thin Solid Films 361–362 (2000) 93.
[27] J. Bougnot, S. Duchemin, M. Savelli, Solar Cells 16 (1986) 221.
[28] P.R. Ram, R. Thangraj, A.K. Sharma, O.P. Agnihotri, Solar Cells 14 (1985) 123.
[29] M.D. Kannan, R. Balasundaraprabhu, S. Jayakumar, P. Ramanathswamy, Solar Energy Mater. Solar Cells 81 (2004) 379.

[30] F.J. Pern, R. Noufi, A. Mason, A. Franz, Thin Solid Films 202 (1991) 299.
[31] A.M. Hermann, M. Mansour, V. Badri, B. Pinkhasov, C. Gonzales, F. Fickett, et al., Thin Solid Films 361–362 (2000) 74.
[32] T. Terasako, Y. Uno, T. Kariya, S. Shirakata, Solar Energy Mater. Solar Cells 90 (2006) 262.
[33] M. Kauk, M. Altosaar, J. Raudoja, A. Jagomagi, M. Danilson, T. Varema, Thin Solid Films 515 (2007) 5880.
[34] P.J. Sebastian, M.E. Calixto, R.N. Bhattacharya, R. Noufi, Sol. Energy Mater. Sol. Cells 59 (1999) 125.
[35] S. Isomura, A. Nagamatsu, K. Shinohara, T. Aono, Solar Cells 16 (1986) 143.
[36] A.G. Chowles, J.H. Neethling, H.V. Niekerk, J.A.A. Engelbrecht, V.J. Watters, Renewable Energy 6 (1995) 613.
[37] A. Yoshida, N. Tanahashi, T. Tanaka, Y. Demizu, Y. Yamamoto, T. Yamaguchi, Solar Energy Mater. Solar Cells 50 (1998) 7.
[38] H.P. Wang, I. Shih, C.H. Champness, Thin Solid Films 361–362 (2000) 494.
[39] S.H. Kwon, S.C. Park, B.T. Ahn, In: 26th IEEE PVSC, Anaheim, 1997, p. 395.
[40] S.H. Kwon, B.T. Ahn, S.K. Kim, K.H. Yoon, J. Song, Thin Solid Films 323 (1998) 265.
[41] S.C. Park, D.Y. Lee, B.T. Ahn, K.H. Yoon, J. Song, Solar Energy Mater. Solar Cells 69 (2001) 99.
[42] G.D. Mooney, A.M. Hermann, J.R. Tuttle, D.S. Albin, R. Noufi, Solar Cells 30 (1991) 69.
[43] J.R. Tuttle, D.S. Albin, R. Noufi, Solar Cells 30 (1991) 21.
[44] J.R. Tuttle, D.S. Albin, R. Noufi, Solar Cells 27 (1989) 231.
[45] E.R. Don, R. Hill, G.J. Russell, Solar Cells 16 (1986) 131.
[46] J.B. Mooney, R.H. Lamoreaux, C.W. Bates Jr., Spray Pyrolysis of $CdS/CuInSe_2$ Cells, (1983) (Final Report Jan.1, 1982-Nov.30).
[47] P. Fons, S. Niki, A. Yamada, H. Oyanagi, J. Appl. Phys. 84 (1998) 6926.
[48] L.-C. Yang, H.Z. Xia, A. Rockett, W.N. Shafarman, R.W. Birkmire, Sol. Energy Mater. Sol. Cells 36 (1995) 445.
[49] E.P. Zaretskaya, V.F. Gremenyuk, V.B. Zalesskii, V.A. Ivanov, I.V. Viktorov, V.I. Kovalevskii, et al., Tech. Phys. 70 (2000) 141.
[50] R. Caballero, C. Guillen, Solar Energy Mater. Solar Cells 86 (2005) 1.
[51] C. Guillen, J. Herrero, Thin Solid Films 387 (2001) 57.
[52] F.O. Adurodija, M.J. Carter, R. Hill, Sol. Energy Mater. Sol. Cells 40 (1996) 359.
[53] J. Zhang, Y. Sun, W. Liu, Q. He, C. Li, Z. Sun, In: 3rd World Conference on Photovoltaic Energy Conversion, 2003, p. 418.
[54] S.D. Kim, H.J. Kim, K.H. Yoon, J. Song, Sol. Energy Mater. Sol. Cells 62 (2002) 357.
[55] W.K. Kim, E.A. Payzant, S. Yoon, T.J. Anderson, J. Cryst. Growth 294 (2006) 231.
[56] A. Gupta, S. Isomura, Sol. Energy Mater. Sol. Cells 53 (1998) 385.
[57] M.E. Calixto, R.N. Bhattacharya, P.J. Sebastian, A.M. Fernandez, S.A. Gamboa, R.N. Nouf, Sol. Energy Mater. Sol. Cells 55 (1998) 23.
[58] C.J. Huang, T.H. Meen, M.Y. Lai, W.R. Chen, Solar Energy Mater. Solar Cells 82 (2004) 553.
[59] C.X. Qiu, I. Shih, Solar Energy Mater. 15 (1987) 219.
[60] A.M. Fernandez, M.E. Calixto, P.J. Sebastian, S.A. Gamboa, A.M. Hermann, R.N. Noufi, Sol. Energy Mater. Sol. Cells 52 (1998) 423.
[61] M. Pattabi, P.J. Sebastian, X. Mathew, R.N. Bhattacharya, Solar Energy Mater. Solar Cells 63 (2000) 315.

[62] H. Sato, T. Hama, E. Niemi, Y. Ichikawa, H. Sakai, In: 23rd IEEE Photovoltaic Specialist Conference, 1993, p. 521.
[63] C. Guillen, J. Herrero, Sol. Energy Mater. Sol. Cells 73 (2002) 141.
[64] V. Alberts, M. Klenk, E. Bucher, Jpn. J. Appl. Phys. 39 (2000) 5776.
[65] S.T. Lakshmikumar, A.C. Rastogi, Appl. Phys. Lett. 66 (1995) 3128.
[66] J. Muller, J. Nowoczin, H. Schmitt, Thin Solid Films 496 (2006) 364.
[67] J.L. Xu, X.F. Yao, J.Y. Feng, Sol. Energy Mater. Sol. Cells 73 (2002) 203.
[68] F.J. Pern, R. Noufi, A. Mason, A. Swartzlander, In: 19th IEEE Photovoltaic Specialist Conference, 1987, p. 1295.
[69] A. Kampmann, J. Rechid, A. Raitzig, S. Wulff, M. Mihhailova, R. Thyen, *et al.*, Mater. Res. Soc. Symp. Proc. 763 (2003) B8.5.1.
[70] J.-F. Guillemoles, P. Cowache, A. Lusson, K. Fezzaa, F. Boisivon, J. Vedel, *et al.*, J. Appl. Phys. 79 (1996) 7273.
[71] F.O. Adurodija, M.J. Carter, R. Hill, In: 1st World Conference on Photovoltaic Energy Conversion, Hawaii, 1994, p. 186.
[72] K.T.L. De Silva, W.A.A. Priyantha, J.K.D.S. Jajanetti, B.D. Chithrani, W. Siripala, K. Blake, *et al.*, Thin Solid Films 382 (2001) 158.
[73] M. Gorska, B. Beaulieu, J.J. Loferski, B. Roessler, J. Beall, Solar Energy Mater. 2 (1980) 343.
[74] N.B. Chaure, J. Young, A.P. Samantilleke, I.M. Dharmadasa, Solar Energy Mater. Solar Cells 81 (2004) 125.
[75] N. Stratieva, E. Tzvetkova, M. Ganchev, K. Kochev, I. Tomov, Solar Energy Mater. Solar Cells 45 (1997) 87.
[76] S. Nakamura, S. Sugawara, A. Hashimoto, A. Yamamoto, Solar Energy Mater. Solar Cells 50 (1998) 25.
[77] M.M. EL-Nahass, H.S. Soliman, D.A. Hendi, K.H.A. Mady, J. Mater. Sci. 27 (1992) 1484.
[78] C. Rincon, C. Bellabarba, J. Gonzalez, G.S. Perez, Solar Cells 16 (1986) 335.
[79] S. Chatraphorn, K. Yoodee, P. Songpongs, C. Chityuttakan, K. Sayavong, S. Wongmanerod, *et al.*, Jpn. J. Appl. Phys. 37 (1998) L269.
[80] M. Rusu, S. Doka, A. Meeder, R. Wurz, E. Strub, J. Rohrich, *et al.*, Thin Solid Films 480–481 (2005) 352.
[81] M. Rusu, P. Gashin, A. Simashkevich, Sol. Energy Mater. Sol. Cells 70 (2001) 175.
[82] G. Orsal, F. Mailly, N. Romain, M.C. Artaud, S. Rushworth, S. Duchemin, Thin Solid Films 361–362 (2000) 135.
[83] F.-J. Haug, M. Krejci, H. Zogg, A.N. Tiwari, M. Kirsch, S. Siebentritt, Thin Solid Films 361–362 (2000) 239.
[84] M.R. Balboul, A. Jasenek, O. Chernykh, U. Rau, H.W. Schock, Thin Solid Films 387 (2001) 74.
[85] D. Fischer, T. Dylla, N. Meyer, M.E. Beck, A. Jager-Waldau, M.Ch. Lux-Steiner, Thin Solid Films 387 (2001) 63.
[86] H. Hallak, D. Albin, R. Noufi, Appl. Phys. Lett. 55 (1989) 981.
[87] T. Kampschulte, A. Bauknecht, U. Blieske, M. Saad, S. Chichibu, M.Ch. Lux-Steiner, In: 26th IEEE PVSC, Anaheim, 1997, p. 391.
[88] D.F. Morron, A. Meeder, U. Bloeck, P. Schubert-Bischoff, N. Pfander, R. Wurz, *et al.*, Thin Solid Films 431–432 (2003) 237.
[89] D. Fischer, N. Meyer, M. Kuczmik, M. Beck, A.-J. Waldau, M.Ch.-L. Steiner, Solar Energy Mater. Solar Cells 67 (2001) 105.
[90] R. Caballero, C. Guillen, Thin Solid Films 403–404 (2002) 107.

[91] I. Martil, J. Santamaria, G.G. Diaz, F.S. Quesada, J. Appl. Phys. 68 (1990) 189.
[92] W. Arndt, H. Dittrich, H.W. Schock, Thin Solid Films 130 (1985) 209.
[93] U. Fiedeler, A. Bauknecht, A. Gerhard, J. Albert, M.Ch. Lux-Steiner, S. Siebetritt, In: 28th IEEE Photovoltaic Specialists Conference, Anchorage, AK, 2000, p. 626.
[94] M. Souilah, X. Rocquefelte, A. Lafond, C. Guillot-Deudon, J.-P. Morniroli, J. Kessler, Thin Solid Films 517 (2009) 2145.
[95] Y. Yan, K.M. Jones, J. AbuShama, M.M. Al-Jassim, R. Noufi, Mater. Res. Soc. Symp. Proc. 668 (2001) H6.10.1.
[96] W.W. Lam, I. Shih, Solar Energy Mater. Solar Cells 50 (1998) 111.
[97] K. Yoshino, H. Yokoyama, K. Maeda, T. Ikari, J. Cryst. Growth 211 (2000) 476.
[98] G. Masse, K. Djessas, F. Guastavino, J. Phys. Chem. Solids 52 (1991) 999.
[99] B. Grzeta-Plenkovic, S. Popovic, B. Celustka, B. Santic, J. Appl. Cryst. 13 (1980) 311.
[100] T. Tinoco, C. Rincon, M. Quintero, G.S. Perez, Phys. Status Solidi A 124 (1991) 427.
[101] A. Kinoshita, M. Fukaya, H. Nakanishi, M. Sugiyama, S.F. Chichibu, Phys. Status Solidi C 3 (2006) 2539.
[102] Y. Yamaguchi, J. Matsufusa, A. Yoshida, J. Appl. Phys. 72 (1992) 5657.
[103] F.B. Dejene, V. Alberts, J. Phys. D Appl. Phys. 38 (2005) 22.
[104] R. Chakrabarti, B. Maiti, S. Chaudhuri, A.K. Pal, Solar Energy Mater. Solar Cells 43 (1996) 237.
[105] A.M. Herman, M. Mansour, V. Badri, B. Pinkhasov, C. Gonzales, F. Fickett, *et al.*, Thin Solid Films 361–362 (2000) 74.
[106] H. Du, C.H. Champness, I. Shih, T. Cheung, Thin Solid Films 480–481 (2005) 42.
[107] H.P. Wang, I. Shih, C.H. Champness, Thin Solid Films 387 (2001) 60.
[108] J.M. Merino, M. Leon, F. Rueda, R. Diaz, Thin Solid Films 361–362 (2000) 22.
[109] C. Amory, J.C. Bernede, E. Halgand, S. Marsillac, Thin Solid Films 431–432 (2003) 22.
[110] S. Marsillac, S. Don, R. Rocheleau, E. Miller, Solar Energy Mater. Solar Cells 82 (2004) 45.
[111] E. Ahmed, A. Zegadi, A.E. Hill, R.D. Pilkington, R.D. Tomlinson, A.A. Dost, *et al.*, Solar Energy Mater. Solar Cells 36 (1995) 227.
[112] A.M. Hermann, C. Gonzalez, P.A. Ramakrishnan, D. Balzar, N. Popa, P. Rice, *et al.*, Solar Energy Mater. Solar Cells 70 (2001) 345.
[113] T. Schlenker, V. Laptev, H.W. Schock, J.H. Werner, Thin Solid Films 480–481 (2005) 29.
[114] L. Zhang, F.D. Jiang, J.Y. Feng, Sol. Energy Mater. Sol. Cells 80 (2003) 483.
[115] T. Yamaguchi, Y. Yamamoto, A. Yoshida, Solar Energy Mater. Solar Cells 67 (2001) 77.
[116] T. Yamaguchi, Y. Yamamoto, T. Tanaka, Y. Demizu, A. Yoshida, Jpn. J. Appl. Phys. 35 (1996) L1618.
[117] J. Kessler, C. Chityuttakan, J. Scholdstrom, L. Stolt, Thin Solid Films 431–432 (2003) 1.
[118] E. Ahmed, R.D. Tomlinson, R.D. Pilkington, A.E. Hill, W. Ahmed, N. Ali, *et al.*, Thin Solid Films 335 (1998) 54.
[119] A.M. Hermann, C. Gonzalez, P.A. Ramakrishnan, D. Balzar, C.H. Marshall, J.N. Hilfiker, *et al.*, Thin Solid Films 387 (2001) 54.
[120] D. Liao, A. Rockett, 28th IEEE Photovoltaic Specialists Conf, Anchorage, AK, 2000 p 446.
[121] O. Lundberg, M. Bodegard, L. Stolt, Thin Solid Films 431–432 (2003) 26.
[122] M. Bodegard, O. Lundberg, J. Malmstrom, L. Stolt, 28th IEEE Photovoltaic Specialists Conf, Anchorage, AK, 2000 p 450.
[123] M.E. Beck, A. Swartzlander-Guest, R. Matson, J. Keane, R. Noufi, Solar Energy Mater. Solar Cells 64 (2000) 135.

[124] F.S. Hasoon, Y. Yan, K.M. Jones, H. Althani, J. Alleman, M.M. Al-Jassim, et al., 28th IEEE Photovoltaic Specialists Conf, Anchorage, AK, 2000, p 513.

[125] T. Satoh, Y. Hashimoto, S.-I. Shimakawa, S. Hayashi, T. Negami, 28th IEEE Photovoltaic Specialists Conf, Anchorage, AK, 2000 p 567.

[126] R. Caballero, C. Guillen, Thin Solid Films 431–432 (2003) 110.

[127] R. Caballero, C. Maffiotte, C. Guillen, Thin Solid Films 474 (2005) 70.

[128] R. Friedfeld, R.P. Raffaelle, J.G. Mantovani, Solar Energy Mater. Solar Cells 58 (1999) 375.

[129] M.A. Contreras, M.J. Romero, R. Noufi, Thin Solid Films 511–512 (2006) 51.

[130] S. Ishizuka, K. Sakurai, A. Yamada, K. Matsubara, H. Shibata, M. Yonemura, et al., In: 4th World Conference on Photovoltaic Energy Conversion, 2006, p. 338.

[131] N.G. Dhere, K.W. Lynn, Solar Energy Mater. Solar Cells 41 (1996) 271.

[132] V.F. Gremenok, E.P. Zaretskaya, V.B. Zalesski, K. Bente, W. Schmitz, R.W. Martin, et al., Solar Energy Mater. Solar Cells 89 (2005) 129.

[133] I.M. Dharmadasa, N.B. Chaure, G.J. Tolan, A.P. Samantilleke, J. Electrochem. Soc. 154 (2007) H466.

[134] A. Ihlal, K. Bouabid, D. Soubane, M. Nya, O. Ait-Taleb-Ali, Y. Amira, et al., Thin Solid Films 515 (2007) 5852.

[135] M. Sugiyama, A. Kinoshita, M. Fukaya, H. Nakanishi, S.F. Chichibu, Thin Solid Films 515 (2007) 5867.

[136] T. Yamaguchi, Y. Yamamoto, T. Tanaka, N. Tanahashi, A. Yoshida, Solar Energy Mater. Solar Cells 50 (1998) 1.

[137] T. Walter, H.W. Schock, Thin Solid Films 224 (1993) 74.

[138] K.H. Kim, M.S. Kim, B.T. Ahn, J.H. Yun, K.H. Yoon, In: 4th World Conference on Photovoltaic Energy Conversion, 2006, p. 575.

[139] R. Caballero, C. Guillen, M.T. Gutierrez, C.A. Kaufmann, Prog. Photovoltaics Res. Appl. 14 (2006) 145.

[140] B. Dmmler, H. Dittrich, R. Menner, H.W. Schock, In: 19th IEEE Photovoltaic Specialist Conference, 1987, p. 1454.

[141] R.N. Bhattacharya, W. Batchelor, J.E. Granata, F. Hasoon, H. Wiesner, K. Ramanathan, et al., Sol. Energy Mater. Sol. Cells 55 (1998) 83.

[142] M. Marudachalam, H. Hichri, R. Klenk, R.W. Birkmire, W.N. Shararman, J.M. Schultz, J. Appl. Phys. 67 (1995) 3978.

[143] I. Dirnstorfer, W. Burkhardt, W. Kriegseis, I. Osterreicher, H. Alves, D.M. Hofmann, et al., Thin Solid Films 361–362 (2000) 400.

[144] I. Dirnstorfer, D.M. Hofmann, D. Meister, B.K. Mwyer, W. Riedl, F. Karg, J. Appl. Phys. 85 (1999) 1423.

[145] M.B. Ard, K. Granath, L. Stolt, Thin Solid Films 361–362 (2000) 9.

[146] D. Rudmann, G. Bilger, M. Kaelin, F.-J. Haug, H. Zogg, A.N. Tiwari, Thin Solid Films 431–432 (2003) 37.

[147] G. Hanna, J. Matheis, V. Laptev, Y. Yamamoto, U. Rau, H.W. Schock, Thin Solid Films 431–432 (2003) 31.

[148] M. Bodegard, O. Lundberg, J. Lu, L. Stolt, Thin Solid Films 431–432 (2003) 46.

[149] M. Lammer, U. Klemm, M. Powalla, Thin Solid Films 387 (2001) 33.

[150] T. Tanaka, Y. Demizu, A. Yoshida, T. Yamaguchi, J. Appl. Phys. 81 (1997) 7619.

[151] T. Tanaka, N. Tanahashi, T. Yamaguchi, A. Yoshida, Solar Energy Mater. Solar Cells 50 (1998) 13.

[152] U.P. Singh, W.N. Shafarman, R.W. Birkmire, Solar Energy Mater. Solar Cells 90 (2006) 623.

[153] T. Delsol, A.P. Samantilleke, N.B. Chaure, P.H. Gardiner, M. Simmonds, I.M. Dharmadasa, Solar Energy Mater. Solar Cells 82 (2004) 587.
[154] J.E. Leisch, R.N. Bhattacharya, G. Teeter, J.A. Turner, Solar Energy Mater. Solar Cells 81 (2004) 249.
[155] E.P. Zaretskaya, V.F. Gremenok, V.B. Zalesski, K. Bente, S. Schorr, S. Zukotynski, Thin Solid Films 515 (2007) 5848.
[156] Y. Tanaka, N. Akema, T. Morishita, D. Okumura, K. Kushiya, In: 17th European Photovoltaic Solar Energy Conference, Munich, 2001, p. 989.
[157] V. Alberts, Thin Solid Films 517 (2009) 2115.
[158] T. Wada, N. Kohara, S. Nishiwaki, T. Negami, Thin Solid Films 387 (2001) 118.
[159] T. Hahn, H. Metzner, B. Plikat, M. Seibt, Thin Solid Films 387 (2001) 83.
[160] J. Eberhardt, J. Cieslak, H. Metzner, Th. Hahn, R. Goldhahn, F. Hudert, et al., Thin Solid Films 517 (2009) 2248.
[161] C. Guillen, J. Herrero, M.T. Gutierrez, F. Briones, Thin Solid Films 480–481 (2005) 19.
[162] S.R. Kodigala, V.S. Raja, J. Mater. Sci. Mater. Electron. 10 (1999) 145.
[163] S.R. Kodigala, V.S. Raja, Mater. Lett. 12 (1991) 67.
[164] M.C. Zouaghi, T.B. Nasrallah, S. Marsillac, J.C. Bernede, S. Belgacem, Thin Solid Films 382 (2001) 39.
[165] P. Guha, D. Das, A.B. Maity, D. Ganguli, S. Chaudhuri, Solar Energy Mater. Solar Cells 80 (2003) 115.
[166] M. Gossla, Th. Hahn, H. Metzner, J. Conrad, U. Geyer, Thin Solid Films 268 (1995) 39.
[167] K. Siemer, J. Klaer, I. Luck, J. Bruns, R. Klenk, D. Braunig, Solar Energy Mater. Solar Cells 67 (2001) 159.
[168] J. Xiao, Y. Xie, R. Tang, Y. Qian, J. Solid State Chem. 161 (2001) 179.
[169] K. Kondo, S. Nakamura, H. Sano, H. Hirasawa, K. Sato, Solar Energy Mater. Solar Cells 49 (1997) 327.
[170] R. Klenk, U. Blieske, V. Dieterle, K. Ellmer, S. Fiechter, I. Hengel, et al., Solar Energy Mater. Solar Cells 49 (1997) 349.
[171] T. Walter, A. Content, K.O. Velthaus, H.W. Schock, Solar Energy Mater. Solar Cells 26 (1992) 357.
[172] J.D. Harris, K.K. Banger, D.A. Scheiman, M.A. Smith, M.H.-C. Jin, A.F. Hepp, Mater. Sci. Eng. B98 (2003) 150.
[173] H.L. Hwang, C.L. Cheng, L.M. Liu, Y.C. Liu, C.Y. Sun, Thin Solid Films 67 (1980) 83.
[174] H. Bouzouita, N. Bouguila, A. Dhouib, Renewable Energy (1998) 1.
[175] R. Scheer, K. Diesner, H.J. Lewerenz, Thin Solid Films 268 (1995) 130.
[176] Y.L. Wu, H.Y. Lin, C.Y. Sun, M.H. Yang, H.L. Hwang, Thin Solid Films 168 (1989) 113.
[177] Y.B. He, T. Kramer, A. Polity, M. Hardt, B.K. Meyer, Thin Solid Films 431–432 (2003) 126.
[178] P. Rajaram, R. Thangaraj, A.K. Sharma, A. Raza, O.P. Agnihotri, Thin Solid Films 100 (1983) 111.
[179] J. Gonzalez-Hernandez, P.M. Gorley, P.P. Horley, O.M. Vartsabyuk, Y.V. Vorobiev, Thin Solid Films 403–404 (2002) 471.
[180] X. Hou, K.-L. Choy, Thin Solid Films 480–481 (2005) 13.
[181] H. Bihri, M. Abd-Lefdil, Thin Solid Films 354 (1999) 5.

[182] M. Krunks, O. Kijatkina, A. Mere, T. Varema, I. Oja, V. Mikli, Solar Energy Mater. Solar Cells 87 (2005) 207.
[183] Y.B. He, A. Polity, R. Gregor, D. Pfisterer, I. Osterreicher, D. Hasselkamp, et al., Physica B 308–310 (2001) 1074.
[184] M. Krunks, O. Kijatkina, H. Rebane, I. Oja, V. Mikli, A. Mere, Thin Solid Films 403–404 (2002) 71.
[185] R. Henninger, J. Klaer, K. Siemer, J. Bruns, D. Braunig, J. Appl. Phys. 89 (2001) 3049.
[186] M. Krunks, O. Bijakina, T. Varema, V. Mikli, E. Mellikov, Thin Solid Films 338 (1999) 125.
[187] M. Ortega-Lopez, A. Morales-Acevedo, Thin Solid Films 330 (1998) 96.
[188] M. Krunks, O. Bijakina, V. Mikli, H. Rebane, T. Varema, M. Altosaar, et al., Solar Energy Mater. Solar Cells 69 (2001) 93.
[189] S. Bandyopadhyaya, S. Chaudhuri, A.K. Pal, Solar Energy Mater. Solar Cells 60 (2000) 323.
[190] T. Terasako, Y. Uno, S. Inoue, T. Kariya, S. Shirakata, Phys. Status Solidi C 3 (2006) 2588.
[191] M. Krunks, V. Mikli, O. Bijakina, E. Mellikov, Thin Solid Films 142 (1999) 356.
[192] A.N. Tiwari, D.K. Pandya, K.L. Chopra, Thin Solid Films 130 (1985) 217.
[193] A. Antony, A.S. Asha, R. Yoosuf, R. Manoj, M.K. Jayaraj, Solar Energy Mater. Solar Cells 81 (2004) 407.
[194] S.P. Grindle, C.W. Smith, S.D. Mittleman, Appl. Phys. Lett. 35 (1979) 24.
[195] Y.B. He, A. Polity, H.R. Alves, I. Osterreicher, W. Kriegseis, D. Pfisterer, et al., Thin Solid Films 403–404 (2002) 62.
[196] Y. Yamamoto, T. Yamaguchi, T. Tanaka, N. Tanahashi, A. Yoshida, Solar Energy Mater. Solar Cells 49 (1997) 399.
[197] R.P. Wijesundera, W. Siripala, Sol. Energy Mater. Sol. Cells 81 (2004) 147.
[198] H. Komaki, K. Yoshino, S. Seto, M. Yoneta, Y. Akaki, T. Ikari, J. Cryst. Growth 236 (2002) 253.
[199] W. Calvet, C. Lehmann, T. Plake, C. Pettenkofer, Thin Solid Films 480–481 (2005) 347.
[200] T. Yukawa, K. Kuwabara, K. Koumoto, Thin Solid Films 286 (1996) 153.
[201] T. Negami, Y. Hashimoto, M. Nishitani, T. Wada, Sol. Energy Mater. Sol. Cells 49 (1997) 343.
[202] T. Watanabe, H. Nakazawa, M. Matsui, H. Ohbo, T. Nakada, Sol. Energy Mater. Sol. Cells 49 (1997) 357.
[203] K. Yoshino, K. Nomoto, A. Kinoshita, T. Ikari, Y. Akaki, T. Yoshitake, J. Mater. Sci. Mater. Electron. 19 (2008) 301.
[204] T. Ohashi, K. Inakoshi, Y. Hashimoto, K. Ito, Sol. Energy Mater. Sol. Cells 50 (1998) 37.
[205] K. Zeaiter, A. Yanuar, C. Llinares, Solar Energy Mater. Solar Cells 61 (2001) 213.
[206] S. Shirakata, T. Terasako, T. Kariya, J. Phys. Chem. Solids 66 (2005) 1970.
[207] T. Yamaguchi, M. Nakashima, A. Yoshida, In: 3rd World Conference on Photovoltaic Energy Conversion, 2003, p. 410.
[208] T. Wada, H. Kinoshita, Thin Solid Films 480–481 (2005) 92.
[209] S.A. Lopez-Rivera, B. Fontal, J.A. Henao, E. Mora, W. Giriat, R. Vargas, Inst. Phys. Conf. Ser. 152 (1998) 175 (11th International Conference on Ternary and Multinary compounds 1997).
[210] J. Bekker, V. Alberts, A.W.R. Leitch, J.R. Botha, Thin Solid Films 431–432 (2003) 116.
[211] S. Kuranouchi, T. Nakazawa, Sol. Energy Mater. Sol. Cells 50 (1998) 31.

5 Optical Properties of I–III–VI$_2$ Compounds

The band structure, band gaps, absorption, transmission, and reflections of I–III–VI$_2$ compounds such as CuInS$_2$, CuInSe$_2$, CuGaSe$_2$, and their solid solutions are emphasized in this chapter. The low temperature photoluminescence studies on the compounds are extensively discussed how the transitions of defect levels take place. On the other hand, what kind of defect levels encourages enhancing efficiency of the cells and discourages to slip the efficiency of the cells. The Raman spectroscopy analysis significantly details structure of the compounds, secondary phases and supports the XRD analysis to confirm the same.

5.1 Band Structure of I–III–VI$_2$ Compounds

A typical band structure of CuInSe$_2$ is given in Figure 5.1 [1]. The lowest conduction band Γ_1 is at the zone center and is approximately isotropic. The valence bands of most of zinc blende crystals, binary analogs of I–III–VI$_2$ compounds are composed of s- and p-like orbitals while the uppermost valance bands of I–III–VI$_2$ are profoundly influenced by the proximity of noble metal d-levels. The anion p-levels and noble metal d-levels hybridize. Two manifestations of this hybridization are [2–4] (i) downward shift of the band gaps relative to their binary analogs and (ii) reduced spin-orbit splitting due to partial cancellation of positive spin-orbit parameter for p-levels and the negative spin-orbit parameter for d-levels. Quantitative estimates of d-like character range from 16 to 45%. The presence of noble metal d-levels in the valence band is confirmed directly from electroreflectance studies [5].

At the point Γ of the Brillouin zone, the triple degeneracy of Γ_{15} is completely lifted due to the simultaneous influences of noncubic crystalline field and spin-orbit interaction. The level splitting is usually described in terms of quasi-cubic model [6]. The band structure and the selection rules for transitions in a chalcopyrite crystal derived from Γ_{15V} to Γ_{1C} energy gap in a zinc blend crystal is shown in Figure 5.2 [7]. It can be seen that the triply degenerate Γ_{15V} in zinc blende splits into nondegenerate Γ_{4V} and doubly degenerate Γ_{5V}. The latter then splits into Γ^5_{6V} and Γ^5_{7V}. Γ_{4V} and Γ_{1C} convert into Γ^4_{7V} and Γ^1_{6C} respectively. Under these conditions, three transitions are possible viz. (i) from Γ^5_{7V} to Γ^1_{6C}, (ii) Γ^5_{6V} to Γ^1_{6C}, and (iii) Γ^4_{7V} to Γ^1_{6C}. The information about crystal field splitting (Δ_{cf}) and spin-orbit splitting (Δ_{so}) can

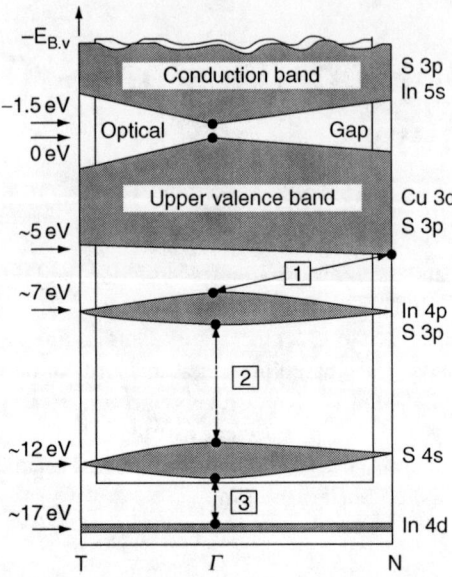

Figure 5.1 Band diagram of CuInSe$_2$ in which 1, 2, 3 are internal gaps and shades indicate sub bands.

be obtained from the optical experimental studies. Electrolyte electroreflectance (EER) investigations on CuInS$_2$ single crystals showed that room temperature energy gaps corresponding to these transitions are equal [8] whilst the EER studies on CuInSe$_2$ single crystals [9] have resulted in the band gaps of 1.04 (A), 1.04 (B), and 1.27 eV (C) corresponding to above three transitions. The observed spin-orbit parameter of 0.23 eV is less than the corresponding splittings of CdSe (0.42 eV) and ZnSe (0.43 eV) that comprise the binary analog Zn$_{0.5}$Cd$_{0.5}$Se of CuInSe$_2$. Recently, Zunger and his coworkers [1,10] presented the first *ab initio* calculation of the electronic structure of CuInSe$_2$, CuInS$_2$, and other ternaries. The calculated structure predicts a direct band gap transition for these materials. The calculated total and local density of states compare well with the published XPS data for CuInSe$_2$ and CuInS$_2$. In CuGaSe$_2$, the transition levels are $A = 1.729$ eV, $B = 1.813$ eV, and $C = 2.016$ eV.

5.1.1 Optical Analysis: Absorption, Transmission, and Reflections

There are several models to determine the absorption coefficient (α) of the absorbing thin films on the nonabsorbing substrates. The sum of absorptance (α_{ab}), transmittance (T), and reflectance (R) yields unity as $\alpha_{ab} + T + R = 1$ from which either $(1-R)/T = (\alpha_{ab} + T)/T$ or $(1+R)/T = (\alpha_{ab} + T - 2)/T$ can be utilized to find absorption coefficient. If two components are known out of three (α_{ab}, T, and R) then it is easy to find out absorption coefficient as well as band gap of interested thin films. It is worth noting that α_{ab}, T, and R are obviously function of wavelength (λ).

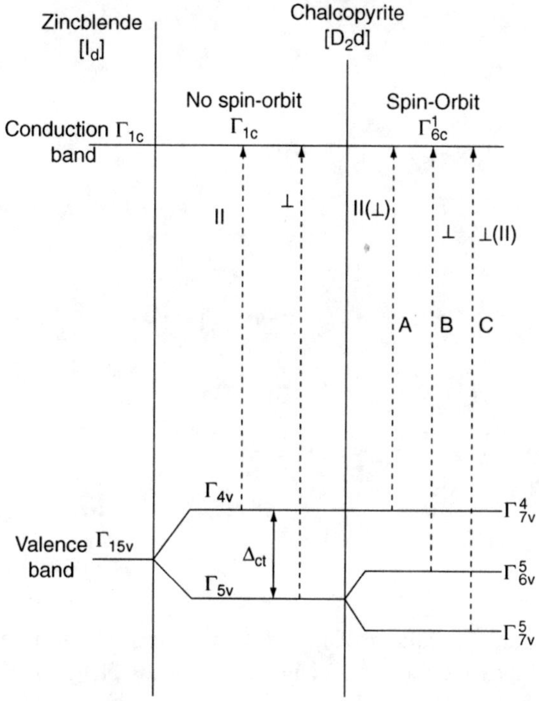

Optical Properties of I–III–VI$_2$ Compounds

Figure 5.2 Band structure and selection rules for the transitions in the chalcopyrite crystal derived from the Γ_{15} to Γ_1 energy gap in a zinc blende crystal.

The absorption coefficient (α) can be obtained using the relation given below provided that R and T of the absorbing film on nonabsorbing substrate are known [11]

$$\frac{1-R}{T} = \frac{1}{2n_s(n^2+k^2)} \left[n\{(n^2+n_s^2+k^2)\sinh\alpha + 2nn_s\cosh\alpha\} \right. \\ \left. + k\{(n^2-n_s^2+k^2)\sin\gamma + 2n_s k\cos\gamma\} \right] \quad (5.1)$$

The absorption coefficient (α) can be calculated from the transmission spectra alone [12,13]. One of the simple models is now given below [14,15]

$$T = \frac{A}{B_1 e^{\alpha d_f} + B_2 e^{-\alpha d_f} + C\cos\gamma + D\sin\gamma}, \quad (5.2)$$

$$\frac{A}{T} - C\cos\gamma - D\sin\gamma = B_1 e^{\alpha d_f} + B_2 e^{-\alpha d_f} = P,$$

$$B_2 e^{-\alpha d_f} = P - B_1 e^{\alpha d_f},$$

$$P^2 = \left(B_1 e^{\alpha d_f} + B_2 e^{-\alpha d_f}\right)^2 = \left(B_1 e^{\alpha d_f} - B_2 e^{-\alpha d_f}\right)^2 + 4B_1 B_2, \quad \sqrt{P^2 - 4B_1 B_2}$$
$$= B_1 e^{\alpha d_f} - B_2 e^{-\alpha d_f} = B_1 e^{\alpha d_f} - P + B_1 e^{\alpha d_f} = 2B_1 e^{\alpha d_f} - P,$$

$$\frac{P + \sqrt{P^2 - 4B_1B_2}}{2B_1} e^{\alpha d_f},$$

$$\alpha = \frac{\ln\left(\frac{P+\sqrt{P^2-4B_1B_2}}{2B_1}\right)}{d_f},$$

$$A = 32 n_0^2 n_s^2 (n^2 + k^2),$$

$$B_{1,2} = \left[(n_0 \pm n)^2 + k^2\right]\left[(n_s^2 + n_0^2)(n_s^2 + n^2 + k^2) \pm 4 n_0 n n_s^2\right],$$

$$C = 2\left[(n_0^2 - n^2 - k^2)(n^2 - n_s^2 + k^2)(n_s^2 + n_0^2) + 8 n_0^2 n_s^2 k^2\right],$$

$$D = 4 n_0 k\left[(n^2 + k^2 - n_s^2)(n_s^2 + n_0^2) - 2 n_s^2 (n_0^2 - n^2 - k^2)\right],$$

$$\gamma = \frac{4\pi n d_f}{\lambda},$$

On the other hand $\alpha = 4\pi k d_f / \lambda$,

where n_o, n_s, and n are the refractive indices of the medium, substrate and thin film, respectively. k and d_f are the extinction coefficient and thickness of the film, respectively. By computing all the values, the absorption coefficient (α), n, and k can be determined with function of λ. The absorption coefficient can also be deduced using a simpler version of Equation (5.3) given below [16].

$$\alpha = \frac{1}{d_f}\ln\left[\frac{(1-R)^2}{2T} + \left\{\frac{(1-R)^4}{4T^2} + R^2\right\}^{1/2}\right], \tag{5.3}$$

where $R = (n-1)^2/(n+1)^2$

Although the Equation (5.3) can be still simplified by assuming $R = 0$ that is, a rough estimation of $\alpha = [\ln(1/T)]/d_f$ provides a large margin of magnitude comparing with others.

There are two kinds of transitions such as direct and indirect in which one is allowed and another one forbidden transition. The relation for the direct optical transition between parabolic bands is given [17]

$$\alpha h\nu = A(h\nu - E_g)^x, \tag{5.4}$$

where A is a constant, $x = 1/2$ for the allowed transition and $x = 3/2$ for forbidden transition. The direct band gap (E_g) of semiconductor can be obtained from the intercept of straight line on $h\nu$ axis in a plot of $(\alpha h\nu)^2$ versus $h\nu$. In the indirect band gap (E_{gi}) transitions, $x = 2$ for allowed transition and $x = 3$ for forbidden transition.

5.1.2 CuInSe$_2$

The CuInSe$_2$ (CIS) has band gap of 0.95–1.05 eV that is slightly lower to match with the solar spectrum to acquire most percentage of photons and its absorption coefficient is in the order of 10^5 cm^{-1} at fundamental absorption region [18]. The variation of transmission with photon energy (hv) and a plot of $(\alpha hv)^2$ versus hv for the stoichiometric CuInSe$_2$ thin films grown onto corning 7059 glass substrates by vacuum evaporation are depicted in Figure 5.3A and B, respectively [19]. The absorption coefficient (α) can be extracted from the transmission spectrum employing above formulae. The sharp transmission spectrum of CuInSe$_2$ thin films prepared by spray pyrolysis technique is observed in high resistance samples (Figure 5.4A) relating to band gap of 1.02 eV, whereas the transmission signal is not in detectable range in low resistance samples due to high carrier absorption [20]. The spectrum of an amorphous 80 nm thick p-CuInSe$_2$ is depicted in Figure 5.4B, from which band gap is calculated to be 1.2 eV [21]. In fact, the band gap is higher in the amorphous CIS than that in the single crystal or polycrystalline due to difference in nature of band structure.

The Cu-rich and nearly stoichiometric In-rich CuInSe$_2$ thin films exhibit band gaps of 0.99 and 1.02 eV, respectively, which are determined from transmission spectra of the layers, as given in Figure 5.5 [22]. Figure 5.6 shows transmission spectra of stoichiometric and Cu-rich CuInSe$_2$ thin films grown onto glass substrates. The poor trasmission is observed in the Cu-rich CuInSe$_2$ thin films due to free carrier absorption [23]. Unlike transmission spectrum alone, the more precise determination of band gap for CuInSe$_2$ thin films can be found out from both transmission and reflectance spectra using Equation (5.1). The band gaps of 1.02 and 0.94 eV are found for CuInSe$_2$ thin films deposited onto glass substrates by evaporation of three constituent elements

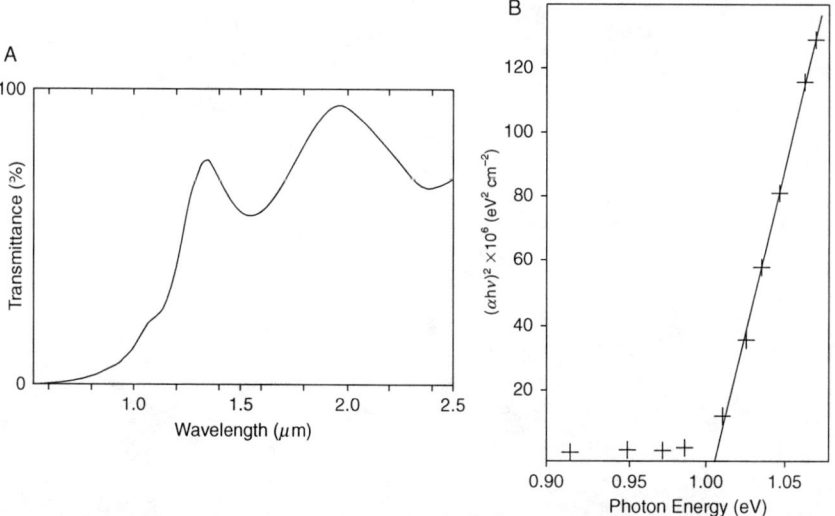

Figure 5.3 (A) Transmission spectrum of CuInSe$_2$ thin film and (B) $(\alpha hv)^2$ versus hv.

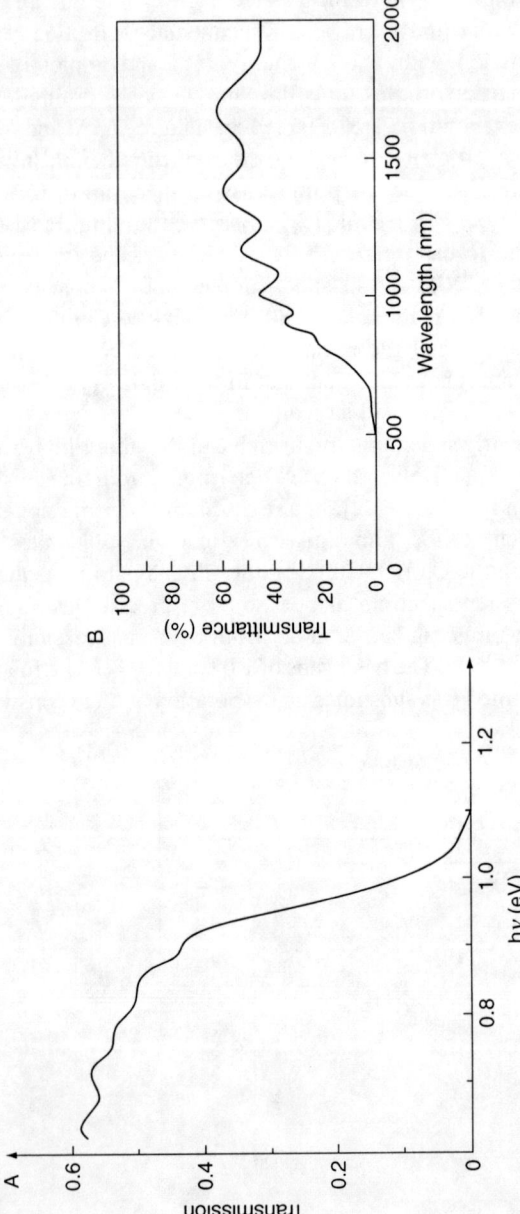

Figure 5.4 (A) Transmission spectrum of CuInSe$_2$ thin films grown by spray pyrolysis technique and (B) Transmission spectrum of an amorphous p-type CuInSe$_2$ thin film.

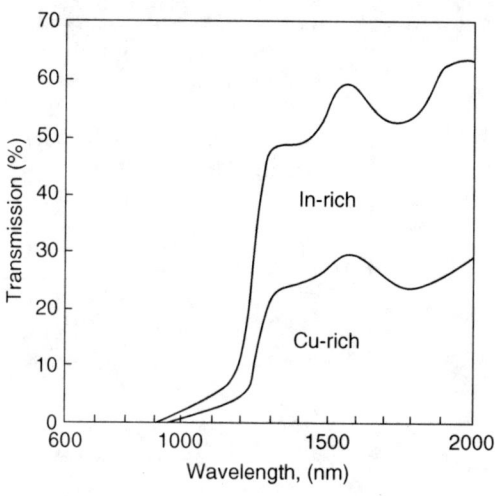

Figure 5.5 Transmission spectra of Cu-rich and nearly stoichiometric In-rich CuInSe$_2$ thin films.

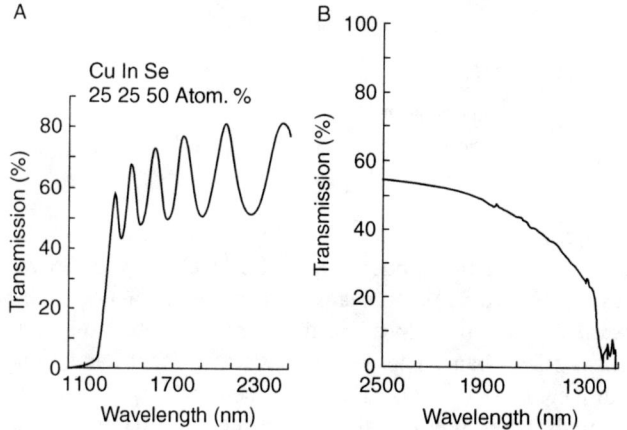

Figure 5.6 Transmission spectra of CuInSe$_2$ thin film: (A) stoichiometric (Cu:In:Se = 25:25:50), and (B) Cu-rich (Cu:In:Se = 42.32:15.1:42.58).

of Cu/In/Se at two different temperatures of 450 and 350 °C, which had Cu:In:Se compositions of 21.9:25.8:52.3 and 24.6:24.9:50.5, respectively (Figure 5.7). On contrast, the latter shows slightly higher tail at below the fundamental absorption region due to excess Cu in the layers. Keeping Se composition of 47.6–48.7% with respect to metal composition of (Cu+ In), the band gap of CIS increases from 1.015, 1.031, 1.043, 1.086, 1.088 to 1.087 eV with decreasing Cu/In ratio from 0.57, 0.53, 0.49, 0.48, 0.44 to 0.43, respectively [24]. After annealing the sphalerite structure films under vacuum at 450 °C for 30 min, the band gap slightly increases and structure changes to chalcopyrite. The CIS films with Cu/In $\geq$ 1.0 show $\rho < 1.0$ Ω-cm whereas the films with Cu/In < 1.0 had $\rho > 10$ Ω-cm [25].

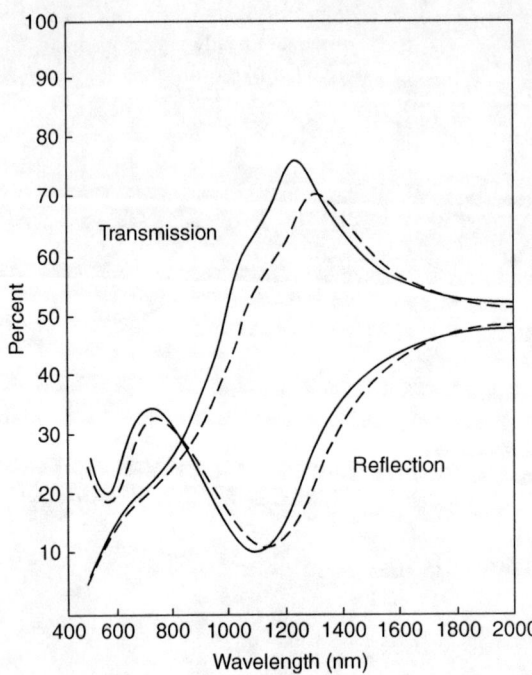

Figure 5.7 Transmission and reflectance spectra of $CuInSe_2$ thin films deposited at: (A) 450 °C, Cu:In:Se = 21.9:25.8:52.3 (solid line) and (B) 350 °C, Cu:In:Se = 24.6:24.9:50.5 (dashed line).

As far as band gap concerned, the Cu-rich $CuInSe_2$ thin films show lower band gap than that of Cu-poor thin films. The Cu 3d levels dictate the valence band in the band structure (Figure 5.1) that means an increase in valence band position takes place to upwards if the Cu is doped in the films but not in the conduction band. Therefore, the optical band gap between valence band and conduction band decreases with increasing Cu or decreasing In content in the compound. The phase change occurs with increasing In in the system such as ordered vacancy compounds (OVC) those have higher band gaps. Bougnot et al., [26] observed that the band gap of spray deposited $CuInSe_2$ varies from 0.9 to 1.03 eV with changing Se atomic ratio with respect to unity of Cu/In in the starting chemical spray solution from 2.5 to 2.8. In another occasion, the band gap increases from 0.97 to 1.12 eV with decreasing Cu/In ratio from 1.1 to 0.8 by keeping Se to Cu/In ratio at 2.5 in the chemical solution. The band gaps of 0.9 and 1.05 eV are reported for the spray deposited $CuInSe_2$ thin films in the literature [27,28]. The band gap of p-$CuInSe_2$ layers deposited from 5% excess Se bulk $CuInSe_2$ decreases from 1.08 to 1.025 eV after annealing layers under vacuum at 300° C due to loss of Se [29]. The $CuInSe_2$ thin films synthesized by two-stage selenization with composition of Cu:In:Se = 24.8:24.9:50.3 reveal band gaps of ∼1.019, 1.044, and 1.050 eV at RT, 78, and 4.2 K, respectively and its absorption coefficient is 1.5×10^5 cm^{-1} at 78 K [30]. The band gap or band edge emission obtained from photoacoustic spectra for $CuInSe_2$ with Cu/In ratio of 1.79 grown onto GaAs (001) follows the Varshini formula,

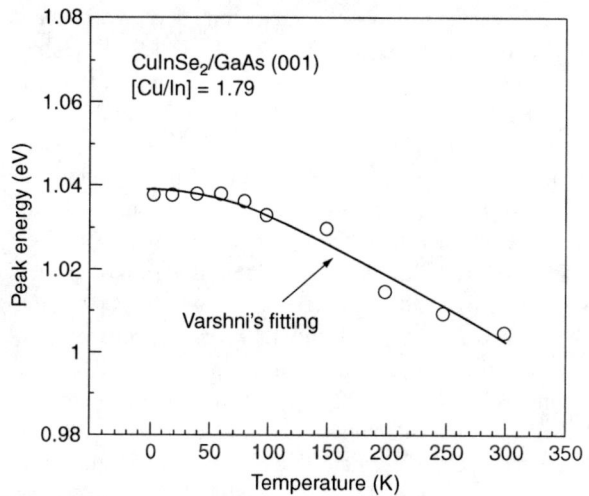

Figure 5.8 Variation of band gap or band edge of CuInSe$_2$ versus temperature.

$$E_g = E_g(0) - \alpha T^2/(\beta + T), \quad (5.5)$$

where $E_g(0)$ is band gap at 0 K, α and β are 3.6×10^{-4} eV/K and 350 K fitting parameters, respectively as shown in Figure 5.8 [31].

After rapid thermal processing (RTP), the films experience slightly Se loss. All the Cu-rich, stoichiometric, and Cu-poor Cu–In–Se precursor layers rapid thermal processed at 700 °C with ramp up rate of 150 °C/s for 30 s show band gap of 1.0 eV and absorption coefficient of 5×10^4 cm^{-1} at fundamental absorption region. The Cu-rich CuInSe$_2$ films contain secondary optical transition due to Cu$_{2-x}$Se phase, in comparison stoichiometric films show sharp optical absorption transition at fundamental absorption region [32]. In order to confirm the presence of Cu$_{2-\delta}$Se phase in the CIS, the CuInSe$_2$ and Cu$_{2-\delta}$Se layers are separately deposited onto transparent glass substrates by thermal evaporation at substrate temperatures of 350–500 and 400 °C, respectively. The Cu$_{2-\delta}$Se is silverish blue in color and exhibits absorption coefficient of 3×10^4 cm^{-1}. The CuInSe$_2$ with composition of Cu:In:Se = 22.2:25.4:52.4 shows sharp absorption in the fundamental absorption region. In the Cu-rich CuInSe$_2$ layers, the dominant subband gap absorption is observed due to secondary phase of Cu$_{2-\delta}$Se. The absorption nature of CuInSe$_2$ coincides with that of Cu$_{2-\delta}$Se by less magnitude. After NaCN treating the Cu-rich CuInSe$_2$ samples, the Cu:In:Se composition of 32.1:12.5:46.3 comes down to 23.6:26.8:49.6. Difference in compositions of 13.2:0:6.6 between etched and as-grown are close to that of Cu$_{2-\delta}$Se. In the optical absorption spectra, the subband gap disappears and sharp absorption occurs in the NaCN etched sample. The ρ value also increases from 10^{-3}–10^0 to 10^0–10^3 Ω-cm supporting elimination off Cu$_{2-\delta}$Se in the sample [33]. The transmission of CuInSe$_2$ thin films decreases at below the band gap with increasing Cu/In ratio from 0.94, 1.05, 1.3 to 2.15 in the layers due to free carrier absorption of secondary phase Cu$_{2-x}$Se, whereby the absorption and emission rates are under equilibrium condition depending on temperature and frequency, according to Kirchhoff's law [34].

The CuInSe$_2$ thin films grown by electrodeposition at potential of −1.0 V and annealed at 350, 450, and 550 °C exhibit band gaps of 0.92, 0.96, and 1.0 eV, respectively, while the films deposited at different potentials of −0.8, −0.9, and −1.0 V, followed by annealing at 550 °C for 2 h show band gaps of 0.978, 0.98, and 1.0 eV, respectively due to difference in composition [35]. The band gap of CuInSe$_2$ deposited onto ITO at −2.2 V versus SCE is 1.1 or 1.2 eV [36,37], whilst the CuInSe$_2$ thin films deposited on Au coated plastic substrates at potential of −1.5 V versus SCE and pH 1.65, followed by annealing under N$_2$ at 150 °C for 1 h show band gap of 1.18 eV and composition of Cu:In:Se = 25.57:25.1:49.42 [38]. The CuInSe$_2$ thin films deposited at lower potential −0.7 eV shows band gap of 0.85 eV. After annealing, followed by etching the CIS films in KCN exhibit band gaps of 1.01 and 1.10 eV, respectively [39]. The band gap of spray deposited In-rich CuInSe$_2$ thin films is 1.22 eV for 0.54 < In/(Cu + In) < 0.67 but increases from 1.22 to 1.36 eV with increasing In/(Cu + In) ratio from 0.67 to 0.78 eV evidencing formation of likely CuIn$_3$Se$_5$ (OVC) [40]. The transmission spectra of CuInSe$_2$ and (CuIn$_3$Se$_5$ + CuInSe$_2$) thin films are given in Figure 5.9. The optical band gaps of CuInSe$_2$ and CuIn$_3$Se$_5$ (OVC) are observed to be 1.01 and 1.24 eV from spectra, respectively [41]. The band gap increases from 1.05 to 1.32 eV with increasing x in the (Cu$_2$Se)(In$_2$Se$_3$)$_{1+x}$ (OVC) compound that is, CuInSe$_2$ (1.05 eV) for $x=0$, CuIn$_2$Se$_{3.5}$ (1.215 eV) for $x=1$, CuIn$_3$Se$_5$ (1.18 eV) $x=2$, and CuIn$_5$Se$_8$ (1.32 eV) $x=4$.

5.1.3 CuGaSe$_2$

The CuGaSe$_2$ thin films are deposited by vacuum evaporation at substrate temperature of 450 and 490 °C with different compositions either Cu-rich or Cu-poor. The band gaps of Cu-poor and stoichiometric CuGaSe$_2$ are 1.72 and 1.66 eV,

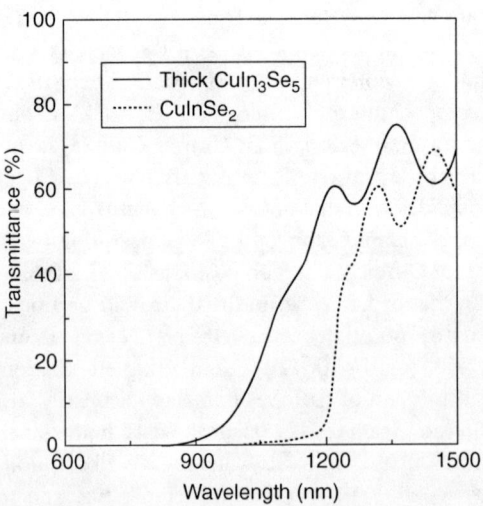

Figure 5.9 Transmission spectra of CuInSe$_2$ and (CuIn$_3$Se$_5$ + CuInSe$_2$) thin films.

respectively. The absorption coefficient of CuGaSe$_2$ is 2×10^5 cm^{-1} at fundamental absorption region of 500 nm. The refractive index (n) of CuGaSe$_2$ is 2.74 at 0.5 eV and 2.97 at 1.53 eV in the single crystals, whereas it is 2.5 at 0.5 eV and 2.73 at 1.53 eV in the thin films. The lower refractive index in the thin films may be due to lower density of thin films [24,42]. The CuGaSe$_2$ layers deposited at substrate temperature of 350–500 °C with Cu/Ga/Se composition of 21.6/26.2/52.2 show sharp absorption in the fundamental absorption region [33]. The transmission spectra of Cu-rich, stoichiometric, and Cu-poor CuGaSe$_2$ thin films are depicted in Figure 5.10. The Ga-rich samples show higher band gap whilst stoichiometric and Cu-rich samples exhibit lower band gaps. At the fundamental absorption region, the optical absorption coefficient is $\sim 10^4$ cm^{-1} [43]. The band gaps of CuGaSe$_2$ thin films are 1.66, 1.68, and 1.69 eV for the compositions of Cu:Ga:Se = 25.2:24.6:50.2, 23.3:25.7:51, and 21.6:26.7:51.7, respectively. The CuGaSe$_2$ thin films used for this investigation are deposited by co-evaporation at substrate temperature of 450 °C and deposition rates of 1.0, 2.6, and 14 Å s^{-1} for Cu, Ga, and Se, respectively and combined growth rate being 8–9 Å s^{-1} [44].

5.1.4 CuInGaSe$_2$

The band gap of CuIn$_{1-x}$Ga$_x$Se$_2$ (CIGS) thin films varies from 1.05 to 1.65 eV with increasing Ga because both CuInSe$_2$ ($x=0$) and CuGaSe$_2$ ($x=1$) are quite soluble solid solutions. The interested band gap of CIGS absorber is 1.5 eV owing to match solar spectrum that can be tailored by varying Ga in the CIGS thin films. In the transmission spectra, the sharp absorption edge at the fundamental absorption region shifts toward lower wavelength with increasing Ga content from 0, 0.1, 0.3, 0.4, 0.69 to 1.0 in the CuIn$_{1-x}$Ga$_x$Se$_2$ films, as shown in Figure 5.11. At the same time, the magnitude of transmittance near the fundamental absorption edge decreases because of decreasing grain sizes due to an increased Ga content and more grain boundaries, which scatter incident light [45]. The similar sharp absorption regions with effect of Ga are found in the transmission spectra for the CuIn$_{1-x}$Ga$_x$Se$_2$ thin films coated onto ITO glass substrates for $x=0$, 0.3, and

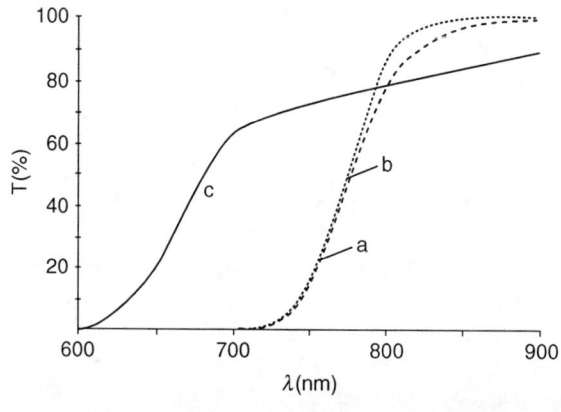

Figure 5.10 Transmission spectra of CuGaSe$_2$ thin films with different compositions; (a, b) stoichiometric, and (c) Cu-poor.

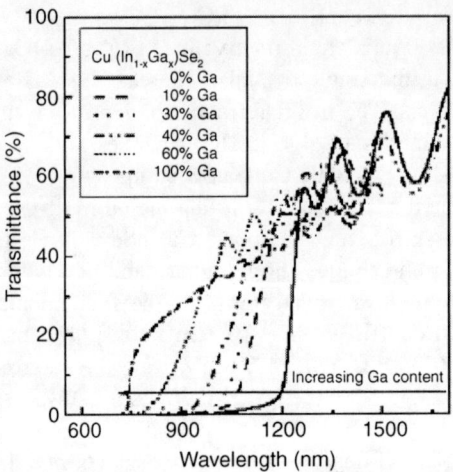

Figure 5.11 Optical transmission spectra of $CuIn_{1-x}Ga_xSe_2$ thin film with different Ga concentrations.

1.0 giving optical band gaps of ∼1.02, 1.18, and 1.63 eV, respectively [46]. There are several reports on the variation of band gap with function of Ga or In composition in the $CuIn_{1-x}Ga_xSe_2$. The $CuIn_{1-x}Ga_xSe_2$ thin films grown by three-stage process exhibit band gaps of 1.0, 1.14, 1.2, 1.4, and 1.6 eV for x values of 0.11, 0.29, 0.46, 0.76, and 1, respectively [47]. The similar trend is observed in the $CuIn_{1-x}Ga_xSe_2$ as $E_g(x) = 1.02 + 0.53x + 0.14x^2$ [48], $E_g(x) = 1.018 + 0.575x + 0.108x^2$ [49], $E_g(x) = 1.006 + 0.473x + 0.161x^2$ [50], and $E_g(x) = 1.018 + 0.575x + 0.108x^2$ eV [51]. The band gap of $Cu(In_{1-x}Ga_x)Se_2$ thin films grown by evaporation also obeys $E_g(x) = 0.998 + 0.291x + 0.43x^2$ eV [52]. In another occasion, the band gap of $CuIn_{1-x}Ga_xSe_2$ thin films grown by evaporation technique follows the relation with x as $E_g(x) = 1.011 + 0.411x + 0.505x^2$ eV. The typical transmission spectra of $CuIn_{1-x}Ga_xSe_2$ thin films for $x = 0.14$ and 0.22 are given in Figure 5.12. A variation of band gap with x from 0 to 0.3 is also given in the inset [53]. The band gaps of $CuIn_{1-x}Ga_xSe_2$ thin films grown by electrodeposition are 1.07, 1.13, and 1.22 eV for Ga/(In + Ga) ratio of 0.16, 0.26, and 0.39, respectively [54]. The precursor Cu $(InGa)Se_2$ thin films with composition of Cu:In:Ga:Se = 24.64:8.65:16.75:49.96 grown by thermal evaporation at room temperature from $Cu(InGa)Se_2$ compound are individually annealed in the presence of Se shots and under vacuum at 500 °C for 1 h denoted as sample-a and sample-b, respectively. The sample-a with composition of Cu:In:Ga:Se = 25.45:8.51:15.43:50.61 and sample-b with composition of 25.58:12.96:14.1:47.36 show band gaps of 1.402 and 1.304 eV, respectively. The samples, which are annealed under vacuum, experience loss of Se and In. Therefore, the band gap is lower in sample-b than that in sample-a [55]. After annealing the layers under maximum Se pressure and inert gas at 300 °C for 2 h, the band gap of $CuIn_{1-x}Ga_xSe_2$ ($x = 0.25$) thin films grown by flash evaporation changes from 1.17 to 1.21 eV [56] and the $CuIn_{0.5}Ga_{0.5}Se_2$ thin films deposited at $T_S = 325$ °C by

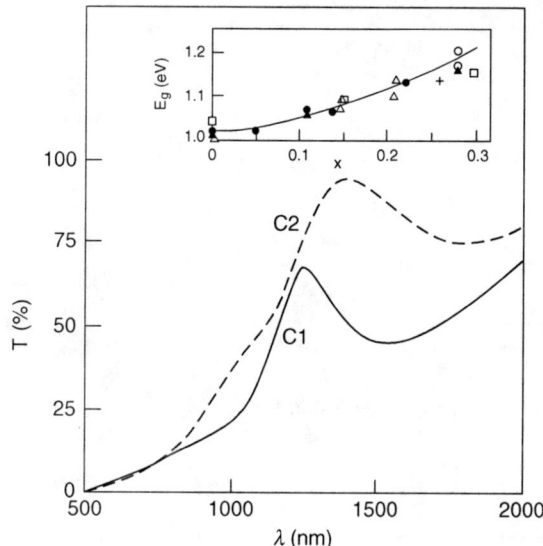

Figure 5.12 The transmission spectra of $CuIn_{1-x}Ga_xSe_2$ thin films for (a) $x=0.14$ (c1) and (b) 0.22 (c2) [53]. (In the inset ●, Δ, +, and □ represent for different samples.)

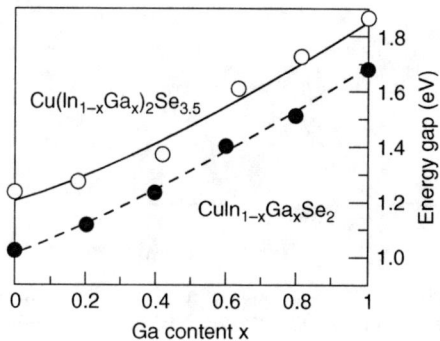

Figure 5.13 The variation of band gap with Ga content in (a) $Cu(In_{1-x}Ga_x)_2Se_{3.5}$ and (b) $Cu(In_{1-x}Ga_x)Se_2$ thin films.

spray pyrolysis show band gap of 1.35 eV [57]. The band gap of the layers varies from 1.07 to 1.68 eV with increasing Ga content from 0.58 to 6.14 at% in the CIGSS layers [58]. The band gaps are slightly higher in the $Cu(In_{1-x}Ga_x)_2Se_{3.5}$ (OVC) than that in the $Cu(In_{1-x}Ga_x)Se_2$ thin films, which sub linearly varies with x, as shown in Figure 5.13. The OVC samples show absorption coefficient of $10^4\,cm^{-1}$ at the fundamental absorption region. The absorption coefficient varies with x in the $Cu(In_{1-x}Ga_x)_2Se_{3.5}$ (OVC), as shown in Figure 5.14 [59]. In summary, the band gap of CIGS single crystal or thin films varies with Ga composition. A change in Se composition makes a shift in band gap of CIGS. The CIGS samples show bowing in band gaps with variation of Ga composition.

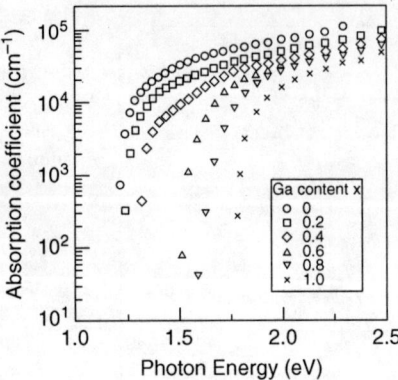

Figure 5.14 Absorption coefficient of $Cu(In_{1-x}Ga_x)_2Se_{3.5}$ thin films.

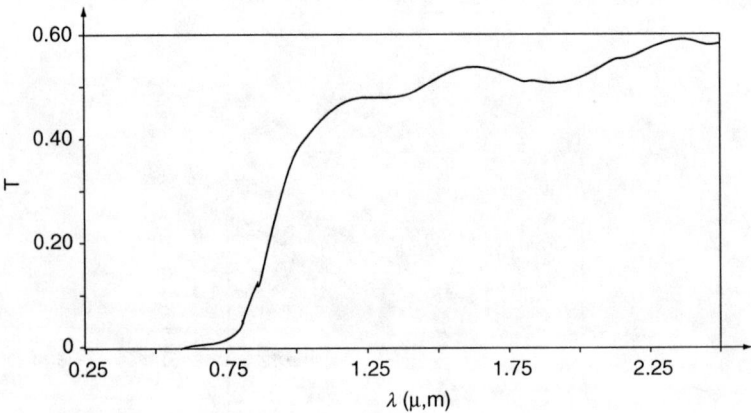

Figure 5.15 Transmission spectrum of $CuInS_2$ thin films deposited at substrate temperature of 300 °C and molar concentration of Cu:In:S = 1:1:3 in the solution by spray pyrolysis.

5.1.5 $CuInS_2$

The $CuInS_2$ is an excellent ideal absorber for thin film solar cells owing to its suitable band gap of 1.53 eV matching with solar spectrum. It's band gap can be tailored from 1.3 to 1.5 eV by varying its intrinsic composition. The transmission spectrum of $CuInS_2$ thin films grown by spray pyrolysis technique is shown in Figure 5.15. The thin films grown at substrate temperature of 300 °C with molar concentration of Cu:In:S = 1:1:3 show band gap of 1.4 eV, which is lower than that of $CuInS_2$ single crystal due to formation of tail state densities at the band edges in particular spray deposited $CuInS_2$ thin films [60]. The thin films grown by sputtering using CuIn alloy and reacting with H_2S gas at substrate temperature of 400 and 500 °C exhibit band gaps of 1.27 and 1.44 eV, respectively. The transmission spectra of films grown at 400 and 500 °C are shown in Figure 5.16 in which the latter shows sharp transition at fundamental absorption region owing to good crystallinity. The films grown at higher temperature had less defects and good crystallinity therefore they show standard band gap [61]. The transmission spectra of $CuInS_2$ thin films grown onto ZnO:Al coated glass substrates by reactive sputtering using Cu_9In_{11} target and H_2S as reactive gas at 200 °C and annealed are shown in Figure 5.17 from which the extracted band gap of 1.49 eV for annealed sample is less than that of ∼ 1.40 eV for as-grown sample [62]. The $CuInS_2$ thin

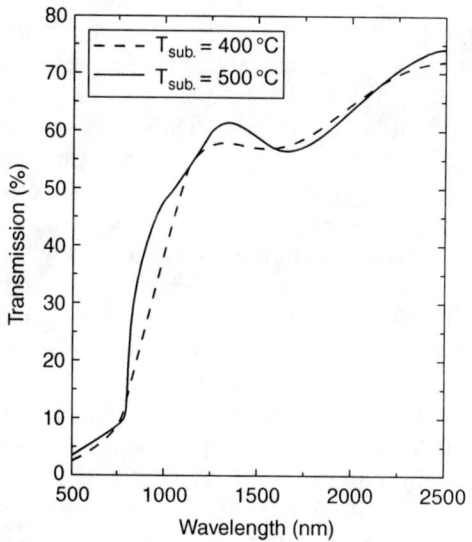

Figure 5.16 Transmission spectra of $CuInS_2$ thin films deposited at substrate temperature of 400 and 500 °C by sputtering.

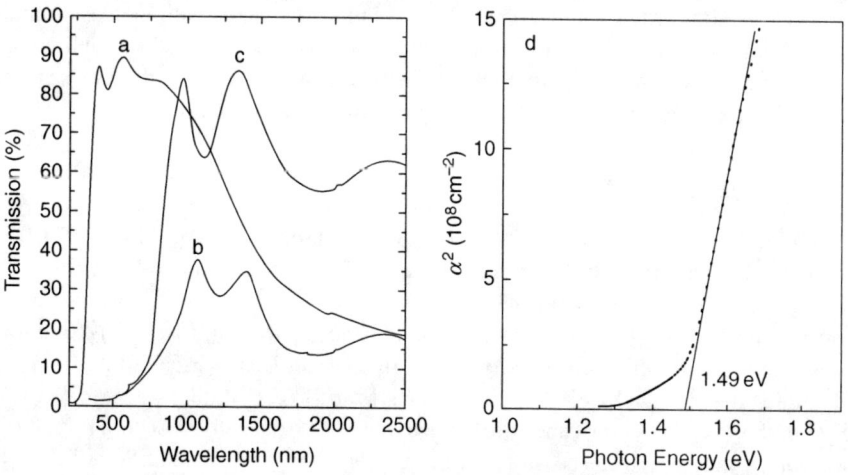

Figure 5.17 Transmission spectra of (A) ZnO, (B) $CuInS_2$/ZnO:Al, and (C) $CuInS_2$/ZnO:Al annealed in vacuum, and (D) α^2 versus $h\nu$ of annealed $CuInS_2$ thin films on ZnO:Al.

films formed by diffusing In into Cu_xS layers grown by CBD shows similar band gap of 1.43 eV [63]. The thin films are grown by single source vacuum evaporation at substrate temperature of 300 °C. After annealing those under sulfur vapor at 200 °C for 8 minutes, the band gap increases from 1.35 to 1.5 eV due to recrystallization, gaining stoichiometric composition, and reduction of grain boundaries [64].

The $CuInS_2$ thin films prepared by sputtering using cold pressed target of $Cu_2S + In_2S_3$ with different ratios of $Cu_2S/In_2S_3 = 4$, 1.5, and 1.0 exhibit band gaps of 1.47, 1.52, and 1.54 eV, respectively and absorption coefficient of $> 10^4$ cm^{-1} at the fundamental absorption region [65]. Similarly the band gaps of 1.48 and 1.53 eV for the Cu/In ratio of 1.3, and 0.99 are found, respectively [66]. The band gap of spray deposited $CuInS_2$ thin films onto SnO_2 covered glass substrates increases from 1.44 to 1.46 eV with increasing deposition substrate temperature from 317 to 377 °C. As the deposition temperature is increased, the Cu content increases and In and S contents decrease in the films. In principle, the losses of In and S are huge at high temperature due to volatile nature of In and S. The oxygen content in the $CuInS_2$ layers also increases because of diffusion of oxygen from the SnO_2 coated glass, which is used as the substrates for the deposition of layers. On the other hand, since spray pyrolysis technique is chemical solution based work, which may also be source for oxygen. The chlorine content is found on far with oxygen in the deposited layers [67]. Unlikely, the band gap of $CuInS_2$ decreases from 1.53 to 1.45 eV with increasing Cu/In ratio from 0.8 to 1.2 in the chemical spray solution due to an increased p-type degeneracy in the sample that means more shallow donors may be suppressed [68]. However, the $CuInS_2$ thin films deposited at $T_S = 380$ °C for Cu/In ratio of 1.2, 1.1, and 1.0 show band gap of 1.45 eV [69]. The band gap varies from 1.31 to 1.35 eV in the spray deposited $CuInS_2$ thin films with varying substrate temperature from 227 to 327 °C in steps of 25 °C but the variation of band gap is not in systematic with function of substrate temperature [70]. The band gap increases from 1.40 to 1.43 eV by incorporation of Ga into $CuInS_2$:Na [71]. The optical absorption coefficient of $CuInS_2$ thin films grown by modulated flux technique increases with increasing thickness of films due to an increase of absorption and light scattering by rough surface into the film, as shown in Figure 5.18. On the otherhand, the band gap decreases from 1.55 to 1.52 eV with increasing thicknesss from 0.2–03 to 0.4 μm [72].

5.1.6 $CuIn(Se_{1-x}S_x)_2$

The spectral analyses are done on $CuIn(Se_{1-x}S_x)_2$ single crystals for $x = 0, 0.3, 0.53, 0.76$, and 1.0 by PL and EER. The band gap of the crystals follows as $E_g(x) = 1.5544x + 1.049(1-x) - bx(1-x)$ eV, where the bowing parameter b is 0.13 and 0.19 for PL and absorption data, respectively [73]. The general form of band gap for $CuIn(Se_{1-x}S_x)_2$ solid solutions linking to electronegativity is written as $E_g(x) = xE_g(CuInS_2) + (1-x)E_g(CuInSe_2) - x(1-x)\Delta y$ eV, where x varies from 0 ($E_g(CuInS_2) = 1.55$ eV) to 1.0 ($E_g(CuInSe_2) = 1.04$ eV) and difference in electronegativities (Δy) between two compounds is 0.04. The simple formula for band gap of $CuIn(Se_{1-x}S_x)_2$ solid solutions can be written as $E_g(x) = 1.04 + 0.47x + 0.04x^2$ eV [74]. The band gap relation for $CuIn(Se_{1-x}S_x)_2$ single crystals follows as $E_g(x) = 1.028 + 0.15x + 0.14x$

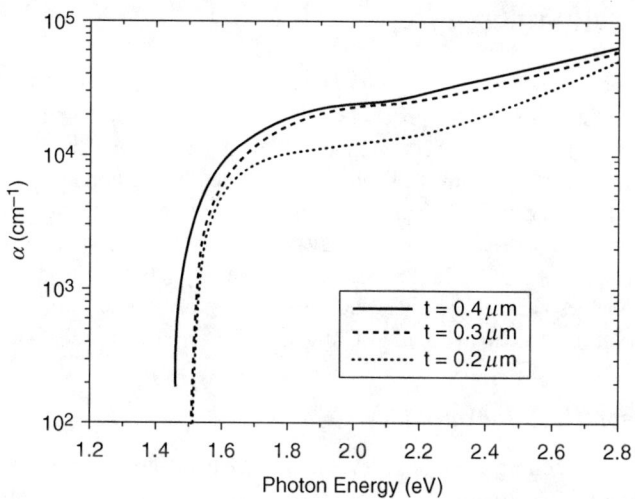

Figure 5.18 Absorption coefficients of CuInS$_2$ thin films with function of photon energy for different thicknesses.

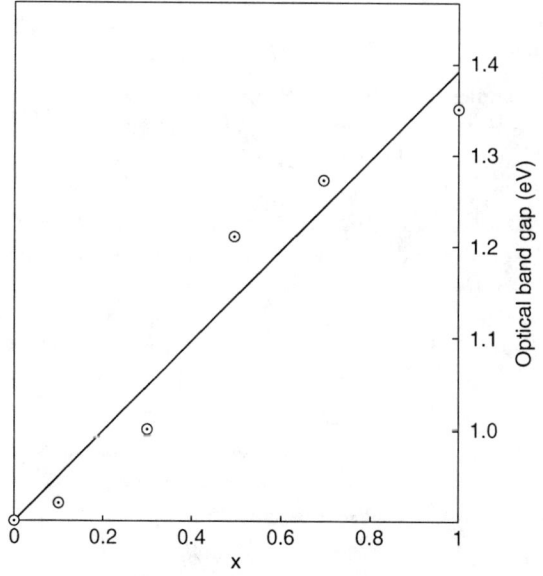

Figure 5.19 Variation of optical band for CuIn(Se$_{1-x}$S$_x$)$_2$ thin films grown by spray pyrolysis technique.

$(x-1)$ eV obtained from the EER spectrum [75]. The band gap of CuIn(Se$_{1-x}$S$_x$)$_2$ thin films grown by spray pyrolysis techniques varies linearly or follows Vegard's law, as shown in Figure 5.19 [76,77]. The similar Vegard's law is observed for the CuIn(Se$_{1-x}$S$_x$)$_2$ system [78].

5.2 Photoluminescence

The photoluminescence is virtually an optical process tool to assess the quality of the material as well as to determine probable defect states. When high energy photons ($h\nu$) greater than the band gap (E_g) of semiconductor hit on it, the optical transition takes place by exciting the electrons from the donor defect states or the conduction band, which recombine with the acceptors available from the acceptor defect states or the valence band by emitting the photons with the energy that is equal to the transition depth between donor and acceptor states in the forbidden gap. We pay attention to discuss about the defect states in the solar cell absorbers such as $CuInSe_2$, $CuGaSe_2$, $CuInS_2$, and their possible solid solutions $Cu(In_{1-x}Ga_x)Se_2$, $CuIn(Se_{1-x}S_x)_2$, and solar cells.

5.2.1 Theoretical Background

The thermal quenching analysis reveals that when the sample temperature is decreased from room to low temperature or vice versa in the photoluminescence experiment, the thermal activation energy of the defect state for the emission peak can be related [79]

$$\frac{I_T}{I_o} = \frac{1}{[1 + C\exp(-\Delta E/k_B T)]}, \tag{5.6}$$

where I_T is the intensity of photoluminescence peak at temperature T, I_o the luminescence intensity at $T = 0$ K, k_B the Boltzmann constant, C a constant or the capture cross section of electron or hole, and ΔE the thermal activation energy of the donor and acceptor.

In the case of participation of donor–acceptor pair (DAP) transition in the semiconductor, the activation energy levels of the donor and acceptor can be obtained by fitting the excitation emission energy ($h\nu$) as [80]

$$h\nu = E_g - (E_D + E_A) + e^2/4\pi r \varepsilon_0 \varepsilon_r, \tag{5.7}$$

where E_g is the energy band gap, E_D and E_A are the donor and acceptor ionization energies, respectively. The last term is related to the Coulombic interaction between the pair; r is the distance between the donor and acceptor involved in the transition, ε_r is the dielectric constant of the material which can be set to $\alpha n_e^{1/3}$, where α is a constant with a value of nearly 2.1×10^{-8} eV-cm and n_e is the carrier concentration of the sample.

The transition from donor to valence band or acceptor to conduction band can be related to [81]

$$h\nu = E_g - E_{D/A} + n_u k_B T, \tag{5.8}$$

where $E_{D/A}$ represents for the activation energy of either donor (D) or acceptor (A) and the remaining functions are already known. Some times in the last term, n_u is either equal to unity or less or higher than unity. In other words, the free to bound transition (FB) can be defined that a free electron recombines with a hole bound to an acceptor that is, FB or a free hole recombine with a donor bound to an electron (DF) in the semiconductor. The intensity of emerging peak can be defined as [82,83]

$$I_{PL}(hv) \propto \{hv - (E_g - E_{A/D})\}^{1/2} \exp\left[\frac{-\{hv - (E_g - E_{A/D})\}}{k_B T}\right], \quad (5.9)$$

where E_g is the band gap, $E_{A/D}$ either acceptor or donor ionization energy level, hv the incident radiation. The FB peak position does not change but its intensity changes proportionally with variation of intensity of incident radiation. On the other hand, the intensity of FB peak is proportional to number of defect states. In Equation (5.7), if the defect level contribution is nil ($E_{A/D}=0$) then Equation (5.9) can be considered to the band to band transition.

The exciton (E_{ex}) is literally generated by the Coulombic interaction between electron and hole [82]. The binding energy of the exciton (E_{ex-be}) can be determined from the relation [84]

$$E_{ex-be} = \frac{m^* e^4}{2(4\pi\varepsilon_r\varepsilon_0 \hbar n_x)^2}, \quad (5.10)$$

where m^* represents the reduced mass as $m^* = m_e m_h/(m_e + m_h)$, m_e and m_h are the electron and hole effective masses, respectively. $\hbar = h/2\pi$, h is the plank constant and n_x the exciton state, which being used as unity for the ground state. If there are multiple excitonic states; first one (E_{x1}) is at ground state ($n_x = 1$), and second one (E_{x2}) at excited state ($n_x = 2$). The relation for the binding energy of the exciton can be written as [85]

$$E_{ex-be} = \frac{4}{3}(E_{ex2} - E_{ex1}). \quad (5.11)$$

The band gap (E_g) of semiconductor can be obtained by adding the binding energy of the exciton (E_{ex-be}) to the excitonic emission energy (E_{ee}) that is, $E_g = E_{ex-be} + E_{ee}$.

5.2.2 Photoluminescence Experimental Setup

A typical photoluminescence experimental setup is depicted in Figure 5.20 [86]. When the laser beam hits surface of the sample, the emitted photons from the sample due to recombination of electron and holes either from defect to defect or valence band to conduction band *etc.*, will be dispersed by the monochromator and detected by the detector. The chopper placed between the sample and laser

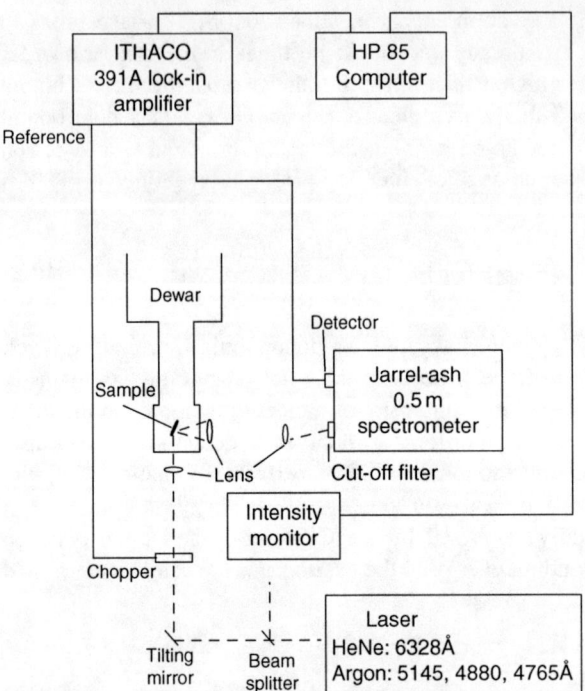

Figure 5.20 Photoluminescence experimental setup.

excitation source is used as a reference signal. The detector feeds the signal to the lock-in-amplifier, which compares the signal with the reference. Finally, the PL signal is recorded with a variation of wavelength. The excitation source such as laser beam has to be chosen in such a way the energy should be higher than the band gap of the semiconductors, for example in the case of $CuInSe_2$ ($E_g = 1.043$ eV), the green laser that is, Ar^+ laser with wavelength of 514.5 nm (2.4 eV) or Kr^+ laser with wavelength of 676.5 nm (1.83 eV) are suitable. In some cases, a 25 mW solid-state laser with a wavelength of 760 nm is also used. The He–Ni laser wavelength of 315 nm (3.94 eV) is right one for higher band gap materials such as GaN (3.2 eV) and ZnSe (3.6 eV). The excitation of He–Ne laser with wavelength of 638 nm, 50 cm monochromator, and Si photodiode, Si–Ge, PbS, InGaAs, *etc.*, as detectors are also employed [87].

5.2.3 $CuInSe_2$

There are 12 possible defect levels identified in the formation of enthalpy energy range between 1.4 and 22.4 eV for $CuInSe_2$. In general, the formation of intrinsic defect levels either in single crystals or thin films depends on the nature of stoichiometry (Cu/In = 1), nonstoichiometry (Cu/In > 1 or Cu/In < 1), and on the recipes of growth of $CuInSe_2$. However, surprisingly, theoretical calculations reveal that V_{Cu} is the most common defect in the $CuInSe_2$ irrespective of Cu to In ratio [88]. The defects, which have higher formation of enthalpies, may have least generation in

Table 5.1 Formation and activation energies of defect levels for CuInSe$_2$ [89–91]

S. No	Intrinsic defect	Formation energy (eV)	Electrical activity	Activation energies (meV)
1	In$_{Cu}$	1.4	Donor	35–45, 10–20
2	Cu$_{In}$	1.5	Acceptor	30, 40–60
3	V$_{Se}$	2.4	Donor/Acceptor	10, 60–80
4	V$_{Cu}$	2.6	Acceptor	40 or 85
5	V$_{In}$	2.8	Acceptor	33, 80, 90, 230, 300–330
6	Cu$_i$	4.4	Donor	21, 55
7	In$_{Se}$	5.0	Donor	–
8	Se$_{In}$	5.5	Acceptor	–
9	Cu$_{Se}$	7.5	Acceptor	230
10	Se$_{Cu}$	7.5	Donor	–
11	In$_i$	9.1	Donor	53, 80
12	Se$_i$	22.4	Acceptor	89–130, 153

the CuInSe$_2$. The enthalpies and activation energies of defect levels are given in Table 5.1. The proper assignment of defect levels can be done with help of experimental results, however some of them are eventually speculation for which there has been continuously debating in the scientific community. If a piece of new evidence is available, the course of assignment of defect level may be changed or confirmed.

1.5 K PL spectrum of p-CuInSe$_2$ single crystals grown by horizontal freezing method exhibits several luminescence peaks, as shown in Figure 5.21A. The emission peaks at 1.0416 and 1.0446 eV so-called A and B free excitons are observed, respectively [92]. There are several fascinating reports on excitons about their values such as $A = 1.0416$ eV and $B = 1.0447$ eV in the p-type CuInSe$_2$ single crystals [91], $A = 1.0424$ eV and $B = 1.0496$ eV in the CuInSe$_2$ nanorods (Figure 5.22) [93], $A = 1.0409$ eV and $B = 1.0444$ eV in the CuInSe$_2$ thin films [30] as well as $A = 1.0398$ eV and $B = 1.0432$ eV in the Cu-rich polycrystalline thin films [94], as shown in Table 5.2. There is a tiny variation in A and B excitons by about 0.3–0.4 meV from film to film, which may be considerable due to variation in composition or stress in the film, etc., 4.2 K PL spectrum of CuInSe$_2$ single crystals at near the band edge emission shows A and B free excitons at 1.0418 and 1.0449 eV for the composition ratio of Cu/In = 0.93, respectively, whereas they are at lower energy levels for the Cu/In ratio of 0.88 or 0.89 because of an increase in the crystal field split that is, $\Delta_{cf} = 3/(2\Delta_{AB})$, where Δ_{AB} is difference between A and B [95]. On the other hand, the full width half maximum (FWHM) of free excitons varies in the range of 0.8–1.0 meV with varying Cu/In ratio. The thermal quenchings of A and B exciton peaks result in binding energies of 17 and 12 meV, respectively closely coinciding with the reported value of 18 meV, as shown in Figure 5.21B [96]. The similar binding energies of 13 and 5 meV for A and B excitons are determined from plots of PL intensity versus inverse temperature for the CuInSe$_2$ nanorods, as shown in inset of Figure 5.22. One can obtain band gap of 1.058 or 1.056 eV for CuInSe$_2$ by adding respective experimental binding energies of excitons to their

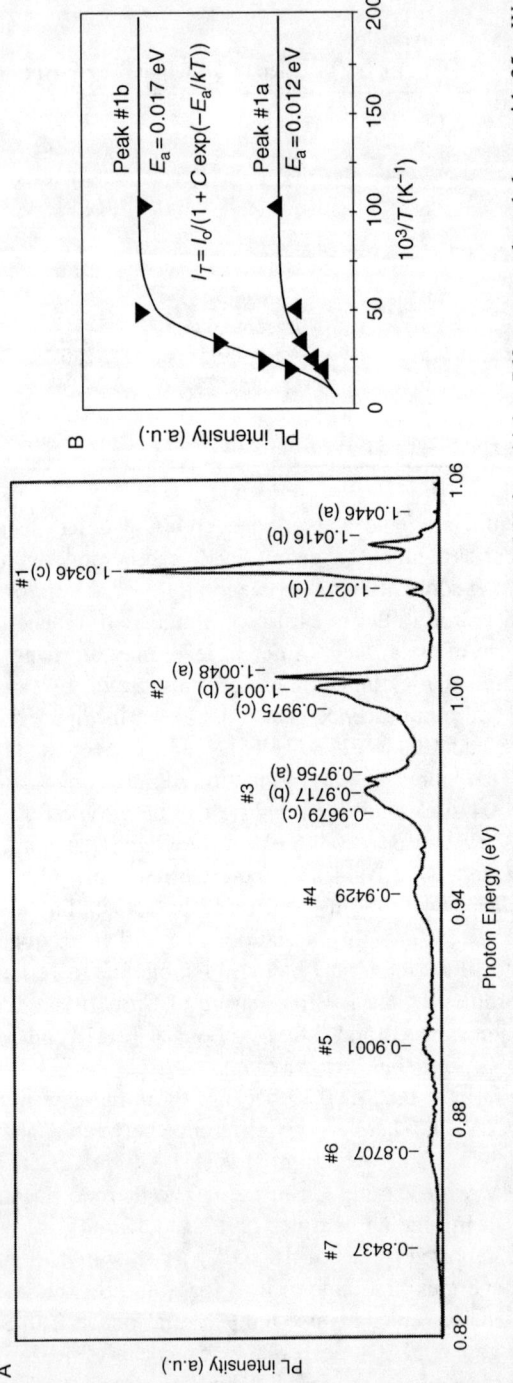

Figure 5.21 (A) Photoluminescence spectrum of CuInSe$_2$ single crystal recorded at 1.5 K for which a 760 nm wavelength laser with 25 mW intensity is used from solid state laser and (B) PL intensity versus inverse temperature for exciton peaks *A* and *B*.

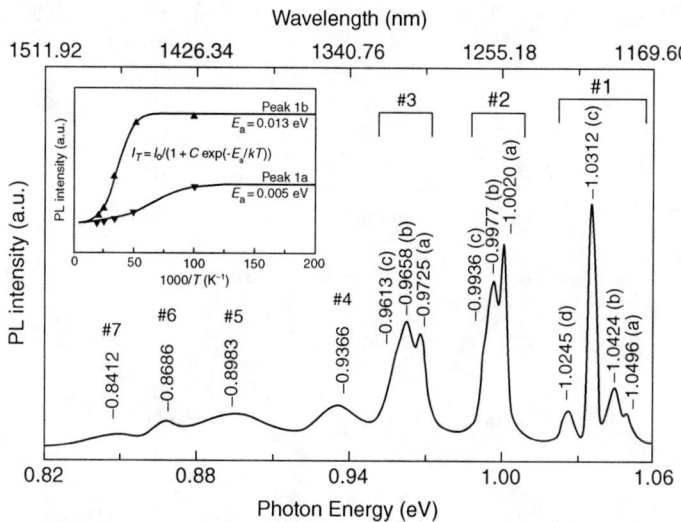

Figure 5.22 Photoluminescence spectrum of CuInSe$_2$ nanorods recorded at 10 K using 760 nm laser.

Table 5.2 Variety of transition energy levels for the CuInSe$_2$

S. No	Single crystal [92]	Nanorods [93]	Thin films [30]	Possible transitions
1	1.0446	1.0496	1.0444	B
2	1.0416	1.0424	1.0409	A
3	1.0346	1.0312	1.0353	BX(VB–In$_{Cu}$/Cu$_i$)
4	1.0048	1.0020	–	BX-LO$_1$
5	0.9975	0.9725	–	BX-LO$_2$
6	1.0277	1.0245	1.0142	BX(VB–Se$_{In}$)
7	1.0012	0.9977	–	(CB–V$_{Cu}$)
8	0.9717	0.9658	0.9716	(CB–Cu$_{In}$)
9	0.9429	0.9366	–	(CB–Cu$_{In}$)-LO$_1$
10	0.9001	0.8983	–	(CB–Se$_i$)
11	0.8707	0.8686	–	(CB–Se$_i$)-LO$_1$
12	0.8437	0.8412	–	(CB–Se$_i$)-LO$_2$
13	–	–	1.0398	A [94]
			1.0432	B
14	–	–	1.0408	A [97]
			1.0446	B

excitonic energies, as shown in Figure 5.23, which matches to the standard E_g value of 1.055 eV [5]. The exciton binding energy calculated from the hydrogenic model using Equation (5.10) is 5.4 meV, where $m_e = 0.09 m_o$, $m_h = 0.73 m_o$, and $\varepsilon_r = 16$. Difference in exciton binding energies between experimental and theoretical or from laboratory to laboratory may be due to variation of effective masses, dielectric

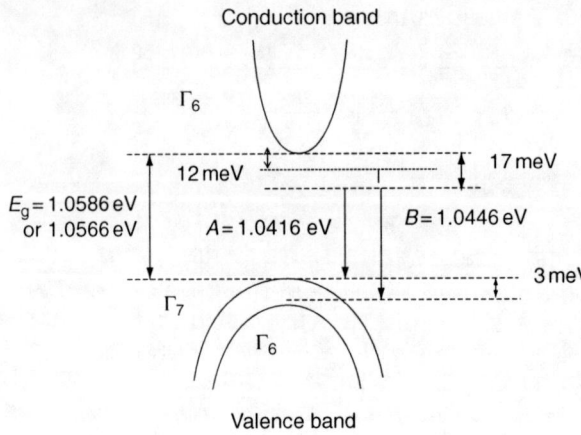

Figure 5.23 Schematic band structure of CuInSe$_2$ at $T = 1.5$ K for A and B excitons by qualitatively.

constants, compositions, other parameters of the CuInSe$_2$ samples, *etc*. The CuInSe$_2$ crystals grown by directional freezing method and annealed under selenium vapor at 792 °C for 200 h exhibit 1.0408 and 1.0446 eV emission lines at 4.2 K corresponding to the ground state ($n_x = 1$) and the first excited state ($n_x = 2$), respectively. The exciton binding energy of 5.1 meV obtained from Equation (5.11) using ground state or first excited state is close to the value obtained from Equation (5.10). The band gap of 1.0459 eV for CuInSe$_2$ at 4.2 K can be estimated by adding binding energy of 5.1 meV to its excitonic ($n_x = 1$) value of 1.0408 eV [97].

The as-grown (110) cleaved facet p-CuInSe$_2$ single crystals with Cu:In:Se = 1:1:1.98 composition show sharp emission at 1.03 eV relating to either free exciton, which is ascribed to transition from Γ^4_{7V} to Γ^1_{6C}, or bound exciton (BX) [98]. The emission peaks at 1.0346 and 1.0277 eV are assigned to BXs and the former one contains LO phonons at 1.0048 and 0.9756 eV with separation energy of nearly 29 meV [92]. Mudryi *et al.*, [30] also found BXs at 1.0353 and 1.0142 eV in the CuInSe$_2$ thin films (Figure 5.24), which may be assigned to (VB–In$_{Cu}$ or Cu$_i$) and (VB–Se$_{In}$), respectively. Table 5.2 shows similar kind of emission lines for the CuInSe$_2$ nanorods [93]. The exciton neutral acceptors $[A_1,X] = 1.0375$ eV, $[A_2,X] = 1.0310$ eV, and neutral donor complex $[D_1,X] = 1.0336$ eV are observed in the etched Cu-rich CuInSe$_2$ thin films, which may be related to (i) free to bound acceptor V$_{In}$ of 24 meV (CB–V$_{In}$), (ii) donor complex of (VB–In$_{Cu}$ or Cu$_i$) or (VB–Se$_{In}$), and (iii) valence band to the donor level Cu$_i$ of 23 meV (VB–Cu$_i$), respectively [94,95]. In Figure 5.21A, the luminescence peak at 1.0012 eV can be assigned to the transition from the conduction band (CB) to acceptor V$_{Cu}$ with activation energy level of 40–50 meV that is FB and the peak at 0.9717 eV is assigned to (CB–Cu$_{In}$) that is also FB and its phonon replica exists at 0.9429 eV. Taking energy gap of 1.058 eV for CuInSe$_2$ the activation energy levels of V$_{Cu}$ and Cu$_{In}$ are 57 and 86 meV for FB transitions of 1.0012 and 0.9717 eV, respectively. The peak at 0.9001 eV is related to FB (CB–Se$_i$) with defect energy level of 138–158 meV containing replicas at 0.8707 and 0.8437 eV with separation energy of ~ 28 meV. The emission line at

Optical Properties of I–III–VI$_2$ Compounds

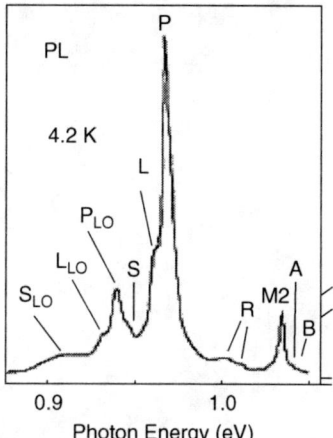

Figure 5.24 PL spectrum of CuInSe$_2$ thin films recorded at 4.2 K; the emission bands exciton A 1.0409 eV, exciton B 1.0444 eV, bound exciton (VB–In$_{Cu}$ or Cu$_i$) M$_2$ 1.0353, bound exciton (VB–Se$_{In}$) R 1.0412, (CB–Cu$_{In}$) P 0.9716, (CB–Se$_{In}$ or V$_{In}$) 0.964 L, and (V$_{Cu}$–V$_{Se}$) S 0.947 eV.

(0.9001 eV) is less effective with variation of excitation intensity unlike DAP, therefore, it can be treated as a FB line [92]. Figure 5.22 shows similar kind of PL spectrum that the peaks at 0.9977 and 0.9658 eV are corresponded to (CB–V$_{Cu}$) and (CB–Cu$_{In}$) that is, (e, A) transitions in the CuInSe$_2$ nanorods, respectively (Table 5.2). The phonon replica of latter is located at 0.9366 eV. The FB (CB–Se$_i$) occurs at 0.8983 eV with activation energy of 156.7 meV and its phonon replicas are resided at 0.8686 and 0.8412 eV [93].

Figure 5.25 shows FB transition that is, (Cu$_{In}$–CB or e, A) at 0.975 eV and several phonon replicas with distribution energy of ~28 meV in the p-type CuInSe$_2$ single crystals, whereas the emission line at 1.00 eV for Cd doped n-type CuInSe$_2$ single crystals is observed in the spectrum [99]. The peaks at 1.0 and 0.975 eV due to transitions of (V$_{Cu}$–CB) and (Cu$_{In}$–CB), respectively, are also found in the cathodeluminescence (CL) spectrum for CuInSe$_2$ single crystals grown by iodine transport method (ITM) recorded at 77 K, as shown in Figure 5.26 [100]. The thermal quenching experimental analysis reveals activation energies of 60 meV (V$_{Cu}$) and 75 meV (Cu$_{In}$) for 1.0 eV and 0.975 eV peaks, respectively, which are close to the calculated values from Equation (5.8). The other emission lines at 1.040 and 1.020 eV due to free exciton and BX (V$_{Cu}$–CB) transitions are observed, respectively. The peaks at 0.94 and 0.91 eV are assigned to DAP transition (V$_{Cu}$–V$_{Se}$) and (CB-Se$_i$), respectively. The latter with activation energy of 135 meV contains phonon lines. The DAP level at 0.95 eV and its phonon replica at 0.925 eV are observed in the CuInSe$_2$ single crystals grown by IT method. The thermal quenching energies 25 and 45 meV of In$_{Cu}$ and V$_{Cu}$ defect levels for 0.985 eV peak are determined in the CuInSe$_2$, respectively [101]. In the case of p-CuInSe$_2$ thin films, the FB transition line occurs at 0.969 eV that is, (Cu$_{In}$-CB) along with LO phonon lines at 0.941

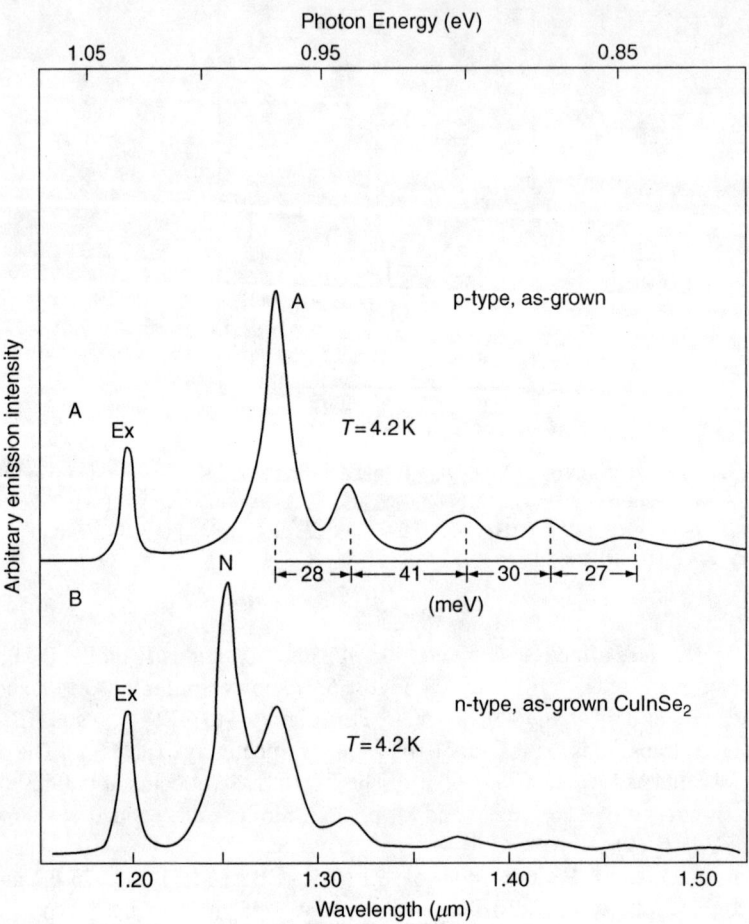

Figure 5.25 PL spectra of (A) *p*- and (B) *n*-type CuInSe$_2$ single crystals recorded at 4.2 K.

and 0.913 eV having dispersion energy of approximately 28 meV. The activation energy of 80 meV is obtained from thermal quenching for 0.969 eV peak closely coinciding with the estimated value of 75 meV from Equation (5.8) ($hv = E_g - E_A + 1/2k_BT$, $E_g = 1.049$ eV). No shift in the 0.969 eV peak position with increasing excitation energy from 0.1 to 10 W/cm^2 is observed. However, a blue shift of 7–9 meV is noticed with increasing temperature from 4.2 to 78 K that is proportional to k_BT [102]. This peak is strongly confirmed to be FB transition from the experimental observation. The same activation energy of 80 meV for FB transition and its phonon line at 0.943 eV are also reported [96]. In most of the cases, 0.943 eV line is DAP, otherwise, it leads as a phonon line to FB transition. Only thermal quenching experimental analysis can confirm whether it is DAP or phonon. In typical case, 4.2 K PL spectrum exhibits only one sharp peak at 0.96 eV for the

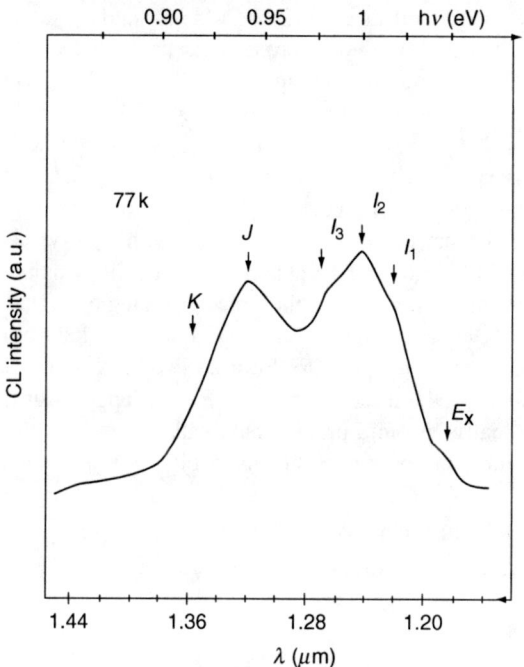

Figure 5.26 Cathodoluminescence spectrum of CuInSe$_2$ crystals, where I_1-1.020, I_2-1.0, I_3-0.975, J-0.94, K-0.91, and E_x-1.040 eV.

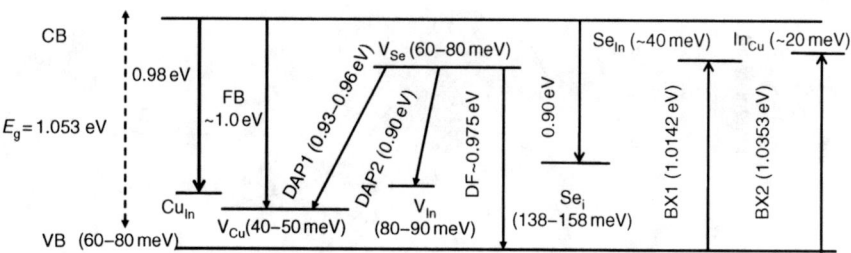

Figure 5.27 Transition levels in the forbidden gap of CuInSe$_2$ (model-1).

p-CuInSe$_2$ single crystals with Cu/In = 1 grown by horizontal Bridgman method, which is a FB transition, whereas two broad peaks at 0.85 and 0.93 eV that is, DAPs are observed in the n-CuInSe$_2$ with Cu/In = 0.76. The former is due to (V$_{Cu}$–V$_{Se}$) in which V$_{Se}$ may be acted as a double donor with activation energy of ~160 meV and the latter being (V$_{Cu}$–In$_{Cu}$), respectively [103]. A schematic diagram of all the possible transition lines in the CuInSe$_2$ is given in Figure 5.27. Obviously, the transition values of (V$_{Cu}$–V$_{Se}$), (Cu$_{In}$–V$_{Se}$), and (V$_{In}$–V$_{Se}$) are very close to each other. The transition from donor to valence band (V$_{Se}$–VB) may be possible in the place of (Cu$_{In}$–CB) transition, since more or less both of Cu$_{In}$ and V$_{Se}$ defect levels having same amount of activation energies.

Lange et al., [98] observed that the natural facet (112) n-CuInSe$_2$ single crystals with composition of Cu:In:Se = 0.99:1:2.08 present a broad band peak at 0.94 eV along with a secondary peak at 0.9 eV, which are due to (V_{Cu}–V_{Se}) and (V_{In}–V_{Se}) transitions, respectively. The possible activation energy levels of 40–50 and 60–70 meV for acceptor (V_{Cu}) and donors (V_{Se}), respectively are due to ~0.94 eV DAP (V_{Cu}–V_{Se}) transition. The grown crystal contains two varieties of faces such as (112) facet n-CuInSe$_2$ on the surface and (110) facet in cleaved or bulk. Variation in the composition between cleaved facet (110) and (112) facet CuInSe$_2$ crystals is due to exchange of indium selenide and Se gas phases while cooling melted charge from high temperature to room temperature in the process of single crystals [98]. The DAP emission band at 0.95 eV in the p-CuInSe$_2$ crystals shifts from ~0.95 to 0.962 eV with increasing sample temperature from 20 to 110 K while at 20 K, which is a dominant emission [104]. In addition, the FB peak appears as a shoulder at ~0.974 eV, which gradually becomes predominant with increasing temperature due to having sufficient thermal energy, as shown in Figure 5.28. In the

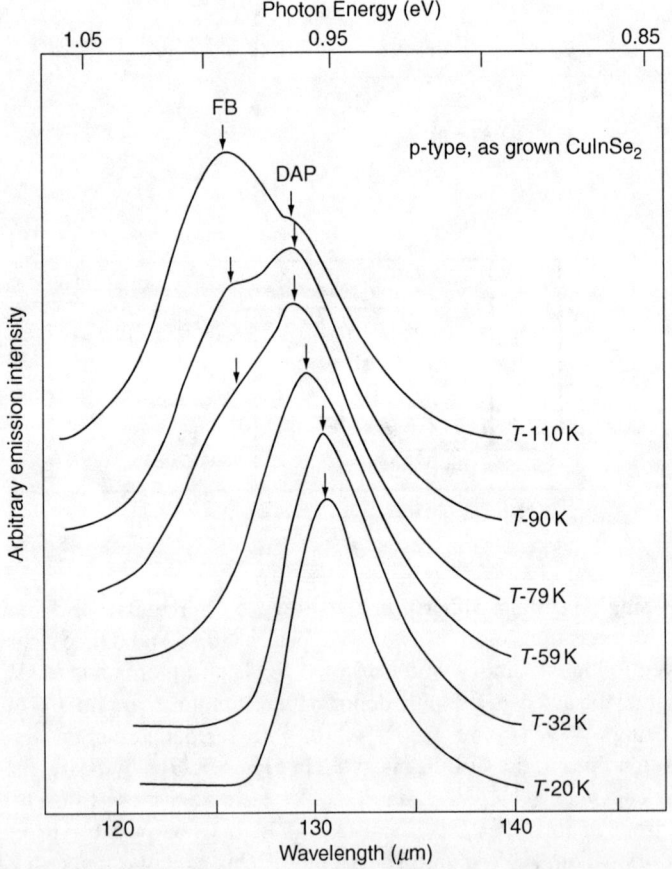

Figure 5.28 PL spectra of p-CuinSe$_2$ single crystals at different temperatures.

CuInSe$_2$ thin films, the similar blue shift at the rate of 3 meV/decade with varying excitation intensity is observed for broad 0.94–0.96 eV DAP (V$_{Cu}$–V$_{Se}$) emission band [102]. The DAP is asymmetric peak, if it is a combination of multiple radiative transitions such as DAP, FB, *etc*. The Columbic interaction energy of DAP increases and an average distance between donor and acceptor decreases that causes to blue shift that is, towards higher energy in the DAP transition with increasing excitation intensity or temperature [105].

The melt-grown CuInSe$_2$ crystals with excess Se or annealed under maximum Se pressure results in *p*-type conductivity and a peak at 1.00 eV in 77 K PL spectrum that is referred as a type-*x*, whereas crystals with excess In or annealed under minimum Se exhibits *n*-type conductivity and a peak at 0.93 eV that is denoted as a sample type-*y*. The peak at 1.00 eV is due to FB transition from the conduction band to the acceptor level of V$_{Cu}$ with activation energy of 40 meV that is, (V$_{Cu}$–CB) in the *p*-type crystals. The peak at 0.93 eV is due to DAP transition that is, from donor to acceptor level (V$_{Cu}$–V$_{Se}$), as shown in Figure 5.29. The type-*y* transition can be converted into type-*x* transition by annealing the *n*-type crystals under maximum Se pressure indicating that the V$_{Se}$ donor levels are suppressed. Either the *p*-type layer is formed onto *n*-type substrate by annealing the *n*-substrate under maximum Se

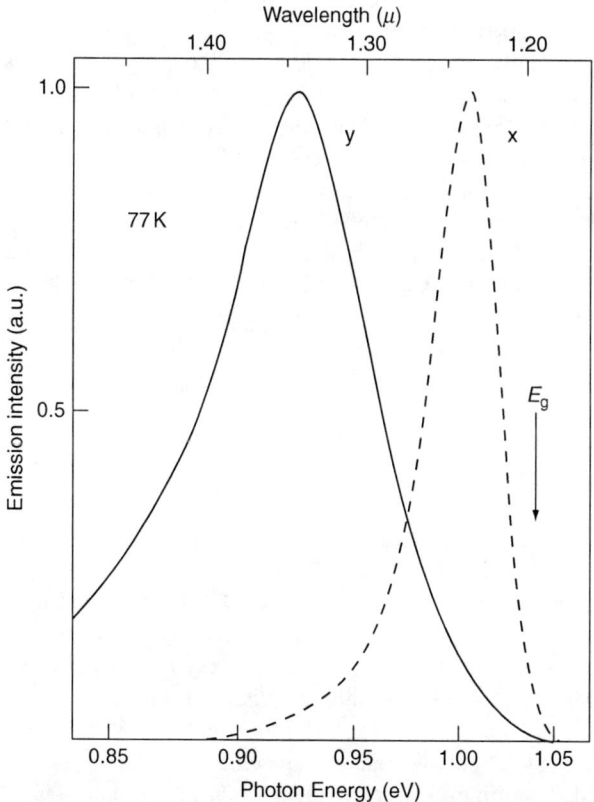

Figure 5.29 (a) PL spectra of CuInSe$_2$ crystals grown with excess Se and annealed in maximum Se vapor at 400–700 °C (*p*-type, type-*x*) and (b) CuInSe$_2$ crystals annealed under minimum Se vapor pressure (*n*-type, type-*y*).

pressure at 500–600 °C for 1–2 min or n-type layer can be formed onto p-type substrate annealing under minimum Se pressure. After annealing at 600 °C under maxium Se pressure the surface of n-type CuInSe$_2$ single crystal gets convert into p-type. The peak at 0.875 eV dominates 0.97 eV peak in the PL spectra at 2 K for the p-type layer on n-type substrate. Although the peak at 0.875 eV blue shifts to 0.93 with increasing the sample temperature from 2 to 77 K, the peak at 0.97 eV resides at the same place irrespective of the temperature. Both of them are more or less at the same intensity. As mentioned earlier, the peaks at 0.93 and 0.97 eV are due to DAP recombination and FB transitions, respectively. The Cd doped samples exhibit similar emission band of type-y such as 0.93 eV emission line, whereas the emission band is red shifted to 0.85 eV for Zn doped samples in 77 K PL spectra. The Cd and Zn create more donor levels in the samples [106].

To grow p- and n-type crystals, slightly an excess selenium and Cd are mixed in the CuInSe$_2$ charge, respectively. The grown crystals are p-type, if both Cd and slightly an excess Se are used. The virgin p-type single crystals show emission band at $\sim$0.945 eV in 4.2 K PL spectrum, which is DAP (V$_{Cu}$–V$_{Se}$) transition. The DAP peak shifts to lower energy of 0.929 eV by 16 meV after annealing the crystals under vacuum at 350 °C for 30 min. The broad emission band further shifts to lower energy of 0.9 eV for Cd doped CuInSe$_2$ crystals, which can also be assigned to DAP (Cd$_{Cu}$–V$_{Se}$). The Cd sits on Cu site or replaces Cu vacancy (V$_{Cu}$) that becoming as antisite (Cd$_{Cu}$) having a defect energy level of 80 meV. 4.2 K PL spectra of stoichiometric p-CuInSe$_2$ crystals show excitonic emission at 1.037 eV, while it is 1.042 eV at 77 K. A small difference of 5 meV between the emission bands with variation of temperature from 77 to 4.2 K is found due to shrinkage in the band gap. The p-type crystals have been annealed under vacuum at different temperatures that is, 330, 400, and 470 °C for 30 min. As the annealing temperature is increased, the exciton peak and other kind of emissions gradually disappear and a broad emission band emerges at 0.925 eV. After annealing under vacuum at 350 °C for 30 min, the Cd implanted p-type CuInSe$_2$ crystals get convert into n-type. On the other hand, all the peaks disappear but a 0.92 eV peak generates. The broad DAP 0.925 eV peak gradually becomes broader and shifts to 0.90 eV with increasing temperature in the PL spectra. The E_A (Cd$_{Cu}$) and E_D (V$_{Se}$) levels for DAP (Cd$_{Cu}$–V$_{Se}$) have activation energies of 85 and 65 meV, respectively [99].

In-rich n-CuInSe$_2$ crystals are annealed under 90%Ar + 10%H$_2$Se at 600 °C for 4 h. The Cu/In ratio changes from 0.92 to 0.87 eV, n_e decreases from 5×10^{17} to 1×10^{16} cm^{-3} and intensities of 0.979, 0.914, and 0.895 eV emission lines gradually decrease with increasing annealing time from 0.5 to 4 h in steps of 0.5 h. Finally, the intensity of 0.979 eV stabilizes at low and 0.914 and 0.895 eV lines are almost suppressed [107]. A decrease in intensities of the peaks such as 0.975, 0.94, and 0.91 eV due to annealing the samples under Se vapor supports that the defects could be related to selenium vacancies (V$_{Se}$). After annealing CuInSe$_2$/CuInS$_2$ crystals under maximum selenium/sulfur at 600 °C for several days, all the emission bands such as 1.0, 0.975 eV FB transition, 0.94 eV donor to acceptor transition and 0.91 eV disappear except 1.020 eV band line in the CL spectra. Again all the emission bands appear if the crystals are annealed under minimum sulfur/selenium vapor pressure at 600 °C for few minutes. This experimental observation literally drives to conclude that 0.975, 0.94, and 0.91 eV peaks could be due to S/Se vacancies [108]. The p-CuInSe$_2$ crystals are grown by three temperature zone horizontal Bridgman method employing different Se vapor pressures. As the PL spectra are recorded at 10 K using excitation energy of 2.41 eV, the DAP peak shifts from 0.922, 0.934, 0.936 to 0.938 eV with increasing Se pressure from 10, 25, 50 to 200 torr, respectively. Due to an increase of Se vapor pressure, the density of V_{Se} defect states decreases. Therefore, DAP position shifts from 0.922 to 0.938 eV that is so-called screening effect [109]. A 0.92 eV peak in the CuInSe$_2$ powder shifts slightly to higher energy after annealing under different Se atmospheres as well as shifts to slightly higher energy with varying excitation intensity from 6 to 160 Wcm^{-2} indicating that the transition might be related to DAP with Se vacancy (V_{Se}) [110]. 4.2 K PL spectrum of CuInSe$_2$ layers formed from the Se sufficient (30–40%) presents a peak at 0.88 eV (1400 nm), which is likely similar to passivation of the surface by oxygen. In fact, the presence of this peak is a symbol of good quality layers for devices. Both the layers with excess Se (50%) and Se deficient (10%) exhibit a peak at 0.98 eV (1270 nm), which may be due to (Se$_{Cu}$–V$_{In}$) and (CB–Cu$_{In}$) or (CB–V$_{In}$), respectively [111]. 4.2 K PL spectra of CuInSe$_2$ thin films prepared at 540 °C show that the near band edge emission (NBE) line at 1.03 eV is labeled to BX emission line. As the period of selenization has been increased from 20, 35 to 45 min, the sharpness of the peaks gradually decreases and finally only one broad peak at 0.969 eV (V$_{Cu}$–CB) presents [102].

There are several p-type single crystals such as a, b, c, d, and e with different Cu, In, and Se compositions as shown in Table 5.3. A typical ~8 K PL spectrum of sample-e is shown in Figure 5.30 in which four emission peaks at ~0.986, 0.971, 0.925, and 0.893 eV are observed. As temperature of sample is increased from 8 to 15–60 K, two additional PL peaks start to appear at 0.946 and 0.935 eV. Above 40 K, intensities of all PL peaks decrease but it is slow for 0.986 eV peak. After Cu thermal diffusion on samples d and e, the intensity of 0.925 eV peak decreases, whereas the intensity of 0.893 eV peak decreases for In thermal diffusion. A rate of decrease of 0.95 eV peak intensity is higher for higher excitation intensity. The intensity (I) of 0.98 eV peak, which is FB transition (CB–V$_{Cu}$), decreases as $I_c < I_e < I_d$ with increasing Cu concentration (Cu) more or less in the sequence of Cu$_c$ > Cu$_e$ > Cu$_d$ for the samples c, e, and d indicating that the Cu diffusion suppresses the copper vacancies (V$_{Cu}$) (Table 5.3). Therefore, the peak at 0.98 eV can be authentically assigned to (CB–V$_{Cu}$) transition beyond the reasonable doubts. The intensity of 0.92 eV peak due to DAP that is, (V$_{Cu}$–V$_{Se}$) decreases in the sequence $I_c < I_e < I_d < I_a < I_b$ for samples c, e, d, a, and b. This peak can be considered to Cu vacancy transition. The experimental study on thermal diffusion of In supports to assign 0.89 eV peak to (V$_{In}$–V$_{Se}$) transition. The peak at 0.935 eV is assigned to (Cu$_{In}$–V$_{Se}$) in the stoichiometric sample-b [112]. The peak at 0.95 eV, which appears in the In-rich samples a, c, and e may be due to (V$_{Cu}$–In$_{Cu}$). A shoulder 0.96–0.97 eV peak in d and e samples may be due to (Cu$_{In}$–In$_{Cu}$). The lowest energy peak presents at 0.82 eV could be related to In$_i$ donor level of 200 meV that supporting the electrical measurement [90,113]. In fact, the experimental results

Table 5.3 Physical parameters of CuInSe$_2$ single crystals

Sample No	Composition Cu:In:Se	ρ (Ω-cm)	μ (cm^2/Vs)	N_a or N_d (cm^{-3})	Type of conductivity	Ref.
a	23.7:26.2:50	0.9	30	2.2×10^{17}	p	[90]
b	24.2:25.6:50.2	0.7	15	6.5×10^{17}	p	
c	23.1:26:50.8	2.1	6	5.0×10^{17}	p	
d	23.7:25.2:51	15	30	1.4×10^{16}	p	
e	22.6:26.6:50.8	4500	10	1.5×10^{14}	p	
f	Cu/In = 1.1, 1.2% Se deficient	–	–	6×10^{15}	p	[114]
f_1	–	–	30–40^a 160–200^b	$6-9 \times 10^{16a}$ 1×10^{14b}	p	[115]
g	Cu/In = 0.97, 0.8% Se deficient	–	–	6×10^{16}	n	[114]
h	Cu/In = 0.85, 2% Se deficient	–	–	2×10^{17}	n	
i	22.79:27.22:49.99	–	50	2×10^{16}	p	[118]
j	22.91:27.07:50.02	–	4	5.8×10^{16}	p	
k	22.97:25.62:51.41	–	–		p	[117]
l	23.77:25.16:51.08	–	–		p	

^{a}RT.
^{b}77 K.

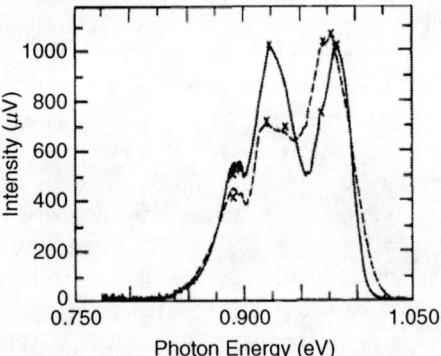

Figure 5.30 7.8 K PL spectra of p-CuInSe$_2$ single crystals.

provide that the Cu vacancies (V$_{Cu}$) decrease with increasing Cu composition as well as In vacancies (V$_{In}$) follow the same foot prints in the samples.

In order to find differences between transition energies in the n- and p-type CuInSe$_2$ single crystals, the crystals with different Se deficiencies and carrier concentrations are studied (Table 5.3) [114]. The p-type single crystal sample-f shows peaks at 1.03, 1.00, and 0.96 eV and the p-type single crystal sample-f_l also provides more or less similar emission lines at 1.04, 0.98, and 0.95 eV [115]. 0.98 and 0.89 eV bands in the n-type CuInSe$_2$ single crystal sample-g and 1.03, 0.98, 0.92, and 0.89 eV lines in the n-type CuInSe$_2$ single crystal sample-h are observed [114]. In the case of In-rich CuInSe$_2$ thin films, the same broad emission lines at 1.1, 0.975, and 0.89 eV are observed at 6 K [116]. The peak at 1.03 eV due to band to band transition, 1.0 or 0.98 eV to FB transition (V$_{Cu}$–CB), 0.96 eV to (Cu$_{In}$–In$_{Cu}$), 0.92 eV to DAP (V$_{In}$–In$_{Cu}$), and 0.89 eV to (V$_{Cu}$–In$_{Se}$) DAP transitions, respectively may be assigned. The transition peaks occur at 0.985–1.0 and 0.975–0.99 eV in the n- and p-type CuInSe$_2$ samples, respectively. On contrast, the peak positions slightly higher in the n-type than in the p-type samples.

The emission lines 0.984 (V$_{Cu}$–In$_{Cu}$), 0.954 (V$_{Cu}$–V$_{Se}$), 0.92 (V$_{In}$–In$_{Cu}$), and 0.89 eV (V$_{Cu}$–In$_{Se}$) in the p-type single crystal sample-i and 0.973 and 0.945 eV (V$_{Cu}$–V$_{Se}$) in the p-type single crystal sample-j are observed, even though both have similar conductivity types but had slightly different compositions, carrier concentrations, and mobilities, as shown in Table 5.3. The p-type CuInSe$_2$ single crystal sample-k with Cu/In = 0.893 or Cu:In:Se = 22.97:25.62:51.41 shows peaks at 0.978 and 0.926 eV, which may be attributed to (CB–V$_{Cu}$ or Se$_{Cu}$) and (Se$_{In}$–In$_{Cu}$), respectively [117], and the emission band at 0.955 eV due to (V$_{Cu}$–V$_{Se}$) appears. Sample-l p-single crystal (IEC sample) with Cu/In = 0.944 or Cu:In:Se = 23.77:25.16:51.08 exhibits a peak at 0.94 eV, which may be due to (V$_{Cu}$–Se$_{Cu}$) transition and 0.978 eV peak due to (CB–V$_{Cu}$) appears, which is already repeated in the sample-k, as shown in Figure 5.31. A pronounced peak at 0.88 eV due to (Cu$_{In}$–In$_{Se}$), a shoulder at 0.901 eV due to (V$_{Cu}$–In$_{Se}$), and another second dominant peak at 0.936 eV are observed for CuInSe$_2$ thin films on far with composition of p-type single crystal sample-l in the PL spectrum. All the peak positions shift to higher energy level with increasing the excitation power from 7 to 14 mW. The same kind of behavior is observed in the cleaved single crystals indicating DAP character. The sample-l polished by mechanically repeats the same peaks, however, the peak positions slightly deviate and intensities also slightly decrease, no dominant peak is observed unlikely shown by virgin sample that indicates formation of nonradiative recombination centers, which obscure the original intensity of the peaks. Due to surface recombination states or surface damage defects, additional peaks at 0.98 and 0.92 eV are found in the polished 2% In excess p-type crystals [118].

The band to band transition peak at 1.03 eV and (CB–V$_{Cu}$) transition peak at 1.0 eV prevail in one sample and do not appear in another sample or vice versa among three p-type single crystals. The intensity of peak at 0.96 eV in the p-type CuInSe$_2$ single crystal sample-m with compositions of 24.87:24.82:50.31 is stronger, whereas the intensity of same peak is moderate in the p-type sample-n with composition of CuIn:Se = 24.2:25.24:50.8 and in the p-type single crystal sample-o with composition of 23.2:25.29:51.5. After annealing n-type crystals under air at 200 °C, a broad DAP emission line at 0.966 eV (Cu$_{In}$–In$_{Cu}$) shifts to higher energy of 0.985 eV. The 0.942 eV peak (DAP) dominates the spectrum for further annealing at 300 °C. After air annealing p-type CuInSe$_2$ crystals at 200 °C, the peak at 0.94 eV shifts to lower energy side, whereas the same peak moves back to 0.955 eV for further air annealing at 450 °C. The p-type crystals show peak at 0.942 eV and three more

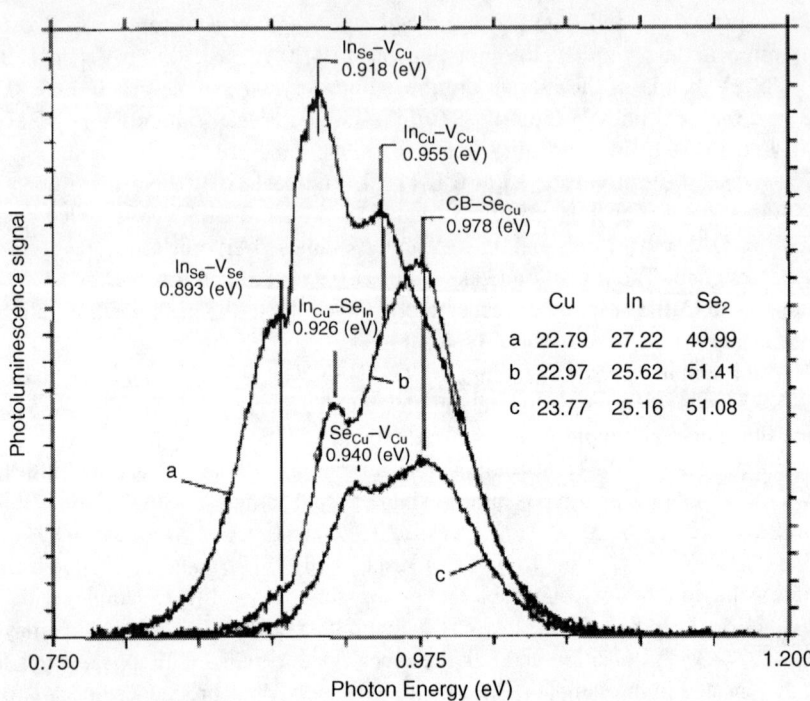

Figure 5.31 PL spectra of CuInSe$_2$ single crystals with three different compositional ratios.

shoulders at 0.907, 0.899, and 0.892 eV. After annealing crystal under selenium vapor at 400 °C, the DAP peak at 0.942 eV (V_{Cu}–V_{Se}) jumps to higher energy side of 0.957 eV with an increased intensity. It could be due to minimum Se pressure that causes to create more density of Se defects (V_{Se}), as observed in earlier. Eventually, the transition length enhances with decreasing defect concentration in the sample due to screening effect. As further increasing annealing temperature to 600 °C, there is not much change but intensities of 0.92, 0.9, and 0.892 eV peaks increase. The Hall mobility increases by annealing the samples under air at 220 °C and under selenium vapor at 450 °C. However, the mobility drastically decreases by annealing the sample under selenium vapor at 600 °C. Typical n-type single crystals had different concentrations and mobilities show wide variation in dominant peak at 0.96 eV [119].

CuIn alloy evaporated onto Mo coated soda-lime glass substrate by the electron beam is selenized under H$_2$Se and Ar atmosphere to form CuInSe$_2$ layers [120]. The CuInSe$_2$ layers with high In concentration (Cu/In = 0.5, Cu:In:Se = 15.7:31.6:52.8) show three peaks at 1.1, 0.975 and 0.89 eV in 60 K PL spectra recorded at 40 mW excitation power. The emission band peaking at 1.1 eV can be assigned to band gap of CuInSe$_2$ because the band gap increases with increasing In content in the layers. A 1.1 eV peak moves toward lower energy with increasing sample temperature in

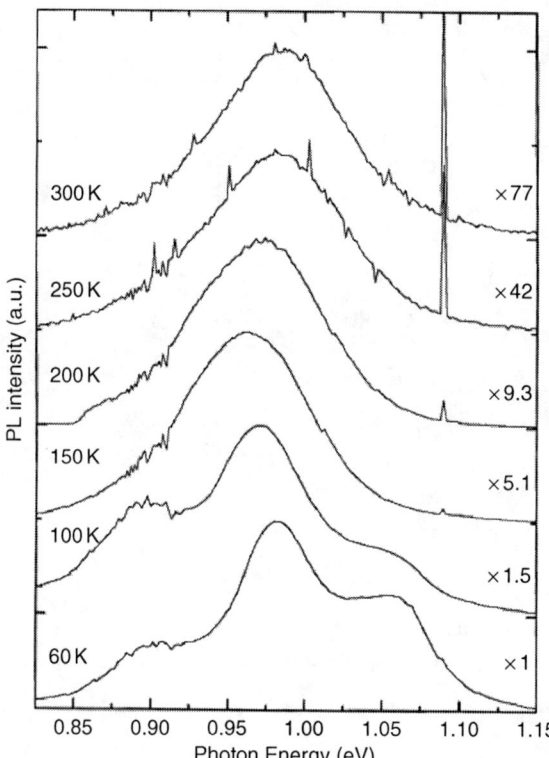

Figure 5.32 Temperature dependent PL spectra of CuInSe$_2$ thin films with Cu/In ratio of 0.5.

the 60–150 K range due to shrinkage of band gap. As further increasing the sample temperature, the peak shifts to higher energy due to thermal excitation of carriers from lower to higher energy state. A variation of peak position with temperature looks like "S" shape character, as shown in Figure 5.32 [121,122]. The peak at 0.975 eV is overlap of DAP and FB transition and another peak at 0.89 eV is also related to DAP. The samples with slightly In-rich (Cu/In = 0.92, Cu:In:Se = 24.5:26.6:48.9) and nearly stoichiometric shows only one emission line at 0.957 eV (Figure 5.33), which shifts to 0.964 eV with increasing excitation power from 0.4 to 40 mW confirming DAP transition (V_{Cu}–V_{Se}), whereas the CuInSe$_2$ thin films show two emission bands at ∼0.86 and ∼0.95 eV due to DAP transitions for the close composition of Cu/In = 0.9 [123]. Taking 1.054 eV as band gap of CuInSe$_2$, using Equation (5.7), the donor and acceptor levels are considered to be (V_{Se}) 60 meV and (V_{Cu}) 40 meV for 0.95 eV emission band, respectively, as mentioned in earlier. The same acceptor (V_{Cu}) participates for 0.86 eV emission band with a different donor (doubly ionized donor V_{Se}) having activation energy level of 160 meV indicating that one acceptor level participates for both emission peaks of 0.95 and 0.86 eV. For Cu-rich (Cu/In = 1.28, Cu:In:Se = 28.6:22.3:49.1) CuInSe$_2$ samples the NBE at 1.036 eV could be due to free exciton, followed by its phonon replica at 1.008 eV. The peaks at 0.993 and 0.971 eV could be due to FBs such

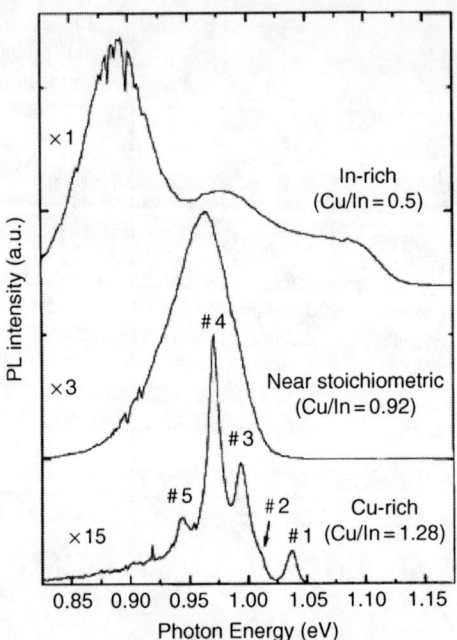

Figure 5.33 6 K PL spectra of CuInSe$_2$ thin films with Cu-rich, stoichiometric, and Cu-poor.

as (CB–V$_{In}$) and (CB–Cu$_{In}$) with ionization energies of 48 and 70 meV, respectively. Since V$_{Cu}$ is a common defect in the CuInSe$_2$ irrespective of Cu to In ratio. Therefore the peak at 0.993 eV may be assigned to (CB–V$_{Cu}$) in the Cu-rich samples. The emission band at 0.942 eV is phonon replica of 0.971 eV peak, as shown in Figure 5.33. In another occasion, 4 K PL spectrum shows that In-rich CuInSe$_2$ films exhibit only one broad peak at about 0.93 eV whereas Cu-rich CuInSe$_2$ thin films, which are fabricated by diffusing Cu into In-rich CuInSe$_2$, as a thin Cu layer is deposited on the sample, followed by annealing at 400 °C for 10 min, show a plethora of emission lines such as peak at 1.03 eV for BX, FB (0.96 eV) along with phonon line at 0.94 eV. The peak at 0.90 eV is due to Se$_i$, followed by three phonon lines, as shown in Figure 5.34. The intensities of PL peaks drastically decrease and peak at 0.90 eV disappears for the thick Cu layer diffused CuInSe$_2$ [124].

Slightly In-rich CuInSe$_2$ films with compositions of Cu:In:Se = 21.40:24.02:54.58 or Cu/In = 0.89 prepared onto GaAs by MBE at substrate temperature of 500 °C, followed by annealing under a selenium beam flux of 1.0×10^{15} atoms/cm^2 at 400 °C for 1 h result in compositions of Cu:In:Se = 20.6:22.79:56.61 or Cu/In = 0.9. The as-grown sample shows peaks at 1.036 (free exciton), 0.97 eV dominant and 0.95 eV, whereas the annealed sample shows peaks at 1.036, 0.995 dominant, and 0.97 eV, as shown in Figure 5.35. The emission line at 0.995 eV can be confidently assigned to (V$_{Cu}$–CB) transition because after annealing the sample under Se over pressure, the density of V$_{Se}$ obviously decreases ultimately density of V$_{Cu}$ increases.

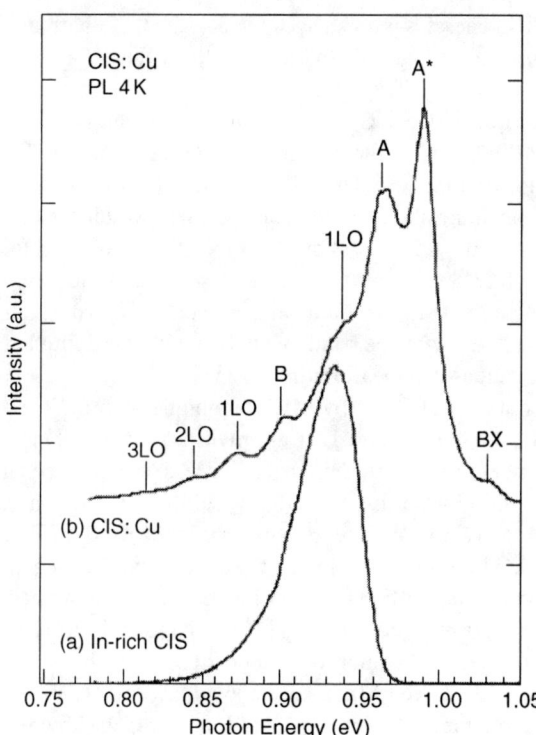

Figure 5.34 4 K PL spectra of Cu-rich and In-rich CuInSe$_2$ thin films.

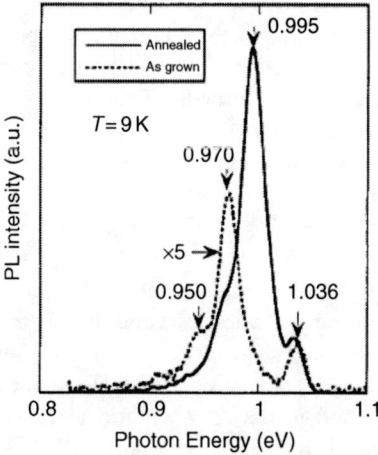

Figure 5.35 PL spectra of as-grown CuInSe$_2$ thin film and annealed under Se at 400 °C for 60 min.

Therefore, 0.995 eV peak (V_{Cu}–CB) generates with sharp intensity. On the otherhand, the peak at 0.95 eV disappears in annealed sample evidencing to assign (V_{Se}–V_{Cu}) transition because of decrease of V_{Se} density. The intensity of 0.97 eV peak increases in the annealed sample that could be due to (Cu_{In}–CB) transition. As annealing Cu-rich samples under Se over pressure at 400 °C for 2 h, the transition at 1.002 eV due to (CB–V_{In}) becomes band to band transition of 1.024 eV. There is no significant difference in the PL spectrum for continuous annealing the sample. The samples, which contain Cu_2Se/Cu_5Se_4 phases determined by XRD, show compositions of Cu:In:Se = 25.04:20.40:54.52 or Cu/In = 1.23 for longer time annealing. After subtracting the secondary phases, the remaining composition would definitely be close to the stoichiometric or chalcopyrite structure that is why the band to band transition is dominant and its intensity is constant for an increased annealing time [125].

The $CuInSe_2$ layers deposited onto (001) GaAs by MBE technique with different composition ratios of Cu/In = 0.81, 1.04, 1.47, and 1.81 are investigated by 2 K PL. The DAP peak at 0.85 eV is a dominant, which gradually becomes less pronounce than other peaks with increasing Cu/In ratio in the In-rich $CuInSe_2$ samples, as shown in Figure 5.36. The exciton emission at 1.036 eV appears with low intensity but FB at 0.97 eV dominates the PL spectra. The phonon replicas of latter appear with equal distribution of ~ 29 meV. The broad emission line (0.85 eV) blue shifts with increasing excitation energy from 0.1 to 300 mW confirming DAP character [126,127]. Nearly stoichiometric Cu-rich $CuInSe_2$ epitaxial layers deposited onto GaAs and GaAs/$In_{0.29}Ga_{0.71}$As epilayers at 500 °C by MBE are labeled as sample-p and sample-q, respectively show a remarkable difference in 2 K PL spectra, as shown in Figure 5.37. The peak at 0.970 eV due to valence band to donor transition (Cu_{In}–CB) dominates the PL spectra, containing phonon line at 0.940 eV in the sample-p. The existence of this peak is indication of large number of point defects in the $CuInSe_2$ layers grown onto GaAs. In addition, free excitons A (1.039 eV) and B (1.045 eV) with weak intensities and BX at 1.03 eV appear. The DAP occurs at 0.913 eV with phonon lines of 0.883 and 0.853 eV, whereas, the intensity of 0.970 eV peak is low in sample-q. The A and B free exciton peaks dominate the spectra and the peaks at 1.03 and 1.043 eV, which are tentatively assigned to exciton-complex, also present. The experimental results indicate that the quality $CuInSe_2$ layers can be obtained when the growth takes place on $In_{0.29}Ga_{0.71}$As buffer layers/GaAs rather than on GaAs substrates [128]. The PL on two different types of $CuInSe_2$ thin films that is, Cu:In:Se = 24.9:24.9:50.2 (sample-r) and 25.6:25.1:49.3 (sample-s), which had carrier concentrations of 5.7×10^{19} and 8.3×10^{19} cm^{-3}, respectively are studied. $A = 1.0374$ and $B = 1.0458$ eV exciton peaks without BX-1.0311 eV are observed in sample-r, whereas BX = 1.0311 eV is the dominant with $A = 1.038$ eV and without B exciton in sample-s. The peaks appear in some cases and disappear in some other cases and a slight difference in peak energy between A in sample-r and sample-s are noticed due to a slight variation in the composition of samples that causes difference in strain relaxation and residual strain in the grown epitaxial layers. The BX (1.0311 eV) appears at 7 K and disappears at ~ 35 K with phonon by separation of 28 meV. The

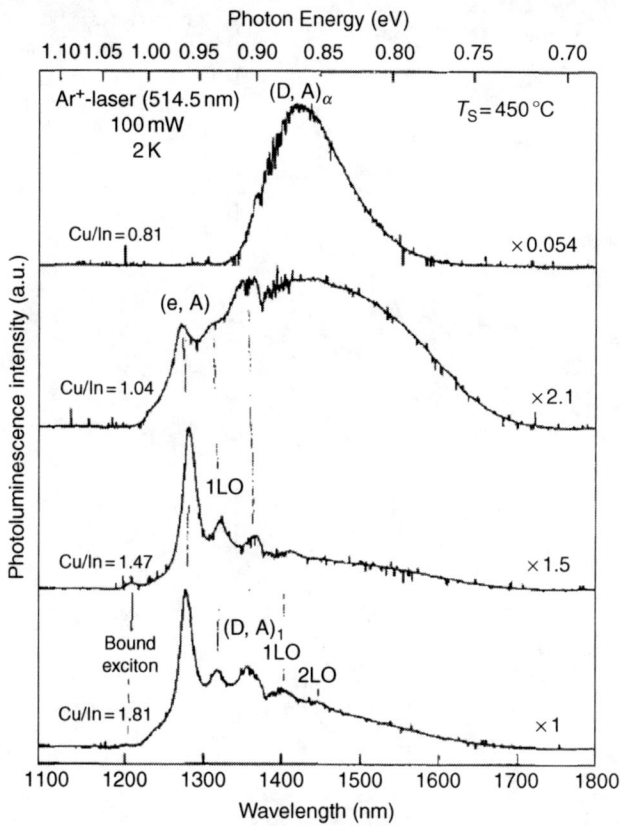

Figure 5.36 PL spectra of CuInSe$_2$ thin films with different composition ratios grown by MBE.

ground state exciton appears between 35 and 102 K in the sample-r. Two DAPs present at ~0.97 and ~0.915 eV with longitudinal optical (LO) phonons [129].

The authors [130] studied the annealing influence on the CuInSe$_2$ thin films, which had composition ratio of Cu/In = 1.08, prepared onto GaAs substrates by MBE technique at 500 °C. 2 K PL spectrum shows FB transition at 0.970 eV, followed by phonon replica at 0.94 eV and DAP at 0.92 eV, followed by its several phonon lines, as shown in Figure 5.38. The sample has been annealed under vacuum at different temperatures such as 260, 350, and 440 °C for 30 min. As annealing temperature is increased from 260 to 440 °C, the FB peak gradually becomes broader and red shifts to lower energy position and other peaks disappear. The overlapping peaks at 0.96–0.97 eV are combination of DAP and FB at excitation intensity of 0.1 mW. As the excitation intensity is increased to 140 mW, the FB peak at 0.970 eV becomes more distinct and DAP peak exists at 0.959 eV. The shallow donor is created during annealing the sample under vacuum, which participates for DAP transition. The sample annealed at 440 °C had a broad peak at ~0.945 eV, which shows blue shift with

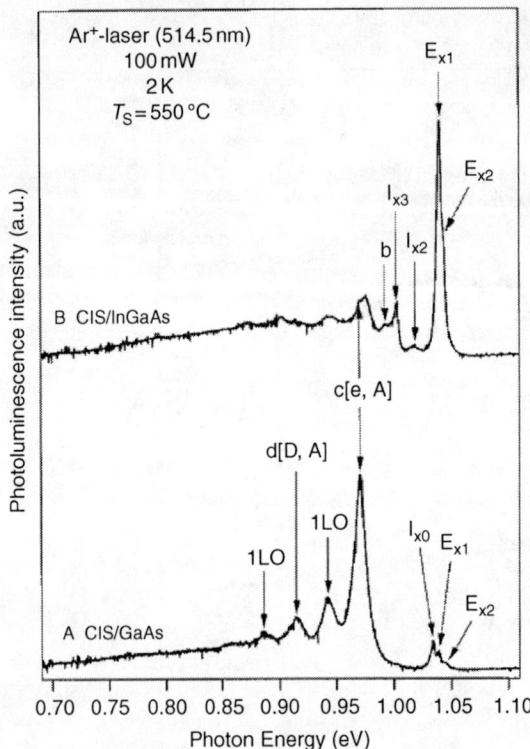

Figure 5.37 PL spectra of CuInSe$_2$ thin films grown onto (A) GaAs and (B) GaInAs epilayer covered GaAs substrates.

increasing excitation intensity containing more compensation, as shown in Figure 5.39. The EPMA analysis reveals Se deficiency in vacuum annealed samples. The p-type CuInSe$_2$ epitaxial layers with excess Cu grown onto GaAs substrates by MBE show CuSe as a secondary phase but PL spectra show no much difference before and after KCN etching for 1 min to remove secondary phase. However, slightly intensity gains in etched samples due to reduction in the reflection losses of the incident Ar-laser. The exciton (1.03–1.04 eV) with one LO phonon replica and broad emission peak at 0.80–0.85 eV due to DAP are observed [131]. Niki *et al.*, [132,133] studied that the p-type CuInSe$_2$ samples with slightly Cu/In ratio of 1.1 grown onto GaAs substrates by MBE technique at 450 and 550 °C emerge several peaks in 2 K PL spectrum. The exciton emission peak at 1.03 eV, FB peak at 0.97 eV and its phonon replica at 0.94 eV are found. The DAP at 0.91 eV along with phonon replicas at 0.88 and 0.85 eV is also observed. The samples with the same conditions deposited at different temperatures such as 450 and 550 °C, which had hole concentrations of 8×10^{19} and 3×10^{19} cm^{-3}, respectively show that the FB is stronger for the former and DAP gains intensity for the latter, as shown in Figure 5.40. No change in the spectra is found except the intensities of 0.97 and 0.94 eV peaks increase with increasing excitation power from 1 mW, 3, 10, 30, 100 to 300 mW. Chichibu [134] prepared n-CuInSe$_2$ epitaxial layers with slightly In- and Se-rich with compositions of Cu:In:

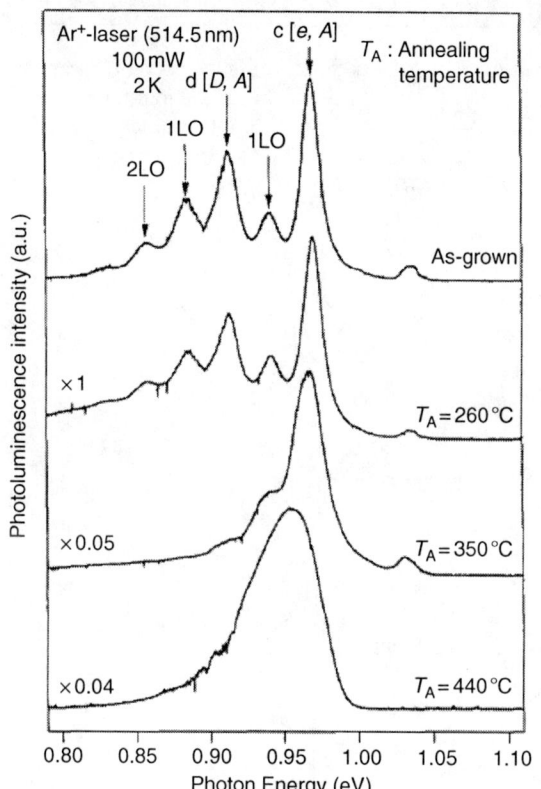

Figure 5.38 PL spectra of CuInSe$_2$ thin films annealed at different temperatures.

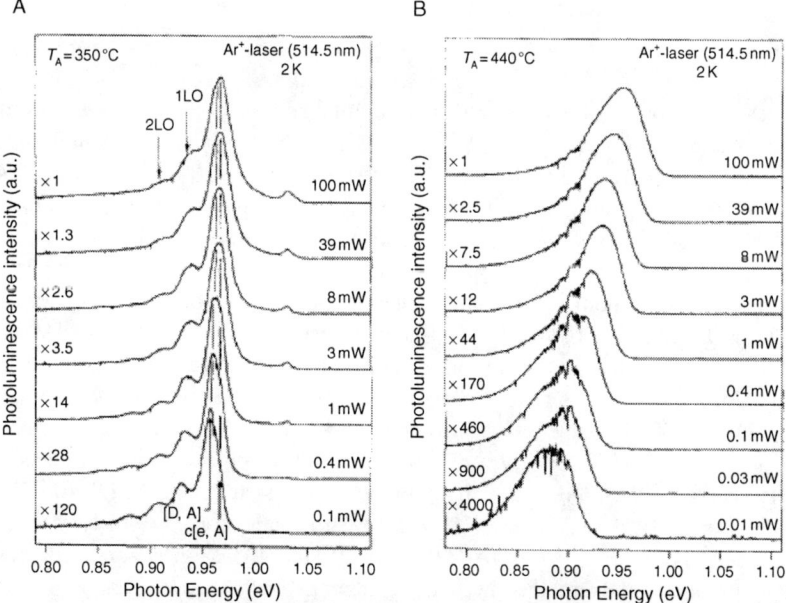

Figure 5.39 PL spectra of CuInSe$_2$ thin films recorded at various excitation intensities (A) sample under vacuum annealed at 350 °C and (B) 440 °C.

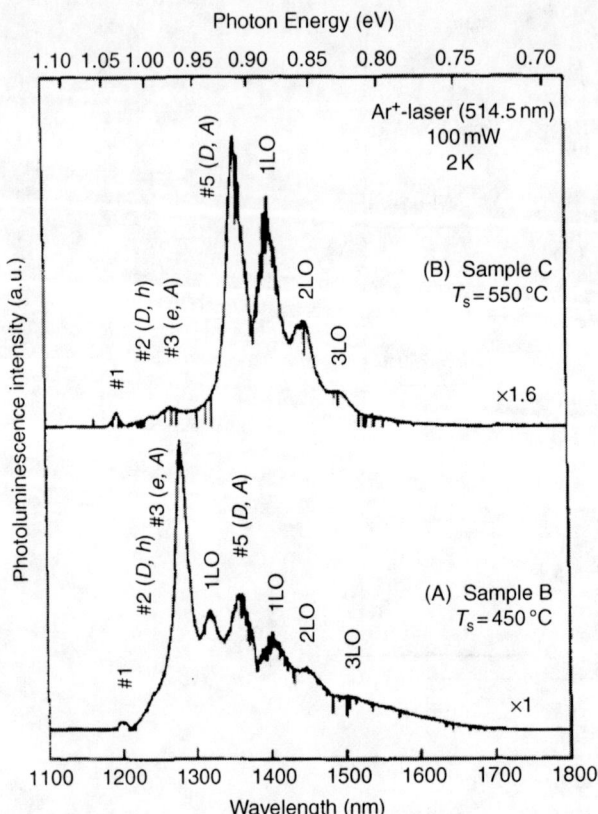

Figure 5.40 PL spectra of CuInSe$_2$ thin films grown at different temperatures (A) 450 °C and (B) 550 °C.

Se = 1:1.02:2.2 or Cu/In = 0.98 onto GaAs, InP, and sapphire substrates by metalorganic vapor phase epitaxy (MOVPE) for which the emission bands at 1.0422 eV (FE), 1.000 eV (FB), and 0.966 eV (FB) are observed in 10 K PL spectra, respectively, as shown in Figure 5.41. It can be concluded that the defect transition is different from substrate to substrate due to formation of different defect structure or different epitaxial growth. The shape of 1.0422 eV peak changes with varying temperature from 10 K to room temperature, hence it could be combination of either of two transitions from BX, FE and band-to-band transition. A blue shift is observed in 1.0422 eV peak with variation of temperature due to the Burstein-Moss effect.

The CuInSe$_2$ thin films grown onto GaAs (100) substrates by MOVPE at substrate temperature of 500 °C have also been studied [135]. In 10 K PL spectra, a broad DAP line appears at 0.96 eV for Cu/In of 0.89. The well-resolved emission lines such as free exciton, DAP$_1$, DAP$_2$, and its phonon replicas at 1.039, 0.991, 0.971, and 0.945 eV, respectively appear with increasing Cu/In ratio from 0.89 to 1.18. The activation energies of 42 meV for DAP$_1$ (0.971 eV) and 56 ± 22 meV for DAP$_2$, respectively, are obtained from the thermal quenching analysis. The DAP$_2$ (0.991 eV) transition is due to transition of donor level of 12 meV to acceptor level of 40 meV,

Optical Properties of I–III–VI$_2$ Compounds

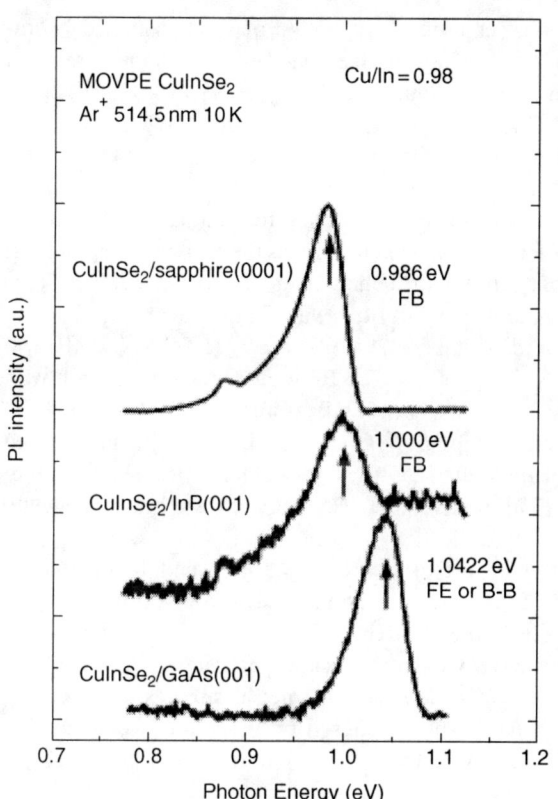

Figure 5.41 10 K PL spectra of 0.8 μm thick CuInSe$_2$ epilayers with Cu/In = 0.98 deposited on different substrates (A) GaAs, (B) InP, and (C) sapphire.

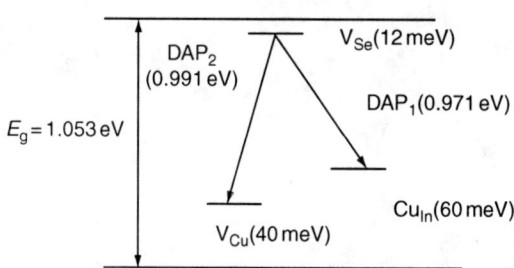

Figure 5.42 Transition levels of CuInSe$_2$ (model-2).

whereas DAP$_1$ (0.971 eV) transition takes place from the same donor level of 12 meV to acceptor level of 60 meV, as shown in Figure 5.42. Note that both DAP$_1$ and DAP$_2$ have a common donor level of 12 meV but different acceptor levels of 40 and 60 meV. The transition levels mentioned in Figure 5.42 are different from Figure 5.27 for the same kind of photoluminescence peaks, which are judged by the thermal quenching experimental analysis *etc*. The emission bands at 1.04 and 0.99 eV are observed in the *n*- and *p*-CuInSe$_2$ crystals grown by synthesis solute diffusion technique [136]. As mentioned earlier, the former is band to band transition and the latter is DAP transition, which is ascribed to donor (V_{Se}) to acceptor level Cu_{In} or V_{In} or V_{Cu} transition. According to Equation (5.7), ($E_A + E_D$) is 50 meV, using the donor level of 10 meV obtained from the Hall measurements, and the remaining activation energy of 40 meV, could be assigned to acceptor level of Cu_{In} or V_{In} or V_{Cu}. The authors [137] studied 10 K PL spectra of $CuInSe_2$ thin films grown onto GaAs substrates with different Cu/In composition ratios of 1.17, 1.01, 1.00, and 0.99. The sample-t with Cu/In ratio of 1.17 shows emission peak at 1.032 eV (#1-a), which is due to exciton recombination. The line at 0.982 eV (#3-c) is assigned to DAP transition, which contains the shallow acceptor level of 6 meV and donor level (V_{Se}) of 60 meV with a phonon replica exists at 0.946 eV (#5-e), as assigned by the authors. Sample-u, $CuInSe_2$ films with Cu/In ratio of 1.01, the peak at 0.992 eV (#2) is contribution of transition from the FB acceptor level V_{Cu} of 40 meV and its phonon signature diminishes at 0.960 eV. The FB transition shows no shift with variation of excitation power. The same FB slightly shifts from 0.992 (#2-b) to 0.980 eV (#2') and its phonon replica also shifts from 0.956 (#4-d) to 0.945 eV (#4') with changing composition from Cu/In = 1.00 (sample-v) to Cu/In = 0.99 (sample-v'), as shown in Figure 5.43. The PL peak recorded at 77 K shifts from 0.975 to 0.86 eV and broadening with increasing Cu/In ratio from 1.29 to 0.52 [138].

The effect of excitation power on the properties of Cu-rich and stoichiometric $CuInSe_2$ samples shows that the DAP peak position shifts from 0.968 to 0.973 eV with shift rate of 2.5 meV/decade in the Cu-rich $CuInSe_2$ samples with increasing excitation power from 2 to 200 mW, whereas the same peak shifts from 0.974 to 0.993 eV with shift rate of 12.5 meV/decade in stoichiometric samples. A difference in shift rate is eventually due to fast response caused by variation of composition that changes the band gap ultimately defects transition length [139]. 7 K PL

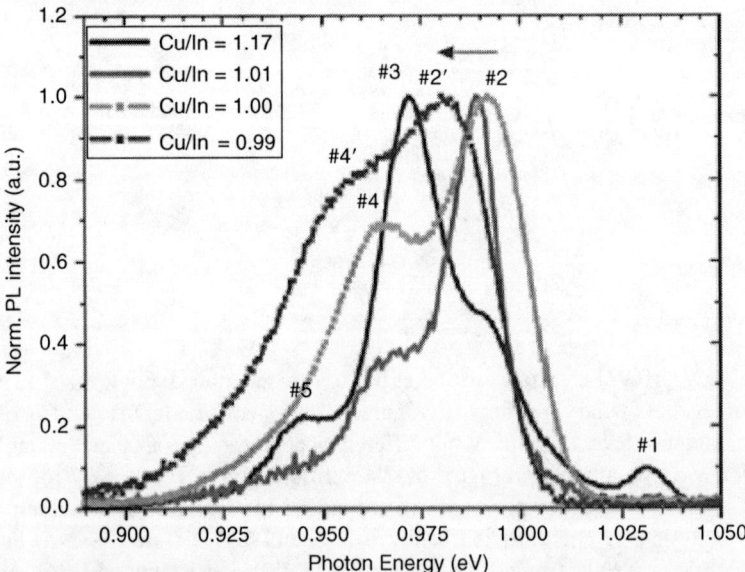

Figure 5.43 10 K PL spectra of $CuInSe_2$ thin films with different composition ratios.

spectrum of n-CuInSe$_2$ crystals shows that the transition at 0.98 eV is assigned to DAP transition in which donor and acceptor levels are probably assigned to V_{Se} and V_{In} with ionization energy levels of 10 and 33 meV, respectively. The other two transitions at 0.99 and 1.013 eV represent to the transition between CB and acceptor level V_{In} (33 meV) and from VB to shallow donor level V_{Se} (10 meV), respectively. The acceptor level of 34 meV can be obtained from the relation

$$E_A = E_{A0} - \beta N_A^{1/3}, \text{ where } E_{A0} = 13.6(m_h*/m_0)\varepsilon_s^2, \qquad (5.12)$$

proportional constant, ($\beta = 2.4 \times 10^{-8}$ eV) which is respecting V_{In} (33 meV) because the electrical analysis on In diffused CuInSe$_2$ indicates the V_{In} vacancies. It is more or less similar to selenization process; the formation of V_{Se} occurs for less Se vapor selenization and it disappears for higher Se vapor selenization. The similar trend is observed on CuInTe$_2$ annealed in low In vapor pressure [140].

Defect levels can also be suppressed by doping of Na into CuInSe$_2$ films [141]. The CuInSe$_2$ films prepared onto corning 7059 substrates by two-stage co-evaporation of Cu/In with Na$_2$Se at substrate temperature of 550 °C exhibit emission lines at 0.92 and 0.875 eV for Cu/In ratio of 0.94 and 0.84, respectively. The Na incorporated CuInSe$_2$ thin films show emission band at 0.91 eV that means Na suppressing the donor defects for irrespective of Cu/In ratios. The (V_{Cu}–V_{Se}) transition is concerned to 0.92 eV DAP peak, whereas the participation of acceptor V_{Cu} and double vacancy donor V_{Se} may be related to 0.875 eV DAP peak. In some cases, no effect by doping of Na is observed. 15 K PL spectra of 0.6 µm thick n-CuInSe$_2$ thin films with Cu/In = 0.84 deposited onto ITO glass substrates shows that the peak at 1.083 eV is due to band edge, its position is little higher than the standard 1.04 eV that may be due to higher percentage of In or stress in the film. The thermal quenching experimental analysis reveals that two peaks at 0.801 and 0.797 eV have activation energies of 26 and 10 meV, respectively. They can be assigned to participation of doubly ionized donors of In$_{Cu}^{2+}$ antisite and singly ionized In$_{Cu}^+$ antisite, respectively. In the band gap, the transitions take place from the deep acceptor level of 0.25–0.28 eV above the valence band to the shallow doubly and singly ionized donor levels below the conduction band. The exist transitions are absolutely DAP, since the peaks show shift in position with variation of excitation intensity [142]. 4.2 K PL spectra of CuInSe$_2$ polycrystalline thin films with different Cu/In composition ratios from 1.34 to 0.58 prepared onto glass substrates by CVT method employing iodine as transport agent exhibit different recombination states. The emission lines at 0.995, 0.950, and 0.930 eV are assigned to (CB–V_{Cu}), (In$_{Cu}$–V_{Cu}), and (Se$_{Cu}$–V_{Cu}), respectively in the Cu-poor CuInSe$_2$ thin films. In the Se-rich, Cu- and In-poor CuInSe$_2$ thin films, the last two emission lines present. In the Cu-rich films 0.995, 0.970, and 0.935 eV can be attributed to (Cu$_{In}$–CB), (V_{In}–CB), and (Cu$_i$–VB), respectively, as shown in Table 5.4. The emission lines at 1.04 and 0.880 eV are assigned to free exciton and (Se$_{Cu}$–Se$_{In}$), respectively. The sample with composition ratio of 19.11:25.14:55.75 shows band gap of 1.15 eV and n-type conductivity indicating formation of CuIn$_3$Se$_5$ phase [143]. 20 K PL spectra illustrate that CuInSe$_2$ with excess In prepared by increasing In ionization voltage from 0 to 4 kV

Table 5.4 Composition and peak positions of CuInSe$_2$ thin films

S. No	Cu:In:Se	Cu/In	Peak positions
1	26.8:20:53.2	1.34	0.935
2	27.6:21.6:50.8	1.28	0.97, 0.995
3	24:19:57	1.26	1.04, 0.93
4	23.7:21.1:55.2	1.12	0.995, 0.88
5	23.1:24.3:52.6	0.95	0.995, 0.955
6	24:26:51	0.92	0.995, 0.93
7	15.7:27.3:57	0.58	0.995, 0.95

with current of 50 mA keeping constant ionization voltage of Cu in the ionized cluster beam technique shows a broad peak at 0.87 eV at below 2 kV. Beyond this voltage there is an additional peak at 0.95 eV resulting by the segregation of CuIn$_2$Se$_{3.5}$. The layers with 0.87 eV emission peak indicates the symbol of quality solar cells [144]. 8 K PL spectra of CuInSe$_2$ films deposited onto GaAs substrates by metal-organic molecular beam epitaxy (MOMBE) shows emission band at 1.03 eV and broad emission at 0.8 eV for the growth time of 4 h, whereas a free exciton or BX at 1.03 eV and sharp emission peak at 0.98 eV due to FB (CB–V$_{Cu}$) occur for longer growth time of 10 h, with a triethyl indium (P$_{TEIn}$) pressure of 2.9×10^{-7} torr, showing quality epilayers [145]. 77 K PL spectra of CuInSe$_2$ single crystal show peaks at 1.03, 0.97, and 0.90 eV, whereas the CuInSe$_2$ thin films deposited from its bulk show peaks at 0.8 and 0.97 eV for the deposition temperatures of 940 and 970/1000 °C, respectively [146].

5.2.4 CuInSe$_2$ Thin Film Solar Cells

The PL studies distinguish the defect level relation between high and low efficiency thin film solar cells well. Two kinds of samples are taken such as sample-w_1: Cu layers are sputtered onto Mo coated glass substrates, followed by sequential evaporation of In/Se at room temperature. The glass/Mo/Cu/In/Se stack is annealed under nitrogen atmosphere at 200–400 °C. The sample-w_2 is prepared by sputtering of Cu and evaporation of In layers, followed by annealing under argon atmosphere and selenization under selenium vapor. The sample-w_1 with slightly Cu-rich composition of Cu:In:Se = 24.4:24.1:51.5 shows PL peak at 0.985 eV (CB–V$_{Cu}$) or (CB–V$_{In}$), which dominates spectrum. In this transition, the acceptor energy level of 40 meV is considered for V$_{In}$. The solar cell made with Cu-rich absorber does not show photovoltaic conversion. However, the cell with absorber containing slightly In-rich compositions of Cu:In:Se = 24.1:25.1:50.8 shows photovoltaic conversion efficiency of 5%, which shows dominant 0.967 eV (V$_{Cu}$–CB) peak and an additional peak at 0.938 eV (V$_{Cu}$–V$_{Se}$) or (Cu$_{In}$–V$_{Se}$). Finally, the luminescence peak at 0.927 eV (V$_{Cu}$–V$_{Se}$) leads the PL spectrum for In-rich CuInSe$_2$ thin films with composition of Cu:In:Se = 24:25.6:50.6 but the cell made with these layers show poor conversion efficiency of 2–3%. The sample-w_2 that is, the Cu/In stack layers annealed under selenium atmosphere for 30 min show peak at 0.965 eV with a small tail but the same stack layers annealed for 4 h exhibit the same luminescence peak (0.965 eV) without tail, whereas there are peaks at 0.985 and 0.965 eV for 1 h annealing. The peaks at 0.985, 0.967–0.965, 0.938, and 0.927 eV are assigned to (CB–V_{In}), (CB–V_{Cu}), (Cu_{In}–In_{Cu}) or (Cu_{In}–V_{Se}), and (V_{Cu}–V_{Se}), respectively. The acceptor level V_{In} is located at 40 meV above the valence band in the forbidden gap for Cu-rich samples [112].

The Cu/In stack layers grown onto corning 7059 annealed under elemental selenium vapor at different temperatures show different compositions of $CuInSe_2$ thin films. Indium composition increases from Cu/In = 0.99, Se/M = 1.01 to Cu/In = 0.72, and Se/M = 1.08 (where M = Cu + In) with increasing Se K-cell temperature from 160 to 200 °C correspondingly the DAP peak at 0.93 eV shifts to 0.85 eV in 4.2 K PL spectrum. The Cu/In/Se stack layers are annealed under N_2 atmosphere at different temperatures for which the PL peak shifts from 0.93 to 0.87 eV with increasing annealing temperature from 330 to 420 °C, respectively. The solar cells made with the $CuInSe_2$ absorber layers grown onto corning 7059 glass substrates and annealed at 350 °C shows efficiency of 6.6%, whereas the efficiency is 2% for the CIS annealed at 420 °C. The cell shows efficiency of 8.5% for the layers grown onto soda-lime float glass and annealed at the same temperature of 420 °C. The PL peak position of $CuInSe_2$ thin film annealed under air at 150 °C for different annealing times shifts from 0.93 eV (V_{Cu}–V_{Se}) or (V_{Cu}–In_{Cu}) to 0.89 eV (V_{Cu}–O_{Se} complex) or (In_{Cu}–Se_{In}). A 5 min annealing time is high enough to drastically shift the peak position [147].

The Cu/In/Se sequential stack layers selenized/annealed without cover glass in a carbon block under N_2 result in compositions of Cu/In = 0.892 and Se/(Cu + In) = 0.965, which shows emission peak at 0.967 eV in 4.2 K PL spectra. The solar cell fabricated with $CuInSe_2$ thin film absorber having emission peak in the range 1200–1400 nm exhibits an efficiency of 2–5%. The stack layers with compositions of Cu/In = 0.848 and Se/M = 1.06 annealed by covering a piece of cover glass on a carbon block show emission peak at 0.939 eV. The cell made with $CuInSe_2$ absorber layers show efficiency of 6.6%. The $CuInSe_2$/CdS/ZnO thin film solar cell made by IEC with similar kind of emission band shows conversion efficiency of 10% for which the absorber is prepared by co-evaporation technique. The peaks at 0.967 and 0.939 eV are related to the transition of (V_{Cu}–CB) and (Cu_{In}–V_{Se}) or (V_{Cu}–V_{Se}), respectively. The selenium vacancy (V_{Se}) with energy level of 70 meV below the conduction band and the copper on indium site (Cu_{In}) with energy level of 40 meV above valence band are occurred. The selenized $CuInSe_2$ thin film with a peak at 1280 nm (0.97 eV) annealed under air at 150 °C for 10 min shows a shift in peak from 1280 (0.97 eV) to 1400 nm (0.89 eV) with broad emission [148]. The $CuInSe_2$ absorber layer, which is employed to >9% efficiency glass/Mo(1 μm)/$CuInSe_2$/CdZnS(0.5 μm)/ZnO(1–2 μm) thin film solar cell yields dominant peak at 0.85 eV (Cu_{In}–In_{Se}) and another low intensity peak at 0.93 eV (V_{Cu}–In_{Cu}), whereas the absorber shows only one dominant peak at 0.95 eV in 4.2 K PL spectra for <1% low efficiency cell. The as-grown CIS thin film solar cell with efficiency of 3% shows double peaks at the same positions of 0.85 and 0.93 eV with low intensity. The efficiency of air anneal thin film solar cell at 200 °C for 2–8 h increases from 3 to >9%. On the other hand, the peak at 0.85 eV pronounces well with unresolved 0.93 eV peak, as compared to that of as-grown cell and $CuInSe_2$ layer. Likely, the $CuInGaSe_2$ absorber with Ga/(In + Ga) = 0.4, which is applied for high efficiency (>9%) thin film solar cells, shows strong emission peak at 1.16 eV. This peak may be due to DAP transition [149]. The peak at 0.85 eV may be due to transition from double donor (V_{Se}) to (V_{Cu}), that is, (V_{Cu}–V_{Se}) or (Cu_{In}–In_{Se}).

Mickelsen et al., [150] studied oxygen annealing effect on the CIS thin films. The CIS/CdS cell shows efficiency of 11%. The $CuInSe_2$ polycrystalline thin film absorbers deposited onto Mo coated alumina substrates by elemental co-evaporation method are employed in the thin film solar cell, which shows a sharp peak at 0.98 eV (1255 nm) in 12 K photoluminescence spectrum. After annealing $CuInSe_2$ thin film under oxygen at 225 °C for 20 min, 0.98 eV peak disappears and broad peaks at 0.94 eV (1318 nm), 0.87 eV (1415 nm), and 0.815 eV (1365 nm) originate. A sharp 0.98 eV peak can be interpreted to the transition from the V_{Cu} acceptor level of 60 meV to the top of the conduction band (FB). It is independent of excitation power indicating assignment of FB is appropriate, as discussed in earlier [151]. The other three peaks could be due to transitions from the three donor levels to the single acceptor level such as (i) V_{Se}(60 meV)–V_{Cu}(40 meV) for 0.94 eV emission peak, (ii) double donor V_{Se} (123 meV)–V_{Cu}(40 meV) for 0.87 eV emission peak, and (iii) donor level V_{Se} (85 meV) and V_{Cu} (40 meV) for 0.915 emission peak. Two different kinds of samples are picked up for testing to find annealing effects such as one with 1340 nm emission peak and another one 1420 nm samples [152]. After annealing the former at 150 and 200 °C for 30 min under air shows peak shift from 1340 to 1420 nm. After annealing the same sample under H_2 at 200 °C for 30 min, the 1420 nm (0.87 eV) peak reverses to 1340 nm (0.93 eV). The 1420 nm peak may be related to the transition from O_{Se}complex to V_{Cu}, there are other possibilities V_{In}–O_{Se} complex, Cu_{In}–O_{Se} complex, but the composition of the films are Cu/In < 1, hence the transition V_{Cu}–O_{Se} complex may be reasonable. The 1340 nm peak shifts to 1420 nm position after annealing the $CuInSe_2$/CdS sample, whereas ZnO coated on $CuInSe_2$/CdS sample annealed under air at 150 °C for 30 min does not show any shift that means the ZnO layer strongly blocks the oxygen interference with the sample. There are two ways to interpret a change in DAP transition (V_{Cu}–V_{Se}) from 0.93 to 0.87 eV. After annealing under air, the sample may be oxygenated that causes to increase the band gap of sample therefore defect level transition length may be proportionally enhanced. Secondly, the Se may be replaced by oxygen in the $CuInSeV_{Se}$ system that turns into $CuInOV_{Se}$ complex therefore the donor level becomes deeper or single donor may be converted into double donor. Since the sulfur and oxygen are the same group (VI) elements in periodic table. Annealing sample under air means oxygen replaces Se atom that is, antisite does not create defect level for the transition unlike (I_{III}) Cu_{In} or In_{Cu} (III_I), etc., however, the participation of oxygen with already existed defect may have reduced the transition length between donor and acceptor by creating deeper donor.

5.2.5 CuGaSe$_2$

The A, B, and C free excitons can be obtained from the low temperature photoconductivity of $CuGaSe_2$. Figure 5.44 shows photoconductivity spectrum of $CuGaSe_2$ (CGS) single crystals grown by ITM recorded at 90 and 300 K [153]. At 90 K, the emission line E_1 (1.83 eV) is from valence band maximum to the conduction band minimum (Γ^4_{7v}–Γ^1_{6c}, i.e., A), a shoulder E_2 (2.0 eV) is from crystal field

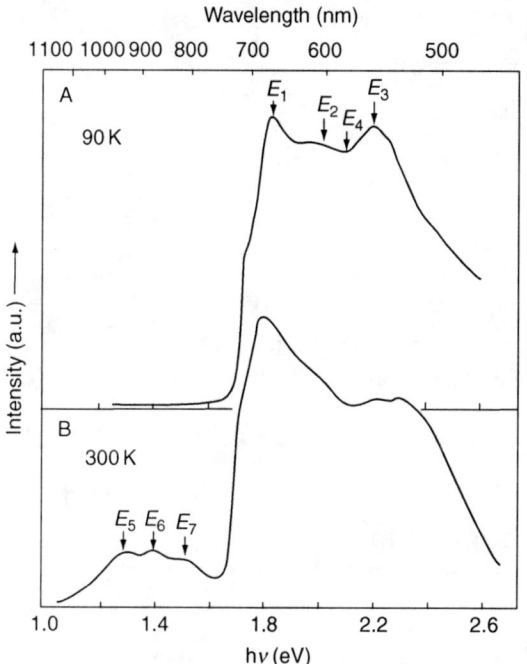

Figure 5.44 Photoconductivity spectra of CuGaSe$_2$ single crystal recorded at (A) 90 K and (B) RT.

splitting state to the conduction band minimum ($\Gamma^5_{6v}-\Gamma^1_{6c}$, i.e., B), another shoulder E_3 (2.21 eV) is from the spin-orbit splitting state to the conduction band minimum ($\Gamma^5_{7v}-\Gamma^1_{6c}$, i.e., C) (Figure 5.2). E_4 (2.13 eV) emission line may be assigned to transition between copper 3d level and valence band. The PL determines E_1 (1.81 eV A) and E_2 (2.02 eV B) emission lines in the Cu-rich and stoichiometric CuGaSe$_2$ thin films [154]. The A, B, and C transitions observed by the photoconductivity are little higher than the standard values of 1.729, 1.813, and 2.016 eV from the reflectivity spectra, respectively [155]. At 300 K, the observed E_5 (1.28 eV) and E_7 (1.55 eV) may be related to deep acceptor centers and E_6 (1.38 eV) is a deep donor due to iodine substitution on selenium. In the case of CuGaSe$_2$ thin films grown onto GaAs, the A and B free excitonic values of 1.713 and 1.71 eV are low due to tensile strain in the films [156]. A several transition lines such as a DAP transition, that is, ($V_{Cu}-V_{Se}$) at 1.584 eV (783 nm), followed by LO phonon lines at 1.56 eV (795 nm) and 1.536 eV (807 nm) with separation of 24 meV and TO phonon line at 1.501 eV (826 nm) with separation of 35 meV are observed in 85 K absorption spectrum (AS) of CuGaSe$_2$ single crystals formed by halogen transport technique, as shown in Figure 5.45 [157]. The similar emission lines at 1.627 eV due to FB transition, that is, (V_{Cu}-CB), DAP at 1.593 due to ($V_{Cu}-V_{Se}$) transition, followed by phonon line at 1.562 eV with separation energy of 31 meV are observed in 10 K PL spectrum of CuGaSe$_2$ epilayers. The value of NBE at 1.709 eV is less than the bulk value of 1.73 eV. A difference between them that is, redshift is due to thermally induced

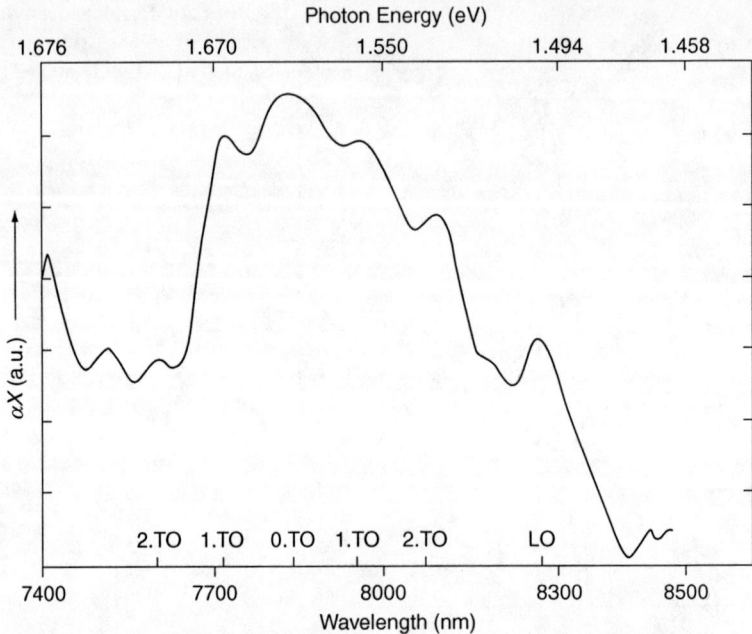

Figure 5.45 Absorption spectrum of CuGaSe$_2$ single crystals recorded at 85 K.

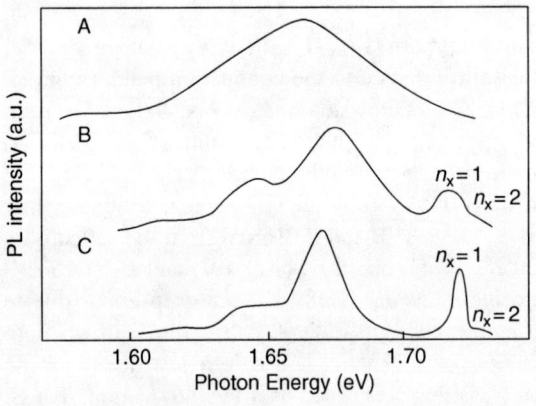

Figure 5.46 PL spectra of CuGaSe$_2$ single crystals recorded at different temperatures (A) 300, (B) 78, and (C) 4.2 K.

strain in the epilayers [158]. There is a very slight difference between DAPs values obtained from AS and PL that could be due to nature of instrumental error.

The emission peaks at 1.7210 and 1.7281 eV in 4.2 K PL spectra recorded on CuGaSe$_2$ crystals grown by directional freezing method (Figure 5.46) are due to the ground state ($n_x = 1$) and first excited state ($n_x = 2$) of the free exciton, respectively. Using Equation (5.10), the exciton binding energy of 9.5 meV can be determined. Adding this to the free excitonic value gives band gap of 1.7305 eV for CuGaSe$_2$ at 4.2 K similar to CuInSe$_2$ [97]. There are several emission peaks emerge in the

CuGaSe$_2$ crystals with composition ratio of Cu/Ga = 1.0 grown by VGF and CVT, as shown in Figure 5.47A, which can be assigned to free exciton (FE-1.720 eV), free to bound acceptor transition (FB-1.678 eV), that is, (CB–V$_{Cu}$), donor to free transition (DF-1.644 eV), that is, (VB–V$_{Se}$), its phonon replica (DF-LO-1.606 eV) and DAP transition (DAP-1.56 eV), that is, (V$_{Cu}$–V$_{Se}$). The depths of defect levels V$_{Cu}$ and V$_{Se}$ are 57 and 102 eV, respectively [156]. 10 K PL shows that the verbatim emission lines such as FE at 1.725 eV, FB at 1.675 eV, and its phonon replica at 1.63 eV and (V$_{Cu}$–Ga$_{Cu}$) defect level at 1.55 eV in the CuGaSe$_2$ single crystals grown by IT method. The thermal quenching energy of 55–60 meV for ~1.678 eV peak is due to V$_{Cu}$ in the CGS [101]. Two BXs at 1.719 and 1.706 eV along with free exciton at 1.723 and three unresolved DAP peaks at 1.466, 1.494, and 1.526 eV in the CVT CuGaSe$_2$ single crystals are also observed [159]. Figure 5.47B shows that the DAP transition triumphs in broad with decreasing Cu/Ga ratio from 1.4 to 0.9 that means the concentrations of selenium vacancies (V$_{Se}$) may be increased in the crystals. The intensities of FB (1.678 eV) and DF (1.644 eV) peaks vary linearly or sublinearly with increasing excitation intensity but no shifts in peak positions are observed. Therefore, assigning FB and DF to 1.678 and 1.644 eV peaks are relevant, respectively [156], whereas the emission band at 1.56 eV shows blue shift with increasing excitation intensity thus the assignment to DAP is virtually proper [160]. Of course, the DAP line position varies toward blue region with variation of excitation intensity and redshifts with variation of temperature [161]. The typical transition levels are shown as model-1 in Figure 5.48. The crystals annealed under vacuum show an increased concentration of donor level of V$_{Se}$ relating to emissions below 1.65 eV. After annealing the crystals under excess Cu, the FB (V$_{Cu}$–CB) emission at 1.675 eV disappears and the DF (V$_{Se}$–VB) emission at 1.645 eV strongly generates in 10 K PL spectrum indicating that copper vacancies (V$_{Cu}$) are suppressed, relatively selenium vacancies (V$_{Se}$) may be enhanced [162]. This experimental observation supports to assign transition levels, as given in Figure 5.48.

The divalancy donor V$_{Se}$ (110 meV) and acceptor V$_{Cu}$ (50 meV) may be the transition centers for DAP (1.56 eV) recombination. The broad emission band at 1.675 eV is due to (V$_{Cu}$–CB) for Cu/Ga = 0.87, which appears with low intensity for Cu/Ga = 0.97 in the CuGaSe$_2$ thin films formed by RTP of stacked elemental layers. In addition, 1.63 (V$_{Se}$–VB), 1.60 (divalency V$_{Se}$–VB), 1.579 (V$_{Cu}$–V$_{Se}$), and 1.55 eV (divalency V$_{Cu}$–V$_{Se}$) peaks present for Cu/Ga = 1.06 in the PL spectra [163]. The FE line position in the 2 K PL spectra gradually shifts from 1.711 to 1.702 eV with increasing Cu/Ga ratio from 1.09, 1.59 to 2.16 in 0.7 μm thick CuGaSe$_2$ epitaxial layers grown onto GaAs (001) substrates at 490 °C by MBE [164]. The quasi DAP transition dominates the spectrum for CGS polycrystalline thin films with Cu/Ga < 1. The films with Cu/Ga > 1 or nearly stoichiometric CGS films show free excitonic emission at 1.726 eV, BX at 1.709 eV, and FB peak at 1.67 eV and its phonon replicas. As Cu content is increased, another DAP at 1.63 eV paramounts 10 K PL spectrum [165]. In some cases, only one peak at 1.61 eV close to 1.59 eV emission band, which demonstrates in the CuGaSe$_2$ thin films, could be due to DAP transition that is, from V$_{Cu}$ to V$_{Se}$ [166,167]. A different group [161] studied 5.2 K PL spectra of CuGaSe$_2$ epilayer grown onto GaAs substrates with varying

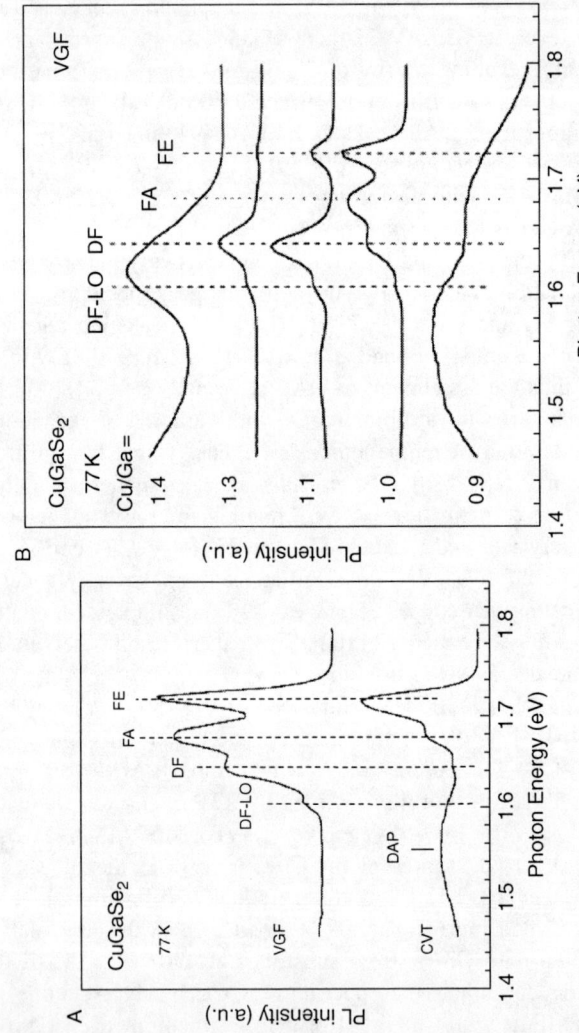

Figure 5.47 (A) 77 K PL spectra of CuGaSe$_2$ crystals grown by different techniques such as VGF and CVT and (B) 77 K PL spectra of CuGaSe$_2$ crystals with different composition ratios of Cu/Ga grown by VGF technique.

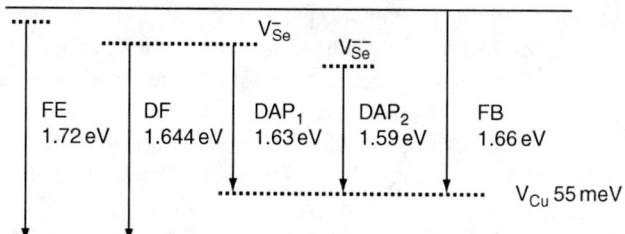

Figure 5.48 The typical radiative recombinations in CuGaSe$_2$ (model-1).

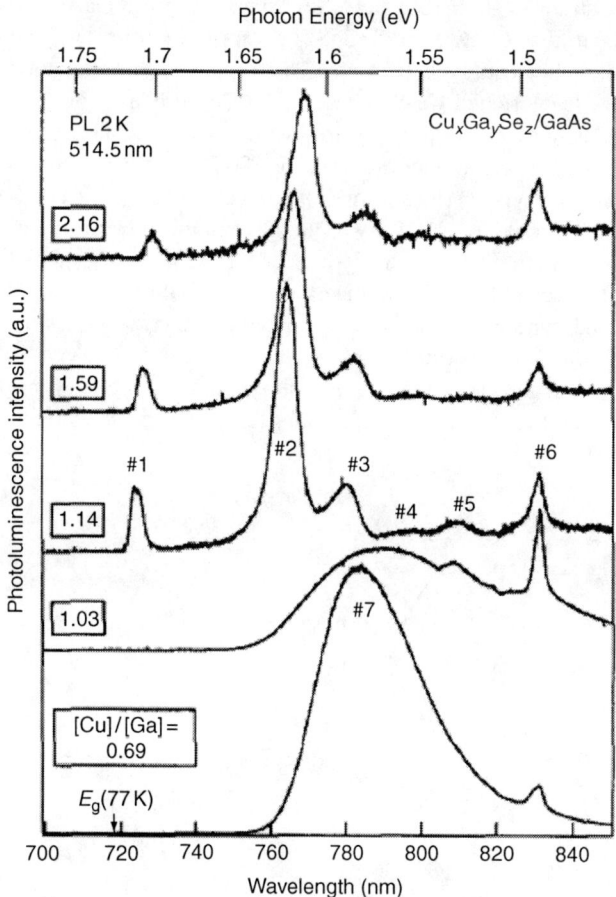

Figure 5.49 PL spectra of Cu$_x$Ga$_y$Se$_z$ thin films with different composition ratios of Cu/Ga grown onto GaAs substrates.

Cu/Ga ratios from 0.69, 1.03, 1.14, 1.59 to 2.16 (Figure 5.49). The excitonic emissions at 1.713 ($n_x = 2$) and 1.710 eV ($n_x = 1$) (#1) as doublets and emission at 1.622 eV (#2) due to DF, that is, (V$_{Se}$–VB) exists for nearly stoichiometric films. FWHM of DF increases with increasing excitation intensity indicating that free carriers participate through the band filling. Its phonon lines present at 1.588 eV (#3), and 1.554 eV (#4) by separation of 34 meV coinciding with the reported value of 34.5 meV from the reflectivity spectrum [168] and DAP line at 1.53 eV. The emission line at 1.58 eV for the composition of Cu/Ga = 0.69 is already attributed to DAP corresponding to transition of (V_{Cu}–V_{Se}). The similar peak (1.58 eV) is observed for the CGS layers with Cu/Ga = 0.98 grown onto ZnO. 1.54 and 1.62 eV peaks for Cu/Ga = 1.82 and 0.59, respectively are observed [169]. Slightly red-shifted exciton emission line at 1.712 eV in the CuGaSe$_2$ layers grown onto GaAs by MBE is sharper than that of 1.720 eV line in VGF crystals. The redshift may be due to tensile strain in the MBE CuGaSe$_2$ layers from the GaAs substrates [156]. After KCN treatment of Cu-rich CuGaSe$_2$ layers the strong DAP peak (1.63 eV) shifts towards higher energy by 8 meV. This is due to the relaxation of CuGaSe$_2$ epitaxial layers from etched Cu$_x$Se phases. The Cu$_x$Se phases in turn create tensile stress in the CuGaSe$_2$ layers [170]. Three varieties of CuGaSe$_2$ crystals grown by iodine vapor transport techniques are used for 100 K PL measurements. One version of CuGaSe$_2$ crystals (type-1) shows emission lines at 1.69 and 1.72 eV so-called A and B excitons and a shoulder at 1.59 eV. Type-2 crystals show emission lines at 1.63 and 1.59 eV. Only one broad peak is noticed at 1.33 eV in type-3 CGS crystals, as shown in Figure 5.50. The raw materials of CuGaSe$_2$ show similar spectra of both type-1 and type-2 except type-3 indicating that the emission line at 1.33 eV might be due to substitution of iodine in selenium. Type-3 CGS crystal annealed under vacuum at 600 °C shows suppression of 1.33 eV peak and generation of 1.69 and 1.63 eV peaks. Again annealing the crystal under iodine vopor

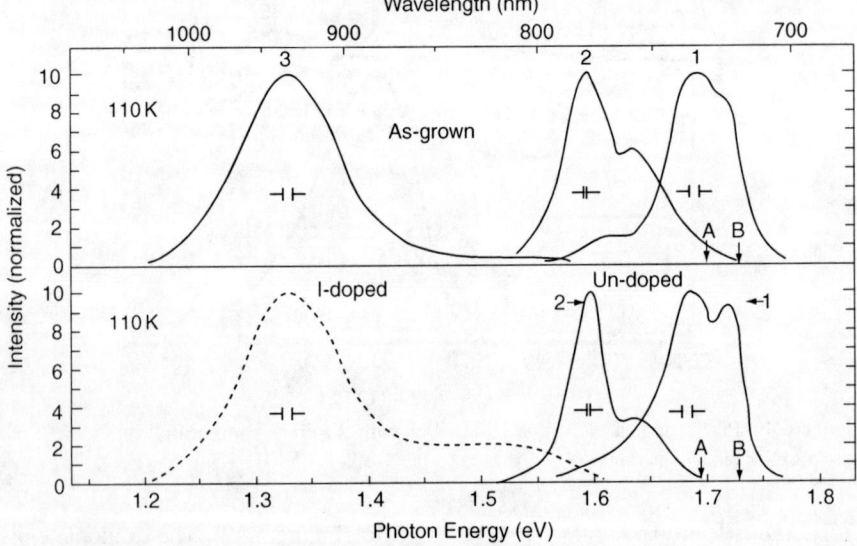

Figure 5.50 PL spectra of CuGaSe$_2$ single crystals grown by iodine vapor transport; type-1, -2, and -3.

at 600 °C it regains 1.33 eV peak and suppression of other peaks such as 1.69 and 1.63 eV. The type-3 crystals annealed under Se vapor at 600 °C show A and B excitons at 1.69 and 1.72 eV, respectively, and 1.33 eV emission line absents. The emission at 1.33 eV is due to substitution of iodine into Se that is, I_{Se} as a deep donor with depth of 0.38 eV. It can be easily confirmed that iodine is responsible for the existence of 1.33 eV emission line [171]. It is already mentioned that the peak at 1.59 eV is due to recombination of DAP and responsible to the transition of (V_{Cu}–V_{Se}), whereby 80 and 55 meV are activation energies of V_{Se} and V_{Cu}, respectively. The experimental observation confirms that the DAP emission at 1.53 eV blue shifts to 1.62 eV with increasing excitation intensity in the $CuGaSe_2$ thin films grown by MOMBE [172].

One set of p-$CuGaSe_2$ crystals show predominant A-exciton emission at 1.72 eV, FB emission line at 1.68 eV, DF at 1.64 eV, and DAP at 1.52 eV, whereas another set of samples present only two emission lines at 1.72 and 1.68 eV with more or less the same intensities. First set of samples annealed under vacuum at 350 °C for 1½ h retains emission line at 1.68 eV as a dominant along with other lines. In addition, a new peak emerges at 1.30 eV and emission peak at 1.52 eV well resolves in the spectrum that could be due to recrystallization. In the second set of annealed samples, emission lines at 1.72 and 1.66 eV remain at the same status without any change in the spectrum [173]. However, after annealing as-grown second set of samples in the presence of Se at 785 °C for 48 h, the free exciton peak at 1.72 eV disappears and the FB (V_{Cu}–CB) peak pronounces well at 1.686 eV, as shown in Figure 5.51. The carrier concentration decreases by 1 order of magnitude by annealing the sample. This observation reveals that the Se compensates Se vacancies and

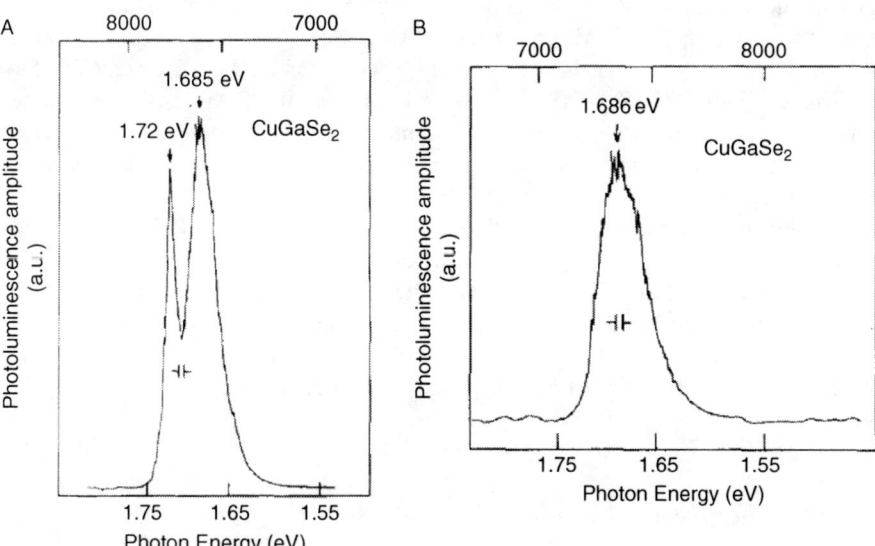

Figure 5.51 PL spectra of $CuGaSe_2$ crystals (A) as-grown and (B) annealed under Se at 785 °C for 48 h.

relatively may enhance copper vacancies [174]. The CL spectrum of CuGaSe$_2$ single crystal shows emission line at 1.59 eV; however, crystals annealed under Ga–Se, Ga and Se atmosphere show emission lines at 1.59, 1.63, and 1.66 eV and their thermal quenching analyses reveal activation energies of 70, 100, and 50 meV, respectively. Subtraction of emission bands 1.63 and 1.66 eV from the energy band gap ($E_g = 1.71$ eV) of CuGaSe$_2$ gives qualitatively nearly equal to the activation energies of the respective emission bands, that is, (i) 1.71–1.63 = 80 meV (V_{Se}) and (ii) 1.71–1.66 = 50 meV (V_{Cu}). The DAP transition (1.59 eV) is the recombination of defect level with activation energy of 80 meV due to V_{Se} donor and the defect level with activation energy of 50 meV due to acceptor level V_{Cu} [175]. Two DAP 1.58 and 1.53 eV emission lines are observed by electroluminescence experiments in the p-CuGaSe$_2$ single crystals [176]. The as-grown single crystals annealed under Se atmosphere show asymmetric emission peak at 1.663 eV, which fits Gaussian-shape that turns into two emission peaks at 1.672 and 1.633 eV with activation energies of 60 and 95 meV corresponding to V_{Cu} and Se$_i$ acceptor levels. Similar kind of activation energies obtained from the Hall measurements are varied from 80 to 90 meV for annealed CGS under excess Se. A broad peak varies from 1.58 to 1.63 eV for CuGaSe$_2$ crystals annealed under Ga atmosphere for which taking V_{Cu} level of 55 meV, a donor level of 110 to 120 eV may be assigned to Ga$_{Cu}$ or Ga$_i$ [162]. The PL studies of CuGaSe$_2$ epitaxial layers grown onto GaAs substrates by MOCVD shows several emission bands such as weak excitonic peak at 1.71 eV, DAP$_1$ at 1.66 eV, DAP$_2$ at 1.63 eV and DAP$_2$-LO at about 1.61 eV. The DAP$_1$ tops the spectrum in the Cu-rich layers, whereas DAP$_2$ leads in stoichiometric layers. There is no remarkable difference observed in the PL spectra for ZnO sputtered on Cu-rich CuGaSe$_2$ absorber layers but the intensity is slightly increased. The DAP$_1$ prominences the spectra for ZnO on moderate Cu-rich CuGaSe$_2$ layers. These observations concluded that the ZnO sputtering on CuGaSe$_2$ does not damage the surface of the absorber layers. No change in the PL spectra for ZnO on Cu-rich CuGaSe$_2$/CdS or Cu-poor CuGaSe$_2$/ZnSe is observed, whereas the DAP$_1$ is the reminisced emission band in the PL spectra for moderately Cu excess samples. An increase in intensity of DAP$_2$ manifests an increase of concentration of shallow acceptor levels [170].

The free to bound acceptor transition (FB$_1$) is separated from the DAP$_1$ transition at temperature ≥ 40 K for Cu$_x$Ga$_y$Se$_2$ epilayers grown onto GaAs (001) by CVD, as shown in Figure 5.52A. The DAP$_1$ leads the PL spectra at low temperature. This kind of combined FB and DAP transitions is occurred at two places as DAP$_1$ + FB$_1$ and DAP$_2$ + FB$_2$ relating to the positions 1.66 + 1.67 and 1.62 + 1.63 eV in the PL spectra, respectively. The FB can be related to Equation (5.8)

$$h\nu_1 = E_g - E_{A/D} + kT, (1.67 = 1.73 - 0.060 + kT).$$

$$h\nu_2 = E_g - E_{A/D} + kT, (1.63 = 1.73 - 0.100 + kT),$$

DAP transition can also be related to Equation (5.7) [177]

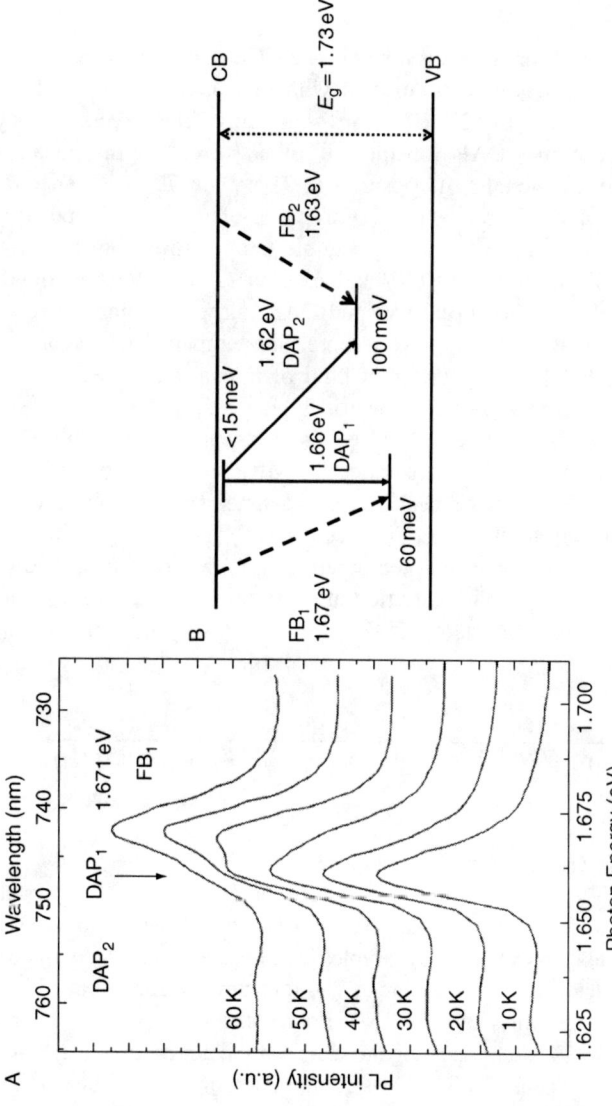

Figure 5.52 (A) PL spectra of CuGaSe$_2$ thin films recorded at different temperatures and (B) Typical transition levels in the CuGaSe$_2$-model-2; participating one donor with two acceptors.

$$hv(DAP_1) = E_g - (E_A + E_D)$$
$$+ e^2/4\pi\varepsilon_0\varepsilon_r r, \{1.66 = 1.73 - (0.060 + 0.010) + 0.003\},$$

$$hv(DAP_2) = E_g - (E_A + E_D)$$
$$+ e^2/4\pi\varepsilon_0\varepsilon_r r, \{1.62 = 1.73 - (0.100 + 0.010) + 0.003\}.$$

A similar kind of observations is noted in the CGS epilayers. The emission line at 1.628 eV relates to FB is remain constant with increasing triethyl gallium (P_{TEGa}) pressure from 2.4×10^{-7} to 12×10^{-7} torr while depositing layers but beyond this pressure, the FB becomes DAP transition. In other words, FB disappears and DAP emerges and A-exciton retains its position (1.71 eV) in 77 K PL spectra of CGS layers grown by MOCVD onto GaAs. Variations of defect level positions in the PL spectrum are observed with varying sample temperature from 8 to 160 K. The DAP_2 (1.625 eV), DAP_1 (1.670 eV) and exciton (1.708 eV) are found at 8 K, whereas DAP_2 becomes FB_2 (1.63 eV) and DAP_1 disappears and A-exciton retains in the same position at 80 K that is, with increasing temperature. The only A-exciton is observed at 160 K [178]. A different kind of transitions of defect levels in the $CuGaSe_2$ are given in model-2, as shown in Figure 5.52B.

The dominant emission line at 1.62 eV is observed for the $CuGaSe_2$ epitaxial layers grown onto GaAs (001) substrates by MBE, which is treated as donor to valence band (V_{Se}–V_B) but the peak can be assigned to DAP based on the effect of temperature or variation of excitation intensity. The peak position shifts from 1.627 to $\sim$1.64 eV with increasing temperature from 6.5 to 94.6 K. The same peak shifts from 1.62 to 1.63 eV with variation of excitation intensity from 1.3 µW to 148 mW. The phonon replica exists at 1.587 eV for DAP. By fitting thermal quenching data to the

$$I = \frac{C_1}{1 + C_2 \exp(-E_D/kT)} + \frac{C_3}{[1 + C_4 \exp(-E_D/kT)][1 + C_5 \exp(-E_A/kT)]}, \tag{5.13}$$

where C_1, C_2, C_3, C_4, C_5, and C_6 are constants.

The obtained E_D and E_A values are 108 and 3.4 meV, which may be probably responsible to the Cu_{Ga} and V_{Se}, respectively. The DAP emission band (1.62 eV) disappears provided that the as-grown $CuGaSe_2$ samples are annealed under Se with combination of Se pressure at 1 torr and Ar pressure at 760 torr atmosphere and at 350 °C for 3 h. The disappeared DAP emission (1.62 eV) arises at the same position along with new emission line at 1.66 eV after annealing the same sample under Ar atmosphere (760 torr) at 380 °C for 2 h. The peak at 1.714 eV is combination of free exciton and BX [179].

There is another defect model proposed that the as-grown $CuGaSe_2$ crystals and annealed under various atmospheres show more or less the same peak positions at 1.58 and 1.55 eV for the point defect pairs of $(V_{Cu}^- + Ga^{2+})^+$ and $2(V_{Cu}^- + Ga^{2+})°$, respectively. However, the position at 1.614 or 1.618 eV disappears in the crystals after annealing under O_2 and Se atmosphere giving impression that the Se vacancy defects are compensated by either oxygen or Se, whereas the peak at 1.618 eV retains at the same position after annealing crystals under H_2, as shown in Table 5.5 [180]. Two emission lines at 1.62 and 1.58 eV appear in the as-grown and moderately resolve in H_2 annealed atmosphere, whereas the peaks disappear and new emission line emerges at 1.57–1.58 eV, after individual annealing under O_2 and Se_2, which may be due to antisite defect of Ga_{Cu} (Figure 5.53). The ESR analysis shows that the ESR peak I_0 at the magnetic field of circa 325 mT in as-grown crystals is as same as in H_2 annealed crystals, whereas I_0 intensity of crystals annealed under O_2 and Se_2 decreases that may be indication of decrease of V_{Se} concentration. The g-factor 2.006 of I_0 is close to the free electrons value of 2.003. The H_2 annealing appears to be increasing density of singly charged point defect pairs $[V_{Cu}^- + Ga_{Cu}^{2+}]$ due to passivation of copper vacancies and maximum amplitude of ESR-I_0 signal indicates cleaning up of singlet paramagnetic centers. The newly generated peak I_1 at circa 321 mT may be due to resonance of Ga–O bands [181].

Table 5.5 Annealing effects on the defect levels by different atmosphere

CGSe crystals	V_{Se}^+ (eV)	$(V_{Cu}^- + Ga^{2+})^+$ (eV)	$2(V_{Cu}^- + Ga^{2+})^\circ$ (eV)
As-grown	1.614	1.58	1.548
Annealed under H_2	1.618	1.57	1.545
Annealed under O_2	disappears	1.58	1.554
Annealed under Se	disappears	1.581	1.557

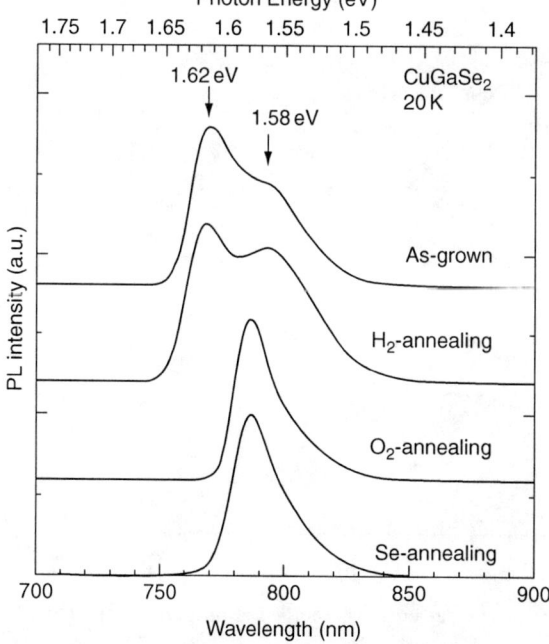

Figure 5.53 PL spectra of as-grown, H_2 and O_2 annealed $CuGaSe_2$ single crystals.

Let us see doping effects in the CuGaSe$_2$ single crystals. Several PL lines at 1.675 (V$_{Cu}$), 1.645 (V$_{Se}$), 1.62 (DAP), 1.52, and 1.30 eV appear in the as-grown CVT CuGaSe$_2$ crystals, which are well pronounced after annealing under vacuum at 400 °C for 15 min. However, B and Ge implanted samples show no emission lines but after rapid annealing, the peaks at 1.675, 1.645, and 1.62 eV appear [182]. If large amount of iodine is used for crystal growth, the emission band at 1.32 eV appears. In such a way the peaks at 1.5 and 1.55 eV for bromine and chlorine doped samples are seen, respectively. The peak positions at 1.57, 1.60, and 1.61 eV for Zn, Mg, and Cd occur at 2–100 K, respectively. Strong secondary emissions at 1.4, 1.42, and 1.44 eV for Zn, Mg, and Cd due to DAP exist at above 100 K, respectively. The estimated activation energies such as 80, 90, 110, and 210 meV are corresponded to Cd$_{Cu}$, Mg$_{Cu}$, Zn$_{Cu}$, and II$_{Ga}$, respectively. Eventually, group II elements doped samples show two DAP lines at 1.35–1.4 eV and 1.6 eV relate to V$_{Cu}$–II$_{Cu}$ and II$_{Cu}$–II$_{Ga}$, respectively. The PL spectrum of a typical Mg-doped CuGaSe$_2$ single crystal is shown in Figure 5.54. They are solely related to the DAP transitions. The Al doped samples show only $\sim$1.675 eV emission line slightly higher than that of undoped samples that means Al is solidified into CuGaSe$_2$ as CuAl$_{1-x}$Ga$_x$Se$_2$ alloy unlikely antisites created by other dopants such as Zn$_{Cu}$ or Zn$_{Ga}$, which is due to transition from V$_{Cu}$ to CB and the activation energy found from the thermal quenching of the PL emission is 55 meV matching with V$_{Cu}$ defect energy [183]. Other kind of dopant effects on CuGaSe$_2$ single crystals are also studied [184,185]. As, P, and Sb doped CuGaSe$_2$ crystals show PL peaks at 1.73, 1.67–1.685, and 1.61–1.63 eV for excitonic transition, FB and DAP transitions, respectively. The activation energies of 45, 50, and 60 meV for As, P, and Sb are obtained from thermal

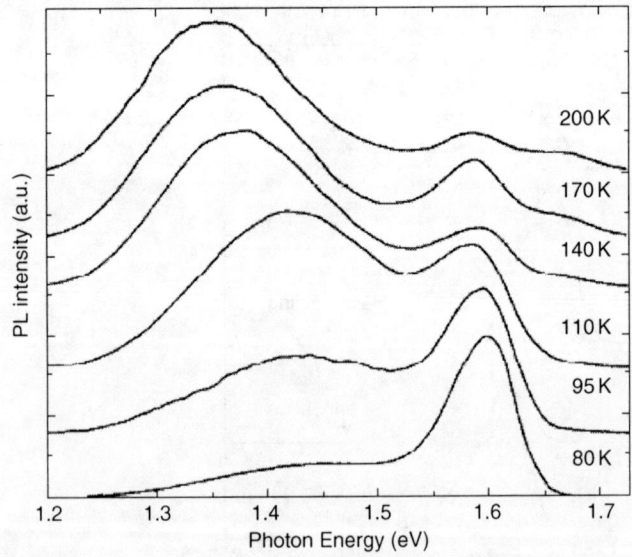

Figure 5.54 PL spectra of Mg-doped CuGaSe$_2$ crystals recorded at different temperatures.

quenching analysis, which replace Se in the CuGaSe$_2$ crystals. After annealing Na doped crystals under air at 200 °C for 5 h, a shift in lower energy for DAP transition (V$_{Cu}$–V$_{Se}$) peak at 1.61–1.63 eV is observed. The intensity of FB (1.67 eV) peak increases because oxygen sits on Se site (O$_{Se}$) and decreases concentration of V$_{Se}$. This kind of effect is high in Na doped crystals. Sn or Ge and Si doped crystals show sharp DAP transitions at 1.35 and 1.425 eV, respectively. The activation energy of 300 meV for Si doped and 400 meV for Sn or Ge doped CuGaSe$_2$ crystals could be due to IV$_{Cu}$ donor. The Ge and Zn co-doped crystals annealed at low temperature show n-type conductivity forming Ge$_{Ga}$ but annealed at high temperature show Zn$_{Ga}$.

In conclusion, the experimental results indicate that the activation energy levels of shallow acceptor V$_{Cu}$ and donor V$_{Se}$ are 45 and 60 meV in the CuInSe$_2$, respectively, whereas they are 60 and 80 meV in the CuGaSe$_2$, respectively. The phonon energies are 27 and 36 meV in the CuInSe$_2$ and CGS, respectively. The donor defect level Ga$_{Cu}$ is deeper in the CuGaSe$_2$ than In$_{Cu}$ in the CuInSe$_2$ therefore it is not easy to create such kind of deeper donors in the CuGaSe$_2$ to fabricate n-CuGaSe$_2$ but n-CuInSe$_2$ can easily be prepared by annealing p-CuInSe$_2$ either under vacuum or in the presence of In [101]. Sometimes it is hard to assign defect levels properly, for example the activation energies of V$_{Cu}$, Cu$_{In}$, and V$_{In}$ are very close to each other but it can be possible based on experiments.

5.2.6 Cu(In$_{1-x}$Ga$_x$)Se$_2$

Figure 5.55 shows 2 K PL spectra of Cu(In$_{1-x}$Ga$_x$)Se$_2$ single crystals with various compositions $x=0.0$, 0.25, 0.50, 0.75, and 1.0 prepared by normal freezing method. The relation between FB transition and variation of composition from $x=0$ to 1.0 in steps of 0.25 in slightly Cu-rich or In/Ga poor single crystals is found to be E_{FB} (eV) $= 1.66 - 0.85(1-x) + 0.17(1-x)^2$ in which bowing parameter is 0.17. The V$_{Cu}$ is one of the probably participating acceptors in FB transition as (V$_{Cu}$–CB) amongst Cu$_{In}$ and Cu$_{Ga}$. Based on the previous studies in the CuInSe$_2$ single crystals, four dominant peaks at 1.042, 1.027, 0.989, and 0.971 eV can be assigned to free exciton (A), donor BX (VB–Se$_{In}$), FB (CB–V$_{Cu}$) and DF (VB–V$_{Se}$) or DAP transition, respectively [186]. PL spectra of CuIn$_{1-x}$Ga$_x$Se$_2$ single crystals grown by THM exhibit single and doublet peaks and their positions shift to higher energy levels with increasing x values from 0, 0.21, 0.42, 0.71 to 0.96, as shown in Figure 5.56. For example the doublets are at 1000 and 1050 nm for $x=0.42$ [187]. Cathode luminescence (CL) spectrum of n-CuGa$_{0.5}$In$_{0.5}$Se$_2$ single crystals grown by melt-grown technique or annealed under vacuum at 600 °C for 1 day shows two emission peaks at 1.32 eV as a strong one (J) and 1.35 eV (I_2+I_3) as a shoulder. As the temperature of sample is decreased to 77 K, almost all the peaks reside at the same position but a new peak at 1.27 eV (K) occurs (Figure 5.57), whereas after annealing the as-grown crystals under Se atmosphere at 700 °C for 3 days turns into p-type, which show emission line at 1.40 eV (I_1) as a distinction and a shoulder at 1.32 eV. The CuIn$_{1-x}$Ga$_x$Se$_2$ crystals with other compositions annealed under Se at 700 °C for 3 days show emission band at 1.23 eV for $x=0.20$, 1.46 eV, and 1.40 eV for $x=0.62$, 1.52, and 1.40 eV for

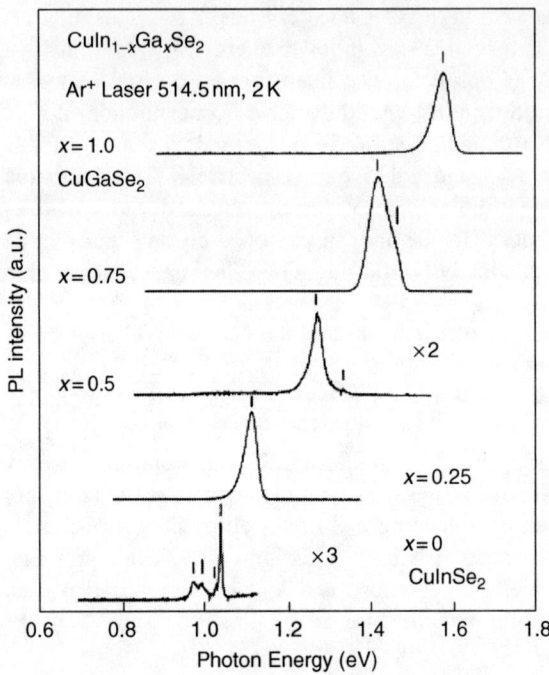

Figure 5.55 2 K PL spectra of $CuIn_{1-x}Ga_xSe_2$ single crystals with different x values.

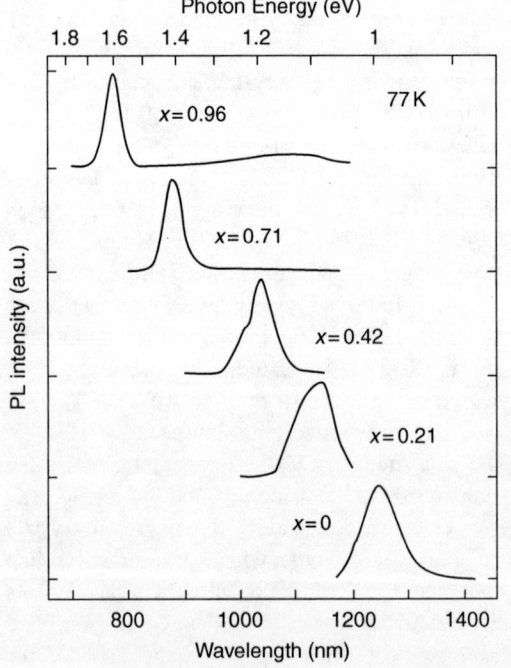

Figure 5.56 77 K PL spectra of $CuIn_{1-x}Ga_xSe_2$ single crystals with different x values.

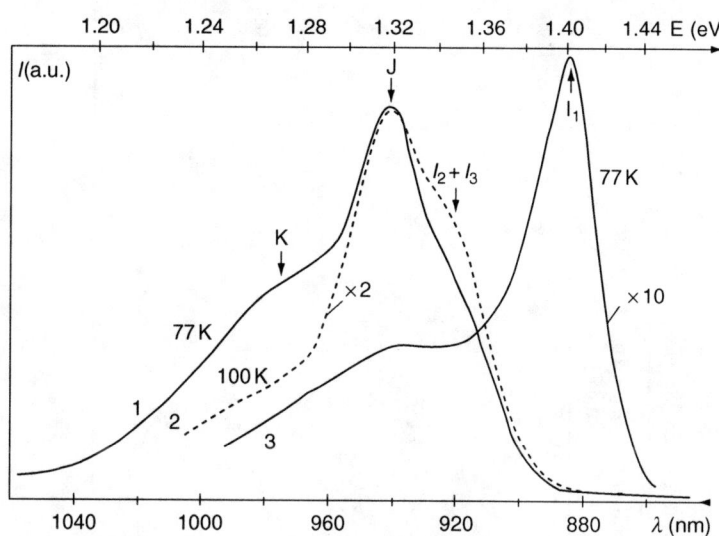

Figure 5.57 Cathodoluminescence of CuIn$_{0.5}$Ga$_{0.5}$Se$_2$ single crystals; (1) as-grown or annealed under vacuum at 600 °C for 1 day recorded at 77 K (*n*-type), (2) recorded at 100 K for the same crystals (*n*-type), and (3) crystals annealed under Se at 700 °C for 3 days (*p*-type).

$x = 0.74$ and 1.68 eV for $x = 1$, respectively. The emissions at 1.40 and 1.35 eV are due to FB (CB–V$_{Cu}$) and free to bound donor (VB–V$_{Se}$) or (Cu$_{In}$–CB), respectively. The activation energies are 50 and 70 meV for 1.4 (I_1) and 1.35 eV (I_2) emission bands, respectively. The thermal quenching analysis reveal two activation regions for 1.32 eV (J) emission band; one at high temperature region (100 meV) and another one at low temperature region (40 meV). The 1.32 eV emission band is due to DAP transition with acceptor level of V$_{Cu}$ but donor assignment is in uncertain. The emission band at 1.40 eV is continuously emerging irrespective of x values in the spectrum therefore it should be due to V$_{Cu}$ because, which is a common defect in the I–III–VI$_2$ system [188].

The CuIn$_{1-x}$Ga$_x$Se$_2$ single crystals with $x = 0.11$, 0.28, and 0.58 grown by Bridgman-Stocbarger high pressure technique show activation energy levels of 37 and 42 meV for 0.95–0.97 and 1.049–1.06 eV peaks, respectively. Two activation energy levels of 51 and 34 meV for 1.195–1.215 eV DAP peak are observed from 80 to 140 K PL measurements by thermal quenching method. The copper vacancy levels of 40 and 50 meV in the CuInSe$_2$ and CuGaSe$_2$ can easily be assigned to DAP (V$_{Cu}$–V$_{Se}$) transition because the DAP peak positions shift in the ranges of 0.95–0.97 eV or 1.049–1.06 eV and 1.195–1.215 eV with varying excitation power [189]. The fascinating results of PL spectra recorded at 7.7 K for various compositions of single crystal and thin films are given in Table 5.6 [190]. A strong similarity is also observed in the Cu(In$_{1-x}$Ga$_x$)Se$_2$ thin films from 4.2 K PL spectra by Kushiya et al. [191]. The peak at 0.925 eV is probably concerned to (V$_{Cu}$–In$_{Cu}$) transition for $x = 0$ in the Cu(In$_{1-x}$Ga$_x$)Se$_2$ thin films. As the Ga content is increased, the PL peak

Table 5.6 Band gaps, PL peak positions, and tentatively assigned defect levels of Cu(In$_{1-x}$Ga$_x$)Se$_2$ single crystal and thin films

Single crystals	E_g (eV)	Peak position (eV)	$(E_g - h\nu)$ (meV)	Assigned transition	Peak position (eV)	$(E_g - h\nu)$ (meV)	Assigned transition	Ref.
$x = 0$	1.03	0.98	50	(In$_{Cu}$–VB)	0.937	95	(In$_{Cu}$–V$_{Cu}$)	[190]
0.1	1.02	1.01	82		0.978	114	(V$_{Cu}$–V$_{Se}$)	
0.18	1.143	1.053	90	(In$_{Cu}$–V$_{Cu}$)	1.036	107	(In$_i$–V$_{Ga}$)	
0.22	1.168	1.067	101	(In$_{Cu}$–V$_{Cu}$)	1.040	128	(In$_i$–V$_{Ga}$)	
0.496	1.4	1.30	100	(Ga$_{Cu}$–V$_{Cu}$)	1.28	120	(In$_i$–V$_{Cu}$)	
1	1.69	1.61	80	(CB–V$_{Se}$)	1.55	140	(Ga$_i$–V$_{Cu}$)	
Thin film								
0		0.938		(V$_{Cu}$–V$_{Se}$)	0.902		(V$_{Cu}$–V$_{Se}$)	[190]
0.28		1.26		(V$_{Cu}$–V$_{Se}$)	1.114			
1		1.54			1.459			
0		0.925		(In$_{Cu}$–V$_{Cu}$)	0.873		(In$_{Se}$–Cu$_{In}$)	[191]
0.161		1.012						
0.399		1.112						
0.571		1.153		(V$_{Cu}$–V$_{Se}$)				
0.748		1.258						
1.0		1.61						

(0.925 eV) position gradually shifts to higher energy, as presented in Table 5.6. The shift could be due to an increased activation energy or formation energy of (V_{Cu}–In_{Cu}). On the other hand, Ga probably replaces In_{Cu} becoming as Ga_{Cu} for (V_{Cu}–Ga_{Cu}) transition. The peak at 0.873 eV may be due to (Cu_{In}–In_{Se}) transition. Of course the peak becomes stronger after annealing the solar cells under air hence it may be related to passivation of the defects by oxygen.

The PL spectrum of Cu(InGa)Se$_2$ shows peak at 0.96 eV, followed by phonon replica at 0.93 eV as well as peak at 0.90 eV separated by three phonon replicas with equal distance of 30 meV. The thermal quenching analysis of 0.90 eV peak shows activation energies of 3.8 and 140 meV at low and high temperature regions, respectively. A slope of $1.1k_B$ from the graph of peak position versus temperature for both 0.96 and 0.90 eV emission peaks is observed. The peak positions shift from 0.961 to 0.963 eV with slope rate of 1.2 meV/decade and from 0.901 to 0.906 eV with slope rate of 1.5 meV/decade with variation of normalized excitation power from 0.01 to 100%. Taking band gap of 1.05 eV for CuInSe$_2$, the obtained $E_{A/D}$ values are 90 and 150 meV for the peaks of 0.96 and 0.90 eV, respectively. Considering 3.8 meV is a common donor, 75 ± 10 and 140 ± 10 meV are acceptors for the peaks, respectively. The DAP peaks at 0.96 and 0.90 eV may be assigned to (Cu_{In}(60 meV)–In_{Cu}(10 meV)) and (Se_i(130 meV)–In_{Cu}(10 meV)), respectively. Since the activation energy of 3.8 meV is lower than the experimental limits of ± 10 meV and the slope rate with respect to excitation power is also low. If the activation energy of 3.8 meV is considered as an artifacts then the peak at 0.9 eV may be assigned to (CB–Se_i) and its thermal activation energy matches well with the acceptor Se_i activation energy [192]. In another approach, the activation energies of donor and acceptor for DAP line at 0.97 eV in air annealed In-rich and Cu-rich CuInGaSe$_2$ may be assigned to V_{Se} of 10 meV and V_{Cu} of 75 meV, respectively. Another line at 0.90 eV may be considered as recombination of shallow donor level V_{Se} of 10 meV and likely doubly ionized acceptor level V_{Cu} with activation energy of 140 meV. At 4.2 K, the In-rich Cu(InGa)Se$_2$ thin films grown by sputtering and evaporation show broad emission band at 0.95 eV. After annealing the sample under air/oxygen at 400 °C for 10 min, the predominant emission line at 0.97 eV likely DAP with four phonon lines by separation of about 28 meV exist. However, there is a red shift in 0.95 eV peak by about 25 meV for the films annealed either under Ar or N$_2$ atmosphere. The Cu-rich Cu(InGa)Se$_2$ thin films show two emission lines at 0.97 and 0.90 eV with phonon replicas, which can be assigned to DAPs (Figure 5.58) [193].

The In-rich co-evaporated CuInSe$_2$ thin films show two emission peaks at 0.86 and 0.95 eV. After air annealing the films at 400 °C for 10 min, the PL spectrum becomes similar to that of Cu-rich CuInSe$_2$ layers but BX emission line at 1.032 eV exists. A 1.1 eV broad peak of CuIn$_{0.72}$Ga$_{0.28}$Se$_2$ thin films is similar to 1.14 eV displayed by annealed In-rich CuInSe$_2$ [194]. Two peaks at 1.16 and 1.01 eV and a shoulder at 0.96 eV corresponding to (V_{Cu}–V_{Se}), (Se_{Cu}–V_{Se}), and (III_{Cu}–V_{III}) transitions, respectively are observed in the 14 K PL spectra of CuIn$_{0.7}$Ga$_{0.3}$Se$_2$. All the peaks show blue shift with increasing intensity of excitation light, indicating DAP defect structure. The 1.16, 1.01, and 0.96 eV peak positions vary nonlinearly with varying $x = 0$ to 1 at 77 K [195]. The similar kind of peaks at 1.08 and 0.94 eV are found in the Cu(InGa)Se$_2$ thin films for composition ratio of Cu:In:Ga:Se = 24.1:23.9:1.4:50.6. The DAP occurs only at higher intensity of excitation laser that might be due to lower Ga gradient at the surface that is, higher

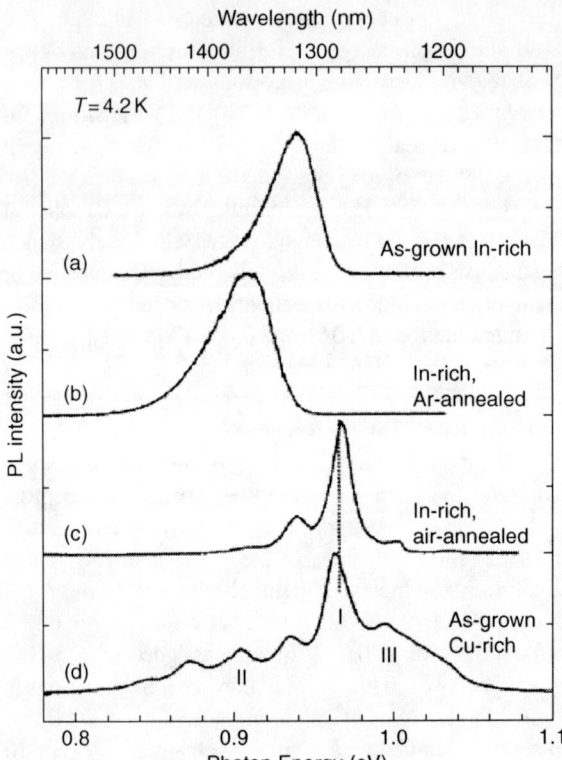

Figure 5.58 PL spectra of as-grown In-rich, Cu-rich, and air annealed CuInGaSe$_2$ thin films.

gradient in depth of the film that create higher band gap of the sample. The laser penetrates deeper into the sample for higher intensity [105]. The E_g changes from 1.045 ($x=0$) to 1.72 eV ($x=1$) in the CuIn$_{1-x}$Ga$_x$Se$_2$ thin films at 77 K. The luminescence is detected from the backside of the Cu(InGa)Se$_2$ layer through the glass substrate is different from that of front side, which shows a broad emission band at 1.02 eV. The Ga content obtained by SIMS analysis on the backside of the layer is $x=0.09$ but the set deposition composition is $x=0.1$ in the typical CIGS sample. In the PL, Cu$_{2-x}$Se effect is nil, however in electrical measurements, after etching in KCN for 3 min the resistance shot up to high [192].

The CIGS layers heated in NH$_4$OH shows two emission lines at 0.97 and 1.05 eV, which have the same acceptor level but different donor levels. After Cd PE treatment the emission line at 1.05 eV quenches and the intensity of 0.97 eV line increases and a new emission segregates at 0.91 eV, which is stronger than other lines. The density of shallow acceptor levels decrease by diffusion of Cd in the place of Cu vacancies that may be caused to quench the emission line of 1.05 eV. A new emission line may be due to formation of new phases such as CdSe or CdIn$_2$Se$_4$ [196]. Figure 5.59 shows PL spectra of Cu(In$_{1-x}$Ga$_x$Se)$_2$ thin films with various values of x. The relation $I \alpha P^k{}_{\text{Ex/DAP}}$ for the photoluminescence of Cu(InGa)Se$_2$ thin films, where I is the emission intensity and P is the excitation power, follows as $k<2$ for

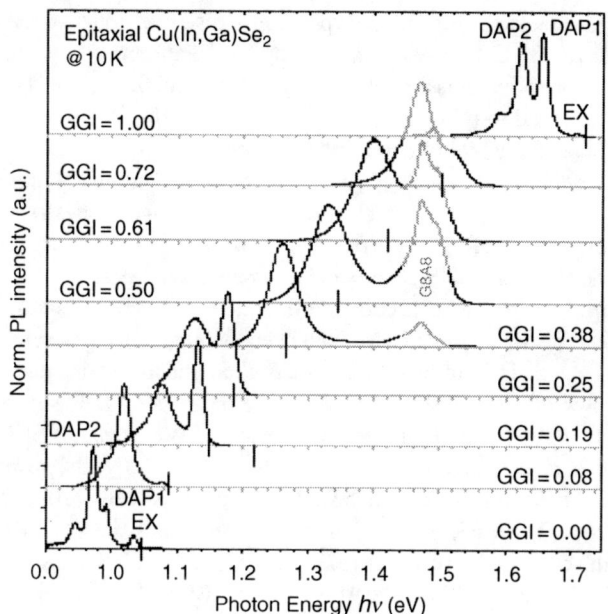

Figure 5.59 PL spectra of $CuIn_xGa_{1-x}Se_2$ thin films for different x values recorded at 10 K Ga/(Ga +In) = GGI.

excitonic peak (Ex) and band to band transition, whereas $k < 1$ for DAP or FBs. One can see the relation between excitonic and DAP transitions in terms of k, $k_{Ex} \alpha 2 k_{DAP}$ for all the compositions of $x = 0$, 0.08, 0.19, 0.25, 0.50, 0.72, 0.84, and 1.00 in the $Cu(In_{1-x}Ga_x)Se_2$ polycrystalline thin films. DAP means either $DAP_1 = 1.62$ or $DAP_2 = 1.64$ eV, and Ex = 1.72 eV for a typical $CuGaSe_2$ thin film. In the DAP transition, the acceptor energy level of 40 meV and donor energy levels of 10 meV for $CuInSe_2$ are observed. They vary to higher value of 60 and 13 meV with increasing x values towards $CuGaSe_2$, respectively [197].

5.2.7 $Cu(In_{1-x}Ga_x)Se_2$ Thin Film Solar Cells

The PL spectra of CIGS with Cu/(In+Ga) = 0.9 and Ga/(In+Ga) = 0.35, CIGS/CdS and CIGS conventional cells are studied. The NBE at 1.18 eV, FBs at 1.12 and 1.01 eV are observed in the CIGS layer at room temperature. After CdS deposition, the intensity of NBE increases by factor of 2–3 due to etching of secondary phases by NH_3. On the other hand, 1.01 eV due to FB (V_{Cu}–CB) disappears and a new 0.97 eV peak appears, which is due to substitution of Cd into V_{Cu} that is, Cd_{Cu}. The intensity of NBE for CIGS cell is higher by two orders of magnitude as compared to that of CIGS thin film. 1.07 (FB_1) and 0.97 eV (FB_2) peaks are observed at 150 K for CIGS thin films with [Ga/(Ga+In) = 0.2]. Two peaks at 1.08 (FB_1) and 1.02 eV (FB_2) are observed for CIGS/CdS heterostructure. The FB_1 and FB_2 disappear but DAP at 1.03 eV appears for CIGS layer, whereas the FB_1 and FB_2 peaks also disappear, but two DAPs at 0.99 and 1.03 eV appear for CIGS/CdS with deccreasing temperature to 8 K [198]. The $CuIn_{0.7}Ga_{0.3}Se_2$ thin films show emission lines at 1.16, 1.01, and 0.96 eV in 14 K PL spectra, which can be assigned to $(V_{Cu}-V_{Se})$, $(Se_{Cu}-V_{Se})$, and $(III_{Cu}-V_{III})$, respectively. Three lines shift to lower energy with increasing temperature [199].

The glass/Mo/Cu(In$_{0.7}$Ga$_{0.3}$)Se$_2$/50nmCdS/50nmZnO/ZnO:Al/50nmNi/1μmAl thin film solar cell in which the absorber is grown by two-stage process shows efficiency of 10.2%, whereas CIS cell shows efficiency of 8.3%. The PL peak at 0.94 eV is common in both the CIGS and CIS absorbers but an additional peak at 1.0 eV is observed in the former [200]. The highest reported efficiency of 16.9% on soda-lime glass/Mo/CuIn$_{1-x}$Ga$_x$Se$_2$/CdS/i-ZnO/n-ZnO:Al has been achieved by conventional method, however, H$_2$O vapor is introduced into MBE chamber to obtain quality CuInGaSe$_2$ layers. The cell with CIGS absorber grown in the presence of H$_2$O shows an increased efficiency of 18.1%. The DAP line at 1.1 eV shows FWHM values of 127.7 and 83.9 meV for 16.9 and 18.1% efficiency CIGS thin film solar cells, respectively (Figure 5.60). It is also sharp for the highest efficiency solar cells [201]. The 14% efficiency glass/Mo/CIGS/CdS/ZnO virgin cells show peaks at 1.07, 1.13, and 1.191 eV, whereas the cells reveal peaks at 1.07 and 1.129 eV in 20 K PL spectra after damp heat testing. The peak at 1.129 or 1.191 eV is the NBE of CIGS layer, which is due to transition of V_{Se} to V_B. The peak at 1.07 eV can be assigned to $(CB-V_{Cu})$ in the OVC formed by defect levels of $(2V_{Cu}^- + In_{Cu}^{2+})$, which is generally occurs on the surface of CIGS layer [202].

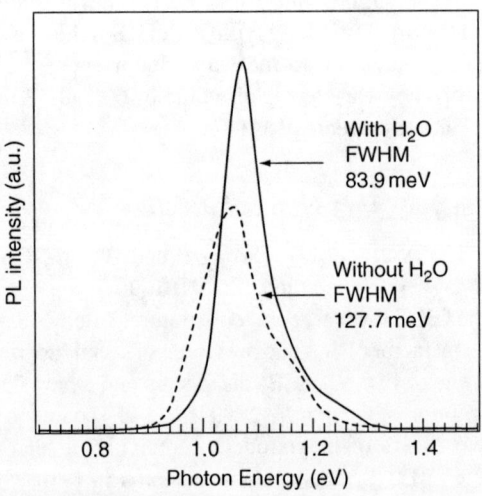

Figure 5.60 PL spectra of CIGS thin films grown for soda-lime glass/Mo/CuIn$_{1-x}$Ga$_x$Se$_2$/CdS/i-ZnO/n-ZnO:Al thin film solar cell.

5.2.8 CuInS$_2$

The free exciton A is the transition from the valence band to the conduction band and the free excitons B and C represent the transitions between the spin-orbit splitting valence band and conduction band (Figure 5.2). The E_{x1} is due to exciton bound to neutral acceptor (V$_{Cu}$) and E_{x2} connects to exciton bound to neutral donor (V$_S$). Figure 5.61 shows that the emission lines such as $A = 1.535$ eV, $B = 1.551$ eV, $E_{x1} = 1.53$ eV, $E_{x2} = 1.525$ eV as a dominant, $E_{x3} = 1.5185$ eV, and donor to valence band (VB–V$_S$) = 1.52 eV are well resolved in the In-rich phosphorus doped p-CuInS$_2$ single crystals, whereas the same emission lines are prevailed on the broad emission band in the virgin In-rich n-CuInS$_2$ crystals. Among these (VB–V$_S$) is a dominant, as shown in Figure 5.62 [203]. The free excitonic emissions $A = 1.5354$ eV and $B = 1.5531$ eV are also observed in the p-CuInS$_2$ single crystals grown by directional freezing method. The similar emission lines $A = 1.536$ eV and $B = 1.554$ eV in the CuInS$_2$ single crystals grown by Bridgman technique are reported in the literature [73,204]. There are two emission lines considered to be as upper (1.537 eV) and lower branch polaritons (1.535 eV) besides A-exciton (~ 1.536 eV) in 8 K PL spectra of CuInS$_2$ crystals grown by traveling heater method (THM), as depicted in Figure 5.63 [205]. A similar kind of emission bands such as $A = 1.535$, $E_{x1} = 1.531$, $E_{x2} = 1.525$ and (V$_S$–VB) = 1.52 eV donor to valence band (DV or FB) in the CuInS$_2$ crystals grown by THM and broad emission peaks at 1.44–1.435, 1.410, and 1.38 eV in the CuInS$_2$ crystals grown by ITM are observed [206,207]. 1.529 (E_{x1}) and 1.526 eV (E_{x2}) are also found in the n-type In-rich CuInS$_2$ crystals in 14 K PL spectrum [208]. 1.53 and 1.44 eV emission lines in undoped [209] and 1.54 and 1.4 eV lines in phosphorus doped p-type CuInS$_2$ single crystals are observed [210]. They are in close to each other.

Figure 5.61 shows that as sample temperature is increased, the free exciton A slowly becomes low profiler and free exciton B presents at 160 K in the In-rich p-CuInS$_2$ single crystals. By this thermal quenching process, the activation or binding energy of 20 meV for free exciton A is obtained. Surprisingly, the binding energy of exciton A is close to the difference of between A and B excitons. The band gap of 1.555 eV at 4.2 K for CuInS$_2$ can be essentially obtained by adding the free exciton binding energy of 20 meV to its A-exciton value of 1.535 eV. 4.2 K PL spectrum shows that the peaks at 1.44 and 1.41 eV in the In-rich and 1.39 and 1.36 eV in the Cu-rich CuInS$_2$ single crystals grown by CVT technique are observed; (a) the activation energies of 35 and 100 meV for donor (V$_S$) and acceptor (V$_{Cu}$), respectively, are assigned to 1.44 eV peak, (b) 72 meV (In$_i$ donor) and 100 meV (V$_{Cu}$) to 1.41 eV emission line, (c) 35 meV (E_D) and 150 meV (E_A) to 1.39 eV peak, and (d) 72 meV (E_D) and 150 meV (E_A) to 1.36 eV peak are observed from thermal quenching analysis [211]. The DAP peak at 1.37 eV is assigned to recombination of donor level (In$_{Cu}$) with activation energy of 145 meV and acceptor level (V$_{Cu}$) with activation energy level of 100 meV transition [208]. The activation energies of 50 and 85 meV obtained by thermal quenching for 1.405 eV peak in the CuInS$_2$ crystals grown by IT technique confirm that the transition is due to DAP from the donor level (V$_S$) to acceptor level (Cu$_{In}$ or V$_{In}$), that is, (Cu$_{In}$–V$_S$ or V$_{In}$–V$_S$) [212]. The PL spectra of some of the as-grown samples with a broad

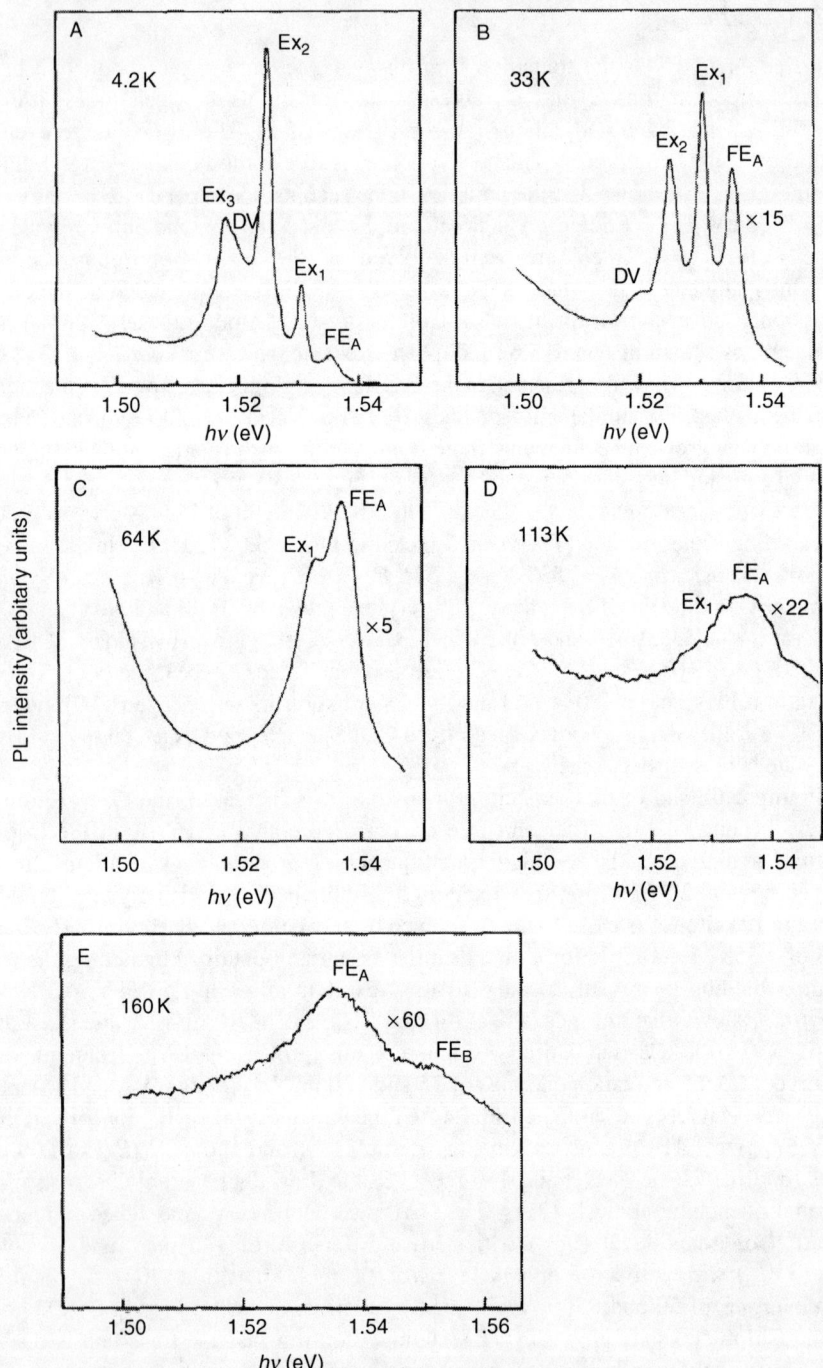

Figure 5.61 PL spectra of In-rich *p*-type CuInS$_2$ single crystals recorded at different temperatures.

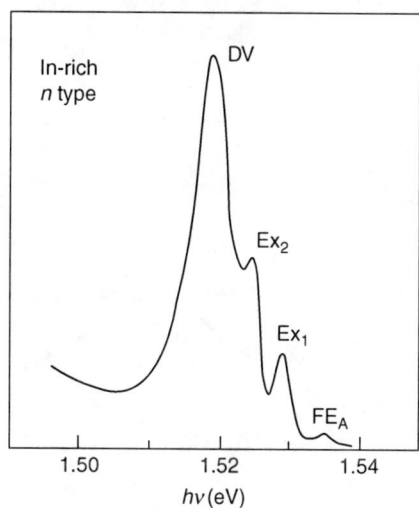

Figure 5.62 PL spectra of In-rich n-type $CuInS_2$ single crystals.

PL peak annealed under Cu and under In are as same as that of as-grown Cu-rich and In-rich $CuInS_2$ samples, respectively. A plot of PL intensity versus inverse temperature that is, thermal quenching (Figure 5.64) gives activation energies $(E_A + E_D)$ of 145 and 90 meV for 1.44 eV peak in the Cu-rich and for 1.39 eV peak in the In-rich $CuInS_2$ crystals, respectively [211]. The emission lines such as 1.52, 1.45, 1.4, 1.35, and 1.28 eV in the $CuInS_2$ thin films deposited by sulfurization of Cu–In alloy [213] and more or less similar lines at 1.53, 1.45, 1.41, and 1.37 eV in the $CuInS_2$ thin films grown by three-source evaporation are observed in 4.2 K PL spectrum [214]. The peaks at 1.52 and 1.45 eV are already assigned to (VB–V_S) and (V_{Cu}–V_S) DAP transitions, respectively. The intensity of 1.45 eV peak is low, if the samples are Cu-rich. The DAP peak at 1.4 eV shifts with variation of temperature, as shown in Figure 5.65, which may be assigned to the recombination between V_S (35 meV) and V_{In} or Cu_{In} (149 meV) because a plot of intensity versus inverse temperature (inset figure) provides two activation energies of 35 and 149 meV for $CuInS_2$ thin film. The DAP transition from donor level In_{Cu} to acceptor level Cu_{In} (Cu_{In}–In_{Cu}) can be assigned to 1.35 eV and the emission line at 1.28 eV could be due to (Cu_i–In_{Cu}). The typical possible transition levels for $CuInS_2$ are depicted in Figure 5.66 [215].

9 K PL spectra of $CuInS_2$ single crystals grown by THM method show donor to valence band (DV), BX (E_{x2}), BX (E_{x1}) and (free exciton A) E_A peaks at 1.520, 1.525, 1.531, 1.535 eV, as shown in Figure 5.67A, respectively. The $CuInS_2$ crystals are annealed under vacuum at 400 °C for 15 min show that the E_{x2} (1.525 eV) quenches, the intensities of other lines also drastically decrease and a new emission line E_{x3} emerges at 1.520 eV, which is due to (V_S–VB) [208,213]. However, the authors [216] mentioned that the E_{x3} (1.520 eV) transition is responsible by an exciton bound to neutral acceptor estimated to be 303 meV and considered to be Cu_{In}. After annealing the crystals under In_2S_3 powder, the PL spectrum remains same except intensity decreases to half and E_{x3} (1.52 eV) peak appears. It concludes that

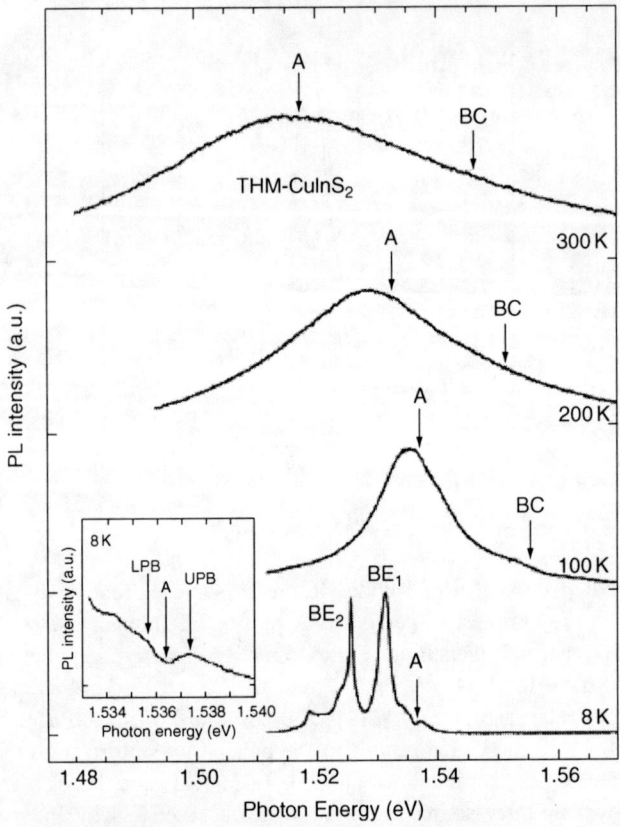

Figure 5.63 PL spectra of CuInS$_2$ single crystal grown by THM (BC represents for B and C excitons, which are one and the same).

annealing the crystals under vacuum allow the dissociation of In and S, whereas annealing under In$_2$S$_3$ blocks the dissipation of In and S from the CuInS$_2$ crystals. The spectra of CuInS$_2$ crystals annealed under In$_2$S$_3$, followed by vacuum annealing show that the E_{x3} peak remains at the same place and E_{x2} (1.525 eV) reappears with low intensity, as compared to that of crystals annealed under vacuum, as shown in Figure 5.67B. The E_{x2} emission may be considered as an exciton bound to neutral acceptor level S interstitial (S_i) of 192 meV. The CuInS$_2$ crystals annealed under sulfur at 700 °C for 40 h show a band at 1.51 eV. If the crystals are annealed more than 50 h, the same emission peak disappears. This experimental observation supports that a lower annealing time creates sulfur vacancies because of having lower saturated sulfur atmosphere, whereas the peak at 1.51 eV disappears for longer time annealing crystal under saturated sulfur atmosphere that nullifies the sulfur vacancies. Therefore the emission band at 1.51 eV can be essentially assigned to (VB–V_S) transition. The role of sulfur annealing on CuInS$_2$ and conductivity type conversion are discussed in Chapter 6. The crystals annealed under (In + S) at 700 °C for 48 h exhibit emission line at 1.45 eV, which is a DAP (V_{Cu}–V_S) recombination.

Optical Properties of I–III–VI$_2$ Compounds

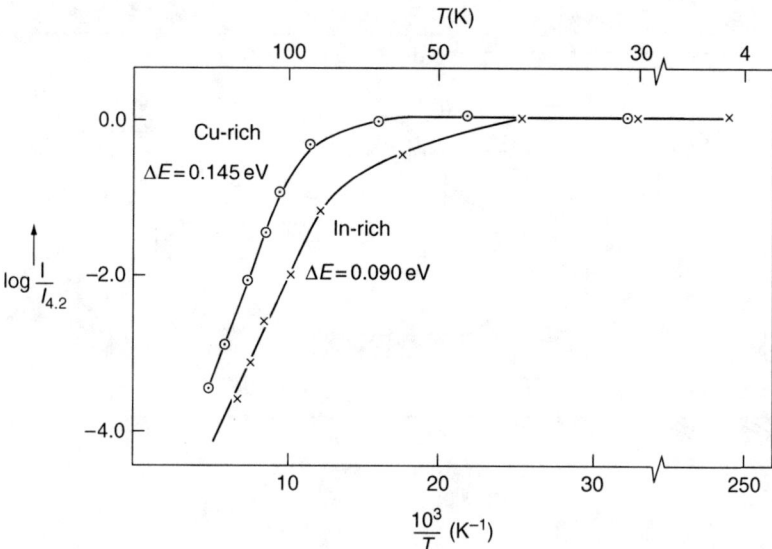

Figure 5.64 PL peaks intensities versus inverse thermal temperature for Cu-rich and In-rich CuInS$_2$ single crystals. Since the low temperature activity is nil, the activation energies could be due to either $(E_A + E_D)$ or E_A.

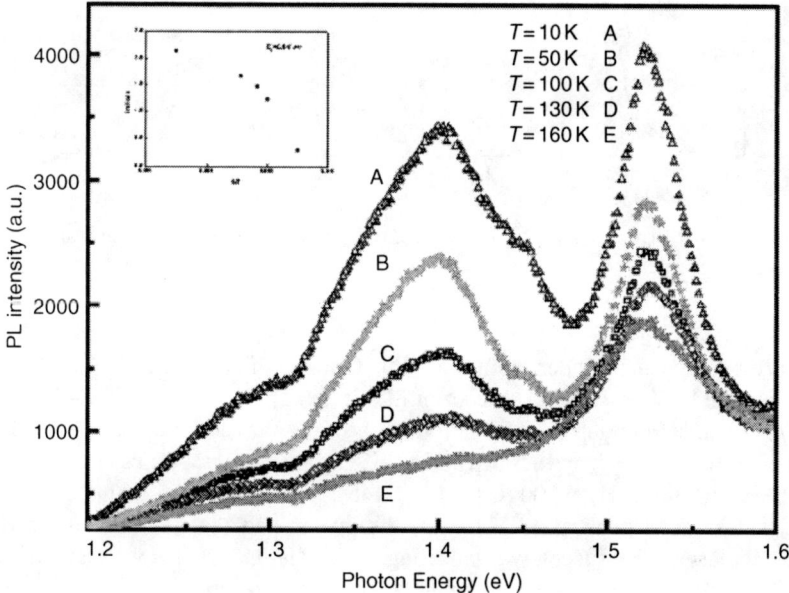

Figure 5.65 PL spectra of CuInS$_2$ thin film with function of temperature and (inset) intensity versus inverse temperature.

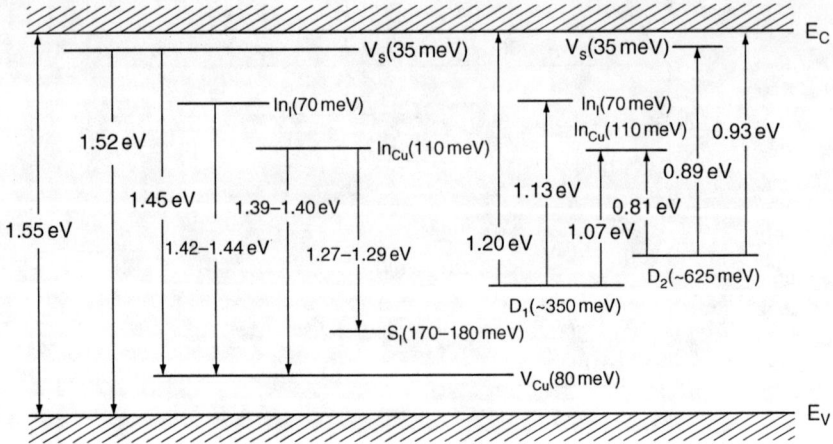

Figure 5.66 A different transition levels of CuInS$_2$.

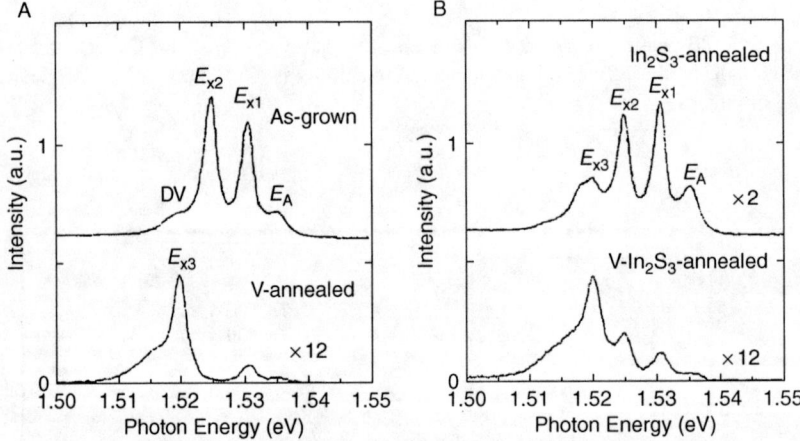

Figure 5.67 9 K PL spectra of (A) as-grown CuInS$_2$ single crystals and vacuum annealed (B) annealed under In$_2$S$_3$, followed by annealing under vacuum.

The crystals annealed under indium at 700 °C for 48 h show two emission lines at 1.39 and 1.34 eV in 77 K CL spectrum [217].

37 K PL spectrum of p-CuInS$_2$ thin films shows a sharp emission band at 1.531 eV, a small line at 1.445 eV, and a broad emission peak at about 1.25–1.3 eV. After annealing the sample under H$_2$ at 200 °C for 1 h, 1.445 eV peak strongly dominates the spectrum and a broad peak at 1.25–1.3 eV disappears and the intensity of 1.531 eV peak decreases. The successive annealing under H$_2$ and O$_2$ shows that the peaks at 1.445 and 1.531 eV are remaining at the same position but their intensities drastically decrease. The similar nature is observed in the CuInS$_2$(3 μm)/CdS(50 nm-CBD)/ZnO(0.4 μm-RF-sputtering) thin film solar cell. There is no change in

the spectrum, if the samples are post-treated either under oxygen or air without H_2 treatment. The H_2 annealing obviously reduces oxygen level at grain boundaries and creates more sulfur vacancies. There is also a possibility that H_2 may be reacted with sulfur to form H_2S then there are chances to obtain more sulfur vacancies. The density of Cu vacancies may be remaining at the same. The thermal quenching analysis of DAP peak at 1.445 eV reveals two kinds of activation energies of 43 and 113 meV at low and high temperature regions, which are close to the previous reports [218]. The peak at 1.521 eV is due to (VB–V_S) transition. The peak at 1.445 eV is already assigned as a combination of donor level (V_S) of 43 meV and acceptor level (V_{Cu}) of 70 meV (DAP), that is, (V_{Cu}–V_S) [219]. The intensity of 1.445 eV peak decreases for multiple successive annealing of $CuInS_2$ thin film under H_2 and O_2, which again gains intensity after successive annealing under H_2, O_2, and H_2, as shown in Figure 5.68. The intensity of peak depends mainly on the final annealing atmosphere. Annealing under H_2 removes oxygen from the sample correspondingly V_S concentration increases, hence 1.445 eV (DAP) peak intensity increases. The peak intensity decreases for annealing $CuInS_2$ thin film under oxygen that oxygen occupies the sulfur vacancy site. The peak shifts to higher energy and intensity increases with increasing excitation power, as shown in Figure 5.69 [218]. Two peaks at 1.49 and 1.42 eV observed in 80 K PL spectrum of $CuInS_2$ thin films grown onto glass substrates by Doctor blade technique are related to excitonic recombination and due to sulfur vacancies (V_S). The intensity of latter enhances after annealing the layers under H_2 but disappears after annealing under O_2. Anneal under H_2 creates sulfur vacancies, whereas anneal under O_2 means suppress sulfur vacancies hence the peak intensity increases in the former and decreases in the latter [220]. 77 K PL of $CuInS_2$ crystals grown by CVT [204] show a broad peak at 1.36 eV due

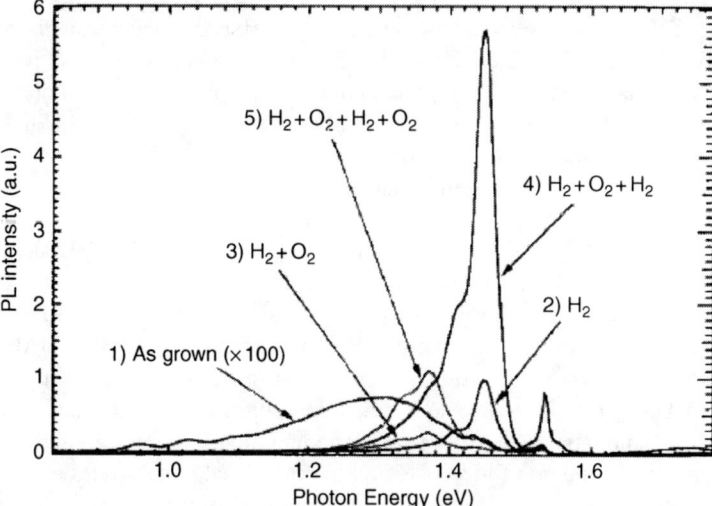

Figure 5.68 PL spectra of as-grown and annealed $CuInS_2$ thin films under hydrogen and oxygen atmospheres.

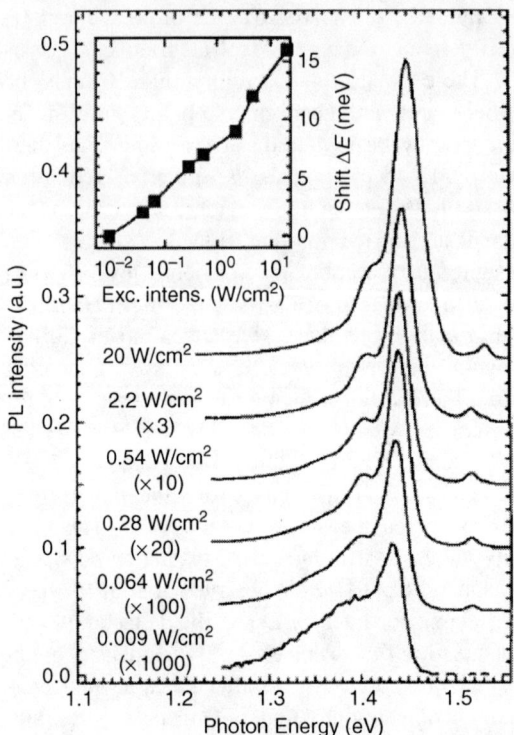

Figure 5.69 PL spectra of $CuInS_2$ thin films annealed under H_2 at 200 °C for 1 h recorded at different excitation energies.

to DAP transition of (Cu_{In}–In_{Cu}) and emission line at 1.24 eV could be due to DAP. After annealing $CuInS_2$ crystals under saturated sulfur atmosphere, the peaks disappear and new peaks generate at 1.44 eV due to DAP transition (V_{Cu}–V_S) [213], followed by 1.40 eV (LO_1), and 1.36 eV (LO_2) phonon lines with separation of ~40 meV and free exciton (A) at 1.535 eV. The other emission bands at 1.446, 1.406, 1.36, and 1.24 eV are also observed in the $CuInS_2$ crystals, as shown in Figure 5.70.

The low energy level hydrogen is implanted into $CuInS_2$ thin films grown onto Si single crystal substrates by MBE, which exhibit several PL lines such as FB-1.485, DAP_1-1.439(3), DAP_2-1.392(3), DAP_3-1.349(3), DAP_4-1.309(3), DAP_5-1.199(5), and DAP-1.034(5) but no exciton line, as shown in Figure 5.71A. After KCN etching, the polycrystalline $CuInS_2$ thin films show excitonic lines at 1.527, DVB-1.465, and DAP-1.435 eV, as shown in Figure 5.72B. The intensities of DAP_1, DAP_2, DAP_3, and FB_1 decrease with increasing temperature from 5 to 65 K. Surprisingly the DAP_2 and DAP_3 disappear due to thermal ionization of the donors and acceptors but DAP_1 remain exists at >80 K [221]. The emission lines at 1.53 and 1.46 eV in $CuInS_2$ polycrystalline thin films coated onto Mo covered Si substrates exist. A bunch of 1.48, 1.44, 1.2, and 1.0 eV emission lines for the $CuInS_2$ epitaxial layers coated onto single crystal Si (111) substrates is observed in 5 K PL spectra. The experimental observation indicates that the formation of defect recombination states are different

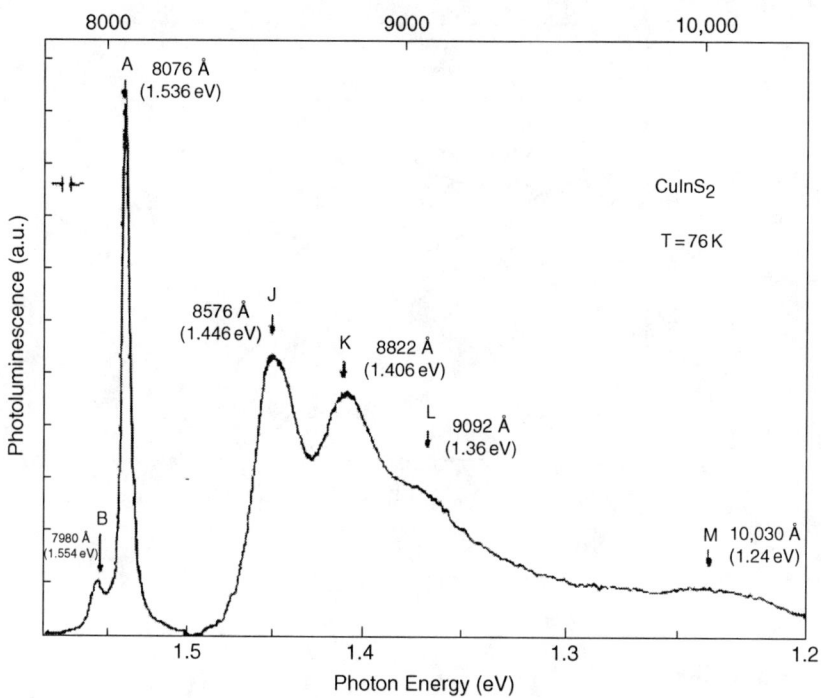

Figure 5.70 PL spectrum of CuInS$_2$ crystals grown by Bridgman technique.

from polycrystalline to epitaxial layers due to the effect of different substrates used that causes different growth direction or formation of defects (Figure 5.72) [222].

The CuInS$_2$ thin films are deposited onto sulfur terminated Si substrates by MBE technique using Cu, In, and S effusion cells at substrate temperature of 527 °C. Lower energy hydrogen is implanted with a kinetic energy of 500 eV per hydrogen molecule and a fluency of 1.0×10^{15} cm^{-2} in the KCN etched CuInS$_2$ layers at room temperature. All the emission lines are stronger in hydrogen implanted CuInS$_2$ thin film than that in the as-grown sample. In addition, extra three DAP lines present in hydrogen implanted CuInS$_2$ thin films. The FB at 1.48 eV with acceptor level (V$_{Cu}$) of 75 meV is appeared in both the samples and the remaining all five peaks are DAP transitions as shown in Figure 5.73. The DAP$_1$ line at 1.436 eV results from the recombination of acceptor level (V$_{Cu}$) and donor level (V$_S$) of 44 meV below the conduction band. The same donor level (V$_S$) of DAP$_1$ participates in recombination with a new acceptor level (V$_{In}$) in the DAP$_2$ (1.395 eV). The (V$_{In}$) acceptor level contains activation energy of 116 meV. There may be possible to occur V$_{In}$ in the Cu-rich samples [211]. The DAP$_3$ at 1.345 eV is due to (Cu$_{In}$–In$_{Cu}$) or (V$_{Cu}$–In$_i$) [208,213] and DAP$_4$ (1.195 eV) transition is currently unknown [225]. The similar kind of transitions at 1.21 and 1.45 eV is observed in the CuInS$_2$ thin films. The transition line at 1.45 eV is also assigned to donor level V$_S$ (43 meV) to V$_{Cu}$ acceptor level (70 meV) [224].

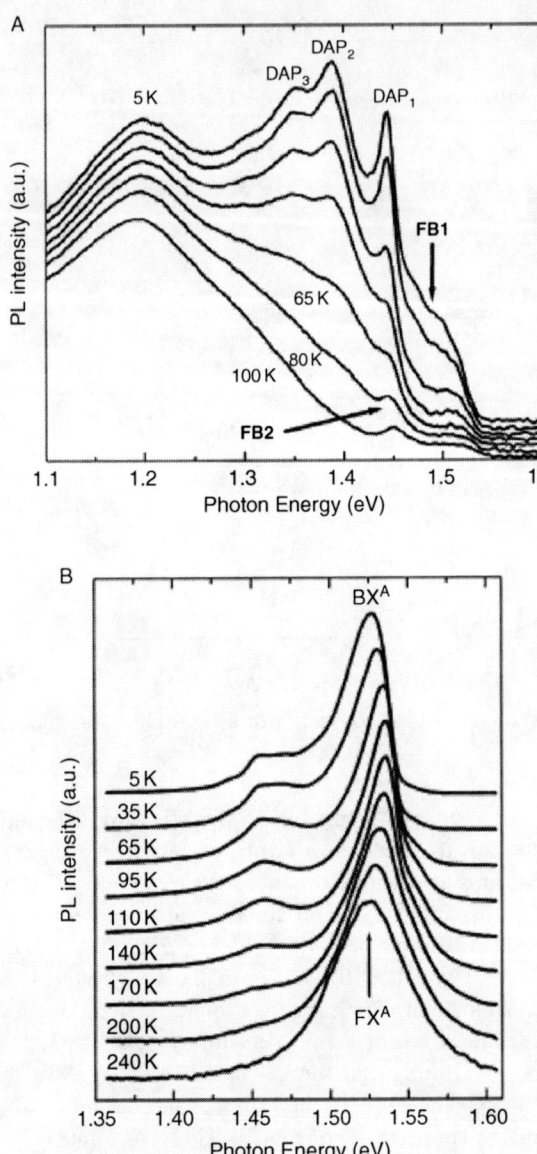

Figure 5.71 (A) PL spectra of hydrogen implanted CuInS$_2$ thin films with function of temperature and (B) PL spectra of KCN etched CuInS$_2$ thin films measured at different temperatures.

The co-evaporated Cu and In layers are sulfurized under 3% H$_2$S and 97% Ar (H$_2$) at high temperature of 500 °C for 3 h to form Cu-rich CuInS$_2$ thin films. The excitonic emission line pronounces at 1.52 eV and DAP at 1.38 eV due to (V$_{In}$–V$_S$) transition in the

Optical Properties of I–III–VI$_2$ Compounds

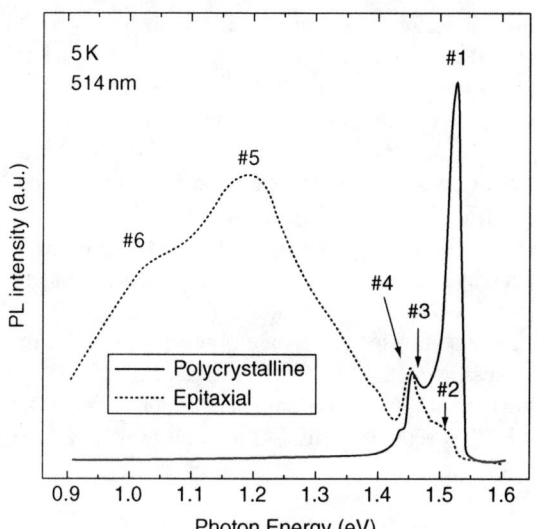

Figure 5.72 5 K PL spectra of CuInS$_2$ thin films grown onto Si substrates and Mo covered Si substrates.

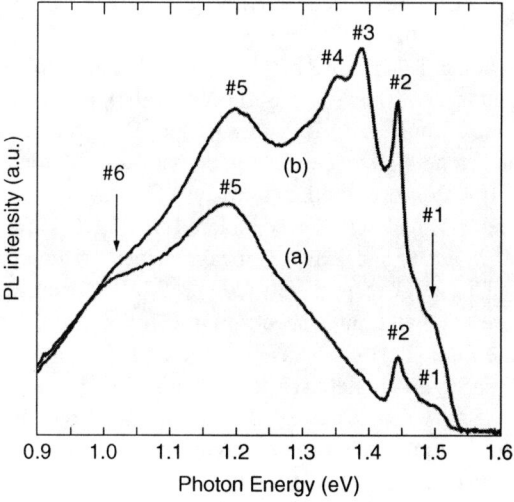

Figure 5.73 PL spectra of (a) as-grown and (b) hydrogen implanted CuInS$_2$ thin films.

Cu-rich layers. The sulfurization of Cu and In layers at low temperature of 350 °C results in In-rich CuInS$_2$ thin film, which shows an additional emission line at 1.44 eV due to DAP recombination of V$_S$ and V$_{Cu}$ that is (V$_{Cu}$–V$_S$) transition. The excitonic band emission disappears after sulfurization of the layers at 350 °C, whereas no emission lines are observed after sulfurization at 600 °C. The intensities of all the emission lines decrease at sulfurization of 375 °C for longer time of 6 h [225]. The authors said that a peak at 1.437 eV for CuInS$_2$ can be assigned to the transition from the free electrons of CB to the acceptor level (V$_{Cu}$) of 110 meV, which can be obtained from Equation (5.10), provided that the emission band does not get shift with variation of excitation energy

[226]. The electrodeposited Cu–In alloy is sulfurized under H_2S at 550 °C for 30 min to form $CuInS_2$ films on Ti substrates. Two emission bands observed at 1.53 and 1.41 eV in 70 K PL spectrum are connected to donor to valence band (D–VB) and DAP, respectively. The DAP peak at 1.41 eV could be due to either In_i (donor level 109 meV) to V_{Cu} (acceptor level 71 meV) [227] or (V_{In}–V_S) transition [213]. The $CuInS_2$ films with Cu/In = 0.99 and S/(Cu + In) = 0.9 grown by sulfurization of Cu and In alloy deposited by vacuum evaporation show peaks at 1.55 (BX), 1.39 (V_{Cu}–In_i) or (V_{In}–V_S), 1.153 and 0.96 eV [228]. 10 K PL spectrum of $CuInS_2$ shows peaks at 1.2, 1.5, and 1.45 eV DAP (V_{Cu}–V_S). The intensity of 1.2 eV peak increases with increasing sulfurization temperature from 550 to 650 °C [229]. The emission peaks at 1.433 and 1.398 eV in 10 K PL spectra due to DAP (V_{Cu}–V_S) and its LO phonon, respectively, are observed in the $CuInS_2$ polycrystalline pellets grown by high pressure method at 700 °C and 22.5 MPa employing Cu_2S and In_2S_3 powders for the growth time of 1 h [87]. The similar emission lines at 1.425 and 1.399 eV are also observed in 14 K PL spectra of p-$CuInS_2$ thin films grown by three-source evaporation [230].

5.2.9 $CuIn(Se_{1-x}S_x)_2$

The n-$CuInS_2$, p-$CuIn(Se_{0.7}S_{0.3})_2$, and p-$CuInSe_2$ crystals grown by Bridgman method are studied by photoluminescence at 15 K with excitation power of 40 mW and wavelength of 351 nm. The excitonic emission at 1.04 eV, FB at 0.973 eV, followed by phonon replica at 0.943 eV with separation of 30 meV are observed in the p-$CuInSe_2$ single crystals. A difference between excitonic and FB is about 70 meV. In the case of p-$CuIn(Se_{0.7}S_{0.3})_2$, more or less the same difference of 72 meV between excitonic emission peak at 1.18 eV and FB at 1.108 eV is observed. In the n-type $CuInS_2$ crystals, three sharp lines at 1.528, 1.523, and 1.517 eV can be considered as E_{x1}, E_{x2}, E_{x3} excitons. A broad band contains 1.457, 1.436, and 1.398 eV peaks, which are due to (V_{Cu}–CB), (V_{Cu}–V_S), and (V_{In}–V_S or Cu_{In}–V_S), respectively [231]. 4.2 K PL spectrum of $CuInSe_2$ single crystals shows more or less the same spectrum of $CuInSe_2$ thin film and the emission lines are A (1.0416 eV), B (1.0499 eV), N (1.002 eV), P (0.971 eV), and K (0.901 eV). The band gap of $CuInSe_2$ is calculated to be 1.049 eV at 4.2 K taking free exciton binding energy value of 7.5 meV. In the case of $CuInS_2$, the emission lines at 1.531 and 1.520 eV are referred to excitons bound to neutral acceptor and neutral donor, respectively. The broad emission line at 1.1–1.3 eV is attributed to DAP recombination of an acceptor V_{Cu} and a donor In_{Cu}. There are sharp lines due to excitons to intrinsic defects in the $CuIn(Se_{1-x}S_x)_2$, between $x=0$ and 1 compositions [73]. In the $CuIn(Se_{1-x}S_x)_2$ alloy system at 4.2 K, $x=1$ that is, n-$CuInS_2$ (Cu/In = 0.89) shows 1.527, 1.473, 1.451, 1.414, and 1.369 eV peaks in which 1.527 eV is a free exciton (A) having binding energy of 16 meV, the peak at 1.451 eV shows a blue shift of 10 meV for variation of excitation power from 1.1 to 114 mW/cm^2. The activation energy of 103 meV for peak at 1.451 eV obtained from the thermal quenching analysis is in good agreement with the reported DAP transition of (V_{Cu}–V_S). The lines at 1.414 and 1.369 eV are due to DAP transitions of (V_{In}–V_S) or (V_{Cu}–In_i) and (Cu_{In}–In_{Cu}), respectively. The emission lines of 1.26, 1.30, 1.33, 1.38, and 1.39 eV are observed for $CuIn(Se_{0.28}S_{0.72})_2$ alloy. The peak at 1.39 eV can be assigned to free exciton because the thermal quenching analysis shows an activation energy of 10 meV that coincides with value of 11 meV obtained by the subtraction of free exciton from the band gap. The calculated binding energy of free exciton is 12 meV from Equation (5.9) [232].

5.2.10 CuGa(SeS)$_2$

The A-exciton (P_{ex}) peak in 77 K PL spectrum occurs nearly at 1.71, 1.82, 2.1, 2.25, 2.5 eV for $x=0$, 0.25, 0.57, 0.8, and 1.0 in the CuGa(Se$_{1-x}$S$_x$)$_2$, respectively. A broader PL peak lies below P_{ex} by 38–80 meV for the sample, which is assigned to the conduction band to the acceptor level Cu$_{Ga}$. A difference between P_1 and P_{ex} varies linearly from 38 to 80 with varying x from 1.85 to 0. This is because of an increase of hole effective mass with increasing x resulting in an increase of ionization energy of the acceptor [233].

5.2.11 Time Resolved Photoluminescence Spectroscopy

The time resolved PL is carried out on CIGS ($x=0.29$) thin film and its 9–16% efficiency thin film solar cell for band to band transistion at RT, as shown in Figure 5.74. A typical pulse laser with wavelength of 636 nm, 88 PS, 17PJ, and repeation rate of 5 MHz is used for TRPL measurement. The generated electron–hole pair concentration of 10^{14}–10^{16} cm^{-3} by one light pulse is lower than the majority carrier of 10^{16}–10^{17} cm^{-3} in the CIGS absorber. The minority carrier life time can be obtained from the PL decay of band to band transition. One can see that the PL rise time of CIGS absorber is as same as CIGS thin film solar cell indicating that the mechanism of photoexcited carriers is one and the same for both the CIGS thin film and CIGS

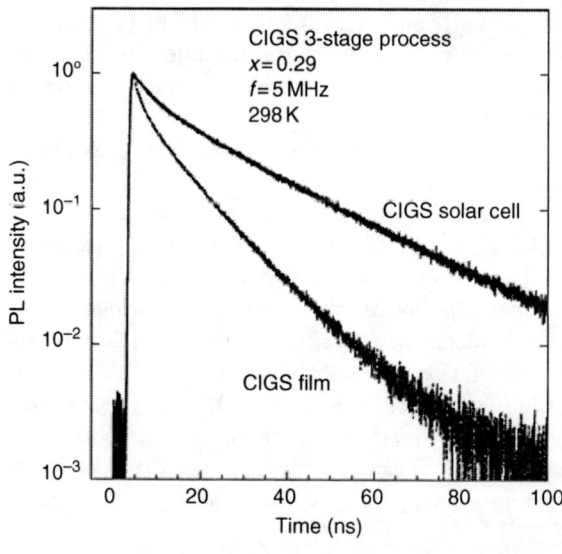

Figure 5.74 Time decay photoluminescence of CIGS thin film and solar cell.

thin film solar cell. The PL decay time is faster in the thin film than in the thin film solar cell revealing that the mechanism is different between them. Both the curves seem to be taking radative recombination mechanism. If there is a nonradiative mechanism, the nature of PL decay curve is asymmetry and response is very slow. The PL decay time is signature of life time of photoexcited minority carrier. The slower decay time in the thin film solar cell may be due to involvement of defect states at interface of CIGS and CdS. In general, Cd diffuses into CIGS and occupies copper place in the CIGS/CdS/ZnO based thin film solar cell [234]. The PL time decay is measured on three CIGS absorber layers, which show efficiencies of 9.3, 10.6, and 12.5% in the thin film solar cells. The minority carrier life time is determined using rate equation that decreases with decreasing efficiency of the cell. The life times of 7.2, 14.8, and 55 ns are observed for the efficiencies of 9.3, 10.6, and 12.5% CIGS thin film solar cells, respectively. The open circuit voltage is higher for higher carrier life time. However, no remarkable change in short circuit current is observed because the generated carriers are swept immediately by the electric field without giving ampoule time to recombine at the junction [235]. The radiative and nonradiative recombinations are observed from the PL decay time for $CuInS_2$ and $CuInSe_2$ [236,237].

5.3 Raman Spectroscopy

The Raman spectroscopy (RS) is one of the marvelous tools to determine single or polytype structures and secondary phases of the materials. The significance difference between RS and XRD is that the RS provides resonance bands for even amorphous materials, whereas X-ray powder diffraction does not. The spectroscopy works on inelastic scattering mechanism that when the laser beam strikes the sample, the atoms will get resonance to create modes in the crystal. If there are several resonance atoms misplaced that is, defects, the Raman spectrum may not give any bands corresponding to those missing atoms, unlike photoluminescence. However, it is exceptional for order vacancy compound (OVC) because its pattern is periodic with resonance atoms. A typical experimental Raman spectroscopy used by several investigators is given in Figure 5.75. The prism filters the laser beam from some other unwanted emission bands. The filtered laser beam is polarized by polarization rotator (Fresnel rhombus) and then focused on the sample through the short distance lens. The scattered light by the sample is collected through the same lens that is so-called back scattering configuration. The polarization analyzer focuses the scattered light onto the entrance of the slit that is close to the M_1 monochromator. The system is combination of M_1, M_2, and M_3 monochromators in which M_1 disperses the spectral lines of scattered light and then which are focused onto the edge slit close to the M_2 to avoid stray light of laser beam that is called subtractive mode. The M_2 monochromator restrains dispersion action of M_1. The M_1 and M_2 obviously avoid the stray light in the main stream. The M_3 monochromator does the spectral dispersion for the liquid nitrogen cooled CCD camera, which records the spectra with variation of frequency. The typical monochromators contain holographic gratings of 1800 lines/mm [238,239].

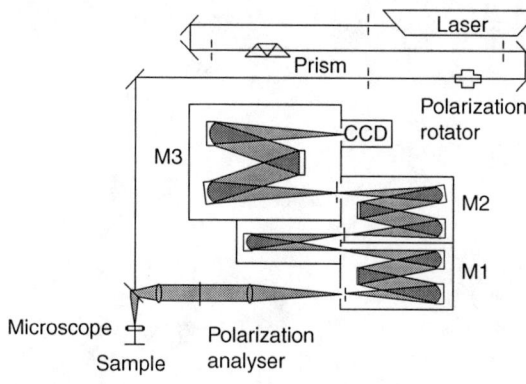

Figure 5.75 A typical Raman spectroscopy experimental setup.

The volume of the chalcopyrite (CH) unit cell for I–III–VI compounds, which had lattice constant of c, is twice ($2a$) of the zincblende (ZB) structure but its Brillouin zone is four times smaller than that of ZB. The ZB contains lattice constants of $a=b=c$ and the symmetric points of $\Gamma(0,0,0)$, $X(0,0,\pm 2\pi/c)$, $W(\pm 2\pi/a, 0, \pm \pi/c)$, and $W(0, \pm 2\pi/b \pm \pi/c)$ in the Brillouin zone. They turn into $\Gamma(0,0,0)$ for CH [240]. In the chalcopyrite unit cell, there are 8 atoms resonating 24 modes in which 21 optical modes transform like $\Gamma 1 + 2\Gamma_2 + 3\Gamma_3 + 3\Gamma_4 + 12\Gamma_5$ and two acoustic phonons transform like $\Gamma_4 + 2\Gamma_5$. All the optical phonons are Raman active except two Γ_2 modes. Γ_4 and Γ_5 modes are infrared active [241]. The atomic vibrations can be obtained from the Cartesian coordinates, as mentioned in Table 5.7 [242]. Figure 5.76 shows that A^I, B^{III} cations, and C^{VI} anions represent as ○, ◎ and ● symbols in the chalcopyrite unit cell, respectively. The A_1 mode is due to the vibration of the anion sublattice. A pair of anions (C^{VI}) vibrates oppositely in the X-axis that means one atom in $-X_3$ and second one in X_4 axis as well as another pair of anions vibrates oppositely in the Y-axis too as Y_1 and $-Y_2$ directions. Finally the total outcome is $(Y_1 - Y_2 - X_3 + X_4)$, as shown in Figure 5.76A. Another example for E_x^5 mode is also given in Figure 5.76B, which follows the resonance directions of $X_1 - Y_1 + X_2 + Y_2$, $-X_1 + Y_1 - X_2 - Y_2$ and $Z_1 - Z_2$ for A^I, B^{III}, and C^{VI} atoms, respectively. Similar way, one can test atomic vibrations for other bands in X, Y, and Z directions, as shown in Table 5.7.

5.3.1 CuInSe₂

A several modes of CuInSe$_2$ and other compounds obtained by either measurements or calculation are summarized in Table 5.8. The chalcopyrite characteristic A_1 mode at 174 cm^{-1} along with 259, 231, 215, 206, and 61 cm^{-1} modes due to B_2 and E in the CuInSe$_2$ crystals grown by vertical Bridgman method are observed by micro-Raman spectroscope. The remaining modes at 124 and 76 cm^{-1} in the spectrum due to B_1 modes are observed, as shown in Figure 5.77 [243]. However, along with the A_1 mode at 175 cm^{-1} and LO mode at 232 cm^{-1} correspond to chalcopyrite (CH) phase, the modes at 52, 186, and 462 cm^{-1} due to CuAu (CA) structure of CuInSe$_2$ thin films grown by MBE onto GaAs (001) 2° off orientation substrates are observed [244].

Table 5.7 Modes and their atomic positions of A^I, B^{III}, and C^{VI} [242]

Mode	A^I	B^{III}	C^{VI}
A_1	0	0	$Y_1-Y_2-X_3+X_4$
A_2^1	0	0	$X_1-X_2+Y_3-Y_4$
A_2^2	0	0	$Z_1+Z_2-Z_3-Z_4$
B_1^1	Z_1-Z_2	$-Z_1+Z_2$	$Y_1-Y_2+X_3-X_4$
B_1^2	Z_1-Z_2	Z_1-Z_2	0
B_1^3	Z_1-Z_2	$-Z_1+Z_2$	$-Y_1+Y_2-X_3+X_4$
B_2^1	Z_1+Z_2	Z_1+Z_2	$-Z_1-Z_2-Z_3-Z_4$
B_2^2	Z_1+Z_2	$-Z_1-Z_2$	$-X_1+X_2+Y_3-Y_4$
B_2^3	Z_1+Z_2	$-Z_1-Z_2$	$X_1-X_2-Y_3+Y_4$
E_x^1	X_1+X_2	X_1+X_2	$-X_1-X_2-X_3-X_4$
E_x^2	$X_1-Y_1+X_2+Y_2$	$-X_1+Y_1-X_2-Y_2$	$-Z_1+Z_2$
E_x^3	Y_1-Y_2	Y_1-Y_2	$X_1+X_2-X_3-X_4$
E_x^4	$X_1+Y_1+X_2-Y_2$	$-X_1-Y_1-X_2+Y_2$	0
E_x^5	$X_1-Y_1+X_2+Y_2$	$-X_1+Y_1-X_2-Y_2$	Z_1-Z_2
E_x^6	Y_1-Y_2	Y_1-Y_2	$-X_1-X_2+X_3+X_4$
E_y^1	Y_1+Y_2	Y_1+Y_2	$-Y_1-Y_2-Y_3-Y_4$
E_y^2	$X_1+Y_1-X_2+Y_2$	$-X_1-Y_1+X_2-Y_2$	Z_3-Z_4
E_y^3	$-X_1+X_2$	$-X_1+X_2$	$-Y_1-Y_2+Y_3+Y_4$
E_y^4	$-X_1+Y_1+X_2+Y_2$	$X_1-Y_1-X_2-Y_2$	0
E_y^5	$X_1+Y_1-X_2+Y_2$	$-X_1-Y_1+X_2-Y_2$	$-Z_3+Z_4$
E_y^6	$-X_1+X_2$	$-X_1+X_2$	$Y_1+Y_2-Y_3-Y_4$

The x and y subscripts of E_x and E_y are indication of E-mode coordinates with x and y dipole moments.

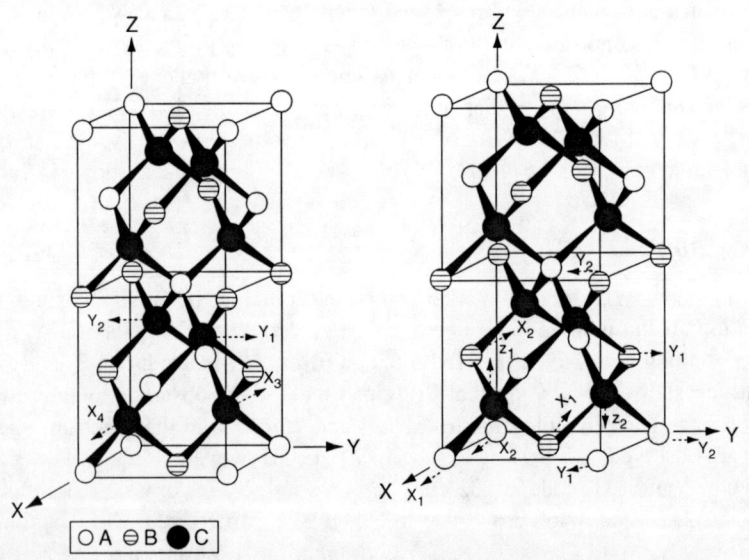

Figure 5.76 (A) Anion vibration directions in the chalcopyrite crystal for A1 mode, and (B) cations, and anions vibration directions in the chalcopyrite crystal for E_x5 mode.

Table 5.8 Phonon modes of $CuInSe_2$, $CuGaSe_2$ and $CuInS_2$

S. No.	$CuInSe_2$ [250] LO/TO (cm^{-1})	$CuGaSe_2$ [251] LO/TO (cm^{-1})	$CuInS_2$ [252] LO/TO (cm^{-1})
A_2^1	Silent	–	Silent
A_2^2	Silent	–	Silent
B_1^1	129	–	Very weak (297)
B_1^2	179	–	157 or 172
B_1^3	67	–	Very weak (110)
B_2^1	233/214	273/257	352/323
B_2^2	192/181	188/177	266/234
B_2^3	65.4/64	96/86	NA/79
A_1	173	186	294
E_1	230/213	277/255	351/322
E_2	212/207	186/180	304/286
E_3	182/179	149/145	239/230
E_4	78/78	109/105	161/160
E_5	61/61	90.9/88.2	124/124
E_6	–	66/64.5	90/90

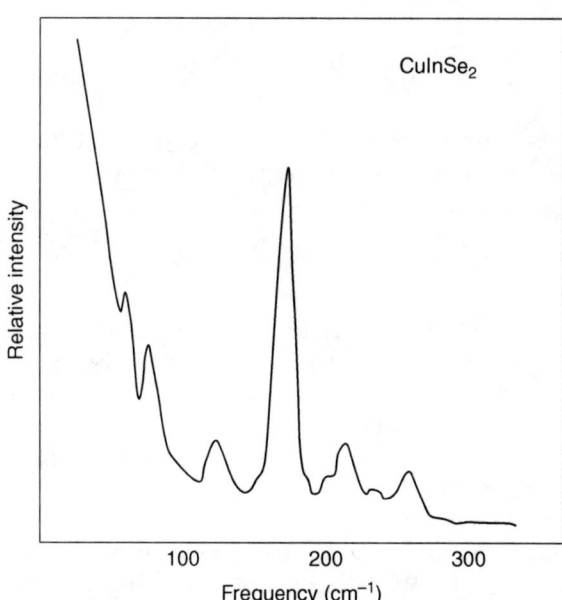

Figure 5.77 Raman spectrum of $CuInSe_2$ single crystal.

The chalcopyrite characteristic peak at 173 cm^{-1} is observed as a stronger one in the $CuInSe_2$ thin films and single crystals for the composition ratio of Cu/In = 0.95 and 0.93, respectively [245]. The good crystallinity in the $CuInSe_2$ nanorods grown by solvothermal technique is evidenced that the FWHM of 9.3 cm^{-1} for A_1 mode at 175.1 cm^{-1} in the CuInSe$_2$ nanorods is lower than that of 12 cm^{-1} in the CuInSe$_2$ thin films [93]. Bodnar *et al.* [246] proposed that the A_1 mode frequency is inversely proportional to the square root of the anion atomic weight. The experimental value of 174 cm^{-1} for A_1 mode can be roughly compared with a simple calculation by Keating model of

$$\omega = (k/M)^{1/2}, \tag{5.14}$$

where k is the force constant and M is the atomic weight of the anion Se. However, the modified formula is reasonably applied by Neumann [247] to determine A_1 frequency (179.8 cm^{-1}) as

$$\omega \approx \sqrt{2(\alpha_A + \alpha_B)/M_{Se}}, \tag{5.15}$$

where $\alpha_A = 32.65$ N m^{-1} and $\alpha_B = 42.565$ N m^{-1} are the two central bond-stretching force constants of A–C and B–C and M_{Se} is the atomic mass of the anion Se [248]. The experimental A_1 value is slightly lower than the calculated one. The first order Raman spectrum shows that the A_1 peak position shifts gradually from 173, 185 to 194 cm^{-1} with increasing pressure (P) from 0.1, 2.7 to 5.4 GPa, respectively in the CuInSe$_2$ single crystals at room temperature, as shown in Figure 5.78. The relation follows as

$$\omega(A_1) = 172.8 + 5.13P - 0.19P^2. \tag{5.16}$$

No Raman peak is observed for further increase of pressure beyond 8 GPa. A change in phonon position with effect of pressure can generally be formulated as

$$\gamma_i = -\frac{\partial \ln \omega_i}{\partial \ln V} = \frac{1}{K_T \omega_i} \frac{\partial \omega_i}{\partial P}, \tag{5.17}$$

where ω_i ($i = A_1$) is a phonon frequency, K_T the isothermal compressibility, and V the crystal volume. The γ_i observed from the relation is 1.46. Taking the sample from room ambient pressure to higher pressure of 9.5 or 13.5 GPa then bringing back to room ambient is so-called a hysteresis of pressure on the sample correspondingly the mode moves from 169.7–169.4 to 173 cm^{-1} and finally stabilizes at its original place in the Raman spectrum [249].

The micro-Raman spectra reveal that the CuInSe$_2$ layers grown onto Mo coated substrates by sequential deposition of In and Cu layers using RF magnetron sputtering and selenized under Ar flow at 350 °C for 5 h, which had well faceted grains with sizes of 2 μm, show A_1 mode at 174 cm^{-1} and mode at 258 cm^{-1} due to secondary phase of Cu$_2$Se/CuSe$_2$. In general, the Cu$_x$Se phase can be wiped off on the surface of CIS by etching sample in KCN solution to have better performance in the thin film solar cells because the Cu$_x$Se phase behaves as a metallic conductor. The details about the Cu$_x$Se phase are given in previous chapter. The Raman spectrum confirms the

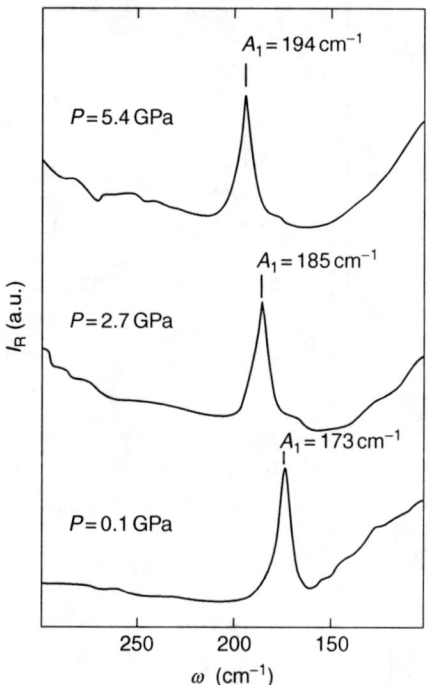

Figure 5.78 Raman spectra of CuInSe$_2$ crystals recorded at different pressures of 0.1, 2.7, and 5.4 GPa.

disappearance of mode at 260 cm^{-1} for Cu$_x$Se phase after etching [253]. After laser annealing CuInSe$_2$ thin film, the A_1 mode disappears and mode at 258 cm^{-1} due to Cu$_x$Se secondary phase becomes stronger, indicating that the phase segregation is high on the surface of CIS that masks CIS. The A_1 mode dominates 258 cm^{-1} mode of secondary Cu$_x$Se phase for Cu/In = 1 in the CuInSe$_2$. The secondary Cu$_x$Se phase mode becomes stronger ultimately the intensity of A_1 mode decreases with increasing Cu/In ratio from 1 to 1.37. The secondary phase mode heavily dominates the spectrum but the A_1 mode gradually disappears with further increasing Cu/In ratio to 4.17. The CuInSe$_2$ layers developed by sequential process in which Cu layer is deposited onto In layer follwed by selenization as usual at 305 °C for 5 h show A_1 mode along with an additional shoulder at 186 cm^{-1}. The shoulder, which may be due to OVC (CuIn$_{2.5}$Se$_4$) phase, is close to the reported value of 184 cm^{-1} for CuIn$_{2.5}$Se$_4$/CuIn$_3$Se$_5$ phase [254], whereas CIS layer grown with In layer onto Cu structure shows only A_1 mode [255]. On contrary, the mode at 186 cm^{-1} could be also related to sphalerite structure of CuInSe$_2$ [256]. In addition to A_1 mode, the peaks at 260 and 183 cm^{-1} are observed for the ratio of Cu/In > 2 in the films. They can be assigned to Cu$_x$Se and CuIn$_{2.5}$Se$_4$ phases, respectively, whereas the chalcopyrite peak disappears and a new peak appears at 150 cm^{-1} probably due to either CuIn$_3$Se$_5$ or In$_2$Se$_3$ for Cu/In < 0.5 [254,257], as depicted in Figure 5.79 [245].

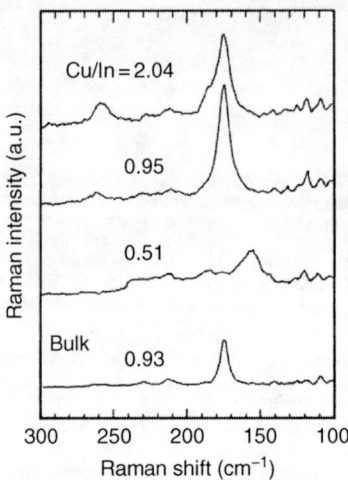

Figure 5.79 Raman spectra of CuInSe$_2$ thin films grown at different Cu/In ratios and stoichiometric single crystal.

Different phases can be identified by varying selenization temperature from low to high while selenizing Cu/In/Cu stack layers in the developing process of CuInSe$_2$ thin films. The stack 1350 Å Cu/6210 Å In/1350 Å Cu thick layers deposited at RT and selenized under Ar/Se pressure of 4×10^{-1} mbar at 199 °C for 1 h show a predominant 263 cm^{-1} mode for CuSe$_2$ phase and a 150 cm^{-1} mode for In–Se bond corresponds to either InSe$_2$ or In$_2$Se$_3$ for the selenization temperature of 215 °C. The sphalerite structure mode at 182 cm^{-1} appears for CuInSe$_2$ for both increased selenization temperatures of 199 and 215 °C. The A_1 mode at 174 cm^{-1} and 2 weak modes at 213 and 230 cm^{-1} accountable for chalcopyrite structure are observed for the selenization temperature of 242 °C. The CuSe$_2$ phase totally disappears for the selenization temperature of 400 °C. The CuInSe$_2$ films deposited at RT by sputtering using Cu/In = 0.79, (Cu + In)/Se = 1.16 at a rate of 0.5 Å/min and annealed under air at 450 °C for 10 min show A_1 at 174 cm^{-1}, E at 213 cm^{-1}, B at 230 cm^{-1}, and unidentified modes at 126 and 144 cm^{-1}. The XRD analysis shows amorphous nature contradictory to the Raman results. The sample may be recrystallized when the laser beam is used for scanning during the Raman measurements that is why the chalcopyrite structure appears in the Raman spectrum [258]. A_1 band at 176 cm^{-1} and a broad peak at 160 cm^{-1}, which may be related to OVC, are observed in the CuInSe$_2$ thin films prepared by electrodeposition with Cu/In = 1.2 in the solution. The modes at 240 and 260 cm^{-1} are also observed for Se and Cu$_x$Se phases in the as-grown precursor layers for $m = 2$Se/(Cu + 3In) = 1.1 but they are well pronounced for $m = 1.7$ indicating that the excess Se is more favorable for generation of secondary phases such as Se and Cu$_x$Se phase [259].

The Raman spectroscopy is useful to determine the quality of the layers for the high efficiency solar cells. Suppose, first 1.5 μm thick Cu-rich CIS layer with Cu/In > 2 is deposited, followed by 0.5 μm thick CuInSe$_2$ layer with Cu/In < 0.5

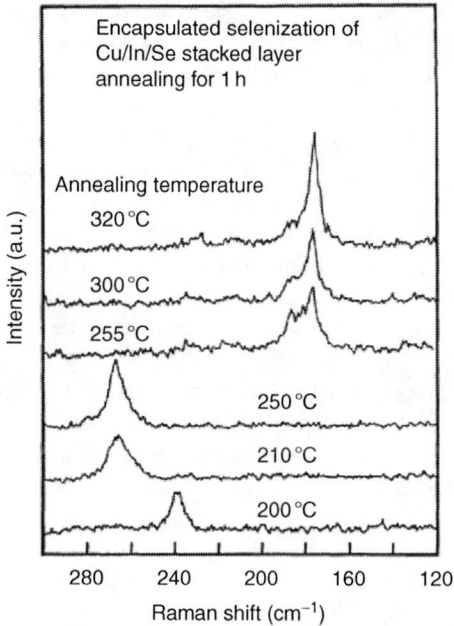

Figure 5.80 Raman spectra of CuInSe$_2$ layers formed by annealing the Cu/In/Se stack at different temperatures for 1 h.

as a bilayer for thin film solar cell. The thin film solar cells with CIS bilayer absorber exhibit high efficiency. In fact the spectrum of bilayer is different from CIS single layers. The Cu and In grown onto Mo coated glass substrates by RF-sputtering and Se by vacuum evaporation with thickness of 0.27, 0.6, and 1.46 μm, respectively to form sequential Cu/In/Se stack layers are annealed in an encapsulated carbon block at different annealing temperatures and times. Figure 5.80 shows Raman peak at 240 cm^{-1}, which is assigned to hexagonal Se for the Cu/In/Se stack layers annealed at below 200 °C, whereas an additional peak at 270 cm^{-1} due to CuSe$_2$ occurs for the layers annealed at or below the temperature of 250 °C. The A_1 mode is found for the stack layers annealed at or above 255 °C and becomes stronger with increasing annealing temperature from 255 to 300 °C, etc. Keeping temperature at 200 °C, the stack layers are annealed at different annealing timings under N$_2$ flow of 300 sccm. The peak at 240 cm^{-1} due to Se phase is found for 1 h annealing and its intensity gradually decreases with increasing annealing time from 1 to 4 h. On the other hand, the intensity of peak at 270 cm^{-1} due to CuSe$_2$ phase gradually increases with increasing annealing time. In order to cross check whether it is CuSe$_2$ phase or not, the CuSe$_2$ is separately developed by annealing Mo/Cu/Se stack with Cu:Se = 1:1 at 300 °C for 1 h, which shows the same mode at 270 cm^{-1}. The XRD analysis also strongly confirms the presence of CuSe$_2$ phase. The glass/Mo/In/Se stack with In:Se = 1:2 annealed at 300 °C shows a peak at 152 cm^{-1} due to β-In$_2$Se$_3$ phase, whereas the same stack annealed at below 250 °C exhibits small intensity peaks at 115 and 210 cm^{-1}. The phases of α-In$_2$Se$_3$ and InSe are detected by XRD. The observations indicate that first CuSe phase forms then In involves to complete the formation of CuInSe$_2$ layer during the course of annealing the stack [257]. On contrast, the Cu$_x$Se$_y$ phase generates at low temperature, whereas In$_x$Se$_y$ phase begins to form at high temperature.

The effect of substrate temperature and pH of chemical solution on CuInSe$_2$ thin films grown by spray pyrolysis are studied by Raman spectroscopy. Keeping pH 1.9, as the substrate temperature is varied from 300 to 360 °C in steps of 20 °C for the deposition of CuInSe$_2$ layers, the dominance of peak at 182 cm^{-1}, which is probably concerned to sphalerite structure, decreases. On the other hand, the chalcopyrite structure characteristic peak at 174 cm^{-1} resembled with sphalerite structure peak dominates the spectra (Figure 5.81A). In the case of pH 4 and $T_s = 300$ °C, the sphalerite peak at 182 cm^{-1} slowly disappears and A_1 pronounces well with increasing Cu/In ratio from 0.8 to 1.3 in steps of 0.1. In addition, other B_2(TO)/E(TO) and B_2(LO)/E(LO) modes are observed irrespective of pH and substrate temperatures within these limits. It is clear that for pH 1.9, Cu/In ratio of 0.8, 0.9, and 1.0 and substrate temperature of 300, 320, and 340 °C, the sphalerite structure occurs, whereas the CuInSe$_2$ layers exhibit chalcopyrite structure for the same pH, higher Cu/In = 1.1, 1.2, and 1.3 and substrate temperature of 360 °C. Secondly, the structure of CuInSe$_2$ is sphalerite for pH 4.0, Cu/In = 0.8, 0.9, and 1.0 and substrate temperature of 300 °C, whereas the sphalerite converts into chalcopyrite structure for

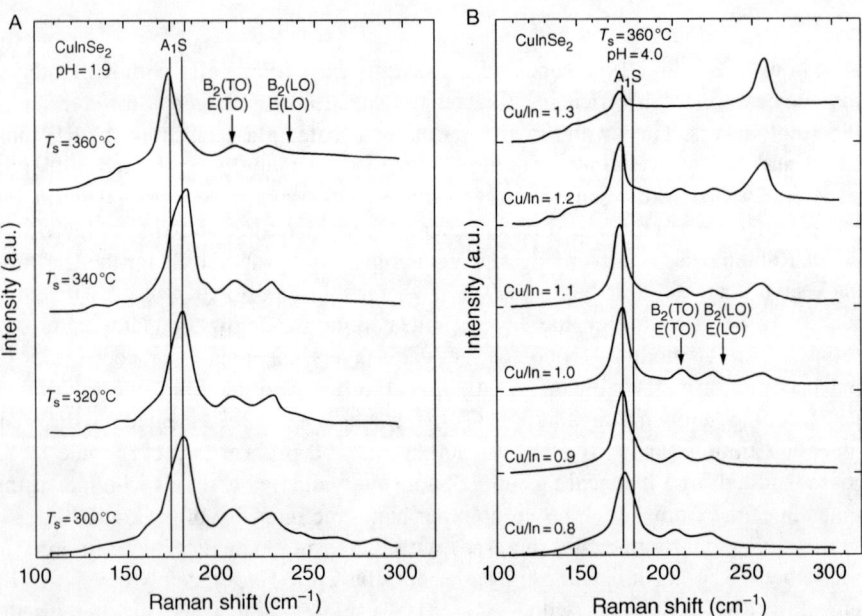

Figure 5.81 Raman spectra of CuInSe$_2$ thin films deposited at (A) different substrate temperatures keeping pH = 1.9 and Cu/In = 1.0 and (B) different Cu/In ratios keeping $T_S = 360$ °C, and pH = 4.0.

above the substrate temperature of 300 °C. The structure is chalcopyrite for pH-4.0, Cu/In = 1.1, 1.2, and 1.3, and substrate temperature of 320, 340, and 360 °C (Figure 5.81B) [256]. The Raman spectroscopy is also useful to assess the degree of sphalerite structure. The polycrystalline $CuInSe_2$ thin films with In/(In+Cu) grown by chemical spray pyrolysis method at $T_S = 360$ and 400 °C show that the A_1 mode at 174 cm^{-1} dominates the Raman spectrum. The $CuInSe_2$ thin films deposited at $T_S = 360$ °C and pH 3.5 for $0.57 \leq$ In/(In+Cu) ≤ 0.64 shows mode at 183 cm^{-1} (S) and a shoulder at 163 cm^{-1} (O), whereas the S peak disappears and O peak dominates the spectrum for the films with In/(Cu+In) = 0.78 deposited at 400 °C. The S mode is probably due to sphalerite structure of $CuInSe_2$. The O peak moves from 163 to 153 cm^{-1} with increasing In/(In+Cu) ratio from 0.64 to 0.78 for both the temperatures. The O peak is an A_1 characteristic mode of OVC of either $CuIn_3Se_5$ or $Cu_2In_4Se_7$ [40]. Both the OVC phases show A_1 characteristic Raman mode at 154 cm^{-1} [260]. It has also been confirmed that the same kind of $CuIn_3Se_5$ phase deposited by the physical vapor deposition (PVD) technique expresses A_1 mode at 153 cm^{-1}. The $CuInSe_2$ thin films grown by PVD show A_1 mode at 177 and 235 cm^{-1} due to Γ_4^7 (Γ_{15}) LO mode. As thickness of the samples is decreased, the A_1 mode slightly moves toward 180 cm^{-1} [261]. Figure 5.82 indicates that a different kind of phases of $CuIn_xSe_y$ layers deposited onto Si substrates by MBE are identified by unpolarized Raman spectroscopy as 174 cm^{-1} (Γ_1) for $CuInSe_2$, 149-151 cm^{-1} for $CuIn_3Se_5$, 152 and 175 cm^{-1} for $CuIn_{1.6}Se_{2.9}$, 152, 175, and 184 cm^{-1} for $CuIn_{2.5}Se_4/CuIn_3Se_5$. Small peaks at 212 and 231 cm^{-1} are also present in all the samples due to E (L) (Γ_4) and E (L) (Γ_5), respectively. The mode represents at 149–152 cm^{-1} due to symmetries of Γ_1, Γ_2, and Γ_3 [254]. The OVC $CuIn_3Se_5$ epitaxial layers grown onto $CuInSe_2$/GaAs indicate that A_1 mode at 153 cm^{-1} is in good agreement with the calculated value of 150 cm^{-1}. The mode at 185 cm^{-1} is probably due to OVC ($CuIn_{2.5}Se_4/CuIn_3Se_5$) bond or sphalerite and the other peaks at 212 and 230 cm^{-1} are due to TO and LO phonon modes of In–Se bond [262].

Let us see the effect of nozzle diameters on the growth of $CuInSe_2$ by Raman studies. The $CuInSe_2$ thin films are fabricated by using ionized cluster beam deposition with different nozzle diameters (D) of 1.7, 2.0, and 3.3 mm and length (L) of 1 mm

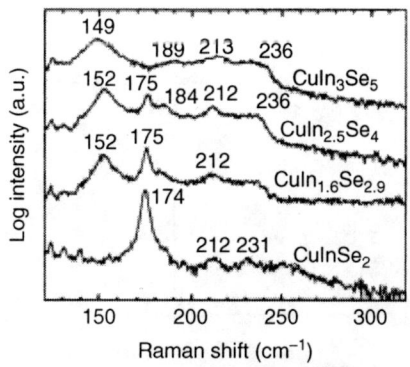

Figure 5.82 Raman spectra of $CuIn_xSe_y$ layers deposited onto Si(111) substrates.

for Cu crucible. $L=1$ mm and $D=1.5$ mm for In and Se crucibles are constantly used in this study. The intensity of 260 cm^{-1} mode increases with increasing Cu nozzle diameter from 1.7, 2 to 3 mm. Several modes at 172, 183, 210, 230, and 260 cm^{-1} are observed for Cu nozzle diameter of 3 mm, as depicted in Figure 5.83. The mode at 260 cm^{-1} disappears after etching the CuInSe$_2$ films by KCN indicating that the mode is related to the Cu–Se bond or Cu$_x$Se phase, as pointed out earlier. The mode at 183 cm^{-1} may be concerned to sphalerite CuInSe$_2$ or CuIn$_{2.4}$Se$_4$. The modes at 210 and 230 cm^{-1} show different intensities in HH and HV polarization configuration of Raman spectra revealing asymmetric E or B_2 modes [253]. The Raman studies on the Cu$_{1-x}$In$_x$Se$_2$ samples with $x=0.4$, 0.44, 0.48, 0.52, 0.55, 0.59, 0.63, and 0.67 indicate that the Cu$_{1-x}$In$_x$Se$_2$ films with $x>0.5$ that is, In-rich samples show emission lines at 153 cm^{-1} due to CuIn$_3$Se$_5$ and A_1 mode at 175 cm^{-1}, whereas the same A_1 appears well and the mode at 260 cm^{-1} due to Cu$_x$Se phase also presents in the Cu-rich samples ($x<0.5$) [263]. The CuInSe$_{2.5}$ layers grown onto Mo coated glass substrates by spray pyrolysis technique using CuI or [Cu(CH$_3$CN)$_4$](BF$_4$)$_2$, InI$_3$ or GaI$_3$ in pyridine and Na$_2$Se in methanol as precursors at 225 °C show A_1-173 cm^{-1}, 212 cm^{-1} (B_2,E), and 262 cm^{-1} (E), which become sharp when the films are annealed under vacuum at 560 °C for 10 min [264]. The Cu–In layers selenized at above 350 °C exhibit

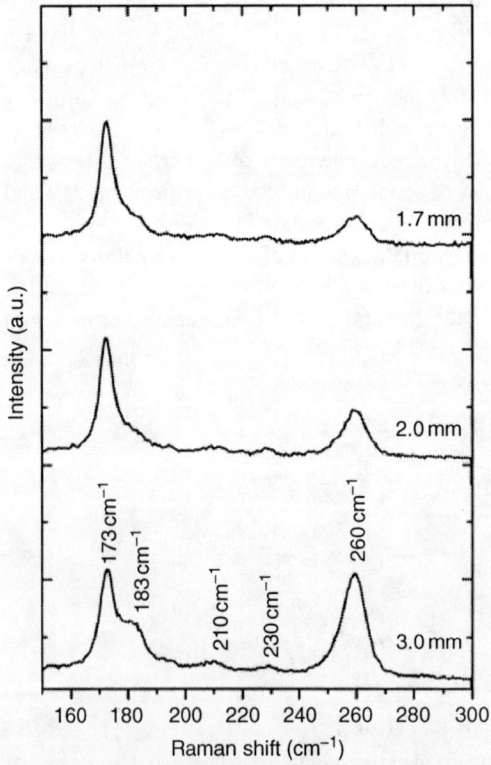

Figure 5.83 Raman spectra of CuInSe$_2$ thin films with different nozzle diameters.

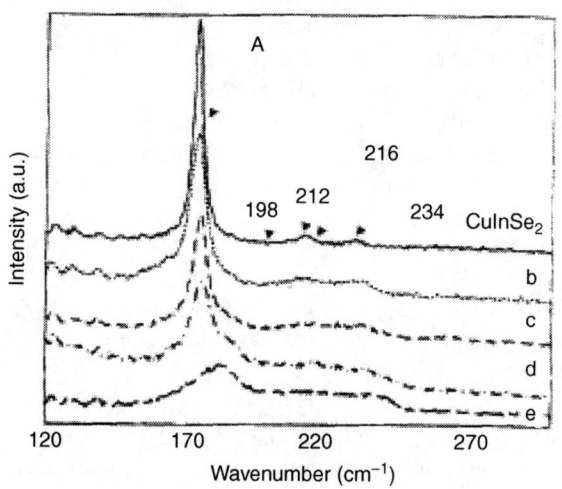

Figure 5.84 Raman spectra of Zn doped CuInSe$_2$ layers with variation of Zn: b) Zn 5.4 at %, c) 9.82, d) 14.26, and e) 20.02.

characteristic A_1 mode whereas 185, 258, and 115 cm^{-1} modes for OVC, CuSe, and InSe are found for below the temperature of 350 °C, respectively [265].

The intensity of the chalcopyrite characteristic mode A_1 gradually decreases with increasing Zn content in the Zn$_{2-2x}$Cu$_x$In$_x$Se$_2$ thin films prepared by two-stage selenization of ZnSe/(Cu+In) precursors, which are sequentially deposited by thermal vacuum evaporation as ZnSe, Cu, and In starting elements at chamber pressure of 5×10^{-6} torr, as shown in Figure 5.84. This observation supports degradation of crystal quality. The mode position also shifts from 174 to 182 cm^{-1} with increasing Zn content from 5.4 to 20 at%. The presence of other modes such as E (L) 212, E (T) 216, and E (L) 230 cm^{-1} is independent of Zn composition [266]. The as-grown p-CuInSe$_2$ single crystals and implanted by hydrogen with dose of 120×10^{16} ions/cm^2 at substrate temperature of 300 °C for 2.19 h prevail the same A_1 mode in the Raman spectra indicating probably not taking lattice damage in the implanted crystals [267]. The spectrum shows two peaks at 175 and 310 cm^{-1} for CuInSe$_2$ and CdZnS in the glass/Mo/CuInSe$_2$/Cd$_{0.8}$Zn$_{0.2}$S/ZnO:Al structure, respectively. Spectrum recorded through the cross section experiences two peaks at 169 and 240 cm^{-1} due to E_{1g} and A_{1g} of 2H-MoSe$_2$ phase. For this analysis the Cu, In, and Se are evaporated by electron beam evaporation to form Cu–In–Se precursor onto Mo coated glass substrates at substrate temperature of 200 °C then annealed under 5% H$_2$Se diluted with Ar at 400 °C. The Cd$_{0.8}$Zn$_{0.2}$S buffer is grown onto CuInSe$_2$ layer by electron beam evaporation, followed by deposition of ZnO:Al using DC magnetron sputtering [268].

5.3.2 CuGaSe$_2$

The Raman spectra recorded at 77 (300) K on CuGaSe$_2$ single crystals show A_1 mode at 188 (187) cm^{-1}, E-278 (274) cm^{-1}, and other modes at 252 (249), 193, 154 (154), 96 (96), 86 (86), and 59 (60) cm^{-1} indicating a slight variation in mode frequency with temperature [269]. The A_1 mode at 185 cm^{-1} in both $X(ZY)\bar{X}$ and $X(ZZ)\bar{X}$ configurations is allowed. However, $B_2(L)$ mode at 195 cm^{-1} and four $E(L)$ modes at 80, 153, 227, and 274 cm^{-1}, which may not be allowed in latter configuration, are present due to resonance effect [270]. The CGS thin films with Cu/Ga = 1.3 grown by physical vapor deposition (PVD) show multiple modes because of its polycrystalline nature. The most intensity mode at E_1 (LO)-277 cm^{-1} due to in phase vibration of Cu and Ga against to Se is observed in all the MOCVD, CVT, and PVD samples. The characteristic A_1-187 cm^{-1}, B_2-82, E_3(LO)-157.5, B_2-199, E_6(TO)-62.5, E_4(TO)-120, E_2(TO)-183, and E_2(LO)-193 cm^{-1} are observed. The mode at 166 cm^{-1} is probably due to second order mode of 82 cm^{-1} mode. Several second order modes 360, 398, 434, 464, 475, and 554 cm^{-1} may be related to B_2^3(LO)E_1(LO), $2B_2^2$(LO), E_3(LO)+E_1(LO), A_1+E_1(LO), B_2^2(LO)+E_1(LO), and $2E_1$(LO), respectively. The E_6(TO)-73, E_5(TO)-80.5, E_5(LO)-83.5, B_2^3(LO)-99, B_1^1-127.5, and E_1(TO)-252 cm^{-1} are also observed. The intensities of A_1, E_1, and B_2 modes are in decending order for Ga-rich CGS samples. On the other hand, peaks become broadening in the same sequence. The A_1 mode observed at 190.5 cm^{-1} for Cu/Ga = 0.75 is higher than that of calculated 188.9 cm^{-1} using modified version of Keating's formula

$$\omega \approx \sqrt{2(x\alpha_A + (2-x)\alpha_B)/M_{Se}}, \qquad (5.18)$$

where Cu/Ga = $x/(2-x)$, α_A = 30.035 N m^{-1}, and α_B = 51.135 N m^{-1} are the two central force constants of A–C and B–C and M_{Se} = 78.96 amu [271]. The measured A_1 mode (187 cm^{-1}) coincides with the calculated one for the stoichiometric CuGaSe$_2$ thin films (Cu/Ga = 1). The Raman modes at 273, 261, 239, 199, 168, and 60 cm^{-1} due to E and B_2 are observed in the CuGaSe$_2$ crystals. The A_1 mode relates to chalcopyrite characteristic signature presents at 184 cm^{-1}. The remaining modes at 116 and 96 cm^{-1} due to B_1 present. The modes at 128, 108, and 90 cm^{-1} due to B_1 are observed where the polarizer plane and analyzer plane are parallel (*aa* or *bb* orientation) to the *a*-axis of CuGaSe$_2$ crystals. There are several 259, 237, 180, 163, and 141 cm^{-1} modes present for *ac* or *bc* that is, polarizer plane parallel to *a*-axis and analyzer parallel to *c*-axis of crystal. The modes 259, 237, and 141 cm^{-1} are related to E modes. The Raman modes at 200 and 185 cm^{-1} are connected to B_2 and A_1 modes for *ab* orientation, respectively. The bands at 273 and 177 cm^{-1} are observed in which 273 cm^{-1} is related to E vibration for *ac* or *bc* orientation. Without dominant orientation, modes at 75 and 69 cm^{-1} are due to E and B_2 are observed, respectively [272]. The CuGaSe$_2$ thin films grown onto Mo coated glass substrates by evaporation technique either ionized or unionized Ga source show 240 cm^{-1} for B_2 or E modes and 210 cm^{-1} for B_2 or E. In addition, Cu–Se phase at 270 cm^{-1} is observed [273]. In another case, the peaks at 274, 185, 82, 31, 21, and 13 cm^{-1} for CuGaSe$_2$ and 45 and 263 cm^{-1} for CuSe in the CuGaSe$_2$ thin films are observed [274]. The A_1 mode shifts from 187 to 189 cm^{-1} with varying Cu/Ga ratio from 1.44 to 0.75 in the Cu$_x$Ga$_y$Se$_2$ thin films, whereas the mode at 187 cm^{-1} resides at constant for stoichiometric and Cu excess Cu$_x$Ga$_y$Se$_2$ thin films. The modes at 193 and 199 cm^{-1} in the CuGaSe$_2$ are related to defect compound of

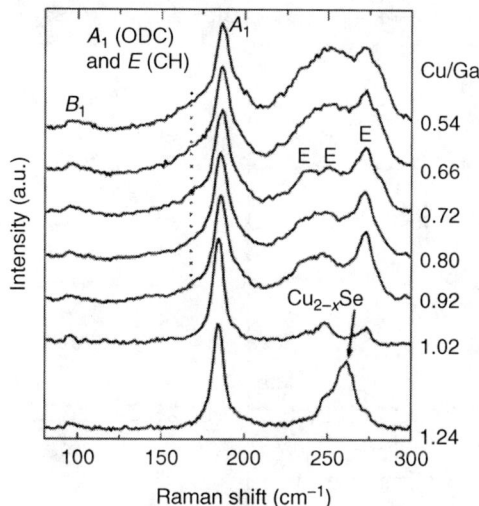

Figure 5.85 Raman spectrum of CuGaSe$_2$ thin films with different Cu/Ga ratios.

CuGaSe$_2$ (OVC) [275]. B_1 band at 96 cm^{-1}, A_1 mode at 185 cm^{-1}, E modes at 235 and 250 cm^{-1}, and E(LO) at 272 cm^{-1} due to in-phase movement of Cu and Ga against Se anions are found in 2 μm thick CGS thin films grown onto glass/0.5 μm Mo by co-evaporation at substrate temperature of 550–560 °C, as shown in Figure 5.85. The Cu$_{2-x}$Se phase mode at 261 cm^{-1} is seen for the Cu/Ga ratio of 1.24. A broad peak at 170 cm^{-1}, which is probably related to CuGa$_3$Se$_5$ (OVC), overlaps with E-mode (168 cm^{-1}) in the Cu-poor CGS films. The frequency of A_1 mode decreases with increasing Cu content due to enhancement of lattice constants [276].

5.3.3 Cu(In$_{1-x}$Ga$_x$)Se$_2$

The A_1-mode of Cu(In$_{1-x}$Ga$_x$)Se$_2$ thin films grown by physical vapor deposition varies linearly from 172.4 to 180.7 cm^{-1} with increasing x from 0.08, 0.19, 0.33, 0.5, 0.6 to 0.82 [275]. The Raman spectra of Cu(In$_{1-x}$Ga$_x$)Se$_2$ single crystals recorded at 10 K shows that the frequency of A_1 mode also varies linearly from 178 to 186 cm^{-1} with varying $x=0$ to 1.0 for the $X(Z,Z)\bar{X}$ configuration [277]. Similar observation is noticed in the CuIn$_{1-x}$Ga$_x$Se$_2$ thin films that the A_1 mode frequency linearly shifts from 174 to 184 cm^{-1} with varying x from 0 to 1. A variation is due to the lower mass of Ga, which occupies In site that creates larger bonding length in the lattice site and lowering the force constants. Therefore, the mode frequency increases with increasing Ga. The additional modes B_2/E-212, B_2/E-230 cm^{-1} for CuInSe$_2$ and E-247, E-269 cm^{-1} for CuGaSe$_2$ thin films are observed. The Raman studies are done at the back side of the CIGS samples, which are on Mo covered glass substrates, indicating MoSe$_2$ features such as A_{1g}-242 cm^{-1}, E_{1g}-169 cm^{-1}, and E_{2g}-288 cm^{-1}, as depicted in Figure 5.86 [278]. The A_1 peak shifts from 174 to 184 cm^{-1} in the spray deposited Cu(In$_{1-x}$Ga$_x$)Se$_2$ thin films with varying x from 0 to 1. The other B_2(TO)-E(TO) and B_2(LO)-E(LO)

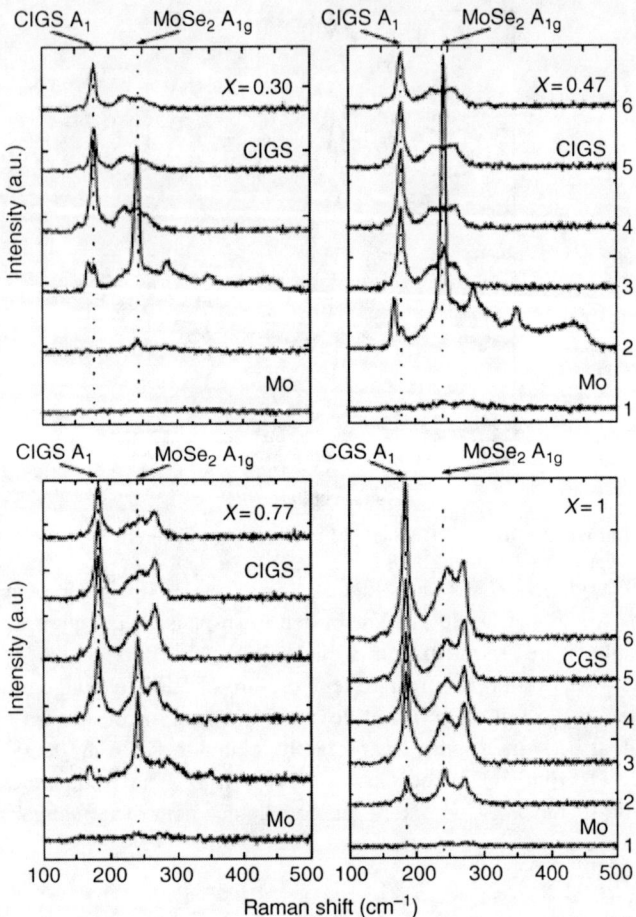

Figure 5.86 Raman spectra of CIGS thin films with different x values, which recorded at the distance of 0.5 μm from the Mo back contact to the top of CIGS absorber in the cross section.

modes at 210 and 230 cm^{-1} are found in the films. The mode at 182 cm^{-1} is observed in the sphalerite CuInSe$_2$ thin films [279]. The A_1 peak position varies from 176, 177, 178 to 179 cm^{-1} with varying Ga/(In+Ga) ratio from 0, 0.205, 0.379 to 0.409 in the CuInGaSe$_2$ thin films, respectively [149].

The chalcopyrite characteristic mode A_1 frequency varies linearly from ~176.7 to 177.8 cm^{-1} with increasing Ga/(Ga+In) ratio from 0.05 to 0.21 in the CuInGaSe$_2$ thin films, as shown in Figure 5.87. This linearity follows the same trend obtained by theoretically from the equation [280]

$$\frac{[(M_A+M_B)/2]^{1/2}}{[(M_X+M_Y)/2]^{1/2}} = \frac{\omega_{A_1}(XYZ_2)}{\omega_{A_1}(ABC_2)}, \tag{5.19}$$

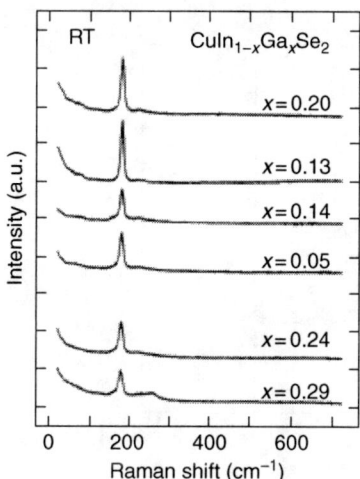

Figure 5.87 Raman spectra of $CuIn_{1-x}Ga_xSe_2$ with variation of x (Exc. Ar^+ laser (2.41 eV) Back scattering).

where XYZ_2 and ABC_2 can be treated as $CuInSe_2$ and $CuGaSe_2$, respectively or vice versa. The intensity of A_1 mode sublinearly decreases with increasing Ga/(Ga+In). The modes E-65.7, B_1-78.6, A_1-177.9, E-211.4, B_2, E-217.1, B_2, E-231.4, and B_2-248.6 cm^{-1} in the $CuInGaSe_2$ thin films for Ga/(Ga+In)=0.13 and Cu/(In+Ga)=0.93 are observed [281].

The A_1 mode at 170 cm^{-1} appears in the $Cu(In_{0.7}Ga_{0.3})Se_2$ thin films, whereas an extra mode at 270 cm^{-1} is observed in the $Cu(In_{0.4}Ga_{0.6})Se_2$ and $CuGaSe_2$ layers grown by co-evaporation of three-stage process, which is due to $Cu_{2-x}Se$ phase. As already mentioned, the electrical measurements confirm the presence of $Cu_{2-x}Se$ secondary phase by exhibiting metallic conductivity. The thin film solar cells made with $Cu(In_{0.4}Ga_{0.6})Se_2$ absorber treated under forming gases of N_2+H_2 5% at 400 °C exhibit efficiency of 11.2% but the cells with as-grown absorber shows efficiency of 7.3%. The mode at 270 cm^{-1} due to $Cu_{2-x}Se$ still exists after the RTP treatment indicating that the phase is not totally eliminated off but partially, as depicted in Figure 5.88 [282]. The $Cu(In_{0.75}Ga_{0.25})Se_2$ films show A_1 mode at 175 cm^{-1}, which becomes stronger as the growth temperature is increased to higher, as shown in Figure 5.89 [283]. The XRD results confirm the same kind of phenomenon that the crystallinity improves with increasing growth temperature. The as-grown $Cu(In_{0.75}Ga_{0.25})Se_2$ thin films annealed under Se at 300 °C, followed by annealing under N_2:H_2=9:1 at the same temperature shows remarkable improvement in the intensity of A_1 peak indicating improved crystallinity of the films [56]. The Cu-treated InGaSe layers annealed at 280 °C, show only one mode at 261 cm^{-1} for Cu_xSe phase, whereas layers annealed at 380 °C exhibit a mode at 177 cm^{-1} for CIGS layer including a mode at 261 cm^{-1} for Cu_xSe evidencing that the CIGS phase begins to form by consuming Cu_xSe phase at latter temperature [284].

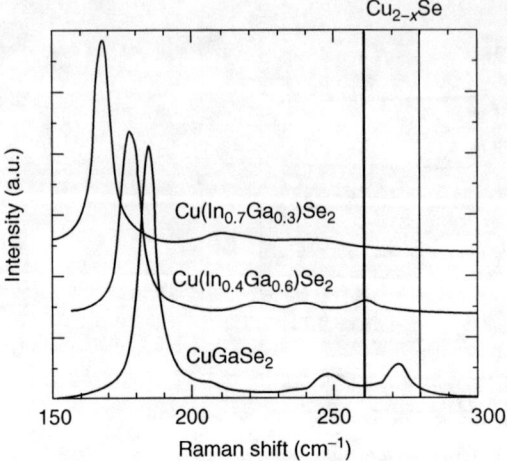

Figure 5.88 Raman peak shift with variation of Ga content in the CIGS.

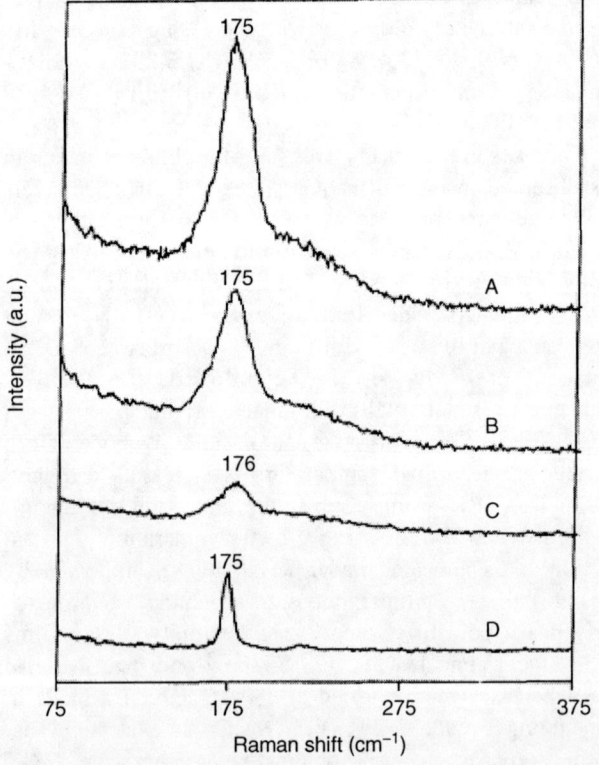

Figure 5.89 Raman spectra of $CuIn_{1-x}Ga_xSe_2$ ($x = 0.25$) thin films grown at different temperatures (A) 200, (B) 100 °C, (C) RT, and (D) single crystal.

The CIGS layers grown by three-stage process using molecular beam deposition show modes at 170 cm^{-1} as chalcopyrite characteristic mode of CIGS and $Cu_{2-x}Se$ phase mode at 270 cm^{-1}. Former mode shifts toward higher wave number with increasing Ga content indicating that the substitution of Ga into In site takes place. On the other hand, the intensity of latter mode also increases with increasing Ga content in the samples. The B_2 mode at 220 cm^{-1} is also observed in the samples. The CIGS samples are annealed using RTA process under forming gases (95% N_2 and 5% H_2). The presence of $Cu_{2-x}Se$ phase continues until to reach the annealing temperature of 300 °C and disappears beyond this temperature that is, at 400 °C for 1 s duration with annealing rate of 450 °C/min indicating that the secondary phase dissolves in the CIGS. The intensity of B_2 mode decreases with increasing annealing duration more than 1 s such as 1 min or 2 min 30 s, indicating probably degradation of sample quality. The metal concentration is high at some of the places on the annealed sample revealing H_2 reacts with Se leaving metal content. Longer annealing time causes to Se deficiency in the layers. The efficiency of typical cells with the $CuIn_{0.4}Ga_{0.6}Se_2$ absorber layers treated under Ar increases from 7.1 to 7.7%, whereas the efficiency reaches to higher value of 11.7 for treating CIGS under forming gases [285]. The A_1 mode frequency decreases from 178.5 to 176 cm^{-1} with varying Cu/(Ga+In) ratio from 0.5 to 1 thereafter increases to 177 cm^{-1} with increasing Cu/(Ga+In) ratio from 1 to 1.2 in 2 μm thick CIGS layers with Ga/(Ga+In)=0.3 grown by in-line co-evaporation, as shown in Figure 5.90. The structure of the sample is disorder below and beyond the stoichiometric ratios of Cu/(Ga+In). Therefore, the Raman mode frequency is low. The CIGS sample contains OVC and CH for 1<Cu/(Ga+In), CH for 1, and CH+$Cu_{2-x}Se$ for Cu/(Ga+In)>1

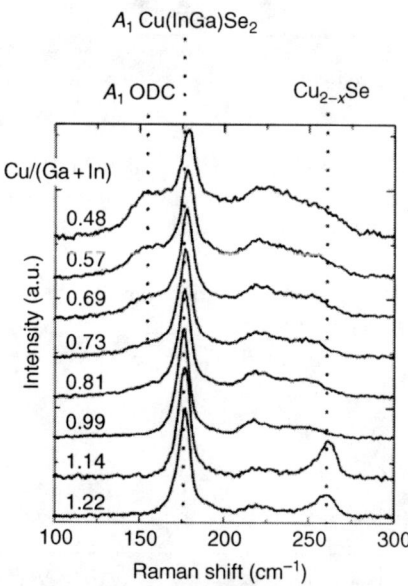

Figure 5.90 Raman spectra of $CuInGaSe_2$ thin films for different Cu/(In+Ga) ratios.

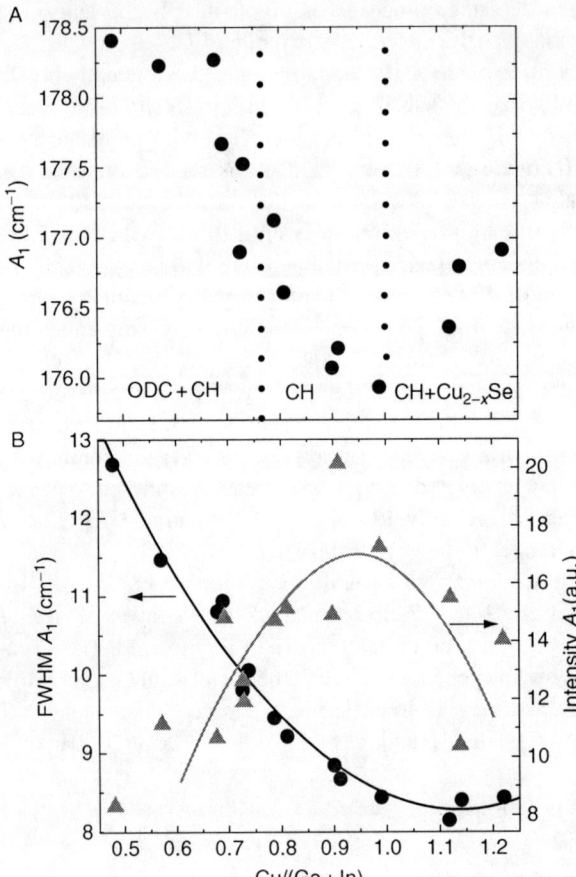

Figure 5.91 (A) Variation of A_1 mode with Cu/(In+Ga) ratio and (B) variation of FWHM and intensity for A_1 mode with Cu/(In+Ga) ratio.

(Figure 5.91A). The intensity of characteristic peak is high at stoichiometric composition, whereas it is low at below or beyond the stoichiometric ratio. The FWHM of A_1 mode is low at the ratio of Cu/(In+Ga) = 1.0, whereas it is high at below or beyond the ratio of 1.0 (Figure 5.94B). A broad peak pertains at 150 cm^{-1} for Cu(InGa)$_2$Se$_3$ or Cu$_2$(InGa)$_4$Se$_7$ phase OVC in the Cu-poor samples $0.8 <$ Cu/(Ga+In). A peak at 260 cm^{-1} and its second order mode at 520 cm^{-1} for Cu$_{2-x}$Se phase in the Cu-rich samples Cu/(In+Ga) > 1 are observed. The B_2/E modes are overlapped at 220 and 250 cm^{-1} in all the samples [286].

5.3.4 CuInS$_2$

The CuInS$_2$ single crystals grown by both the traveling heater and iodine vapor transport methods show more or less the same kind of features in the Raman spectra recorded at 9 K. The modes 345–347, 315, 295, 263–267, 156, 140–141, 76, and 60 cm^{-1} are corresponded to $B_2(L)/E(L) = E_5$, $E(L) = E_4$, A_1, $B_2(L)/E(L) = E_3$,

$2B_2'$, $E(L) = E_2$, $B_2(L) = B_2'$, and $E(L) = E_1$, respectively. The A_1 mode dominates spectrum [206]. The CuInS$_2$ layers deposited by reactive sputtering of Cu, In, and H$_2$S + Ar exhibit active Raman mode at 292 cm^{-1} for chalcopyrite lattice due to sulfur sublattice vibration, which presents in all the samples irrespective of growth conditions. The characteristic mode (A_1') at 305 cm^{-1} confirms the signature of CuAu (CA) ordering defect structure, whereas only one 292 cm^{-1} mode presents and mode at 305 cm^{-1} disappears in the CuInS$_2$ thin films grown by sequential sputtering [287]. The 60 and 305 cm^{-1} modes in the CuInS$_2$, 185 and 233 cm^{-1} in the CuInSe$_2$ are signatures of CA. As the In/(In+Cu) ratio is increased from 0.47, 0.55 to 0.62 in the CuInS$_2$, the intensities of 60 and 305 cm^{-1} modes increase, indicating that CA phase becomes stronger with ratio [288,289]. After KCN treating CuInS$_2$ thin films prepared by RF-ion plating technique, different modes at 240, 266, 292, 307, 320, and 340 cm^{-1} present, which are assigned to E, B_2, A_1, CA (A_1'), E, and B_2, respectively. As shown in Figure 5.92, the spectrum is recorded in horizontal–horizontal (HH) and horizontal–vertical (HV) configuration using a polarizer and a $\lambda/2$ plate that does not show much difference between (HH) and (HV) configurations indicating that the modes are in symmetry. The Raman modes at 173 (A_1), 183 (A_1'), 210, and 230 cm^{-1} for CuInSe$_2$ thin films are also observed. According to formula proposed by Matsushita et al. [280] the frequency ratio of A_1 $(ABC_2)/A_1'(XYC_2) = 292/307$ cm^{-1} = 0.951 for CuInS$_2$ to CA-CuInS$_2$ and 173/183 cm^{-1} = 0.945 for CuInSe$_2$ to CA-CuInSe$_2$. The ratio of CuInS$_2(A_1)$/CuInSe$_2(A_1)$ is 292/173 = 1.69 and CA-CuInS$_2(A_1')$/CA-CuInSe$_2(A_1')$ is 1.68. The observations indicate that the mean atomic weight of cations in the A_1' mode is lower than that in A_1 mode of stoichiometric compounds [290]. The A_1 is calculated to be 285 cm^{-1} for CuInS$_2$ using the formula (Equation (5.15)) and force constants $\alpha_A = 28.4$ N m^{-1} and $\alpha_B = 48.475$ N m^{-1}, which is slightly less than the measured value of 292 cm^{-1}.

The A_1 mode at 290 cm^{-1} and CA mode at 305 cm^{-1} are detected for CuInS$_2$, which is grown by sulfurizing Cu–In alloy onto Mo coated glass substrates under H$_2$S and elemental sulfur at 450 °C. The dominance of CA mode at 305 cm^{-1} gradually decreases

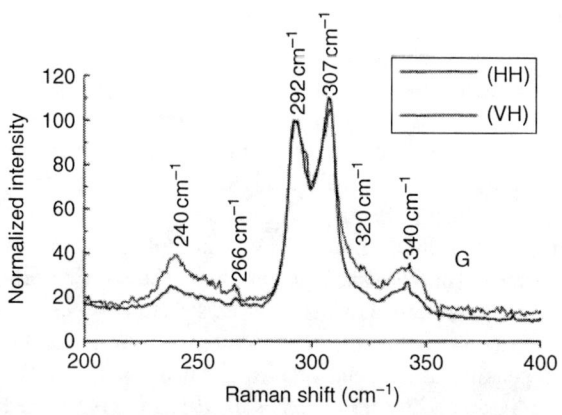

Figure 5.92 Raman spectra of CuInS$_2$ thin films recorded in HH and VH configurations.

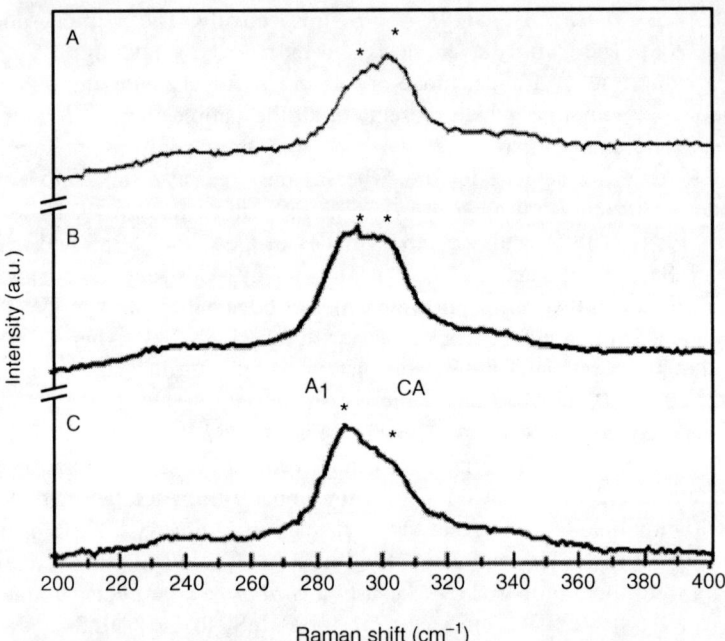

Figure 5.93 CA ordering percentage decreases with increasing sulfur pressure during the course of sulfurization in the Raman spectra of CuInS$_2$ thin films; (a) 1 atm, (b) >1 atm, and (c) >1 atm.

with increasing sulfur pressure probably from 1 atm to higher as shown in Figure 5.93. The CA mode is found to disappear for a partial incorporation of oxygen into the sulfurization chamber. The CA mode is more dominant under H$_2$S sulfurization than that under elemental sulfurization [213]. In the spectra, the intensity ratio of chalcopyrite peak 288 cm^{-1} to CA phase 298 cm^{-1} peak increases with decreasing Cu/In ratio from 0.66, 0.75, 0.87, 1.1 to 1.29 for the CuInS$_2$ thin film deposition conditions of pH 3.5 and $T_S = 360\,°C$ [291]. The CA ordering phase mode at 305 cm^{-1} dominates the same phase mode at 60 cm^{-1} in the Cu-poor (Cu/In < 1) CuInS$_2$ thin films, whereas the chalcopyrite mode A_1 at 290 cm^{-1} pronounces well in the Cu-rich CuInS$_2$ thin films. The similar kind of observation is noticed by X-ray diffraction analysis that (0002) peak is stronger in the Cu-rich CuInS$_2$ films than that in the Cu-poor CuInS$_2$ films [292]. Figure 5.94 shows that the chalcopyrite characteristic A_1 mode is found either as a broader band with low intensity or disappears in the Cu-poor CuInS$_2$ films, whereas the same mode is stronger in the Cu-rich films. The A_1 mode along with the CA mode at 305 cm^{-1} appears in the Cu-rich samples, however it is also close to the modes of In$_2$Se$_3$ [293]. The CuInS$_2$ thin films grown by ILGAR technique contribute two kinds of features on its surface; one is darker and another one being brighter. The Raman spectra reveal that the darker region is solely concerned to chalcopyrite structure by peaking chalcopyrite characteristic band at 292 cm^{-1} while the brighter area contributes combination of CH and CA structures peaking at 292 and 305 cm^{-1}, respectively [294]. The Raman modes at 292, 305, 340, and 472 cm^{-1} are observed in the crystallized

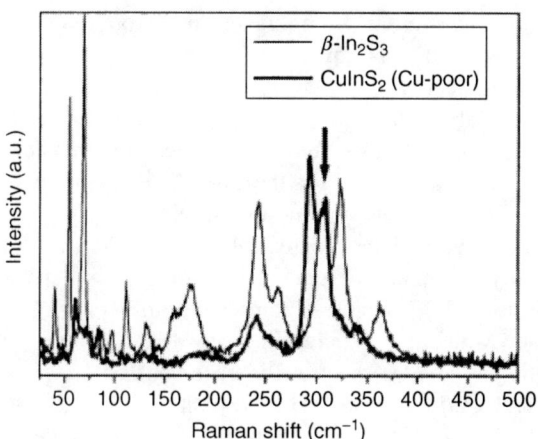

Figure 5.94 Raman spectra of Cu-poor CuInS$_2$ and In$_2$S$_3$ thin films.

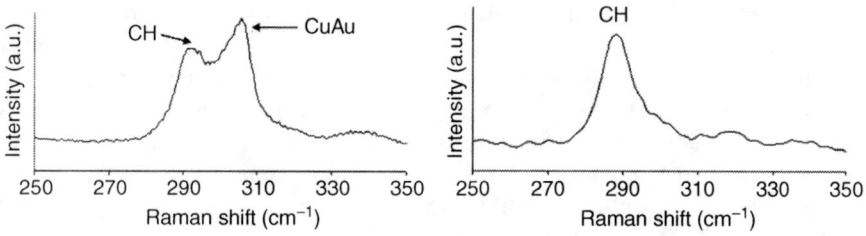

Figure 5.95 Raman spectra of CuInS$_2$ thin films after annealing at (A) 500 °C and (B) 700 °C.

CuInS$_2$ powder grown by chemical solution route at different pHs. As pH is increased from lower to higher value of 11.8, 472 cm^{-1} mode due to SO$_4^{-2}$ phase disappears. The other modes such as 292, 305, and 340 cm^{-1} are signatures of common characteristic A_1, CA, and B_2, respectively. The CuInS$_2$ powder made at lower pH of 9.0 shows B_2 mode at 265 cm^{-1} [220].

The CuInS$_2$ thin films deposited onto ITO glass substrates by RF-sputtering with 70 W, Ar 1.3×10^{-2} mbar at RT using Cu$_2$S–In$_2$S$_3$ mixed and cold pressed target. The target shows CH phase peak at 290 cm^{-1} as well as one at 327 cm^{-1}. The as-grown p-type thin films show broad peaks at 280 and 340 cm^{-1}. The former broad peak is obviously resemblance of CH and CA phases. After annealing the samples at 400 °C, no change in peak is observed. The CH and CA at 290 and 305 cm^{-1} phases, respectively are well resolved with increasing annealing temperature to 500 °C. The CH phase only retains and CA phase disappears for further increasing annealing temperature to 700 °C, as shown in Figure 5.95 [295]. The intensity of CA phase A_1' mode at 300 cm^{-1} is greater than that of CH phase A_1 mode at 290 cm^{-1} in the CuInS$_2$ thin films grown by spray pyrolysis technique at 370 °C. The CH content increases in the samples, when they are heat treated under H$_2$S atmosphere at 525 °C rather than treated under the same atmosphere at 450 °C.

The FWHM of CH phase A_1 mode decreases from higher to 12 or 13 cm^{-1} for Cu/In = 1.0 or 1.1, respectively when the as-grown samples are annealed under H$_2$S at 525 °C indicating an improved quality of the layers. In the Cu-rich samples, the mode at 348 cm^{-1} disappears probably relates to CuIn$_5$S$_8$ phase. In addition, the modes at 321 and 338 cm^{-1} could be related to B_2^1 and E_{LO}^1 modes of the chalcopyrite [296]. Lines at 300, 345, and ~480 cm^{-1} for the growth temperature of 350 °C are observed in the CuInS$_2$ thin films with Cu/In ratio of 0.9 grown by reactive magnetron sputtering, which can be assigned to resemblance of A_1 and CA, CuIn$_5$S$_8$ and CuS, respectively. Only one mode presents at 300 cm^{-1} with increasing growth temperature from 350 to 420 °C, which is combination of A_1 and CA. The A_1 mode dominates CA mode with further increasing growth temperature to 500 °C, as shown in Figure 5.96. The similar spectrum is observed for the same growth temperature of 500 °C and Cu/In ratio of 1.1 [297]. In addition to 340 cm^{-1} the modes present at 325, and 360 cm^{-1} for CuIn$_5$S$_8$ phase with depth profiling in Cu-poor CuInS$_2$ films [298]. The mode appears at 472 cm^{-1} is related to CuS hexagonal phase in the Cu-rich Cu$_{1-x}$In$_x$S$_2$ thin films, whereas the mode at 309 cm^{-1} is concerned to β-In$_2$S$_3$ phase in the In-rich samples. The A_1 mode at 291 cm^{-1} presents well in both the Cu-rich and In-rich samples [263]. 1 μm thick Cu/In alloy sputtered having a Cu/In = 1.8, which is suitable for quality layers, is ramped in elemental sulfur to sulfurize or form CuInS$_2$ compound at different temperatures and different ramping times. The CuInS$_2$ layers are made by ramping up to 300 °C (type-*a*), ramping up to 500 °C within 50 s (type-*b*), after ramping up to 500 °C and kept at the same temperature for 1/2 min (type-*c*), for 1 min (type-*d*) and for 2 min (type-*e*). The type-*b* CuInS$_2$ thin film shows A_1 mode at 290 cm^{-1} along with CA

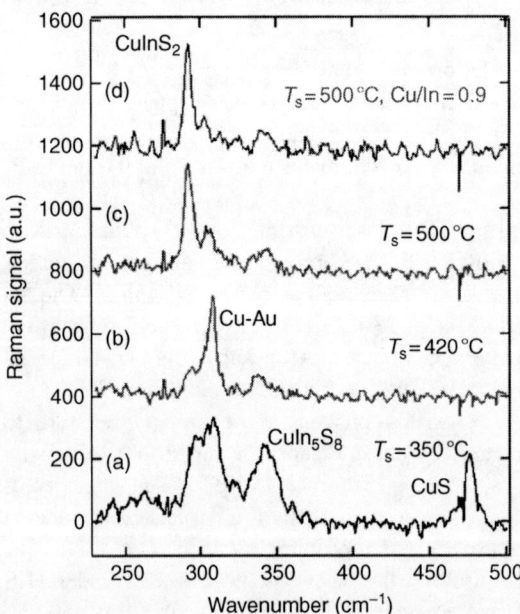

Figure 5.96 Raman spectra of CuInS$_2$ thin films grown by reactive magnetron sputtering at different temperatures with constant Cu/In ratio of 0.9.

mode at 305 cm^{-1}. These two modes are well resolved in type-c samples, whereas type-d and -e samples show only CuS phase modes that could be due to loss of In. After eliminating Cu$_x$S phase by sputter etching, the A_1 mode and lower intensity CA mode appear in all the samples, as shown in Figure 5.97 [299]. The Raman signals appear at 55 and 470 cm^{-1} for the growth temperature of 180 °C relate to CuS phase. The mode at 290 cm^{-1} due to CuInS$_2$ phase is observed for the growth temperature of 205 °C. On the other hand, 305 and 60 cm^{-1} mode frequencies are concerned to CA phase. The binary phase disappears for the growth temperature of 240 °C. At 290 °C, β-In$_2$S$_3$ phase (150 cm^{-1}) appears, but which disappears at 440 °C. Up to the end of the experiment the CuInS$_2$ and CuAu phases appear. While cooling down the sample, the CuIn$_5$S$_8$ phase appears. Below 180 °C, the CuInS$_2$ phase is not found. In the 160–215 °C ranges, CuS phase appears. As temperature is increased, the CuS phase disappears. Between 270 and 480 °C, β-In$_2$S$_3$ phase appears. While cooling down to 240 °C, chalcopyrite and CuAu phases present, whereas β-In$_2$S$_3$ phase and CuIn$_5$S$_8$ appear at 225 °C. In the case of Na doping CuS phase appears and disappears earlier stage. The order phase and chalcopyrite phase appear at later stage. The appearance and disappearance of β-In$_2$S$_3$ phase occur at later stage that is, at high temperature. The observation of β-In$_2$S$_3$ phase occurs at higher temperature and InS appearance at lower temperature indicating

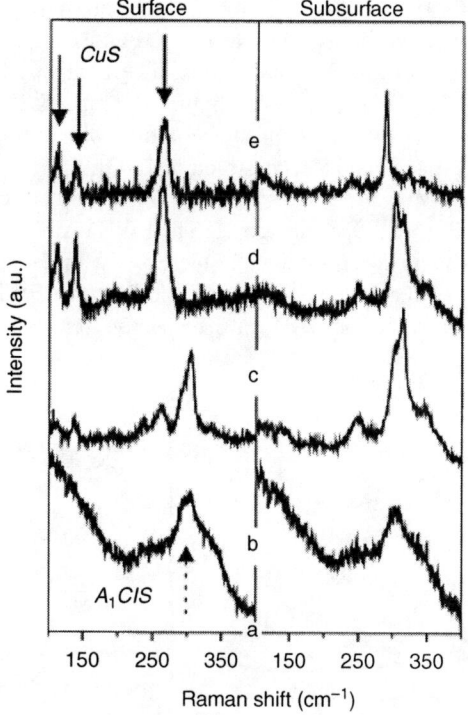

Figure 5.97 Raman spectra of samples a, b, c, d, and e recorded at surface and subsurfaces.

that In segregation toward surface takes place at higher temperature. Raman spectra of Na incorporation during the growth of $CuInS_2$ give us different results indicating that growth reaction path is different for Na doped $CuInS_2$ thin films [300].

In some of the as-grown $CuInS_2$ thin films, the modes at 62, 142, 267 and A_1 mode at 300 cm^{-1} are observed. In addition, a strong mode at 474 cm^{-1} is due to CuS phase, which is confirmed by recording Raman spectra on CuS single crystal. After KCN etching the thin film sample, all the modes except A_1 mode are eliminated indicating that CuS is etched off [275]. The CuS made at 470 cm^{-1} along with emission bands at 280, and 337 corresponding to A_1, and B_2 are observed in the etched $CuInS_2$ layers grown onto conducting ITO glass substrates (10 Ω/sq) by electrodeposition, respectively. The similar characteristics are also observed for the $CuInS_2$ layers deposited onto Ti substrates. The mode at 470 cm^{-1} for secondary CuS phase disappears after KCN etching (Figure 5.98). The XRD measurements support that the CuS phase is no longer exist. In addition, the intensities of XRD reflections of $CuInS_2$ increase in the etched $CuInS_2$ samples [301]. Cu_xS phase mode at 474 cm^{-1} and modes at 294 and 306 cm^{-1} for CH and CA, respectively as a resemblance peak are found in the $CuInS_2$ for which the Cu/In layers are deposited by vacuum evaporation, followed by sulfurization in the $CuInS_2$ for which H_2S at 500 °C for 30 min is employed. After electrochemical etching the sample, the intensity of 474 cm^{-1} mode decreases but still exists [302]. As Zn concentration is increased from 0.1 to 10 at% in the $CuInS_2$, the mode at 340 cm^{-1} gradually pronounces, which can be assigned to $E_1(LO)$ and $B_2^1(LO)$ modes caused by a combined movement of cations and anions. The chalcopyrite characteristic A_1 frequency at ~290 cm^{-1} disappears for higher doping of 25 at% Zn concentration. This indicates that the chalcopyrite structure of $CuInS_2$ disintegrates into zincblende structure that is, sphalerite structure [303].

The quality of glass/Mo/$CuInS_2$/CdS/ZnO thin film solar cells can be assessed using Raman and photoluminescence spectroscopes. The FWHM of chalcopyrite characteristic A_1 mode in the Raman spectra gradually increases with decreasing efficiency and other PV parameters of the cell, as given in Table 5.9. The FWHMs of PL peaks at 1.53 and 1.45 eV for excitonic and DAP peaks, respectively, also gradually increase with

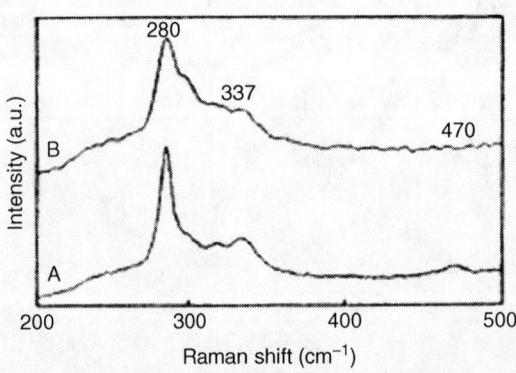

Figure 5.98 Raman spectra of (A) as-grown and (B) etched $CuInS_2$ in KCN.

Table 5.9 Correlation between FWHM of A_1 Raman mode and solar cell parameters

Solar cell	V_{oc} (mV)	J_{sc} (mA/cm^2)	FF (%)	η (%)	FWHM (cm^{-1})
A	728	21.5	71.1	11.1	3.27
B	585	21.4	36.4	4.6	3.83
C	363	16	25.2	1.5	4.15

decreasing efficiencies of the solar cells [304]. It is worthwhile to note that the same kind of trend is observed in the CuInS$_2$ solar cells with different open circuit voltages of (a) 752, (b) 715, and (c) 685 mV. The FWHM of about 2.7–3.8 cm^{-1} for A_1 mode 292 cm^{-1} is lower for high V_{oc} samples, whereas it is higher (>3.8 cm^{-1}) for lower V_{oc} samples [305]. The FWHM of A_1 mode increases with decreasing V_{oc} from 710 to 688 mV. On the other hand, the efficiency also decreases from 10.7 to 9% in the CuInS$_2$ cells indicating an increase of defect density [306].

5.3.5 CuIn(Se$_{1-x}$S$_x$)$_2$

Figure 5.99 shows A_1 mode at 174 cm^{-1} and a weak mode at 200 cm^{-1} in the CuIn(Se$_{1-x}$S$_x$)$_2$ thin films grown by spray pyrolysis technique. The latter is assigned to mixed mode of B_2 and E. The CuInS$_2$ thin films show A_1 mode at 300 cm^{-1} and a weak mode at 220 cm^{-1}. The A_1 mode linearly shifts from 295 to 287 cm^{-1} with varying x from 1.0 to 0.35 in the CuIn(Se$_{1-x}$S$_x$)$_2$. On the other hand, the intensity of peak also increases with decreasing x [307]. Trigonal structured Se-240 cm^{-1}, Cu$_x$Se modes at 43 and 260 cm^{-1} are found in 1.6 µm thick nanocrystalline Cu–In–Se precursor layers with composition ratio of In/Cu=0.9 and Se/(Cu+In)=1.2 deposited onto 0.6 µm thick Mo coated glass substrates by electrodeposition using one-step method. In addition, OVC mode at 160 cm^{-1} and CuInSe$_2$ mode at 176 cm^{-1} are found. The mode value of 176 cm^{-1} is higher than the standard chalcopyrite characteristic peak of 173 cm^{-1} that may be due to compressive stress in the layers or nonstoichiometry. The Raman mode at 173 cm^{-1} is the intensity one at some places on the sample, whereas the 240 cm^{-1} mode is intensity one at some other places indicating inhomogeineity in the sample. The precursor layers sulfurized under S vapor using RTP exhibit modes at ~290 cm^{-1} for S–S and ~200 cm^{-1} for Se–Se vibrations. The intensities of S–S and Se–Se vibration modes increase and decrease with increasing S content in the CuIn(SSe)$_2$ system by RTP, respectively. This observation obiviously indicates an increased alloying of S in the CuIn(SSe)$_2$ system. The CuIn(SSe)$_2$ layers sputter etched at different timings show that the intensity of A_1 mode of CuInS$_2$ gradually decreases with etching. After 40 min etching, the CuIn$_5$S$_8$ (OVC) broad mode at 337 cm^{-1} disappears and MoS$_2$ modes at 375 and 407 cm^{-1} start to appear [308]. Changes in the CuInSe$_2$ thin films with Cu/In=1.1 and Se/(Cu+In)=1.3 grown by single step electrodeposition in an acidic media, followed by annealing in the presence of sulfur atmosphere at 500 °C for different timings 0–90 min using RTP system are studied by Raman spectroscope. The electrodeposited CuInSe$_2$ thin films annealed under sulfur

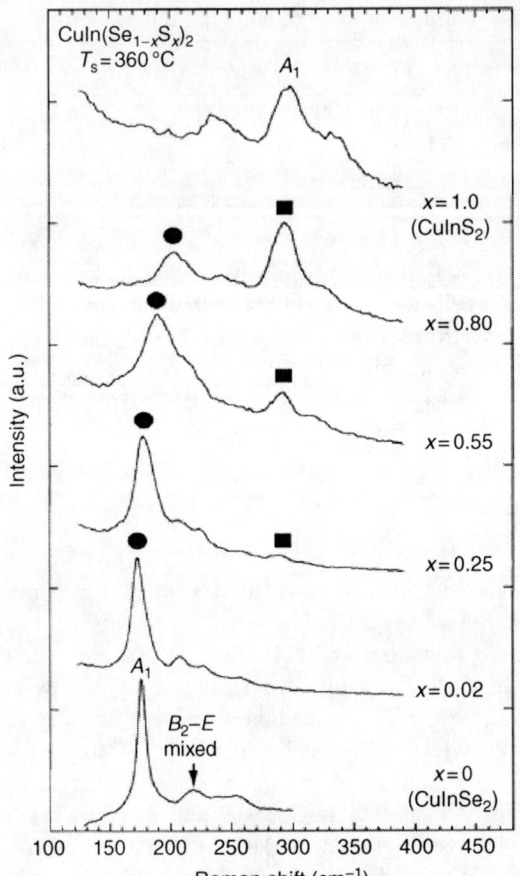

Figure 5.99 Raman spectra of CuIn(Se$_{1-x}$S$_x$)$_2$ thin films with different x values deposited at 360 °C.

atmosphere up to 10 min show modes at 290, 173, and 340 cm^{-1} for S–S, Se–Se, and CuIn$_5$S$_8$, respectively. In addition, 380 and 465 cm^{-1} are found for Cu(S,Se) phases. The secondary Cu(S,Se) phases disappear with further increasing annealing time to 20 min that means those are consumed by CuIn(SeS)$_2$ system, as shown in Figure 5.100. The Raman frequency modes of MoS$_2$ and CuIn$_5$S$_8$ are observed at the cross section of the layers. The latter exists throughout the layer no matter whether at the surface or bottom of the layer. The broad mode at 305 cm^{-1} for CA phase overlapped with SS peak slowly disappears with increasing annealing time, as shown in Figure 5.101. The cells made with electrodeposited and annealed CuIn(SSe) layers reveal that the cell V_{oc} increases from 450 to 625 mV with increasing annealing time from 0 to 20 min thereafter slightly decreases to 600 mV with further increasing annealing time to 90 min. The J_{sc} also follows more or less the similar pattern from 12.5, 13.75 to 15.5 mA/cm^2 but FF monotonically increases. It can be concluded that there are two regimes based on the experimental results.

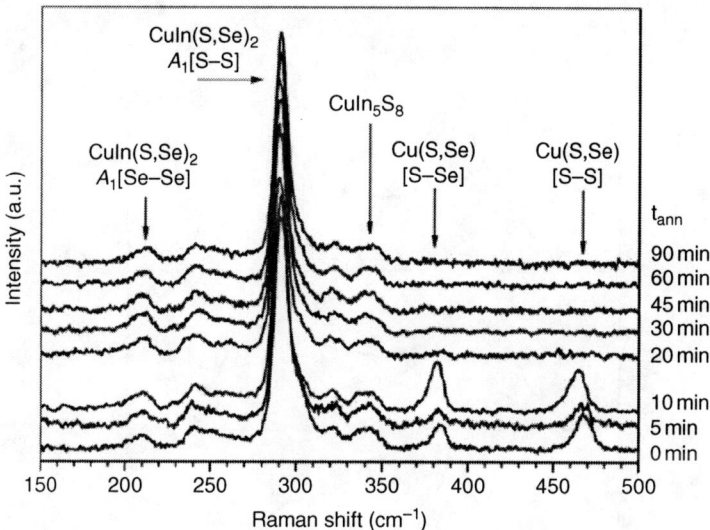

Figure 5.100 Raman spectra of CuIn(SSe)$_2$ thin films fabricated at 500 °C by sulfurization of electrodeposited CuInSe$_2$ precursor for different timings.

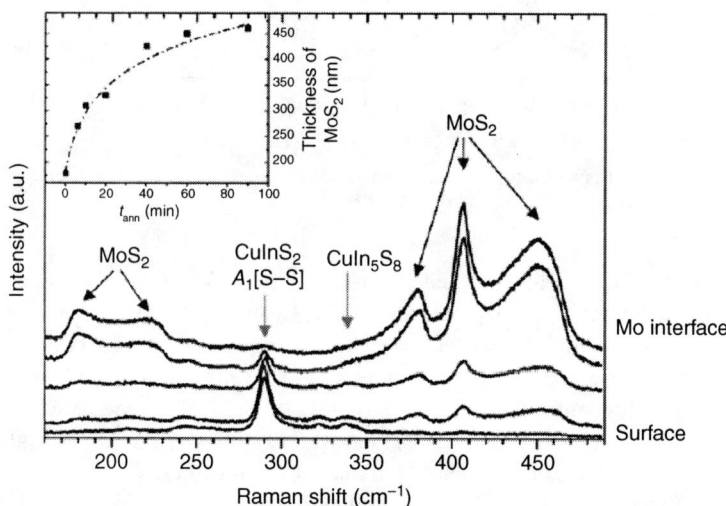

Figure 5.101 Raman spectra of cross-sectional CuIn(SSe)$_2$ thin film fabricated at 500 °C for 90 min.

One is from 0 to 20 min and another one from 20 to 90 min for these annealing conditions. The S–S peak shifts from 290.75 to 289.5 cm^{-1} with increasing annealing time. The cell shows efficiency of 5.5% for the optimum conditions [309]. The

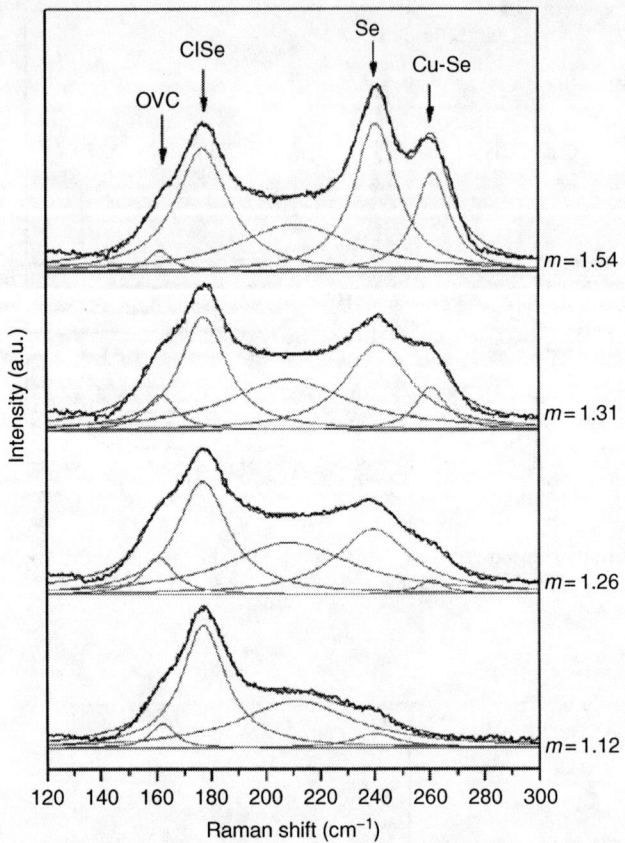

Figure 5.102 Raman spectra of $CuInSe_2$ thin films grown at different $m = 2Se/(Cu+3In)$ ratios.

modes at 160, 177, 240, and 260 cm^{-1} correspond to OVC, $CuInSe_2$, Se, and CuSe phases in 2 μm thick $CuInSe_2$ thin films with Cu/In = 1.2 grown onto Mo coated glass substrates by single step electrodeposition technique are observed, as shown in Figure 5.102. The Se and CuSe modes dominate the spectrum with increasing $m = 2Se/(Cu+3In)$ value from 1.12, 1.26, 1.31 to 1.51. A mode at 210 cm^{-1} is due to overlap of A_1 and four Raman modes of $B_1^2 + B_2^2 + 2E^{(3,4)}$. The glass/Mo/$CuIn(SeS)_2$/CdS/ZnO cell with NaCN treated CIS absorber shows $V_{oc} = 640$ mV and $J_{sc} = 19$ mA/cm^2 for $m = 1.5$. The cell performance degrades beyond $m = 1.5$. The cell exhibits efficiency of 6.5% for $m = 1.3$–1.4 and Cu/In = 1.1–1.2 [310].

5.3.6 Cu(InGa)(SeS)$_2$

The Raman spectra of $Cu(InGa)(SeS)_2$ thin films prepared by two-stage technique is analyzed; CuInGa alloy grown by DC sputtering onto Mo coated glass substrates, followed by selenization and sulferization under H$_2$Se and H$_2$S gases at above

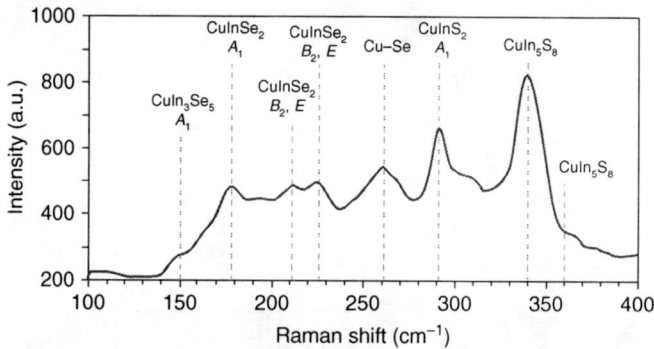

Figure 5.103 Raman spectrum of CIGSS thin films grown onto Mo covered glass substrates.

450 and 500 °C, respectively, show several modes, as shown in Figure 5.103. The mode at 176 cm^{-1} is due to A_1 chalcopyrite characteristic mode of CuInSe$_2$, as already mentioned. The modes at 212 and 226 cm^{-1} are due to B_2 and E by vibrations of all the atoms, respectively. A shoulder at 150 cm^{-1} for A_1 mode in the spectrum is due to CuIn$_3$Se$_5$ phase [40,260]. The modes at 291 and 261 cm^{-1} are also already assigned to A_1 chalcopyrite mode of CuInS$_2$ and Cu$_x$Se phase, respectively. The other two modes at 340 and 360 cm^{-1} may be due to CuIn$_5$S$_8$ phase. Based on the analysis, the phases such as CuInSe$_2$, CuInS$_2$, Cu$_x$Se and other CuIn$_3$Se$_5$ and CuIn$_5$Se$_8$ present may not be true because the resonance peaks from Ga–Se, In–Se, In–S, etc., may be overlapping each other in the Cu(InGa)(SSe)$_2$ system [311]. The Cu(GaIn)(SSe)$_2$ thin films with different S/Se composition ratios of 0, 0.06, 0.25, 0.9, and 1.2, which are originally obtained from line intensity ratios from XPS analysis, reveal that the intensities of A_1-178 cm^{-1} mode as well as other modes E-212 cm^{-1} and B_2-230 cm^{-1} due to CuInSe$_2$ gradually decrease with increasing sulfur to selenium ratio, as shown in Figure 5.104. The modes at 260 and 294 cm^{-1} are corresponded to overlap of E/B_2 modes and A_1 mode of CuInS$_2$ for higher sulfur composition, respectively. The two modes at ~185 and ~285 cm^{-1} are related to vibrations of neither CuInSe$_2$ nor CuInS$_2$. The A_1 mode is symmetric, nonpolar and its intensity is stronger than that of others [312]. The intensities of A_1 mode at 294 cm^{-1} and CA mode at 307 cm^{-1} for CuInS$_2$ increase and on the other hand the intensity of A_1 mode at 176 cm^{-1} for CuInSe$_2$ decreases with increasing ratio of S/(S+Se) from 0.14, 0.19, 0.28, and 1.0 in the CuInGa(SSe)$_2$ thin films, as shown in Figure 5.105. The contribution of 307 cm^{-1} mode is higher, if the samples are grown at lower temperature [58].

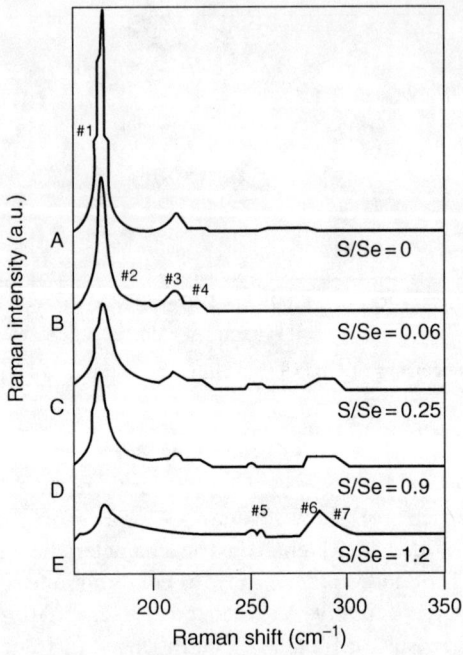

Figure 5.104 Raman spectra of Cu(GaIn)(SSe)$_2$ thin films with variation of S/Se (#1–178, #2–?, #3–212, #4–230, #5–250, #6–294 cm^{-1}, #7–A' CuInS$_2$).

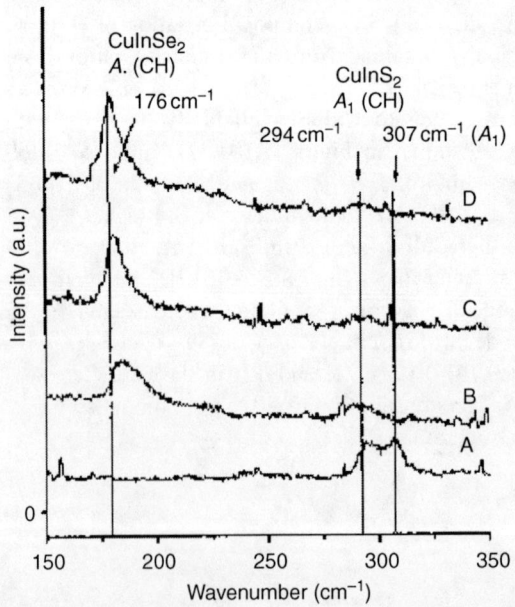

Figure 5.105 Raman spectra of CIGSS thin films (A) S/(S+Se) = 1, (B) 0.28, (C) 0.19, and (D) 0.14.

References

[1] J.E. Jaffe, A. Zunger, Phys. Rev. B 28 (1983) 5822.
[2] J.L. Shay, H.M. Kasper, Phys. Rev. Lett. 29 (1972) 1162.
[3] J.L. Shay, B. Tell, H.M. Kasper, L.M. Schiavone, Phys. Rev. B 5 (1972) 5003.
[4] J.L. Shay, B. Tell, Surf. Sci. 37 (1973) 748.
[5] J.L. Shay, J.H. Wernick, Ternary Chalcopyrite Semiconductors: Growth, Electronic Properties and Applications, Pergamon Press, New York, 1975.
[6] J.E. Rowe, J.L. Shay, Phys. Rev. B 13 (1971) 451.
[7] J.L. Shay, E. Buchler, J.H. Wernick, Phys. Rev. Lett. 24 (1970) 1301.
[7] B. Tell, J.L. Shay, H.M. Kasper, Phys. Rev. B 4 (1971) 2463.
[9] J.L. Shay, B. Tell, H.M. Kasper, L.M. Schiavone, Phys. Rev. B 7 (1973) 4485.
[10] P. Bendt, A. Zunger, Phys. Rev. B 26 (1982) 3114.
[11] J.R. Tuttle, D. Albin, R.J. Matson, R. Noufi, O.S. Heavens, Optical Properties of Thin Solid Films, Butterworth, London, (1955); S.G. Tomlin, Br. J. Appl. Phys. (J. Phys. D), 1 (1968) 1667.
[12] R. Swanepoel, J. Phys. E: Sci. Instrum. 16 (1983) 1214.
[13] A.E. Rakhshani, J. Appl. Phys. 81 (1997) 7988.
[14] H. Neumann, W. Horig, E. Reccius, H. Sobotta, B. Schumann, G. Kuhn, Thin Solid Films 61 (1979) 13.
[15] S.R. Kodigala, V.S. Raja, A.K. Bhatnagar, R.D. Tomlinson, R.D. Pilkington, A.E. Hill, et al., Semicond. Sci. Technol. 15 (2000) 676.
[16] J. Gonzalez-Hernandez, P.M. Gorley, P.P. Horley, O.M. Vartsabyuk, Y.V. Vorobiev, Thin Solid Films, 403–404 (2002) 471, Y.I. Uhanov, Optical Properties of Semiconductors, Nauka, Moscow, 1977.
[17] A.L. Fahrenbruch, R.H. Bube, Fundamentals of Solar Cells, Academic Press, New York, 1983.
[18] R. Caballero, C. Guillen, Solar Energy Mater. Solar Cells 86 (2005) 1.
[19] A. Knowles, H. Oumous, M. Carter, R. Hill, Semicond. Sci. Technol. 3 (1988) 1143.
[20] M. Gorska, R. Beaulieu, J.J. Loferski, B. Roessler, J. Beall, Solar Energy Mater. 2 (1980) 343.
[21] H. Sakata, H. Ogawa, Solar Energy Mater. Solar Cells 63 (2000) 259.
[22] F.O. Adurodija, M.J. Carter, R. Hill, Solar Energy Mater. Solar Cells 40 (1996) 359.
[23] R. Noufi, R. Axton, D. Cahen, S.K. Deb, 17th IEEE Photovoltaic Specialist Conference, (1984), p. 927.
[24] J.R. Tuttle, D. Albin, J. Goral, C. Kennedy, R. Noufi, Solar Cells 24 (1988) 67; G.F. Yuksel, B.M. Basol, H. Safak, H. Karabbiyik, Appl. Phys. A 73 (2001) 387.
[25] J.R. Tuttle, R. Noufi, R.G. Dhere, 19th IEEE Photovoltaic Specialist Conference, (1987), p. 1494.
[26] J. Bougnot, S. Duchemin, M. Savelli, Solar Cells 16 (1986) 221.
[27] B. Pamplin, R.S. Feigelson, Thin Solid Films 60 (1979) 141.
[28] O.P. Agnihotri, P. Rajaram, R. Thangaraj, A.K. Sharma, A. Raturi, Thin Solid Films 102 (1983) 291.
[29] N.M. Shah, J.R. Ray, K.J. Patel, V.A. Kheraj, M.S. Desai, C.J. Panchal, et al., Thin Solid Films 517 (2009) 3639.
[30] A.V. Mudryi, V.F. Gremenok, I.A. Victorov, V.B. Zalesski, F.V. Kurdesov, V.I. Kovalevski, et al., Thin Solid Films 431–432 (2003) 193.

[31] K. Yoshino, T. Shimizu, A. Fukuyama, K. Maeda, P.J. Fons, A. Yamada, et al., Solar Energy Mater. Solar Cells 50 (1998) 127.
[32] G.D. Mooney, A.M. Hermann, J.R. Tuttle, D.S. Albin, R. Noufi, Solar Cells 30 (1991) 69.
[33] J.R. Tuttle, D.S. Albin, R. Noufi, Solar Cells 27 (1989) 231.
[34] N. Kohara, T. Negami, M. Nishitani, T. Wada, Jpn. J. Appl. Phys. 34 (1995) L1141.
[35] N. Stratieva, E. Tzvetkova, M. Ganchev, K. Kochev, I. Tornov, Solar Energy Mater. Solar Cells 45 (1997) 87.
[36] R.P. Raffaelle, H. Forsell, T. Potdevin, R. Friedfeld, J.G. Mantovani, S.G. Bailey, et al., Solar Energy Mater. Solar Cells 57 (1999) 167.
[37] M. Pattabi, P.J. Sebastian, X. Mathew, R.N. Bhattacharya, Solar Energy Mater. Solar Cells 63 (2000) 315.
[38] C.J. Huang, T.H. Meen, M.Y. Lai, W.R. Chen, Solar Energy Mater. Solar Cells 82 (2004) 553.
[39] C. Guillen, J. Herrero, Solar Energy Mater. Solar Cells 43 (1996) 47.
[40] T. Terasako, Y. Uno, T. Kariya, S. Shirakata, Solar Energy Mater. Solar Cells 90 (2006) 262.
[41] S.H. Kwon, B.T. Ahn, S.K. Kim, K.H. Yoon, J. Song, Thin Solid Films 323 (1998) 265.
[42] I. Martil, J. Santamaria, G. Gonzalez, F. Sanchez, J. Appl. Phys. 68 (1990) 189.
[43] W. Arndt, H. Dittrich, H.W. Schock, Thin Solid Films 130 (1985) 209.
[44] R. Noufi, R. Powell, C. Herrington, T. Coutts, Solar Cells 17 (1986) 303.
[45] S.H. Kwon, D.Y. Lee, B.T. Ahn, J. Korean Phys. Soc. 39 (2001) 655.
[46] T. Tokado, T. Nakada, 3rd World Conference on Phtovoltaics Energy Conversion, (2003), 2PD366.
[47] G.H. Bauer, R. Bruggemann, S. Tardon, S. Vignoli, R. Kniese, Thin Solid Films 480–481 (2005) 410.
[48] W.S. Chen, J.M. Stewart, B.J. Stanbery, W.E. Devaney, R.A. Mickelsen, 19th IEEE Photovoltaic Specialist Conference, (1987) p. 1446.
[49] B. Dmmler, H. Dittrich, R. Menner, H.W. Schock, 19th IEEE Photovoltaic Specialist Conference, (1987), p. 1454.
[50] T.F. Ciszek, R. Bacewicz, J.R. Durrant, S.K. Deb, D. Dunlay, 19th IEEE Photovoltaic Specialist Conference, (1987), p. 1448.
[51] B. Dimmler, H. Dittrich, R. Menner, H.W. Schock, 19th IEEE Photovoltaic Specialist Conference, (1987), p. 1454.
[52] F.B. Dejene, V. Alberts, J. Phys. D: Appl. Phys. 38 (2005) 22.
[53] R. Chakrabarti, B. Maiti, S. Chaudhuri, A.K. Pal, Solar Energy Mater. Solar Cells 43 (1996) 237.
[54] R.N. Bhattacharya, W. Batchelor, J.E. Granata, F. Hasoon, H. Wiesner, K. Ramanathan, et al., Solar Energy Mater. Solar Cells 55 (1998) 83.
[55] T. Yamaguchi, Y. Yamamoto, T. Tanaka, Y. Demizu, A. Yoshida, Jpn. J. Appl. Phys. 35 (1996) L1618.
[56] E. Ahmed, A. Zegadi, A.E. Hill, R.D. Pilkington, R.D. Tomlinson, A.A. Dost, et al., Solar Energy Mater. Solar Cells 36 (1995) 227.
[57] K.T.R. Reddy, R.B.V. Chalapathy, Solar Energy Mater. Solar Cells 50 (1998) 19.
[58] E.P. Zaretskaya, V.F. Gremenok, V.B. Zalesski, K. Bente, S. Schorr, S. Zukotynski, Thin Solid Films 515 (2007) 5848.
[59] T. Tanaka, N. Tanahashi, T. Yamaguchi, A. Yoshida, Solar Energy Mater. Solar Cells 50 (1998) 13.

[60] H. Bouzouita, N. Bouguila, A. Dhouib, Renewable Energy (1998) 1.
[61] Y.B. He, T. Kramer, A. Polity, M. Hardt, B.K. Meyer, Thin Solid Films 431–432 (2003) 126.
[62] Y.B. He, W. Kriegseis, T. Kramer, A. Polity, M. Hardt, B. Szyszka, et al., J. Phys. Chem. Solids 64 (2003) 2075.
[63] S. Bini, K. Bindu, M. Lakshmi, C.S. Kartha, K.P. Vijajakumar, Y. Kashiwaba, et al., Renewable Energy 20 (2000) 405.
[64] M. Abaab, M. Kanzari, B. Rezig, M. Brunel, Solar Energy Mater. Solar Cells 59 (1999) 299.
[65] Y. Yamamoto, T. Yamaguchi, T. Tanaka, N. Tanahashi, A. Yoshida, Solar Energy Mater. Solar Cells 49 (1997) 399.
[66] G.-C. Park, H.D. Chung, C.-D. Kim, H.-R. Park, W.-J. Jeong, J.-U. Kim, et al., Solar Energy Mater. Solar Cells 49 (1997) 365.
[67] M.C. Zouaghi, T.B. Nasrallah, S. Marsillac, J.C. Bernede, S. Belgacem, Thin Solid Films 382 (2001) 39.
[68] M. Ortega-Lopez, A. Morales-Acevedo, Thin Solid Films 330 (1998) 96.
[69] M. Krunks, O. Bijakina, T. Varema, V. Mikli, E. Mellikov, Thin Solid Films 338 (1999) 125.
[70] S.R. Kodigala, V.S. Raja, Mater. Lett. 12 (1991) 67.
[71] T. Abe, S. Kohiki, K. Fukuzaki, M. Oku, T. Watanabe, Appl. Surf. Sci. 174 (2001) 40.
[72] C. Guillen, J. Herrero, M.T. Gutierrez, F. Briones, Thin Solid Films 480–481 (2005) 19.
[73] A.V. Mudryi, I.A. Victorov, V.F. Gremenok, A.I. Patuk, I.A. Shakin, M.V. Yakushev, Thin Solid Films 431–432 (2003) 197.
[74] K. Zeaiter, Y. Llinares, C. Llinares, Solar Energy Mater. Solar Cells 59 (1999) 299.
[75] H. Neff, P. Lange, M.L. Fearheiley, K.J. Backmann, Appl. Phys. Lett. 47 (1985) 1089.
[76] S.R. Kodigala, V.S. Raja, Thin Solid Films 208 (1992) 247.
[77] S.R. Kodigala, V.S. Raja, Thin Solid Films 207 (1992) L6.
[78] I.V. Bodnar, B.V. Korzun, A.I. Lukomskii, Physica Status Solidi (B) 105 (1981) K143.
[79] P.J. Dean, Phys. Rev. 157 (1967) 655.
[80] F. Williams, Physica Status Solidi 25 (1968) 493.
[81] J.I. Pankov, Optical Processes in Semiconductors, Prentice-Hall, Englewood Cliffs, NJ, 1971.
[82] S. Siebentritt, U. Rau, Wide-Gap Chalcopyrites, Springer 2006.
[83] E.M. Eagles, J. Phys. Chem. Solids 16 (1960) 76.
[84] J.H. Davies, The Physics of Low-Dimensional Semiconductors, An Introduction, Cambridge University Press 1998.
[85] P.J. Dean, J.D. Cuthbert, D.G. Thomas, R.T. Lynch, Phys. Rev. Lett. 18 (1967) 122.
[86] J.R. Sites, R.E. Hollingsworth, Solar Cells 21 (1987) 379.
[87] H. Komaki, K. Yoshino, S. Seto, M. Yoneta, Y. Akaki, T. Ikari, J. Crystal Growth 236 (2002) 253.
[88] S.B. Zhang, S.-H. Wei, A. Zunger, Phys. Rev. Lett. 78 (1997) 4059.
[89] S.M. Wasim, Solar Cells 16 (1986) 289.
[90] G. Dagan, F.-A. Elfotouh, D.J. Dunlay, R.J. Matson, D. Cahen, Chem. Mater. 2 (1990) 286.
[91] M.V. Yakushev, Y. Feofanov, R.W. Martin, R.D. Tomlinson, A.V. Mudryi, J. Phys. Chem. Solids 64 (2003) 2005.
[92] S. Chatraphorn, K. Yoodee, P. Songpongs, C. Chityuttakan, K. Sayavong, S. Wongmanerod, et al., Jpn. J. Appl. Phys. 37 (1998) L269.

[93] Y. Yang, Y. Chen, J. Phys. Chem. B 110 (2006) 17370.
[94] J.H. Schon, V. Alberts, E. Bucher, J. Appl. Phys. 81 (1997) 2799.
[95] M.V. Yakushev, A.V. Mudryi, Y. Feofanov, R.D. Tomlinson, Thin Solid Films 431–432 (2003) 190.
[96] S. Zott, K. Leo, M. Ruck, H.-W. Schock, J. Appl. Phys. 82 (1997) 356.
[97] A.V. Mudryi, I.V. Bodnar, V.F. Gremenok, I.A. Victorov, A.I. Patuk, I.A. Shakin, Solar Energy Mater. Solar Cells 53 (1998) 247.
[98] P. Lange, H. Neff, M. Fearheiley, K.J. Bachmann, Phys. Rev. B 31 (1985) 4074.
[99] P.W. Yu, J. Appl. Phys. 47 (1976) 677.
[100] G. Masse, E. Redjai, J. Appl. Phys. 56 (1984) 1154.
[101] J.H. Schon, E. Bucher, Solar Energy Mater. Solar Cells 57 (1999) 229.
[102] M.V. Yakushev, A.V. Mudryi, V.F. Gremenok, E.P. Zaretskaya, V.B. Zalesski, Y. Feofanov, *et al.*, Thin Solid Films 451–452 (2004) 133.
[103] K. Zeaiter, A. Yanuar, C. Llinares, Solar Energy Mater. Solar Cells 70 (2001) 213.
[104] P.W. Yu, Solid State Commun. 18 (1976) 395.
[105] V. Alberts, S. Zweigart, J.H. Schon, H.W. Schock, E. Bucher, Jpn. J. Appl. Phys. 36 (1997) 5033.
[106] P. Migliorato, J.L. Shay, H.M. Kasper, S. Wagner, J. Appl. Phys. 46 (1975) 1777.
[107] F.-A. Elfotouh, L.L. Kazmerski, D.J. Dunlavy, Solar Cells 14 (1985) 197.
[108] G. Masse, J. Appl. Phys. 68 (1990) 2206.
[109] H. Matsushita, T. Suzuki, S. Endo, T. Irie, Jpn. J. Appl. Phys. 34 (1995) 3774.
[110] J. Krustok, J. Madasson, K. Hjelt, J. Mater. Sci. Lett. 13 (1994) 1570.
[111] K. Kushiya, A. Shimizu, A. Yamada, M. Konagai, Jpn. J. Appl. Phys. 34 (1995) 54.
[112] M. Tanda, S. Manaka, J.R.E. Marin, K. Kushiya, H. Sano, A. Yamada, *et al.*, Jpn. J. Appl. Phys. 31 (1992) L753M. Tanda, S. Manaka, J.R.E. Marin, K. Kushiya, H. Sano, A. Yamada, M. Konagai, K. Takahashi, 22nd IEEE Photovoltaic Specialist Conference, (1991) p. 1169.
[113] T. Irie, S. Endo, S. Kimura, Jpn. J. Appl. Phys. 18 (1979) 1303, H. Neumann, E. Nowak, G. Kuhn, Cryst. Res. Technol. 16 (1981) 1369.
[114] F.A.- Elfotouh, D.J. Dunlavy, D. Cahen, R. Noufi, L.L. Kazmerski, K.J. Backmann, Prog. Crystal Growth Charact. 10 (1984) 365.
[115] H. Miyake, H. Ohtake, K. Sugiyama, J. Crystal Growth 156 (1995) 404.
[116] V. Alberts, R. Herberholz, T. Walter, H.W. Schock, J. Phys. D: Appl. Phys. 30 (1997) 2156.
[117] F.A. Elfotouh, H. Moutinho, A. Bakry, T.J. Coutts, L.L. Kazmerski, Solar Cells 30 (1991) 151.
[118] F.A.- Elfotouh, L.L. Kazmerski, H.R. Moutinho, J.M. Wissel, R.G. Dhere, A.J. Nelson, *et al.*, J. Vac. Sci. A 9 (1991) 554.
[119] F.A. Elfotouh, D.J. Dunlavy, T.J. Coutts, Solar Cells 27 (1989) 237.
[120] J.H. Schon, V. Alberts, E. Bucher, Thin Solid Films 301 (1997) 115.
[121] S.R. Kodigala, Y.K. Su, S.J. Chang, B. Kerr, H.P. Liu, I.G. Chen, Appl. Phys. Lett. 84 (2004) 3307.
[122] S.R. Kodigala, D. Huang, M.A. Reshchikov, F. Yun, H. Morkoc, J. Jasinski, *et al.*, J. Mater. Sci. Mater. Electron. 14 (2003) 233.
[123] S. Zott, K. Leo, M. Ruck, H.-W. Schock, Appl. Phys. Lett. 68 (1996) 1144.
[124] O. Ka, H. Alves, I. Dirnstorfer, T. Christmann, B.K. Meyer Thin, Solid Films 361–362 (2000) 263.
[125] B.H. Tseng, S.-B. Lin, K.-C. Hsieh, H.-L. Hwang, J. Crystal Growth 150 (1995) 1206.

[126] S. Niki, Y. Makita, A. Yamada, A. Obara, S. Misawa, O. Igarashi, et al., Jpn. J. Appl. Phys. 32 (Suppl. 32–33) (1993) 161.
[127] S. Niki, Y. Makita, A. Yamada, O. Hellman, P.J. Fons, A. Obara, et al., J. Crystal Growth 150 (1995) 1201.
[128] S. Niki, P.J. Fons, A. Yamada, T. Kurafuji, S. Chichibu, H. Nakanishi, et al., Appl. Phys. Lett. 69 (1996) 647.
[129] S. Niki, H. Shibata, P.J. Fons, A. Yamada, A. Obara, Y. Makita, et al., Appl. Phys. Lett. 67 (1995) 1289.
[130] S. Niki, I. Kim, P.J. Fons, H. Shibata, A. Yamada, H. Oyanagi, et al., Solar Energy Mater. Solar Cells 49 (1997) 319.
[131] S. Niki, P.J. Fons, A. Yamada, Y. Lacroix, H. Shibata, H. Oyanagi, et al., Appl. Phys. Lett. 74 (1999) 1630.
[132] S. Niki, Y. Makita, A. Yamada, A. Obara, O. Igarashi, S. Misawa, et al., Solar Energy Mater. Solar Cells 35 (1994) 141.
[133] S. Niki, Y. Makita, A. Yamada, A. Obara, S. Misawa, O. Igarashi, et al., Jpn. J. Appl. Phys. 33 (1994) L500.
[134] S. Chichibu, Appl. Phys. Lett. 70 (1997) 1840.
[135] S. Siebentritt, N. Rega, A. Zajogin, M.C. Lux-Steiner, Physica Status Solidi (C) 1 (2004) 2304.
[136] H. Matsushita, S.-I. Ai, A. Katsui, J. Crystal Growth 224 (2001) 95.
[137] N. Rega, S. Siebentritt, I. Beckers, J. Beckmann, J. Albert, M. Lux-Steiner, J. Crystal Growth 248 (2003) 169.
[138] E.P. Zaretskaya, V.F. Gremenyuk, V.B. Zalesskii, V.A. Ivanov, I.V. Viktorov, V. I. Kovalevskii, et al., Tech. Phys. 70 (2000) 141.
[139] N. Rega, S. Siebentritt, I. Beckers, J. Beckmann, J. Albert, M. Lux-Steiner, Thin Solid Films 431–432 (2003) 186.
[140] C. Rincon, J. Gonzalez, G.S. Perez, J. Appl. Phys. 54 (1983) 6634.
[141] R. Kimura, T. Nakada, P. Fons, A. Yamada, S. Niki, T. Matsuzawa, et al., Solar Energy Mater. Solar Cells 67 (2001) 289.
[142] R. Jayakrishnan, K.G. Deepa, C.S. Kartha, K.P. Vijajakumar, J. Appl. Phys. 100 (2006) 046104.
[143] A. Zouauui, M. Lachab, M.L. Hidalgo, A. Chaffa, C. Llinares, N. Kesri, Thin Solid Films 339 (1999) 10.
[144] T. Uahiki, A. Ueno, T. Yano, H. Sano, H. Usui, K. Sato, Jpn. J. Appl. Phys. 32 (Suppl. 32–33) (1993) 103.
[145] S. Shirakata, S. Yudate, T. Terasako, S. Isomura, Jpn. J. Appl. Phys. 37 (1998) L1033.
[146] S. Isomura, S. Shirakata, T. Abe, Solar Energy Mater. 22 (1991) 223.
[147] K. Kushiya, A. Yamada, H. Hakuma, H. Sano, M. Konagai, Jpn. J. Appl. Phys. 32 (Suppl. 32–33) (1993) 54.
[148] M. Tanda, S. Manaka, A. Yamada, M. Konagai, K. Takahashi, Jpn. J. Appl. Phys. 32 (1993) 1913.
[149] K. Urabe, T. Hama, M. Roy, H. Sato, H. Fujisawa, M. Ohsawa, et al., 22nd IEEE Photovoltaic Specialist Conference, (1991), p. 1082.
[150] R.A. Mickelsen, W.S. Chen, Y.R. Hsiao, V.E. Lowe, IEEE Trans. Electron. Devices ED-31 (1984) 542.
[151] M. Nishitani, M. Ikeda, T. Negami, S. Kohoki, N. Kohara, M. Terauchi, et al., Solar Energy Mater. Solar Cells 35 (1994) 203.
[152] K. Kushiya, H. Hakuma, H. Sano, A. Yamada, M. Konagai, Solar Energy Mater. Solar Cells 35 (1994) 223.

[153] O.C. Cantser, L.L. Kulyuk, T.D. Shemyakova, A.V. Siminel, V.E. Tezlevan, Jpn. J. Phys 32((Suppl. 32–3) (1993) 630.
[154] N. Esser, W. Richte, S. Siebentrilt, M.Ch. Lux-Steiner, J. Appl. Phys. 94 (2003) 4341.
[155] A.V. Matveev, V.E. Grachev, V.V. Sobolev, V.E. Tazlavan, Physica Status Solidi (B) (1996) K7.
[156] K. Yoshino, M. Sugiyama, D. Maruoka, S.F. Chichibu, H. Komaki, K. Umeda, et al., Physica B 302–303 (2001) 357.
[157] N.V. Joshi, A.J. Mejias, R.W. Echeverria, Solid State Commun. 55 (1985) 933.
[158] T. Kampschulte, A. Bauknecht, U. Blieske, M. Saad, S. Chichibu, M.Ch. Lux-Steiner, 26th IEEE PVSC, Anaheim, (1997) p. 391.
[159] A. Meeder, D.F. Marron, V. Tezlevan, E. Arushanov, A. Rumberg, T.-S. Niedrig, et al., Thin Solid Films 431–432 (2003) 214.
[160] A. Bauknecht, S. Siebentrit, J. Albert, M.Ch. Lux-Steiner, J. Appl. Phys. 89 (2001) 4391.
[161] A. Yamada, Y. Makita, S. Niki, A. Obara, P. Fons, H. Shibata, Microelectron. J. 27 (1996) 53.
[162] J.H. Schon, O. Schenker, H. Riazi-Nejad, K. Friemelt, Ch. Kloc, E. Bucher, Physica Status Solidi A 161 (1997) 301.
[163] M. Klenk, O. Schenker, E. Bucher, Thin Solid Films 361–362 (2000) 229.
[164] K. Yoshino, N. Mitani, T. Ikari, P.J. Fons, S. Niki, A. Yamada, Solar Energy Mater. Solar Cells 67 (2001) 173.
[165] A. Meeder, D.F. Marron, V. Chu, J.P. Conde, A.-J. Waldau, A. Rumberg, et al., Thin Solid Films 403–404 (2002) 495.
[166] M. Rusu, P. Gashin, A. Simashkevich, Solar Energy Mater. Solar Cells 70 (2001) 175.
[167] D. Fischer, N. Meyer, M. Kuczmik, M. Beck, A.-J. Waldau, M.Ch.-L. Steiner, Solar Energy Mater. Solar Cells 67 (2001) 105.
[168] I.V. Bodnar, A.G. Karoza, G.F. Smirnova, Physica Status Solidi (B) 84 (1977) K65.
[169] G. Orsal, F. Mailly, N. Romain, M.C. Artaud, S. Rushworth, S. Duchemin, Thin Solid Films 361–362 (2000) 135.
[170] U. Fiedeler, J. Albert, S. Siebentritt, M.Ch. Lux-Steiner, 17th European Photovoltaic Solar Energy Conference 22–26 October, (2001), pp. 1143–1146.
[171] M. Susaki, T. Miyauchi, H. Horinaka, N. Yamamoto, Jpn. J. Appl. Phys. 17 (1978) 1555.
[172] S. Shirakata, K. Morita, S. Isomura, Jpn. J. Appl. Phys. 33 (1994) L739.
[173] M.P. Vecchi, J. Ramos, W. Giriat, Solid-State Electron. 21 (1978) 1609.
[174] J. Stankiewicz, W. Giriat, J. Ramos, M.P. Vecchi, Solar Energy Mater. 1 (1979) 369.
[175] G. Masse, N. Lahlou, N. Yamamoto, J. Appl. Phys. 51 (1980) 4981.
[176] C. Paorici, N. Romeo, G. Sberveglieri, L. Tarricone, J. Lumin. 15 (1977) 101.
[177] A. Bauknecht, S. Siebentritt, J. Albert, M.Ch. Lux-Steiner, J. Appl. Phys. 89 (2001) 4391.
[178] S. Shirakata, K. Tamura, S. Isomura, Jpn. J. Appl. Phys. 35 (1996) L531.
[179] A. Yamada, P. Fons, S. Niki, H. Shibata, A. Obara, Y. Makita, et al., J. Appl. Phys. 81 (1997) 2794.
[180] G.A. Medvedkin, T. Nishi, Y. Katsumata, H. Miyake, K. Sato, Solar Energy Mater. Solar Cells 75 (2003) 135.
[181] T. Nishi, Y. Katsumata, K. Sato, H. Miyake, Solar Energy Mater. Solar Cells 67 (2007) 273.
[182] J.H. Schon, E. Arushanov, L.L. Kulyuk, A. Micu, D. Shaban, V. Tezlevan, et al., J. Appl. Phys. 84 (1998) 1274.

[183] J.H. Schon, H. Riazi-Nejad, Ch. Kloc, F.P. Baumgartner, E. Bucher, J. Lumin. 72–74 (1997) 118.
[184] J.H. Schon, J. Phys. D: Appl. Phys. 33 (2000) 286.
[185] J.H. Schon, F.P. Baumgartner, E. Arushanov, H.-R. Nejad, Ch. Kloc, E. Bucher, J. Appl. Phys. 79 (1996) 6961.
[186] K. Yoshino, H. Yokoyama, K. Maeda, T. Ikari, J. Crystal Growth 211 (2000) 476.
[187] H. Miyake, T. Haginoya, K. Sugiyama, Solar Energy Mater. Solar Cells 50 (1998) 51.
[188] G. Masse, K. Djessas, F. Guastavina, J. Phys. Chem. Solids 52 (1991) 999.
[189] R. Bacewicz, A. Dzierzega, R. Trykozko, Jpn. J. Appl. Phys. 32 (Suppl. 32–3) (1993) 194.
[190] F.-A. Elfotouh, D.J. Dunlavy, L.L. Kazmerski, D. Albin, K.J. Bachman, R. Menner, J. Vac. Sci. Tech. 6 (1988) 1515.
[191] K. Kushiya, Y. Ohtake, A. Yamada, M. Konagai, Jpn. J. Appl. Phys. 33 (1994) 6599.
[192] Mt. Wagner, I. Dirnstorfer, D.M. Hofmann, M.D. Lampert, F. Karg, B.K. Meyer, Physics Status Solidi (A) 167 (1998) 131.
[193] I. Dirnstorfer, D.M. Hofmann, D. Meister, B.K. Meyer, W. Riedl, F. Karg, J. Appl. Phys. 85 (1999) 1423.
[194] I. Dirnstorfer, W. Burhardt, W. Kriegseis, I. Osterreicher, H. Alves, D.M. Hofmann, et al., Thin Solid Films 361–362 (2000) 400.
[195] M. Sugiyama, A. Kinoshita, M. Fukaya, H. Nakanishi, S.F. Chichibu, Thin Solid Films 515 (2007) 5867.
[196] K. Ramanathan, R.N. Bhattacharya, J. Grantala, J. Webb, D. Niles, M.A. Contreras, et al., 26th IEEE PVSC, Anaheim, (1997), p. 319.
[197] N. Rega, S. Siebentritt, J. Albert, S. Nishiwaki, A. Zajogin, M.Ch. Lux-Steiner, et al., Thin Solid Films 480–481 (2005) 286.
[198] S. Shirakata, K. Ohkubo, Y. Ishii, T. Nakada, Solar Energy Mater. Solar Cells 93 (2009) 988.
[199] A. Kinoshita, M. Fukaya, H. Nakanishi, M. Sugiyama, S.F. Chichibu, Physics Status Solidi (C) 3 (2006) 2539.
[200] F.B. Dejene, Solar Energy Mater. Solar Cells 93 (2009) 577.
[201] S. Ishizuka, K. Sakurai, A. Yamada, H. Shibata, K. Matsubara, M. Yonemura, et al., 20th European Photovoltaic Solar Energy Conference, (2005), p. 1740.
[202] G.A. Medvedkin, E.I. Terukov, Y. Hasegawa, K. Hirose, K. Sato, Solar Energy Mater. Solar Cells 75 (2003) 127.
[203] J.J.M. Binsma, L.J. Giling, J. Bloem, J. Lumin. 27 (1982) 55.
[204] M.P. Vecchi, J. Ramos, J. Appl. Phys. 52 (1981) 2958.
[205] S. Shirakata, II. Miyake, J. Phys. Chem. Solids 64 (2003) 2021.
[206] K. Wakita, H. Hirooka, S. Yasuda, F. Fujita, N. Yamamoto, J. Appl. Phys. 83 (1998) 443.
[207] K. Wakita, G. Hu, N. Nakayama, D. Shoji, Jpn. J. Appl. Phys. 41 (2002) 3356.
[208] H.Y. Ueng, H.L. Hwang, J. Phys. Chem. Solids 50 (1989) 1297.
[209] N. Yamamoto, J. Ogihara, H. Horinaka, Jpn. J. Appl. Phys. 29 (1990) 650.
[210] H.Y. Ueng, H.L. Hwang, J. Appl. Phys. 62 (1987) 434.
[211] J.J.M. Binsma, L.J. Giling, J. Bloem, J. Lumin. 27 (1982) 35.
[212] N. Lahlou, G. Masse, J. Appl. Phys. 52 (1982) 978.
[213] M. Nanu, J. Schoonman, A. Goossens, Thin Solid Films 451–452 (2004) 193.
[214] A. Amara, W. Rezaiki, A. Ferdi, A. Hendaoui, A. Drici, M. Guerioune, et al., Solar Energy Mater. Solar Cells 91 (2007) 1916.
[215] H.J. Lewerenz, N. Dietz, J. Appl. Phys. 73 (1993) 4975.

[216] K. Wakita, M. Matsuo, G. Hu, M. Iwai, N. Yamamoto, Thin Solid Films 431–432 (2003) 184.
[217] G. Masse, N. Lahlou, C. Butti, J. Phys. Chem. Solids 42 (1981) 449.
[218] K. Topper, J. Krauser, J. Bruns, R. Scheer, A. Weidinger, D. Brauning, Solar Energy Mater. Solar Cells 49 (1997) 383.
[219] K. Topper, J. Bruns, R. Scheer, M. Weber, A. Weidinger, D. Braunig, Appl. Phys. Lett. 71 (1997) 482.
[220] P. Guha, D. Das, A.B. Maity, D. Ganguli, S. Chaudhuri, Solar Energy Mater. Solar Cells 80 (2003) 115.
[221] J. Eberhardt, K. Schulz, H. Metzner, J. Cieslak, Th. Hahn, U. Reislohner, et al., Thin Solid Films 515 (2007) 6147.
[222] J. Eberhardt, J. Cieslak, H. Metzner, Th. Hahn, R. Goldhahn, F. Hudert, et al., Thin Solid Films 517 (2009) 2248.
[223] J. Eberhardt, H. Metzner, R. Goldhahn, F. Hudert, U. Reislohner, C. Hulsen, et al., Thin Solid Films 480–481 (2005) 415.
[224] Y.B. He, A. Polity, H.R. Alves, I. Osterreicher, W. Kriegseis, D. Pfisterer, et al., Thin Solid Films 403–404 (2002) 62.
[225] M. Gossla, H. Metzner, H.-E. Mahnke, Thin Solid Films 387 (2001) 77.
[226] T.M. Hsu, J.S. Lee, H.L. Hwang, J. Appl. Phys. 68 (1990) 283.
[227] R. Garuthara, R. Wijesundara, W. Siripala, Solar Energy Mater. Solar Cells 79 (2003) 331.
[228] S. Bandyopadhyaya, S. Chaudari, A.K. Pal, Solar Energy Mater. Solar Cells 60 (2000) 323.
[229] K. Siemer, J. Klaer, I. Luck, J. Bruns, R. Klenk, D. Braunig, Solar Energy Mater. Solar Cells 67 (2001) 159.
[230] Y.L. Wu, H.Y. Lin, C.Y. Sun, M.H. Yang, H.L. Hwang, Thin Solid Films 168 (1989) 113.
[231] B. Eisener, D. Wolf, G. Muller, Thin Solid Films 361–362 (2000) 126.
[232] K. Zeaiter, Y. Llinares, C. Llinares, Solar Energy Mater. Solar Cells 61 (2000) 313.
[233] S. Shirakata, A. Ogawa, S. Isomura, T. Kariya, Jpn. J. Appl. Phys. 32 (Suppl. 32–33) (1993) 94.
[234] S. Shirakata, T. Nakada, Mater. Res. Soc. Symp. Proc. 1012 (2007) Y04–Y06.
[235] B. Ohnesorge, R. Weigand, G. Bacher, A. Forchel, W. Riedl, F.H. Karg, Appl. Phys. Lett. 73 (1998) 1224.
[236] K. Puech, S. Zott, K. Leo, M. Ruckh, H.-W. Schock, Appl. Phys. Lett. 69 (1996) 3375.
[237] K. Wakita, K. Nishi, Y. Ohta, N. Nakayama, Appl. Phys. Lett. 80 (2002) 3316.
[238] N. Esser, J. Geurts, Raman spectroscopy, in: G. Baur, W. Richter (Eds.), Optical Characterization of Epitaxial Semiconductor Layers, Springer, 1996 Chapter 4.
[239] T. Riedel, Technical University, Berlin, 1992, Ph.D. Thesis.
[240] J.P. van der Ziel, A.E. Meixner, H.M. Kasper, J.A. Ditzenberger, Phys. Rev. B 9 (1974) 4286.
[241] J.N. Gan, J. Tauc, V.G. Lambrecht Jr., M. Robbins, Phys. Rev. B 13 (1976) 3610.
[242] E. Rudigier, Ph.D. Thesis, Faculty of Physics, der Philips Universitat Marberg (2004).
[243] C. Rincon, F.J. Ramirez, J. Appl. Phys. 72 (1992) 4321.
[244] B.J. Stanbery, S. Kincal, S. Kim, T.J. Anderson, O.D. Crisalle, S.P. Ahrenkiel, et al., 28th IEEE Photovoltaic Specialists Conference, Anchorage, AK, (2000), p. 440.
[245] S. Yamanka, M. Konagai, K. Takahashi, Jpn. J. Appl. Phys. 28 (1989) L1337.
[246] I.V. Bodnar, L.V. Golubev, V.G. Plotnichenko, E.A. Smolyaninov, Physica Status Solidi B 105 (1981) K111.

[247] H. Neumann, Helv. Phys. Acta. 58 (1985) 337.
[248] V. Kumar, D. Chandra, Physica Status Solidi B 212 (1999) 37.
[249] J. Gonzalez, M. Quintero, C. Rincon, Phys. Rev. B 45 (1992) 7022.
[250] H. Neumann, R.D. Tomlinson, W. Kissinger, N. Avgerinos, Physica Status Solidi B 118 (1983) K51.
[251] A.M. Andriesh, N.N. Syrbu, M.S. Iovu, V.E. Tazlavan, Physica Status Solidi B 187 (1995) 83.
[252] W.H. Koschel, M. Bettini, Physica Status Solidi B 72 (1975) 729.
[253] K. Kondo, H. Sano, K. Sato, Thin Solid Films 326 (1998) 83.
[254] A.N. Tiwari, M. Krejci, F.-J. Huang, H. Zogg, Thin Solid Films 361–362 (2000) 41.
[255] J.H. Park, I.S. Yang, H.Y. Cho, Appl. Phys. A 58 (1994) 125.
[256] S. Shirakata, H. Kubo, C. Hamaguchi, S. Isomura, Jap. J. Appl. Phys. 36 (1997) L1394.
[257] S. Yamanaka, M. Tanda, N. Nakada, A. Yamada, M. Konagai, K. Takahashi, Jpn. J. Appl. Phys. 30 (1991) 442.
[258] A.F. Cunha, M.M.P. Azevedo, R.J.O. Ferrao, A.A.C.S. Lourenco, C. Boemare, Mater. Res. Soc. Symp. Proc. 668 (2001) H8.18.1.
[259] J. Alvarez-Garcia, X. Fontane, V. Izquierdo-Roca, A. Perez-Rodriguez, J.R. Morante, E. Saucedo, et al., Mater. Res. Soc. Proc. 1165 (2009) M02–M03.
[260] S. Nomura, S. Ouchi, S. Endo, Jpn. J. Appl. Phys. 36 (1997) L1075.
[261] J.H. Ely, T.R. Ohno, T.E. Furtak, A.J. Nelson, Thin Solid Films 371 (2000) 36.
[262] A.N. Tiwari, S. Blunier, M. Filzmoser, H. Zogg, D. Schmid, H.W. Schock, Appl. Phys. Lett. 65 (1994) 3347.
[263] G. Morell, R.S. Katiyar, S.Z. Weisz, T. Walter, H.W. Schock, I. Balberg, Appl. Phys. Lett. 69 (1997) 987.
[264] D.L. Schulz, C.J. Curtis, A. Cram, J.L. Alleman, A. Mason, R.J. Matson, et al., 26th IEEE PVSC, Anaheim, (1997) p. 483.
[265] S.D. Kim, H.J. Kim, K.H. Yoon, J. Song, Solar Energy Mater. Solar Cells 62 (2002) 357.
[266] E.P. Zaretskaya, V.F. Gremenok, V.B. Zalesski, O.V. Ermakov, W. Schmitz, K. Bente, et al., 19th European Photovoltaic Solar Energy Conference 7–11 June, (2004), pp. 1776–1779.
[267] K. Otte, G. Lippold, D. Hirsch, R.K. Gebhardt, T. Chasse, Appl. Surf. Sci. 179 (2001) 203.
[268] R. Takei, H. Tanino, S. Chichibu, H. Nakanishi, J. Appl. Phys. 79 (1996) 2793.
[269] I.V. Bodnar, L.V. Golubev, V.G. Plotnichenko, E.A. Smolyaninova, Physica Status Solidi B 105 (1980) K111.
[270] K. Wakita, T. Miyazaki, Y. Kikuno, S. Takata, N. Yamamoto, Jpn. J. Appl. Phys. 38 (1999) 664.
[271] C. Xue, D. Papadimitriou, N. Esser, J. Phys. D: Appl. Phys. 37 (2004) 2267.
[272] F.J. Ramirez, C. Rincon, Solid State Commun. 84 (1992) 551.
[273] A. Yamada, H. Miyazaki, T. Miyake, Y. Chiba, M. Konagai, 4th World Conference on Photovoltaic Solar Energy Conversion (IEEE), (2006) p. 343.
[274] C. Xue, D. Papadimitriou, Y.S. Raptis, W. Richter, N. Esser, S. Siebentritt, et al., J. Appl. Phys. 96 (2004) 1963.
[275] D. Papadimitriou, N. Esser, C. Xue, Physica Status Solidi B 242 (2005) 2633.
[276] W. Witte, R. Kniese, M. Powalla, Mater. Res. Soc. Symp. Proc. 1165 (2009) M05–M20.
[277] H. Tanino, H. Deai, H. Nakanishi, Jpn. J. Appl. Phys. 32 (Suppl. 32–33) (1993) 436.

[278] W. Witt, R. Kniese, A. Eicke, M. Powalla, 4th World Conference on Photovoltaic Solar Energy Conversion (IEEE). (2006), p. 553.
[279] S. Shirakata, Y. Kannaka, H. Hasegawa, T. Kariya, S. Isomura, Jpn. J. Appl. Phys. 38 (1999) 4997.
[280] H. Matsushita, S. Endo, T. Irie, Jpn. J. Appl. Phys. 31 (1992) 18.
[281] S. Roy, P. Guha, S.N. Kundu, H. Hanzawa, S. Chaudhuri, A.K. Pal, Mater. Chem. Phys. 73 (2002) 24.
[282] H. Miyazaki, R. Mikami, A. Yamada, M. Konagai, J. Phys. Chem. Solids 64 (2003) 2055.
[283] E. Ahmed, R.D. Tomlinson, R.D. Pilkington, A.E. Hill, W. Ahmed, N. Ali, et al., Thin Solid Films 335 (1998) 54.
[284] C.J. Hibberd, M. Ganchev, M. Kaelin, K. Emits, A.N. Tiwari, 33rd IEEE Photovoltaic Specialist Conference, (2005), 368_08050850533.
[285] A. Yamada, H. Miyazaki, R. Mikami, M. Konagai, 3rd World Conference on Photovoltaic Energy Conversion, (2003), S40B123-2859.
[286] W. Witte, R. Kniese, M. Powalla, Thin Solid Films 517 (2008) 867.
[287] T. Unold, J. Hinze, K. Ellmer, 19th European Photovoltaic Solar Energy Conference 7-11 June, (2004), pp 1917–1920.
[288] J.A. Garcia, Characterization of $CuInS_2$ films for solar cell applications by Raman spectroscopy, (2002), University of Baracelona, Ph.D. Thesis.
[289] J. Alvarez-Garcia, B. Barcones, A. Perez-Rodriguez, A. Romano-Rodriguez, J.R. Morante, A. Janotti, et al., Phys. Rev. B 71 (2005) 54303.
[290] K. Kondo, S. Nakamura, K. Sato, Jpn. J. Appl. Phys. 37 (1998) 5728.
[291] T. Terasako, Y. Uno, S. Inoue, T. Kariya, S. Shirakata, Physica Status Solidi C 3 (2006) 2588.
[292] J.-A. Garcia, A.-P. Rodgiguez, B. Barcones, A.-R. Rodriguez, A. Janotti, S.-H. Wei, et al., Appl. Phys. Lett. 80 (2002) 562.
[293] J.-A. Garcia, A.-P. Rodriguez, A.-R. Rodriguez, T. Jawhari, J.R. Morante, R. Scheer, et al., Thin Solid Films 387 (2001) 216.
[294] C. Camus, N.A. Allsop, T. Kohler, M. Kruger, S.E. Gledhill, J. Klaer, et al., 33rd IEEE Photovoltaic Specialist Conference, (2005) 339-08050850525.
[295] R. Cayzac, F. Boulch, M. Bendahan, P. Lauque, P. Knauth, Mater. Sci. Eng. B 157 (2009) 66.
[296] I. Oja, M. Nanu, A. Katerski, M. Krunks, A. Mere, J. Raudoja, et al., Thin Solid Films 480–481 (2005) 82.
[297] T. Unold, T. Enzenhofer, K. Ellmer, Mater. Res. Soc. Symp. Proc. 865 (2005), F16.5.1.
[298] J.-A. Garcia, J.-M. Ruzafa, A.-P. Rodriguez, A.-R. Rodriguez, J.R. Morante, R. Scheer, Thin Solid Films 361–362 (2000) 208.
[299] L.-C. Barrio, A.-P. Rodriguez, J.-A. Garcia, A.-R. Rodriguez, B. Barcones, J.R. Morante, et al., Vacuum 63 (2001) 315.
[300] E. Rudigier, Ch. Pietzker, M. Wimbor, I. Luck, J. Klaer, R. Scheer, et al., Thin Solid Films 431–432 (2003) 110.
[301] S.D. Sartale, A. Ennaoui, M.Ch. Lux-Steiner, 19th European Photovoltaic Solar Energy Conference 7–11 June 2004, (2004) pp. 1988–1991.
[302] S. Nakamura, Physica Status Solidi C 8 (2006) 2564.
[303] T. Enzenhofer, T. Unold, R. Scheer, H.-W. Schock, 20th European Photovoltaic Solar Energy Conference, Barcelona, (2005) p. 1751.
[304] E. Rudigier, T. Enzenhofer, R. Scheer, Thin Solid Films 480–481 (2005) 327.
[305] E. Rudigier, I. Luck, R. Scheer, Appl. Phys. Lett. 82 (2003) 43700.

[306] E. Rudigier, J. Alvarez-Garcia, I. Luck, J. Klaer, R. Scheer, J. Phys. Chem. Solids 64 (2003) 1977.
[307] S. Shirakata, T. Terasako, T. Kariya, J. Phys. Chem. Solids 66 (2005) 1970.
[308] V.-I. Roca, J.-A. Garcia, A.-P. Rodriguez, L.-C. Barrio, A.-R. Rodriguez, J.R. Morante, *et al.*, IEEE, Electron Devices, Spanish Conference, (2007) p. 146.
[309] V. Izquierdo-Roca, X. Fontane, L. Calvo-Barrio, A. Perez-Rodriguez, J.R. Morante, J. Alvarez-Garcia, *et al.*, Thin Solid Films 517 (2009) 2264.
[310] V. Izquierdo-Roca, X. Fontane, J. Alvarez-Garcia, L. Calvo-Barrio, A. Perez-Rodriguez, J.R. Morante, *et al.*, Thin Solid Films 517 (2009) 2163.
[311] T. Delasol, A.P. Samantilleke, N.B. Chaure, P.H. Gardiner, M. Simmonds, I. M. Dharmadasa Solar, Energy Mater. Solar Cells 82 (2004) 587.
[312] E. Rudigier, J. Palm, R. Scheer, E. Rudigier, J. Palm, R. Scheer, 19th European Photovoltaic Solar Energy Conference 7–11 June, (2004), pp. 1913–1916.

6 Electrical Properties of I–III–VI$_2$ Compounds

The electrical conductivities of CuInSe$_2$, CuGaSe$_2$, CuInS$_2$, and their solid solutions such as Cu(In$_{1-x}$Ga$_x$)Se$_2$ and CuIn(Se$_{1-x}$S$_x$)$_2$ play an immense role in the thin film or single crystal solar cells to control efficiencies of the cells. The electrical transport properties of the compounds at RT and low temperature with effect of intrinsic doping are systematically discussed. The conductivity of the sample can easily be varied at large extent by varying its intrinsic composition ratios without making extrinsic doping. The n- and p-type conductivities can be made in the compounds. However, it is little difficult to make n-type CuGaSe$_2$ rather than p-type because the creation of Ga$_{Cu}$ sites in the CGS is not easy task like In$_{Cu}$ in the CuInSe$_2$, which are responsible for n-type conductivity.

6.1 Conductivity of CuInSe$_2$

The CuInSe$_2$ (CIS) is one of the potential candidates as absorber for thin film solar cells owing to its n-, intrinsic, and p-type conductivities and profound chemical and radiation stability. The composition of CuInSe$_2$ plays a great role in the electrical properties of the sample. The molecularity deviation $\Delta m = (Cu/In) - 1$ and nonstoichiometry $\Delta S = [2Se/(Cu+3In)] - 1$ decide type of conductivity. If $\Delta m > 0$ (Cu-rich) and $\Delta S > 0$ (Se-rich), the samples show p-type conductivity, otherwise the samples exhibit n-type conductivity for $\Delta S < 0$ (Se-poor) and Δm either negative (In-rich) or positive. In some cases both n- and p-type conductivities are observed for $\Delta m > 0$ and $\Delta S < 0$ conditions. The above results are indeed confirmed by studying more than hundreds of CuInSe$_2$ single crystals. Figure 6.1 shows that how n- and p-type conductivities depend on the compositions of near stoichiometry CuInSe$_2$ single crystals [1]. More than or 25 atomic percent of Cu in the CuInSe$_2$ leads to strongly p-type while less than that is intrinsic or n-type conductivity. The stoichiometry deviation ΔS is positive; the films are likely p-type indicating participation of metal vacancies and excess Se. The intrinsic or n-type conductivity occurs, if ΔS is negative [2]. The CuInSe$_2$ or CuInS$_2$ samples with excess selenium or sulfur are obviously p-type, whereas n-type samples contain selenium or sulfur deficient and excess In. The CuInSe$_2$, CuGaSe$_2$ and CuInS$_2$ single crystals annealed under minimum selenium or sulfur at 600–800 °C for 24 h show n-type, whereas samples show p-type conductivity for maximum selenium or sulfur. The electrical properties of CuInSe$_2$,

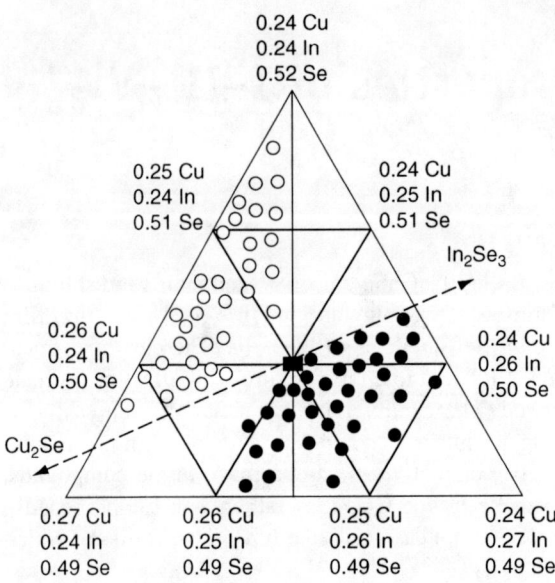

Figure 6.1 Distinguishion of n- (●) and p-type (○) conductivity based on composition of near stoichiometry (■) CuInSe$_2$ single crystals.

Table 6.1 Electrical properties of CuInS$_2$, CuInSe$_2$, and CuGaSe$_2$ single crystals

S.No	Sample	Type	ρ (Ω-cm)	n_e or p (cm^{-3})	μ(cm^2/Vs)	Annealed under	Ref.
1	CuInS$_2$	p	5	1×10^{17}	15	Max. S	[3]
2		n	1	3×10^{16}	200	Min. S	
3	CuInSe$_2$	p	0.5	1×10^{18}	10	Max. Se	
4		n	0.05	4×10^{17}	320	Min. Se	
5	CuGaSe$_2$	p	0.05	5×10^{18}	20	–	

CuGaSe$_2$ and CuInS$_2$ single crystals annealed under the vicinity of maximum or minimum S/Se are given in Table 6.1 [3]. The CuInSe$_2$ crystals prepared by three zone temperature horizontal Bridgman method under Se pressure below 10 torr show n-type and CuIn as a secondary phase, whereas crystals grown under higher Se pressure show p-type single crystal. The Hall mobility of p-CuInSe$_2$ single crystals decreases from ~40 to 10 cm^2/Vs and carrier concentration increases from ~4×10^{15} to 8×10^{17} cm^{-3} with increasing Se pressure from 8 to 200 torr [4].

The surface smoothness of the CIS turns into roughness when moving from n- to p-type CIS. The In-rich n-type CuInSe$_2$ thin films are silver/grey, whereas p-type Cu-rich sample is bluish black in color. The carrier concentration and mobility of CIS sample vary from 10^{13} to 10^{20} cm^{-3} and from low to 150 cm^2/Vs, respectively [5]. Three typical CIS thin films with composition ratios of Cu:In:Se = 25.2:24.9:49.9, 24:26.5:49.5, and 21.1:29.3:49.6 show resistivities of 0.1,

500, and $>10^4$ Ω-cm, respectively. It is clear that the Cu-excess CIS layers show p-type conductivity, whereas In-excess CIS layer is semi-insulating. The Cu vacancies (V_{Cu}) and Cu_{In} participate as acceptors for p-type conductivity, whereas Se vacancies, In interstitials (In_i), and In_{Cu} act as donors for n-type conductivity in the CIS layers [6].

6.1.1 p-Type Conductivity of $CuInSe_2$

After etching the as-grown 0.6–0.8 μm thick p-$CuInSe_2$ thin film grown by MBE in 5–10 wt% KCN solution for 1 min at RT, the composition ratio Cu/In and hole concentration of the sample decrease from 2.1–2.6 to 0.92–0.93 and from 1×10^{19} to 7.5×10^{16} cm^{-3}, respectively and the resistivity increases from 9.5×10^{-3} to 3.6 Ω-cm due to removal off $Cu_{2-x}Se$. The $Cu_{2-x}Se$ is a p-type semi-metallic compound, which is one of the predominant secondary phases in the as-grown $CuInSe_2$ thin films. Therefore, the CIS sample had always high hole concentration and low resistivity, if it is in the sample. The AFM and optical analyses support the presence of $Cu_{2-x}Se$ in the CIS samples, as mentioned in Chapter 3 [7]. A variation of composition with effect of etching on as-grown p-$CuInSe_2$ thin films is observed. All the time the etched samples indeed show higher resistivity and lower carrier concentrations due to elimination off $Cu_{2-x}Se$ secondary phase, as shown in Table 6.2 [8].

The resistivities of CIS layers can be varied from 10^{-2} to 10^3 Ω-cm by varying deposition conditions. A decrease in resistivity at high and low temperature regions in the temperature dependent resistivity plots may be participation of lattice, optical phonon and impurity scattering mechanisms, respectively. A change in mobility also occurs at the same temperature regions in the temperature dependent mobility plots. The resistivities of typical $CuInSe_2$ thin films with composition ratios of Cu:In:Se = 24.6:24.9:50.5 and 23.5:25.9:50 increase from 1×10^2 to 0.6×10^4 Ω-cm and from 5×10^2 to 7×10^5 Ω-cm, respectively with decreasing temperature from 300 to 100 K in the temperature dependent resistivity measurements. A change in conductivity from p- to n-type occurs at certain temperature in the samples [9]. The resistivity of typical p-$CuinSe_2$ films deposited by vacuum evaporation (5×10^{-3} torr) is 100–60 Ω-cm, whereas the resistivity decreases to 10^{-2} Ω-cm for the films deposited under oxygen pressure of 2×10^{-1} torr [10]. After annealing stack under air for 1 h in the temperature range of 200–350 °C, the resistivity of stacked In/Se/Cu layers grown by vacuum evaporation technique decreases from 10^4 to 10^1 Ω-cm. This could be due to by oxidation of $CuInSe_2$ compound [11].

Table 6.2 Electrical parameters of as-grown and etched $CuInSe_2$ thin films

S.No	Sample	Cu:In:Se	Cu/In	ρ (Ω-cm)	p or n_e (cm^{-3})	μ (cm^2/Vs)
1	p-Cu-rich CIS	31.3:19.4:49.3	1.1	3×10^{-3}	3×10^{19}	32
2	Etched p-CIS	22.1:25.4:52.5	0.87	37×10^2	2×10^{13}	3.4×10^2
3	Cu-rich p-CIS	31.9:22.6:43.5	1.4	3×10^{-2}	1×10^{20}	7.6
4	Etched p-CIS	18.4:23.4:58.2	0.79	2×10^2	3×10^{16}	6.8
5	In-rich n-CIS	24:27:49	0.89	2×10^5	8×10^{14}	64
6	n-$CuIn_3Se_5$	10:32:54.9:3.1:Na	0.31	2	7×10^{17}	3.4

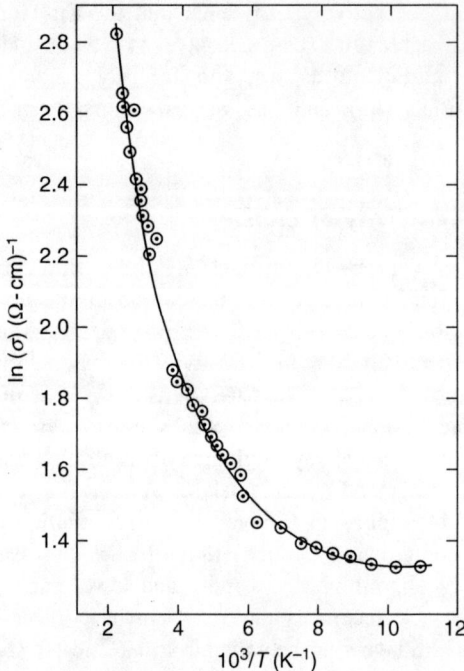

Figure 6.2 Variation of conductivity with inverse temperature for spray deposited CuInSe$_2$ thin films.

The activation energies of 30 and 70 meV at low and high temperature regimes are determined for the p-CuInSe$_2$ thin films grown by spray pyrolysis (Figure 6.2), respectively [12] from Arrhenius plot of $\ln(\sigma)$ versus $1/T$ in the temperature range of 473–90 K using the Pertritz model [13]

$$\sigma = \sigma_o \exp(-\Delta E/k_B T). \tag{6.1}$$

More or less the similar activation energies of 40 and 80 meV at low and high temperature regimes, respectively, are also obtained for the p-CuInSe$_2$ thin films, which had mobilities of 10–15 cm^2/Vs and acceptor concentration in the range of 10^{16}–10^{17} cm^{-3} [14]. The activation energies of 42–44 and 92–94 meV in the temperature range of 80–400 K for the p-CuInSe$_2$ layers grown by hybrid process of sputtering of Cu–In and thermal evaporation of Se are found to V$_{Cu}$ and V$_{Se}$, respectively [15]. The PL study supports the electrical data that the PL peak recorded at 77 K shifts from 0.975 to 0.86 eV and broadening with decreasing Cu/In ratio from 1.29 to 0.52, which (0.975–0.86 eV) is assigned to DAP transition of (V$_{Cu}$–V$_{Se}$) indicating participation of V$_{Cu}$ and V$_{Se}$ defect levels. The similar activation energies of 40 and 85 meV are observed for the single crystals from the PL measurements, as mentioned in Chapter 5. The authors [16] suggested that the activation energies of 12 and 135 meV for stoichiometric $CuInSe_2$ thin films at low and high temperature regions, respectively may be attributed to potential barriers of the grain boundaries. Unlike other reports, lower activation energy of 30 meV is observed in the high temperature range of $\sim 300\text{–}400$ K for 0.2 μm thick p-$CuInSe_2$ thin films with carrier concentration of 10^{17}–10^{18} cm^{-3} grown onto glass substrates by flash evaporation technique using $CuInSe_2$ powder at substrate temperature of 250°C [17]. Two slope regions even with low activation energies of <10 meV are found from the resistivity dependent plot for the p-CIS thin films grown onto Mo and glass substrates by flash evaporation (Figure 6.3) [18]. The reason for lower activation energy could be due to higher carrier concentration, causing degeneracy. The degeneracy is observed for the carrier concentration higher than 10^{17} cm^{-3} in the $CuInSe_2$ thin films [19].

The high activation energies of 0.85 and 0.4–1.01 eV for the growth rate of 3.5 and 0.1 nm/s, respectively are found for the p-$CuInSe_2$ thin films prepared by single source vacuum evaporation [20]. The activation energies of 301 and ~ 56 meV for 80 nm and ≥ 160 nm thick amorphous p-$CuInSe_2$ thin films are observed, respectively [21]. The $\ln(\sigma)$ versus $1/T$ plots (Figure 6.4) give activation energies of 100–170 and 350–600 meV at low and high temperature regions for 15 μm thick p- or n-$CuInSe_2$ layers grown by close-spaced vapor technique (CSVT), as given in Table 6.3 [22]. The electrical properties of p- and n-$CuInSe_2$ layers grown by DC/RF-sputtering of Cu–In and thermal evaporation of Se as a hybrid process, followed by annealing under vacuum at 500 °C for 1 and 2 h are shown in Table 6.3 [23]. As the grain sizes of p-$CuInSe_2$ thin films increase from 0.1 to 0.5–1.0 μm, the mobility increases from 0.4 to 6.1 cm^2/Vs. The Cu-rich p-CIS sputtered thin films show low resistivity, mobility, and high carrier concentration, as compared to that of near stoichiometric CIS films (Table 6.3) [24].

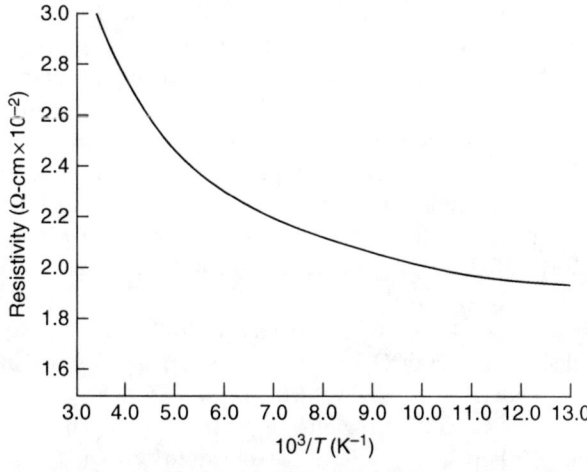

Figure 6.3 Temperature dependent resistivity of p-$CuInSe_2$ thin films grown by flash evaporation.

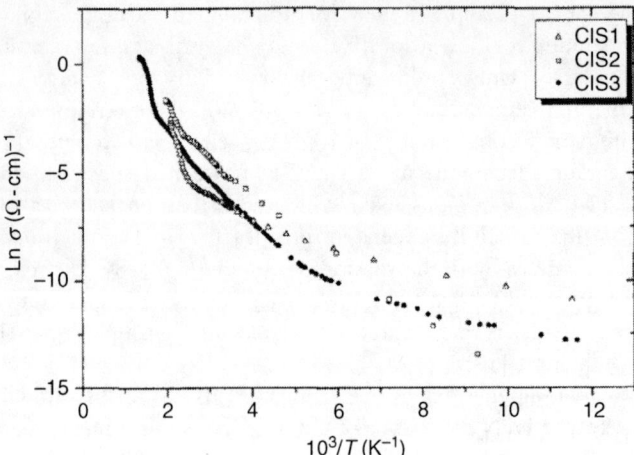

Figure 6.4 Arrhenius plot of ln σ versus $1/T$ for p- and n-CIS thin films; p-CIS1, p-CIS2, and n-CIS3.

The p-CuInSe$_2$ thin film with composition ratio of Cu:In:Se = 23.5:27:49.5, in which In is little higher than the one using for devices, shows resistivity of 10^4–10^5 Ω·cm at room temperature. The activation energies of 0.25 and 0.21 eV for before and after annealing the samples under air atmosphere at 200 °C for 20 min are obtained from the temperature (300–125 K) dependent resistivity measurements. Again after annealing the sample under air high activation energy decreases to 0.17 eV and subsequently annealing under H$_2$ regains 0.23 eV [25]. The resistivity of p-CuInSe$_2$ thin films decreases and then increases to higher quasi level upon reduction and oxidation, respectively. The typical virgin CIS sample shows activation energies of 156 and 193 meV at low and high temperatures, respectively. After air baking the samples, the activation energies decrease to lower and those increase for hydrazine treatment and again decrease for air baking, as shown in Table 6.3. The same trend is observed in resistivities for the same kind of treatment. A change in conductivity is negligible in the stoichiometric samples with variation of temperature, whereas Cu-poor and Se-poor CuInSe$_2$ samples exhibit change in resistivities with variation of temperature due to higher defect concentration indicating that they are temperature dependent. Upon air annealing, the oxygen reacts with In to form In–O hence the donor (In$_{Cu}$) concentrations decrease and V$_{Cu}$ concentration relatively increases that means p-type conductivity increases in the annealed CuInSe$_2$ sample [6]. An activation energy of 0.25 eV may be responsible to In$_{Cu}$ defect levels with a different valancy.

The Arrhenius plots of log(σ) versus $1/T$ for (a) as-grown p-CuInSe$_2$ thin films by CBD technique, (b) air annealed those at 200 °C for 1 h, and (c) vacuum annealed at 200 °C for 1 h reveal activation energies of (a) 40.5, 70.72, (b) 40.5, 103, and (c) 70.29, 414.5 meV at low and high temperature regions, respectively [26]. The activation energies of ~40 and ~70 meV are already assigned to V$_{Cu}$ and V$_{Se}$,

Table 6.3 Electrical parameters of CuInSe$_2$ layers

S.No	Sample	Cu:In:Se	ΔS	σ (Ω-cm)$^{-1}$	Reduction/Oxidation	μ (cm^2/Vs)	n_e or p (cm^{-3})	Activation energy (meV) LT	HT	Ref.
1	p-CIS	–	–	33	–	4	9×10^{19}	2	8	[18]
2	n-CIS	–	–	4.5	–	2	7×10^{18}	–	–	
3	p-CIS (CVT)	–	–	1.3×10^{-3}	–	7.5	1.4×10^{16}	104	560	[22]
4	p-CIS (CVT)	–	–	3.7×10^{-3}	–	80	4.1×10^{14}	168	351	
5	n-CIS (CVT)	–	–	1.1×10^{-3}	–	18	3.1×10^{14}	154	592	
6	n-CIS	20.7:27:52.3	0.028	6.6	–	1.89	2.12×10^{19}	–	–	[23]
7	p-CIS	23.9:19:51.9	0.075	2.08	–	2.3×10^4	5.37×10^{18}	–	–	
8	p-CIS	30.2:25.8:45.9	–	0.3ρ	–	0.1	5×10^{19}	–	–	[24]
9	p-CIS	24.4:26.6:49	–	0.9ρ	–	6.1	1×10^{18}	–	–	
10	p-CIS	25.2:24.9:49.9	–	10	5–10	–	–	–	–	[6]
11	p-CIS	24:26.5:49.5	–	2×10^{-3}	0.001–0.02	–	–	–	–	
12	p-CIS	21.1:29.3:49.6	–	$>10^4$	10^{-4}–2×10^{-4}	–	–	–	–	
13	Virgin-CIS	–	–	–	–	–	–	156	193	
14	Trt1	–	–	–	–	–	–	71	106	
15	Trt2	–	–	–	–	–	–	94	145	
16	Trt3	–	–	–	–	–	–	63	117	

Trt1-air bake at 200 °C for 30 min, Trt2-dipped in Hydrazine for 3 min and Trt3–Trt1 follows after Trt2; LT = temperature, HT = temperature.

Table 6.4 Annealing effects on CuInSe$_2$ thin films

S.No	As-grown/annealed[a] (°C)	ρ (Ω-cm)	μ (cm^2/Vs)	$p \times 10^{17}$ (cm^{-3})	Activation energy (eV)		Ref.
					Low temp.	High temp.	
1	473	0.59	144	7.2	82	302	[28]
2	573[a]	0.15	14	2.9	42	154	
3	523	3.5	141	0.12	56	175	
4	573[a]	0.3	23	8.2	22	80	
5	573	18.8	165	0.002	46	152	
6	573[a]	0.02	42	58	12	67	

respectively. The higher activation energy of 414.5 meV could be due to Fe^{2+} on In site [27], whereas 100–170 meV may be due to double donor V$_{Se}$, as mentioned in Chapter 5. The CuInSe$_2$ thin films are grown by evaporation using CIS melt charge with 5% excess of Se. The resistivity, mobility, carrier concentration and activation energies of p-CuInSe$_2$ thin films deposited at different substrate temperatures and their annealing effects are given in Table 6.4. The activation energies of 12–82 and 67–302 meV are obtained at low and high temperature regimes, respectively, from the temperature dependent resistivity measurements in the temperature range of $\sim$294–385 K for the as-grown and annealed p-CIS samples [28].

The activation energies can be deduced not only from temperature dependent resistivity plots but also from temperature dependent carrier concentration plots using the electroneutral equation [29]

$$p(p+N_D)/(N_A - N_D - p) = (N_V/g)\exp(-E_A/k_B T),$$

where density of states

$$N_V = 2\left[2\pi m_h^* k_B T/h^2\right]^{3/2}, \tag{6.2}$$

p is the free carrier concentration, N_A the acceptor concentration, N_D the donor concentration, g the degeneracy factor, E_A the activation energy, k_B the Boltzmann constant, m_h^* effective mass, and h the Plank's constant. Until to obtain a straight line from a plot of $\ln[p(p+N_D)/\{T^{3/2}(N_A - N_D - p)\}]$ against $1/T$, the N_A and N_D values should be iterated. The slope and intercept are equal to be E_A/k_B and $\ln[2/g(2\pi m_h^* k_B/h^2)^{3/2}]$, respectively [30]. The activation energies of 93 and 420 meV for the p-CuInSe$_2$ thin films grown onto semi-insulating GaAs substrates at 547 °C are determined from the plot of $pT^{-3/2}$ versus $1/T$ (Figure 6.5). The other samples grown at 497 and 597 °C consist of activation energies of 380 and 88 meV, respectively [31]. As mentioned in earlier, the V$_{Se}$ is the responsible for the activation energy of 88 or 93 meV and the deeper

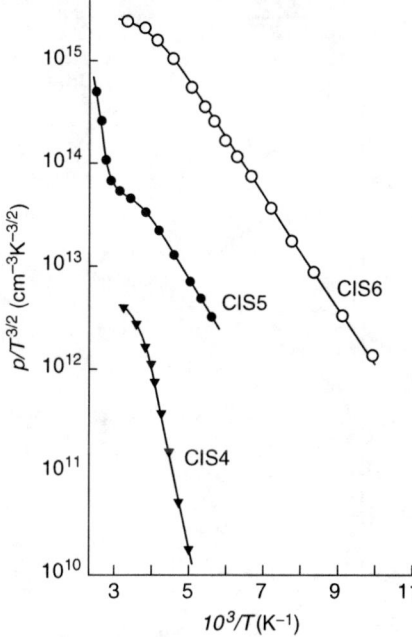

Figure 6.5 Arrhenius plot of $pT^{-3/2}$ versus $1/T$ for p-CuInSe$_2$ thin films deposited at different substrate temperatures of 497 °C (sample-CIS4), 547 °C (CIS5), and 597 °C (CIS6). The CIS4 shows 380 meV, CIS5 shows 93 and 420 meV, and sample-CIS6 exhibits 88 meV.

acceptor level of 380 or 420 meV could be due to Fe^{2+} on In site. The acceptor activation energy (E_A) of 34 meV for the p-CuInSe$_2$ can be related to the carrier concentration as equation (5.12), where $\beta \approx 2.4 \times 10^{-8}$ eV-cm and $N_A = 9 \times 10^{15}$ cm^{-3}. The activation energy (E_{A0}) at the dilute limit of acceptor concentration can also be obtained from the hydrogenic model (eq. 5.12), where $\varepsilon_o = 16$, $m_h^* = 0.73 m_0$. The activation energy of 34 meV is ascribed to the copper vacancy (V$_{Cu}$) or V$_{In}$ [32]. The temperature dependent carrier concentration $p(T)$ plot incurs acceptor activation energy of 12 ± 4 meV. The calculated acceptor activation energy is 21 meV from Equation (5.12), using $p = 3.5 \times 10^{17}$ cm^{-3} for p-CuInSe$_2$ single crystals ($\mu_p = 5$ cm^2/Vs). It is evident that the acceptor activation energy decreases with increasing carrier concentration due to screening effect, which could be related to standard acceptor level (V$_{Cu}$) of 40 meV [33]. The similar activation energy of 12 meV is deduced from a plot of carrier concentration versus inverse T [27].

If the crystals are annealed under air, the oxidation eventually takes place only on the surface of the crystals to certain depth but not into deeper. In order to avoid ambiguity and to oxidize entire crystals, the grown crystals are powdered and annealed under air at different temperatures of 200, 250, and 300 °C and again regrown to study their electrical properties, which are correspondingly labeled as

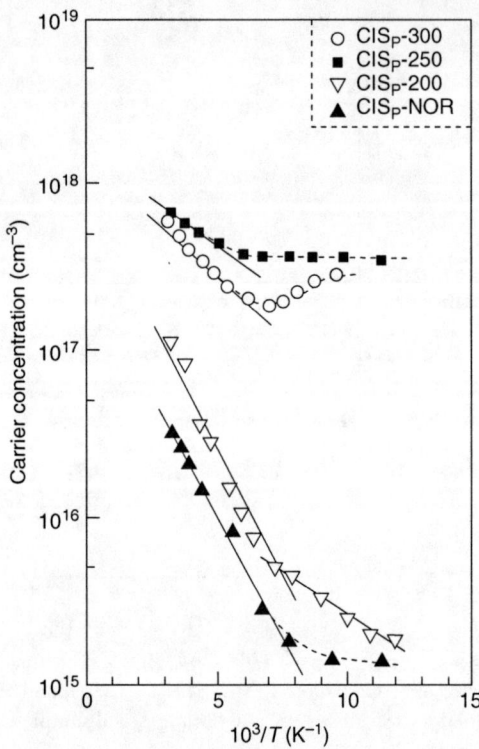

Figure 6.6 Carrier concentration versus inverse temperature of oxinated p-CuInSe$_2$ single crystals.

CISp-200, CISp-250, and CISp-300. All the crystals show p-type conductivity. The temperature dependent carrier concentrations of oxidized CIS crystals are shown in Figure 6.6. The activation energy of 50 meV is found in all the air annealed CISp-200, CISp-250, and CISp-300 samples, which may be due to V_{Cu} defect, additional activation energy of 120 meV is observed only in the CISp-200 sample and in the as-grown CIS single crystals (CISp-NOR). The n-type CISn-NOR sample can be converted into p-type by annealing under air at 300 °C [34]. The temperature dependent experimental and theoretical carrier concentrations of CuInSe$_2$, CuGaSe$_2$, and CuIn$_{0.5}$Ga$_{0.5}$Se$_2$ single crystals grown by CVT technique are shown in Figure 6.7. The theoretical carrier concentrations of the samples are obtained using two acceptors and one donor model [35]. The activation energies of 20 or 28 meV for acceptor levels are obtained from the plots of Hall coefficients versus inverse temperature for the p-CIS single in the crystals [36]. The variation of Hall coefficient with temperature is also given in Figure 6.8 [22].

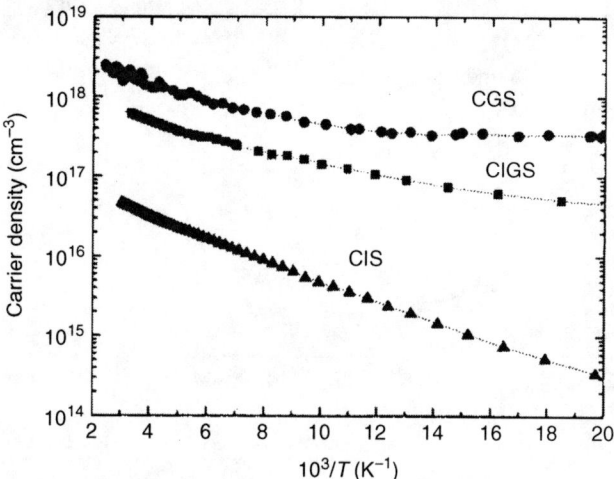

Figure 6.7 Temperature dependent carrier concentration of CIS, CGS, and CIGS (points-experimental, dotted lines-theoretical fitting).

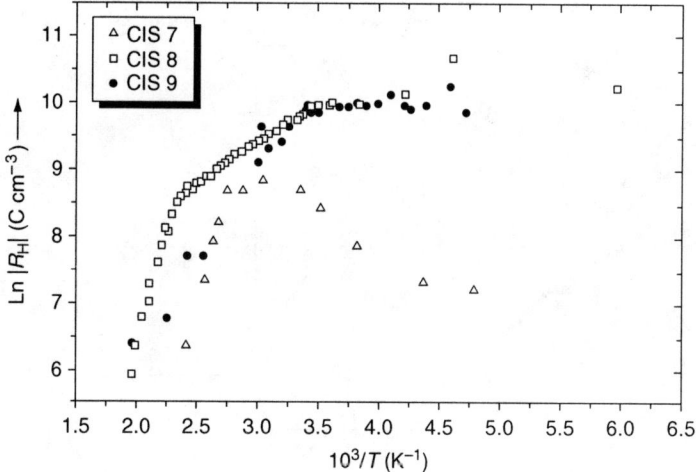

Figure 6.8 Arrhenius plot of $\ln(R_H)$ versus $1/T$ for p- and n-CIS thin films; p-CIS7, p-CIS8, and n-CIS9.

6.1.2 n-Type Conductivity of CuInSe$_2$

The intrinsic defects such as In$_{Cu}$, V$_{Se}$, and In$_i$ are the main donors in the n-type CuInSe$_2$, as already pointed out. The activation energies of 3 and 30 meV at low and high temperature regions, respectively for the n-CuInSe$_2$ thin films are found from the Arrhenius plot of $\ln(\sigma)$ versus $1/T$, as shown in Figure 6.9. The n-CuInSe$_2$ thin films used for conductivity measurements are prepared using Cu/In bilayer

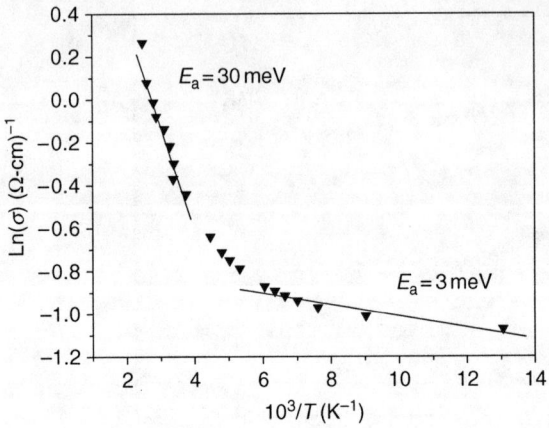

Figure 6.9 Arrhenius plot of ln σ versus inverse temperature for the n-CuInSe$_2$ thin film.

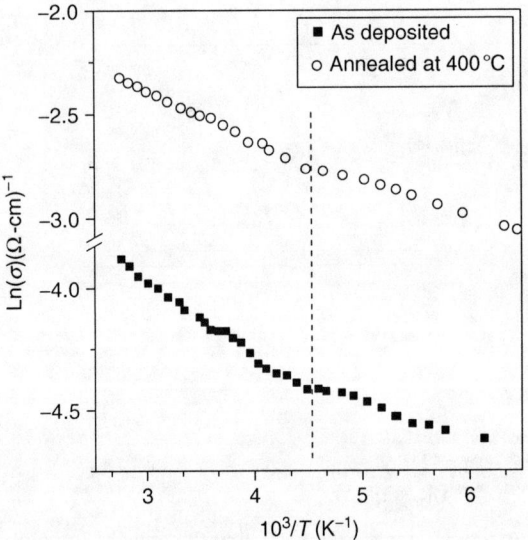

Figure 6.10 Conductivity versus inverse temperature of n-CuInSe$_2$ thin films grown by electrodeposition.

process by thermal evaporation onto Si (100) substrates at RT and annealed at 150 °C on which Se is evaporated from graphite effusion source then again annealed under Argon flow at 450 °C for 15 min. After annealing the combined 120, 240, and 540 nm thick Cu, In, and Se layers become 0.8 μm thick n-CuInSe$_2$ layer, which consists of composition ratio of Cu:In:Se = 25:29:46 [37]. In another case the as-grown electrodeposited n-CuInSe$_2$ thin films exhibit activation energies of 29 and 12 meV at high and low temperature regions from the temperature dependent resistivity measurements, respectively. After annealing at 400 °C under vacuum, the CIS layers show activation energy of 20 meV (Figure 6.10) [38]. Unlike the activation

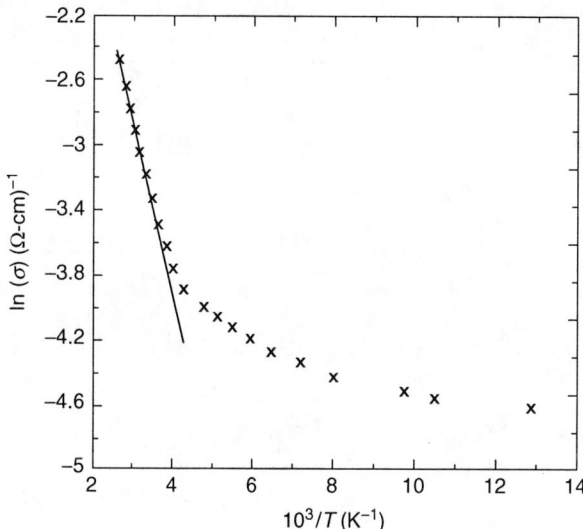

Figure 6.11 Conductivity versus inverse temperature of n-CuInSe$_2$ thin films.

energies of 46, 82, and 154 meV are observed for the n-CuInSe$_2$ thin films grown by electrodeposition technique from $\ln(\rho)$ versus $1/T$ plots in the temperature range of 300–600 K (three temperature regions). The resistivity of 10^1–10^4 Ω-cm for thin films is in the temperature range of 300–600 K [39]. As shown in Figure 6.11, The n-type CuInSe$_2$ thin film samples with Cu:In:Se = 27.1:24.6:48.3 and 24.5:30.1:45.5 show activation energies of 91 and 71 meV, respectively, which are assigned to thermally activated conduction across the grain boundaries by the authors [40]. The electrical conductivities of n- and or p-CuInSe$_2$ thin films vary from 2.15×10^{-3} to 6.14×10^{-3} and from 1.6×10^{-2} to 1.6×10^{-1} $(\Omega\text{-cm})^{-1}$ with varying Cu/In ratio from 0.509 to 1.633, respectively. The electrical conductivity versus $1/T$ for different composition ratios of Cu/In is shown in Figure 6.12. As shown in Figure 6.13, even though the carrier concentration is low in the n-type CIS samples, the magnitude of n-type conductivity is high due to higher mobility as compared to that of p-type CIS films. It is well-known fact that the electron mass is less than the hole mass therefore the mobility is higher in the n-type films [41]. Keeping lower Se composition, the p-CuInSe$_2$ thin films show low conductivity for higher Cu/In ratio while the n-CuInSe$_2$ show higher conductivity for lower Cu/In ratio. Keeping higher Se composition, the conductivity of p-type CuInSe$_2$ thin films is higher for higher Cu/In ratio, whereas the conductivity is lower in the n-type CuInSe$_2$ thin films for lower Cu/In ratio [42]. No detectable change in the resistivity is observed for the n-CuInSe$_2$ single crystals after 20 keV Li implantation with fluence of 3×10^{14} cm^{-2}. However, the resistivity of p-CuInSe$_2$ single crystals increases from 0.7 to 29 Ω-cm. On the other hand in both cases no type conversion is observed [43].

Two activation energy levels of 9.5 and 194 meV at lower and higher temperature regions from the temperature dependent Hall coefficient in the temperature

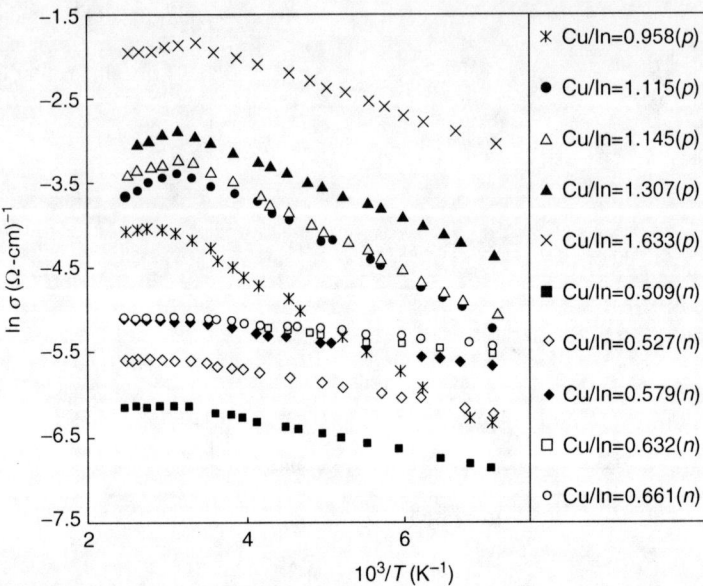

Figure 6.12 Conductivity versus inverse temperature of CuInSe$_2$ thin films for different compositions.

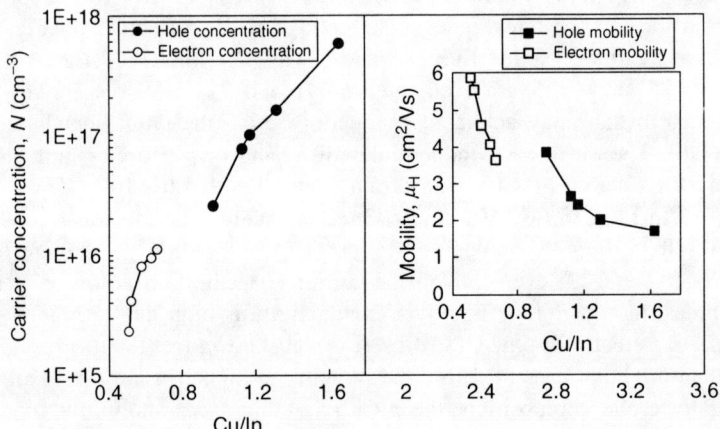

Figure 6.13 Variation of carrier concentration with Cu/In ratio and variation of mobility with Cu/In ratio for CuInSe$_2$.

range of 77–300 K are determined by employing the relation $R_H \alpha \exp(\Delta E/kT)$ for the n-CuInSe$_2$ thin films. Former can be assigned to donor level and latter due to acceptor level. The n-CuInSe$_2$ thin films grown onto mica or glass substrates by evaporation using 2 at% In-excess charge at substrate temperature of 350 °C exhibit mobilities of 3.4×10^3 and 20 cm^2/Vs at 77 and 300 K, respectively. The room temperature carrier concentration of the layers is 1.36×10^{15} cm^{-3} [44].

The $N_D = 9.5 \times 10^{16}$ cm^{-3}, $N_A = 3.7 \times 10^{16}$ cm^{-3}, and $E_D = 21$ meV are extracted from a plot of temperature dependent carrier concentration for the n-CuInSe$_2$ thin film grown onto GaAs (001) by RF-sputtering using the electroneutral relation

$$n_e(n_e + N_A)/(N_D - N_A - n_e) = (N_c/g)\exp(-E_D/k_B T),$$

where

$$N_c = 2\left[2\pi m_c^* k_B T/h^2\right]^{3/2}. \tag{6.3}$$

A lower activation energy of 21 meV may be due to participation of In$_{Cu}$ donor level. The defect densities of 1×10^{17} and 5.7×10^{17} cm^{-3} are observed for the irradiation fluencies of 3×10^{13} and 1×10^{14} cm^{-2}, respectively. An activation energy of 95 meV is observed for irradiated CuInSe$_2$ thin films, which could be due to V$_{Se}$. A slope increases with increasing proton irradiation fluence indicating an increase of density of defect levels (Figure 6.14). The irradiation studies reveal that the CuInSe$_2$ thin films are harder for proton irradiation than III–V compounds [45,46]. No remarkable change in the carrier concentration of 1×10^{17}–5×10^{17} cm^{-3} for the n-CIS epitaxial layer on GaAs substrates is observed with variation of temperature indicating that shallow donor levels may be the responsible for n-type conduction [31]. The hydrogenic model $E_{A0} = 13.6(m_e^*/m_0)/\varepsilon_s^2$ exploits an activation energy of 5–7 meV for donor level, where $m_e^* = 0.09m_0$ and $\varepsilon_s = 16$, which is close to the value of 11 meV obtained from the Schottky junction studies [47]. All typical n-CuInSe$_2$ single crystals show donor level of ≤ 5 meV except one sample shows additional donor level at 80 meV that means two donor levels may be contributed n-type conductivity [48]. The activation energies of 12 and 180 meV for donors are deduced from the plots of Hall coefficient versus inverse temperature for the

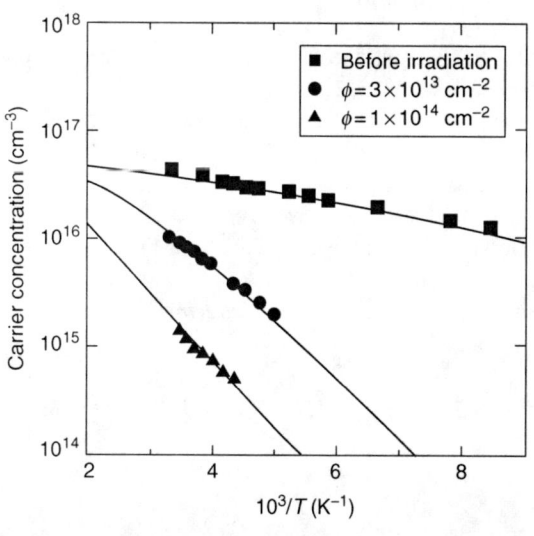

Figure 6.14 Variation of carrier concentration with inverse temperature for as-grown, proton irradiation with fluence of 3×10^{13} and 1×10^{14} cm^{-2} CuInSe$_2$ thin films.

n-type CIS single crystals in the low and high temperature regions [36]. In a different case the temperature dependent carrier concentration reveals activation energy of 5.8 meV, which matches well with the value of 5–7 meV obtained by hydrogenic approximation for the n-type CuInSe$_2$ single crystals [49].

6.1.3 Scattering Mechanisms

A various kinds of scattering mechanisms participate to lead variation of mobility with temperature in the semiconductors. The acoustic, polar, nonpolar optical, and impurity scattering mechanisms are dominants in the CuInSe$_2$, CuGaSe$_2$, and CuInS$_2$ compounds.

(a) *Acoustic scattering* (μ_{ac}): Variation of mobility due to acoustical mode scattering with temperature can be related to

$$\mu_{AC} = \frac{\sqrt{8\pi}\, e\hbar^4 C_{11}}{3m^{*5/2}(k_B T)^{3/2} E_I^2} \tag{6.4}$$

where $\hbar$ is the reduced Plank constant, m^* the effective mass of the carriers, k_B the Boltzmann constant, C_{11} the average longitudinal elastic constant, and E_I the shift of the edge of the conduction or valence band with temperature per unit dilation, that is, $V(dT/dV)(dE_g/dT)$, which can be obtained with aid of change in the band gap with temperature (dE_g/dT) and the volume expansion coefficient $(1/V)(dV/dT)$. In the case of p-type semiconductor, E_I or E_{AC} is the valence band deformation potential, C_{11} can be estimated as $C_{11} = \rho_d u^2$, where ρ_d is the density and u_u is the velocity of sound

$$u = \frac{k\theta_D}{\hbar}\left(\frac{V}{6\pi^2}\right)^{1/3}; \quad \theta_D = CM^{-1/3}\rho_d^{-1/6}\kappa^{-1/2}, \tag{6.5}$$

where C is constant, M is the molecular weight and κ the hardness of the semiconductor [50]. The dE_g/dT of CuInSe$_2$ varies from -1.1×10^{-4} to -3×10^{-4} eV K^{-1} [51] and its average value can be taken. The volume expansion coefficient of 9.6×10^{-6} K^{-1} for the CuInSe$_2$ can be derived from the temperature dependent lattice constants [52]. Using the data, E_I of CuInSe$_2$ is calculated to be 21.4 eV. We have reported $dE_g/dT = 4.872 \times 10^{-4}$ eV K^{-1}, $1/V(dV/dT) = 12 \times 10^{-6}$ K^{-1}, and $E_I = 40.6$ eV for CdS [53]. However, E_I value of ~ 75 eV in the literature is larger than what we have calculated for CdS [30].

(b) *The combination of acoustical and nonpolar optical scattering* ($\mu_{AC,NOP}$): The expression for $\mu_{AC,NOP}$ can be related to

$$\mu_{AC,NPO} = \mu_{AC} S(\theta, \eta, T); \quad S(\theta, \eta, T) = [1 + (\theta/T)\eta H/\exp(\theta/T) - D]^{-1} \text{ and } \eta = \left(\frac{E_{NPO}}{E_{AC}}\right)^2, \tag{6.6}$$

where θ is the optical phonon characteristic temperature, E_{NPO} is the optical nonpolar deformation potential, where H and D are constants, supposing $H=1.34$ and $D=0.914$, if $\eta=4$ [36,54].

(c) *Polar optical phonon scattering* (μ_{PO}): The equation for μ_{PO} can be written as

$$\mu_{PO} = \frac{T^{1/2}}{\theta}\left(\frac{1}{\varepsilon_\infty} - \frac{1}{\varepsilon_s}\right)^{-1} \times \left(\frac{m_0}{m^*}\right)^{3/2}(e^z - 1)G(z), \qquad (6.7)$$

$Z=\theta/T$ and $G(Z)$ is a tabulated function, which may be followed as $0.48\exp(0.18\theta/T)$ in the temperature range of 120–300 K [54].

(d) *Ionized impurity scattering* (μ_I): The Brooks–Herring expression can be used to calculate μ_I approximately for nondegenerate semiconductor. The ionized impurity scattering is temperature dependent or dominant one at low temperature, which is proportional to $T^{3/2}$ [55].

$$\begin{aligned}\mu_I &= \frac{2^{7/2}(4\pi\varepsilon_s\varepsilon_0)^2(k_BT)^{3/2}}{\pi^{3/2}e^3N_i m_e^{*1/2}f(x)}; \quad f(x) = \ln(1+x) - \frac{x}{(1+x)}; \quad x\\ &= \frac{6(4\pi\varepsilon_s\varepsilon_0)m_e^*(k_BT)^2}{\pi e^2\hbar^2 n_s}.\end{aligned} \qquad (6.8)$$

The condition is that $x \gg 1$, where

$$n_s = n_e + (N_D - N_A - n_e)(n_e + N_A)/N_D \quad \text{and} \quad N_I = n_e + 2N_A \qquad (6.8a)$$

is the concentration of ionized defects, N_D and N_A are the donor and acceptor concentrations. N_D can be extracted using the relation $r_s = (3/(4\pi N_D))^{1/3}$, $r_s \approx 3a_0$, and $a_0 = \varepsilon(m_0/m^*)a_H$ the effective Bohr radius and $a_H = 0.053$ nm. On the other hand, N_D and N_A can be derived from a plot of $\ln[n_e(n_e+N_A)/\{T^{3/2}(N_D-N_A-n_e)\}]$ against $1/T$ by iteration method for *n*-type semiconductor. If it is *p*-type semiconductor, *p* related terms replace *n*-type terms in Equations (6.8) and (6.8a). The parameters used to estimate the mobilities for $CuInSe_2$, $CuGaSe_2$, and $CuInS_2$ are given in Table 6.5.

Table 6.5 Physical parameters of $CuInSe_2$, $CuGaSe_2$, and $CuInS_2$ compounds for mobility evaluation

S.No	Compound	E_{AC} H (eV)	u (cm/s)	d_g (g/cm³)	ε_s	θ_c (K)	k^{-1} (kg/cm³)	$(1/\varepsilon_\alpha - 1/\varepsilon_s)^{-1}$	Ref.
1	$CuInSe_2$	4 7	2.18×10^5	5.77	16	395	–	–	[54,56]
2	$CuGaSe_2$	4 6.4	2.47×10^5	5.57	13.6	245	4.7×10^{-5}	–	[57]
3	$CuGaSe_2$	– 2.65	3.02×10^5	5.57	11	294	–	29.3	[58]
4	$CuInS_2$	4 8.4	4.97×10^5	4.74	10.2	470	–	13.7	[59]

The total mobility of the sample can be approximated by Mathiessen's formula

$$\frac{1}{\mu} \approx \frac{1}{\mu_I} + \frac{1}{\mu_{AC}} + \frac{1}{\mu_{PO}} + \frac{1}{\mu_{NOPO}} \tag{6.9}$$

6.1.3.1 Mobility of CuInSe$_2$

Among the several scattering mechanisms the ionized impurity scattering (μ_I), polar or nonpolar optical scattering (μ_{PO}), and lattice scattering (μ_{AC}) mechanisms are predominant in the temperature dependent Hall mobility of p- or n-type CuInSe$_2$ single crystals. The peak mobility of 1250 cm^2/Vs at 250 K is observed in the n-CuInSe$_2$ single crystals [56,60]. The n-type CuInSe$_2$ single crystals had carrier concentrations in the range of 2×10^{15}–3×10^{17} cm^3 and mobilities in the range of 50–360 cm^2/Vs in which one of the typical crystals with size of $5 \times 5 \times 0.5$ mm^3 exhibits $n_e = 4.8 \times 10^{15}$ cm^{-3}, $\mu = 360$ cm^2/Vs, $N_I = 1 \times 10^{17}$ cm^{-3}, and activation energy (ΔE) = 0.25 eV. The temperature dependent mobility for the n-type CuInSe$_2$ single crystal is recorded in the temperature range of 80–700 K. As shown in Figure 6.15, the impurity scattering is dominant one below the temperature of 300 K, whereas the lattice scattering is dominant one above 350 K but nonpolar optical scattering is also considered as inclusion in it [61]. The similar combinations of acoustical and nonpolar optical scattering ($\mu_{AC,NOP}$) and polar optical scattering (μ_{PO}) are dominant scattering mechanisms at high temperature, whereas the ionized impurity scattering (μ_I) is dominant one at low temperature in the temperature dependent mobility of n- and p-type CuInSe$_2$ single crystals grown by normal freezing method. The typical p-type CIS single crystals had $p = 9.3 \times 10^{16}$ cm^{-3}, $\mu = 18$ cm^2/Vs, and $N_I = 8.4 \times 10^{18}$ cm^{-3} as well as n-type samples contain $\mu \sim 600$ cm^2/Vs and $\rho = 1$ Ω/sq [36]. The domination of polar (μ_{PO}) scattering or combination of polar and acoustic scatterings ($1/\mu_{OP} + 1/\mu_{AC}$) is observed at high temperature in some cases [49].

Wasim et al. [56] disregarded nonpolar optical phonon scattering to the total mobility for n-type CuInSe$_2$ because their opinion is that the S-like wave functions do not couple to nonpolar optical modes in the CIS samples. After annealing under Se ambient at 600 °C the as-grown n-type CuInSe$_2$ single crystals become p-type, which consists of $p = 2.6 \times 10^{14}$ cm^{-3}, higher mobility $\mu = 485$ cm^2/Vs and activation energy $E_A = 98$ meV. No correlation between the annealing process under Se/temperature and electrical parameters such as hole concentration or mobility is found. The mobility is low in some of the samples that could be due to higher compensation of N_D and N_A. The nonpolar optical scattering in the first order of the phonon wave vector follows $\mu_P \alpha T^{-2.35 \pm 0.1}$ for the p-type CuInSe$_2$ single crystals, which had $p = 5 \times 10^{18}$ cm^{-3}, $\mu = 63$ cm^2/Vs, and $E_A = 65$ meV. The as-grown p-type CuInSe$_2$ single crystals can be converted into n-type by diffusing In into it. The temperature dependent electron mobility of n-CuInSe$_2$ single crystals ($n = 4 \times 10^{16}$ cm^{-3} at RT) follows the relation $\mu_I \alpha T^{1.5}$ at low temperature of below 140 K and $\mu_{AC,NPO} \alpha T^{-2.5}$ at high temperature of above 240 K [27]. The temperature dependent mobility studies are carried out on (112) oriented n- (0.5 at% excess In) and

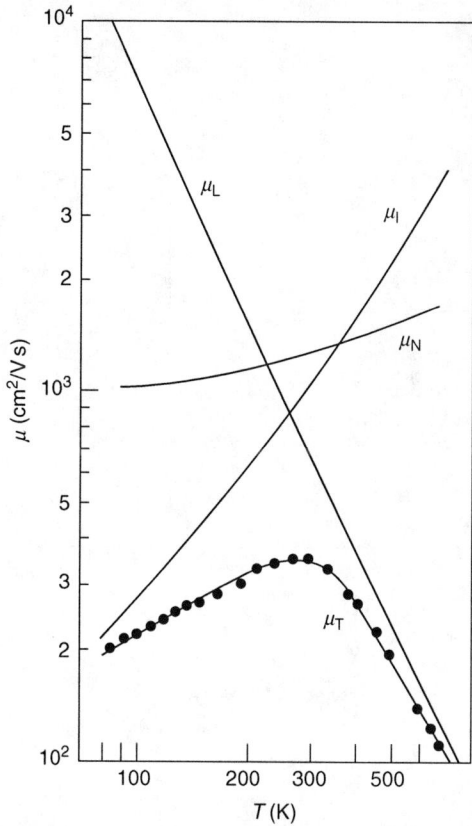

Figure 6.15 Temperature dependent Hall mobility of n-CuInSe$_2$ single crystals.

p-CuInSe$_2$ (0.3 at% excess Se) single crystals. As shown in Figure 6.16, in the inset (b), the slopes from Hall factor (R) versus $1/T$ incur activation energies of 55 ± 3 and 22 ± 3 meV probably relating to V$_{Cu}$ and V$_{In}$ for the p-CuInSe$_2$, respectively and 10 meV relating to V$_{Se}$ for n-CuInSe$_2$ single crystals. A combination of acoustical and nonpolar optical phonon scattering ($\mu_{AC,NPO} \sim \mu°_{AC,NPO} T^{-3/2}$) dominates in the temperature range of $T > 250$ K and the ionized impurity scattering ($\mu_I \sim \mu°_I T^{3/2}$) dominates in the temperature region $150 < T < 180$ K for the n-CuInSe$_2$ single crystals. The hoping conduction mechanism $\mu_{HOP} \sim \mu_{HOP} \exp(-T_o/T)^{1/4}$ is also predominant at < 150 K. Both ionized impurity and acoustical-nonpolar optical phonon scatterings are dominants in the temperature region of 180–250 K. In the case of p-CuInSe$_2$ single crystals, the ionized impurity scattering is dominant in the temperature range of $80 < T < 120$ K and the acoustic-nonpolar optical phonon scattering is dominant one for $T > 180$ K [62]. Both n- and p-type CuInSe$_2$ thin films grown by thermal evaporation show domination of ionized impurity scattering at ≤ 300 K, whereas lattice vibration scattering dominates at $T > 300$ K. As expected the magnitude of mobilities is

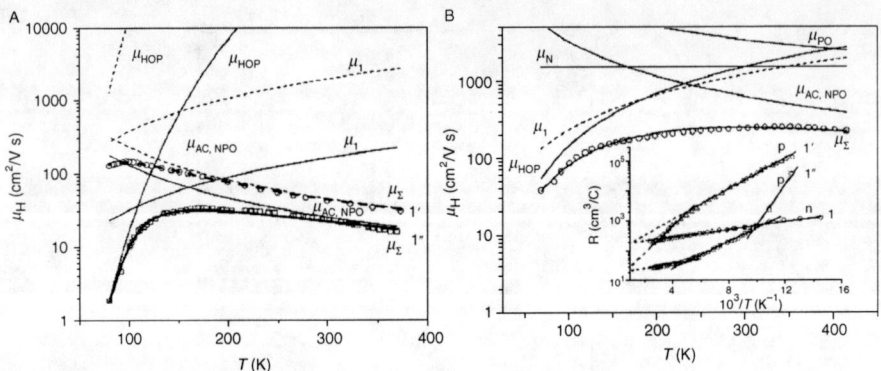

Figure 6.16 Temperature dependent hole mobility of (A) p-type (1′, 1″) and (B) n-type CuInSe$_2$ single crystals (samples 1).

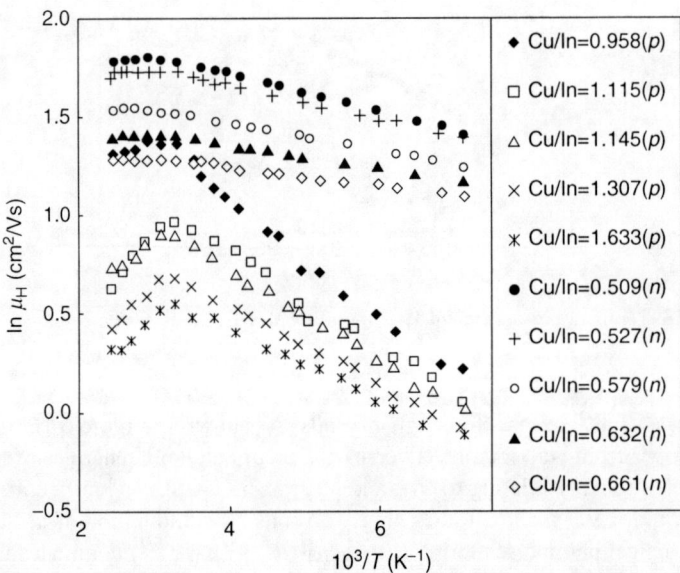

Figure 6.17 Hall mobility versus inverse temperature of CuInSe$_2$ thin films for different compositions.

higher in the n-type samples than in the p-type samples [63]. As shown in Figure 6.17, the nature of temperature dependent mobility curves for the CuInSe$_2$ thin films varies with Cu/In ratios. A variation of mobility with temperature is more curve nature in the p-CuInSe$_2$ samples than in the n-CuInSe$_2$ thin films indicating that the scattering mechanisms are stronger in the p-type samples than that in the n-CuInSe$_2$ [41].

6.2 Conductivity of CuGaSe$_2$

The CuGaSe$_2$ thin films deposited onto glass and or alumina substrates at the deposition temperature of 400–490 °C show resistivities of 10^5–10^6 and 10^0–10^1 Ω-cm for $45 < \text{Cu}/(\text{Cu}+\text{Ga}) < 49$ and $50 < \text{Cu}/(\text{Cu}+\text{Ga}) < 52$ ratios, respectively. The resistivity sharply decreases from 10^5–10^6 to 10^0–10^1 Ω-cm with changing composition of Cu/(Cu+Ga) from 49 to 50 [9]. The Hall measurements on Cu-rich and Cu-poor CuGaSe$_2$ thin films grown onto semi-insulating GaAs (001) substrates by MOVPE technique using AIXTRON AIX200 system are carried out. In the case of Cu-rich CuGaSe$_2$ thin films, the samples are etched in KCN prior to conduct Hall measurements. As shown in Figure 6.18, the conductivity, hole concentration and mobility decrease with decreasing Cu/Ga ratio in the samples due to an increase of defect concentration. The hole concentration is higher in the Cu-rich CuGaSe$_2$ thin films than in the Ga-rich samples at room temperature due to an increase of shallow acceptor levels. The conductivity is low in the Cu-poor layers, as expected [64]. The activation energies of 90 and 300 meV for the CuGaSe$_2$ thin films are obtained from low and high temperature regions in the plot of log σ versus $1/T$ (Figure 6.19) [65], former can be attributed to V_{Se}. Santamaria et al. [66] obtained an similar activation energy value of 80 meV from a plot of log characteristic relaxation frequency (v) versus $1/T$ for CuInSe$_2$/CdS thin film solar cells, which is assigned to V_{Se}. As shown in Figure 6.20, the conductivity is high in the stoichiometric CuGaSe$_2$ thin films (♦, ▲) grown by MOVPE due to low defect concentration but its response with temperature is slow, whereas the conductivity of high Cu-rich (●, ■) and moderate Cu-rich (▼) CuGaSe$_2$ thin films respond well with varying temperature due to high defect density [67]. An activation energy (E_A) of 30 ± 3 meV for the CuGaSe$_2$ single crystals is determined from the experimental temperature dependent hole concentration using Equation (6.2) assuming $N_A - N_D = 3.45 \times 10^{16}$ cm^{-3}, the estimated other values are $m_{vd} = 1.2 m_o$, $\beta = 2$, and $N_D = 10^{18}$ cm^{-3}. The CuGaSe$_2$ sample is highly compensated, that is, $p << N_D \approx N_A$ [68]. Using known hole concentration and employing the best fit parameters of $N_A - N_D = 3.8 \times 10^{18}$ cm^{-3} and $N_D = 4.4 \times 10^{18}$ cm^{-3}, higher activation energy of 388 meV and higher effective mass of $3.1 m_o$ for the CuGaSe$_2$ single crystals are obtained from the slope and intercepts of plots of ln $[p(p+N_D)/\{T^{3/2}(N_A-N_D-p)\}]$ versus $1000/T$, as shown in Figure 6.21 [57]. The obtained effective mass value seems to be higher or abnormal.

6.2.1 Mobility of CuGaSe$_2$

In the high Cu-rich (●, ■) and moderate Cu-rich (▼) CuGaSe$_2$ thin films grown by MOVPE, the optical phonon scattering ($\mu \alpha T^{-2}$) dominates at temperature > 200 K. At temperature < 180 K, the impurity scattering ($\mu \alpha T^{3/2}$) dominates in the moderate Cu-rich CuGaSe$_2$ thin films. In the case of near stoichiometric (♦,▲) CuGaSe$_2$ thin films, the mobility sharply falls off it could be due to domination of hopping conduction mechanism, as shown in Figure 6.22 [67]. A plot of temperature dependence Hall mobility for the typical (a) CuGaSe$_2$ single crystals grown by Bridgman method (b) CGS thin films by flash evaporation, and (c) CGS thin films by single source evaporation are

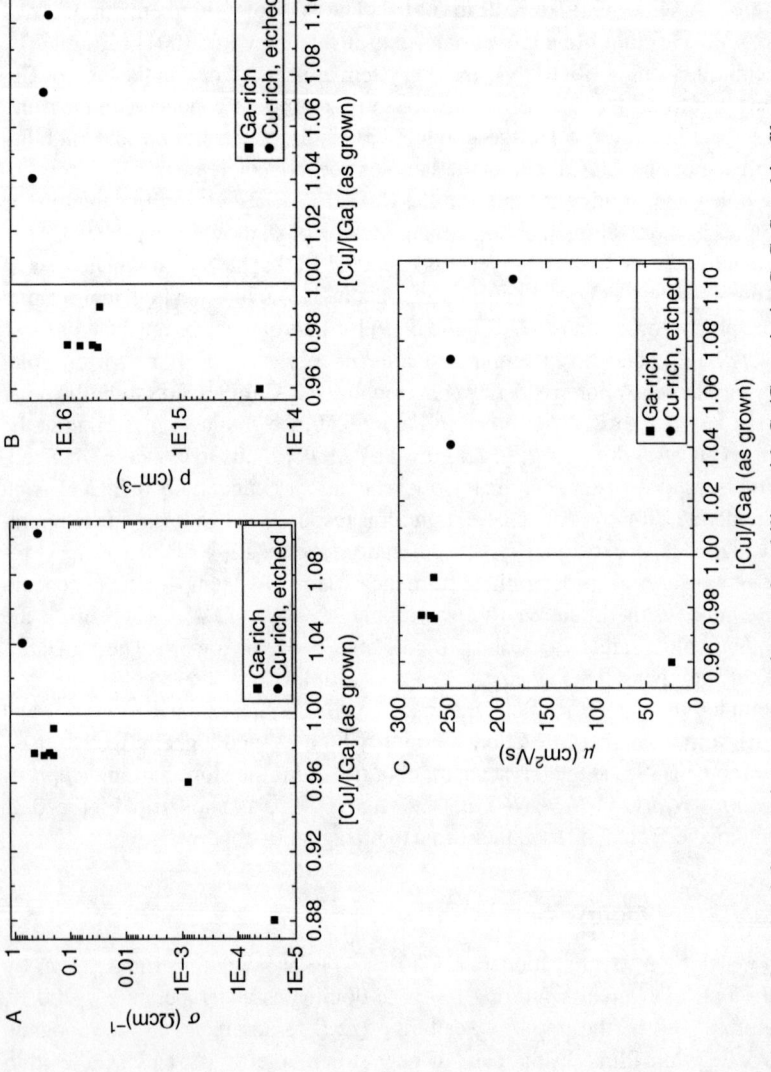

Figure 6.18 Variation of conductivity, carrier concentration, and mobility with Cu/Ga ratio in CuGaSe$_2$ thin films.

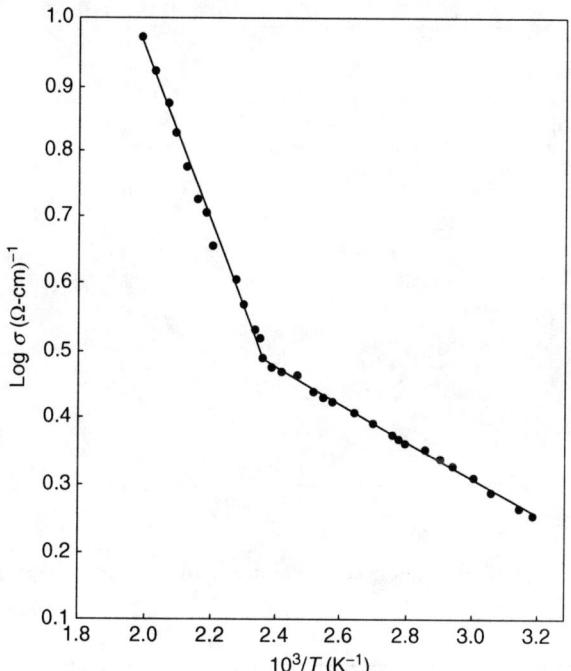

Figure 6.19 A plot of log conductivity versus inverse temperature for CuGaSe$_2$ thin films.

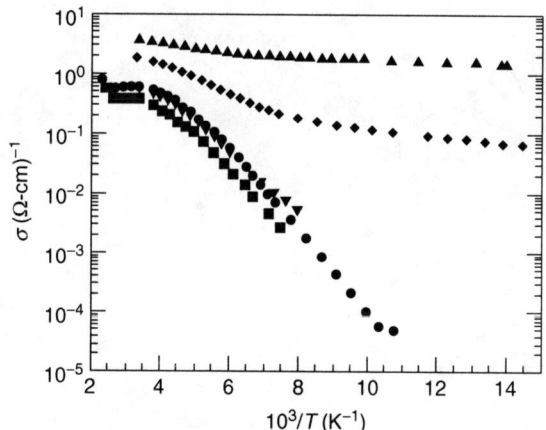

Figure 6.20 Conductivity versus inverse temperature for CuGaSe$_2$ thin films (high Cu-rich (●, ■), moderate Cu-rich (▼), and near stoichiometric ◆,▲).

depicted in Figure 6.23. A change in mobility for Bridgman CuGaSe$_2$ single crystals follows as $\mu \alpha T^{-1.6 \sim -3/2}$, that is, lattice scattering dominates. A change in mobility for flash evaporated CGS films, which contain carrier concentration of 1.8×10^{18} cm^{-3}, is due to $\mu \alpha T^{0.7 \sim 1/2}$ in the temperature range of 160–400 K that means the mobility follows as

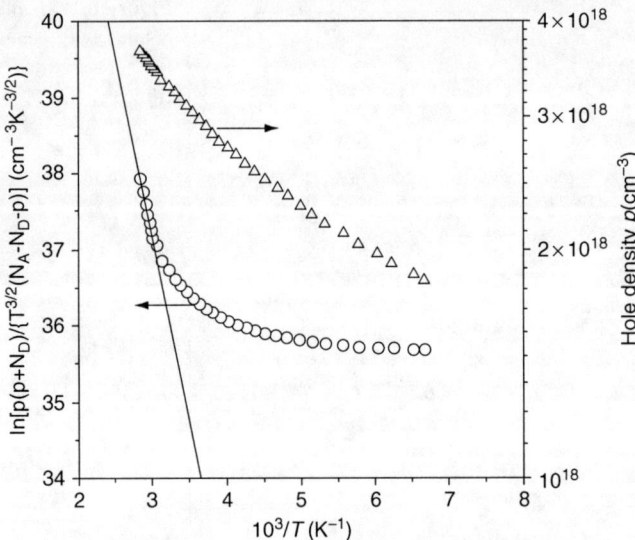

Figure 6.21 $\ln[p(p+N_D)/\{T^{3/2}(N_A\text{-}N_D\text{-}p)\}]$ and hole concentration versus $10^3/T$ of CuGaSe$_2$ single crystal.

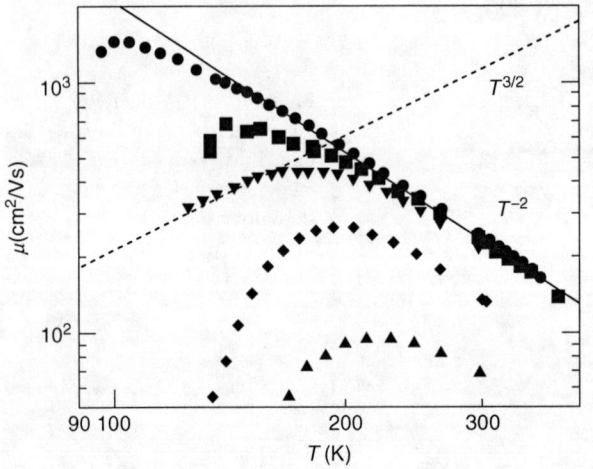

Figure 6.22 Mobility versus temperature for CuGaSe$_2$ thin films (high Cu-rich (●, ■), moderate Cu-rich (▼), and near stoichiometric (♦, ▲)).

$$\mu = q\lambda_p/(2kTm^*)^{1/2} \quad (6.10)$$

where λ_p is the mean free path of carriers. In the case of CGS thin films with $p = 1.6 \times 10^{21}$ cm^{-3} grown by single source evaporation, the mobility follows as $\mu = f(T)$.

The electrical parameters of single crystals, flash, and single source evaporated CuGaSe$_2$ are given in Table 6.6 [69]. The temperature dependent mobility at high

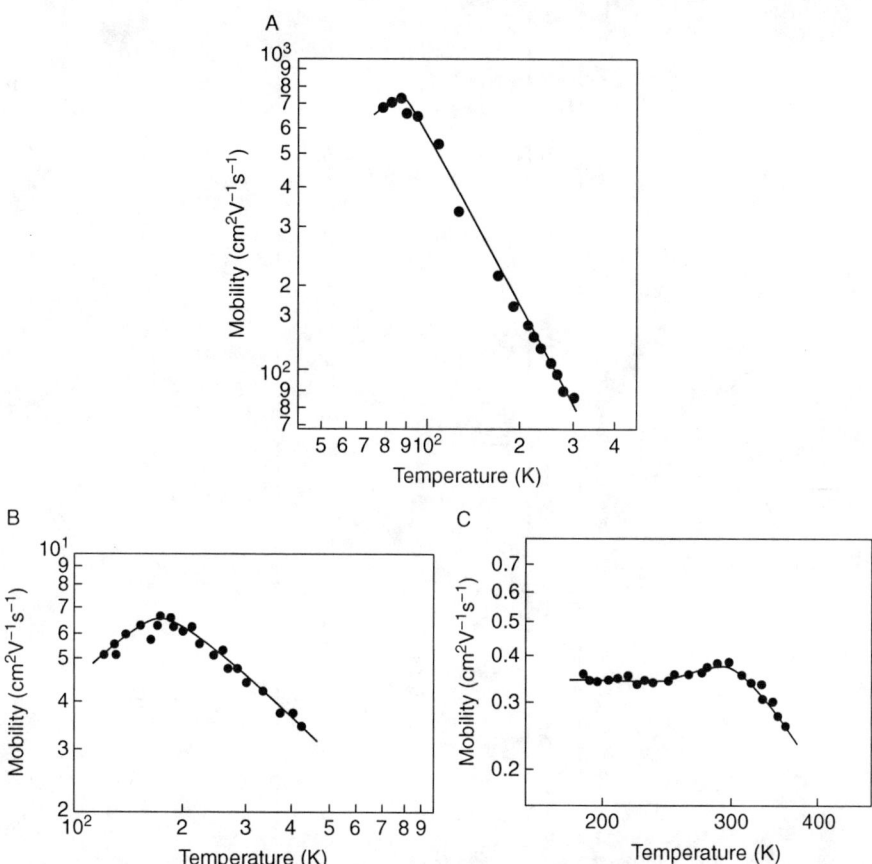

Figure 6.23 Temperature dependent Hall mobility of CuGaSe$_2$ single crystal grown by Bridgman method ($p = 0.8$–1.5×10^{17} cm^{-3}), (B) flash evaporated CuGaSe$_2$ ($p = 1.8 \times 10^{18}$ cm^{-3}), and (C) single source evaporated CuGaSe$_2$ thin films ($p = 1.6 \times 10^{21}$ cm^{-3}).

temperature indicates that the nonpolar optical scattering in the first order of the phonon wave vector is the dominant scattering mechanism ($\mu_{NPO} \alpha T^{-2.1}$) in the CGS crystals from the University of Salford, as shown in Figure 6.24 [68]. In another case the authors found that in the temperature dependent mobility of p-CuGaSe$_2$ single crystals the μ_I and $\mu_{AC,NPO}$ are the dominant scattering mechanisms at low and high temperatures, respectively [57]. As expected the mobility in the CuGaSe$_2$ polycrystalline layers is lower than that in the Cu-rich CuGaSe$_2$ epitaxial layers by one order of magnitude at room temperature due to higher concentration of defect levels. Unlike taking assumptions for scattering formulas, the authors have derived equations for different scattering mechanisms based on quantum mechanisms and Boltzmann transport equations those are employed to interpret experimental mobilities for more accuracy. The hole effective mass (m^*) of 1.0 is

Table 6.6 Electrical parameters of CuGaSe$_2$ single crystals and thin films

S.No	CuGaSe$_2$ process	ρ (Ω-cm)	μ (cm^2/Vs)	$p \times 10^{17}$ (cm^{-3})	N_D	N_A	N_D/N_A	Ref.
1	Bridgman	0.61–1.04	68.4–89	0.8–1.5	1.5×10^{17}	1.17×10^{18}	0.13	[69]
2	Iodine trans. (IT)	0.03	60.3	35	–	–	–	
3	Single source	0.01	0.35	16,000	–	–	–	
4	Flash evaporation	0.42	4.1–24	1.5–18	4.3×10^{20}	4.53×10^{20}	0.95	[70]
5	As-grown (IT) (RT)	0.4	60	3	–	–	–	
6	(76 K)	4	140	0.1	–	–	–	
7	350 °C, 1-1/2 h (RT)	0.1	40	30	–	–	–	
8	(76 K)	0.8	45	1.6	–	–	–	

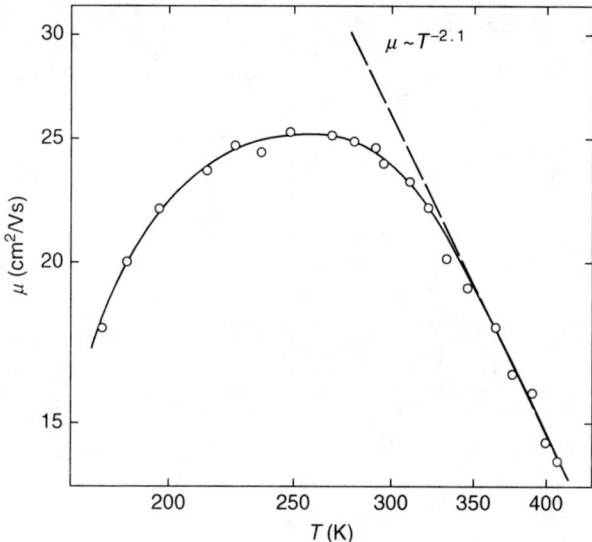

Figure 6.24 Temperature dependent Hall mobility of CuGaSe$_2$ single crystals.

chosen rather than low value of 0.4 or higher, *etc*. The polar optical scattering mechanism depends on the chosen value of hole effective mass. If it is low, the role of polar optical scattering mechanism dominates high. The authors found that the theoretically obtained mobilities are low for optical phonon scattering and high for neutral defect scattering as compared to that of experimentally observed mobilities in the CGS compound. On the other hand, the ionic bonding character is low in the CGS therefore the polar optical scattering mechanism is not playing a role within the experimental temperature ranges therefore it may be excluded but thorough investigation is still needed. The experimental and theoretical mobilities fit for various scattering mechanisms are shown in Figure 6.25. The hopping conduction is clearly seen at below 200 K that means a decrease in mobility follows sharp linearity in the polycrystalline layers. The grain boundary scattering is also dominant parameter in the polycrystalline CGS samples [58]. The similar hopping conduction is observed in the *n*-CIS polycrystalline samples at below 50 K [56]. Over all a sharp variation in total mobility with response to temperature is less in the polycrystalline samples than in the epitaxial layers that could be due to weak scattering mechanisms.

6.3 Conductivity of CuInGaSe$_2$

The typical CIGS thin films from NREL and Shell Solar Inc., are characterized by Hall measurement before and after non melting laser annealing by 248 nm KrF laser with different energy densities and pulses. Upon annealing, the samples show

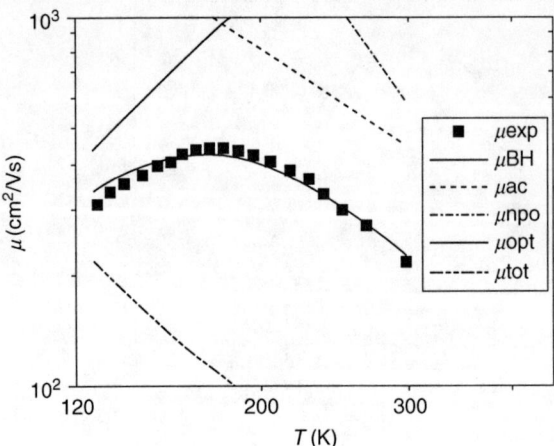

Figure 6.25 Temperature dependent mobilities of CuGaSe$_2$ epitaxial layer.

variation in hole concentration and mobility. Particularly in the CIGS sample-1, mobility increases, as shown in Table 6.7 [71]. The Cu-rich CuInGaSe$_2$ Cu/(In+Ga)=0.97 thin films with carrier concentration of 10^{20} cm^{-3} prepared by sputtering and evaporation technique show resistances of 1000 and 500 Ω at RT and 4.2 K, respectively. A variation of resistance with temperature shows metallic behavior. After etching the samples in KCN solution to remove Cu$_{2-x}$Se phase, they show low activation energies of 20 and 40 meV at low and high temperature regions, respectively. The resistance decreases with increasing temperature. A variation of resistance with temperature responses well [72]. The co-electrodeposited CuInGaSe$_2$ thin films, onto Mo coated soda-lime glass substrates exhibit n-type conductivity. Table 6.7 indicates that the carrier concentration increases and simultaneously the mobility decreases with increasing Cu/(In+Ga) ratio [73].

The Cu and In thin films by e-beam evaporation and Ga by thermal evaporation in the sequence of In/Ga/Cu/In are prepared. Some of the metallic layers under Se vapor and others under combination of selenium and Ar atmosphere are selenized to obtain CuInGaSe$_2$ thin films. The activation energies of 34 and 50 meV are obtained from plots of ln(σ) versus $1/T$ for the CuInGaSe$_2$ thin films grown by former and latter processes, respectively as shown in Figure 6.26 [74,75]. As shown in Figure 6.27, the resistivity decreases with increasing x in the as-grown CIGS thin films from 0 to 1 range. In comparison the films annealed under Se shows that the resistivity is lower for lower x values to some extent then it follows the pattern of as-grown samples [76]. Figure 6.28 shows that the activation energies of 65–100 and 275–315 meV at low and high temperature regimes, respectively, are obtained from the temperature dependent conductivity [77]. Note that they could be due to V$_{Se}$ and Fe^{2+} on In site, respectively, as already assigned. The resistivity (ρ) of CIGS layers on SLG increases from 10^1 to 5×10^5 Ω-cm with decreasing deposition substrate temperature in the range 550–450 °C, whereas no remarkable variation in ρ is observed with variation of deposition substrate temperature below 450 °C indicating that the Na diffusion takes

Table 6.7 Electrical properties of CIGS thin films

S.No	Cu:In:Ga:Se	ρ (Ω-cm)		n_e or p (10^{16} cm^{-3})		μ (cm^2/Vs)		NLA annealing (mJ/cm^2, pulses)	Type	Ref.
		Before	After	Before	After	Before	After			
1	–	133	37.3	0.53	0.445	8.89	37.6	20,10	p	[71]
2	–	235	86.4	2.9	2.43	0.93	2.98	20,20	p	
3	–	94	2.67	4.3	1.8	1.54	6.1	40,10	p	
4	–	148	4.64	7.1	3.3	0.6	2.8	40,20	p	
5	13.65:23.85:7.7:54.8	0.020	–	80.26	–	386	–	–	n	[73]
6	13.27:17.19:9.69:59.25	0.019	–	100.7	–	323	–	–	n	
7	21.92:18.2:7.76:52.12	0.017	–	115.8	–	311	–	–	n	
8	27.21:20.72:9.77:42.3	0.015	–	148.7	–	271	–	–	n	

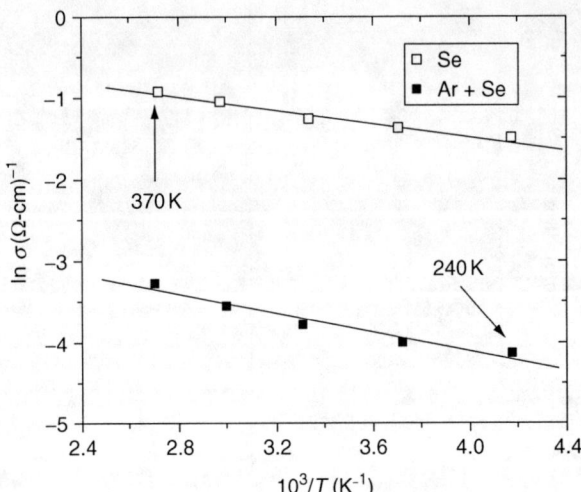

Figure 6.26 ln σ versus $1/T$ of CIGS thin films grown by two different processes.

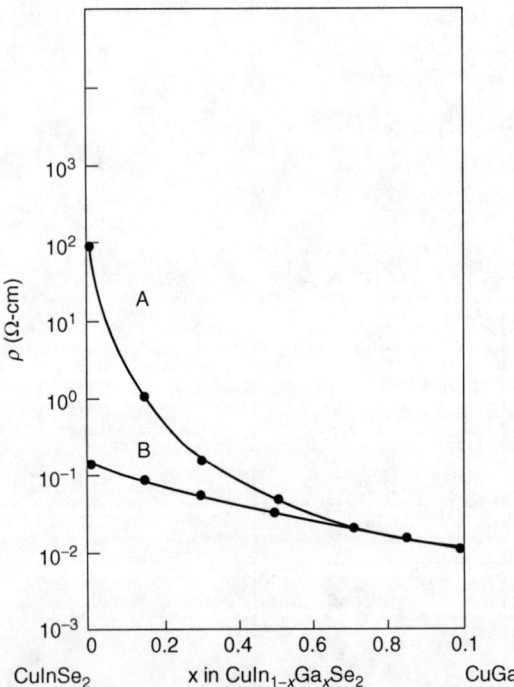

Figure 6.27 Variation of resistivity with composition in the CIGS thin films; (A) as-grown and (B) annealed.

place only at above the deposition temperature of 450 °C. The resistivity of 10^5 Ω-cm for the CIGS layers on Na free glass substrates is all the time high for the high deposition substrate temperatures. The diffusion of Na into CIGS enhances carrier concentration that causes to decrease the resistivity of the samples for high substrate temperature [78].

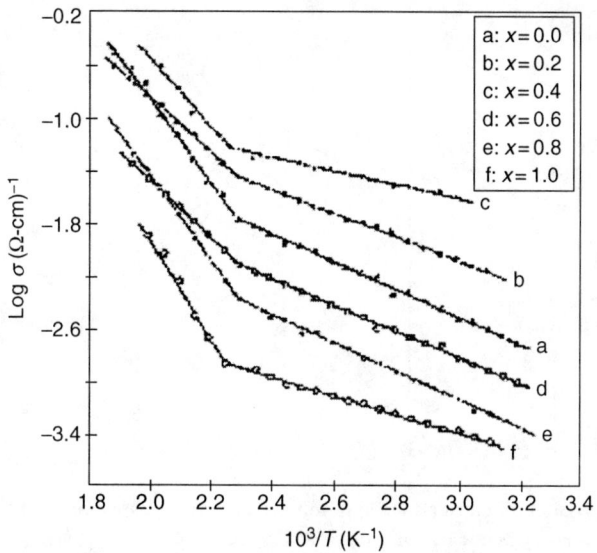

Figure 6.28 Temperature dependent conductivity versus inverse temperature for the $CuIn_{1-x}Ga_xSe_2$ thin films grown by spray pyrolysis.

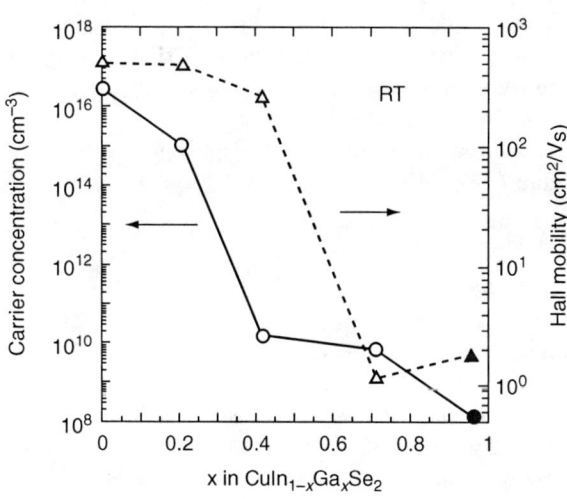

Figure 6.29 Variation of carrier concentration and Hall mobility for the $CuIn_{1-x}Ga_xSe_2$ crystals with function of x.

6.3.1 Mobility of CuInGaSe$_2$

As shown in Figure 6.29, the carrier concentration and Hall mobility decrease with increasing x in THM $CuIn_{1-x}Ga_xSe_2$ crystals and the conductivity type changes from n- to p-type indicating that the shallow donors are compensated by the acceptors. All the n-type crystals convert to p-type after annealing under Se maximum pressure at 600 °C for 24 h and their hole concentration is more than

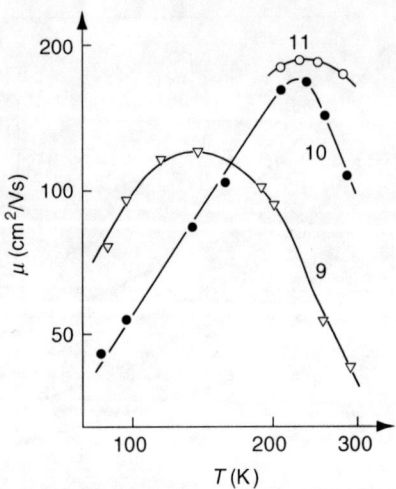

Figure 6.30 Variation of carrier concentration versus inverse temperature for the CuInGaSe$_2$, and (B) temperature dependent mobility.

10^{17} cm^{-3} [79]. The CuIn$_{0.7}$Ga$_{0.3}$Se$_2$ thin films prepared by flash evaporation onto GaAs (111) substrates at different substrate temperatures of 552, 587, and 597 °C show $p = 8.2 \times 10^{18}$ and $\mu = 42$ for sample 9, $p = 2.4 \times 10^{19}$ and $\mu = 170$ for sample 11, and $p = 1.2 \times 10^{19}$ cm^{-3} and $\mu = 103$ cm^2/Vs for sample 10, respectively. The samples show ionized impurity scattering $\mu \alpha T^{1.4-1.6}$ at low temperature region and the nonpolar optical phonon scattering in the first order of the phonon wave vector $\mu \alpha T^{-2.4-2.55}$ at high temperature region, irrespective of deposition temperatures but mobility factor differs (Figure 6.30) [80].

6.4 Conductivity of CuInS$_2$

In-rich CuInS$_2$ thin films exhibit n-type, whereas S-rich films show p-type conductivity indicating that the sample is p-type for $\Delta S > 0$ while sample is n-type for $\Delta S < 0$ [59]. The type of conductivity changes from n to p with increasing Cu to In ratio from 0.6 to 1.5 in the CuInS$_2$ single crystals grown by hot-press method under high pressure either in the Cu-rich or In-rich conditions at 700 °C for 1 h, which contain a small quantity of CuIn$_5$S$_8$ secondary phase. A change in carrier concentration and n to p type occur with increasing Cu to In ratio from 0.6 to 1.5, as shown in Figure 6.31. The authors suggested that the defect levels of V$_S$ and In$_i$ for n-type and V$_{In}$ and Cu$_{In}$ for p-type conductivity are responsible [81]. As shown in Figure 6.32, the resistivity of CuInS$_2$ thin films decreases from 10^6 to 10^{-2} Ω-cm with increasing Cu content in the layers. The Cu percentage is accounted in the chemical solution that may not be same in the layers [82]. The resistivity of CuInS$_2$ thin films deposited by spray pyrolysis at 360 °C and pH 3.5 also decreases from 10^3 to ~ 0.3 Ω-cm with increasing Cu/In

Electrical Properties of I–III–VI$_2$ Compounds

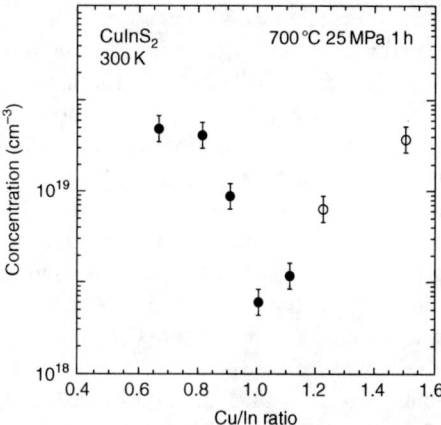

Figure 6.31 Variation of carrier concentration with Cu/In ratio for the CuInS$_2$ (solid dots, n-type; open dots, p-type).

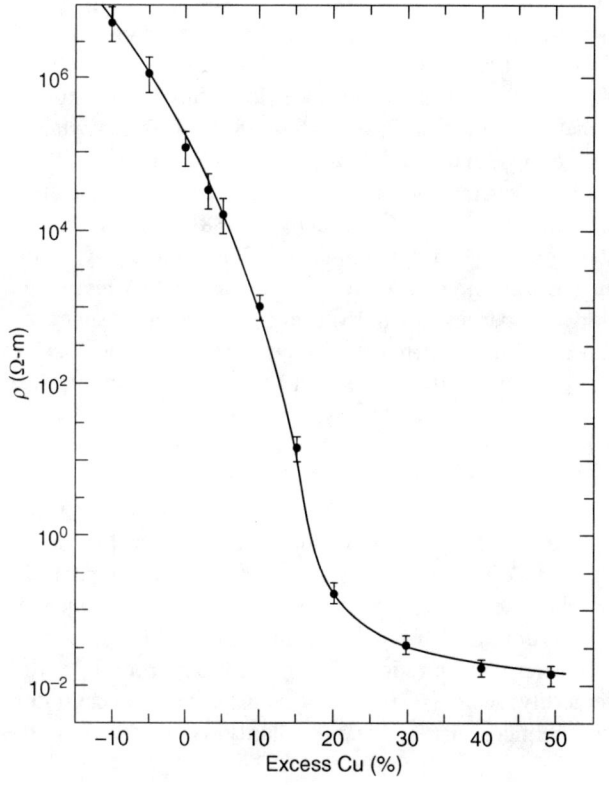

Figure 6.32 Variation of resistivity with Cu/In ratio for the sprayed CuInS$_2$ thin films.

ratio from 0.6 to 1.3, which is higher by 3 orders of magnitude for the Cu/In ratio of 0.66–0.87 than the one with Cu/In ratio of 0.87–1.29. The resistivity is in the range of 0.2–5 Ω-cm for the ratio of Cu/In = 0.87–1.29. The lower resistivity of the films is due to an increase of acceptor (V_{Cu} or Cu_{In}) concentration and decrease of In_{Cu} donor concentration with increasing Cu/In ratio in the range of 0.87–1.29. On the other hand Cu_xS phase is also observed in the Cu-rich $CuInS_2$ that causes to reduce resistivity in the films. In the Cu/In composition 0.66–0.87 range the In_2S_3 secondary phase and intrinsic defect levels may be dominant to have higher resistivity [83]. After electrochemically etching the $CuInS_2$ thin films in KCl solution, which are prepared by sulfurization of CuIn under H_2S atmosphere at 500 °C for 30 min, the Cu/In and S/(Cu + In) composition ratios change from 1.05 to 0.97 and from 0.97 to 0.79, respectively. Ultimately the resistivity of the films increases from 3.8 to 1.77×10^1 Ω-cm. In some cases, after etching Cu-rich $CuInS_2$ thin films, the Cu/In and S/(Cu + In) ratios change from 1.96 to 0.8 and from 1.7 to 0.78, respectively but the resistivities are more or less at the same level of 4.7 and 7.1 Ω-cm in the as-grown and etched samples, respectively [84]. The typical as-grown $CuInS_2$ thin films by three source evaporation show resistivities of 0.01–10 Ω-cm, electron and hole mobilities of 21.84 and 1.42 cm^2/Vs, respectively [85]. The resistivity of p-$CuInS_2$ thin films grown by spray pyrolysis technique is given in Table 6.8 [86].

The p-$CuInS_2$ thin films prepared by successive ionic layer absorption and reaction method with different Cu/In ratios are given in Table 6.8. As the Cu/In ratio is increased, the resistivity of p-$CuInS_2$ thin films decreases. It could be due to formation of Cu_{In}, V_{In}, and semi-metallic nature of $Cu_{2-x}S$ phase [87]. The resistivities of $CuInS_2$ films are 0.14–0.57, 26–500, and >4500 Ω-cm for Cu/In = 1.38, 1.18, and 0.94, respectively. The mobility, carrier concentration and band gap of the films are 0.1–1.0 cm^2/Vs, 10^{16} cm^{-3}, and 1.55 eV, respectively [88]. Surprisingly the typical $CuInS_2$ single crystals grown by Bridgman method show n- and p-type conductivities for Cu/In > 1 and Cu/In < 1, respectively, as opposing to general rule (Table 6.8) [89]. The electrical properties of $CuInS_2$ thin films grown by sputtering are given in Table 6.8. With aging in air from 0 to 50 days the resistivity of layers gradually decreases from 2×10^3 to 2×10^{-1} Ω-cm and type of conductivity also changes from n to p-type. The surface of $CuInS_2$ covered with Cu_xS phase is oxidized during exposure to air that forms ultra thin In_2O_3 layer. Even though n-In_2O_3 layer is formed, still existing islands of Cu_xS phase actively participate to contribute p-type conduction. In order to prepare $CuInS_2$ thin films, the Cu–In layers are sputtered using power of 200 W and sulfurized under H_2S gas with flow rate of 25 sccm at 400 °C [90]. The p-$CuInS_2$ thin films with Cu/In = 1.38 prepared by sputtering and sulfurization show resistivity of 6.4×10^{-2} Ω-cm. After annealing it remains more or less at the same level but the mobility slightly decreases from 0.3 to 0.17 cm^2/Vs and the carrier concentration slightly increases from 3.2×10^{20} to 4.0×10^{20} cm^{-3}. The mobility decreases with increasing carrier concentration is a general pattern in the semiconductors. The similarity is observed in the p-$CuInS_2$ films with Cu/In = 1.52 [91].

The $CuInS_2$ thin films grown by spray pyrolysis technique with different compositions show activation energies (ΔE or E_a) in the range between 0.11 and 0.5 eV.

Table 6.8 Effect of composition on the electrical parameters of $CuInS_2$

S.No	Process	S/(In+Cu)	Cu/In	Type	ΔS	p or n_e (cm^{-3})	ρ (Ω-cm)	μ (cm^2/Vs)	Ref.
1	–	1.0	0.67	n	−0.09	2.64×10^{18}	–	–	[59]
2	–	0.92	0.96	n	−0.091	1.99×10^{16}	–	–	
3	–	0.90	0.99	n	−0.104	3.37×10^{14}	–	–	
4	–	–	24:27:48	–	–	–	18	–	[86]
5	–	–	26:29:45	–	–	–	0.7	–	
6	–	–	27:29:44	–	–	–	0.2	–	
7	–	–	24:29:47	–	–	–	40	–	
8	–	–	27:30:43	–	–	–	17	–	
9	–	–	27:29:44	–	–	–	18	–	
10	–	–	29:28:43	–	–	–	7.3	–	
11	–	–	1	p	0.01	1.2×10^{16}	80.5	4.92	[87]
12	–	–	1.25	p	0.02	5.24×10^{16}	18.9	6.44	
13	–	–	1.5	p	0.04	1.45×10^{16}	1.13	3.41	
14	–	–	1.75	p	0.14	2.4×10^{16}	5.9×10^{-2}	10.3	
15	–	–	2	p	0.36	8.6×10^{18}	3.4×10^{-3}	50.3	
16	Fresh	–	–	n	–	3×10^{14}–10^{15}	2×10^3–10^4	2–20	[90]
17	Aged in air	–	–	p	–	10^{19}–10^{20}	0.1–2	0.2–0.7	
18	KCN etched	–	–	n	–	4×10^{13}–3×10^{15}	6×10^2–2.5×10^4	2–100	
19	Single crystal	–	25.9:24.6:49.5	n	–	–	10^6	–	[89]
20	Single crystal	–	19.2:27.3:53.5	p	–	–	1–10	–	
21	n–$CuInS_2$	–	–	–	–	3.82×10^{16}	–	–	[92]
22	n–$CuInS_2$	–	–	–	–	3.43×10^{16}	–	–	

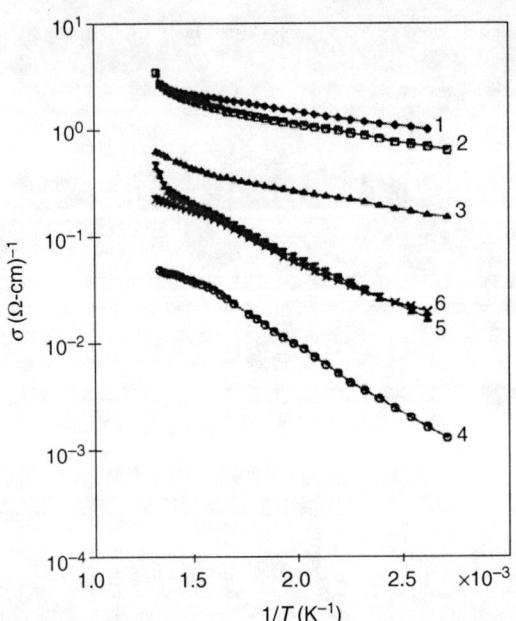

Figure 6.33 Plots of conductivity versus $1/T$ for the CuInS$_2$ thin films.

Table 6.9 Compositional dependent resistivities and activation energies of CuInS$_2$ thin films

S.No	Cu:In:S	ρ (Ω-cm)	ΔE (eV)	Ref.
1	1:1:2	–	0.11	[93]
2	1:1:2.2	–	0.14	
3	1:1:2.4	–	0.14	
4	1:1:2.8	–	0.50	
5	1:1:3	–	0.37	
6	1:1:3.2	–	0.31	
7	In/(In+Cu)=0.51	2×10^5	0.092	[94]
8	0.51	1.1×10^3	0.075	
9	0.498	1×10^2	0.025	

As shown in Figure 6.33, the plots of conductivity versus $1/T$ show that the activation energy increases from 0.11 to 0.50 eV with increasing sulfur to metal ratio from 2 to 3.2, as depicted in Table 6.9 [93]. Two activation energy regimes are observed in the n-type CuInS$_2$ thin films for three different composition ratios of Cu/In = 0.67 (●), 0.96 (○) and 0.99 (■) (Figure 6.34) [59].

Employing experimental $\mu_m = 170$ cm^2/Vs, $n_m = 5.3 \times 10^{12}$ cm^{-3} at $T_m = 295$ K and $\varepsilon_s = 11$ and $m_n = 0.16 m_o$, the ionized impurity concentration (N_I) is found to be 1.05×10^{18} cm^{-3}, $\mu_m = \mu$, $n_m = n_e$. Based on the relation $N_I = 2N_A + n_e$, the

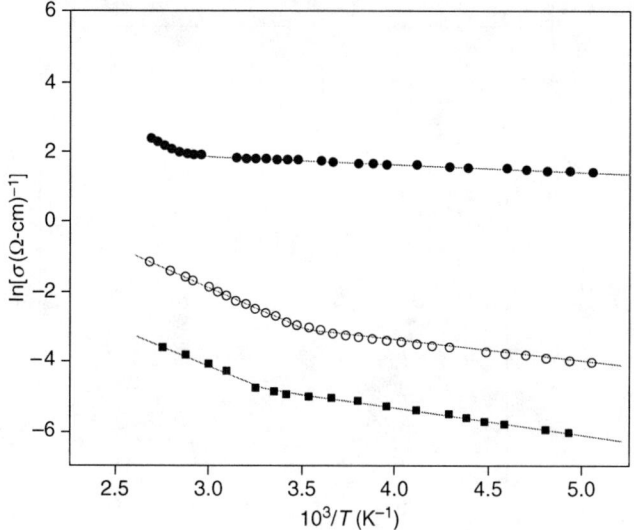

Figure 6.34 Temperature dependent conductivity of n-CuInS$_2$ thin films.

Table 6.10 Electrical parameters of n-CuInS$_2$

S.No	μ (cm²/Vs)	n_e (cm^{-3})	N_I (cm^{-3})	N_A (cm^{-3})	T_m (K)	Ref.
1	170	5.3×10^{12}	1.05×10^{18}	5.27×10^{17}	295	[95]
2	95	5×10^{15}	4×10^{18}	2×10^{18}	310	[98]
3	–	–	6×10^{18}	3×10^{18}	–	[99]

acceptor concentration (N_A) is calculated to be 5.27×10^{17} cm^{-3}. Two donors–one acceptor model is applied for partly compensated semiconductor using the relation

$$\frac{ND_1}{1 + g_1 \frac{N_c}{n_e} \exp(-ED_1/k_B T)} + \frac{ND_2}{1 + g_2 \frac{N_c}{n_e} \exp(-ED_2/k_B T)} = n_e + N_A, \quad \text{where } N_c = (2\pi m_o k_B T)^{3/2}. \tag{6.11}$$

Using $g_1 = g_2 = 2$ and $N_A = 5.27 \times 10^{17}$, the best fitted ED$_1$ and ED$_2$ are estimated to be 24 and 350 meV [95]. The typical defect concentrations of CuInS$_2$ are given in Table 6.10. The temperature dependent conductivity and carrier concentration of p-CuInS$_2$ thin films are shown in Figure 6.35A and B. The acceptor, donor, and ionized impurity concentrations ($N_I = N_A + N_D$) are estimated to be 8.2×10^{16}, 3.8×10^{16} and 1.2×10^{17} cm^{-3} from the electroneutral Equation (6.2) using $g = 2$, respectively [54]. An activation energy (E_A) of p-CuInS$_2$ single crystals is calculated to be 0.15 eV from

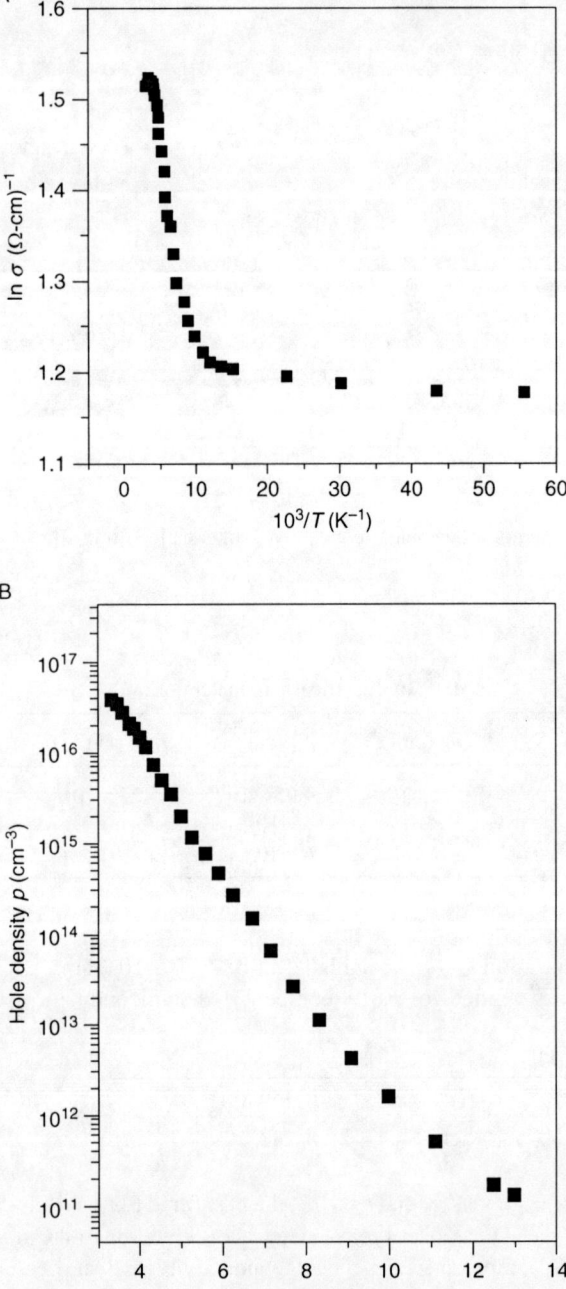

Figure 6.35 (A) Temperature dependent conductivity of $CuInS_2$ and (B) Temperature dependent hole concentration of $CuInS_2$ thin film.

Electrical Properties of I–III–VI$_2$ Compounds 357

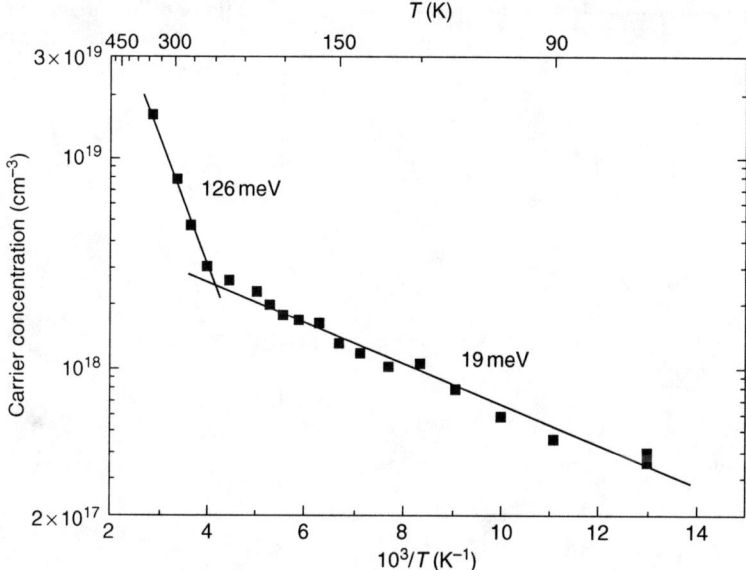

Figure 6.36 Variation of carrier concentration versus inverse temperature for the CuInS$_2$ thin films.

a plot of log $pT^{-3/2}$ versus $1/T$ by iterating $N_D = 2.2 \times 10^{18}$ and $N_A = 2.8 \times 10^{18}$ cm^{-3} ($N_A/N_D = 1.28$) for the best fit [96]. The Cu-rich CuInS$_2$ thin films grown by sulfurization of Cu–In sputtered layers, which contain combination of Cu$_2$S and CuS secondary phases, show carrier concentration of 3.55×10^{17}–1.58×10^{19} cm^{-3}, mobility of 4.15–1.15 cm^2/Vs and resistivity of 4.24–0.35 Ω-cm in the temperature range of 77–350 K. The temperature dependent carrier concentration shows activation energies of 19 and 126 meV at low and high temperature regions for the Cu-rich CuInS$_2$ thin films, respectively (Figure 6.36). The CuS is a metallic compound, which had carrier concentration of 10^{22} cm^{-3} while Cu$_2$S is a semiconductor [97]. It is hard to assign these activation energies to the particular defect levels since secondary phases exist in the samples. They may be interfered with the results. The activation energies approximately in the range of 20–50 and 125–150 meV could be due to V$_{Cu}$ and V$_S$, respectively (see Chapter 5). A linear plot of $\ln(\sigma T^{1/2})$ versus $1000/T$ for CuInS$_2$ thin films is indication of grain boundary participation in the layers (Figure 6.37) [54].

6.4.1 Mobility of CuInS$_2$

The temperature dependent mobility of p-CuInS$_2$ single crystals grown by iodine vapor transport and melt grown with variation of temperature are methods shown in Figure 6.38 and 6.39 [54]. The c-axis is perpendicular to the grown lengthy bar sample. The total mobility of the sample is combination of impurity (μ_I), acoustical, nonpolar ($\mu_{AC,NPO}$), and polar scatterings (μ_{PO}) [99]. Both the In-rich and Cu-rich CuInS$_2$ single crystals grown by CVT method show that the impurity scattering is the dominant scattering, as shown in Figure 6.40 [96]. Variation of mobility with temperature

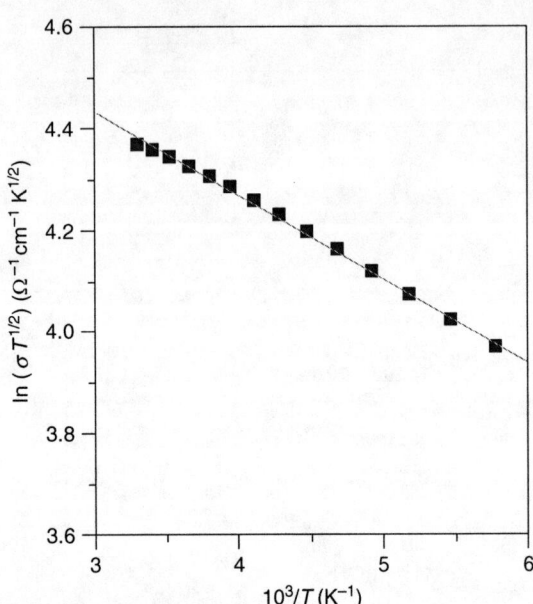

Figure 6.37 Variation of $\ln(\sigma T^{1/2})$ with $10^3/T$ for the CuInS$_2$ thin films.

for the p-CuInS$_2$ thin films containing Cu/In = 0.97 and S/(Cu + In) = 1.15 is given in Figure 6.41. The total mobility of the sample is combination of impurity (μ_I), acoustical, nonpolar ($\mu_{AC,NPO}$), and polar scatterings (μ_{PO}). In the ln(T) range of 5.3–5.7 (high temperature region), the nonpolar and polar scatterings are dominants, whereas at low temperature region, that is, < ln(T) = 5.3, the impurity scattering is dominant one. For this analysis the CuInS$_2$ thin films are grown onto soda-lime glass substrates by sulfurization of Cu–In alloy deposited by thermal evaporation in which solid sulfur is heated for sulfurization whilst keeping Cu–In alloy in graphite box that can be fitted into quartz tube furnace [59]. The n-CuInS$_2$ single crystals with excess of 1% In are grown by Bridgman method using 5N purity Cu, In, and Se. On the polished and cleaned crystals, the In layer is evaporated for contacts. The Hall mobility measurements are carried out in the temperature range of 90–300 K. $\mu \alpha T^{1.5}$ for $T < 250$ K and $\mu \alpha T^{-1.5}$ for $T > 320$ K are observed, whereas at $T_m = 295$ K, the peak or maximum mobility is contribution of both mechanisms [95].

6.5 Conductivity of CuIn(Se$_{1-x}$S$_x$)$_2$

As shown in Figure 6.42, the resistivity of 10^3 Ω-cm for CuIn(Se$_{1-x}$S$_x$)$_2$ is constant with increasing Cu/In sputtering power ratio from 0.3 to 0.7. In this range, the CuIn$_5$(SSe)$_8$ and In-rich CuIn(SSe)$_2$ thin films are formed. As expected, with increasing Cu content, the resistivity decreases from 10^3 to 10^{-1} Ω-cm. Keeping In sputtering power of 60 W at constant, the Cu sputtering power is increased from 15 to 55 in steps of 5 W to form Cu–In alloy. The grown Cu–In alloy by sputtering

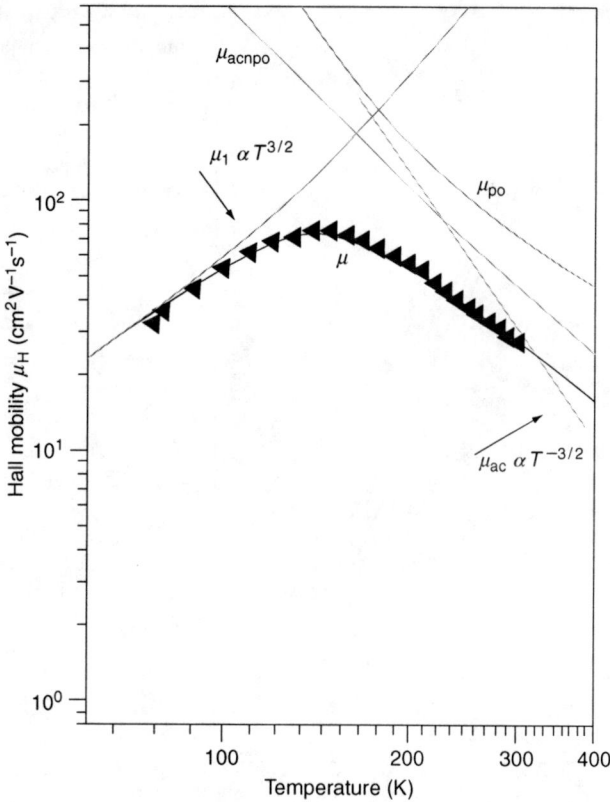

Figure 6.38 Temperature dependent Hall mobility of p-CuInS$_2$ single crystals.

is sulfurized and selenized by thermal process in the furnace at 250 °C, followed by processing at 500 °C to form CuIn(SSe)$_2$ thin films [100]. The conductivity, hole concentration and Hall mobility of CuInSe$_2$ (two samples) and CuIn(Se$_{0.7}$S$_{0.3}$)$_2$ single crystals (two samples) are shown in Figures 6.43 and 6.44. Two pieces of crystals are selected from each crystal bowl: one is from middle and another one from end of the crystal bowl for electrical measurements. The theoretically estimated hole concentrations using two acceptors one donor model are also given along with experimental values in the graphs [101]. The electrical resistivities of CuIn(Se$_{1-x}$S$_x$)$_2$ thin films prepared by spray pyrolysis technique measured in the temperature range of 170–420 K are shown in Figure 6.45A and B, which incur activation energies of 10–25 and 13–140 meV at low and high temperature regions, as depicted in Table 6.11 [102]. The CuIn(Se$_{1-x}$S$_x$)$_2$ thin films with composition (Cu:In:S:Se = 36.91:19.77:17.60:25.72) reveal activation energies of 42 and 743 meV at low and high temperature regions, respectively. The films show conductivity of 9.39×10^{-16} (Ω-cm)$^{-1}$ and thermoelectric power of 4.91 μV/K at 384 K. The samples show n-type conductivity at RT but exhibits positive thermoelectric power above 325 K, indicating a change in conductivity from n- to p-type [103].

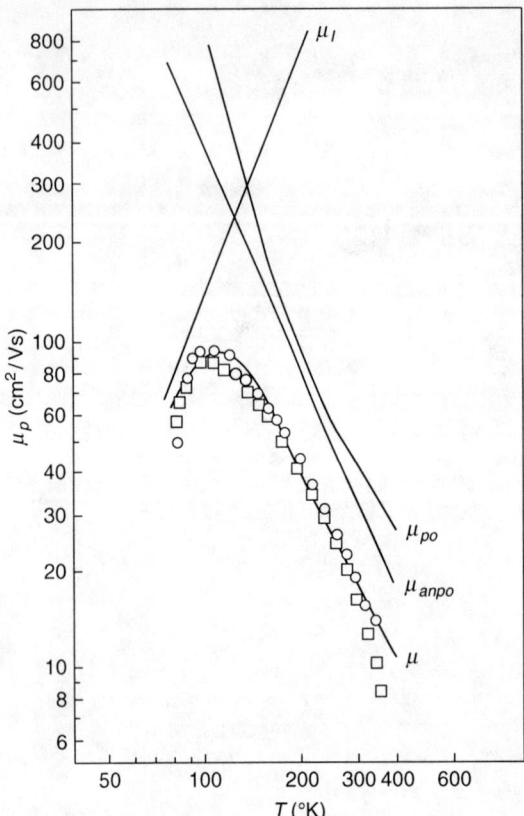

Figure 6.39 Variation of mobility with temperature for p-CuInS$_2$ single crystals grown by melt grown method in which the dominant mechanisms are μ_I, μ_{PO} and μ_{AC}, μ_{NPO} (circle bigger sample squares small samples cut from bigger sample).

In conclusion, sometimes the k factor in the scattering mechanisms of mobilities $\mu \alpha T^k$ does not exactly depend on either impurity ($k = 3/2$) or lattice scattering ($k = -3/2$) but its value deviates to lower or higher due to variation of composition or spherical wave functions, *etc.*, in the samples. The temperature dependent mobility from sample to sample responses early or late for impurity scattering and falls off early or late for lattice scattering in the measured temperature range due to variation in carrier concentration, composition, *etc.* The resistivity and activation energy of the samples eventually depend on composition of the samples, growth parameters, post annealing treatments, *etc.*

6.6 Hopping Conduction

When a transition takes place from impurity band to either conduction or valence band is so-called hopping conduction for which the Mott parameters quantify disorderness of the semiconductors. The derivation for Mott variable range hopping conduction can be written as

Electrical Properties of I–III–VI$_2$ Compounds

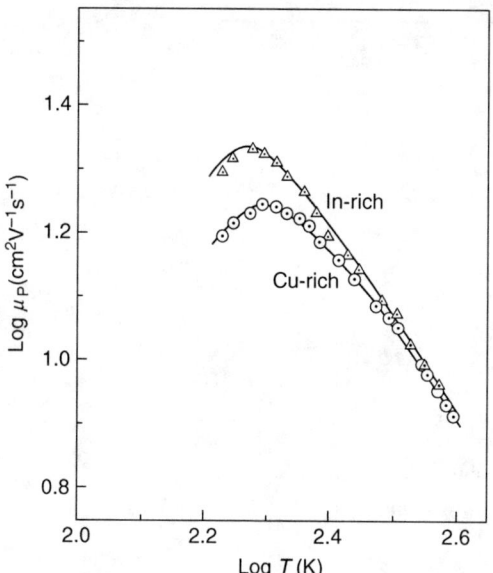

Figure 6.40 Temperature dependent mobilities of In-rich and Cu-rich CuInS$_2$ single crystals.

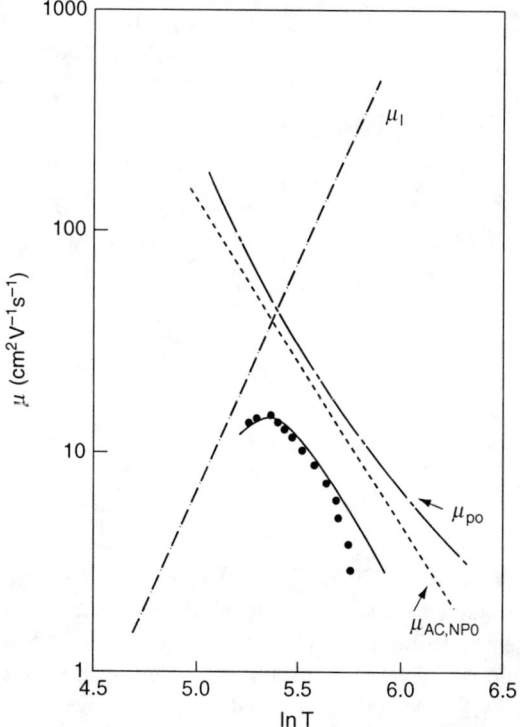

Figure 6.41 Variation of mobility with temperature for the CuInS$_2$ thin films grown by evaporation in which the dominant mechanisms are μ_I, μ_{PO}, and $\mu_{AC,NPO}$.

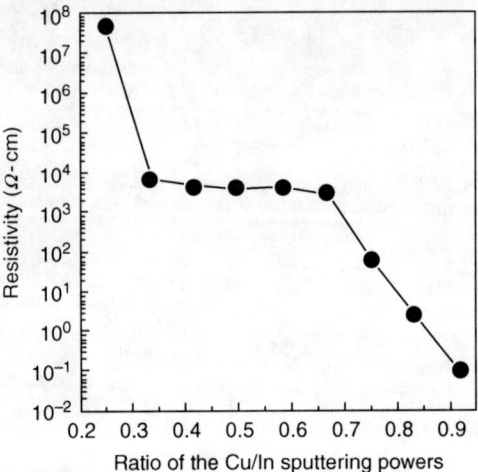

Figure 6.42 Resistivity versus ratio of Cu/In sputtering powers for CuIn(SSe)$_2$ layers.

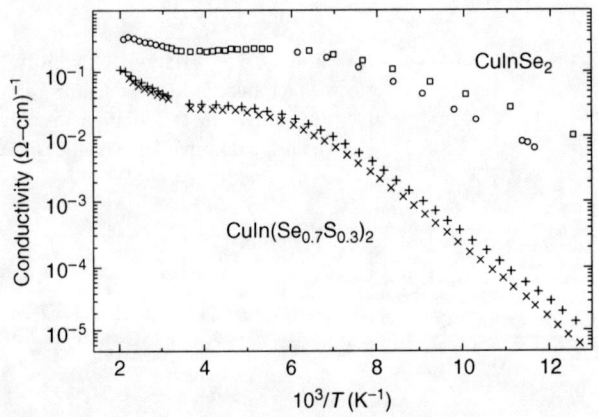

Figure 6.43 Temperature dependent conductivity of CuInSe$_2$ and CuIn(Se$_{0.7}$S$_{0.3}$)$_2$ single crystals.

$$\sigma = \frac{\sigma_0}{T^{1/2}} \exp\left[-\left(\frac{T_0}{T}\right)^{1/4}\right], \qquad (6.12)$$

the degree of disorder (T_0) can be related as $T_0 = \lambda_C \alpha_d^3 / k_B N(E_F)$, where $N(E_F)$, λ_C, α_d, and k_B are density of localized states near the Fermi level, dimensionless constant, the decay constant of the wave function of the localized states near the Fermi level, and Boltzmann constant, respectively. The pre-exponential factor can be considered as $\sigma_0 = 3e^2 v_{ph}[N(E_F)/8\pi\alpha_d k_B]^{1/2}$, where Debye frequency $v_{ph} \sim 3.3 \times 10^{12}$ Hz, The hopping range distance R and energy W can be written as [104].

Electrical Properties of I–III–VI$_2$ Compounds

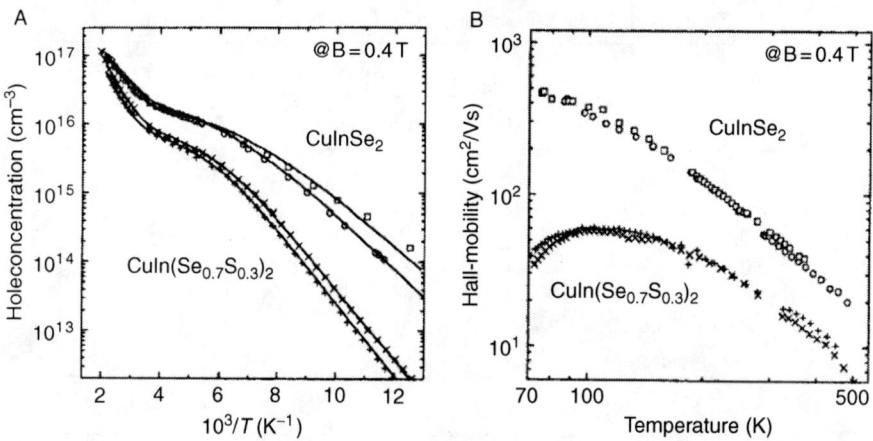

Figure 6.44 Temperature dependent hole concentration and mobility of CuInSe$_2$ and CuIn(Se$_{0.7}$S$_{0.3}$)$_2$ single crystals.

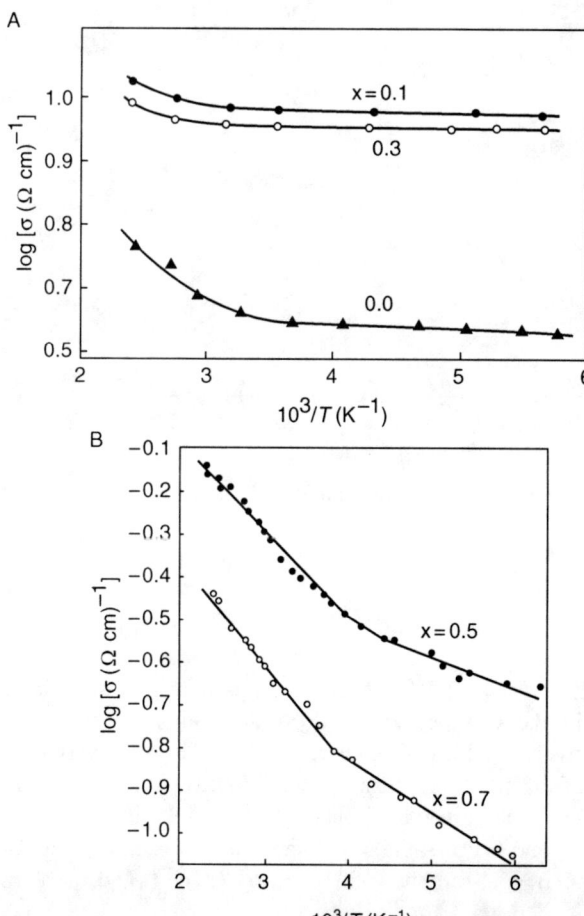

Figure 6.45 (A, B) Electrical conductivity of CuIn(Se$_{1-x}$S$_x$)$_2$ thin films.

Table 6.11 Electrical parameters of $CuIn(Se_{1-x}S_x)_2$ thin films

S.No	x	ρ (Ω-cm) at RT	Activation energies (eV)		Ref.
			Low temp.	High temp.	
1	1.0	150	–	140	[102]
2	0.7	4.7	25	50	
3	0.5	2.0	10	44	
4	0.3	0.1	–	16	
5	0.1	0.1	–	13	
6	0.0	0.2	–	33	
7	0.0	–	12	135	[16]

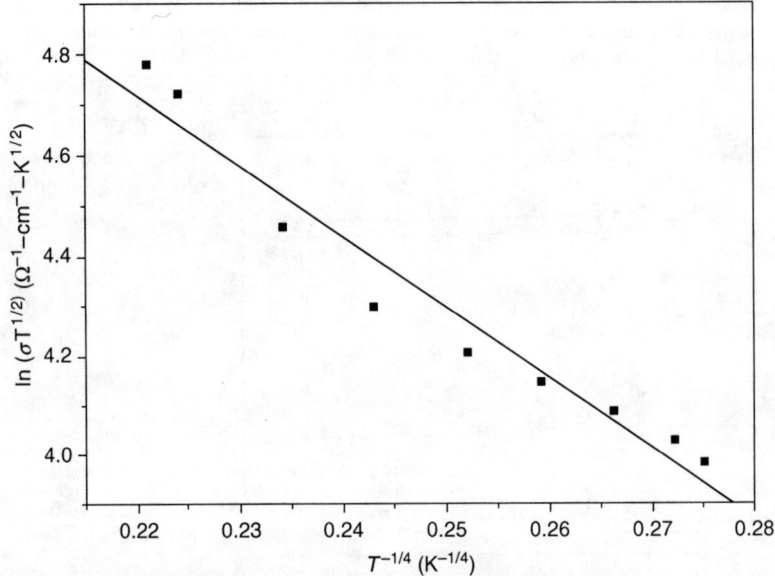

Figure 6.46 Variation of $\ln(\sigma T^{1/2})$ with $T^{-1/4}$ for the $CuInSe_2$ thin films.

$$R = \left[\frac{9}{8\pi\alpha_d k_B T N(E_F)}\right]^{1/4} \quad and \quad W = \frac{3}{4\pi R^3 N(E_F)}. \tag{6.13}$$

A linear variation of $\ln(\sigma T^{1/2})$ versus $T^{-1/4}$ plot indicates participation of variable range hopping conduction in the $CuInSe_2$ and $CuInS_2$ thin films, as shown in Figures 6.46 and 6.47, respectively [54,105]. Keeping $\lambda_C = 18$, all the parameters T_o/T, $\alpha_d R$, W, and $N(E_F)$ estimated from the experimental results (Fig. 6.48) for sample n-$CuInS_2$ (Cu:In:Se = 27.1:24.6:48.3) are given in Table 6.12. The similar plots of CIGS layers with Cu/(In + Ga) = 1.24 and Ga/(In + Ga) = 0.33 and another sample with Cu/(In + Ga) = 0.99 and Ga/(In + Ga) = 0.40 are shown in Figure 6.49. The T_0 and σ_0 values

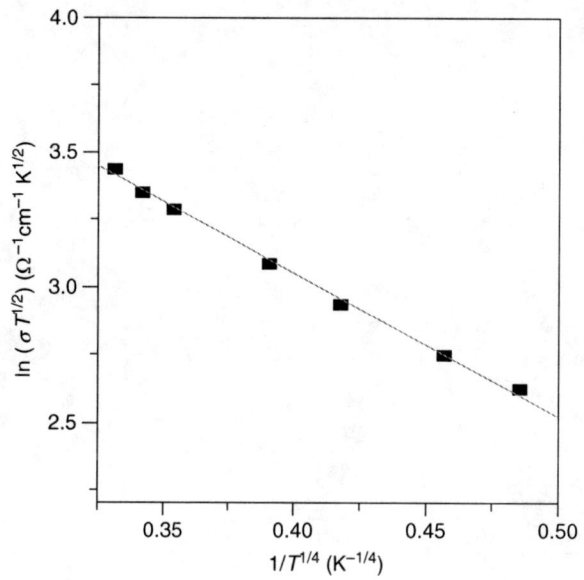

Figure 6.47 Arrhenius plot of $\ln(\sigma T^{1/2})$ versus $T^{-1/4}$ for the $CuInS_2$ thin films.

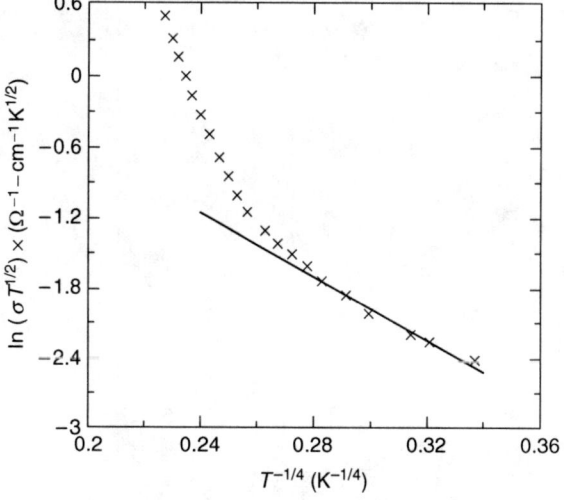

Figure 6.48 Arrhenius plot of $\ln(\sigma T^{1/2})$ versus $T^{-1/4}$ for the n-$CuInSe_2$ thin films.

extracted from a plot of $\ln[\sigma T^{1/2}]$ versus $T^{-1/4}$ are given in Table 6.12. The disorder (T_0) is less in the CIGS layers prepared by selenization under combination of Se and Ar than the one prepared under Se atmosphere. The former with composition ratio of Cu:In:Ga:Se = 27.4:17.7:10.2:44.8 and the latter with Cu:In:Ga:Se = 24.4:15.7:9:51 had activation energies of 34 and 50 meV and the band gaps of 1.1 and 1.19 eV, respectively [74,75]. The mobility due to hoping carriers can be expressed as

Table 6.12 Hopping conduction parameters of CuInSe$_2$, CuInS$_2$, CuGaSe$_2$, and CuInGaSe$_2$

S.No	Compound (Cu:In:Se)	v_{ph} (Hz)	$\alpha_d R$	W (eV)	$N(E_F)$ eV^{-1} cm^{-3}	σ_o	T_o	Ref.
1	p-CIS	8.23×10^{12}	576	6.65	9×10^{21}	–	–	[105]
2	1% Cu-excess CIS	–	1.02	5.7×10^{-3}	4.98×10^{17}	–	–	
3	p-CIS	–	4.36	19×10^{-3}	4.1×10^{17}	–	–	[107]
4	n-CIS	–	10	19×10^{-3}	4.0×10^{18}	–	–	
5	p-CIS	3.3×10^{12}	3.3	19.13×10^{-3}	2.8×10^{23}	–	6.6×10^2	[108]
6	p-CIS	–	–	–	6.2×10^{17}	–	–	[16]
7	p-CIS	–	–	–	–	–	12.8×10^6	[22]
8	p-CIS	–	–	–	–	–	22.1×10^6	
9	n-CIS	–	–	–	–	–	11.6×10^6	
10	n-CIS (27.1:24.6:48.3)	–	–	–	–	7.99	3.35×10^4	[40]
11	n-CIS (24.5:30.1:45.5)	–	–	–	–	4.32	1.99×10^4	
12	n-CIS	–	–	–	1.5×10^{21}	–	–	[38]
13	Annealed n-CIS	–	–	–	1×10^{21}	–	–	
14	CIGS processed under Se	–	–	–	–	64.82	20,641	[74,75]
15	CIGS processed under Se + Ar	–	–	–	–	2.39	9605	
16	p-CuInS$_2$	–	–	–	4.1×10^{23}	–	4.5×10^2	[108]
17	p-CuInS$_2$ thin film	3.3×10^{12}	3.56	26.79	7.8×10^{15}	–	$T_o/T = 7.7$	[54]
18	n-CuInS$_2$ single crystal	–	–	–	–	–	163	[92]
19	n-CuInS$_2$ single crystal	–	–	–	–	–	1154	

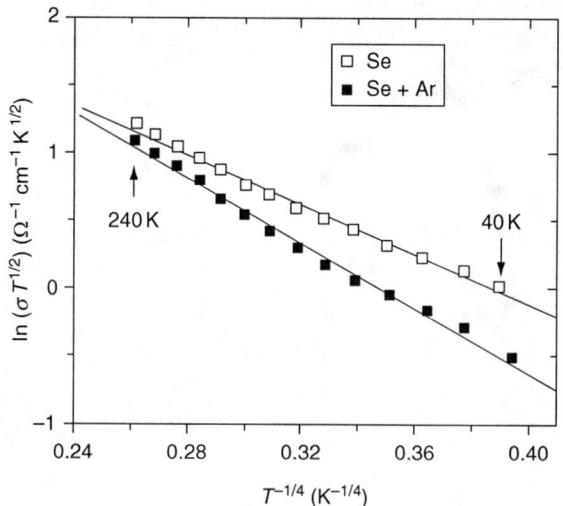

Figure 6.49 Variation of $\ln(\sigma T^{1/2})$ with $T^{-1/4}$ for the CIGS thin films.

$$\mu_{hop} = \mu_{hop}^0 \exp\left(\frac{\varepsilon_{hop}}{k_B T}\right)^{1/4}; \quad \mu_{hop}^0 = C f_{N_A} \exp\left(N_A^{1/3}\right), \quad (6.14)$$

where ε_{hop} is the characteristic hopping conduction energy (1 meV), f_{N_A} is the fraction of unionized shallow acceptors, N_A is the shallow acceptor density, and C is a constant. The parameters $N_A = 2 \times 10^{16}$ cm^{-3}, $p = 1.3 \times 10^{17}$ cm^{-3}, $\mu = 234$ cm^2/Vs, $E_{NPO} = 4$, and $E_{AC} = 2$ are used to fit the temperature dependent experimental mobilities for sample CIGST1 with Cu/(In+Ga) composition ratio of 0.93 and Ga/(In+Ga) ratio of 0.031 while the parameters $N_A = 1 \times 10^{15}$ cm^{-3}, $\mu = 135$ cm^2/Vs, $E_{AC} = 2$, and $E_{NPO} = 4.5$ are used to fit the data for the sample CIGST2 with Cu/(In+Ga) = 0.88 and Ga/(Ga+In) = 0.20 in hopping mobility Equation (6.14). The mobilities drastically decrease at low temperature due to participation of hopping conduction in both CIST1 and CIST2, as shown in Figure 6.50. There is no remarkable change in temperature dependent mobility by adding Ga in the CIS system. The neutral defect impurity and optical polar scatterings in the CIGS samples are disregarded to the total mobility fitting data due to having higher mobilities than experimental mobilities. The E_{AC} and E_{NPO} are considered to be 1.5–2.0 and 4–5 eV for I–III–VI$_2$ compounds, respectively [106].

6.7 Thermoelectric Power

The thermoelectric power devices play a major role in the semiconductor swiching devices. The thermoelectric power (S) expression for nondegenerate semiconductors can be written as

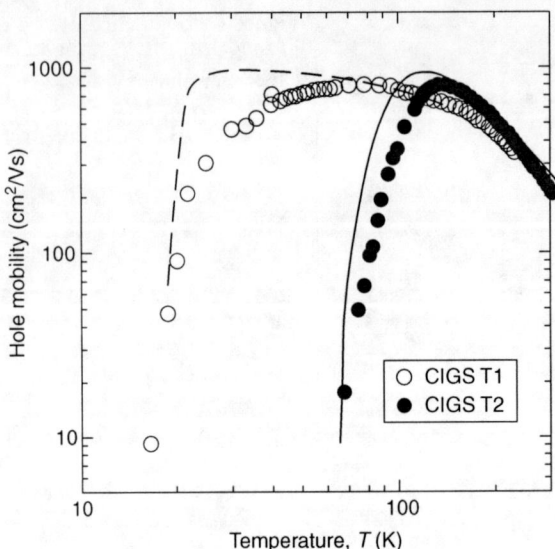

Figure 6.50 Temperature dependent mobility of CIGS samples (lines-theoretical fit, and dots- experimental).

$$S = \left(\frac{k_B}{e}\right)\left[\frac{5}{2} + S' + \ln\left(\frac{N_V}{p}\right)\right];$$
$$N_V = \frac{2(2\pi m_h^* k_B T)^{3/2}}{h^3} \text{ or } 2.5 \times 10^{19}(m_h^*/m)^{3/2}(T/300)^{3/2}, \quad (6.15)$$

where N_v is the density of states and $k_B/e = 86.2$ μV/K. $S' = -1/2$ suits well for the lattice scattering mechanism, which is the dominant one at high temperature region, whereas the ionized impurity scattering is the predominant at low temperature region therefore $S' = 3/2$ can be predicted [109]. Both n- and p-CuInSe$_2$ single crystals exhibit similar thermoelectric power of ∼17–45 μV/K in the temperature range of 80–300 K. Taking the hole concentration (p) of 10^{19} cm^{-3}, the m_h^* is estimated to be $0.09m_o$ at 300 K by considering participation of ionized impurity scattering mechanism, which is lower than the standard $m_h^* = 1.3m_o$. This could be due to higher carrier concentration of the sample [51,110]. A linear behavior is observed in the plots of thermoelectric power (S) versus $1/T$ for the p-CuIn(Se$_{1-x}$S$_x$)$_2$ layers (Figure 6.51A). The hole concentrations of 2×10^{20} and 0.8×10^{20} cm^{-3} for the CuInS$_2$ and CuInSe$_2$ thin films, respectively are determined from the thermoelectric power measurements using Equation (6.15). The hole effective masses of $1.3m_o$ and $0.73m_o$ for CuInS$_2$ and CuInSe$_2$ are taken for this analysis, respectively. The thing is that the obtained higher carrier concentration indicates degeneracy behavior [102]. A variation of thermoelectric power with temperature is depicted in Figure 6.51B, which shows two regimes. At room temperature, the measured thermoelectric powers of CuInSe$_2$ and CuInS$_2$ are 325 and 335 μV/K, respectively [108].

Electrical Properties of I–III–VI$_2$ Compounds

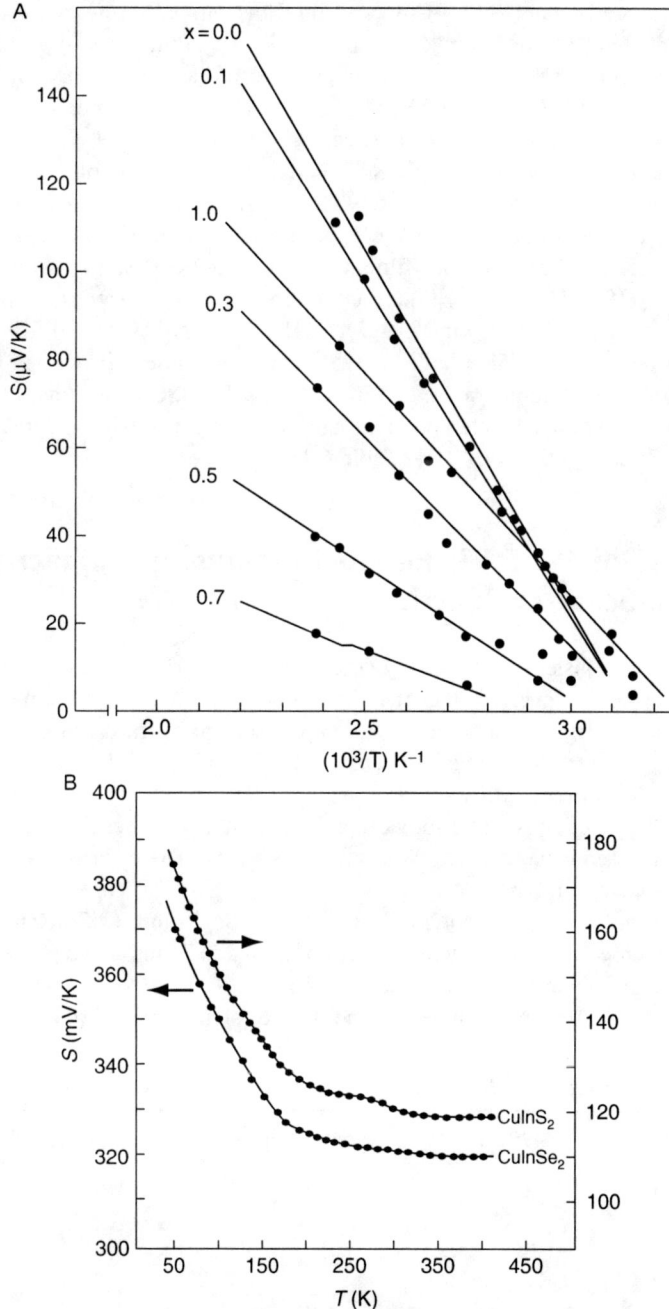

Figure 6.51 (A) Variation in thermoelectronic power of p-CuIn(Se$_{1-x}$S$_x$) thin films with inverse temperature and (B) Temperature dependent thermoelectric power of CuInS$_2$ and CuInSe$_2$ thin films.

The Seebeck coefficient (S) of n-CuInSe$_2$ thin films varies from -75 to -225 μV/K with varying temperature from 50 to 330 K [111]. Akin the p-type amorphous CuInSe$_2$ thin films exhibit thermoelectric power of 20 and 220 μV/K for the thickness of 80–160 and 240 nm at 390 K, respectively [20]. The thermoelectric power of typical n-CuInSe$_2$ thin films varies between -100 and -400 μV/K, whereas the p-CuInSe$_2$ thin films show 10–120 μV/K in the temperature range of 300–475 K. The electrical parameters such as $p = 1.7 \times 10^{19}$, $N_V = 1.5 \times 10^{19}$ cm^{-3}, and $\mu = 2.3$ cm^2/Vs ($\sigma = eN_c\mu$) are estimated from the experimental thermoelectric power of 35.8 μV/K at 315 K for the p-CuInSe$_2$ films ($m_h^* = 0.73 m_o$) deposited at substrate temperature of 100 °C using Equation (6.15). Similar way the parameters $n_e = 7.2 \times 10^{17}$, $N_c = 6 \times 10^{16}$ cm^{-3}, and $\mu = 7.1$ cm^2/Vs are evaluated from the thermoelectric power of -258 μV/K at 315 K for n-CuInSe$_2$ films ($m_e^* = 0.09 m_o$) deposited at substrate temperature of 150 °C. The p-CuInSe$_2$ thin films show activation energy of less than 10 meV while n-CuInSe$_2$ thin films exhibit activation energies of 80–85 and 140 meV [107] (Table 6.13).

6.8 DLTS of I–III–VI$_2$ Heterostructures, Homojunctions, and Schottky Junctions

The deep level transient spectroscopy (DLTS) is a powerful tool to precisely determine type of defect level whether it is hole or electron trap in the semiconductors. The activation energy, density and capture cross section of traps can be determined. When the p–n junction is under reverse bias, the voltage pulses are periodically employed. Hence the majority carrier traps are filled and the space charge width is reduced. If the junction is under forward bias, the minority carrier traps are filled then the trap centers emit the carriers with time constant τ_e. The capacitance transient is related to a change in space charge by traps [112,113]. The capacitance transients of 9% efficiency glass/1μm Mo/2μm bilayer p-CuInSe$_2$/2μm n-CdS:In/ZnO thin film solar cell recorded at different temperatures of 210, 270.2, and 315.5 K are shown in Figure 6.52, for which the reverse bias voltage $V_r = -0.5$ V, pulse height of 0.45 V, and width of 3 ms are employed. The DLTS signal $\Delta C = C(t_2) - C(t_1)$ versus

Table 6.13 Thermoelectric power of CuInSe$_2$ and CuInS$_2$

S.No	Sample	S (μV/K)	p or n_e (cm^{-3})	N_V or N_C (cm^{-3})	Ref.
1	p-CIS	35.8	1.7×10^{19}	1.5×10^{19}	[107]
2	n-CIS	-258	7.2×10^{17}	6×10^{16}	
3	p-CIS	325	–	–	[108]
4	a-p-CIS	20 and 220	–	–	[21]
5	p-CIS	~200	1.4×10^{16}	–	[22]
6	p-CIS	~5	4.1×10^{14}	–	
7	n-CIS	~-700	3.1×10^{14}	–	
8	p-CuInS$_2$	335	–	–	

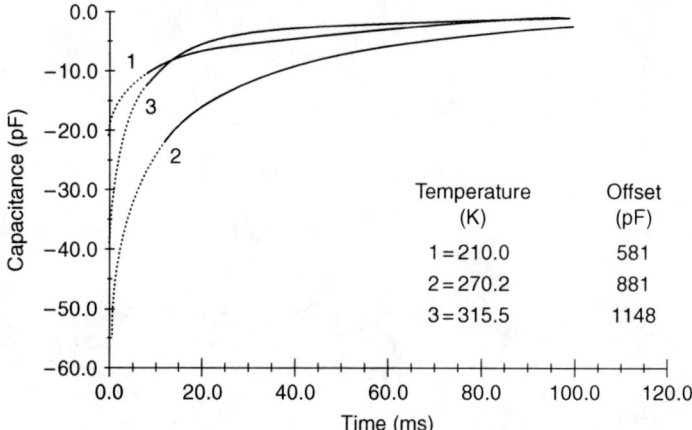

Figure 6.52 Capacitance transients of 9% efficiency glass/1μm Mo/2μm bilayer CuInSe$_2$/2μm CdS:In/ZnO thin film solar cell recorded at different temperatures of 210, 270.2, and 315.5 K.

temperature ($1/T$) (Figure 6.53A) can be drawn from Figure 6.52, where $C(t_2)$ and $C(t_1)$ are capacitances at timings t_2 and t_1, respectively. For example in Figure 6.53A, curve-1 is drawn by taking t_1 and t_2 timings of 2 and 10 ms at different temperatures, respectively. Similar fashion, other curves such as curve-2, -3, and -4 can be drawn by setting some constant as $t_2/t_1 = 5$ with different t_1 and t_2 combinations at different temperatures. The formula for emission rate (e_{max}) from the trap can be written as e_{max} or $(1/\tau) = \ln(t_1/t_2)/(t_1 - t_2)$ and at the same time emission rate (e) can be mentioned

$$e = \text{constant}\, T^2 \exp(-\Delta E/kT) \tag{6.16}$$

The emission rate can be related to hole trap

$$e_p = \sigma_p v_{th} N_v \exp\left(-\frac{\Delta E}{kT}\right), \tag{6.17}$$

where $v_{th} N_v$ depends on T^2, v_{th} is thermal velocity, σ_p is the hole capture cross-section and N_v is the effective density of the states in the valence band. The activation energy (ΔE) of trap can be obtained from the slope of $\ln(\tau^{-1} T^{-2})$ versus $1/T$ plot, as shown in Figure 6.53B. The negative peak is indication of hole trap for p-type absorber and its activation energy and trap density are $E_v + 0.85$ eV and 5×10^{14} cm^{-3}, respectively. The DLTS peak height depends on the density of trap. The cell junction is made negative bias by employing minority carrier injection, that is, applying pulse height of 1.2 V to be greater than $V_r = -0.8$ V and pulse width of 5 ms. The DLTS spectra recorded at various temperatures generate positive peak at around the neighborhood of 140 K (Figure is not shown). The observed activation energy and density of electron trap are $E_c - 0.35$ eV and 0.5×10^{14} cm^{-3}, respectively. A positive peak (negative peak) is signature of electron trap (hole trap) for

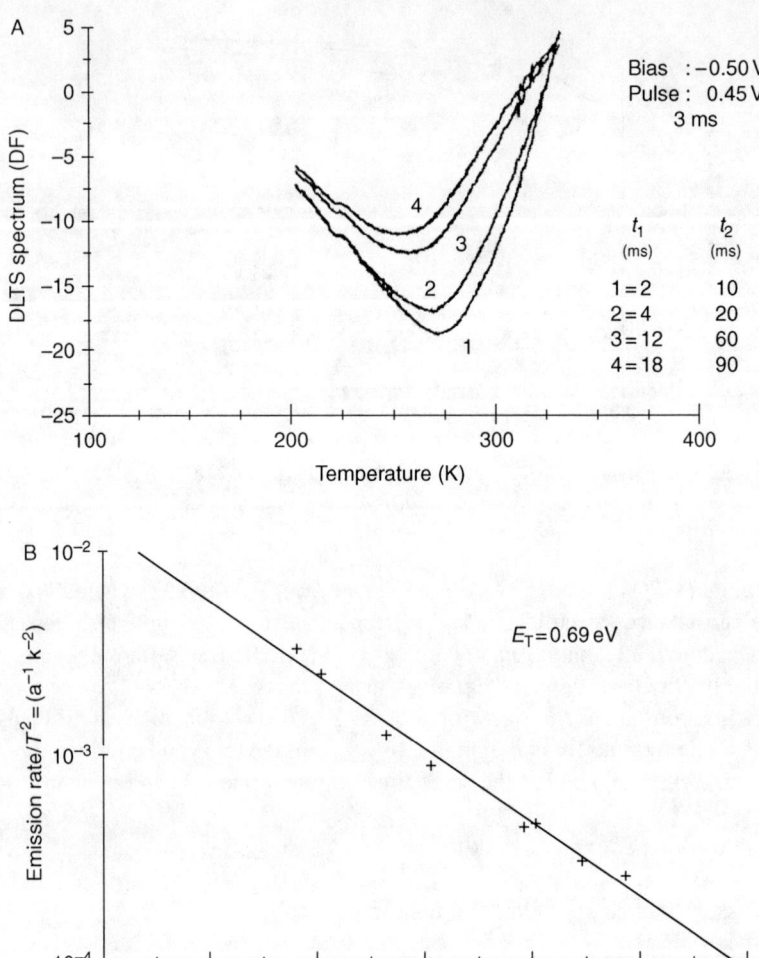

Figure 6.53 (A) DLTS spectra of 9% efficiency glass/1μm Mo/2μm bilayer CuInSe$_2$/2μm CdS:In/ZnO thin film solar cell and (B) A plot of emission rate versus $1/T$ deduced from the DLTS spectrum for 9% efficiency glass/1μm Mo/2μm bilayer CuInSe$_2$/2μm CdS:In/ZnO thin film solar cell.

p-CIS/n-CdS heterojunction. The 9% efficiency cell exhibits hole trap level at $E_v + 0.85$ eV from the valence band and electron trap level at $E_c - 0.35$ eV from the conduction band. The hole trap density can also be deduced using relation

$$N_T \approx 2N_A \Delta C / C_o \tag{6.18}$$

for $N_T < N_A$, where N_A is the net hole concentration of the absorber layer. A change in capacitance (ΔC) is proportional to the density of traps from the trap level, where C_o is the junction capacitance recorded at the DLTS peak temperature while applying a reverse bias [114]. A broad negative DLTS peak is observed for the p-CuInSe$_2$/n-CdZnS heterostructure at around 270 K for the conditions of $V_r = -3$ V, pulse height $= 2.9$ V and width $= 2$ ms, which is probably combination of two hole traps with activation energies of 423 and 498 meV [115]. An electron trap level $E_c - 0.135$ eV is also observed in the p-CIS/n-CdS cells [116].

Prior to deposition of 50 nm thick CdS layer on 1–2 mm thick p-CuInSe$_2$ single crystals with carrier concentrations of $1 \times 10^{16} - 1 \times 10^{17}$ cm^{-3} grown by Bridgman method, which are mechanically polished and etched in 0.5%Br-methanol for 1 min and annealed under Ar at 200–400 °C for 4 h, followed by deposition of 20 Ω/sq ZnO layer by RF magnetron sputtering. Au on back side of p-CuInSe$_2$ and Al on front side of ZnO are grown by vacuum evaporation to obtain complete cell configuration of Au/p-CuInSe$_2$/CdS/n^+ZnO/Al. The cells with total area of 2.5–8 cm^2 show efficiency of 10.3%. The DLTS spectra of 10.3% efficiency single crystal Au/p-CuInSe$_2$/CdS/n^+ZnO/Al cell are measured at different rate windows, as shown in Figure 6.54. The hole trap level at 0.24 eV from the valence band, that is, $E_v + 0.24$ eV with density of 10^{14} cm^{-3} is observed but no minority trap level is detected. Some of the samples from different ingots show variation in trap level position from 0.14 to 0.23 eV. 4–6% efficiency single crystal p-Cu(In$_{0.7}$Ga$_{0.3}$)Se$_2$/CdS/ZnO cells annealed under Ar show both majority and minority carrier traps at $E_v + 0.33$ with trap density of 2×10^{13} cm^{-3} and from $E_c - 0.45$ to $E_c - 0.55$ eV, respectively, as shown in Figure 6.55. The minority carrier trap level is close to mid gap of CIGS [117]. The DLTS spectra are carried out on the different

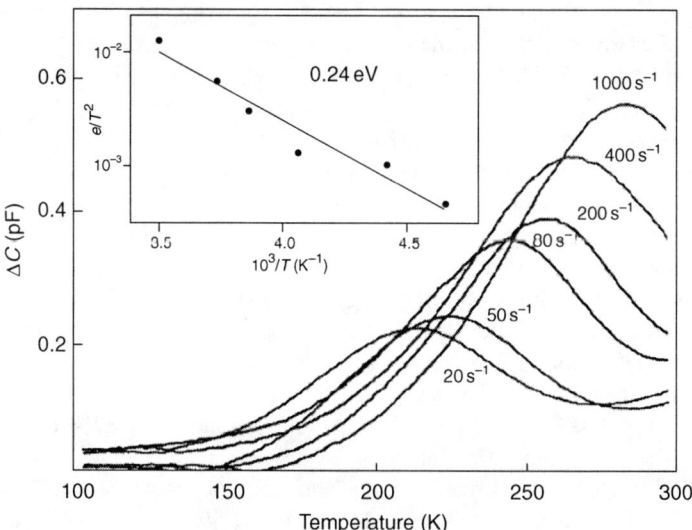

Figure 6.54 DLTS spectra of 10.3% efficiency single crystal Au/p-CuInSe$_2$/CdS/n^+-ZnO/Al cell.

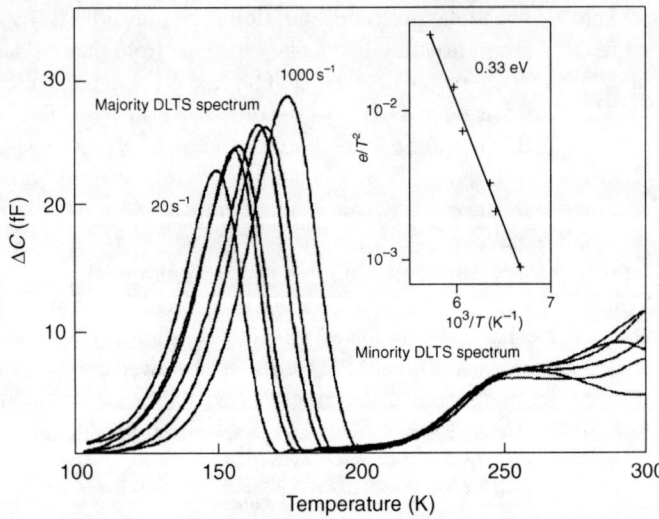

Figure 6.55 (A) DLTS spectra of 4–6% efficiency $Cu(In_{0.7}Ga_{0.3})Se_2/CdS/ZnO$ cells at different rate windows and inset is plot of emission rate versus $1/T$.

efficiency cells employing reverse bias $V_r = -0.4$ V, filling pulse amplitude of 0.4 V with time period of 1 ms and the rate window of 332 s. The glass/Mo/CuInSe$_2$/CdS/ZnO thin film solar cells consist of lower efficiencies of 7–8% exhibit hole trap defect level $\sim E_v + 0.079$ eV, whereas cells with higher efficiencies of 9–11% contain trap levels in the range of between $E_v + 0.226$ and $E_v + 0.160$ eV. The trap levels $E_v + 0.226.5$ and $E_v + 0.186$ eV are observed for 11.3 and $\sim 9\%$ efficiency cells, respectively. In fact, higher efficiency cells show trap level with higher activation energies. However, one of the lowest efficiency of 5% cells had higher energy defect level at 0.1605 eV. This could be due to higher sheet resistance of ZnO layer used in the cell. In the samples, no electron traps are observed. The cells with CuInSe$_2$ layers are mostly fabricated by annealing Cu/In/Se stack under N$_2$ show lower efficiency, whereas the cells made with CIS absorber, which is processed with Cu/In stack annealed under Se vapor at low pressure, show higher efficiency. Onto CIS, 50 nm thick CdS by CBD method, 10 Ω/sq ZnO by MOCVD, and Al–Ni grids are sequentially grown to make the cell structure. The generation of defects mainly depends on the process method of CIS layer [118]. The DLTS measurements are carried out on different efficiencies cells such as 5, 13, and 19.2% obtained from the different laboratories such as the University of Florida, EVP and NREL. For 5% efficiency cell $V_r = -0.7$ V and trap filling pulse height $= 0.4$ V, for 13% efficiency cell $V_r = -0.1$ V and pulse height $= 0.3$ V, and for 19.2% efficiency cell $V_r = -0.5$ V and filling pulse height $= 0.4$ V, whereas for minority trap level detection, filling pulse height $= 0.7$ V are employed. The filling pulse width of 10 ms is maintained for all the experiments. The DLTS spectra of 5% efficiency CIS/CdS cell is shown in Figure 6.56A and B, which exhibits two trap levels such as one minority trap level at $E_1 = E_c - 0.52$ eV with $N_t = 1.3 \times 10^{12}$ cm^{-3} and another majority trap level at

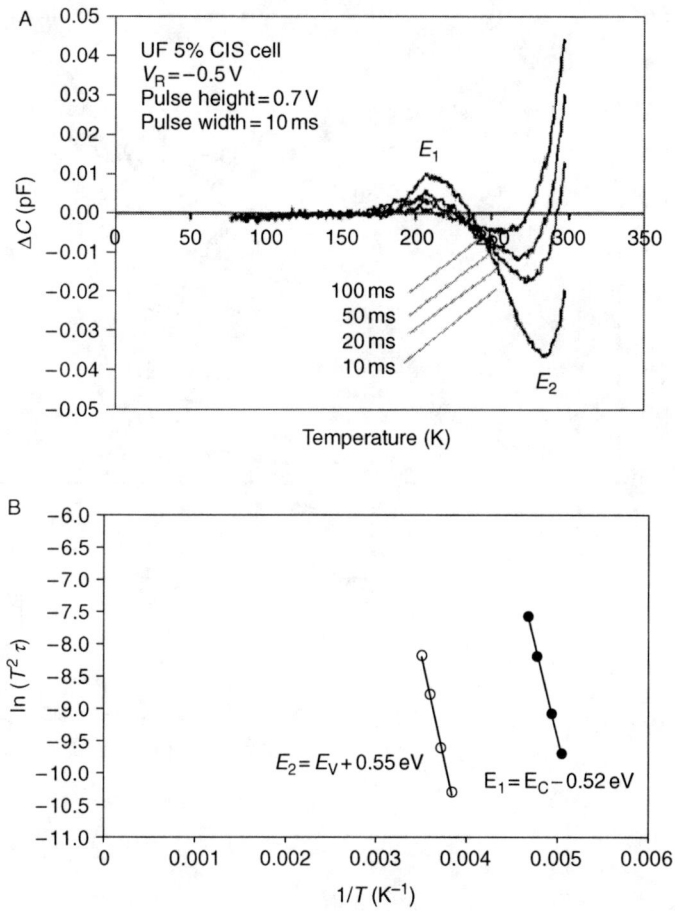

Figure 6.56 (A) DLTS spectra of 5% efficiency CIGS/CdS thin film solar cell and (B) Plots of emission rate versus $1/T$ for 5% efficiency CIGS/CdS cell.

$E_2 = E_v + 0.55$ eV with $N_t = 4.6 \times 10^{12}$ cm^{-3}. The minority carrier traps can be observed by employing minority carrier injection. An optical method such as a laser beam with wavelength of 532 nm is applied for minority carrier injection instead of filling pulse in order to find out minority trap levels such as $E_c - 0.16$ eV in the sample. A 13% efficiency cell, which possesses hole density of 3×10^{15} cm^{-3} for absorber, shows hole trap at $E_a = E_v + 0.94$ eV with $N_t = 6.5 \times 10^{13}$ cm^{-3} from DLTS spectrum, as shown in Figure 6.57A and B. 19.2% efficiency cell shows $E_1 = E_c - 0.07$ (V$_{Se}$) with $N_t = 4.2 \times 10^{13}$ cm^{-3} and another minority trap at $E_1 = E_c - 0.14$ eV for filling pulse amplitude of 0.7 V (Figure 6.58A and B). The V$_{Se}$ and In$_{Cu}$ are the donors, whereas Cu$_{In}$ is the acceptor [119].

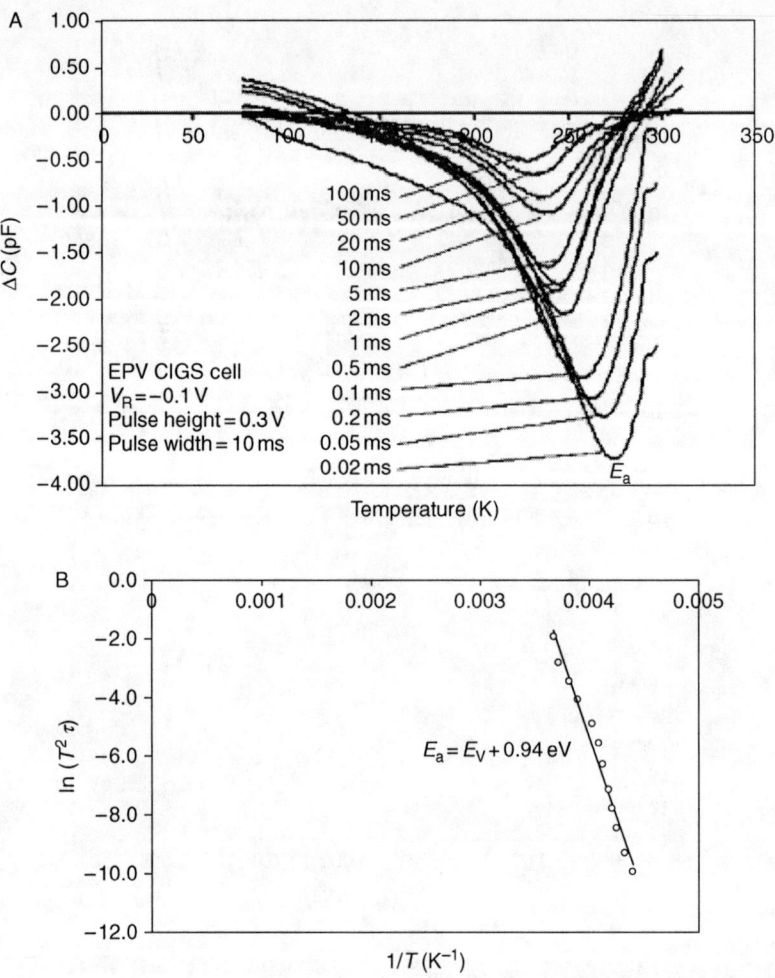

Figure 6.57 (A) DLTS spectra of 13% efficiency CIGS thin film solar cell and (B) Variation of emission rate with inverse temperature for 13% efficiency CIGS thin film solar cell.

The CuInSe$_2$ based homojunctions are made that first 2 μm thick In or Bi layers deposited on p-CuInSe$_2$ single crystals and annealed under N$_2$ flow at 200 °C to convert partially n^+-CuInSe$_2$ layer on it then Ag epoxy is applied to the backside of the crystal for ohmic contact. Al layer simply deposited on cleaned p-CuInSe$_2$ single crystals is annealed under N$_2$ at 200 °C for 10 min to form Schottky junctions. In the DLTS experiments the rate window from 50 to 100 s^{-1}, pulse height of -0.5 V and duration of 10 ms are employed and the temperature range is 80–340 K. Two hole trap levels at 250 ± 15 and 529 ± 15 meV are observed in homojunctions and in some of the Schottky junctions too. The trap density is higher by one order of magnitude in some

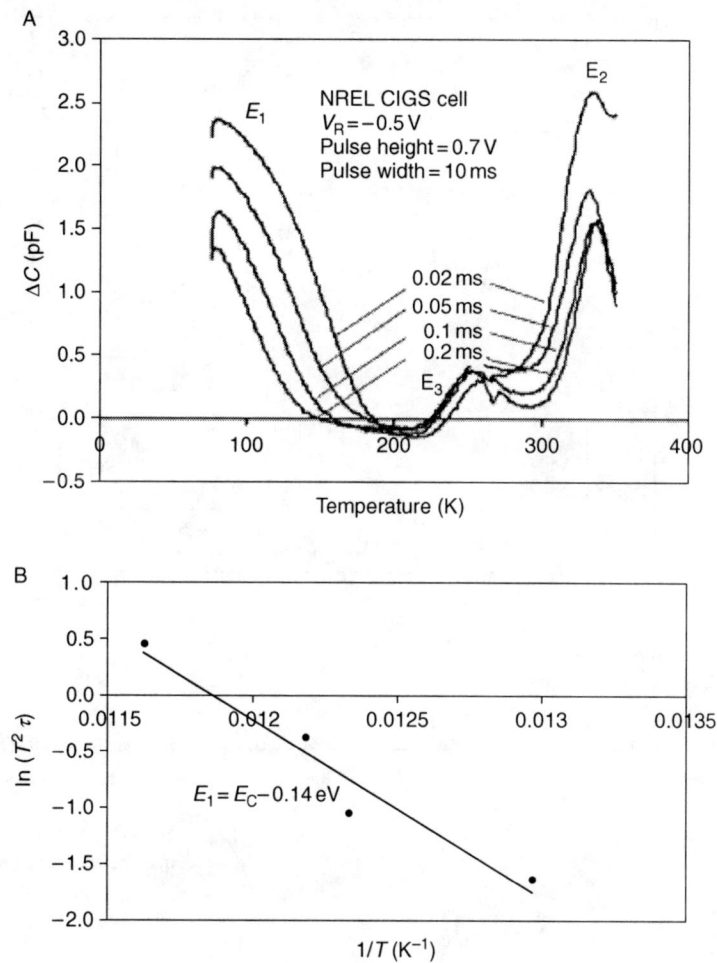

Figure 6.58 (A) DLTS spectra of 19.2% efficiency CIGS thin film solar cell and (B) Variation of emission rate with inverse temperature for 19.2% efficiency CIGS thin film solar cell.

of the Schottky junctions than in the homojunctions ($N_T = 10^{13}$ cm^{-3}). It could be due to mechanical damage of the crystal while polishing [120]. The similar hole trap levels at 234, 493 meV in the single crystal based p-CuInSe$_2$ (Cu:In:Se = 23.:25.47:51.23)/CdZnS heterostructure (Figure 6.59) and an hole trap at 282 meV in the p-CuInSe$_2$/Al Schottky junctions are observed, whereas in the thin film the p-CuInSe$_2$/CdZnS heterostructure, an deep hole trap level at 530 meV is determined (Figure 6.60). In this case, the trap density is 2–3 orders of magnitude higher in the single crystal than in the thin film [121]. The Schottky Al/p-CuInSe$_2$ thin film shows deep levels at 234 and 481 meV, whereas after Se heat treatment the junction shows only one defect level at 115 meV by suppressing other defect levels. The trap level is probably due to Se$_{Cu}$ [122]. In the

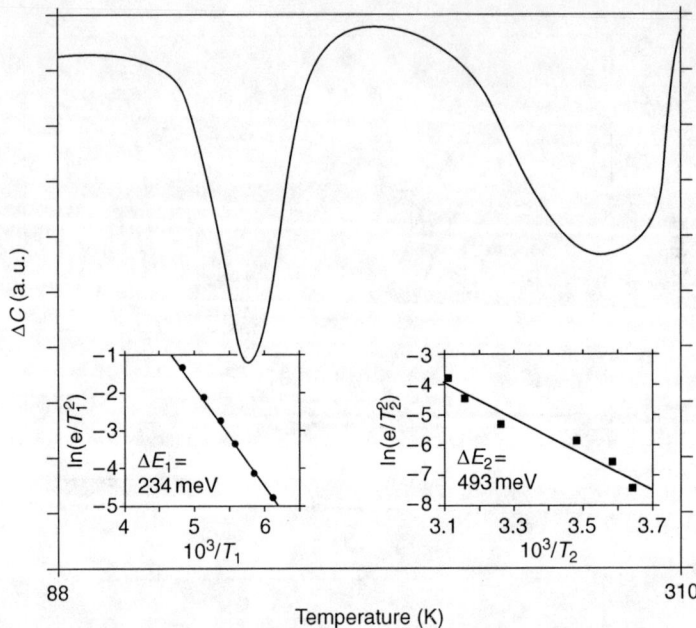

Figure 6.59 DLTS spectrum of CIS/CdS single crystal solar cell and plots of emission rate versus inverse temperature.

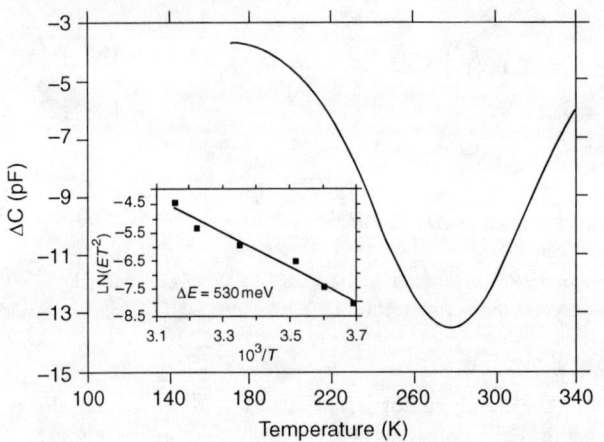

Figure 6.60 DLTS spectrum of CIS/CdS thin film solar cell and a plot of emission rate versus inverse temperature.

n-CuInSe$_2$/Au Schottky junction, trap levels 105 meV ($N_T = 1.3 \times 10^{13}$ cm^{-3}) and 365 meV (8.8×10^{12} cm^{-3}) are observed, whereas in another same kind of Schottky diode, the trap levels with little higher activation energies of 135 meV ($N_t = 6.84 \times 10^{12}$ cm^{-3}) and 395 meV ($N_t = 1.1 \times 10^{13}$ cm^{-3}) are determined. It is found that the deep levels reduce open circuit voltage (V_{oc}) of the device [123]. The as-grown n-CuInSe$_2$ thin film/Au and electron irradiated n-CuInSe$_2$ film/Au Schottky junctions show electron trap levels at $E_c - 0.35$ eV with $\sigma_n = 6 \times 10^{-15}$ cm^2 and $E_c - 0.22$ eV with $\sigma_n = 1 \times 10^{-17}$ cm^2, respectively. The negative DLTS signal peak in the n-CuInSe$_2$ sample indicates electron trap level. The n-CuInSe$_2$ thin films used in this experiment consist of composition ratios of Cu/In = 0.94 and Se/(Cu + In) = 1.05. The n-CuInSe$_2$ thin film is irradiated with electron beam fluence of 5×10^{17} cm^{-2} and acceleration energy of 3 MeV. The electron concentration of n-CuInSe$_2$ decreases from 0.9–2 $\times 10^{17}$ to 7×10^{16} cm^3 upon irradiation [124].

Two varieties of Schottky junctions such as Pt/p-CuInSe$_2$/In and In/n-CuInSe$_2$/Pt are made in which p-CuInSe$_2$/In and n-CuInSe$_2$/Pt are rectifying junctions. For DLTS measurements, $V_r = -2$ V and majority carrier pulse = 2 V and width = 0.1–0.5 ms are employed. The hole trap activation energy of 0.26 eV and capture cross section of 4×10^{-14} cm^2 are observed in the p-CuInSe$_2$. In the case of n-CuInSe$_2$, two electron trap levels at $E_1 = E_c - 0.35$ and $E_2 = E_c - 0.57$ eV corresponding capture cross sections of 5×10^{-14} and 3×10^{-14} cm^2 are also determined [125]. The Schottky diodes with area of 2.6×10^{-3} cm^2 prepared by depositing 100 nm Au on p-CuInSe$_2$ single crystals grown by Bridgman method as ohmic contact on one side and 200 nm Al layer on the other side as rectifying contact. The carrier concentration (p), defect concentration (N_t) and activation energies E_a are given in Table 6.14 [126]. The electron trap position changes from $E_c - 0.1$ ($N_t = 0.2 \times 10^{13}$ cm^{-3}) to $E_c - 0.22$ eV ($N_t = 6.5 \times 10^{13}$ cm^{-3}) with changing the filling pulse duration from 20 μs to 10 s, that is, band like trap. This observation indicates that the trap location depends on the Fermi level. The hole trap $E_v + 0.23$ eV is also observed along with band like trap in the high efficiency CIS cells [127].

The DLTS spectra of CIGS cell are recorded using reverse bias of 0.4 V and filling pulse amplitude of 1 V with a width of 10 ms. The parameters used in the DLTS experiments make cell in forward bias of 0.6 V and causes minority carrier injection. The filling pulse allows free carriers to flow into the depletion region and fill charge-trapping defects. A negative peak at 160 K and a positive peak at ~220 K are observed for the rate window of 0.5 ms in the DLTS spectrum. An Arrhenius plot of $\ln(T^{-2}\tau^{-1})$ versus $1/T$ gives activation energies of 0.42 and 0.17 eV, as shown in Figure 6.61, which are related to donor like trap $E_c - 0.42$ eV and acceptor

Table 6.14 Electrical parameters of p-CIS Schottky diodes obtained by DLTS

S.No	p (cm^{-3})	N_t (cm^{-3})	E_a (meV)	Defects	Ref.
1	6.4×10^{17}	2.8×10^{16}	87	In$_{Se}$	127
2	1.4×10^{18}	3.5×10^{13}	166		
3	2.1×10^{18}	6.1×10^{16}	276		
4	1.9×10^{17}	4.6×10^{15}	39	In$_{Cu}$	
5	2.1×10^{17}	1.3×10^{16}	16		
6	–	2.9×10^{16}	191		
7	6.3×10^{16}	2.5×10^{15}	172		
8	3.2×10^{17}	3.9×10^{16}	30	V$_{Cu}$	
9	1.5×10^{17}	9.5×10^{15}	92	In$_{Se}$	

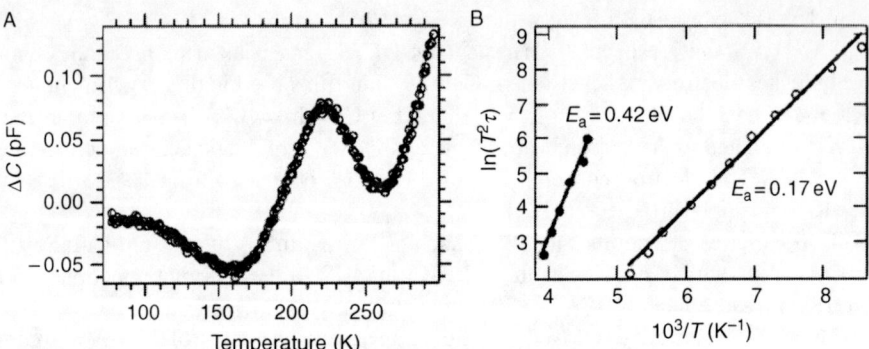

Figure 6.61 DLTS spectrum of CIGS cell give two trap states at 0.42 and 0.17 eV.

like trap $E_v + 0.17$ eV, respectively [128]. The majority carrier traps and minority carrier traps are observed in (1) CIGSS/ZnO, (2) CIGSS/ZnO, (3) CIGSS/CdS, (4) CIGSS/ZnS(10 nm), and (5) CIGSS/ZnS (20 nm) samples, as given in Table 6.15. The performance of ZnO is efficient in sample-1 CIGSS/ZnO and poor in sample-2 CIGSS/ZnO, respectively. The reverse biases $V_r = -0.1$ and $V_r = -0.4$ V are applied for sample-1 and sample-2, respectively. The rate window can be varied from 0.02 to 100 ms in steps of 12. The filling pulse of 0 V and width of 10 ms are employed. The DLTS spectra of these five samples for electron and hole traps are shown in Figures 6.62A, B and 6.63A, B, respectively [129]. The DLTS spectra of different samples are measured employing reverse bias in the range from -1.2 to -0.2 V and pulse width of 1 ms, as depicted in Figure 6.64. The spectra are measured at emission rate of 465.2 ms. The trap levels and their densities of conventional CIGS cells (samples 9–12) made by different processes are given in Table 6.15. The H_1 ($E_v + 0.26$) and E_1 ($E_c - 0.1$) traps are observed in all the samples but the E_2 ($E_c - 0.83$) trap is only one found in the low efficiency cells [130].

The CuInGaSe$_2$ thin films are made by co-evaporation of three stage process to have from Cu-rich to In-rich range in such a way the samples are removed from the deposition chamber with different timings in between beginning to ending period of 3rd stage. The conventional CIGS/CdS/ZnO cells (samples a–d$_1$) made with different absorber compositions as shown in Table 6.16. The activation energies of electron and hole traps and their densities obtained from the DLTS are also given in Table 6.16. It can be concluded that In-rich Cu/(In+Ga) = 0.93–0.9 samples show donor like defect levels, whereas Cu-rich Cu/(In+Ga) = 1.24–0.95 samples possess acceptor like defect levels [131]. The minority and majority carrier traps at 305 and 400 meV, respectively are observed in the CIGS cells. The minority carrier trap disappears and the density of majority carrier trap decreases in the Na doped cells [132].

The p-CuInS$_2$/n-CdS based thin film solar cells with different efficiencies (samples 1–4) and trap levels are given in Table 6.17. The hole trap $E_v + 0.806$ eV is close to half of the band gap of CuInS$_2$, the origin of the trap may be from the

Table 6.15 Activation energies and densities of electron and hole traps for CIGS thin film solar cells

S.No	Cell-efficiency	Hole trap			Electron trap		Ref.
		E_a (eV)	N_t (cm^{-3})	σ_{ht} (cm^2)	E_a (eV)	N_t (cm^{-3})	
1	CIGSS/ZnO-12%	$E_v + 0.85$	2×10^{13}	2.4×10^{-15}	$E_c - 0.12$	1×10^{14}	[129]
2	CIGSS/ZnO-7%	$E_v + 0.73$	4×10^{12}	3.1×10^{-17}	$E_c - 0.07$ (V$_{Se}$)	8×10^{14}	
3	CIGSS/CdS-13%	$E_v + 1.33$	3×10^{12}	4.6×10^{-9}	$E_c - 0.05$ (V$_{Se}$)	1×10^{14}	
4	CIGSS/ZnS-12%	$E_v + 0.81$	5×10^{12}	1.4×10^{-15}	$E_c - 0.83$	2×10^{14}	
5	CIGSS/ZnS-6%	–	–	–	$E_c - 0.26$ (In$_{Cu}$)	1×10^{12}	
6	CIGS/CdS	$E_v + 0.12$	2×10^{14}	–	$E_c - 0.1$	2.1×10^{14}	[130]
7	CIGS/CdS	$E_v + 0.28$	1.1×10^{13}	–	$E_c - 0.12$	7.2×10^{13}	[131]
8	CIGS/CdS	–	–	–	$E_c - 0.25$	2.1×10^{12}	
9	PVD-18.5%	$E_v + 0.26$	2.7×10^{14}	–	$E_c - 0.1$	2.1×10^{14}	[130]
10	EP-15.4%	$E_v + 0.27$	1.2×10^{13}	–	$E_c - 0.09$	8.4×10^{14}	
11	AP-13.4%	$E_v + 0.26$	1.5×10^{14}	–	$E_c - 0.09$	1.2×10^{15}	
12	AP-12.4%	$E_v + 0.26$	3.2×10^{14}	–	$E_c - 0.83$	2.1×10^{14}	

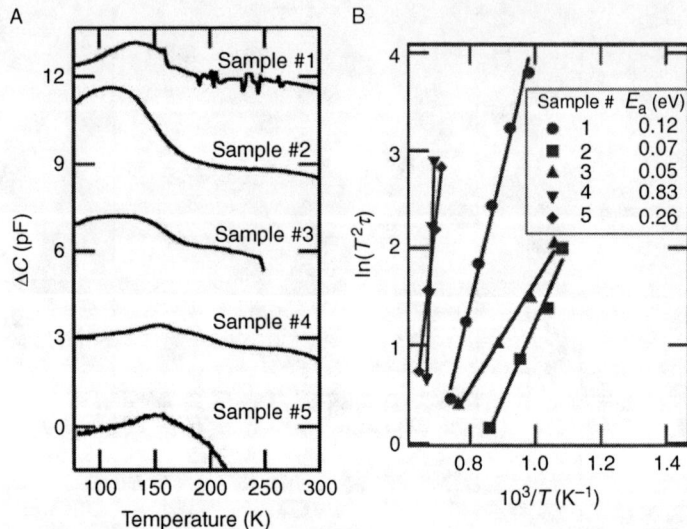

Figure 6.62 DLTS spectra of different samples (1) CIGS/ZnO (efficient), (2) CIGS/ZnO (poor), (3) CIGSS/CdS, (4) CIGSS/ZnS(10 nm), and (5) CIGSS/ZnS (20 nm) and (B) their Arrhenius plots at low temperature.

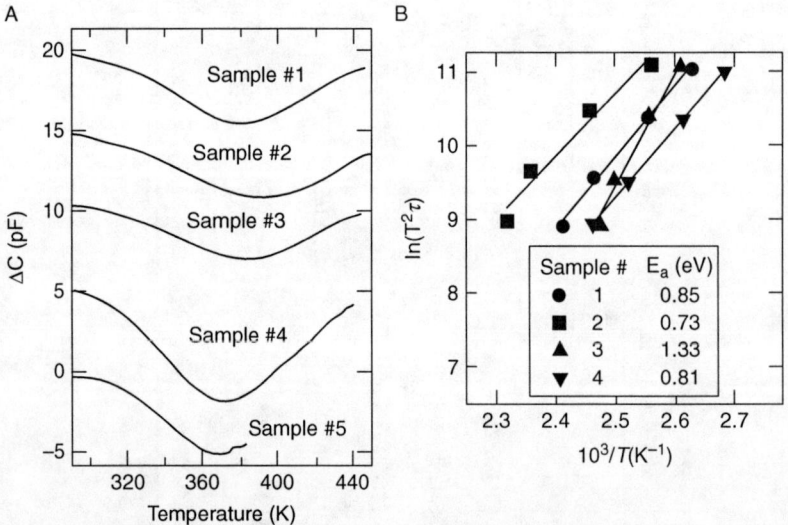

Figure 6.63 DLTS spectra of different samples (1) CIGS/ZnO (efficient), (2) CIGS/ZnO (poor), (3) CIGSS/CdS, (4) CIGSS/ZnS (10 nm), and (5) CIGSS/ZnS (20 nm) and (B) their Arrhenius plots at high temperature.

Electrical Properties of I–III–VI$_2$ Compounds

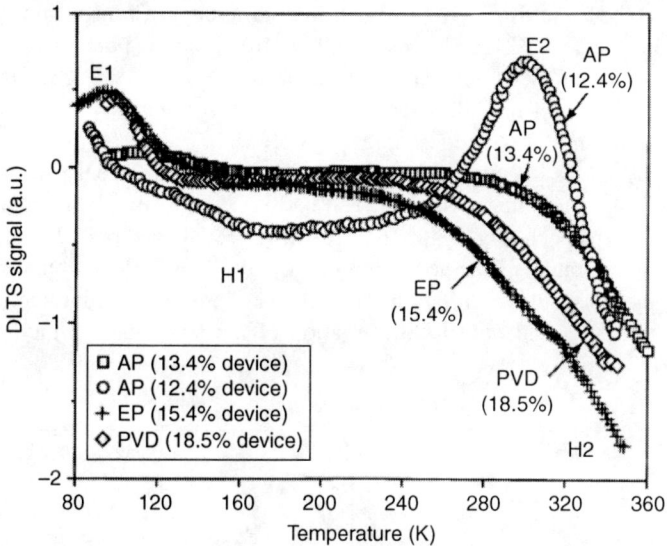

Figure 6.64 DLTS spectra of CIGS thin film solar cells with different efficiencies.

Table 6.16 Effect of variation of composition on the defect levels in the CIGS cells

S.No	Sample	Cu/(In+Ga)	Ga/(In+Ga)	Trap	E_a (eV)	N_t (cm^{-3})	p (cm^{-3})
1	CIGS-a	1.24	0.19	HT	$E_v+0.12$	2×10^{14}	4.8×10^{16}
2	CIGS-a1	1.02	0.19	–	–	–	–
3	CIGS-b	0.95	0.24	HT	$E_v+0.28$	1.1×10^{13}	7.5×10^{15}
4	CIGS-c	0.93	0.26	ET	$E_c-0.47$	6.6×10^{11}	4.8×10^{15}
5	CIGS-d	0.9	0.28	ET	$E_c-0.12$	7.2×10^{13}	7.9×10^{15}
				ET	$E_c-0.25$	2.1×10^{12}	–
6	CIGS-d1	0.9	0.28	ET	$E_c-0.63$	1.4×10^{13}	1.4×10^{16}

Table 6.17 Trap level and efficiencies of CuInS$_2$ cells

S.No	η (%)	E_a (eV)	Origin of traps	Ref.
1	11.4	$E_c-0.259$	Bulk	[133]
2	10.4	$E_v+0.806$	Interface?	
3	10.2	$E_v+0.241$	Interface	
4		$E_v+0.354$	Bulk	
5	8.6	$E_v+0.218$	–	
6		$E_v+0.321$	–	

interface of the heterostructure. In the low efficiency $CuInS_2$ thin films, two $E_v+0.218$ and $E_v+0.321$ eV traps might probably cause poor efficiency [133]. Unlike conventional cells, the DLTS studies are done on 9% efficiency n-type absorber based Cu tape/n-$CuInS_2$/p-CuI/ZnO:Al cells using $V_r=-1$ V and an AC signal amplitude of 0.1 V and frequency of 1 MHz in the temperature range of 4–305 K. The DLTS spectra of the cell recorded at two different period widths of 5.12 and 512 ms show negative peak (H_1) and positive peak (E_1) at approximately 125 and 260 K, respectively. Figure 6.65A shows DLTS spectra of E_1 trap recorded at different temperatures. The activation energy of 583.4 meV is obtained from the Arrhenius plot, as shown in Figure 6.65B. The pre-exponential factor (K_T) of 4.82×10^6 s^{-1} K^{-2} and defect concentration of 2×10^{16} cm^{-3}, which is close to

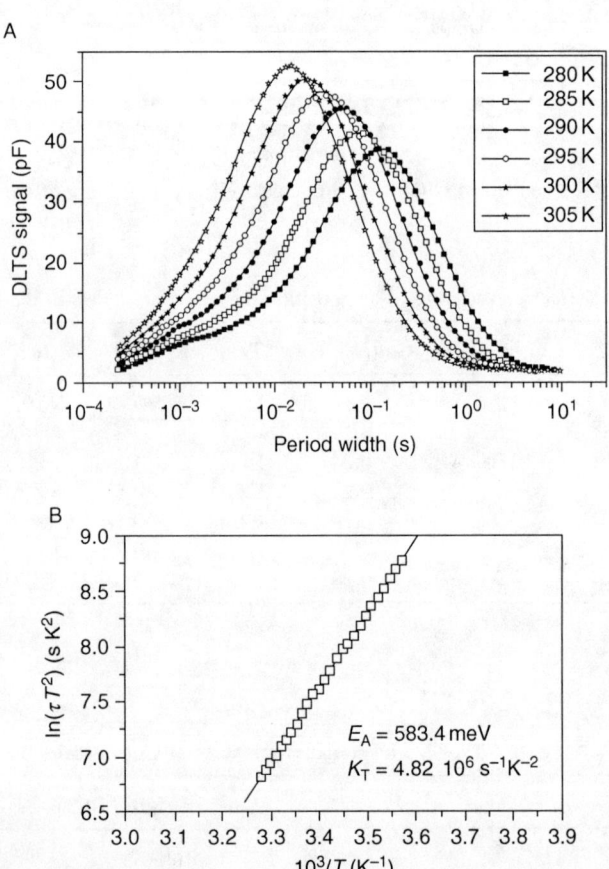

Figure 6.65 (A) DLTS spectra of Cu tape/n-$CuInS_2$/p-CuI/ZnO:Al cell recorded at different temperatures and (B) A plot of emission rates versus inverse temperature for E_1 trap in the Cu tape/n-$CuInS_2$/p-CuI/ZnO:Al cell.

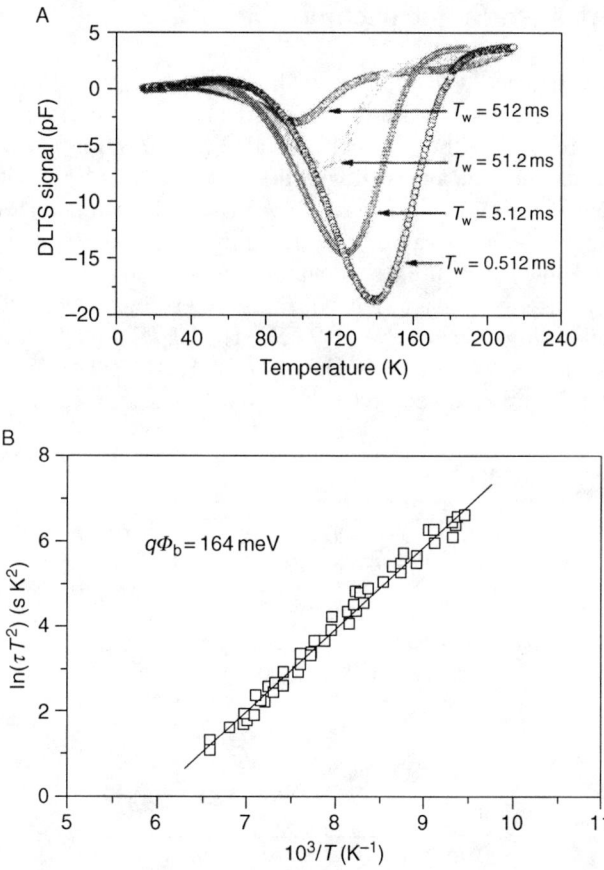

Figure 6.66 (A) DLTS spectra of H_1 trap in the Cu tape/n-CuInS$_2$/p-CuI/ZnO:Al cell recorded at different rate windows and (B) A plot of emission rate versus inverse temperature for H_1 trap in the Cu tape/n-CuInS$_2$/p-CuI/ZnO:Al cell.

the background concentration of 8×10^{16} cm^{-3}, are also determined. The H_1 is recorded at different period widths with conditions of $V_r = -1$ V, $V_{pulse} = 0$ V, and pulse duration = 1 ms, as shown in Figure 6.66A. The activation energy of 164 meV is obtained from the Arrhenius plot, as shown in Figure 6.66B. In the DLTS spectrum, the negative peak of Cu tape/n-CuInS$_2$/p-CuI/ZnO:Al cell is eventually indication of minority trap but the set experimental conditions obviously cannot provide the forward voltage to the cell to create minority trap hence the assign of negative peak to minority trap can easily be excluded. The authors assigned H_1 trap to barrier energy, which occurs when the solar cell is under reverse bias. The energy barrier is probably located between Cu–In alloy and CuInS$_2$ layer [134].

6.9 Admittance Spectroscopy

A change in capacitance with variation of frequency is so called admittance spectroscopy, which is recorded at different temperatures from 160 to 260 K for glass/Mo/CIGS/CdS/ZnO thin film solar cell annealed at 330 K for 1 h with zero bias (sample-x) and another one annealed with the same conditions but with -2 V bias voltage (sample-y), as shown in Figure 6.67A. The activation energies of 0.24 and 0.34 eV for sample-y and -x can be obtained from the emission rate versus inverse temperature, as shown in Figure 6.67B but the determination of sign for trap level is unavailable. The as-grown glass/Mo/CIGS/CdS/ZnO and glass/Mo/CIGS/ZnO cells also show similar activation energy of 0.24 eV. The DLTS, RDLTS, optical pulse of 1350 nm biasing RDLTS spectra of glass/Mo/CIGS/CdS/ZnO thin film solar cell annealed at 330 K are recorded under dark at 120 K show negative peak instead

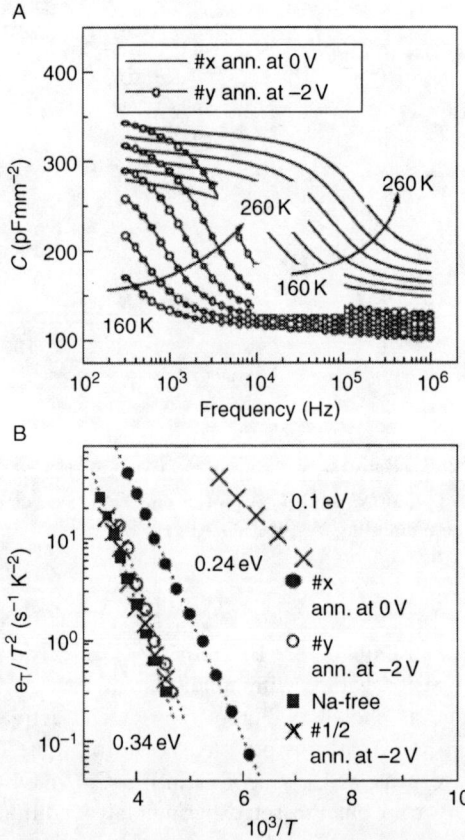

Figure 6.67 (A) Admittance spectra of annealed CIGS/CdS/ZnO ($x=0.2$) thin film solar cell at 0 and -2 V bias recorded in the temperature range from 160 to 260 K and (B) emission rates versus inverse temperature of trap levels derived from admittance spectra.

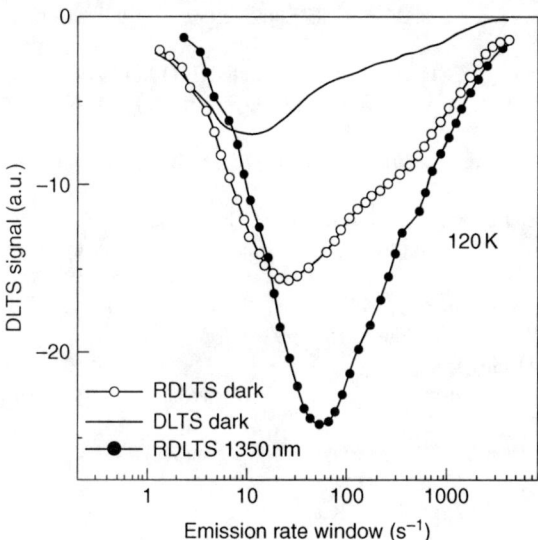

Figure 6.68 DLTS signal of CIGS/CdS/ZnO ($x=0.2$) thin film solar cell recorded at different conditions.

of positive peak for minority trap level. This is commonly assigned to minority trap in annealed sample because the trap level arises from the inversion region of the junction but much more investigation is needed in this direction for confirmation. In the measurements, a positive pulse (1 V) has been employed on -1 V quiescent bias for DLTS, whereas a negative pulse (-0.5 V) has been superimposed on 0 V quiescent bias for RDLTS. The intensity of trap level (0.24 eV) is in the descending order from optical pulse biased RDLTS to RDLTS and DLTS, as shown in Figure 6.68. The trap level N_1 with different energies of 0.24 and 0.34 eV are assigned to In_{Cu} donor levels with different valancies [135].

In conclusion, the activation energy of trap level in the range from $E_c - 0.24$ to $E_c - 0.34$ eV can be assigned to In_{Cu}, whereas the activation energy of trap level in the range $E_c - 0.07$ to $E_c - 0.05$ eV is considered to be V_{Se} defect. There are several interface, bulk and barrier defects found, as mentioned in the text or tables. More investigations are needed in this direction to assign high energy defect levels properly based on their activation energies.

References

[1] H. Neumann, R.D. Tomlinson, Solar Cells 28 (1990) 301.
[2] A. Catalano, 1st World Conference on Photovoltaic Energy Conversion, Hawaii, (1994) p. 52.
[3] B. Tell, J.L. Shay, H.M. Kasper, J. Appl. Phys. 43 (1972) 2469.
[4] H. Matsushita, T. Suzuki, S. Endo, T. Irie, Jpn. J. Appl. Phys. 34 (1995) 3774.

[5] R. Noufi, R. Axton, D. Cahen, S.K. Deb, 17th IEEE Photovoltaic Specialist Conference, (1984) p. 927.
[6] R. Noufi, R.J. Matson, R.C. Powell, C. Herrington, Solar Cells 16 (1986) 479.
[7] S. NiKi, P.J. Fons, A. Yamada, Y. Lacroix, H. Shibata, H. Oyanagi, et al., Appl. Phys. Lett. 74 (1999) 1630.
[8] R.R. Philip, B. Pradeep, G.S. Okram, V. Ganesan, Semicond. Sci. Technol. 19 (2004) 798.
[9] J. Tuttle, D. Albin, J. Goral, C. Kennedy, R. Noufi, Solar Cells 24 (1988) 67.
[10] E.P. Zaretskaya, V.F. Gremenok, I.V. Viktorov, I.V. Bodnar, J. Appl. Spectrosc. 61 (1994) 785.
[11] A. Ashour, J. Mater. Sci. Mater. Electron. 17 (2006) 625.
[12] Y.D. Tembhurkar, J.P. Hirde, Thin Solid Films 215 (1992) 65.
[13] R.L. Pertritz, Phys. Rev. 104 (1956) 1508.
[14] J.-F. Guillemoles, P. Cowache, A. Lusson, K. Fezzaa, F. Boisivon, J. Vedel, et al., J. Appl. Phys. 79 (1996) 7273.
[15] E.P. Zaretskaya, V.F. Gremenyuk, V.B. Zalesskii, V.A. Ivanov, I.V. Viktorov, V.I. Kovalevskii, et al., Tech. Phys. 70 (2000) 141.
[16] V.K. Gandotra, K.V. Ferdinand, C. Jagadish, A. Kumar, P.C. Mathur, Physica Status Solidi (A) 98 (1986) 595.
[17] N.M. Shah, J.R. Ray, V.A. Kheraj, M.S. Desai, C.J. Panchal, B. Rehani, J. Mater. Sci. 44 (2009) 316.
[18] R.D.L. Kristensen, S.N. Sahu, D. Haneman, Solar Energy Mater. 17 (1988) 329.
[19] A. Rockett, R.W. Birkmire, J. Appl. Phys. 70 (1991) R81.
[20] M.M. El-Nahass, H.S. Soliman, D.A. Hendi, Kh.A. Mady, J. Mater. Sci. 27 (1992) 1484.
[21] H. Sakata, H. Ogawa, Solar Energy Mater. Solar Cells 63 (2000) 259.
[22] O. Tesson, M. Morsli, A. Bonnet, V. Jousseaume, L. Cattin, G. Masse, Opt. Mater. 9 (1998) 511.
[23] Y.-J. Kim, H.-H. Yang, W.-J. Jeon, J.-Y. Park, G.-C. Park, C.-D. Kim, et al., Physica Status Solidi (C) 3 (2006) 2601.
[24] J. Piekoszewski, J.J. Loferski, R. Beaulieu, J. Beall, B. Roessler, J. Shewchun, Solar Energy Mater. 2 (1980) 363.
[25] R. Vaidhyanathan, R. Noufi, T. Datta, R.C. Powell, S.K. Deb, 18th IEEE Photovoltaic Specialist Conference, (1985) p. 1054.
[26] N.A. Zeenath, P.K.V. Pillai, K. Bindu, M. Lakshmy, K.P. Vijayakumar, J. Mater. Sci. 35 (2000) 2619.
[27] H. Neumann, E. Nowak, G. Kuhn, Cryst. Res. Technol. 16 (1981) 1369.
[28] N.M. Shah, J.R. Ray, K.J. Patel, V.A. Kheraj, M.S. Desai, C.J. Panchal, et al., Thin Solid Films 517 (2009) 3639.
[29] J.S. Blakemore, Semiconductor Statistics, Pergamon Press, New York, 1962.
[30] S.S. Devlin, Transport properties, in: M. Aven, J.S. Prener (Eds.), Physics and Chemistry of II-VI Compounds, North-Holland Publishing Company, Amsterdam, 1967, p. 551, Chapter 11.
[31] B. Schumann, C. Georgi, A. Temple, G. Kuhn, N.V. Nam, H. Neumann, et al., Thin Solid Films 52 (1978) 45.
[32] C. Rincon, J. Gonzalez, G.S. Perez, J. Appl. Phys. 54 (1983) 6634.
[33] H. Neumann, R.D. Tomlinson, E. Nowak, N. Avgerinos, Physica Status Solidi (A) 56 (1979) K137.
[34] H. Matsushita, S. Endo, T. Irie, Jpn. J. Appl. Phys. 31 (1992) 2687.

[35] J.H. Schon, Ch. Kloc, E. Bucher, Thin Solid Films 361–362 (2000) 411.
[36] T. Irie, S. Endo, S. Kimura, Jap. J. Appl. Phys. 18 (1979) 1303.
[37] O. Aissaoui, S. Mehdaoui, L. Bechiri, M. Benabdeslem, N. Benslim, A. Amara, et al., J. Phys. D: Appl. Phys. 40 (2007) 5663.
[38] K. Bouabid, A. Ihlal, A. Manar, A. Outzourhit, E.L. Ameziane, Thin Solid Films 488 (2005) 62.
[39] A. Ashour, A.A. Akl, A.A. Ramadan, K.A. El-Hady, J. Mater. Sci. Mater. Electron. 16 (2005) 599.
[40] J. Schmidt, H.H. Roscher, R. Labusch, Thin Solid Films 251 (1994) 116.
[41] S.M.F. Hasan, M.A. Subhan, Kh.M. Mannan, Opt. Mater. 14 (2000) 329.
[42] S. Isomura, S. Shirakata, T. Abe, Solar Energy Mater. 22 (1991) 223.
[43] D. Fink, G. Lipppold, M.V. Yakushev, R.D. Tomlinson, R.D. Pilkington, Solar Energy Mater. Solar Cells 59 (1999) 217.
[44] A. Kumar, A.L. Dawar, P.K. Shishodia, G. Chauhan, P.C. Mathur, J. Mater. Sci. 28 (1993) 35.
[45] H.-S. Lee, H. Okada, A. Wakahara, A. Yoshima, T. Ohshima, H. Itoh, et al., Solar Energy Mater. Solar Cells 75 (2003) 57.
[46] M. Yamaguchi, J. Appl. Phys. 78 (1995) 1467.
[47] J. Parkes, R.D. Tomlinson, M.J. Hampshire, Solid State Electron. 16 (1973) 773.
[48] H. Neumann, R.D. Tomlinson, N. Avgerinos, E. Nowak, Physica Status Solidi (A) 75 (1983) K199.
[49] S.M. Wasim, A. Noguera, Physica Status Solidi (A) 82 (1984) 553.
[50] J. Bardeen, W. Shockley, Phys. Rev. 80 (1950) 72.
[51] S.M. Wasim, Solar Cells 16 (1986) 289.
[52] Ternary Compounds, Organic Semiconductors, Springer (2000), H. Dittrich, N. Karl, S. Kuck, H. W. Schock.
[53] S.R. Kodigala, J. Mater. Sci. Mater. Electron. 10 (1999) 291.
[54] A. Amara, W. Rezaiki, A. Ferdi, A. Hendaoui, A. Drici, M. Guerioune, et al., Solar Energy Mater. Solar Cells 91 (2007) 1916.
[55] H. Brooks, Adv. Electron. Electron Phys. 7 (1955) 85.
[56] L. Essaleh, S.M. Wasim, J. Galibert, J. Appl. Phys. 90 (2001) 3993.
[57] A. Amara, W. Rezaiki, A. Ferdi, A. Hendaoui, A. Drici, M. Guerioune, et al., Physica Status Solidi (a) 204 (2007) 1138.
[58] S. Siebentritt, Thin Solid Films 480–481 (2005) 312.
[59] S. Bandyopadhyaya, S. Chaudari, A.K. Pal, Solar Energy Mater. Solar Cells 60 (2000) 323.
[60] L. Essaleh, S.M. Wasim, in: I.A. Lukyanchuk, D. Mezzane (Eds.), Smart Materials for Energy, Communications and Security, Springer Science+Business Media, BV, 2008.
[61] H. Neumann, N.V. Nam, H.-J. Hobler, G. Kuhn, Solid State Commun. 25 (1978) 899.
[62] P.M. Gorley, V.V. Khomyak, Y.V. Vorobiev, J.G.- Hernandez, P.P. Horley, O.-O. Galochkina, Solar Energy 82 (2008) 100.
[63] M.R.-A. Magomedow, D.K. Amirkhanova, S.M. Ismailov, P.P. Khokhlachev, R.Z. Zubairuev, Tech. Phys. 42 (1997) 282.
[64] S. Siebentritt, A. Bauknecht, A. Gerhard, U. Fiedeler, T. Kampschulte, S. Schuler, et al., Solar Energy Mater. Solar Cells 67 (2001) 129.
[65] J.L. Annapurna, K.V. Reddy, Indian J. Pure Appl. Phys. 24 (1986) 283.
[66] J. Santamaria, G.G. Diaz, E. Iborra, I. Martil, F.S.- Quesada, J. Appl. Phys. 65 (1989) 3236.

[67] S. Siebentritt, A. Gerhard, S. Brehme, M.Ch. Lux-Steiner, Mater. Res. Soc. Symp. Proc. 668 (2001) H4.4.1.
[68] L. Mandel, R.D. Tomlinson, M.J. Hampshire, H. Neumann, Solid State Commun. 32 (1979) 201.
[69] M. Rusu, P. Gashin, A. Simashkevich, Solar Energy Mater. Solar Cells 70 (2001) 175.
[70] M.P. Vecchi, J. Ramos, W. Giriat, Solid State Electron. 21 (1978) 1609.
[71] X. Wang, S.S. Li, C.H. Huang, S. Rawal, J.M. Howard, V. Craciun, *et al.*, Solar Energy Mater. Solar Cells 88 (2005) 65.
[72] M.T. Wagner, I. Dirnstorfer, D.M. Hofmann, M.D. Lampert, F. Karg, B.K. Meyer, Physica Status Solidi 167 (1998) 131.
[73] X. Donglin, X. Man, L. Jianzhuang, Z. Xiujian, J. Mater. Sci. 41 (2006) 1875.
[74] R. Caballero, C. Maffiotte, C. Guillen, Thin Solid Films 474 (2005) 70.
[75] R. Caballero, C. Guillen, Thin Solid Films 431–432 (2003) 200.
[76] C. Paorici, L. Zanotti, N. Romeo, G. Sberveglieri, L. Tarricone, Solar Energy Mater. 1 (1979) 3.
[77] K.T.R. Reddy, R.B.V. Chalapathy, 11th International Conference on Multinary and Ternary Compounds, (1997) 317.
[78] M. Lammer, U. Klemm, M. Powalla, Thin Solid Films 387 (2001) 33.
[79] H. Miyake, T. Haginoya, K. Sugiyama, Solar Energy Mater. Solar Cells 50 (1998) 51.
[80] B. Schumann, H. Neumann, A. Tempel, G. Kuhn, E. Nowak, Krist. Tech. 15 (1980) 71.
[81] K. Yoshino, K. Nomoto, A. Kinoshita, T. Ikari, Y. Akaki, T. Yoshitake, J. Mater. Sci. Mater. Electron. 19 (2008) 301.
[82] A.N. Tiwari, D.K. Pandya, K.L. Chopra, Thin Solid Films 130 (1985) 217.
[83] T. Terasako, Y. Uno, S. Inoue, T. Kariya, S. Shirakata, Physica Status Solidi (C) 3 (2006) 2588.
[84] S. Nakamura, Physica Status Solidi 3 (2006) 2564.
[85] Y.L. Wu, H.Y. Lin, C.Y. Sun, M.H. Yang, H.L. Hwang, Thin Solid Films 168 (1989) 113.
[86] M. Gorska, R. Beaulieu, J.J. Loferski, B. Roessler, Solar Energy Mater. 1 (1979) 313.
[87] Y. Shi, Z. Jin, C. Li, H. An, J. Qiu, Appl. Surf. Sci. 252 (2006) 373.
[88] S.P. Grindle, C.W. Smith, S.D. Mittleman, Appl. Phys. Lett. 35 (1979) 24.
[89] N. Satoh, K. Abe, K. Wakita, K. Mochizuki, Physica Status Solidi (C) 3 (2006) 630.
[90] Y.B. He, T. Kramer, I. Osterreicher, A. Polity, B.K. Meyer, M. Hardt, Semicond. Sci. Technol. 20 (2005) 685.
[91] Y. Ogawa, S. Uenishi, K. Tohyama, K. Ito, Solar Energy Mater. Solar Cells 35 (1994) 157.
[92] L. Essaleh, S.M. Wasim, J. Galibert, J. Leotin, P. Perrier, S. Askenazy, Phil. Mag. B 65 (1992) 843.
[93] H. Onnagawa, K. Miyashita, Jpn. J. Appl. Phys. 23 (1984) 965.
[94] R. Scheer, M. Alt, I. Luck, R. Schieck, H.J. Lewerenz, Mater. Res. Soc. Symp. Proc. 426 (1996) 309.
[95] D. Cybulski, A. Opanowicz, Cryst. Res. Technol. 32 (1997) 813.
[96] J.J.M. Binsma, J. Phys. Chem. Solids 44 (1983) 237.
[97] Y.B. He, A. Polity, R. Gregor, D. Pfisterer, I. Osterreicher, D. Hasselkamp, *et al.*, Physica B 308–310 (2001) 1074.
[98] H.Y. Ueng, H.L. Hwang, J. Phys. Chem. Solids 51 (1990) 1.
[99] D.C. Look, J.C. Manthuruthil, J. Phys. Chem. Solids 37 (1976) 173.
[100] F.O. Adurodija, J. Song, I.O. Asia, K.H. Yoon, Solar Energy Mater. Solar Cells 58 (1999) 287.

[101] B. Eisener, D. Wolf, G. Muller, Thin Solid Films 361–362 (2000) 126.
[102] S.R. Kodigala, V.S. Raja, Mater. Lett. 20 (1994) 29.
[103] R.H. Bari, L.A. Patil, A. Soni, G.S. Okram, Bull. Mater. Sci. 30 (2007) 135.
[104] N.F. Mott, E.A. Davis, Electronic Process in Non-Crystalline Materials, Clarendon, Oxford, 1971.
[105] S.R. Kodigala, V.S. Raja, A.K. Bhatnagar, F.S. Juang, S.J. Chang, Y.K. Su, Mater. Lett. 45 (2000) 251.
[106] D.J. Schroeder, J.L. Hernandez, G.D. Berry, A. Rockett, J. Appl. Phys. 83 (1998) 1519.
[107] H. Sakata, N. Nakao, Physica Status Solidi (a) 161 (1997) 379.
[108] A.M.A.E. Soud, H.A. Zayed, L.I. Soliman, Thin Solid Films 229 (1993) 232.
[109] P.S. Kireev, Semiconductor Physics, Mir Publishers, Mascow, 1978.
[110] G.P.S. Porras, S.M. Wasim, Physica Status Solidi (A) 59 (1980) K175.
[111] A. Amara, A. Drici, M. Guerioune, Physica Status Solidi (A) 195 (2003) 405.
[112] D.V. Lang, J. Appl. Phys. 45 (1974) 3014.
[113] R. Herberholz, M. Igalson, H.W. Schock, J. Appl. Phys. 83 (1998) 318.
[114] N. Christoforou, J.D. Leslie, Solar Cells 26 (1989) 197.
[115] T.R. Hanak, A.M. Bakry, D.J. Dunlavy, F. Abou-Elfotouh, R.K. Ahrenkiel, M.L. Timmons, Solar Cells 27 (1989) 347.
[116] R.K. Ahrenkiel, Solar Cells 16 (1986) 549.
[117] W.W. Lam, L.S. Yip, J.E. Greenspan, I. Shih, Solar Energy Mater. Solar Cells 50 (1998) 57.
[118] S. Kuranouchi, M. Konagai, Jpn. J. Appl. Phys. 34 (1995) 2350.
[119] L.L. Kerr, S.S. Li, S.W. Johnston, T.J. Anderson, O.D. Crisalle, W.K. Kim, et al., Solid State Electron. 48 (2004) 1579.
[120] I. Shih and A. I. Li, 22nd IEEE Photovoltaic Speacialist Conference 1991, p1100, Las Vegas, Nevada.
[121] F.A. Abou-Flfotouh, H. Moutinho, A. Bakry, T.J. Coutts, L.L. Kazmerski, Solar Cells 30 (1991) 151.
[122] H.R. Moutinho, D.J. Dunlavy, L. Kazmerski, R.K. Ahrenkiel, F.A. Abou-Elfotouh, 23rd IEEE Photovoltaic Specialist Conference, Louisville, KY, (1993) p572.
[123] F.A. Abou-Elfotouh, L. Kazmerski, A.M. Bakry, A. Al-Douri, 21st IEEE Photovoltaic Specialist Conference, (1990) p. 541.
[124] H. Okada, N. Fujita, H.-S. Lee, A. Wakahara, A. Yoshida, T. Ohshima, et al., J. Electron. Mater. 32 (2003) 15.
[125] M. Igalson, R. Bacewicz, 11th EC Photovoltaic Solar Energy Conference, Montreax, Switzerland, (1992) p. 874.
[126] A.M. Bakry, Λ.M. Elnaggar, J. Mater. Sci. Mater. Electron. 7 (1996) 191.
[127] J.A.M. Abushama, S. Johnston, R. Noufi, J. Phy. Chem. Solids 66 (2005) 1855.
[128] J.A.M. AbuShama, S. Jonston, T. Moriarty, G. Teeter, K. Ramanathan, R. Noufi, Prog. Photovolt.: Res. Appl. 12 (2004) 39.
[129] S.N. Kundu, S. Johnston, L.C. Olsen, Thin Solid Films 515 (2006) 2625.
[130] R.N. Bhattacharya, A. Balcioglu, K. Ramanathan, Thin Solid Films 384 (2001) 65.
[131] J. Abushama, S. Johnston, R. Ahrenkiel, R. Noufi, 29th IEEE Photovoltaic Specialist Conference, New Orleans, Louisiana, (2002) p. 740.
[132] B.M. Keyes, F. Hasoon, P. Dipp, A. Balcioglu, F. Abulfotuh, 26th IEEE PVSC, Anaheim, (1997) p. 479.
[133] K. Siemer, J. Klaer, I. Luck, D. Braunig, Thin Solid Films 387 (2001) 222.
[134] J.V. Gheluwe, P. Clauws, Thin Solid Films 515 (2007) 6256.
[135] M. Igalson, M. Edoff, Thin Solid Films 480–481 (2005) 322.

7 Fabrication and Properties of Window Layers For Thin Film Solar Cells

In this chapter, the fabrication of different n-type window or buffer layers such as CdS, ZnS, InS, *etc.*, and their surface, structural, optical, and electrical properties are illustrated. In addition, the preparation and characteristics of transparent n-type conducting layers such as ZnO, ITO, and FTO layers are demonstrated for CIGS thin film solar cell applications.

7.1 CdS Window Layer

The n-CdS thin films are extensively employed as window layers with many p-type absorbers such as Cu_2S, CdTe, InP, and $CuInGaSe_2$ in the thin film solar cells. The p-$CuInGaSe_2$/CdS and p-CdTe/CdS thin film solar cells exhibit world record efficiencies of 20.3 and 15.8%, respectively for which the CdS is prepared by simple chemical bath deposition (CBD) [1,2]. The CdS based thin film solar cell with low cost, cheap, and new prospective Cu_2ZnSnS_4 absorber shows efficiency of 7–10% [3,4]. The CdS is a wide band gap (2.42 eV) semiconductor and had reasonable lattice mismatch with CIGS. In order to improve open circuit voltage of CIGS/CdS thin film solar cell, the band gap of CdS layer has to be widen by adding Zn into it that is, $Cd_{1-y}Zn_yS_y$. On the other hand, the lattice mismatch can also be reduced by choosing proper y and x values in the $Cd_{1-y}Zn_yS$ and $CuIn_{1-x}Ga_xSe_2$, respectively. Both the band gap and lattice mismatches are tradable to obtain high efficiency from the CIGS/CdZnS thin film solar cells.

7.1.1 Growth Process of CdS Thin Films

The CdS layers have been prepared by several techniques such as spray, CBD, sputtering, vacuum evaporation, MOCVD, e-beam, *etc.* Among these CBD is a simple, easy, and inexpensive method. The quality of CBD-CdS layer depends on temperature, solution concentrations, type of solutes, stirring, pH of solutions, *etc.* The CdS thin films are grown by dipping CIGS coated glass substrates or bare substrates into the stirring chemical solutions of 1 mM CdX_2, 5 mM NH_4X, 100 ml H_2O, and 7.5 mM thiourea at bath temperature of 60 °C, where X=Cl, SO_4, (COOH), I, *etc.* The CdS layers grown at below bath temperature of 60 °C are in inferior quality. The chemical reaction follows as $Cd(NH_3)_4^{2+} + SC(NH_2)_2 + 4OH^- \rightarrow CdS + 6NH_3 + CO_3^{2-} + 2H_2O$. Another version of chemical reaction is also proposed as $Cd(NH_3)_4^{2+} + SC(NH_2)_2 + 2OH^- \rightarrow CdS + 4NH_3 + CN_2H_2 + 2H_2O$ [5]. Apparently the deposition of CdS layer lasts about 20–30 min to obtain required thickness of about 30–50 nm for thin film solar cell. The CdS thin film growth takes place ion by ion as heterogeneous that is advantageous process to enhance film thickness. At the beginning of deposition of CdS layers, the heterogeneous reaction may be dominant. After completion of deposition of CdS layers, the colloidal precipitation is observed underneath of waste chemical bath solution and surroundings of the glass beaker. The quantity of yellow precipitation increases with time indicating that the homogenous reaction may have well taken place by forming colloids in the solution. The homogeneous reaction may be dominant over the heterogeneous reaction with time or at the end. At the starting, the chemical solution is light green in color that slowly becomes yellow evidencing that homogeneous reaction may have gradually dominated heterogeneous reaction with time, as shown in Figure 7.1 [6,7]. The growth of Zn(S,O) buffer layers on CIGSS is considered as combination of heterogeneous and homogeneous process. First the ZnS layer grows as a heterogeneous reaction and then $Zn(OH)_2$ may be released from $Zn[(NH_3)_n]^{2+}$ and S^{2-} from thiourea that proceeds as homogeneous reaction [8]. The CBD ZnS buffer is grown using chemical solutions of $ZnSO_4 \cdot 7H_2O$, $(NH_4)_2SO_4$, thiourea, ammonium hydroxide, and hydrazine hydrate at 80 °C [9].

The CdS thin films are grown onto 7059 corning glass substrates by using chemical solutions of 1.4 mM CdI_2, NH_3, and 0.14 M thiourea. By adding NH_3 solution to the chemical bath, the pH of solution can be increased from 11.5, 11.8 to 11.9. The CdS layers show grain sizes of 200–500 Å for pH 11.5. The grain sizes of 200 and 800–1000 Å and voids are observed for 11.8, whereas grain sizes of 100–300 Å are found for pH 11.9 but coverage by grains on the substrate is little poor. The RMS and S/Cd ratios are 4.983, 11.873, 26.275 nm and 0.96, 0.99, and 0.96 for pH 11.5, 11.8, and 11.9, respectively. In this case pH 11.9 is used for obtaining CIGS cells with high efficiency of 17% [10]. At high bath temperature of 90 °C and low solution pH of 9.5, the CdS layers are also grown by CBD. The

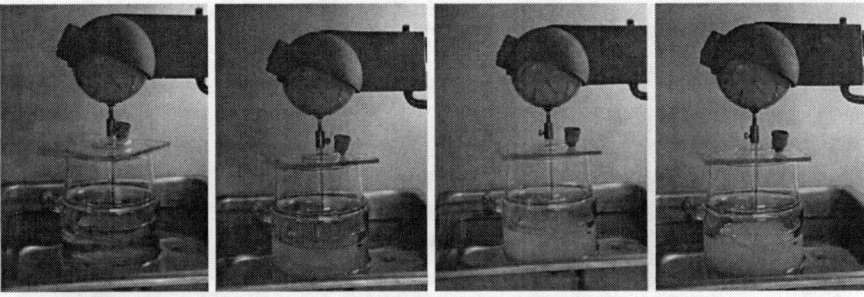

Figure 7.1 Chemical bath deposition experiment of CdS thin films; change in color of chemical solution with time.

CdZnS thin films are also deposited by CBD method using chemical solutions of 0.015 M $Cd(COOH)_2$, 0.01 M $Zn(COOH)_2$, 0.05 M thiourea, 0.6 M NH_3, and 0.1 M $NH_2(COOH)$ at bath temperature of 75 °C. A 40 nm thick In is deposited onto CdZnS film and annealed at 150–550 °C under air, in order to dope In. The In doped CdZnS exhibits low resistivity of 0.3 Ω-cm [11]. Unlike acetates, using chloride salts such as 1.2 mM$CdCl_2 \cdot H_2O$, 1.39 mMNH_4Cl, 11.9 mM$(NH_2)_2CS$, 0.627 mM$ZnCl_2$, and 52.7 mMNH_3 and deposition temperature of 80–85 °C and pH 7, the CdZnS films are also prepared by CBD method. Composition ratio of $y = Zn/(Cd+Zn)$ is varied from 0, 0.3, 0.5 to 0.7 in the solution in order to study the effect of Zn on the deposited CdZnS layers. The color of the layers changes from yellow to whitish-yellow with increasing y values. On the other hand, the adherence to substrates becomes poor for higher y values [12]. To study surface properties, the CdS thin films are deposited by closed space sublimation (CSS) onto either SnO_2 coated glass or SnO_2 coated silicon substrates.

The CdS thin films are also formed by simple and cheap spray pyrolysis technique for thin film solar cell applications. The chemical solution of 0.025 M $CdCl_2$ and thiourea is sprayed onto hot glass substrates at substrate temperature of 270–470 °C where it decomposes by providing desired film. The proposed endothermic reaction is $CdCl_2 + (NH_2)_2CS + 2H_2O \rightarrow CdS + 2NH_4Cl\uparrow + CO_2\uparrow$. The Cd to S ratio is kept to be 1:1.5 in the chemical solution. The In doped CdS films are prepared by adding indium chloride ($InCl_3$) to the chemical solution. The doping concentration of In 3 at% with respect to Cd in the solution is maintained. The spray rate of chemical solution is 7 ml/min [13]. The CdS films are also grown onto ITO coated borosilicate glass substrates by chemical vapor deposition (MOCVD) using dimethyl cadmium (DMCd), diethyl sulfide (DES) at substrate temperature of 350 °C, with atomic ratio of S/Cd = 4, deposition time of 90 min and total flow rates of 30 or 60 sccm to obtain thickness of 0.3 μm. The precursors used in this experiment are kept at zero centigrade [14]. In order to avoid splattering, the CdS pellet is used for evaporation of CdS thin films, which is annealed under vacuum at 450 °C for 30 min. By employing substrate temperature of 150 °C the CdS layers are formed onto glass substrates [15].

7.1.2 Surface Analysis of CdS Thin Films

The AFM images of CdS layers grown by CBD and CSS are depicted in Figure 7.2. The RMS and grain sizes are 9–10 and 81 nm for CBD-CdS, whereas they are in higher value of 17–20 and 136–142 nm for CSS-CdS thin films [16,17]. In order to grow CdS thin film on glass substrates under influence of magnetic field, the uniform magnetic field at constant has been applied to bath container during the film growth, which contains chemical solutions of $CdCl_2$, KOH, NH_4NO_3, CS$(NH_2)_2$, and pH 10. Presumably the CdS layers formed in the presence of magnetic field cause to create sulfur vacancies. In all aspects of atomic weight and radius, Cd is heavier than S, therefore, the drift on S is higher that may be caused to have sulfur deficiency in the CdS thin films. The CdS layers grown under magnetic field show the preferential parallel columnar growth with respect to the substrate due to

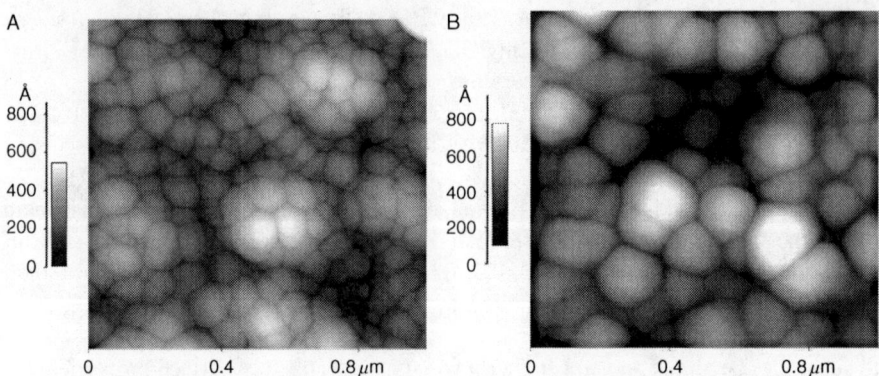

Figure 7.2 AFM images of CdS layers grown by (A) CBD and (B) closed space chemical vapor deposition.

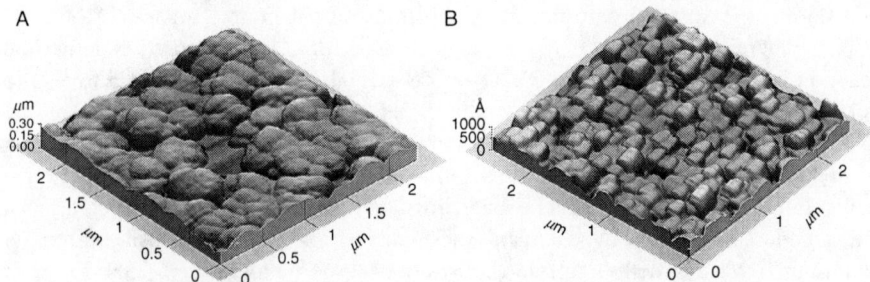

Figure 7.3 AFM images of CdS thin films grown by CBD; (A) without and (B) with magnetic field.

stronger Lorentz's force. The layers grown without magnetic field consist of no preferred growth, as shown in Figure 7.3 [18]. The 0.1 at% In doped CdS films deposited at different substrate temperatures of 300, 375, and 400 °C by spray pyrolysis technique using In(COOH)$_3$ source for In doping show agglomerates of grains look like cauliflower, whereas the layers deposited at 450 °C had same shapes but with developed grains, as shown in Figure 7.4. The SEM analysis reveals that the layers grown at low temperature of 350–375 °C contain smooth surface and pores, whereas the layers grown at high temperature of 400–450 °C show no pores and continuous structure shapes [19]. On the other hand the Cd to S ratio increases with increasing substrate temperature [22,23]. In the XPS spectra of PLD CdS, the Cd3d$_{5/2}$ or Cd3d$_{3/2}$ position decreases and S2P position increases with increasing CdS deposition substrate temperature from 100 to 450 °C due to a change in electrostatic charging or composition ratio, as shown in Figure 7.5. The AFM images of FTO and CBD-CdS layers on FTO show RMS values of 22.5 and 18.8 nm, respectively (Figure 7.6). The grain sizes of the layers increase from 0.34 to 0.42 μm for

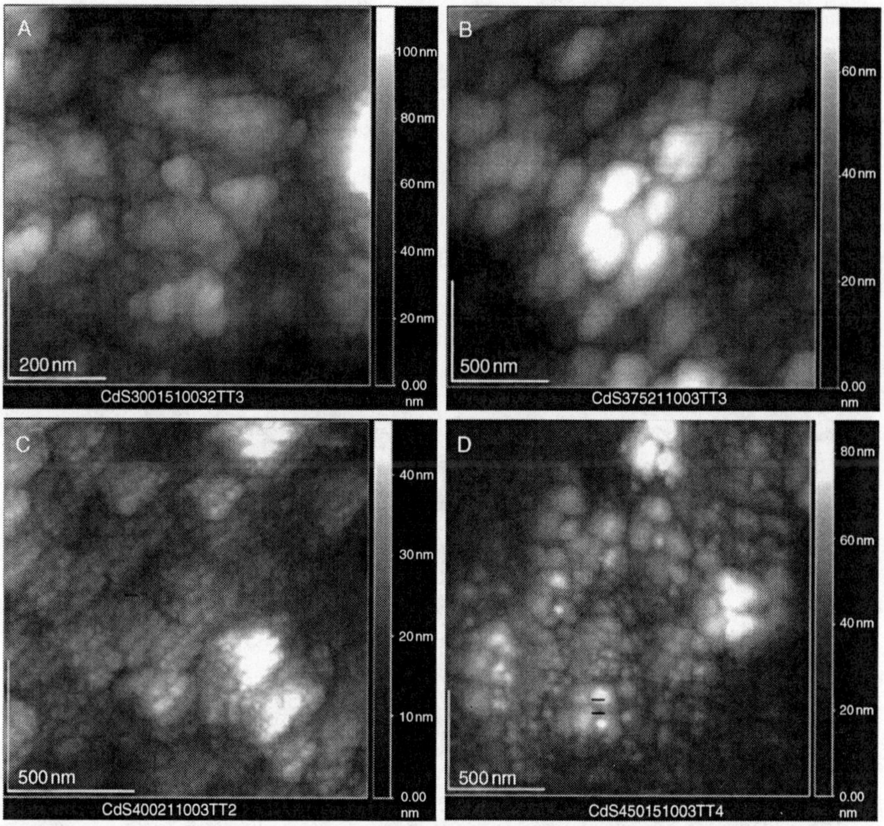

Fabrication and Properties of Window Layers For Thin Film Solar Cells 397

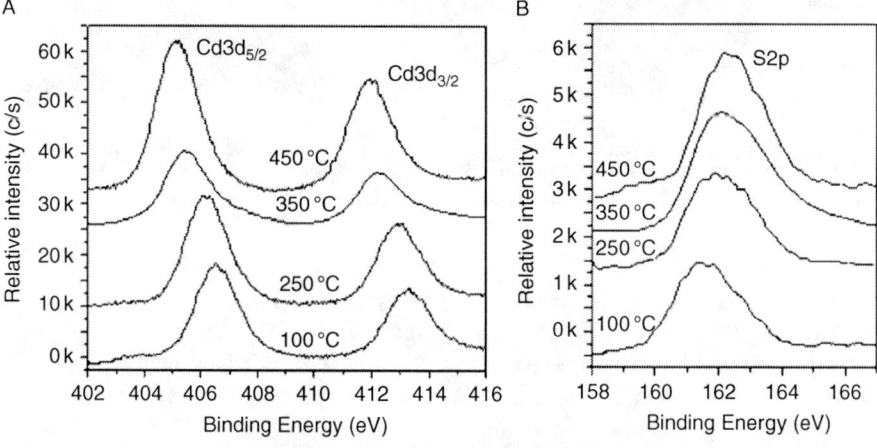

Figure 7.4 AFM images of In doped CdS thin films grown by spray pyrolysis technique at different substrate temperatures of (A) 300 (B) 375, (C) 400, and (D) 450 °C.

Figure 7.5 XPS spectra of Cd and S core levels in the CdS thin films.

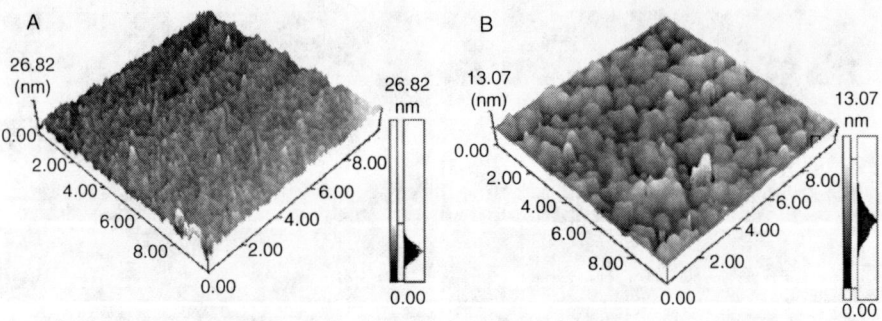

Figure 7.6 AFM images of (A) FTO and (B) CdS grown on FTO.

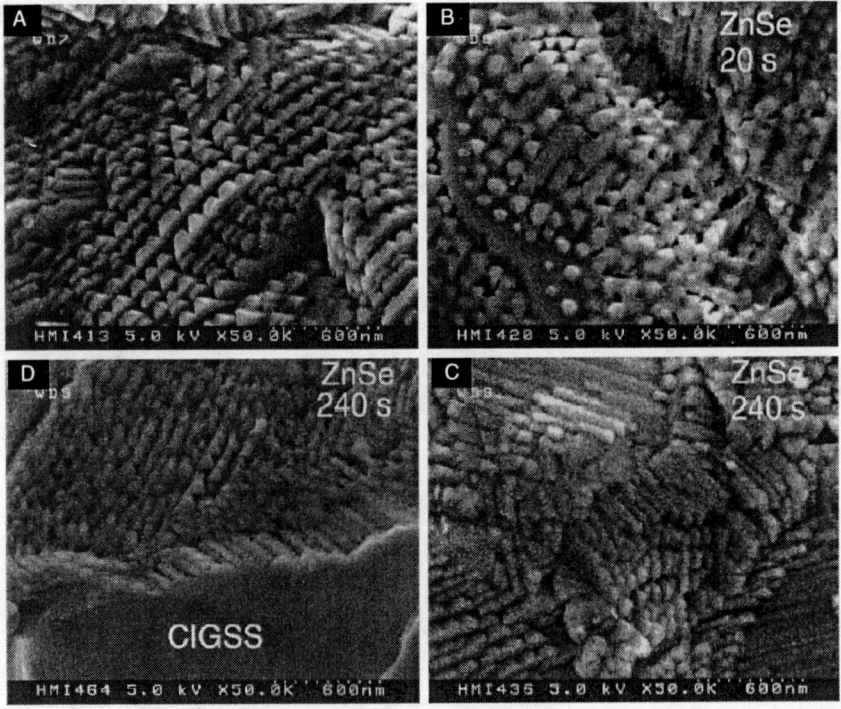

Figure 7.7 SEM of (ZnSe-OH) grown on CIGS by CBD.

annealing treatment [20]. In the case of ZnSe by CBD, a change in growth of Zn(Se, OH) on CIGSS can be observed with increasing growth time from the SEM, as shown in Figure 7.7. After the growth of 20 s, the Zn(Se,OH) growth pattern is as same as surface of CIGSS (Figure 7.7A and B) that looks like a two dimensional growth covering laterally all over the surface of CIGSS then the three dimensional growth takes place (Figure 7.7C and D) [21].

7.1.3 Structural Analysis of CdS Thin Films

7.1.3.1 TEM

The CdS exhibits hexagonal and cubic structures or mixed. The former is preferable for thin film solar cell applications, despite of higher lattice mismatch of 1.2% with CIS as compared to 0.7% of cubic CdS owing to stableness. The lattice parameters a and c of hexagonal structure (Figure 7.8) can be derived using simple equation

$$\frac{1}{d^2} = \frac{3}{4}\frac{(h^2+k^2)}{a^2} + \frac{l^2}{c^2}, \tag{7.1}$$

where d is interplanar lattice spacing and (hkl) is miller index. The electron diffraction of CdS films grown by CBD shows ring patterns, as shown in Figure 7.9. The CdS films are annealed under $CdCl_2$ vapor at 450 °C for 5 min in order to reduce planar defects and improve crystallinity. The annealed CdS layers show discontinued and number of spots indicating well-developed crystalline nature. The discontinued rings with number of spots are observed in the as-grown CSS CdS layer without any post-treatment indicating well-developed crystalline structure. No difference between as-grown and $CdCl_2$ treated CSS-CdS is observed [16]. The CBD-CdS thin film grown by us shows continuous ring patterns without spotty patterns. However, after annealing thin films by laser, spotty ring pattern occurs, as shown in Figure 7.10 [24]. As shown in Table 7.1, the (111), (220), (222), (400), and (422) are the reflections of cubic phase CdS. The standard d values of (111),

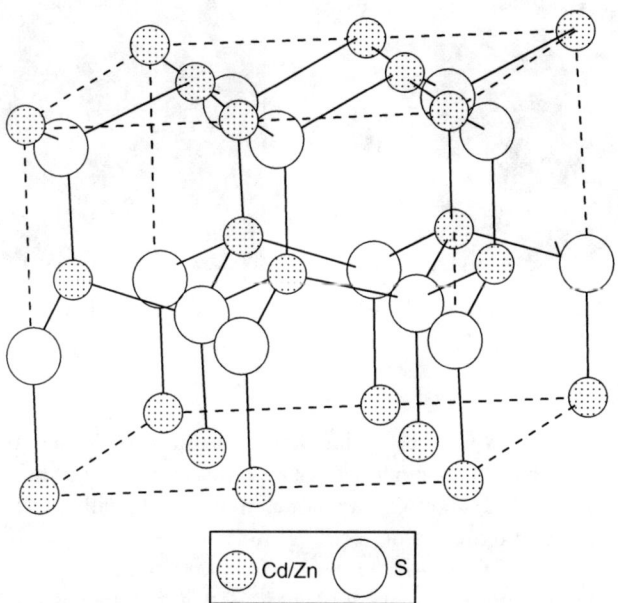

Figure 7.8 Hexagonal structure of CdS.

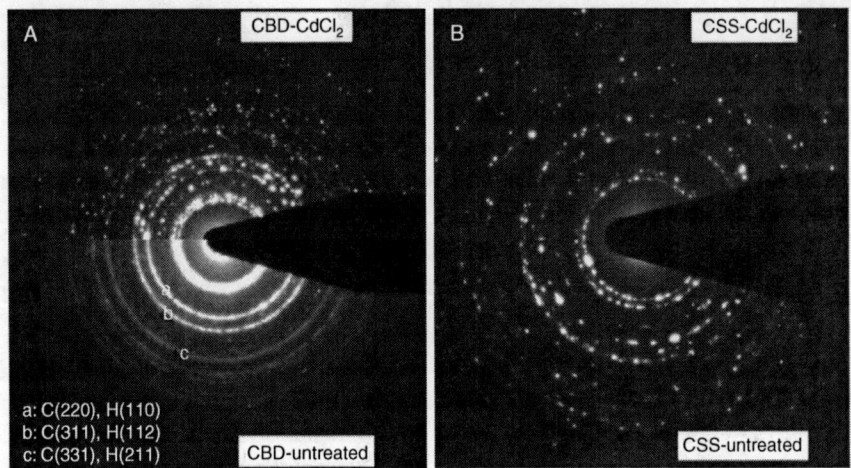

Figure 7.9 (A) Selected area diffraction (SAED) pattern of as-grown CBD-CdS layer (bottom) and annealed under $CdCl_2$ (top), (B) as-grown CSS-CdS layer (bottom) and annealed under $CdCl_2$ (top).

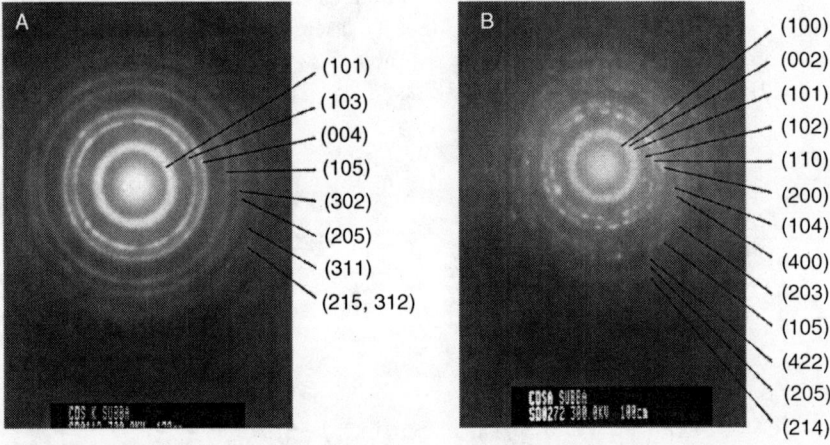

Figure 7.10 SAED pattern of as-grown and annealed CdS layer by laser under vacuum.

(220), and (222) are very closely coinciding with those of hexagonal reflections. Therefore, it is difficult to conclude whether the phase is cubic or hexagonal. However, the d values of (400) and (422) are not at all matching with those of hexagonal. Hence the presence of these two reflections either in the XRD or SAED easily confirms cubic structure of CdS. All the new diffraction patterns appear except few in the SAED pattern of annealed samples. The SAED pattern reveals that the presence of (400), and (422) reflections is indication of cubic phase of CdS. On the other

Table 7.1 SAED pattern of as-grown, laser annealed CBD-CdS layer, and JCPDS data of hexagonal and cubic CdS

S.No.	CdS hexagonal d (nm)	JCPDS (hkl)	CdS cubic d (nm)	JCPDS (hkl)	SAED (As) d (nm)	SAED (Ann) d (nm)	XRD d (nm)	XRD[a] d (nm)
1	0.3583	(100)	–	–	–	0.3672	–	0.3555
2	0.3357	(002)	0.336	(111)	–	0.3385	0.328	0.3326
3	0.3160	(101)	–	–	0.3190	0.3098	–	0.3132
4	0.2450	(102)	–	–	–	0.2542	–	0.2439
5	0.2068	(110)	0.2058	(220)	–	0.2155	–	0.2058
6	0.1898	(103)	–	–	0.1944	–	–	–
7	0.1791	(200)	–	–	–	0.1836	–	0.1784
8	0.1761	(112)	–	–	–	–	–	0.1757
9	0.1731	(201)	–	–	–	–	–	0.1725
10	0.1679	(004)	0.1680	(222)	0.1660	–	–	0.1677
11	0.1581	(202)	–	–	–	–	–	0.1578
12	0.1520	(104)	–	–	–	0.1525	–	–
13	–	–	0.1453	(400)	–	0.1471	–	–
14	0.1398	(203)	–	–	–	0.1416	–	0.1394
15	0.13536	(210)	–	–	–	–	–	0.1353
16	0.13271	(211)	–	–	–	–	–	0.1324
17	0.13032	(114)	–	–	–	–	–	0.1301
18	0.12572	(105)	–	–	0.1260	0.1239	–	0.1256
19	0.11940	(300)	–	–	–	–	–	0.1194
20	0.11585	(213)	–	–	–	–	–	0.1159
21	–	–	0.1186	(422)	–	0.1180	–	–
22	0.11249	(302)	–	–	0.1132	–	–	0.11304
23	0.10743	(205)	–	–	0.1091	0.1078	–	–
24	0.10540	(214)	–	–	–	0.1044	–	–
25	0.09827	(311)	–	–	0.0983	–	–	–
26	0.095533	(215,312)	–	–	0.0942	–	–	–

[a]Powder CdS; As = As-grown; Ann = annealed.

hand huge number of hexagonal planes such as (101), (002), (100), (103), (105), *etc.*, present in the pattern of annealed sample (Figure 7.10) [24]. Therefore, it can be concluded that the annealed CBD-CdS is mixed phase of cubic and hexagonal. Surprisingly, the as-grown CBD-CdS shows hexagonal structure. The same is observed by optical transmission spectra.

The SAED patterns of CBD-CdS on CIGS represent combined patterns of CdS and CIGS, as shown in Figure 7.11 in which CdS shows hexagonal structure. The planes of CBD-CdS such as $\{01\bar{1}0\}$ and $\{10\bar{1}0\}$ coincide with the CIGS planes of $\{112\}$ and $\{132\}$, respectively. On the other hand, plane $\{0001\}$ of PVD-CdS also matches with $\{112\}$ of CIGS in the sense of lattice mismatch. The pattern is

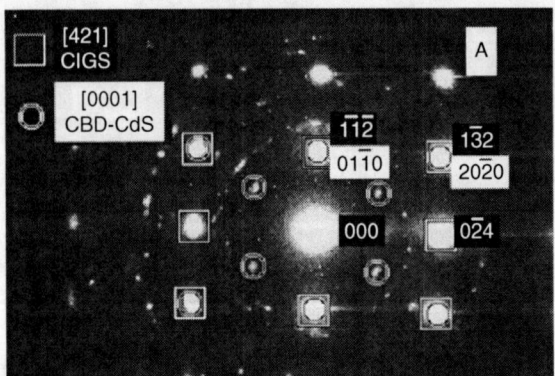
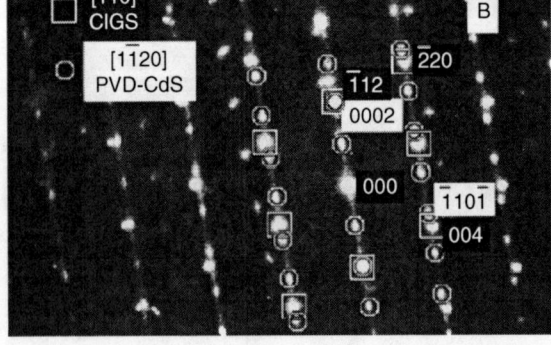

Figure 7.11 SAED pattern of (A) CBD-CdS on CIGS and (B) PVD-CdS on CIGS.

coherent in the CBD-CdS/CIGS, whereas it is incoherent in the PVD-CdS/CIGS [25]. Figure 7.12 shows (0001) plane position of CdS in hexagonal crystal and (112) plane position in tetragonal (chalcopyrite) structure. In general, the CdS grown by CBD is combination of cubic and hexagonal structure. Despite having large number of stacking faults, the CIGS cell with hexagonal CBD-CdS shows high efficiency due to having larger band gap of 2.58 eV comparing with band gap of 2.38 eV for cubic CdS. The SAED pattern of CBD-CdS grown on CIGS reveal that the d spacings of 3.36 and 2.06 Å are close to both cubic CdS (111) and (220) reflections as well as to hexagonal CdS (002) and (110), respectively. The (111) or (002) is parallel to (112) plane of CIGS. Therefore, the orientation of CdS depends on the structure of CIGS, that is, epitaxial growth [26]. The SAED pattern of CdS reveals that the grown CdS is epitaxial rather than amorphous, that is, $\{111\}_{cubic-CdS}//\{112\}_{CIGS}$. The HRTEM or lattice image of CBD-CdS along the zone axis of $\langle 110 \rangle$ shows {ABCABC} type stacking for cubic {111} plane. The $d_{(111)}$ spacing of 3.4 Å, ABC stacking and parallel lattice springes of CdS and CIGS are shown in Figure 7.13A and B, respectively [27]. The high-resolution transmission electron micrograph is recorded on (002) oriented lattice fringes and the d

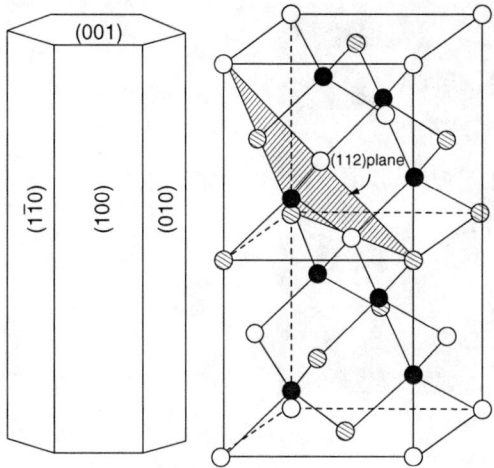

Figure 7.12 (A) Typical planes in hexagonal structure and (B) (112) plane in chalcopyrite structure (○ Cu, ◎ In and ● Se).

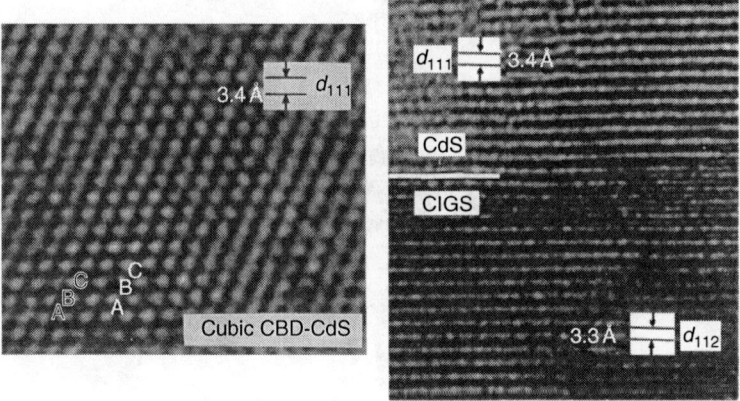

Figure 7.13 (A) ABC stacking defects of CdS and (B) parallel lattice springes of CdS and CIGS.

spacing is measured to be 0.35 nm, as shown in Figure 7.14. The similar TEM image reveals that the $d_{(10\bar{1}0)}$ of 0.37 nm and $d_{(112)}$ of 0.26 nm are observed for CdS and CIGS, respectively, as given in Figure 7.15A. One can see higher defect density in the PVD-CdS/CIGS than in the CBD-CdS/CIGS, as shown in Figure 7.15B.

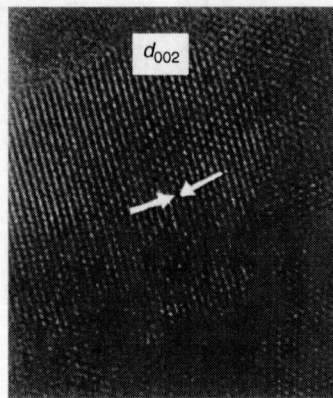

Figure 7.14 TEM of $d_{(002)}$ spacing of annealed CBD-CdS thin film.

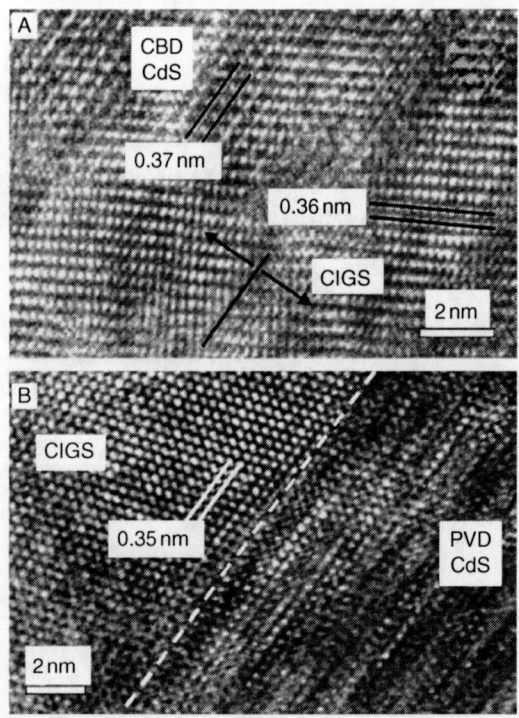

Figure 7.15 HRTEM micrograph of (A) CBD-CdS/CIGS and (B) PVD-CdS/CIGS heterostructures.

7.1.3.2 XRD

The XRD pattern of hexagonal CdS can theoretically be constructed using model, which is elaborately illustrated in Chapter 4. The structure factor of CdS can be obtained as

$$F_{(hkl)} = f_{Cd}\sum e^{2\pi i(hx+ky+lz)} + f_S \sum e^{2\pi i(hx+ky+lz)}. \tag{7.2}$$

The Cd and S atomic positions in two molecule hexagonal unit of zincite (ZnO) structure are at Cd: 000, 1/3 2/3 1/2 and S: 00u, 1/3 2/3u + 1/2 and $u = 0.375$. After substituting atomic positions in Equation (7.2), the structure factor is turn out to be

$$F_{(hkl)} = \left[1 + e^{2\pi i\left(\frac{h}{3}+\frac{2k}{3}+\frac{l}{2}\right)}\right]\left[f_{Cd} + f_S e^{2\pi i l u}\right], \tag{7.3}$$

the $|F_{(hkl)}|^2$ can be obtained by substituting (hkl) values for each plane in Equation (7.3), which are given in Table 7.2. The intensities of diffraction peaks are calculated by substituting structure factor and other parameters in Equation (4.5), which nearly coincide with the experimental values, as listed in Table 7.2. The theoretically constructed XRD pattern of hexagonal CdS is depicted in Figure 7.16. The intensities of some of the experimental peaks for CdS by spray are in low comparing with theoretical values that could be due to influenced by growth conditions of CdS thin films [13].

Many researchers observed that a diffraction angle at $\sim 26.5°$ is either related to cubic (111) or hexagonal (002) structure. It is difficult to precisely judge the

Table 7.2 X-ray intensity data of CdS, CdS:In (3 at%), and annealed CdS:In (3 at%) thin films

(hkl)	d (nm)	Cal. CdS $|F_{(hkl)}|^2$	I/I_0	JCPDS 6-314 I/I_0	exp. CdS	I/I_0 CdS:In	Annealed CdS:In
(100)	0.3581	$(f_{Cd}+f_S)^2$	59	75	36	62	17
(002)	0.3374	$4(f^2_{Cd}+f^2_S)$	45	59	18	46	5
(101)	0.3163	$3[(f_{Cd}-f_S\backslash 2)^2+(f_S\backslash 2)^2]$	100	100	100	100	73
(102)	0.2456	$(f^2_{Cd}+f^2_S)$	27	25	18	31	7
(110)	0.20674	$4(f_{Cd}+f_S)^2$	54	57	55	58	14
(103)	0.19040	$3[(f_{Cd}+f_S\backslash 2)^2+(f_S\backslash 2)^2]$	56	42	24	31	5
(200)	0.17903	$(f_{Cd}+f_S)^2$	8	17	12	7	–
(112)	0.17629	$4(f^2_{Cd}+f^2_S)$	43	45	76	58	100
(201)	0.17305	$3[(f_{Cd}-f_S\backslash 2)^2+(f_S\backslash 2)^2]$	19	18	8	9	4
(004)	0.16873	$4(f_{Cd}-f_S)^2$	3	4	–	–	–
(202)	0.15921	$(f^2_{Cd}+f^2_S)$	8	7	–	–	5
(104)	0.15263	$(f_{Cd}-f_S)^2$	4	2	–	–	–
(203)	0.14001	$3[(f_{Cd}+f_S\backslash 2)^2+(f_S\backslash 2)^2]$	20	15	–	13	5
(210)	0.13534	$(f_{Cd}+f_S)^2$	3	6	–	–	–
(211)	0.13270	$3[(f_{Cd}-f_S\backslash 2)^2+(f_S\backslash 2)^2]$	8	11	–	9	5
(114)	0.13072	$4(f_{Cd}-f_S)^2$	7	7	–	–	–
(105)	0.12630	$3[(f_{Cd}+f_S\backslash 2)^2+(f_S\backslash 2)^2]$	13	11	–	–	4
(204)	0.12279	$(f_{Cd}-f_S)^2$	1	1	–	–	–
(300)	0.11940	$4(f_{Cd}+f_S)^2$	8	8	–	–	–
(213)	0.11597	$3[(f_{Cd}+f_S\backslash 2)^2+(f_S\backslash 2)^2]$	9	12	–	–	12

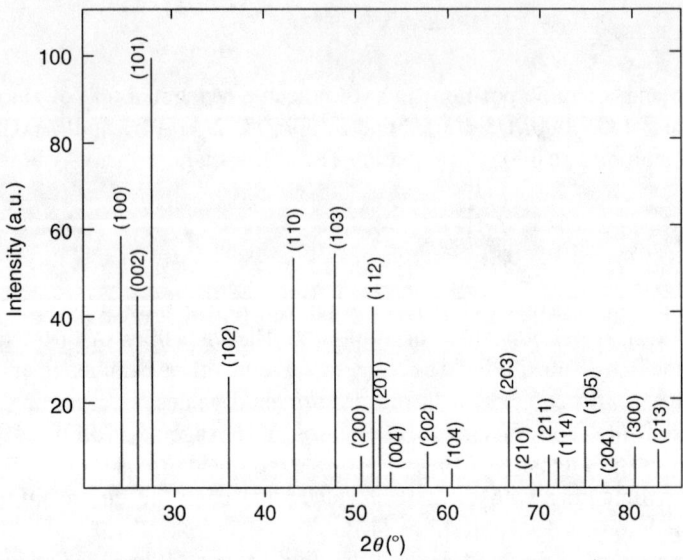

Figure 7.16 Theoretically constructed XRD pattern of CdS.

structure unless a lot of diffraction peaks occur [28]. The XRD patterns of our CBD and powder CdS are shown in Figure 7.17 in which the lonely X-ray reflection peak occurs at $2\theta = 27.04°$ relating to either (002) hexagonal or (111) cubic phase for the CdS (CBD) films, whereas the powder pattern had multiple reflections. After laser annealing CBD-CdS thin film, the intensity of (002) or (111) peak at $2\theta = 27.15°$ increases but not sharply in the XRD spectrum. The grain size of CdS (CBD) thin films increases from 4.55 to 5.3 nm, which is determined by Scherrer formula. Some of the diffraction reflections such as (103), (004), (105), (302), (205), etc., present in the SAED pattern but absent in the XRD spectrum for our CBD-CdS thin film. This anomaly could be due to a difference in the sample thickness, which is < 100 Å for the SAED pattern and $\gg 100$ Å for the XRD pattern. In the thinner samples, the orientation of the planes is random comparing with the results of thicker samples [24]. The (111) peak in the XRD spectrum shows higher intensity for CdS films grown by CBD under magnetic field than one grown without magnetic field [18].

The structure of CdS layers grown by CBD changes from cubic to hexagonal by decreasing solution pH from 10.5 to 8 in the chemical solutions of 2.4 mM $CdCl_2$, 28 mM NH_4Cl, and 57 mM thiourea at bath temperature of 70 °C. The hexagonal phase can also be obtained after annealing CBD-CdS thin film under $CdCl_2$ with irrespective of structure of the films, growth conditions and substrates. The films show grain sizes of 90 nm and lattice parameters of $a = 4.14$ and $c = 6.72$ Å [29]. The CdS thin films with Cd/S = 0.92 or 1 grown by CBD at bath temperature of 75 °C and solution pH of 10.5 show cubic structure giving multiple diffraction peaks (111), (200), (220), and (311) at diffraction angles of 26.5, 30.8, 43.9, and 52.1°, respectively, whereas the CdS films show same cubic structure with extra reflection

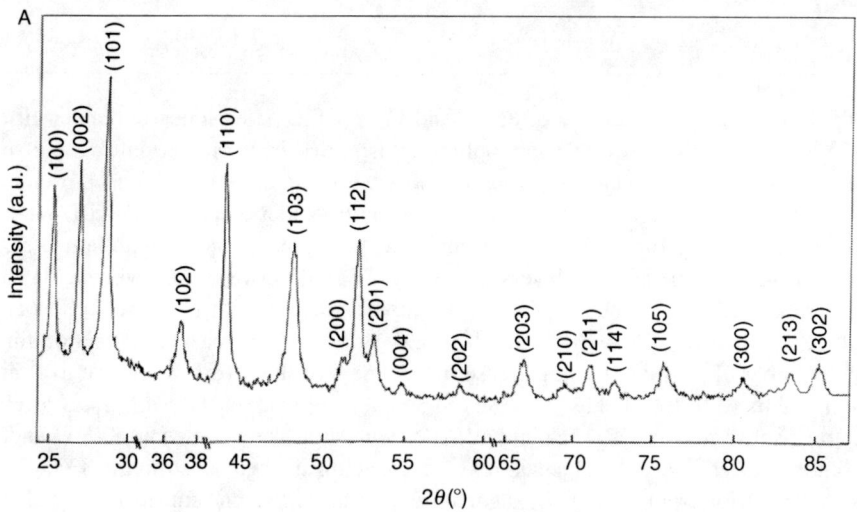
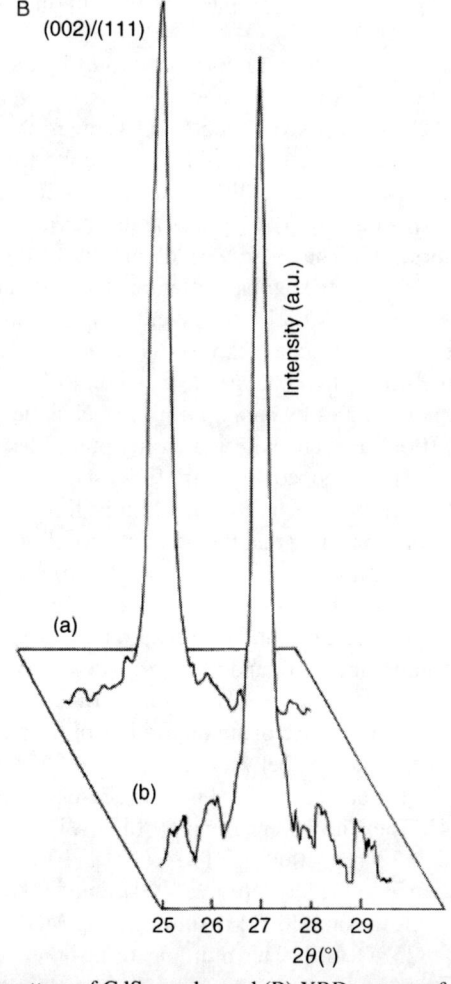

Figure 7.17 (A) XRD pattern of CdS powder and (B) XRD pattern of (a) CdS grown by CBD and (b) laser annealed.

of (200) for bath temperature of 85 °C and Cd/S = 1.2. After annealing under air at 450 °C for 15 min, the films show polymorphism, that is, combination of cubic and hexagonal [30]. However, the phase of CdS layer changes from cubic to hexagonal and the d spacing slightly increases upon annealing under S_2 [31]. The XRD analysis reveals that (100) and (110) peaks appear as hexagonal structure with lattice constant of 4.08 Å for the CdS layers grown by CBD at above 60 °C, whereas layers grown at above 40 °C show presumably cubic structure with (111) reflection and lattice constant of 5.77 Å [6]. The (002) peak shifts with increasing deposition time and number of dips due to an increase of thickness of the layers that cause to strain in the films [32]. (002)/(111) is observed as intensity peak in the CdS layers grown onto Si substrates by CBD technique. After annealing the samples at 500 °C under vacuum, several hexagonal peaks exist evidencing hexagonal structure [33]. The phase transition from cubic to hexagonal for CdS thin films on semi-insulating GaAs substrates is observed after annealing under argon at 400 °C [34]. The structure also changes from cubic to hexagonal in the sputtered CdS thin films with increasing sputter power from 50, 75 to 150 W [35].

The CdS:In thin films deposited onto glass substrates by electron beam evaporation at deposition temperature of 60 °C show multiple peaks such as (002), (100), (101), (110), and (112) in the XRD spectrum, whereas the films show strong (002) orientation for the temperature of 100 °C or above [36]. (100), (002), and (101) are the dominant peaks at low diffraction angles for the CdS and CdS:In (3 at%) thin films grown by spray pyrolysis. Among these, the (101) is the intense peak, as shown in Figure 7.18. The as-grown CdS and In or Al doped CdS thin films grown by spray pyrolysis show hexagonal structure. The d spacing of different (hkl) for CdS, CdS:In, and annealed CdS:In are given in Table 7.3. After annealing In doped CdS films, the orientation changes from (101) to (112) due to recrystallization [13]. The preferred orientation changes from (002) to (101) irradiating samples by laser while depositing samples by spray pyrolysis technique [37]. The orientation of CdS changes from (002) to (100) with increasing pulsed laser deposition density from 2 to 4 J/cm^2 [38]. The In doped CdS thin films show inferior quality comparing with un-doped CdS layers. The In doped CdS thin films grown onto glass substrates by vacuum evaporation show that the orientation changes from (002) to (110) with increasing In concentration from 3×10^{18} to 1×10^{21} cm^{-3} in the CdS thin films [15]. In addition, several peaks exist due to an increase of polycrystalline nature. The degree of preferred orientation of the crystallites depends on the cation/anion ratio, substrate temperature and other growth recipes [39–43]. In the case of Ga doped CdS, the $d_{(111/002)}$ spacing is minimum for the Ga/Cd ratio of 0.034 due to substitution of Ga into Cd because of smaller radius of Ga comparing with Cd but d spacing increases for the ratio of between 0.034 and 0.06 and it crossover that of un-doped CBD-CdS for beyond ratio of 0.06 due to interstitials and an increase of strain in the films [44]. The CdS layer grown by MOCVD technique shows hexagonal structure evidencing by reflections of (002), (101), (102), (103), and (105) [14]. The hexagonal structure with ordered (002), (004), and (006) reflections are also found in the CdS films grown onto Si (111) substrates by MOCVD technique at substrate temperature of 325 °C [45]. The multiple reflections such as (002), (102),

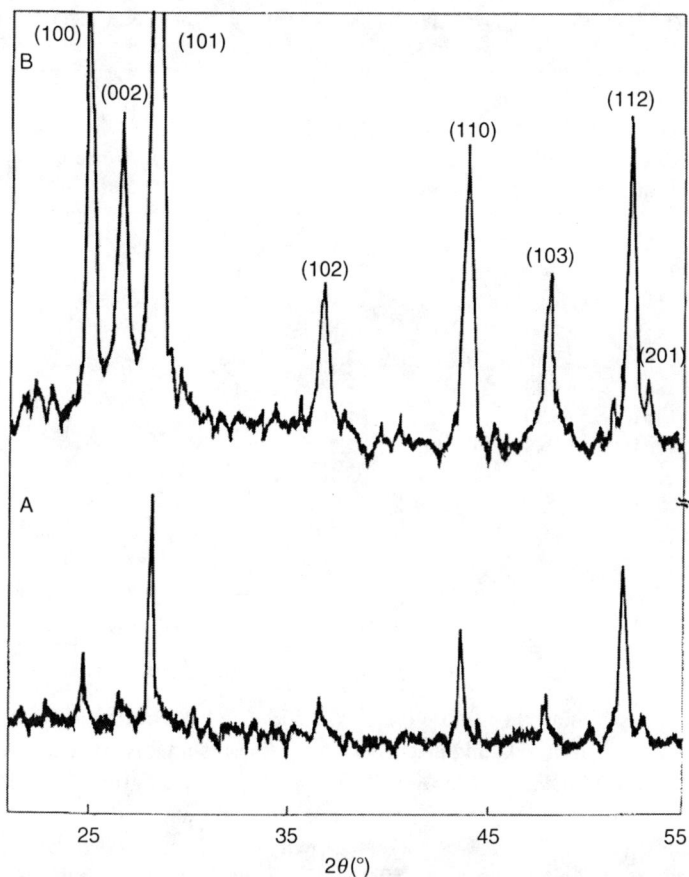

Figure 7.18 XRD pattern of (A) CdS and (B) In doped CdS thin films grown by spray pyrolysis technique.

(110), (103), and (004) present with increasing substrate temperature from 120, 200 to 350 °C evidencing authentic hexagonal structure [46].

The CdS thin films are deposited onto glass substrates by vacuum evaporation at substrate temperature of 200 °C and evaporation source temperature of 850 °C under vacuum pressure of 3×10^{-2} Pa. During deposition of CdS layers white light is illuminated on the sample in the range of 50–150 mW/cm^2 with an increment of 50 mW/cm^2. The CdS layers show (100), (002), and (101) peaks in the XRD spectrum without light illumination. With increasing light intensity during growth of CdS layers the intensity of (002) peak increases [47]. The as-deposited CdS thin films by sputtering show amorphous. 70 nm thick CdCl$_2$ layer is coated onto CdS and annealed under air at 500 °C for 20 min show hexagonal structure. However, a CdO secondary phase presents for thicker CdCl$_2$ layer. The orthorhombic structure is observed for the films annealed below 500 °C [48]. The CdS thin films grown onto glass substrates by laser ablation at 200 °C show combined cubic and

Table 7.3 X-ray diffraction data of spray-deposited CdS, CdS:In (3 at %) and annealed CdS:In (3 at %) thin films

(hkl)	JCPDS 6-314 d (nm)	CdS 2θ (°)	d (nm)	CdS: In (3 at%) 2θ (°)	d (nm)	CdS: In (3 at%) Annealed 2θ (°)	d (nm)
(100)	0.3583	24.83	0.359	24.93	0.357	24.90	0.358
(002)	0.3357	26.40	0.338	26.60	0.335	26.35	0.338
(101)	0.3160	28.13	0.317	28.30	0.315	28.35	0.315
(102)	0.2450	36.60	0.246	36.81	0.244	36.65	0.245
(110)	0.2068	43.66	0.207	43.70	0.207	43.70	0.207
(103)	0.1898	48.00	0.190	48.04	0.189	47.95	0.190
(200)	0.1971	50.00	0.182	51.20	0.178	–	–
(112)	0.1761	51.83	0.176	52.18	0.175	51.90	0.176
(201)	0.1731	53.00	0.173	53.20	0.172	52.90	0.173
(202)	0.1581	–	–	–	–	58.35	0.158
(203)	0.1398	–	–	67.60	0.139	66.75	0.140
(211)	0.13271	–	–	71.40	0.132	70.75	0.133
(105)	0.12572	–	–	–	–	75.45	0.126
(213)	0.11585	–	–	–	–	83.10	0.116

hexagonal phases that the cubic phase is evidenced by reflections of (111) at 26.542° and (222) at 56.64° and hexagonal phase is presumably witnessed by reflections of (103) at 48.01°, and (105) at 75.56°, whereas layers grown at 400 °C show hexagonal structure by presenting reflections of (002) and (004) only [49]. (002) reflection is stronger in the CdS layers grown onto Si substrates rather than the one grown onto glass substrates [50]. (002) is the intensity peak for the CdS thin films deposited onto Si substrates by femtosecond pulsed Ti:sapphire laser at 100–450 °C, whereas films grown by nanosecond laser shows (002) as the preferred orientation. In the former case, the laser contains high energy that creates different dynamic plasma plume. Therefore, the species could not find sufficient time to align in preferred growth direction of (002) [20].

The $Cd_{1-y}Zn_yS$ thin films are deposited by evaporation of CdS and ZnS powders in separate crucibles at substrate temperature of 150 °C under pressure of 10^{-3} Pa. The grown $Cd_{1-y}Zn_yS$ samples show hexagonal structure for $y \leq 0.65$, whereas the samples show cubic structure for $y \geq 0.85$. The (002), (101), (103), (112), and (004) peaks appear in the hexagonal samples, whereas (111) peak appears in cubic structure samples. Variation of lattice parameters a and c with varying y is linear in the $Cd_{1-y}Zn_yS$. The roughness increases with increasing Zn content in the samples [51]. In another occasion, the XRD pattern of CBD $Cd_{1-y}Zn_yS$ thin films shows (100), (002), (101), (102), (110), (103), (112), and (004) reflections for $y=0$. As y is increased to 0.3 from the intensities of peaks decrease and the films become amorphous with further increasing y. Among all y values, the film growth is faster for $y=0.3$. The diffraction angle increases toward higher angle with increasing Zn content in the layers [12]. The lattice parameters a and c of $Cd_{1-y}Zn_yS$ layer like paste formed on glass substrates using powder and acetone vary linearly with y, as shown in Figure 7.19. The lattice parameters obey Vegard's law $a = 4.136 - 0.34y$ and $c = 6.705 - 0.46y$. The hexagonal structure is observed for $0 < y < 0.85$ range, whereas the cubic structure is found above 0.85 [52]. The XRD patterns show that the intensities of peaks are high in the CdZnS:In (6 at%) than in the CdZnS:Al (6 at%). This could be due to deterioration of crystal structure by Al doping in the CdZnS layers, as shown in Figure 7.20 [53].

7.1.4 Determination of Thickness by Optical Method

Thickness of the films can be calculated from spectral transmittance curve if multiple interference maxima and minima exit. The thickness (d_f) of film can be written as [54].

$$d_f = \frac{M\lambda_1\lambda_2}{2[n(\lambda_1)\lambda_2 - n(\lambda_2)\lambda_1]}, \quad (7.4)$$

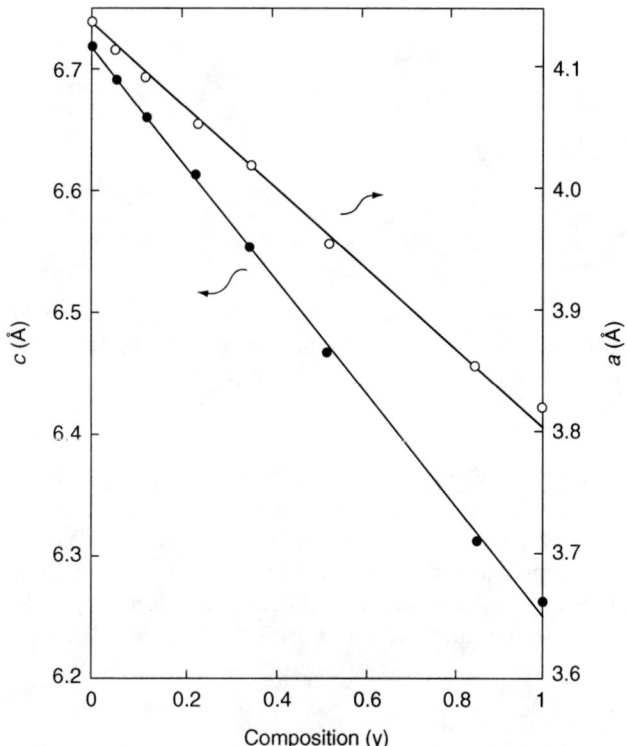

Figure 7.19 Variation of lattice constants a and c with function of composition y.

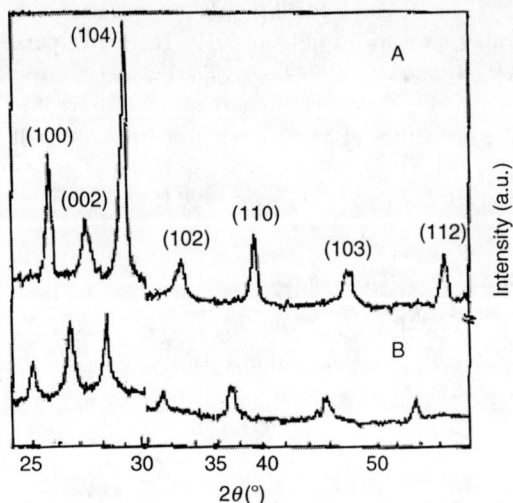

Figure 7.20 XRD patterns of (A) CdZnS:In and (B) CdZnS:Al thin films grown by spray pyrolysis technique.

where M is the number of oscillations between two extrema at wavelengths λ_1 and λ_2. $n(\lambda_1)$ and $n(\lambda_2)$ are the refractive indices of the film. The refractive index (n) of the film can be calculated from the equation

$$n = \left[N + \left(N^2 - n_0^2 n_1^2 \right)^{1/2} \right]^{1/2},$$

where

$$N = \frac{n_0^2 + n_1^2}{2} + 2n_0 n_1 \frac{T_{\max} - T_{\min}}{T_{\max} T_{\min}}, \qquad (7.5)$$

n_0 and n_1 are the refractive indices of the air and glass. $T_{\max}$ is the maximum transmittance at any particular wavelength (λ) on the interference pattern and the $T_{\min}$ is the minimum transmittance at the same 'λ' on the envelope drawn to maxima and minima.

7.1.4.1 Burstein-Moss shift

The band gap of semiconductor increases with increasing doping concentration (n_e) because the electrons occupy the lowest states in the conduction band in the form of electron gas for n-type. Therefore, the optical transition takes place from valence band to upper states of conduction, as shown in Figure 7.21 because the excited electron is not allowed to occupy lower states of conduction band according to Pauli principal. The band gap broadening can be achieved using the Burstein-Moss shift relation

$$\Delta E_g = \frac{h^2}{8m^*} \left(\frac{2}{\pi} \right)^{2/3} n_e^{2/3}, \qquad (7.6)$$

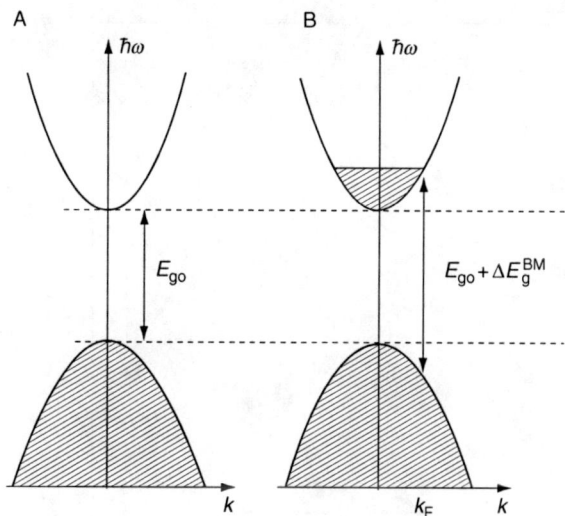

Figure 7.21 Schematic band diagram of semiconductor (A) before and (B) after Burstein-Moss effect.

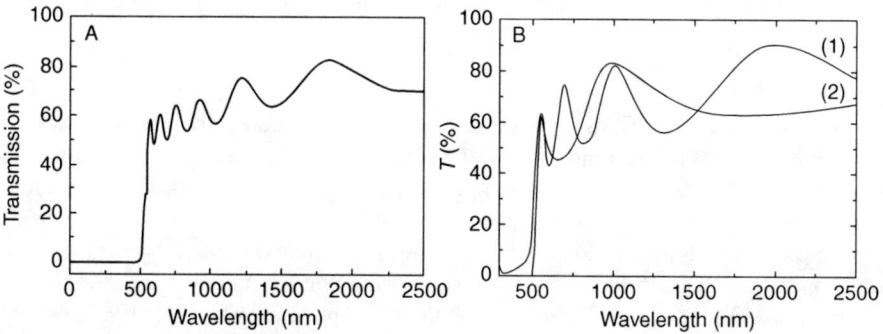

Figure 7.22 Transmission spectra of CdS thin film grown by (A) CBD method and (B) (1) as-grown CBD and (2) annealed.

where h is Plank's constant, m^* is the electron effective mass. In some cases, the band gap decreases with increasing doping due to many body effects on the conduction and valence band. On the other hand, the strain and imperfection of the crystals play a role [55].

7.1.5 Optical Properties of CdS Thin Films

An average thickness of a typical CBD-CdS layer is determined to be 903 nm using Equations (7.4) and (7.5) from the transmission spectrum, as shown in Figure 7.22A. The band gap of 2.48 eV for as-grown CBD-CdS thin films is determined from its transmission spectrum. After annealing the sample under vacuum, the band gap decreases from 2.48 to 2.34 eV [56]. Another report reveals that the band gap of

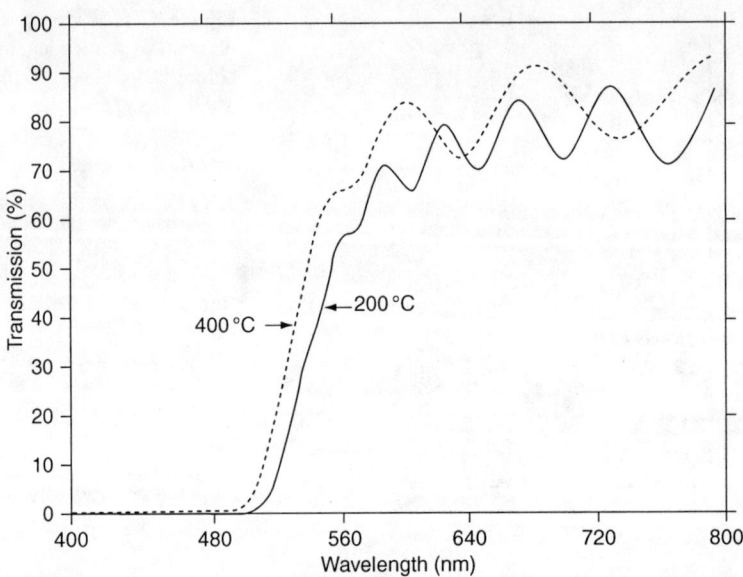

Figure 7.23 Transmission spectra of CdS thin films grown by PLD at substrate temperature of 200 and 400 °C.

2.4 eV for CdS thin films deposited onto glass substrates by CBD method is obtained from the transmission spectra (Figure 7.22B). The band gap increases from 2.4 to 2.62 eV after annealing the samples by laser. An increased band gap matches well with the standard hexagonal phase. In addition, sub band gap of 2.35 eV is observed in the same sample that could be due to cubic phase of CdS revealing that the annealed CBD-CdS had polymorphism, that is, combination of cubic and hexagonal phases. The XRD analysis confirms the same phenomenon [24]. The band gap of typical CdS thin films increases from 2.41 (515) to 2.48 eV (500 nm) with increasing substrate temperature from 200 to 400 °C and color of the sample changes from light yellowish orange to yellowish brown, as shown in Figure 7.23 due to improvement in crystallinity [49]. The CdS layers grown under influence of 0.077 T magnetic field at bath temperatures of 85 and 65 °C show band gaps of 2.48 and 2.45 eV, respectively, whereas thin films grown without influence of magnetic field exhibit lower band gap of 2.35 eV. It could be due to nonstoichiometry of CdS thin films, that is, sulfur deficiency in the CdS grown in the presence of magnetic field [18]. The band gaps of 2.51 and 2.42 eV are found for the CdS films deposited by CBD method using $CdCl_2$, NH_4Cl, ammonia, thiocarbanide, pH of the solution 10 and Cd/S = 2.0 and by spray pyrolysis technique using $CdCl_2$ and thiocarbanide solutions (Cd:S = 1:2 and $T_S = 400$ °C), respectively. After annealing the samples under vacuum, band gaps of 2.47 and 2.46 eV exist for former and latter, respectively [57]. The band gap of CdS thin film decreases from 2.47, 2.44 to 2.38 eV with increasing annealing temperature from RT, 200 to 400 °C, respectively [58]. After annealing the sample under air at 450 °C for 20 min, the band gap of

CdS layers also decreases from 2.42 to 2.25 eV [22]. Similarly, the band gap decreases from 2.51 to 2.42 eV after annealing CBD-CdS films under vacuum at 450 °C for 5 min. The reason is that the as-grown CBD-CdS shows higher band gap due to quantum size confine effect. The grain sizes increase upon annealing, hence no more quantum size confinement therefore the band gap decreases [59]. On the other hand, there may be sulfur loss in the films. The band gap of CBD-CdS thin films with grain sizes of 20 nm decreases from 2.54 to 2.44 eV with increasing surface roughness (RMS) from 4.4 to 24.5 nm that may be due to change in structure from one phase to another [60]. However, the band gap increases from 2.38 to 2.41 eV with increasing substrate temperature from RT to 120 °C or higher due to improvement in crystallinity. The refractive index of film is in the range of 2.19 to 2.39 at 1000 nm [46]. By B doping ($H_3BO_3/Cd(Ac)_2 = 0.01$) into CdS thin films, the band gap of CdS thin films also increases from 2.38 to 2.41 eV due to the Burstein-Moss effect and the resistivity decreases from 2×10^4 to 2 Ω-cm [61].

The reflectivity spectra of un-doped and In doped CdS thin films are shown in Figure 7.24 in which the transitions exist at 2.5 (E_0), 4.85 (E_A), and 5.4 eV (E_B) corresponding to Γ_{9v}-Γ_{7c}, Γ_{5v}-Γ_{3c}, and U_{4v}-U_{3c} transitions, respectively. The thicknesses of CdS, CdS (In 1 at%), and CdS (In 5 at%) are found to be 164, 273, and 464 nm from reflectance spectra using Equation (7.4), respectively [62]. One can see the well-developed Fabry–Perot fringes due to internal reflections in the CdS (In 5 at%) sample comparing with other samples. The similar results are noticed in the CdS on quartz rather than on glass substrates. It could be due to difference in growth pattern by nature of substrates [63]. The band gap increases from 2.42 to 2.6 eV with increasing In concentration from 2×10^{18} to 2×10^{19} cm^{-3} in the CdS thin films due to the Burstein-Moss shift. Thereafter, the band gap decreases to $\sim$2.49 eV with further increasing In concentration to 1×10^{21} cm^{-3}, as shown in Figure 7.25. A decrease in band gap for higher concentration of In may be due to formation of secondary phase of $CdIn_2S_4$ that causes to have less In to the majority of CdS phase or many body effects [15]. The band gap of CdS thin films decreases from 2.42 to 2.26 eV with increasing Ga/Cd ratio from 0 to 0.017 and then increases to 2.32 eV with further increasing Ga/Cd ratio of 0.06. The reason is that the donor concentration increases with increasing doping donor levels, becoming degenerate that causes to extend the conduction band into band gap. Therefore, band gap decreases. An increase in band gap may be due to lattice strain caused by an increased grain sizes of Ga doped CdS thin films [44].

The band gap varies from 2.39 to 3.53 eV with increasing y value in the $Zn_yCd_{1-y}S$. It obeys the formula

$$E_g(y) = E_{gCdS} + (E_{gZnS} - E_{gCdS} - b)y + by^2, \qquad (7.7)$$

where $b = 0.32$. A variation of band gap with varying y is given in Figure 7.26 [51]. The band gap of $Cd_{1-y}Zn_yS$ thin films grown by CBD technique varies from 2.43, 2.46, 2.56, 2.61, 2.65 to 2.66 eV with varying from $y = 0$, 0.1, 0.2, 0.3, 0.4 to 0.5, respectively [64]. Similarly the band gap of $Cd_{1-y}Zn_yS_y$ (CBD) layers increases from 2.4, 2.55 to 2.7 eV with increasing y values from 0, 0.3 to 0.5, respectively.

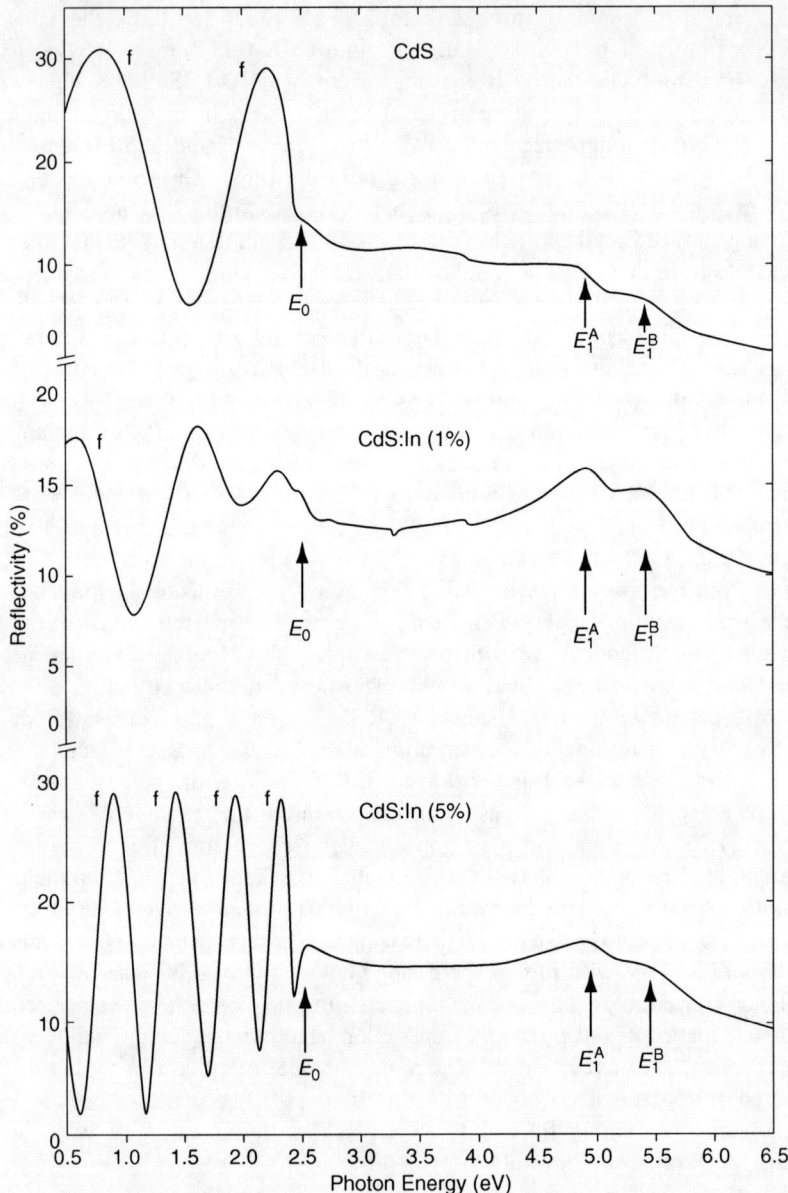

Figure 7.24 Reflectivity spectra of CdS and CdS:In thin films.

The transmission of the layers is 80% above the wavelength of 600 nm for $y=0.3$ [12]. A linear variation in band gap $E_g = (2.43 + 1.28y)$ eV is observed with varying y in the typical $Cd_{1-y}Zn_yS$ single crystals [52].

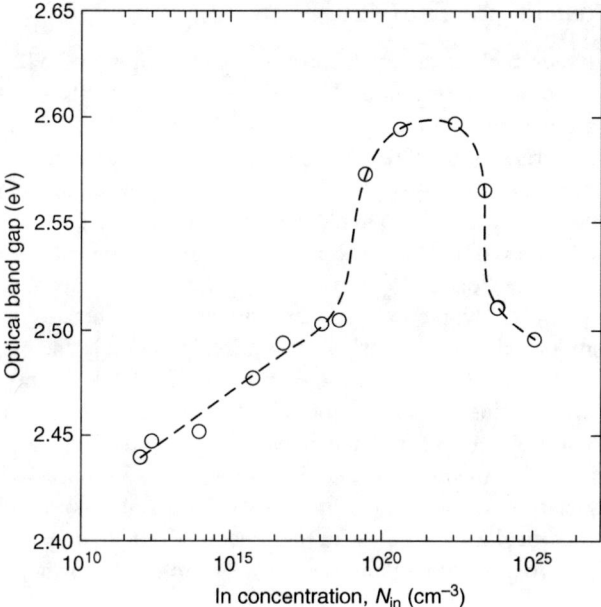

Figure 7.25 Variation of optical band gap with function of In concentration.

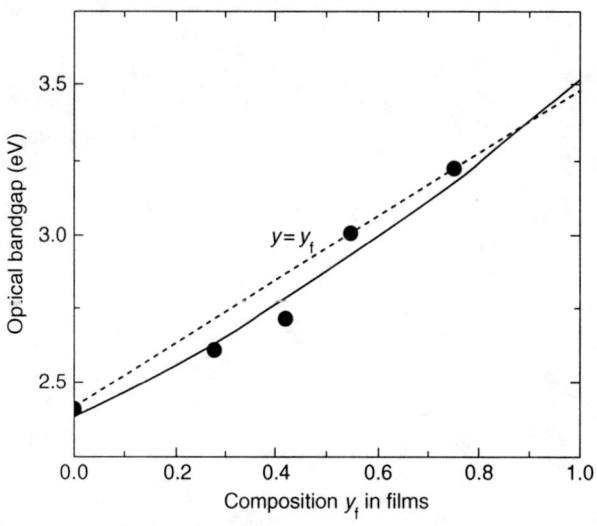

Figure 7.26 Variation of optical band gap with function of Zn composition in the $Cd_{1-y}Zn_yS$ thin films (solid line-experimental fit and dotted line-theoretical fit).

7.1.6 Photoluminescence of CdS

The bound excitons such as neutral donor to bound exciton at 2.542 eV (I_2 or B_1), neutral acceptor to bound exciton at 2.532 eV (I_1 or B_2), DAPs at 2.487 (D_1), and 2.42 eV (D_2) are observed in the CdS layers grown onto Si substrates by MOCVD at substrate temperatures of 300, 280, and 270 °C, whereas the layers grown at 325 °C show only DAP at 2.487 eV (Figure 7.27) [65]. The similar neutral donor to bound exciton (I_2) peak at 2.545 eV, neutral acceptor to bound exciton (I_1) at 2.532 eV, and free exciton (Γ_6) at 2.553 eV are reported in the bulk CdS [66]. As the PL temperature of CdS epilayer grown onto CdTe substrates by CVD is increased from 10 to 50 K, the I_1 line disappears due to dissociation of bound exciton and intensity of I_2 line decreases, as shown in Figure 7.28. The binding energies of 6.3 and 17.2 meV for I_2 and I_1 emission lines at 2.534 (489.4 nm) and 2.545 eV (487.3 nm) are determined, respectively. The free excitonic emission (FE) line at 2.551 eV (486 nm) along with phonon lines FE-1LO and FE-2LO exist [67]. The I_2 and I_1 lines at 2.543 (487.6) and 2.537 eV (488.8 nm) are observed, respectively in 4.2 K PL spectrum of CdS thin films grown onto BaF_2 substrates by hot-wall epitaxy technique at 480–550 °C, which is recorded employing emission line of Kr + laser with 476 nm with intensity of 0.1 W/cm^2, as shown in Figure 7.29.

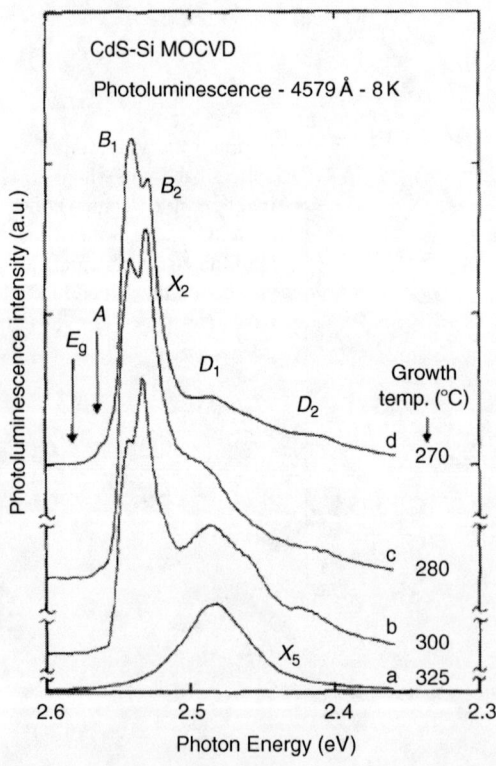

Figure 7.27 PL spectra of CdS thin films grown onto Si at different substrate temperatures.

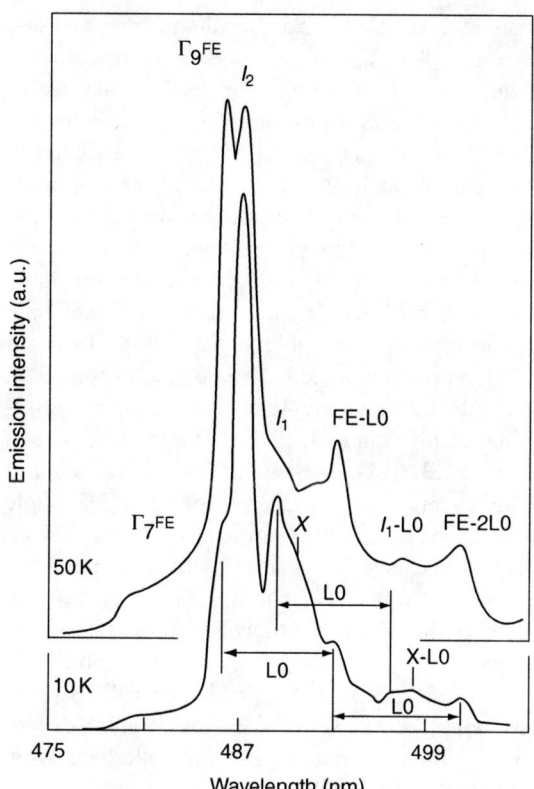

Figure 7.28 Excitonic emission of CdS epilayer grown on CdTe substrates.

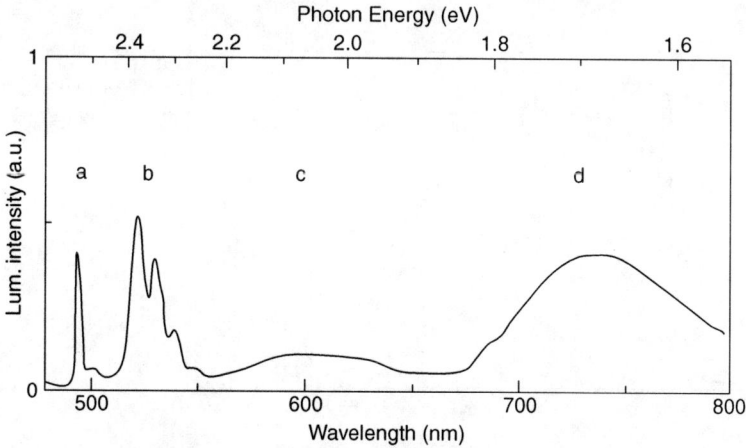

Figure 7.29 4.2 K PL spectrum of CdS thin films.

The high-energy spectrum line exists at 2.420 eV (512.5 nm) due to free-to-bound transition. The yellow emission line at 2.033 eV (610 nm) with activation energy of 0.51 eV is assigned to transition from Cd_i to valence band. The intensity of 2.0–2.1 eV peak decreases after annealing CdS films under $CdCl_2$ revealing a decrease in density of Cd_i [68]. Therefore peak at 2.0–2.1 eV can be assigned to cadmium interstitial (Cd_i). The transition line at 1.699 eV (730 nm) may be considered as a transition from Cd_i to Cl_S. The low energy spectrum shows DAP (zero phonon line) at 2.396 eV (517.6 nm), accompanied by phonon lines with separation of 36 meV [69]. A similar DAP is observed at 2.42 eV with phonon lines separated by 36 meV in the CdS single crystals in 80 K PL spectrum, as shown in Figure 7.30. The yellow emission (2.04 eV) due to Cd_i defect presents [70]. The PL peaks at 2.4 and 1.7 eV are observed at RT for 70 nm crystal size CdS grown by gas evaporation. As sample temperature is decreased to 10 K, the yellow luminescence strongly pronounces. The green emission line well resolves into G_1 (2.45 eV) and G_2 (2.42 eV) luminescence peaks, as shown in Figure 7.31 [71]. The similar G_1 at 2.54 (488 nm) and G_2 at 2.37 eV (522 nm) peaks are observed in 4 K CL spectrum for CdS single crystals. The yellow and red emission bands at 2.08 (594 nm) and 1.69 eV (734 nm) due to Cd interstitials (Cd_i) and sulfur vacancies (V_S) are observed, respectively [72]. The G_1 and G_2 emission bands shift from 2.431 and 2.324 eV to 2.464 and 2.341 eV in 77 K CL spectra by F doping into CdS thin films grown by RF sputtering using 3% CHF_3 and 97% Ar gases, respectively (Figure 7.32). A blue shift occurs in band to band transition at 2.518, 2.5, and 2.48 eV for CdS, CdS:In (1%), and CdS:In (5%) thin films with decreasing samples temperature due to shrinkage of band gap, respectively. The intensity ratio of green to yellow emission is high with increasing In content in the CdS [62]. The Y_1, Y_2, and R at 2.15, 2.0, and 1.81 eV are found in un-doped CdS, whereas Y and R are observed at 2.05 and 1.84 eV in the CdS:F thin films. The intensities of yellow and red emission bands decrease with F doping into CdS thin films [73]. The CdS thin films with

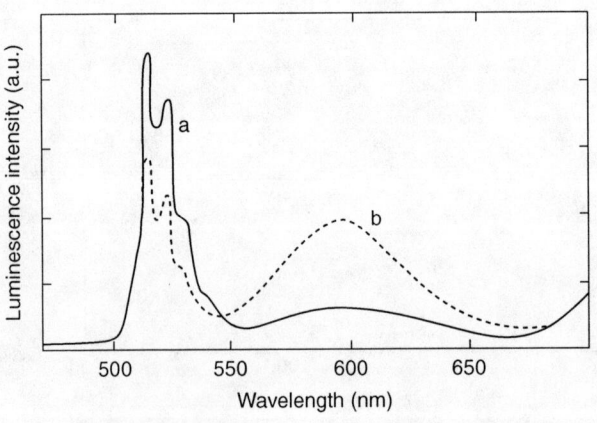

Figure 7.30 80 K PL spectra of two different CdS single crystals.

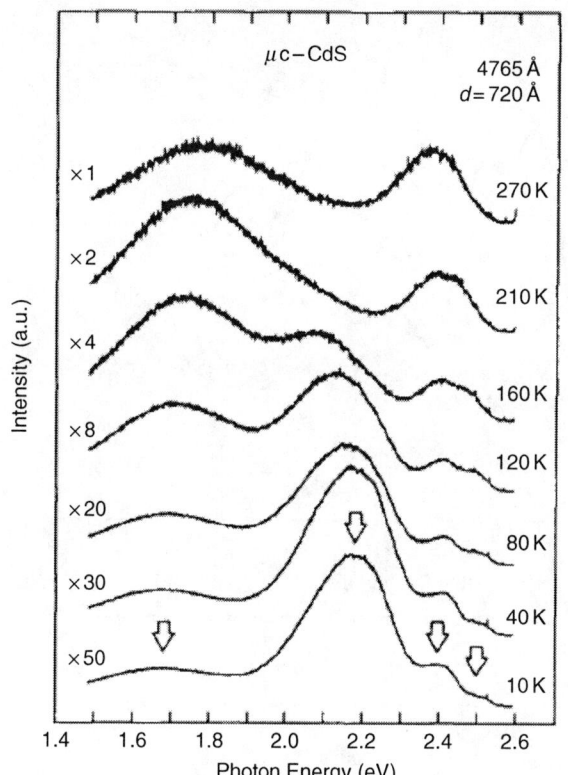

Figure 7.31 PL spectra of CdS single crystal with function of temperature.

different S/Cd ratios from 3 to 8 grown by CBD method show PL lines at 2.35 and 1.7 eV due to S_i and V_S, respectively [68], whereas the nanoparticle CdS shows band to band transition at 3.0 eV and defect to band transition at 2.56 eV [74]. The CdS quantum dots grown by colloidal method show a strong 2.2 eV band and a weak band to band transition. However, after photo etching and treating with Cd(OH)$_2$, the band to band transition dominates spectrum indicating that the nonradiative transition centers dominate surface in the as-grown samples [75]. In doped CdS crystals with carrier concentration of 1×10^{17}–5×10^{18} cm^{-3} reveal that the green emission becomes broader and recombination dynamics is faster. The bound exciton complex at 2.546 eV due to In_{Cd} is observed [76]. A several peaks at 2.43 (510), 2.1 (590) [77], 2.61 (475), 1.80 (690) [78], 2.40 (517), and 1.69 eV (735 nm) [79] are found in nanostructured CdS. A different possible transition levels in the CdS such as (i) transition from Cd_i (or I^+_{Cd} donor) to V_B (2.012 eV), (ii) transition from S-vacancy (V_S) to VB (2.175 eV), (iii) transition from S interstitial (I^-_S) to the conduction band (2.275 eV), (iv) donor–acceptor pair (DAP) transition (2.349 eV), (iv) neutral donor to bound exciton (D^0X) (2.455 eV) are shown in Figure 7.33 [24,80,81].

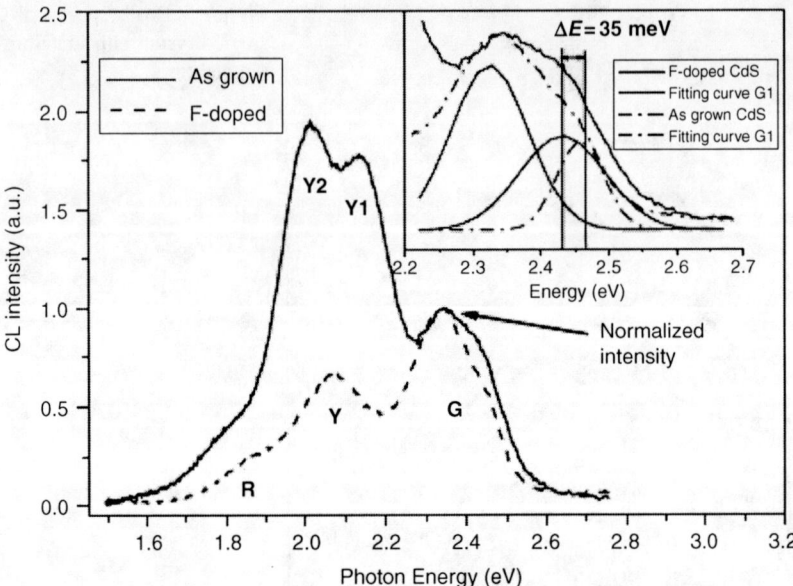

Figure 7.32 77 K CL spectra of CdS and CdS:F thin films grown by RF sputtering.

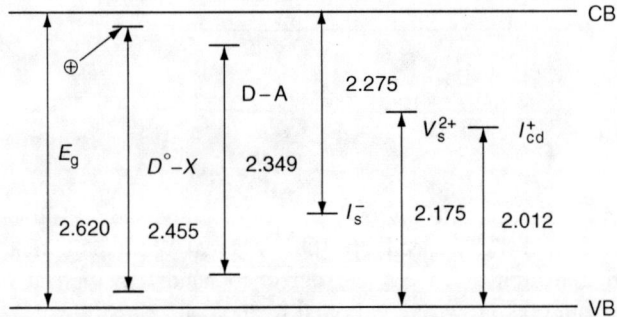

Figure 7.33 Different transition levels in the CdS.

7.1.7 Raman Spectroscopy of CdS

The CdS is wurtzite structure with two formula unit or four atoms per unit cell. The CdS contains nine optical branches $\Gamma = A_1 + 2E_1 + 2E_2 + 2B_1$, the doubly degenerated E_1 is Raman and infrared active but B_1 is inactive. The low and high frequency modes of E_2 are denoted as E_2^1 and E_2^2 for simplicity, respectively. The Raman modes E_2^1, E_2^2, $E_1(TO)$, $A_1(TO)$, $E_1(LO)$, and $A_1(LO)$ at 44, 252, 235, 228, 305, and 305 cm^{-1} are observed in the bulk CdS, respectively (Table 7.4). The multiphonons at 97, 207, 328, 347, 364, 556, and 604 cm^{-1} are also recorded in the bulk [82]. Typical Raman spectra of CdS nanowires and bulk are shown in Figure 7.34 [77]. The 1LO (301 cm^{-1}) and 2LO (601 cm^{-1}) phonon lines show red shift with

Table 7.4 Optical modes of CdS observed by Raman spectroscopy

S.No	Mode	Frequency (cm^{-1})	A_1(LO) Multiphonons	Frequency (cm^{-1})
1	E^1_2	44	1LO	302–306
2	E^2_2	252	2LO	607
3	E_1(TO)	235	3LO	910
4	A_1(TO)	228	4LO	1214
5	E_1(LO)	305	5LO	1520
6	A_1(LO)	305	6LO	1819
7	–	–	7LO	2118
6	–	–	8LO	2417
7	–	–	9LO	2716

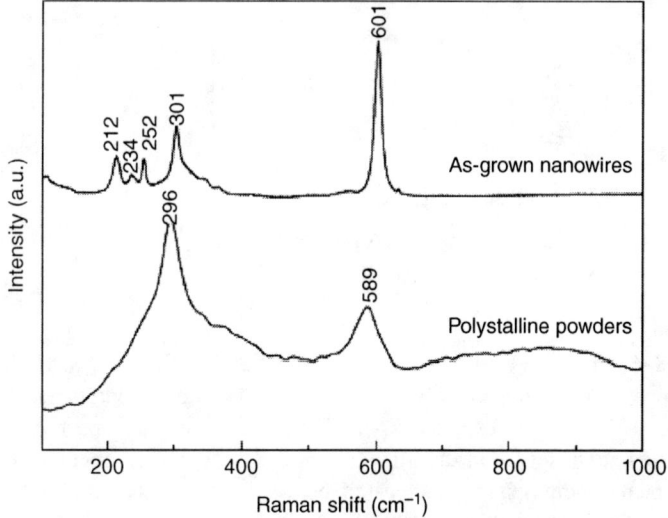

Figure 7.34 Raman spectra of CdS nanowires and bulk.

respect to bulk, it could be due to quantum size effect [24]. A several multiphonons from 1LO to 9LO are observed in the CdS (Table 7.4) [83]. In the Raman spectra, the 1LO, 2LO, 3LO, and 4LO modes at 303, 600, 899, and 1190 cm^{-1} are observed for continuous dip coated CdS films, whereas they are slightly deviated from their positions and the last one absent due to a small dispersion of LO mode phonon wave vector in the single and multiple dipped CdS thin films, as shown Figure 7.35. An asymmetry in peak is due to high density of stacking faults [28]. The similar 1LO and 2LO phonons at 300 and 600 cm^{-1} are observed in the CdS thin film and CdS/TiO$_2$ composite materials by exciting 514.5 nm (2.42 eV) laser in the Raman spectroscopy [84,85]. However, they are weaker in the electrochemically deposited quantum dot CdS thin films than in the bulk CdS [86]. Similar LO modes at 300,

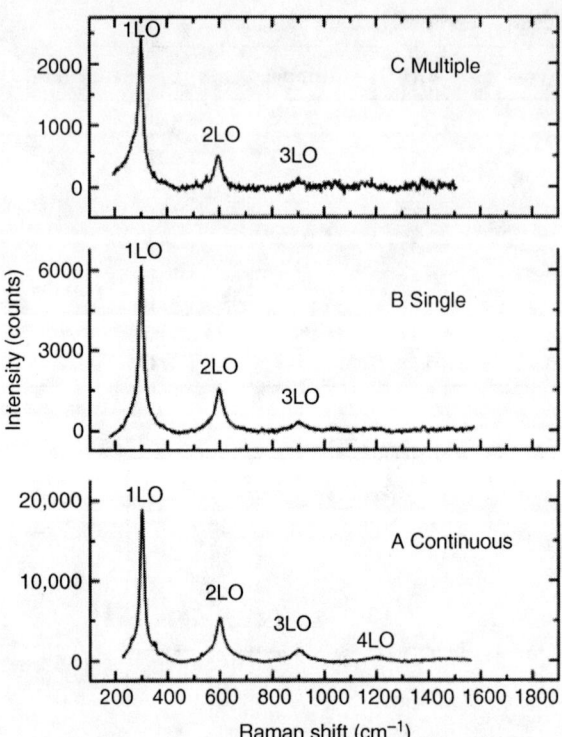

Figure 7.35 Raman spectra of CdS thin films grown by CBD technique.

601, and 846 cm^{-1} are observed in the CBD-CdS thin films on Si [87]. The CdS layers grown onto Si substrates by MOVCD at substrate temperatures of 325, 300, 280, and 270 °C show A_1(LO) and $E^2{}_2$ at 303 and 257 cm^{-1}, respectively, but some extra features are observed in the films grown at 325 °C [65]. The intensities of 1LO and 2LO peaks increase with increasing boron doping concentration (B/Cd) from 0.01 and 0.1 in the CdS thin films [88], whereas the Raman spectrum of CBD-CdS films with Ga/Cd ratio of 6.8×10^{-2} shows that the A_1(LO)/E_1(LO) at 302 cm^{-1} due to either hexagonal or LO cubic, TO and E_2 at 246 and 260.5 cm^{-1} are found, respectively [44]. 1LO and 2LO modes at 300 and 605 cm^{-1} present in the CdS films deposited at 200 °C by PLD, whereas the films grown at 400 °C show five LO modes strongly confirming hexagonal structure of the CdS films deposited at 200 °C by PLD [49]. In addition, 1TO and 2TO phonon lines at 214 and 428 cm^{-1} are observed, respectively [89].

7.1.8 Electrical Properties of CdS Thin Films

In general, the as-grown CBD-CdS layers show high resistivity (ρ) in the range of 10^{8-10}–10^{6-8} Ω-cm [6]. There are several methods to reduce resistivities by tailoring growth recipes and annealing procedures. The resistivity of CdS films grown under influence of magnetic field is higher by three orders of magnitude than the one grown without magnetic field. They are 20 and 2×10^5 Ω-cm with and without magnetic field, respectively [18]. The reason is that the films grown under influence of magnetic field increase sulfur vacancies that lead to increase conductivity of the layers. The resistivity of CdS layers grown onto 10^{-3} Ω-cm ITO layer at $T_S = 400\,°C$ by spray pyrolysis technique decreases from 4.8×10^5 to 1.0×10^5 Ω-cm by changing composition ratio from Cd:S = 1:1 to 1:2 in the chemical solution. The resistivity further decreases from 1.0×10^5 to 5.6×10^4 Ω-cm with increasing deposition temperature from 400 to 420 °C for keeping Cd:S = 1:2 in the solution. The CdS layer grown onto glass substrates shows low resistivity of 17–55 Ω-cm for all the deposition conditions. After annealing CdS films, the resistivity decreases to lower by approximately 1 order of magnitude. The resistivity of CdS layers grown onto ITO at 85 °C by CBD decreases from 2.5×10^4 to 1.0×10^3 Ω-cm after annealing those, whereas it increases from 3.8×10^5 to 4.4×10^5 Ω-cm for CdS on glass [57]. It could be due to Indoping from ITO and impurities from glass substrates informer and latter, respectively. The CdS layers with S/Cd = 0.96 grown by CBD method using 1 mM $CdSO_4$, 10 mM thiourea, 2 M NH_3 chemical solutions at bath temperature of 70 °C, and pH 11–12 for 20 min show dark and light resistivities of 1.5×10^8 and 1.4×10^3 Ω-cm, respectively. They decrease from 1.5×10^8 to 7.3×10^5 and from 1.1×10^5 to 2.2×10^2 Ω-cm, respectively after annealing the samples under air at 400 °C. In addition, the S/Cd ratio in the CdS layers decreases from 0.96 to 0.88 [58]. Similar resistivities of 7.3×10^3–3.63×10^5 and 3.42–2.57×10^2 Ω-cm under dark and light conditions are observed for CBD-CdS thin films, respectively [90]. After annealing CBD-CdS samples under S_2 atmosphere at 250 °C the resistivity decreases from 2.3×10^7 to 8.2×10^5 Ω-cm and continue to decrease to 11 Ω-cm for further annealing under $H_2 + In$ at 350 °C [31]. The typical as-deposited CBD-CdS layers show resistivity in the range of 1.08×10^4–1.34×10^4 Ω-cm. After annealing those at 450 °C for 20 minutes, no striking difference is observed in the resistivity [22]. The CdS films grown by spray pyrolysis technique at substrate temperature of 400 °C show low resistivities of 17 and 60 Ω-cm for composition ratio of Cd:S = 1:1 and 1:2, which decrease to 0.3 and 9.1 Ω-cm after annealing films under vacuum at 450 °C for 5 min, respectively due to loss of S, whereas the resistivity of CBD-CdS films slightly increases from 3.8×10^5 to 4.4×10^5 Ω-cm due to chemisorption of oxygen [57]. The resistivity decreases from 0.59×10^5 to 0.4×10^3 Ω-cm with increasing Cd:S ratio in the spray-deposited CdS thin films [91]. In the case of doped CdS thin films, the resistivity of CdS:In thin films decreases from 0.09 to ~ 0.01 Ω-cm with increasing substrate temperature from 60 to 200 °C but resistivity increases to 0.1 Ω-cm for further increasing substrate temperature to 400 °C [36]. It could be due to re-evaporation of In or S for high temperature. The CdS:In (1%) (Cd:S = 1.5:1) thin films grown onto mica sheet by evaporation show resistivity of 10^{-2} Ω-cm and mobility of 30 cm^2/Vs [92,93]. The carrier concentration increases from 4×10^{16} to 2.96×10^{19} cm^{-3} with increasing Ga/Cd from 0.0 to 0.034 in the CdS:Ga. On the other hand, the dark resistivity of CdS decreases from 1.03×10^2 to 1.2×10^{-2} Ω-cm due to an increased carrier concentration [44].

The resistivity of vacuum evaporated CdS thin films decreases from 10^3 to 5×10^{-3} Ω-cm by doping In concentration of 3×10^{20} cm^{-3}. The resistivity starts to increase for further increasing In concentration [15]. The carrier concentration of CdS thin films can be estimated using the Burstein-Moss shift relation. For example, the PL spectra recorded for three typical CdS thin film samples so-called x, y, and z deposited by PLD using different power densities of 2, 3, and 4 J/cm^2 and wavelength of 1064 nm show band edge emissions at 2.45, 2.478, and 2.493 eV. Shifts in band edges (ΔE) are found to be 28 and 43 meV for sample y and z with respect to sample x, respectively. The carrier concentrations (n_e) for y and z samples are determined to be 1.9×10^{18} and 3.6×10^{18} cm^{-3} with respect to sample x by taking $m_e^* = 0.2 m_o$. The results can be cross checked employing their dark conductivity (σ) of 1 Ω-cm^{-1} and mobility (μ) of 1–2 cm^2/Vs using simple relation $\mu = \sigma/ne$, which gives $n_e \sim 3 \times 10^{18}$ cm^{-3} [38]. As mentioned in earlier, the resistivity and mobility of spray-deposited typical CdS thin films ($n_e = 1.82 \times 10^{11}$ cm^{-3}) on ITO glass substrates decrease from 1.48×10^8 to 4.3×10^2 Ω-cm and from 2.42×10^3 to 42 cm^2/Vs, after annealing the sample at 200 °C [94]. The donor and acceptor energies of 34 and 858 meV, respectively are calculated using the relation (Equation 5.12) by taking $m_e^* = 0.2 m_o$, $m_h^* = 5 m_o$, and $\varepsilon = 8.9$ [95]. The thermally simulated current experiments on CBD-CdS thin films reveal activation energies of 0.3, 0.45, and 0.51 eV for using Cd acetate and 0.04 and 0.33 eV for CdCl$_2$ chemical solutions in CBD method [96]. The CdS single crystals from Eagle Picher company show $E_1 = 0.014$ eV ($n_e = 1.5 \times 10^{17}$ cm^{-3}) in the range 200–50 K and $E_2 = 0.007$ eV ($n_e = 7 \times 10^{16}$ cm^{-3}) in the temperature range of 50–10 K in a plot of n_e versus $1/T$ [97].

The theoretically estimated mobilities for CdS due to lattice as well as ionized impurity scattering are two orders of magnitude higher than the experimental mobilities of CdS:In, as shown in Figure 7.36. The CdS:In thin films exhibit room temperature experimental conductivity of 0.045 (Ω-cm)$^{-1}$ and peak mobility of 3 cm^2/Vs. The low experimental mobilities of CdS:In thin films might be due to its polycrystalline nature with large grain boundaries, partially degeneracy, *etc*. It is clear that the lattice scattering is predominant at high temperature, whereas the scattering due to impurities is dominant at low temperature. At low temperature, the phonon population is depleted by cooling and the ionized impurity scattering controls electron mean free path so that the mobility decreases with decreasing temperature [98]. Woodbury [99] observed that the theoretical fitting is well done in some of the samples but the fitting deviates at large due to domination by defect pairing rather than ionized impurity scattering in some of the samples. The grain boundary and deffect scatterings are the dominant mechanism in the indium doped CdS thin films [100]. The mobility of CdS thin films grown onto glass and Si by vacuum evaporation at substrate temperature of 150 °C decreases with increasing temperature, as shown in Figure 7.37. A decrease in mobility starts above 100 K due to acoustical (lattice) scattering that follows as $\mu_H \alpha \mu_o T^{-1.52}$. The conductivity and mobility are 1.85×10^{-1}(Ω-cm)$^{-1}$ and 1.25×10^2 cm^2/Vs at 77 K, whereas they are 1.34 and 34.5 cm^2/Vs at 300 K, respectively [50]. The mobilities of as-grown CdS$^{\parallel}$ and CdS$^{\perp}$ single crystals from Eagle Picher Co measured by employing 4KG field with function of temperature in the range 10–300 K are shown in Figure 7.38. The latter sample shows higher mobility than former due to anisotropy. Two typical samples are cut from lengthy boule sample; one is parallel to c-axis and another one perpendicular to c-axis. They are electron irradiated with energy

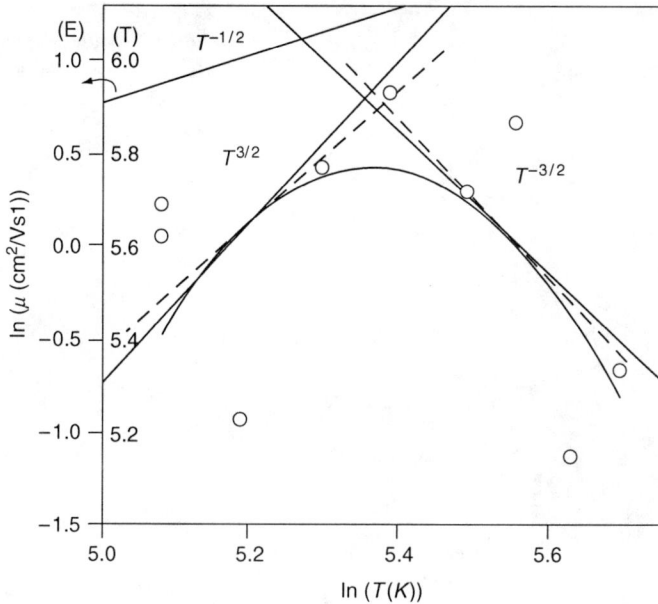

Figure 7.36 Variation of mobility with function of temperature for indium doped CdS thin films by spray.

of 10 MeV at 120 K and fluence of 2.3×10^{16} electrons/cm^2 and annealed at 50 °C for 1 h and again isothermally annealed at 97 °C for different timings of 10, 50, 200, 500, and 1000 min. The electron irradiation creates acceptor concentration and suppresses donor concentration in the sample. Above 40 K, the lattice scattering dominates mobility ($\mu_H \alpha \mu_o T^{-3/2}$). The sample Hall mobility is nearly equal to the one that governs by the lattice scattering. In particular, in the irradiated samples, the ionized impurity scattering dominates in the temperature range of 40–20 K ($\mu_H \alpha \mu_I T^{3/2}$), the impurity band conduction is suppressed that falls very sharply at low temperature, which is discussed in Chapter 6. In the lattice scattering mechanism formula, the acoustic phonon deformation potential (E_I) value of 7.5 fits well rather than 3.3 in both irradiated CdS$^\parallel$ and CdS$^\perp$ samples indicating that an increased defect levels. The piezoelectric constants are 0.207 and 0.152 for CdS$^\parallel$ and CdS$^\perp$ samples, respectively. The various scattering mechanisms are theoretically fitted to the experimental temperature dependent mobilities, as shown in Figure 7.39 for which the parameters used are given in Table 7.5. The $N_d = 2.35 \times 10^{15}$, $N_a = 1.0 \times 10^{13}$cm^{-3}, and $E_d = 31$ meV may be used in the impurity scattering analysis [101]. In some of the crystals, longitudinal optical scattering (μ_{op}) is dominant at above 50 K temperature regions, whereas piezoelectric scattering (μ_p) is dominant below 50 K, that is, 1.8–50 K, as shown in Figure 7.40 [102]. The crystals show mobility of 240 cm^2/Vs in the RT-200 K, peak mobility of 1700 cm^2/Vs at 50 K and mobility falls sharply to 1 cm^2/Vs with decreasing temperature to liquid temperature. The mobility

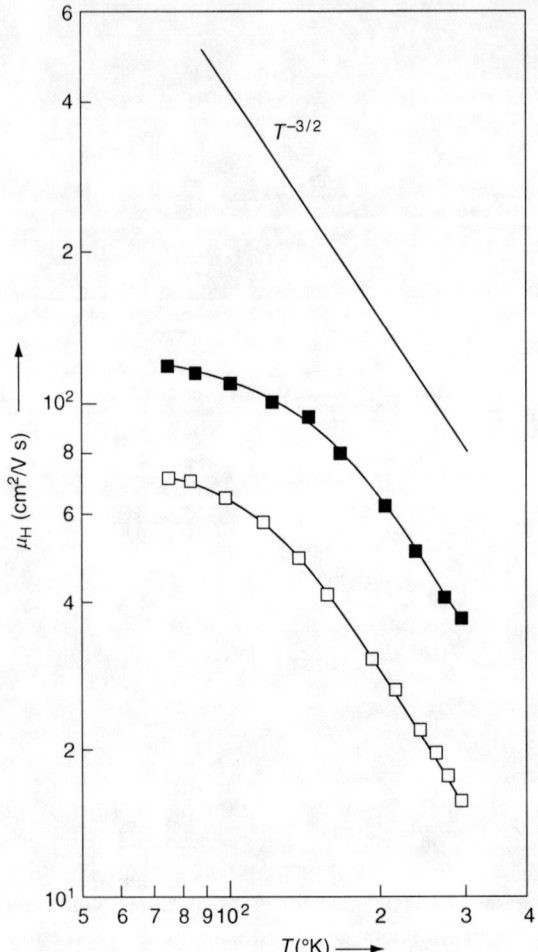

Figure 7.37 Variation of mobility with temperature for CdS thin films on glass (□), Si (■), and solid line-theoretical fit.

follows $\mu \alpha T^{-3/2}$ or $\mu_H = 6.4 \times 10^4 T^{-3/2}$, that is, lattice scattering dominates in the 200–50 K [97].

The resistivity increases from 27 to 2.7×10^7 Ω-cm with increasing y from 0 to 0.35 in the $Cd_{1-y}Zn_yS$, which is very high beyond $y = 0.35$ [51]. In another case, the resistivity of $Cd_{1-y}Zn_yS$ thin films varies from 1.1×10^2, 2.7×10^2, 4.1×10^2, 6.7×10^3, 6.9×10^4 to 2.7×10^5 Ω-cm with varying x from 0, 1, 2, 3, 4 to 5, respectively [64]. Similarly it increases from 10^2 to 7×10^3 Ω-cm with increasing y from 0 to 0.5 [12]. The conductivity also varies from 7, 16 to 10^{-3} Ω-cm^{-1} with varying y from 0, 0.05 to 0.45, respectively, whereas the samples become semi-insulating for $y > 0.45$. The electron mobility of CdZnS crystals decreases from 235 to 5 cm^2/Vs with increasing y from 0.15 to 0.22 [52].

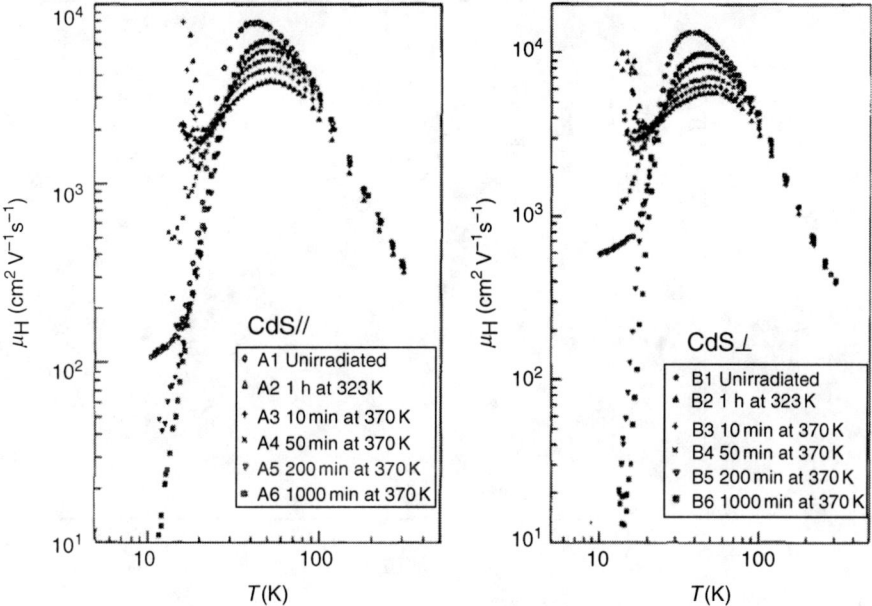

Figure 7.38 Variation of mobility with function of temperature; (A) CdS$^{\|}$ and (B) CdS$^{\perp}$.

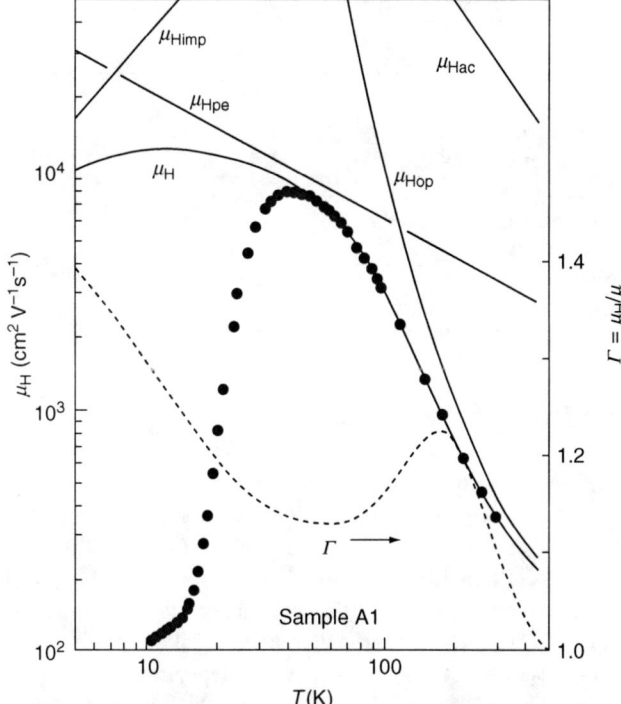

Figure 7.39 The various models such as impurity (μ_i), polar (μ_{po}), acoustic (μ_{ac}), optical scattering mechanisms (μ_{op}) fitted for sample-A$_1$ (CdS$^{\|}$), and weak field Hall factor (Γ).

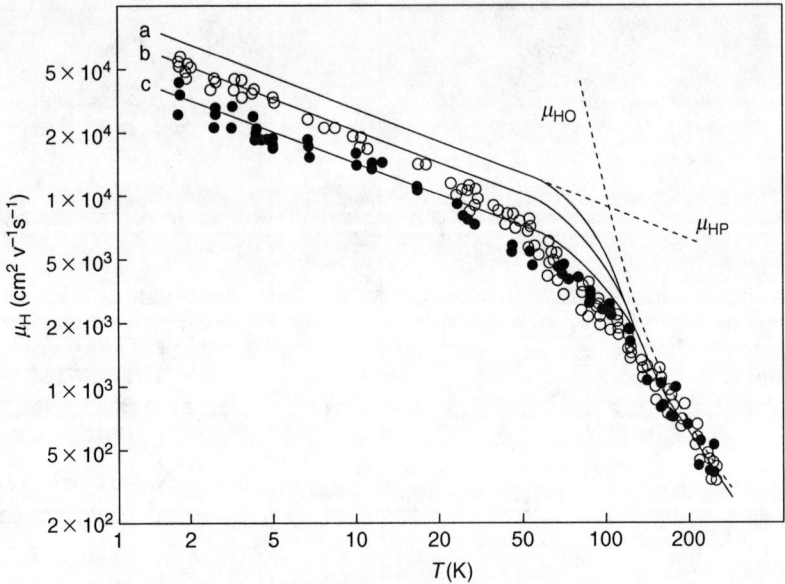

Figure 7.40 Mobility versus temperature for CdS single crystals.

Table 7.5 Physical constants of CdS [101]

Parameters	CdS$^{\|}$	CdS$^{\perp}$
Static dielectric constant ($\varepsilon_0/\varepsilon$)	9.1	8.58
High frequency dielectric constant ($\varepsilon_\infty/\varepsilon$)	5.31	5.26
Optic phonon temperature top (K)	428	428
Longitudinal elastic constant C_λ (N m^{-2})	8.47×10^{10}	8.47×10^{10}
Acoustic phonon deformation potential E_I (eV)	3.3	3.3
Piezoelectric constant (P)	0.207	0.207
Band gap E_g (eV)	2.52	2.52
Density of states effective mass m^*/m	0.208	0.208

7.2 InS or In$_2$S$_3$ Thin Films

The Indium sulfide is one of the nontoxic friendly buffer layers for thin film solar cells, which consists of different structures such as α-In$_2$S$_3$ (cubic), β-In$_2$S$_3$ (tetragonal or spinal-type), and γ-In$_2$S$_3$. The band gap varies from 2.0 to 2.7 eV for the structures. The β-In$_2$S$_3$ thin films are prepared by spray pyrolysis technique using chemical solutions of InCl$_3$, thiourea 400 ml, and spray rate 20 ml/min at substrate temperature of 300 °C. The chemical reaction takes place as $2\text{InCl}_3 + 3\text{CS}(\text{NH}_2)_2 + 6\text{H}_2\text{O} \rightarrow \text{In}_2\text{S}_3 + 3\text{CO}_2 + 6\text{NH}_4\text{Cl}$ while depositing chemical solution onto heated substrates. The oxygen and chlorine are the contaminants in the spray-deposited films [103]. The chemical solutions of 0.02 M $InCl_3$, 0.01 M HCl, 0.3 M acetic acid, and 0.5 M CH_3CSNH_3 thioacetomide are employed to deposit InS films at bath temperature of 70 °C [104]. The trimethyl-indium and t-butyl-thiol (tBuSH) are also used to prepare InS using closed coupled showerhead AIXTRON AG MOCVD system under pressure of 450 mbar at substrate temperature of 300–550 °C [105]. Unlike using individual In and S evaporation, the In_2S_3 powder is evaporated onto soda-lime glass substrates at substrate temperature of 50 °C and source temperature of 720 °C under base pressure of 5×10^{-5} mbar to form β-In_2S_3 films [106].

The InS layers grown at substrate temperature of 350 °C with growth rate of 3 nm/min onto CIGS by MOCVD layers show tetrahedral growth, as shown in Figure 7.41. The VI/III ratio influences growth of InS layers [105]. The CBD InS film is a combination of In_2O_3 evidencing by presence of 404 eV peak correspond to In_2O_3 phase in the XPS spectrum [104]. The XRD pattern of 200 nm thick In_2S_3 layer grown by atomic layer chemical vapor deposition (ALCVD) technique is depicted in Figure 7.42 revealing tetragonal structure, that is, β-In_2S_3 [107]. The preferred orientation of (220) is observed for In/S ratio of 2/8 in the sprayed In_2S_3 layers [103]. The highly (103) orientation is observed for the In_2S_3 films deposited at 350 °C by evaporation of In and S individually. In addition other reflections such as (206), (10 15), and (40 12) are also observed. The In/S ratio of 0.7 for evaporated films is close to the In_2S_3 [108]. The In_2S_3 buffer layer with thickness of 50 nm deposited onto CIGS absorber layer using pulse sequence as H_2S/N_2/indium acetylacetonate ($In(acac)_3$)/N_2:1500/1100/2500/3000 ms at 210 °C shows band gap of 2.7 eV [107]. The band gaps of 2.52–2.54 eV for 68 and 120 nm thick In_2S_3 layers are observed. The InS films are prepared by CBD method using chemical solutions of thioacetamide (CH_3CSNH_2), $InCl_3$, HCl, and CH_3COOH under acidic media conditions at bath temperature of 50 °C. The S/In ratio of 0.65 and oxygen impurities are found in the films. The band gap of the films decrease from 2.73, 2.58 to 2.46 eV with increasing thickness of InS layers from 85, 231 to 470 nm, respectively [108]. The spray-deposited β-In_2S_3 films show band gap of 2.64 eV [103]. However, CBD InS films show lower band gap of 2.2 eV [104].

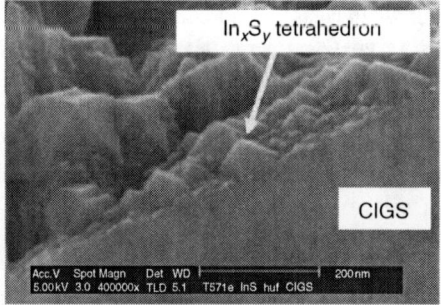

Figure 7.41 SEM cross section of In_xS_y/CIGS heterostructure: the rough surface traps light and reduces reflections.

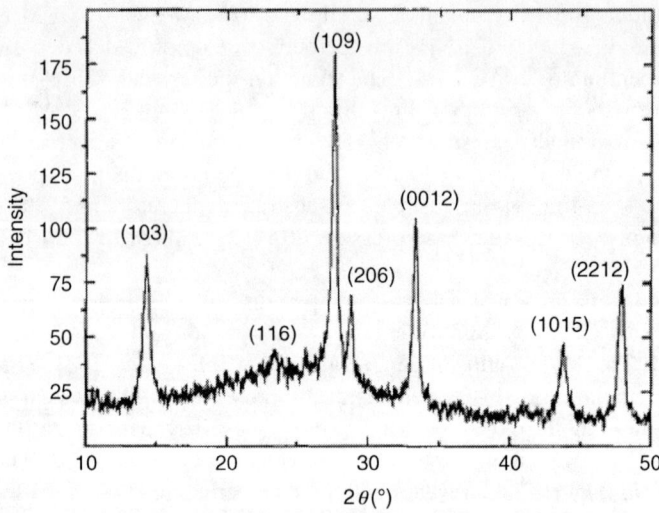

Figure 7.42 XRD pattern of tetragonal β-In$_2$S$_3$ thin films grown by ALD.

7.3 ZnO

High transmittance and high conductive metal oxide thin films are so-called transparent conducting oxide (TCO) thin films, which are essential for thin film solar cell applications. The ZnO is one of the promising TCO candidates among In$_2$O$_3$:Sn (ITO), SnO$_2$:F (FTO), CdIn$_2$O$_4$ etc., for photovoltaic cells, ultraviolet light emitting diodes, laser diodes, detectors, SAW devices, gas sensors, flat panel display, electrochromic devices, piezo-optic, and piezo-electric devices. The ZnO is a wide band gap (3.2 eV) semiconductor and had high exciton binding energy of 60 meV at room temperature. The ZnO exhibits good optoelectronic properties such as high transparency (>80%) and electrical conductivity ($> 10^3$ Ω-cm^{-1}), those are important characteristics for optoelectronic device applications. Maximum carrier concentration of 10^{21} cm^{-3} and mobility of 50 cm^2/Vs are so far reported for ZnO thin films [109]. The quality ZnO layers can be deposited at low substrate temperatures comparing with other TCO layers. In the case of indium tin oxide (ITO) layer, the ITO has to be deposited at high temperature in order to obtain optimum characteristics for devices that create adverse affects on the thin film solar cell structure. The fluorine doped tin oxide (FTO) contains low chemical stability at high temperature that causes fluorine to diffuse into device layers from FTO. The Zn is highly abundant in the earth crust, therefore, it's cost effective to prepare ZnO. The i-ZnO and ZnO:Al are the part of bilayer for thin film solar cells.

7.3.1 Growth of ZnO Thin Films

There are several potential techniques to grow ZnO thin films for thin film solar cells or optoelectronic devices such as MOCVD, spray, sol gel, spin coating, vacuum evaporation, MBE, *etc.* Among these RF or DC sputtering technique is one of the advantageous techniques that the layers can be deposited at low substrate temperature, good adhesion of films to the substrates, high deposition rates, uniform thickness, and long-term stability. The reactive gas mixtures and elements can be combined to make compound thin film in the reactive sputtering. The materials with different vapor pressures can be grown and scalable largely ($3 \times 6 \, m^2$) [110]. A schematic of typical conventional sputtering chamber is shown in Figure 7.43 [111]. In order to avoid splattering or to reduce bombardment on the substrates by sputter particles, substrates are kept at side to deposit ZnO thin films in the chamber. Typical 4 in. diameter sintered ZnO or ZnO:Al (2.5 wt%) targets are positioned in face to face by separation of 8 cm or less, as shown in Figure 7.44. On the other hand, employing off-axis configuration, a single target can be used. Typical distance between substrate and target is 1.5 cm and the height between target and substrate is 2.5 cm [112]. The Ar and oxygen partial pressures of 0.3 and 0.6 Pa are kept, respectively. The typical parameters such as RF power of 150–200 W, chamber working pressure of 5 mtorr, substrate to target distance of 5–6 cm, substrate rotation speed of 144 rpm and substrate temperature of 100–275 °C are also employed to grow ZnO:Al thin films [113–115]. There is another typical recipe that the ZnO with 2 wt% Al_2O_3 target with diameter of 12.5 cm, the distance between target and substrate of 9 cm, working pressure of 1.33 Pa and RF power of 40 W are employed to grow ZnO:Al thin films [116]. A 12 cm dia ZnO(Al_2O_3 2 wt%) target is used to grow AZO thin films by DC magnetron sputtering with DC power of 80 W and the chamber pressure of 0.26–12 Pa [117]. The Zn–Al film deposition rates can easily be controlled in reactive sputtering technique to deposit either ZnO or ZnO:Al thin

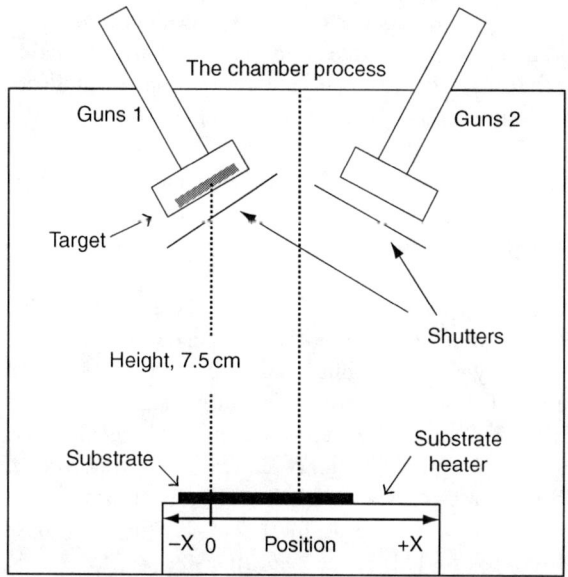

Figure 7.43 A typical conventional sputtering chamber.

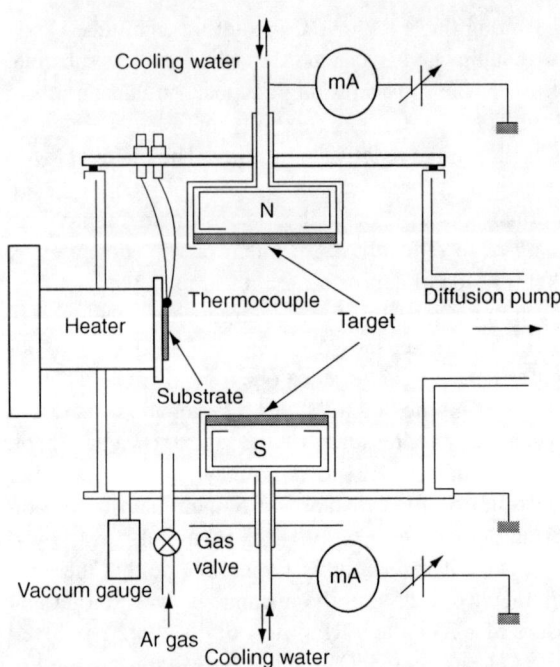

Figure 7.44 Typical sputtering system for deposition of ZnO thin films.

films. The fabrication of Zn–Al target is simpler than ZnO:Al. The chemical composition on the surface of target gets easily modify in latter [118]. Zn–Al alloy target yields higher deposition rate and precise control of thickness rather than using ZnO:Al target. A typical facing target sputtering experimental set up is shown in Figure 7.45 in which ZnO:Al and Zn targets are connected up and down configurations and substrate being kept aside. The oxygen flow rate ($O_2/(Ar + O_2)$) is maintained to be 0.1–0.3. The typical distances from target to target and between target and substrate are kept at 10 cm and working pressure being 1 mtorr [119]. The surface of target gets oxidation for lower RF power of 120 W. The oxidation occurs due to chemisorptions of particles and particle–particle collisions. The density of sputter particles is high for higher sputtering power for which the oxygen concentration is low therefore the density of metallic particles are high on the substrate owing to low defect levels due to higher growth rates [120].

The typical Al doped ZnO target with density of 5.3 g/cm^3 has been prepared that the ZnO and Al_2O_3 powders are mixed using agate mortar and pestle then the powder is pressed at 800 MPa followed by sintering either under air at 900 °C for 10 h or under an Ar atmosphere at 1300 °C for 6 h. In general the target size depends on the configuration of system [112,118]. The similar report reveals that the ZnO disc with 10 mm diameter and 2 mm thick is formed by uniaxial pressing ZnO powder at 100 MPa, followed by the cold iso-static press at 200 MPa. The disc is sintered at 600 °C for 2 h and at 1200 °C for 4 h to condensate the target. Finally, the target is attached to the holder by the epoxy resin [114]. The ZnO + Ga_2O_3 powder is used

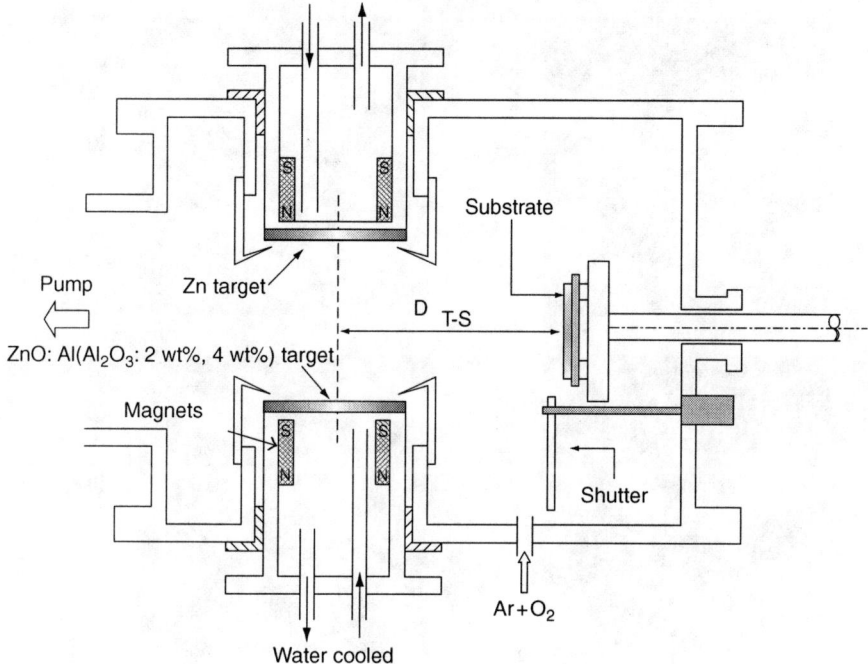

Figure 7.45 A typical sputtering experimental set up to grow ZnO thin films.

for target with typical size of 12.6 cm dia and 4.2 mm thick disc sintered under air from 800 to 900 °C for 1 h to form target [121].

The pulsed laser deposition technique is alternative viable approach to deposit ZnO or ZnO:Al thin films because the bombardment of high-energy sputter particle damages surface of frontier layer in the sputtering that degrades thin film solar cell. The KrF excimer laser ($\lambda = 248$ nm, $\tau = 25$ or 30 ns, 5 or 10 Hz, 1 J/cm^2) is employed to ablate ZnO target in the presence of oxygen at 200 mtorr [122,123]. A third harmonic of Q-switched Nd:YAG laser with wavelength of 355 nm, 10 Hz, and 6 nS with fluence of 2 J/cm^2 is used to ablate ZnO or ZnO:Al pellets at substrate temperature of 400 °C with working pressure of 10^{-3} torr [124].

7.3.2 Surface and Structural Properties of ZnO Thin Films

The deposition recipes control grain sizes that the ZnO thin films grown by RF sputtering show smaller grain sizes comparing with the films grown by DC sputtering probably due to bombardment by low energetic particles on substrate and formation of low density of nucleation sites [125]. The AFM shows that the films deposited at RT, 200, 300, and 600 °C have grain sizes of 50–70, 40–80 smaller +150–200 larger, 60–100, 250+60–100 nm, respectively, as shown in Figure 7.46. The layers deposited at RT show densely packed grains, whereas layers grown at high temperature show large grain sizes but with voids. However, the layers grown at 200 °C

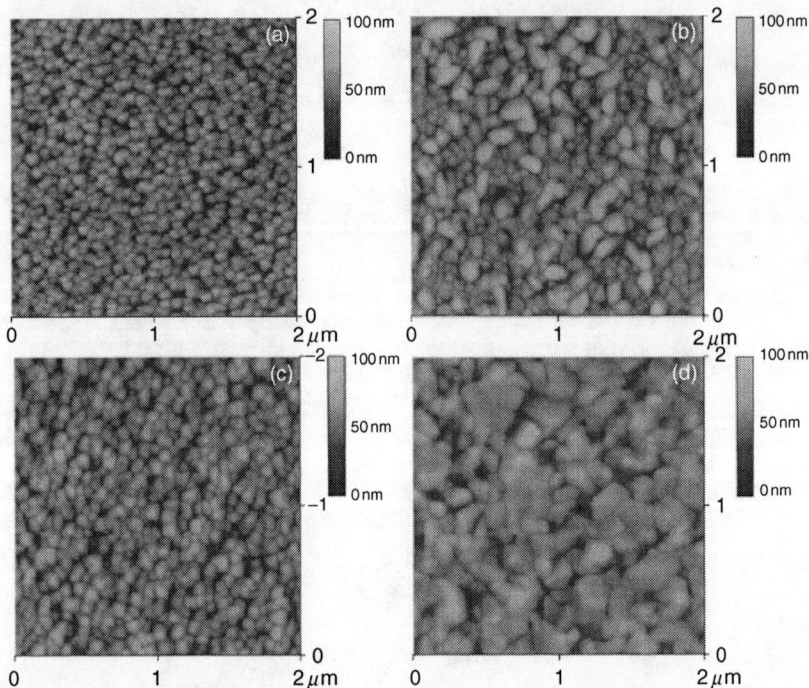

Figure 7.46 AFM images of ZnO thin films grown at different substrate temperatures of (A) RT, (B) 200, (C) 300, and (D) 600 °C.

show two kinds of grain sizes due to influence of Zn adsorption where it contains higher vapor pressure [126]. The grain sizes change from round shape to wedge like structure with increasing Al doping from 0, 0.5, 0.75 to 2 wt% into 3 μm thick ZnO thin films grown by sputtering technique at substrate temperature of 350 °C under pressure of 12 Pa. On the other hand average grain sizes decrease from 1000 to 500 nm with increasing Al doping from 0 to 2 wt% (Figure 7.47) [127]. The similar report reveals that the grain sizes of ZnO layers decrease from 160, 64, 47 to 41 nm with increasing Al_2O_3 dopant concentration from 0, 1, 2 to 4 wt% [128]. However, the grain sizes of ZnO:Al increase from 40 to 45–65 nm with increasing bias voltage from zero to 40 V in the sputtering system. The grains slightly tilt with respect to substrate for the higher deposition bias voltage [129]. The grain sizes also increase from 60 to 105 Å with decreasing working pressure from 1.33 to 0.04 Pa [130]. The ZnO and ZnO:Al thin films grown onto glass substrates by reactive RF sputtering show that the as-grown ZnO thin films have cluster type surface, whereas Al doped ZnO thin films have ordered cone-like shape, as shown in Figure 7.48 [131].

The ZnO has hexagonal (wurtzite) structure with lattice constants of $a = 3.25$ Å and $c = 5.20$ Å (JCPDS). The SAED pattern reveals that (002) is the intensity diffraction ring for un-doped ZnO thin films, whereas the Al doped ZnO thin films show that (002) and (101) are at the same intensities [113]. Obviously, (002) is a

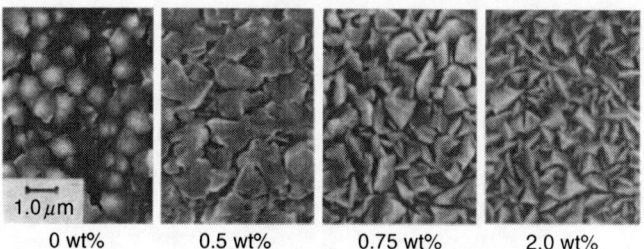

Figure 7.47 SEM of ZnO layers with a function of Al doping concentration.

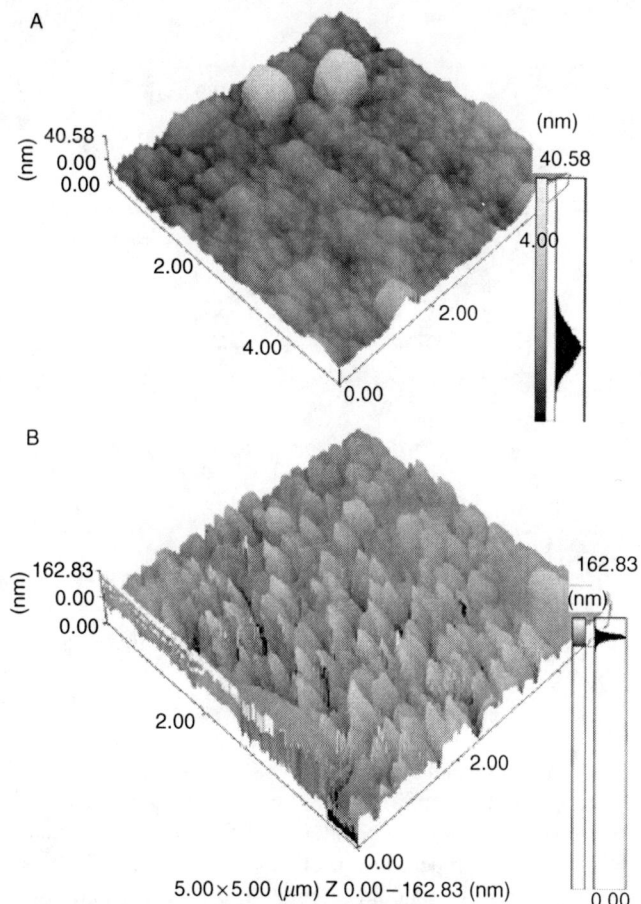

Figure 7.48 AFM images of (A) ZnO and (B) ZnO:Al thin films.

common preferred orientation but the standard powder X-ray diffraction shows (101) as an intensity peak (Figure 7.49) [132]. The XRD analysis reveals that the (002) orientation is stronger for the films grown onto Al_2O_3 buffer layer rather than

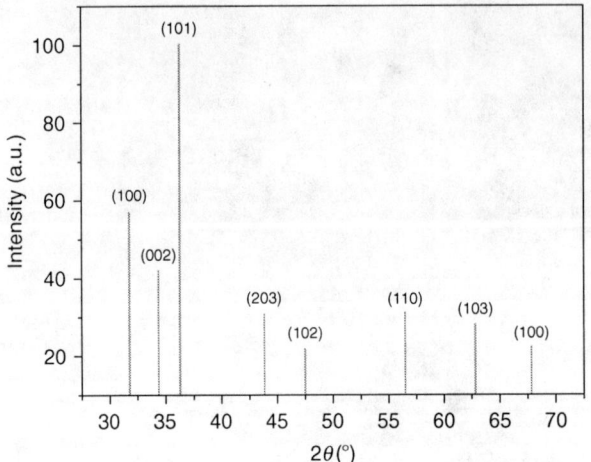

Figure 7.49 Standard XRD pattern of ZnO thin films.

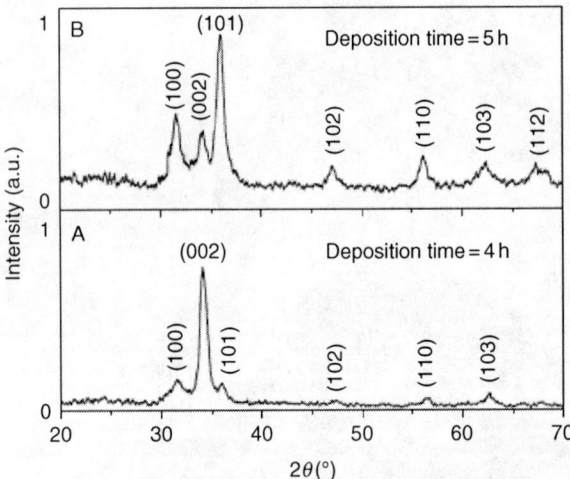

Figure 7.50 XRD pattern of ZnO:Al thin films grown by RF sputtering with two different thicknesses.

layers grown onto bare PET substrates [129]. (002) is sharper for ZnO:Al thin films grown onto glass substrates at 350 °C than the one deposited at 100 °C [133]. The ZnO layers deposited onto glass substrates by DC sputtering at chamber pressure of 0.4 mbar and substrate temperature of 60 °C for 4 h shows (002) as the intensity peak, whereas the films show random orientation for 5 h deposition time due to an increase of roughness of the surface, the incoming species would have less freedom to seek the minimum energy site, that is, adatom mobility decreases therefore the growth becomes random, as shown in Figure 7.50 [134]. (002) is the preferred orientation for 100% Ar gas, which gradually decreases with increasing H_2/Ar ratio from 0.3, 0.8 to 2% [109]. The orientation of ZnO:Al thin films grown onto glass substrates from Zn and ZnO:Al_2O_3 2 wt% targets changes from (002) to (110) with changing oxygen flow rate from 0.1 to 0.3 indicating that the required surface energy is available for the growth of (110) orientation plane [119]. Eventually, a low surface energy is good enough for the growth of (002) orientation comparing with other planes. Keeping working pressure of 1.7×10^{-2} mbar and argon gas flow rate of 10 sccm, the intensity of (002) preferred orientation for AZO thin films grown by RF magnetron sputtering technique from ZnO:Al (ZnO:Al$_2$O$_3$ = 98:2) target increases gradually with increasing RF power from 150 to 200 W [135]. In addition, the (002) orientation of thin films becomes stronger with increasing RF power from 50 to 200 W [136]. The intensity of (002) peak increases with increasing substrate temperature from RT to 250 °C thereafter decreases for the ZnO:Al thin films grown onto quartz substrates by DC reactive sputtering technique using Zn–Al target. The diffraction angle of (002) peak increases from 34.02 to 34.43° with increasing substrate temperature from RT to 250 °C thereafter decreases to 34.41° with further increasing substrate temperature of 350 °C [137,138]. Angle of (002) peak shifts from 34.18 to 34.44° with increasing substrate temperature from RT to 400 °C for the ZnO thin films grown onto quartz substrates by reactive RF magnetron sputtering using 3 in. Zn target, Ar flow rate of 24 sccm, O$_2$ flow rate of 6 sccm, working pressure of 1×10^{-2} mbar, and RF power of 400 W [126]. A change in angle with increasing substrate is due to influence of compressive stress in the films.

The ZnO thin films grown by reactive RF magnetron sputtering using metallic Zn target and Ar–O$_2$ reactive gases under chamber pressure of 5×10^{-4} Pa show that the d spacing of (002) orientation decreases from 2.641 to 2.622 Å with increasing Ar concentration from 25% Ar in Ar + O$_2$ to higher in the deposition chamber [139]. The Zn–Al alloy target with Zn:Al = 98.5:1.5 wt% is employed for the deposition of ZnO:Al thin films onto 1737 corning glass substrates by Hallow cathode sputtering technique using working chamber pressure of 50–80 Pa and sputtering power of 1500 W (5.9 W/cm^2). The orientation of ZnO:Al layers grown at substrate temperature of 200 °C strongly depends on O$_2$ flow, which is varied from 0 to 50 sccm while keeping Ar flow constantly at 3000 sccm. The ZnO and Zn multiphases appear for O$_2$ flow rate of 10 sccm, (002) oriented ZnO films generate for O$_2$ flow rate of 20 sccm, whereas stronger (002) oriented ZnO layers occur for O$_2$ flow rate of 50 sccm [138]. The lattice constants of pellet are $a = 3.243$ and $c = 5.202$ Å, which coincide with JCPDS data. The pellet shows that (100), (002), and (101) peaks are the most intensity peaks in the XRD spectrum. In addition extra ZnAl$_2$O$_4$ features are observed by means of existing (220), (311), (422), (511), and (440) peaks. The ZnO:Al films are deposited by RF sputtering with power density of 3 W/cm^2 at 70 °C and the distance between substrate and target is kept to be 7 cm. After annealing ZnO:Al films at 400 °C, no secondary phases such as Al, Zn, and ZnAl$_2$O$_4$ are observed. The as-deposited and annealed ZnO films show (002) preferred orientation [118]. The ZnO:Al thin films with columnar structure are prepared onto float glass substrates by high rate reactive mid frequency (MF) magnetron sputtering employing working gas pressure of 400 MPa and discharge power of 3.3 kV. The Zn–Al (1–2 wt% Al) metallic targets are used and O$_2$ being as reactive gas. The ZnO:Al layers deposited by sputtering show peaks at (002) and (101) for substrate temperature of 50 °C, whereas only (002) intensity peak presents at 34.5° for 150 °C and the sequence of (002) and (101) is again repeated for substrate temperature of 300 °C in the XRD spectrum [140]. The ZnO:Al(2 wt %) thin films deposited onto (12$\bar{1}$0) sapphire substrates by RF sputtering show (002) orientation for the deposition temperature of 350 °C, whereas the films contain (002) and (101) peaks for the growth temperature of 400 °C [116]. The ZnO:Al thin films grown onto amorphous Si substrates at 150 °C under pressure of 10 mtorr by RF magnetron sputtering show only (002) and (004) reflections, whereas the target consists of multiple reflections [112].

The diffraction angle of (002) shifts from 34.86 to 34.36° with increasing Al dopant concentration from 0 to 4 wt% for the ZnO films deposited onto 1737 corning glass substrates by RF magnetron sputtering at RT employing chamber pressure of 1×10^{-3} Pa, working pressure of 0.34 Pa, argon gas flow rate of 10 sccm, and RF power of 80 W, as shown in Figure 7.51 indicating an increase of lattice constants of ZnO [128]. Similarly, the diffraction angle of ZnO gradually decreases from 34.45, 34.39, 34.22 to 34.20° with increasing Al content from 0, 2, 3 to 5 wt%, respectively. The intensity of (002) peak also decreases with increasing Al content in the ZnO layers [127]. The lattice constants of ZnO should be decreased or the diffraction angle of (002) should be increased with increasing Al concentration, since radius of 0.53 Å for Al^{3+} is smaller than of 0.72 Å for Zn^{2+}. However, the experimental results are in quite opposite. This anomalous behavior may be due to internal stress in the films. The stress is caused by difference in the thermal expansion coefficients of 7×10^{-6} and 4.6×10^{-6} °C^{-1} for ZnO and glass, respectively [124].

As expected, the diffraction angle of (002) preferred orientation for AZO films deposited by RF using deposition conditions of $T_S = 150$ °C, $P_{RF} = 150$ W, and $P_W = 2$ mtorr shifts by 0.4° and 1.0° toward higher angle with increasing Al content from 0 to 3 and 5 wt%, respectively [141]. The Al(OH)$_3$ has been used as a dopant instead of Al$_2$O$_3$. The diffraction angle of (002) slightly increases with increasing doping concentration from 0, 2 to 4 wt% Al(OH)$_3$ in the films grown onto glass substrates by RF at substrate temperature of 250 °C using ZnO + Al(OH)$_3$ target. In these cases stress may be less in the films. On the other hand, the intensity of (002) peak decreases with increasing Al doping [142]. The ZnO target shows several

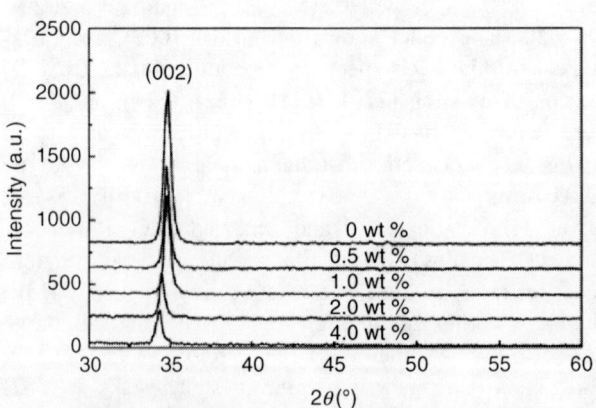

Figure 7.51 XRD pattern of ZnO layers with different dopant concentrations of Al$_2$O$_3$ wt%.

diffraction reflections such as (100), (002), (101), (102), (110), (103), (200), (112), and (201), whereas the ZnO films deposited onto sapphire (0001) substrates by KrF laser ablation technique in the presence of O_2 atmosphere of 50 Pa, 30 Hz, 3 J/cm^2 at 400 °C show (002) as the preferred one along with other reflections and the layers deposited onto polyimide at substrate temperature of 300 °C show (002) peak as slightly intensity peak [143]. The (002) is the intensity peak for In 1–2 at% doping, whereas (101) becomes intensity peak for beyond doping concentration of In 2 at% for i-ZnO/ZnO:In layers [144].

7.3.3 Optical Properties of ZnO Thin Films

7.3.3.1 Reflectance and transmittance

The conduction band with symmetry of Γ_7 generates from 4s levels of Zn in the band structure of ZnO. Three valence bands A, B, C with symmetries Γ_7, Γ_9, and Γ_7 originate from 2p states of O, respectively. The crystal field and spin-orbit splitting cause to provide three two-fold degenerate valence bands such as A, B, and C, as shown in Figure 7.52 [145]. In other words, the A, B, and C transitions can be mentioned as (i) Γ^V_9–Γ^C_7, (ii) Γ^V_7 (upper band)–Γ^C_7, and (iii) Γ^V_7 (lower band)–Γ^C_7, respectively. The valence band splitting is shown in envelope mode (Figure 7.52B). The $\Delta E_{AB}=4.9$ meV, $\Delta E_{BC}=43.7$ meV, and $E_g=3.4376$ eV are observed for ZnO at 4.2 K [146,147], whereas CdS at 1.6 K shows $\Delta E_{AB}=16$ meV, $\Delta E_{BC}=57$ meV, and $E_g=2.582$ eV [148]. The $A=3.376$, $B=3.397$, and $C=3.435$ eV are observed from 13 K reflectance spectra of $+c$, $-c$, and m plane ZnO, as shown in Figure 7.53 [149]. The similar $A=3.3782$, $B=3.3856$ eV and their excitonic states ($n_x=2$) at 3.4216 and 3.4273 eV, respectively, are also observed from 10 K reflectance spectra [150]. $A=3.3773$, 3.4221 eV for $n_x=2$, 3.4303 eV for $n_x=3$ and $B=3.3895$, 3.4325 eV for $n_x=2$ are reported from PL studies as well as from absorption spectra [151,152].

The transmission of AZO thin films at infrared region decreases with increasing substrate temperature from RT to 230 °C (Figure 7.54A). The Haacke figure of merit $\Phi_{TC}=T^{10}/R_\square=55\times 10^{-3}\,\Omega^{-1}$, $R_\square=4.59\,\Omega/\text{sq}$, and $\rho=1.2\times 10^{-3}\,\Omega\text{-cm}$ are observed for the films deposited at substrate temperature of 230 °C [153]. The similar property is observed that over all transmission of AZO samples decreases with increasing substrate temperature from 50 to 150 °C except at 300 °C, as shown in Figure 7.54B [140]. In some cases, the ZnO:Al layers with (002) preferred orientation deposited at 200 °C show higher transmission comparing with the films grown at 50 °C due to improvement in crystallinity [114]. The transmission of ZnO:Al thin films increases with increasing substrate temperature from 150, 250 to 350 °C. On the other hand, the band gap of ZnO:2 wt%Al thin films decreases from 3.77 to 3.32 eV with increasing substrate temperature from RT to 500 °C in steps of 100 °C (Figure 7.55) [122]. Similarly, the band gap decreases from 3.31, 3.30, 3.28, 3.27 to 3.26 eV with increasing substrate temperature from 90, 140, 230, 300 to 350 °C, respectively due to an increased crystallinity [133]. The band gap of ZnO:Al (2 wt%) thin films increases from 3.40 to ~3.7 eV with increasing

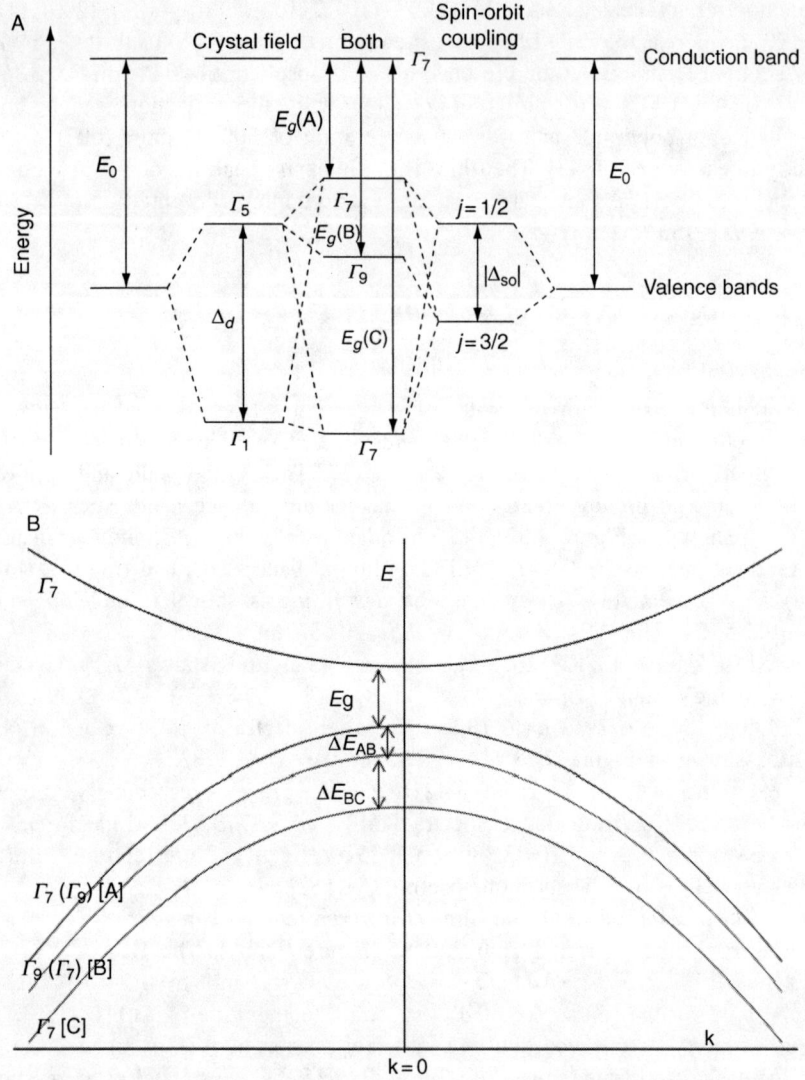

Figure 7.52 (A) Band splitting of ZnO by the influence of crystal field and spin-orbit splittings and (B) Valence band splitting due to crystal field and spin-orbit splitting, shown in envelope modes.

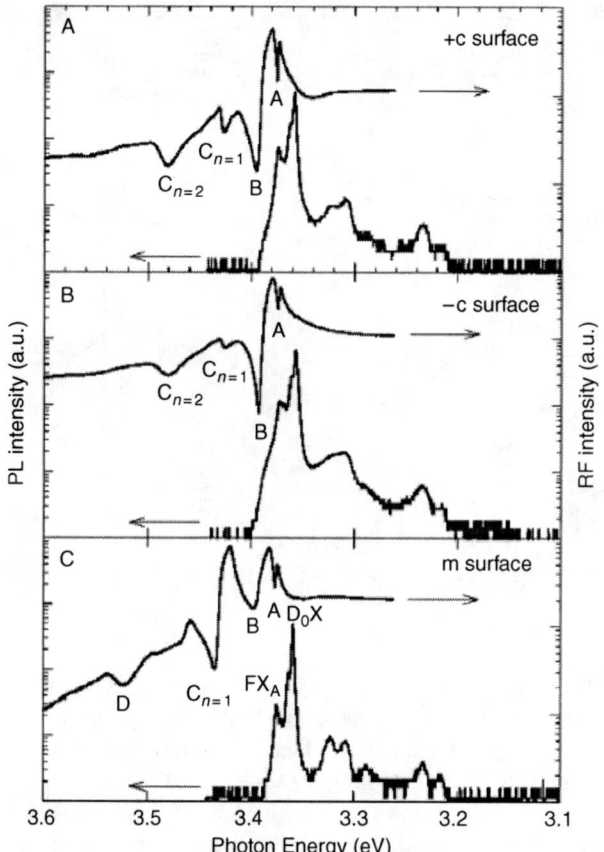

Figure 7.53 13 K reflectance and PL spectra of +C, −C, and M plane ZnO single crystal.

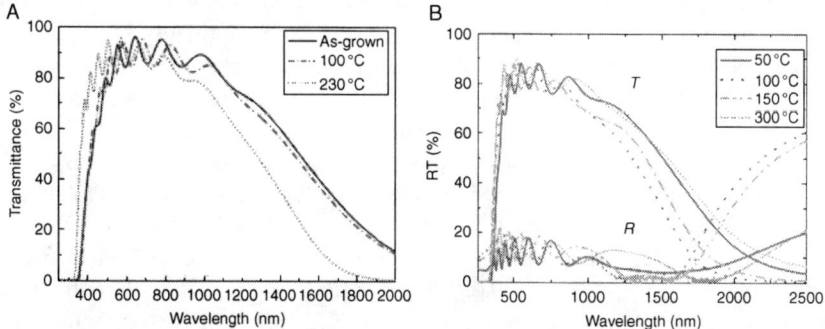

Figure 7.54 (A) Transmittance spectra of AZO films and (B) Transmittance and reflectance spectra of ZnO:Al layers grown onto float glass by reactive sputtering at different temperatures.

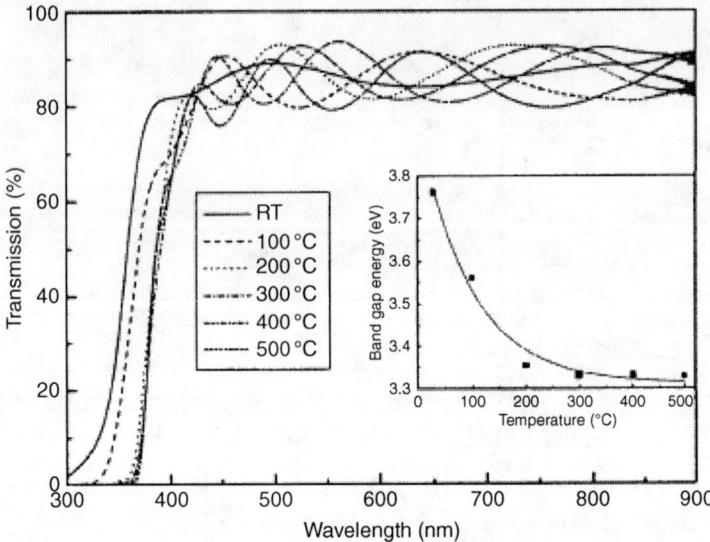

Figure 7.55 Transmission spectra of ZnO:Al thin films with function of substrate temperature.

substrate temperature from RT to 230 °C due to an increase of carrier concentration. In general, the amorphous, quasi-crystalline samples show higher band gap as compared to that of crystalline sample. The transmission of ZnO:Al thin films decreases with increasing RF power from 100, 200, 300 to 400 W [154]. Akin the transmission of ZnO decreases with increasing RF value, as shown in Figure 7.56. The reason is that the carrier concentration increases with increasing RF power. The band gap also increases from 3.31, 3.35, 3.37, 3.45, 3.4 to 3.51 eV with increasing RF power from 50, 100, 125, 150, 175 to 200 W for $P_w = 1$ Pa and $T_S = 250$ °C in lieu of an increase of carrier concentration [136].

The overall transmittance decreases with increasing Al doping into ZnO films due to an increase of doping concentration that deteriorate quality of crystal. The transmittance of 78% and least sheet resistance of 2 Ω/sq for dopant concentration of 0.75 wt% Al_2O_3 are determined [127]. The band gap increases from 3.32, 3.325, 3.37, 3.375 to 3.45 eV with increasing Al concentration from 1, 2, 3, 5 to 7%, that is, a blue shift in the band gap is observed with increasing carrier concentration in the films [133]. A similar blue shift in the band gap from 3.3 to 3.44 eV is observed with increasing Al content from 0, 2, 4, 6, 8 to 10 wt% $Al(OH)_3$ [142]. Another report reveals that the band gap changes from 3.3, 3.52, 3.54 to 3.47 eV with increasing Al_2O_3 wt% from 0.0, 1.9, 4 to 6.2 at% [141]. The authors found that the band gap of ZnO thin films varies from 3.4, 3.55, 3.79 to 3.9 with increasing Al doping concentration from 0, 0.95, 1.4 to 2.14 at% correspondingly the carrier concentration varies from 0.95×10^{20}, 1.8×10^{20}, 2.6×10^{20} to 4.5×10^{20} cm^{-3} [55]. With increasing Al doping concentration from 0, 2, 3 to 5% the band gap increases from 3.27 to 3.67 eV [124] and in another case it varies from 3.32 to 3.83 eV with

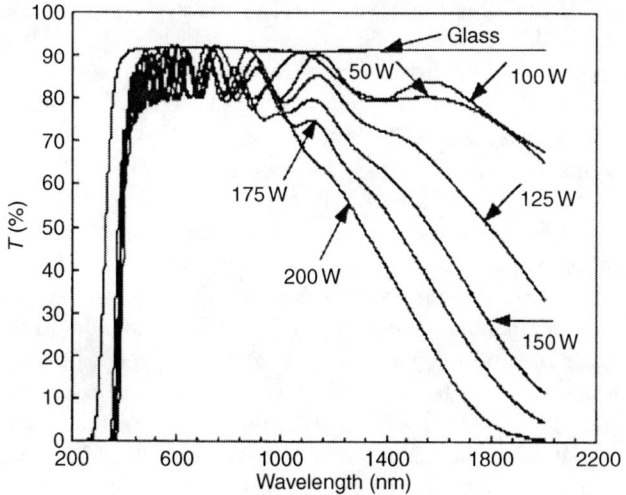

Figure 7.56 A decrease of transmission with increasing RF power for ZnO thin films.

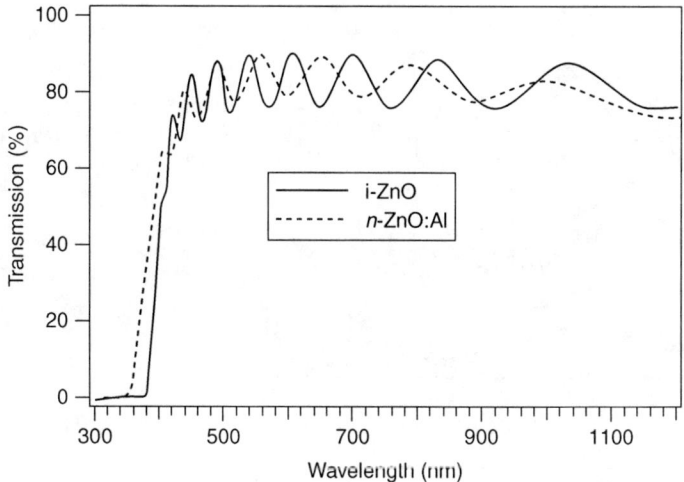

Figure 7.57 Transmission spectra of ZnO and Al doped ZnO thin films.

increasing doping concentration of Al_2O_3 from 0 to 4 wt% [128]. By doping 3 mol% Al into ZnO the band gap quietly increases from 3.3 to 3.375 eV [155]. The transition starts early in the Al doped ZnO thin films comparing with un-doped ZnO layers due to an increased band gap, as shown in Figure 7.57 [156]. With increasing In 1.5 at% doping the band gap of spray-deposited ZnO thin films also increases from 3.3 to 3.34 eV [157]. Obviously, all the observations indicate that an increase in the band gap of ZnO thin films with increasing carrier concentration is so-called the Burstein-Moss shift [158].

The band gap of ZnO:Al layers decreases from 3.81 to 3.63 eV with increasing O_2 flow rate from 25, 30, 35 to 50 sccm due to a decrease of carrier concentration, that is, the Burstein-Moss effect. The transmittance and resistivity are 80% and 5.2–6.4 × 10^{-4} Ω-cm, respectively for O_2 flow of 25–50 sccm deposited at substrate temperature of 200 °C [138]. The O_2 partial pressure from 0 to 3%, working pressure of 1.3 Pa, DC bias voltage of −200 V, and the weak RF power of 10–20 W are applied to deposit ZnO:Al thin films. The transmission is higher at infrared region for higher concentration of O_2. After annealing the samples under vacuum (1.3 × 10^{-3} Pa) at temperature of 400 °C for 4 h, the transmission decreases to lower due to an increase of carrier concentration, as shown in Figure 7.58. Anneal creates proper substitution of Al. Therefore, the resistivity too decreases from 1 × 10^2 to 5 × 10^{-3}–1 × 10^{-2} Ω-cm [159]. The ZnO:Ga thin films grown onto glass substrates by DC arc ion plating method at substrate temperature of 200 °C using ZnO mixed with 3–4 wt% Ga_2O_3. The transmission increases with increasing O_2 flow rate from 0 to 30 sccm in steps of 5 sccm due to a decrease of free carrier absorption, as shown in Figure 7.59 [160]. The as-deposited ZnO:Al 1 wt% films by both RF and DC show lower transmission at longer wavelength region due to free carrier

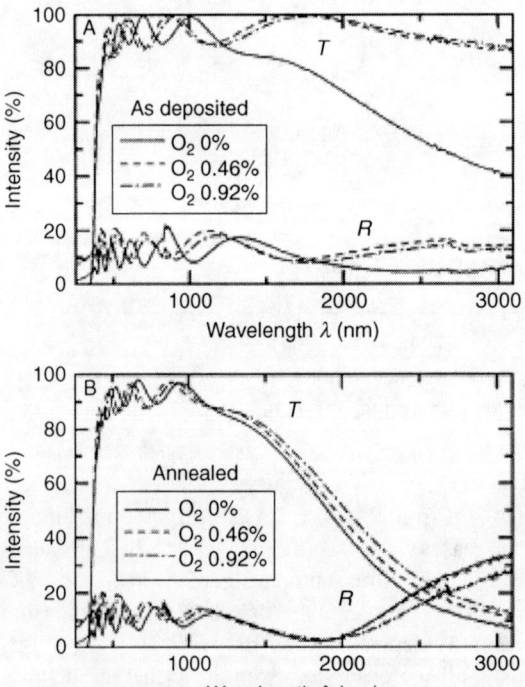

Figure 7.58 Transmission spectra of ZnO thin films with function of (A) O_2 pressure and (B) annealing.

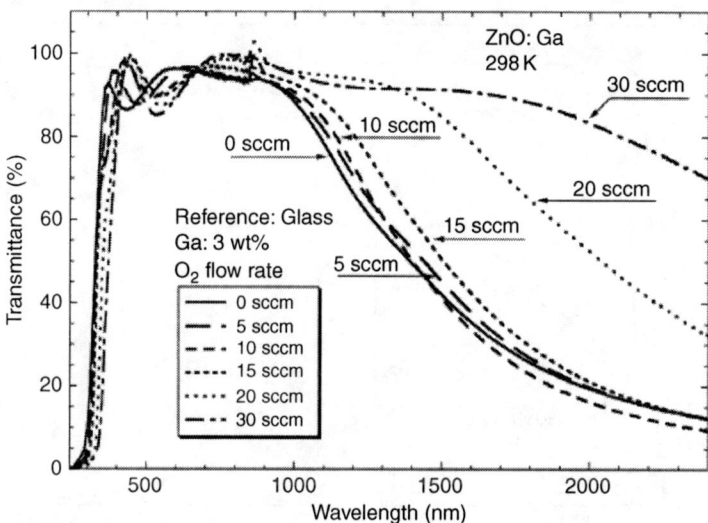

Figure 7.59 Transmission spectra of ZnO:Ga thin films grown at different O_2 flow rates of 0, 5, 10, 15, 20, and 30 sccm.

absorption. After annealing the films under air at 450 °C for 1 h, the transmission increases to higher level due to reduction in carrier concentration of samples. A reduction is caused by oxygen doping into air annealed films. The transmission spectrum shows that the optical transition at band edge red shifts in annealed samples, as shown in Figure 7.60. There is a difference in transmission spectra between similar resistivity samples grown by DC and RF technique [125].

The ZnO:Al 1.0% and ZnO:Ga 2.7% thin films have nearly same minimum resistivity of 2.5×10^{-4} Ω-cm. However, former shows higher transmission of 83% due to having slightly lower carrier concentration of 5.7×10^{20} cm^{-3} and higher mobility of 44 cm^2/Vs, as shown in Figure 7.61. The CIGS cells with ZnO:Al and ZnO:Ga layers show efficiencies of 11.9 and 10.8%, respectively [123]. The former shows higher efficiency that could be due to having higher transmittance by ZnO:Al. As expected, the transmission of ZnO:Sb thin films gradually decreases with increasing Sb doping concentration in the films due to free carrier absorption, as shown in Figure 7.62 [125]. The band gap increases from 3.24 to 3.64 eV with increasing carrier concentration but not exactly following the Burstein-Moss shift that may be due to shrinkage of band gap, as shown in Figure 7.63 [137]. The transmission of ZnO:In thin films grown by spray pyrolysis slightly increases with increasing In concentration from 0 to 5 at% in steps of 1 at% [144].

The ZnO layers with (002) orientation are grown onto (100) Si substrates by electron beam evaporation with electron beam energy of 5.5 keV in the presence of partial O_2 atmosphere using ZnO as a source at substrate temperature of 100 °C. The band gap of 3.47 eV, refractive indices ($n = 1.7$–1.95) and dielectric functions of ZnO thin films can be obtained by ellipsometry. The real and imaginary parts of the dielectric coefficients of ZnO thin films with function of wavelength are

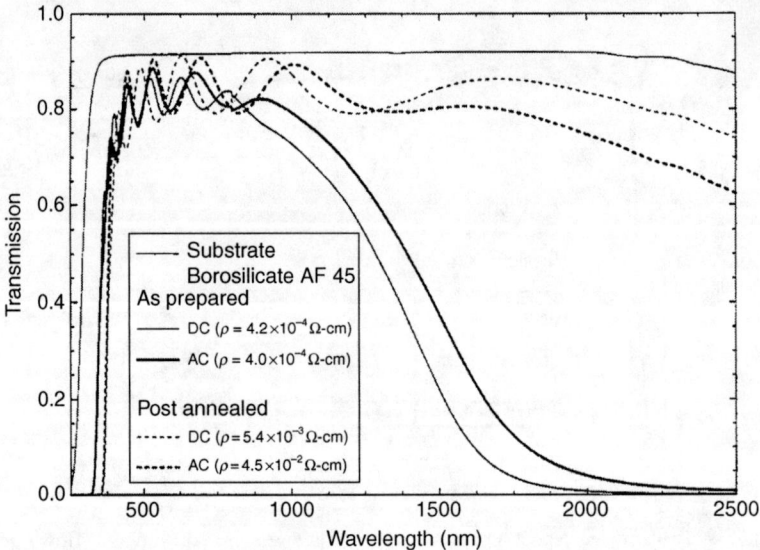

Figure 7.60 Variation of transmission spectra of ZnO:Al thin films with function of deposition techniques such as DC and RF sputtering and annealing effects. The ZnO:Al thin films with thickness of 480–640 nm annealed under air at 450 °C for 1 h.

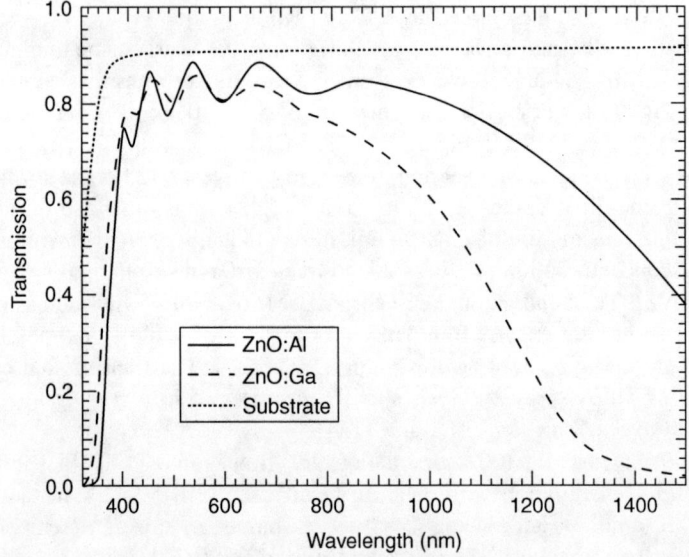

Figure 7.61 Transmission spectra of ZnO:Al and ZnO:Ga thin films grown by pulsed laser technique.

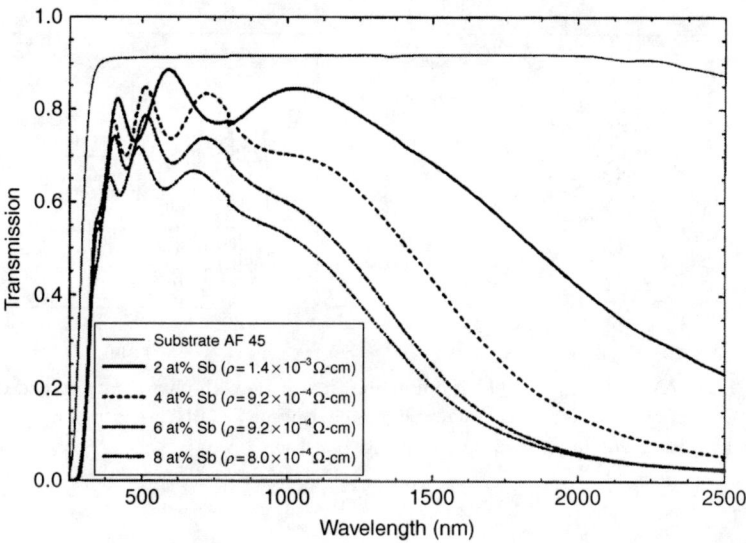

Figure 7.62 Variation of transmission spectra with function of Sb doping into 270–370 nm thick ZnO thin films grown by RF sputtering.

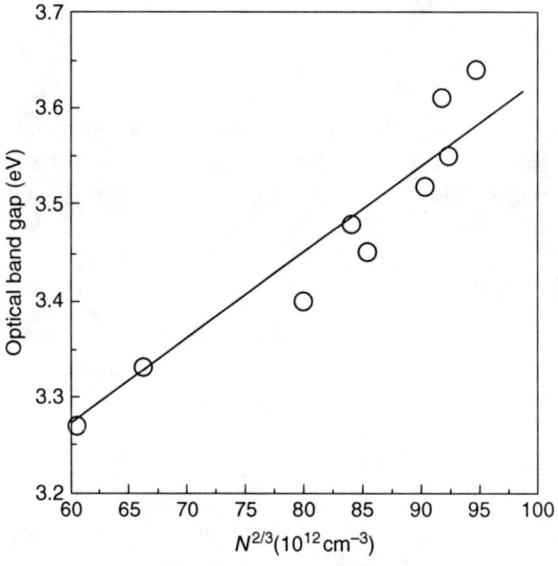

Figure 7.63 Variation of band gap with carrier concentration ($N^{2/3}$).

measured using the phase modulated ellipsometer (Figure 7.64). The n and ε are high in annealed samples. The dielectric function (ε) can be related to $\varepsilon = \varepsilon_r + i\varepsilon_i$,

$$\varepsilon = \sin\Phi\left[1 + \tan^2\Phi\left(\frac{1-\rho}{1+\rho}\right)^2\right], \tag{7.8}$$

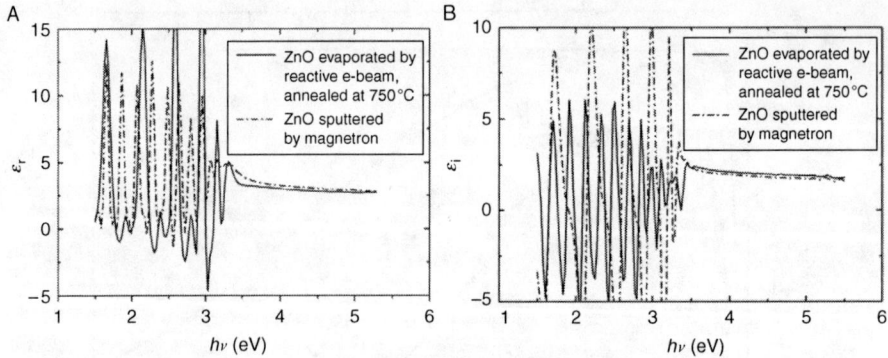

Figure 7.64 (A) Variation of dielectric function (real part) with photon energy and (B) dielectric function (imaginary part) with photon energy for ZnO layers grown by reactive E-beam and annealed, and grown by sputtering.

where complex reflectance ratio $\rho = r_p/r_s$, r_p, and r_s are linear polarized light components in perpendicular and parallel, respectively, $\varepsilon_r = n^2 - k^2$; $\varepsilon_i = 2nk$; if $n(\lambda_1) = n(\lambda_2)$, Equation (7.4) turns out to be

$$2nd_f = \frac{\lambda_1 \lambda_2}{\lambda_1 - \lambda_2}, \tag{7.9}$$

where λ_1 and λ_2 are the wavelengths between two successive interference envelopes and d_f thickness of thin film. The thickness of ZnO thin films grown by E-beam evaporation and sputtering technique determined from Equation (7.9) are 0.5 and 1.6 μm, respectively [161].

7.3.3.2 Photoluminescence of ZnO

The free excitons A (FXA) and B (FXB), ionized donor to bound exciton (D^+X), neutral donor to bound exciton (D^0X), acceptor to bound exciton (A^0X), free electron to neural acceptor (eA^0), two-electron transitions (TES), *etc.*, are common emission lines in the ZnO. The typical ranges for existence of $3.3769 < FXA < 3.4930$, $3.36499 < D^+X < 3.3769$, $3.3609 < D^0X < 3.36499$, $3.3541 < A^0X < 3.3613$, $3.3333 <$ deep centers < 3.3541, $3.3155 <$ two-electron transitions < 3.3333 and 3.3155 eV $<$ phonon replicas are scaled [162]. 4.2 K PL spectrum shows recombination lines of ionized donor to bound exciton (D^+X) such as I_0, I_1, and I_2 lines at 3.3726, 3.3718, and ~ 3.3665 eV for ZnO homoepitaxial layers, respectively. The emission lines of neutral donor to bound excitons (D^0X) such as I_{6a}(Al), I_8(Ga), and I_9(In) at 3.360, 3.359, and 3.356 eV are also observed (Figure 7.65), respectively. The intensity ratio of I_0/I_1 and I_{6a}/I_8 is found to be 3. The localization energy $E_{loc} = E_{FXA} - E_{DX}$, where DX represents either D^+X or D^0X, which are established according to Haynes rule in terms of binding energies as

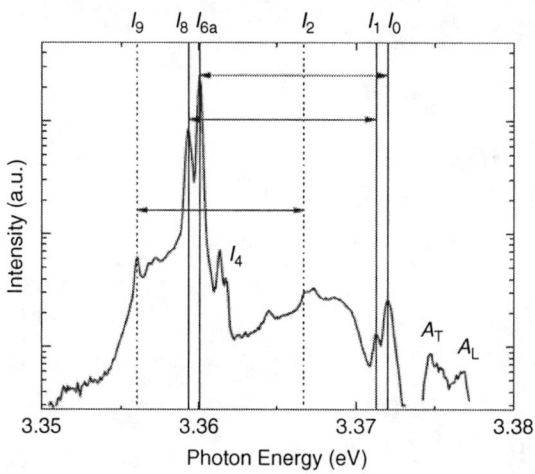

Figure 7.65 4.2 K PL spectrum of homoepitaxial ZnO.

$$E_{\text{loc}} = 0.50 E_D - 23\,\text{meV} \quad \text{and} \quad E_{\text{loc}} = 0.37 E_D - 4.2\,\text{meV} \qquad (7.10)$$

for ionized donors and neutral donors, respectively. The localization energies (E_{loc}) and donor binding energies of I_0, I_1, and I_2 emission lines as well as I_{6a}, I_8, and I_9 peaks are given in Table 7.6 [163]. The I_0, I_2, and I_3 at 3.3724, 3.3686, and 3.3702 eV are also reported in the ZnO single crystals, respectively [164]. 2 K PL spectra of ZnO thin films grown onto (11$\bar{2}$0) sapphire substrates by PLD under different oxygen partial pressures of (A) 8×10^{-2}, (B) 4×10^{-2}, (C) 2×10^{-2}, and (D) 2×10^{-3} mbar show emission lines such as neutral Al-donor bound exciton (I_6) at 3.3605 eV, neutral Zn_i-donor bound exciton (I_{3a}) at 3.3636 eV, ionized Al-donor bound exciton (I_0) at 3.3719, and free exciton (FXA) at 3.3754 eV (Figure 7.66A). The activation or localization energies of 10.9 and 15 meV are obtained for I_{3a} and I_6 peaks by thermal quenching experiment for ZnO thin film grown with oxygen partial pressure of 2×10^{-3} mbar (Figure 7.66B). They match very well with the calculated localization energy (E_{loc}) values of 10.9 and 14.8 meV, respectively. The donor binding energies of 40.2 and 50.9 meV for (I_{3a}) and (I_6) can be obtained from the Haynes formula (Equation (7.10)), respectively [165]. The bound exciton emission lines pertain at 3.356, 3.360, and 3.367 eV for un-doped ZnO thin film in 10 K PL spectrum. After Ga doping into ZnO thin film ($n_e = 5.7 \times 10^{18}$ cm^{-3}), a new emission line exists at 3.358 eV marked as I_{Ga}. On the other hand, line at 3.360 disappears and intensities of other two lines at 3.356 and 3.367 eV reduce [166]. The Ga doped ZnO nanorods on Si substrates show neutral donor to bound exciton ($D^0 X$) line or (I_8) at 3.364 eV and DAP at 3.228 eV, followed by several phonon lines in 8 K PL spectra, whereas (eA^0) and FXA lines exist at RT, as shown in Figure 7.67A. The thermal quenching analysis for ($D^0 X$) line reveals activation energies of 18 and 6 meV (Figure 7.67B). The former is due to participation of bound exciton localization energy (E_{loc}). Thus, the binding energy (E_D) of

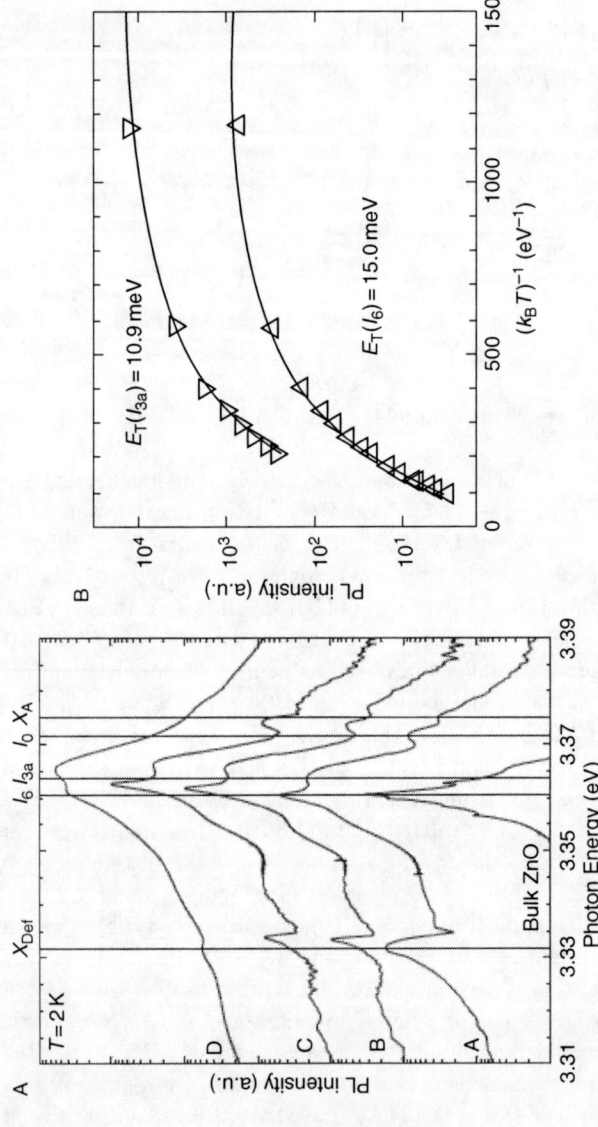

Figure 7.66 (A) PL spectra of ZnO epitaxial layer grown by sputtering and (B) PL intensity versus $(k_B T)^{-1}$ for I_{3a} and I_6 peaks.

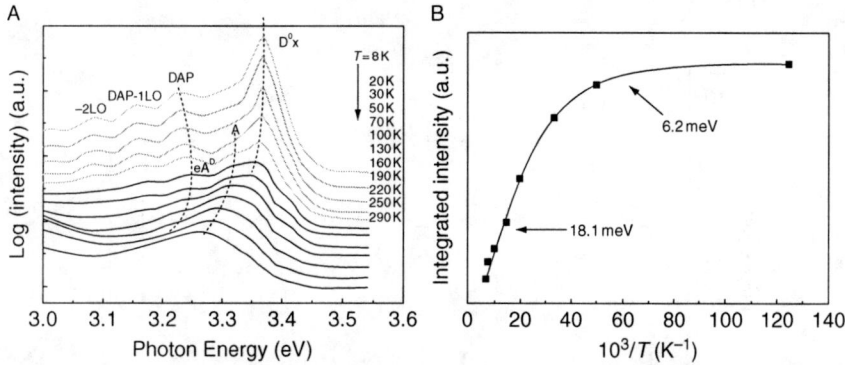

Figure 7.67 (A) Low temperature PL spectra of ZnO nanorods grown onto Si substrates and (B) PL intensity versus $1/T$ curve for D^0X line.

Table 7.6 Binding energies of (D^+X) and (D^0X) lines calculated from Equation (7.10)

Emission line	Peak position (eV)	Loc (meV)	E_D (meV)	Ref.
FXA	3.376	–	–	[163]
I_0	3.3726	3.4	53	
I_1	3.3718	4.1	54	
I_2	~3.3665	8.5	63	
I_{3a}	3.3636	10.9	40.2	[165]
I_{6a}(Al)	3.360	15.5	53	
I_8(Ga)	3.359	16.1	55	
I_9(In)	3.356	19.2	63	

60 meV for neutral donor to bound exciton is determined, which coincides well with the calculated value, as depicted in Table 7.6. An activation energy of 6 meV is considered to be capture cross section of the carriers [167]. The ZnO single crystals show I_6 emission line at 3.3618 eV (3688.46 Å), accompanied by LO phonon line at 3.2785 eV (3688.46LO). In addition, its excited line ($n_x = 2$) exists at 3.3223 eV (3732.35 Å), followed by LO phonon line at 3.2479 eV (3732.35 Å-LO). After In doping into crystals, all the lines disappear but new line (I_9) appears at 3.3580 eV (3692.64 Å), followed by LO phonon line at 3.2872 eV (3692.64 Å-LO) [168]. However, the virgin sample shows I_9 at 3.352 eV, followed by phonon lines at 3.28 and 3.21 eV, I_7 at 3.36 and FX at 3.38 eV in 2 K PL spectrum [169].

The two-electron satellite (TES) line pertains when the neutral donor electron is left in an excited (2s, 2p, *etc.*,) state in the exciton recombination process, or the two-electron satellite line is down shifted from (D^0X) by 30–50 meV [170]. In other words, the TES recombination is either final state of the donor 1s state (D^0_iX) or the 2s/2p state (TES_i) due to transition from an exciton bound to a neutral donor. The energy difference between 1s and 2s/2p states is equal to 3/4 of donor binding energy (E_D) [171]. The neutral donor bound exciton $(D^0_{1a}X)$, $(D^0_{2a}X)$, and

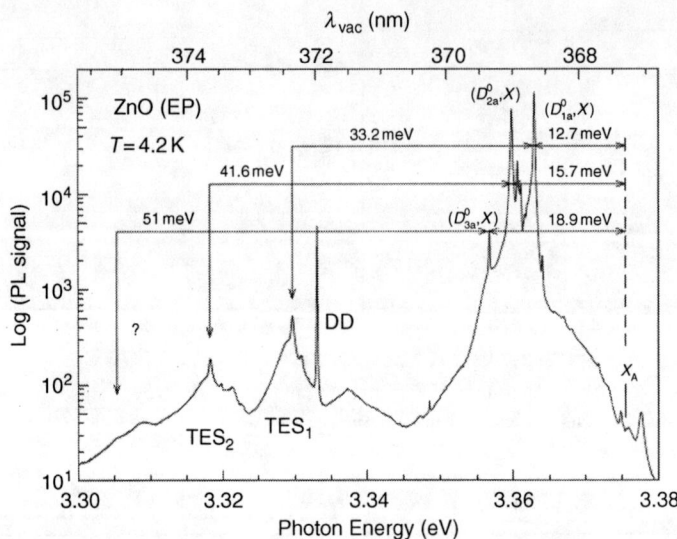

Figure 7.68 PL spectrum of ZnO bulk showing donor-bound exciton lines and their TES lines.

($D^0_{3a}X$) lines at 3.362(8), 3.359(7), and 3.356(6) eV and their two-electron states such as $TES_{(D^0_{1a}X)}$, $TES_{(D^0_{2a}X)}$, and $TES_{(D^0_{3a}X)}$ lines are separated by donor single electron energies (ΔE) of 33.3, 41.5, and 51 meV, respectively, as shown in Figure 7.68. The binding energies of 47.3, 55.5, and 65 meV for ($D^0_{1a}X$), ($D^0_{2a}X$), and ($D^0_{3a}X$), respectively can be obtained by adding donor single electron energies to $1/4 E_{D,EMT}$ (since $E_{D,EMT} = R* = Ry \frac{m_e^*}{m_0} \frac{1}{\epsilon_0^2}$, $Ry = 13.6$ eV, $\epsilon_0 = 7.93$, $m_e^* = 0.26\, m_0$), where $E_{D,EMT}$ calculated to be 56 meV. The doublet ($D^0_{1a}X$) at 3.3639 eV is separated from its ($D^0_{1b}X$) partner by 1.1 eV, as shown in Figure 7.69 [170]. The similar neutral donor bound excitons such as D^0_1X, D^0_2X, and D^0_3X at 3.364, 3.362, and 3.36 eV are observed, respectively, as shown in Figure 7.70. In addition, the acceptor bound exciton (A^0_1X) and free exciton (FXA) emissions are found at 3.358 and 3.377 eV, respectively. After annealing the sample under N_2 atmosphere from 600 to 800 °C for 30 min, the line at 3.364 eV disappears, whereas other lines such as A^0_1X, D^0_2X, and D^0_3X are quite stable. On the other hand, the $TES_{(D^0_2X)}$ and $TES_{(D^0_3X)}$ lines are moderately resolved. The binding energies are calculated to be 55, 52, and 43 meV for the $TES_{(D^0_3X)}$, $TES_{(D^0_2X)}$, and $TES_{(D^0_1X)}$ lines at 3.32, 3.323, 3.332 eV, respectively. They follow proportional to the bound exciton localization energies ($E_{Fx} - E_{D^0X}$) known as Haynes rule [171]. In another case, the neutral donor to bound exciton (D^0_1X) and (D^0_2X) lines at 3.3636 and 3.3614 eV and their corresponding TES at 3.3220 and 3.3189 eV are observed in the PL spectrum, respectively. The (D^0_1X) and (D^0_2X) have binding energies of 55.5 and 56.7 meV, respectively. After annealing the sample at 800 °C the TES at 3.3137 ($n_x = 2$) and 3.3058 eV ($n_x = 3$) occur for D^0X emission line (3.357 eV) [172]. There are several 2S and 2P states in TES transitions, as shown in

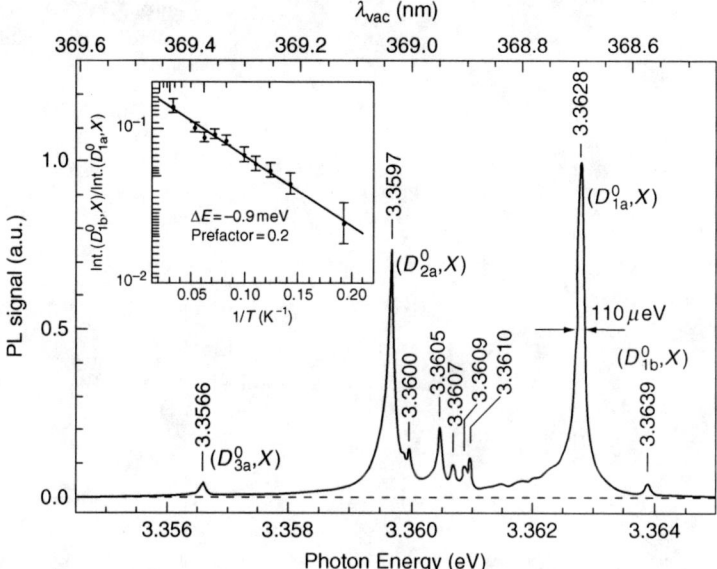

Figure 7.69 4.5 K PL spectrum shows bound exciton region.

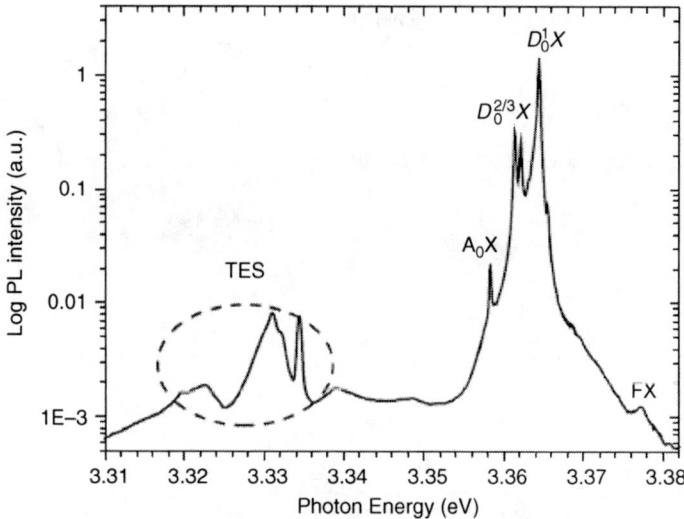

Figure 7.70 Near band edge PL spectra of ZnO bulk recorded at liquid helium temperature.

Figure 7.71 [146]. The CL spectra show that the free electron to neutral acceptor transition (e,A^0) at 3.314 eV is due to a complex defect by segregation of basal plane stacking faults, followed by phonon replicas (Figure 7.72). D^{0}X at 3.35–3.40 eV dominates spectra at low temperature. The acceptor activation energy of 130 meV for (e,A^0) is

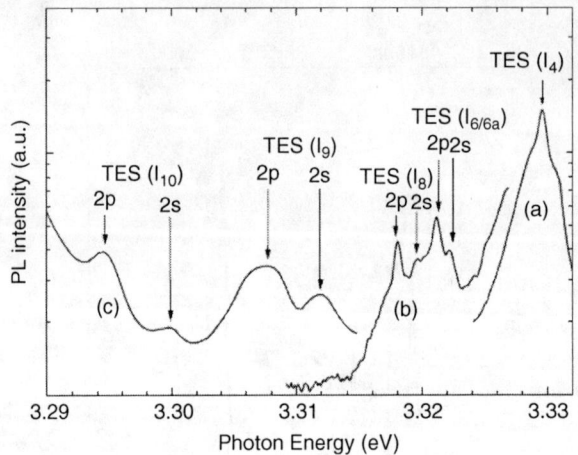

Figure 7.71 Two-electron satellite transitions of donor bound excitons.

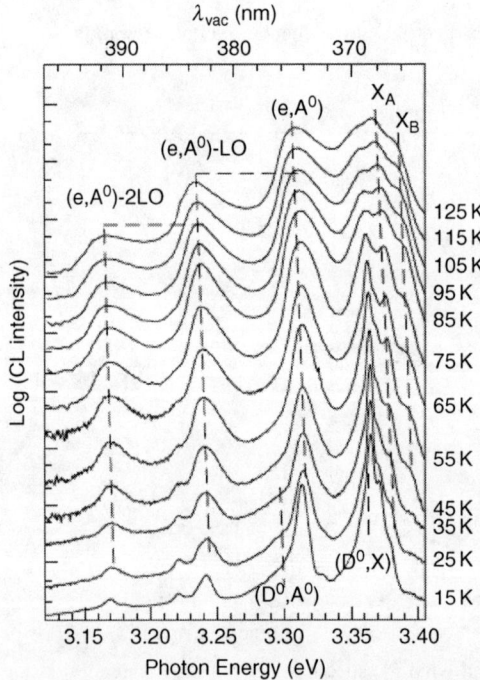

Figure 7.72 CL spectra of ZnO single crystal (FXA = X_A and FXB = X_B).

obtained from a relation (Equation (5.8)), which is close to that of phosphorus, antimony, arsenic, and nitrogen [173]. The TES of donor bound exciton (D^0X) is located less than by 30 meV from bound excitons. As shown in Figure 7.73, the DAP transition

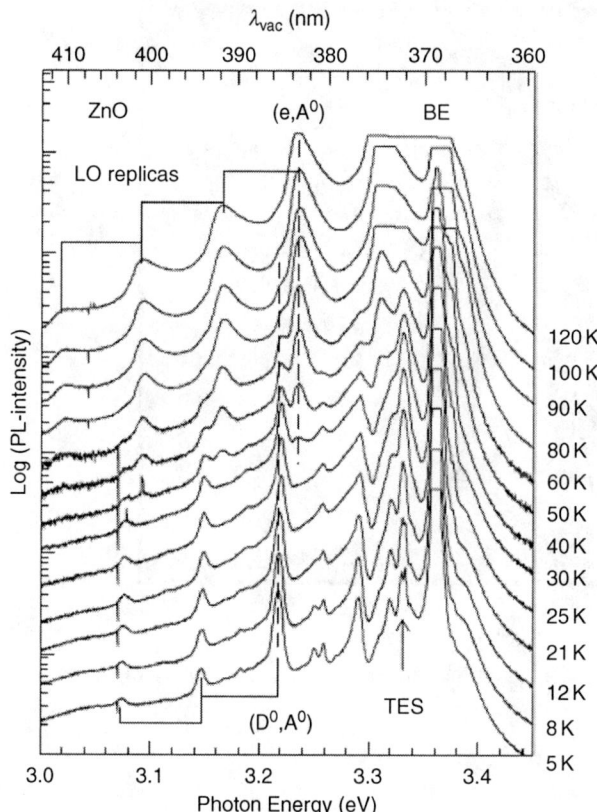

Figure 7.73 PL spectra of ZnO single crystal.

exists at 3.220 eV, followed by several phonon lines at low temperature. The intensity of DAP gradually decreases and free electron to neutral acceptor transition (e,A^0) line at 3.236 eV emerges and gains intensity with increasing temperature. The donor activation energy of 30 meV (E_D) for DAP can be determined using relation

$$E_D = [E(e, A^0) - k_B T/2] - E_{DAP} - \alpha N_D^{1/3}, \quad (7.11)$$

where N_D is the donor concentration and $\alpha = \sqrt[3]{\frac{4\pi}{3}} \frac{e^2}{4\pi\epsilon\epsilon_0} = 2.7 \times 10^{-8}$ eV cm. Similarly $E_A = 195 \pm 10$ meV can be obtained from Equation (5.7) using $E_g = 3.438$ eV [174]. The ZnO thin films on SiO_2 fiber mat show peaks at 3.375, 3.360, and 3.44 eV relate to free exciton (FXA), neutral donor to bound exciton (D^0X) and neutral acceptor to bound exciton (A^0X) in 80 K PL spectrum, respectively. The FXA has 1LO, 2LO, and 3LO phonon lines at 3.305, 3.234, and 3.167 eV, respectively (Figure 7.74) [175]. 80 K PL spectrum of (0002) highly oriented ZnO thin films grown by electrodeposition also exhibits FXA at 3.374 eV as a shoulder to D^0X (3.345 eV), followed by FXA-1LO, 2LO, and 3LO phonon replicas at 3.307,

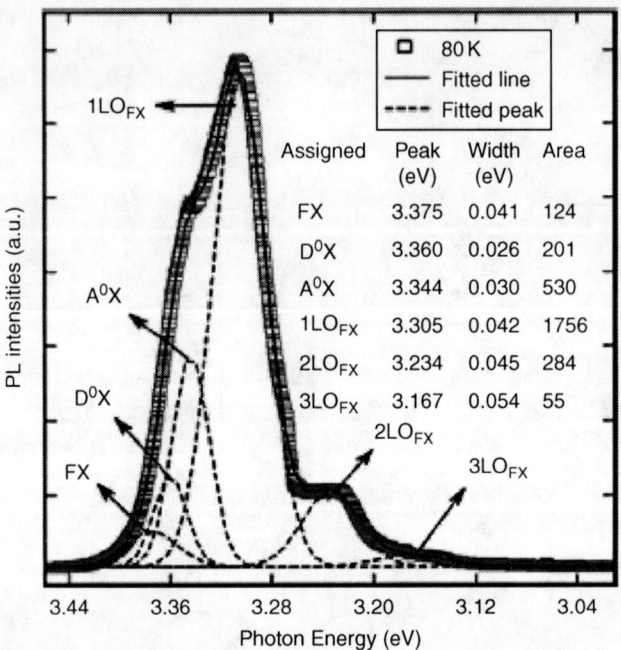

Figure 7.74 80 K PL spectra of ZnO thin films grown onto SiO$_2$ fiber mat by vapor transport method at 600 °C.

3.326, and 3.167 eV, respectively. The neutral acceptor to bound exciton (A^{0}X) at 3.345 eV is also a shoulder to D^{0}X [176].

A variation of exciton FXA position with temperature is fitted with several models such as Varshni's equation (5.5),

modified version of Varshni's equation

$$E_g(T) = E(0) - \alpha T^4/(\beta + T)^3 \tag{7.12}$$

and semi-empirical equation equal to Bose–Einstein

$$E(T) = E(0) - \alpha \Theta/[\exp(\Theta/T) - 1], \tag{7.13}$$

where the Debye temperature (Θ_D) is the effective phonon temperature ($\Theta_D = 3\Theta/2$). The authors claim that the last one shows best fit and all the fitting parameters are given in Figure 7.75 [177]. The free exciton (FXA) at 3.377 eV, followed by LO phonon lines and BX at 3.36 eV occur in 4.8 K PL spectrum. The BX and

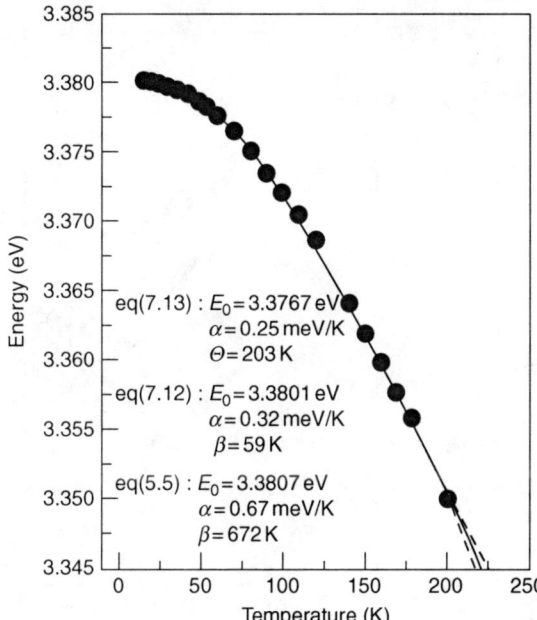

Figure 7.75 Variation of free exciton energy versus temperature.

FXA lines are in equal intensities. As the sample temperature is increased from low temperature to RT, all the peaks disappear but a peak at 3.263 eV exists. Employing Varshni's equation (5.5) for E_{FXA}, the fitting parameters are found to be $E_{FXA}(0) =$ 3.380 eV, $\alpha = 0.8.2 \pm 0.3$ meV/K, and $\beta = 700 \pm 30$ K [178]. Temperature dependent PL spectra of N doped ZnO thin films are shown in Figure 7.76A. Using Equation (7.12), the Varshni parameters α and β can be obtained for different kinds of peaks from the plots of peak positions versus temperatures, as shown in Figure 7.76B [179]. 80 K PL spectrum of ZnO nanobelts shows donor to bound exciton (I_4) at 3.350 eV, acceptor to bound exciton (I_a) at 3.300 eV, followed by number of phonon replicas. A shoulder to I_4, that is, free exciton (FXAB) at 3.365 eV is shown in Figure 7.77. The α parameters are found to be 1.3 ± 0.03, 1.5 ± 0.08, and 1.0 ± 0.03 meV/K for free exciton, donor to bound exciton and acceptor to bound exciton by using $E_{FX}(0) = 3.37$, $I_4(0) = 3.36$, $I_a(0) = 3.31$, and $\beta = 920$ K [180]. For free exciton FXA the fitting parameters are found to be $E(0) = 3.379$ eV, $U = -5.04 \times 10^{-5}$ eV/K$^{1.01}$, $S = 1.01$, $V = -1.84 \times 10^{-4}$ eV/K, and $\Theta = 398.4$ K in the temperature range of 3.2–300 K using

$$E = E_0 + UT^s + V\Theta[\coth(\Theta/2\,T) - 1] \qquad (7.14)$$

relation, where UT^s is lattice dilatation and $V\Theta[\coth(\Theta/2T) - 1]$ is electron–phonon interaction [181]. As shown in Figure 7.78A, FXA at 3.375 eV, followed by phonon lines and donor to bound exciton (BX) at 3.360 eV for ZnO tetrapod-like microrods shift with increasing temperature from 10 to 170 K. The first phonon line (3.310 eV) is separated from FXA by 65 meV. The FXA-1LO line follows the relation with

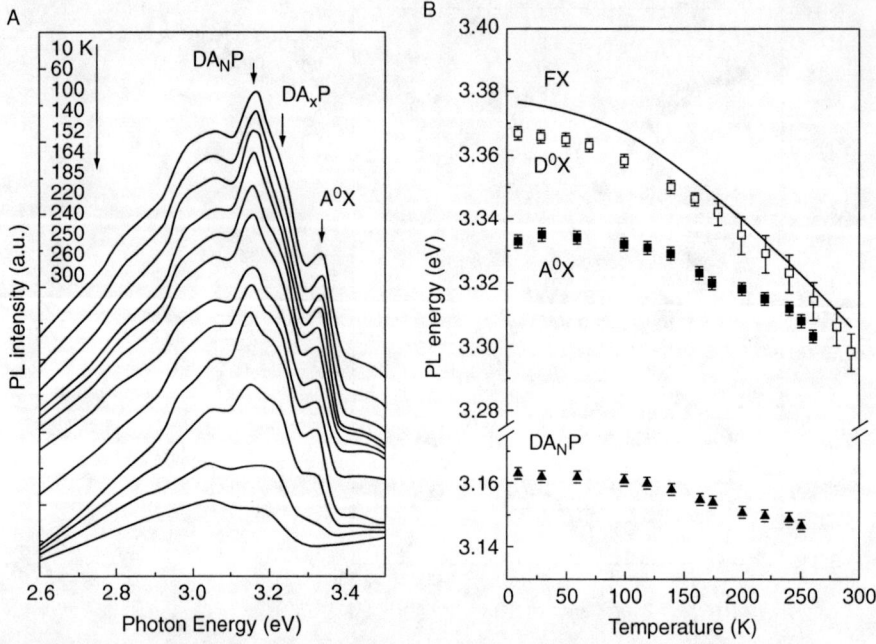

Figure 7.76 (A) PL spectra of N doped ZnO thin film grown by DC reactive sputtering technique and (B) peak energy versus temperature for FX, D^0X, A^0X, and DAP lines, the D^0X line is observed in the PL of n-type sample that is not shown here.

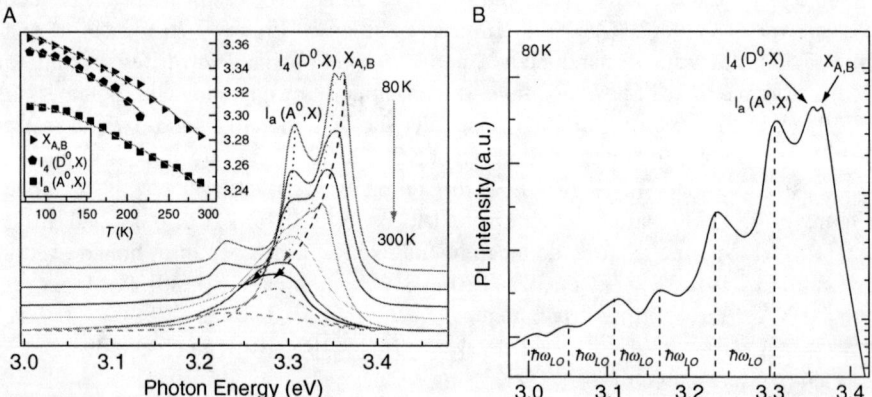

Figure 7.77 (A) PL spectra of ZnO nanobelts with variation of temperature and (B) 80 K PL spectra of ZnO nanobelts.

variation of temperature as $E_{FXA}(1LO) = E_{FXA} - \hbar\omega + 3/2k_BT$, whereas second LO phonon line follows as

$$E_{FXA}(2LO) = E_{FXA} - 2\hbar\omega + 1/2k_BT. \tag{7.15}$$

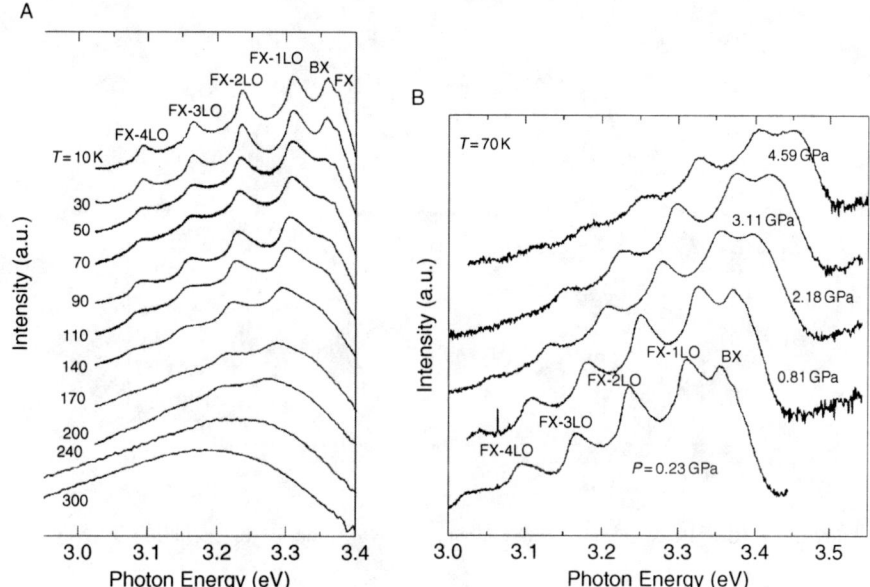

Figure 7.78 PL spectra of ZnO nanorods recorded at (A) different temperatures and (B) pressures.

In the case of FXA, $E_0 = 3.377 \pm 0.001$ eV, $U = -(4.5 \pm 0.1) \times 10^{-5}$ eV/K, $V = -(2.5 \pm 0.1) \times 10^{-4}$ eV/K, $S = 1.2 \pm 0.2$, and $\Theta = 401 \pm 20$ K parameters are extracted from Equation (7.15). The separation (ΔE) between FXA and FXA (1LO) is given by

$$\Delta E = \hbar \omega_{LO} - (p + 1/2) k_B T, \qquad (7.16)$$

where $p = 1/2$ but it varies from 0 to 1. As shown in Figure 7.78B, the BX, FX, and its phonon peak positions shift with varying pressure from 0.23 to 4.59 GPa, the relation follows as

$$E(P) = E_0 + \alpha P, \qquad (7.17)$$

where α is the pressure coefficient and E_0 is the emission line energy at $P = 0$ GPa [182]. The PL intensity of ZnO nanocrystals increases with increasing excitation intensity and follows the relation $I \alpha I_0^k$ where k is found to be ∼0.87 and ∼0.85 for free exciton (∼3.377 eV) and A line (3.315 eV, probably DAP), respectively. DX lines at ∼3.364 and ∼3.368 eV are concerned to ionized donor to bound exciton and neutral donor to bound exciton, respectively. The FX-1LO, A-1LO, and FX-2LO lines occur at 3.30, 3.245, and 3.218 eV, respectively (Figure 7.79) [183].

The phonon lines of BE gradually develop, whereas the phonon lines of FXA slowly diminish with decreasing ZnO bulk sample temperature in the range of 70–12 K, as shown in Figure 7.80. The phonon replicas of bound exciton disappear

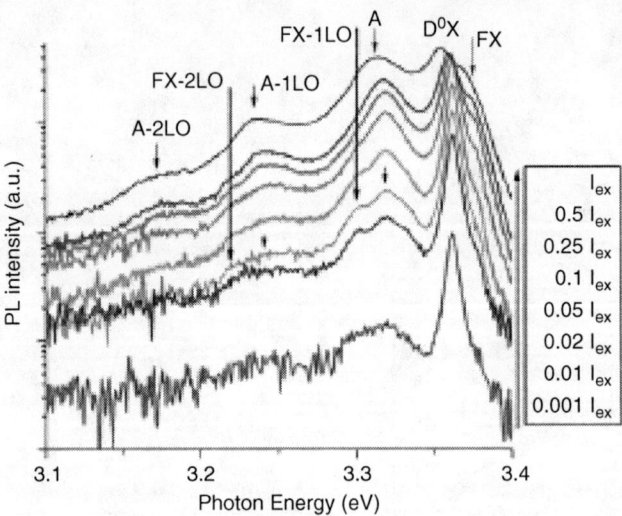

Figure 7.79 10 K PL spectra of ZnO nanocrystals with function of excitation intensity.

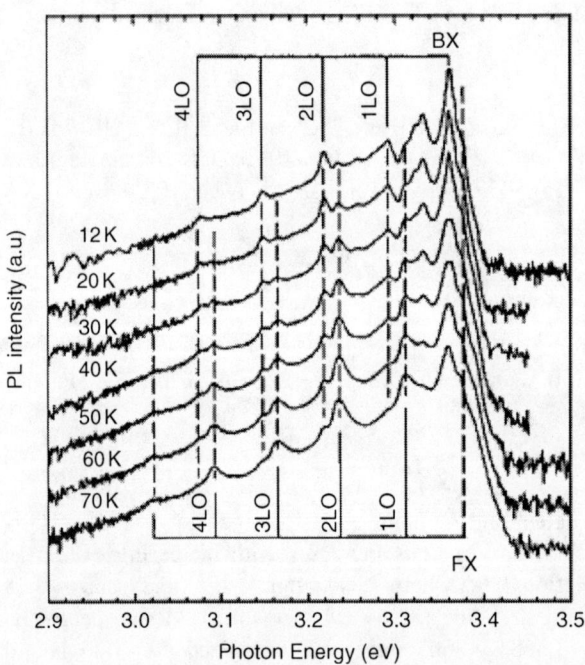

Figure 7.80 PL spectra of ZnO recorded at different temperatures.

above 70 K due to change in the ground states of free and bound excitons by thermal energy and rapid thermal ionization of bound excitons. The expression for the exciton and phonon emissions can be written as

$$E_m = E_0 - m\hbar\omega_{LO} + \Delta E, \tag{7.18}$$

where E_m is the peak position, E_0 is the exciton energy at $K=0$, m is an integer and $\hbar\omega_{LO}$ is phonon energy of 72 meV, and $\Delta E = \hbar^2 K^2/2\,M$ is the kinetic energy of free exciton [150]. A number of emission lines such as X, Y, and Z exist at 3.324, 3.307, and 3.287 eV, which are speculated as free-to-bound emissions and their phonon emissions being pertain, as shown in Figure 7.81 [149]. A several FX-LO phonons at 3.31, 3.235, 3.164, 3.092, and 3.020 eV separated by 72 meV are observed. The neutral donor to bound exciton (D^0X) at 3.355 eV, followed by shoulder line at 3.369 eV due to ionized donor to bound exciton (D^+X) present in 10 K PL spectrum for ZnO nanocrystals fabricated by irradiating laser beam on ZnO powder, as shown in Figure 7.82 [184]. The similar free exciton (FX) at 3.370 eV, followed by phonon emission lines at 3.299, 3.223, 3.147, and 3.071 eV separated by 71 meV in 80 K PL spectrum of ZnO thin films are found (Figure 7.83) [185]. The similar free exciton at 3.3696 eV (368 nm), separated by multiple phonons at 3.3155 (374), 3.2376 (383), and 3.1633 eV (392 nm) in 77 K PL spectrum for the ZnO ceramic sintered at 950 °C without pressing pellet are observed (Figure 7.84). In addition, donor to bound exciton sharp peak at 3.3604 eV (369 nm) is overlapped with free exciton [186].

20 K PL spectrum of ZnO layers grown by RF sputtering under $O_2 + 0.1\%N_2$ atmosphere shows neutral donor to bound exciton (D^0X) at 3.362 eV, TES of D^0X and DAP at 3.273 eV with multiple phonon lines by separation of 73 meV (Zn/O = 1.8), as shown in Figure 7.85. By taking donor activation energy of 40 meV and $N_D = 5 \times 10^{16}$ cm^{-3}, the acceptor activation energy of 135 meV can be obtained for DAP peak at 3.273 eV. The yellow band (2.3 eV) shifts to green emission (2.4 eV) for the layers grown at O_2-rich conditions [187]. A change in 3.3913 eV

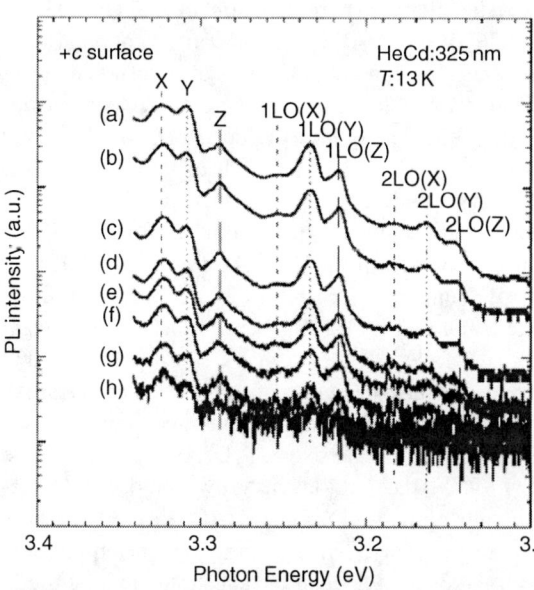

Figure 7.81 13 K PL spectra of +c ZnO single crystal with function of different excitation intensities of (A) 1.8, (B) 1.4, (C) 1.0, (D) 0.8, (E) 0.6, (F) 0.4, (G) 0.3, and (H) 0.2 W/cm^2.

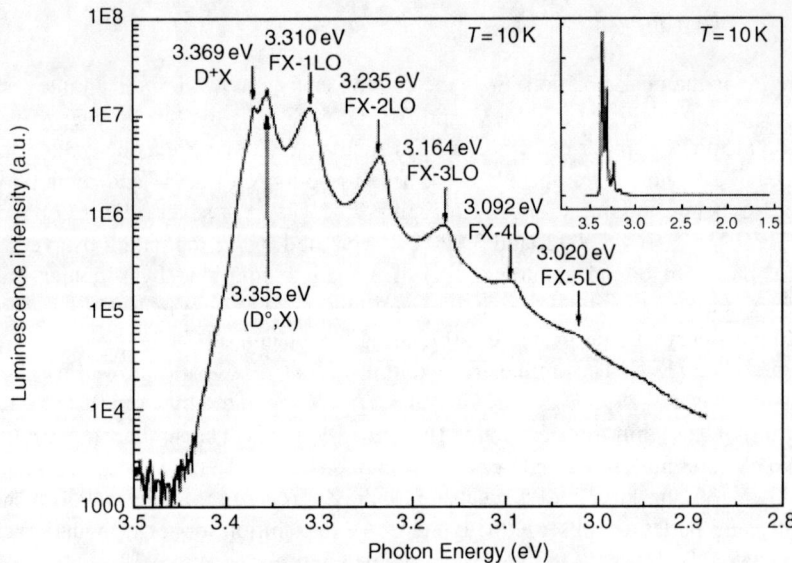

Figure 7.82 10 K PL spectrum of ZnO nanocrystals formed by irradiating ZnO powder.

(A), 3.3774 eV (B) excitons, and DAP with variation of temperature is shown in Figure 7.86. Shifts in exciton emissions with temperature could follow models (eq. 5.5, 7.12–7.14). A broad peak at RT may be due to responsible by surface states of ZnO single crystals [188]. The exciton emission line is stronger in the ZnO on GaN buffered sapphire substrates rather than on bare sapphire in 8 K CL spectra. In addition, DX bound exciton at 3.36 eV and DAP at 3.32 eV and its phonon lines at 3.24 and 3.18 eV exist and green emission band at 2.43 eV is also observed [189].

The positions of various defect levels calculated by full potential linear muffin-tin orbital are depicted in Figure 7.87 [190]. The O face Li doped ZnO single crystal wafers annealed in Zn-rich atmosphere at 1050 °C, in O-rich at 900 °C and ZnO wafer show emission lines at 2.53, 2.35, and 2.17 eV, respectively. The emission lines, deep band emission lines (DBE) show shift from 2.53 to 2.47 eV by ~ -0.018 eV for V_O relate emission and from 2.35 to 2.44 eV by $\sim +0.078$ eV for V_{Zn} emission line during thermal quenching from RT to 27 K. The V_{Zn} follows temperature dependent band gap of ZnO, that is, Varshni's formula, whereas negative shift by -0.018 eV for V_O does not follow. The V_{Zn} may be probably related to free to bound transition or DAP and V_O may be related to color center transition [191]. The donor to bound exciton (I_4) at 3.3628 eV and weaker acceptor to bound exciton (I_8) at 3.3597 eV exist. The donor to bound exciton (D^0X) and free exciton (FXA) are found at 3.36 and 3.374 eV for the ZnO layers grown onto (0001) sapphire substrates by MBE. The phonon replicas (D^0X)-1LO, (D^0X)-2LO for D^0X are observed at 3.288 and 3.214 eV, respectively. The atomic oxygen beam pressure of 5×10^{-6} mbar, atomic Zn pressure of 3×10^{-6} mbar and substrate temperature of

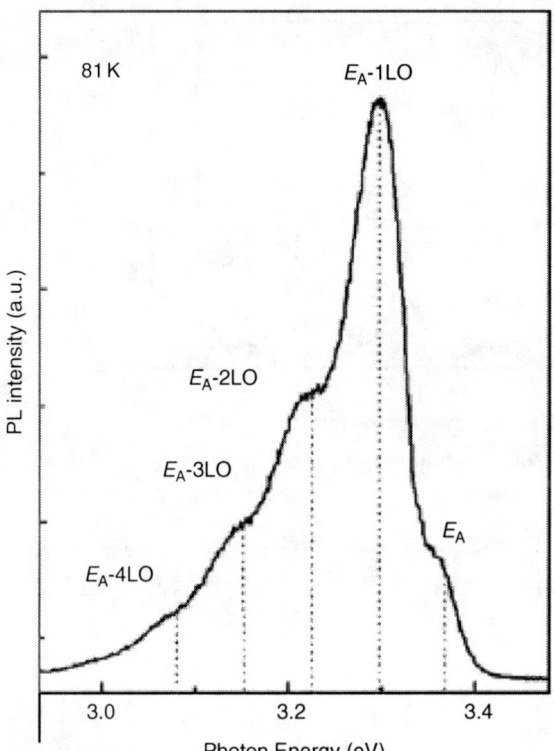

Figure 7.83 PL spectrum of ZnO thin film grown onto Si substrates by CVD.

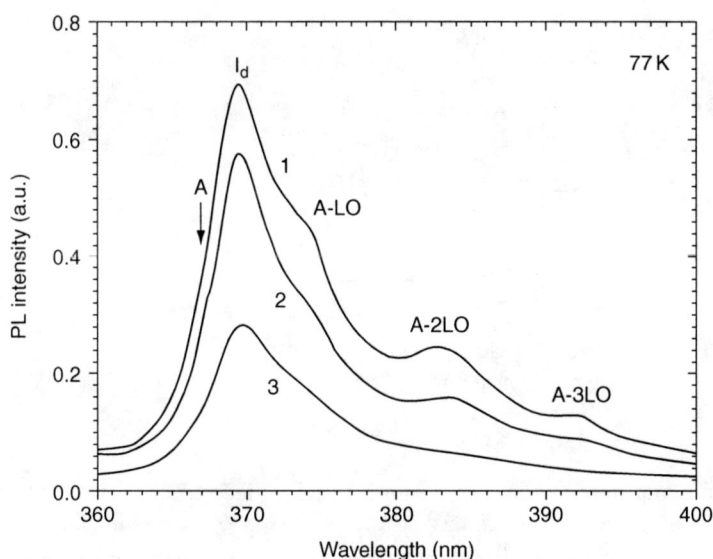

Figure 7.84 PL spectra of ZnO without pressing pellet.

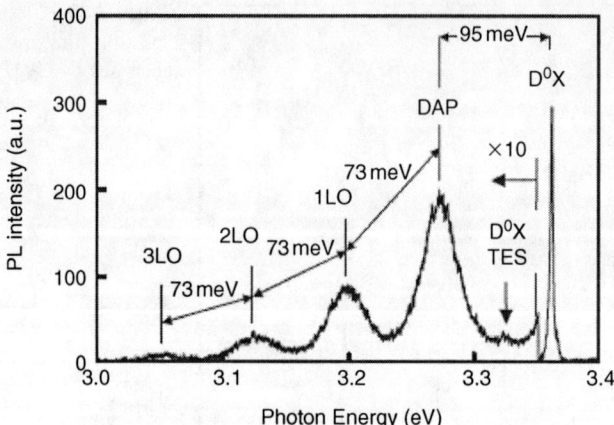

Figure 7.85 20 K PL spectrum of ZnO thin film grown by RF sputtering.

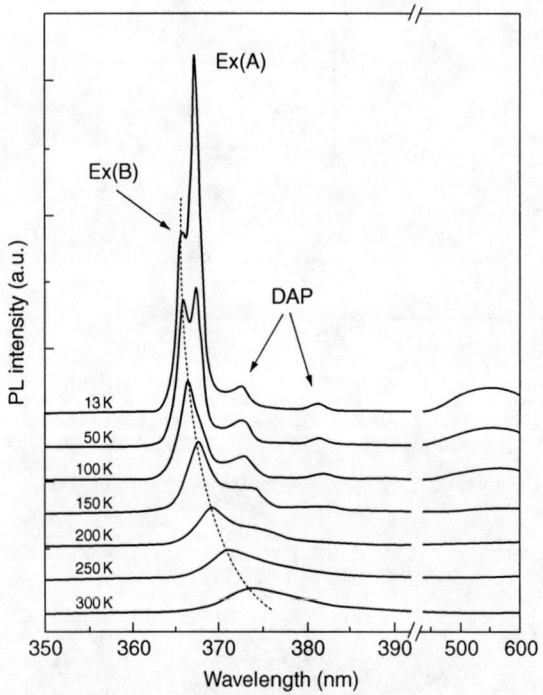

Figure 7.86 PL spectra of ZnO single crystal grown by hydrothermal method.

450 °C are employed to grow ZnO thin films. The deep level emission at RT shifts from 2.04, 2.18, 2.21 to 2.24 eV with increasing Zn pressure from 2×10^{-7}, 5×10^{-7}, 1×10^{-6} to 2×10^{-6} mbar, respectively. The green emission (2.4 eV) is due to recombination of donor V_O and acceptor V_{Zn} with activation energy of

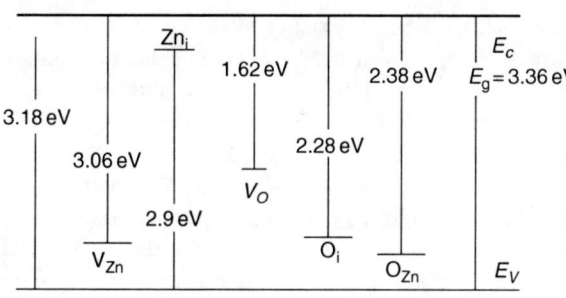

Figure 7.87 Defect levels in the ZnO.

190 meV [192]. The PL studies reveal that the deep level emission line at 2.43 eV dominates the spectrum after annealing the sample in the presence of oxygen indicating responsible for V_{Zn} [193]. The intensity of green emission decreases with increasing 5 mol% Al doping into ZnO thin films. The green emission could be also responsible for formation of oxygen vacancies. On the other hand zinc vacancies (V_{Zn}) may be reduced by Al doping. The Al doped ZnO films used for PL studies are prepared by using $Zn(CH_3COO)_2 \cdot H_2O + Al(NO_3)_3 \cdot 9H_2O$ chemical solution by spin coating at substrate temperature of 350 °C and annealed at 850 °C [155]. The neutral donor to bound exciton (D^0X) at 3.355 eV and DAP peak at 3.307 eV, accompanied by phonon lines at 3.235 and 3.163 eV occur in 12 K PL spectra of ZnO ceramics either annealed by two step or one step processes at different temperatures [194], whereas the DAP line is observed at 3.104 eV, followed by phonon lines with separation of 72 meV for the sample grown by PLD with oxygen partial pressure of 2×10^{-3} mbar. The activation energy of 40.2 meV is obtained for DAP line (3.104 eV) from thermal quenching analysis, which is equal to activation energy of Zn_i-donor level. By using donor activation energy level of 40.2 meV, the activation energy of 320 eV for E_A can be obtained from Equation (5.7). The acceptor could be due to V_{Zn} among O_i and O_{Zn}, which coincides with the literature value of 260–390 meV [165]. The un-doped ZnO thin films exhibit three donor to bound excitons (DX) at 3.3667, 3.3624, and 3.3587 eV, whereas the lightly phosphorus (P) doped ZnO thin films show a violet luminescence (VL) at 3.1099 eV, followed by several phonon lines with separation of 72 meV. The VL line is confirmed as DAP that the line shows a blue shift by 14 meV with increasing laser excitation intensity. Using the donor activation energy of 42 meV obtained by thermal quenching analysis, the acceptor activation energy is determined to be 339 meV from Equation (5.7), which may be considered as $P_{Zn}-2V_{Zn}$ complex [195,196]. The theoretical analysis supports the defect nature of $(P_{Zn}-2V_{Zn})^{2-}$ for the activation energy of 305 meV [197].

The ZnO layers grown onto polyimide substrates by PLD show that the PL intensity ratio of band edge emission to yellow band increases with increasing substrate temperature from RT to 300 °C in steps of 100 °C indicating improvement in crystal quality [143]. No PL features are observed for the ZnO layers grown onto different substrates of sapphire, mica, and fluorite by laser ablation of ZnO pellets using Nd-YAG laser ($\lambda = 354.7$ nm, 10 Hz, 8 ns pulse) at substrate temperature of 400 °C and working pressure of 2×10^{-4} mbar. However, after annealing the ZnO layers, donor to bound exciton at 3.359–3.366 eV exists containing LO phonon by separation energy of 72 meV, which changes from sample to sample due to variation of compressed stress, as shown in Figure 7.88 [198]. 10 K PL spectra of ZnO on GaAs and sapphire substrates show that peaks at 3.3645 and 3.3583 eV are assigned to neutral donor to bound exciton and its excited state. A broader peak at 3.3086 eV is due to neutral acceptor to bound exciton. A shoulder at 3.2938 eV may be due to phonon line of 3.3645 eV peak. There are slightly deviations in peak positions comparing with substrate to substrate due to different stress or strains in the grown ZnO films (Figure 7.89) [199]. The intensity of band edge emission also decreases with increasing Al doping [133]. A blue shift in PL peak from ~3.320 to 3.400 eV occurs with increasing Ga doping from 0 to 2%, as shown in Figure 7.90. The un-doped ZnO thin films show NBE peak at 3.3 eV and yellow luminescence at 2.2 eV. The peak at 3.2 and yellow luminescence peak at 2.2 eV pronounce well in PL spectra with increasing oxygen flow rate from 30, 40 to 50 sccm [160]. The ZnO

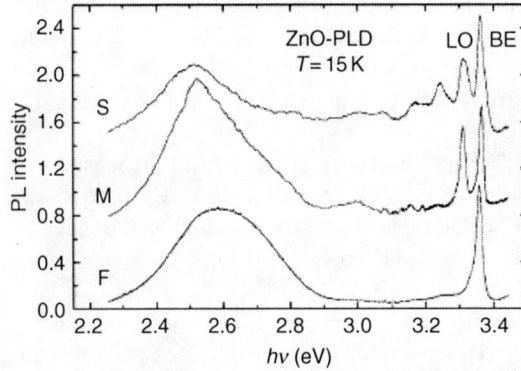

Figure 7.88 15 K PL spectra of ZnO layers grown onto sapphire (S), mica (M), and fluorite (F) substrates.

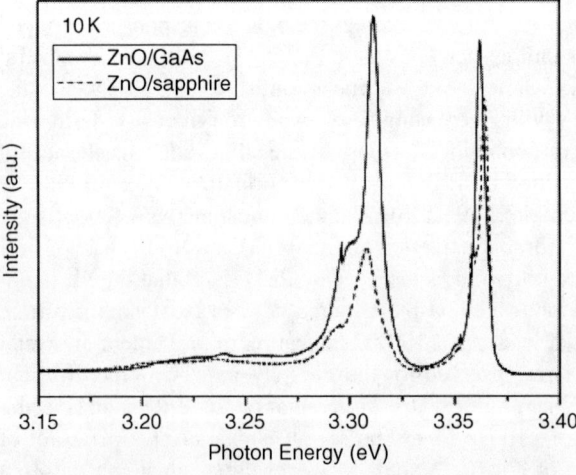

Figure 7.89 PL spectra of ZnO layers grown onto GaAs and sapphire substrates.

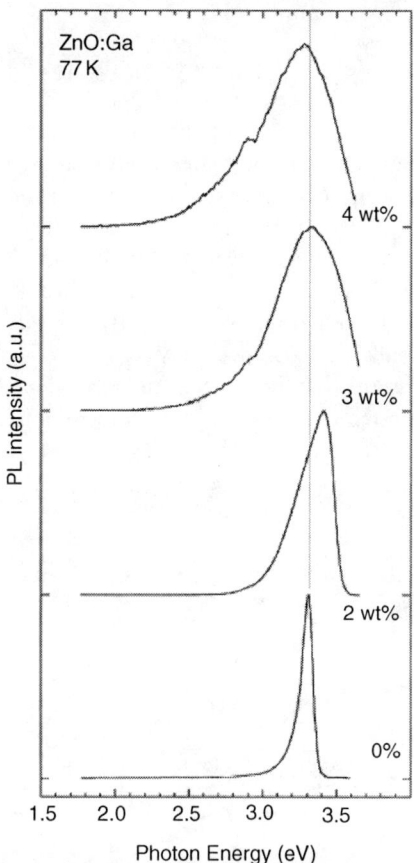

Figure 7.90 77 K PL spectra of ZnO:Ga thin films grown by DC arc ion plating method at 200 °C recorded for different Ga doping concentrations from 0, 2, 3 to 4 wt%.

nanocrystal sample annealed below 600 °C does not show change in deep level emission (2.4 eV), which is responsible to oxygen vacancy, whereas the intensity of emission line dominates the spectrum for the sample annealed under air at above 600 °C. The reason is that the oxygen desorption occurs for high temperature annealing, whereas probably oxygen adsorption takes place for low temperature annealing [200]. The ZnO layers grown onto Si (100) substrates by PECVD using $Zn(C_2H_5)_2:CO_2 = 4:5$, RF power of 35 W and gas flow rate of 4 sccm show free exciton at 3.26 eV in the PL spectra. By adding free exciton binding energy of 60 meV to free exciton provides band gap of 3.32 eV [201]. The neutral donor to bound exciton, A, B excitons, followed by exciton phonon at 3.3608, 3.377, 3.384, and 3.45 eV are observed in 5 K PL spectrum, respectively [202]. The FXA, BE, FB emission lines nearly at 3.374, 3.354, and 3.312 eV in 80 K PL spectrum are observed for ZnO nanowires on Si, respectively in which the BE dominates the spectrum and FB line contains LO lines with separation of ∼60 meV [203].

7.3.3.3 Raman Spectroscopy of ZnO

The wurtzite ZnO consists of C^4_{6v} space group with two formulae units per primitive cell. The group theory reveals the optical phonons such as $1A_1+1E_1+2E_2+2B_1$ at Γ point of Brillouin zones in which the polar A_1, E_1, and nonpolar two E_2 modes (low and high frequency) are Raman active but the B_1 modes are silent. The A_1 and E_1 modes divide into LO and TO modes due to association of macroscopic electric field with LO phonons. The polar E_2 and B_1 modes are IR inactive. The Raman frequency versus zone boundary shows locations of different modes in the band structure, as shown in Figure 7.91 [204]. The A_1(TO), E_1(TO), E^2_2, E^1_2, A_1(LO), and E_1(LO) at 380, 407, 437, 101, 574, and 583 cm^{-1} are observed in the bulk ZnO, respectively (Table 7.7) [205]. The E^2_2 mode (436.20 cm^{-1}) is wurtzite characteristic mode, whereas A_1(LO) mode at 576.2 cm^{-1} could be due to either V_O, Zn_i, or defect complex [188]. The E^1_2(98),

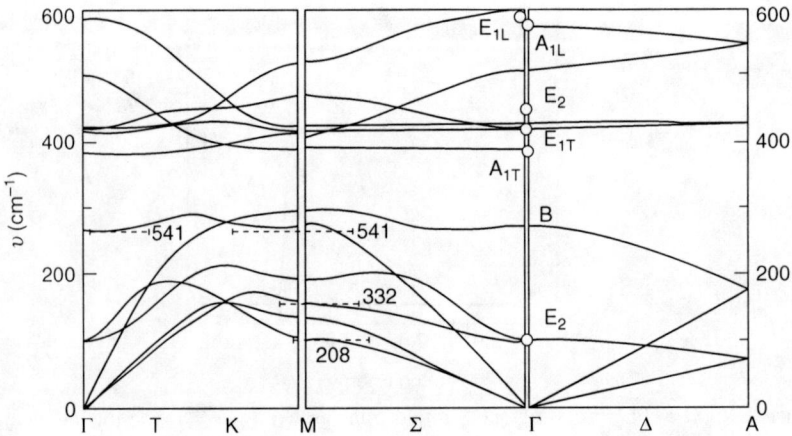

Figure 7.91 Raman mode versus different zones.

Table 7.7 Raman modes of ZnO

Mode	Bulk [240]	Bulk [205]	Bulk [207]	Nanorods [241]	Thin film [216]	ZnO thin film [227]	ZnO:Al (2.5%) thin film
A_1(TO)	380	380	393	383	380(2)	379	382
E_1(TO)	413	407	–	–	409(2)	–	–
E_2-high (E^2_2)	444	437	464	438	438(1)	434	434
E_2-low (E^1_2)	101	101	131	–	102(1)	–	–
A_1(LO)	579	574	–	–	–	544	539
E_1(LO)	591	583	584	583	587(2)	–	558

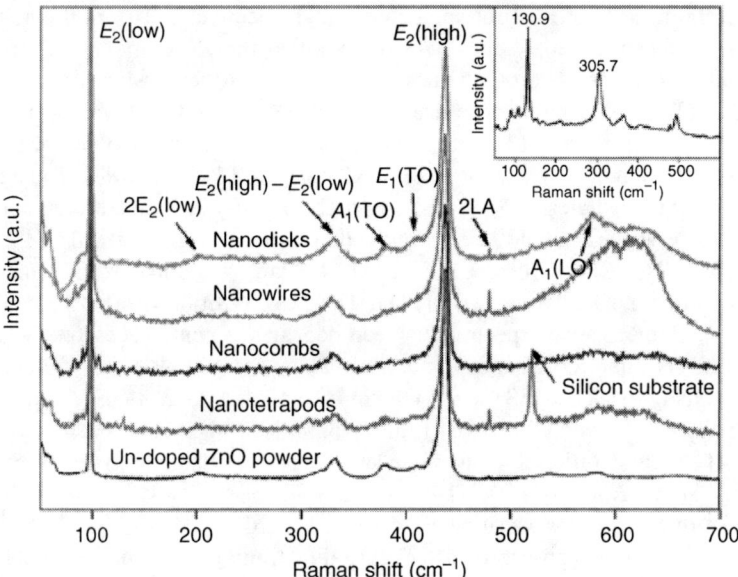

Figure 7.92 Raman spectra of different type of ZnO structures.

E^2_2(438), A_1(TO)(380), A_1(LO)(574), and E_1(TO)(410 cm^{-1}) are observed in the nanostructures (Figure 7.92). The second order modes at 203, 334, and 483 cm^{-1} are assigned to $2E^1_2$, $E^2_2 - E^1_2$, and 2LA, respectively. The 2LA mode is stronger in In doped ZnO nanostructures due to change in polarizability [206]. The peaks at 205 and 332 cm^{-1} in the ZnO nanostructures grown by low temperature hydrothermal technique are assigned to 2-TA(M) and 2-E_2(M) modes, respectively, which arise from M point zone boundary. The similar 2-TA(M)-205, $2E_2$(M)-331, and 2LA (M)-539 cm^{-1} phonons are also observed in the bulk ZnO, whereas in some cases only 2TA (M) mode at 209 cm^{-1} is noticed in the nanoparticles [207]. The Raman modes at 398, (E_2) 440, and 569 cm^{-1} are observed in the ZnO thin films deposited onto quartz substrates at different temperatures from RT to 600 °C. The Raman mode at 398 cm^{-1} is combination of A_1(TO) and E_1(TO). The mode at 569 cm^{-1} is also mixture of A_1(LO) and E_1(LO). The combined modes are not well resolved that could be due to stress in the films because of difference in thermal expansion coefficients (α) of ZnO and quartz substrate, which are $\alpha_{ZnO} = 4 \times 10^{-6}$ K^{-1} and $\alpha_{quartz} = 0.5 \times 10^{-6}$ K^{-1} [126].

The A_1(TO)-380, E_1(TO)-407, E^2_2-437, and E^1_2-101 modes exist in the ZnO nanobelts except E_1(LO)-583 cm^{-1} comparing with bulk. The modes present at 221 and 553 cm^{-1} may be due to oxygen vacancies. The B_2 silent mode presents at 276 cm^{-1} in the nanobelts due to electric field induction Raman mode. The second order modes are found at 205 and 334 cm^{-1} in both nanobelt and powder ZnO samples [180,208]. In fact, the A_1 modes are Raman active, whereas E_1 modes are Raman inactive in the zinc terminated face, therefore the E_1 mode is absent. The lattice vibrations of A_1 and E_1 modes are parallel and perpendicular to the c-axis of Zn terminated (0001) ZnO single crystal, respectively. The multiphonons at 330.54, 516.62, and 650.94 cm^{-1} are observed in the Zn terminated face (0001) ZnO bulk grown by hydrothermal process. The DAP emission line at 3.3271 (372.65), followed by phonon replica at 3.2505 eV (381.43 nm) with separation of ~70 meV are observed in 13 K PL spectrum. The phonon energy of 70 meV closely matches with Raman peak measured at 576.2 cm^{-1} (69.1 meV) [188]. The nanorods grown onto Si substrates at 500 °C shows high intensity E_2 mode at 436 cm^{-1} and low intensity A_1(TO) at 580 cm^{-1} and other modes at E_1(TO) 408, A_1(TO) at 380 cm^{-1} [209,210], whereas $E^2{}_2$ mode at 433 cm^{-1} dominates spectrum in the mushroom-like ZnO microcrystals [211]. The peak position lies between 574 and 570 cm^{-1}, if the Raman spectrum is recorded at different places on the sample due to different nature of nanostructures [212]. A sharp peak E_2 at 437 cm^{-1} and small intensity peaks at 331 and 378 cm^{-1} assigned to $E^2{}_2 - E^1{}_2$ and A_1(TO), respectively are observed but E_1(LO) is absent in the nano needle-type ZnO [213,214]. Second order Raman mode at 339 cm^{-1} arising from zone-boundary phonons $3E^2{}_2 - E^1{}_2$, A_1(TO)-388 cm^{-1}, $E^1{}_2$-98 and $E^2{}_2$-449, and E_1(LO) at 561 cm^{-1} due to oxygen vacancy are observed [215].

The A_1(LO) is not observed in the ZnO thin films grown onto c-sapphire substrate by pulsed laser deposition because the scattering cross section of A_1(LO) mode is smaller than of that A_1(TO) mode due to contribution of destructive interference between the Frohlich interaction and the deformation potential to the LO scattering in the ZnO [216]. The A_1(TO), A_1(LO), $E^2{}_2$ at 376, 532, and 438 cm^{-1} are observed in the nanobelt ZnO but E_1(TO) at 407 cm^{-1} is missed in the Raman spectrum that could be due to asymmetry nature of crystals. Obviously, the E_1(LO) at 579 cm^{-1} is forbidden in the back scattering configuration but presents here that may be due to the relaxation of q-selection rule. The other modes such as 335, 661, 780, 1073, and 1149 cm^{-1} are also observed [217]. The E_2 at 438 and A_1(LO) at 579 cm^{-1} occur with blue shift respect to bulk due to slipping symmetry and contribution of piezoelectric effect in the nanocrystals [200].

The wurtzite characteristic mode (E_2) at 437 cm^{-1} becomes stronger upon annealing samples indicating improvement in crystallinity. The A_1(LO) and A_1(2LO) phonon replicas at 556 and 1110 cm^{-1} are observed along with A_1(TO) 378 and E_1(TO) 409 cm^{-1} modes in the nanorods, as shown in Figure 7.93 [218]. After annealing the films under air at 500 °C for 30 min, an increase in intensity of $E^2{}_2$ mode (435 cm^{-1}) is general phenomenon, whereas an increase in intensity for 580 cm^{-1} mode could be due to local vibrational modes (LVMs) of nitrogen on oxygen site [219]. However eventually, the intensity of A_1(LO)-570 cm^{-1} mode responsible to oxygen vacancy decreases, after annealing the sample under air at 200–800 °C for 120 min [220,221]. A decrease in intensity for former mode may be due to reduction in oxygen vacancies. After annealing ZnO thin films at higher temperature of above 400–600 °C, the intensity of E_2-440 cm^{-1} [222] or E_2-437.6 cm^{-1} [193] increases. Similarly, after irradiating ZnO powder by laser, the intensities of E_2 modes at 99 and 437 cm^{-1} increase indicating improvement in crystallinity [184]. However, the E_2 mode shifts from 438.1 to 437.4 cm^{-1} for the ZnO thin films on Si after annealing at high temperature of 800 °C. This could be

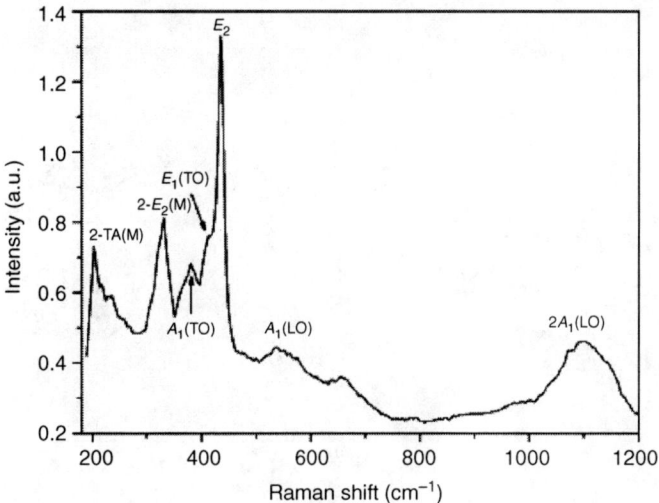

Figure 7.93 Raman spectrum of ZnO nanorods.

due to relaxation of layers from compressive stress [223]. The intensity of $E^2{}_2$-434.42 cm^{-1} mode increases, whereas the intensity of A_1(LO)-573.62 cm^{-1} decreases with increasing O$_2$/Zn flow rates during the deposition process, indicating improvement in the ZnO crystallinity [224]. The intensity of 556 cm^{-1} mode decreases with increasing substrate temperature from RT to 230 °C for the ZnO films grown onto Si substrates by filter cathode vacuum arc method. Finally, the peak disappears with further increasing the substrate temperature to 430 °C. Obviously, the density of Zn$_i$ is high for the layers grown at low substrate temperature. This experimental observation indicates that this peak could be due to either Zn$_i$ or oxygen vacancy. A broad peak at 180–250 cm^{-1} appears only in the films grown at RT [225]. In general the E_2-439, E_1(TO)-410, and A_1(TO)-379 cm^{-1} are observed in the bulk ZnO, whereas no TO phonons are observed in the quantum dots. The peaks in the film show red shift with respect to bulk due to defects and impurities [226].

The modes at 144 and 153 cm^{-1} are assigned to lattice vibration, whereas the modes at 211, 243, and 258 cm^{-1} are due to deformation in the ZnO thin films grown onto glass substrates by dip and dry method. However, they disappear in Al doped ZnO thin films, as shown in Figure 7.94. The nonpolar phonon mode $E^2{}_2$ at 436 cm^{-1} becomes stronger, whereas the intensity of A_1(TO) at 380 cm^{-1} decreases with increasing Al doping indicating decrease of polar character of ZnO [227]. Variation of Raman modes for sintered ZnO:Al (1 wt%) with temperature is shown in Figure 7.95. Negative shifts of 0.0051 and 0.0191 cm^{-1}/K for $E^1{}_2$ (99 cm^{-1}) and $E^2{}_2$ (438 cm^{-1}), respectively are observed with increasing sample temperature from 293 to 973 K due to variation of band gap of sample. The similar shift -0.0185 cm^{-1} for former mode is reported for single crystal in the literature [228]. An increase in intensities is observed for 99 (E_2), 208 (2E_2), 332

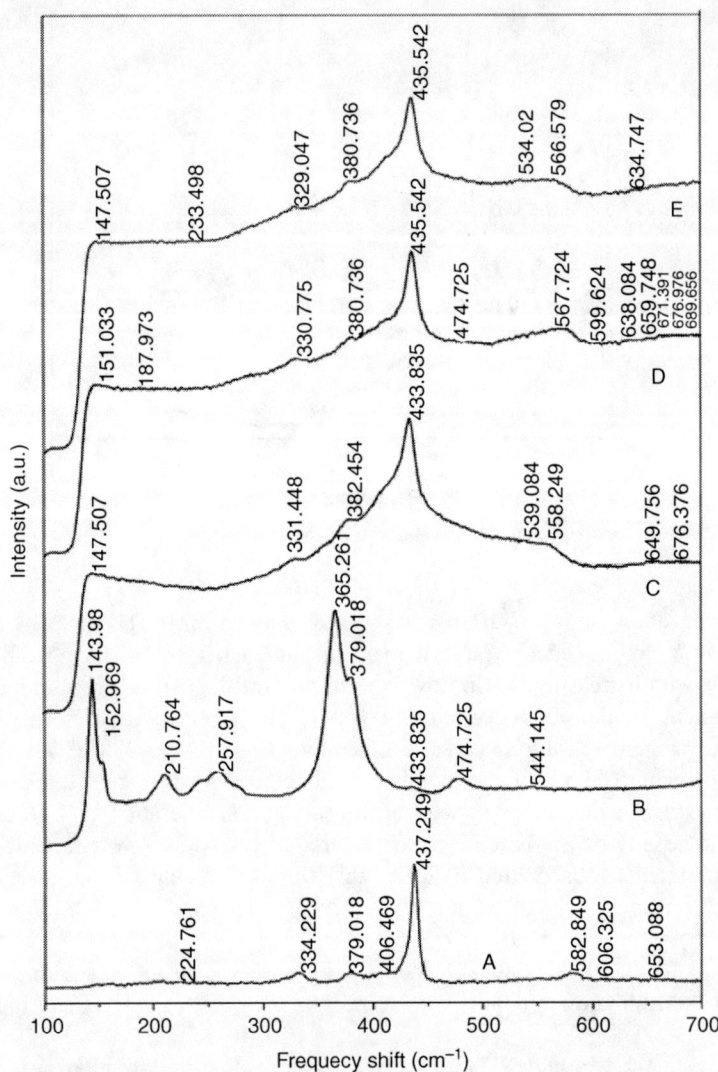

Figure 7.94 Raman spectra of (A) ZnO bulk, (B) ZnO, (C) ZnO:Al(2.5), (D) ZnO:Al(5.0), and (E) ZnO:Al(7.5 wt%).

(multiphonon), and 1153 cm^{-1} (2LO) with respect to 437 (E_2) mode above 573 K. After annealing Al doped ZnO thin films, the intensity of 579 cm^{-1} mode decreases. On the other hand, the conductivity also increases in the samples. Shifts of -0.123 and -0.494 cm^{-1}/kbar are observed for 99 and 435 cm^{-1} modes with increasing pressure from 1 atm to 101 kbar, respectively [229]. The intensities of E^1_2-100 and E^2_2-438 cm^{-1} modes increase after annealing ZnO:0.5% Al at 600 °C, whereas those decrease for ZnO:Al3at% indicating that the quality of the films is high for lower doping level of Al but the quality degrades for higher doping

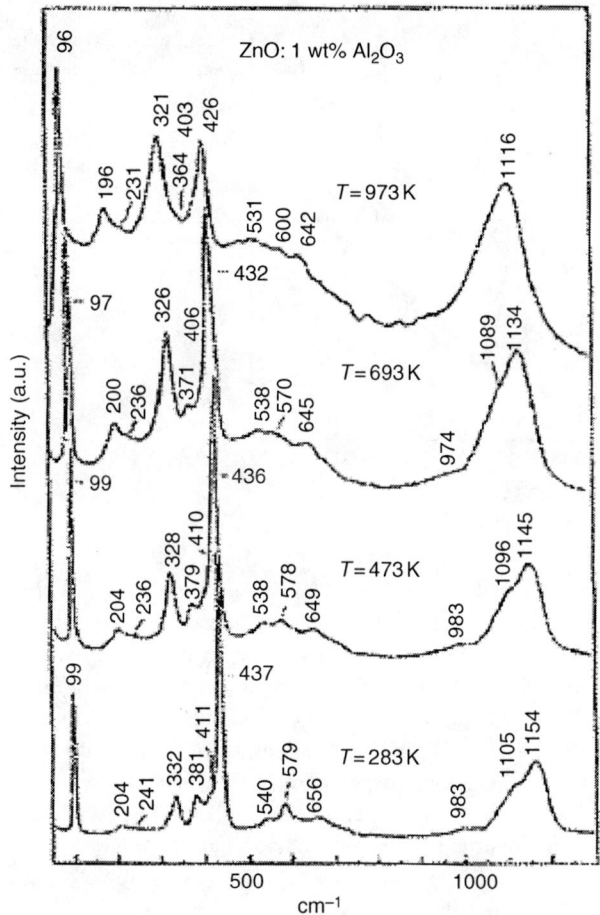

Figure 7.95 Raman spectra of sintered ZnO:Al 1 wt% measured at different temperatures.

concentration of Al [230]. The mode shifts from 437 to 468 cm^{-1} by doping Al into ZnO thin films. In some cases, the mode at 570 cm^{-1} dominates Raman spectrum irrespective of doping into ZnO thin films. An asymmetric peak at 578 cm^{-1} turns into symmetric peak and intensity of 437 cm^{-1} increases with increasing oxygen pressure during deposition of layers. The intensity of 578 cm^{-1} mode increases with increasing excitation intensity [231]. The intensities of LVMs such as 275, 510, 582, and 643 cm^{-1} increase with increasing N concentration in the ZnO samples grown by CVD [232]. In the case of N doped samples, the peak at 275 or 279 cm^{-1} is assigned to N complex [233,234]. However, it is established that the mode at 279 cm^{-1} is not related to N because it appears in all Fe, Sb, Al, Ga, and Li doped ZnO thin films, therefore, it could be due to intrinsic host lattice defects [235].

The multiphonons at 331 and 1050 cm^{-1} along with E^2_2 436, A_1(TO) 374, A_1(LO)$-E_1$(LO) at 565 cm^{-1} are observed in the ZnO nanocrystals [236]. The

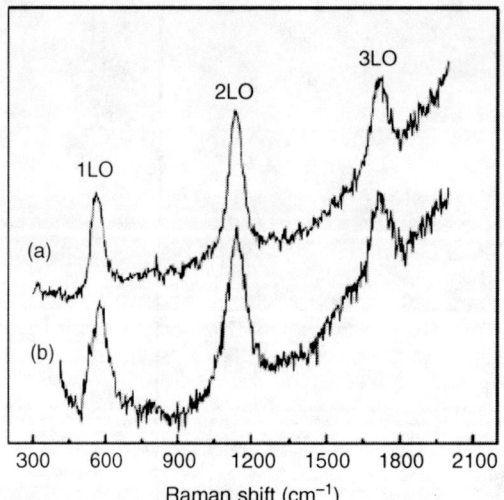

Figure 7.96 Raman spectra of ZnO nanorods; (A) one step and (B) two step reaction hydrothermal.

similar multiphonons at 563, 1130, and 1710 cm^{-1} are also observed in the nanorods like ZnO in which smaller size nanorod samples show blue shift, that is, toward lower frequency (Figure 7.96). It could be due to heating by laser beam because smaller size nanorods have more air gaps that causes lower thermal conductivity [237]. The similar 1LO, 2LO, 3LO, 4LO, 5LO, 6LO, 7LO, and 8LO phonons at 585, 1165, 1749, 2343, 2928, 3520, 4101, and 4678 cm^{-1} are observed for ZnO in the X(ZX)Y configuration, respectively [238]. The ZnO nanorods grown by chemical solution method show number of A_1(LO) phonon lines, as shown in Figure 7.97. The incident and propagation waves are parallel to the c-axis of nanorods therefore the detected phonon q wave vector is parallel to the c-axis, that is, contribution by A_1(LO) [239].

7.3.4 Electrical Properties of ZnO Thin Films

The low resistivity or sheet resistance, high mobility, and transmittance of ZnO thin films are potentially important for solar cell applications. The zinc interstitials (Zn_i) and oxygen vacancies (V_O) are responsible for n-type conductivity in the ZnO thin films. However, the intrinsic states could not afford required conductivity for thin film solar cell applications. Therefore, extrinsic dopings such as trivalent In, Ga, Al, *etc.*, are amenable to dope into Zn^{2+} state to promote higher n-type conductivity. Keeping in mind, the radius of external doping should be smaller than recipient. On the other hand, it should be in a position to create a free electron for conduction. The deposition parameters have strong influence on the electrical properties of ZnO or doped ZnO thin films. As expected the resistivity varies exponentially with varying thickness below the critical thickness of 300 nm [142]. Therefore, the thickness of >300 nm for ZnO is preferable for high conductivity.

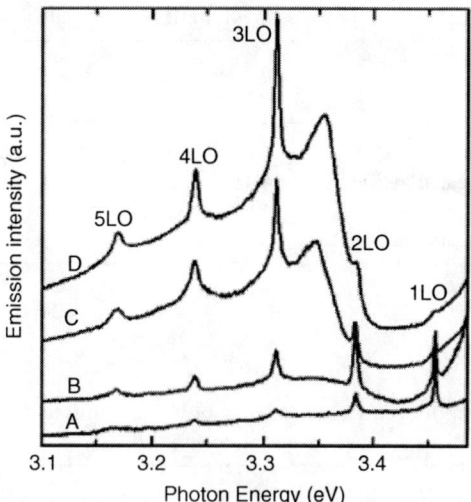

Figure 7.97 Raman spectra of ZnO nanorods grown by chemical solution method; (A) as grown and (B) annealed at 400, (C) 650, and (D)750 °C.

The three typical un-doped ZnO thin film samples grown by evaporation of Zn from effusion cell at source temperature of 540 °C using glow discharge in the presence of oxygen by maintaining base chamber pressure of 3×10^{-3} mbar show different resistivities, mobilities, and carrier concentrations of (i) 9.4×10^2, 10.4, 6.3×10^{14}, (ii) 4.2×10^{-3}, 14.7, 1.1×10^{20} cm^{-3}, and (iii) 8.5×10^{-4} Ω-cm, 9.8 cm^2/Vs, 7.5×10^{20} cm^{-3}, respectively [242]. The resistivity and carrier concentration of un-doped ZnO thin films vary from 4.2×10^{-3} to 9.6×10^{-1} Ω-cm and from 1×10^{20} to 1.2×10^{18} cm^{-3} with varying O$_2$ flow rate from 0 to 40 sccm during deposition of layers, respectively [160]. After annealing spray-deposited ZnO thin films under H$_2$ at different temperatures from 200 to 340 °C in steps of ~20 °C, the conductivity, carrier concentration and mobility increase from 0.0012 to 6.6 Ω-cm^{-1}, from 0.15×10^{18} to 1.47×10^{18} cm^{-3} and from 0.1 to 28 cm^2/Vs, respectively [243], whereas the resistivity increases from 1.6×10^{-2} to 1.3×10^1 Ω-cm, after annealing ZnO thin films under air at 550 °C for 1 h and continue to increase to 10^9 Ω-cm for further annealing under air at 750 °C for 1 h [161]. The ZnO thin films grown by reactive PLD at substrate temperature of 200 °C under oxygen pressure of 20 and 40 Pa show resistivities of 3.26 and 4.99 Ω-cm, respectively. After annealing the films under air at 300 °C for 70 min, the carrier concentration slightly reduces, resistivity and mobility increase due to reduction of oxygen vacancies and zinc interstitials Zn$_i$, as shown in Table 7.8 [244].

The (002) oriented ZnO:Al (1.2 wt%) (AZO) thin films grown onto glass substrates by DC (0.21 Pa) and RF sputtering (0.15 Pa) at substrate temperature of 300 °C show sheet resistances of 6.6 and 7.5 Ω/sq, respectively. The DC sputtered ZnO:Al thin films show slightly lower carrier concentration of 2.6×10^{20} cm^{-3} and

Table 7.8 Change in electrical parameters with effect of annealing under O_2 atmosphere

O_2 pressure (Pa)	Process	$\rho \times 10^{-3}$ (Ω-cm)	$n_e \times 10^{19}$ (cm^{-3})	μ (cm^2/Vs)	Ref.
20	As-deposited	3.26	7.24	26.43	[244]
20	Annealed at 300 °C	4.46	4.57	30.67	
40	As-deposited	4.99	1.45	86.37	
40	Annealed at 300 °C	6.05	1.13	91.10	

Table 7.9 Comparison of electrical parameters of ZnO:Al thin films grown by DC and RF sputtering technique

S.No	Sputtering	$R_\square$ (Ω/sq)	ρ (Ω-cm)	n_e (cm^{-3})	μ (cm^2/Vs)	d_f (nm)
1	DC	6.6	4.2×10^{-4}	2.6×10^{20}	57	640
2	DC annealed	70	3.7×10^{-3}	6.3×10^{19}	27	530
3	RF	7.5	4.0×10^{-3}	4.9×10^{20}	32	530
4	RF annealed	270	1.3×10^{-2}	4.6×10^{19}	11	480

higher mobility of 57 cm^2/Vs comparing with those of 4.9×10^{20} cm^{-3} and 32 cm^2/Vs for the one grown by RF sputtering, as shown in Table 7.9. The ZnO:Al thin films with thickness of 500 nm deposited by RF sputtering using RF power of 180 W show minimum resistivity of 1.4×10^{-4} Ω-cm and transmittance of 95% in the visible region. However, the high resistive films can be deposited even from Al doped target by suppressing oxygen vacancies and Al dopings. For example, the highest resistivity of 8.1×10^9 Ω-cm for ZnO:Al thin films is achieved by introducing 10% oxygen content into the chamber. Variation in resistivity, mobility and carrier concentration with varying RF power from 60 to 300 W is given in Figure 7.98. The lowest resistivity, highest carrier concentration and mobility are observed for optimum RF power of 180 W [245]. The resistivity of ZnO:Al thin films grown onto (11$\bar{2}$0) sapphire substrates by RF sputtering technique using ZnO mixed 2 wt% Al_2O_3 target with 12.5 cm diameter at substrate temperature of 200 °C and chamber pressure of 1.33 Pa slightly increases from 1.4×10^{-4} to 3×10^{-4} Ω-cm and carrier concentration and mobility slightly decrease from 1.3×10^{21} to 1.1×10^{21} cm^{-3} and from 34 to 27 cm^2/Vs, respectively with increasing growth rate from 1.2 to 22.3 nm/min. The growth rate can be increased by increasing RF power from 25 to 170 W [246]. On contrary, in some cases the resistivity decreases from 2.7×10^{-1} to 3.6×10^{-2} Ω-cm, carrier concentration and mobility increase from 9.3×10^{19} to 2.3×10^{20} cm^{-3} and from 0.25 to 0.74 cm^2/Vs with increasing RF power from 150 to 200 W [135]. The lowest resistivity of 2.7×10^{-4} Ω-cm is obtained for the optimum deposition parameters of 0.2Pa, P_{RF}=100W, T_S = 250 °C

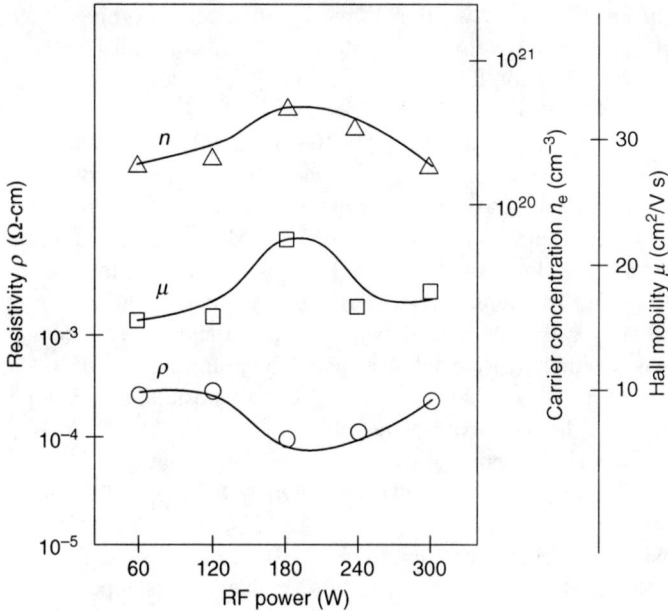

Figure 7.98 Variation of resistivity, carrier concentration and mobility with function of RF power.

and ZnO:Al$_2$O$_3$ 2wt% target. The growth rate of AZO thin films is 1.2-1.5 times higher for the layers grown at working gas pressure of 0.2Pa than the one grown at 1.0Pa [136].

The recipes such as substrate temperature, chamber working pressure, Ar flow rate, oxygen flow rate, RF or DC power, the distance between target and substrate *etc.*, govern the electrical and optical properties of AZO thin films. The optimum resistivity of 4.23×10^{-4} Ω-cm, carrier concentration of 9.21×10^{20}, mobility of 16 cm^2/Vs and transmission of 80% are observed for typical DC reactive sputtered ZnO:Al (1.5wt%) thin films grown at 250 °C [137]. Similar optimum resistivity of 3×10^{-4} Ω-cm, carrier concentration of 6×10^{20} cm^{-3} and mobility of ~30 cm^2/Vs are determined for ZnO:Al (1wt%) thin films grown onto corning 7059 glass substrates by PLD at substrate temperature of 200 °C [247]. In typical case the higher resistivity (ρ) of $3-5 \times 10^{-3}$ Ω-cm (R$_\square$=30–60 Ω/sq, n_e=4×10^{19} cm^{-3} and μ=7cm^2/Vs) is measured for 0.8μm thick ZnO:Al thin films grown by RF sputtering technique at RT [248]. 560–750 nm thick ZnO:Al (1–2wt%) layers grown onto glass substrates at substrate temperature of 150 °C by reactive sputtering of Zn-Al target in the presence of oxygen show minimum resistivity, higher carrier concentration and mobility of 4.6×10^{-4} Ω-cm, 5.8×10^{20}cm^{-3} and 24 cm^2/Vs, respectively. The resistivity decreases from 4.1×10^{-3} to 4.6×10^{-4} Ω-cm with increasing substrate temperature from 50 to 150 °C thereafter increases with further increasing substrate temperature from 150 to 300 °C [249]. In the case of PLD, the resistivity of ZnO:Al (1.5wt%) thin films deposited onto corning 7059 glass substrates slightly increases from 3×10^{-4} to 6×10^{-4} Ω-cm with increasing substrate temperature from 200 (or RT) to 550 °C [123]. The carrier concentration varies from 2×10^{20}, 4.25×10^{20}, 5.5×10^{20}, 5×10^{20}, 3×10^{20} to 2.1×10^{20} cm^{-3} with increasing substrate temperature from 50, 100, 150, 200, 250 to 300 °C, respectively indicating that the carrier concentration increases up to growth temperature of 150 °C thereafter decreases with further increasing temperature up to 300 °C. A variation in carrier concentration or mobility of ZnO:Al thin films with function of substrate temperature follows opposite trend of resistivity, as expected, as shown in Figure 7.99 [140]. The similar pattern is observed that the minimum resistivity is 1.5×10^{-4} Ω-cm for ZnO:Al thin films deposited at substrate temperature of 150 °C, which increases with increasing substrate temperature from 150 to 400 °C. The carrier concentration of 8×10^{20}cm^{-3} is minimum for the deposition temperature of 300 °C [116]. Note that the resistivity varies from $\sim2.4\times10^{-4}$, 2.35×10^{-4}, 2.1×10^{-4}, 1.9×10^{-4} to 3.5×10^{-4} Ω-cm with varying temperature from RT, 70, 170, 270 to 370 °C for the ZnO:Al thin films deposited using ZnO:Al$_2$O$_3$ 2.5wt% [130]. As noted earlier, the resistivity decreases from $\sim5.75\times10^{-4}$, 5.5×10^{-4}, 4.25×10^{-4} to 3.18×10^{-4} Ω-cm with increasing substrate temperature from 50, 100, 150 to 200 °C thereafter increases from 4.5×10^{-4} to 8.0×10^{-4} Ω-cm with further increasing temperature from 250 to 300 °C for the ZnO:Al thin films grown by DC magnetron sputtering at growth rate of 30 nm/min. A similar opposite pattern is observed in the carrier concentration with varying substrate temperature that the carrier concentration varies from $\sim3.9\times10^{20}$, 4.0×10^{20}, 4.1×10^{20}, 4.6×10^{20}, 4.4×10^{20} to 2.9×10^{20} with varying temperature from 50, 100, 150, 200 to 300 °C [114]. On overall a decrease in resistivity with increasing temperature from RT to 150 °C is due to an increase of effective Al doping concentration

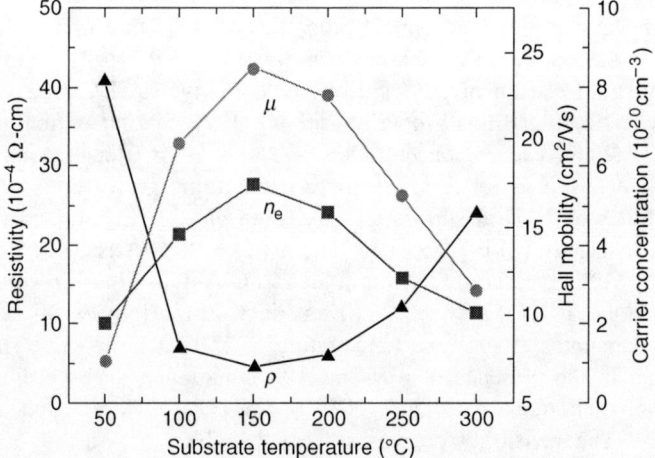

Figure 7.99 Variation of resistivity and mobility with function of substrate temperature for ZnO:Al thin films.

and an improved crystallinity, whereas an increase in resistivity for higher deposition temperature (>150 °C) is due to desorption of Zn caused by high vapor pressure of Zn (up and down phenomenon) [140]. The same principle may be applicable for increasing RF or DC sputtering powers. In some cases, the lowest resistivity of $2-5 \times 10^{-4}$ Ω-cm ($R_\square = 2$ Ω/sq) is observed for 500 nm thick AZO (Al_2O_3 2wt%) layer grown at substrate temperature of 350 °C and working pressure of 12 Pa, whereas the layers grown below the substrate temperature of 350 °C show inferior quality [117]. A single digit sheet resistance of 4 Ω/sq, transmission of 91% and figure of merit 2.5 Ω^{-1} are measured for PLD ZnO:Al (2 wt%) thin films grown at substrate temperature of 300 °C under oxygen ambient pressure of 1 mtorr for optimization conditions. The grown layers show higher sheet resistance for > or < 2wt% Al [250]. In general the optimum substrate temperature lies at 150–200 °C in some cases, which deviates to higher that could be dictated by some other growth parameters. The mobility increases from 20 to 35 cm^2/Vs with increasing substrate temperature from 50 to 200 °C and thereafter decreases to 5 cm^2/Vs with further increasing substrate temperature to 400 °C [116]. As verbatim, the mobility of AZO films decreases from 16 to 3.2 cm^2/Vs with increasing substrate temperature from 250 to 350 °C (up and down phenomenon) [137]. On the other hand, higher Al doping takes place at high temperature that will cause to have neutral impurity, grain boundary scattering *etc*.

The ZnO:Al thin films grown from Zn-ZnO:Al (Al_2O_3 2wt%) targets at room temperature for oxygen flow rate ratio of 0.2 {O_2 flow rate/(O_2 flow rate+Ar flow rate)} show lower resistivity of 10^{-4} Ω-cm [110]. The high resistivity of ZnO thin films decreases from 3×10^{-2} to 6×10^{-4} Ω-cm with increasing Al doping from 0 to 2at.% into it ($n_e = 8 \times 10^{20}$ cm^{-3}) [124]. As expected, the resistivity decreases from 3.8×10^{-2} to 5.7×10^{-4} Ω-cm with increasing doping concentration from 0 to 2 wt% Al_2O_3 in DC planar magnetron sputtered ZnO thin films (Table 7.10) [127]. Similarly the resistivity decreases from 3×10^{-3} to 4×10^{-4} Ω-cm with increasing

Table 7.10 Electrical parameters of ZnO and ZnO:Al thin films

S.No.	H_2/Ar (%)	Al_2O_3 wt%	$R_\square$ (Ω/sq)	$\rho \times 10^{-4}$ (Ω-cm)	$n_e \times 10^{20}$ (cm^{-3})	μ (cm^2/Vs)	Ref.
1	0.3	0	101	–	0.33	48	[109]
2	0.3	0.5	12.3	–	3.4	36	
3	0.3	1.0	7.2	–	5.5	32	
4	100Ar	2	9.2	–	5.9	25	
5	–	0	–	380	0.16	–	[127]
6	–	0.5	–	7.8	1.8	–	
7	–	0.75	–	4.9	3.0	–	
8	–	1.0	–	5.3	3.1	–	
9	–	2.0	–	5.7	5.6	–	
10	–	0	–	$>3 \times 10^5$	$<1 \times 10^{14}$	2–3	[156]
11	–	NA	–	10	2.0	18–19	

doping level from 0 to 4at.% Al in the ZnO films [141]. The typical ZnO:Al (2wt%) thin films grown by RF sputtering show resistivity of 6.25×10^{-3} Ω-cm, carrier concentration of 1.6×10^{19} and transmittance of 90% for the deposition chamber pressure of 1.6×10^{-2} mbar and RF power of 150W [251], If the resistivity (ρ) is decreased to lower value of 1.9×10^{-4} Ω-cm, automatically the transmittance (T) of sample decreases to 85% for typical optimum conditions of T_S=300 °C, argon pressure of 0.27 Pa, argon flow rates of 3 sccm, RF power of 100 W and deposition time of 120 min revealing facts that both ρ and T parameters are tradable [130]. In some cases even for low Al doping of 0.5% the low resistivity and reasonable mobility can be achieved that depends on configuration of deposition system and recipes. For example the ZnO:Al films grown by RF sputtering show low resistivity of 3.7×10^{-4} Ω-cm and mobility of 44 cm^2/Vs for low Al doping concentration of 0.5%. The resistivity decreases by factor of 2 with increasing doping concentration from 0.5 to 2% as well as transmittance [252]. The lowest resistivity of 9.8×10^{-2} Ω-cm is observed for doping concentration of 6 wt% Al(OH)$_3$ instead of Al$_2$O$_3$ whereas the resistivity increases with increasing doping concentration beyond 6 wt% in the films grown at RT [142].

The carrier concentration increases from 1.6×10^{19} to 5.6×10^{20} cm^{-3} with increasing doping concentration from 0 to 2 wt% Al$_2$O$_3$ in the ZnO thin films (Table 7.10) [127]. The carrier concentration of AZO slightly increases from 3×10^{20} to 4.8×10^{20} cm^{-3} and mobility decreases from 57 to 18 cm^2/Vs with increasing Al wt % from 1 to 2 in the target for the films grown by DC sputtering. However, the carrier concentration slightly decreases from 4.8×10^{20} to 1.4×10^{20} cm^{-3} with further increasing to 4wt% Al. A decrease in carrier concentration with increasing Al concentration at higher doping is mainly due to formation of grain boundaries, decrease of zinc interstitials Zn$_i$, oxygen vacancies, trapping of Al, existence of secondary phases etc [125]. Keeping deposition conditions of T_S=150 °C, P_{RF}=150 W and P_w=2 mTorr, the resistivity, carrier concentration and mobility vary with varying Al$_2$O$_3$ concentration in the ZnO:Al thin films from 0, 1, 3 to 5 wt% corresponding Al at.% from 0, 2, 4 to 6.2. The carrier concentration of ZnO thin films increases from 8.7×10^{19} to 8×10^{20} cm^{-3} with increasing doping from 0 to 4 at.% Al, whereas the mobility starts to decrease from 25.3 to minimum 1.3 cm^2/Vs with increasing doping from 0 to 5at.% Al, as shown in Fig. 7.100 [141]. Similarly the mobility decreases from 24, 13, 12.5 to 8 with increasing Al doping from 0, 2, 3 to 5 at.% in the ZnO thin films grown by laser ablation [124]. A decrease in mobility with increasing Al doping concentration may be due to domination of grain boundary, neutral impurity scatterings, reduction in mean free path of carriers etc.

Let us see a role of H$_2$ in the ZnO:Al thin films during deposition. The carrier concentration of ZnO:Al thin films increases from 1×10^{20} to 8×10^{20} cm^{-3} with increasing Al content from 0.1 to 2 wt% for optimum H$_2$/Ar ratio of 0.3% and T_S=200 °C. The mobility decreases for higher Al$_2$O$_3$ and H$_2$ content that could be due to domination of grain boundary scattering, etc. The mobility and carrier concentration of the films deposited at T_S=200 °C and H$_2$/Ar=0.3% decrease from 50 to 1 cm^2/Vs and from 10^{19} to 10^{17} cm^{-3} with increasing annealing temperature from RT to 400 °C due to destruction or rearrangement of structure [253]. Keeping H$_2$/Ar gas ratio of 0.3%, the carrier concentration of ZnO thin films grown by RF

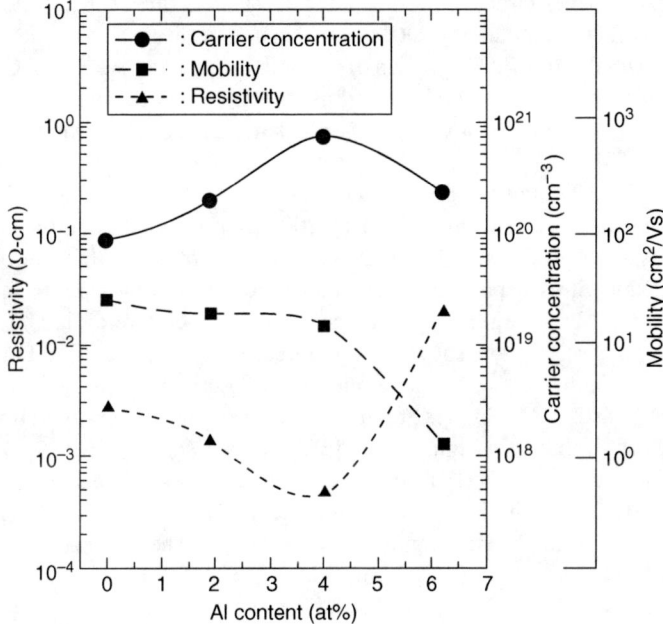

Figure 7.100 Variation of resistivity, carrier concentration and mobility with function of Al concentration in the ZnO:Al thin films.

sputtering increases from 3×10^{19} to 3.4×10^{20} cm^{-3} and mobility decreases from 48 to 36 cm^2/Vs by doping Al$_2$O$_3$ 0.5 wt% into ZnO target. The highest carrier concentration of $\sim 10^{21}$ cm^{-3} is observed for the dopant concentration of Al$_2$O$_3$ 2 wt% and mobility of ~ 28 cm^2/Vs for the typical films deposited at substrate temperature of 150 °C and sputtering power of 72 W. The carrier concentration increases and mobility decreases with increasing Al concentration in the ZnO films, as shown in Table 7.10 [109].

In the thin film solar cells, the i-ZnO layer is first deposited on CIGS layer in order to prevent damage of surface and to avoid segregation of defect levels in the CIGS. The i-ZnO and ZnO:Al layers are successively grown as a bilayer configuration by sputtering technique using ZnO and ZnO:Al 2.5 wt% targets for which argon gas is diluted with a small quantity of 2.5% oxygen for i-ZnO layer deposition, whereas only Ar gas is used for the deposition of ZnO:Al layers. The ZnO films show resistivity of 1×10^{-3} Ω-cm for working pressure of 2 mtorr. The minimum resistivity of 4×10^{-4} Ω-cm and mobility of 33 cm^2/Vs are achieved for ZnO:Al films grown at substrate temperature of 100 °C [254]. The ZnO bilayer (i-ZnO/ZnO:Al) is employed in the CIGS thin film solar cells in which the thickness of i-ZnO layer is critical. The CIGS cells show efficiency of 16 and 10.6% for the thickness of 70 and 20 nm i-ZnO layer, respectively. The resistivities of 1 μm thick i-ZnO and ZnO:Al are $> 3 \times 10^5$ and 1×10^{-3} Ω-cm, respectively (Table 7.10) [156].

The Zn:Al(2 wt%) target is used to deposit ZnO:Al thin films on polyethylene terephalate (PET) substrates by DC reactive magnetron sputtering under working pressure of 0.6 Pa and 7.5% oxygen ratio with respect to Ar and DC power of 80 W. While depositing ZnO:Al thin films, the substrate temperature is found to be 80 °C due to bombardment of high-energy particles. The ZnO:Al thin films show resistivity of 8.4×10^{-4} Ω-cm and transmittance of 80% [129]. The ZnO:B thin films grown by DC sputtering using ZnO target and 2.5% diborane (B_2H_6) with respect to Ar at substrate temperature of 200 °C show resistivity of 4×10^{-4} Ω-cm, mobility of 60 cm^2/Vs and carrier concentration of 2.6×10^{20} cm^{-3}, whereas the ZnO:Al thin films deposited by DC magnetron sputtering using 4 in. diameter ZnO and 2 wt% Al_2O_3 target at optimum substrate temperature of 350 °C and working pressure of 2×10^{-2} torr show minimum resistivity of 3.6×10^{-4} Ω-cm, carrier concentration of 5.6×10^{20} cm^{-3} and mobility of 37 cm^2/Vs. The ZnO:Al thin films show higher resistivity for higher growth temperature due to re-evaporation of interstitial Zn (Zn_i), as shown in Figure 7.101 [255]. 1 μm thick ZnO:Al layers are grown onto glass substrates by CVD at substrate temperature of 350 °C and at 60 torr. The flow rates of $Zn(C_5H_7O_2)_2$, $Al(C_5H_7O_2)_2$, and H_2O precursors by the carrier gas of N_2 are 400, 140, and 20 cm^3 min^{-1} are employed, respectively. The ZnO:Al thin films by CVD also show ($\rho = 3.4 \times 10^{-3}$ Ω-cm, and $R_\square = 32$ Ω/sq, $n_e = 3.4 \times 10^{20}$ cm^{-3}) similar to sputtering [256,257]. $R_\square = 10.4$ Ω/sq, $\rho = 1.27 \times 10^{-3}$ Ω-cm, $n_e = 1.3 \times 10^{20}$ cm^{-3}, and mobility = 37.8 cm^2/Vs are found for 1.22 μm thick ZnO:B thin films grown by MOCVD using diethlyzinc, H_2O, and B_2H_6 gas. The thin film solar cells made with ZnO:B thin films exhibit efficiency of 12% [258].

Ga doping in the ZnO thin films is one of the suitable dopings. The resistivity decreases from 2.58×10^{-2} to 3.9×10^{-4} Ω-cm, carrier concentration and Hall mobility increase from 6.5×10^{19} to 1.12×10^{21} cm^{-3} and from 3.7 to 12.8 cm^2/Vs, respectively with increasing RF power from 50 to 200 W for (002) oriented ZnO:Ga thin films deposited by RF sputtering using ZnO and 3 wt% Ga_2O_3 mixed 3 inch sintered

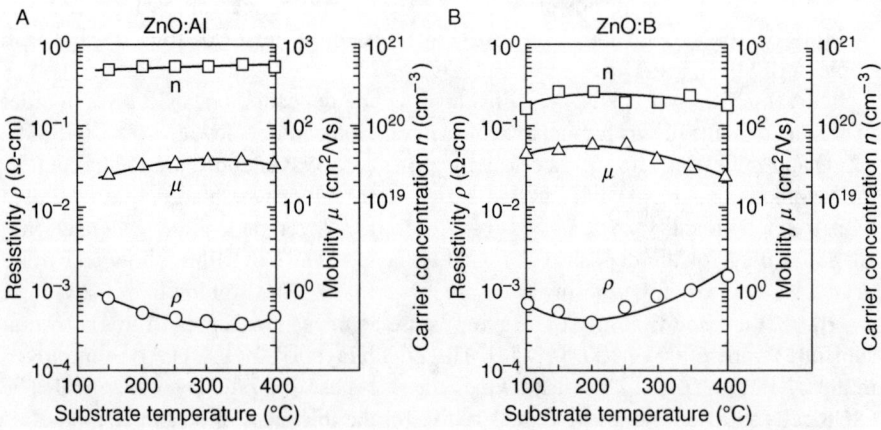

Figure 7.101 Variation of resistivity, carrier concentration and mobility with increasing substrate temperature for ZnO:Al and ZnO:B thin films grown by RF sputtering.

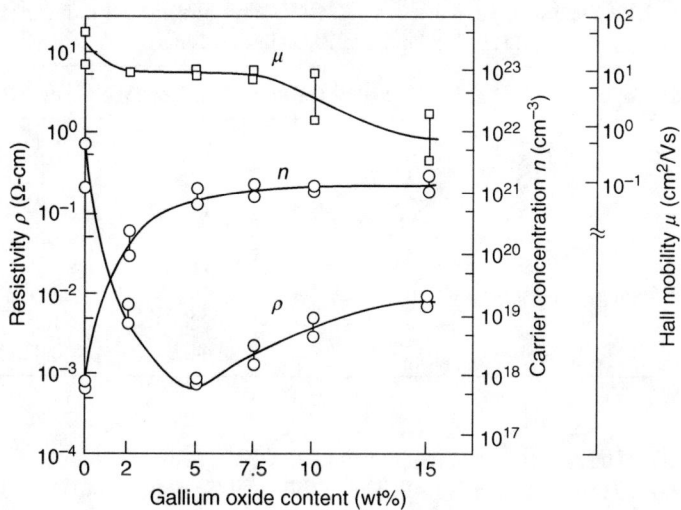

Figure 7.102 Variation of resistivity, carrier concentration and mobility of ZnO:Ga thin films grown by RF sputtering.

target. A single digit sheet resistance of 4.6 Ω/sq ($\rho = 3.9 \times 10^{-4}$ Ω-cm) for 800 nm thick samples is observed due to an increase of crystalline quality, reduction in grain boundaries. On the other hand, the concentration of zinc interstitial (Zn_i) decreases and substitution of Al in Zn increases. A variation of mobility is nil in the ZnO:Ga thin films with temperature due to degeneracy of the layers, however, there is a slight variation in mobility at high temperature that could be due to grain boundary scattering [259]. The carrier concentrations of 7.5×10^{17} and 1×10^{21} cm^{-3} are found for un-doped and 5 wt% Ga_2O_3 doped ZnO thin films grown onto glass substrates by RF sputtering at RF power of 0.42–2.1 W cm^{-2} and argon pressure of 5–80 mtorr, respectively for optimum deposition conditions of 5 mtorr and 0.84 W/cm^2, as shown in Figure 7.102. The mobility decreases from 30 to 7 cm^2/Vs with increasing Ga concentration from 0 to 2 wt%. The Ga_2O_3 weight percent of 2, 5, and 10 are close to 1.86, 4.65, and 8.84 wt% of Ga in the ZnO:Ga thin films, respectively. The AES analysis confirms the same Ga wt% in the ZnO:Ga thin films. The un-doped and 5 wt% Ga_2O_3 doped ZnO thin films show band gaps of 3.28 and 3.59 eV, respectively [121].

The ZnO:Ga thin films evaporated by DC arc-discharge plasma method employing ZnO:Ga pellets with Ga_2O_3 3 wt%, 100 A current, chamber working pressure of $4-9 \times 10^{-9}$ torr, Ar flow rate of 30 and O_2 gas flow rate of 0–17.5 sccm show the lowest resistivity of $2-3 \times 10^{-4}$ Ω-cm, carrier concentration of 8.6×10^{20} cm^{-3}, mobility of 27 cm^2/Vs and transmittance of above 90% in the visible region. The carrier concentration decreases from 8.6×10^{20} to 5×10^{20} cm^{-3} and mobility increases from 27 to 29 cm^2/Vs with increasing oxygen partial pressure from 0.2×10^{-4} to 3×10^{-4} torr [260]. The ZnO:Ga 4 wt% thin films for O_2 flow rate of 10 sccm show similar resistivity of 2.6×10^{-4} Ω-cm and carrier concentration of $10^{19}-10^{21}$ cm^{-3} and mobility of 25 cm^2/Vs but the highest mobility of 32 cm^2/Vs is observed for

Table 7.11 Electrical parameters of i-ZnO/ZnO:In thin films with thickness of 1120–1380 nm and ZnO:Ga thin films

S.No	In at%	$R_\square$ (Ω/sq)	$\rho \times 10^{-3}$ (Ω-cm)	μ (cm^2/Vs)	$n_e \times 10^{20}$ (cm^{-3})
1	1	43.7	5.7	8.3	1.33
2	2	28.5	3.3	11.9	1.61
3	3	22.4	2.9	12.5	1.71
4	4	31.1	3.5	9.7	1.85
5	5	24.3	3.6	11.2	1.56
6	Ga doping	–	0.5	15	3
7		–	3	38	1.3
8		–	20	44.5	0.033

2 wt% Ga$_2$O$_3$. The resistivity increases from 2.6×10^{-4} to 10^{-2} Ω-cm with increasing O$_2$ flow from 10 to 15–30 sccm [160]. The mobility of Ga doped ZnO thin films grown by CVD method gradually decreases with increasing Ga doping as shown in Table 7.11 [261]. In the case of In doping the ZnO:In 3 at% thin films show minimum resistivity of 2.9×10^{-3} Ω-cm and mobility of 2.5 cm^2/Vs, as given in Table 7.11 [144].

The electrical parameters of ZnO:Al thin films can be varied by annealing under vacuum, oxygen and H$_2$ ambient, *etc.* The resistivity of AZO thin films can be decreased or increased by choosing suitable anneal conditions. The resistivity of ZnO:Al (2 wt%) thin films decreases from 2.9×10^{-2} to 2.76×10^{-4} Ω-cm, the mobility and carrier concentration increase from $\sim$7 to 16.5 cm^2/Vs and from 2.5×10^{19} to 1.4×10^{21} cm^{-3} with increasing annealing temperature from RT to 500 °C under vacuum for 2 h. The ZnO:Al thin films used for this investigation are prepared using 2×10^{-1} Ω-cm resistive ZnO:Al pellets [118]. The films had high resistivity of 1×10^{2} Ω-cm for O$_2$ pressure of 0.5%, which decreases to 5×10^{-3}–1×10^{-2} Ω-cm after annealing under vacuum at 1.3×10^{-3} Pa and at temperature of 400 °C for 4 h. The annealed samples show carrier concentration of 2×10^{20} cm^{-3} and mobility of 3–6 cm^2/Vs [159]. Unlike after annealing ZnO:Al (1 wt% Al$_2$O$_3$) thin films under vacuum at pressure of 1×10^{-4} Pa, and 400–500 °C the resistivity and sheet resistance increase from 3×10^{-4} to 1×10^{-3} Ω-cm and from 4 to 12 Ω/sq and carrier concentration decreases from $4.1–4.8 \times 10^{20}$ to 1.5×10^{20} cm^{-3} but mobilities stay in the range of 42 to 48 cm^2/Vs [262]. It could be due to low doping effect after annealing ZnO:Al thin films under N$_2$ and air at 400 °C, the resistivity increases from 2.76×10^{-4} to 7.12×10^{-3} and from 2.76×10^{-4} to 6.14×10^{-2} Ω-cm, respectively due to absorption of oxygen. The ZnO and Al are co-sputtered using working pressure of 0.67 Pa and Ar gas. After annealing ZnO:Al thin films under H$_2$ atmosphere at 270 °C for 60 min, the resistivity of sample decreases from 4.8×10^{-3} to 8.3×10^{-4} Ω-cm. The carrier concentration and mobility increase from 2.11×10^{20} to 8.86×10^{20} cm^{-3} and from 6.09 to 9.7 cm^2/Vs, respectively [158]. The ZnO:Al (5–6 at%) thin films grown by ion beam sputtering technique using 4.92% H$_2$ in Argon and RF power of 60 W. The as-deposited ZnO:Al thin films show the lowest resistivity of 1.2×10^{-3} Ω-cm ($\mu = 12$ cm^2/Vs and $n_e = 5 \times 10^{20}$ cm^{-3}), which is higher

Table 7.12 Effect of annealing on doped ZnO thin films.

Process	ρ (Ω-cm)	μ (cm^2/Vs)	n_e (cm^{-3})	E_g (eV)
As-deposited ZnO:Al	8.3×10^{-2}	2	–	3.5
Annealed under H$_2$ at 400 °C	9.12×10^{-3}	1.47	4.68×10^{20}	3.52
As-deposited ZnO:In	8.3×10^{-2}	2	3×10^{19}	3.35
Annealed under H$_2$ at 400 °C	3.72×10^{-3}	12.62	1.33×10^{20}	3.32

comparing with the reported values but after annealing films under 10.1% H$_2$/N$_2$ at 350–400 °C, the resistivity tremendously decreases from 1.2×10^{-3} to 5×10^{-4} Ω-cm [263] due to desorption of oxygen from the grain boundaries [120]. The resistivities of ZnO:Al (3 wt%) and ZnO:In (2 wt%) thin films grown onto glass substrates by DC reactive sputtering technique decrease after annealing under H$_2$ atmosphere at 400 °C, as shown in Table 7.12 [264].

The ZnO:Al thin films grown by sputtering technique using ZnO:Al$_2$O$_3$ 2 wt% target, Ar flow rate 20 sccm, working pressure of 0.40 Pa are studied by damp heat process conditions of 85 °C and 85% humidity for 1000 h. After damp heat process the carrier concentration, mobility, and transmission decrease from 2×10^{20} to 1×10^{20} cm^{-3}, from 27 to 22 cm^2/Vs, and from 90 to 88–89% for 320 nm thick ZnO:Al layers due to penetration of oxygen and water molecules into grain boundaries that cause to enhance grain boundary scattering. A decrease in quality is severe in thin layers comparing with thick layers because the thick layers contain dense surface and large grain sizes therefore those block penetration of impurities. After the damp heat test the sheet resistance increases from 18 to 25 Ω/sq. The (002) orientation is stronger in the thicker samples than in the thin samples [265].

The mobility of 2 in. diameter and 1 mm thick ZnO single crystal (sample A) grown by vapor phase method varies with varying temperature, as shown in Figure 7.103. The peak mobility of >1900 cm^2/Vs for single crystal is observed at low temperature. The combination of various scattering mechanisms such as (i) optical mode lattice scattering, (ii) acoustic mode deformation potential, (iii) acoustic mode piezoelectric potential, and (iv) ionized impurity scattering are employed using Debye temperature (θ_D) of 837 K, deformation potential (E_I) of 15, piezoelectric coupling coefficient of 0.21, $\varepsilon_0 = 8.12$, $\varepsilon_\infty = 3.72$, $N_a = 2 \times 10^{15}$ cm^{-3}, and $m^* = 0.318 m_o$ to fit theoretical curve with mobility curve [266]. The experimental carrier concentration versus inverse temperature for the samples denoted as A, B, and C grown by seeded vapor phase method are simulated with data of $N_{d1} = 9 \times 10^{15}$ cm^{-3}, $E_{d1} = 31$ meV; $N_{d2} = 1 \times 10^{17}$ cm^{-3}, $E_{d2} = 61$ meV, $N_a = 1 \times 10^{15}$ cm^{-3} for sample A, $N_{d2} = 4 \times 10^{16}$ cm^{-3}, $E_{d2} = 63$ meV, and $N_a = 1 \times 10^{16}$ cm^{-3} for sample B and $N_{d3} = 7 \times 10^{15}$ cm^{-3}, $E_{d3} = 340$ meV, and $N_a = 1 \times 10^{15}$ cm^{-3} for sample C. The activation energy of 30 meV is assigned to H shallow donar level, 60 meV due to either Al$_{Zn}$ or Cl$_O$ and higher activation energy of 340 meV is due to deep donor level of O$_{Zn}$ [267]. The ZnO samples with $\rho = 5$–200 Ω-cm $\mu = 125$ cm^2/Vs, and

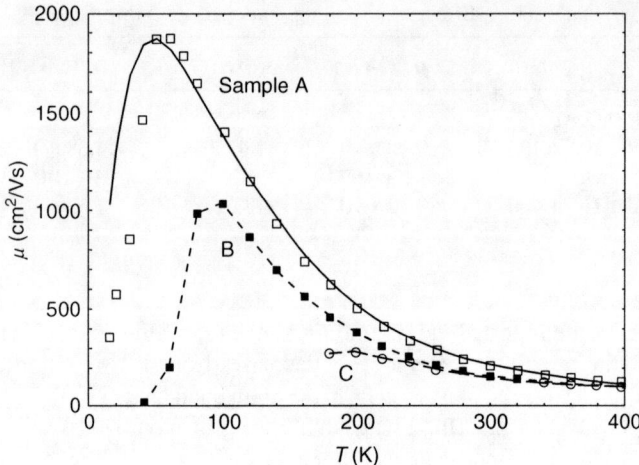

Figure 7.103 Variation of mobility with temperature.

$n_e = 3 \times 10^{17}$ cm^{-3} grown by hydrothermal method show donor levels of 4–48, 60–66, and 300 meV by temperature dependent conductivity measurements [268]. In conclusion, the sheet resistance and transmittance should be in single digit and 85–90% for thin film solar cell applications, respectively.

7.4 ITO

The ITO thin films provide high performance electrical and optical properties comparing with other TCOs but it needs high growth temperature. The ITO is body centered cubic structure containing 80 atoms per unit cell and its band gap is in the range of 3.5–4.3 eV. It is a degenerate semiconductor and its least resistivity is 2×10^{-4} Ω-cm. The dopant concentration of Sn and oxygen vacancies determine conductivity of ITO layers. The grain sizes of PLD ITO determined by SEM are 200 nm for the growth temperature of 200–400 °C. Keeping substrate temperature at 200 °C, the surface is dense with round shape grains of 50–80 nm diameter for the layers grown at $P_{O_2} = 0.1$ Pa, whereas the surface is less homogeneous with different shaped grains of 80–100 nm diameter for $P_{O_2} = 6.6$ Pa [269]. The XRD analysis of ITO films deposited by spray pyrolysis technique at two different substrate temperatures of 520 and 480 °C reveals lattice constants of 10.16 and 10.09 Å, respectively, which are close to standard value of 10.118 Å (Figure 7.104). The ITO layers deposited at higher substrate temperature shows (400) preferred orientation and SnO$_2$ phase along with noisy signals due to lattice distortion, whereas the layers deposited at low temperature of 480 °C exhibit (400) and (222) more or less at same intensities [270]. The intensity of (222) preferred orientation increases with increasing thickness of sputtered ITO layer from 174 to 450 nm and the other peaks are also generated [271]. (400) is a preferred orientation in the near stoichiometric

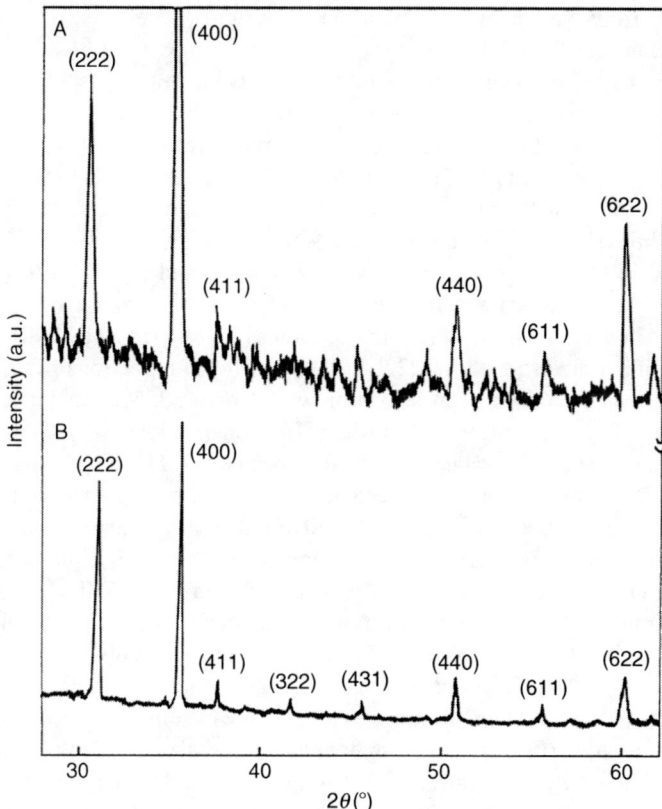

Figure 7.104 XRD pattern of ITO coated onto glass substrates at substrate temperatures of (A) 520 and (B) 480 °C.

or low oxygen content 700 nm thick samples, whereas (222) is the intensity peak in the oxygen rich ITO films prepared by sputtering using 90 wt% In_2O_3 and 10 wt% SnO_2 target, chamber pressure of 4×10^{-1} Pa, sputter power of 2 W/cm^2 at room temperature [272]. Alike the ITO films grown by CVD technique using 0.2 M In (III)-acetylacetonate and 5 mol% Sn(IV)-bis-acetylacetonate-dibromide dissolved in acetylacetone at substrate temperature of 600 °C show preferred orientation of (400) [273]. The SAED patterns reveal that the ITO films deposited onto glass substrates at RT by PLD using 90 wt% In_2O_3 + 10 wt% SnO_2 target under oxygen pressure of 1.3 Pa shows amorphous structure, whereas films deposited at 200 or 400 °C show polycrystalline structure. The ITO layers grown at RT and 400 °C show RMS value of 0.5 and 2 nm, respectively. The intensity of (222) increases with increasing substrate temperature from 175 to 300 °C thereafter decreases for $P_{O2} = 1.3$ Pa. Keeping substrate temperature of 200 °C, the (222) is dominant one for Pa = 1.3, whereas the intensity of (222) peak decreases above or below of oxygen pressure [269].

The ITO films grown by laser ablation shows (310) preferred orientation with lattice constant of 10.53 Å [274]. In the case of sputtering, the films deposited onto glass substrates by microwave enhanced DC reactive magnetron sputtering with In:Sn = 9:1 at low pressure shows amorphous, whereas films deposited at oxygen partial pressure $P_{O_2} = 1.2 \times 10^{-3}$ mbar and total pressure PT = 5×10^{-3} mbar show (222) preferred orientation [275]. The physical parameters of ITO layers with thickness of 252 nm deposited onto glass substrates at RT and $T_S = 300$ °C by DC reactive sputtering using 90 W bias voltage of -100 V are given in Table 7.13. 10 wt% SnO_2 doped In_2O_3 target is used to deposit ITO layers and $P_{O_2 a} = P_{O_2}/(P_{O_2} + P_{Ar}) = 21.2\%$, argon and oxygen and total pressure of 1.1×10^{-1} Pa are used [276]. The layers deposited at RT show amorphous, whereas layers deposited at 180 °C show polycrystalline structure having (222), (400), and (622) reflections [277]. The ITO layers are grown by pulsed laser deposition with 3 J/cm^2, 355 nm and chamber pressure of $1-6 \times 10^{-2}$ torr and 90% In_2O_3 and 10% SnO_2 target.

The transmittance, reflectance, and absorptance of ITO films are shown in Figure 7.105. The transmittance is lower at higher wavelength region in the thicker samples (160 nm) than in the thinner samples (140 nm) due to free carrier absorption [273]. As expected the transmission is higher at infrared region in slightly high sheet resistance samples than in low sheet resistance samples, as shown in Figure 7.106 [272]. The band gap varies from 3.38 to 3.68 eV with varying carrier concentration. In fact, the intrinsic In_2O_3 exhibits band gap of 3.53 eV [276] but ITO grown by CVD shows little higher band gap of 3.9 eV [273]. The resistivity of ITO films by PLD increases from 2×10^{-4} to 1.4×10^{-3} Ω-cm with increasing O_2 pressure from 0 to 7 Pa [269]. The ITO layers grown by PLD at 250 °C under oxygen pressure of 10 mtorr show similar resistivity of $1-3 \times 10^{-4}$ Ω-cm [273,274]. Akin the layers grown onto glass substrates at RT and 180 °C show resistivity of 8×10^{-4} and 5.34×10^{-4} Ω-cm, respectively [277]. The layers deposited onto polycarbonate (PC) substrates by laser ablation with energy of 230 mJ, 20 Hz, using In_2O_3 95 wt% + SnO_2 5 wt% grown at RT using $P_{O_2} = 1.3$ Pa show the highest mobility of 38 cm^2/Vs [278].

7.5 FTO

The fluorine doped SnO_2 (FTO) thin films are high resistance to against heat, strong adhesion to glass, and good chemical stability. The applications of FTO are high in thin film solar cells, flat panel displays, electrochromic windows, heat reflectors, and gas sensors. The SnO_2:F with F/Sn atomic ratio of 1.0 in the chemical solution deposited by spray pyrolysis technique at substrate temperature of 450 °C shows (200) preferred orientation. The grown films show tetragonal structure with lattice constants of $a = 4.75$ and $c = 3.197$ Å [282]. The theoretically constructed XRD pattern of SnO_2 shows (110) preferred orientation, whereas experimental XRD pattern shows (200) preferred orientation, as shown in Figures 7.107 and 7.108, respectively. This could be due to effect of deposition parameters. The similar (200) preferred orientation is observed for un-doped and F doped SnO_2 thin films grown by

Table 7.13 Physical parameters of ITO layers

S.No.	T_S (°C)	Sn/In	T	$\Phi(\lambda_{1550\text{ nm}})$ ($10^{-3}\ \Omega^{-1}$)	n	k	E_g (eV)	μ (cm^2/Vs)	$R_\square$ (Ω/sq)	$n_e \times 10^{20}$ (cm^{-3})	$\rho \times 10^{-3}$ (Ω-cm)	Ref.
1	520	0.023	0.8–0.9	4	2.1–1.1	0.6–0.01	3.46	18.5	26.6	2.3	1.5	[270]
2	480	0.25	0.8–0.95	0.5	–	–	3.4	6.1	194	0.94	10.9	
3	498	0.023	0.8–0.9	22	–	–	3.5–3.8	16.1	10.9	7.4	0.52	[279–281]
4	435	0.023	0.8–0.9	0.3	–	–	3.5–3.8	3.0	420	4	5.2	
5	–	–	0.85	4.1	1.79	0.03	3.84	–	209	–	–	[275]
6	330	–	0.98	6.8	–	–	3.68	23.2	146.3	1.1	2.4	[276]
7	RT	–	0.975	4.1	–	–	3.54	33.7	238.6	0.19	9.7	
8	RT	–	–	–	–	–	–	38	–	6.5	0.25	[278]

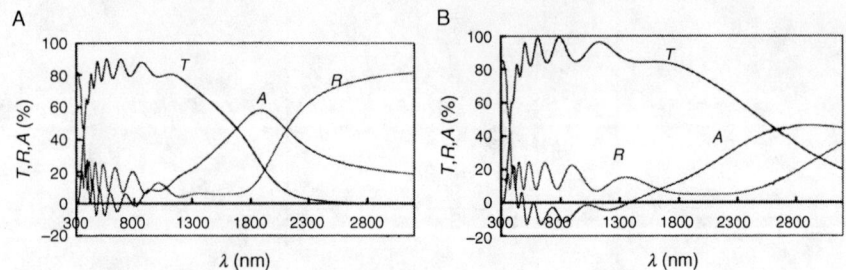

Figure 7.105 Transmittance, reflectance, and absorptance of ITO films grown by CVD with different thicknesses of (A) 160 and (B) 140 nm.

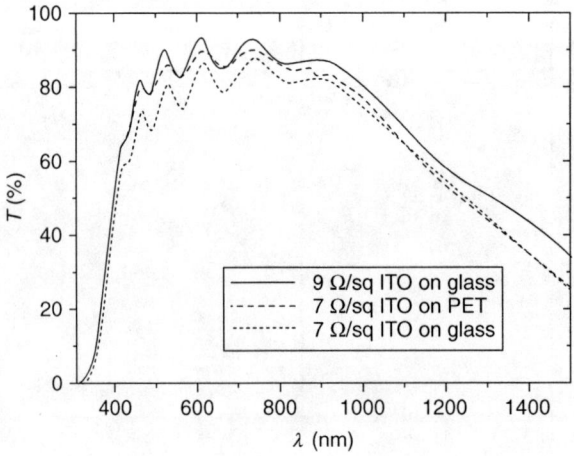

Figure 7.106 Transmission spectra of ITO thin films on glass and polyethylene terephtalate (PET) substrates. There is a difference in transmission at infrared region from sample to sample due to slightly difference in sheet resistance.

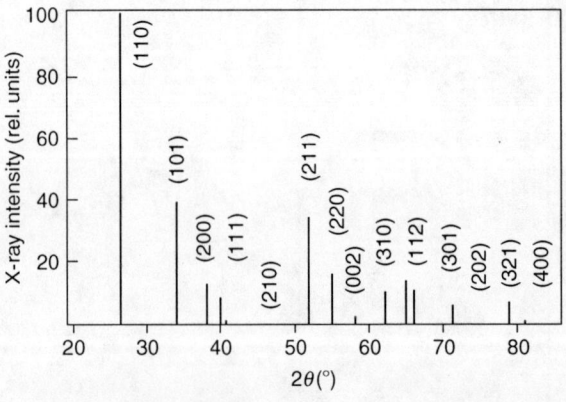

Figure 7.107 Theoretically constructed XRD pattern of SnO_2.

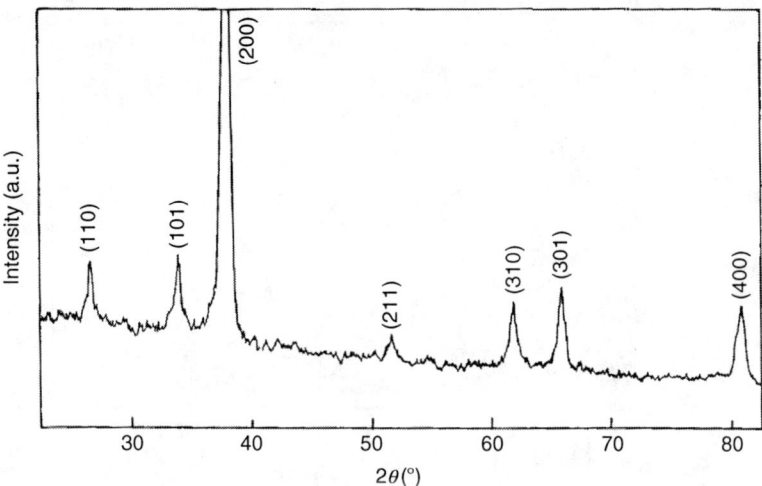

Figure 7.108 XRD pattern of fluorine doped SnO$_2$ thin films grown by spray pyrolysis.

spray pyrolysis [283]. The films deposited by CVD at lower substrate temperature of <450 °C shows orthorhombic and tetragonal structure [284]. The FTO layers by spray show similar (200) preferred orientation and lattice constants of $a=4.761$ and $c=3.225$ Å [285]. The films deposited at substrate temperature of 500 °C shows (200) as the intensity peak, whereas the films deposited below the substrate temperature of 500 °C show (211) and (114) as intensity peaks [286].

The optical band gap of SnO$_2$ thin films grown onto quartz substrates by CVD at substrate temperature of 500 °C increases from 4.03 to 4.17–4.29 eV with increasing fluorine concentration in the films. The FTO films are grown using SnCl$_2$ for which oxygen is used as transport agent, whereas N$_2$ is used as a transport agent for trifluoroacetic acid (15 vol%) for fluorine doping [287]. The transmittance and reflectance spectra of SnO$_2$:F thin films grown by spray pyrolysis technique using 3 wt% HF dopant source with function of wavelength are shown in Figure 7.109. The transmittance of FTO films decreases with increasing thickness in the visible and infrared regions. On the other hand, the reflectance increases with increasing thickness in the infrared region, that is, $\lambda > \lambda_p$ (plasma wavelength). The λ_p is independent of thickness of layers, which is 1500 and 1900 nm for FTO and ITO layers. The λ_p can be calculated using the relation

$$\lambda_p = 2\pi c_o \left(\frac{n_e e^2}{\varepsilon_0 \varepsilon_L m_{\text{eff}}} \right)^{-0.5}, \tag{7.19}$$

where c_o is the velocity of light, ε_o is permittivity in vacuum, m_{eff} effective mass of electron for conductive layer. λ_p is found to be 1.3 µm for ITO layers taking $n_e = 3.2 \times 10^{20}$ cm^{-3}, $m_{\text{eff}} = 0.35 m_o$, $\varepsilon_L = 3$ [288].

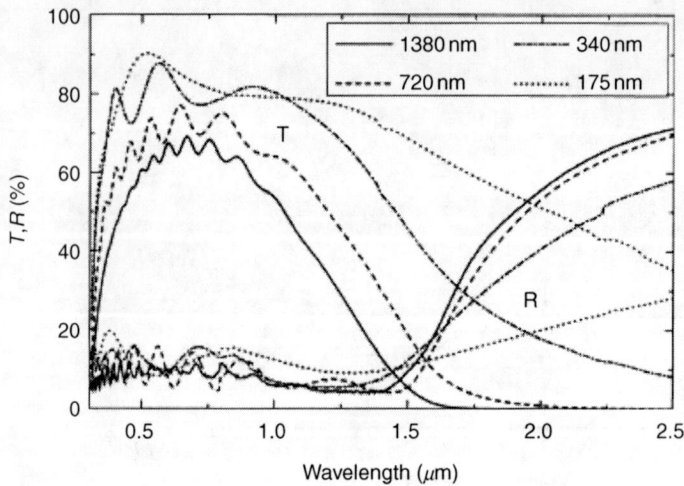

Figure 7.109 Transmittance and reflectance spectra of FTO thin films with different thicknesses grown by spray pyrolysis technique (3 wt% HF doped).

The mobility and carrier concentration vary with varying orientation of the films. The highest mobility and carrier concentration are obtained for the orientation of (110), whereas they are the lowest for the orientation of (301) [289,290]. The F doped SnO_2 thin films by spray exhibit mobility and carrier concentration of 16 cm^2/Vs and 3×10^{20} cm^{-3} at room temperature, respectively [282]. The resistivity of SnO_2 by spray pyrolysis technique using $SnCl_2 H_2O$ and NH_4F chemical solution decreases from 3×10^{-2} to 6.2×10^{-4} Ω-cm with increasing fluorine content up to 50 mol% and the mobility increases from 11 to 18 cm^2/Vs with increasing fluorine content up to 10 mol% [286]. The mobility decreases with increasing carrier concentration in the degenerate semiconductor for which the relation can be written as

$$\mu \approx \left(\frac{4e}{h}\right)\left(\frac{\pi}{3}\right)^{1/3} n_e^{-2/3}. \tag{7.20}$$

The sheet resistance of 400 nm thick FTO is 15 Ω/sq. the resistivity of 500 nm thick FTO layers is 3.5×10^{-4} Ω-cm and mobility increases from 10 to 25 cm^2/Vs with thickness [288]. The FTO layers are grown onto glass substrates by spray pyrolysis technique at substrate temperature of 400 °C using chemical solution of $SnCl_2 \cdot 2H_2O$, HCl, methanol, NH_4F, and distilled water. The carrier concentration increases from 5.7×10^{20} to 13.2×10^{20} cm^{-3} and mobility decreases from 9.6 to 7 cm^2/Vs with increasing from 4 to 10 wt% NH_4F in the solution. The least sheet resistance of 5.7 Ω/sq and resistivity of 2.5×10^{-3} Ω-cm are observed for 10 wt% NH_4F [285].

References

[1] M. Powella, Zentrum fur Sonnenenergie- und Wasserstoff-Forschung Baden-Wurttemberg, Germany. *www.pv-tech.org*, August 29th 2010.
[2] J. Britt, C. Ferekides, Appl. Phys. Lett. 62 (1993) 2851.
[3] H. Katagiri, K. Jimbo, S. Yamada, T. Kamimura, W.S. Maw, T. Fukano, *et al.*, Appl. Phys. Express 1 (2008) 041201.
[4] IBM, private communication (2010).
[5] D. Lincot, B. Mokili, J. P. Badiali, P. Mandin, private communication.
[6] T. Nakashi, K. Ito, Solar Energy Mater. Solar Cells 35 (1994) 171.
[7] U.S. Ketipearachchi, D.W. Lane, K.D. Rogers, J.D. Painter, M.A. Cousins, Cranfield University, Swindon, UK, private communication (2010).
[8] R.S.- Araoz, D.A.-. Ras, T.P. Neisser, K. Wilchelmi, M.Ch.L-. Steiner, A. Ennaoui, Thin Solid Films 517 (2009) 2300.
[9] S. Kundu, L.C. Olsen, 29th IEEE Photovoltaic Specialist Conference, (2002), p. 648.
[10] Y. Hashimoto, N. Kohara, T. Negami, N. Nishitani, T. Wada, Solar Energy Mater. Solar Cells 50 (1998) 71.
[11] J.-H. Lee, W.-C. Song, J.-S. Yi, Y.-S. Yoo, Solar Energy Mater. Solar Cells 75 (2003) 227.
[12] J. Song, S.S. Li, S. Yoon, W.K. Kim, J. Kim, J. Chen, V. Craciun, T.J. Anderson, O.D. Crisalle, F. Ren, 31st IEEE Photovoltaic Specialists Conference, (2005) p. 449.
[13] S.R. Kodigala, V.S. Raja, M. Sharon, J. Mater. Sci. Mater. Electron. 9 (1998) 261.
[14] H. Uda, H. Yonezawa, Y. Ohtsubo, M. Kosaka, H. Sonomura, Solar Energy Mater. Solar Cells 75 (2003) 219.
[15] S.Y. Kim, D.S. Kim, B.T. Ahn, H.B. Im, Thin Solid Films 229 (1993) 227.
[16] H.R. Moutinho, D. Albin, Y. Yan, R.G. Dhere, X. Li, C. Perkins, *et al.*, Thin Solid Films 436 (2003) 176.
[17] H.R. Moutinho, D. Albin, Y. Yan, R.G. Dhere, C. Perkins, X. Li, M.M. Al-Jassim, Mater. Res. Soc. Symp. Proc. 668 (2001) H7.4.1.
[18] O. Vigil, Y. Rodriguez, O.Z.- Angel, V.V.- Lopez, A.M.- Acevedo, J.G.V.- Luna, Thin Solid Films 322 (1998) 329.
[19] D.R. Acosta, C.R. Magana, A.I. Martinez, A. Maldonado, Solar Energy Mater. Solar Cells 82 (2004) 11.
[20] N.B. Chaure, S. Bordas, A.P. Samantilleke, S.N. Chaure, J. Haigh, I.M. Dharmadasa, Thin Solid Films 437 (2003) 10.
[21] A. Ennaoui, S. Siebentritt, M.Ch. Lux-Steiner, W. Riedl, F. Karg, Solar Energy Mater. Solar Cells 67 (2001) 31.
[22] X.L. Tong, D.S. Jiang, M.Z. Luo, L. Liu, L.C. Xiong, Mater. Sci. Eng. B 136 (2007) 62.
[23] M.T.S. Nair, P.K. Nair, F.L.A. Zingaro, E.A. Meyers, J. Appl. Phys. 75 (1994) 1557.
[24] S.R. Kodigala, R.D. Pilkington, A.E. Hill, R.D. Tomlinson, A.K. Bhatnagar, Mater. Chem. Phys. 68 (2001) 22.
[25] D. Abou-Ras, G. Kostorz, A. Romeo, D. Rudmann, A.N. Tiwari, Thin Solid Films 480–481 (2005) 118.
[26] T. Wada, Sol. Energy Mater. Sol. Cells 49 (1997) 249.
[27] T. Nakada, Thin Solid Films 361–362 (2000) 346.
[28] I.O. Oladeji, L. Chow, J.R. Liu, W.K. Chu, A.N.P. Bustamante, C. Fredricksen, *et al.*, Thin Solid Films 359 (2000) 154.

[29] U.S. Ketipearachchi, D.W. Lane, K.D. Rogers, J.D. Painter, M.A. Cousins, Mater. Res. Soc. Symp. Proc. 836 (2005) L5.35.1.
[30] A. Cortes, H. Gomez, R.E. Marotti, G. Riveros, E.A. Dalchiele, Solar Energy Mater. Solar Cells 82 (2004) 21.
[31] O. de Melo, L. Hernandez, O.Z.- Angel, R.L.- Morales, M. Becerril, E. Vasco, Appl. Phys. Lett. 65 (1994) 1278.
[32] C.D.G. Lazos, E. Rosendo, M. Ortega, A.I. Oliva, O. Tapia, T. Diaz, et al., Mater. Sci. Eng. B 165 (2009) 74.
[33] D. Kaushik, R.R. Singh, M. Sharma, D.K. Gupta, N.P. Lalla, R.K. Pandey, Thin Solid Films 515 (2007) 7070.
[34] S. Mishra, A. Ingale, U.N. Roy, A. Gupta, Thin Solid Films 516 (2007) 91.
[35] B.-S. Moon, J.-H. Lee, H. Jung, Thin Solid Films 511 (2006) 299.
[36] A. Kuroyanagi, T. Suda, Thin Solid Films 176 (1989) 247.
[37] M. Kalafi, H. Bidadi, S. Sobhanian, A.I. Bairamov, V.M. Salmanov, Thin Solid Films 265 (1995) 119.
[38] B. Ullrich, H. Sakai, Y. Segawa, Thin Solid Films 385 (2001) 220.
[39] R.R. Chamberlin, J.S. Skarman, J. Electrochem. Soc. 113 (1966) 86.
[40] B.K. Gupta, O.P. Agnihotri, Phil. Mag. 37 (1978) 631, B.K. Gupta, O.P. Agnihotri, Solid State Commun. 23 (1977) 295.
[41] Y.Y. Ma, R.H. Bube, J. Electrochem. Soc. 124 (1977) 1430.
[42] R. S. Feigelson, Ph.D. Dissertation, Stanford University (1974).
[43] A. Mzerd, D. Sayah, I.J. Saunders, B.K. Jones, Phys. Stat. Sol. (a) 119 (1990) 487.
[44] H. Khallaf, G. Chai, O. Lupan, L. Chow, S. Park, A. Schulte, Appl. Surface Sci. 255 (2009) 4129.
[45] J.L. Boone, S.A. Howard, D.D. Martin, Thin Solid Films 176 (1989) 143.
[46] U. Pal, R.S.- Gonzalez, G.M.- Montes, M.G.- Jimenez, M.A. Vidal, Sh. Torres, Thin Solid Films 305 (1997) 345.
[47] E. Bacaksiz, V. Novruzov, H. Karal, E. Yanmaz, M. Altunbas, A.I. Kopya, J. Phys. D Appl. Phys. 34 (2001) 3109.
[48] J.-H. Lee, D.-J. Lee, Thin Solid Films 515 (2007) 6055.
[49] O. Trujillo, R. Moss, K.D. Vuong, D.H. Lee, R. Noble, D. Finnigan, et al., Thin Solid Films 290 (1996) 13.
[50] A.L. Diwar, P.K. Shishodia, G. Chauhan, Thin Solid Films 201 (1991) L1.
[51] J.-H. Lee, W.-C. Song, J.-S. Yi, K.-J. Yang, W.-D. Han, J. Hwang, Thin Solid Films 431–432 (2003) 349.
[52] M.K.B. Saidin, G.J. Russell, A.W. Brinkman, J. Woods, J. Crystal Growth 101 (1990) 844.
[53] S.R. Kodigala, Ph.D. Thesis, Sri Venkateswara University (1992).
[54] J.C. Manifacier, M. De Murcia, J.P. Fillard, Thin Solid Films 41 (1977) 127.
[55] B.E. Sernelius, K.-F. Berggren, Z.-C. Jin, I. Hamberg, C.G. Granqvist, Phys. Rev. B 37 (1988) 10244.
[56] H. Oumous, H. Hadiri, Thin Solid Films 386 (2001) 87.
[57] J. Hiie, T. Dedova, V. Valdna, K. Muska, Thin Solid Films 511–512 (2006) 443.
[58] C. Guillen, M.A. Martinez, J. Herrero, Thin Solid Films 335 (1998) 37.
[59] S.R. Kodigala, A.K. Bhatnagar, R.D. Pilkington, A.E. Hill, R.D. Tomlinson, J. Mater. Sci. Mater. Electron. 11 (2000) 269.
[60] M.D. Archbold, D.P. Halliday, K. Durose, T.P.A. Hase, D.-S. Boyle, K. Govender, 31st IEEE Photovoltaic Specialists Conference, (2005) 476.
[61] J.-H. Lee, Y.-S. Yi, K.-J. Yang, J.-H. Park, R.-D. Oh, Thin Solid Films 431 (2003) 344.

[62] G. Perna, V. Capozzi, M. Ambrico, V. Augelli, T. Ligonzo, A. Minafra, et al., Thin Solid Films 453 (2004) 187.
[63] B. Ullrich, H. Sakai, N.M. Dushkina, H. Ezumi, S. Keitoku, T. Kobayashi, Mater. Sci. Eng. B 47 (1997) 187.
[64] A. Antony, M.K. Jayaraj, 20th European Photovoltaic Solar Energy Conference, (2005) p. 1870.
[65] A.K. Berry, P.M. Amirtharaj, J.-T. Du, J.L. Boone, D.D. Martin, Thin Solid Films 219 (1992) 153.
[66] Y. Endoh, Y. Kawakami, T. Taguchi, A. Hiraki, Jpn. J. Appl. Phys. 27 (1988) L2199.
[67] N. Lovergine, R. Cingolani, A.M. Mancini, M. Ferrara, J. Crystal Growth 118 (1992) 304.
[68] J.A-. Hernandez, J.S-. Hernandez, N.X-. Quiebras, R.M-. Perez, O.V-. Galan, G.C-. Puente, M.C-. Garcia, Solar Energy Mater. Solar Cells 90 (2006) 2305.
[69] J. Humenberger, G. Linnert, K. Lischka, Thin Solid Films 121 (1984) 75.
[70] N.V. Klimova, N.E. Korsunskaya, I.V. Markevich, G.S. Pekar, A.F. Singaevsky, Mater. Sci. Eng. B 34 (1995) 12.
[71] M. Agata, H. Kurase, S. Hayashi, K. Yamamoto, Solid State Commun. 76 (1990) 1061.
[72] K. Mam, K. Durose, D.P. Halliday, A. Szczerbakow, Thin Solid Films 480–481 (2005) 236.
[73] A. Podesta, N. Armani, G. Salviati, N. Romeo, A. Bosio, M. Prato, Thin Solid Films 511 (2006) 448.
[74] T.R. Ravindran, A.K. Arora, B. Balamurugan, B.R. Mehta, Nanostructured Mater. 11 (1999) 603.
[75] D.G. Kim, K. Tomihira, S. Okahara, M. Nakayama, J. Crystal Growth 310 (2008) 4244.
[76] Ch. Fricke, R. Heitz, R. Lummer, V. Kutzer, A. Hoffmann, I. Broser, et al., J. Crystal Growth 138 (1994) 815.
[77] X.L. Fu, L.H. Li, W.H. Tang, Solid State Commun. 138 (2006) 139.
[78] J. Butty, N. Peyghambarian, Y.H. Kao, J.D. Mackenzie, Appl. Phys. Lett. 69 (1996) 3224.
[79] W.F. Liu, C.G. Jin, C. Jia, L.Z. Yao, W.L. Cai, X.G. Li, Chem. Lett. 33 (2004) 228.
[80] O. Vigil, L. Riech, M.-G. Rocha, O.A.- Angel, J. Vac. Sci. Technol. A 15 (1997) 2282.
[81] H.A.- Calderon, R.L.- Morales, O.A.- Angel, I.G.M.- Alvarez, L. Banos, J. Vac. Sci. Technol. A 14 (1996) 2480.
[82] B. Tell, T.C. Damen, S.P.S. Porto, Phys. Rev. 144 (1966) 771.
[83] J.F. Scott, R.C.C. Leite, T.C. Damen, Phys. Rev. 188 (1969) 1285.
[84] X.W. Wang, F. Spitulnik, B. Campell, R. Noble, R.P. Hapanowicz, R.A. Condrate Sr., et al., Thin Solid Films 218 (1992) 157.
[85] P. Corio, J.C. Tristao, F. Magalhaes, M.T.S. Sansiviero, Modern Topics in Raman Spectroscopy, IQUSP, Sao Paulo, Brazil (2006).
[86] A. Balandin, K.L. Wang, N. Kouklin, S. Bandyopadhyay, Appl. Phys. Lett. 76 (2000) 137.
[87] Z.R. Khan, M. Zulfequar, M.S. Khan, Mater. Sci. Eng. B 174 (2010) 145.
[88] J. Lee, Thin Solid Films 451–452 (2004) 170.
[89] T.L. Tong, D.S. Jian, Z.M. Liu, M.Z. Luo, Y. Li, P.X. Lu, et al., Thin Solid Films 516 (2008) 2003.
[90] J.N.X.- Quiebras, G.C.- Puente, J.A.- Hernandez, G.S.- Rodriguez, A.A.-C. Readigos, Solar Energy Mater. Solar Cells 82 (2004) 263.
[91] B. Su, K.L. Choy, Thin Solid Films 359 (2000) 160.

[92] N. Romeo, G. Sberveglieri, L. Tarricone, Thin Solid Films 43 (1977) L15.
[93] N. Romeo, G. Sberveglieri, L. Tarricone, Thin Solid Films 55 (1978) 413.
[94] D. Cha, S. Kim, N.K. Huang, Mater. Sci. Eng. B 106 (2004) 63.
[95] B. Ullrich, H. Ezumi, S. Keitoku, T. Kobayashi, Mater. Sci. Eng. B 35 (1995) 117.
[96] L. Pintilie, E. Pentia, I. Pintilie, D. Petre, Mater. Sci. Eng. B 44 (1997) 403.
[97] M. Itakura, H. Toyoda, J. Phys. Soc. Jpn. 18 (1963) 150.
[98] S.R. Kodigala, J. Mater. Sci. Mater. Electron. 10 (1999) 291.
[99] H.H. Woodbury, Phys. Rev. B 9 (1974) 5188.
[100] J.C. Joshi, B.K. Sachar, P. Kumar, Thin Solid Films 88 (1982) 189.
[101] K. Morimoto, M. Kitagawa, T. Yoshida, J. Crys. Growth 59 (1982) 254.
[102] H. Fujita, K. Kobayashi, T. Kawai, K. Shiga, J. Phys. Soc. Jpn. 20 (1965) 109.
[103] T.T. John, K.P. Vijayakumar, C.S. Kartha, Y. Kashiwaba, T. Abe, 3rd World Conference on Photovoltaic Energy Conversion, (2003), 1pc324-155.
[104] B. Asenjo, C. Sanz, C. Guillen, A.M. Chaparro, J. Herrero, M.T. Gutierrez, 20th European Photovoltaic Solar Energy Conference, (2005) p. 1866.
[105] S. Spiering, L. Burkert, D. Hariskos, M. Powalla, B. Dimmler, C. Giesen, et al., Thin Solid Films 517 (2009) 2328.
[106] P. Pistor, R. Caballero, D. Hariskos, V.I.- Roca, R. Wachter, S. Schorr, et al., Solar Energy Mater. Solar Cells 93 (2009) 148.
[107] S. Spiering, D. Hariskos, M. Powalla, N. Naghavi, D. Lincot, Thin Solid Films 431–432 (2003) 359.
[108] J.F. Trigo, B. Asenjo, J. Herrero, M.T. Gutierrez, Solar Energy Mater. Solar Cells 92 (2008) 1145.
[109] J.N. Duenow, T.A. Gessert, D.M. Wood, T.M. Barnes, M. Young, B. To, T.J. Coutts, J. Vac. Sci. Tech. A 25 (2007) 955.
[110] K. Ellmer, J. Phys. D Appl. Phys. 33 (2000) R17.
[111] L.A.- Silva, M.S. Tomar, O.P.- Perez, S.P. Singh, S.J.- Rosas, Mater. Res. Soc. Symp. Proc. 1165 (2009), 1165-M08-38.
[112] M.K. Jayaraj, A. Antony, M. Ramachandran, Bull. Mater. Sci. 25 (2002) 227.
[113] C. Wang, P. Zhang, J. Yue, Y. Zhang, L. Zheng, Physica B 403 (2008) 2235.
[114] W. Li, Y. Sun, Y. Wang, H. Cai, F. Liu, Q. He, Solar Energy Mater. Solar Cells 91 (2007) 659.
[115] Z.-C. Jin, I. Hamberg, C.G. Granqvist, J. Appl. Phys. 64 (1988) 5117.
[116] Y. Igasaki, H. Saito, J. Appl. Phys. 69 (1991) 2190.
[117] H. Sato, T. Minami, S. Takata, T. Mouri, N. Ogawa, Thin Solid Films 220 (1992) 327.
[118] G.J. Fang, D. Li, B.-L. Yao, Thin Solid Films 418 (2002) 156.
[119] M.J. Keum, J.S. Yang, I.H. Son, S.K. Shin, H.W. Choi, W.S. Lee, et al., Mater. Res. Soc. Symp. Proc. 730 (2002) V5.12.1.
[120] J.F. Chang, H.L. Wang, M.H. Hon, J. Crystal Growth 211 (2000) 93.
[121] B.H. Choi, H.B. Im, J.S. Song, K.H. Yoon, Thin Solid Films 93 (1990) 712.
[122] F.K. Shan, Y.S. Yu, Thin Solid Films 435 (2003) 174.
[123] K. Matsubara, P. Fons, K. Iwata, A. Yamada, K. Sakurai, H. Tampo, et al., Thin Solid Films 431 (2003) 369.
[124] R.K. Shukla, A. Srivastava, A. Srivastava, K.C. Dubey, J. Crystal Growth 294 (2006) 427.
[125] S. Jager, B. Szyszka, J. Szczyrbowski, G. Brauer, Surface and Coatings Technology 98 (1998) 1304.
[126] S. Singh, R.S. Srinivasa, S.S. Major, Thin Solid Films 515 (2007) 8718.

[127] H. Sato, T. Minami, Y. Tamura, S. Takata, T. Mouri, N. Ogawa, Thin Solid Films 286 (1994) 86.
[128] S.N. Bai, T.Y. Tseng, Thin Solid Films 515 (2006) 872.
[129] Z.L. Pei, X.B. Zhang, G.P. Zhang, J. Gong, C. Sun, R.F. Huang, et al., Thin Solid Films 497 (2006) 20.
[130] J. Yoo, J. Lee, S. Kim, K. Yaoon, I.J. Park, S.K. Dhungel, et al., Thin Solid Films 480–481 (2005) 213.
[131] J.J. Ding, S.Y. Ma, H.X. Chen, X.F. Shi, T.T. Zhou, L.M. Mao, Physica B 404 (2009) 2439.
[132] V.D. Falcao, M.E.L. Sabini, D.O. Miranda, A.S.A.C. Diniz, J.R.T. Branco, 33rd IEEE Photovoltaic Specialist Conference, (2008), 442-080511140518.
[133] H.W. Lee, S.P. Lau, Y.G. Wang, K.Y. Tse, H.H. Hng, B.K. Tay, J. Crystal Growth 268 (2004) 596.
[134] A.N. Banerjee, C.K. Ghosh, K.K. Chattopadhyay, H. Minoura, A.K. Sarkar, A. Akiba, et al., Thin Solid Films 496 (2006) 112.
[135] E. Fortunato, P. Nunes, A. Marques, D. Costa, H. Aguas, I. Ferreira, et al., Mater. Res. Soc. Symp. Proc. 685E (2001) D5.10.1.
[136] D. Song, P. Widenborg, W. Chin, A.G. Aberle, Solar Energy Mater. Solar Cells 73 (2002) 1.
[137] C. Meng, P. Zhilliang, W. Xi, S. Cao, W. Lishi, Mater. Res. Soc. Symp. Proc. 666 (2001) F1.2.1.
[138] H. Takeda, Y. Sato, Y. Iwabuchi, M. Yoshikawa, Y. Shigesato, Thin Solid Films 517 (2009) 3048.
[139] O. Takai, M. Futsuhara, G. Shimizu, C.P. Lungu, J. Nozue, Thin Solid Films 318 (1998) 117.
[140] V. Sittinger, B. Szyszka, R.J. Hong, W. Werner, M. Ruske, A. Lopp, 3rd World Conference on Photovoltaic Energy Conversion, (2003), 2PD355.
[141] K.C. Park, D.Y. Ma, K.H. Kim, Thin Solid Films 305 (1997) 201.
[142] S.H. Jeong, J.W. Lee, S.B. Lee, J.H. Boo, Thin Solid Films 435 (2003) 78.
[143] M. Matsumura, R.P. Camata, Thin Solid Films 476 (2005) 317.
[144] J. Wienke, A.S. Booij, Thin Solid Films 516 (2008) 4508.
[145] C. Klingshirn, Physica Status Solidi B 71 (1975) 547.
[146] B.K. Meyer, H. Alves, D.M. Hofmann, W. Kriegseis, D. Forster, F. Bertram, et al., Physica Status Solidi B 241 (2004) 231.
[147] V. Srikant, D.R. Clarke, J. Appl. Phys. 81 (1997) 6357.
[148] D.G. Thomas, J.J. Hopfield, Phys. Rev. 128 (1962) 2135.
[149] M. Yoneta, K. Yoshino, M. Ohishi, H. Saito, Physica B 376 (2006) 745.
[150] W. Shan, W. Walukiewicz, J.W. Ager III, K.M. Yu, H.B. Yuan, H.P. Xin, et al., Appl. Phys. Lett. 86 (2005) 191911.
[151] D.C. Reynolds, D.C. Look, B. Jogai, C.W. Litton, G. Cantwell, W.C. Harsch, Phys. Rev. B 60 (1999) 2340.
[152] Y.S. Park, C.W. Litton, T.C. Collins, D.C. Reynolds, Phys. Rev. 143 (1966) 512.
[153] J.-H. Park, K.-J. Ahn, K. Park, S.-I. Na, H.-K. Kim, J. Phys. D Appl. Phys. 43 (2010) 115101.
[154] S.-Y. Kuo, W.-T. Lin, L.-B. Chang, M.-J. Jeng, Y.-T. Lu, S.-C. Hu, Mater. Res. Soc. Symp. Proc. 1201 (2010) 1, 1201-H05-32.
[155] S.-Y. Kuo, W.-C. Chen, F.-I. Lai, C.-P. Cheng, H.-C. Kuo, S.-C. Wang, et al., J. Crystal Growth 287 (2006) 78.

[156] S. Ishizuka, K. Sakurai, A. Yamada, K. Matsubara, P. Fons, K. Iwata, et al., Solar Energy Mater. Solar Cells 87 (2005) 541.
[157] P.S. Reddy, G.R. Chetty, S. Uthanna, B.S. Naidu, P.J. Reddy, Solid State Commun. 77 (1991) 899.
[158] B.-Y. Oh, M.-C. Jeong, D.-S. Kim, W. Lee, J.-M. Myoung, J. Crystal Growth 281 (2005) 475.
[159] T. Tsuji, M. Hirohashi, Appl. Surface Sci. 157 (2000) 47.
[160] S. Shirakata, T. Sakemi, K. Awai, T. Yamamoto, Thin Solid Films 445 (2003) 278.
[161] R.A. Asmar, G. Ferblantier, F. Mailly, P.G.- Borrut, A. Foucaran, Thin Solid Films 473 (2005) 49.
[162] S.A. Studenikin, M. Cocivera, W. Kellner, H. Pascher, J. Luminescence 91 (2000) 223.
[163] B.K. Meyer, J. Sann, S. Lautenschlager, M.R. Wagner, A. Hoffmann, Phys. Rev. B 76 (2007) 184120.
[164] A. Teke, U. Ozgur, S. Dogan, X. Gu, H. Morkoc, B. Nemeth, et al., Phys. Rev. B 70 (2004) 195207.
[165] C.P. Dietrich, M. Lange, G. Benndorf, H.V. Wenckstern, M. Grundmann, Solid State Commun. 150 (2010) 379.
[166] H.J. Ko, Y.F. Chen, S.K. Hong, H. Wenisch, T. Yao, D.C. Look, Appl. Phys. Lett. 77 (2000) 3761.
[167] L. Zhu, J. Li, Z. Ye, H. He, X. Chen, B. Zhao, Opt. Mater. 31 (2008) 237.
[168] D.C. Reynolds, T.C. Collins, Phys. Rev. 185 (1969) 1099.
[169] D.M. Bagnall, Y.F. Chen, M.Y. Shen, Z. Zhu, T. Goto, T. Yao, J. Crystal Growth 184–185 (1998) 605.
[170] A. Schildknecht, R. Sauer, K. Thonke, Physica B 340 (2003) 205.
[171] H. Alves, D. Pfisterer, A. Zeuner, T. Riemann, J. Christen, D.M. Hofmann, et al., Opt. Mater. 23 (2003) 33.
[172] D.C. Reynolds, D.C. Look, B. Jogai, C.W. Litton, T.C. Collins, W. Harsch, et al., Phys. Rev. B 57 (1998) 12151.
[173] M. Schirra, R. Schneider, A. Reiser, G.M. Prinz, M. Feneberg, J. Biskupek, et al., Physica B 401 (2007) 362.
[174] K. Thonke, Th. Gruber, N. Teofilov, R. Schonfelder, A. Waag, R. Sauer, Physica B 308 (2001) 945.
[175] X.H. Li, C.L. Shao, Y.C. Liu, X.T. Zhang, S.K. Hark, Mater. Lett. 62 (2008) 2088.
[176] Y.L. Liu, Y.C. Liu, W. Feng, J.Y. Zhang, Y.M. Lu, D.Z. Shen, et al., J. Chem. Phys. 122 (2005) 174703.
[177] C. Boemare, T. Monteiro, M.J. Soares, J.G. Guilherme, E. Alves, Physica B 308 (2001) 985.
[178] L. Wang, N.C. Giles, J. Appl. Phys. 94 (2003) 973.
[179] H. Tang, Z. Ye, H. He, Opt. Mater. 30 (2008) 1422.
[180] S.J. Chen, G.R. Wang, Y.C. Liu, J. Luminescence 129 (2009) 340.
[181] D.W. Hamby, D.A. Lucca, M.J. Klopfstein, G. Cantwell, J. Appl. Phys. 93 (2003) 3214.
[182] F.H. Su, W.J. Wang, K. Ding, G. Hua, Y. Liu, A.G. Joly, et al., J. Phys. Chem. Solids 67 (2006) 2376.
[183] S.S. Kurbanov, T.W. Kang, J. Luminescence 130 (2010) 767.
[184] W. Cao, W. Du, J. Luminescence 124 (2007) 260.
[185] Z.Y. Xiao, Y.C. Liu, D.X. Zhao, J.Y. Zhang, Y.M. Lu, D.Z. Shen, et al., J. Luminescence 122–123 (2007) 822.
[186] I.V. Markevich, V.I. Kushnirenko, Solid State Commun. 149 (2009) 866.

[187] S. Yamauchi, Y. Goto, T. Hariu, J. Crystal Growth 260 (2004) 1.
[188] C.J. Youn, T.S. Jeong, M.S. Han, J.H. Kim, J. Crystal Growth 261 (2004) 526.
[189] R.D. Vispute, V. Talyansky, S. Choopun, R.P. Sharma, T. Venkatesan, M. He, et al., Appl. Phys. Lett. 73 (1998) 348.
[190] Z. Fang, Y. Wang, D. Xu, Y. Tan, X. Liu, Opt. Mater. 26 (2004) 239.
[191] P. Klason, T.M. Borseth, Q.X. Zhao, B.G. Svensson, A.Y. Kuznetsov, P.J. Bergman, et al., Solid State Commun. 145 (2008) 321.
[192] Y.W. Heo, D.P. Norton, S.J. Pearton, J. Appl. Phys. 98 (2005) 073502.
[193] X. Yang, G. Du, X. Wang, J. Wang, B. Liu, Y. Zhang, et al., J. Crystal Growth 252 (2003) 275.
[194] T.-B. Hur, G.S. Jeen, Y.-H. Hwang, H.-K. Kim, J. Appl. Phys. 94 (2003) 5787.
[195] A. Allenic, X.Q. Pan, Y. Che, Z.D. Hu, B. Liu, Appl. Phys. Lett. 92 (2008) 022107.
[196] H.V. Wenckstern, G. Benndorf, S. Heitsch, J. Sann, M. Brandt, H. Schmidt, et al., Appl. Phys. A 88 (2007) 125.
[197] W.J. Lee, J. Kang, K.J. Chang, Phys. Rev. B 73 (2006) 024117.
[198] J.A. Sans, A. Segura, M. Mollar, B. Mari, Thin Solid Films 453–454 (2004) 251.
[199] K.-H. Bang, D.-K. Hwang, M.-C. Jeong, K.-S. Sohn, J.-M. Myoung, Solid State Commun. 126 (2003) 623.
[200] X.Q. Meng, D.Z. Shen, J.Y. Zhang, D.X. Zhao, Y.M. Lu, L. Dong, et al., Solid State Commun. 135 (2005) 179.
[201] B.S. Li, Y.C. Liu, Z.Z. Zhi, D.Z. Shen, Y.M. Lu, J.Y. Zhang, et al., Thin solid Films 414 (2002) 170.
[202] T. Makino, C.H. Chia, N.T. Tauan, Y. Segawa, N. Kawasaki, A. Ohtomo, et al., Appl. Phys. Let. 76 (2000) 3549.
[203] Q.X. Zhao, M. Willander, R.E. Morjan, Q.-H. Hu, E.E.B. Campbell, Appl. Phys. Lett. 83 (2003) 165.
[204] J.M. Calleja, M. Cardona, Phys. Rev. B 16 (1977) 3753.
[205] T.C. Damen, S.P.S. Porto, B. Tell, Phys. Rev. 142 (1966) 570.
[206] J. Zhao, X. Yan, Y. Yang, Y. Huang, Y. Zhang, Mater. Lett. 64 (2010) 569.
[207] M. Rajalakshmi, A.K. Arora, B.S. Bendre, S. Mahamuni, J. Appl. Phys. 87 (2000) 2445.
[208] K. McGuire, Z.W. Pan, Z.L. Wang, D. Milkie, J. Menendez, A.M. Rao, J. Nanosci. Nanotech. 2 (2002) 1.
[209] H. Tang, L. Zhu, Z. Ye, H. He, Y. Zhang, M. Zhi, et al., Mater. Lett. 61 (2007) 1170.
[210] Z. Peng, G. Dai, P. Chen, Q. Zhang, Q. Wan, B. Zou, Mater. Lett. 64 (2010) 898.
[211] H. Niu, Q. Yang, F. Yu, K. Tang, Y. Xie, Mater. Lett. 61 (2007) 137.
[212] X.D. Guo, R.X. Li, Y. Hang, Z.Z. Xu, B.K. Yu, H.L. Ma, et al., Mater. Lett. 61 (2007) 4583.
[213] A. Umar, S.H. Kim, J.H. Kim, Y.K. Park, Y.B. Hahn, Mater. Lett. 61 (2007) 4954.
[214] A. Umar, S.H. Kim, J.H. Kim, Y.B. Hahn, Mater. Lett. 62 (2008) 167.
[215] A. Khan, S.N. Khan, W.M. Jadwisienczak, M.E. Kodesch, Mater. Lett. 63 (2009) 2019.
[216] N. Ashkenov, B.N. Mbenkum, C. Bundesmann, V. Riede, M. Lorenz, D. Spemann, et al., J. Appl. Phys. 93 (2003) 126.
[217] Y. Tong, Y. Liu, C. Shao, Y. Liu, C. Xu, J. Zhang, et al., J. Phys. Chem. B 110 (2006) 14714.
[218] U. Pal, J.G. Serrano, P. Santiago, G. Xiong, K.B. Ucer, R.T. Williams, Opt. Mater. 29 (2006) 65.
[219] Y.F. Mei, G.G. Siu, K.Y.Fu. Ricky, P.K. Chu, Z.M. Li, Z.K. Tang, Appl. Suf. Sci. 252 (2006) 2973.

[220] J.-B. Lee, M.-H. Lee, H.-J. Lee, J.-S. Park, Mater. Res. Symp. Proc. 720 (2002) H313.
[221] G.J. Exarhos, S.K. Sharma, Thin Solid Films 270 (1995) 27.
[222] T. Kumpika, W. Thongsuwan, P. Singjai, Surf. Interface Anal. 39 (2007) 58.
[223] Y.-C. Lee, S.-Y. Hu, W. Water, Y.-S. Huang, M.-D. Yang, J.-L. Shen, et al., Solid State Commun. 143 (2007) 250.
[224] M. Yan, D.G.- Tong, Y.S.- Ren, Y.T.- Peng, Y.H.- Yun, Y.X.- Tian, et al., Chem. Phys. Lett. 20 (2003) 1155.
[225] X.L. Xu, S.P. Lau, J.S. Chen, G.Y. Chen, B.K. Tay, J. Crystal Growth 223 (2001) 201.
[226] K.A. Alim, V.A. Fonoberov, A.A. Balandin, Appl. Phys. Lett. 86 (2005) 053103.
[227] D. Behera, B.A. Acharya, J. Luminescence 128 (2008) 1577.
[228] D.G. Mead, G.R. Wilkinson, J. Raman Spectrosc. 6 (1977) 123.
[229] R.K. Sharma, G.J. Exarhos, Solid State Phenomena 55 (1997) 32.
[230] O. Lupan, L. Chow, S. Shishiyanu, E. Monaico, T. Shishiyanu, V. Sontea, et al., Mater. Res. Bull. 44 (2009) 63.
[231] M. Tzolov, N. Tzenov, D.D.- Malinovska, M. Kalitzova, C. Pizzuto, G. Vitali, et al., Thin Solid Films 379 (2000) 28.
[232] A. Kaschner, U. Haboeck, M. Strassburg, M. Strassburg, G. Kaczmarczyk, A. Haffmann, et al., Appl. Phys. Lett. 80 (2002) 1909.
[233] S. Eisermann, A. Kronenberger, M. Dietrich, S. Petznick, A. Laufer, A. Polity, et al., Thin Solid Films 518 (2009) 1099.
[234] L. Wu, Z. Gao, E. Zhang, H. Gao, H. Li, X. Zhang, J. Luminescence 130 (2010) 334.
[235] C. Bundesmann, N. Ashkenov, M. Schubert, D. Spemann, T. Butz, E.M. Kaidashev, et al., Appl. Phys. Lett. 83f (2003) 1974.
[236] S.K. Panda, C. Jacob, Bull. Mater. Sci. 32 (2009) 493.
[237] Y. Tong, L. Dong, Y. Liu, D. Zhao, J. Zhang, Y. Lu, et al., Mater. Lett. 61 (2007) 3578.
[238] J.F. Scott, Phys. Rev. B 2 (1970) 1209.
[239] V.V. Ursaki, O.I. Lupan, L. Chow, I.M. Tiginyanu, V.V. Zalamai, Solid State Commun. 143 (2007) 437.
[240] C.A. Arguello, D.L. Rousseau, S.P.S. Porto, Phys. Rev. 181 (1969) 1351.
[241] L. Fan, H. Song, L. Yu, Z. Liu, L. Yang, G. Pan, et al., Opt. Mater. 29 (2007) 532.
[242] G. Gordillo, C. Calderon, Solar Energy Mater. Solar Cells 69 (2001) 251.
[243] S.A. Studenikin, N. Golego, M. Cocivera, J. Appl. Phys. 87 (2000) 2413.
[244] M. Stamataki, I. Fasaki, G. Tsonos, D. Tsamakis, M. Kompitsas, Thin Solid Films 518 (2009) 1326.
[245] W.-J. Jeong, G.-C. Park, Solar Energy Mater. Solar Cells 65 (2001) 37.
[246] Y. Igasaki, H. Saito, J. Appl. Phys. 70 (1991) 3613.
[247] M.A. Martinez, J. Herrero, M.T. Gutierrez, Solar Energy Mater. Solar Cells 45 (1997) 75.
[248] M. Kumar, R.M. Mehra, A. Wakahara, M. Ishida, A. Yoshida, Thin Solid Films 484 (2005) 174.
[249] K. Matsubra, P. Fons, K. Iwata, A. Yamada, K. Sakurai, H. Tampo, et al., Thin Solid Films 431–432 (2003) 369.
[250] A.V. Singh, R.M. Mehra, A. Wakahara, A. Yoshida, 3rd World Conference on Photovoltaic Energy Conversion, (2003), 2PD370.
[251] P. Nunes, D. Costa, E. Fortunato, R. Martins, Vacuum 64 (2002) 293.
[252] C. Agashe, O. Kluth, G. Schope, H. Siekmann, J. Hupkes, B. Rech, Thin Solid Films 442 (2003) 167.

[253] J.N. Duenow, T.A. Gessert, D.M. Wood, A.C. Dillin, T.J. Coutts, J. Vac. Sci. Tech. 26 (2008) 692.
[254] J.C. Lee, K.H. Kang, S.K. Kim, K.H. Yoon, I.J. Park, J. Song, Solar Energy Mater. Solar Cells 64 (2000) 185.
[255] T. Nakada, N. Murakami, A. Kunioka, Mater. Res. Soc. Proc. 426 (1996) 411.
[256] H. Sato, T. Minami, T. Miyata, S. Takata, M. Ishii, Thin Solid Films 246 (1994) 65.
[257] T. Minami, H. Sato, H. Sonohara, S. Takata, T. Miyata, I. Fukuda, Thin Solid Films 253 (1994) 14.
[258] B. Sang, Y. Nagoya, K. Kushiya, O. Yamase, Solar Energy Mater. Solar Cells 75 (2003) 179.
[259] X. Yu, J. Ma, F. Ji, Y. Wang, X. Zhang, C. Cheng, et al., J. Crystal Growth 274 (2005) 474.
[260] T. Yamamoto, T. Sakemi, K. Awai, S. Shirakata, Thin Solid Films 451–452 (2003) 439.
[261] B.M. Ataev, A.M. Bagamadova, A.M. Djabrailov, V.V. Mamedov, R.A. Rabadanov, Thin Solid Films 260 (1995) 19.
[262] M. Berginski, J. Hupkes, W. Reetz, B. Rech, M. Wutting, Thin Solid Films 516 (2008) 5836.
[263] M. Ruth, J. Tuttle, J. Goral, R. Noufi, J. Crystal Growth 96 (1989) 363.
[264] S. Ghosh, A. Sarkar, S. Bhattacharya, S. Chaudhuri, A.K. Pal, J. Crystal Growth 108 (1991) 534.
[265] W. Lin, R. Ma, J. Xue, B. Kang, Solar Energy Mater. Solar Cells 91 (2007) 1902.
[266] D.C. Look, D.C. Reynolds, J.R. Sizelove, R.L. Jones, C.W. Litton, G. Cantwell, et al., Solid State Commun. 105 (1998) 399.
[267] D.C. Look, Mater. Sci. Eng. B 80 (2001) 383.
[268] R. Schifano, E.V. Monakhov, L. Vines, B.G. Svensson, W. Mtangi, F.D. Auret, J. Appl. Phys. 106 (2009) 043706.
[269] G. Giusti, L. Tian, I.P. Jones, J.S. Abell, J. Bowen, Thin Solid Films 518 (2009) 1140.
[270] S.R. Kodigala, V.S. Raja, A.K. Bhatnagar, R.D. Tomlinson, R.D. Pilkington, A.E. Hill, et al., Semicond. Sci. Technol. 15 (2000) 676.
[271] J. Herrero, C. Guillen, Thin Solid Films 451–452 (2004) 630.
[272] C. Guillen, J. Herrero, Thin Solid Films 480–481 (2005) 129.
[273] K. Maki, N. Komiya, A. Suzuki, Thin Solid Films 445 (2003) 224.
[274] C. Coutal, A. Azema, J.-C. Roustan, Thin Solid Films 288 (1996) 248.
[275] L.-J. Meng, E. Crossan, A. Voronov, F. Placido, Thin Solid Films 422 (2002) 80.
[276] V. Teixeira, H.N. Cui, L.J. Meng, E. Fortunato, R. Martins, Thin Solid Films 420 (2002) 70.
[277] C. Viespe, I. Nicolae, C. Sima, C. Grigoriu, R. Medianu, Thin Solid Films 515 (2009) 8771.
[278] H. Izumi, T. Ishihara, H. Yoshioka, M. Motoyama, Thin Solid Films 411 (2002) 32.
[279] J.C. Manfacier, L. Szepessy, J.F. Bresse, M. Perotin, R. Stuck, Mater. Res. Bull. 14 (1979) 163.
[280] J.C. Manfacier, L. Szepessy, Appl. Phys. Lett. 31 (1977) 459.
[281] J.C. Manfacier, J.P. Fillard, J.M. Bind, Thin Solid Films 77 (1981) 67.
[282] S.R. Kodigala, V.S. Raja, Appl. Surf. Sci. 253 (2006) 1451.
[283] A. Chitra, B.R. Marathe, M.G. Takewale, V.G. Bhide, Solar Energy Mater. 17 (1988) 99.
[284] U. Kaplan, A. Ben-Shalom, R.L. Boxman, S. Goldsmith, U. Rosenbery, M. Nathan, Thin Solid Films 253 (1994) 1.

[285] B. Thangaraju, Thin Solid Films 402 (2002) 71.
[286] C.-C. Lin, M.-C. Chiang, Y.-W. Chen, Thin Solid Films 518 (2009) 1241.
[287] S.R. Reddy, A.K. Mallik, S.R. Jawalekar, Thin Solid Films 143 (1986) 113.
[288] H. Bisht, H.-T. Eun, A. Mehrtens, M.A. Aegerter, Thin Solid Films 351 (1999) 109.
[289] M. Bender, W. Seelig, C. Daube, H. Frankenberger, B. Ocker, J. Stollenwerk, Thin Solid Films 326 (1998) 72.
[290] A.K. Kulakarni, T. Lim, M. Khan, K.H. Schulz, J. Vac. Technol. A 16 (1998) 1639.

8 Cu(In$_{1-x}$Ga$_x$)Se$_2$ and CuIn(Se$_{1-x}$S$_x$)$_2$ Thin Film Solar Cells

The quaternary Cu(In$_{1-x}$Ga$_x$)Se$_2$ and CuIn(Se$_{1-x}$S$_x$)$_2$ ($x=0$ to 1) compounds based thin film solar cells perish well with high efficiencies. Therefore, it is important to understand the basic Physics behind the thin film solar cells. In this chapter, on the first hand the fabrication and characterizations of CuInSe$_2$, CuGaSe$_2$, and their alloy based cells are well illustrated. On the other hand, the work on CuInS$_2$ and CuIn(Se$_{1-x}$S$_x$)$_2$ alloy based cells are also exploited. The role of each layer in the thin film solar cell is thoroughly discussed. The band diagrams of the heterojunction solar cells with different buffer layers and working functions are illustrated by citing several examples. The high efficiency is observed in only vacuum processed cells those are not cost-effective to the mankind. In order to reduce cost, the low cost techniques are essential to build thin film solar cells. Hence, the cells made by low cost fabrication techniques are given preference to discuss even though their efficiencies are low. The fabrication of modules, how the efficiency decreases with increasing module size and solutions for the problems are fruitfully demonstrated in this chapter.

8.1 Basic Theory of CuInSe$_2$ Based Solar Cells

The theoretical simulation of p-CuInSe$_2$/n-CdS/ZnO thin film solar cell demonstrates $V_{oc}=656$ mV, $J_{sc}=45.2$ mA/cm^2, FF$=0.837$, and $\eta=24.8$–28.1% for which physical parameters of CuInSe$_2$ (CIS) such as $E_g=1.02$ eV, $N_A=5\times10^{17}$ cm^{-3}, $\tau_{rad}=10^{-7}$ s, and $\mu_n=400$ cm^2/s and ZnO parameters such as $E_g=3.3$ eV and $N_D=2\times10^{18}$ cm^{-3} are employed [1,2]. The theoretical efficiency is always higher than the experimental efficiency of 14.5–15.4% in CIS cells [3,4]. In order to improve the experimental efficiency of the cells, the physical parameters of the participating layers in the thin film solar cells have to be tailored. For example, the role of p-CIS layer is very critical in the conventional CIS thin film solar cells. The Cu/In ratio, carrier concentration, resistivity, thickness of CIS absorber, phases, *etc.*, play important roles. The Cu/In ratio and carrier concentration of CIS layers should be in the range of 0.8–0.9, that is, nonstoichiometric and 10^{16}–10^{17} cm^{-3}, respectively [5]. The best devices can be obtained with either discrete or mixed phases of OVC or In-rich CIS and CIS [6]. It is an essential to have the depletion layer in the CIS to enhance collection of minority carriers, therefore, the resistivity of CIS to be maintained around neighborhood of 100 Ω-cm. According to Beer's law, the thickness of the absorber should be sufficient to absorb most of the photons from the solar spectrum, when they impinge on the solar cells. Ninety percent of photons can be absorbed by the thickness of >1 μm, therefore 1.5–2.5 μm thick absorber is enough [7].

The role of window layer is also critical in the thin film solar cells. The typical resistivity of n-CdS window layer is considered to be 10^{-3} Ω-cm because the Fermi level should be close to the conduction band [8]. The thickness and carrier concentration of window layer should also be optimized otherwise the barrier height turns into high for higher thickness that causes to obstruct electron flow from absorber to window, hence the short circuit current decreases. The open circuit voltage and fill factor decrease for lower thickness of window layer, therefore the typical optimized thickness of CdS is ~80 nm [9]. The junction barrier height is high for a typical lower carrier concentration (p) of 10^{15} cm^{-3} for absorber, which acts as a second junction to block carrier transportation, whereas no such kind of higher barrier junction is observed for (p) 5×10^{16} cm^{-3} in the CIS thin film solar cells. The band diagram of p-CIS/n-CdS shifts to down with increasing carrier concentration from 10^{15} to 5×10^{16} cm^{-3}, as shown in Fig. 8.1A. It is learnt that there is a kink for lower carrier concentration of 10^{15} cm^{-3} in the calculated I–V plots (Fig. 8.1B) [10]. The positive value of conduction band offset or discontinuity (ΔE_C) creates a spike, which obstructs the flow of photogenerated carriers from the absorber to the buffer layer. According to the simulation work [11,12], $0 \leq \Delta E_C \leq 0.4$ is acceptable to obtain the regionable efficiency from the solar cells. If $\Delta E_C < 0$ cliff arises in the vicinity of band structure, as shown in Fig. 8.2. The cliff indeed enhances interface recombinations which reduce open circuit voltage in the solar cells.

The XPS scans are done on Mo/p-CIS/n-CuIn$_3$Se$_5$(OVC), glass/CdS and n-OVC/CdS in the range of -4 to 22 eV, as shown in Fig. 8.3A and similar scans are also recorded on Mo/p-CIS, glass/CdS, and p-CIS/CdS stacked layers, as shown in Fig. 8.3B to obtain band offsets. The valence band offset $\Delta E_V = \Delta E_{V\text{-In4d}} - \Delta E_{V\text{-Cd4d}} - \Delta E_{CL}$, where $\Delta E_{V\text{-In4d}}$ ($18 - 0.5 = 17.5$ eV) and $\Delta E_{V\text{-Cd4d}}$ ($11.7 - 1.4 = 10.3$ eV) are binding energies of In4d and Cd4d in n-OVC and CdS from the valence band

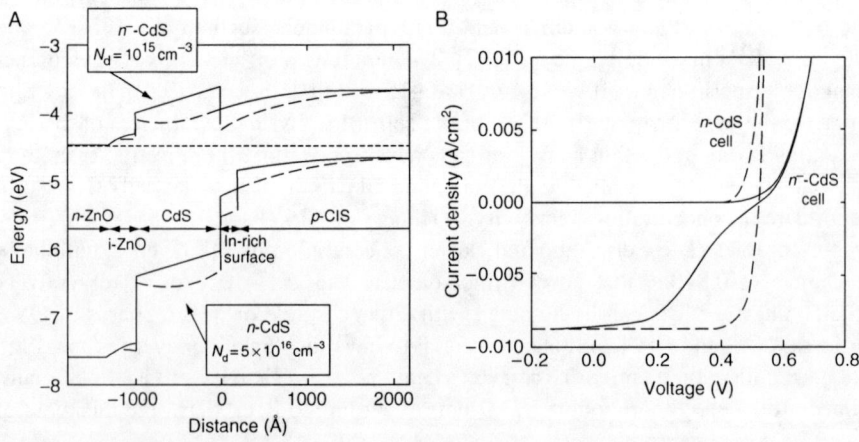

Figure 8.1 (A) Simulated band diagram of CIS/CdS/ZnO heterojunction and (B) I–V curves.

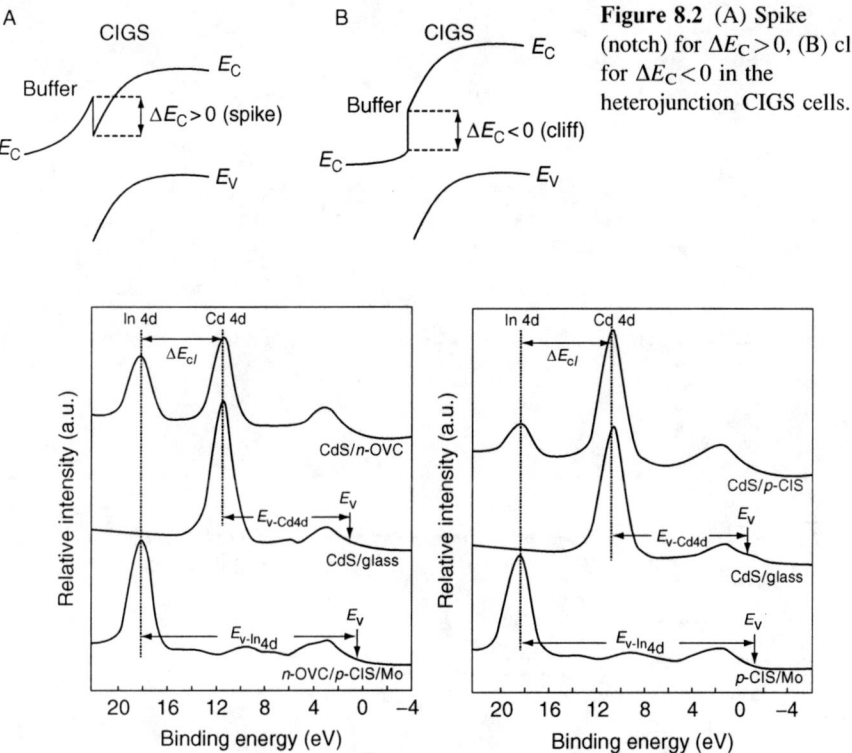

Figure 8.2 (A) Spike (notch) for $\Delta E_C > 0$, (B) cliff for $\Delta E_C < 0$ in the heterojunction CIGS cells.

Figure 8.3 XPS scans of (A) Mo/p-CIS/n-OVC, glass/CdS, n-OVC/CdS, and (B) Mo/p-CIS, glass/CdS, p-CIS/CdS junctions.

maximum, respectively, and ΔE_{CL} is the difference between In4d (n-OVC) and Cd4d (CdS), that is $18 - 11.7 = 8.3$ eV. In other words, $\Delta E_{CL} = E^{(OVC)CuIn_3Se_5/CdS}(Cd4d) - E^{(OVC)CuIn_3Se_5/CdS}(In4d)$ is the difference between the kinetic energies of the respective core level electrons measured for different thicknesses of CdS layer on the (OVC) CuIn$_3$Se$_5$. Finally, the valence band and conduction band offsets of CuIn$_3$Se$_5$/CdS are determined to be $\Delta E_V = 17.5 - 10.3 - 8.3 = 0.9$ eV and $\Delta E_C = E_{g2} - E_{g1} - \Delta E_V = 2.46 - 1.23 - 0.9 = 0.33$ eV, respectively. In the case of p-CIS/n-CdS heterojunction, the ΔE_V is 0.4 eV and $\Delta E_C = 2.46 - 1.04 - 0.4 = 1.02$ eV. The band line up of CuInSe$_2$/CdS and CuIn$_3$Se$_5$/CdS can be constructed using experimental results of XPS, as shown in Fig. 8.4 [13]. The band diagram can also be prepared using standard physical parameters of CuInSe$_2$ and CdS, which are available from Table 1.3. The conduction band offset ($\Delta E_c = \chi_2 - \chi_1$) and valence band offsets $\Delta E_v = (E_{g2} - E_{g1}) - \Delta E_c$ are depicted in Fig. 8.5, the suffixes 1 and 2 represent for window and absorber, respectively, and final form of band diagram for p-CIS/n-CdS/ZnO is also given in Fig. 8.6 [14].

The current–voltage (J–V) relation of solar cell under light illumination can be written in the form of expression by combination of forward diode current and light generated current (J_L) in the opposite direction, as mentioned in chapter 1[15–17].

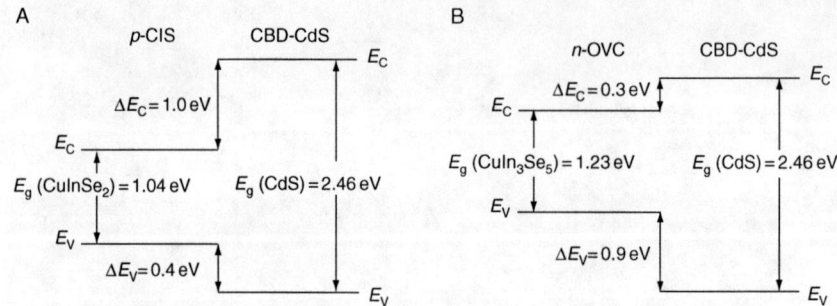

Figure 8.4 Band diagram of (A) p-CuInSe$_2$/CBD-CdS, and (B) n-OVC/CBD-CdS.

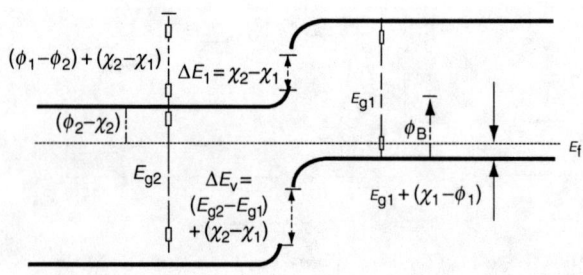

Figure 8.5 Band diagram of CuInSe$_2$/CdS.

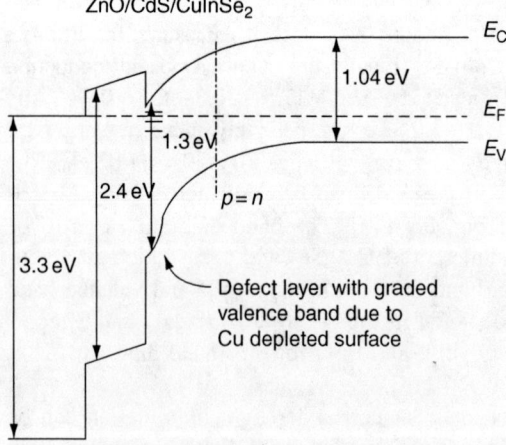

Figure 8.6 Band diagram of CuInSe$_2$/CdS/ZnO qualitatively.

$$J = J_o \left[\exp \frac{q}{Ak_BT}(V - JR_s) - 1 \right] + \frac{V - R_s J}{R_{sh}} - J_L \tag{8.1}$$

where

$$J_o = J_{oo} \exp\left(\frac{-E_a}{Ak_BT}\right) \tag{8.2}$$

The current density (J) in the forward direction for p–n junction

$$J = J_o \exp\left(\frac{qV}{Ak_BT}\right) = J_{oo} \exp\left(\frac{-E_a}{Ak_BT}\right) \exp\left(\frac{qV}{Ak_BT}\right) \tag{8.3}$$

The open circuit voltage (V_{oc}) can be arrived from Equations (8.1) and (8.2) by considering $J = 0$ and $J_L = J_{sc}$

$$V_{oc} = \frac{E_a}{q} + \frac{Ak_BT}{q} \ln\left(\frac{J_{sc}}{J_{oo}}\right) \tag{8.4}$$

A plot of V_{oc} versus temperature (T) for solar cell results in a straight line, by extending it to $T = 0$ K inevitably contributes the activation energy (E_a). In this case, A, J_{sc}, and J_{oo} in Equation (8.4) are considered to be independent of temperature. In the tunneling process by introducing diode quality factor into saturation current exponent in Equation (8.3), the expression can be derived as [18]

$$A \ln(J_o) = A \ln(J_{oo}) - \frac{E_a}{k_BT} \tag{8.5}$$

The activation energy E_a can be obtained from a plot of $A \ln(J_o)$ versus inverse temperature ($1/T$), where R_s is the series resistance, R_{sh} the shunt resistance, A the diode quality factor, k_B the Boltzmann constant, q the electron charge, J_o the saturation current density, J_{oo} the weakly temperature-dependent prefactor, T the temperature, and $R_s = 0$ and $R_{sh} = \infty$ are for ideal solar cell. In order to arrive at V_{oc} versus T plots, the J–V measurements are done under light illumination at different temperatures varying from 393 to 232 K in steps of 40 K, as shown in Fig. 8.7A. The activation energy (E_a or ϕ) of 1.0 eV is obtained from V_{oc} versus T plot (Fig. 8.7B), in some cases which could be close to either band gap of CIS or less. Figure 8.7C yields same activation energy of 0.99 eV from slopes of J_o versus $1/T$ plots [19]. The diode quality factor (A) of 2 is deduced from the J–V measurements. The diode quality factor and activation energy evidence that the Shockley–Read–Hall (SRH) recombination is the dominant one rather than interface states recombination in the space charge region via mid gap states in the CIS cells, since $A < 2$ and E_a is close to band gap of CIS. On the other hand, (i) if the recombination dictates current in the space charge region through the midgap states, the saturation current $(J_o) \alpha \exp(-E_g/2k_BT)$, and (ii) if (diode factor) $A = 1$,

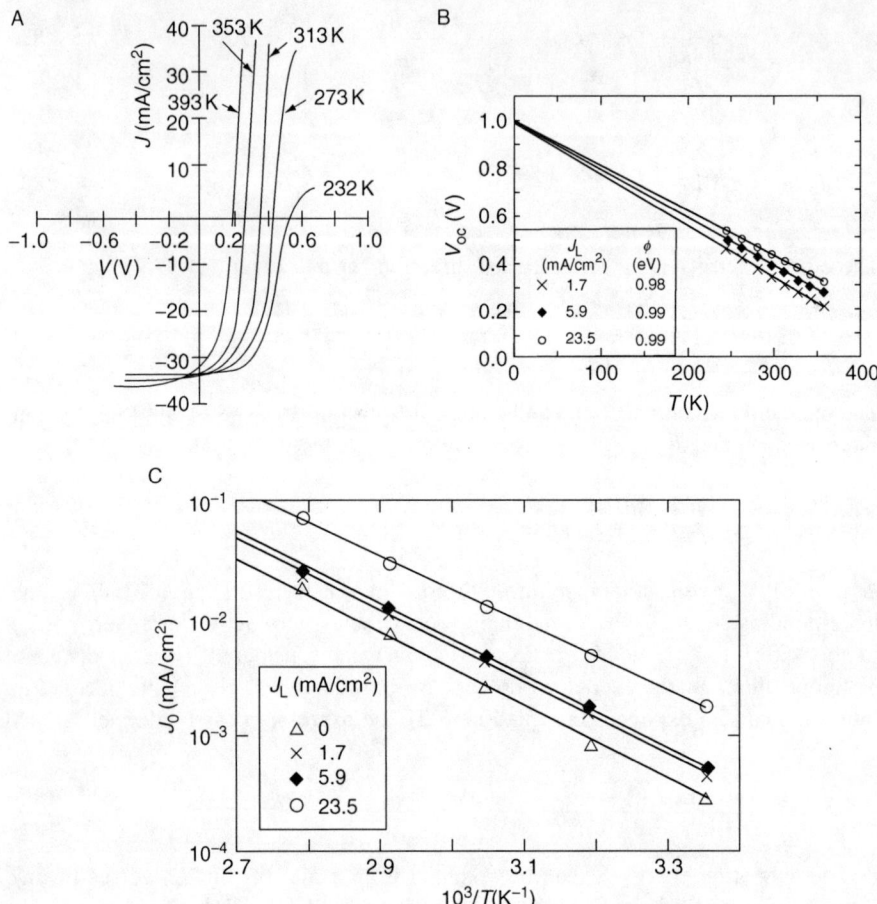

Figure 8.7 (A) *I–V* curves at different temperatures under illumination, and (B) V_{oc} versus temperature at different light illuminations, and (C) Plot of J_o versus $1/T$ at different illuminations for CIS cell.

the recombination takes place in the quasi-neutral region of the absorber, the saturation current $(J_o) \alpha \exp(-E_g/k_BT)$. In third case, the above both recombinations participate ($1 < A < 2$), the thermally activated saturation current $(J_o) \alpha \exp(-\Delta E_g/k_BT)$, where ΔE_g lies in the range of $E_g/2 < \Delta E < E_g$. Therefore, the recombination may be treated thermal recombination [20]. In the SLG/Mo/CIS/CdS/ZnO cell, the dark *I–V* curves are recorded at different temperatures varying from 303 to 101 K, as shown in Fig. 8.8 [21]. The curve nature deviates from its originality with varying temperature that is blocking behavior or roll over due to influence of second diode. The degree of deviation mainly depends on the participation of an increased acceptor like states in the CIS/CdS junction, which decreases fill factor (FF) and slightly increases the open circuit voltage (V_{oc}). In this case, the role of donor-

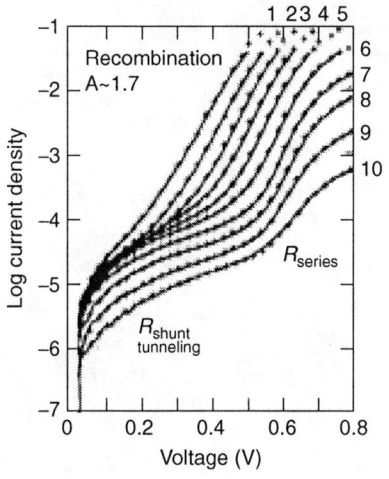

Figure 8.8 *I–V* curves of 12.4% efficiency CIS cells recorded under dark at different temperatures.

like-states is negligible because the Fermi level persists near to the conduction band that causes to neutralize donor-like-states. The tunneling mechanism dominates and the series resistance increases with decreasing temperature [22]. The expression for J_{oo} can be written as

$$J_{oo} = \frac{q}{2} w \sigma V_{th} N_t (N_C N_V)^{1/2} \qquad (8.6)$$

where w is the depletion width (0.1–1.0 μm), σ the electron capture cross section (10^{-14} cm^{-2}), V_{th} the thermal velocity (10^7 cm/s), N_c and N_v the effective density of states (10^{19} cm^{-3}) in the conduction and valence bands, respectively, and N_t is the midgap density of states (10^{14} cm^{-3}). At $T \geq 333$ K, $J_{oo} \sim 4 \times 10^6$ A/cm^2, $J_L = 32.5$ mA/cm^2, and $R_s = 0.1$ Ω-cm^2 for a typical CIS/CdS junction. J_o is not a constant but varies with temperature [16].

8.1.1 CIS Based Single Crystal Solar Cells

Shay *et al.*, [23] developed single crystal based solar cells at the Bell laboratories in 1970s. The *p*-CuInSe$_2$ single crystals grown by melt–grown technique with carrier concentration of 10^{16} cm^{-3} are annealed under Se atmosphere at 600 °C for 24 h. They are polished and etched in aqua regia then on which 5–10 μm thick CdS layer is deposited by evaporation. Indium grids and Au contact on backside of CIS crystal are developed. The Au/*p*-CIS/*n*-CdS/In/SiO$_2$ cell with an area of 0.79 mm^2 exhibits efficiency of 12% under clear sunny day of 92 mW/cm^2. Prior to prepare solar cells, the polished *p*-CuInSe$_2$ single crystals are processed by annealing under Ar at 350 °C. 0.8 mm^2 active area solar cells with processed CIS substrates contribute efficiency of 10.3%, whereas cells without processed substrates exhibit lower efficiency of 8% (Table 8.1). Instead of evaporated CdS, a 50 nm thick CdS by

Table 8.1 Photovoltaic (PV) parameters of single crystal and thin film solar cells

Single crystal (SC)/Cu:In:Se	Cu/In	V_{oc} (mV)	J_{sc} (mA/cm^2)	FF (%)	η (%)	A	J_o (μA/cm^2)	R_s (Ω-cm^2)	R_{sh} (kΩ-cm^2)	Ref.
SC	–	500	~39	60	12	–	–	–	–	[23]
SC	–	470	32.8	67	10.3	1.87	3.75	–	–	[25]
SC	–	500	43	58	12.5	–	–	–	–	[26]
SC (112)	–	410	44.7	50	8.5	–	–	–	–	[27]
SC (101)	–	370	46.8	40	8.3	–	–	–	–	
		340	31	54	5.7	2	–	–	–	[29]
		385	33.77	62.06	8.08	1.54	–	1	0.43	[30]
		395	34.5	65.3	8.9	–	–	3.59	8.06	[31]
23.9:25.4:50.7	0.94	370	24.8	60.1	5.5	–	–	–	–	[32]
24.8:26.1:49.4	0.95	280	35.8	58.7	6	–	–	–	–	
24.6:24.1:51.3	1.02	310	36	47	5.2	–	–	–	–	
24.3:24.5:51.2	0.99	378	35.8	61.6	9.3	–	–	–	–	[33]
23.6:24.7:51.7	0.96	412	34.8	66.5	10.7	–	–	–	–	
	1 or 0.9	500	–	60	9.5	–	–	–	–	[34]
23.5:24.5:52	0.96	513	40.4	71.6	14.8	–	–	–	–	[35]
	0.85	491	41.1	71.9	14.5	–	–	–	–	[4]
23.7:26.9:49.4	0.88	348	32.9	55	7.2	–	–	–	–	[36]
22.3:27.7:50	–	–	–	3.89	–	–	–	–	–	
As-grown	–	326	26.7	60.4	6.1	–	3	–	–	[38]
185 °C[a]	–	413	30	66	8.2	–	0.8	–	–	
225 °C[a]	–	315	14.4	35.2	1.5	–	1000	–	–	
200 °C[a]	–	370	32.7	52	8.3	–	–	–	–	
As-grown	–	385.6	32.48	–	7.72	0.7	1.4	3.33	28	[39]
Illuminated	–	379	31.5	–	7.25	1.0	1.4	2.4	22	[40]
As-grown	–	352	31.9	58	6.5	–	–	–	–	[5]
Reduction	–	152	28.7	41	1.8	–	–	–	–	
Oxidation	–	424	38.4	65	10	–	–	–	–	

[a] Air annealed.

chemical bath deposition (CBD) is covered onto 1–2 mm thick p-CuInSe$_2$ single crystals grown by Bridgman technique. The CIS cell with a ~80 nm thick CdS grown by CBD gives better short circuit current, which decreases with increasing thickness because of an increase of absorption in the CdS [9]. Using simple expression, the doping concentration and depletion width can be obtained from the plots of $1/C^2$ versus V.

$$\frac{1}{C^2} = \frac{2}{\varepsilon q N_A}(V_{bi} - V) \qquad (8.7a)$$

the slope

$$\frac{d(C^{-2})}{dV} = -\frac{2}{\varepsilon q N_A} \qquad (8.7b)$$

where N_A is the acceptor density, the depletion width [24]

$$w = \frac{\varepsilon}{C} = \varepsilon\sqrt{C^{-2}} \qquad (8.7c)$$

In another version, A 1–2 μm thick In$_2$O$_3$ doped ZnO layer is RF sputtered onto CuInSe$_2$/CdS stack. Au on the backside of CuInSe$_2$ and Al grids on the front side of ZnO layer are deposited for ohmic contacts. The $1/C^2$ versus V graphs are straight lines for the Au/CIS/CdS/ZnO:In$_2$O$_3$/Al cells with processed substrates, which have carrier concentration (p) of 1.8×10^{17} cm^{-3}, whereas solar cells without processed substrates those visualize as curve nature, indicating higher concentration of deep defect levels [25]. Prior to anneal under Ar at 350 °C for 2 h, the p-CuInSe$_2$ single crystals are cleaned in bromine and methanol for 1 min. Au and In grids are formed on the backside of p-CuInSe$_2$ single crystals and on the front side of ZnO layer, respectively. The Au/CuInSe$_2$/CdS/ZnO:In/In solar cell with an active area of 0.126 cm^2 and an aperture area of 0.136 cm^2 demonstrates the highest efficiencies of 12.5% (Table 8.1). Surprisingly, the cell efficiency decreases with passing days or weeks. The diffusion length (L_n) of 0.3 μm for minority carrier can be obtained from the relation $L_n = -\varepsilon_0\varepsilon_r A_s/C_i$, where ε_0 and ε_r are permittivity of vacuum and dielectric constant of CuInSe$_2$, respectively, and A_a is an area of the cell. The intercept C_l is deduced from the plots $\Delta I = I_{ill} - I_{dark}$ versus $1/C_p$ known as photocurrent–capacitance method. The sample shows series resistance of 11 Ω-cm^2 under dark conditions. In general, the C–V measurements are carried out at intermediate frequencies (~10 kHz) to calculate carrier concentration and other parameters of solar cells amid the noise signal ratio is high at lower frequency (Hz), whereas the recognition of change in capacitance with bias is difficult at higher frequency (MHz). The acceptor concentration (N_A) of 7×10^{16} cm^{-3} at bias voltage of -1.5 V for the cell is calculated from $1/C^2$ versus V plot (Fig. 8.9) [26]. Instead of standard TCO, such as ZnO, the CIS single crystal/CdS (CBD) solar cell with CdO as a conducting layer demonstrates an efficiency of 8.5%, as shown in Table 8.1. 0.1 μm thick CdS by CBD, followed by 0.7 μm thick CdO by reactive sputtering technique

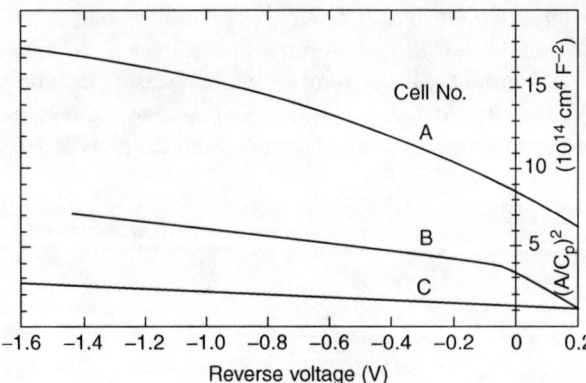

Figure 8.9 $1/C^2$ versus V for 6 (A), 10.7 (B), and 12.5% efficiency (C) cells. They have carrier concentrations of 5.6×10^{16}, 13×10^{16}, and 7×10^{16} cm^{-3} and diffusion lengths (L_n) of 0.4, 0.2, and 0.3 µm, respectively.

are deposited onto 2–3 mm thick CIS single crystals. Finally Au contact layer and grids on the backside of the crystal as well as on the front side of CdO layer are developed, respectively. The Au/CIS/CdS(CBD)/CdO/Au cells with (112) crystallographic plane CIS single crystals show higher efficiency than that of (101) based CIS cells. It may be related to differences in the resistivities and density of defect levels [27].

8.1.2 CIS Based Thin Film Solar Cells

Kazmerski *et al.*, [28] first fabricated glass/Mo(0.35 µm)/Au(0.06–0.1 µm)/5–6 µm CIS/6 µmCdS/Al thin film solar cell employing CIS absorber with conversion efficiency (η) up to 6% in which CuInSe$_2$ by two source (CuInSe$_2$ + Se) and CdS by evaporation are prepared under vacuum at substrate temperatures of 250 and 225 °C in the same chamber, respectively. The Boeing group [29] vigorously worked to develop efficient cells by employing bilayer process. The Cu and In elements are independently evaporated from separate Ta crucibles but they are mixed at one point before reaching to substrate and Se is evaporated from separate boat. In order to avoid Se reaction with Au, the CuInSe$_2$ layer deposited at low substrate temperature of 350 °C has resistivity of 500 Ω-cm and grain sizes of <1 µm and second CuInSe$_2$ layer with 100 Ω-cm is deposited at 450 °C, which has grain sizes of 1 µm then the CuInSe$_2$ sample is cooled down to 150 °C to deposit CdS layer. Indium evaporation is introduced after half of the deposition of 2–4 µm thick CdS layer to dope 1.5%In, which shows resistivity of 10 Ω-cm. Finally, Al grids are developed to made full pledged cell structure. Near the junction, the growth of CIS and CdS layers is controlled in such a way to have resistivities of 1000 and 20 Ω-cm, respectively. After annealing CIS cell under vacuum at pressure of 10^{-2} torr and temperature of 100 °C for 5 min, the efficiency of cell increases from 3.4 to 4.8%. The efficiency enhances to 5.7% for further annealing the sample at 90 °C for 5 min. The acceptor density near the junction decreases from 1.2×10^{15} to 5.4×10^{14} cm^{-3}. Similarly, first 1.5–2.0 µm thick CIS layers at 350, followed by layers at 425 °C are deposited by co-evaporation onto 1.5–2 µm thick Mo metalized glass or alumina substrates as bilayer configuration. A 0.5 μm thick CdS layer with high resistivity followed by 1.5 μm thick 1%In doped CdS layer with sheet resistance of 10–20 Ω/□ in the second phase are evaporated onto CIS by vacuum evaporation from Knudsen cell at 150 °C. Finally the In or In–Ag grids are covered onto glass/Mo/CIS/CdS/CdS:In. After annealing at 150 °C for 10 min or to 1 h, the glass/Mo/CIS/CdS/CdS:In/In cell exhibits efficiency of 8% [30].

Alike the CIS bilayer consists that first 2.5 μm thick Cu-rich CIS layer at 350 °C for 40 min and second 0.8 μm thick CIS layer at 450 °C for 17 min are deposited onto glass/Mo. A CdS bilayer consists that 0.8 μm thick CdS with resistivity of 10^2–10^3 Ω-cm, followed by 2.5 μm CdS thick layer with resistivity of 10^{-3} Ω-cm at growth rate of 150 Å/s is deposited onto glass/Mo/CIS by vacuum evaporation at substrate temperature of 200 °C. The as-grown glass/Mo/CIS/CdS/Al cells do not show any sign of photovoltaic response but after annealing those under air at 200 °C for 20–30 min exhibit efficiency of 8.9%. The CIS layer processed by bilayer method can be distinguished into three regions as bottom, overlap region and top regions, which compose compositions of Cu:In:Se = 19.1:28.5:52.4, 29.3:23.7:47, and 38.3:17.8:43.9, respectively [31]. A typical 1–2 μm thick In doped CdS layer is deposited onto 3–5 μm thick CuInSe$_2$ composite or bilayer by co-evaporation of CdS and In materials at substrate temperature of 200 °C, which has resistivity of 10^{-2}–10^{-3} Ω-cm, followed by deposition of 250 nm thick ITO layer with resistivity of 10^{-3} Ω-cm. Finally 0.5–1 μm thick Ni grids are coated. The Alumina/Mo/CIS/CdS(Co-Ev.)/ITO/Ni cells show low V_{oc} due to higher In content at the junction. The efficiency of cells varies in between 5 and 6% for the variation of Cu, In, and Se compositions of the absorber layer in the narrow window range. In particular, the short circuit current increases and open circuit voltage decreases with changing Cu to In ratio from 0.94 to 1.02, as shown in Table 8.1 [32]. Instead of feedback controlled co-evaporation of Cu, In, and Se, Knudsen type source cells are used to deposit CuInSe$_2$ thin films. Two micrometers thick CuInSe$_2$ thin film with Cu/In ratio of 1 is deposited at substrate temperature of 350 °C on which 1 μm thick CIS second layer with Cu/In = 0.6 at substrate temperature of 450 °C is grown. The Se flux is maintained three times higher than combined metal flux of Cu and In during the deposition. Indium doped CdS layer with resistivity of 10^{-3} Ω-cm is deposited onto CuInSe$_2$ bilayer. The entire grown stack is annealed under air at 200 °C for 48 h. If the CdS bilayer is used, the annealing time should be shortening to 1 h. The fabricated typical 12 cells lead as array and each cell contains an area of 3×3 mm^2. The CIS/CdS/CdS:In cell shows efficiency of 10% for the CIS composition of Cu-23-27% and In-25-28%. The quantitative analysis reveals difference in efficiency with effect of Cu composition, as shown in Table 8.1 [33].

The CIS cells made with CIS absorber, which contains Cu/In = 1.0 or 0.9, show efficiency of 9.5%. The cells with Cu/In < 0.7 have low open circuit voltage, short circuit current and FF because the p–n junction is not abruptly formed to efficiently separate electron–hole pairs. The higher sheet resistance causes lower FF [34]. Akin the CuInSe$_2$ absorbers with composition of Cu:In:Se = 23.5:24.5:52 are evaporated by maintaining Se evaporation of three times higher than the metals composition. In order to avoid Se depletion on the surface of CIS layer, the Se flux has

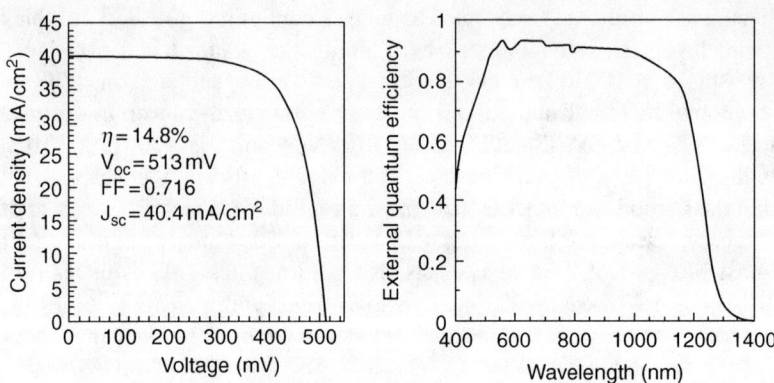

Figure 8.10 *I–V* curve, and quantum efficiency of CIS/CdS thin film solar cell.

been continued while cooling the sample until to at the end of the deposition. 0.8 mmSLG/1.5 μmMo/2.5 μmCuInSe$_2$/10 nmCdS/0.5 μmZnO/0.12 μmMgF$_2$ cell efficiency increases from 14.1 to 14.8% by replacing evaporated CdS with CBD-CdS. Figure 8.10 shows *I–V* curve and quantum efficiency (QE) of CIS cells [35]. The QE can be defined as

$$\mathrm{QE} = (1 - R)\exp(\alpha_{\mathrm{ZnO}} t_{\mathrm{ZnO}})\exp(-\alpha_{\mathrm{CdS}} t_{\mathrm{CdS}}) \frac{1 - \exp(-\alpha_{\mathrm{CIS}} w_{\mathrm{CIS}})}{1 + \alpha_{\mathrm{CIS}} L_{\mathrm{CIS}}} \tag{8.8}$$

where R is the front reflection loss, α the absorption coefficient function of wavelength, t thickness, w the depletion width (0.4 μm), and L the minority carrier diffusion length (0.5 μm). The V_{oc} versus composition ratio of Cu/(In + Ga) shows platue region in the composition range of 0.9–1.0, whereas below or beyond this range, the V_{oc} decreases. The platue region (0.85–1.0) of short circuit current (J_{sc}) versus composition ratio plot is wider than that of V_{oc} versus composition plot [21].

As the In composition is slightly increased, the efficiency of glass/Mo/CISbilayer/1.8 μmCd$_{0.93}$Zn$_{0.07}$S:In/0.19 μmITO/Ni cell with an active area of 0.08 cm^2 decreases from 7.2 to 3.9%. In the 7.2% efficiency cells, the CIS with 3 μm thick Cu-rich and 0.6 μm thick In-rich bilayer has global composition of Cu:In:Se = 23.7:26.9:49.4, whereas the CIS with 3 μm thick Cu-rich and 0.9 μm thick In-rich bilayer has total composition of Cu:In:Se = 22.3:27.7:50 in 3.8% efficiency cells. The similar results are reported on 10% efficiency CIS cells, which contains total composition of Cu:In:Se = 23.5:27:49.5 in the CIS bilayer. In the 3.5 μm bilayer, 2.5 μm thick bottom, and 0.8 μm thick top CIS layers consist composition of Cu:In:Se = 25.4:24.9:49.7 and Cu:In:Se = 20.5:30.2:49.3, respectively [5]. The cells made with either extreme Cu-rich or In-rich CIS layers are shorted. The Cu-rich (Cu:In:Se = 26:22:52) and In-rich CIS layers (Cu:In:Se = 19:29:52) show band gaps of 0.95 and 1.07 eV, respectively. The Cu-rich or stoichiometry, In-rich CIS single, and CIS bilayers show resistivities of 1000, 0.04, and 100 Ω-cm, respectively. After annealing CIS bilayer under air at 200 °C for 8 h, the band gap changes from 0.95 to 0.93 eV. The efficiency of cell increases from 2.4 to 7.2%, after annealing as-grown cell under air at 200 °C for 12 h. Anneal the cells under air causes to reduce the defect states that oxygen occupies isovalent Se. On the other hand, the oxygen also occupies grain boundaries that pursue to reduce defect density [36]. The oxidation increases acceptor concentration (N_A) causing change in the space charge region width or electric field width close to CdS. The oxidation decreases density of recombination centers and interface states, which are sources for charge carriers to recombine with minority carriers [5]. The oxygen diffuses into CIS through window layer in air annealed CIS cells. On the other hand, the oxygen also reacts with CdS to form $CdSO_4$ [37]. The Al_2O_3/Mo/low ρ (10^{-1} Ω-cm) CIS/high ρ (10^3–10^5 Ω-cm) CIS/0.5 μm high ρ (10^1 Ω-cm) CdS/low ρ (10^{-3} Ω-cm) CdS:In/Al grid cell is formed by depositing CdS bilayer onto CIS bilayer. The cells annealed up to 150 °C show a small increase in spectral response uniformly but the response sharply decreases for exceeding annealing temperature of 185 °C. Similarly, the efficiency of cell increases from 6.1 to 8.2% after annealing the cell at 185 °C thereafter decreases to 1.5% for further annealing at 225 °C. An increase in efficiency is probably due to a change in electrical properties of CIS bilayer and CdS/CIS junction. Degradation starts in high temperature annealed cells that could be due to interdiffusion of S and Se. On the other hand, In diffuses into junction below the detection limits of AES. After annealing the CIS bilayer at 200 °C, the carrier concentration enhances from 10^{14} to 10^{15} cm^{-3} [38]. The (112) oriented CIS films with Cu/In = 0.7 are deposited onto Mo coated glass substrates at 450 °C, followed by deposition of 0.5 μm thick CdS films by vacuum evaporation at 200 °C and 2 μm thick ZnO:Al layer with 4 Ω/□ by DC sputtering. After air annealing at 200 °C for 5 h, the best glass/Mo/CIS/CdS(Ev·)/ZnO:Al cells without antireflection coating yield efficiency of 8.3% [39]. The efficiency of glass/0.3 μmMo/2.0 μm low p-CuInSe$_2$/0.8 μm high p-CuInSe$_2$/0.8 μm high n-CdS/low n-CdS:In3%/grids/SiO$_2$ AR-coating cell under light illumination of 100 mW/cm^2 decreases from 7.7 to 7.2% with increasing illumination time from 0 to 31 h [40].

The glass/Mo/CIS/50 nmCdS/50 nmZnO/350 nmZnO:Al/50 nmNi/300 nmAl cell with active area of 0.441 cm^2 shows the highest reported efficiency of 14.5% (Table 8.1) in which the structure of CIS is of course grown by three-stage process. It is obvious from the SEM analysis that the surface of CIS film used in the high efficiency cell contains structure like steps and ridges. The CIS layer grown by three-stage process has less Cu content on the surface of the layer. Secondly, the grain sizes are small from the surface to depth of 0.7 μm. The XRD analysis reveals that the intensity ratio of (220,204)/(112) is 1.9 in the bulk CIS, whereas at the surface being <0.8. On the other hand, (220,204) shows texture structure in the bulk but random orientation at the surface. The (220,204) oriented CIS sample contains steps and ridges. The CuInSe$_2$ thin films with Cu/In of 0.85 used for the cells exhibit band gap of 0.97 eV [4]. After dipping the CIS cell in oxygen scavenger like hydrazine, the efficiency of as-grown CIS cell decreases from 6.5 to 1.8%. After annealing under air at 200 °C for 3 min, the efficiency of cell restores to 10%, as shown in Table 8.1 [5].

The low efficiency cells are one of the prime objectives to understand Physics behind the photovoltaic cell in order to improve efficiency of cells. The CuInSe$_2$ thin films are deposited onto alumina substrates by sputtering at substrate temperature of 500 °C on which without breaking vacuum, 5–15 µm thick In doped CdS films with resistivity of 0.3–5 Ω-cm are sputtered at substrate temperature of 220 °C, followed by evaporation of In metal contacts. The Alumina/CIS/CdS:In (Sput.) cell with an active area of 0.02 cm^2 shows low efficiency of 0.5%. The poor efficiency may be mainly due to causing of inter-diffusion of elements between CdS and CuInSe$_2$ that degrades the device performance. The high resistivity of CdS may be another reason. A longer time of 1 and ½ h is required for the deposition of CdS at T_s of 200 °C onto CIS by sputtering that may also cause to change in composition of CIS and CdS layers ultimately the quality of the layers degrades. Another version of cells is also developed, that a 10 µm thick CdS layer with resistivity of 0.04–0.1 Ω-cm is deposited onto Alumina/Au-Cr/CuInSe$_2$ stack at substrate temperature of 220 °C by vacuum evaporation with breaking process, followed by deposition of In metal contacts. After vacuum annealing at 200–220 °C for 1 h, the alumina/Au–Cr/CuInSe$_2$/CdS:In(Vac.)/In cell with an active area of 0.015 cm^2 accomplishes efficiency of 2.1%. The dark and light illuminated I–V curves of 0.5 and 2.1% efficiency cells are depicted in Fig. 8.11. The I–V curves of 0.5% low efficiency cells recorded under dark and light illumination conditions are crossed over each other indicating domination of secondary junction in the cells. A small 0.2 cm^2 area CIS cell developed with quality CIS absorber shows reasonable efficiency of 5.1% (Table 8.2) [41].

Similar crossover nature is observed in back wall configured spray deposited glass/ITO(2–5 Ω/□)/Cd$_{0.85}$Zn$_{0.15}$S:In(5 mol%)/CIS/Au thin film solar cells, as shown in Fig. 8.12A. All spray deposited cells with an active area of 0.4 cm^2 exhibit low efficiency of 3.2% in which the CdZnS layer is grown onto glass/ITO at substrate temperature of 350 °C by spray pyrolysis. The reason for low efficiency is due to higher series resistance and lower shunt resistance of the cells. The sheet resistance of 50 Ω-cm for CdZnS used in the cells is absolutely high causing higher series resistance of the cell. On the other hand, the band edge tails are eventually high in the spray deposited CIS layer. Even though the CIS absorber used in the cells is made by bilayer process such as first 0.5 µm thick CIS layer with composition ratio of Cu:In:Se = 1:1:4 is deposited onto glass/ITO/CdZnS at substrate temperature of 300 °C, followed by deposition of second 1.5 µm thick CIS layer with composition of 1.2:1:4. The first and second layers contain resistivities of 100 and 0.1–1 Ω-cm, respectively. Note that the composition of CIS layers mentioned here is in chemical solution that may be different in the grown film because during the growth of the film in fact the In and Se losses take place. In order to protect interdiffusion of CdZnS and CIS, the growth temperature of CIS is lowered to 300 °C [42,43]. This is a common practice in back wall configured cells. As anticipated the CuInSe$_2$/50 nmCdS(CBD)/ZnO/ZnO:Al cells made with very In-rich CIS absorber yield efficiency of 1.2%. As mentioned in earlier, the I–V curves measured under dark and light illumination conditions are in crossover position for the low efficiency cells unlike standard parallel I–V curves (Fig. 8.12B). The CIS absorber used in the cells is prepared that the Cu/In/In alloy prepared by E-beam

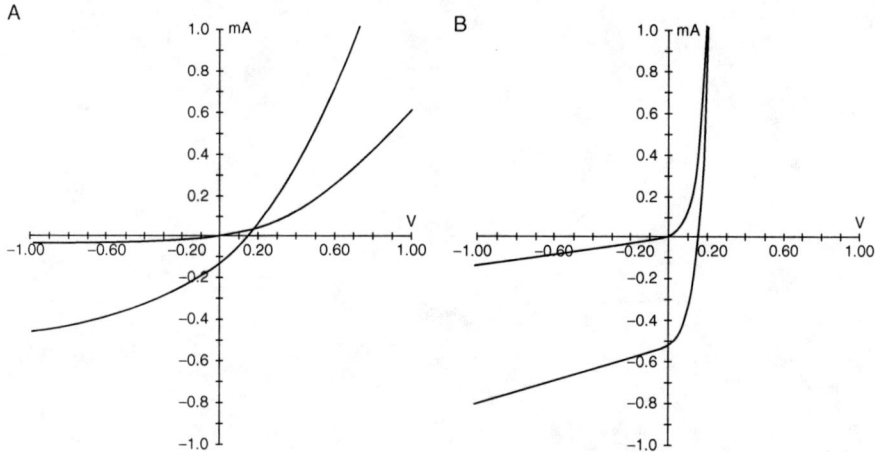

Figure 8.11 *I–V* curves of alumina/CIS/CdS:In under dark and light illumination conditions; (A) 0.5% efficiency cell, and (B) 2.1% efficiency cell.

Table 8.2 PV parameters of CIS thin film solar cells

Process/area of cells (cm^2)	V_{oc} (mV)	J_{sc} (mA/cm^2)	FF (%)	η (%)	Ref.
0.015	160	9.8	22	0.5	[41]
0.015	160	33	40	2.1	
0.2	450	22	51	5.1	
0.4	305	12.8	32	3.2	[42]
–	320.8	5.3 mA(I_{sc})	36	1.2	[44]
0.09	405	34.44	59.7	8.3	[45]
0.25	400	27	55	6.5	[46]
CIS (PLD)	381	36.9	60	8.5	[47]

evaporation is selenized under $H_2Se + Ar$ atmosphere in such a way the thickness of Cu and In layers are adjusted to have In composition of 28–30 at% in the CIS layers. The CuInSe$_2$ layers grown by selenization process have composition of Cu:In:Se = 15.6:31.5:52.8, that is, low Cu content [44]. In this case, the higher In content CIS causes to poor junction formation. In fact, high In-rich CIS samples have higher resistivity that does not meet junction conditions for efficient cells. If pose a question, does second junction participate in the high efficiency cells? The answer is yes but the *I–V* curve of high efficiency cell either dominates or suppresses second junction, therefore the second junction no longer visible.

8.1.3 CIS Cells with OVC Buffer

The glass/Mo/CuInSe$_2$/CuIn$_3$Se$_5$/CdS/i-ZnO/ZnO:Al/Ag thin film solar cell exhibits efficiency of 5.4% for optimum OVC thickness of 150 nm. If the thickness of OVC

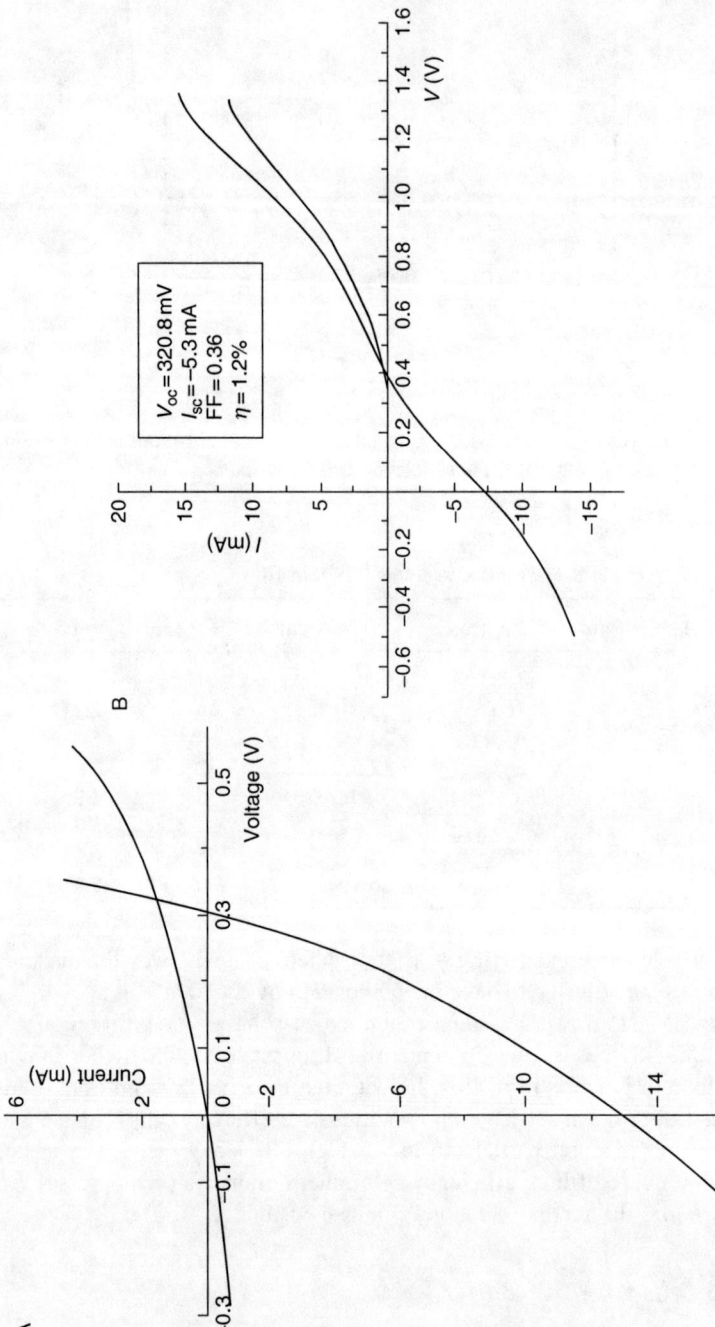

Figure 8.12 (A) *I–V* curves of low efficiency CIS thin film solar cells with crossover nature under dark, and illumination, and (B) *I–V* curves of low efficiency CIS thin film solar cell under dark, and illumination.

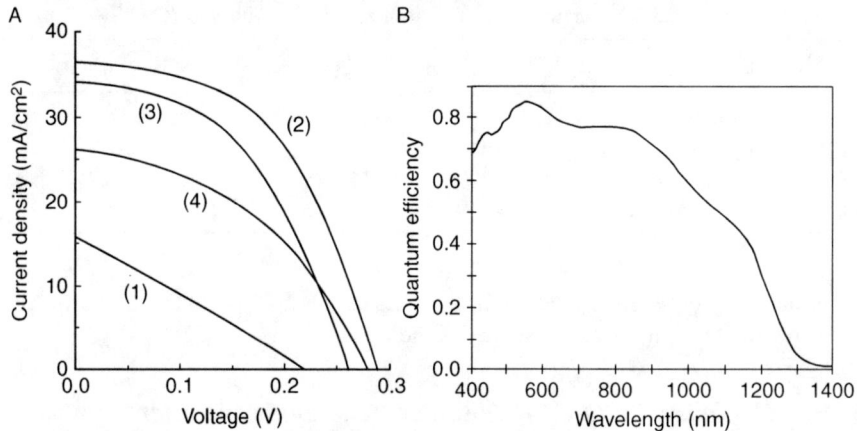

Figure 8.13 I–V curves of (A) CIS cells (1) without CuIn$_3$Se$_5$, (2) with 150, (3) 200, and (4) 500 nm, and (B) QE of CIS cells with CuIn$_3$Se$_5$ buffer.

layer is less than or greater than 150 nm, the efficiency of the cell slips into lower, as shown in Fig. 8.13A. The absorber layer is deposited by using Cu$_2$Se and In$_2$Se$_3$ binary compounds. They are sequentially deposited at room temperature by keeping In$_x$Se layer thickness of 1 μm at constant and thickness of Cu$_2$Se layer is varied from 0.2 to 0.5 μm. The deposited Cu$_2$Se/In$_x$Se double layers are annealed under Se vapor at 550 °C for 1 h to form CuInSe$_2$ layer on which CuIn$_3$Se$_5$ layer is grown by evaporation of In$_2$Se$_3$ and Se compounds at the same temperature for 10 min. The CuInSe$_2$ thin film along with γ-In$_2$Se$_3$ phase is formed for <0.35 μm thick Cu$_2$Se layer, whereas single phase CuInSe$_2$ thin film is appeared for >0.4 μm thick Cu$_2$Se layer [48]. Instead of Cu$_2$Se binary compound, Cu and Se are used along with In$_2$Se$_3$ to grow OVC on the surface of CuInSe$_2$ thin films. A 2000 Å thick Cu layer is first deposited onto Mo coated glass substrates by thermal evaporation at room temperature on which In$_2$Se$_3$ is evaporated at substrate temperature of 500 °C, followed by evaporation of Se at the same temperature. The evaporation cycles of In$_2$Se$_3$ and Se are repeated at 500 °C. The XRD analysis reveals that split in (116, 312), and (204, 220) is observed, promoting chalcopyrite structure. In addition, the other low intensity reflections such as (110), (202), and (114) are found to assign CuIn$_2$Se$_{3.5}$ (OVC). The surface composition is close to the CuIn$_3$Se$_5$. The net composition ratio of CuInSe$_2$ layer is Cu:In:Se = 22.5:27.5:50. The glass/Mo/CIS/n-OVC/(CBD)CdS/ZnO thin film solar cell yields efficiency of 10.8%. In the quantum efficiency measurements, a part of the red response is lost in the wavelength region of 900–1300 nm, as compared to that of conventional solar cells, judging higher content of In in the CIS (Fig. 8.13B) [49]. The p-CuInSe$_2$/CdS and p-CuInSe$_2$/n-OVC/CdS cells are prepared using In-rich CuInSe$_2$ films etched in NH$_3$ solution to remove In on the surface of the layers and without etched layers, respectively. The In-rich CuInSe$_2$ layers form OVC as CuIn$_3$Se$_5$ on the surface of p-CuInSe$_2$. The cells with OVC show higher efficiency of 8.7% than that of conventional CIS cells (6.9%) for the same CIS absorber because the OVC reduces conduction band

Table 8.3 PV parameters of thin film solar cells with n-OVC CIS

Type of cell/active area (cm^2)	V_{oc} (mV)	J_{sc} (mA/cm^2)	FF (%)	η (%)	Ref.
p-CIS/n-OVC/n-CdS/i-ZnO/ZnO:Al	286	36	52	5.4	[48]
p-CIS/n-OVC/n-CdS/ZnO:Al	431	39.2	64	10.8	[49]
p-CIS/n-OVC/n-CdS/ZnO:Al (0.317)	390	34.4	65	8.7	[13]
p-CIS/n-CdS/ZnO:Al (0.272)	360	32.8	58	6.9	

offset hence it forms good junction in the cells (Table 8.3) [13]. The low efficiency could be due to high series resistance of the cell.

8.1.4 CIS Cells with CuInSe$_2$ Absorber Made by Selenization

The CuInSe$_2$ is one of the potential components in the thin film solar cells therefore the growth process determines quality of CIS layers for which selenization process can be adopted to lead efficient cells. This process reduces voids and improves the good mechanical adhesion between CIS and Mo [50]. The Cu, In, and Se are evaporated by thermal evaporation and annealed under either vacuum or Se atmosphere at 500 °C for 20 min at pressure of 10^{-2} torr. First a 200 nm thick In layer is deposited for strong adhesion of the layers on the substrates, that is, to avoid delaminating of the films from the substrate, followed by deposition of 450 nm Cu or Se and 690 nm In layers. The thickness of Se layer is maintained twice of the metal layers. Annealing under high vapor pressure of Se provides quite well formation of CuInSe$_2$ thin films. Smooth surface layers are observed when the films are annealed in the graphite box. The typical CIS/CdS/ZnO cells demonstrate efficiency of 6.5% (Table 8.4) [51]. Unlike previous process, the Cu and In layers are sequentially deposited under selenium vapor onto Mo substrates by three source evaporation to form CuInSe precursor, which are selenized by two steps; first selenization is done under selenium vapor at 250 °C for 30 min, followed by second selenization at 450 °C for 1 h to incur CuInSe$_2$ layers. The PL studies reveal that the layers formed by this method result in PL peaks at 0.988 and 0.926 eV for composition ratio of Cu/In = 1 and 0.93 in the CuInSe$_2$ layers, respectively. The cells made with CIS layers exhibit higher photovoltaic conversion efficiency as compared to that of cells with slightly In excess CuInSe$_2$ films formed by co-evaporation with 0.87 eV PL peal. 2 μm thick CuInSe$_2$ thin films, 0.5 μm thick CdS by evaporation and 2 μm thick ZnO:Al layers by sputtering are successively grown onto metalized glass to develop glass/Mo/CIS/CdS/ZnO:Al solar cell, which yields efficiency of 6.02% [50]. The CIS cells are also fabricated by hybrid process that the Se is evaporated onto 180 alternate 600 nm thick Cu and In layers grown by sputtering (SP), followed by annealing under Se/Ar atmosphere using tubular furnace at 500 °C for 30 min. Surprisingly, the CIS layers with carrier concentration (p) of 4.33×10^{15} cm^{-3} grown by hybrid process of sputtering and selenization do not exhibit MoSe$_2$ phase. The CIS layers with Cu/In = 0.94 and band gap of 1.01 eV are employed in the CIS/CBD-CdS/ZnO/ZnO:Al cells, which with an area of 0.51 cm^2 show efficiency of 8.3% [52].

The cells made with In doped CdS instead of TCO layer show efficiency of 7–8%. The sheet resistance of doped CdS layer is maintained close to that of TCO. The glass/Mo/CIS/CdS/CdS:In cells with areas of 1.04 and 0.071 cm^2 exhibit efficiencies of 7.3 and 8.43%, respectively. 200 nm Cu and 440 nm In sequentially electrodeposited onto Mo metalized glass are selenized under H$_2$Se and Ar at 400 °C to obtain 2 µm thick CuInSe$_2$ thin films, which show (112) preferred orientation at diffraction angle of 26.7°, band gap of 1.03 eV and $\rho = 5$–500 Ω-cm. 0.6 µm thick CdS is grown onto CIS, followed by vacuum evaporation of 1.2 µm thick CdS:In with $R_\square = 25$ $\Omega/\square$ as a CdS/CdS:In double layer structure [53]. The CIS absorber for thin film cell is made by bilayer process. 0.2 µm thick Cu and 0.45–0.5 µm thick In layers are sequentially deposited onto 2 µm thick Mo metalized 7059 Corning glass. The Mo layer with $R_\square = 0.1$ $\Omega/\square$ is grown by electron beam evaporation. The CIS bilayer process is done as the bottom Cu–In alloy layer with Cu to In ratio of 1:1 is first selenized, followed by selenization of top alloy layer with Cu/In ratio of 0.7 at 400 °C using H$_2$Se in tubular furnace whereby the thickness of second CIS layer is maintained to be 0.8 µm. A 0.5 µm thick CdS and 0.8 µm thick CdS:In with sheet resistance of 30 $\Omega/\square$ are evaporated at 200 °C and growth rate of 60 Å/s. The sheet resistance of sputtered ZnO/ITO (TCO) is 20 $\Omega/\square$. The glass/Mo/CIS/CdS(Ev·)/ZnO(MOCVD) cells yield low efficiency of 8.6% in which the CdS is grown at growth rate of 50–80 Å/s and ZnO layers have resistivities of 100–500 and 2×10^{-3} Ω-cm, respectively. The glass/Mo/CIS/CdS (CBD, $\rho > 10^5$ Ω-cm)/ZnO/ITO cell with an active area of 0.075 cm^2 exhibits efficiency of 10.89%, which decreases to 10.74% with increasing cell area to 1×1 cm^2. The strategic difference between CdS by evaporation and CBD is found that the former has very sharp absorption line between nearly 500 and 600 nm, whereas latter consists of sublinear absorption line between nearly 300 and 525 nm in the transmission spectra. A typical 1 cm^2 area cell shows best efficiency of 12.4%, which has series resistances of 0.2 and 0.3 Ω-cm^2 under dark and light illuminations, respectively. 100 nm thick CdS layer by CBD at 60 °C for 5 min, 800 nm thick ZnO, and 200 nm thick ITO layer with sheet resistance of 20 $\Omega/\square$ are successively deposited onto CuInSe$_2$ layers by sputtering technique in order to complete cell structure. The CIS films with Cu/In composition ratio of 1.0 and nearly stoichiometric show grain size of 1 µm, whereas films with Cu/In < 1.0 have small grain sizes. The CIS layers with Cu/In ratio of 0.9 consist of (112) orientation chalcopyrite structure. The resistivities of CIS films are 0.1–10^4 and 50–500 Ω-cm for the composition ratio of Cu/In = 1.1–0.7 and 0.9, respectively and the carrier concentration (p) is 3×10^{16} cm^{-3} for the ratio of 0.9 [54–56].

The CIS cells with not only CIS bilayer but also single layer have shown efficiency of more than 10% but the thing is that the CIS layers have to be etched by KCN. Prior to selenize Cu–In alloy evaporated by electron beam evaporation with Cu/In ratio of 0.95 under 1.7% H$_2$Se with O$_2$/H$_2$Se = 0.004 at 400 °C for 1 h, the CuIn layers are preselenized under flow of Ar/O$_2$ at 400 °C for 10 min to form CuInSe$_2$ single layers. After etching as-grown CIS layers in KCN, the composition of CIS changes from Cu:In:Se = 25.4:25:49.6 to 24.6:25.9:49.5. The cells fabricated with etched CIS layers and evaporated CdS window layer demonstrate efficiency of 10.5%, whereas

Table 8.4 PV parameters of CIS thin film solar cells with CIS absorber grown by selenization

Cell area (cm^2)/process	V_{oc} (mV)	J_{sc} (mA/cm^2)	FF (%)	η (%)	A	R_s (Ω-cm^2)	R_{sh} (kΩ-cm^2)	Ref.
0.25	400	27	55	6.5	–	–	–	[51]
–	372	30.1	53	6.02	–	–	–	[50]
0.51	443	35.4	53	8.3	1.8	2.3	0.25	[52]
0.1/(CBD-CdS)	462.7	35.36	66.59	10.89	–	–	–	[54]
0.1/(Ev.CdS)	450	25	63	7	–	–	–	
0.075	462.7	35.36	66.59	10.89	–	–	–	
1	483.2	35.6	66.65	11.5	1.7	–	–	
0.08	460	38	71.9	12.6	–	–	–	[58]

the cells with unetched CIS layers contribute no conversion efficiency. The efficiency of cells increases from 10.5 to 11% when the cells are made with post annealed CIS layers under Ar at 400 °C for 30 min. On contrary, the device with CdS grown by CBD technique annealed under air at 200 °C for 2.15 h yields lower efficiency of 1.5–7.1%. The cells with absorber layer having composition of Cu:In:Se = 25.8:25.1:49.1 also show efficiency of 8.4–11%. The efficiency of the cells does not reach to high unless oxygen is introduced [57]. The performance of cells not only depends on selenization temperature but also on the substrate temperature of Cu–In stack. It is concluded that the favorable temperature for the growth of Cu–In stack is 200 °C. The CuInSe$_2$ thin films prepared by selenization of Cu–In–Se stack layers, which are formed by co-evaporation at 200 °C show better results in all aspects rather than done at 25 °C. The CIS films formed by selenization using Cu–In–Se stack at 25 °C show amorphous nature. Se/(In + Cu) ratio of 0.4 and deposition temperature of 200 °C are optimum conditions for the preparation of Cu–In–Se stack precursor to obtain high efficiency cells. The glass/Mo/CuInSe$_2$/CdS/ZnO solar cell with an active area of 0.08 cm^2 yields efficiency of 12.6% (Table 8.4) [58]. Even though the short circuit current density obtained from quantum efficiency (QE) under low intensity light illumination is three to four times higher than that of I–V measurements carried out under light illumination of 100 mW/cm^2. In the case of I–V measurements, the current collection dramatically decreases under full spectrum light illumination. This is commonly anticipated trend because of deterioration of space charge region due to low carrier concentration of $N_A \approx 10^{14}$ cm^{-3} for the sample. The breakdown electric field is not sufficient to separate generated electron–hole pairs. On the other hand in the In-rich layers, the planar defects such as microtwins and stacking faults are commonly present[44].

8.1.5 CIS Cells with CdZnS Window

The main stream of alloying Zn with CdS is to decrease lattice mismatch with CIS that contribute to reduce stacking faults and point defects at interface of CIS and CdS. On the other hand, the band gap of CdS increases with Zn doping that allows more photons to the junction in the blue region. The preferable doping range of Zn is 10–20%, beyond this range the performance of cell disintegrates due to slipping optimum junction conditions. The CuInSe$_2$ thin films onto Mo or Cu–Mo metalized glass substrates are prepared by magnetron sputtering of Cu and In and evaporation of Se. The Cu-rich then In-rich CIS are processed at 450 and 400 °C, respectively as a bilayer. 1.7 μm thick Cd$_{0.9}$Zn$_{0.1}$S with sheet resistance of 50 Ω/□, and 180 nm thick ITO layer/Ni bus bars are subsequently deposited onto CIS absorber by evaporation and sputtering, respectively. The glass/Cu–Mo/CIS/Cd$_{0.9}$Zn$_{0.1}$S/ITO/Ni cell with an active area of 0.0745 cm^2 demonstrates efficiency of 10% (Table 8.5). The adhesion test by Scotch-brand tape shows that the CuInSe$_2$ is more adhesive on Cu–Mo than on Mo layer [59]. The 7059 corning glass/2 μmMo/2 μmCuInSe$_2$/ 1.5 μmCd$_{0.9}$Zn$_{0.1}$S/0.2 μmITO/Ni thin film solar cell with an active area of 0.08 cm^2 yields efficiency of 10.1%. The Cd$_{0.9}$Zn$_{0.1}$S:In layers with sheet resistance of 35 Ω/□ are grown at growth rate of 0.8 μm/min onto evaporated CIS layer by evaporation at substrate temperature of 200 °C, followed by sputtering of ITO and Ni. A difference in V_{oc} between two identical cells is 23 mV but short circuit current density is same. Lower value of diode quality factor ($A = 1.6$) is indication for the recombination of Shockley–Read–Hall (SRH). The diode factor increases from 1.6 to 1.75 under light illumination. The barrier height of 0.99 eV is obtained from the plots of open circuit voltage versus temperature [19]. Unlike on metalized glass substrates, the Cu and In are deposited onto GaAs substrates by magnetron sputtering technique, followed by evaporation of Se to form CuInSe$_2$ layer, which has composition of Cu:In:Se = 24:26:50. 1.7 μm thick Cd$_{0.9}$Zn$_{0.1}$S layer with sheet resistance of 50 Ω/□, 180 nm thick ITO layer by sputtering and Ni bus bars are successively deposited onto GaAs/CIS to form GaAs/CIS/CdZnS/ITO cell, which yields efficiency of 8.4%. The CIS layers deposited by bilayer process as first Cu-rich then In-rich CIS layers show pits and small islands or ripples, respectively. The voids between interface of GaAs substrate and CIS are observed as to Mo/CIS [60]. The glass/Ta/pattern1.5 μmMo/0.1 Ω□/CuInSe$_2$/Zn$_{0.13}$Cd$_{0.87}$S/Algrids/ MgF$_2$/SiO$_x$ monolithic cells over the large area of 91 cm^2 as a submodule show differences in PV parameters before and after AR coating, as shown in Table 8.5. The CdZnS is evaporated onto CIS by E-beam evaporation using sintered CdZnS bulk with 13%Zn. This technique has disadvantage of spitting of source material and high resistance in grown CdZnS. The CuInSe$_2$ thin films are prepared by feedback controlled co-evaporation of Cu, In, and Se. Two In evaporation cells stay besides of Cu evaporation cell in order to incur uniform growth of CuInSe$_2$ in the evaporation chamber. The compositions of Cu and In layers are sensed by electron impact emission spectrometer (EIES) sensors and separately by Se sensor for Se. The substrates are heated by quartz lamps rather than using resistive heater to maintain uniform of the temperature [61].

The CIS layers are deposited by co-evaporation of constituent elements of Cu, In, and Se under vacuum of 10^{-4} Pa. The deposition is started at substrate temperature of 360 °C and 15 step processes have been carried out between 420 and 520 °C in the 30 min duration. In the first 15 min time, the temperature is ramped up from 420 to 520 °C as stair case steps then down to 420 °C in the remaining period of 15 min. In the first stage, Cu to In ratio is maintained to rich, whereas in the second stage ratio is poor. The evaporation ratio of Se to metals composition is maintained higher than two to three times. Onto CIS layer with composition of Cu:In:Se = 24:25:51, a 50 nm thick CdZnS (Zn-15%), 150 nm thick CdS, and 1.5 μm thick Ga-doped CdS layers as a combined window are sequentially deposited at substrate temperature of 200 °C resulting in sheet resistance of 15 Ω/□. The CdS is also deposited by CBD method using different source materials of CdI_2, $CdCl_2$, and $CdSO_4$. A 50 nm thick ZnO with conductivity of $< 10^{-4}$ $(\Omega\text{-cm})^{-1}$ is deposited and 1 μm thick Al doped ZnO layer with conductivity of 700 $(\Omega\text{-cm})^{-1}$ is covered by RF sputtering. In the case of CBD-CdS, as the thickness of CdS is increased, the J_{sc} decreases but V_{oc} and fill factor increase to some level then remain constant. The reason is that the absorption of light by CdS increases with increasing thickness of layer. The optimum thickness of CdS buffer layer for the best cell is considered to be 80 nm. The standard CIS/$Cd_{0.85}Zn_{0.15}S$/CdS/CdS:Ga/i-ZnO/ZnO:Al cell annealed under air at 200 °C for 40 min shows efficiency of 8.8%, whereas the other buffered CIS/CdS(CBD)/i-ZnO/ZnO:Al cells are annealed for 2 min only. An increased blue response by the CBD-CdS layers causes high efficiency in the cells. The cell with CBD-CdS buffer layer formed by source of CdI_2 exhibits little higher conversion efficiency (11.7%) than that of one (11.5%) made with CdS using source of $CdSO_4$ or standard cell. The cell with MgF_2 coating with an active area of 0.315 cm^2 shows higher efficiency of 12.8%, as shown in Table 8.5 [62]. The alumina or glass/1.5 μmMo/3–4 μm $CuInSe_2$/CdZnS/ 2 μmAl grid cell demonstrates efficiency of 11%, whereas cell with CdS experiences low efficiency of 8.7%. The p-$CuInSe_2$ absorber layer used in the cells contains sheet resistance of 1 kΩ/□, carrier concentration (p) of 10^{14}–10^{15} cm^{-3} and mobility (μ) of 3–9 cm^2/Vs. The Cu, In, and Se are evaporated at substrate temperature of 350–450 °C to obtain bilayer and cooled down to 200 °C to deposit CdS or CdZnS (Zn-20 at%). Obviously, the Cu-rich $CuInSe_2$ layer shows larger grain sizes and rough surface while In-rich layer shows mirror like smooth surface. After ¾ part of evaporation of CdS or CdZnS layer from Cd(Zn)S compound, indium is started to evaporate to obtain 1–3 at % In doping in the top layers. Finally, the cell is annealed under air at 200–225 °C for 20–60 min. The sheet resistance of mixed CdZnS layer is brought down from 20, 100 Ω-cm to 50 Ω/□. A 950 Å thick SiO_x layer is evaporated by E-beam [63,64]. In another occasion, 2–2.5 μm thick Mo layers with sheet resistance of 0.08 Ω/□ deposited onto SLG by E-beam on which Cu and In layers are grown by vacuum co-evaporation. The Cu–In stack layers are selenized under 95%Ar and 5%H_2Se atmosphere at 400 °C for 1 h to convert $CuInSe_2$ layers. Onto CIS layers, 1.1 μm thick CdZnS (15%Zn) with 175 Ω/□, and 200 nm thick ITO layer having sheet resistance of 20 Ω/□ are sequentially deposited by evaporation and sputtering, respectively. In the CIS cells, $CuInSe_2$ thin films with Cu/In ratio of 0.92 have resistivity of ∼100 Ω-cm. The SLG/Mo/$CuInSe_2$/CdZnS/ITO cells show efficiency of 6.96% under light illumination of 87.5 mW/cm^2 (Table 8.5) [65].

8.1.6 CuInSe₂ Cells with Different Types of Windows or Buffers

The conduction band (0.84 eV) and valence band offsets (0.78 eV) of CIS/ZnSe can be determined using similar approach of CIS/CdS heterojunction by XPS

Table 8.5 Photovoltaic parameters of CIS/CdZnS cells with different Z_n compositions

Window	V_{oc} (mV)	J_{sc} (mA cm^{-2})	FF (%)	η (%)	$J_0 \times 10^{-3}$ (mAcm^{-2})	ϕ_b (eV)	A	R_s (Ω-cm^2)	R_{sh} (Ω-cm^2)	Ref.
$Cd_{0.90}Zn_{0.10}S$	426	35.3	66.5	10	–	–	–	–	–	[59]
$Cd_{0.90}Zn_{0.10}S$	440	33.8	67.6	10.1	–	0.99	1.6	–	–	[19]
$Cd_{0.90}Zn_{0.10}S$	417	33.8	67.2	9.3	4	0.91	1.5	0.6	20	
$Cd_{0.90}Zn_{0.10}S$	392	33.9	63.1	8.4	–	–	–	–	–	[60]
$Cd_{0.87}Zn_{0.13}S$	1.68V	29.82	63.91	7.99	–	–	–	–	–	[61]
$Cd_{0.87}Zn_{0.13}S$	1.73 V	35.32	62.61	9.56	–	–	–	–	–	
$Cd_{0.85}Zn_{0.15}S$	400	32.4	68	8.8	–	–	–	–	–	[62]
$CdS(CdSO_4)$	424	38.5	70	11.5	–	–	–	–	–	
$CdS (CdI_2)$	450	38	68	11.7	–	–	–	–	–	
CdS/MgF_2	453	41	69	12.8	–	–	–	–	–	
$Cd_{0.80}Zn_{0.20}S$	436	38.6	65.3	11	–	–	1.22	1.2	600	[63]
CdS	423	32.5	63	8.67	–	–	–	–	–	
$Cd_{0.85}Zn_{0.15}S$	332	32.24	56.91	6.96	–	–	–	–	–	[65]

analysis. The XPS analysis reveals that the $CuInSe_2/ZnSe$ heterojunction has $\Delta E_{V\text{-}In4d} = 18.55 - 0.8 = 17.75$ eV and $\Delta E_{V\text{-}Zn3d} = 10.65 - 1.4 = 9.25$ eV, the ΔE_{CL} is difference between In4d and Zn3d, that is, $18.55 - 10.65 = 7.92$ eV but taken an average value of 7.72 eV, $\Delta E_V = \Delta E_{V\text{-}In4d} - \Delta E_{V\text{-}Cd4d} - \Delta E_{CL}$, that is, $\Delta E_v = 17.75 - 9.25 - 7.72 = 0.78$ eV and $\Delta E_c = 2.67 - 1.05 - 0.78 = 0.84$ eV [66]. The electron affinity of 5.5 eV for ZnSe can be determined using band offset parameters from XPS analysis.

The glass/Mo/$CuInSe_2$/ZnSe/ZnO cells show efficiency of 8.5%. The ZnSe layer is grown by DC magnetron reactive sputtering technique using Zn target and H_2Se as a reactive gas at Zn flux of 305 Å/min and Zn target power of 94.4 W [67]. The ZnSe buffer layer formed onto glass/Mo/CIS (Siemens) at substrate temperature of 200–250 °C by MOCVD shows (111) line in the XRD spectrum indicating cubic structure, on which the ZnO films with resistivity of 0.05–6 Ω-cm is also grown by the same technique. The glass/Mo/CIS/ZnSe/ZnO cell with an active area of 0.0616 cm^2 shows efficiency of 14.1% [68]. The $CuInSe_2$ solar cell with In_2Se_3 buffer layer is found to show lower efficiency (8.3%) than that of reference cell (9.2%), as shown in Table 8.6. The CIS thin films are grown onto Mo coated glass substrates by two-stage process as the Cu is first deposited onto Mo and subsequently In_xSe_y layers by evaporation. In the second stage, the stacked Cu/In_xSe_y system is annealed at 450 °C under Se vapor. The In_xSe_y buffer layer ($E_g = 1.75$ eV) with thickness of 80 nm is formed onto CIS by evaporation using In_2Se_3 powder at source temperature of 850 °C. To complete the cell configuration, ZnO is finally deposited by reactive evaporation, which had conductivity of 8×10^{-4} Ω-cm and $T > 80\%$ [69,70]. The CIS cells with CBD $In(OH)_xS_y$ buffer exhibit efficiency of 9.5% as compared to that of CdS control cell (11.9%) [71]. The superstrate glass/ZnO:Al/In_xSe_y(6 mgNa_2S)/CIS/Au cell had an active area of 0.1 cm^2 shows efficiency of 7.5% while the cell shows low efficiency for above or below optimum doping of 6 mg Na_2S in the CIS layer (Table 8.6). First 2 μm thick ZnO:Al layer with $R_\square$ of 10 $\Omega/\square$ is deposited onto glass substrate at deposition temperature of 540 °C by RF magnetron sputtering on which 0.2–0.6 μm thick In_xSe_y buffer layer and 2 μm thick CIS layer are successively grown by vacuum co-evaporation at substrate temperature of 350 °C. 6 mg Na_2S is added to the CIS that reflects to 5 at% Na in the layer. The deposition of CIS at above 400 or below 350 °C, the cell slips to poor performance. The XRD spectra of CIS films show that as the amount of

Table 8.6 Photovoltaic parameters of CIS cells with different buffer layers

Cell	V_{oc} (mV)	J_{sc} (mA/cm^2)	FF (%)	η(%)	Ref.
CIS/In_xSe_y/ZnO	445	30.8	60	8.3	[69,70]
CIS/CdS/ZnO (Ref.)	430	34	63	9.2	
CIS/$In(OH)_3$-In_2S_3	446	34.5	62	9.5	[71]
CIS/CdS/ZnO (Ref.)	460	37.5	69	11.9	
Glass/ZnO:Al/In_xSe_y/CIS/Au	430	29.4	60	7.5	[72]

Na_2S is increased from 0, 6 to 12 mg in the CIS film while performing deposition, the intensity of preferred orientation of (112) for CIS films increases. On the other hand, small intensity peaks such as (211), (400), (332), *etc.*, pronounce well [72].

8.1.7 CIS Cells with CIS Absorber Grown by Electrodeposition

The $CuInSe_2$ thin films onto Mo coated glass substrates are grown by single step electrodeposition (ED) using typical 250 ml chemical solution with combinations of 2.6 mM $CuCl_2$, 9.6 mM $InCl_3$, 5.5 mM H_2SeO_3, and 240 mM LiCl at pH 3. The potentials of −0.476 and −0.576 V have been applied during first stage deposition time of 20 min and second stage deposition time of 50 min, respectively. Ag/AgCl and Pt are used as a reference and a counter electrode for the deposition of CIS layers, respectively. The grown CIS layers are annealed under Se atmosphere with a combination of 10%H_2 and 90%N_2 maintaining pressure of 10 mbar at 550 °C for 30 min. The grain sizes of grown and annealed layers are more or less 50 nm in diameters. The compositional changes are observed in annealed 1.7 μm thick ED CIS layers that the Cu/In ratio increases from 0.72 to 0.79 and Se/(Cu + In) decreases from 1.07 to 0.96 due to losses of Se and In, in addition $MoSe_2$ phase is observed. The CIS thin films with Cu/In = 0.79 and carrier concentration of 1.85×10^{16} cm^{-3} show band gap of 0.99 eV. The resistivity of ED CIS samples can be tailored from 0.1 to 100 Ω-cm by annealing the layers under Se. The carrier concentration of *p*-CIS layers grown by electrodeposition is in the order of 10^{16}–10^{17} cm^{-3}. The Mo/CIS/CdS/ZnO cells with electrodeposited absorber show efficiency of 6.5%. The typical *I–V* curves of 6.3% efficiency glass/Mo/1.5 μmCuInSe$_2$/50 nmCdS(CBD)/1 μmZnO thin film solar cells are shown in Fig. 8.14A (Table 8.7). The $CuInSe_2$ thin films used in the cells are annealed under Ar instead of Se vapor at 300–420 °C for 20 min and CdS layers with resistivity of 5×10^4 Ω-cm are deposited by chemical bath deposition technique. Under light illumination, the resistivity of CdS decreases by two orders of magnitude [52,73,74]. The typical glass/Mo/CIS/CdS/i-ZnO/ZnO/Al-Ni cell with an area of 0.06 cm^2 shows efficiency of 8.8%. 32 cells over the area of 5×5 cm^2 substrate show an average efficiency of 4.5%. The *C–V* measurements reveal doping concentration of 1×10^{16} cm^{-3}, space charge region width of 250 nm, and barrier height of 0.75 eV for 8.8% efficiency cells taking junction between CIS and CdS. The $MoSe_2$ phase is determined in the annealed CIS layers by XRD [75].

The $CuInSe_2$ thin films deposited onto Mo coated glass substrates by electrodepostion technique, followed by vacuum deposition of CdS:In layer using CdS and In_2S_3 as co-evaporation sources at substrate temperature of 180–200 °C. The grown CdS:In (2 wt%) layer has $n = 10^{20}$–10^{21} cm^{-3}, $\mu = 20$ cm^2/Vs and transmission of 80%. After air annealing at 200 °C for 1 h, the glass/Mo/CuInSe$_2$/CdS:In cell with an active area of 0.71 cm^2 shows efficiency of 4%. 20–50 nm thick and 2–3 μm thick CdS layers are essential for thin film solar cells, if they are grown by CBD and evaporation techniques, respectively. The acceptor concentration of 1.2–5×10^{16} cm^{-3} for *p*-CuInSe$_2$ is observed by CV measurements. In some cases, prior to deposition of CdS, the electrodeposited CIS layers are annealed between 300 and 420 °C. The cell with an area of 0.11 cm^2 shows efficiency of 7%. During

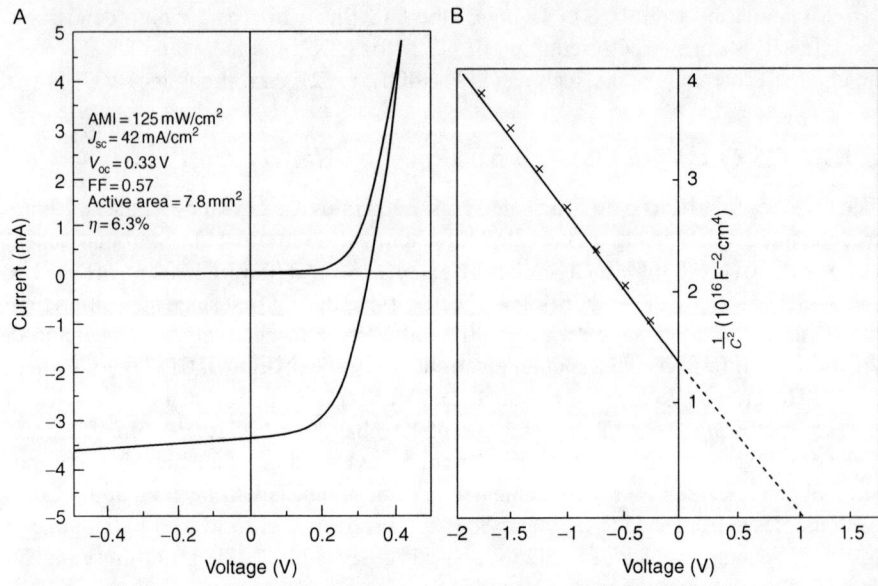

Figure 8.14 (A) *I–V* curves of 6.3% efficiency CIS cells with electrodeposited CIS absorber under dark and light illumination, and (B) $1/C^2$ versus of V glass/SnO$_2$:F(1 μm)/*p*-CuInSe$_2$(0.450 μm)/*n*-CdS(1.2 μm)/1 μm thick In grid cell.

Table 8.7 Photovoltaic parameters of CIS cells with CIS absorbers grown by electrodeposition

Cell area (cm^2)	V_{oc} (mV)	J_{sc} (mA/cm^2)	FF (%)	η (%)	A	R_s (Ω-cm^2)	R_p (kΩ-cm^2)	J_o (μA cm^{-2})	Ref.
0.51	395	31.2	52	6.7	2	2	0.29	12	[52]
0.078	330	42	57	6.3	–	–	–	–	[74]
0.06	416	32.7	65	8.8	1.6	0.8	–	1	[75]
0.71	300	30	50	4	–	–	–	–	[76]
0.11	360	35	56	7	–	–	–	–	[77]
A	394	32.1	62.3	7.9	–	–	–	–	[78]
B	410	36	63.5	9.4	–	–	–	–	
CdS	403	33.4	49	7.7	–	–	–	–	[79]
ZnS	340	35.4	57	6.1	–	–	–	–	
1.04	408	26.76	66.5	7.3	–	–	–	–	[53]
0.071	395	30.7	69.4	8.43	–	–	–	–	
0.1	365	12	61	3.1	–	–	–	–	[81]

annealing, In and Se losses occur that makes slightly Cu-rich at the surface comparing with bulk layer [76,77]. The Cl impurity in the ED CIS layer lowers efficiency of cells. Two kinds of approaches have been used to deposit CuInSe$_2$ layers onto

Mo coated glass substrates by electrodeposition. In the first case, Cl related salts such as 0.025 M $CuCl_2$, 0.025 M $InCl_3$, and 0.025 M H_2SeO_3 solutions are used for the deposition of layers and maintained pH 1.5. Pt gauge is used as counter, SCE as reference and Mo coated glass as working electrode. The In and Se are added to the electrodeposited CIS layers at substrate temperature of 550 °C by physical vapor deposition (PVD) to tailor composition of $CuInSe_2$ that can fit to thin film solar cell. The entire glass/Mo/$CuInSe_2$/CdS/ZnO/MgF_2 cell is annealed under air at 200 °C for 15 min, which is considered as cell-A. In the second approach, sulfur related salts such as 0.025 M $CuSO_4$, 0.025 M $In_2(SO_4)_3$, and 0.025 M H_2SeO_3 chemical solutions with pH 1.5 are used instead of chloride salts to grow $CuInSe_2$ layers. However, the grown layers are found to be $CuIn_2Se_{3.5}$, In_2Se_3 and metallic In and Cu, which are annealed under Ar at 250 and/or 450 °C for 1.5 min. The Cu, In, and Se are added to the electrodeposited layers by PVD method to monitor the composition of $CuInSe_2$ layers. The cells grown by this method is denoted as cell-A. The cell-B shows little higher efficiency (9.4%) than that of cell-A (7.9%). A difference in efficiency could be due to effect of Cl contamination in the cell-A [78]. The CIS cells fabricated with CBD ZnS buffer shows lower efficiency (6.1%) than that of one with CBD-CdS buffer (7.7%), as given in Table 8.7 [79].

The CIS absorbers are fabricated by electrodeposition using one-step method at different potentials of 1.0, 0.75, and 0.6 V to form n–i–p $CuInSe_2$ layer structure onto glass/SnO_2:F (FTO)/CdS CBD stack followed by annealing under Se at 450 °C for 10 min to form CIS. The CIS layers deposited at individual potentials show different phases such as CuSe, In_2Se_3, and In_2O_3 along with $CuInSe_2$ phase. The back wall configured glass/FTO/CdS/CIS cells with KCN etched CIS layers show $V_{oc}=410$ mV, $J_{sc}=36$ mA/cm^2, and FF=45% [80]. Alike the $CuInSe_2$ layers with thickness of 450 nm are deposited onto 15 Ω/□ sheet resistance FTO coated glass substrates by electrodeposition using copper citrate (0.2 M), indium citrate (0.2 M), and sodium selenosulphite (0.1 M) at room temperature unlike deposition at 50–70 °C. One deposition run consists of only 120 nm thick $CuInSe_2$ layers, in order to incur 450 nm thick layers, multiple depositions being done. CdS layers are also deposited onto glass/FTO/CIS by the same technique. The stack of glass/SnO_2:F(1 μm)/$CuInSe_2$(0.450 μm)/CdS(1.2 μm) is annealed under air at 200 °C for 7 h and dipped in stannic chloride solution for few seconds then dried at 100 °C for 5 min. After the deposition of 1 μm thick In grids on the stack, again the samples are annealed under vacuum at 150 °C for 1 h. All the electrodeposited cells with an active area of 0.1 cm^2 demonstrate low efficiency of 3.1% under light illumination of 85 mW/cm^2. The performance of the cells may be viable by increasing thickness of absorber and decreasing thickness of window layer. The acceptor density of 1.25×10^{16} cm^{-3} is observed in all the electrodeposited cells from CV measurements (Fig. 8.14B). The cells with Br etched CIS layers explore that the V_{oc} slightly increases or remains at the same level, whereas the J_{sc} decreases from 33.6–34.4 to 31.4 mA/cm^2 and the fill factor also follows decreasing from 63.2–67.2 to 60.6% in the cell because of reduction of CIS thickness [81,82].

8.1.8 Effect of Conducting Layer on CIS Thin Film Solar Cells

In the thin film solar cells, the role of conducting layer is predominant to pioneer efficient cells. The conducting layers, such as ZnO, ITO, FTO, CdO, *etc.*, are successfully applied in the thin film solar cells over the buffer layer as an electrode, which also acts as window for the solar cell to allow solar radiation as well as partial antireflector. 0.86 μm thick ZnO layer deposited by LPCVD has sheet resistance ($R_\square$) of 9 Ω/□, resistivity of 7.7×10^{-4} Ω-cm and transmission of 88.4%, which are preferable for better efficiency solar cells, whereas 1.45 and 0.04 μm thick ZnO layers show $R_\square = 4$ and 28–32 Ω/□, transmission of 82.3 and 90.4%, respectively. The data indicate that the physical parameters of 0.86 μm thick ZnO layer stand in between 0.04 and 1.45 μm thick ZnO layers. Therefore, the thickness of ZnO can be trade off suitably to have optimum sheet resistance and transmission to obtaining high efficiency thin film solar cells. The LPCVD ZnO layers deposited at substrate temperature of 180 °C is preferred one in terms of transmission otherwise beyond or below the temperatures, the layers have lower or higher transmissions, respectively those inversely influence the sheet resistance of the samples. The glass/Mo/CIS/CdS(CBD)/ZnO:B/Ni–Al cell with an active area of 0.074 cm^2 shows efficiency of 5.5% for which the CdS with thickness of 110 nm is covered onto CIS by CBD technique, followed by 1.5 μm thick ZnO layer with sheet resistance of 4–10 Ω/□. The LPCVD ZnO is grown at growth rate of 10–30 Å/s employing diethylzinc and diborane gas sources at substrate temperature of 200 °C. An improved version of glass/Mo/CIS/CdS(CBD)/ZnO:B (LPCVD) cell with an active area of 0.18 cm^2 shows efficiency of 10.23%. The Cu, In, and Se are sputtered to form CuInSe$_2$ thin film onto glass/Mo magnetron sputtering from individual targets at substrate temperature of 300–500 °C, chamber pressure of 1.5–2 mtorr, and growth rate of 3–4 Å/s [83,84]. In some of the cases, the oxide layers are used as *n*-type layer in the *p-n* junction solar cells. The SIMS analysis reveals that In loss is observed when ZnO layer is grown onto CuInSe$_2$ by sputtering rather than by CVD. The glass/Mo/CIS/CdS/ZnO cells with ZnO grown by sputtering and CVD show efficiencies of 10.4 and 9.2%, respectively (Table 8.8) [85]. The ZnO:Al and ZnO:B films are prepared onto glass/Mo/CIS/CdS structure at 350 and 200 °C by DC magnetron sputtering, respectively. The featureless and textured surface morphology are observed in former and latter, respectively. The quality of ZnO:Al thin films degrades, if they are deposited at less than substrate temperature of 350 °C. The Al and B doped ZnO films show mobility of 37 and 60 cm^2/Vs, carrier concentration of 5.6×10^{20} and 2.6×10^{20} cm^{-3}, transmission 29 and 83% at wavelength of 1.2 μm, respectively. The CIS cells with ZnO:B window layer contribute higher short circuit current as compared to one with ZnO:Al because of quality of ZnO:B layers and higher transmission at near infrared regions that allows photons to have higher current (Table 8.8) [86].

In the CIS/CdS/ZnO(O$_2$)/ZnO:Al solar cell (type-1), i-ZnO layer is grown by RF sputtering using ZnO target under the O$_2$:Ar = 1:3, which had resistivity of 1 Ω-cm. 400 nm thick Al doped ZnO (Al$_2$O$_3$ 2 wt%) layer with sheet resistance of 20 Ω/□ and mobility (μ) of 20 cm^2/Vs is grown without using O$_2$ during sputtering. The ZnO and ZnO:Al layers without using O$_2$ are deposited for CIS/ZnO/ZnO:Al cells denoted as type-2. In comparison, with inclusion of CdS buffer, the open circuit

Table 8.8 PV parameters of CIS thin film solar cells with effect of conducting layers

S.No.	Configuration of cell	V_{oc} (mV)	J_{sc} (mA/cm^2)	FF (%)	η (%)	Ref.
–	CIS/CdS(CBD)/ZnO:B (LPCVD)	298	32.9	56.5	5.5	[83]
–	CIS/CdS(CBD)/ZnO:B (LPCVD)	427.1	37.41	64.1	10.23	[84]
–	CIS/CdS/ZnO (Sput.)	418	38.4	65	10.4	[85]
–	CIS/CdS/ZnO (CVD)	387	38.7	61.2	9.2	
–	CIS/CdS/ITO	343	31.9	56.2	6.2	
–	Mo/CIS/CdS/ZnO:B	431	34.4	71	10.5	[86]
–	Mo/CIS/CdS/ZnO:Al	445	31.4	70	9.79	
1	CIS/CdS/ZnO(O$_2$)/ZnO:Al	460	38	71	12.4	[87]
2	CIS/ZnO/ZnO:Al	398	39	68	10.5	
3	CIS/CdS/ZnO:Al	430	37.7	77	11.8	
4	CIS/ZnO/ZnO:Al	375	40.3	65	9.8	
5	CIS/ZnO(O$_2$)/ZnO:Al	390	34.6	40	5.4	
6	CIS/ZnO:Al	220	40	52	4.6	
–	CIS/i-ZnO(MOCVD)/ZnO/Ni-Ag	492	34.26	66	11.1	[89]
–	CIS/ZnO(CBD)/n-ZnO(MOCVD)	420	35.5	65	9.7	[90]
–	Mo/CIS/CdS/ZnO/ITO	426	35.6	69.1	10.5	[91]
–	Mo/CIS/CdS/ZnO/ITO	340	33.5	53	6	
–	Mo/CIGS/CdS/ZnO/ITO	493	33.7	70.5	11.7	

voltage increases. In the case of type-4, -5, and -6 cells, the efficiency decreases in the sequence because of (a) no CdS used in the cell 4, (b) O$_2$ is used during sputtering of ZnO that causes to increase resistance of ZnO layer in the cell 5, and (c) in the cell 6, no CdS and no i-ZnO layer used those cause to decrease efficiency of the cell (Table 8.8). The i-ZnO layer reduces leakage current in the cells. The CdS buffer grown by CBD protects CIS absorber from the strong energy sputtering particles and diffusion of oxygen. Note that the post annealing under air at 200 °C is not essential for thicker CdS buffer layers. The efficiency of type 1 cell increases from day to day for ZnO sputtered without oxygen [87]. The CuInSe$_2$ based thin film solar cells can be fabricated without using CdS buffer but CIS layer should be chemically treated by either Zn or Cd. The CuInSe$_2$/CdS/ZnO standard thin film solar cells demonstrate efficiency of 12.78%, whereas Zn and Cd treated CuInSe$_2$/CdS/ZnO cells show little lower efficiencies of 11.56 and 11.46%, respectively but higher currents. Dipping the absorber layer either in Cd or Zn salts with combination of ammonium hydroxide at different times and temperatures is known as chemically treated in this context. Both the chemically treated cells show little higher currents, as compared to that of standard CuInSe$_2$/CdS/ZnO cell. This is due to conversion of absorber surface into n-type by incorporating Zn and Cd. The XPS depth profile also confirms signatures of Zn and Cd in the layers. The CuInGaSe$_2$/CdS/ZnO, CuInGaSe$_2$/Cd treated/ZnO, and CuInGaSe$_2$/Zn treated/ZnO thin film solar cells yield efficiencies of 13.6, 11.4, and 5.3%, respectively. The cells with CIGS absorbers treated by chemicals also exhibit little higher current.

As is observed in the CIS cells [88]. The glass/Mo/CIS/i-ZnO(MOCVD)/ZnO/Ni-Ag thin film solar cell without buffer layer exhibits efficiency of 11.3% (Table 8.8). The CIS absorber is graded with Ga, and S, and i-ZnO layer is formed by MOCVD technique with thickness of 200 to 400 Å exhibiting resistivities of 1000–10,000 Ω-cm. The glass/Mo/CIS/i-ZnO structure is annealed under ethyliodide at 225 °C to reduce resistivity of intrinsic ZnO layer on which conducting ZnO layer with sheet resistance of 10–20 Ω/□ is deposited. The photovoltaic parameters of 11% efficiency glass/Mo/CIS/i-ZnO(MOCVD)/ZnO/Ni-Ag cell are $R_s = 0.28$, 1.28 Ω-cm^2, $R_{sh} = 1500$, 4.67×10^7 Ω-cm^2, $A = 2.1$, 1.5, $J_o = 8.34 \times 10^{-4}$ and 1.2×10^{-7} mA/cm^2 under dark and light illumination, respectively [89]. 0.95 cm^2 area CIS/ZnO (CBD)/n-ZnO(MOCVD) thin film solar cell with 10 nm thick ZnO buffer shows efficiency of 9.7%. The QE measurements reveal that the spectral response starts early at shorter wavelength in the CIS/ZnO(CBD)/ZnO than that in the CIS/CdS/ZnO cells because of difference in band gaps between ZnO (3.4 eV) and CdS (2.4 eV). The similar characteristics are observed in their transmission curves [90]. The efficiency of typical Mo/CIS/CdS/ZnO/ITO cell decreases from 10.5 to 6% by sacking CdS off from the cell. The efficiency of Mo/CIGS/CdS/ZnO/ITO cell increases from 10.5 to 11.7% by adding Ga (Ga/III = 0.125, III = In + Ga) to the CIS layer. The CIS bilayer is prepared by multisource evaporation for cells [91].

8.1.9 Back Wall Configuration/Superstrate CIS Cells

In the superstrate thin film solar cells the CIS absorber is deposited onto CdS window layer (glass/ITO/ZnO/CdS) (Fig. 8.15A) or TCO unlike substrate type thin film solar cells therefore optically, structurally and electrically quality window layer is very essential in order to develop better absorber. In the preparation process of superstrate thin film solar cells, the absorber should be prepared at low substrate temperature of less than ~450 °C otherwise the interdiffusion takes place between CdS and CuInSe$_2$. 0.25 μm thick ZnO films are sputtered on ITO coated glass substrates, followed by evaporation of 0.25 μm thick CdS layers at $T_s = 150$ °C. The glass/ITO/ZnO/CdS films are annealed with CdS:CdCl$_2$ powder at different temperatures of 520, 540, and 570 °C under N$_2$ atmosphere for several hours to obtain large grain sizes of 1.7, 2.1, and 2.7 μm, respectively. The as-grown CdS layers have (002) preferred orientation, whereas annealed one shows (103) orientation. The CIS films deposited onto as-grown and annealed CdS layers at 325 °C show (112) and (204)/(220) orientations, respectively. The efficiency of 6.7% is measured for the glass/ITO/ZnO/CdS/CIS/Au cells made with annealed CdS layers (Table 8.9). As shown in Fig. 8.15B, the quantum efficiency is higher in the cells made with annealed CdS than that one with as-grown CdS amid the quality of CIS layers is probably improved on annealed CdS layer. In general, the diffusion length is longer in the quality CIS cells. The photoresponse of cells with annealed CdS layer starts early in the shorter wavelength region rather than one with unannealed CdS because the annealed CdS layer contains wider band gap comparing with as-grown CdS [92]. The CuInSe$_2$ based superstrate solar cells prepared on glass substrates show improved efficiency of 8.1% in which 2 μm thick ZnO:Al layer with sheet resistance of 2 Ω/□ and transmittance of 85% at 600 nm is first deposited onto

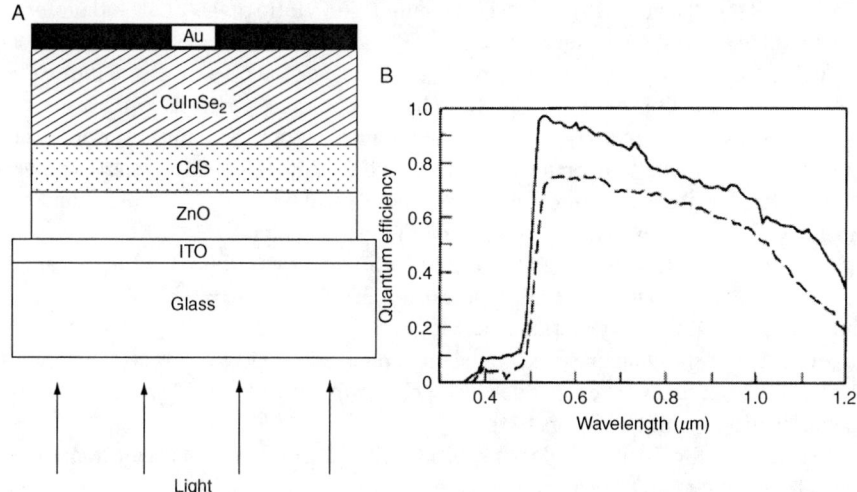

Figure 8.15 (A) Schematic diagram of superstrate type CIS thin film solar cell, and (B) Quantum efficiency of CIS cells made on without (dashed line) and with annealed CdS window (solid line).

Table 8.9 Photovoltaic parameters of back wall configuration/superstrate cells

Superstrate cell configuration	V_{oc} (mV)	J_{sc} (mA/cm^2)	FF (%)	η (%)	Ref.
Glass/ITO/ZnO/CdS(Ev.)/CIS/Au	357	32.2	58.2	6.71	[92]
Glass/ZnO:Al/CdS(CBD)/CuInSe$_2$/Au	350	39.8	58	8.1	[93]
Glass/ITO/ZnO/CdS(Ev.)/CuInSe$_2$/Au	300	31	53	4.9	[94]
Glass/TCO/CdS(CBD)/CuInSe$_2$/Au	220	32.86	29	2.06	[95]

glass then 300 nm CdS layer is deposited by several times using CBD in order to develop thick CdS. In this case, the thickness of CdS per run is 40 nm. The CdS layer also grown by vacuum evaporation on similar kind of ZnO:Al coated substrates at 150–200 °C. The transmission curve of CBD-CdS layer begins early at shorter wavelength region as compared to that of vacuum evaporated CdS layer. The reason is that CBD-CdS is combination of Cd(OH)$_2$, and CdO compounds. Therefore, band gap of CBD-CdS is higher than that of evaporated CdS. On the other hand, CBD-CdS layer contains less pinhole density. Finally, 2–2.5 μm thick CuInSe$_2$ layer is deposited onto CdS at substrate temperature of 450 °C by vacuum evaporation. In the CuInSe$_2$ deposition process, at starting point, the Cu-rich CuInSe$_2$ then In-rich CuInSe$_2$ layer is formed. If the order is changed, the In diffuses into CdS and forms CuCd$_2$InSe$_4$ layer at interface of CdS that will invariably short the device. Secondly, CuIn$_2$Se$_{3.5}$ can be formed for below Cu/In = 0.8. Cu also diffuses into CdS for heavy Cu-rich CuInSe$_2$ layers that make high resistance CdS, which is also not suitable for device. The CuInSe$_2$

thin film solar cells with CIS absorber containing Cu/In ratio of 0.9–0.98 and preferred orientation of (112) show better results. The XRD analysis reveals that multiphases such as $CuCd_2InSe_4$ and $Cu_{2-x}Se$ for Cu/In = 1.20, $CuIn_2Se_{3.5}$ and In_6Se_7 for Cu/In = 0.81 present in grown CIS layers. The $CuInSe_2$ phase is commonly present with irrespective of Cu/In ratios in the grown layers. The performance of solar cells for lower or beyond the optimum substrate temperature of 450 °C and Cu/In = 0.9 deteriorates. The layers show triangular grain structure for the optimum conditions of $T_s = 450$ °C and Cu/In = 0.98. Small grains and poor crystallinity are observed by SEM analysis for $T_s = 350$ °C. The efficiency of glass/ZnO:Al/CdS/$CuInSe_2$(Cu/In = 0.98)/Au solar cell with an area of 3.5×3 mm^2 is 8.1%, whereas it is 2.7% for Cu/In = 0.81. The cells made with evaporated CdS show inferior efficiency of 5.4%. As shown in Fig. 8.16, the photoresponse at both shorter and longer wavelength regions is wider for CBD-CdS thin film solar cells comparing with evaporated CdS cells owing to wider band gap and probably no interdiffusion with CIS layers [93].

The back wall configured glass/ITO/ZnO/CdS/$CuInSe_2$/Au solar cell is prepared that 0.8 μm thick ZnO layer by sputtering, 0.3 μm thick CdS by hot wall vacuum evaporation, CIS by three source evaporation and 0.1 μm Au by electron beam evaporation are successively grown onto ITO glass to complete the cell structure. The CIS layers contain composition ratios of Cu/In = 0.9–1 and Se/(Cu + In) = 1, indicating nearly stoichiometric or slightly In-rich. The layers prepared at substrate temperature of 250 °C shows less (112) preferred orientation, whereas layers grown at 300 and 350 °C have (112) preferred orientation with intensity ratio of 9–10 for (112)/(220) peaks. The chalcopyrite characteristic peaks such as (211) and (103) present for the CIS layers grown at 300 and 350 °C, whereas they absent for the layers grown at 250 °C. Figure 8.17 shows that the reverse saturation current is lower in the cells with CIS layers prepared at 300 °C than that in the cells with layers deposited at 350 °C indicating quality diode characteristics. 0.02 cm^2 active area cells annealed under air at 300 °C for 3 h show efficiency of 4.9% [94].

In some cases, the cells show low efficiency due to presence of minute pores between CdS and CIS interface, if everything is fine. The $CuInSe_2$ films are grown onto CBD-CdS covered TCO glass substrates at substrate temperature of 320 °C by

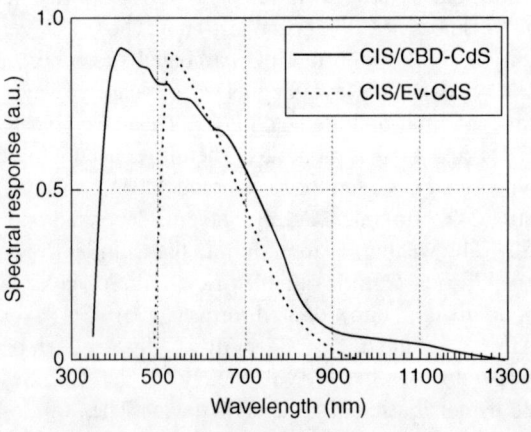

Figure 8.16 Quantum efficiency of CIS superstate cells with different process a) CBD-CdS, and b) Ev-CdS window layers.

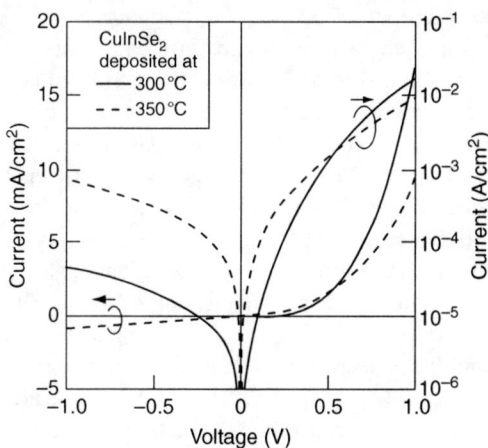

Figure 8.17 I-V curves (forward and reverse bias) of cells formed with CIS layers deposited at a) 300 °C, and b) 350 °C, prior to Au deposition, the cells annealed under air at 300 °C for 3 hours.

evaporating Cu, In, and Se using ionized cluster beam at source temperatures of 1320, 950, and 300 °C, respectively under the pressure of 1×10^{-5} torr. If the substrate temperature of 300 °C is exceeded, CdS and CuInSe$_2$ may be interdiffused. The PL peaks at 0.83 and 0.91 eV are observed in 20K PL spectra for efficient solar cells in which the CIS absorber has Cu/In ratio of 0.88. The preferred orientation of (112) for CuInSe$_2$ and low intensity peak due to In$_6$Se$_7$ are observed by XRD analysis. The SEM and cross section find leaf shape morphology and columnar with densely packed structure. The glass/TCO/CdS/CuInSe$_2$ superstrate cells with an active area of 0.07 cm^2 yield efficiency of 2.06% (Table 8.9) [95]. There is a drawback to make cells with superstrate configuration because the lattice mismatch between ZnO and CIS is large that creates interface states in the junction. In the interface recombination process, the open circuit voltage of cell decreases with increasing acceptor concentration N_A. The diode quality factor (A) can be related to $A = 1 + (N_A/N_D)$, where $N_D > N_A$ for asymmetrical doping [15].

8.1.10 CuInSe$_2$ Modules

1 ft^2 SLG/CIS/CdS/ZnO (MOCVD) module developed by conventional monolithic integration with aperture area of 816 cm^2 exhibits efficiency of 5% and power output of 4 W, whereas small area modules show efficiency of 7.4%. The preparation of cells begins for a typical module that a 50–500 Å thick Te layer evaporated onto 1–2 μm thick Mo covered soda-lime glass substrates by DC sputtering on which 0.65–0.7 μm thick In/Cu layers with Cu to In ratio of 0.9–1.0 sequentially deposited by sputtering technique are selenized using H$_2$Se at 400 °C to form CuInSe$_2$ absorber layer. The CdS layer and 1.5–2 μm thick ZnO layer with sheet resistance of 10 Ω/□ are successively grown onto CuInSe$_2$ by CBD and MOCVD, respectively [96]. The CIS based monolithic circuit size of nearly 0.4 m^2 (32.2 × 128.6 cm^2) is fabricated with total solar cells of 53. There are two kinds of modules, one version of modules shows efficiency of 10.4% with power output of 39 W and another one

exhibits efficiency of 8.5% with power output of 33 W. The mapping of V_{oc} is done all over the places that means on each cell for higher efficiency module. The V_{oc} mapping of higher efficiency module shows uniform open circuit voltage of 464 mV/cell, whereas low efficiency module shows open circuit voltage of 360 mV/cell at some places, particularly at the edges of one side. The lower efficiency suffers due to poor adhesion of CIS to the Mo. The best cell yields efficiency of 14.1%, whereas 0.1 m^2 unencapsulated module with 55 cells and an average 0.4 m^2 module with 53 cells exhibit efficiencies of 11.2 and 8.5%, respectively. In comparison to the best cell, module or monolithic circuit experiences lower efficiency due to the power loss by distribution of resistance, for example ZnO with sheet resistance of 5.5 $\Omega/\square$ and Mo with 0.3 $\Omega/\square$ and interconnects with 0.5 mΩ-cm^2 show loss power 7, 0.5, and 0.5%, respectively [97]. Smaller size module (55 monolithic cells), which contains an active area of 897 cm^2 and aperture area of 938 cm^2 show efficiencies of 11.7, and 11.2%, respectively. The larger size module (53 cells) over the area of 3900 cm^2 shows efficiency of 9.1%, as shown in Table 8.10. In comparison, the larger size module has lower cell V_{oc} and FF due to lower junction quality and lower ZnO uniformity. In the case of larger size module, the cell spacing is 0.557 cm, the Mo isolation spacing of 18 µm is done by irradiating 1.06 µm wavelength Nd:YAG laser. The ZnO/Mo interconnect is made through a 61 µm wide. The typical sheet resistances of Mo and ZnO are 0.3 and 4–5 $\Omega/\square$, respectively. The interconnect contact resistance is 1 m-Ω-cm^2. An active area glass/Mo/2 µmCIS(15%Ga)/50 nmCdS/1.5 µmZnO cell shows efficiency of 14.1% [21]. In the module design, the front contact of the conducting layer sheet resistance should be low without having lower transmittance and at the same time the low segment width offers lower contact resistivity. The efficiency of submodules increases from 11.1 to about 13% by reducing segment/cell width from 0.635 to 0.623 cm in which the conducting layer contains sheet resistance of 15 $\Omega/\square$. The best cell efficiency is 14.4% revealing huge difference in the efficiency between CIS cells and modules. In the case of Si solar cells, difference is very narrow [98].

The inhomogeinity is one of the major problems over the large area CIS coatings that degrade efficiency of the cell. The Cu/In ratio varies from 0.76 to 0.98 at different places over the CIS substrate size of 400 cm^2 or $8'' \times 8''$. The grain sizes

Table 8.10 PV parameters of CIS modules and active area cells

Module	900 cm^2	3900 cm^2	Active area of cell (cm^2)	Ref.
Area (cm^2)	938	3916	–	[21]
V_{oc}/cell (V)	0.464	0.443	0.508	
J_{sc} (mA/cm^2)	39.3	37	41	
V_{oc} (V)	25.5	23.5	–	
I_{sc} (A)	0.639	0.598	–	
η (%)	11.2	9.1	14.1	
Power (W)	10.5	35.8	–	
FF (%)	–	–	67.7	

Table 8.11 PV parameters of cells and modules

Process/area of cell (cm^2)	V_{oc} (mV)	J_{sc} (mA/cm^2)	FF (%)	η (%)	J_o (μA/cm^2)	A	R_s (Ω-cm^2)	R_{sh} (kΩ-cm^2)	Ref.
Active area	431	37.41	64.1	10.2	–	–	–	–	[99]
Sub-module	368	37.2	53.6	7.3	–	–	–	–	
0.61	424	33.5	66	9.4	1.5	1.7	–	5	[100]
0.4233^a	343.3	38.32	50.33	6.62	–	–	–	–	[101]
0.4402^b	426.5	36.11	64.27	9.9	–	–	–	–	

^{a}Flexible substrate.
^{b}SLG substrate.

of CuInSe$_2$ increase with decreasing In composition in the CIS. The Cu/In ratio changes with increasing Se deposition rate during the course of growth. An increase of Se favors to form In$_2$Se$_3$ causing to loss In in the sequential evaporation of Cu, In, and Se elements. The elements are heat-treated under an inert gas unlike co-evaporation. The active area cells and submodules exhibit efficiencies of 10.3 and 7.2%, respectively. The photovoltaic parameters of active area and 25 cm^2 submodule glass/CIS/CdS/ZnO/grid are given in Table 8.11 [99]. The ARCO glass/Mo/2 μmCuInSe$_2$/50 nmCdS/1.5 μmZnO (10 Ω/□) module shows efficiency of 7.3%. 16 individual cells, each with 4 cm^2 is prepared on 10 × 10 cm^2 glass substrates. Some of the active area cells show efficiency of 9.4% [100]. The CIS/CdS/ZnO/MgF$_2$ cells developed on (30 × 30 cm^2 size) Mo coated flexible and glass substrates with absorber containing Cu/In ratio of 0.99 ± 0.04 deposited by sequential process show efficiencies of 6.62 and 9.9%, respectively [101].

8.2 CuGaSe$_2$ Solar Cells

In the band diagram of CGS/CdS heterojunction, by taking valence band offset $\Delta E_V = 0.9$ eV, the conduction band offset is found to be 0.8 eV ($\Delta E_C = E_{g2} - E_{g1} - \Delta E_V = 2.42 - 1.72 - 0.9 = 0.8$ eV), as shown in Fig. 8.18 [102].

The CGS single crystal based homojunction cells are made that the CuGaSe$_2$ crystals are grown by chemical vapor transport (CVD) method using iodine as a transport agent. The CuGaSe$_2$ crystals implanted by Ge with a dose of 10^{15} cm^{-2} and annealed at 400 °C in the presence of Zn result in n-type with electron densities of 5×10^{16} cm^{-3} and mobilities of 150 cm^2/Vs. In order to develop 100–200 nm thick p-type CuGaSe$_2$ on n-type CGS single crystal substrate, a thin Sb layer is diffused into it. The Au fingers and Ag layer on the p-type and n-type sides of CuGaSe$_2$ crystals are formed, respectively. The efficiency of 9.4% is recorded on the Au/p-CuGaSe$_2$/n-CuGaSe$_2$/Ag homojunction solar cell with an active area of 5.8 mm^2. The extrapolation of open circuit voltage line toward 0 K in the open circuit voltage (V_{oc}) versus temperature (T) graph for homojunction cell results in an activation energy of 1.6 eV,

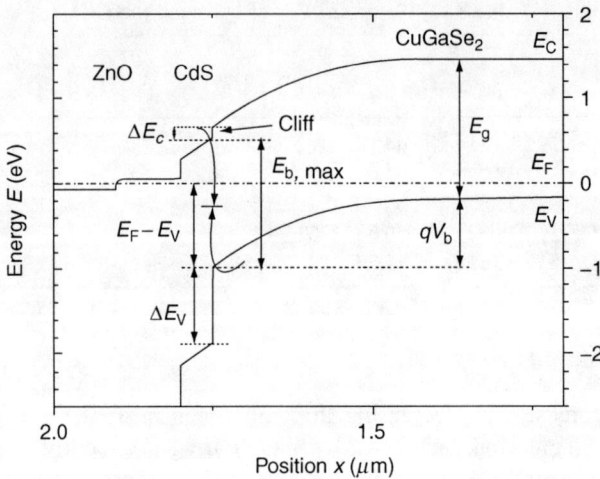

Figure 8.18 Band diagram of CGS/CdS/ZnO thin film solar cell.

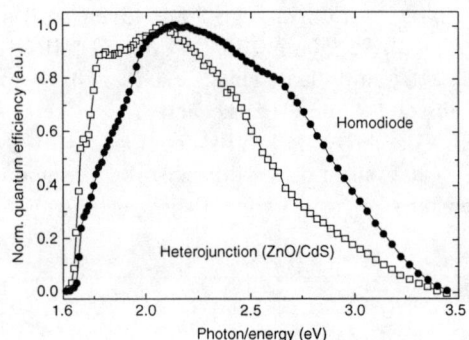

Figure 8.19 Quantum efficiency of p-CGS/n-CGS homojunction and p-CGS/n-CdS heterojunction solar cells.

which is slightly less than the band gap of 1.72 eV for CGS at 0 K that could be due to the position of Fermi energy level, which lies slightly below the conduction band of n-CGS. Figure 8.19 shows QEs of p-CGS/n-CGS homojunction ($\eta = 9.4\%$) and CGS/CdS/ZnO heterojunction ($\eta = 7.4\%$) that the photoresponse is high at shorter wavelength region in the heterojunction due to the absorption of window layer, whereas photoresponse is low at higher wavelength region in the homojunction that could be due to less defect density in the CGS layers [103]. After annealing single crystal based Au/p-CuGaSe$_2$/CdS(CBD)/ZnO(Sput.)/In/MgF$_2$ heterojunction solar cells under vacuum at 200 °C, the efficiency increases from 6.0 to 6.7%. Despite of an increase in efficiency, the series resistance slightly increases from 2.3 to 15.9 Ω-cm^2 and shunt resistance decreases from 3.8×10^5 to 2.5×10^5 Ω-cm^2 in the cells. If the present trend is in reverse direction, the results may be encouragable to improve efficiency of the cell. In fact both as-grown and annealed cells show crossover nature in *I–V* curves under dark and light illumination, as shown in Fig. 8.20. The higher diode quality factor of 3.8 indicates that the tunneling transport mechanism in the cells is indeed dominant one. The quantum efficiency is slightly higher at longer wavelength region in annealed cell (6.7%) than that in the as-grown cell (6.0%) because the carrier concentration decreases due to interdiffusion of Cd and Cu or S interstitials that will cause to increase the depletion width in the annealed one. The extrapolation of V_{oc} versus T graphs toward 0 K temperature gives barrier height of 1.56–1.69 eV for the heterojunction cells that is as same as homojunction cells (Fig. 8.21) [104]. The improved version single crystal CGS based heterojunction solar cell shows efficiency of 9.7% but the diode quality factor (*A*) determined to be higher value of 5.9 is due to

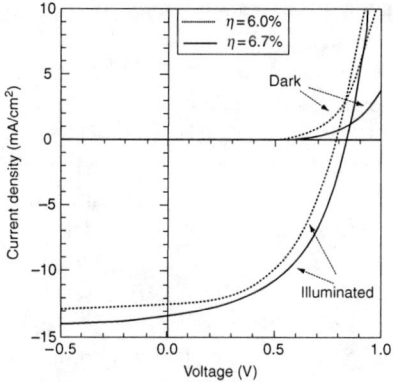

Figure 8.20 *I–V* curves of Au/*p*-CuGaSe$_2$/CdS(CBD)/ZnO(Sput)/In/MgF$_2$ thin film solar cells under dark, and light illumination.

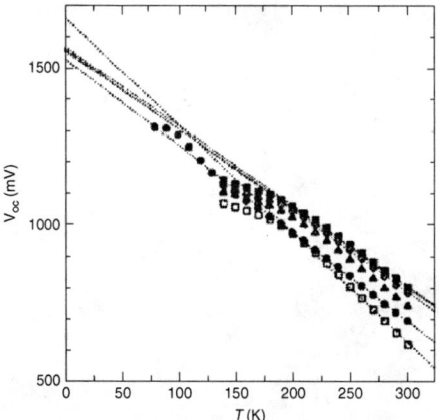

Figure 8.21 Open circuit voltage (V_{oc}) with function of temperature (*T*).

domination of tunneling mechanism rather than SRH. Prior to develop solar cell, the single crystal CGS is polished and treated with Br and methanol solutions, which have carrier concentration of 10^{17} cm^{-3} and mobility of 40 cm^2/Vs. The CdS and ZnO with resistivities of 500 and 3×10^{-3} Ω-cm are sequentially deposited onto CGS crystals by CBD and RF sputtering, respectively. The extracted minority carrier diffusion length and depletion widths for the Au/CGS/CdS/ZnO/MgF$_2$/In single crystal cells annealed at 200 °C are 0.32 and 0.7 μm, respectively. As shown in Fig. 8.22, the quantum efficiency spectrum of Au/CGS/CdS/ZnO/MgF$_2$/In cell reveals three band gaps of 1.68, 1.75, and 1.96 eV for CGS. In addition, the band gaps of 2.48 and 3.3 eV for CdS and ZnO, respectively are observed [105].

The SLG/Mo/CGS/CdS/ZnO:Ga thin film solar cells show conversion efficiency of 8.7% for which CGS layers with Ga/Cu = 1.12 are prepared by close-spaced chemical vapor transport technique. Figure 8.23A shows plots of I–V at different temperatures under dark and light illumination from which $A\{\ln(J_o)\}$ versus $1/T$ drawn curve gives slope as an activation energy of 1.52 eV (Fig. 8.23B). More or less the similar activation energy of 1.50 eV is obtained from the plots of $J_{sc} - V_{oc}$ at different temperatures for the cells. On the other hand, the diode quality factor of (1.6–1.8) is less than 2 and independent of temperature therefore the recombination at the interface can be assigned to SRH [102]. The CGS layers used in the SLG/Mo/CGS/CdS/i-ZnO/ZnO:Ga solar cells are prepared by CVD technique employing Cu$_2$Se and Ga$_2$Se$_3$ solid sources as precursors transporting by HI and HCl gases, respectively. The source temperature of 600 °C and growth temperature of 500 °C for 32–4 h are considered as first stage, whereas the latter precursor is transported to the substrate at $T_s = 532$ °C by HCl and HI gases in second stage. The layers formed with first stage had Cu$_{2-x}$Se as secondary phase, whereas for the combined process of first and second stages, the layers certainly contain slightly Ga-rich phases. The precautionary measurements have to be taken to incur Ga-rich layers in the first stage process. Table 8.12 shows efficiencies of 4.1 and 5.9% for SLG/Mo/CGS/CdS/i-ZnO/ZnO:Ga cells with CGS absorber grown by first and second

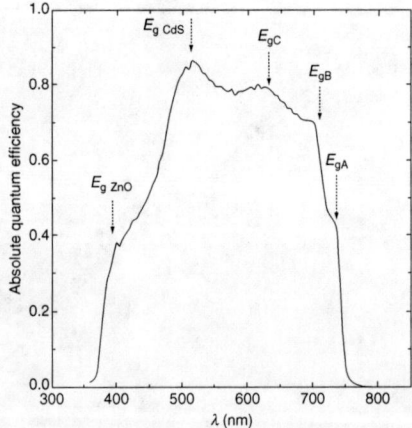

Figure 8.22 Quantum efficiency of single crystal based CGS/CdS solar cell.

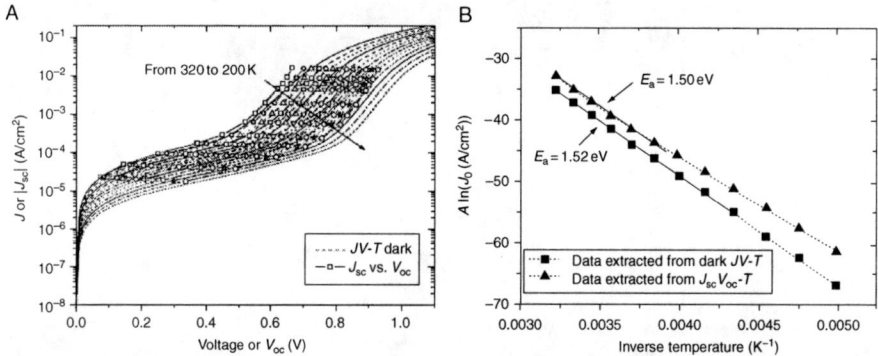

Figure 8.23 (A) *I–V* curves with variation of temperature for CGS cells, and (B) Plot of *A* ln (J_o) versus inverse temperature for SLG/Mo/CGS/CdS/ZnO:Ga cells.

stage processes, respectively. The quantum efficiencies of CuGaSe$_2$ thin film solar cells made with two varieties of processes are given in Fig. 8.24A in which the CuGaSe$_2$ band gaps of 1.68, 1.74, and 1.93 eV, CdS-2.4 and ZnO-3.24 eV are observed for the cells with CGS absorber grown by single stage process ($\eta = 5.9\%$), whereas the band transition at 1.68 eV is missing in the cells with two-stage processed CGS absorber ($\eta = 4.1\%$). The observed band transitions of CGS thin film solar cell from QE (Fig. 8.24A) are as verbatim as CGS single crystal cells (Fig. 8.22). The spectral response is apparently higher in two stage samples than that in single stage samples. Secondly the cutoff wavelength is longer in 5.9% efficiency (two stages) samples than in 4.1% efficiency (single stage) samples. The investigation ensures that the defect density may be less in former. Figure 8.24B contributes that the reverse saturation current density and diode quality factors are low in 5.9% efficiency cells than in 4.1% efficiency cells indicating relatively less defect density in former. The diode quality factor more >2 substantially assures the domination of tunneling mechanism in the low efficiency samples ($\eta = 4.1\%$). As predicted, the low series and high shunt resistances are observed in 5.9% efficiency cells as compared to that in low efficiency cells. The EBIC analysis presents narrow depletion width or space charge region width ($w = 0.13$ μm) in low efficiency cells than in high efficiency cells ($w = 0.37$ μm) that may be prompted tunneling mechanism in former. However, the minority carrier diffusion lengths (L_e) are more or less similar, they are 0.91 and 0.87 μm for low and high efficiency cells, respectively [106].

The best glass/Mo/CGS/CdS/ZnO/ZnO:Al/Ni-Al/MgF$_2$ cell with an active area of 0.442 cm^2 shows efficiency of 9.53% [107]. For which the CGS films are deposited by three-stage process using vacuum evaporation; in the first stage, Ga is evaporated with deposition rate of 4 Å/s at substrate temperature of 450 °C. In the second stage, the Cu is evaporated with growth rate of 3 Å/s at substrate temperature of 600 °C while the temperature is decreased by 2 °C in order to make Cu-rich films at the end of the second stage. In the final stage, temperature is increased to normal temperature of 600 °C by 2 °C to obtain Ga-rich films. The Se flux is maintained to 30 Å/s during the entire three-stage process. The SEM and XRD analyses reveal that

Table 8.12 PV parameters of single crystal and thin film based CuGaSe$_2$ solar cells

CGS cell configuration	V_{oc} (mV)	J_{sc} (mA/cm^2)	FF (%)	η (%)	J_o (A/cm^2)	A	R_s (Ω-cm^2)	R_{sh} (kΩ-cm^2)	Ref.
CGS/CdS/ZnO/ZnO:Ga	743	17.7	65.8	8.7	2.2×10^{-7}	1.8	1.3	4.5	[102]
p-CGS/n-CGS[a]	945	13.6	73	9.4	–	–	–	–	[103]
p-CGS/n-CdS/ZnO	837	13.4	50	6.7	1.9×10^{-4}	3.8	15.9	2.5×10^2	[104]
p-CGS[a]/n-CdS	946	15.5	66.5	9.7	8.3×10^{-3}	5.9	28	–	[105]
SLG/Mo/CGS/CdS/ZnO/ZnO:Ga[b]	804	12.9	57	5.9	1.3×10^{-9}	2.13	1.83	15.4	[106]
SLG/Mo/CGS/CdS/ZnO/ZnO:Ga[c]	684	11.7	51	4.1	9.8×10^{-7}	3.42	0.881	0.481	
CGS/CdS/ZnO/ZnO:Al	905	14.88	70.8	9.53	7×10^{-7}	2.1	1.4	1.396	[107]
SLG/Mo/CuGaSe$_2$/CdS	580	9.6	46	2.6	–	–	–	–	[109]
SLG/Mo/CGS/CdS/ZnO/ZnO:Al	574	11.9	57.9	4	–	–	–	–	[110]
p-GaAS/p-CGS/n-CdS	590	8.14	44	2	1.5×10^{-3}	2.8	10^{-7}	0.183	[111]
p-GaAS/p-CGS/n-ZnSe	555	15.5	38.2	3.3	4.2×10^{-4}	2.9	1.4×10^2	1.3×10^2	
CGS/CdS/ZnO	853	10.1	45.8	4	–	–	–	16	[112]
p-CGS/n-CdZnS	820	14.6	65	6.2	–	–	–	–	[113]

[a]Single crystal.
[b]1st stage-Ga-rich.
[c]Combined process.

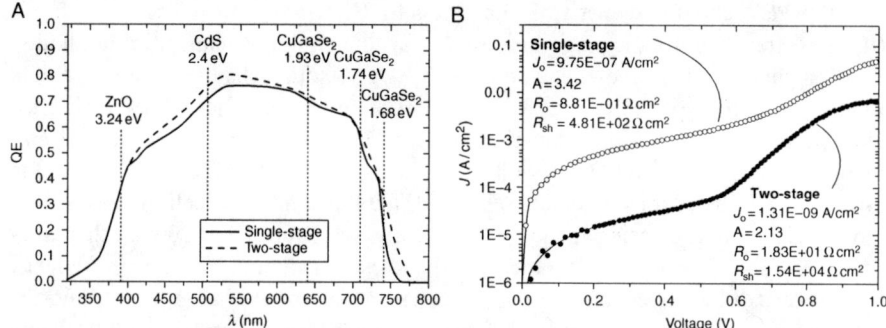

Figure 8.24 (A) The QE of CuGaSe$_2$ thin film solar cells, and (B) I–V curves of cells under dark.

the CGS films show columnar structure, preferred orientation of (204,220) and intensity ratios of (204,220)/112 = 11.4. The carrier concentration (p) of CuGaSe$_2$ layers is 2×10^{15} cm^{-3}. Let us study the low efficiency CGS cells. The CGS/80–100 nm CdS/200 nm ZnO(60–80 Ω/□)/Ni–Al cells experience FF = 0.66, $R_s > 8$ Ω-cm^2, and $R_{sh} < 1000$ Ω-cm^2. The cells with CdS layer deposited by CBD at growth temperature of 85 °C show superior spectral response as compared to that of cells with CdS grown at lower temperature of 60 °C. The CuGaSe$_2$ thin film absorber used in the cells prepared by the same kind of three-stage process at different substrate temperatures of 600, 630 and 660 °C, at first stage the substrate temperature is kept at 500 °C and the Se ratio is varied in between of $2 <$ Se/(Cu + Ga) < 4. The (110) preferred orientation occurs for Se to metals ratio of >3.5. Either random or (112) preferred orientation develops for the ratio of <3, whereas secondary phase of GaSe generates for <2.5. The grain sizes of the samples increase with increasing substrate temperature [108]. The CuGaSe$_2$ thin films deposited by vacuum evaporation with evaporation rates 1.0, 2.6, and 14 Å/s of constitute elements such as Cu, Ga, and Se. 2.8 μm thick CGS layer used in the cells have composition ratio of Cu:Ga:Se = 24.2:24.7:51.1 that is, slightly Ga-rich with resistivity of 3×10^3 Ω-cm, which can be varied from 10^{-6} to 10^6 Ω-cm by changing composition. The band gap of layers can also be varied from 1.66 to 1.69 eV. The EBIC signal of 2.6% efficiency thin film Mo/CuGaSe$_2$/CdS solar cells with an area of 1 cm^2 stays at heterojunction [109]. The CuGaSe$_2$ thin films flash evaporated at 350 °C are treated by RTP under Se atmosphere at 550 °C on which CdS and ZnO/ZnO:Al layers are sequentially made by CBD and RF sputtering, respectively. Finally, the SLG/Mo/CGS/CdS(CBD)/ZnO/ZnO:Al (RF Sput.) cell annealed under vacuum and light soaked shows efficiency of 4% [110].

Instead of traditionally using metalized glass substrates the p-GaAs substrates with carrier concentration (p) $> 10^{18}$ cm^{-3} are employed for the growth of solar cells. The p-GaAs/CGS(epitaxial)/ZnSe/ZnO/Al–Ni and p-GaAs/CGS(polycryst.)/CdS/ZnO/Al–Ni cells show efficiencies of 3.3 and 2%, respectively (Table 8.12). The cells with ZnSe show lower reverse saturation current and higher shunt resistance as compared to that of CdS based cells. The CuGaSe$_2$ and ZnSe thin films by MOVPE technique whilst CdS thin films by CBD are grown. The diode factor (A) is greater than 2 indicating participation of tunneling mechanism in the diodes. The *I–V* curves of CuGaSe$_2$/ZnSe and CuGaSe$_2$/CdS solar cells under dark and light illumination are depicted in Fig. 8.25 [111]. The shunt resistance of 4% efficiency glass/Mo/CuGaSe$_2$/CdS/ZnO thin film solar cells with an active area of 0.51 cm^2 increases from 6 k to 16 kΩcm^2 with increasing thickness of CBD-CdS from 30 to 60 nm [112]. The efficiency of CGS/CdZnS/ZnO thin film solar cell increases from 2.6 to 4.5% with increasing thickness of CdZnS layer from 50 to 200 nm thereafter decreases to 4.3% with further increasing thickness of 300 nm. The lower efficiency for thicker CdZnS layer may be due to extension of space charge region into the CdZnS causing to decrease electric field in the junction. Therefore, the optimum thick CdS or CdZnS is required to promote higher efficiency in the cells. The cell shows better efficiency for CdZnS layer with conductivity of $>10^{-2}$ Ω-cm^{-1}, whereas the efficiency slips to below 3% for lower conductivity of $<10^{-2}$ Ω-cm^{-1}. The Zn concentration of 30–50% is used in the CdZnS layer. The cell shows efficiency of below 3% without ZnO intrinsic layer, whereas the same cell yields efficiency of >5% for incorporation of 500 nm thick i-ZnO layer. The efficiency of best CGS/CdZnS/ZnO solar cell is 6.2%. The CGS layers with excess Cu deposited at substrate temperature of 650 °C with Se/metal ratio of 3 either single or double ramp processes are etched in either NaCN or KCN for thin film solar applications. Despite of excess Ga in the CGS layers, the layers show *p*-type conductivity, unlike CuInSe$_2$ [113]. In conclusion, the important points have to be considered to increase efficiency of CGS solar cells such as (1) slightly Ga-rich CGS absorber layer, (2) Na doping into CGS layer to enhance its quality, (3) the growth rate and bath temperature of CdS, and (4) the annealing of device under air at moderate temperature.

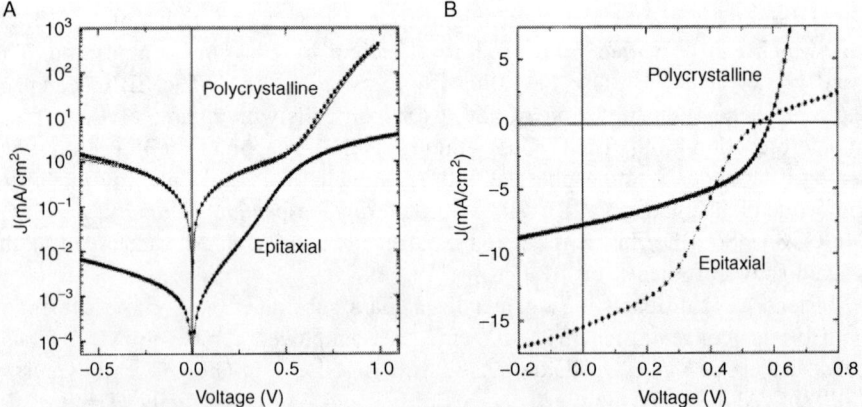

Figure 8.25 *I–V* curves of CuGaSe$_2$/ZnSe epitaxial, and CuGaSe$_2$/CdS polycrystalline thin film solar cells under; (A) dark, and (B) light illumination.

8.3 Cu(In$_{1-x}$Ga$_x$)Se$_2$ Thin Film Solar Cells

The Cu(In$_{1-x}$Ga$_x$)Se$_2$ (CIGS) based thin film solar cells have earned special interest among the thin film solar cells because their efficiency has been significantly reaching to 20%. There are several important merits in CIGS to reach high efficiency: (i) the band gap of CIGS can be varied from 1.04 to 1.68 eV in such a way by varying Ga composition to obtain required band gap that meets the solar spectrum to absorb most of the photons, (ii) the thermal expansion coefficient of CIGS matches with soda-lime glass (SLG), (iii) in order to make abrupt junction with window layer, the carrier concentration and resistivity of CIGS can be varied by controlling its intrinsic composition without using extrinsic dopants, (iv) the band gap of absorber layer decides how much quantity of thickness to be needed for the absorber in the thin film solar cells. 1.5–2.5 μm thick CIGS layer is enough to form CIGS based thin film solar cells because of its direct band gap, whereas in the case of Si based solar cells, the Si needs thicker layers about 250 μm owing to indirect band gap. In the veteran CIS thin film solar cells, 1–20 μm heavy thick CdS window layer grown by evaporation has been essentially employed to mask pin holes and defects to some extent in it. In the modern CIGS cells, a very thin CBD-CdS layer about 30–100 nm as a buffer replaces evaporated CdS probably to have minimum pin holes, and (v) low cost soda-lime glass substrate acts as a source for diffusion of Na into CIGS but the diffusion is random and uncontrollable for self-doping.

The demerits are that if the Ga concentration is reached beyond $\sim x = 0.26$ in the CuIn$_{1-x}$Ga$_x$Se$_2$ ($E_g = 1.2$ eV) thin film solar cells, the efficiency of cell decreases because of an increased defect density by Ga. Even though the optimum optical band gap of 1.55 eV can be reached out in the CuIn$_{1-x}$Ga$_x$Se$_2$ by increasing x from 0 to ~ 0.8 to meet the solar spectrum but the fact is that the cell efficiency is low as compared to one with band gap of 1.2 eV. In the preparation of CIGS thin film solar cells, the vacuum process has to be broken in order to deposit CdS layer by chemical bath deposition. If the deposition of CdS is by evaporation then the in-line vacuum process would be in continuous for the preparation of CIGS modules without breaking vacuum process. The Cd is a toxic therefore a new full pledged nontoxic window has to be invented that is under process. Of course a control of precise composition and the single phase in the CIGS system is somewhat difficult, because which consists of a lot of elements.

The glass/Mo/CuIn$_{1-x}$Ga$_x$Se$_2$(2.5–2.75 μm)/CdS(50–60 nm)/i-ZnO(90 nm)/ZnO: Al(120 nm)/Ni–Al/MgF$_2$(100 nm) thin film solar cells (Fig. 8.26) show the world record efficiency of 19.5–20.3%. The CIGS absorber layer used in the hallmark efficiency cells virtually contains $0.88 < $ Cu/(In+Ga) < 0.95, $x \sim 0.3$ and band gap of 1.21–1.14 eV. The National Renewable Energy Laboratory (NREL) group found that $0.69 \leq$ Cu/(In+Ga) ≥ 0.98 and $0.21 \leq$ Ga/(Ga+In) ≥ 0.38 are also indeed the favorable compositions to obtaining high efficiency (19–19.5%) CIGS thin film solar cells (Table 8.13) [114,115]. The CIGS absorber layer has been developed onto Mo sputtered glass substrates by three-stage process that first (InGa)$_2$Se$_3$ precursor layer is deposited by co-evaporation and in the second stage, on which Cu

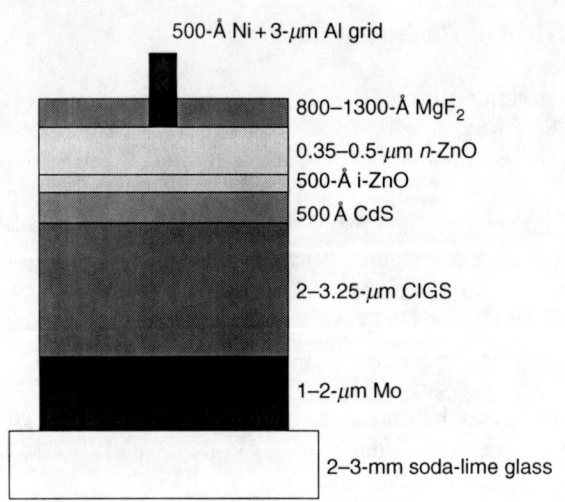

Figure 8.26 Schematic diagram of a typical high efficiency CIGS thin film solar cell.

Table 8.13 PV of CIGS solar cells with an active area of <0.5 cm^2

Cell No./area (cm^2)	V_{oc} (mV)	J_{sc} (mA/cm^2)	FF (%)	η (%)	J_o (A/cm^2)	A	R_s (Ω-cm^2)	R_{sh} (kΩ-cm^2)	Ref.
A_{01} (0.449)	678	35.22	78.65	18.8	–	1.5	0.2	10	[114]
A_{02} (0.414)	674	33.98	77.24	17.7	–	1.6	0.3	2.8	
A_{03} (0.410)	693	35.34	79.40	19.5	–	–	–	–	
(220 CIGS)	693	33	75	17	–	–	–	–	[118]
(112 CIGS)	640	33	74	15.5	–	–	–	–	
A_1	701	34.60	79.65	19.3	5×10^{-11}	1.35	–	–	[119]
A_2	704	34.33	79.48	19.2	6×10^{-11}	1.36	–	–	
A_3	703	34.08	79.23	19.0	6×10^{-11}	1.35	–	–	
A_4	740	31.72	78.47	18.4	4×10^{-10}	1.57	–	–	
A_5	737	31.66	78.98	18.2	5×10^{-10}	1.62	–	–	
A_6	689	35.71	78.12	19.2	N/A	N/A	–	–	
B_1	649	36.1	75.1	17.6	7×10^{-9}	1.5	0.5	7.4	[121]
B_2	674	35.4	77.4	18.5	6×10^{-10}	1.4	0.5	4.8	
Normal	664	29.24	78.04	15.1	–	–	–	–	[124]
Double	652	33.18	77.44	16.8	–	–	–	–	
	608	36.6	75.7	16.9	–	–	–	–	[123]

and Se are deposited and substrate temperature is changed in order to change Cu-poor to Cu-rich layer. In, Ga, and Se are also co-evaporated in the third stage to have slightly Cu-poor Cu(InGa)Se$_2$ layer with band gap of 1.15 eV for the high efficiency cells. The CdS layer is deposited onto SLG/MO/CIGS by chemical bath deposition (CBD) using chemical solutions of 1.5 mM CdSO$_4$, 1.5 M NH$_4$OH, and 7.5 mM thiourea at 60 °C for 15 min. The ZnO bilayer with sheet resistance ($R_\square$)

of 65–70 Ω/□, which consists of 90 nm thick ZnO and 120 nm thick Al doped ZnO, is deposited onto SLG/Mo/CIGS/CdS stack using intrinsic and ZnO:Al$_2$O$_3$(2 wt%) targets, respectively in the presence of Ar/O$_2$ as sputtering gases by sputtering technique. The Ni/Al grids are developed onto ZnO layers by electron beam (E-beam) evaporation and finally 100 nm thick MgF$_2$ as an antireflection coating is deposited by the same technique. The laboratory SLG/Mo/CIGS/CdS/i-ZnO/ZnO:Al/Ni-Al/MgF$_2$ thin film solar cells denoted as A_1 and A_4 with efficiencies of 19.3 and 18.4% had composition ratios of Ga/(In + Ga) = 0.26 and 0.31, respectively. The large grain size CIGS thin films are very essential for extremely high efficiency cells, whereas the smaller grain size samples are well enough for moderate efficiency ($\sim$15%) cells. The cross sections of scanning electron micrographs (SEM) for high efficiency CIGS cells grown on flexible and glass substrates are shown in Fig. 8.27A–C. The nanocolumnar structure can be clearly seen in the Mo contact layer. The large grain columnar structure of CIGS absorber is observed in the high efficiency cells developed on flexible substrates [116,117]. The top and bottom halves of CIGS layer consist of grain sizes of 1 and 0.2–0.5 μm, respectively [115].

The XRD analysis reveals that the intensity or texture ratio of (220)/(112) is 15.4 for CIGS absorber employed in the high efficiency (19.5%) cells, whereas it is 1.94 in the low efficiency cells, indicating that (204 or 220) is the preferred orientation. The CIGS thin films can be grown with either (112) or (220) orientation as dominant one by tailoring growth conditions. The In–Ga–Se precursors are deposited at different Se/(In + Ga) flux ratios varying from 6.7 to 11.9 at 350 °C, followed by deposition of CIGS thin films by keeping Se/Cu ratio of 60 at 550 °C. The highly preferred (112) orientation can be obtained for Se/(In + Ga) flux ratio of <7.1 and (220) becomes preferred one for >7.6. The SLG/Mo/2.2 μm CIGS/CdS(50 nm)/ZnO (MOCVD) solar cells made with (220) oriented CIGS thin films show higher efficiency of 17 than 15.5% for (112) oriented CIGS thin film solar cells (Table 8.13) because (220) oriented CIGS with Ga/(In + Ga) = 0.28 and Cu/(In + Ga) = 0.82 layers allow more Cd diffusion into their system while demonstrating surface treatment by Cd, which is practiced by dipping the samples into combination of ammonia (1.5 M) and CdSO$_4$ (3 mM) solutions at 65 °C for 15 min or deposition of CdS. The surface treated CIGS samples are annealed under air at 200 °C for 15 min to properly dope Cd into the layers. The solar cells show higher efficiency for (220/112) ratio of 2, whereas the efficiency decreases for further increased ratio [118].

Table 8.13 indicates that the solar cell with lower diode quality factor (A) and reverse saturation current (J_o) exhibit higher efficiency or vice versa. The cells exhibit series resistance (R_s) of 0.2–0.3 Ω-cm^2 and shunt resistance (R_{sh}) of 10 kΩ-cm^2, whereas the $R_s = 0$ and $R_p = \infty$ are the realistic parameters for ideal solar cells. The parallel resistance (R_p or R_{sh}) of the solar cell is governed by localized leakage currents, which reduces the parallel resistance and degrades efficiency of the solar cell. The thin film solar cells with efficiencies of 19.3 and 18.2% have band gaps of 1.15 and 1.21 eV, respectively. On the other hand, it is clear that the cells with higher band gap absorbers have higher reverse saturation current. For example, the reverse saturation current increases from 5×10^{-11} to 5×10^{-10} A/cm^2 with increasing band gap from $\sim$1.15 to 1.21 eV. The simulation analysis presents that the diode quality factor increases with increasing band gap of the absorber [114,119,120]. The typical high efficiency cells have shown lower short circuit current and higher open circuit voltage as compared to those of lower efficiency cells (Table 8.13). However, Negama *et al.*, [121] found lower open circuit voltage and higher short circuit current in the high efficiency CIGS cells in contrast to the same efficiency of NREL cells. The typical glass/Mo/CIGS (1.9 μm)/CdS(0.1 μm)/ZnO(0.1 μm)/ITO(0.1 μm)/MgF$_2$(0.1 μm)/Au(0.12 μm) solar cells with an area of 0.96 cm^2 show efficiencies of 17.6–18.5%. The CIGS absorber layer, CdS, ZnO/ITO, MgF$_2$, and Au for the cells are sequentially deposited by three-stage process, CBD, sputtering, and E-beam evaporation, respectively. The reverse saturation current J_o is also lower in the high efficiency solar cells ($\eta = 18.5\%$) because of decreased recombination in the depletion region. During CdS deposition, Cd diffuses into *p*-CIGS to form *n*-type Cd doped CIGS surface that

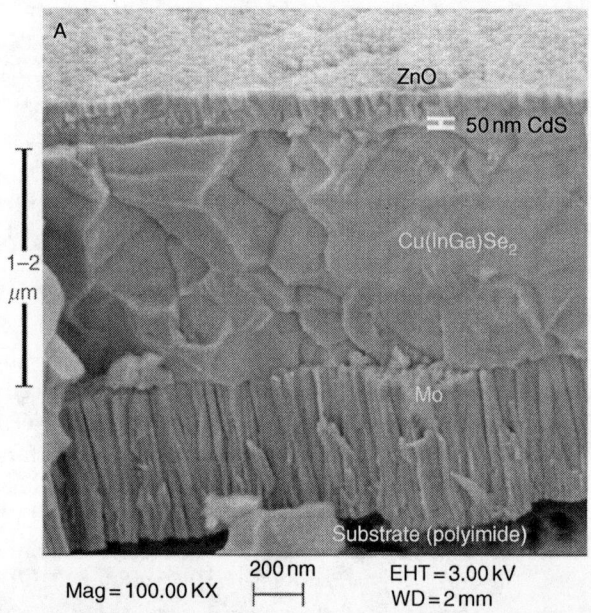

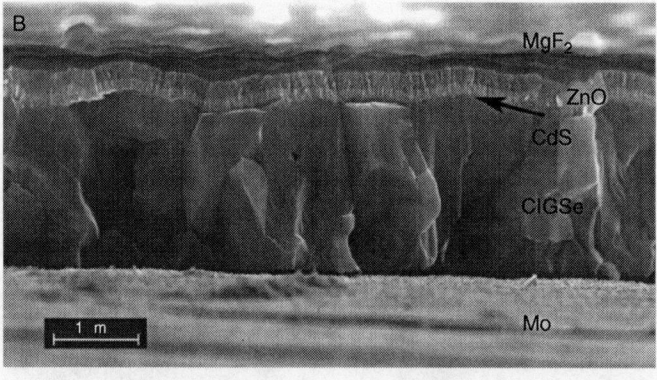

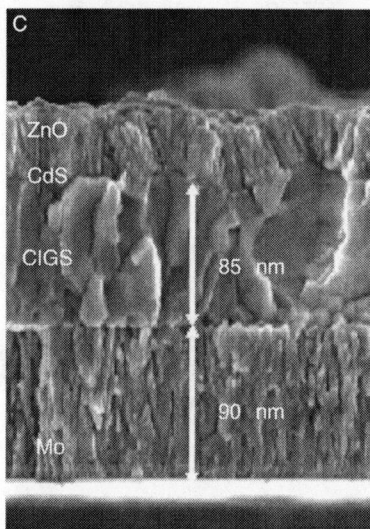

Figure 8.27 (A) SEM cross section of 14.1% efficiency CIGS cell developed on polyimide substrate without antireflection coating, (B) SEM cross section of 16.4% efficiency CIGS cell developed on Ti foil, and (C) SEM cross section of 14.2% efficiency CIGS cell developed on glass.

results in to form buried p–n homojunction as p-CIGS/n-CIGS:Cd. The uniformly formed buried junction reduces the recombination at the interface of p–n junction. The typical I–V curves of 17.6 (B_1) and 18.5% efficiency (B_2) CIGS cells are shown in Fig. 8.28. Slightly higher V_{oc} and lower J_{sc} are observed in cell B_2 than that in B_1.

A small difference in efficiency between two identical high efficiency solar cells can indeed be distinguished by taking their device parameters obtained from I–V measurements. The series resistances of 0.3 and 0.2 Ω-cm^2, shunt resistances of 3.8 k and 10 k Ω-cm^2, diode quality factors of 1.6 and 1.5 are obtained for 17.7 (A_{02}) and 18.8% (A_{01}) efficiency solar cells in which the absorbers show band gaps of 1.14 and 1.12 eV, respectively. On contrast, the best efficiency thin film solar cells always exhibit lower diode factor, series resistances and higher shunt resistance. Perhaps, the higher efficiency cell contains reduced space charge recombination in space charge region. The electron beam induced current (EBIC) analysis reveals that the collection profile lengths of 1.5 and <0.2 μm in the best solar cells with and without CdS buffer layer are observed, respectively. A larger minority carrier diffusion length of absorber layers is obviously advantage for the record efficiency cells. The efficiency of 0.462 cm^2 active area cell decreases from higher to 15%, if CdS buffer layer is removed off in the cell [122]. 1.1 cm^2 area CIGS cells with CIGS absorber layers deposited by three-stage process contain intensity ratio of 4 for (220)/(112) prevailing efficiency of 16.9%, whereas the same cells over the large area of 14 cm^2 show lower efficiency of 13.5% [123]. The high efficiency of laboratory cell decreases from 18–19 to 13.7% while transforming technology from small to large area of 43.89 cm^2 module [122].

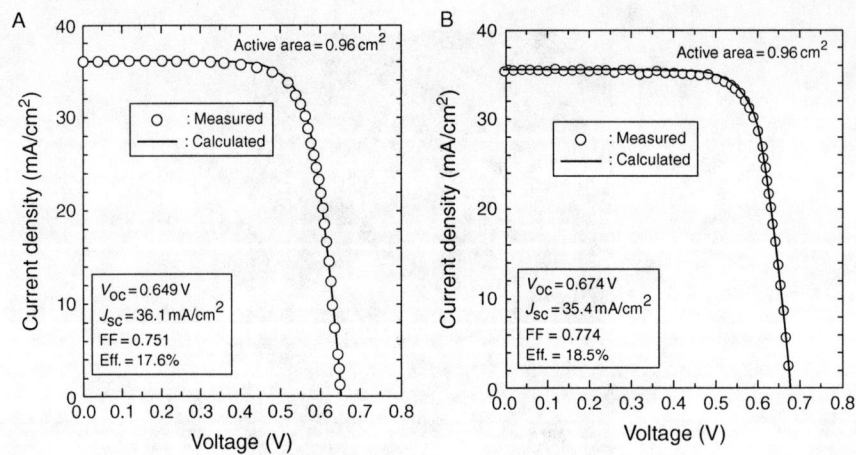

Figure 8.28 $I-V$ curves of high efficiency CIGS thin film solar cells under light illumination; (A) cell-B_1, and (B) cell-B_2.

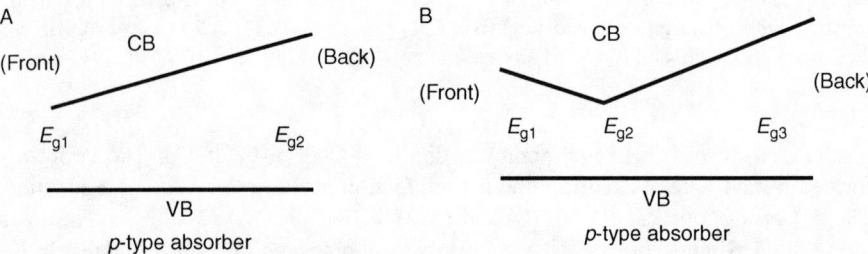

Figure 8.29 Schematic diagram of (A) normal, and (B) double grading band gap profiles in the CIGS layers for solar cell applications.

There are two kinds of band gap profiling gradings such as normal and double adopted from Si technology (Fig. 8.29); the band gap gradually decreases from back to front in the former, whereas the intermediate band gap (E_{g2}) material is sandwitched between low band gap (E_{g1}) front surface layer and high band gap (E_{g3}) back surface layer in the latter. In the normal grading absorber, the excited electron from valence band to conduction band rolls down to lower potential edge that means toward space charge region, thus the recombination at back surface is potentially avoided and the current by minority carriers increases in the solar cell. The back surface field (BSF) due to Ga grading dormantly reduces reverse saturation current. Double grading is well-suited phenomena to absorb more photons from different kinds of spectral regions such as blue and red, which are assigned to Eg_3 and Eg_1 band gaps. Therefore, the solar cell current increases by absorbing most of the photons. On contrast to normal grading, the double grading exhibits fruitful results in the solar cells, the typical examples are given in Table 8.13. The band gaps of both the normal (1.1 eV) and double (1.09 eV) graded CIGS layers are one and the same. The quantum efficiency analysis supports that the spectral response moves toward higher wavelength side for the solar cells with double grading absorber layers as compared to that of normal grading. It is also evidenced by other analysis that the Cu(InGa)Se$_2$ sample with global atomic composition of Cu:In:Ga:Se = 23.18:16.87:9.03:50.92 recorded at a typical voltage of 20 kV by EPMA shows resolving peaks for (112) reflection at 2θ diffraction angle of 27 and 27.3° corresponding to 25 and 75% Ga in the normal profiling grading Cu(InGa)Se$_2$ thin films that means the back and front surface layers contain Cu(In$_{0.25}$Ga$_{0.75}$)Se$_2$ and Cu(In$_{0.75}$Ga$_{0.25}$)Se$_2$ phases in the XRD spectrum, respectively [124,125]. In the graded absorbers, all the time Ga composition is higher at back surface of the absorber as compared to that at front surface in the solar cells, as evidenced by AES depth profile [122].

Figure 8.30 shows external quantum efficiency of 19.3 and 18.2% solar energy conversion efficiencies for CIGS thin film solar cells [119]. The quantum efficiency analysis distinguishes that the quantum efficiency of the best cell enhances toward longer wavelength region because of (i) less grain boundaries in the quality CIGS absorber and (ii) improved higher transmission and lower absorption caused by low free carrier concentration of TCO such as ZnO layer. The situation at shorter wavelength is one and the same in both the 19.3 and 18.2% efficiency cells. In comparison to CdS buffered cells, the collection efficiency starts sharply at ~350 nm in the CdS free buffer solar cells because higher energy photons are freely allowed into the cell thus the current gains to higher [122]. A striking difference between CuInSe$_2$ and CuInGaSe$_2$ based cells is observed by means of

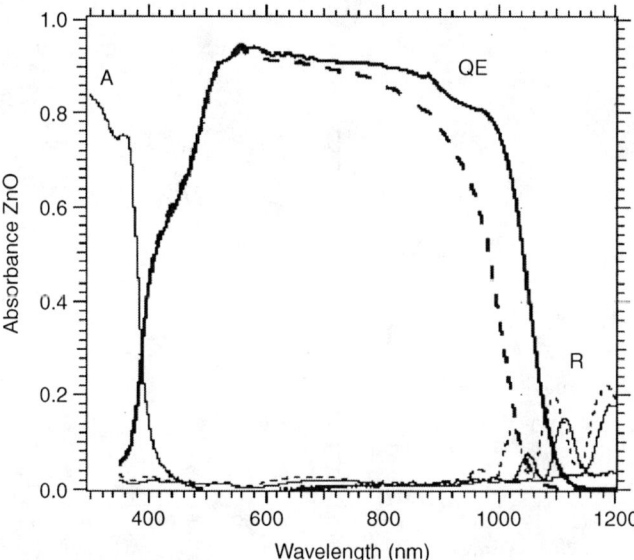

Figure 8.30 External quantum efficiency and reflectance of CIGS thin film solar cells; $\eta = 19.3$ (solid line), and 18.2% (dashed line).

conversion and quantum efficiencies for the same deposition conditions. The typical SLG/Mo/CuInSe$_2$/CdS/ZnO/MgF$_2$ and SLG/Mo/CuInGaSe$_2$/CdS/ZnO/MgF$_2$ thin film based solar cells prepared by standard techniques exhibit efficiencies of 15.4 and 16.9%, respectively. As shown J–V and quantum efficiency curves in Fig. 8.31, a difference of 0.126 V between open circuit voltages of CuInSe$_2$ ($V_{oc}=0.515$ V) and CuInGaSe$_2$ ($V_{oc}=0.641$ V) based solar cells is coinciding with a difference of 0.14 V between cutoff wavelength regions in the quantum efficiencies. The cells used for this investigation are prepared by standard method that a 0.8 μm thick Mo layer with sheet resistance of 0.2–0.3 Ω/□ is deposited onto 0.8 mm thick soda-lime float glass substrates on which 3 μm thick either CuInSe$_2$ or Cu(InGa)Se$_2$ layer is deposited by three-stage process at 550 °C. The Cu/(In + Ga) ratio slightly less than unity, Ga to In ratio of 3:1 and the Se mass ratio three times higher than the stoichiometric of CuInSe$_2$ or Cu(InGa)Se$_2$. A variation in composition is found ±4% over the area of 5 × 5 cm^2 substrates. The CuInSe$_2$ layers deposited onto Mo coated soda-lime glass substrates show pronounced (112) orientation as compared to that of layers on Mo/borosilicate glass, whereas it is very well pronounced on the bare glass substrates. The CdS buffer layer by CBD, 50 nm ZnO layer with high sheet resistance under Ar:1%O$_2$ and 300 nm ZnO:Al layer with sheet resistance of 20 Ω/□ under Ar by sputtering are successively grown onto CIGS or CIS absorber layer. No difference is noticed either using ITO or ZnO as window layers in the solar cells. After depositing Ni as an intermediate layer, a 2 μm thick Al grid is deposited with a probe distance of 1 mm. Finally, MgF$_2$ as antireflection coating is covered on the thin film solar cells. The CIGS cells are annealed under air at 200 °C for 2–10 min, however longer annealing time is necessary for CuInSe$_2$ cells [3]. Unlike three source evaporation, (112) oriented 1.8 μm thick CIGS layers with grain size of 0.5–1 μm are deposited onto 1 μm thick Mo coated glass substrates by electron beam evaporation using In$_2$Se$_3$, Cu, and

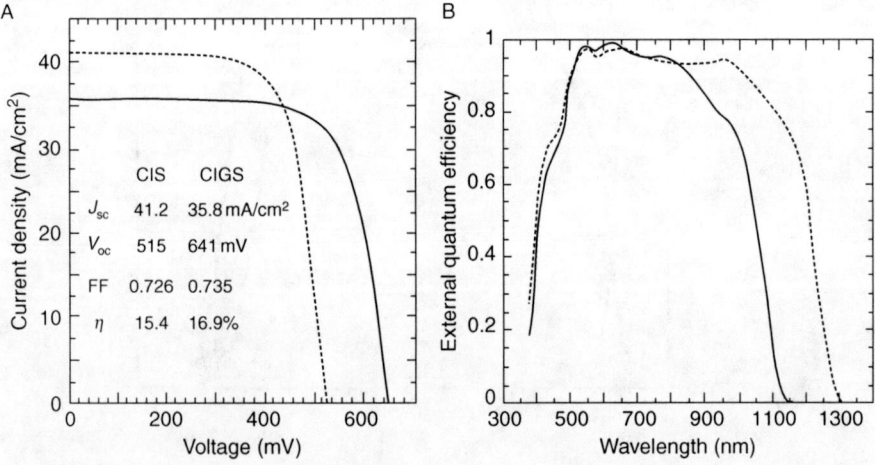

Figure 8.31 (A) I–V characteristics and (B) quantum efficiencies of CuInSe$_2$ and CuInGaSe$_2$ thin film solar cells.

Ga precursor sources. The glass/Mo/CIGS/CdS/ZnO/ZnO:Al cell exhibits efficiency of 14.5% for which 60 nm CdS is grown by sputtering at substrate temperature of 200 °C or by CBD. 100 nm thick ZnO and 2 μm thick ZnO:Al layers are deposited onto glass/Mo/CIGS/CdS stack by sputtering technique. No remarkable difference is found in the cells with CdS buffer grown by either of techniques in this case. All the cells show uniform efficiency over the 1 in.^{2} substrate [126].

The authors [127] fabricated cells on CIGS or CIGSS absorbers grown by different processes received from two different laboratories and four industries, that is, NREL (sample-a), EPV (b), Global Solar (c), ISET (d), SSI (e), and IEC (f). The CdS buffer layer with 60 nm by CBD, 50 nm i-ZnO with resistivity of 1–10 Ω-cm, and 150 nm thick ITO with sheet resistance of 25 Ω/□ by RF sputtering are sequentially covered onto CIGS absorber layers. Finally, 50 nm Ni and 2 μm Al grids are deposited onto the thin film solar cells. Six cells on 2.2 × 2.5 cm^2 substrate are formed and an each cell contains an area of 0.47 cm^2. Sample-a is grown by three-stage process onto Mo coated glass substrates. The hybrid process such as sputtering and evaporation is adopted to prepare sample-b for which Cu layer is first deposited onto Mo coated glass substrates by sputtering and selenized, followed by evaporation of In and Ga in the presence of Se vapor. Sample-c is grown that first In and Ga then Cu are evaporated under Se flux as three-stage process to make CIGS. The ISET has developed CIGS thin film that 10–15 μm thick metal oxide particles based wet ink is coated by nonvacuum knife technique onto metalized glass substrate and annealed under N$_2$ and H$_2$ atmosphere to incur uniform Cu–In–Ga–O alloy layer then annealed under H$_2$Se and H$_2$S atmosphere to convert oxides into 1.2–1.5 μm thick CuInGaSSe$_2$ layers, which are denoted as sample-d. The Shell Solar Industry (SSI) has grown first CuInGa layers by sputtering using Cu–Ga and In targets then selenosulfised under H$_2$Se and H$_2$S atmosphere. The CIGSS samples formed by this type of technique are considered as sample-e. The IEC group has grown type-f samples by two-stage process to form CIGS samples maintaining Cu/(In + Ga) > 1 at initial stage and finally Cu/(In + Ga) = 0.8–0.9 at substrate temperature of 550 °C. The low efficiency samples-d and e have low fill factor, as given in Table 8.14. The studies indicate that the cells made at outside the laboratories show slightly lower efficiencies as compared to their original efficiencies. The cells prepared by SSI show efficiency of 13.3%, whereas the cells formed on the same (SSI) CIGSS absorber by NREL shows efficiency of 7.9%. The reason for low efficiency is that the surface of CIGSS may be well passivated by SSI for their cells to have higher efficiencies. On the other hand, changes in the absorbers induce while building cell. The ISET Company has fabricated glass/Mo/CIGS/CBD-CdS/MOCVD-ZnO cells (d_1) with an active area of 0.084 cm^2 using CIGS absorber by ink based process shows efficiency of 13.6% on soda-lime glass substrates, whereas cells (d_2) on flexible Mo foil, upilex foil, SS foil, and titanium substrates, exhibit efficiency of 9–13% (Table 8.14). 18 monolithic cells made over the large area of 10 × 9 cm^2 show an average efficiency of 10% [128].

By introducing n-Cu(InGa)$_2$Se$_{3.5}$ (OVC) into the CIGS cells, the buried junction can be formed to enhance the efficiency of the cells. Prior to deposition of 0.8 μm thick CuInGaSe$_2$ layer by evaporation from bulk CuInGaSe$_2$, 0.2 μm thick In layer is deposited onto Mo coated glass substrates. Again In layer is deposited with

Table 8.14 CIGS cells made on CIGS absorbers obtained from different laboratories

Sample	$p \times 10^{15}$ (cm^{-3})	μ (cm^2/Vs)	$\rho \times 10^2$ (Ω-cm)	V_{oc} (mV)	J_{sc} (mA/cm^2)	FF (%)	η (%)	R_{oc} (Ω-cm^2)	G_{sc} (mS/cm^2)	Ref.
a	5.3	8.9	1.3	645	33.6	72.8	15.5	1.6	2	[127]
b	0.26	21	12	507	32.2	58.4	9.2	2.8	7	
c	15	9	0.5	530	33.8	67.9	12.2	1.7	1	
d	8.4	0.4	20	444	34.6	55.8	8.6	2.4	11	
e	–	–	–	559	24.5	61.9	8.5	2.9	5	
f	–	–	–	629	32.4	74.9	15.3	1.7	1	
d$_1$/glass	–	–	–	535	35.18	72	13.6	–	–	[128]
d$_2$/Mo foil	–	–	–	491	38.39	73	13	–	–	

R_{oc} can be derived from dV/dJ at $J = 0$ that is combination of series resistance and a diode term and G_{sc} is shunt conductance that is, dJ/dV at $V = 0$.

different thicknesses from 0 to 1.0 μm onto Mo/In/CuInGaSe$_2$. Finally, the entire glass/Mo/In/CIGS/In stack is annealed under Se vapor at 550 °C for 1 h. The ordered vacancy compound such as Cu(InGa)$_2$Se$_{3.5}$ is formed with top 0.75 μm thick In or higher whilst the stoichiometric Cu(InGa)Se$_2$ layer is formed without top In layer coating. The XRD reflections such as (110), (202), and (114) confirm the formation of OVC. The photovoltaic parameters of the cell decrease with increasing In thickness from 0 to 1 μm, however the V_{oc} increases for 0.75 μm thick In layer or more due to a change in band gap of layers. The Cu(InGa)$_2$Se$_{3.5}$ phase starts to form at 0.75 μm thick In layer. The band gaps of 1.23 and 1.06 eV for Cu(InGa)$_2$Se$_{3.5}$ and CuInGaSe$_2$ are determined, respectively. The best glass/Mo/p-Cu(InGa)Se$_2$/n-OVC/n-CdS/i-ZnO/ZnO:Al thin film solar cell exhibits efficiency of 9.56%. The conventional glass/Mo/CIGS/CdS/ZnO thin film solar cells show higher efficiency due to having OVC [129]. The simulation work reveals that the V_{oc} increases and FF decreases with increasing acceptor concentration in the CIGS/n-Cu(InGa)$_3$Se$_5$(OVC)50 nm/CdS cell amid the negative space charge at CdS/OVC or in the OVC increases. The built-in-electric field in the space charge region of CIGS layer lowers that causes to decrease FF in the cell [22].

8.3.1 Effect of CIGS Composition

The (In$_{1-x}$Ga$_x$)$_2$Se$_3$ precursor has been grown onto heated glass/Mo substrates by thermal evaporation of In, Ga, and Se on which Cu layer is deposited by magnetron sputtering at stationary deposition as a hybrid process. The entire glass/Mo/(In$_{1-x}$Ga$_x$)$_2$Se$_3$/Cu stack is selenized under selenium atmosphere at different substrate temperatures without using toxic H$_2$Se gas. The Cu/(In$_{1-x}$Ga$_x$)$_2$Se$_3$ precursor with composition ratio of Cu/(In+Ga) = 1.27 and selenization temperature of 550 °C are optimal to obtain high efficiency cells. After selenization of precursor, the composition ratio of Cu/(In+Ga) = 1.27 becomes 0.72 in the film. The cells made with CIGS layer, which is processed by Cu-poor Cu(In$_{1-x}$Ga$_x$)$_2$Se$_3$ precursor with Cu/(In+Ga) = 0.87 and low selenization temperatures perform inferior efficiency. The efficiency of cell decreases from 10.9 to 8% with reducing selenization temperature from 550 to 500 °C. The efficiency increases from 10.9 to 12% with rising the selenization temperature from 550 to 580 °C. In the case of selenization of Cu-poor precursor, the efficiency does not depend on the selenization temperature. The glass/Mo/CIGS/CdS/i-ZnO/n-ZnO cells with CIGS layer containing Ga/(In+Ga) = 0.28, 0.31 and Cu/III = 0.72, and 0.63 show efficiencies of 10.9 and 8.4%, respectively, as shown in Table 8.15 [130]. The V_{oc} increases from 668, 688 to 702 mV and the FF decreases from 77, 76 to 74% with increasing Cu/(In+Ga) ratio from 0.88, 0.92 to 0.96 in 2 mm thick float glass/Mo/CIGS/CdS(CBD)/i-ZnO/ZnO:Ga/Ni-Al cells. The cells show efficiencies of 16.6, 17.5, and 16.9% for Cu/(In+Ga) ratios of 0.88, 0.92, and 0.96, respectively. The carrier concentration increases with increasing Cu content in the films hence the band gap of the films and V_{oc} increase. The lower V_{oc} and FF for lower Cu/(In+Ga) ratio may be due to presence of Cu$_x$Se, which is not totally nullified by the less Cu composition. The Cu content in the films can be reduced by shortening deposition time of third stage process. Prior to continue

Table 8.15 Effect of Ga composition on the properties of photovoltaic parameters

x	E_g(eV)	E_a (eV)	V_{oc} (mV)	J_{sc} (mA/cm^2)	FF (%)	η (%)	Ref.
0.3	–	–	499	26.9	62.2	8.4	[130]
0.28	–	–	558	26.9	72.7	10.9	
Std.	–	1.11	569	32.3	73.5	13.5	
Pilot	–	–	554	30.7	70.6	12	
0	–	–	411	37.9	67.5	10.5	[133]
0.07–0.12	–	–	457	37.4	68.7	11.7	
0.0	–	–	454	39	69.5	12.3	[134]
0.25	–	–	526	36.5	69.3	13.3	
0.40	–	–	454	35.1	63.3	10.1	
0.20	–	–	518	34.1	72.8	13.1	[135]
0.26	–	–	545.8	36.71	68.38	13.7	
0.48	1.297	–	724	28.9	75	15.7	[136]
0.52	1.325	–	746	28.1	73	15.2	
0.52/MgF$_2$	1.325	–	730	31.8	73	16.9	
0.3 Meas.	–	–	606	33.2	68.8	13.9	[137]
0.3 Cal.	–	–	635	33.5	71	15.1	
0.5–0.7 Meas.	–	–	732	18.5	74.4	10.1	
0.5–0.7 Cal.	–	–	736	18.8	76.6	10.6	
0	0.98	0.99	431	34.7	67.5	11.1	[139]
0.27	1.11	1.19	590	30.2	76	14.8	
0.43	1.22	1.28	675	28	74.9	14.3	
0.49	1.26	1.39	715	26.9	76.2	14.6	
0.63	1.39	1.44	795	22.7	75.9	12.2	
1.0	1.67	1.44	846	15.3	67.3	7.5	

third stage, that is, at the end of second stage, the Cu_xSe and $CuInGaSe_2$ phases exist, which shunt the device causing to decrease FF. The diode quality factor (A) increases from 1.4 to 1.8 with increasing Cu/In ratio that might be due to higher defect levels in the space charge region [131]. In some cases, the cells with Cu/(In+Ga)=0.88 show higher efficiency than the one with 0.90. Two kinds of CIGS layers are evaporated onto glass/0.4 μm Mo at substrate temperature of 500 °C to have Cu/(In+Ga)= 0.88 and Ga/(In+Ga)=0.41 denoted as g_1 and 0.9 and 0.4 as g_2. The substrate temperature is lowered by 100 °C at beginning of the CIGS deposition for g_1 layers, whereas the substrate temperature of 500 °C has constantly been maintained throughout the deposition for g_2. The layers with Cu/(Ga+In) ratio of 0.88 consist of randomly oriented small grains at the bottom and large grains at the top of the layers. The preferred orientation with (112)/(204,220) ratio of 13 and large grains are observed for the Cu/(Ga+In) ratio of 0.90 in the layers. The glass/Mo/CIGS/CdS/ZnO cells formed with g_1 and g_2 absorbers yield efficiencies of 14.5 and 12.2%, respectively [132]. The growth process and composition influence efficiency of the cells. The V_{oc} gradually increases from ~411 to 457 mV with increasing Cu/(In+Ga) ratio from 0.86 to 0.95 and then sharply decreases to ~300 mV with further increasing Cu/(In+Ga) from ~0.95 to 1.08 for Ga/(In+Ga) ratios of 0.01–0.12. The efficiency of CIS cell increases from 10.5 to 11.7% by incorporation of Ga into CIS because of an increased open circuit voltage but the short circuit current density remain lies at more or less the same value of 37.4 mA/cm^2. The advantage of Ga incorporation is that the Ga may have obstructed the shunting causing by Cu$_x$Se phase to the back contact of Mo [133]. ~0.2 cm^2 active area glass/Mo/CIGS/CdS/ZnO (MOCVD) thin film solar cells for which CIGS layers are prepared by two-stage process exhibit efficiencies of 12.3, 13.3, and 10.1% with different Ga/(In + Ga) composition ratios of 0, 0.25, and 0.4, respectively (Table 8.15). Prior to deposition of 2 μm thick boron doped ZnO layer with resistivity of 1.5×10^{-3} Ω-cm at 130 °C by MOCVD technique, the glass/Mo/CIGS/CdS (CBD) is annealed under air at 200 °C for 30 min [134]. The CuIn$_{1-x}$Ga$_x$Se$_2$ bilayer grown by multisource evaporation onto Mo coated alumina substrates for which Cu-rich CIGS layer is grown at 450–500 °C as first layer, followed by deposition of second layer at 500–550 °C as a Cu-poor. The grain sizes of the layers can be increased from 1 to 5 μm provided that the final substrate temperature of 530 °C is applied. The CdZnS layer by CBD, 50 nm ZnO layer with high sheet resistance of 65 Ω/□, and 300 nm ZnO layer with low sheet resistance of 25 Ω/□ by RF magnetron sputtering and Ni/Al grids by vacuum evaporation are successively deposited onto Alumina/Mo/CIGS stack. The efficiency of 0.979 cm^2 area alumina/Mo/CIGS/CdZnS(CBD)/i-ZnO/ZnO/Ni–Al thin film solar cells increases from 13.1 to 13.7% with increasing Ga concentration in the CIGS absorber from 5.4 (Cu:In:Ga:Se = 23.4:21.2:5.4:50) to 6.99% (Cu:In:Ga: Se = 23.38:19.7:6.99:49.93). As expected, the cell open circuit voltage increases with increasing Ga concentration. Despite of an increased Ga composition, the cell current increases due to improvement of window layers. Eventually, the cell current should be in low with increasing Ga because the defect density increases with increasing Ga [135].

Let us see beyond optimum Ga composition of $x \sim 0.3$ corresponding $E_g = 1.15$ eV in the CIGS cells. The efficiency of glass/Mo/CIGS/50 nm thick CdS/70 nm i-ZnO/ 1 μm thick ZnO:Al cell decreases from 15.7 to 15.2% with increasing Ga/(In + Ga) ratio from 0.48 to 0.52 due to an increased defect concentration and grain boundaries, as depicted in Table 8.15. As expected, the SIMS analysis reveals that the Ga content is high at the substrate and gradually decreases to certain point with increasing thickness then follows the reverse trend toward the surface of the layer. The CIGS layers for this cell are grown by three stage process using MBE at substrate temperature of 350 °C for first stage and 550 °C for second and third stages [136]. The measured (calculated) efficiencies of 13.9 (15.1) and 10.1 (10.6%) for different Ga contents of 0.3 and 0.5–0.7 in the CIGS cells are observed, respectively. The 13.9% efficiency cell shows R_s and R_{sh} of 0.2 Ωcm^2 and 1.7 kΩcm^2, respectively. The latter is quite low comparing with high efficiency cells [137]. After deposition of 20 nm thick CdS layer onto CIGS, the CIGS/CdS stack is annealed at 200 °C for 30 min, followed by deposition of 2 μm thick ZnO by MOCVD. The efficiency, J_{sc} and FF of glass/Mo/CIGS/CdS/ZnO(MOCVD) cell with an active area of 0.16 cm^2 decrease from 14.9 to 8%, from 33.8 to 22 mA/cm^2 and from 73.3 to 52% but V_{oc} increases from 600 to 700 mV with increasing Ga content from 40 ($E_g = 1.3$ eV) to 70%, respectively. The CIGS absorber used in the cell is grown by co-evaporation of Cu,

In, Ga, and Se at the substrate temperature of 150–200 °C for 30 min then ramping up to 500 °C in 5 min. The grown CIGS precursor layers are selenized at 500 °C for 1 h under Se beam flux intensity of 1×10^{-5} torr. In the precursor preparation stage, Se content is limited to 30–40% to prevent peeling of CIS from Mo. The In loss occurs by means of forming In_2Se during selenization. On the surface of the layers to some extent, the Ga grading is spontaneously observed in the CIGS films. Porous round-grains, fine grains in conical form and some of them as columnar structure in the cross section are observed. Big chucks are also observed in the graded layers. This may be due to reaction of In_2Se and stable of InSe phase. The device with 70% of Ga performs poor efficiency due to low deposition substrate temperature of 500 °C. The thermal energy is insufficient to fully diffuse Ga into thin films [138]. The efficiency of typical glass/Mo/$CuIn_{1-x}Ga_xSe_2$/CdS/ZnO/ZnO:Al cell increases from 11.1 to 14% with increasing x from 0 to 0.49 thereafter decreases to 7.5% with continuously increasing x to 1.0. The activation energies of CIGS cells obtained from V_{oc}–T plots coincide with their band gaps except for $x=1$, as given in Table 8.15 [139].

The efficiency of typical SLG/Mo/CIGS/CdS/ZnO/ZnO:Al cell increases from 5.2 to 7.3% by using etched CIGS absorber. The efficiency also increases from 9 to 9.7% for changing the deposition sequence from Mo/In/Ga/Cu/In to Mo/Ga/In/Ga/Cu/In that means an extra 110 Å thick Ga layer is inserted on top of Mo. 9.7 and 9.0% efficiency cells with 1.7 μm thick CIGS absorber etched in KCN had compositions of Cu:In:Ga:Se = 23.08:17.9:6.62:52.4 (Se/M = 1.1) and Cu:In:Ga:Se = 22.98:17.48:6.68:52.85 (Se/M = 1.12), respectively. An increase in efficiency may be due to increase of back surface field in the cell by inserting Ga [140]. In the typical case the efficiency of glass/Mo/CIGS/CdS/ZnO/ZnO:B cell increases from 12.5 to 13.5% and then decreases to 10.7% with increasing x values from 0.161, 0.445 to 0.815 in the $Cu(In_{1-x}Ga_x)Se_2$ absorber (Fig. 8.32A). The V_{oc} increases ultimately J_{sc} decreases with increasing x values (Table 8.16). Up to 50% of Ga contained CIGS sample shows moderate efficiency beyond that the efficiency

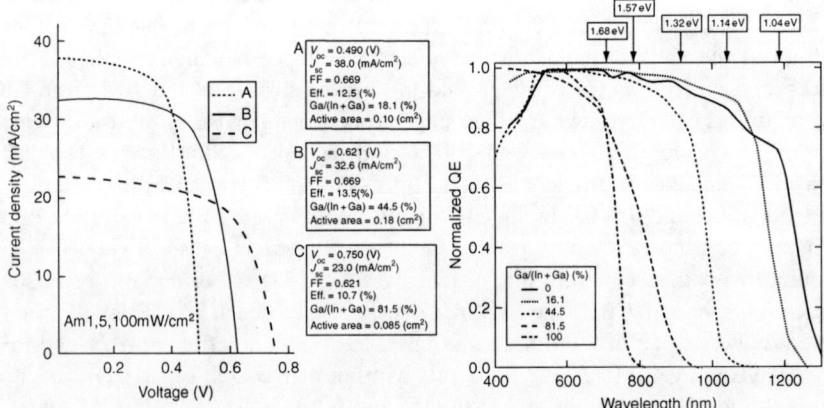

Figure 8.32 (A) I–V curves, and (B) quantum efficiencies of $CuIn_{1-x}Ga_xSe_2$ thin film solar cells with varying x.

Table 8.16 PV parameters of CIGS cells with effect of Ga composition

x	E_g (eV)/Seq.	E_a (eV)/ area (cm²)	V_{oc} (mV)	J_{sc} (mA/cm²)	FF (%)	η (%)	J_o (mA/cm²)	A	Ref.
0.27	Seq.1	–	462.2	32.8	64.1	9.7	–	–	[140]
0.28	Seq.2	Area (cm²)	449.9	33.2	60.6	9.0	–	–	
0.161	1.14	0.100	490	38	66.9	12.5	–	–	[141]
0.445	1.32	0.180	621	32.6	66.9	13.5	–	–	
0.815	1.57	0.065	750	23	62.1	10.7	–	–	
0.681[a]	NA	0.126	609	24.6	60.6	9.1	–	–	
0.25	–	–	560	36	67	13.4	–	–	[143]
0.38	–	E_a (eV)	639	31.9	74.3	15.1	–	–	
0.00	–	0.602	426	35.6	69.1	10.5	7.96×10^{-7}	1.63	[144]
0.13	–	0.614	494	34.2	70.5	11.9	4.29×10^{-7}	1.73	
0.21	–	0.643	581	33.2	70.8	13.7	3.82×10^{-8}	1.84	
0.60	–	0.657	691	23.1	67.6	10.8	6.89×10^{-9}	1.93	

[a]ZnSe, Seq.1-Mo/Ga/In/Ga/Cu/In and Seq.2-Mo/In/Ga/Cu/In.

decreases revealing that the higher substrate temperature is required to deposit over 50% of Ga contented CIGS sample. The quality layers can be obtained at higher substrate temperature of 600 °C or above but the drawback is that the glass substrates cannot withstand beyond 550 °C but withstand at mild temperature. The band gap of $Cu(In_{1-x}Ga_x)Se_2$ absorbers varies from 1.04, 1.14, 1.32, 1.57 to 1.68 eV with varying x from 0, 0.161, 0.445, 0.815 to 1.0, respectively, which are derived from the cutoff wavelengths at red regions in the spectral response curves (Fig. 8.32B) [141]. At longer wavelength sloping in quantum efficiency of typical CIGS cells reveals band gaps of 1.23 eV ($\eta = 11.7\%$) and 1.39 eV (9.4%) for CIGS absorber layers corresponding Ga percentages of 55 and 75 but the expected band gaps are 1.38 and 1.52 eV, respectively. Difference in band gaps may be due to some grading in Ga composition and losses. The efficiency is low for higher concentration of Ga because of formation of multiple phases and an increased density of defects [142].

For example, the efficiency of glass/Mo/CIGS/CdS/ZnO/Ni/Al/MgF$_2$ cell increases from 13.4 to 15.1% with increasing Ga/(In+Ga) ratio from 0.25 to 0.38. The open circuit voltage (V_{oc}) increases and short circuit current (J_{sc}) decreases with increasing Ga. The cells demonstrate efficiency of ~15% for all the compositions of Ga/(In+Ga) = 0.27–0.43 correspondingly the band gap of CIGS layers varies from 1.16 to 1.27 eV. The cells made with CIGS layers containing Ga/(In+Ga) = 0.25 and single phase show efficiency of 13.4%, whereas cells with multiphase layers exhibit low efficiency. The CIGS layers used in the cells are prepared that the Cu–In–Ga multilayer films deposited by sputtering with Cu/(In+Ga) = 0.9, and Ga/(In+Ga) ratio from 0 to 0.75 are selenized under H$_2$Se at 450 °C for 90 min, followed by annealing at 500–600 °C under Ar for 90 min. The XRD analysis reveals that both the selenized layers and layers annealed at 500 °C for 90 min show mixed phases of CuInSe$_2$, CuGaSe$_2$, and CuInGaSe$_2$, however the layers annealed at 600 °C exhibit single phase of CuInGaSe$_2$. The CIGS layers prepared by four elemental sources using two-step process; the single phase CIGS is obtained by first Cu-rich CIGS layers deposited at substrate temperature of 450 °C, followed by continuing deposition of In, Ga, and Se except Cu and increasing substrate temperature to 600 °C. [143]. The efficiency of typical glass/Mo/CIGS/CdS/ZnO/ITO cells increases from 10.5 to 13.7% thereafter decreases to 10.8% with increasing Ga content from 0, 0.21 to 0.6. The same kind of trend follows in V_{oc} and reverse trend in J_{sc}. On contrast, as the Ga content is increased from 0, to 0.21, the hole concentration increases from $\sim 10^{14}$ to 10^{15} cm^{-3} thereafter saturates for further increase of Ga content. The CIGS layers by bilayer process, 40–60 nm thick CdS by CBD, 0.3 µm ZnO and 0.5 µm thick ITO by RF and MgF$_2$ by E-beam evaporation are successively administered for the cells [144].

The glass/Mo/CIGS/CdS/ZnO/ZnO:Al/Ni-Al cells with electrodeposited absorber containing Ga/(In+Ga)=0.24 prevails efficiency of 14.1%. Some of the samples show efficiencies such as 12.2, 13.4, 13.7, and 13.5% for the compositions of Ga/(In+Ga)=0.16, 0.25, 0.26, and 0.29, respectively, which are around the neighborhood of good samples. The cell efficiency increases with increasing Ga content due to an increase of V_{oc}, J_{sc}, and band gap of the CIGS layers. The shunt resistance of ED cells is as low as PVD cells indicating more leakage current. The lower quantum efficiency of ED solar cells is due to difference in morphology and distribution of Ga in the samples those may cause to have lower red response at higher wavelength range as compared to that of PVD cells. The fabrication of cell is as follows: Cu–In–Ga–Se precursor layers deposited onto 1 µm thick Mo using CuCl$_2$ or 0.02 M Cu(NO$_3$)$_2$·6H$_2$O, 0.08 M InCl$_3$, 0.024 M H$_2$SeO$_3$, and GaCl$_3$ or 0.08 M Ga(NO$_3$)$_2$ at potentials of −1.0 V by electrodeposition. Some of the samples are grown by superimposition of 3 V DC and 3.5 V AC at 18.1 kHz. The 0.7 M LiCl chemical is added to the chemical solution to control pH of 2 and contribute to enhance conductivity of the solution. The as-grown layers show composition of Cu:In:Ga:Se = 42.1:15:0.6:42.3 for a typical sample and consisting phases of CuInGaSe$_2$ and Cu$_2$Se. The In, Ga, and Se are additionally deposited onto precursor layers by PVD technique at substrate temperature of 550 °C resulting in composition of Cu:In:Ga:Se = 24.6:19.5:6.2:49.8 and Ga/(In+Ga)=0.24. The AES analysis reveals that the Ga concentration is higher on the surface of CuInGaSe$_2$ thin films, as compared to that in the bulk. 50 nm CdS, 50 nm ZnO/350 nm ZnO:Al and Ni/Al grids by CBD, sputtering, and E-beam evaporation are sequentially grown onto CuInGaSe$_2$ thin films. The CuInGaSe precursor layer is prepared by chemical bath deposition using Cu(NO$_3$)$_2$, In(SO$_3$NH$_2$)$_3$, Ga(NO$_3$)$_3$, Na$_2$SeSO$_3$, triethanolamine, and NH$_4$OH/NaOH. The results on CBD cells are also presented in Table 8.17 [145,146].

The thin film solar cells with high In-rich CuIn$_{1-x}$Ga$_x$Se$_2$ absorbers, which have compositions of Cu:In:Ga:Se = 18.9:28.1:2.5:50.5 and mixed phases of CuIn$_{1-x}$Ga$_x$Se$_2$ and CuIn$_2$Se$_{3.5}$, show poor performance such as V_{oc} of 300 mV and J_{sc} of 1 mA, whereas cells with moderately In-rich CuIn$_{1-x}$Ga$_x$Se$_2$ thin films with composition of Cu:In:Ga:Se = 19.3:27.6:1.3:51.8 demonstrate V_{oc} of 331 mV, J_{sc} of 18 mA/cm^2 and low fill factor of 28%. The high Cu deficiency absorber layers in the thin film

Table 8.17 PV parameters of CIGS cells made by different techniques ED, PVD, and CBD.

Ga/(In+Ga)/Tech.	Area (cm²)	V_{oc} (mV)	J_{sc} (mA/cm²)	FF (%)	η (%)	R_s (Ω-cm²)	R_{sh} (kΩ-cm²)	A	$p \times 10^{16}$ (cm⁻³)	E_g (eV)	w (μm)
0.16/ED	0.414	521	34.93	68.2	12.2	1	1	1.7	1	1.07	0.45
0.24/ED	–	656	29.03	74.1	14.1	0.3	1.4	1.8	–	1.2	0.45
0.26/ED	0.42	602	31.7	69.4	13.4	0.7	0.75	1.8	2	1.13	0.3
0.39/ED	0.419	689	27.65	71.6	13.7	0.9	1	2.1	1	1.22	0.3
0.25/PVD	0.414	674	34	77.2	17.7	0.3	8	1.5	3	1.13	0.4
0.19/CBD	0.43	480	32.15	47.1	7.27	–	–	–	–	–	–

solar cells contribute high series resistance causing poor performance of the cells. The $CuIn_{1-x}Ga_xSe_2$/CdZnS:In/ITO/Al cell with CIGS absorber containing composition of Cu:In:Ga:Se = 23.5:25.5:1.5:49.5 and low pit density show efficiency of 8.2%, whereas cells with CdS shows efficiency of 5.8% in which the first precursor layer has low In content. The cells have high series resistance of 11.2–29.7 Ω-cm^2 and low parallel resistance of 398–1200 Ω-cm^2 indicating inferior efficiency of the cells. The CIGS/CdS/ZnO/Ni/Al thin film solar cell with composition of Cu:In:Ga:Se = 24.25:22.21:4.4:49.14 yields efficiency of 9.02%. The reproducible results can be obtained for the selenization temperature of 550–560 °C. Unlike vacuum evaporation, the $CuInGaSe_2$ thin films employed in the cells are prepared onto Mo coated glass substrates by magnetron sputtering technique using Cu–Ga(22 at%) alloy and In targets; 2210 Å thick CuGa layer, 981 Å thick In, and 736 Å thick CuGa are sequentially grown, followed by selenization at 450–560 °C as a first precursor layer. Second selenization is done to convert $Cu_{2-x}Se$ using Cu-poor process; the first selenized layers are heated *in situ* at 80–90 °C for 10 min, on which 2943 Å In and 736 Å CuGa layers are grown, followed by selenization at 450–560 °C to obtain 2.5 μm thick $CuIn_{1-x}Ga_xSe_2$ layers. The studies indicate that slightly (In + Ga) rich over Cu composition in the CIGS layers demonstrates reasonable results [147–149]. The authors indentified efficiency of 15.5% for Mo/CIGS/CdS/ZnO/ITO cells that Cu/(In + Ga) ratio less than unity is essential for CIGS absorber [150].

Under low irradiations, the standard solar cells are not at all suitable. The cells need to be made with high shunt or parallel resistance (R_p) that can be achieved from the high sheet resistance of absorber layer. If the light irradiation is < 1 mW/cm^2, the parallel resistance should be maintained at > 100 kΩ-cm^2. In order to achieve high R_p, the possible Cu content is 18% in the CIGS. The conversion efficiency is 6% under light irradiation of 0.1 mW/cm^2 for the cells with low Cu content CIGS absorber. The J_o and A slightly increase when the cells move from light to dark due to an increase of parasitic resistance caused by photoconductivity behavior of absorber layer. The SEM shows that ordered columnar growth with grain sizes of 2–4 μm is perpendicular to the Mo back contact in the Cu-rich absorbers, whereas they are in the range of 0.5 μm and 1–1.5 μm in the Cu-poor and standard absorbers, respectively (Fig. 8.33). The resistivity of absorber layer increases from 12 to 115 Ω-cm with decreasing Cu content from 23.3 to 18 at%. Variation of solar cell parameters with effect of Cu content recorded under light illumination of 0.1 mW/cm^2 is presented in Table 8.18 [151]. If the irradiation is below 1 mW/cm^2, the efficiency of cell becomes very worse, that is, < 2–3%. Hence the efficiency of the cell has to be increased by enhancing R_p value more than 100 kΩ-cm^2 that is critical parameter for low irradiance cases. It may be possible by tailoring the Cu composition in the CIGS absorber layer from 21–23 to 18–19 at%. The $R_\square$ of TCO such as ZnO:Al influences less at low light illumination hence its thickness can also be reduced from 1 to $\sim$0.5 μm to have higher transmittance. An increased CdS thickness strengthens hope to have higher shunt resistance but it works adversely by absorbing more photons. The modified cells have to be annealed under air at 200 °C for 15–20 min that influences to reduce the shunt across p_1 scribing area by $\sim$1 order of magnitude in the monolithic cells. The final device on the

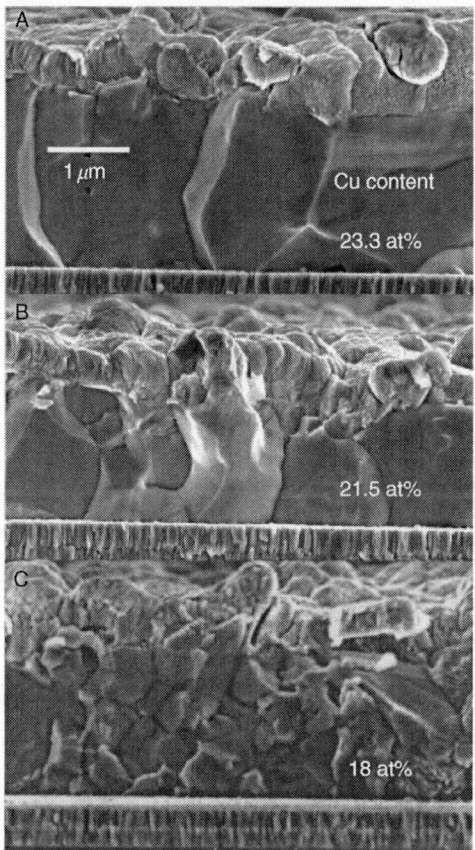

Figure 8.33 SEM cross section of CIGS cells with different Cu compositions.

area of 17.55 cm^2 performs $V_{oc}=4.96$ V, $I_{sc}=40.6$ μA, FF$=58.7\%$, $\eta=5.8\%$, and $P_{max}=118.2$ μW under light illumination of 0.11 mW/cm^2 [152]. 10.3% efficiency SLG/Mo/CIGS/CdS/i-ZnO/ZnO:Al(28±7 Ω/□)/Ni–Al cells show that the R_s, G_s, and A are 0.7 Ω-cm^2, 1 ms/cm^2, and 2.2 under dark conditions while they are $R_s=0.1$ Ω-cm^2 and $A=2.5$ under light illumination [153].

8.3.2 Effect of Substrate Temperature and Thickness of Absorber

The substrate temperature determines quality of CIGS layers at the beginning of deposition of In–Ga–Se precursor layers. The cells with CuInGaSe$_2$ films grown by elemental evaporation at substrate temperature of 550 °C show efficiency of 13.5% for the composition of Cu/(In+Ga)$=0.9$ and Ga/(In+Ga)$=0.3$. The Cu/(In+Ga) ratio is substrate temperature dependent, if the temperature (T_S) is below 400 °C, whereas the ratio is independent of temperature for above the temperature of 400 °C at constant elemental flux. There are several grain boundaries in the layers

Table 8.18 The effect of Cu composition on the properties of photovoltaic parameters

Cu:In:Ga:Se	V_{oc} (mV)	J_{sc} (mA/cm²)	FF (%)	η (%)	J_o (A/cm²) (10⁻⁹)	A	R_s (kΩ-cm²)	R_{sh} (kΩ-cm²)	Ref.
18.9:28.1:2.5:50.5	300	1	–	–	–	–	–	–	[147]
19.3:27.6:1.3:51.8	331	18	28	–	–	–	–	–	[148]
23.5:25.5:1.5:49.5[a]	377	34.8	62.5	8.2	–	–	–	–	[149]
23.5:25.5:1.5:49.5	345–404	30.8–34.2	41.6–49.1	5.8	–	–	0.011–0.03	0.398–1.2	
24.25:22.21:4.4:49.14	451.8	34.5	57.87	9.02	–	–	–	–	
23.3Cu at%	93.9	.0278	25.4	0.7	4.8	1.6/1.7	0.4/0.6	1.9/3.5	[151]
21.5Cu at%	256	.0254	31.6	2.3	31	1.7/1.9	0.6/1	1.7/11.8	
18Cu at%	405	.0236	57.1	6	1	1.5/1.5	0.6/0.8	1.9/142	

[a]Buffer CdZnS:In.

for the growth temperature of 400 °C, whereas the large grain growth is observed for the growth temperature of 550 and 600 °C. This is mainly due to an increased diffusion of Na concentration. The efficiency of typical cells decreases from 12.8 to 7% with decreasing substrate temperature from 400 to 350 °C. The efficiency also decreases from 13.5, 11.7 to 9.9% with decreasing thickness of CIGS layers from 2.5, 1.2 to 1.0 μm [154]. The glass/Mo/CIGS/CdS/i-ZnO/n-ZnO/Au solar cells fabricated with CIGS for which $(InGa)_2Se_3$ (IGS) layers are deposited at 150 and 325 °C show efficiencies of 9.3 and 11.7%, respectively, as shown in Table 8.19. As for as J–V curve concerned (Fig. 8.34A) the reverse saturation current densities are 3.0×10^{-6} and 4.8×10^{-7} A/cm^2 for the cells with the IGS layers fabricated at 150 and 325 °C in the first stage, respectively. The latter, which has low reverse saturation current, can be attributed to the less recombination centers in the space charge region due to larger grain sizes, whereas the high reverse saturation currents are due to interface recombination. The spectral response of cells in the range from 380 to 1200 nm indicates that the cells formed with IGS layers at higher temperature of 325 °C shows higher quantum efficiency, as compared to that of the cells formed with IGS layer at lower

Table 8.19 Variation of photovoltaic parameters with effect of growth temperature

Growth temp. (°C)	V_{oc} (mV)	J_{sc} (mA/cm^2)	FF (%)	η (%)	A	$J_o \times 10^{-6}$ (mA/cm^2)	CIGS $p \times 10^{15}$ (cm^{-3})	Ref.
150	529	28.15	62.5	9.3	2.13	3.0	1.8	[155]
325	550	31.25	68.2	11.7	1.87	0.48	6.8	
Optimum	634	38.4	75.6	17.5	–	–	–	[156]
350	661	31.9	73.6	15.5	–	–	–	
200	614	32.6	71.2	14.2	–	–	–	
100	615	31.8	67.9	13.3	–	–	–	

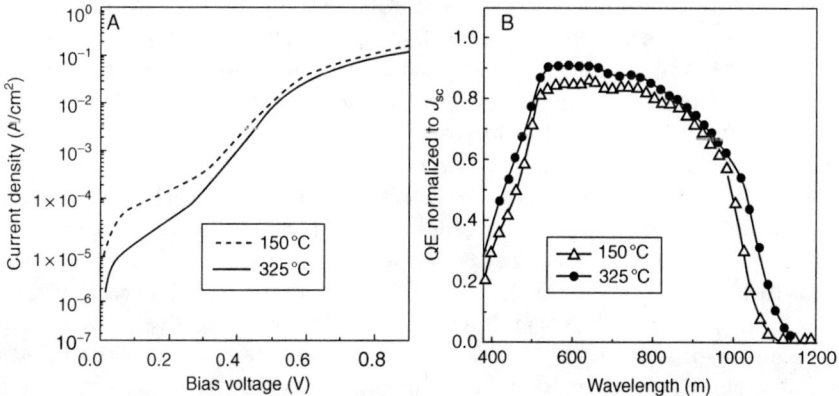

Figure 8.34 (A) I–V curves under dark, and (B) QE of CIGS thin film solar cells made with first stage temperatures of 150 and 325 °C.

temperature of 150 °C. The higher quantum efficiency above 520 nm is attributed to higher diffusion length of the junction because of larger grain sizes, whereas above 1000 nm it is due to higher band gap and an increased grain sizes (Fig. 8.34B). The CIGS absorber used in the cells is prepared that the binary compounds such as In_2Se_3, Ga_2Se_3, and Se are evaporated at 150 or 325 °C to form $(InGa)_2Se_3$ with thickness of 1 μm, followed by evaporation of Cu_2Se at 500 °C to form Cu-rich $Cu(InGa)Se_2$ films, which are annealed under Se atmosphere at 500 °C for 10 min then In_2Se_3, Ga_2Se_3, and Se are evaporated by a small amount to avoid formation of $Cu_{2-x}Se$ [155]. The efficiency of typical glass/Mo/CIGS/CdS/ZnO/ITO/MgF_2 cells decreases from 15.5, 14.2 to 13.3% with decreasing substrate temperatures of In–Ga–Se precursor layers from optimum 350, 200 to 100 °C. This is mainly due to lack of single phase in the CIGS layers and undeveloped grains. The XRD analysis also supports the same thing by exhibiting CuSe phases in the layers [156].

The Cu and In layers deposited onto Mo coated substrates by sputtering technique are selenized by vaporization of elemental selenium to convert into $CuInSe_2$ thin films. After finishing half of the selenization process in the deposition process, Ga is obviously introduced at the substrate temperature of 370 °C by evaporation. However, if the substrate temperature (T_S) of 420 °C is employed instead of a standard temperature of 370 °C, the surface layers contain $CuIn_3Se_5$ (OVC) compound due to slow diffusion of Ga. The active area glass/CIGS/CdS/ZnO cells show efficiency of 11.1% for Ga introducing at T_S of 370 °C, whereas the efficiency of 9.5% is observed for Ga introducing at T_S of 420 °C. The solar cells with absorber layers either $CuInSe_2$ or $CuInGaSe_2$ formed by the binary compounds in the sequence of InSe/CuSe, CuSe/InSe/CuSe, InSe ends up with Cu, CuSe terminates with In, CuSe/InSe, GaSe/CuSe/InSe, GaSe/CuSe terminates with In, InSe/CuSe/InSe, and CuSe/(In+Ga)Se show efficiencies in the 5.5–9.1% range. In the binary layers deposition, first deposition of either InSe or GaSe makes good adhesion, whereas CuSe less adhesive to the substrates [157,158].

Ramanathan et al., [159] studied that the effect of purity, thickness of CIGS absorber, and its deposition temperature on the efficiency of the solar cells. As the thickness of absorber in the glass/Mo/CIGS/CdS(40 nm)/ZnO(50 nm)/ZnO:Al (400 nm) cell is decreased from 1 to 0.4 μm, the cell efficiency decreases from 16.2 to 9.1% (Table 8.20). The absorption co-efficient of the absorber determines magnitude of thickness required for high efficiency cells. The efficiency of cells decreases from 16.2 to 15.5% for low purity elements, the same decreases to 13.7% for higher deposition rates and continues to 12.95% for lower substrate temperature. With decreasing thickness of absorber in the glass/Mo/CIGS/CdS(40 nm)/ZnO(50 nm)/ZnO:Al(400 nm) solar cell, the cell efficiency gradually decreases; the efficiency becomes worse particularly at below 0.4 μm. The efficiency decreases from 16.1 to 15% and from 12.1 to 4.8% with decreasing thickness of CIGS layer from 1.8 to 1.0 μm and from 0.6 to 0.15 μm, respectively. The reason for decrease of efficiency with thickness is that the cell gets shunted for very thin absorber. On the other hand, the absorber follows Beer's law. The shunt resistance of the cell also decreases with decreasing absorber thickness. Secondly, the generated carriers have higher probability to recombine at back contact. The XRD

Table 8.20 PV parameters of CIGS cells with effect of temperature, thickness, and purity of elements

Thickness of CIGS (μm)/or T_s	V_{oc} (mV)	J_{sc} (mA/cm^2)	FF (%)	η (%)
1.0[a]	654	31.6	78.3	16.2
1.0[b]	699	30.6	75.4	16.1
0.75[b]	652	26.0	74	12.5
0.5[b]	607	23.9	60	8.7
0.4[b]	565	21.3	75.7	9.1
T_s = 450 °C (1.4 μm)	585	31.1	70.6	12.9
High purity (99.999%)	671	33.1	77.3	17.2
Low purity (99.9%)	624	34.5	72.1	15.5
High deposition rates	572	32.7	73	13.7

[a]Three-stage.
[b]Co-deposition.

peak is broaden for CIGS layer with Ga/(Ga + In) = 0.3 and 0.5 and with 160 nm thick CuGaSe$_2$ bottom layer, indicating inhomogeneous of the CIGS composition in the layers [160].

8.3.3 Incorporation of Sulfur into the CIGS Thin Film Solar Cells

The band gap of CIGSS absorber increases by alloying sulfur with CIGS. On the other hand, the surface of CIGS is passivated. The formation of CIGSS layer on the surface of CIGS reduces the trap states. In order to form CuInGa(SSe)$_2$ (CIGSS), In$_2$S$_3$ is deposited onto three-stage processed CIGS by evaporation at 580 °C. In another method the CIGS is annealed under sulfur vapor at 580 °C or sulfurization in RTP for several minutes. The glass/Mo/absorber/90 nmCdS(CBD)/ZnO:Al (0.6 μm)/Al thin film solar cell with CIGS and CIGSS absorber yield efficiencies of 12.9–14.5 and 16%, respectively (Fig. 8.35). The signature of CIGSS along with CIGS is observed by means of (112) diffraction peak in the XRD spectrum at low grazing incidence angle of 0.1° (Fig. 8.36A). The AES analysis shows that the concentration of S is higher at surface than that in bulk CIGSS, as shown in Fig. 8.36B [161]. The XRF analysis reveals that the composition ratio of S/(Se + S) varies from 4.8 to 7.9% over the area of 60 × 90 cm^2 CIGS module. The XRD peak position at 26.8° very slightly deviates from pilot line 60 × 90 cm^2 CIGSS module to laboratory10 × 10 cm^2 module [162]. The lowest spectral response curve is projected by the glass/Mo/homogeneous CIGSS/CdS(50 nm)/ZnO cells in which the absorber has higher band gap of 1.43 eV. The spectral response of graded CIGSS is closely similar to standard CIGS, whereas the response of homogeneous CIGSS is inferior and falls sharply very early at longer wavelength region, as shown in Fig. 8.37. In general, the slope at longer wavelength is responsible by the combination of free carrier absorption and absorption in the quasi-neutral region. It is clear from the Auger depth profile analysis that the concentration of sulfur is higher at front and backsides

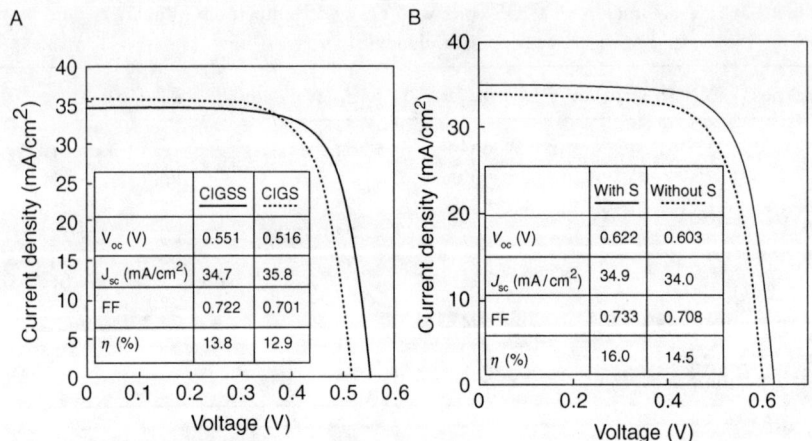

Figure 8.35 CIGSS absorber used in the cells made with (A) In_2S_3, and (B) S vapor.

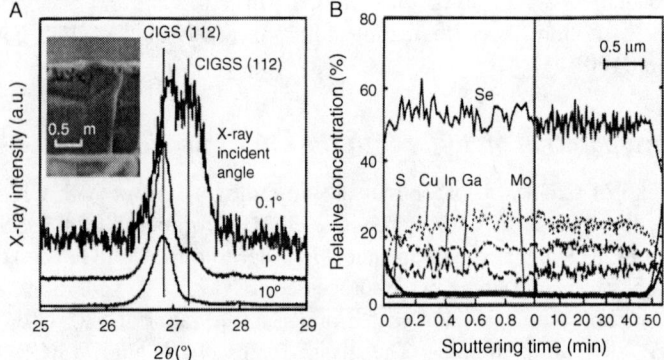

Figure 8.36 (A) XRD pattern of CIGSS layer and (B) AES depth profile of CIGSS; left from In_2S_3, right from S vapor.

in the graded CIGSS absorber, which is used in 15.1% efficiency cells (Table 8.21). On the other hand, Ga concentration is also high at back, as anticipated. By incorporation of Ga into CIS adhesion condition improves. The spatial variation in optical beam induced current (OBIC) density is uniform in the graded devices, whereas which deviates in the normal devices. The efficiency of the cell with high sulfur concentrated CIGSS layer is sensitive to Cu/(Ga+In) ratios below 0.95 and above 1.0, whereas which is not sensitive in the graded cells. InS and $CuInS_2$ phases may be formed in the high sulfur content samples. The efficiency of cell is degraded for above 50% of sulfur. Sulfur rich at the front surface of CIGS causes to have offset in the valence band that lower the recombination of majority carriers in the space charge region, whereas minority carriers do not have much resistance by minimal conduction band offset [163].

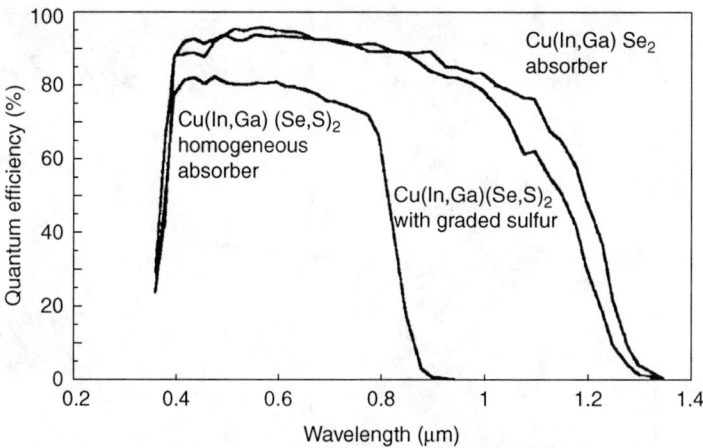

Figure 8.37 QE of CIGS, homogeneous, and graded CIGSS thin film solar cells.

Table 8.21 PV parameters of CIGSS thin film solar cells

Cell	ΔE_a (eV)	E_g (eV)	V_{oc} (mV)	J_{sc} (mA/cm^2)	FF (%)	η (%)	Ref.
Cu(InGa)(SSe)$_2$	–	1.43	728	23.3	65.6	11.1	[163]
Cu(InGa)(SSe)$_2$	–	–	534	39.2	71.8	15.1	
Cu(InGa)Se$_2$	–	0.95	508	41	67.7	14.1	
CuIn(Se$_{0.59}$S$_{0.41}$)$_2$	0.90	1.15	–	–	–	–	[166]
CuIn(Se$_{0.23}$S$_{0.77}$)$_2$	0.89	1.43	–	–	–	–	[167]
Cu(In$_{0.75}$Ga$_{0.25}$)(Se$_{0.73}$S$_{0.27}$)$_2$	1.25	1.22	–	–	–	–	
Cu(In$_{0.69}$Ga$_{0.31}$)(Se$_{0.46}$S$_{0.54}$)$_2$	1.40	1.49	–	–	–	–	
Cu$_{0.9}$In$_{0.69}$Ga$_{0.29}$Se$_{1.96}$S$_{0.13}$	–	>1.2	590	24.6	71	10.1	[171]

Gossla and Shafarmann [164] studied sulfur based CIGS thin film solar cells. The glass/Mo/CIGS/CdS/i-ZnO/ITO/Ni-Al cells with absorber Ga/(In+Ga)=0.73 demonstrate efficiency of 9.4%, whereas cells with sulfur S/(S+Se)=0.26 show lower efficiency of 7.4% but the V_{oc} increases from 808 to 854 mV. The CIGSS absorber layers formed by rapid thermal annealing of CIGS stacked elemental layers deposited onto Mo coated glass substrates in the presence of S vapor, followed by Cd^{2+}/NH$_3$ treatment. 20 nm thick ZnO window extension layer (WEL) by ion layer gas reaction method (ILGAR) at 100 °C, 100 nm i-ZnO and 400 nm ZnO:Ga by RF sputtering and Ni/Al grids are successively deposited to finalize device structure. After light soaking under light illumination of 100 mW/cm^2 for 30 min, 0.5 cm^2 glass/Mo/CIGSS/ZnO-WEL/i-ZnO/ZnO:Ga/Ni–Al and glass/Mo/CIGSS/ 40 nmCBD-CdS/i-ZnO/ZnO:Ga/Ni–Al cells demonstrate efficiencies of 13.8 and 13.4%, respectively then damp heating (DH) test at 85 °C and 85% relative humidity for 100 h results in decrease an efficiency of 9.3 and 10.9%, respectively. The *I–V*

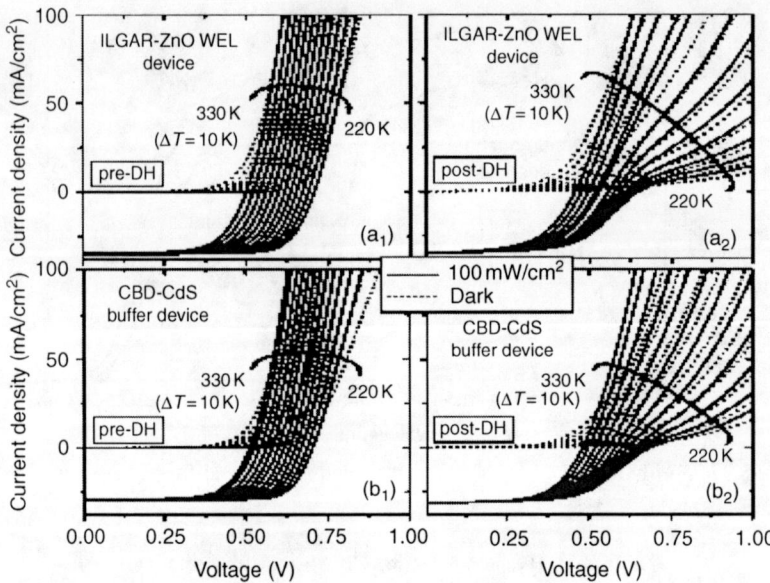

Figure 8.38 *I–V* curves of CIGS cells with variation of temperature.

characteristics of these cells with effect of temperature are shown in Fig. 8.38. The slopes of Arrhenius plots of $A \ln(J_o)$ versus $1/T$ derived from *I–V* measurements with function of temperature give activation energies of 1.12 and 1.10, which are similar to the extrapolation of V_{oc} toward 0 K results in 1.11 and 1.12 eV for ZnO-WEL and CdS based thin film solar cells, respectively. The activation energies are responded by the band gaps of CIGSS absorbers in the as-grown thin film solar cells indicating thermally activated recombination process. After DH process, the obtained activation energies (E_A) of 0.97 and 0.98 eV for WEL-ZnO and CdS based CIGSS cells are less than the band gap of 1.07 eV for CIGSS but a change in diode quality factor is low; therefore, the recombination could be due to thermally activated interface recombination. The recombination takes place at heterojunction interface unlike in bulk for pre-DH samples. After DH, the diffusion length increases and decreases in WEL ZnO based, and CdS control cells, respectively. On the other hand, there is a kink in both the temperature-dependent *I–V* curves of CdS and WEL ZnO based cells due to an increased acceptor like defect states [165].

The plots of V_{oc} versus T (200–375 K) (Fig. 8.39) for SLG/Mo/2 μm CIGSS etched/CdS(50 nm)/ZnO(300 nm) cell result in straight lines and their extensions toward zero temperature give activation energies of 1.25 and 1.4 eV, which coincide with the band gaps of 1.22 and 1.49 eV for $Cu(In_{0.75}Ga_{0.25})(Se_{0.73}S_{0.27})_2$ and $Cu(In_{0.69}Ga_{0.31})(Se_{0.46}S_{0.54})_2$, respectively [166,167]. The activation energies of other systems are also given in Table 8.21, which indicate that the activation energies of Cu-rich absorbers depend on their band gaps, whereas those are independent of band gaps in the Cu-poor absorbers. The CIGSS cells with CIGSS absorber containing composition of Cu:In:Ga:Se:S = 23:25:2:44:5 shows efficiency of 11%, which

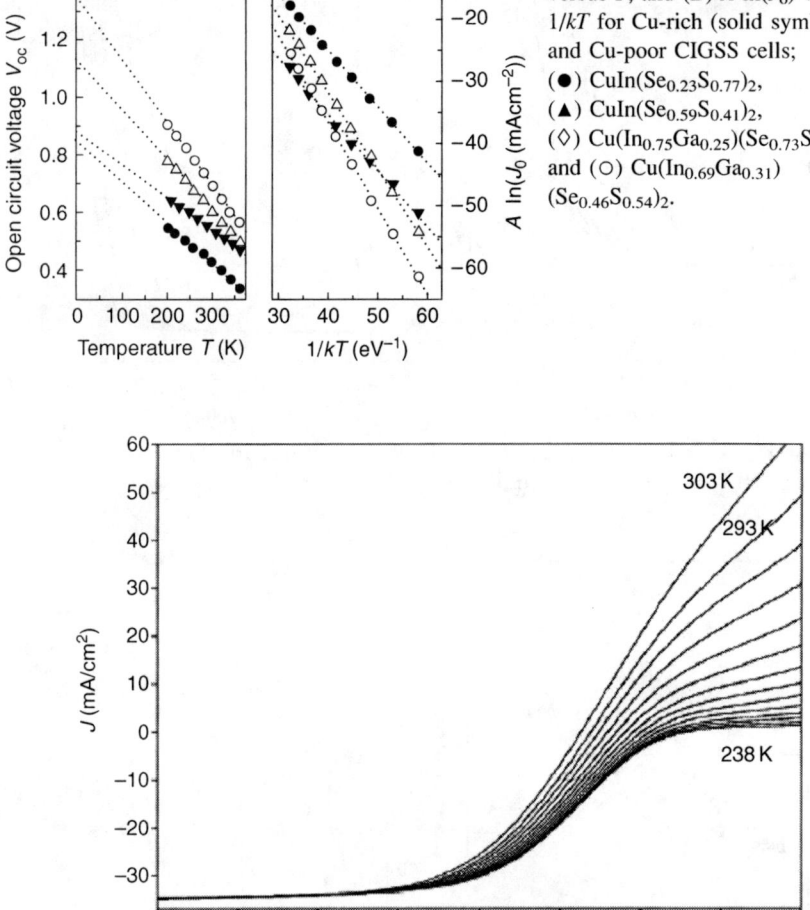

Figure 8.39 Plots of (A) V_{oc} versus T, and (B) $A \ln(J_o)$ versus $1/kT$ for Cu-rich (solid symbols) and Cu-poor CIGSS cells; (●) $CuIn(Se_{0.23}S_{0.77})_2$, (▲) $CuIn(Se_{0.59}S_{0.41})_2$, (◇) $Cu(In_{0.75}Ga_{0.25})(Se_{0.73}S_{0.27})_2$, and (○) $Cu(In_{0.69}Ga_{0.31})(Se_{0.46}S_{0.54})_2$.

Figure 8.40 Temperature-dependent I–V curves of glass/Mo/CIGSS/CdS/ZnO:i/ZnO:Al/Cr/Ag/MgF$_2$ thin film solar cell.

are tested at different temperatures from 220 to 330 K by means of I–V measurements, as shown in Fig. 8.40. The kink, that is, turning current (J_t) is found in I–V curves at current density of 6 mA/cm^2 for the temperature of 273 K. The roll over effect is seen in I–V curves with variation of temperature. The barrier height (ϕ_b or E_a) can be obtained from the expression (Equation (8.4) or (8.5)). The barrier height of 0.44 eV at 273 K is experimentally determined [168]. The sulfur and Ga are found higher level at rear place of 12.78% efficiency glass/Mo/CIGSS/CdS/i-ZnO/ZnO:Al/Cr/Ag/MgF$_2$ cells. No Ga but sulfur is found little lower at front than that at rear side as evidenced by AES depth profile revealing double grading band gap [169]. Figure 8.41 shows that the open circuit voltage increases with decreasing

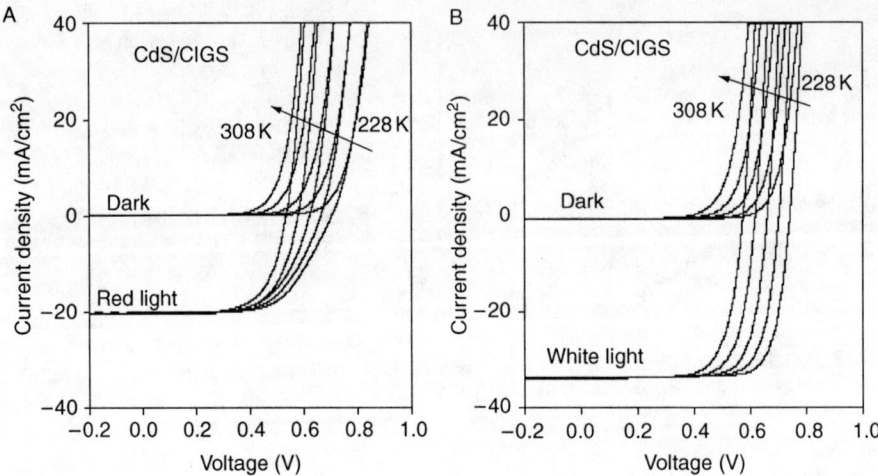

Figure 8.41 (A) *I–V* curves of CIGS cells at different temperatures; (A) under red light illumination, and (B) under white light illumination.

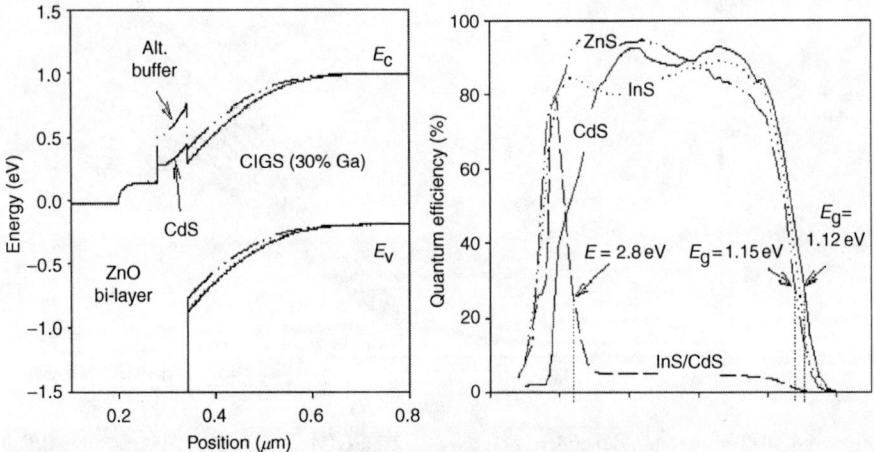

Figure 8.42 (A) Band structure of CIGS/CdS, and (B) QE of CIGS/CdS, CIGS/InS, and CIGS/ZnS.

temperature for CIGS/CdS cells similar to CIGSS/CdS cells. The reason is that the acceptor concentration increases with temperature. Only fewer electrons have thermal energy to cross the barrier. There may be other factors such as window, *etc*. In the simulated band structure, the conduction band offset for buffers such as ZnS, InS, *etc*., is higher than that of CdS, as shown in Fig. 8.42A. The cutoff wavelengths at long wavelength region in the quantum efficiency spectra of CIGS cells with different buffer layers are in between 1.12 and 1.15 eV, as shown in Fig. 8.42B [170]. The $Cu_{0.9}In_{0.69}Ga_{0.29}Se_{1.96}S_{0.13}$/50 nmCdS/70 nm i-ZnO/200 nm ITO (50–60 $\Omega/\square$)

cells made with CIGSS absorber by nonvacuum hydrazine process show efficiency of 10.1%. The cells are annealed under O_2 flow at 180 °C for 15 min and light soaked at 1 sun conditions for 24 h [171].

8.3.4 Effect of Na on the CIGS Thin Film Solar Cells

The silica is used as a barrier to block the Na self diffusion from the soda-lime glass substrates into CIGS in order to control uniform Na in the absorber by doping. A 30 nm thick NaF layer deposited onto CIGS layers grown by three-stage process is annealed at 400 °C for 20 min, as a part of post-treatment (PT) or Na doping into CIGS. The CIGS cells formed on SLG/Mo reveal that the efficiency increases from 0, 11.1, 12.2 to 14.2% with increasing substrate temperature for CIGS layer from 370, 400, 500, to 580 °C, respectively, whereas the efficiency of post treated cells increases from 11.1, 13.3 to 13%, and thereafter decreases to 10.6% with increasing substrate temperature from 370, 400, 500 to 580 °C, respectively. The results indicate that the Na doped cells for which the CIGS is prepared at high temperature show inferior efficiency due to slow down of intermixing of In and Ga by Na. In fact, the cells with Na doping show high efficiency for the moderate growth temperatures of CIGS. The cells on SS/Mo show efficiencies of 12.6 and 8% with and without Na doping, respectively [172]. Na occupies Cu site in the CIGS layers. The efficiency is high (11.7%) in the typical cells with intentionally Na doped absorber layers, whereas the efficiency is low (8.8%) in one with Na free absorber layers, irrespective of Na free and Na containing substrates (Table 8.22) [173]. The SLG/Mo/CIGS/CdS/ZnO cells made with typical $CuIn_{0.75}Ga_{0.25}Se$ thin films, which have different Cu/(In+Ga) ratios up to seven ranges from 0.95 to 0.85 show efficiency from 11.2 to 9.3% without NaF and from 14.2 to 12.1% with 180 nm thick NaF, as shown in Table 8.22. The efficiency of cell increases with Na doping due to an increased carrier concentration from 3×10^{15} to 1×10^{16} cm^{-3} in the CIGS samples that enhances open circuit voltage. On the other hand, the junction width decreases from 0.9 to 0.6 μm with Na doping [174]. The efficiency increases from 5 to 9.4% by adding Na even though in the typical low efficiency CIGS thin film solar cells. The intensity of Na signal positioned at 1072 eV in the XPS increases with increasing Na concentration in the films indicating that doping is confined [175].

The efficiency of glass/Mo/CIGS/CdS/ZnO/ITO/Ni–Al cell increases from 12.6 to 12.8% with Na doping, whereas the efficiency of similar cell on Ti also increases from 10 to 12.1% with doping for which CIGS is deposited at substrate temperature of 550 °C, as shown in Table 8.22. The similar kind of improvement is observed in the cells with CIGS absorber grown at low substrate temperature of 350 °C with Na doping. The compositions such as Cu:In:Ga:Se = 23.5:18.3:7.8:50.4, Cu/(In+Ga) = 0.9, and Ga/(Ga+In) = 0.3 for the CIGS layers deposited at substrate temperature of 550 °C and Cu:In:Ga:Se = 20.6:20.6:8.8:50, Cu/(In+Ga) = 0.7, and Ga/(Ga+In) = 0.3 for the CIGS layers deposited at 350 °C are observed. In comparison, at high temperature the In and Ga composition percentages are little lower. This could be due to role played by Na blocking intermixing of In and Ga, as mentioned earlier [176]. The effect of substrate temperature and Na doping on the properties of CIGS layers with compositions

Table 8.22 Effect of Na doping on CIGS thin film solar cells

Substrate or Cu/(In+Ga)	T_s (°C)	Na Deposition	V_{oc} (mV)	J_{sc} (mA/cm^2)	FF (%)	η (%)	Ref.
SS/Mo	–	No	523	29.5	52.5	8.1	[172]
SS/Mo	–	Yes	605	29	71.8	12.6	
SLG	400	No	550	11.6	68.8	8.8	[173]
SLG	400	Yes	620	12.8	73.3	11.7	
SLG/Mo, 0.97	–	No	528	30.3	–	11.2	[174]
SLG/Mo, 0.97	–	Yes	621	31	–	14.2	
SLG/Mo, 0.85	–	No	514	30	–	10.2	
SLG/Mo, 0.85	–	Yes	589	28.4	–	12.1	
SLG/Mo	550	Yes	590	31.2	69.6	12.8	[176]
SLG/Mo	550	No	580	31.3	69.4	12.6	
Ti/Mo	550	Yes	560	31.7	68.3	12.1	
Ti/Mo	550	No	530	31.0	60.8	10.0	
SLG/Mo	350	Yes	500	18.4	59.8	4.9	
SLG/Mo	350	No	510	16.7	55.8	4.8	
Ti/Mo	350	Yes	450	12.1	48.3	2.6	
Ti/Mo	350	No	420	11.0	48.0	2.2	
SS/Mo		Yes	609	34.5	72.9	15.3	[178]
SS/Mo		No	457	28.7	61.8	8.1	
SS/Mo		Best cell	727	37.2	72.7	17	
7059 corning-glass		CIGS/Nil	520	31.2	67.2	10.9	[179]
7059 corning-glass		CIGS/NaF	602	30.8	72.7	13.5	
7059 corning-glass		CIGS/KF	551	32.6	66	11.9	

of Cu-22 at% and Ga/(Ga+In) = 32–34% prepared by in-line process of Cu, In, Ga, and Se are studied. The intensity ratio of (112)/(204,220) is 9.5, 1.6, and 0.7 for the CIGS layers deposited onto glass substrates at substrate temperature of 550, 420, 420 °C+Na co-evaporation, respectively in the XRD, whereas JCPDS shows ratio of 2.2 for $CuIn_{0.7}Ga_{0.3}Se_2$. The sheet resistances of 4×10^3, 3×10^5, and 5×10^3 kΩ/□ for CIGS layers grown without buffer layer, whereas the sheet resistances of 5×10^5, 9×10^5, and 6×10^3 kΩ/□ for CIGS with Al_2O_3 buffer layer grown onto glass at substrate temperatures of 550, 420, 420 °C+Na co-evaporation are determined, respectively, indicating that the sheet resistance of CIGS layers is lower for Na doping. On the other hand, the sheet resistance is lower without barrier layer due to diffusion of Na or impurities from the glass substrates. The grain sizes of CIGS layers grown onto Mo covered glass substrates at 550, 420, and 420 °C+Na co-evaporation are 1, 0.5, and 0.5 μm, respectively. The efficiencies of the cells are 13.8 and 7.6% for CIGS grown at 550 and 420 °C, respectively encouraging that the efficiency rises to higher for higher growth temperature. The efficiencies are 7.6 and 9.6% for the layers grown at 420 and 420 °C+Na co-evaporation, respectively, indicating that Na doping enhances efficiency to higher. At longer wavelength the quantum efficiency of cells with CIGS layers grown at

420 °C + moderate Na co-evaporation is higher than the one with CIGS grown with high Na incorporation. The spectral response is higher for the cells with CIGS layers grown at higher temperature of 550 °C comparing with the one grown at lower temperatures, as shown in Fig. 8.43A [177].

The efficiency of typical SS/Mo/CIGS/CdS/ZnO/ITO/MgF$_2$ cell increases from 8.1 to 15.3% with Na doping (Table 8.22). Na compound is sputtered during the deposition of CIGS layers to dope Na into the CIGS layers. The carrier concentration of CIGS layers increases from 2.73×10^{16} to 7.73×10^{16} cm^{-3} with Na doping. The best CIGS cells with an area of 0.96 cm^2 prepared on flexible stainless steel substrates exhibit record efficiency of 17% close to glass substrate based cell [178]. Prior to deposition of Cu(InGa)Se$_2$ thin films, NaF, CsF, and KF thin films with thickness of 200 Å (impurity molar concentration 3%) are deposited onto 7059 corning glass substrates. The conductivity of CuInGaSe$_2$ layer with Ga/(In+Ga)=0.4 is $\sim 1 \times 10^{-1}$, 5×10^{-3}, 1×10^{-4}, 1×10^{-5} Ω-cm^{-1} for Na, K, doping free, and Cs doping, respectively, indicating that the conductivity is high for Na doping. The efficiency of cell is also high (Table 8.22). The grain sizes of CIGS layers increase with Na doping. The conductivities of typical Na doped and undoped CuInSe$_2$ samples for Cu/In=0.9 are $\sim 10^{-2}$ and 6×10^{-5} Ω-cm^{-1}, respectively. In the XRD, the diffraction angle moves toward lower angle side with Na doping that means the unit cell volume increases from 372.92 to 374.25 Å^3 with doping. The similar observation is noticed in CuNaInSe$_2$ samples that the unit cell increases from 385.75 to 388.49 Å^3 with Na doping. The secondary phase diffractions at 12.37, 24.85, and 37.63° in the XRD are observed for CuNa(InGa)Se$_2$ with Ga/(Ga+In)=0.3 [179]. The efficiency of 10–13.5% is reported for CIGS cells with Na doping using Na$_2$Se source for the wide composition range of Cu/(In+Ga)= 0.51–0.96. Otherwise the composition range of high efficiency cells is limited to the narrow range of 0.85–0.95. The glass/SiO$_x$/Mo/CIGS:Na(Co-Ev)/60 nmCdS(CBD)/ 1 μmZnO:Al(RF Sput.) cell with Cu/(In+Ga)=0.56 yields efficiency of 13.5% [180].

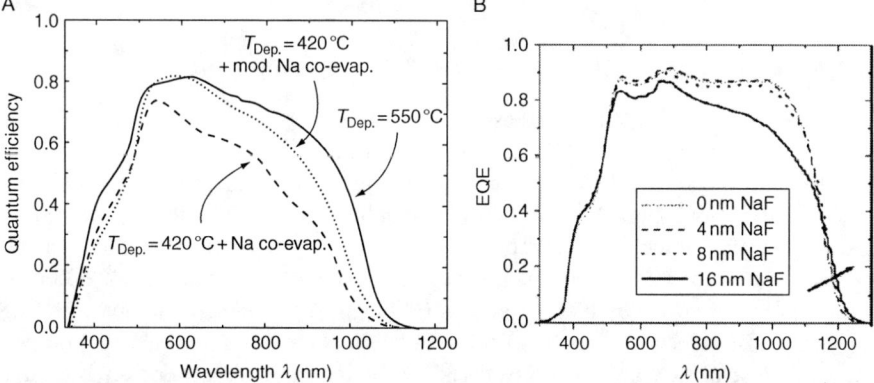

Figure 8.43 (A) Quantum efficiencies of CIGS cells made with CIGS absorber at different temperatures and Na incorporation, and (B) Quantum efficiencies of CIGS cells with CIGS absorber made with different NaF concentrations.

A 100°Å thick NaF layer is optimal for both polymide and Ti substrates; however the layer thickness can be increased up to 200Å without obtaining changes in J_{sc} and fill factor except open circuit voltage for latter. The absorber layer easily peels off from the substrate by further increasing NaF thickness to 400 Å. The deposition temperature of CIGS layers is limited to 450 and 550 °C for polymide and glass substrates, respectively because the thermal expansion coefficient is high about 20–40 and 8 ppm/°C for former and latter, respectively. The thermal expansion coefficient of glass is as close as CIGS. Antireflection coated 0.5 cm^2 area cells on glass and polymide substrates with 30 Å NaF precursor buffer layer show efficiencies of 14.6 and 10.8%, respectively [181]. Unlike direct Na doping into CIGS, the doping is carried out through Mo layer that can participate into CIGS. In the Alumina/Mo:Na/Mo, the Na doped Mo layer thickness is in the range of 100 to 600 nm and the total thickness of Mo:Na (bottom) and Mo (top) is 1100 nm. The bottom and top layers are deposited at 10 and 3 mtorr pressure, respectively in order to have good adhesion to the substrate. The alumina/Na doped 100 nm Mo/Mo layer(1000 nm)/CIGS/60 nmCdS/50 nmZnO/500 nmZnO cells with an active area of 0.45 cm^2 show maximum efficiency of 13.34%, whereas the cell without Na doping shows lower efficiency of about 9%. The efficiency of cell decreases to $\sim$12% with increasing thickness of Mo:Na layer (bottom) from 100 to 600 nm that means with increasing Na doping [182]. As shown in Fig. 8.43B, the spectral response of PI/Mo/CIGS/CdS/i-ZnO/ZnO:Al cell is low for Na doping, whereas the response is high in undoped samples at red region. The reason is that the current collection is low in the Na doped samples due to an increased hole concentration that reduces the space charge region width. The acceptor concentration of cell varies from 1.5×10^{14} to 1.5×10^{15} cm^{-3} with increasing thickness of NaF from 0 to 16 nm [183]. The merit is that the optimum Na doping not only increases the carrier concentration and grain sizes but also reduces the defect levels. The demerit is that the cell performance deteriorates for higher Na doping. The dark spots are found in the CIGS layers causing by diffusion of Na through the Mo from the glass substrates, which create pinholes in the absorber [132].

8.3.5 Role of CdS and CdZnS in the CIGS Thin Film Cells

The band gap of $Cd_{1-y}Zn_yS$ increases from 2.4 to 3.8 eV with increasing 'y' from 0 to 1. The Zn doping into CdS shares better lattice match with CGS or CIGS. On the other hand, it allows blue region photons to the absorber layer at shorter wavelength region. Thus, V_{oc} and J_{sc} of solar cell enhance. The effect of CdS buffer grown by different techniques such as physical vapor deposition (PVD) and chemical bath deposition (CBD) has been studied on the CIGS solar cells. The CIGS cells with CdS buffer grown by PVD offer lower efficiency, as compared to one with CdS buffer by CBD. Even though the CdS layers by PVD have large grain sizes comparing with those of CdS buffer by CBD. The reason is that the defect density is high in the PVD CdS layers, hence the efficiency of the cells is low (Table 8.23). A 70 nm thick CdS buffer used in the cells is grown by PVD at substrate temperature of 50 °C for 30 min under pressure of 10^{-8} mbar. In general, the structure of CdS

Table 8.23 PV parameters of CdZnS based cells

Buffer/$Cd_{1-y}Zn_yS$	Absorber	V_{oc} (mV)	J_{sc} (mA/cm²)	FF (%)	η (%)	J_o (nA/cm²)	A	R_s (Ω-cm²)	R_{sh} (kΩ-cm²)	Ref.
CdS (PVD)	CIGS	569	26.1	71	10.5	–	–	–	–	[184]
CdS (CBD)	CIGS	690	26.1	76.7	14.6	–	–	–	–	
CdS	CIGSS	587.7	32.9	73.1	14.1	0.51	1.31	0.88	9.54	[187]
Cd^{2+}/NH_3/ILGAR-ZnO	CIGSS	580	35.1	73.7	15	1.27	1.36	0.55	6.99	
ILGAR-ZnO	CIGSS	519	34.1	61	10.8	7.16	1.45	1.27	2.38	[189]
CdZnS(CBD)	CIGS	555	34.2	65.7	12.5	–	–	–	–	[188]
$y = 0.19$	CIS	419	37.8	68.4	10.5	–	–	–	–	
0.19	CIGS	503	32.4	61.4	10	–	–	–	–	
0.3	CGS[a]	613	15.34	56.7	5.33					[190]
0.3	CGS	394	12.22	35.96	1.73					
0.3	CIGS[a]	501	34.89	66.81	11.67					
0.3	CIGS[a]	526	35.6	69.53	13.02					
0.2	CIGS[a]	420	28.04	51.47	6.06					
CdS	CIGS	482	31.22	65.1	9.80					
CdS	CIS	366	33.7	64	9.3					[191]
CdS	$x = 0.56$[b]	522	15.6	30	2.9					
0.25	CGS	845	11.6	50	5.8					
0.35	CGS	680	11.21	35	2.71					[192]
0.12	$x = 0.50$[b]	540	23.81	52	6.72					
0.12	$x = 0.25$[b]	510	30.37	66	10.06					

[a]KCN treated.
[b]$CuIn_{1-x}Ga_xSe_2$.

thin films grown via CBD technique onto CIGS depends on the surface of the CIGS layer and its deposition parameters. The CIGS {112} plane matches with CdS {0001} for CdS grown by PVD process, whereas the CIGS {112} and CIGS {132} planes match with CdS {01$\bar{1}$0} plane for the CdS processed by CBD. The Cu or Cd diffuses from CIGS into CdS or vice versa. The ionic radii of 0.97 and 0.96 Å for Cd and Cu are close to each other. Figure 8.44 shows that CBD-CdS with CIGS cells exhibit better efficiency than that of the cells with PVD CdS [184]. As shown in Fig. 8.45, the quantum efficiency is low in thick CdS layer based cells as compared to that in thinner CdS based cells due to higher absorption by thicker CdS layer at shorter wavelength region [185]. Prior to deposition of ZnO layer, the Shell Solar CIGSS absorber is treated with Cd^{2+}/NH_3 solution whereby the solution is prepared by dissolving $CdSO_4$ 1.5 mM in aqueous NH_3. The absorbers are dipped into Cd^{2+}/NH_3 solution at room temperature, heated to 80 °C for 10 min, and cleaned in distilled water then dried on which 25 ILGAR deposition cycles are done at 100 °C to form ZnO layer. The ZnO layers grown onto Cd^{2+}/NH_3 treated CIGSS absorbers by ILGAR shows flat layer with isolated crystallites. The CIGSS layers treated by Cd^{2+}/NH_3 solution may have one monolayer of CdS but it is not in detectable range, whereas layers treated with concentrated 15 mM Cd^{2+}/NH_3 solution contain one CdS monolayer along with fast growth of Cd $(OH)_2$ layer. In the CBD-CdS thin film, Cd-$3d_{5/2}$ peak at 405.3 eV is close to standard CdS-405.3 eV, CdO-405.2 eV, and Cd-OH-405 eV. In addition, $S2p_{3/2}$ peak at 161.5 eV is observed. The oxygen peaks at 529.49 and 531.09 eV are close to standard CdO-529.2 and CdOH-530.9–532 eV, respectively. This investigation confirms that CdS is combination of CdO and CdOH species [186]. The ILGAR ZnO layer on SLG/Mo/CIGSS/Cd^{2+}/NH_3treated CIGSS layers shows (002) peak at diffraction angle of 34.4° ($d = 3.2$ Å) and has small crystallites and several cluster type regions.

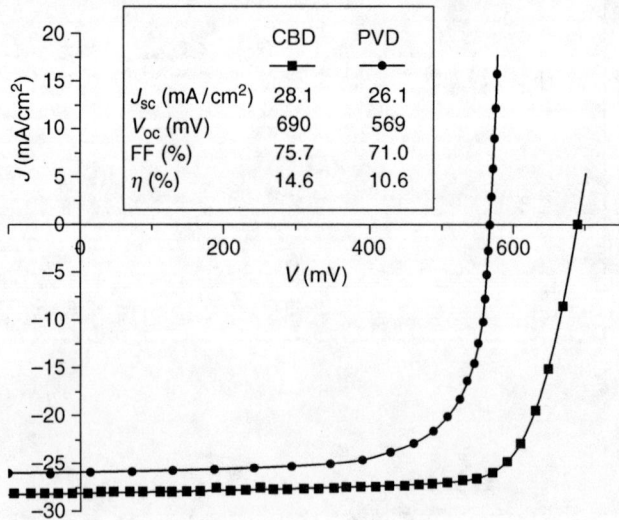

Figure 8.44 *I–V* curves of CIGS/CdS(CBD) and CIGS/CdS(PVD) thin film solar cells.

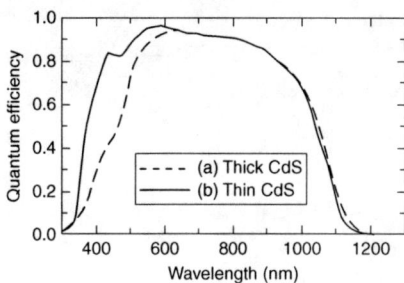

Figure 8.45 Quantum efficiencies of CIGS cells with two different thicknesses of CdS.

On top of the ILGAR ZnO structure, 100 nm i-ZnO, followed by 400 nm ZnO:Ga layers are deposited by RF sputtering to finalize SLG/Mo/CIGSS/Cd^{2+}/NH$_3$treated/ZnOILGAR/i-ZnO100 nm/ZnO:Ga400 nm cell. The SLG/Mo/CIGSS/Cd^{2+}/NH$_3$ treated/ILGAR ZnO/100 nm i-ZnO/400 nm ZnO:Ga cell with an area of 0.5 cm^2 shows higher efficiency of 15% as compared to the efficiency of 14.1% for CIGSS/CdS reference cell, whereas cell without Cd^{2+}/NH$_3$ treated absorber shows lower efficiency of 10.8% (Table 8.23). The R_s of cell is low in the case of Cd^{2+}/NH$_3$ treated CIGSS/ILGAR ZnO based cells due to formation of CdS monolayer [187].

The CuInSe$_2$ or CuInGaSe$_2$ (Ga/(In + Ga) = 0.4 and E_g = 1.2 eV) absorber layers for the glass/Mo(1 μm)/CuInSe$_2$ or CuInGaSe$_2$/CdZnS(0.5 μm)/ZnO(1–2 μm) thin film solar cell are prepared by vacuum evaporation of constituents elements of Cu, In, Ga, and Se. In the beginning, the Cu-rich absorber layer with thickness of 2.5–3.0 μm is deposited out of the total thickness of 3.5–4.5 μm. The remaining thick layer is filled with In or In + Ga rich CIS or CIGS layer to complete bilayer structure. The controlled Cu/(In + G) or Cu/In ratio in between 0.9 and 1.0 is essential for the efficient solar cells correspondingly its resistivity varies in between 10^4 and 10^{-1} Ω-cm. The band gap and resistivity of Cd$_{0.81}$Zn$_{0.19}$S layer prepared by electron beam evaporation are nearly 2.5 eV and 10^2 Ω-cm, respectively. The Zn concentration of 0.19 is optimal for efficiency cells, whereas below or above the optimum concentration of Zn, the open circuit voltage and fill factor of the cell deteriorate. The 7059glass/Mo/CIGS/CdZnS/1–2 μm ZnO cells with an area of 2 × 4mm^2 annealed under air at 200 °C for 2–8 h demonstrate efficiency of 10%, as shown in Table 8.23. The quantum efficiencies of CuInSe$_2$/CdZnS (η = 10.5%) and CuInGaSe$_2$/CdZnS (η = 10% and E_g = 1.2 eV) cells are shown in Fig. 8.46. The cutoff wavelengths in both cells can be seen at ∼1300 and ∼1100 nm for CIS and CIGS absorbers in the cells and those are presumably equal to their band gaps, respectively [188]. The alumina/Mo-1250 nm/CuIn$_{0.72}$Ga$_{0.28}$Se$_2$(2300–1200 nm bilayer)/Cd$_{0.8}$Zn$_{0.2}$S 16 nm/ZnO (high ρ) 90 nm/ZnO (low ρ) 560 nm/Ni 50 nm–Al 5000 nm cells exhibit efficiency of 12.5%. The CIGS layers for the cells are prepared by thermal vacuum co-evaporation technique using bilayer process with composition ratios of Cu:In:Ga:Se = 23.6:19.4:7.4:49.4. A 40–50 nm thick Cd$_{0.8}$Zn$_{0.2}$S layer, which has resistivity in the order of 5 × 10^5 Ω-cm, are grown by CBD process using chemical solutions

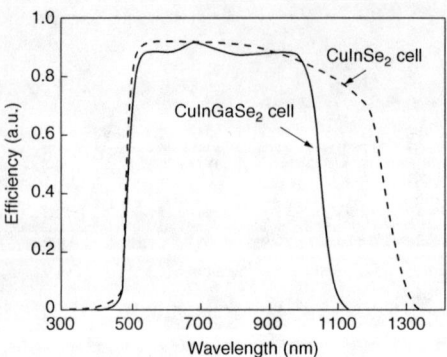

Figure 8.46 Quantum efficiencies of CIGS and CIS cells.

of 0.01 M $CdCl_2$ + $ZnCl_2$, 0.026 M NH_4Cl, 0.26 M NH_4OH, and 0.083 M thiourea at bath temperature of 85 °C for 30 min. A 10 nm CdZnS thick layer grown by this technique covers well on even rough surfaces of CIGS. First high sheet resistance ZnO:Al is deposited in the presence of oxygen, followed by depositon of low sheet resistance AZO under only Ar gas conditions by RF sputtering. A difference in quantum efficiencies of $CuInGaSe_2$/CdZnS/ZnO and $CuInSe_2$/CdZnS thin film solar cells is shown in Fig. 8.47. The spectral response starts early at short wavelength region in the former sample due to response of ZnO that is close to its band gap [189].

Note that in the CIGS/$Cd_{1-y}Zn_yS$ and CGS/$Cd_{1-y}Zn_yS$ solar cells $y = 0.2$ and 0.3 are optimal to incur better efficiencies, respectively. The CIGS/$Cd_{1-y}Zn_yS$ cell with $y = 0.2$ and 0.3 exhibit efficiencies of 13.2 and 11.7%, respectively. The photovoltaic parameters of glass/Mo/(CIGS or CGS)/40 nm $Cd_{1-y}Zn_yS$/i-ZnO/n^+ ZnO/ 50 nm Ni–300 nm Al cells with effect of KCN etching on absorber layers and Zn compositions are given in Table 8.23. The surface of the absorber layers etched in KCN solution gives better photovoltaic performance [190]. The efficiency of cell varies with different x and y values of CIGS and ZnCdS layers. Variation of x and y values in the CIGS and ZnCdS causes to change electron affinities of the absorber and window layer resulting in reduced minority carrier diffusion length in the absorber [191]. The CIGS ($x = 0.25$)/CdZnS($y = 0.12$) cells annealed under oxygen ambient at 225 °C for 5–20 min contribute higher efficiency of 10.5% as compared to that of other combinations of x and y. The cells prepared that $CuIn_{1-x}Ga_xSe_2$ thin films are deposited onto Mo coated glass substrate by co-evaporation using bilayer process. As first 2 μm thick Cu-rich CIGS layer at substrate temperature of 400–450 °C, 1 μm thick In-rich CIGS layer at slightly higher substrate temperature of 500–550 °C, 3 μm thick $Zn_{0.12}Cd_{0.88}S$ layer at substrate temperature of 200 °C and Al grid are successively evaporated [192].

8.3.6 ZnS Buffer Based CIGS Thin Film Solar Cells

The reason to choose ZnS buffer layer in the place of CdS is owing to wide band gap of 3.8 eV and nontoxic. The NREL research group developed Zn(S,O,OH)

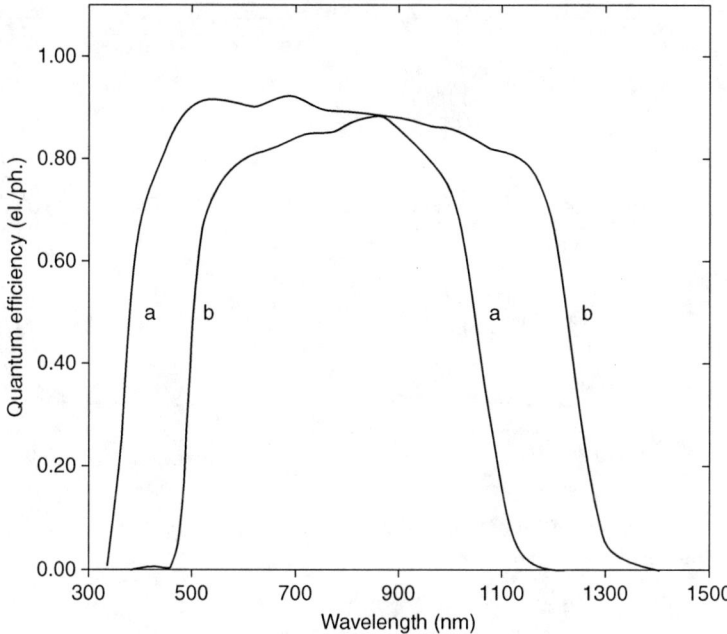

Figure 8.47 Quantum efficiencies of (A) CuInGaSe$_2$/CdZnS/ZnO, and (B) CuInSe$_2$/CdZnS thin film solar cells; the spectral response starts early in former sample at shorter wavelength due to participation of ZnO.

buffer based CIGS thin film solar cells with solar energy conversion efficiency of 18.6%. The Zn(S,O,OH) prepared on CIGS by CBD technique using ZnSO$_4$ (0.1–0.3 M), ammonia (5–8 M) and thiourea (0.4–0.8 M) at 80 °C for 15 min has composition ratio of Zn:S:O = 48:28:24. The Zn(S,O,OH) layers deposited onto 2.5 μm thick CIGS absorbers with Cu/(In+Ga)~0.88 and Ga/(In+Ga)~0.3 ratios and band gap of 1.12 eV are annealed under air at 200 °C for 15 min. The ZnO bilayers with sheet resistance of 60–80 Ω/□, Ni/Al grids and 100 nm MgF$_2$ antireflection coatings are successively deposited. The absorber used for solar cell prepared by conventional three-stage process has (220) preferred orientation. The SEM analysis reveals that the surface morphology of CIGS is combination of steps and ridges character growth unlike normal triangular smooth surfaces of (112) oriented absorber layers. The I–V characteristics of 18.6% efficiency glass/Mo/CIGS/CBD-ZnS/i-ZnO/ZnO:Al/Ni-Al/MgF$_2$ thin film solar cells along with CdS reference cell are studied at different temperatures from 240 to 320 K range under dark and light illuminations, as shown in Fig. 8.48 from which it can be concluded that the efficiency, open circuit voltage and fill factor increase linearly with decreasing temperature, however no variation in short circuit current with effect of temperature is noticed. As the temperature is reduced to lower, an increase in efficiency for ZnS based solar cells is slightly faster than that in the CdS based reference solar cells. As expected, the short circuit current is also higher in the ZnS based thin film solar cells than that in the

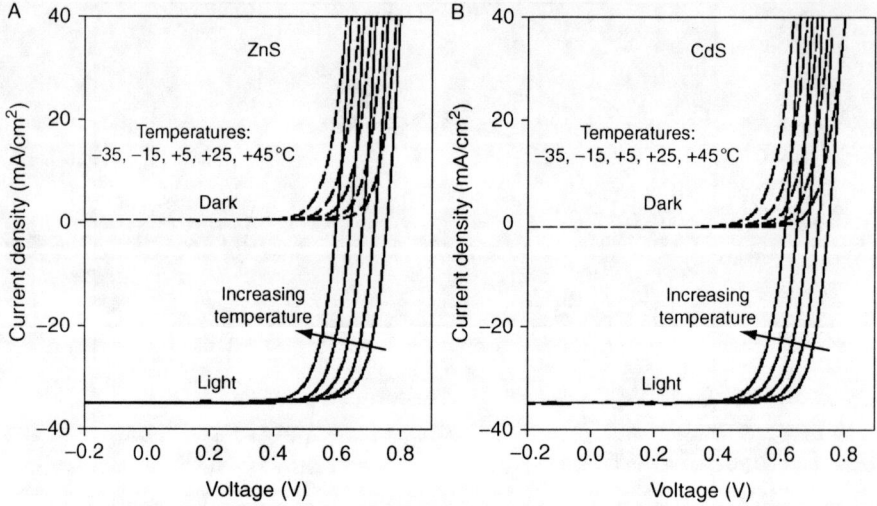

Figure 8.48 *I–V* curves of ZnS (18.6%), and CdS buffer based thin film solar cells under dark, and illumination at different temperatures.

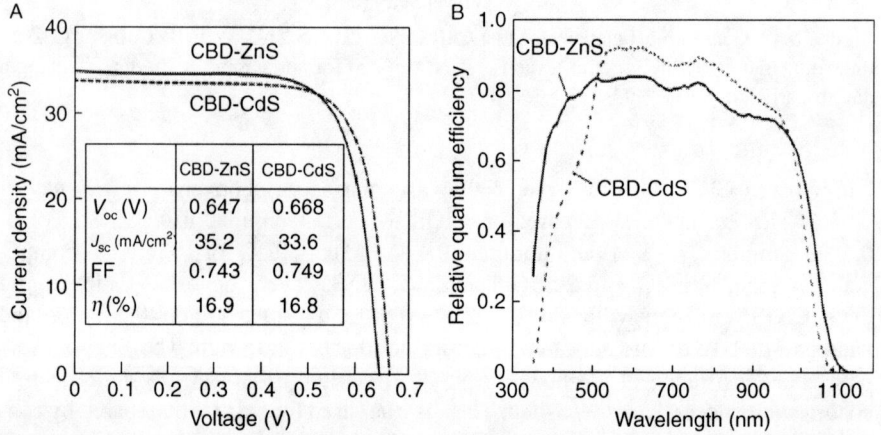

Figure 8.49 *I–V* curves, and quantum efficiencies of ZnS, and CdS based CIGS thin film solar cells.

CdS based cells. The V_{oc} increases with decreasing temperature due to an increase of carrier concentration that eventually causing to increase band gap of CIGS [193]. The similar efficiencies of 16.8 and 16.9% for SLG/Mo/CIGS/CdS(90 nm)CBD/ZnO (70 nm)/ZnO:Al/MgF$_2$ and SLG/Mo/CIGS/ZnS(100 nm)CBD/ZnO:Al/MgF$_2$ cells with an area of 0.2 cm^2 are observed, respectively (Fig. 8.49A) (Table 8.24). A slight difference in the quantum efficiencies between typical ZnS and CdS based cells is observed at longer wavelength range of $\sim$1100 nm due to deviation of Ga composition of Ga/III = 0.34, and 0.36 in the cells, respectively (Fig. 8.49B) [194].

Table 8.24 PV parameters of ZnS buffer based CIGS cells

Buffer/cell area (cm^2)	V_{oc} (mV)	J_{sc} (mA/cm^2)	FF (%)	η (%)	$J_o \times 10^{-6}$ (A/cm^2)	A	R_s (Ω-cm^2)	R_{sh} (kΩ-cm^2)	Ref.
ZnS/0.402	660.7	36.105	78.16	18.6	–	–	–	–	[193]
CdS/0.402	648	34.98	76.21	17.25	–	–	–	–	
CdS/0.2	668	33.6	74.9	16.8	–	–	–	–	[194]
ZnS/0.2	647	35.2	74.3	16.9	–	–	–	–	
ZnS/0.15 cm^2	671	34	77.6	17.7	–	–	–	–	[195]
ZnS/0.15 cm2	671	34.7	77.6	18.1	–	–	–	–	
ZnS without ZnO	649	34.8	70.6	16.0	–	–	–	–	
CdS	NA	NA	69	14.9	0.021	1.61	1.9	60	[198]
ZnS	552	39.35	62	13.3	4.2	1.73	3.1	16	
ZnS(O,OH) USCBD	600	35.4	71.2	15.1	–	–	–	–	[197]
ZnS(O,OH) CBD	586	35	64.7	13.4	–	–	–	–	
Zn(O,OH)$_x$ (0.95 cm^2)	420	35.5	65	9.7	–	–	–	–	[199]
ZnS(O,OH)$_x$ (3.2 cm^2)	523	36.5	66.9	12.8	–	–	–	–	
ZnS	618	32.4	69.3	13.9	–	1.8	0.7	–	[200]
CdS	660	31.3	72.4	15	–	1.6	0.3	–	
CIGS(NREL)/ZnS(O,OH)[a]	678	35.74	72	17.4	–	–	–	–	[202]
CIGSS(SS)/ZnS(O,OH)[b]	581.8	34.74	71.2	14.4	–	–	–	–	
Zn(O,S)10%	642	34.3	74.4	18.4	–	–	–	–	[203]
CdS	624	33.3	75.8	15.8	–	–	–	–	

[a]Double layer.
[b]Single layer.

The undoped ZnO and 0.6 μm ZnO:Al layers are deposited onto glass/Mo/2–2.5 μm CIGS/100 nm ZnS CBD stack by RF sputtering. The Al doped ZnO layer has sheet resistance of 10 Ω/□ and 85% transmission in the 500–1200 nm range. Al grids and 0.1 μm thick MgF$_2$ antireflection coating are sequentially deposited onto stack. 0.2 cm^2 active area glass/Mo/CIGS/ZnS/ZnO:Al/Al/MgF$_2$ cell without ZnO layer exhibits little lower efficiency of 16%. The glass/Mo/CIGS/ZnS/ZnO/ZnO:Al/Al/ MgF$_2$ cell with an active area of 0.15 cm^2 for 130–150 nm thick ZnS buffer exhibits efficiency of 17.7%, as shown in Table 8.24. The best cell, which shows efficiency of 18.1%, contains 130 nm thick ZnS buffer and CIGS absorber layer by three-stage process using MBE with compositions of Cu:In:Ga:Se = 23.2:14.0:9.5:53.3, Cu/In = 0.99, Ga/In = 0.40, Se/(In + Ga) = 1.14, and band gap of 1.3 eV. However, the spectral response stands at 1.19 eV due to Ga grading in the layers [195]. The I–V characteristics of ZnS based CIGS solar cells with efficiency of 18.1% from the Aoyama Gakuin University and CdS based CIGS cell with efficiency of 18.8% from the NREL reveal shunt resistance of 800 and 3900 Ω-cm^2 and diode quality factor of 1.55 and 1.5, respectively [196]. As anticipated, the analysis exploits slightly higher shunt resistance and lower diode factor in the slightly high efficiency cells. 2 μm thick CIGS thin films with Cu/(In + Ga) = 0.94, Ga/(In + Ga) = 0.35, and Se/metals = 1.0 by three-stage process at 550 °C and ZnS(O, OH) layers by ultrasonic vibration chemical bath deposition (UVCBD) are successively deposited onto SLG/Mo. The CIGS cells with ZnS by UVCBD show higher efficiency of 15.1% than that of one with ZnS(O, OH) prepared by conventional CBD method (13.4%) on the same CIGS absorber (Table 8.24) [197].

In another version of 13.3% low efficiency glass/Mo/CIGS/ZnS/ZnO:Al (MOCVD)/Ag typical cells, the efficiency is little lower than that of CdS control cell (14.9%). As thickness of ZnS is increased from 10 to 12 nm to higher, the short circuit current density decreases from 38 to 25 mA/cm^2 due to an increased barrier height at the interface that causes to block the current flow. The cell shows lower efficiency below the optimum thickness of 10–12 nm. In comparison with CdS reference cells, the ZnS based cells show higher diode quality factor, sheet resistance, reverse saturation current density, and lower shunt resistance probably due to larger conduction band offset in the heterojunction [198]. Two kinds of buffers such as Zn(O,OH)$_x$ and Zn(S,O,OH)$_x$ can be grown by CBD method onto CIGS absorber. The SLG/Mo/CIGS/Zn(O,OH)$_x$ stack is annealed under air at 200 °C for 15 min, whereas the SLG/Mo/CIGS/Zn(S,O,OH)$_x$ stack is annealed under vacuum at 150–220 °C for 15–45 min. If the stack is annealed at 300 °C or above, the structure gets destroyed. The annealing may have resulted in either abrupt p–n junction or in between p–i–n and p–n junction of CIGS/CBD-ZnS. On top of the annealed SLG/Mo/CIGS/Zn(S,O,OH)$_x$ or Zn(O,OH)$_x$ stack, a 0.8 μm thick ZnO:Al is grown by DC sputtering technique in order to complete the cell structure. Without adding NH$_2$CS (0.2–2 M) to the chemical solutions of Zn sulfate or Zn acetate (0.04 M) and ammonia 25% at 60–80 °C, Zn(O,OH)$_x$ can be formed. After constant irradiation of Zn(O,OH)$_x$ and Zn(S,O,OH) buffers based thin film solar cells with active areas of 0.95 and 3.2 cm^2 under 1-sun (1.5 AM, 100 mW/cm^2) light illumination conditions for 60 and 150 min show efficiencies of 9.7 and 12.8%, respectively, as shown in Table 8.24. Beyond the irradiation timings the characteristics of cell degrade. The quantum efficiency of $Zn(S,O,OH)_x$ based solar cells starts at early wavelength of 350 nm as compared to that of CdS based solar cells indicating participation of larger band gap (3.2 eV), which allows higher energy photons to the absorber that generates higher current in the solar cell. The CIGS absorber used in the films is prepared by two-stage method; first Cu–Ga and In layers are deposited by DC sputtering employing respective sputtering targets, followed by selenization of metals using H_2Se gas at 550 °C [199].

The glass/Mo/2 µm thick CIGS ($E_g = 1.2$ eV)/ZnS(CBD)/ZnO/ITO cells demonstrate higher current in I–V curve (Fig. 8.50A) because at shorter wavelength the absorption by ZnS is low therefore the spectral response is high comparing with CdS (Fig. 8.50B). The ZnS growth by CBD is slow comparing with CdS because Cd tetramine is unstable comparing with Zn tetramine $[Zn(NH_3)_4]^{2+}$ in the chemical bath solution. The PV parameters of typical cells with CdS and ZnS buffers are given in Table 8.24 [200]. At the end of three-stage process of CIGS layers, Zn layer is deposited on it at 300 °C for 10–1200 s in the same vacuum chamber. In the CIGS absorber layers, the Zn concentration of 10^{21} atoms/cm^3 is penetrated to the depth of 0.4 µm. The cells with and without Zn doping show efficiencies of 11.9 and 13.4%, respectively. The high resistive doping layer acts as a semi-insulator, which compensates the acceptors and suppresses the shunting between p-type CIGS and n-type window layer. The best glass/Mo/CIGS:Zn/Zn(O,S) (CBD)/ZnO thin film solar cells exhibit efficiency of 14.8% [201]. The SLG/Mo/CIGSS or CIGS/ZnS(CBD)/i-ZnO (110 nm)/ZnO:Al(400 nm)/Ni–Al/MgF$_2$ cells with Shell Solar CIGSS/ZnS(O,OH) or NREL CIGS/ZnS(O,OH) stacks annealed under air at 200 °C for 10 min show efficiencies of 14.4 and 17.4% for single (15 nm) and double (30 nm) layer of ZnS(O,OH), respectively. The QE of glass SLG/Mo/CIGS/ZnS-CBD/i-ZnO (110 nm)/ZnO:Al(400 nm)/Ni–Al/MgF$_2$ cells decreases with increasing thickness of ZnS layer due to an increase of absorption by ZnS, as shown in Fig. 8.51 [202].

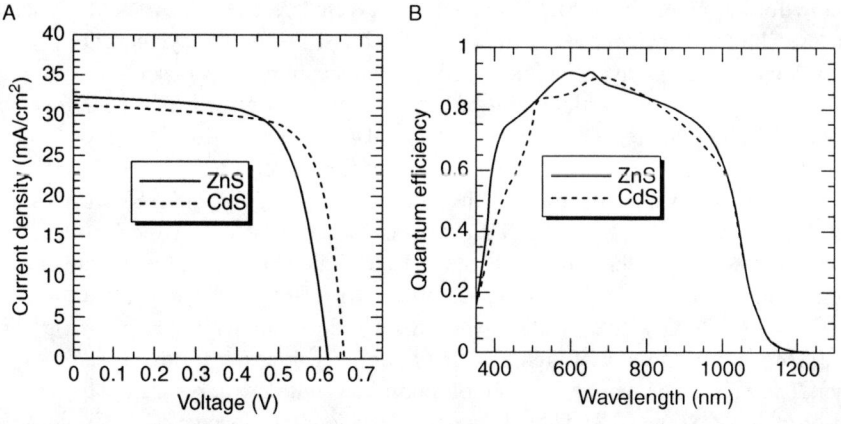

Figure 8.50 (A) I–V curves, and (B) quantum efficiencies of ZnS, and CdS based CIGS thin film solar cells.

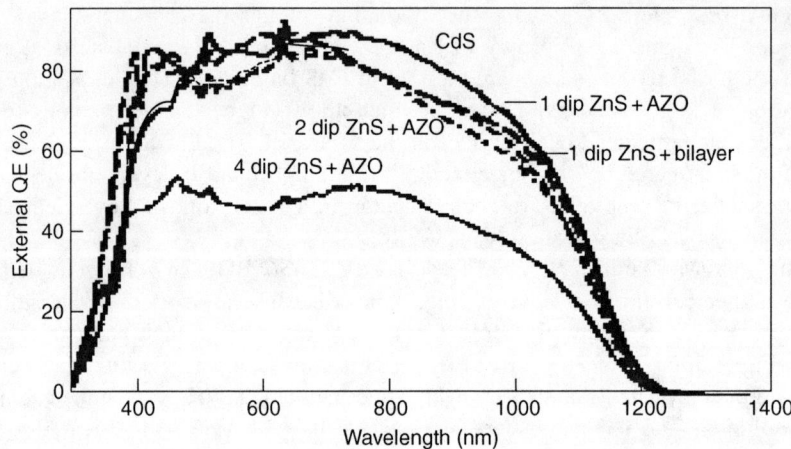

Figure 8.51 Quantum efficiency of SLG/Mo/CIGS/ZnS-CBD/i-ZnO(110 nm)/ZnO:Al (400 nm)/Ni–Al/MgF$_2$ cell with different buffer thicknesses, and CdS control sample.

The Zn(O,S) buffer is grown onto CIGS with Ga/(In+Ga)=0.3 by ALD in Microchemistry F-120 reactor employing pulsing sequence of 200/400/200/400 ms for DEZ/N$_2$/H$_2$O/H$_2$S at substrate temperature of 120 °C. In this context, the Zn (O,S) 20% means 1H$_2$S is allowed in 4H$_2$O sequences. In the device, the thicker buffer layer blocks photocurrent. If sulfur concentration is low that means O rich layer lowers V_{oc}. On the other hand, S rich layer blocks photocurrent, the precise control of composition and thickness is needed to obtain the best performance of cells. The best results are gained with Zn(O,S) 10%, whereas Zn(O,S) 5% shows low V_{oc}. 100, 300, and 500 cycles, with Zn(O,S)20% and 33% show low efficiencies but few shows zero activities. On 100 cycle ZnS layer, the protective 1000 cycle ZnO layer deposited by ALD shows efficiency of 11.4%. The performance of cell is low for below or above of 100 cycled ZnS layers because of low shunt resistance or low fill factors, whereas the efficiency of cell is 16% for 300 cycled Zn(O,S) 20% with protective ZnO layer. An increased sulfur concentration is observed at interface of CIGS layer due to participation of higher nucleation growth of ZnS rather than ZnO at the beginning. The efficiencies of 18.4 and 15.8% for SLG/Mo/CIGS/Zn (O,S)/ZnO/ZnO:Al, and CdS reference cells are observed, respectively [203].

The XPS analysis confirms that the thickness of ALCVD Zn(O,S) layer used in the CIGS cells is 1–2 nm. The S/Zn ratio is found to be 0.7 in the Zn(O,S) 20% layer. The sulfur concentration is higher in the films than that in the precursor reaction quantity. This may be due to desorption of H$_2$O by the reaction process of ZnO + H$_2$S $\Leftrightarrow$ H$_2$O + ZnS. In the layers, S/Zn ratio is higher for an increased H$_2$S and S/Zn is also low for an increased H$_2$O but latter does not follow as same as former. The ZnS is hexagonal structure or amorphous with nanocrystals for the composition ratio of S/Zn $\sim$ 0.7, whereas the structure is either hexagonal or cubic for S/Zn < 0.5. The volume of hexagonal unit cell increases with increasing S in the O-rich side [203]. The ZnS(O,OH) layer grown by CBD for glass/1 μmMo/

$Cu_{0.9}In_{0.72}Ga_{0.28}Se_2$/ZnS(O,OH)/150 nmZnO:Al/Ni–Al/MgF$_2$ thin film solar cell is combination of ZnO and Zn(OH)$_2$ in which O1s peaks at 529.8 and 531.2 eV represent by ZnO and Zn(OH)$_2$, respectively. ZnS(O,OH) layer by CBD on CIGS is annealed at 200 °C under air for 10 min to obtain abrupt junction. Interestingly, the CIGS cell without i-ZnO layer shows efficiency of 18.5%. On the same absorber, the reference cell shows efficiency of 17.25%, as already mentioned [204]. The similar oxygen peaks at 530.6–530.1 and 531.6 eV correspond to oxygen signatures of standard ZnO (530.4 eV) and Zn(OH)$_2$ (530.9–532 eV) are reported, respectively. Zn2p$_{3/2}$-1022 eV is possibly related to standard ZnS (1022 eV) or ZnO (1021.8 eV) [186]. Nakada et al., [195] also observed the same binding energies at 530.1 and 531.7 eV for oxides and hydroxides, respectively. All these investigations reveal that the ZnS layer deposited by CBD is not only ZnS but also combinations of ZnO and Zn(OH)$_2$. The ZnS layers with thickness of 15 nm deposited onto Shell Solar (SS) CIGSS absorber layers by CBD are in cubic structure confirming broad (111) peak.

8.3.7 ZnSe and InSe Window Based CIGS Cells

Choosing ZnSe as a buffer layer is advantage owing to suitable band gap of 2.67 eV and low lattice mismatch with CIGS as compared to that of CdS. The Zn(Se,OH) thin films grown onto CIGSS by CBD technique are studied using X-ray photoelectron spectroscopy employing radiation of $Mgk_\alpha = 1253.6$ eV. The $\Delta E_{V\text{-In4d}}$ and $\Delta E_{V\text{-Zn3d}}$ are binding energies of In4d and Zn3d in the CIGSS and Zn(Se,OH) with reference to the valence band maximum, which are found to be 17.1 and 8.8 eV, respectively and the ΔE_{CL} is 7.7 eV, as shown in Fig. 8.52. From all the values ΔE_V is found to be 0.6 eV, as mentioned in earlier, and ΔE_C can also be obtained by substituting (CIGSS) $E_{g1} = 1.04$ and Zn(Se,OH) $E_{g2} = 2.9$ eV in $\Delta E_C = E_{g2} - E_{g1} - \Delta E_V$, that is, 1.26 eV [205].

7 nm thick Zn(Se,OH)$_x$ buffer layers with grain sizes of 20–25 Å on NH$_2$–NH$_2$·H$_2$O treated Shell Solar CIGSS absorber layers are deposited at 70 °C by CBD.

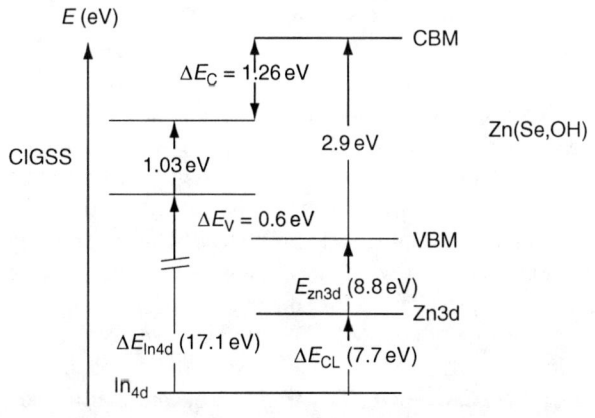

Figure 8.52 Band diagram of CIGS/ZnSe is designed based on XPS results.

A 110 nm thick ZnO, 400 nm ZnO:Al, Ni–Al grids, and 120 nm MgF_2 antireflection coatings are subsequently grown to have glass/Mo/CIGSS/Zn(Se,OH)$_x$/i-ZnO/ZnO:Al/Ni–Al cell. 0.5 cm^2 area glass/Mo/CuInGa(S,Se)$_2$/Zn(OH)$_2$/Zn(Se,OH)/i-ZnO (100 nm)/ZnO:Ga(100 nm)/Al–Ni thin film solar cells show efficiency of 14.4% close to the one with CdS buffer based solar cells (14.6%), whereas the glass/Mo/CuInGa(S,Se)$_2$/Zn(OH)$_2$/ZnO cell without Zn(Se,OH) buffer shows lower efficiency of 10.7%. The cells with CBD Zn(Se,OH)$_x$ show better performance than that of MOCVD ZnSe based cell. The Zn(OH)$_2$ buffer can be deposited onto the absorber using combination of $ZnSO_4$ (0.8 M), $(NH_2)_2$(25%), and NH_3 (25%) solutions at 70 °C. Without MgF_2 antireflection coating 0.6 cm^2 area CIGSS/Zn(Se,OH)/i-ZnO/ZnO:Ga cell shows slightly lower efficiency of 13.67% as compared to one with antireflection coating. The efficiency of CIGSS/ZnO solar cells increases from 5.1 to 12.1%, if Zn-treated CIGSS absorber is used in the cells. Similarly the efficiency of 0.5 cm^2 area glass/Mo/CIGSS/Zn(Se,OH)/ZnO cells improves efficiency from 12.7 to 14.5% for Zn-treated CIGSS absorber, as shown in Table 8.25.

The XPS analysis on the Zn(OH)$_2$ based buffer layer reveals that the oxygen core levels of 531.5–532, 530–530.8, and 533–534 eV are related to zinc hydroxides, zinc oxides, and H_2O, respectively. By adding $SeC(NH_2)_2$ (0.08 M) and Na_2SO_3 (0.08 M) to the above solutions, the Zn(Se,OH)$_x$ can be prepared. The better results are obtained for thin film solar cells, if Zn(S,OH) and Zn(Se,OH) layers are grown at growth temperatures of 50 and 70 °C, respectively, otherwise the efficiency slips to lower. The undoped ZnO layer with thickness of 100 nm and Ga-doped ZnO layer with thickness of 400 nm deposited by sputtering are employed for the cells. In the XPS spectra, the Cu2p-932.4 eV peak from CIGSS absorber is observed but after Zn treatment its intensity drastically decreases. The Cu peak disappears, as a result of deposition of 10 nm Zn(Se,OH) by CBD method. The hydrogen depth profile studies on the Zn(Se,OH) films shows that the concentration of hydrogen increases up to depth of $\sim$7 nm then steeply decreases with moving into depth indicating that the concentration of hydrogen is higher at or near the surface of ZnSe layer. The Raman spectra reveal that the peak positions at 177, 258, and 290 cm^{-1} due to ZnSe, CuInSe$_2$, and CuInS$_2$ are observed in the glass/Mo/CIGSS/Zn(Se,OH) structure layers, confirming the formation of ZnSe layers as a part of compound in Zn(Se,OH) [206–209].

The efficiency of 0.548 cm^2 active area CIGS/10 nmZnSe/i-ZnO/ZnO:Ga cell improves from 11 to 15.1% by replacing Shell Solar (SS) CIGS with CIGSS absorber (SS). The I–V and quantum efficiencies of CIGSS/10 nmZnSe/i-ZnO/ZnO:Ga based cell and CdS reference cell are given in Fig. 8.53A and B. The J_{sc} is higher in the ZnSe based cells than that in the CdS reference cells due to higher band gap of ZnSe. The ZnSe based cells also show higher quantum efficiency than that of CdS reference cell in the entire wavelength range. The ZnSe thin films for cells are grown onto CIGSS layer by horizontal AIX200 MOCVD reactor using ditertiarybutylselenide and dimethyl-zinctriethylamine at substrate temperature of 280 °C, Se/Zn ratio of 2.7, chamber pressure of 300 mbar and H_2 flow of 3.5 l/min for 3 min. The similar results on glass/Mo/CIGS/30 nm ZnSe (Vac.Ev at 320 °C) cell annealed under air at 200 °C for 30 min are observed. The band gaps of 2.42 and 2.67 eV for CdS and ZnSe can be obtained from the quantum efficiency

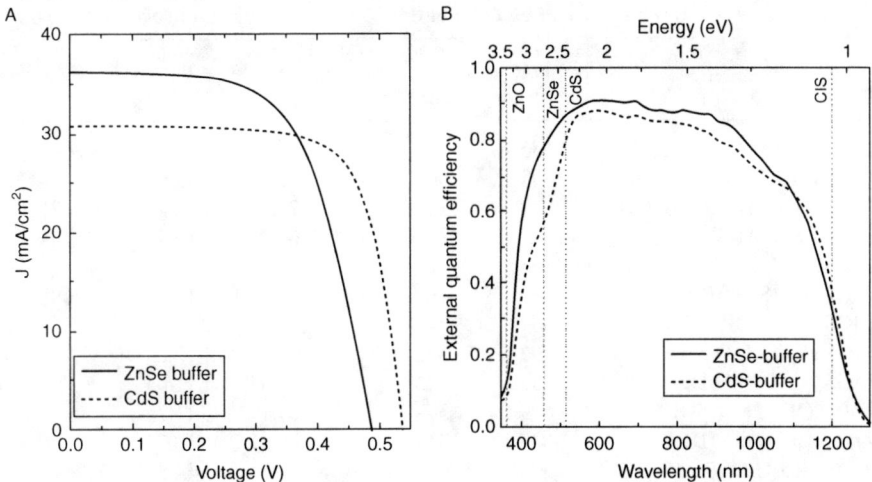

Figure 8.53 (A) Light illuminated I–V curves of ZnSe, and CdS based cells, and (B) quantum efficiencies of ZnSe, and CdS cells.

measurements, respectively [141]. The CIGSS cells do not show any difference in PV parameters by light soaking for 30 min. However, the damp heat test on the cells shows a decrease in efficiency. After light soaking the efficiency increases in damp heat cells but not reaching to virgin efficiency [210,211]. A 30 nm thick ZnSe layer, which contains 900 ZnSe monolayers, is deposited onto glass/Mo/2–3 μm thick CIGS layer by Zn and Se Knudsen cells as an atomic layer deposition at 250 °C. The shutters have been sequentially opening 7 s for Zn and closing 7 s for Se or vice versa in the interval of 1 s. In this process, one monolayer of ZnSe is deposited per cycle. The ZnSe grown layers show cubic structure with (111) preferred orientation at 2θ diffraction angle of 27.22°. The final glass/Mo/CIGS/ZnSe/ZnO thin film solar cells with an area of 0.172 cm^2 exhibit efficiency of 11.6%. The same thick ZnSe buffer layer deposited by vacuum evaporation at 300 °C for 2 min is incorporated in the cells, which shows lower efficiency of 9.1% as compared to one with ZnSe grown by ALD (Table 8.25). The reason for lower efficiency is that while depositing ZnSe on CIGS, several recombination centers are generated on the surface of CIGS confirming by PL studies. The donor–acceptor pair emission lines are due to participation of In$_{Cu}$ donor to V$_{Cu}$ acceptor or Ga$_{Cu}$ donor to V$_{Cu}$ acceptor transitions in the CIGS/CdS structure, whereas no such emission lines are observed in the CIGS/ZnSe. This may be due to formation of Zn$_{Cu}$ antisites, which suppress V$_{Cu}$ therefore no possibility of action of DAP. Secondly, the evaporation temperature of 320 °C for ZnSe is high that causes to reevaporation of Se leading to form V$_{Se}$ [212,213].

The efficiency of 0.172 cm^2 active area typical CIGS solar cells with 10 nm thick ZnSe layer grown by ALCVD technique improves from 5 to 11.6% after irradiating under 1 sun light for 1 h. The efficiency, V_{oc} and FF increase, however no changes in J_{sc}. After keeping the cell under dark conditions for 30 min, the efficiency of cell decreases to 5%. The forward bias is applied to the cell at different voltage levels to

Table 8.25 PV parameters of ZnSe buffer based CIGS thin film solar cells

Buffer	V_{oc} (mV)	J_{sc} (mA/cm^2)	FF (%)	η (%)	Ref.
CIGSS/Zn(Se,OH)$_x$ (CBD)	535.1	36.1	70.76	13.7	[206,210]
ZnSe (MOCVD)	487.7	36.3	62.2	11.03	
CIGSS/Zn(OH)$_2$/Zn(Se,OH)	583	33.9	72.9	14.4	[207]
CIGSS/Zn(OH)$_2$	498	32.1	66.7	10.7	
CIGSS/CdS	588	33.7	73.7	14.6	
CIGSS/Zn(Se,OH) (CBD)	570	36.6	69	14.2	[208]
CIGSS/Zn(S,OH) (CBD)	569.4	34.9	71.3	14.2	
CIGSS/CdS/ZnO(MOCVD)[a]	592.6	34.58	68.12	12.69	
CIGSS/CdS/ZnO(DC-SP)[a]	638.6	32.2	67.95	11.64	
Zn(Se,OH)/ZnO(MOCVD)[a]	609.1	38.47	63.19	11.70	
Zn(Se,OH)/ZnO (DC-SP)[a]	635.3	31.40	64.61	10.74	
Zn-treated CIGSS/ZnSe	570	35.2	72.3	14.5	[209]
CdS	573	32.35	73.12	13.6	
CIGSS/ZnSe(MOCVD)	552	39	70	15.1	[211]
CIGS/ZnSe(ALD)	502	35.2	65.4	11.6	[212,213]
CIGS/ZnSe(Co-Ev.)	609	24.6	60.6	9.1	

[a]20 cm^2 module is connected by 11 or 12 interconnected cells.

test cell performance under the dark conditions; the efficiency of cell increases with pre-bias then after holding 10 min without bias under the same dark condition the cell performance reverses to normal. This is mainly due to carrier injection phenomenon. The ZnIn$_x$Se$_y$ (ZIS) and In$_x$Se$_y$ layers with <50 nm are prepared by co-evaporation of individual elements onto 2 mm thick soda-lime glass/Mo(1 μm)/CuInGaSe$_2$(2 μm) at substrate temperature of 300 °C with Se/Zn and Se/In ratios of 10 and 15, respectively. Prior to deposition of ZnO layers by MOCVD, glass/Mo/CIGS/ZnInSe and glass/Mo/CIGS/InSe are annealed under air at 200 °C, which show efficiencies of 12.7 and 13%, respectively. There is no remarkable change in the characteristics of device after light illumination. The Auger depth profiles of CIGS/In$_x$Se$_y$ and CIGS/ZnIn$_x$Se$_y$ up to 100 nm reveal that no Cu diffusion into buffer layers is observed. The Raman spectra confirm the characteristic mode of 170 cm^{-1} for CIGS and probably mode at 150 cm^{-1} for γ-In$_2$Se$_3$ defect wurtzite structure in the In$_x$Se$_y$/CIGS structure, whereas no Raman signal is observed for ZnIn$_x$Se$_y$ buffer in the ZnIn$_x$Se$_y$/CIGS structure [214]. Figure 8.54 shows I–V and quantum efficiency of CIGS/ZnInSe cells that the higher efficiency can be achieved for the cells by improving the quality of CIGS layer and deposition conditions of ZnInSe buffer. The efficiency of 15.1% for ZnInSe (Zn/In = 1.5) based cell is close to efficiency of 15.9% for CdS reference cell, as shown in Table 8.26. Some times it is hard to explain about the QE results because the low 13.9% efficiency cell shows its QE as close to high efficiency CdS reference cell, as shown in Fig. 8.55A. Auger depth profile on CIGS/ZnInSe indicates that Cu diffusion takes place into the interface from CIGS layer that may have impeded doping of Zn from ZnInSe buffer

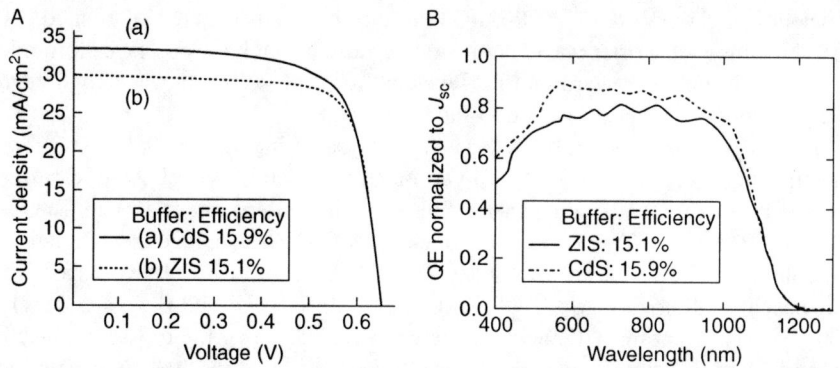

Figure 8.54 (A) *I–V* and (B) quantum efficiency of CIGS cells with CdS, and ZnInSe buffer layers.

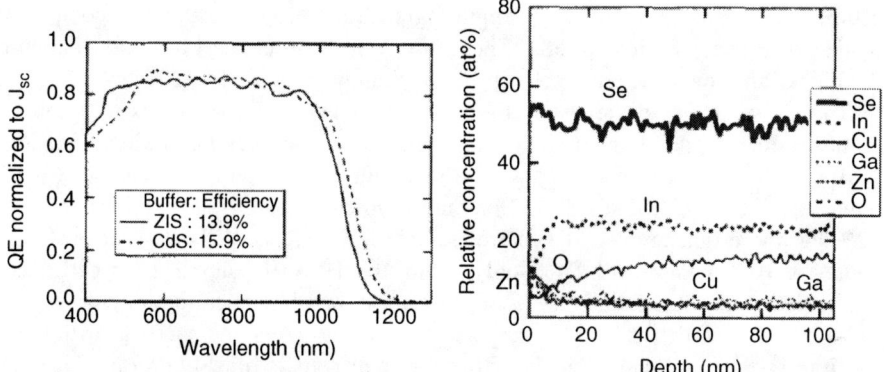

Figure 8.55 (A) Quantum efficiencies of CIGS/CdS, and CIGS/ZIS cells, and (B) Auger depth profile of CIGS/ZIS cell.

layers while growing buffer layers at 550 °C, as depicted in Fig. 8.55B. The AES analysis confirms thickness of 10 nm for ZnInSe buffer layer. The $ZnIn_xSe_y$, ZnSe, and In_xSe_y buffer layers show band gaps of 2.0, 2.65, and 2.1 eV from optical studies. The XRD analysis reveals that the (112), (111), and (006) peaks in the 2θ range between 26 and 28° for $ZnIn_2Se_4$, ZnSe, and γ-In_2Se_3 confirming defect chalcopyrite, cubic, and wurtzite structure, respectively. Finally, the Raman modes at 135 and 240 cm^{-1} for $ZnIn_2Se_4$ and 150 cm^{-1} for In_2Se_3 in the Raman spectra support the XRD analysis. The cells with $ZnIn_xSe_y$, ZnSe, and In_xSe_y buffer layers show efficiency of more than 10% but cells with $ZnIn_xSe_y$ buffer deposited at 550 °C shows higher efficiency of 15.1% [215,216]. The cells with combination of (112) oriented CIGS layer with $ZnIn_xSe_y$ buffer layer deposited by co-evaporation of Zn, In, and Se show efficiency of 15.1%. A little improvement in efficiency to 15.3% is observed for the cells, which contain (220) oriented CIGS layer (Table 8.26). More importantly, an improvement in V_{oc} takes place from 652 to 682 mV. The quantum efficiency of glass/Mo/CIGS/CdS/ZnO thin film solar cells with (220) oriented CIGS absorber is low at shorter wavelength side because of thick n-CdSe layer formation, whereas a shift in QE toward shorter wavelength at 1100–1200 nm wavelength region is due to wider band gap of CIGS, as shown in Fig. 8.56 [118].

The Zn is evaporated onto CIGS layers with beam intensity of 0.7×10^{-7} torr at substrate temperature of 300 °C, as a part of Zn treatment on which ZnInSe buffer is grown at the same temperature then a bilayer with combination of 100 nm ZnO and 1500 nm B doped n-ZnO layers is formed by MOCVD technique to complete cell structure. The resistivities of i-ZnO and ZnO:B are 0.1 and 10^{-3} Ω-cm, respectively. The post annealing glass/Mo/CIGS-Zn-treated/ZnInSe/ZnO/ZnO:B cells under air at 200 °C reveals that the efficiency of the cell decreases over the crossing annealing time of 30 min, whereas a decrease in efficiency occurs over the time of 60 min in the Zn untreated samples. The efficiency is low (2.3%) for higher Zn beam intensity of 4.7×10^{-7} torr, whereas the efficiencies of 10.7 and 9.6% are observed for Zn beam intensities of 1×10^{-7} and 0.2×10^{-7} torr, respectively. The Zn treatment is done for the duration of 210 s. As anticipated, the QE is low in the case of low efficiency cells in the entire wavelength region from 400 to 1200 nm as compared to that of higher efficiency cells. The cutoff wavelengths are 1140 and 1080 nm for higher and low efficiency cells in the IR region, respectively [217].

The CIGS layers are deposited by novel linear evaporation sources unlike four points source evaporation or three-stage process method for glass/Mo/CIGS/CdS (CBD)/ZnO:Al-(Sput) thin film solar cells in the pilot line program. The cells with either ZnInSe or pilot line CdS buffer shows more or less the similar efficiency of 12% but the efficiency of CdS control sample is 16%. The cells show lower efficiency of 10% without ZnInSe or only with ZnO. The efficiency of solar cell drastically decreases to 3.13 and 6.59% for the buffer growth temperature of below 100 or beyond 200 °C that is, 315 °C, respectively indicating that the buffer growth temperature is also important to incur efficiency solar cells [218]. SLG/Mo/CIGS/ZnO

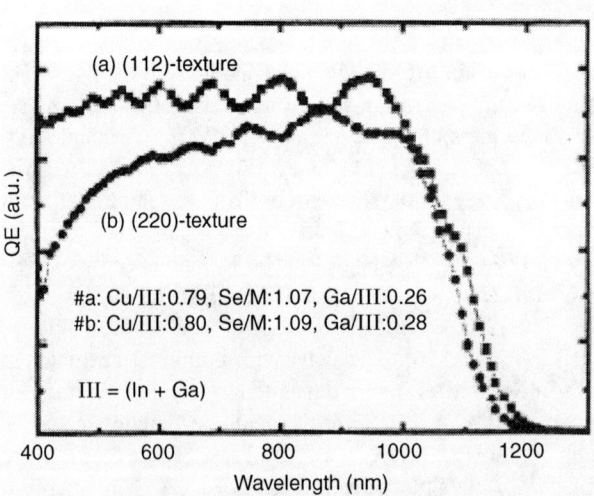

Figure 8.56 Quantum efficiencies of CIGS cells with different orientations of CIGS absorbers.

(ALD)/ZnO:B solar cells with an area of 0.184 cm^2, in which 1–1.5 μm thick CIGS absorber has CIGSS surface of 60–80 nm, made by Showa Shell Sekiyu KK yield efficiency of 13.9%. The resistivity of i-ZnO layers grown at 165 °C by ALCVD technique is nearly 10^3 Ω-cm. Prior to deposition of i-ZnO layer, the CIGSS absorber layer is heat-treated at 285 °C in order to remove any unwanted stuff such as In$_x$S$_y$ on the surface of it. ZnIn$_x$Se$_y$, ZnO, and ZnO:B layers are sequentially grown onto CIGSS by evaporation, ALD, and MOCVD, respectively to complete the SLG/MO/CIGSS/ZnIn$_x$Se$_y$/ZnO(ALD)/ZnO:B(10 Ω/□)(MOCVD) cell. By adding ZnIn$_x$Se$_y$ layer to SLG/Mo/CIGS/ZnO(ALD)/ZnO:B, the efficiency of cell increases from 13.9 to 14.4% (Table 8.26). After illumination of 60 min that is, light soaking, the cell efficiency decreases from 14.4 to 14.1%, correspondingly V_{oc} reduces from 598 to 596 mV, J_{sc} decreases from 33.6 to 33.1 mA/cm^2 but FF remains at the same level of 71.4%. The phenomenon of light soaking effect is that the efficiency of cell increases when the cell is illuminated under light for certain time and then reverses to virgin efficiency for keeping the cell under dark for certain time [219].

8.3.8 InS Window Based CIGS Cells

The InS buffer replaces CdS in the CIGS cells owing to friendliness to environment. The band gap of In$_x$S$_y$ varies from 2.2 to 2.8 eV depending on its composition. 30–40 nm thick In$_2$S$_3$ buffer layers prepared by atomic layer chemical vapor deposition (ALCVD) at 160 °C have been employed in the CIGS solar cells, which realize efficiency of 12.1% (Table 8.27). The XPS/UPS studies of In$_2$S$_3$ grown onto CIGS at 160 °C reveal that by taking band gap (E_g) of 2.1±0.05 eV for In$_2$S$_3$ and 1.15±0.05 eV for CIGS; $\Delta E_v = -1.2\pm 0.2$ eV (-1.4 ± 0.2 eV for Na free), and $\Delta V_c = -0.25\pm 0.2$ eV (-0.45 ± 0.2 eV for Na free) are the band offset values in

Table 8.26 PV parameters of ZnSe, InSe, and ZnIn$_x$Se$_y$ based CIGS cells

Buffer/cell area	V_{oc} (mV)	J_{sc} (mA/cm^2)	FF (%)	η (%)	Ref.
ZnSe (ALD)	502	35.2	65.4	11.6	[214]
In$_x$Se$_y$ (Co-Ev.at 300 °C)	595	30.4	72	13	
ZnIn$_x$Se$_y$ (Co-Ev.at 300 °C)	579	29.2	75.2	12.7	
ZnIn$_x$Se$_y$ (Co-Ev.at 550 °C)	652	30.4	78.3	15.1	[215]
CdS (CBD)	648	33.7	73.1	15.9	
(220 orient. CIGS)/ZnIn$_x$Se$_y$	682	29.2	76.5	15.3	[118]
(112 orient. CIGS)/ZnIn$_x$Se$_y$	652	30.4	76.2	15.1	
ZnInSe	562	30.8	67.2	11.64	[218]
ZnO only	541	31.8	59.9	10.29	
CdS-control	610	35.8	74.8	18.33	
CdS pilot line	581	30.1	68.7	12	
ZnO only	510	36.9	73.6	13.9	[219]
ZnInSe	598	33.6	71.4	14.4	

ZnO only indicates without ZnInSe.

Table 8.27 PV parameters of InS based CIGS cells

Buffer thickness/ T_s/ substrate	Buffer layer	V_{oc} (mV)	J_{sc} (mA/cm^2)	FF (%)	η (%)	w (nm)	Ref.
	In$_x$S$_y$	660	26.9	68.4	12.1	–	[220]
	CdS	622	28.3	78	13.7	–	
	In situ In$_x$S$_y$[a]	609	30.6	63.8	11.9	–	[221]
	Ex situ In$_x$S$_y$[a]	608	29.5	64.6	11.6	–	
	CBD-CdS	648	32.6	76.1	16.1	–	
	In situ In$_x$S$_y$[b]	665	31.3	71.1	14.8	–	
	Ex situ In$_x$S$_y$[b]	612	32.6	66.2	13.2	–	
	CBD-CdS	630	31.7	76.2	15.2	–	
	In$_x$S$_y$ (0.5 cm^2)	723	28.8	72.3	13.2	–	[223]
	(10 × 10 cm^2)	632	27.9	63.6	11.2	–	
	CdS (0.5 cm^2)	668	28.4	76.1	14.4	–	
	(10 × 10 cm^2)	661	28	66.1	12.3	–	
30 nm	β-In$_2$S$_3$:Na	300	23.9	36	2.6	–	[225]
50 nm	β-In$_2$S$_3$:Na	610	20.8	60	7.6	–	
100 nm	β-In$_2$S$_3$:Na	661	20	62	8.2	–	
–	In$_2$S$_3$ spray	502	34.7	71	12.4	–	[227]
–	CdS (reference)	521	31.1	65	10.5	–	
–	CBD-CdS	610	28	60	10.2	–	
–	In$_2$S$_3$	572	41.9	64.2	15.1	–	[228]
130 °C	In$_2$S$_3$	686	31	72	12.4	–	[229]
200 °C	In$_2$S$_3$	620	27	45	8.3	–	
–	In$_x$(OH,S)$_y$	632	32.4	73	14.9	–	[231]
–	CdS	595	34.9	73	15.2	–	
Corning-7059	In$_x$(OH,S)$_y$	590	34.9	72	14.8	~200	[232]
Soda lime	In$_x$(OH,S)$_y$	630	32.4	73	14.9	~150	
Corning-7059	CdS	536	34.3	72	13.2	~1000	
Soda lime	CdS	595	34.9	73	15.2	300–500	
–	CdS	510	30.5	63.8	9.99	–	[233]
–	In$_x$(OH,S)$_y$	570	29.1	44.6	7.39	–	
–	In(OH)$_3$:Zn^{2+}	575	32.1	75.8	14	–	[234]
–	In(OH)$_3$	469	30.7	67	9.6	–	
–	In(OH)$_x$S$_y$	525	27.3	59.7	8.6	–	[186]
–	ZnS	520	27.7	68.4	9.8	–	
–	CdS	530	31.4	68.3	11.1	–	
–	In$_2$S$_3$	557	32.6	72.4	13.33	$A = 1.68$	[20]
–	ZnMgO	515	34	69	12.09	$A = 1.89$	
–	CdS	564	35.2	69.8	13.87	$A = 1.98$	
–	In$_2$S$_3$	665	31.5	78	18.4	–	[235]

[a]In and S co-evaporation.
[b]Evaporation from InS bulk compound.

the band structure of In$_2$S$_3$/CIGS heterostructure. This negative offset causes to support interface recombination resulting in poor photocurrent collection [220]. As stated in earlier, the optimum value of ΔV_c is in between 0 and 0.4 eV for thin film solar cells. The band diagrams of CIGS/CdS and CIGS/In$_x$S$_y$ are theoretically simulated employing band gap of 1.14 eV and hole concentration of 10^{16} cm^{-3} for CIGS absorber, as shown in Fig. 8.57. The surface CIGS layer is Cu-poor with respect to bulk CIGS layer in the high efficiency CIGS/CdS based cells. The Cu-poor CIGS layer contains n-Cu(InGa)$_3$Se$_5$ phase that is strongly inverted. As pointed out earlier, the Cd diffuses into CIGS while depositing CdS by CBD method that also makes inversion layer (n-type). The valence band is pulled down (solid line) from its original place (dashed line) due to inversion layer. The Fermi level in the band structure should be close to the conduction band in order to prevent interface recombinations. The admittance spectroscopy reveals that the distance between the Fermi level and the conduction band is 0.1–0.2 eV. The barrier for holes (Φ_b^P) can be arrived as $\Phi_b^P = E_g - \Delta E_F$, which is substantially sufficient to block recombination of photogenerated electrons with holes. Therefore, the recombination in the space charge region is dominant. The donor-like defects are favorable in In$_x$(OH, S)$_y$ and Zn$_x$(OH,S)$_y$ buffer layers to keep Fermi level close to the conduction band akin CdS in the band structure. However, acceptor like defect density in them leads along with positive states from the absorber. Therefore, the Fermi level slips away from the conduction band. The barrier for electrons (Φ_b^n) increases and holes (Φ_b^P) decreases. Hence, the photogenerated electrons in the absorber recombine with holes rather than separation as interface recombination [221,222].

After air annealing SLG/Mo/CIGS/30–80 nmIn$_x$S$_y$/RF-ZnO(50 nm)/DC-ZnO:Al (400 nm)/Ni/Al thin film solar cells with an active area of 0.5 cm^2 at 200 °C for 10 min, they show efficiency of 13.3%, whereas the modules over the large area of 10×10 cm^2 without air annealing show efficiency of 11.2% (Table 8.27). The In$_x$S$_y$ thin films prepared by sputtering technique using In$_2$S$_3$ ceramic target at different temperatures of 120, 230, and 320 °C show band gaps of 1.88, 2.04, and 2.16 eV, respectively [223]. However, the cells with In$_2$S$_3$ buffer formed by reactive

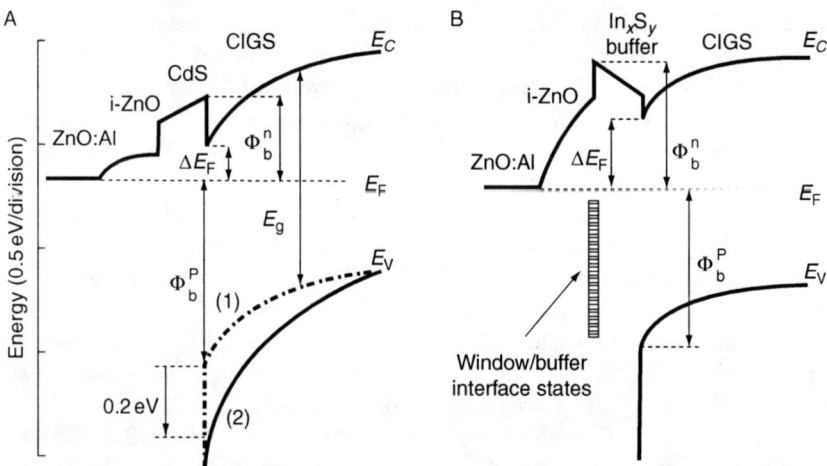

Figure 8.57 Band diagram of (A) CIGS/CdS, and (B) CIGS/InS heterostructures.

sputtering of In target and H_2S:Ar = 1:1 gas exhibits little lower efficiency of 11.1% due to lower current collection [224]. The efficiency of glass/Mo/ZSW-CIGS/In_2S_3-ALCVD/ZnO/ZnO:Al solar cell gradually increases from 8.8, 8.7, 10.4, 12.1 to 12.6% with increasing growth temperature of In_2S_3 buffer layer from 140, 160, 180, 200 to 220 °C, respectively. The Na content is observed on the surface of In_2S_3 buffer due to diffusion, which is virtually higher with an increased deposition temperature. On the other hand, the Cu diffusion into In_2S_3 buffer layers from CIGS takes place that creates OVC layer on the surface of CIGS layer. As a result of OVC formation, the n-OVC/p-CIGS buried homojunction increases efficiency of the cell, however there is a possibility of forming of $CuIn_5S_8$. The cells annealed under Ar for 15 min at varies temperatures within the window limits of 200–400 °C reveal that they can tolerate up to annealing temperature of 200–250 °C thereafter all the PV parameters of them deteriorate that means cells lose thermal stability. The glass/Mo/CIGS/β-In_2S_3:Na-PVD/i-ZnO/ITO/Ni-Al thin film solar cells realize efficiency of 8.2%, which is low comparing with efficiency of 10.2% for the CdS reference solar cell. The efficiency of the cells increases from 2.6, 7.6 to 8.2% with increasing thickness of β-In_2S_3:Na buffer layer from 30, 50 to 100 nm (Table 8.27). The lower V_{oc} and FF cause to have lower efficiency in the cells due to lower buffer thickness. The In and NaF then sulfur are evaporated at 200 °C to form 100 nm thick β-In_2S_3:Na buffer with Na/In ratio of 0.12 and band gap of 2.8 eV on the CIGS thin films with Cu/(In + Ga) = 0.28. The band gap of Na doped β-In_2S_3, that is, $In_{21.33-x}Na_xS_{32}$ thin films varies from 2.15 to 2.9 eV with varying x from 0 to 0.8 [225,226].

The SLG/Mo/CIGSS/In_2S_3/i-ZnO/ZnO:Al/Ni/Al cell shows efficiency of 12.4% as compared to that of 10.5% for CdS reference cell, as shown in Fig. 8.58. The In_2S_3 buffer layer used in this cell is prepared by ultrasonic spray pyrolysis at substrate temperature of 200 °C for 15 min using chemical solutions of $InCl_3$, $(NH_2)_2CS$, and methanol. The measured short circuit currents are 33.1, 31.3, and 32 mA/cm^2 for the cell with indium sulfide buffers having different In:S composition ratios of 1:4, 1:3, and 1:1 in the chemical solution, respectively. The buffers could be InS, In_xS_y, and In_2S_3 for the chemical composition of In:S = 1:1, 1:3, and 1:4, respectively. The short circuit current is relatively low in the In_xS_y based cells. It could be due to nonstoichiometry of In_xS_y or undeveloped buffer. The indium sulfide layer may be partially amorphous nature for the ratio of 1:4. The QE of In_2S_3 based cell is higher at shorter wavelength region with respect to reference cell (Fig. 8.59) [227]. The best CIGSS/In_2S_3/ZnO/ZnO:Al/Ni/Al cells show efficiency of 15.1%, $R_s = \sim 0.55$ Ω-cm^2 and $R_{sh} = \sim 400$ Ω-cm^2 for which the In_2S_3 layer is grown by spray-ILGAR deposition onto CIGSS Shell Solar absorber layers. $InCl_3$ is dissolved in ethanol and transported by N_2 as a carrier gas with flow rate of 75 ml/s by ultrasonic spray and reacted with H_2S gas to form In_2S_3 layer at substrate temperature of 175–250 °C. The growth rate is 3.3 nm/cycle and total cycles of about 7 [228]. The typical glass/Mo(0.5 μm)/ASC-CIGS(1.5 μm)/50 nm PVD-In_2S_3 cells annealed at 200 °C for 1 min offer efficiency between 8 and 12%. On contradictory to previous results, Table 8.27 reveals that the cells with In_2S_3 buffer grown at higher substrate temperature of 200 °C gives lower efficiency of 6.3% than that of

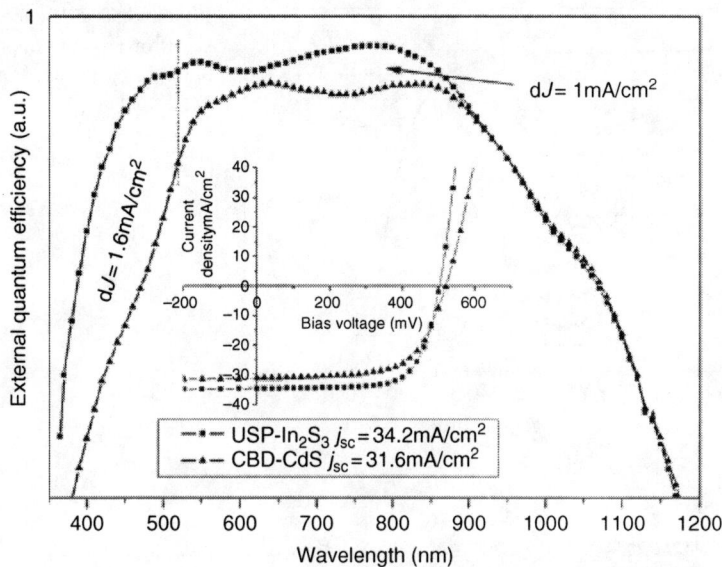

Figure 8.58 Quantum efficiencies, and *I–V* curves of CIGS/In$_2$S$_3$, and CdS reference cells.

12.4% for the cells formed with In$_2$S$_3$ grown at 130 °C (Fig. 8.59). The reason may be due to migration of Cu into the buffer that forms [In$_{16}$]$_{Oh}$[In$_{5.33-x-}$Cu$_{3x}$L$_{2.66-2x}$]$_{Th}$S$_{32}$ phase for $0 \leq x \leq 1.33$ or CuIn$_5$S$_8$ for $x = 1.33$, where L, Oh, and Th are cation vacancy, octahedral and tetrahedral sites, respectively. Na occupies Cu site, if it is doped. The indium sulfide thin films used as a buffer layer in the CIGS cells are grown by vacuum co-evaporation of In and S from tungsten and pyrex crucibles, respectively at two different substrate temperatures of 130 and 200 °C under the pressure of 10^{-3} Pa over the large area of 12.5×12.5 cm^2 CIGS substrates. The XPS analysis reveals interdiffusion of Se and S in the layers [229]. The optimum conditions from system to system may vary.

0.48 cm^2 area glass/Mo(1.2 μm)/Cu(InGa)Se$_2$(1.6–2 μm)/In$_x$S$_y$/i-ZnO(50 nm)/ZnO:Al(300 nm) solar cells with In$_x$S$_y$ buffer layer shows efficiency of 14.8%. The SEM cross section of solar cell is shown in Fig. 8.60A. One can see very thin pin pointed In$_x$S$_y$ buffer layer. The In$_x$S$_y$ layer is deposited by InS bulk compound or In and S individual elements as sources by co-evaporation technique, which show band gaps of 2.5 and 2.4 eV, respectively. In general, Cu diffusion into InS is observed from CuInGaSe$_2$/InS interface while preparing InS by ALCVD deposition technique, however by evaporation technique such kind of diffusion is not observed. The InS by evaporation of bulk compound shows better results than that of co-evaporation of In and S in terms of efficiency, as shown in Table 8.27. The buffer thickness $30 \leq d \leq 90$ nm and substrate temperature $23 \leq T_s \leq 120$ °C for InS compound evaporation, whereas $30 \leq d \leq 100$ nm and $120 \leq T_s \leq 340$ °C for In and S co-evaporation are suitable parameters to have better efficiency solar cells. The cells show no difference in photovoltaic parameters after annealing them either under vacuum

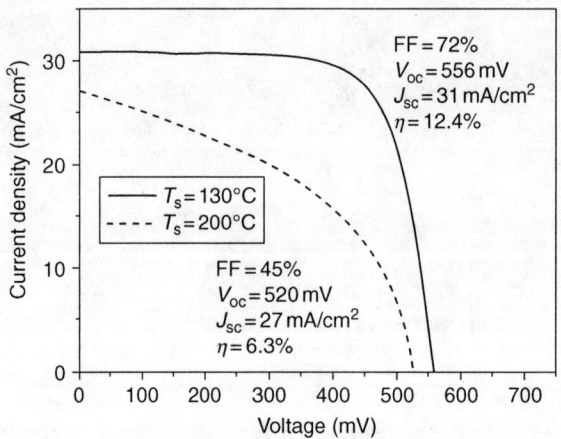

Figure 8.59 *I–V* curves with effect of growth temperatures of In_2S_3 buffer used in the CIGS cells.

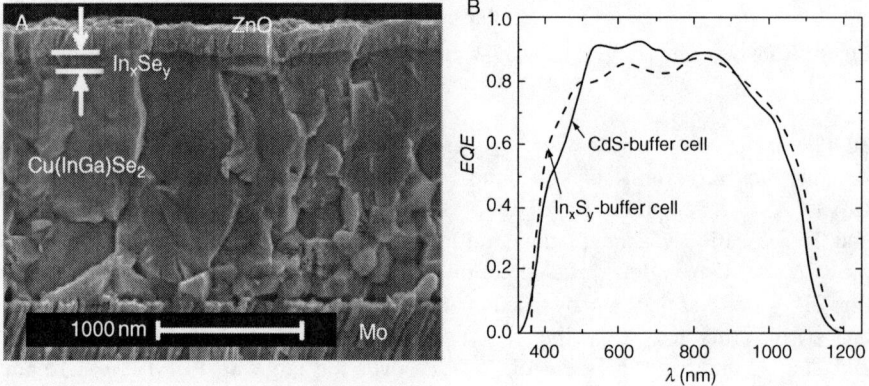

Figure 8.60 *I–V* curves with effect of growth temperatures of In_2S_3 buffer used in the CIGS cells.

or air. The In_xS_y buffered solar cells contribute slightly higher quantum collection efficiency in blue wavelength regime that results in higher J_{sc} as compared to that of CdS buffer layer, as shown in Fig. 8.60B. *In situ* In_xS_y with thickness of 90 nm is deposited using InS bulk compound as a source at substrate temperature (T_s) of 120 °C and *ex situ* In_xS_y with thickness of 40 nm is also deposited at substrate temperature (T_s) of 20 °C. The cells with *in situ* and *ex situ* In_xS_y buffer layers formed using bulk InS are annealed under air at 200 °C for 24 and 12 min, respectively. 50 nm thick In_xS_y layer is co-evaporated for *in situ* and *ex situ* process at substrate temperature of 320 °C using In and S sources for solar cells. The corresponding cells are annealed under air at 300 °C for 25 min and 320 °C for 20 min, respectively. The *in situ* and *ex situ* denote that the InS is grown onto CIGS without breaking vacuum and the CIGS layer is exposed to air prior to deposition of InS, respectively [221].

The CIGS (Cu:In:Ga:Se = 23.1:18:6:52.9) cells with 30 nm thick $In_x(OOH)_yS_z$ buffer layer yield efficiency of 10% for which the $In_x(O,OH)_yS_z$ buffer layer deposited by CBD method using chemical solutions 0.03 M $InCl_3$ and 0.1 M CH_3CSNH_2 at bath temperature of 68 °C and pH of 2 for 15–30 min. The $In_x(O,OH)_yS_z$ buffer layer shows In_2S_3 (111), (220), and In(O,OH) (011), (200), and (211) phases confirmed by XRD analysis. The XPS analysis reveals that $In3d_{5/2}$ and $In3d_{3/2}$ symmetric peaks appeared at 445.2 and 452.8 eV, respectively close to the In signals of In_2S_3, In_2O_3, and $In(OH)_3$ compounds. Well-connected cucumber shape fine grain structure is found in the buffer by SEM analysis [230]. The CIGS solar cells with CdS and $In_x(OH,S)_y$ buffer layers formed by CBD exhibit efficiencies of 15.2 and 14.9%, respectively (Table 8.27). The lower red response in QE may be due to smaller space charge region width (w). An increased N_A contributes to smaller space charge region width, which is inversely proportional to the square root of N_A [231]. As shown in Fig. 8.61, the as-grown CIGS (Ga/In + Ga = 0.18)/$In_x(OH,S)_y$/ ZnO cells show poor performance, however air anneal at 200 °C for 2 min, followed by light soaking they show efficiency of 15.75%. Fig. 8.62 shows I–V curves and QE of CIGS/CdS/ZnO and CIGS/$In_x(OH,S)_y$/ZnO cells made on different substrates of corning 7059 and soda-lime glass. The efficiencies of the cells are little higher in both the $In_x(OH,S)_y$ and CdS buffered cells made on soda-lime glass substrates comparing with the cells made on corning 7059. This could be due to self diffusion of Na from SLG. On the other hand, as anticipated the V_{oc} is little higher in the $In_x(OH,S)_y$ buffer based cells because of its wider band gap. The same phenomena is observed in the QE that the spectral response is higher in shorter wavelength region for $In_x(OH,S)_y$ buffered based cells irrespective of type of substrates. Overall the spectral response is higher for CdS based cells at longer wavelength regions [232].

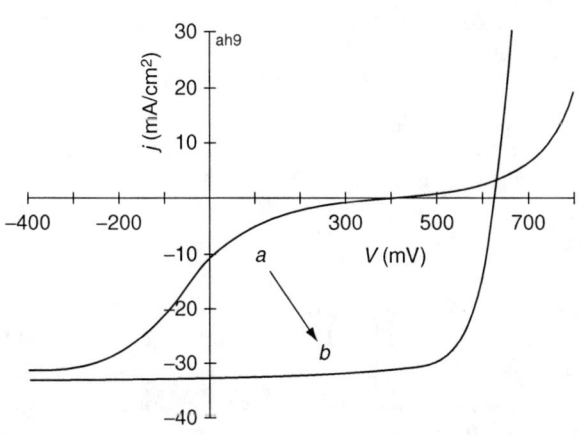

Figure 8.61 I–V curves of $CuInGaSe_2$/$In_x(OH,S)_y$/ZnO thin film solar cells under illumination; (a) as-grown, and (b) annealed, and light soaking.

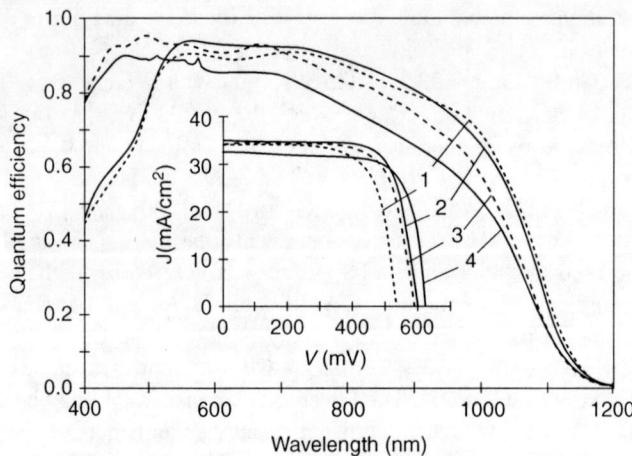

Figure 8.62 Quantum efficiencies and *I–V* curves of CIGS cells with different buffers and substrates (A) corning7059/CIGS/CdS/ZnO, (B) soda-lime/CIGS/CdS/ZnO, (C) corning7059/CIGS/In$_x$(OH,S)$_y$/ZnO, and (D) soda-lime/CIGS/In$_x$(OH,S)$_y$/ZnO.

The In$_x$(OH,S)$_y$ buffer layer by CBD technique on CIGS with Cu/(In + Ga) = 0.8–0.9 and Ga/(In + Ga) = 0.3 is formed. The crossover nature is less in 0.47 cm^2 area CIGS/In$_x$(OH,S)$_y$/ZnO-bilayer/Ni–Al than that in the CIGS/CdS/ZnO-bilayer/Ni–Al cells in their *I–V* curves under dark and light illumination conditions. After annealing In$_x$(OH,S)$_y$ and CdS buffer based CIGS cells under air at 200 °C for 20 and 5 min show efficiencies of 7.4 and 10%, respectively (Table 8.27). On contrast to the CdS based solar cells, the V_{oc} is higher in In$_x$(OH,S)$_y$ for the same CIGS absorber layer, it is mainly due to an increased carrier concentration in the absorber that causes to reduce the depletion width in the junction. An increase in carrier concentration is due to wider band gap of In$_x$(OH,S)$_y$ buffer that allows more radiation. On the other hand, there may be a formation of OVC compound likely Cu(InGa)$_2$Se$_3$ phase at the interface between buffer and absorber layers. Auger depth profile of CIGS/In$_x$(OH,S)$_y$, as shown in Fig. 8.63 reveal that the atomic concentrations of S, O, In, and Ga are high at surface of the sample. The atomic concentrations of Se and Cu are found to increase with etching. The XPS analysis of In$_x$(OH,S)$_y$ reveals that S/O and In/S ratios are 1 and 1.75, respectively. The oxygen peaks at 530.2 and 531.6 eV are related to bonding of In–O and In–OH, respectively. The optical band gap of 2.54 eV for In$_x$(OH,S)$_y$ is observed. The In$_x$(OH,S)$_y$ buffer layer employed in the CIGS cells can be deposited by CBD method using InCl$_3$ (5 mmol/l) and either CH$_3$CSNH$_2$ (0.15 M) or 0.15 M thioacetamide solutions with pH of 1.8 at bath temperature of 70 °C for 20 min [233].

The CIGS samples prepared by MBE three-stage process are immersed into combined 0.01 M ZnCl$_2$, 0.01 M InCl$_3$·4H$_2$O and 0.15 M thiourea solutions and brought to 60 °C from room temperature keeping for 30 min in order to deposit In(OH)$_3$ buffer layer. The pH of 3.6 for chemical bath solution is favorable for more incorporation of Zn^{2+} into In(OH)$_3$. The XRD analysis reveals that the 2θ peak at 22.2°

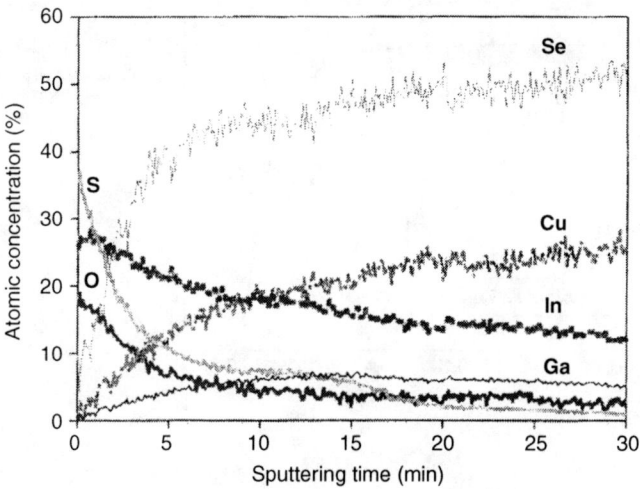

Figure 8.63 Auger depth profile of $In_x(OH,S)_y$ grown on CIGS layer.

is due to (200) reflection of $In(OH)_3$ crystal. After deposition of buffer layer onto CIGS, the layers are annealed under air at 200 °C for 20 min. The ZnO:B window layer by MOCVD and Al grids are sequentially deposited onto SLG/Mo/CIGS/CBD-$In(OH)_3$:Zn^{2+}. 0.2 cm^2 area SLG/Mo/CIGS/CBD-$In(OH)_3$:Zn^{2+}/ZnO:B(MOCVD) thin film solar cells show higher efficiency of 14% than that of one with In(OH)$_3$ buffer layer (Fig. 8.64). Zn doping into buffer results in higher efficiency because of an increased band gap of buffer that contribute to enhance V_{oc}, J_{sc}, and FF (Table 8.27) [234]. The XPS of $In(OH)_xS_y$ buffer layer grown onto CIGS by CBD revelas that the S2p$_{3/2}$-161.24 eV and weak peak at 168.96 eV are probably related to indium sulfate (169.171 eV), which disappears after sputtering layer for 4 min, indicating that indium sulfate phase is formed on the surface of buffer. Two oxygen peaks such as one at 531.4 eV (InOH) as a stronger and another peak at 529.5 eV (InO) as a weaker peak are present. After sputtering former becomes weak and latter leads stronger indicating that InOH concentration is higher on the surface of the as-grown layer. In some cases, in comparison with CdS buffer layer, the CIGS cells with $In(OH)_xS_y$ buffer layer show higher open circuit voltage because of an increased carrier concentration in the CIGSS absorber. The CIGSS/$In(OH)_xS_y$/ZnO(MOCVD), CIGSS/ZnS/ZnO(MOCVD), and CIGSS/CdS/ZnO(MOCVD) cells with different buffer layers experience efficiencies of 8.6, 9.8, and 11.1%, respectively (Table 8.27) [186]. The thicker buffer layers cause to higher series resistances and poor fill factor. As shown in Fig. 8.65A, plots of saturation current versus inverse temperature give activation energies of 0.6, 0.54, and 0. 52 eV for the glass/Mo/CIGSS/In_2S_3/ZnO, ZnMgO, and CdS based cells, respectively, in which CIGSS absorber is made by Avancis company. In the present samples, the recombination may be due to thermal

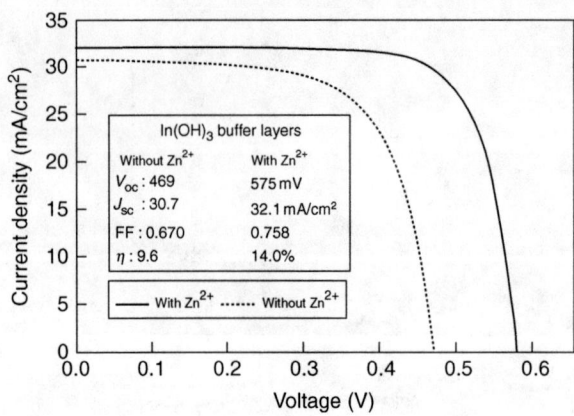

Figure 8.64 *I–V* curves of In(OH)$_3$ buffer based CIGS cell under illumination.

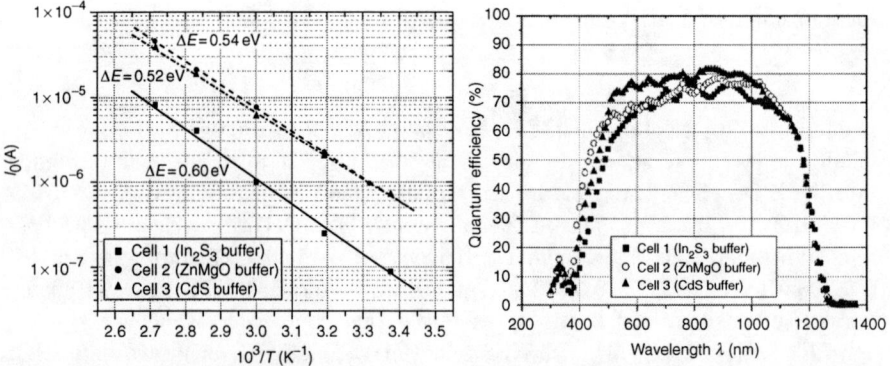

Figure 8.65 (A) Short circuit current versus inverse temperature of CIGS cells with different buffer layers, and (B) quantum efficiencies of CIGS cells with different buffer layers.

recombination. The spectral responses of the cells are more or less the same for all the cells but at shorter wavelength about (500 nm), which are close to the band gaps of buffer layers whereas at longer wavelength region the spectral transition region at 1265 nm (0.98 eV) which is close to the band gap (1.1 eV) of CIGSS ($p = 10^{16}$ cm^{-3}), as shown in Fig. 8.65B. The reason could be due to sulfur grading in the sample whereby the transition takes place [20]. By employing ALCVD In$_2$S$_3$ buffer, an improved efficiency of 18.4% is significantly achieved from 0.1 cm^2 area CIGS/In$_2$S$_3$(31 nm)/i-ZnO(100 nm)/ZnO:Al(0.5 μm) thin film solar cells. The In$_2$S$_3$ layer grown by ALCVD is a beneficial due to wider band gap than that of 2.0 V of common β-In$_2$S$_3$ compound. The efficiency of solar cells drastically decreases with decreasing buffer thickness as well as with increasing the deposition temperature, for example, the efficiency of cell is 4% for 18 nm thick In$_2$S$_3$ buffer layer grown at 270 °C [235].

8.3.9 Effect of Conducting Layer on the CIGS Cells

The ZnO layer is one of the important materials as window and conducting layer to achieve high efficiency in the glass/Mo/CuInGaSe$_2$/CdS/i-ZnO/n-ZnO thin film solar cells. The typical RF deposition conditions for the growth of (002) oriented ZnO are 2 mtorr and substrate temperature of 100 °C and RF power of 120 W. In order to prepare ZnO by RF sputtering method, one with high resistance and another one with low resistance Al doped ZnO targets are used for i-ZnO and ZnO:Al· (n-ZnO) layers, respectively. Prior to deposition of doped ZnO layer (ZnO:Al$_2$O$_3$ 2.5 wt%), the insulating i-ZnO is traditionally deposited onto glass/Mo/CuInGaSe$_2$/CdS stack, followed by deposition of Al doped ZnO with thickness of 300 nm. Insulating ZnO is probably protecting the CdS window layer from the higher energy ZnO sputter particles over the period of deposition time. On the other hand, it acts as a seed layer for the growth of doped ZnO layer. An average transmission and resistivity of ZnO:Al are 85% and 4×10^{-4} Ω-cm, respectively. The resistivity of i-ZnO varies from 9.2×10^1, 8.4×10^2 to 2.4×10^6 Ω-cm with varying O$_2$ content from 0, 1 to 1.4%. During the deposition process, the Ar flow is kept to be at constant. The typical solar cell with an active area of 0.18 cm^2 shows efficiency of 14.5% along with $R_s = 0.36$ Ω-cm^2 and $R_p = 632.9$ Ω-cm^2 for the combination of i-ZnO (RF) and ZnO:Al (RF) [236]. The sheet resistance ($R_\square$) of typical as-grown ZnO layer on glass by RF sputtering increases from 50 to >500 kΩ/□ for annealing under O$_2$ at 225 °C otherwise it decreases to >7 kΩ/□ for annealing under 10% H$_2$ and 90% Ar. The resistivity (ρ) varies from 4×10^{-3} Ω-cm in the first quarter of the layer to 2.6×10^{-4} Ω-cm in the last quarter of the layer because of preconditioning done under oxygen before each run [189].

The efficiency of CuInGaSe$_2$ solar cells is low unless the composition of Ga is set to 0.3 in the CIGS ($E_g = 1.15$ eV). Beyond the Ga=0.3, the efficiency decreases with increasing Ga content. However, there is a special case for Ga=0.5 and $E_g = 1.3$ eV, glass/Mo(0.8 μm)/CIGS(1.8 μm)/CdS(50 nm)/i-ZnO(70 nm)/ZnO:Al (1 μm) solar cells show efficiency of 16% by choosing right thickness of i-ZnO layer. In this structure, onto Mo coated soda-lime glass substrates with size of nine 3×3 cm^2, the CIGS layers are deposited by three-stage process using a molecular beam epitaxy (MBE) technique. A 70 nm thick i-ZnO is an optimal for thin film solar cells to obtain higher efficiency of 16%. As the thickness of i-ZnO is decreased from 70 to 20 nm, the efficiency of cell sharply decreases from 16 to 10.6% and other PV parameters also decrease. The fill factor of cell decreases from 66.6 to 58.2% with increasing thickness of i-ZnO from 110 to 180 nm indicates deterioration of performance of solar cell. The i-ZnO and n-ZnO:Al layers are prepared by RF sputtering technique with intrinsic ZnO and ZnO:Al$_2$O$_3$ targets, respectively. To deposit layers, the typical deposition parameters of chamber pressure of 5×10^{-1} Pa, power of 50 W, and the distance of 5 cm between target and substrate and substrate temperature of 150 °C are employed. The undoped and doped layers show $\rho = 3 \times 10^5$ and 1×10^{-3} Ω-cm, $\mu = 2$–3 and 18–19 cm^2/Vs, $n_e = 1 \times 10^{14}$ and 2×10^{20} cm^{-3}, respectively. A 70 nm thick i-ZnO layer is determined to be optimal for good performance of cells. The V_{oc} is low due to shunting in the junction whether a 20 nm thick i-ZnO layer is used or not. If i-ZnO layer thickness is greater than 70 nm, the short circuit current density decreases because of an increase in series resistance of the solar cells. The fill factor becomes worse due to an increase of leakage current for less than 70 nm thickness. The efficiency of cell increases from 15.2 to 16.9% by adding MgF_2 anti-reflection coating [136,237]. In the case of glass/Mo/CIGS/i-ZnO/ZnO: B thin film solar cells yield efficiency of 11.7% without CdS buffer. The i-ZnO layer used in the cells is grown by atomic layer deposition using diethylzinc and H_2O whereby the transportation of precursors is allowed as ON for 2 s and OFF for 8 s. The cell with a 70 nm thick i-ZnO buffer layer with resistivity of 3.8×10^3 Ω-cm shows better efficiency of 11.7%, whereas cell with a 80 nm thick i-ZnO layer containing with lower resistivity of 8×10^{-2} Ω-cm experiences lower efficiency of 8%. The efficiency of cell decreases with decreasing resistivity of i-ZnO layer, as shown in Table 8.28 [238].

The doped ZnO layer is also imminent component as a TCO in the cells to obtain high efficiency cells. The carrier concentration (n) and band gaps (E_g) increase from 1×10^{20} to 8×10^{20} cm^{-3} and from 3.2 to 3.8 eV with increasing Al_2O_3 wt% from

Table 8.28 Variation of PV parameters with effect of ZnO layers

Cell	d_f (nm) i-ZnO	ρ (Ω-cm) i-ZnO	V_{oc} (mV)	J_{sc} (mA/cm^2)	FF (%)	η (%)	Ref.
i-ZnO (RF)/ZnO:Al (RF)	–	–	581.5	43.88	71.38	14.48	[236]
i-ZnO(70 nm)/ZnO:Al	70	–	713	29.6	76.1	16	[237]
i-ZnO(20 nm)/ZnO:Al	20	–	676	29.2	53.7	10.6	
Glass/Mo/CIGS/i-ZnO/ZnO:B	80	8×10^{-2}	399	36.1	54.5	7.9	[238]
Glass/Mo/CIGS/i-ZnO/ZnO:B	80	2.1	440	36.7	58	9.4	
Glass/Mo/CIGS/i-ZnO/ZnO:B	70	3.8×10^3	502	35.5	65.7	11.7	
ZnO:Al			614	35.9	76.1	16.7	[240]
ZnO:B			603	37	75.1	16.8	
ZnO:B-new cell			645	36.8	76	18	
CdS(CBD)/ZnO(125 °C)/ZnO:Al(125 °C)			584	35	68	14	[241]
CdS(CBD)/ZnO:Al(150 °C)/ITO(15Ω/□)			563	33	71	13.2	
ZnO(150 °C)/ZnO:Al(175 °C)			542	35.8	56.8	11	
ZnO (CBD)			550	35	70	14	[242]
ZnO (CBD)			557	35.5	–	14.3	
Traditional cell			554	30.7	706	12	[244]
CdS/ZnO/In$_2$O$_3$:Ti			565	28.4	66.8	10.7	
Std. large area- 3453.9 cm^2, 71 cells[a]			38.49 V	1.198 A	58.4	7.52	
15 cell minimodule/cell			553	28.15	62.5	9.74	
Total output			8.3 V	82.7 mA	–	–	
With MgF$_2$			616	33.9	73	15.2	[144]
Without MgF$_2$			605	32	73.5	14.2	

[a]V_{max} = 26.95 V and I_{max} = 0.9645 A.

0.05 to 2 in the ZnO target, respectively. The Al doped ZnO thin film shows mobility of 50 cm^2/Vs for doping concentration of 0.02–0.2 wt% Al_2O_3. The 2 mm thick SLG/1 μmMo/2.5 μmCIGS/CdS/ZnS/0.1 wt%Al_2O_3 cell shows efficiency of 18.1%. The QE of this cell is higher or close to that of world record efficiency of 19.5% cell. The transmission of ZnO:Al decreases with increasing Al doping at infrared region in the transmission spectra, as shown in Fig. 8.66A due to an increase of free carrier concentration. In other words, the free carrier absorption increases at infrared region with increasing carrier concentration in the ZnO layer by Al doping. 0.1 wt% Al_2O_3 doping in the ZnO is optimal [239]. In glass/Mo/CIGS/CdS(50 nm)/i-ZnO (70 nm)/ZnO:Al/MgF$_2$/Al grid cell, ZnO:Al conducting layer is replaced by ZnO:B thus the cell exhibits higher efficiency of 18% because of an increased J_{sc} due to allowing infrared radiation, which is partly blocked by ZnO:Al layer. The ZnO:B, which has higher optical transmission in the infrared region, is deposited by reactive RF magnetron sputtering using B_2–H_6–Ar gas mixtures with power of 250 W. Al_2O_3 2 wt% doped ZnO target is applied with system pressure of 8×10^{-3} torr for ZnO:Al layer. As shown in Fig. 8.66B, the transmission of ZnO:Al films is lower than that of ZnO:B at infrared region in the transmission spectrum, indicating that the free carrier absorption is higher in the ZnO:Al than that in the ZnO:B thin films. The CIGS (Ga/(In + Ga) = 0.25 and Cu/(In + Ga) = 0.91) cells with ZnO:B exhibit higher efficiency of 18% than the one with ZnO:Al (16.7%), as shown in Fig. 8.67A due to higher quantum efficiency. The ZnO:B with higher transmission at infrared region offers higher quantum efficiency at the same wavelength region in the cell as compared to one with ZnO:Al (Fig. 8.67B) [240].

A change in substrate temperature on the growth of ZnO layer by ALE method influences efficiency of the cell. The sheet resistance of ZnO layer varies from 400, 120, 40 to 35 Ω/□ with varying substrate temperature from 125, 150, 175 to 200 °C, respectively. The optimum growth temperature of ALE ZnO and ALE ZnO:Al is 125 °C for the glass/Mo/CIGS/50 nmCdS(CBD)/50 nmZnO(125 °C)/400–500 nmZnO:Al(125 °C) cells, which show efficiency of 14%, whereas the cell shows inferior performance for higher growth temperatures. If ITO layer replaces

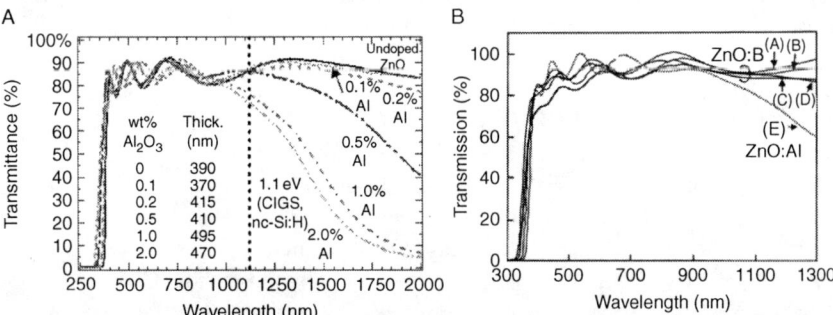

Figure 8.66 (A) Transmission spectra of ZnO layers with different Al concentrations and (B) with different B concentrations.

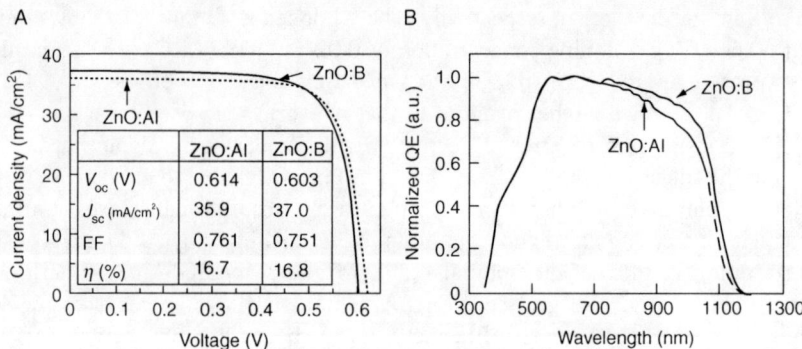

Figure 8.67 (A) *I–V* curves, and (B) quantum efficiencies of CIGS cells with different ZnO layers.

ZnO:Al layer for which the optimum growth temperature of i-ZnO layer is 150 °C. The glass/Mo/CIGS/CdS/i-ZnO(150 °C)/ITO shows efficiency of 13.2%. 0.15 μmCdS/1.35 μmCdS:Ga stack by evaporation is also used instead of CdS (CBD) layer in the cells as a reference cell. If the CdS buffer is not used in the cells, the growth temperatures for i-ZnO and ZnO:Al should be higher as 150 and 175 °C, respectively. The typical glass/Mo/CIGS/i-ZnO(150 °C)/ZnO:Al(150 °C) cell exhibits efficiency of 11%. This study reveals that the growth temperatures vary with varying buffer configurations in the cells to reach higher efficiency. This is mainly due to the formation of the right junction that requires proper sheet resistance of ZnO layers and crystallinity. The dimethyl zinc or diethyl zinc, trimethyl aluminium, and H_2O are employed for the growth of ZnO by ALE [241].

The CIGS solar cells made on CIGS grown by the Showa Shell Sekiyu KK covered with ZnO buffer by CBD method show efficiency of approximately 14%. The set conditions for CBD deposition are $Zn(CH_3COO)_2$(0.01 M), NH_3(0.038 M), growth time of 9 min, and bath temperature of 50 °C. The efficiency of ZnO (CBD) based cells is improved from 9 to 14.3% by using diluted chemical solutions. The grown ZnO layers show hexagonal structure with (100), (002), and (101) reflections at 2θ of 32, 34.5, and 36° in the XRD spectrum, respectively. The optical band gap of 3.4 eV is observed from the transmission spectra, which is higher than that of CdS. The CBD ZnO layers show (002) as a preferred orientation, and resistivity of 9.8×10^8 Ω-cm for the growth of 45 min at solution pH of 8.5. The glass/Mo/ 1.5 μmCIGSS/50 nmZnO(CBD)/1 μmZnO:B(MOCVD)/Al cells with ZnO (CBD) consist of moderate efficiency of 14.3% [242,243].

In the standard glass/Mo/CIGS/CdS/i-ZnO/n-In_2O_3:Ti cell structure, the transparent conducting n-In_2O_3:Ti layer grown by reactive environment hall cathode sputtering (RE-HCS) is inserted in the place of traditional n-ZnO, which shows efficiency of 10.7% (Fig. 8.68A), whereas the cell with ZnO, ZnO:B, and without CdS buffer shows efficiency of less than 10%. The resistivity, sheet resistance and mobility of In_2O_3:Ti layers are 1.8×10^{-4} Ω-cm, $R_\square = 3.3$ Ω/□, and 80 cm^2/V, respectively. The transmission, absorption, and reflectance spectra of n-In_2O_3:Ti ($R_\square = 7.1$ Ω/□) and ZnO:Al

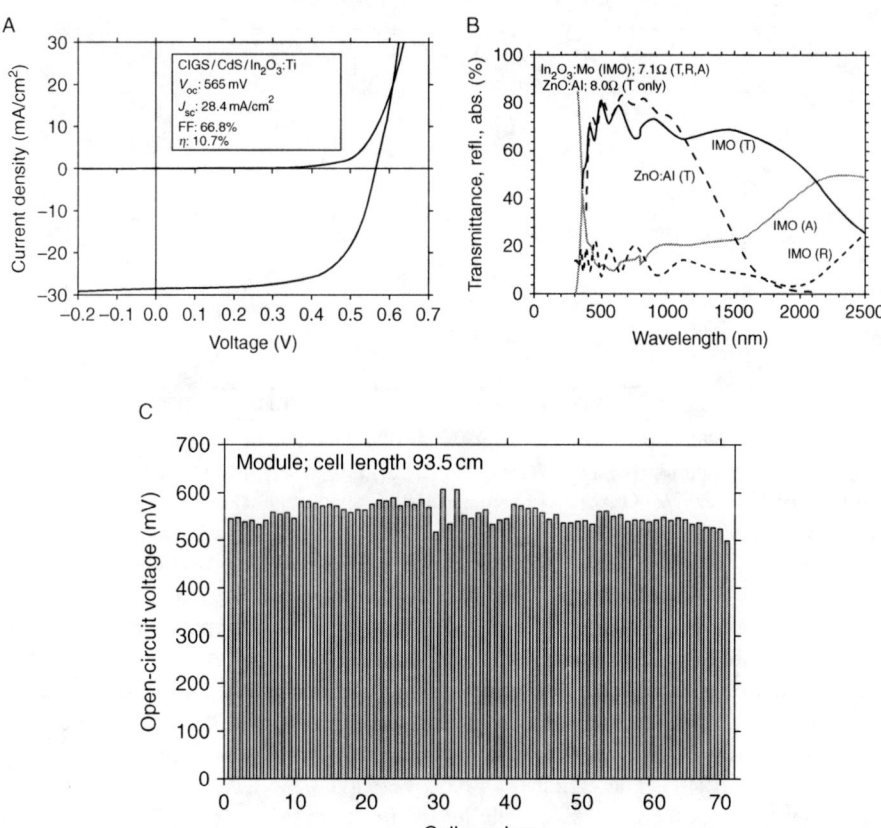

Figure 8.68 ((A) Dark and light *I–V* characteristics of CIGS cells with Ti doped In_2O_3 layer as TCO and (B) Transmission, absorption, and reflectance spectra of IMO and ZnO:Al layers. (C) Mapping of V_{oc} versus number cells made with CIGS.

($R_\square = 8$ $\Omega/\square$) are given in Fig. 8.68B, which has more or less the same sheet resistance of In_2O_3:Mo. However, the free carrier absorption is higher in the ZnO:Al than that in the In_2O_3:Mo at longer wavelength regions. Therefore, the transmission is lower in the ZnO:Al sample. A variation of voltage from cell to cell as a mapping for number of cells is given in Fig. 8.68C. The modeling indicates that power losses are 14 and 21% for the TCO sheet resistances of 15 and 30 $\Omega/\square$, respectively indicating that the losses are higher for higher sheet resistances that cause to decrease fill factor. It is noticed that the sheet resistance of ZnO:Al is higher on CdS than that on glass. The cell without i-ZnO exhibits low efficiency of about 7% [130,244]. In the glass/Mo/CIGS/Zn(O,S, OH)$_x$/i-ZnO/ZnO:Al thin film solar cells, the ZnO role is also decisive. ZnO and doped ZnO layers can be prepared either by RF or DC sputtering technique. Figure 8.69 shows difference between quantum efficiencies of the cells for which ZnO layers are prepared

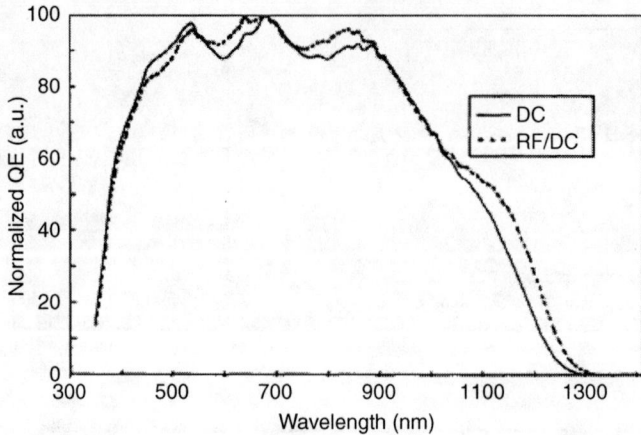

Figure 8.69 Quantum efficiency of glass/Mo/CIGS/Zn(O,S,OH)$_x$/i-ZnO/ZnO:Al thin film solar cells; (A) i-ZnO/ZnO:Al layers prepared by DC sputtering, and (B) the same layers prepared by RF/DC sputtering.

by DC and RF–DC sputtering techniques. The quantum efficiency of cells with RF–DC sputtered ZnO layers is higher than the one with i-ZnO/ZnO:Al sputtered by DC amid the CIGS is not extensively damaged in former case. If DC sputtering is used for deposition of both ZnO layers, the performance of cell is not that much impressive because the energetic plasma charge by DC sputtering damages the absorber layer and creates defects therefore i-ZnO layer is preferred to deposit by RF sputtering rather than by DC sputtering [245]. The efficiency of the best glass/Mo/CIGS/CdS/ZnO/ITO cell increases from 14.2 to 15.2% by adding 0.1 μm thick MgF$_2$ antireflection coating (refractive index = 1.32) [144]. The cells with sputtered ZnO over the area of 11.1 and 1.85 cm^2 show efficiencies of 12 and 13.2%, respectively, whereas cells with CVD ZnO show efficiency of 12.6% [175]. The glass/Mo/CIGS/i-ZnO/ZnO:Al cell without CdS buffer shows efficiency of 13.4%, whereas the efficiency of cells with CdS buffer is 14.6% and the cell without i-ZnO contribute efficiency of 8% but Al diffusion into CIGS layers is observed [246].

8.3.10 CIGS Cells with ZnMgO Buffer

The $Zn_{1-x}Mg_xO$ is also one of the alternative buffers to replace CdS in the CIGS thin film solar cells to avoid toxic Cd for friendly environment. The preparation techniques of $Zn_{1-x}Mg_xO$ material are eventually vacuum process such as sputtering, ALD, *etc.*, that is advantage to continue in-line process of CIGS thin film solar cells unlike CBD process. The band gap of $Zn_{1-x}Mg_xO$ window layer prepared by sputtering technique using MgO and ZnO targets varies from 3.24, 3.31, 3.38 to 3.60 eV with varying x from 0, 0.03, 0.06 to 0.17. The valence band offset (VBO) position more or less remain stays at 2.3, 2.3, 2.3, and 2.2 eV but the conduction band offset (CBO) position shifts from −0.16 to −0.09, −0.02, and 0.30 with x in the band diagram of $Zn_{1-x}Mg_xO$. The valence band offset shifts to the vacuum level by 0.1 eV in the $Zn_{0.83}Mg_{0.17}O$. The ZnMgO is a suitable window layer compared with CdS owing to wider band gap that allows higher energy photons at blue region hence J_{sc} increases. The V_{oc} is higher for higher conduction band offset (CBO) of over 0.25 eV as compared to lower CBO of 0.2 eV ultimately fill factor also increases in the glass/Mo/CIGS/$Zn_{1-x}Mg_xO$/ITO cells. The simulation work shows that the V_{oc} decreases with varying CBO from 0 to –0.6 eV then it becomes steady over the value of 0. The FF also decreases then constant between 0.0 and 0.4 eV thereafter decreases. The J_{sc} is constant in the CBO range of from –0.6 to 0.4 eV and decreases beyond the value of 0.4 eV. In the heterojunction, the cliff or notch (spike) occurs when the conduction band of window layer is below or above that of CIGS, respectively (Figure 8.2). The cliff leads to decrease V_{oc} and FF because an increase of recombination of majority carriers through the defects at the interface of window and absorber. The notch acts as a barrier toward photogenerated electrons in the CIGS and its influence is negligible when the CBO is below 0.4 eV. The notch reduces recombination of majority carriers through the defects at the interface. The Mo, CIGS, $Zn_{1-x}Mg_xO$, and ITO with thicknesses of 0.8, 2.0, 0.1, and 0.1 μm respectively are successively grown onto glass substrates to made glass/Mo/CIGS/$Zn_{1-x}Mg_xO$/ITO thin film solar cell. The cell with an active area of 0.96 cm^2 demonstrates efficiency of 13.2% [247].

50 nm thick ZnMgO buffer as well as 1 μm thick ZnO:B grown by MOCVD technique are used in the glass/Mo/CIGS/$Zn_{0.91}Mg_{0.09}O$/ZnO:B/Al-grid cell, which shows efficiency of 6.5% with lower V_{oc} of 388 mV and FF of 51.8% [248]. The glass/Mo/CIGS($Eg=1.15$ eV)/CdS/$Zn_{1-x}Mg_xO$/ITO thin film solar cell shows higher efficiency (15%), V_{oc}, J_{sc}, and fill factor as compared to those of glass/Mo/CIGS($E_g=1.15$ eV)/CdS/ZnO/ITO reference thin film solar cells ($\eta=12$%) due to reduction of interface states in the CIGS/CdS. The conduction band offset in the CIGS/CdS is lower therefore it needs to be increased to enhance V_{oc} consequently the interface recombination states decrease in the case of higher band gap (>1.15 eV) CIGS absorbers. The typical I–V curves of glass/Mo/CIGS ($E_g=1.15$ eV)/CdS/$Zn_{1-x}Mg_xO$/ITO ($\eta=11$%) and glass/Mo/CIGS($E_g=1.15$eV)/CdS/ZnO/ITO ($\eta=9$%) cells are shown in Fig. 8.70 [249]. The incorporation of $Zn_{0.85}Mg_{0.15}O$ layer instead of ZnO:Ga in the glass/Mo/CIGSS/CdS(CBD)/ZnO:Ga cells increases open circuit voltage from 581 to 586 mV by a small amount without much change in the short circuit current because of an increase in conduction band position toward vacuum level. The synonym that is, from 427 to 514 mV is observed moving from glass/Mo/CIGSS/ZnO to glass/Mo/CIGSS/ZnMgO cell, as presented in Table 8.29. The CIGSS layers used in the cells are etched by KCN in order to remove secondary phases on the surface. The conductivities of 4 and 20 Ω-cm^{-1} for ZnO ($E_g=3.24$ eV) and ZnMgO ($E_g=3.51$ eV) are observed, respectively [250]. The glass/Mo/CIGS/ZnMgO/i-ZnO/ZnO:Al and glass/Mo/CIGS/ZnMgO/ZnO:Al cells show efficiencies of 11.5 and 10%, respectively, as shown in Table 8.29 [251].

The higher efficiency of 16.2% is observed for ZnMgO based cells as compared to that of 15.4% for CdS based cells. The I–V curves and quantum efficiencies of SLG/Mo/CIGS/buffer/ZnO/ZnO:Al based cells, where buffer is ZnMgO or CdS, are depicted in Fig. 8.71A and B. As anticipated, the spectral response at shorter

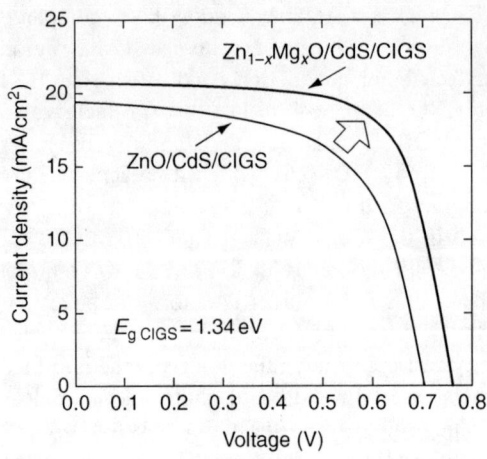

Figure 8.70 *I–V* curves of CIGS cells with ($\eta=11\%$) and without Mg doping in the ZnO layer ($\eta=9\%$).

Table 8.29 PV parameters of ZnMgO based CIGS thin film solar cells

Cell configuration	V_{oc} (mV)	J_{sc} (mA/cm^2)	FF (%)	η (%)	Ref.
Glass/Mo/CIGS/ $Zn_{1-x}Mg_xO$/ITO	572	33.8	68.2	13.2	[247]
Glass/Mo/CIGS/$Zn_{0.91}Mg_{0.09}O$/ZnO:B or Al	388	32.7	51.8	6.5	[248]
Glass/Mo/CIGSS/CdS/ZnO	581	30.8	69.9	12.5	[250]
Glass/Mo/CIGSS/CdS/ZnMgO	586	30.6	68	12.1	
Glass/Mo/CIGSS/ZnO	427	29	50.8	8.3	
Glass/Mo/CIGSS/ZnMgO	514	28.2	57.5	8.32	
Glass/Mo/CIGSS/ZnMgO (best cell)	521	33.6	56	9.7	
CIGS/ZnMgO/i-ZnO/ZnO:Al	583	29.4	67.2	11.5	[251]
CIGS/ZnMgO/ZnO:Al	551	29.3	62.3	10.1	
CIGS/ZnMgO/i-ZnO/ZnO:Al	640	33.7	75.1	16.2	[252]
CIGS/CdS/i-ZnO/ZnO:Al	527	32.2	76.5	15.4	
Glass/Mo/CIGS/CIGS:Zn/$Zn_{0.9}Mg_{0.1}O$/ITO	587	40.2	68.9	16.2	[253]
Glass/Mo/CIGS/10 nmCdS/$Zn_{0.9}Mg_{0.1}O$/ITO	593	34.8	69.7	14.4	[185]
Glass/Mo/CIGS/60 nmCdS/$Zn_{0.9}Mg_{0.1}O$/ITO	580	32.3	65.9	12.3	
Glass/Mo/CIGS/CdPE/$Zn_{0.9}Mg_{0.1}O$/ITO	576	34.6	54.8	10.9	
Glass/Mo/CIGS/CdS/$Zn_{0.9}Mg_{0.1}O$/ITO Module (10×10 cm^2) 18 cells	10.8 V	137mA	68.9	12.6	

wavelength region is higher in the ZnMgO based cells than in the CdS based cells due to wider band gap of ZnMgO. The current and voltages are higher in former. In the CIGS/ZnMgO cells, the Ga/(In+Ga) ratio of 0.4 and Ga grading are maintained. The ZnMgO layers are grown by ALD technique using diethylzinc, H$_2$O, and cyclopentadienylmagnesium. The typical deposition sequences of 600/400/400/400 ms and 1200/400/400/400 ms for DEZ/N$_2$/H$_2$O/N$_2$ and MgCp$_2$/N$_2$/H$_2$O are maintained, respectively, whereas the optimized process is one MgCp$_2$/H$_2$O

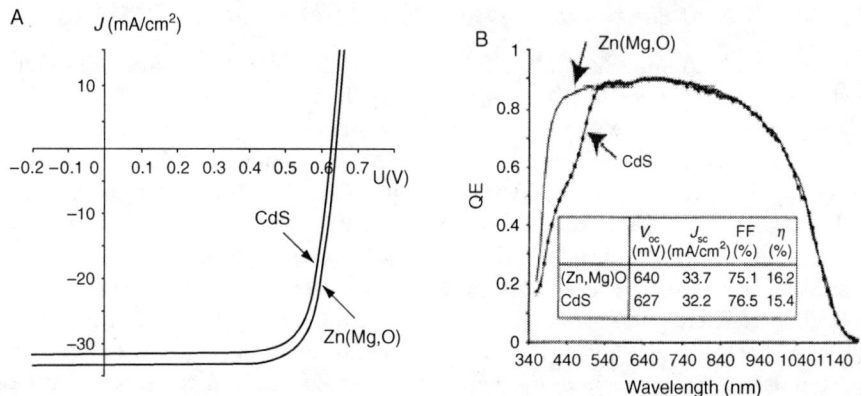

Figure 8.71 (A) Light *I–V* curves, and (B) QE of CIGS/CdS, and CIGS/ZnMgO thin film solar cells.

and six DEZn/H$_2$O cycles that is, 1:6 and the deposition temperature is 105–135 °C for ZnMgO. The band gaps of ZnMgO are 3.6, >3.9, and 3.75 eV for 1:6, 1:2, and 1:3 process, respectively. The cell loses J_{sc}, V_{oc}, and FF to lower due to change in properties of ZnMgO for the precursor cycle ratios of 1:2 and 1:3. The same pattern follows for the deposition temperatures of 150 and 180 °C for ZnMgO, even though the ZnMgO layers have the same band gap of 3.6 eV. It could be due to damage of CIGS/ZnMgO interface [252]. Unlike using bare CIGS absorber in the cells, Zn doped CIGS buffer is sand witched between CIGS and Zn$_{0.9}$Mg$_{0.1}$O layers. Zn is incorporated into CIGS by sequential deposition of Zn and CIGS at 300 °C on the surface of CIGS. The larger Zn exposing time of 15 min for CIGS exhibits higher efficiency that means stronger buried homojunction is formed. The high resistance Zn doped CIGS that is, *n*-type layer replaces shunt between *p*-CIGS and *n*-window layers. The glass/Mo/*p*-CIGS/*n*-CIGS:Zn/*n*-Zn$_{0.9}$Mg$_{0.1}$O/ITO cell yields efficiency of 16.2% [253].

The glass/Mo/CIGS/CdS/Zn$_{0.9}$Mg$_{0.1}$O/ITO submodules are developed over the area of 10 × 10 cm^2 and active area of 81.4 cm^2 for which CdS, 100 nm thick ZnMgO (E_g = 3.45 eV)/ITO (In$_2$O$_3$:Sn) and Ag–NiCr metal grids are sequentially deposited by CBD, RF-magnetron sputtering and electron beam evaporation, respectively. The QE response of the cells with thin CdS starts at earlier wavelength than that of one with thick CdS layers because of less absorption by thin CdS that enhances the higher collection of photons. Therefore, the cells with thin CdS show higher efficiency due to higher current collection. The cells with free CdS buffer and CIGS absorber treated with cadmium partial electrolyte demonstrate lower efficiency and fill factor as compared to that of other cells amid the CIGS is damaged by the sputtering species during the deposition of ZnMgO. In fact, the module efficiency increases by inclusion of antireflection coating MgF$_2$. The best cell and submodule show efficiencies of 14.4 and 12.6%, respectively (Table 8.29). The submodule consists of 18 cells connected in series over the area of 10 × 10 cm^2 and its active area is 81.36 cm^2 [185].

8.3.11 Role of Electrolyte Treated CIGS Absorbers in the CIGS Cells

The surface of 2.5 μm thick $CuIn_{0.7}Ga_{0.3}Se_2$ thin film layers with carrier concentration of $1-3 \times 10^{16} cm^{-3}$ and Cu/(In+Ga) less than unity grown by three source evaporation are treated with either Cd or Zn partial electrolytes (PE) prior to deposition of CdS buffer layer. The CIGS sample is immersed in the CdS CBD solution excluding thiourea at 75 °C for 20 min. The glass/Mo/CIGS/CdS/ZnO/ZnO:Al solar cell with Cd-PE treated CIGS shows higher quantum efficiency at both sides of shorter and higher wavelength regions as compared to that of cell with as-grown absorber because the treated one suppresses the absorption by CdS and enhances collection efficiency at shorter wavelength, as shown in Fig. 8.72. A difference in QE at higher wavelength is due to difference in Ga composition between both treated and untreated absorbers in this case. The reverse saturation currents under dark and light illumination are two and one order of magnitudes higher in the Cd treated samples than that in the conventional cells, respectively because of higher interface recombinations or different kind of mechanisms. The Cd fills Cu vacancies that suppresses the acceptor levels causing reduction in carrier concentration and enhance in depletion width (w). On the other hand, the absorber surface treated by Cd forms n-type layer. The cells with Cd-PE absorber show higher J_{sc} but lower V_{oc} and FF than that of cells with untreated CIGS. The efficiencies of typical glass/SiO_x/Mo/CIGS/CdPE/CdS/ZnO/ZnO:Al with Cd partial electrolyte (Cd PE) treated absorber and glass/SiO_x/Mo/CIGS/CdPE/ZnO/ZnO:Al cell with as-grown absorber are 11.9 and 11.1%, respectively, as given in Table 8.30 [254,255]. A 3000 and 7200 Å thick Ga and In layers are deposited onto electrodeposited CIGS layers (ED) by PVD method in order to compensate lack of composition in the CIGS in such a way to

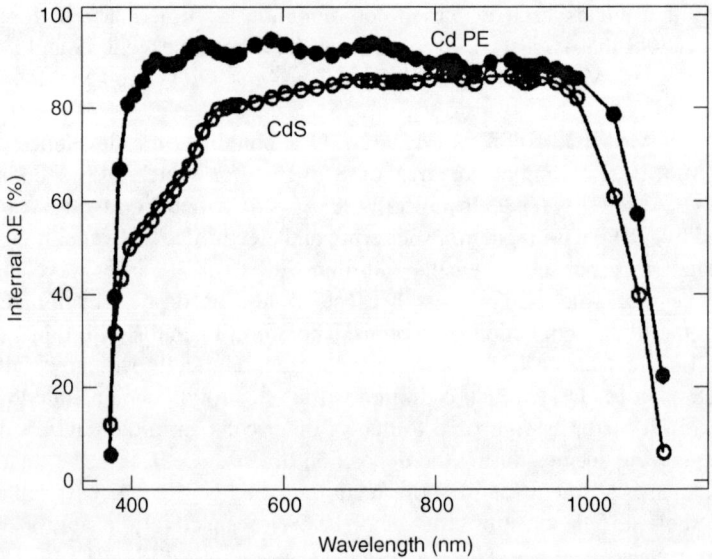

Figure 8.72 QE of CIGS cells with as-grown CIGS, and Cd-PE treated CIGS absorber.

Table 8.30 Effect of surface treatment on CIGS cells

Treatment/buffer	V_{oc} (mV)	J_{sc} (mA/cm^2)	FF (%)	η (%)	J_o (nA/cm^2)	A	R_s (Ω·cm^2)	R_{sh} (kΩ·cm^2)	Ref.
Cd PE	636	34.6	72	15.7	100	1.9	–	–	[254]
Zn PE	558	38.3	70	14.2	–	–	–	–	
Cd layer	564	30.7	68.8	11.9	–	1.6	1.3	1	[255]
CdS PE	521	32.8	65	11.1	–	2.2	0.4	0.6	[256]
CIGS-ED	666	30.51	75.56	15.4	–	–	–	–	
CIGS-EI	686.6	29.29	66.87	13.4	–	–	–	–	
As-grown CIGS	452	30.48	56.8	7.8	–	–	–	–	[257]
CIGS-(In-S)	631	32.16	71.7	14.6	–	–	–	–	
As-grown CIGS	555	31.2	61.3	10.6	–	–	–	–	
CIGS-CuInS$_2$ buffer	605	32.2	71.5	13.9	–	–	–	–	
CIGS-(In-S)	641	35	76.1	17.1	–	–	–	–	[258]
As-grown CIGS	611	38.3	73.5	18.3	–	–	–	–	
Br etched	657	21.8	57	8.1	–	–	–	–	[259]
Br+CN etched	660	24	73	11.6	–	–	–	–	
Cd^{2+}/ZnO WEL	586.7	34.6	72.1	14.6	0.67	1.30	0.680	7.220	[260]
Cd^{2+}/no buffer	537	32.1	68.8	11.9	4.26	1.35	0.802	1.540	
No/no buffer	367.8	30.8	36.8	6.50	1460	1.96	0.890	1.310	
CdS reference	587.7	32.9	73.1	14.1	0.12	1.19	1.190	9.940	

obtain final $CuIn_{0.72}Ga_{0.47}Se_{2.05}$ absorber layer for accommodating in the thin film solar cells. In similar fashion, 2600 Å thick Ga and 5620 Å thick In layers are deposited onto CIGS layer grown by electroless (EL) technique to form $CuIn_{0.85}Ga_{0.32}Se_{2.22}$. The band gaps of both $CuIn_{0.72}Ga_{0.47}Se_{2.05}$ and $CuIn_{0.85}Ga_{0.32}Se_{2.22}$ thin films are determined to be 1.2 and 1.11 eV, respectively. On these absorber layers, 50 nm thick CdS by CBD, 50 nm thick i-ZnO, 350 nm thick ZnO:Al by RF, and Ni/Al grids by e-beam evaporation are subsequently formed to build the cell structure. The cells with CIGS absorber by ED and EL show efficiencies of 15.4 and 13.4%, respectively (Table 8.30). The Cd free buffer layers are also formed by soaking the PVD CIGS absorber layers into $ZnCl_2$ solution. The Zn-treated CIGS annealed under air at 200 °C for 1 h is etched by HCl then cleaned with distilled water. The remaining process of i-ZnO, ZnO:Al, and grids is traditionally done on the glass/Mo/CIGS/CdS stack to fulfill the solar cell structure, which shows efficiency of 14.2% [256].

The three-stage processed CIGS absorber surface is treated by In–S in order to improve the solar cell efficiency. Prior to deposition of CdS by CBD, the CIGS sample is In–S treated by dipping into $InCl_3$ (5 mM) and CH_3CSNH_2 (0.1 M) combined solution at pH of 1.95 for 10 s. 0.96 cm^2 area glass/Mo/CIGS/CdS/ZnO(RF)/In_2O_3:Sn(RF) cells with and without In–S treated CIGS absorber show efficiencies of 14.6 and 7.8%, respectively. The $CuInS_2$ thin buffer layer with thickness of 80 Å is deposited onto CIGS absorber by PVD prior to deposition of CdS. The thin film solar cell without $CuInS_2$ buffer shows efficiency of 10.6%, whereas the cell with $CuInS_2$ buffer layer shows efficiency of 13.9% [257]. The glass/Mo/CIGS/CdS/ZnO/ZnO:Al solar cell fabricated with CIGS absorber dipped in $InCl_3$ (0.005 M) and CH_3CSNH_2-thioacetamide (0.1 M)) solution at 80 °C for 10 s shows better efficiency than the one with untreated absorber, as shown in Table 8.30. The reason is that the In–S treated sample reduces surface recombination. Presumably, the modified surface contains an 80 Å thick $CuInS_2$ buffer layer, which has higher band gap causing reduction of the conduction band offset that enhances V_{oc} in the solar cells. The cells with In–S treated CIGS absorber have better results among Ga–S, Al–S, and Y–S. On the other hand, after the In–S treatment, the intensities of In-3d and Cu-2p peaks increase in the XPS spectra that supports eventually formation of $CuIn(SSe)_2$ phase on the surface of CIGS absorber. The composition ratios measured by XPS show that Cu/(In+Ga) and Ga/(In+Ga) are 0.98 and 0.24, respectively [258]. The cells made with CIGS absorber etched with Br solution shows lower efficiency comparing with standard cell. However, the absorber is etched with cyanides after Br etching shows fruitful results as shown in Table 8.30. The Br etches surface of CIGS that leads to decrease the thickness and roughness. The absorber used in this investigation shows composition of Cu:In:Ga:Se = 21.7:17.4:9.2:51.7 [259].

The CIGSS absorbers treated in $CdSO_4 + NH_3 + H_2O$ solution at 80 °C for 10 min then cleaned with deionized water are dipped in 10 mM $Zn(ClO_4)_2$ solution. The cleaned and dried samples are treated again in $NH_3 + H_2O$ gas mixtures at 155 °C for 1 min. The deposition cycles of ZnO are continued several times until to incur required thick ZnO WEL layers. The results on glass/Mo/CIGSSe/30 nmZnO-WEL/i-ZnO/ZnO:Ga-sputtered/Ni–Al cell with an active area of

0.486 cm² from 1 in² area module, which had 8 cells, are given with and without treated samples in Table 8.30 [260]. The cells formed with absorber treated by Cd^{2+} solution and with ZnO WEL buffer provide good results, indicating abrupt junction formed between absorber and window layer. Secondly, the ZnO WEL layer masks pinholes in the absorber, which cause shunt in the cells. In principle, good cells or diodes always taste lower diode quality factor, reverse saturation current, series resistance, and higher shunt resistance. The glass/Mo/2 µmCIGS/CSD-CdS/ZnO:Al(20 Ω/□)/Ni–Al cells with double layered CdS (CSD) yield an efficiency of 13.2%, which is little lower than that of standard cell efficiency of 15.4%. The CdS layer is grown by chemical surface deposition (CSD) instead of chemical bath deposition where the chemical solution is flown onto heated CIGS (Ga/(In+Ga) = 0.15) thin film substrates to form thin CdS layer for the cells [261].

8.3.12 Effect of Annealing on the CIGS Cells

The soda-lime glass/Mo/CIGS/CdS/ZnO/Ni–Al cells formed on NREL CIGS samples are annealed using rapid thermal annealing process (RTA) at different temperatures such as 100, 200, and 300 °C for 30 s with ramp rate of 60 °C/s under nitrogen ambient. The Table 8.31 indicates that after RTA process the efficiency of type-N_1 cell increases from 9.52 to 15.77%, however type-N_2 cells, which enjoy higher efficiencies do not show much effect with RTA. The solar cells made on the CIGS

Table 8.31 Effect of annealing on different efficiencies cells

Cell	Annealing temp. (°C)	V_{oc} (mV)	J_{sc} (mA/cm²)	FF (%)	η (%)
N_1 (NREL)	Nil	628	31.66	47.88	9.52
	100	633	34.30	56.81	12.32
	200	652	34.85	68.43	15.55
	300	627	35.39	71.05	15.77
N_2 (NREL)	Nil	652	32.97	68.52	14.73
	100	656	35.47	71.10	16.55
	200	657	35.11	72.14	16.65
	300	623	38.35	70.48	15.96
R_1 (EPV)	Nil	500	24.74	61.45	8.19
	100 °C,He,30s	503	24.68	67.60	8.35
R_2 (EPV)	Nil	430	25.56	45.91	5.03
	150 °C,He,30s	509	28.85	51.90	7.62
R_3 (SSI)	Nil	455	27.49	52.69	6.591
	300 °C,air,60s	471	31.43	49.01	7.259
R_4 (SSI)	Nil	471	26.55	57.48	7.196
	300 °C,air,120s	487	30.51	49.96	7.422
R_5 (SSI)	Nil	456	28.75	52.14	6.975
	350 °C,air,60s	471	33.94	48.58	7.759
R_6 (SSI)	Nil	465	27.67	48.00	6.177
	350 °C,air,120	422	30.41	39.36	5.051

grown by EPV and SSI reveal that the annealing time of 1 min is enough to reach higher efficiency, if it is more than 1 min the cells lose their V_{oc}, J_{sc}, and fill factors. The performance of R_1 and R_2 cells increases for lower annealing temperatures. All the PV parameters of R_3, R_4, and R_5 cells increase except the fill factor [262].

The as-grown or control glass/Mo/CIGS/50 nmCdS(CBD)/ZnO/Ni–Al thin film solar cell and annealed by KrF laser with 248 nm, energy density of 20 mJ/cm^2 and 10 pulses show hole concentration of 0.53×10^{16} and 0.445×10^{16} cm^{-3}, hall mobility of 8.89 and 37.6 cm^2/Vs and resistivity of 133 and 37.3 Ω-cm, respectively. After annealing at 30 mJ/cm^2 and 5 pulses, the glass/Mo/CIGS/CdS/ZnO thin film solar cells with an area of 0.429 cm^2 show better results or optimal as compared to different annealing parameters of 30–10, 40–5, 40–10, 50 mJ/cm^2–10 pulses. The efficiency of cell increases from 7.69 to 13.41% after annealing at 30 mJ/cm^2 and 5 pulses. On the other hand, the shape of dark J–V curve splits into two different curve natures, as shown in Fig. 8.73 from which two different diode factors (A) of ∼2 and ∼4.3 can be derived, whereas the as-grown or control cell shows only one diode factor of ∼4.1. In the as-grown sample (A ∼4.3), the recombination is through the surface defects, whereas after annealing the sample, a decrease in dark current density is due to reduction in surface defect density. On the other hand, the recombination is dominated by bulk defects in the space charge region ($A = 2$). The annealed sample also shows higher quantum efficiency than that of control sample [263]. One square foot area CIGS films are prepared onto Mo coated glass substrates by sputtering technique at room temperature using independent elements, as sequential deposition of Cu, In, Ga, followed by Se thermal evaporation at high temperature. The type-a CIGS cells in which CIGS layers had Cu/(In + Ga) = 0.87, Ga/(In + Ga) = 0.4, and Se/(Cu + In + Ga) = 0.88 (type-a) annealed at 560 °C for 1 h show higher efficiency of 10%. The type-b CIGS cells in which CIGS layers

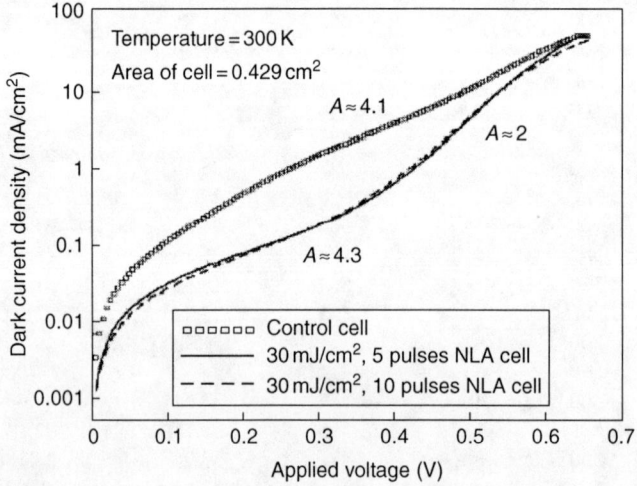

Figure 8.73 Dark I–V curves of glass/Mo/CIGS/50 nmCdS(CBD)/ZnO/Ni–Al thin film solar cell, and annealed by laser.

had Cu/(In+Ga)=0.74, Ga/(In+Ga)=0.49, and Se/(Cu+In+Ga)=0.92 (type-b) annealed at 580 °C for 4 h show efficiency of 7.3%, as shown in Table 8.32. The low efficiency in type-b kind of cells is because of annealing the layers at higher temperature destroys the crystal quality and rearranges to form amorphous impurity phases in the layers [264]. The CIGS/CdS heterojunction annealing at 200 °C under air does not change remarkably band line up or built-in voltage but decreases recombination centers [265].

8.3.13 Superstrate CIGS Thin Film Solar Cells

The superstrate solar cells are potential candidates for tandem cell applications in the shorter wavelength region. On other hand, it is probably easy to encapsulate that reduces the manufacturing cost. The configuration of superstrate cells is nothing but reverse process of conventional thin film solar cells. The superstrate SLG/2 μmZnO:Al/0.2 μmZnO/2–2.5 μmCIGS/Au solar cells also demonstrate efficiency of 10.2% in which first 2 μm thick ZnO:Al layer with sheet resistance of 10 Ω/□ and optical transmission of 80% in the 500 to 1000 nm range is deposited onto soda-lime glass substrate by RF-magnetron sputtering at 500 °C on which 0.1 μm ZnO with high sheet resistance of 10^8 Ω/□ in the presence of Ar:O_2 ratio of 1:1 at pressure of 1 Pa is grown. 2 μm thick Cu(InGa)Se_2 absorber layer is deposited onto ZnO coated glass substrates at 500 °C by co-evaporation and finally 0.5 μm thick Au contacts are formed to complete superstrate structure. The evaporation of Na_2S is done in the beginning stage of Cu(InGa)Se_2 growth in order to enhance the carrier concentration and grain sizes of the absorber. Using the conditions of Ga/(In+Ga)=0.15 and 12 g of Na_2S evaporation during deposition of CIGS layers, the as fabricated cell with an active area of 0.10 cm^2 shows efficiency of 0.9%, which is light soaked under illumination of 100 mW/cm^2 and bias of 1.0 V for 30 min showing hallmark improvement in the efficiency of 10.2%. The cells show efficiencies of 6.9, 10.2, 9.7, 5.7, and 5.2% for different ratios of Ga/(In+Ga) 0, 0.15, 0.18, 0.30, and 0.57 and doping concentrations of Na in the form of 12 mg Na_2S, respectively. The open circuit voltage increases simultaneously short circuit current decreases with increasing Ga content in the CIGS layers as same as substrate type cells. The reason is that V_{oc} increases with increasing Ga due to an increase of band gap. The J_{sc} decreases due to an increase of defect levels, grain boundaries, *etc.*, with increasing Ga content. In the present case, the light soaking effect is reversible under light and dark cycles, whereas in some cases the soaking effect is not reversible in substrate type cells. This could be due to deep trap states in the ZnO and interface states at ZnO/CIGS. The cells with CIGS layers grown at either <450 or >500 °C show poor performance that is due to diffusion of Zn/O into CIGS layers at high temperature and undeveloped CIGS phase at low temperature [266].

The improved version 2 μm thick ZnO:Al layers with properties of sheet resistance of 2.4 Ω/□, resistivity of 4.8×10^{-4} Ω-cm, mobility of 25 cm^2/Vs, carrier concentration of 5.1×10^{20} cm^{-3} and optical transmission of 80% in the wavelength range of 500–800 nm are prepared onto SiO_x coated glass at substrate temperature of 540 °C on which 0.2 μm thick ZnO layer is deposited at room temperature,

Table 8.32 Effect of annealing on CIGS cells by laser and thermal

Cell	NLA (30 mJ/cm^2) anneal temp.	V_{oc} (mV)	J_{sc} (mA/cm^2)	FF (%)	η (%)	V_m (mV)	J_m (mA/cm^2)	A	J_0 (mA/cm^2)	R_s (Ω-cm^2)	Ref.
D$_{01}$	–	528	29.78	48.86	7.69	365	21.37	4.13	3.22(10^{-3})	10.17	[263]
D$_{02}$	5 pulses	577	34.24	67.88	13.41	458	28.99	1.98	1.1(10^{-3})	14.06	
D$_{03}$	10 pulses	572	32.00	66.78	12.22	453	26.88	1.96	1.06(10^{-3})	15.94	
CIGS(*b*)	580 °C/4 h	527	31	NA	7.3	–	–	–	–	–	[264]
CIGS(*a*)	560 °C/1 h	625	25	NA	10	–	–	–	–	–	

sputtering pressure of 2×10^{-2} torr and the sputter gases Ar:O_2 ratio of 1:1. Onto SLG/SiO$_x$/ZnO:Al/ZnO, the CuInGaSe$_2$ thin films are prepared by co-evaporation at substrate temperature of 550 °C keeping Se to metal ratio of 3 to have stoichiometric layers. Na$_2$S is evaporated during beginning of CIGS layer growth. The composition ratio of CIGS thin films is Cu:In:Ga:Se = 13.5:17.4:13.8:55.3, Cu/(In+Ga) = 0.43, and Ga/(In+Ga) = 0.44. The Cu(InGa)$_3$Se$_5$ phase is observed for excess Ga deposition on the surface of CIGS films and sometimes ZnSe$_2$O$_5$ is also found for above the substrate temperature of 550 °C. 30 mg of Na$_2$S equal to 2%Na in the CIGS film is observed as optimum source to gain high efficiency thin film solar cells. The superstrate SLG/2 μmZnO:Al/0.2 μmZnO/2–2.5 μmCIGS/Au thin film solar cell exhibits efficiency of 12.8%. The cell performance degrades below or above the optimum thickness of Na$_2$S. The Cu(InGa)$_3$Se$_5$ phase is observed for the Cu/(In+Ga) composition ratio of 0.44 that causes maximum efficiency in the cells. In the superstrate cells, the quantum efficiency starts at early wavelength of 350 nm (Fig. 8.74) that quitely signs by band gap of 3.2 eV for ZnO, whereas cells with CdS have the cutoff wavelength at circa 518 nm equal to band gap (2.4 eV) of CdS [267].

In order to prepare CIGS based superstrate solar cells, first ZnO:Al layer with $n_e = 2.5 \times 10^{20}$ cm^{-3}, $\mu = 20$ cm^2/Vs, and $\rho = 1.8 \times 10^{-3}$ Ω-cm is deposited onto glass substrate then 200 nm undoped ZnO buffer is grown on which CIGS layer is deposited by modified three-stage process in which (InGa)$_2$Se$_3$ layer and Cu-rich then Cu-poor regimes are formed. The ZnO layer inhibits the diffusion of Na into CIGS, however at the interface insulating Ga$_2$O$_3$ layer is observed rather than buried homojunction of ZnCIGS or conductivity type inversion. The Au grids are used as ohmic contacts for p-CIGS. The glass/ZnO:Al/200 nm ZnO/p-CIGS/Au superstrate cells express the light soaking effect under 1 sun conditions and its efficiency

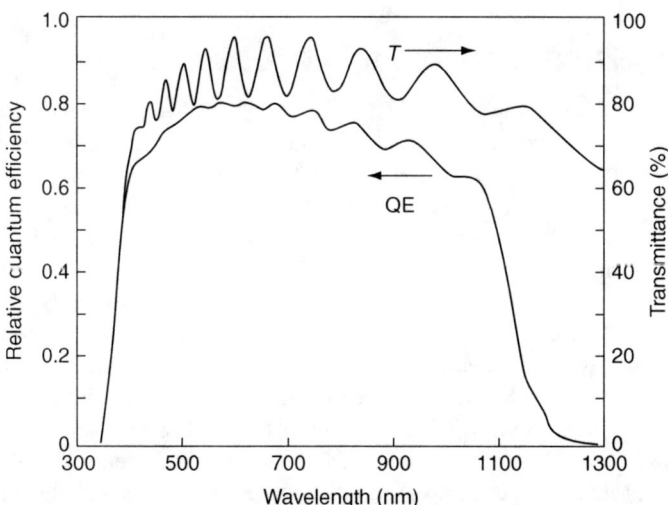

Figure 8.74 Quantum efficiency of SLG/2 μmZnO:Al/0.2 μmZnO/2–2.5 μmCIGS/Au superstrate cell, and transmission of ZnO layer.

Table 8.33 Effect of annealing on CIGS cells

Superstrate cells	V_{oc} (mV)	J_{sc} (mA/cm^2)	FF (%)	$J_o \times 10^{-7}$ (A/cm^2)	A	η (%)	R_s (Ω-cm^2)	R_p (kΩ-cm^2)	Ref.
As-grown	100	22	4	–	–	0.9	–	–	[266]
Annealed	476	37.6	57	–	–	10.2	–	–	
As-grown	601	31.32	68	–	–	12.8	–	–	[267]
As-grown	530	34	64	1.2	1.2	11.2	–	–	[268]
As-grown	240	32	36	8.3	1.2	2.8	10	13	[269]
Light soaking	450	34	64	1.2	1.6	10.2	1.3	240	

increases from poor performance to 11.2%. In other words, V_{oc}, J_{sc} increase from 300 to 530 mV and from 32 to 34 mA/cm^2, respectively. The diode quality factor and reverse saturation currents also increase from 1.2 to 1.6 and from 1.2×10^{-7} to 6.3×10^{-7} mA/cm^2, respectively. The cells without light soaking show crossover nature and low efficiency. The C–V measurements reveal that upon light soaking, the acceptor concentration of the cells increases from 1.0×10^{14} to 1.5×10^{15} cm^{-3} [268]. The photovoltaic parameters with ($\eta = 10.2\%$) and without ($\eta = 2.8\%$) effect of light soaking on a typical superstrate cell are given in Table 8.33 [269]. The CIGS and CGS layers are grown onto front and back side of SnO$_2$:F coated glass substrates by three-stage process at substrate temperature of 605 and 610 °C, respectively. As usual CdS and other conducting ZnO layers are developed on both sides. The cells on front and backside of glass substrates can be treated as tandem cells. The CIGS and CGS cells demonstrate poor efficiencies of 2.45 and 3.28%, respectively. This could be due to degradation of quality of SnO$_2$:F layers employed at high temperatures for the deposition of absorbers [270].

8.3.14 Bifacial CIGS Thin Film Solar Cells

The transparent conducting oxide (TCO) layer replaces Mo as a back contact in the CIGS thin film solar cells, which is beneficial for the tandem solar cells owing to its transparency, if it is used as a front side TCO in the solar cell. The cell can be used as a bifacial cell for the light illumination from the rear and front sides, as shown in Fig. 8.75. The CIGS layers are grown onto ITO coated glass substrates by three-stage process at optimum substrate temperature of 520 °C that is less than standard process temperature. As conventional CdS, i-ZnO, and ZnO:Al layers are sequentially deposited onto glass/ITO/CIGS stack by CBD and sputtering methods to complete cell structure, respectively. To deposit CdS layer, at bath temperature of 80 °C, CdSO$_4$ (0.16 M), ammonia (7.5 M), and thiourea (0.6 M) chemical solutions are employed. The glass/ITO/CuIn$_{0.7}$Ga$_{0.3}$Se$_2$/CdS/ZnO/ZnO:Al and glass/ITO/CuGaSe$_2$/CdS/ZnO/ZnO:Al solar cells with an active area of 0.2 cm^2 show efficiencies of 9.6 and 4%, respectively (Table 8.34). An improved 0.2 cm^2 area solar cells without antireflection coating for CIGS optimum growth temperature of 520 °C shows efficiency of 15.2%, whereas the cell efficiency decreases to lower value of 11.8 and 9.6% for higher and

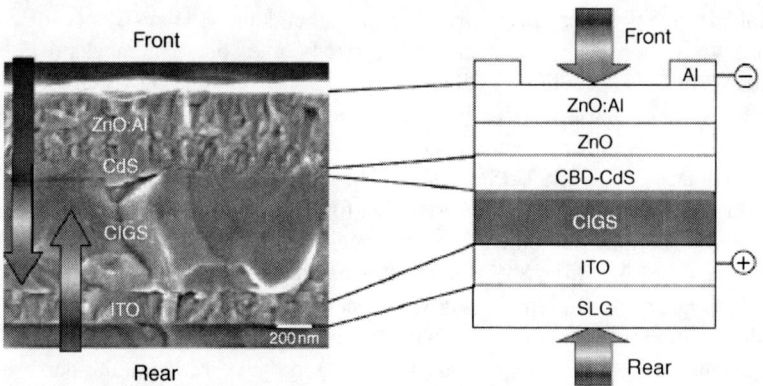

Figure 8.75 SEM cross section of bifacial glass/ ITO/CIGS/CdS/ZnO/ZnO:Al thin film solar cell.

Table 8.34 CIGS ells made on different back contact layers of ITO, Mo, ZnO:Al, and SnO$_2$

Back contact/absorber	T_S (°C)	V_{oc} (mV)	J_{sc} (mA/cm^2)	FF (%)	η (%)	Ref.
ITO/CIGS	520	651	34.4	68.2	15.2	[272]
ZnO:Al/CIGS	520	267	0.45	26.9	0.04	
ITO/CIGS	550	564	33.7	51.8	9.6	
Mo/ITO/CIGS	550	546	34.5	67.8	12.8	
SnO$_2$/CGS	500	528	14.0	44.6	3.3	
ITO/CGS	520	673	13.4	43.8	4.0	
ITO/CIGS	–	602	35.2	65.2	13.8	[274]
Mo/CIGS	–	565	36.8	64.3	13.3	
SnO$_2$/CIGS	500	608	36.1	62.6	13.7	

lower growth temperatures of 550 and 500 °C, respectively. The lower substrate temperature of 500 °C enroutes to undeveloped grain structure of CIGS onto ITO. The higher substrate temperature is needed to increase crystallinity but the cell efficiency decreases even though CIGS had improved crystallinity. The Ga$_2$O$_3$ is developed at the interface amid the Ga reacts with oxygen obtaining from ITO at the high temperature. A parasite p–n junction is developed between n-Ga$_2$O$_3$ and p-CIGS, which reduces the performance of the cell by opposing current flow. The thin film solar cells formed with CIGS absorbers deposited at 520 and 550 °C show series resistances of 15 and 51 Ω-cm^2, respectively [271].

The XPS analysis concludes that the peak position at binding energy of 424.9 eV for Ga LMM is related to Ga$_2$O$_3$. In order to mitigate secondary junction problem, a 70 nm thin Mo layer is sand witched between ITO and CIGS layer then the cell efficiency increases to 12.8%. Why does not Mo layer provide any adverse effect? Because the Mo layer reacts fast with Se to form p-type MoSe$_2$ with band gap of 1.4 eV that is driving force not to exit another p–n junction. In the case of SnO$_2$:F back contact in the solar cells, the cell efficiency is 13.7% for the CIGS substrate temperature of 500 °C. If the CIGS growth temperature is increased to 520 °C or higher, the cell efficiency deteriorates fast. On the other hand, the resistance of SnO_2:F increases to higher because the fluorine dopant bubbles out. The mechanism behind the process of SnO_2 is entirely different from the ITO sample. The ZnO:Al back contact in the solar cells follows the same phenomenon of ITO solar cells. The CGS cells with back contacts of SnO_2:F and ITO show single digit efficiency [272]. The typical soda-lime glass/ITO/CIGS/CdS/ZnO/ZnO:Al solar cell with Ga/(In + Ga) = 0.3 and 0.95 shows efficiencies of 12.6 and 7.4% for front and back light illuminations, respectively. An apparent total efficiency of cell is 20% for simultaneous light illumination. The transmission of 85% and sheet resistance of 10 $\Omega/\square$ are observed for ITO layer. Rear side light illumination shows lower efficiency as compared to that of front side illumination that is mainly due to insufficient quality of crystalline formed in the CIGS because of high Ga content at ITO side. On the other hand, the narrow space charge region may be formed at CIGS/CdS that encourages photogenerated carriers to recombine in the CIGS layer before entering into space charge region [273]. The CIGS (E_g = 1.26 eV or 990 nm) and CGS (E_g = 1.68 eV or 740 nm) layers are grown onto SnO_2 coated glass substrates by three-stage process with composition ratios of Cu:In:Ga:Se = 23.20:16.54:9.20:51.07, Ga/(In + Ga) = 0.36, Cu/(In + Ga) = 0.9, Se/metal = 1.04, and Cu:Ga:Se = 23.90:25.6:50.6, Cu/Ga = 0.93, Se/metal = 1.02, respectively. The SLG/SnO_2/CIGS/CdS(CBD)/ZnO/ZnO:Al and SLG/SnO_2/CGS/CdS/ZnO/ZnO:Al cells show efficiencies of 4 and 13.3% and some of them show low efficiencies of 8.7 and 8.4%. In the case of SnO_2 back contact, the best cell performance is obtained for the CIGS deposition temperature of 500–520 °C, whereas the PV parameters of cell deteriorate for higher deposition temperature of 550 °C due to out diffusion of F from SnO_2:F and formation of Ga_2O_3 [274]. The different substrate temperatures are used for the growth of CIGS layers on the different conducting layers those influence PV parameters of solar cell.

8.3.15 EBIC Analysis of CIGS Cells

The electron beam induced current (EBIC) analysis is one of the tools to assess the junction properties of thin film solar cells. In the cleaved glass/Mo/CIS/CdS/ZnO/ZnO:Al/Al thin film solar cell, one electrical contact from the bottom Mo layer and another one from the top Al layer are connected to the amplifier for EBIC measurements. The cross section of the sample is normally scanned by the electron beam using scanning electron microscope. The typical beam voltage, current, and beam rapture diameter are 5 kV, 30 pA, and 0.25 μm, respectively. The output signal of amplifier that is, electron beam induced current is recorded with a function of beam scanning distance. The combined EBIC scan line and SEM of 10.5% efficiency glass/Mo/CIS/CdS/ZnO/ITO thin film solar cell annealed under air at 200 °C is shown in Fig. 8.76. In general, the EBIC signal generates in the CIGS absorber very close to the $p-n$ junction [91,275]. Thousands of electron–hole pairs are generated and separated by the junction electric field when the electrons impinge. The charge collection takes place throughout the absorber layer in the cells. The EBIC scan lines of Ga/(In + Ga) = 0% and 30% are one and the same and show a slow development toward heterojunction, whereas the scan lines of

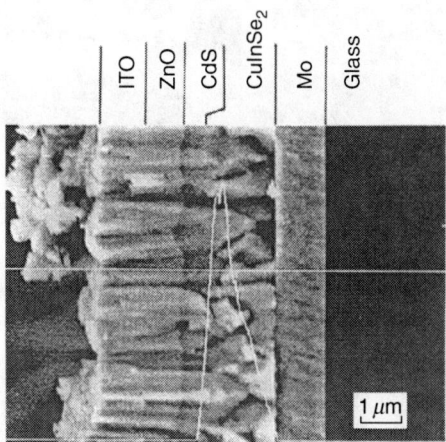

Figure 8.76 EBIC scan of 10.5% efficiency air annealed (200 °C) glass/Mo/CIS/CdS/ZnO/ITO thin film solar cell.

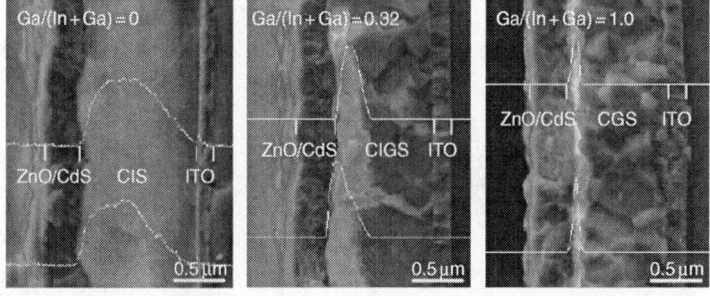

Figure 8.77 Superimposed EBIC and SEM profiles of CIGS thin film solar cells (A) Ga/(In+Ga)=0, (B) 0.32, and 1.0.

Ga/(In+Ga) = 50, 75 and 100% are sharp indicating abrupt p–n junction at near the heteroface. In comparison, the evolution between Ga/(In+Ga) = 30 and 50% is in between p–i–n and abrupt p–n junction. In other words it is clear that as Ga content is increased from Ga/(In+Ga) = 0, 0.32 to 1.0, the space charge region gradually shifts toward interface of CdS/CIGS. The EBIC line correspondingly shifts toward CdS/CIGS interface and becomes sharper with increasing Ga content indicating that the heterojunction becomes stronger with increasing Ga content in the samples [276]. The EBIC scans and superimposed SEM of bifacial CIGS cells are shown in Fig. 8.77. The space charge region and diffusion length widths are found to be 0.4 and 0.6 μm, respectively for the typical Ga/(In+Ga) composition of 0.32. The space charge region is obviously wider in the CIS cells comparing with CIGS due to lack of back surface field. On the other hand the acceptor density of CIGS increases with increasing Ga content in the cells. Thus the sample virtually turns into more p-type [273]. The EBIC line lies at interface of CdS/CIGS junction in the as-grown cells. After annealing the cell under vacuum at high temperature of 500 °C its efficiency decreases

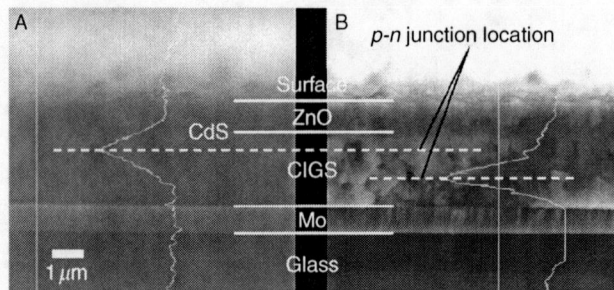

Figure 8.78 EBIC line scans and SEM of (A) as-grown, and (B) annealed under vacuum at 500 °C.

from 13 to 2% correspondingly the EBIC line shifts into CIGS absorber from CdS/CIGS interface that means opposite direction. The backward direction shift in EBIC line indicates that the p-n junction shifts into CIGS layer from interface, as shown in Fig. 8.78 [277]. The EBIC analysis on 11 and 2% efficiency CIGS/ZnInSe cells demonstrates that the EBIC peak position locates at 200 and 500 nm from the interface of CIGS/ZnInSe, respectively indicating that the junction is faraway from the interface of CIGS/ZnInSe in latter due to higher Zn in the CIGS. The formation of p–n homojunction caused by Zn doping is confirmed and the junction is located in the CIGS [217].

8.3.16 CIGS Cells Made on the Variety of Flexible Substrates

There is a need to transfer conventional glass substrate based CIGS solar cells onto flexible substrates in order to reduce cost, easy to install on the buildings, vehicles, satellites, *etc.*, because of flexibility. On the other hand, the cells on the flexible substrates are lightweight and easy to roll out. By considering low cost and thermal expansion coefficient of the substrates, so far stainless steel (SS), Cr steel, PI, Ti, Cu, and Al substrates have been used to develop CIGS thin film solar cells. The V_{oc} of standard Mo/CIGS/CdS/ZnO/ITO cells on the SS is less than that of cells on the glass substrates. The reason is that the carrier concentration of CIGS layers on the SS is less than that of those on the glass substrates. On contrast, the grain sizes of CIGS layers are smaller in the layers grown on SS than those on the glass substrates. However, the XRD results reveal that there is no much difference in the CIGS layers grown either on Mo covered SS or glass substrates/Mo. The depth profile reveals that the Cu concentration is constant and Ga grading is observed in the CIGS on both the SS and glass substrates. The concentration of Ga decreases from top to bottom up to depth of 0.3 μm thereafter increases. The CIGS layer used in the cells contains compositions of Cu/(In+Ga)=0.9, Ga/(In+Ga)=0.3, and Cu:In:Ga:Se=23.5:18.6:6.9:51. Either 300 nm SiO_2 or 200 nm Al_2O_3 insulating layer is sandwiched between SS substrates and Mo coating. <2 μm thick CIGS layers by three-stage process at 500 °C in-line evaporation system, 0.1 μm thick CdS by CBD, 0.1 μm thick ZnO and ITO layers by sputtering are sequentially deposited for the flexible solar cells. The best 0.96 cm^2 area SS/Mo/CIGS/CdS/ZnO/ITO thin film solar cells with and without SiO_2 insulating layer show more or less similar efficiencies of 12.4%, whereas the efficiency of the same structure with Na doping on soda-lime glass substrate shows higher efficiency of 13.6%. In the case of low efficiency cells, the efficiency is slightly better in the insulating buffered cells (11%) comparing with noninsulating buffer layer cells (10.8%) [278,279].

The Cu, In, and Ga are sequentially deposited onto Mo coated copper, steel, and stainless steel substrates by electrodeposition, followed by evaporation of Se and Na compounds. The metal foil/Mo/Cu–In–Ga–Se–Na compound stack is heat-treated by RTP to form CIGS. The best efficiencies of CIGS/CdS(CBD)/ZnO(SP) cells formed on different substrates Mo, SS, steel, and Cu are 8.8, 10.4, 9.5, and 9%, respectively. The efficiencies of the cells formed over the area of 1, 13.3, and 36 cm^2 SS substrates are 10.4, 6.6, and 8.7% for roll to roll process, respectively. Unlike standard flexible substrate/Mo/CIGS/CdS/ZnO cells, the multi metal layer stack is sand witched between flexible substrate and Mo layer. On the cleaned Cu substrate, which is cleaned by ethanol, ultrasonic bathing with cleaning agent and cathodic H_2 degreasing, a 20 µm thick Ni by electrodeposition, 200 nm Cr, 100 nm Ta, 200 nm TaN, and 400 nm Mo by sputtering are formed. The Cu foil/multi metal stack/CIGS/CdS(CBD)/ZnO(SP) cell with an area of 1 cm^2 demonstrates efficiency of 9.02% [280,281]. In order to block diffusion of Fe into CIGS thin film cells from SS, a 2 µm thick ZnO layer is sand witched between SS and Mo layer in the SS/2 µmZnO/0.89 µmMo/CIGS/CdS/i-ZnO/ZnO:Al cell, which exhibits efficiency of 10%. No improvement is observed beyond 2 µm thickness of ZnO layer in the cells, whereas the efficiencies of cells are 9.94 and 6.63% for 1 µm thick ZnO layer and without ZnO layer, respectively [282]. The cells formed on Fe/Ni alloy substrate without Cr barrier layer underneath of Mo show poor performance by showing efficiency of 1.7% otherwise 9.1%. The SIMS analysis reveals that the Ni and Fe are found in the CIGS absorber layer indicating their diffusion from substrates [181]. The efficiency of 0.25 cm^2 area 0.5–1 µmMo/2 µmCIGS/50 nmCdS(CBD)/50 nmi-ZnO(RF)/1 µmZnO:Al(DC) typical cell decreases from 6, 4.3 to 3.1% with increasing number of cells from 4, 5 to 10, respectively in the conventional module size of 10×10 cm^2 made on the Cr steel substrates, whereas the average efficiency of 10 cells made on the polyimide substrates is 4.5%. The efficiencies of best cells made on the substrates of Kovar, Ti, Cr, and polyimide are 12.8, 12.5, 11.2 and 7.1%, respectively. Ten cells made on 10×10 cm^2 glass substrates with 30 nm In_2S_3 buffer grown by ALCVD show efficiency of 11.2%. The efficiency of standard CIGS 10×10 cm^2 modules varies with changing Ga concentration in the cells [283].

27.1 cm^2 roof tiled Mo/CIGS/CdS/i-ZnO/ZnO:Ga cell is developed on the flexible Ti substrates, which exhibits an efficiency of 15%, whereas the integrated module, which is combination of four cells over the large area of 108.4 cm^2 shows an efficiency of 10.5%. Both Ti and float glass substrate based solar cells show one and the same efficiency of 16.7%, as shown in Table 8.35. The cells with an efficiency of greater than 10% show parallel resistance of < 100 Ω-cm^2 [284]. The CIGS is developed onto 30 µm thick Ti/Si_3N_4 by DC sputtering using Cu75% + Ga25% and In targets, followed by Se evaporation and sulfurization by RTP with

Table 8.35 PV parameters of CIGS cells made on different flexible substrates

Substrate/Area(cm^2)/cells	E_g (eV)	V_{oc} (mV)	J_{sc} (mA/cm^2)	FF (%)	η (%)	Ref.
SS	–	555	31.2	70.2	12.2	[278]
Cu foil/multimetal stack	–	407.18	35.06	63.16	9.02	[281]
SS/7715/126	–	50.53	2006	56.5	7.43	
PI/0.68/1	–	0.5	36.6	62.3	11.3	
PI/383/55	–	24.21	1723	43.4	4.73	
(10 × 10 cm^2), Ga/(Ga+In)=0.3	1.37	791.32	20.4	69.2	11.1	[283]
(10 × 10 cm^2), Ga/(Ga+In)=0.6	1.13	646.5	30	66.7	12.8	
Ti/0.5/1	–	640	34.5	75.6	16.7	[284]
Ti/27.1/-	–	640	33.9	69.3	15	
Float glass/0.5/1	–	661	33.8	74.6	16.7	
Al foil	–	434	30.7	49.3	6.6	[286]
PI	–	612.7	30.62	68.2	12.8	[287]
Ti	–	647	29.5	72.3	13.8	
50 μm thick PI	–	639	23.3	73.5	11	[288]
25 μm thick Ti	–	651	28.5	74.1	13.8	
150 μm thick SS	–	585	27.7	71	12.6	
Soda lime	–	606	33.2	74.3	14.9	[291]
Polyimide	–	507	34.3	68.3	11.5	
Ti/46/10	–	5.8	24.3	49	6.8	[293]
Ti/467/39	–	14.9	24	41	3.8	
PI/46/10	–	5.2	25.3	56.7	7.5	
PI/422/37	–	18.7	25.1	43.5	5.2	
PI/45/10 (In$_2$S$_3$ buffer)	–	4.9	26.2	49.2	8.4	

ramp of 10 °C/s. Finally on the Ti/Si$_3$N$_4$/CIGS stack, CdS by CBD, ZnO by sputtering and Ni–Al grids are sequentially deposited. Without antireflection coating, eighteen Ti/Si$_3$N$_4$/CIGSS/CdS/ZnO cells on 8 × 8 cm^2 submodule and each one contains an area of 1.4 cm^2 shows efficiency of 12.4%. The best cells on Ti and glass yield efficiencies of 13.9 and 15.3%, respectively. They are 11.7 and 13.6% over the large area of 24.7 and 24.4 cm^2, respectively [285]. The CIGS layers are grown onto 100 μm thick Al foil substrates by three-stage evaporation of Cu, In, Ga, and Se at substrate temperature of 550 °C, whereas the layers deposited at low substrate temperature of 400 °C do not show fruitful results due to lack of substrate temperature for proper interdiffusion of In and Ga in the CIGS layers. The Al/CIGS/CdS/i-ZnO/ZnO:Al(200 nm)/Ni–Al cells with an area of 0.3 cm^2 show efficiency of 6.6% without Na and antireflection coating. By adding Al$_2$(SO$_4$)$_3$ in CBD bath for the preparation of CdS layers, the efficiency of the cells increases slightly that may be due to different reaction path [286].

The CIGS solar cells prepared on polymide (PI) substrates yield good efficiency that a thin NaCl layer is deposited onto 3 × 3 cm^2 area glass substrates by evaporation, followed by coating of 20 μm thick polymide by spin coating, which is heated under air at 400 °C. The CIGS layers onto polymide by co-evaporation at 450 °C, CdS by

CBD, ZnO, and ZnO:Al layers are subsequently deposited to build the cell structure. The glass/NaCl/polymer/Mo/CIGS/CdS/ZnO/ZnO:Al cell with an active area of 0.13 cm^2 shows efficiency of 15.8%, whereas the best Mo/CIGS/CdS/ZnO/ZnO:Al cell on the 20–25 µm thick polymer sheet exhibits efficiency of 12.8%. The cell can easily be detached from the glass substrates by dissolving the NaCl into water [287]. Similar way, the Wurth solar company improved efficiency of solar cell module (60 × 120 cm^2) made on PI substrates as 9–10% in 2002, 10–11% in 2003, and 11–11.6% in 2004. The modules with the same size deliver power of 85 Wp and 16,000 modules produce power of 1.3 kWp/a. In the role to role process, 32 µm thick SiO_x by PECVD, 0.6 µm Mo by DC sputtering, CIGS layers by co-evaporation and 10 nm thick NaF, 40–50 nm CdS by CBD, i-ZnO by RF, and 0.3 or 1 µm thick ZnO:Al by DC sputtering are successively grown onto metal foils such as SS and Ti with surface roughness of ≤ 1.5 µm. The substrate temperature of 450 °C is chosen instead of 550 °C for PI because of keeping in mind the substrate withstanding temperature. The standard monolithic technique is applied for the modules prepared on glass substrates, whereas the same technique does not work for polyimide (PI) substrate to fabricate the cells therefore the photolithographic technique is employed as a suitable method. By applying photolithographic technique P_1 scribing is done on Mo covered PI substrates for which a weak laser (409 nm) is illuminated on positive photoresistance and P_2 pattern is done after deposition of CIGS without damaging back contact of Mo. Finally, P_3 scribing is done on CdS/ZnO/ZnO:Al layers. The efficiency of the cells formed on PI is apparently low as compared to that of one on the SS because of lower substrate temperature that is employed to deposit CIGS. The cells with large area of 46 and 467 cm^2 on Ti foil show efficiency of 6.8 and 3.8%, respectively. With ZnOS (S/Zn = 0.3) by ALD deposition, the CIS module size of 30 × 30 cm^2 produces efficiency of 12.7%, whereas the CIS module with ZnOS by CBD exhibits efficiency of about 10%. The polyimide substrates based CIS cells with Na doping exhibit efficiency of 11.2, 7.5, and 5.2% over the areas of 0.5, 10 × 10, and 20 × 30 cm^2, respectively. The CIS cells on Ti/SiO_x exhibit efficiency of 6.8 and 3.8% over the area of 7 × 8 and 20 × 30 cm^2, respectively and the best cell efficiency on Ti substrates is 13.8%. The CIGS cells with ZnS buffer by CBD either over the area of 27.1 cm^2 and 25 µm thick Ti or Cr steel foil show efficiency of 10.3% [288–290].

1.4 cm^2 area cells without any barrier layer on PI substrates show efficiency of 11.3% [285]. The CIGS layer on glass/Mo is well adherent, whereas the same on polymide shows poor adherent after subsequent deposition of CdS. The CIGS films grown at 400 °C shows (112) orientation, whereas the same films grown at high temperature shows random orientation. Initially, Ga–In–Se films at 350 °C and final CIGS layers are grown over the area of 5 ft × 5 in. polymide substrates at 400–450 °C in which the Cu-rich then Cu-poor growth is maintained. The substrates web moves with speed of 0.75 in./min for about 20 min. The CIGS layers used for the cell structure contain composition of Cu:In:Ga:Se = 24:19.7:7.5:48.7, Cu/(In + Ga) = 0.88, and Ga/(Ga + In) = 0.28. The efficiency of the best cells increases from 11.5 to 12.1% by adding 1200 Å MgF_2 antireflection coating. The growth temperature of CIGS is 450 °C for polyimide substrates, whereas it is 525 °C for glass substrates. In the pilot line, the I–V parameters of PI/Mo/CIGS/CdS/ZnO/ITO/Ni–Al devices with an area of 0.56 cm^2 are given in Table 8.35. The efficiencies of cells in the module at different places of center and edges show 10 and 8.4–8.8%, respectively [291]. The efficiency increases for precise control of Na in the CIGS layers, otherwise the quality of device deteriorates with higher concentration of Na. The Mo on PI provides cracking and extends to longer length due to large difference in thermal expansion coefficients of PI and Mo [292]. The efficiencies of conventional CIGS cells formed on different flexible substrates are given in Table 8.35. In another case, the best efficiency of 5.2% is achieved for the cells made over the large area of 20×30 cm^2 size and 50 μm thick polyimide substrates. In addition, the cells with In_2S_3 buffer grown by ALD technique on polyimide are developed [293].

8.3.17 Role of Mo Layer in the CIGS Cells

The effect of Mo bed layer on the properties of CIGS thin film solar cells has been studied. The Mo layer is preferable among Ni, Al, and Au metallic contact layers because which withstands corrosions against from Se and S, whereas Al or Au diffuses fast into CIGS absorber. Ni is good ohmic contact on already grown n- and p-type CIS layers with wide range of resistivities but the drawback is that Ni as a substrate layer reacts with CIGS. The thermal expansion co-efficient of 5×10^{-6} K^{-1} for Mo is close to 9.5×10^{-6} K^{-1} of CIS [30]. The Mo layers deposited at low Ar pressure have dense and low resistivity because of compressed stress. The stress changes from compression to tensile with increasing pressure and the structure of the films becomes porous microstructure. The resistivity of Mo layers is one order of magnitude higher than bulk (5.5×10^{-6} Ω-cm) [294]. The efficiency of cells with SiO_2 is lower than one without SiO_2 layer. For example, the glass/SiO_2/0.8 μmMo/CIGS (1.8 μm)/CdS/ZnO/ITO and glass/Mo/CIGS/CdS/ZnO/ITO thin film solar cells demonstrate efficiencies of 9 and 14%, respectively. The reason is that the SiO_2 barrier layer blocks diffusion of Na from glass substrate into CIGS layers. The formation of $MoSe_2$ is less favorable with SiO_2 barrier. The Na diffuses into CIGS without SiO_2 layer that enhances to form $MoSe_2$ layer between Mo and CIGS. The $MoSe_2$ layer is ohmic to the CIGS whereas Mo/CIGS is more or less Schottky type, evidenced by I–V measurements. The XRD analysis confirms the formation of $MoSe_2$ by the presence of (100) and (110) reflections [295].

The surface properties of CIGS mainly depend on the growth pressure of Mo. In general, the Mo layers are deposited by DC sputtering. The CIGS layers grown onto glass/Mo exhibit porous, fibrous, elongated grains and contain closed intercolumnar gaps for the Mo growth pressure of 5 mtorr, whereas the CIGS layer had above structure with micro cracks and with column boundary voids for the Mo deposition pressure of 0.8 mtorr. Another set of CIGS layers show lamellar-type structure, tightly packed and faceted grains for the Mo sputter pressure of 0.8 mtorr. Some other CIGS films have less faceted, very dense and smooth structure for the Mo layers sputter pressure of 5 mtorr. The structure changes from sample to sample. The CIGS layers show rough surfaces and faceted grains for the Mo deposition pressure of 8 mtorr. The CIGS cells show higher efficiency of 16% for the Mo deposition optimum pressure of 5 mtorr, whereas the efficiency of the cell degrades below or above the optimum pressure, as

Table 8.36 Influence of Mo on CIGS cells, and cell with electrodeposited CIGS layers

Process	V_{oc} (mV)	J_{sc} (mA/cm^2)	FF (%)	η (%)	R_s (Ω·cm^2)	Ref.
0.8 mtorr (Ar)	624	30.9	74.6	14.36	–	[296]
5 mtorr (Ar)	671	32.6	73.2	15.99	–	
Mo selen. temp. 450 °C	585	29.3	68.9	11.8	–	[299]
Mo selen. temp. 580 °C	559	27.1	60.8	9.8	–	
No selen. temp 580 °C	314	21.6	29.6	2.0	–	
Mo	601.2	31.3	73.3	13.8	0.37	[298]
ZnO:Al + 10 nm MoSe$_2$	613.8	33.5	65.4	13.4	2.9	
Electro-dep.+PVD CIGS	550	26.55	67.9	9.87	–	[300]
Electro-dep. only	550	26.55	67.9	9.87	–	[301]
Electro-dep. only	300	31.7	45.6	4.35	–	[303]

mentioned in Table 8.36 [296]. The 0.8 μm thick Mo layer onto glass is deposited by RF-magnetron sputtering under Ar at chamber pressure of 2×10^{-3} and another layer at 8×10^{-3} torr. The CIGS/CdS/ZnO/ITO/MgF$_2$ thin film solar cells are prepared on Mo coated glass substrates in which 50 nm thick CdS by CBD, 0.3 μm thick ZnO, and 0.5 μm ITO by RF sputtering and MgF$_2$ by electron beam evaporation are sequentially grown. The CIGS thin film solar cells with composition of Cu:In:Ga:Se = 22.92:17.15:7.61:52.32, Cu/(In+Ga) = 0.93, and G/(In+Ga) = 0.31 formed for Mo layer growth pressure of 2×10^{-3} torr show efficiency of 15%, whereas the CIGS cells with Cu:In:Ga:Se = 23.78:19.24:5.26:51.72, Cu/(In+Ga) = 0.97, and Ga/(In+Ga) = 0.21 fabricated for the Mo growth pressure of 8×10^{-3} torr show efficiency of 15.4%, indicating that Mo layer grown at lower pressure shows thinner interfacial layer, which leads to dense grain structure and is less reactive with Se. Therefore, the growth pressure of 8×10^{-3} torr is preferable than 2×10^{-3} torr for Mo [297]. The results on SLG/ZnO:Al/Mo/CIGS/CdS/i-ZnO/ZnO:Al and SLG/Mo/CIGS/CdS/i-ZnO/ZnO:Al cells with back contacts of Mo and ZnO:Al are given in Table 8.36. They show efficiencies of 13.4 and 13.8%, respectively indicating that the cells with ZnO:Al layer and less thick Mo layer show less ohmic contact due to higher sheet resistance and less quantity of MoSe$_2$. At least 15 nm thick Mo layer is essential between CIGS and ZnO:Al layers to obtain an ohmic contact. The sheet resistances of Mo/CIGS/Au, ZnO:Al/Mo/NaF/CIGS/Au, and n-ZnO:Al/p-CIGS/Au are 0.045, 0.6, and 27.2 kΩ-cm^2, respectively. The last one shows rectifying p–n diode characteristics [298].

The hexagonal phase MoSe$_2$ (100), MoSe$_2$ (110), and Mo (110) reflections are observed in the CIGS cells by XRD analysis [129]. The p-type MoSe$_2$ is amenable to the cells in terms of electrical. 175 nm thick Mo layer deposited onto Si by DC magnetron sputtering is selenized at 580 °C for 60 min. The RBS analysis confirms the existence of 300 nm thick MoSe$_2$ layer and 115 nm thick Mo layer. 60 nm thick Mo and 300 nm MoSe$_2$ are observed from TEM analysis. The MoSe$_2$-(10$\bar{1}$0) and MoSe$_2$-(11$\bar{2}$0) reflections and Mo (110) from XRD are found. The SAD pattern shows (0002) and (0004) reflections. Both XRD and TEM analyses reveal the hexagonal structure. The lattice constants of MoSe$_2$ are $a = 3.28$ and $c = 12.9$ Å.

The BCC structured Mo converts into hexagonal $MoSe_2$ layer while selenization. The c-axis of $MoSe_2$ layer is perpendicular to the substrate for the selenization temperature of 550 °C and parallel to the higher substrate temperature of 580 °C. The selenization time between 10 and 40 min offers 10 nm thick $MoSe_2$ [299].

8.3.18 CIGS Thin Film Solar Cells with CIGS Absorbers Grown by Electrodeposition

The Cu–In–Se precursor layers onto Mo coated glass substrates are deposited by electrodeposition using 12 mM $CuCl_2$, 25 mM $InCl_3$, 25 mM H_2SeO_3 chemical solution, pH 1.5 and −0.5 V at RT on which 398 nm thick In, 214 nm thick Ga, and Se are added by PVD at substrate temperature of 550 °C to make adequate compositional CIGS layer that fit to solar cell, followed by selenization under Se vapor while cooling the sample. Upon annealing the as-grown electrodeposited CIS layer with composition of Cu:In:Se = 39.18:17.46:43.36 under Se vapor does not show much difference in composition. As usual using standard procedure, CdS, ZnO, ZnO:Al, Ni–Al, and MgF_2 are successively grown onto CIGS substrate to construct cell configuration. The glass/Cr/1 μmMo/CIGS/0.5 μmCdS/50 nmZnO/350 nmZnO:Al/Ni–Al/100 nm MgF_2 cells with an active area of 0.41 cm^2 demonstrate efficiency of 10%. The XRD analysis demonstrates that the as-grown layer shows mixed phases of CuSe and $CuInSe_2$. After selenization of precursor in tubular furnace at 550 °C, only $CuInSe_2$ exists by suppressing secondary phase. The composition analysis categorizes no much difference in composition between as-grown (Cu:In:Se = 39.18:17.46:43.36) and selenized CIS samples (Cu:In:Se = 39.5:16.64:43.85). The CIGS films with composition of Cu:In:Ga:Se = 28.77:21.05:1.84:48.34 can be prepared by adding 0.2839 M $GaCl_3$ solution to the CIS precursor solutions, increasing electrodeposition potential from –0.5 to –0.6 V and keeping the same pH. The broad and less sharp peaks are found in the CIGS layers due to distortion of In sites by Ga, in addition Cu_2Se (311) phase is observed. However, the sharp standard reflections are observed from XRD measurements for CIS films. The cells with CIGS layers grown by electrodeposition also show same efficiency of 9.87% (Table 8.36). The acceptor density of 1.4×10^{17} cm^{-3} and flat band potential of –0.35 V for CIS thin films are determined [300–302]. The CuInGa metallic layers are deposited onto Mo substratates by electrodeposition using CuCl, $InCl_3$, $Ga(NO_3)_3 \cdot 7H_2O$, KSCN solutions, and pH 5. CuInGa layers annealed under Se vapor at substrate temperature of 560 °C lead to 2–2.5 μm thick $CuInGaSe_2$ thin films. The Mo/CIGS/CdS/i-ZnO/ZnO:Al cells demonstrate efficiency of 4.35% [303].

8.3.19 CIGS Monolithic Cells and Modules

By modifying design parameters of cell, the efficiency of CIGS/CdS/ZnO/ZnO:Al/Ni–Al–Ni/MgF_2 cell can be increased from 17.8 to 19.3% that the finger width is decreased from 50–100 to 5–10 μm by 1 order of magnitude and height is increased from 2–3 to 6–8 μm ultimately the finger numbers increase from 3 to 10. The thickness of ZnO:Al is decreased from 300 to 100 nm. The thicknesses of Mo and CIGS are increased from 1.2 to 1.5 μm, and from 1.5–1.8 to 2.8 μm, respectively and other parameters are remaining at the same. The cell dimensions are changed from 11×4.4 to 7.6×6.58 mm^2 [114]. The efficiency of CIGS cell can also be increased from 18.7 to 21.5% with increasing light concentration from 1.67 to 14.05 suns. However, the layers may be delaminated due to lower thermal conductivity of substrates. It may be possible to overcome by using thick conducting substrates like SS [304]. The CIGS/CdS/ZnO/ITO standard cells interconnected submodules with an active area of 80.8 cm^2 prepared by in-line process show efficiency of 10.5%. However, each segment cell with an area of 4.5 cm^2 shows higher efficiency of 14%. The in-line process system is capable to deposit 10×10 cm^2 area for small scale cells, as shown in Fig. 2.6 [305]. The laboratory scale deposition process is used for small size modules, whereas the deposition process is different for large scale modules such as the sources are kept on top of the substrates, as shown in Fig. 8.79. The CIGS layer is deposited over the large area of 60×120 cm^2 and 2–4 mm thick float glass substrates in-line process by evaporation of respective pure elements as a pilot line. The flatness of the substrates is an important factor to gain uniform temperature on the surface of substrates, which are rolled by supporting rigid body. A module with combination of 42 circuits or cells consists of an area of 729 cm^2 that reflects into an active area of 651 cm^2, whereas a cell contains an area of 0.5 cm^2 that turns into an active area of 0.46 cm^2. The glass/Mo/CIGS/CdS/iZnO/ZnO:Al module and cell exhibit efficiencies of 10.5 and 13.2%, respectively as given in Table 8.37. The series and shunt resistances of 3.5 and 62 Ω-cm^2 for modules and 0.8 and 4.5 kΩ-cm^2 for cells are determined, respectively. The thicknesses and sheet resistances of ZnO layers are 1 μm and 12 Ω/□ for modules and 0.35 μm and 60 Ω/□ for cells, respectively. In the modules, the low efficiency is due to nonactive area of the patterning and inhomogeneity. On the other hand, the internal connectors occupy 10% area of the modules. A substrate size of 30×30 cm^2 is used to deposit CIGS in-line process for the fabrication of module area of 729 cm^2. The efficiency decreases from 10.5 to 8% with increasing size of module from 30×30 to 120×60 cm^2. The CIGS module fabricated by the ZSW Company has ten cells and active area of 48.2 cm^2 in which each cell contains width of 8.4 mm. Another module size of 10×10 cm^2 with cell

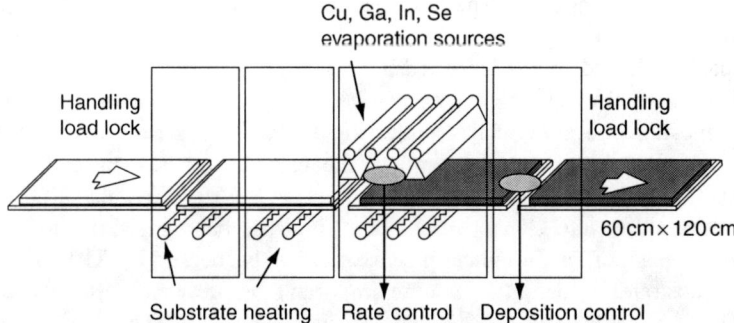

Figure 8.79 In-line process of CIGS film by co-evaporation over the large area substrates.

Table 8.37 PV parameters of monolithic cells and modules

Area or Ga/(Ga+In)	V_{oc} (mV)	J_{sc} (mA/cm^2)	FF (%)	η (%)	R_s (Ω-cm^2)	R_p (kΩ-cm^2)	P_{max} (W)	Ref.
80.8 cm^2-18 cells	11.3 V	119 mA	63.1	10.5	–	–	–	[305]
4.5 cm^2 cell	626/cell	26.5	–	14	–	–	–	
0.26	602	138	62.1	10.7	19	–	–	[307]
0.32	646	138	60.4	11.1	20	–	–	
0.37	692	128	67.9	12.5	20	–	–	
0.45	718	119	66.7	11.8	18	–	–	
0.6	788	96	69.3	10.9	14	–	–	
WS (0.3)	606	182	59.4	10.3	16	–	–	
Cell	614	29.8	72.1	13.2	0.8	4.5	–	[308]
Module (42 cells)	597	25.5	68.8	10.5	3.5	0.062	–	
	654	33.3	76.5	16.7	–	–	–	[132]
As-grown	607	39.7	64.7	11.1	–	–	–	[312]
Irradiated	611	40.7	67	11.8	–	–	–	
1.1 cm^2/(AM1.0)	666.8	40.83	78.3	15.2	–	–	–	
900 × 4 cm^2	31.17 V	2.153 A	68	13.4	–	–	48.3	[314]
3600 cm^2	555/cell	34.13	67.4	12.8	–	–	–	
Module 1 (42 cells)	23.7	1.49	61	9.37	0.035	0.1999	21.55	[315]
After 4 months	24.01	1.51	62	9.73	0.052	0.029	22.38	
Module 2 (42 cells)	23.37	1.41	59	8.51	0.032	0.02074	19.56	
After 4 months	23.84	1.44	60	9.02	0.041	0.02038	20.75	

width of 7 mm and aperture area of 82.3 cm^2 made by Wurth solar company with different Ga/(In + Ga) (GGI) ratios are given in Table 8.37 [306–308].

In the pilot line process, a size of 60 × 90 cm^2 float glass substrate with thickness of 3 mm has been employed from substrate washing stage to absorber deposition point then the size of substrate is split into 30 × 30 cm^2 for further process of other layers to make modules. Figure 8.80 shows process development of flow chart of monolithically integrated CIGS thin film solar cells. The SiN is first deposited onto glass substrate as an alkali barrier on which Mo layer by DC sputtering is deposited by vertical in-line sputtering. The laser patterning (P_1) is carried out. The Na compound is sputtered on top of the Mo to require Na doping in the CIGSS layers. On top of that, CuGa and In are deposited by DC magnetron sputtering then Se is evaporated by thermal evaporation with thick layer that can make to obtain required final composition because while conducting RTP, Se gets re-evaporation due to thermal power. In this context thicker layer means an excess Se is evaporated during the deposition of layers. No Se vapor is used during RTA process but sulfur vapor is used to dope sulfur into CIGS that causes to increase band gap. Prior to sulfurization, the composition of CuInGaSe precursor layers is checked by XRF. The homogeneous substrate temperature is more important to prevent substrate crack and wrap. The acceptable error limit of temperature is ±10 °C. The S/(Se + S) ratio is controlled by the Raman, XRF, and PL to have adequate quality CIGSS layers,

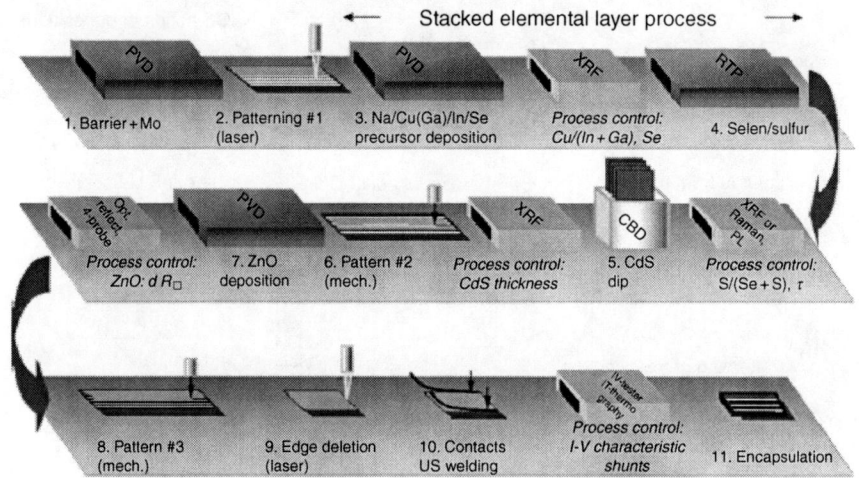

Figure 8.80 Process flow of monolithically integrated thin film solar cells.

which are dipped into CBD bath to deposit CdS whereby the XRF is used to monitor composition of CdS buffer layer. On glass/SiN/Mo/CIGSS/CdS structure, mechanical pattern (P_2) is done and 75 nm thick ZnO and 600–800 nm thick n-ZnO layers are successively deposited by RF and DC sputtering techniques. The thickness and sheet resistances of ZnO are measured by reflectometry and four point probes method, respectively. The mechanical pattern (P_3) is done on ZnO layers. The air flow exhausts ablated particulates while patterning the layers. The edge deletation and contacts are also done on the samples. The I–V and thermagraphy are done on 53 circuits or 30×30 cm^2 panels for testing purpose. Finally the modules are encapsuled in order to protect from atmosphere [309]. The typical monolithic circuit pattern is shown in Fig. 8.81A in which each cell with typical area of 1×1 cm^2 or more is made and the cells are internally connected in series. One can see that P_1 pattern on Mo is carried out by pulsed laser and the P_2 and P_3 mechanical patterns are sequentially done on the absorber and ZnO, respectively, as pointed out in earlier [116]. As shown in Fig. 8.81B, a typical module made on 4×2 ft^2 size has number of monolithically interconnected cells. The module is covered with ethyl vinyl acetate (EVA) sheet and then glass plates are used to encapsulate [310].

The efficiency of 13.1% is observed in the typical CIGS submodule size of 12.5×12.5 cm^2 with ZnO layer with sheet resistance of 5–10 $\Omega/\square$, whereas the best 0.5 cm^2 area cell demonstrates efficiency of 16.7% [132]. Monolithically integrated submodules with aperture area of 144.9 cm^2 have 20 cells connected in series and each cell or segment with a width of 0.55 cm shows efficiency of 9.7%. The Ga concentration is high at back because of fast formation of CIS, as compared to that of CIGS. The In/Cu/Ga precursor is selenized under H_2Se atmosphere at 400–450 °C in such a way to obtain Cu/(In + Ga) = 0.9 and 20% Ga for thin film solar cells [142]. In monolithic circuit or submodules, the post-absorber formation felicitates higher efficiency of 11.7% as compared to that of conventional (10.9%) and

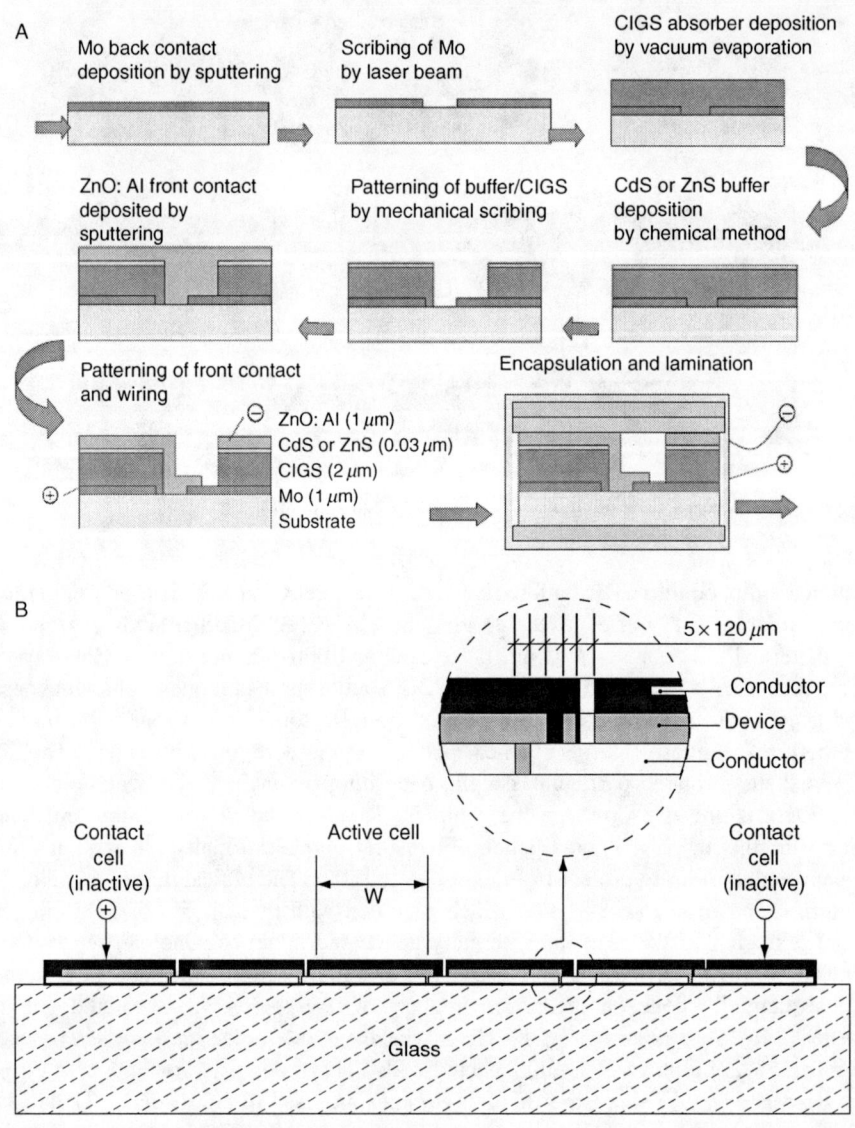

Figure 8.81 (A) Fabrication of CIGS monolithic circuit for module and (B) Final schematic cross section of module with combination of number of monolithic cells.

post-device fabrication (10.3%) because of reduction in defect formation [311]. 15–150 μm thick metal foil/0.5–2 μm Mo/2.5 μm CIGS/30 nmCdS/500 nmZnO/2–5 μm SiO_x thin film solar cell structure is made for extra terrestrial applications. The CIGS layer is made by two-stage process in which Cu-rich CIGS layers are deposited by evaporation technique to have large grain sizes then titrated to incur Cu-poor CIGS layers in the surface by evaporating InGaSe layers. Prior to deposition of absorber, Na contained 0.2 μm thick CIGS layer is deposited. The SiO_x is used to act as a radiation protector. The samples are irradiated with 10^{15} fluence of 1 MeV electrons and the typical 24 cm^2 area cell results are given in Table 8.37. The best cell shows efficiencies of 16.9 and 15.2% at AM1.5 and AM0, respectively. The CIGS thin film solar cell without cover glass delivers specific power of 1430 W/kg, power density of 210 W/m^2 and cost is 30–50\$/W, whereas it is \$130/W for Si and \$260/W for III–V three junction cells per life period [312]. After damp heating at 85 °C and humidity of 85% for 144 h, the free carrier concentration decreases from 5.5×10^{15} to 2.5×10^{15} cm^{-3} and the space charge region width increases from 300 to 600 nm [313]. The modules show stable efficiency, after testing under standard conditions of 85 °C, 1000 h and 200 cycles [314]. Two modules M_1 (9.4%) and M_2 (8.5%), which contain 42 monolithic cells, are tested at the Nelson Mandela Metropolitan University, Port Elizabeth, South Africa in outdoor. The modules have been kept on dual axis solar tracker at latitude of 34°S in order to obtain 27% more radiation than normal from the Sun. The efficiencies of modules slightly increase under the irradiation of Sun over the period of 4 months (Table 8.37) [315]. The glass/CIGS/CdS/ZnO submodules with aperture area of 202.3 cm^2 with combination of 25 cells show an efficiency of 9%, $V_{oc} = 11.53$ V (0.461 V/cell), $I_{sc} = 0.2482$ A ($J_{sc} = 1.227$ mA/cm^2), $I_{max} = 0.2147$ A, $V_{max} = 8.483$ V, and FF $= 0.6364$ [157,158].

8.3.19.1 ZnS Buffer Based CIGS Modules

The CuInGa alloy layers have been grown by sputtering using In and Cu–Ga targets, which are selenized by H_2Se gas to form $CuInGaSe_2$ layers. Using the same analogy of chemical bath deposition of CdS, the Zn(O,S,OH) buffer by CBD is employed in the CIGSS solar cells. The performance of the module is improved using ZnO (MOCVD) and by light soaking as a part of post annealing. The typical patterning is shown in Fig. 8.82 for 51.7 cm^2 area SLG/CIGSS/Zn(O,S,OH)/ZnO minimodules, which show $V_{oc} = 6$ V, $J_{sc} = 3.1$ cm^2, FF $= 0.6825$, and $\eta = 14.19\%$ [316]. The dead area in the module is a part of loss factor for low efficiency of the cell. The glass/Mo/CuInGaSe$_2$/Zn(O,S,OH)$_x$/i-ZnO(50 nm) based 30×30 cm^2 modules are studied with effect of doping in the ZnO that is, ZnO:B and ZnO:Ga grown by MOCVD and

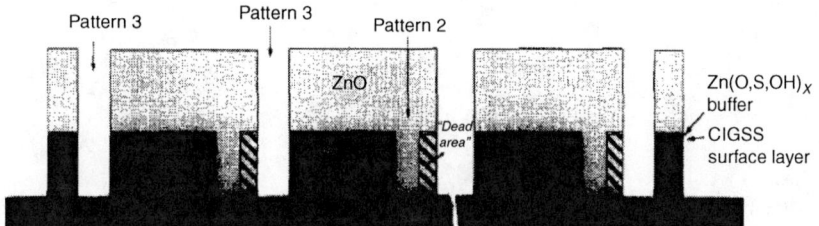

Figure 8.82 Patterning of SLG/CIGSS/Zn(O,S,OH)/ZnO minimodules.

sputtering techniques, respectively. The modules with ZnO:B (MOCVD) shows higher efficiency owing to well enough transparency of 90%, sheet resistance of 10 Ω/□ and high electron mobility of 37.7 cm^2/Vs, as already discussed. The CIGS module shows efficiencies of 12 and 11.67% with 100 nm i-ZnO and without i-ZnO layer, as shown in Table 8.38. The i-ZnO forms proper contact between Zn(S,O,OH)$_2$ and doped ZnO layer. In the case of ZnO sputtered layers, the thicker ZnO layer is indeed needed to improve the fill factor. The ZnO:B window layer is coated onto Zn(O,S,OH) by MOCVD technique is little high resistive at the beginning because lack of B doping in the beginning growth of ZnO and less textured surface then the grown layers normally show low sheet resistance. The sufficient doping and textured surface provide good optical and electrical properties of ZnO [317,318].

As shown in Fig. 8.83A, the glass/Mo/CIGS/Zn(O,S,OH)$_x$/ZnO-doped solar cells with ZnO:B (MOCVD) show higher short circuit current and fill factors than that of cells with ZnO:Al/ZnO:Ga grown by sputtering for an active area of 864 cm^2 with 50 cells. Prior to deposition of 300 nm thick ZnO:Ga by DC sputtering, 60 nm thick ZnO:Ga layer is grown by RF sputtering in order to protect Zn(O,S,OH)$_x$ buffer from higher energy particles produced by DC sputtering. In the sputtering process to grow ZnO:Ga, the sequence is used as RF/DC/DC sputtering corresponding to monolayer/trilayer/trilayer, which shows better results. Ga-doped ZnO (ZnO:Ga) layers deposited using ZnO with 5.7 wt% Ga$_2$O$_3$ target proves lower transmission at the longer wavelength region because of higher free carrier absorption comparing with the one grown by ZnO-3.4 wt% Ga$_2$O$_3$ target. In comparison, ZnO:B grown by MOCVD exhibits higher transmission than that of one grown by sputtering using ZnO-5.7 wt% Ga$_2$O$_3$ (Fig. 8.83B), even though both layers show more or less similar sheet resistance ($R_\square$) of 6.8 and 7.9 Ω/□, respectively. The lower transmission at longer wavelength in sputtered ZnO due to free carrier absorption may be caused to have

Table 8.38 PV parameters of ZnS and InS based cells and modules

ZnS buffer based cells and modules/cell area (cm^2)	V_{oc} (mV)	J_{sc} (mA/cm^2)	FF (%)	η (%)	Ref.
ZnO:Ga (SP)	584	29.9	61.8	10.82	[317]
(100 nm ZnO)/ZnO:B (MOCVD)	584	33.32	61.9	12.03	
Without ZnO	576	33.53	60.4	11.67	
ZnO:B (861.3 cm^2)	630	30.9	66.7	13	
ZnO:5.7%Ga (855 cm^2)	614	31	64.2	12.2	
855 cm^2	34.25 V	0.543 A	70	15.22	[324]
Single cell	601	36.18	70		
10 × 10 cm^2	667	31.6	72	15.2	[327]
0.5 cm^2	661	35.1	74.9	17.3	
50 cm^2	550	31.8	60	10.7	[245]
3.2 cm^2	570	35.2	70	14.1	
InS based modules					
30 × 30 cm^2	125.6 V	I_{sc}=0.463 A	67.9	10.4	[330]
30 × 30 cm^2 (typical cell)	680	–	70	12.9	[334]

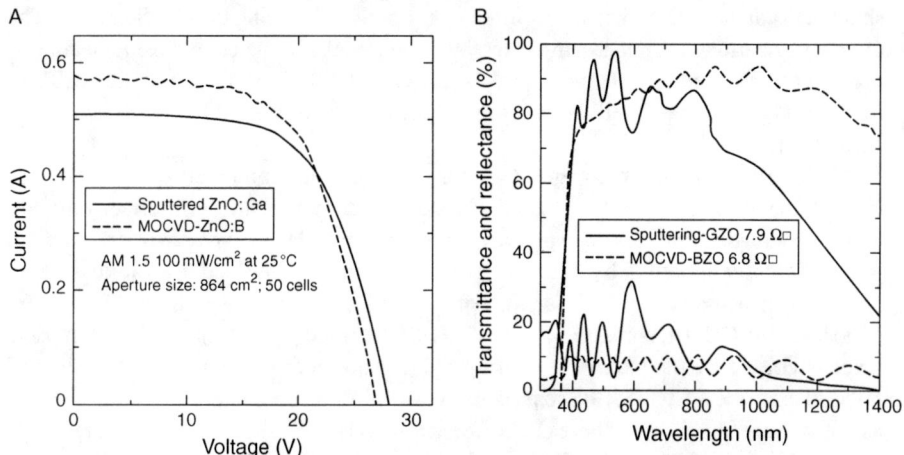

Figure 8.83 A) *I–V* characteristics of glass/Mo/CIGS/Zn(O,S,OH)$_x$/ZnO-doped cells with ZnO:Ga and ZnO:B grown by sputtering and MOCVD techniques, respectively, and (B) Transmission, and reflectance spectra of ZnO:Ga and ZnO:B layers.

lower short circuit current in the cells. On the other hand, the higher reflectance at shorter wavelength also causes to have lower efficiency of the cells. The mobilities of 18 and 35 cm^2/Vs, carrier concentrations of 4.69×10^{20} and 1.45×10^{20} cm^{-3} are found for sputtered 800 nm thick ZnO and 1600 nm thick MOCVD ZnO layers, respectively [319,320]. By avoiding H$_2$Se toxic gas, a new approach has been employed to replace with elemental Se. Several Cu(85%)+Ga(15%) and In layers are deposited onto Mo coated 3 mm thick glass substrates by magnetron sputtering on which Se layer is deposited by evaporation, finally the entire layers are selenized and sulfurized at the heating rate of 10 °C/s to form Cu(InGa)SSe$_2$ layers. The CdS and ZnO layers are formed onto CIGSS by CBD and sputtering techniques, respectively. The 30×30 cm^2 module efficiency is improved to 12.9 from 12.2% by changing sputter composition from Cu$_{85}$Ga$_{15}$ to Cu$_{75}$Ga$_{25}$. 60×90 cm^2 area cells deliver efficiency of 12.8% and power of 65 Wp. The 30×30 cm^2 module with Zn(S,OH) buffer shows efficiency of 12%, whereas 8×8 cm^2 area cells on Ti-foil substrates show efficiency of 12.4% and 1.4 cm^2 area cell exhibits efficiency 13.9% [321]. The chemical solution applied for the growth of Zn(O,S,OH) buffer is used six or more times in the process of buffer deposition that reduce the production cost. The glass/Mo/Cu(InGa)SSe$_2$/Zn(O,S,OH)-CBD/ZnO/ZnO:B solar cells made over the large area of 900 and 3600 cm^2 exhibit efficiency of 14.3 and 12.8%, respectively. The mosaic module with 900 cm^2 $\times$ 4 circuits that is, 3600 cm^2 shows efficiency of 13.4%. 288 number, 30×120 cm^2 sized mosaic modules are tested after exposing to the Sun for 53 days, which yield better performance by improving fill factor. The glass/SiO$_x$/Mo/CIGSS/Zn(O,S,OH)-CBD/ZnO(MOCVD)/ZnO:B(MOCVD) module with an active area of 855 cm^2 or 30×30 cm^2 have 57 interconnected cells show an improved efficiency of 15.22% and delivers maximum power of 13.01 Wp. In such a way, the cell open circuit voltage is tailored in between 630 and 650 mV/cell to avoiding shunt paths in the cell, which are nullified by properly sulfurizing the CIGSS layers. The over sulfurization of CIGSS samples creates shunt paths in the cells. Post annealing is done in order to reduce OH radical in the Zn(O,S,OH) buffer layer that reduces R_s of the cell. On the other hand to increase R_{sh} in the cells, the ZnO layer is grown by MOCVD [322–324].

The Shell Solar Inc. company, Germany established similar efficiency modules that CIGSS thin films on Mo coated glass substrate patterned by laser scribing are fabricated in-line process by sputtering Cu, In, and Ga layers and Na doping on which Se is evaporated by thermal evaporation and sulfurized by rapid thermal annealing process using H_2S on top either CdS or Zn(S,O,OH) buffer layers are deposited by CBD technique and patterns are formed by mechanically then ZnO layer is coated by magnetron sputtering again patterned mechanically. 30×30 and 60×90 cm^2 area CIGSS/CdS based modules show efficiencies of 11 and 13.1% as the best, respectively, whereas Cd free Zn(S,OH) buffered 30×30 cm^2 area cells yield efficiency of 11.9% [325]. Pilot line process is adopted to fabricate large area solar cell modules for which the deposition method as follows; SiN and Mo are sequentially deposited onto float glass substrates on which Cu, Ga, In, and Na compound are grown by DC magnetron sputtering followed by thermal evaporation of Se. The entire layers are sulfurized under sulfur in RTP reactor. The CdS by CBD and ZnO by magnetron sputtering are sequentially formed. The Mo is patterned by a near infrared pulse laser, absorber, and ZnO are patterned by mechanical scribing. The Shell Solar company steadily improved its panel (30×30 cm^2) efficiency from year to year as 10.5% (2002), 11% (2003), 11.5% (2004), and 12.3% (2005). A 60×90 cm^2 module with ZnO (MOCVD) shows efficiency of 13.1%, whereas 1.4 cm^2 cell and 30×30 cm^2 modules with Zn(S,OH) buffer layer show efficiencies of 13.2 and 10.5%, respectively. Another buffer layer (Zn,Mg)O grown by RF sputtering with 30 at%Mg–ZnO target is also used in the place of CdS for CIGS solar cells, which show efficiencies of 11.7 and 9.1% for 1.4 and 30×30 cm^2 areas, respectively [326]. The ZSW company improved quality of CIGS layers by modifying in-line process adding extra evaporation sources of Cu, In, Ga, and Se to in one single run by controlling evaporation rates and temperatures. On the other hand, Mg is doped into ZnO layer for improvement. The SEM (Fig. 8.84) shows ZnS on CIGS as a mat with small grains. The fill factor of cells decreases if the cells kept under dark and again increases upon light illumination [327]. The transmission of the chemical solution is monitored by the Si based sensor. The optimized conditions to incur best Zn(SO) buffer layers are NH_3 30%, $ZnSO_4$, $(NH_2)_2CS$, bath temperature of 75 °C, and the dipped sample to be removed when the chemical bath solution turns turbidity with transmission of 20%. The SLG/CIGSS/Zn(SO)/i-ZnO/ZnO: Al cell is light soaking for 1 h in order to improve the efficiency. The light soaked cell with 0.5 cm^2 shows higher efficiency of 13.5% than that of as-grown cell. 30×30 cm^2 area module shows efficiency of 10–5–12.5% [328].

In the pilot line process, the homogeneity of the layers has been controlled by atomic absorption spectroscopy. The XRF is used to find out compositions of the layers except light elements such as Se and S, which are detected by the Raman spectroscopy, for example, the A_1 modes of Cu(InGa)SeS$_2$ system appear at 180

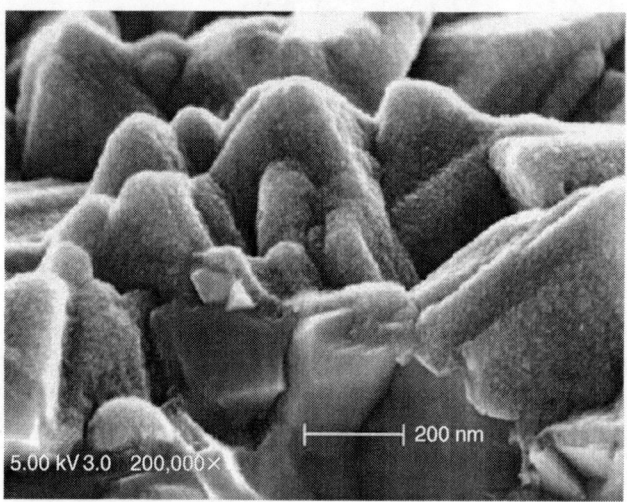

Figure 8.84 SEM of ZnS layer covered on CIGS, and the CIGS surface look like a pyramidal shapes.

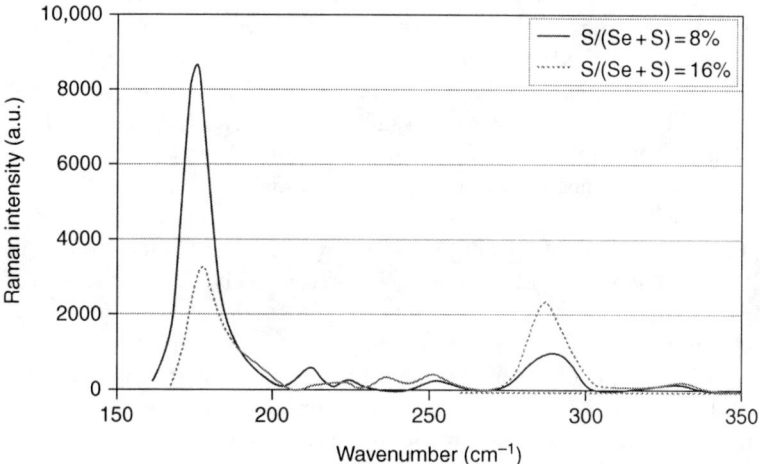

Figure 8.85 Raman spectroscopy of CIGSS layer for two different sulfur compositions.

and 290 cm^{-1}. The intensity ratio of Se–Se (180 cm^{-1})/S–S (290 cm^{-1}) vibration peaks reveals composition ratio of Se to S as shown in Fig. 8.85 [309]. The PL checks minority carrier lifetime and the optical detector in reflection mode from the interference fringes measures the thickness of Se and ZnO layers. The four points probe method evaluates sheet resistance of ZnO layer. The developed 60×90 cm^2 module is tested by infrared cameras, which show difference in thermal image if there is a variation in temperature that occurs due to resistive losses caused by microshunts. The damp heat test reveals 1% efficiency loss [329].

8.3.19.2 InS Based CIGS Modules

The nontoxic In_2S_3 buffer for CIGS modules is grown by atomic layer chemical vapor deposition (ALCVD or ALD) method using an ASM microchemistry F-450 reactor at substrate temperature of 210 °C. The indium acetylacetonate In $(CH_3COCHCOCH_3)_3$ and hydrogen sulphide used in this process are the precursors. A large module size of 30×30 cm^2 with In_2S_3 buffer shows efficiency of 10.4%, whereas the efficiency is 11.2% in small modules (10×10 cm^2). The cells indeed do not exhibit light soaking effect. The QE shifts to higher wavelength by little with increasing thickness of In_2S_3 buffer from 15 to 45 nm in the float glass/CIGS/In_2S_3-ALCVD/i-ZnO/ZnO:Al cells due to absorption of photons, as already mentioned [330]. An improved version In_2S_3 buffer based large area 30×30 cm^2 module with an active area of 714 cm^2 and combination of 42 solar cells shows efficiency of 12.9%. The In_2S_3 buffer layer with thickness of 50 nm is deposited onto CIGS absorber layer using pulse sequence as H_2S/N_2/indium acetylacetonate ($In(acac)_3$)/N_2:1500/1100/2500/3000 ms at 210 °C, followed by i-ZnO and ZnO:Al layers using sputtering. The SIMS analysis reveals that the In_2S_3 buffer layer contains 44.1%–In, 48.3%–S, 2.1%–Na, 1.2%–Cu, 6%–O, and 0.3%–Se. The Na self-doping occurs into film from the glass substrate, the Cu and Se interdiffuse into In_2S_3 buffer from the absorber layer and the oxygen is due to adsorption from the atmosphere while keeping the grown sample in out door for some time. In the case of 1 nm In_2S_3 buffer layer deposited at the same conditions, the composition is entirely different as Ga—4.9%, Na–7.9%, Cu–5.7%, O–19.4%, In–20.7%, S–0%, and Se–41.5%. These anomalous results are because of thinner In_2S_3 layer causing to add composition of bottom layer by XPS whereby the penetration depth of X-rays is high. The thin film solar cells annealed at 200 °C for 1 h, which have In_2S_3 and i-ZnO layers deposited at substrate temperature of 160 °C and RT, respectively shows efficiency of $\sim$11%, whereas the efficiency increases from 11 to 14% for In_2S_3 and i-ZnO growth temperatures of 210 °C and RT, respectively. The efficiency further increases from 14 to 15% as the growth temperature of ZnO buffer is further increased from RT to 200 °C while keeping deposition temperature of 210 °C for In_2S_3 buffer. Once the efficiency is reached to higher, the post annealing effect does not contribute its role much to enhance efficiency. On the same absorber, the CdS and In_2S_3 buffered solar cell modules show similar efficiencies, as shown in Fig. 8.86. The SEM cross section of CIGS/In_2S_3 cell is also shown in Fig. 8.86A [331–333]. The stability of 15×30 cm^2 CIGS modules with In_2S_3 buffer has been tested before and after the storage in the dark for 8 months under atmospheric conditions. They show the same results but slightly lower FF by about 1–2%. The modules are kept under light illumination of 100 mW/cm^2 (AM 1.5) at 25 °C for 1 h. During first 15 min light illumination, the PV parameters slightly decrease due to warming up the panel. The dry stability test is lightly problematic to the panel, whereas the damp heat test is harmful to the panel. The module passes damp heat test, which kept at 85 °C for 1000 h at humidity of 85% is so-called damp heat test. The modules are also tested by thermal cycling method varying temperature from –40 to 85 °C per cycle with speed of 4 h per cycle and total cycles of 200. The test

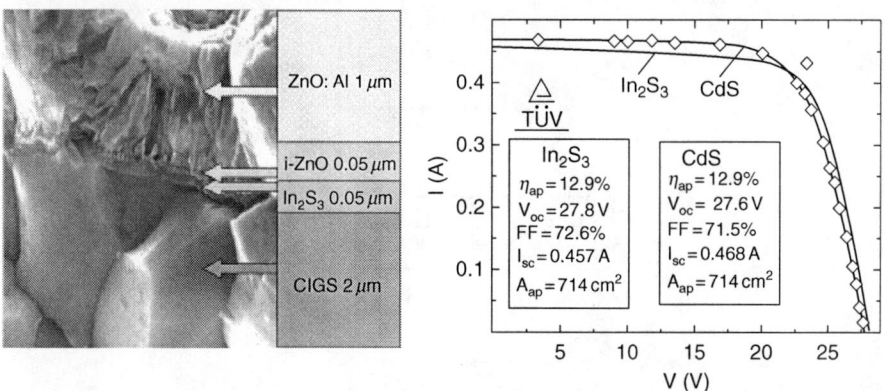

Figure 8.86 (A) Cross-sectional scanning electron micrograph of CIGS thin film solar cell, and (B) *I–V* curves of CdS and In_2S_3 based CIGS modules under illumination.

results show stable conditions of modules. According to IEC61646 standard condition, the module efficiency should not decrease below 95% of the original efficiency. In the outdoor, the 30×30 cm^2 panel, which consists of 42 cells in series, has been tested in every 4 months over the period of 16 months, which is stable in terms of *I–V* parameters. The efficiency of a typical cell is 12.9% [334].

8.4 CuInS$_2$ Thin Film Solar Cells

A typical schematic diagram of CuInS$_2$ single layer based thin film solar cell with encapsulation is shown in Fig. 8.87. The simulation work on the CuInS$_2$ thin film solar cells reveals theoretical efficiency of 20.6% by using parameters $L_h = 1.6$ μm, $L_e = 2.5$ μm, and $V_{oc} = 1.01$ V, midgap defect density of 2×10^{15} cm^{-3}, carrier lifetime of 50 ns and band gap of 1.5 eV [335]. The band diagram of CuInS$_2$/CdS/ZnO is shown in Fig. 8.88 in which $\Delta E_C = 4.5(\chi_{CdS}) - 4.1(\chi_{CIS}) = 0.4$ eV, and $\Delta E_V = E_{gCdS} - E_{gCIS} - \Delta E_C = 0.47$ eV. The glass/Mo/Cu-rich CuInS$_2$(KCN etched)/CdS/n^+ZnO cells with an active area of 0.4 cm^2 show efficiency of 10.2%, whereas cells made with KCN etched In-rich CuInS$_2$ show low efficiency of 1.5% (Table 8.39). The results virtually indicate that Cu-rich CuInS$_2$ growth is always favorable to obtain quality layer for thin film solar cell applications. After etching absorbers those are in turn suitable for solar cell. The CuInS$_2$ absorber used in the thin film solar cells is prepared by co-evaporation of Cu, In, and S at substrate temperature of 550 °C. As-grown 3 μm thick Cu-rich CuInS$_2$ thin films etched in KCN to remove CuS phase show slightly In-rich composition of Cu $= 24.3 \pm 0.5$%, In $= 24.7$, S $= 51.0 \pm 1$%, and hole concentration of 10^{17} cm^{-3}. On the other hand, the In-rich CuInS$_2$ thin films etched in KCN solution show lower hole concentration of 10^{13} cm^{-3} [336]. In typical case an efficiency of 11.1% is reported on 0.48 cm^2 float glass/Mo/KCN etched CuInS$_2$/CdS 50 nm/ZnO based thin film solar cells (Table 8.39) and the best cell exhibits the highest efficiency of 12.5%. The CuInS$_2$

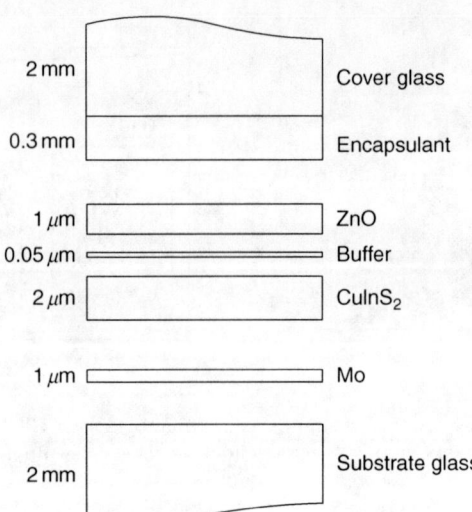

Figure 8.87 Schematic diagram of typical CuInS$_2$/buffer/ZnO with encapsulated cover glass, where buffer is either CdS or ZnS, *etc*.

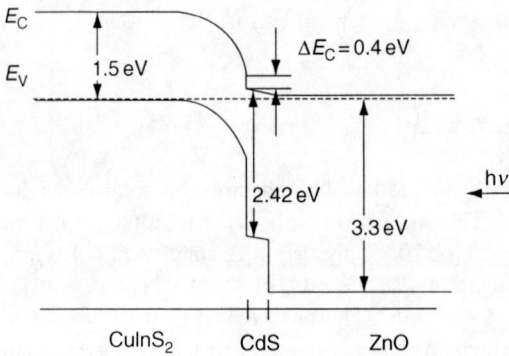

Figure 8.88 Band diagram of CuInS$_2$/CdS/n^+ZnO solar cell.

thin films employed as absorber in the cells are prepared by DC magnetron sputtering technique; first thick Cu layer then In layer are deposited, followed by sulfurization under sulfur vapor using elemental sulfur as a source at 500–550 °C. The etched sample shows composition of Cu/In = 1.0, as determined by EDAX [337]. The SLG/Mo/0.7 µm etched CuInS$_2$(PVD)/CdS(CBD)/i-ZnO/n-ZnO/Ni-Al cells with bilayer zinc oxide show little improvement in efficiency (6.7%), as compared to that of cells with single ZnO:F layer (5.3%). The i-ZnO and ZnO:F (*n*-ZnO) thin films applied in the cells grown by sputtering had sheet resistance of 40 and 200 Ω/□, respectively. [338].

Three stack Cu(100–130 nm)/In(170–200 nm) alternate layers with total thickness of 1 µm deposited either onto Mo coated glass substrates or 0.2 mm thick Pt sheet are sulfurized under 5%H$_2$S + Ar at 550 °C for 4 h to made CuInS$_2$ thin films with

Table 8.39 PV parameters of $CuInS_2$ thin film solar cells

Cell configuration	V_{oc} (mV)	J_{sc} (mA/cm^2)	FF (%)	η (%)	A	$J_o \times 10^{-10}$ (A/cm^2)	Ref.
Simulation	1010	–	–	20.6	–	–	[335]
$CuInS_2$/CdS(CBD)/i-ZnO/n-ZnO	710	24	70	11.4	–	–	
In-rich $CuInS_2$/CdS/n^+-ZnO (0.4 cm^2)	520	–	30	1.5	–	–	[336]
Cu-rich $CuInS_2$/CdS/n^+-ZnO (0.4 cm^2)	697	21.5	69	10.2	1.5	2	
SLG/Mo/$CuInS_2$/CdS50 nm/ZnO (0.48 cm^2)	728	21.42	70.9	11.1	–	–	[337]
SLG/Mo/$CuInS_2$(PVD)/CdS/ZnO:F/Al	477	39.7	38	5.3	–	–	[338]
SLG/Mo/$CuInS_2$(PVD)/CdS/i-ZnO/n-ZnO/Ni–Al	465	42.5	46	6.7	–	–	

different concentrations of Cu/In ratio from 1 to 2.1. After etching in 10%KCN solution for 3 min, no difference is found in the $CuInS_2$ thin film by XRD analysis but the composition changes to less than unity. The etched $CuInS_2$ absorber used in the cells contain resisivity of 58–550 Ω-cm, Hall mobility of 0.5–5 cm^2/Vs and hole concentration of 4×10^{16}–3×10^{17} cm^{-3}, respectively, whereas as-grown $CuInS_2$ layers show resisivity of 0.4–5.7 Ω-cm, Hall mobility of 0.1–1 cm^2/Vs and hole concentration of 2×10^{19}–2×10^{20} cm^{-3}. The cells made with etched samples show fair results because of removal off Cu_xS secondary phase on the surface, which is not found in the XRD spectrum that may be found by either in grazing incidence angle X-ray diffraction measurements or XPS, *etc*. The efficiencies of 1.7–1.5 mm^2 area Pt/$CuInS_2$/CdS (CBD)/ZnO(RF) cells with and without KCN etched $CuInS_2$ thin film absorber are 5.8 and 3.9%, respectively (Table 8.40). The annealing effect is nil on the cells if the etched absorber is used. The low efficiency is due to (i) high dark current of 10^{-4} mA/cm^2, (ii) low fill factor due to high series resistance, and (iii) the large interface recombination velocity reduces open circuit voltage. Figure 8.89 shows quantum efficiencies of cells with and without KCN etched absorber. Difference in efficiencies is mainly due to difference between short circuit currents of cells with etched and unetched absorbers. In the red spectral region, the cutoff region is associated to $CuInS_2$ band gap of 1.5 eV. On the other hand, the slopes are obviously witnessed by band gaps of CdS (2.4 eV) and ZnO (3.2 eV) in the shorter wavelength region [339]. The spectral response is sharper in the Cu-rich $CuInS_2$ thin film solar cells (Cu/In = 1.4 and 1.8) than in the Cu-poor $CuInS_2$ thin film solar cells (Cu/In = 1.0 and 1.2) at longer wavelength region [337]. The quantum efficiency response is lower in the cells fabricated on ceramic substrates than the one on glass substrates. This could be due to inferior quality of the $CuInS_2$ layers on ceramic substrates. The bubbling and cracks are seen on the ceramic substrates upon using for the deposition of $CuInS_2$ thin films due to effect of substrate temperature. The ceramic/Mo/$CuInS_2$/CdS/ZnO cells exhibit efficiency of 10.4% for which the Cu/In layers evaporated onto Mo coated ceramic substrates are sulfurized under sulfur vapor and under 5% H_2S/Ar

Table 8.40 CuInS$_2$ based thin film solar cells with different processes

Cell configuration	V_{oc} (mV)	J_{sc} (mA/cm^2)	FF (%)	η (%)	Ref.
Pt/CuInS$_2$(Virgin)/CdS(CBD)/In$_2$O$_3$	570	18.4	42	3.9	[339]
Pt/CuInS$_2$(KCN etched)/CdS(CBD)/In$_2$O$_3$	663	19.3	45	5.8	
Ceramic/CuInS$_2$/CdS/ZnO	740	10.1	70	10.4	[340]
Pt/CuInS$_2$/CdS(CBD)/In$_2$O$_3$	717	22.1	61	9.7	[341]
Pt/CuInS$_2$/CdS(CBD)/In$_2$O$_3$	700	24	63	10.5	[342]
Glass/Mo/CuInS$_2$/CdS(CBD)/In$_2$O$_3$	680	23.3	64.7	10.3	
Glass/Mo/CuInS$_2$:Ga/CdS(CBD)/In$_2$O$_3$	717	22.2	66	10.5	
CuInS$_2$(spray)/CdS/ZnO	705	22.7	68.5	10.9	[343]
	390	9.37	39	1.4	[344]
Glass/Mo/CuInS$_2$/CdS/ZnO/ITO	702	16.2	65.7	7.5	[345]
Glass/Mo/CuInS$_2$/CdS/ZnO (0.5 cm^2 ref. cell)	726	22	69	11.1	[346]
SLG/Mo/MgO/CuIn(Zn or Mg 0.5 at%)S$_2$	760	17.8	53	7.3	
SLG/Mo/CuInS$_2$(Mg 1 at%)/ZnO	794	18.8	69	9.7	
SLG/Mo/ZnO/CuIn(Zn or Mg 0.25 at%)S$_2$	792	19.2	69	10.5	
SLG/Mo/CuInS$_2$(Mg0.5 at%)/ZnO	807	19.8	68	10.9	

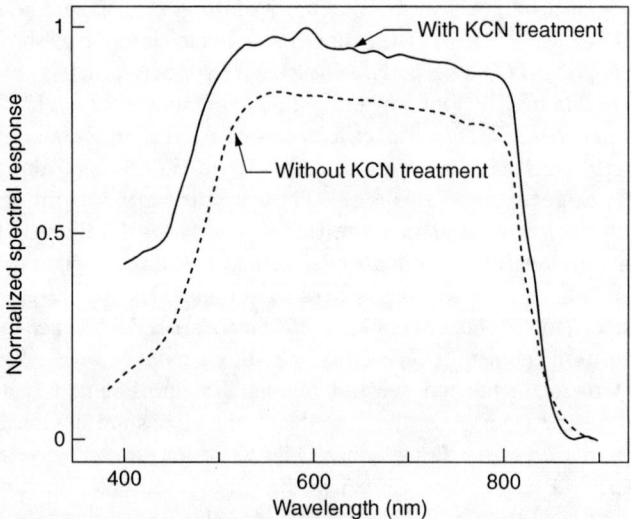

Figure 8.89 Quantum efficiency of CuInS$_2$/CdS/ZnO cells with, and without KCN treatment.

mixture gases at 500–550 °C to form CuInS$_2$ thin film absorber. The CuInS$_2$ layers show (112) preferred orientation [340].

The cells with an area of 1.7 mm^2 formed on Pt are mechanically well adherent but not on Mo coated glass substrates. The XPS analysis reveals that the concentration of In is higher than Cu at interface of Pt/CuInS$_2$. The In may be reacted with Pt that causes to make strong adhesion. The efficiency of cells steadily improves from 5.8, 9.7 to 10.5%. The cells with different configurations and KCN etched absorber exhibit more or less similar efficiencies: (a) Pt/1.8 µm thick CuInS$_2$/80–129 nmCdS(CBD)/150 nmIn$_2$O$_3$(Sput.) cells exhibit efficiency of 10.5%, (b) the cells with 0.9 µm thick CuInS$_2$ absorber layer and Cu/In ratio of 0.91 made on glass/Mo show efficiency of 10.3%, and (c) the cells formed with 27 nm thick Ga layer sand witched between CuInS$_2$ and Mo/substrates show efficiency of 10.5% (Table.8.40). An introduction of Ga layer between CuInS$_2$ and Mo strengthens adhesion of CuInS$_2$ well to the substrate. The cell parameters drastically decrease with increasing Ga thickness beyond 27 nm that is, from 27, 50, 90 to 130 nm. The FWHM of (112) peak decreases with increasing Cu/In ratio from ~1.3 to1.7 in the XRD spectrum. After KCN etching, the film composition ratio of Cu/In changes from 1.3–1.5 (Cu:In:S = 33:16:51) to 0.91 (Cu:In:S = 21:23:56) nonstoichiometry. Eventually after etching CuInS$_2$ layers in KCN solution, Cu/In composition ratio lowers to less than unity irrespective of Cu/In ratio in the Cu-rich CuInS$_2$ layers. The FWHM of (112) peak for CuInS$_2$ also decreases from 76.7 ± 0.5 to 72.2 ± 0.5″ evidencing as one of the probable tools to confirm elimination of CuS phase on the surface of the layers, which masks large grain sized CuInS$_2$ thin films [341,342].

The electroreflectance studies confirm the presence of CuS secondary phase by exhibiting band gaps of 1.42 and 1.47 eV in the as-grown CuInS$_2$ thin films. In addition the band gap of 1.5–1.52 eV for CuInS$_2$ is observed [337]. The Cu2p$_{3/2}$ and In3d$_{5/2}$ broad bands at 932.4 and 446 eV due to Cu$_2$O and In$_x$O$_y$ phases are observed in the as-grown CuInS$_2$ thin films, respectively. In fact, the binding energies of 932.0 ± 0.05 and 444.5 eV for Cu2p$_{3/2}$ and In3d$_{5/2}$ represent to pure Cu and In, respectively. After KCN etching the sample the broad bands disappear in the CuInS$_2$ thin films. The CuInS$_2$ thin films grown by co-evaporation are annealed under vacuum at 400 °C, followed by sulfurization under H$_2$S gas at 300 or 400 °C for 32 h. (004)/(200) XRD reflection peak appears at 32° for CuInS$_2$ and additional phases such as CuS secondary phase peaks appear at 29.3 and 31.8°. The CuS phase forms an island on the surface of the CuInS$_2$ layers. The CuInS$_2$/CdS/ZnO thin film solar cells with etched CuInS$_2$ absorber (Cu/In = 1) show efficiency of 10.9% [343]. Unlike reactive sputtering, two varieties of Cu–In–S targets are made by sintering Cu$_2$S and In$_2$S$_3$ powders under 5% H$_2$S/Ar and under Ar at 850 °C as target-A and target-B, which contain S/(Cu + In) ratios unity and much less than unity (0.5), respectively. The CuInS$_2$ layers are prepared by sputtering at substrate temperature of 250 °C, chamber pressure of 30 mtorr and RF power of 0.9 W/cm^2 from both targets. The layers obtained from target-A and -B are annealed under 5%H$_2$S/ Ar at 700 °C. They show (112) FWHMs of 0.27 and 0.17°, respectively. The cells with CuInS$_2$ (type-B) layer shows better performance than that of one with absorber (type-A). The typical 1.6 µm thick CuInS$_2$/0.25 µmCdS (CBD)/1 µm ITO cells show efficiency of 1.41%. This observation indicates that the nonstoichiometric layers annealing under H$_2$S give better results that means the sulfur diffusion occurs fairly in nonstoichiometric layers rather than stoichiometric layers. The spectral response is low at higher wavelength region in the low efficiency cells due to lower diffusion length [344]. Unlike sulfur based compounds, 1 μm thick Cu–In–O precursor films prepared onto Mo coated glass substrates by sputtering using $Cu_2In_2O_5$ powder target and $5\%O_2+Ar$ as sputtering gases. The precursor layers sulfurized under H_2S gas with a constant pressure of 56 torr and optimum H_2 pressure of 150 torr at substrate temperature of 550 °C for 1 h are used to form $CuInS_2$ thin films. After etching the layers in 10 wt% KCN solution, the composition ratio of Cu/In changes from 1.2 to 0.9 by eliminating CuS phase. The In_2O_3 phase is observed while sulfurizing the precursors, if only H_2S, Ar, and without H_2 gases are used. As shown in Fig. 8.90A, the photovoltaic efficiency of glass/Mo/$CuInS_2$/ CdS (50–100 nm)/ZnO(0.2 μm)/ITO(0.3 μm) thin film solar cells is 7.5%. The photovoltaic performance decreases if H_2 pressure is kept below 150 torr while growing $CuInS_2$ thin films. The similar way, the spectral response decreases with decreasing H_2 pressure from 150 torr to below, as shown in Fig. 8.90B [345]. The glass/Mo/ $CuInS_2$/CdS/ZnO(0.5 cm^2) thin film solar cells, in which $CuInS_2$ layers are prepared by sequential process of Cu and In sputtering and doped with either Zn or Mg. The entire CuIn:Zn or Mg precursor layers are sulfurized by evaporation of elemental sulfur. The Zn or Mg doping increases open circuit voltage irrespective of doping concentration, as shown in Table 8.40 [346].

The CuIn thin films grown by electrodeposition (ED) using $CuCl_2$ and $InCl_3$ solutions in the potential range from −0.8 to −1.0 V versus Ag/AgCl and carbon electrode as a counter electrode. The chemical solution pH is in the range of 3.22–1.74. The grown Cu–In precursor layers onto titanium (Ti) substrates are annealed under sulfur atmosphere at 500 °C for 1 h to obtain $CuInS_2$ films. The Cu/In ratio varies from 0.5 to 1.5 with varying HCl concentration in the solution. The Cu-rich films show single phase $CuInS_2$, whereas In-rich films show $CuInS_2$ and In_2O_3 phases. The as-grown $CuInS_2$ thin films contain composition of Cu/In = 1.79 and S/(Cu + In) = 0.92.

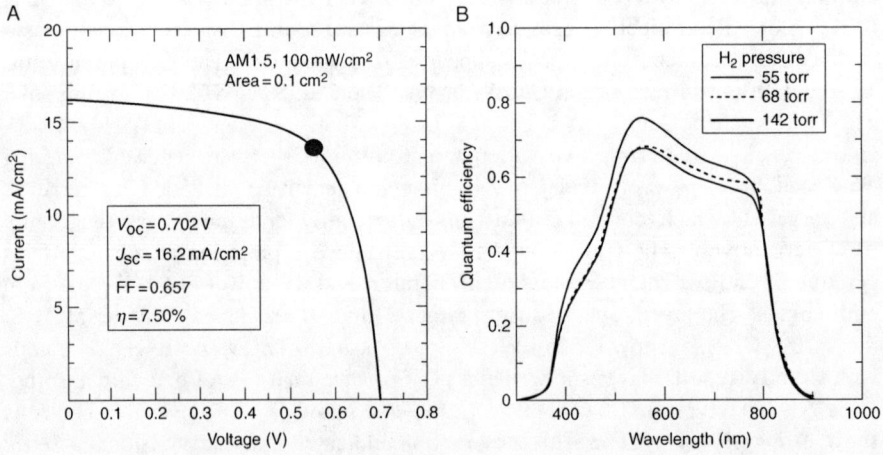

Figure 8.90 (A) *I–V* curve of $CuInS_2$ thin film solar cell, and (B) quantum efficiencies of $CuInS_2$ thin film solar cell with effect of H_2 pressure while depositing $CuInS_2$ thin films.

The Ti/CuInS$_2$(ED)/CdS(CBD)/ZnO:Al/Ag-past cells with etched CuInS$_2$ absorber demonstrate lower efficiency of 1.3% [347]. The glass/Mo/CuInS$_2$/CdS(CBD)/1 μm ZnO:F(Sput.)/Al cells with an area of 0.5 cm^2 exhibits low efficiency of 0.68% in which the CuInS$_2$ layers prepared by CVD method using single precursor of (PPh$_3$)$_2$CuIn(Set)$_4$ at substrate temperature of 390 °C. Prior to prepare cells, the absorber layer is etched in 1.5 M KCN solution for 1 min [348]. A set of cells is fabricated with CuInS$_2$ absorber grown by aerosol-assisted chemical vapor deposition (AACVD). The CuInS$_2$ layers are grown using single source precursor of (PPh$_3$)$_2$–CuIn(Set)$_4$ dissolving in toluene by spraying (AACVD) onto heated substrates at evaporation zone temperature of 120 °C and hot zone temperature of 395 °C in the reactor. The cells with CuInS$_2$ layer by AACVD witnesses poor performance, as presented in Table 8.41 [338]. The dark brown color CuInS$_2$ thin films with (112) preferred orientation grown by electrostatic spray assisted vapor deposition (ESAVD) using precursor copper nitrate, indium nitrate, thiourea, H$_2$O, and alcohol show band gap in the range from 1.41 to 1.48 eV for the variation of Cu/In ratio from 1.3 to 0.8. The surface is dense, smooth, and containing large grain sizes for Cu/In = 1.0. The CdS layer is also prepared by the same ESAVD technique using CdCl$_2$ and thiourea solutions. The back wall configured ITO/CdS(300 nm)/CuInS$_2$(2 μm) cells also exhibit low efficiency of 0.65% [349].

The photovoltaic parameters of CuInS$_2$/CdS/ZnO:In/ZnO cells, which consist of CuInS$_2$ absorber layers grown by spray pyrolysis with different ratios of Cu/In are given in Table 8.42. The cell efficiency is 1% for Cu/In ratio between 0.9 or unity, whereas the quality of the cells deteriorates for above the ratio of unity. The barrier

Table 8.41 Low efficiency CuInS$_2$ based thin film solar cells

Configuration of cell	V_{oc} (mV)	J_{sc} (mA/cm^2)	FF (%)	η (%)	Ref.
Ti/CuInS$_2$/CdS(CBD)/ZnO:Al/Ag-past	284	17.87	26.4	1.3	[347]
Glass/Mo/CuInS$_2$/CdS(CBD)/ZnO:F	304	5.25	29	0.68	[348]
ITO/CdS (300 nm)/CuInS$_2$ (2 μm)	205	10.4	30	0.65	[349]
SLG/Mo/CuInS$_2$(AACVD)/CdS/ZnO:F/Al	309	12.5	37	1	[338]
SLG/Mo/CuInS$_2$(AACVD)/CdS/i-ZnO/ITO/Ni-Al	412	7.2	45	1	

Table 8.42 PV parameters of spray deposited CuInS$_2$ thin film solar cells

Cu/In	V_{oc} (mV)	J_{sc} (mA/cm^2)	FF (%)	η (%)	R_s (Ω-cm^2)	p (cm^{-3})	ϕ_b (eV)	Ref.
0.98	336	0.26	27	–	3k	1.8×10^{17}	0.973	[350]
1.09	443	5.5	37	–	270	2.1×10^{17}	1.065	
1.17	380	3.4	38	–	220	8.7×10^{16}	0.994	
Bilayer	440	8.25	27.8	2.66	–	–	–	[42]

heights of the cells can be obtained from intercepts of the plots of V_{oc} versus T, as shown in Fig. 8.91 [350]. The growth process of spray pyrolysis is borrowed from the fabrication of CuInSe$_2$ thin film solar cells in order to develop all spray deposited CuInS$_2$ thin film solar cells. The glass/ITO/CdZnS/CuInS$_2$/Au thin film solar cells show efficiency of 2.7%. The explanation given for low efficiency spray deposited CuInSe$_2$ cells suits well to all spray deposited CuInS$_2$ based thin film solar cells [42].

As witnessed, the segregation of CuS phase enhances grain sizes and carrier concentration in the Cu-rich CuInS$_2$ layers. In order to avoid generation of CuS phase and toxic process of KCN etching, the Cu-poor CuInS$_2$ growth is chosen. However, the growth process of Cu-poor CuInS$_2$ realizes small grain sizes and low carrier concentration. In order to mitigate crises, Na doping is applied. The incorporation of Na in the absorber layers leads to an increase carrier concentration and enhance grain sizes of layers. On the other hand, the band gap of CuInS$_2$ can be increased by incorporation of Ga into it that causes to increase V_{oc} in the cells. The V_{oc} increases from 760 to 802 mV with increasing Ga from 0.0 to 0.18%. 0.1 cm^2 area SLG/Ti/Mo/CuInS$_2$:Na/80 nmCdS/10 nmZnO/500 nmITO and CuInGaS$_2$:Na based thin film solar cells without KCN etched absorber exhibit efficiency of 10.6 and 11.2%, respectively (Table 8.43). The CuInS$_2$ thin film absorber used in the cells is prepared by hybrid sputtering technique. In this process, SiO$_2$ is first deposited onto soda-lime glass substrates as an alkaline barrier layer on which Ti (500 nm)/Mo (500 nm) layers are grown. The Cu/In stacked layers by RF sputtering, Na$_2$O$_2$ layer by evaporation from effusion cell and In–S layer by reactive sputtering with H$_2$S/Ar gas are sequentially formed. They are annealed under 5%H$_2$S/Ar atmosphere at 500 °C for 1 h to make 2 μm thick Na doped CuInS$_2$ layers with Cu/In = 0.9 and Na/(Cu + In) = 0.02. The XRD analysis reveals well resolved peaks of (204) and

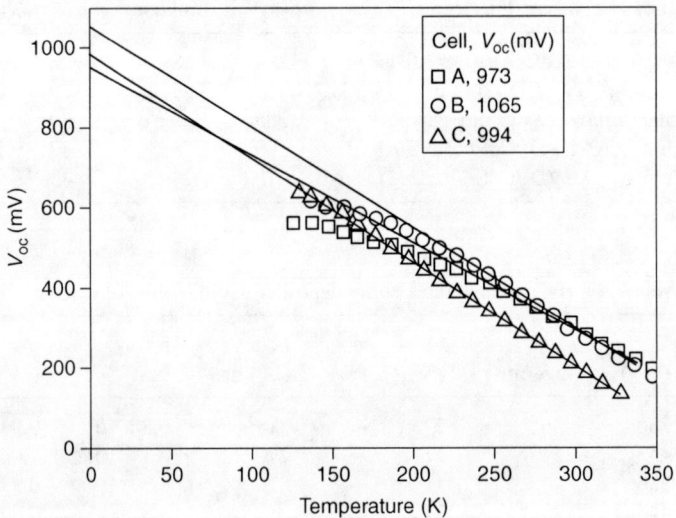

Figure 8.91 Plots of open circuit voltage (V_{oc}) versus temperature (T).

Table 8.43 PV parameters of $CuInS_2$ based thin film solar cells

Cell configuration	V_{oc} (mV)	J_{sc} (mA/cm^2)	FF (%)	η (%)	R_s (Ω-cm^2)	Lattice mismatch %	Ref.
SLG/Ti/Mo/CuInS$_2$:Na/CdS/ZnO/ITO	760	20.2	68.8	10.6	–	–	[351]
SLG/Ti/Mo/CuIn$_{0.82}$Ga$_{0.18}$S$_2$:Na/CdS/ZnO/ITO	802	20.9	66.7	11.2	–	–	
Glass/Mo/CuInS$_2$/CdS/ZnO:Al/ITO (without Na)	235	18.7	27	1.18	–	–	[352]
Glass/Mo/CuInS$_2$/CdS/ZnO:Al/ITO (with Na)	665	18.9	52.6	6.61	–	–	
Conventional	461	26.9	68.5	5.66	5.1	3.2	[353]
SLG/Mo/lowρ/highρ0.2 μmCuInS$_2$/highρ/lowρ CdS/In	580	30.6	69.7	8.25	4.3	2.8	
0.5 cm^2	729	21.8	71.7	11.4	–	–	[355]
5 × 5 cm^2	5.09 V	19.1	65.9	9.2	–	–	

(220), indicating chalcopyrite structure and $NaInS_2$ phase in the $CuInS_2$ layer, whereas no such kind of split is occurred in the $CuInGaS_2$. In the case of $CuInGaS_2$ layer formation, Ga layer is deposited by evaporation onto InS then Cu layer and the rest of the process is as usual [351]. The glass/Mo/CuInS$_2$/CdS/ZnO:Al/ITO cell with an active area of 0.06 cm^2 yields low efficiency of 1%, whereas the efficiency of cell increases from 1 to 6% with Na doping and without KCN treatment. Na occupies Cu site to form $NaInS_2$ phase, which increases with decreasing Cu/In ratio but no Na is observed for Cu-poor layers. In the In-rich films, the $CuIn_5S_8$ phase increases but $NaInS_2$ phase remains at the same level. The Cu/In = 0.9 is optimal for high efficiency $CuInS_2$ cells. For this analysis the Cu–In–S or Cu/InS mixture precursors prepared onto Mo coated glass substrates by reactive sputtering with H_2S gas are annealed under 5%H_2S gas + Ar for 2 h at 550 °C to incur $CuInS_2$ layers. Na is incorporated into $CuInS_2$ by evaporating Na_2S whereby the optimum thickness is 200–1000 Å [352].

A schematic bilayer based $CuInS_2$ cell structure (glass/Mo/lowρ/highρ0.2 μm CuInS$_2$/highρ/lowρ CdS/In grids) is depicted in Fig. 8.92. The bilayer process allows to avoid segregation of CuS and toxic process of KCN etching. The bilayer and single layer structured cells show efficiencies of 8.25 and 5.66%, respectively. The bilayer structure, which has band gap grading, not only suppresses CuS phase but also probably strengthens to form back surface field and junction electric field those enhance efficiency of cell. Unlike co-evaporation, the S/In/Cu stacked layers sequentially deposited with excess sulfur onto Mo coated glass substrates are heat-treated at substrate temperature of 250 °C for 1 h to convert chalcopyrite $CuInS_2$ thin films. The intensities of XRD peaks decrease for the stack layers heat-treated at 300 °C, whereas layers treated at 200 °C for 2 h show multiphases. The band gaps of the layers are 1.48 and 1.53 eV for Cu/In ratio of 1.3 and 0.99, respectively. The p-type conductivity has

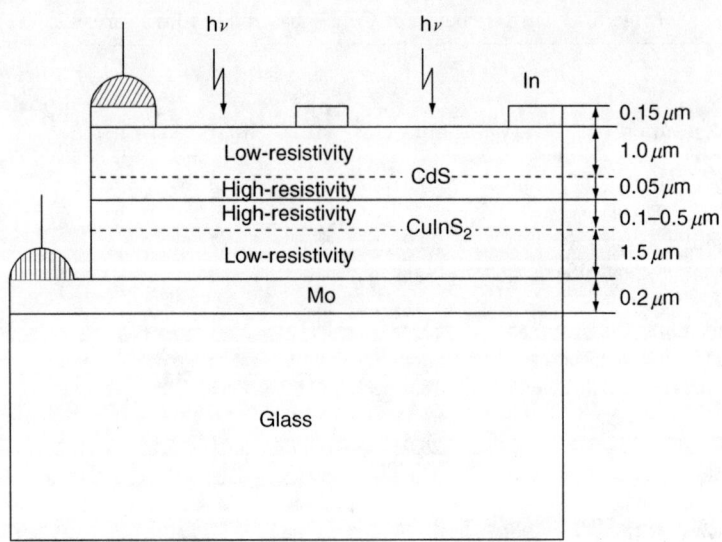

Figure 8.92 CuInS$_2$ cell with double layer absorber and window.

started to appear for Cu/In ratio of 0.99 and continue beyond the value, whereas the conductivity becomes n-type below the ratio of 0.99. The resistivities of CuInS$_2$ thin films are 1.608×10^2 and 5.587×10^{-2} Ω-cm for Cu/In ratios of 0.99 and 1.3, respectively. The bilayers may be chosen in the Cu/In range of 0.99–1.3. The typical cubic n-CdS doped with In 1.5 at% and undoped n-CdS layers grown by electron beam evaporation at substrate temperature of 150 °C show resistivities of 4×10^{-3} and 10^3 Ω-cm, respectively. The band gaps of 2.49 and 2.43 eV are observed for CdS:In (0.5 at%) and CdS layers, respectively. The transmission of CdS layers is in the range of 80%. The performance of cell degrades for below or above In doping percentage of one in CdS, which is optimal. On the other hand, 0.2 μm thick CdS layer is optimal for high resistive CuInS$_2$ thin films in bilayer cell structure. The results of single layer CuInS$_2$ ($\rho = 5.59 \times 10^{-2}$ Ω-cm)/CdS:In(1%) and bilayer cells are presented in Table 8.43 [353].

The Cu/In sputtered with Cu/In = 1.4 layers are sulfurized by heating in elemental sulfur atmosphere at 500 °C to form CuInS$_2$ and etched in KCN then chemically treated the surface. After mechanical scribing, ZnO is deposited by sputtering and again mechanically scribed. The glass/Mo/CuInS$_2$/ZnO module size of 125×65 cm^2 shows efficiency of 8.5% [354]. The fabrication of CuInS$_2$ based thin film solar cell minimodule; 0.4–1 μm thick Mo coated glass substrates are patterned by Nd:YAG laser with wavelength of 1064 nm as P$_1$ pattern. Cu and In layers sputtered to have Cu/In = 1.8 that become CuIn$_2$ and Cu at room temperature and the Knudsen cell supplies sulfur vapor with beam pressure of 1×10^{-2} Pa. The sulfurization can be done at substrate temperature of 500–600 °C to form CuInS$_2$ but not optimal. The CuInS$_2$ layers formed by this process show sheet resistance of 1×10^6 $\Omega/\square$, carrier concentration of 1×10^{16} cm^{-3}, and band gap of 1.52 eV.

The encouraging Cu/In ratio is between 1.4 and 1.8. The etching rate of CuS in KCN is 5 orders of magnitude higher than that of absorber. The CdS buffer layer with $n_e = 1 \times 10^{19}$ cm^{-3} is formed by CBD then mechanical scribing P$_2$ is done. The i-ZnO and conducting n-ZnO layers with $n = 1 \times 10^{20}$ cm^{-3} and sheet resistance of 9 Ω/□ are formed by sputtering technique. Finally scribing P$_3$ separates cells. 5×5 cm^2 minimodule contains 7 cells and each optimum cell width is 6 mm. The minimodules are encapsulated by laminating glass on the front side of the cells. They show efficiency of 10.3%. The thermally activated recombination is in the $1 \leq A \leq 2$ range and the tunneling enhanced recombination occurs for $A > 2$. The best CuInS$_2$/CdS(CBD)/i-ZnO/n-ZnO cells show efficiency of 11.4% for which FWHM of A$_1$ mode in the Raman spectra is < 3.7 cm^{-1}. Beyond this value the cell properties deteriorate. In the CuInS$_2$ minimodule fabrication, 11 process steps take place, whereas the total process steps are twenty in the Si minimodules [335].

In order to grow CuInS$_2$ thin films, the CuIn sputtered layers are sulfurized by elemental sulfur vapor which is done by two different ways one is conventional thermal process at ramp of 2–3 °C/s and holding at 600 °C for 3 min. In another approach the layers are sulfurized by rapid thermal processing at ramp of 10 °C/s and holding at 580 °C for 8 min. The CdS buffer layer by CBD, 0.1 μm thick ZnO, and 0.4–1.6 μm thick Ga-doped ZnO by sputtering results in sheet resistance of 3–15 Ω/□ are successively grown onto CuInS$_2$ layers. The SLG/Mo/CuInS$_2$(2–3 μm)/CdS/ZnO(0.5–1.7 μm)/ZnO:Ga(0.4–1.6 μm) cells with 0.5 cm^2 shows efficiency of 11.4%, whereas cells on 5×5 cm^2 area that is, minimodules with seven series show efficiency of 9.7%. The efficiency of large area cells is lower because of poor fill factor that is caused by high series resistance. The cells with 4 mm width and 0.4 mm interconnect show the best efficiency of 9.5% for ZnO thickness of 900 nm. On the other hand, cells with 7 mm width and 0.8 mm interconnects need ZnO with thickness of 1300 nm for the best efficiency of 9.5%, which had seven series connections. The cells tested under air or nitrogen at 85 °C for 2000 h show no degradation. A typical I–V curve of 9.66% efficiency CuInS$_2$ based minimodule is shown in Fig. 8.93 [355–358].

8.4.1 CuInS$_2$ Cells with Variety of Buffer Layers

After annealing under air at 200 °C for 5 min and light soaking for 10 min the glass/Mo/CuInS$_2$/ZnS/i-ZnO/ZnO:Ga/Ni/Al solar cell with an area of 0.5 cm^2 shows efficiency of 10.4% (Table 8.44). The effect of light soaking is not predominant on air anneal device. The CuInS$_2$ absorber layers are prepared by sequential sputtering of Cu and In then reacted with elemental sulfur by rapid thermal processing. The ZnS buffer layer is prepared onto KCN etched CuInS$_2$ layer using *base media* that contains zinc sulfate (0.15 M), thiourea (0.6 M), ammonia (1.1 M), and pH of 10–11 by CBD technique at 75 °C for 15 min, followed by deposition of ZnO by RF sputtering technique [359]. The ZnS layers prepared onto CuInS$_2$ thin film by CBD method at 70 °C in *acidic media* using chemical solutions of 0.5 M thioacetamide, 0.01 M ZnSO$_4$, 0.3 M acetic acid, and 0.01 M HCl for about 40 min. Prior to deposition of ZnS, the CuInS$_2$ thin film is etched in 0.1 M KCN solution at 40 °C for 2 min. The grown

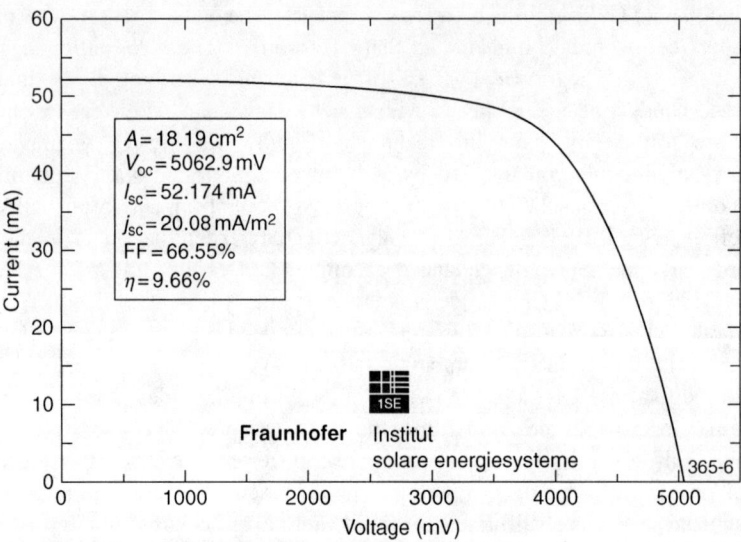

Figure 8.93 $I–V$ curve of glass/Mo/CuInS$_2$/CdS/i-ZnO/ZnO:Ga/Ni–Al minimodule.

Table 8.44 CuInS$_2$ based thin film solar cell with variety of buffer layers

Cell configuration/buffer	V_{oc} (mV)	J_{sc} (mA/cm^2)	FF (%)	η (%)	Ref.
ZnS by base media	700	22.5	65.8	10.4	[359]
CdS (Ref. cell)	708	21.4	68.1	10.3	
ZnS by acidic media	613	19.6	44.3	5.3	[360]
CdS	699	11	66	10.2	
Glass/Mo/CuInS$_2$/ZnSe/n-ZnO	594	21	58	7.1	[361]
Glass/Mo/CuInS$_2$/n-ZnO	220	16.8	36	1.3	
Glass/Mo/CuInS$_2$/CdS/n-ZnO	686	19.1	67	8.7	
CuInS$_2$/ZnO(CBD)	360	19.64	55	3.8	[362]
CuInS$_2$/CdS(CBD)	676	19.65	65	8.6	
Glass/ZnO/InS(CBD)/CuInS$_2$ (SP)	456	14.6	43	2.9	[363]
Glass/ZnO/CdS(SP)/CuInS$_2$ (SP)	443	6.7	37	1.0	
In(OH)$_x$S$_y$	685	10.7	61.8	9.1	[364]
In$_x$(OH,O)$_y$	490	10.6	49.4	5.2	
In(OH,S)	735	23.2	67	11.4	[365]
CdS	714	22.3	68	10.8	

ZnS layers have mixed phases of cubic and hexagonal as well as mixed with oxygen, CNS and OH components. The grown ZnS can be generally written as Zn(O,SOH) and its band gap is close to 3.6 eV. The glass/Mo/CuInS$_2$/ZnS 64 nm/ZnO cell exhibits efficiency of 5.3%, $R_s = 5.67$ Ω-cm^2 and $R_p = 343$ Ω-cm^2 (Table 8.44). The thickness of ZnS layer can be increased by depositing longer time but the contents of oxygen and Zn increase. The composition of buffer layer is Zn-44.86% and S-55.14% for optimum cells. As the thickness of buffer layer is increased, the efficiency of cell slips to lower due to decrease of short circuit current even though there is a small increase in V_{oc} [360]. The thing is that the cells made with ZnS by acidic bath show lower efficiency.

The glass/Mo/CuInS$_2$/ZnSe/n-ZnO/Ni–Al solar cells with ZnSe grown by MOCVD show efficiency of 7.1%, which is less than that of CdS reference cell (8.7%). The same cell without ZnSe buffer shows lower efficiency of 1.3%. The ZnSe buffer is developed by MOCVD using DMZN-TEN and DtBSe precursors at optimum substrate temperature of 250 °C for 6 min. The quality of CuInS$_2$ layer degrades beyond this substrate temperature therefore the efficiency of the cell decreases. The efficiency of cell also slightly decreases with light soaking [361]. Figure 8.94 shows I–V and quantum efficiency of CuInS$_2$/CdS and CuInS$_2$/ZnO thin film solar cells. The efficiency of 3.8% for CuInS$_2$/ZnO thin film solar cells is less than that of CdS based cell (8.6%) due to higher series resistance. Even though the cutoff region in the spectral response curve starts at early in ultraviolet region for ZnO based cells. The ZnO layer for the cell is grown by CBD method employing ZnSO$_4$, H$_2$O, and NH$_3$ solutions at bath temperature of 65 °C. The XRD analysis reveals that ZnO is hexagonal structure evidencing diffraction reflections of (101), (002), and (100) and lattice constants of $a = 3.249$ and $c = 5.205$ Å [362].

The spray deposited (SP) CuInS$_2$ thin films on buffer covered on ZnO:Ti coated glass substrates are used as thin film solar cells. The CuInS$_2$ layers are prepared by spray using chemical solution containing of CuCl, InCl$_3$, and SC(NH$_2$)$_2$. The back wall configured glass/ZnO/InS(CBD)/CuInS$_2$ cells with spray deposited CuInS$_2$ thin film absorber show efficiency of 2.9% but cells with all spray deposited CuInS$_2$ and CdS buffer layers being 1%, as shown in Table 8.44. In spray pyrolysis, there may be diffusion of elements from one to another layer or vice versa while growing layers at high temperature that causing lower efficiency in the cells. It is obvious that the CuInS$_2$ thin films grown by spray pyrolysis have band tailing that causes to make optical transition between defect to valence band or conduction band to defect or vice versa unlike direct band-to-band transition. The hoping conduction in the sprayed CuInS$_2$ thin films confirms band-tailing nature [363]. The In$_x$(OH,O)$_y$ layer first forms

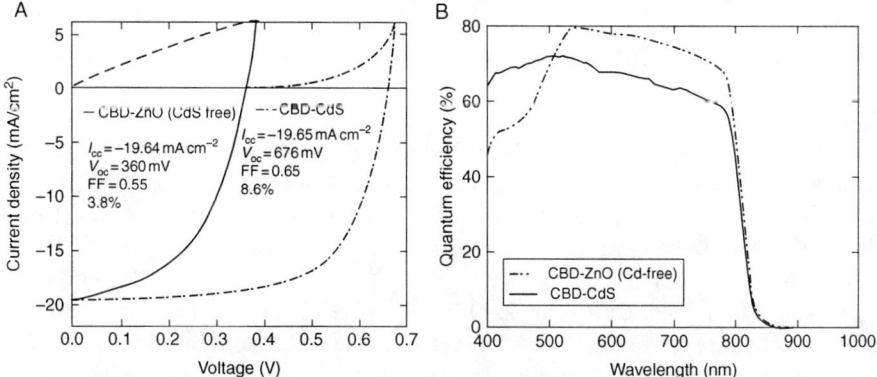

Figure 8.94 (A) I–V and (B) QE curves of CuInS$_2$/CdS and CuInS$_2$/ZnO.

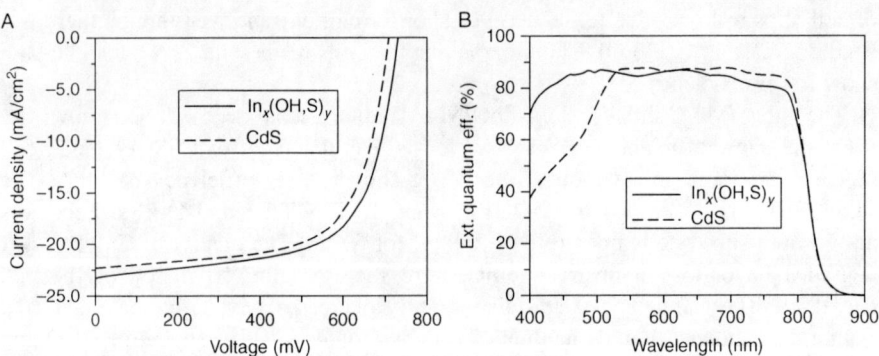

Figure 8.95 (A) *I–V* curves, and (B) QE of CuInS$_2$/In$_x$(OH,S)$_y$ and CuInS$_2$/CdS cells.

over the time of 10 min and then the In(OH)$_x$S$_y$ layer starts to form those are confirmed by elastic recoil detection analysis. The latter film gives indirect band gap of 2.2 eV and its elemental composition is In—36.5, S—26.7, O—14.3, H—16.7, and C—5.9 at%. The KCN etched CuInS$_2$/In based buffer/ZnO/ZnO:Al/Ni–Al thin film solar cells with In(OH)$_x$S$_y$ and In$_x$(OH,O)$_y$ buffer layers yield efficiencies of 9.1 and 5.2%, respectively (Table 8.44) [364]. 3 μm thick CuInS$_2$ thin films grown by thermal vacuum co-evaporation with excess sulfur atmosphere of 15 times with respect to metals onto Mo coated glass substrates at substrate temperature of 630 °C. The KCN etched CuInS$_2$ sample is treated by In^{3+} in In$_2$(SO$_4$)$_3$/NH$_3$ solution on which In(OH,S) (CBD), 90 nm i-ZnO, 300 nm ZnO:Al, and Ni/Al grids are sequentially deposited. Figure 8.95 shows *I–V* curves and QE of glass/Mo/CuInS$_2$/In(OH,S)/i-ZnO/ZnO:Al/Ni–Al and reference cells. The In(OH,S) based cell shows efficiency of 11.4% greater than that of the reference cell (10.8%), as shown in Table 8.44. As anticipated, the quantum efficiency of In(OH,S) based cell is higher in ultraviolet region, as compared to that of CdS reference cell [365].

8.4.2 n-CuInS$_2$ Based Thin Film Solar Cells

The *p*-CuI is grown by successive ionic layer adsorption and reaction (SILAR) technique onto Cu tape/*n*-CuInS$_2$ thin films and then treated with iodine solution by means of dipping in iodine solution for 5 s. The solution contains 630 mg of iodine in 100 ml ethanol. The CuI thin film shows γ-phase zincblende structure with (111) preferred orientation. The surface structure of CuI thin film by SEM looks like fibrous. After iodine treatment it becomes an increased grain structure. Prior to complete cell structure, the photovoltage of Cu-tape/*n*-CuInS$_2$/*p*-CuI junction with function of wavelength is measured, as shown in Fig. 8.96. The band gaps of 3.0 and 1.45 eV for *p*-CuI and *n*-CuInS$_2$ at shorter and longer wavelength regions can be observed. The efficiency of Cu-tape/*n*-CuInS$_2$/*p*-CuI/i-ZnO/ZnO:Ga/Ni–Al cell increases from 2.5 to 4% with iodine treatment (Table 8.45) [366]. The cells without CuI layer show more or less similar currents of one with CuI but lower voltage. The *I–V* curves of Cu-tape/*n*-CuInS$_2$/i-ZnO/ZnO:Ga/Ni–Al are as same as standard *I–V* curves indicating there must be *p–n*

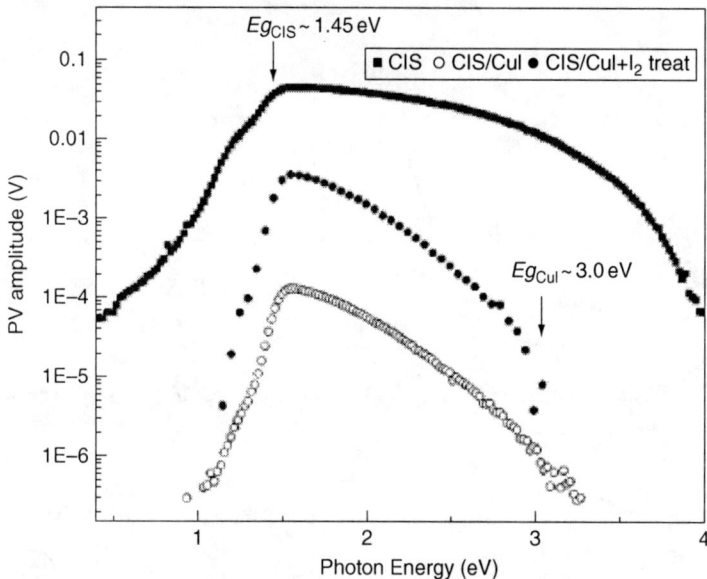

Figure 8.96 Photovoltage versus wavelength as a spectral response of n-CuInS$_2$/p-CuI stack.

Table 8.45 PV parameters of n-type CuInS$_2$ based thin film solar cells

Treatment/cell config. Area	V_{oc} (mV)	J_{sc} (mA/cm^2)	FF (%)	η (%)	Ref.
Without Iodine	440	14.55	38	2.5	[366]
With Iodine	520	15	53	4	
Cu/CuInS$_2$/CuIn$_5$S$_8$/CuInS$_2$/CuI/ZnO/ZnO:Al (3.7 cm^2)	657	20.9	66.1	9.1	[369]
Cu/CuInS$_2$/CuIn$_5$S$_8$/CuInS$_2$/CuI/ZnO/ZnO:Al (21 cm^2)	3.7 V	67.2	60	7.1	

homojucnction in the cells (Fig. 8.97A). By adding CuI to the cell structure, the recombination mechanism probably is tunneling amid it demonstrates higher A values, whereas the cells without CuI that is, with ZnO perform Shockley-Read-Hall (SRH) recombination in the p–n junction. The QE of cells with CuI and without CuI is shown in Fig. 8.97B. The cutoff regions at longer wavelength region reveal CuInS$_2$ (Eg = 1.45 eV) and CuIn$_5$S$_8$ (1.31 eV) phases [367].

The fabrication of Cu tape based CuInS$_2$ minimodules by role to role is shown in Fig. 8.98. Unlike CuInS$_2$ deposition on Cu-tape, 80–100 μm thick and quasi-endless tape itself is used as a Cu source for CuInS$_2$. Prior to deposition of 0.7 μm thick In layer onto 1 cm width Cu-tape by electrodeposition, which is chemically cleaned then sulfurization of Cu–In is carried out in the presence of sulfur vapor at substrate temperature of 580 °C (750 K) whereby CuIn$_5$S$_8$ layer forms in which first CuIn alloy forms.

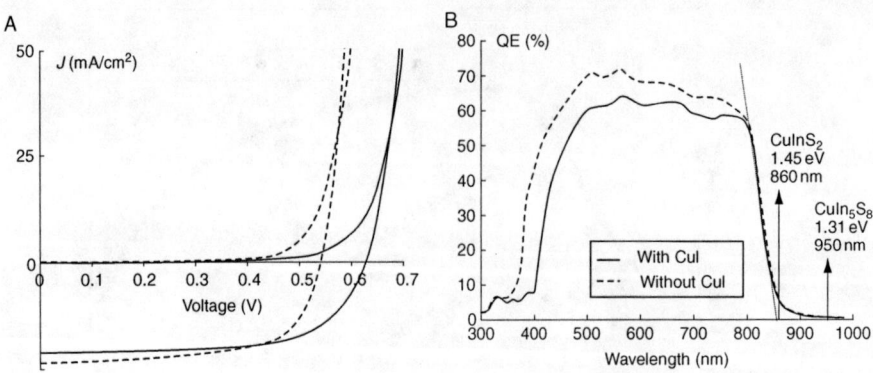

Figure 8.97 (A) *I–V* curves, and (B) QE of Cu tape/CuInS$_2$/CuI/ZnO/*n*-ZnO cell with CuI, and without CuI.

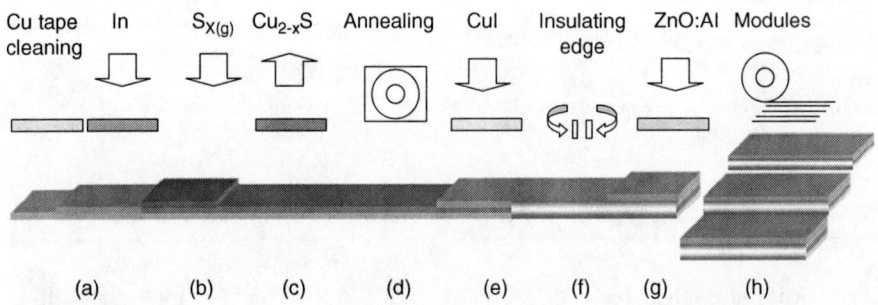

Figure 8.98 Schematic process of Cu tape/CuInS$_2$/CuI/ZnO/*n*-ZnO minimodules.

The sulfurization is continued until to form CuInS$_2$ + Cu$_{2-x}$S. The probable reaction paths are Cu + 5In + 4S$_2$ → CuIn$_5$S$_8$ and 4Cu + CuIn$_5$S$_8$ → CuInS$_2$ + 2Cu$_{2-x}$S + InS↑ at 480 and >480 °C, respectively. After cooling down reel to room temperature, the absorber layer is cleaned by KCN solution to remove Cu$_{2-x}$S phase then the reel is annealed under vacuum for 30 min at reasonable temperature. A 70 nm thick *p*-CuI is deposited by spray pyrolysis technique at substrate temperature of 80 °C using chemical solution of 0.4 g of CuI dissolved in 80 ml acetonitril. In order to make roof tiles, the glassy insulating layer using nanomere solution masks edges of the tape. 100 nm thick insulating TCO layer is deposited, followed by 1 μm thick ZnO:Al (2 at%) conducting layer by DC sputtering at substrate temperature of 165 °C. Six solar cells with 22 cm length are combined as roof tile pattern occupying 1 mm width to encapsulate with plastic foils. The output voltage and currents depend on number of strips and its width, respectively. In this deposition process, the tape velocity is 6.5 m/min that gives totally 4 m^2/h. Computer controls the substrate temperature, gas flow, and pressure of the chamber, *etc*. The CuI had band gap of 3.0 eV, Hall mobility of 4 cm^2/Vs and carrier concentration of 10^{20} cm^{-3}. The typical equipment used for roll-to-roll process is shown in Fig. 8.99 [368]. The Cu/CuInS$_2$/CuIn$_5$S$_8$/CuInS$_2$/

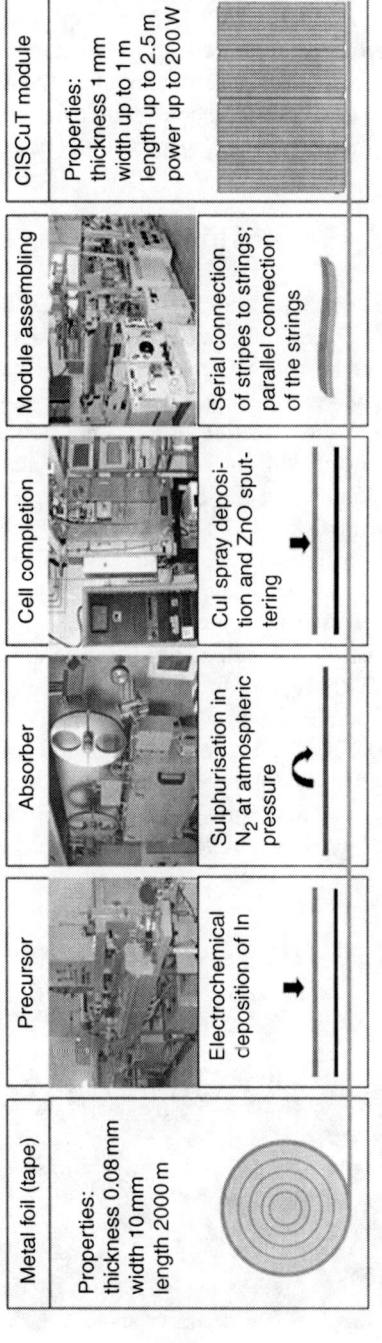

Figure 8.99 Role to role process of CIS/Cu tape based modules.

CuI/ZnO/ZnO:Al cell with areas of 3.7 and 22 cm^2 show efficiencies of 7.1 and 9.1%, respectively. In the absorber close to surface, the CuInS$_2$ and CuIn$_5$S$_8$ phases had compositions of Cu:In:S = 25.5:25:49.9 and 8.5:35:56.5, respectively [369]. There is a less probability to reach high efficiency in the n-CuInS$_2$ based thin film solar cells because the hole mass is higher than that of the electron therefore the hole mobility or conductivity is low, as discussed in Chapter 6. The hole participates as a minority carrier in the n-CuInS$_2$, unlike electron transport in the p-type CuInS$_2$ based thin film solar cells.

8.5 CuIn(Se$_{1-x}$S$_x$)$_2$ Based Thin Film Solar Cells

The corning glass/(CuInSe$_{1-x}$S$_x$)$_2$/CdS(CBD)/30 nmZnO/120 nmITO cell exhibit efficiency of 7.41% and series resistance of 1.67 Ω-cm^2 in which the CuIn(Se$_{1-x}$S$_x$)$_2$ (CISS) absorber has been grown by solution based hydrazine method employing spin coating. The remaining part of the layers is grown by standard conventional method. The SEM cross section of CISS cell is shown in Fig. 8.100 that the grown CISS absorber layer with S/(S + Se) = 0.13 contains band gap of 1.1 eV and small grain sizes of 45 nm. The Cu$_2$S and In$_2$Se$_3$ with excess Se powders are separately dissolved in hydrazine (N$_2$H$_2$) and stirred for several days and filtered. The CISS layers are spin coated onto Mo coated glass substrates using these solutions, followed by annealing under N$_2$ atmosphere at 350 °C for 30 min. The thickness of CISS layers is increased by repeating the same process for several times. The S/(Se + S) can be varied by manipulating combinations of Cu$_2$S and In$_2$Se$_3$ solutions [370]. The Cu (3100 Å) and In (5900 Å) thick layers deposited onto Mo coated glass or quartz substrates by vacuum evaporation and sulfurized under Ar containing 5%H$_2$S at 550 °C for 2 h to obtain 2 µm thick CuInS$_2$ layer on which 5 µm thick Se is evaporated at RT followed by annealing at 550 °C for 10 min under vacuum at chamber pressure of 10^{-6} torr. The Se composition in the CuIn(Se$_{1-x}$S$_x$)$_2$ is varied by adding additional Se and/or monitoring annealing conditions. No

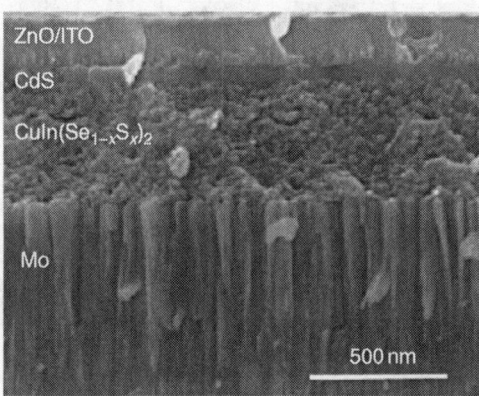

Figure 8.100 SEM cross section of CISS/CdS/ZnO/ITO thin film solar cell.

secondary phases are observed in the films by XRD analysis. As-grown CuIn$(Se_{1-x}S_x)_2$ thin films with $x=0.6$ are etched with KCN to sack Cu_xS or Cu_xSe phases off to use in the solar cells. After KCN etching as-grown absorber, the Cu/In ratio turns out from 1.1 to 0.9. The as-grown layers show lower resistivity of 1 Ω-cm because of influence of Cu_xS and Cu_xSe phases. The CuIn$(Se_{1-x}S_x)_2$/80 nmCdS(CBD)/150 nmIn$_2$O$_3$(RF sputtering) cell with an area of 1.8 mm^2 exhibits efficiency of 8.15% (Fig. 8.101A). It is learnt from the XPS analysis that the x varies from 0.75 to 0.4 from top to bottom of the CISS layer. The similar observation is noticed from the quantum efficiency measurements for CISS ($x=0.6$) cell that the spectral response changes from 850 to 1000 nm that is close to the change in band gaps from 1.38 (900 nm) to 1.2 eV (1030 nm) corresponding to the x from 0.75 to 0.4 in the CuIn$(Se_{1-x}S_x)_2$ thin films, as shown in (Fig. 8.101B) [371].

The sulfur composition in the quaternary CuIn$(Se_{1-x}S_x)_2$ layers can be varied in the interest range of x from 0 to 1 for the Cu-rich layers, whereas it is difficult to achieve required range of x in the In-rich layers with irrespective of Se/S ratio because of nonparticipation of higher sulfur concentration into the nonstoichiometric (In-rich) films thus the precise controllable quaternary system could not be achieved. In the Cu-rich samples, the participation of Cu_xSe or Cu_xS nucleates to accommodate more sulfur in the system. The glass/Mo/Cu-rich CuInS$_2$/Cd$_{0.9}$Zn$_{0.1}$S:Ga/ZnO:Al and glass/Mo/In-richCuIn$(Se_{0.58}S_{0.42})_2$/CdS(CBD)/ZnO:Al cells with an area of 23 mm^2 exhibit efficiencies of 6.1 and 10.1%, respectively (Table 8.46). The I–V curve of glass/Mo/In-rich CuIn$(Se_{0.58}S_{0.42})_2$(Cu/In = 0.8)/CdS(CBD)/ZnO:Al is shown in Fig. 8.102A. No photoactivity is observed in the high In-rich CuInS$_2$ based thin film solar cells. The spectral response curves of Cu- and In-rich CuIn$(Se_{1-x}S_x)_2$/80 nmCdS(CBD)/ZnO:Al thin film solar cells are depicted in Fig. 8.102B. Surprisingly, the responses show band gaps of 1.1 and 1.36 eV for In-rich and Cu-rich CuIn$(Se_{1-x}S_x)_2$ thin films, respectively. As opposing that the In-rich samples obviously exhibit higher band gap than that of Cu-rich samples. In the present case, the

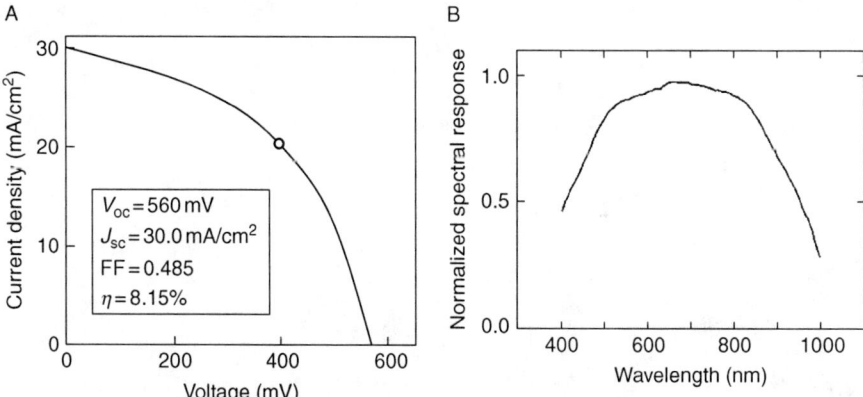

Figure 8.101 (A) I–V curve and (B) quantum efficiency of CuIn$(Se_{1-x}S_x)_2$/80 nm CdS (CBD)/150 nm In$_2$O$_3$ thin film solar cell ($x=0.6$).

Table 8.46 Photovoltaic parameters of CuIn(Se$_{1-x}$S$_x$)$_2$ thin film solar cells

Cell configuration	V_{oc} (mV)	J_{sc} mA/cm^2	FF (%)	η (%)	Ref.
Glass/Mo/(CuInSe$_{1-x}$S$_x$)$_2$/CdS(CBD)/30 nmZnO/ 120 nmITO	500	28.04	53	7.41	[370]
Glass/Mo/CuIn(Se$_{0.4}$S$_{0.6}$)$_2$/80 nmCdS(CBD)/ 150 nmIn$_2$O$_3$(RF)	560	30	48.5	8.15	[371]
Glass/Mo/In-richCuIn(Se$_{0.58}$S$_{0.42}$)$_2$/CdS(CBD)/ ZnO:Al	492	31.2	66	10.1	[372]
Glass/Mo/CuInS$_2$/Zn$_{0.1}$Cd$_{0.9}$S:Ga/ZnO:Al	549	19.8	56	6.1	
Glass/Mo/CuIn(Se$_{1-x}$S$_x$)$_2$/CdS/ZnO (0.1 cm^2)	722	23.8	67	11.5	[373]
Glass/Mo/CuIn(Se$_{1-x}$S$_x$)$_2$/CdS/ZnO (0.1 cm^2)	718	22.3	66	10.6	[374]
Glass/Mo/CuIn(Se$_{1-x}$S$_x$)$_2$/CdS/ZnO (0.1 cm^2)	692	21	68	9.9	
15 × 15 cm^2 module[a]	528	16.7	49	5.7	
Corning 7059 glass/ITO/Cd$_{0.95}$Zn$_{0.05}$S:In/CuIn (Se$_{0.5}$S$_{0.5}$)$_2$/Au	325	10.3	33	1.1	[375]

[a]Each cell area 0.4 cm^2.

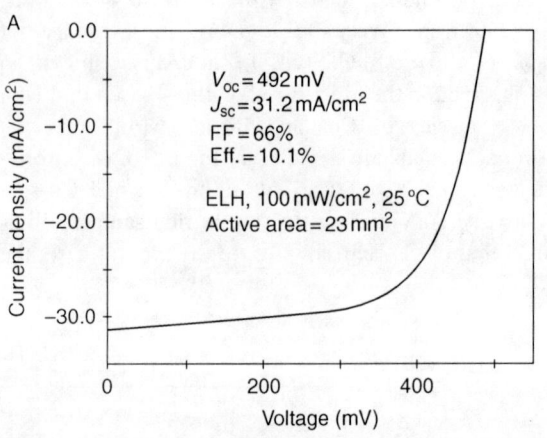

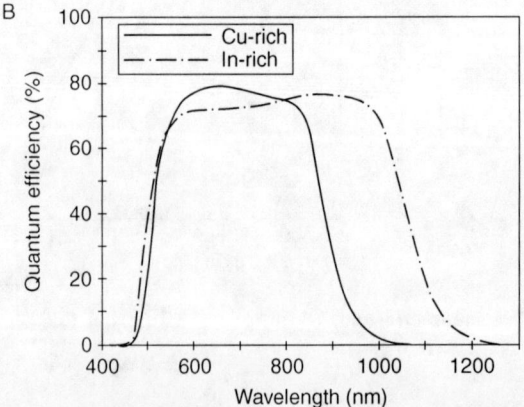

Figure 8.102 (A) I–V curve of glass/Mo/In-rich CuIn (Se$_{0.58}$S$_{0.42}$)$_2$(Cu/In = 0.8)/CdS (CBD)/ZnO:Al, and (B) quantum efficiency of Cu- and In-rich glass/Mo/CuIn (Se$_{1-x}$S$_x$)$_2$/80 nm CdS(CBD)/ ZnO:Al cells.

In-rich samples might have not allowed to participate more sulfur in the system. As mentioned in earlier, it is difficult to dope more sulfur in the In-rich quaternary $CuIn(Se_{1-x}S_x)_2$ samples. The band gap increases with increasing sulfur content in the quaternary samples that may be the reason the Cu-rich samples showing higher band gap than that of In-rich $CuIn(Se_{1-x}S_x)_2$ system. The $CuIn(Se_{1-x}S_x)_2$ absorber bilayer is prepared by vacuum co-evaporation of at substrate temperature of 400–500 °C for Cu-rich bottom layer. After 2/3 growth, the In-rich CISS layer is grown by increasing substrate temperature of 100 °C and the (S+Se)/metal ratio being three times [372].

Unlike annealing CIS under S/inert gas, the sulfur or its compound is directly deposited onto CIS layer then recrystallized by rapid thermal processing to form $CuIn(Se_{1-x}S_x)_2$, which is etched to remove CuS or Cu_2Se. The final absorber contains x from 0.2 to 0.4 having band gap of between 1.1 and 1.2 eV. The XRD analysis reveals chalcopyrite structure and SEM cross section indicates formation of Mo$(SSe)_2$ layer between Mo and $CuIn(Se_{1-x}S_x)_2$ layers. The CdS buffer layer is coated onto glass/Mo/$CuIn(Se_{1-x}S_x)_2$ by conventional CBD method. The final glass/Mo/$CuIn(Se_{1-x}S_x)_2$/CdS/ZnO thin film solar cells with an active area of 0.1 cm^2 show efficiency of 11.5%, whereas 30×30 cm^2 modules exhibit efficiency of about 8%. Single route electrodeposited $CuIn(SeS)_2$ solar cells with an small area of 0.1 cm^2 shows efficiency of 10.6%, whereas 15×15 cm^2 large area cells show maximum efficiency of 5.7% (Table 8.46). The band gap of CISS with $S/(S+Se)=95\%$ is 1.47 eV. The spectral responses of glass/Mo/$CuIn(Se_{1-x}S_x)_2$/CdS/ZnO thin film solar cells with different efficiencies of 10.5 and 9.9% are depicted in Fig. 8.103 [373,374].

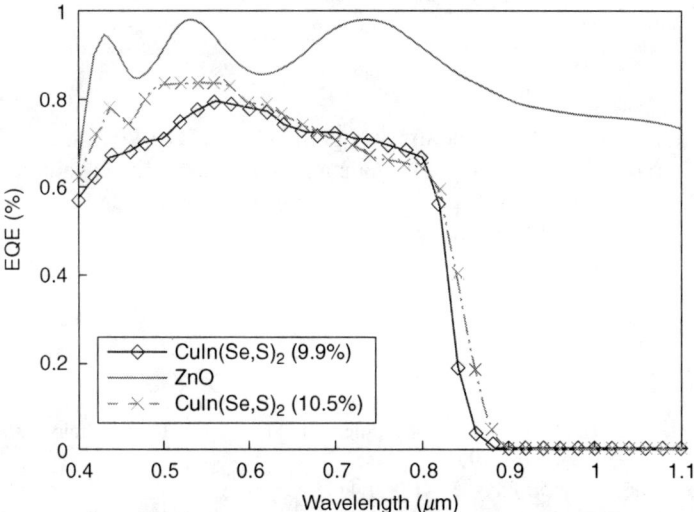

Figure 8.103 Quantum efficiencies of 10.5, and 9.9% efficieny CISS thin film solar cells, and transmission spectra of ZnO.

Figure 8.104A Monolithic array of thin film solar cells; number of rows 9, each row contains 18 cells, total cells 162 over the area of 11×11 cm^2, and each cell area is 0.5 cm^2, and average efficiency of 7%. Fine grids are visible.

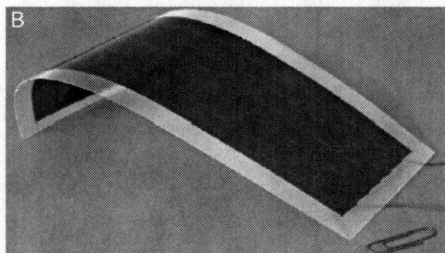

Figure 8.104B CuInS$_2$ based roof tile minimodules.

All spray deposited back wall configured glass/0.5 μmITO/1 μm n-Cd$_{0.95}$Zn$_{0.05}$S:In (6 at%)/2 μm p-CuIn(S$_{0.5}$Se$_{0.5}$)$_2$ heterojunction thin film solar cell exhibits efficiency of 1.1% [375]. Small area (15×15 cm^2) module configuration and roof tile module are shown in Fig. 8.104A and B [369,373].

References

[1] R. Menner, H.W. Schock, Proceedings of 11th E. C. Photovoltaic Solar Energy Conference, Montreux, (1992), p. 834.
[2] C.H. Henry, J. Appl. Phys. 51 (1980) 4494.
[3] J. Hedstrom, H. Ohlsen, M. Bodegard, A. Kylner, L. Stolt, D. Hariskos, M. Ruck, H.W. Schock, Proceedings of 23rd IEEE Photovoltaic Specialists Conference, (1993), p. 364.

[4] J.A.M. AbuShama, S. Johnston, T. Moriarty, G. Teeter, K. Ramanathan, R. Noufi, Progress in Photovoltaics: Research and Applications 12 (2004) 39.
[5] R. Noufi, R.J. Matson, R.C. Powell, C. Herrington, Solar Cells 16 (1986) 479.
[6] J.R. Tuttle, D.S. Albin, R. Noufi, Solar Cells 30 (1991) 21.
[7] A. Fahrenbruch, R.H. Bube, Fundamentals of Solar Cells, Academic Press, New York, 1983.
[8] R.W. Birkmire, R.B. Hall, J.E. Phillips, 17th IEEE Photovoltaic Specialist Conference, (1984), p. 882.
[9] J.S. Britt, S. Wiedeman, U. Schoop, D. Verebelyi, 33rd IEEE Photovoltaic Specialist Conference, (2008) 670-080515190533.
[10] I.L. Eisgruber, J.E. Granata, J.R. Sites, J. Hou, J. Kessler, Solar Energy Mater. Solar Cells 53 (1998) 367.
[11] T. Minemoto, T. Matsui, H. Takakura, Y. Hamakawa, T. Negami, Y. Hashimoto, T. Uenoyama, M. Kitagawa, Solar Energy Mater. Solar Cells 67 (2001) 83.
[12] M. Burgelman, P. Nollet, S. Degrave, Thin Solid Films 361–362 (2000) 527.
[13] Y. Okano, T. Nakada, A. Kunioka, Solar Energy Mater. Solar Cells 50 (1998) 105.
[14] S.A.A. Kuhaimi, Solar Energy Mater. Solar Cells 52 (1998) 69.
[15] I. Hengel, A. Neisser, R. Klenk, M.Ch. Lux-Steiner, Thin Solid Films 361–362 (2000) 458.
[16] M. Roy, S. Damaskinos, J.E. Phillips, Proceedings of the 20th IEEE Photovoltaic Specialist Conference, (1988), p. 1618.
[17] S.R. Kodigala, I. Bhat, T.P. Chow, J.K. Kim, E.F. Schubert, D. Johnstone, S. Akarca-Biyikli, J. Appl. Phys. 98 (2005) 106108.
[18] V. Nadenau, U. Rau, A. Jasenek, H.W. Schock, J. Appl. Phys. 87 (2000) 584.
[19] W.N. Shafarman, J.E. Phillips, Proceedings of 22nd IEEE Photovoltaic Specialists Conference, (1991), p. 934.
[20] J. Marlein, K. Decock, M. Burgelman, Thin Solid Films 517 (2009) 2353.
[21] K.W. Mitchell, C. Eberspacher, J.R. Emer, K.L. Pauls, D. Pier, IEEE Trans. Electron Devices ED-37 (1990) 410.
[22] M. Topic, F. Smole, J. Furlan, Solar Energy Materials and Solar Cells 49 (1997) 311.
[23] J.L. Shay, S. Wagner, H.M. Kasper, Appl. Phys. Lett. 27 (1975) 89.
[24] H. Tavakolian, J.R. Sites, Proceedings of 18th IEEE Photovoltaic Specialist Conference, (1985), p. 1065.
[25] L.S. Yip, I. Shih, 1st World Conference on Photovoltaic Energy Conversion, (1994), p. 210.
[26] H. Du, C.H. Champness, I. Shih, Thin Solid Films 480–481 (2005) 37.
[27] Z.A. Shukri, C.H. Champness, 23rd IEEE Photovoltaic Specialist Conference, (1993), p. 603.
[28] L.L. Kazmerski, F.R. White, G.K. Morgan, Appl. Phys. Lett. 29 (1976) 268.
[29] R.A. Mickelsen, W.S. Chen, Appl. Phys. Lett. 36 (1980) 371.
[30] V.K. Kapur, P. Singh, U.V. Choudary, P.M. Uno, L. Elyash, S. Meisel, 17th IEEE Photovoltaic Specialist Conference, (1984), p. 777.
[31] R. Noufi, R. Axton, D. Cahen, S.K. Deb, 17th IEEE Photovoltaic Specialist Conference, (1984), p. 927.
[32] T.C. Lammasson, H. Talieh, J.D. Meakin, J.A. Thornton, 19th IEEE Photovoltaic Specialist Conference, (1987), p. 1285.
[33] R.E. Rocheleau, J.D. Meakin, R.W. Birkmire, 19th IEEE Photovoltaic Specialist Conference, (1987), p. 972.

[34] M. Kauk, M. Altosaar, J. Raudoja, A. Jagomagi, M. Danilson, T. Varema, Thin Solid Films 515 (2007) 5880.
[35] L. Stolt, J. Hedstrom, J. Kessler, M. Ruckh, K.-O. Velthaus, H.-W. Schock, Appl. Phys. Lett. 62 (1993) 597.
[36] H. Talieh, A. Rocket, Solar Cell 27 (1989) 321.
[37] L.L. Kazmerski, P.E. Russell, O. Jamjoum, P.J. Ireland, R.A. Mickelsen, W.S. Chen, K.J. Bachmann, 16th IEEE Photovoltaic Specialist Conference, (1982), p. 786.
[38] R. Noufi, P. Souza, C. Osterwald, Solar Cells 15 (1985) 87.
[39] T. Nakada, K. Migita, A. Kunioka, 23rd IEEE Photovoltaic Specialist Conference, (1993), p. 560.
[40] A.Y. Yahia, N.M. Shaalan, M.S. Shaalan, O.M. Jamjoum, 19th IEEE Photovoltaic Specialist Conference, (1987), p. 222.
[41] J. Piekoszewski, J.J. Loferski, R. Beaulieu, J. Beall, B. Roessler, J. Shewchun, Solar Energy Mater. 2 (1980) 363.
[42] P.R. Ram, R. Thangaraj, A.K. Sharma, O.P. Agnihotri, Solar Cells 14 (1985) 123.
[43] M.S. Thomar, F.J. Garcia, Thin Solid Films 90 (1982) 419.
[44] V. Alberts, R. Herberholz, T. Walter, H.W. Schock, J. Phys. D: Appl. Phys. 30 (1997) 2156.
[45] B.M. Basol, V.K. Kapur, A. Halani, C. Leidholm, A. Minnick, 11th E. C. Photovoltaic Solar Energy Conference Montreux, (1992), p. 803.
[46] F.O. Adurodija, M.J. Carter, R. Hill, Solar Energy Mater. Solar Cells 40 (1996) 359.
[47] H. Dittrich, M. Klose, M. Brieger, R. Schaffle, H.W. Schock, 23rd IEEE Photovoltaic Specialist Conference, (1993), p. 617.
[48] S.C. Park, D.Y. Lee, B.T. Ahn, K.H. Yoon, J. Song, Solar Energy Mater. Solar Cells 69 (2001) 99.
[49] J. Kessler, H.W. Schock, 11th E. C. Photovoltaic Solar Energy Conference Montreux, (1992), p. 838.
[50] T. Nakada, T. Ichen, T. Ochi, A. Kunioka, 11th E. C. Photovoltaic Solar Energy Conference Montreux, (1992), p. 794.
[51] F.O. Adurodija, M.J. Carter, R. Hill, 1st World Conference on Photovoltaic Energy Conversion, Hawaii, (1994), p. 186.
[52] P.J. Dale, A.P. Samantilleke, G. Zoppi, I. Forbes, L.M. Peter, J. Phys. D: Appl. Phys. 41 (2008) 085105.
[53] V.K. Kapur, B.M. Basol, E.S. Tseng, Solar Cells 21 (1987) 65.
[54] B.M. Basol, V.K. Kapur, R.C. Kullberg, Solar Cells 27 (1989) 299.
[55] B.M. Basol, V.K. Kapur, IEEE Trans. Electron. Devices 37 (1990) 418.
[56] B.M. Basol, V.K. Kapur, A. Halani, 22nd IEEE Photovoltaic Specialists Conference, (1991), p. 893.
[57] S. Verma, T.W.F. Russell, R.W. Birkmire, 23rd IEEE Photovoltaic Specialist Conference, (1993), p. 431.
[58] H. Sato, T. Hama, E. Niemi, Y. Ichikawa, H. Sakai, 23rd IEEE Photovoltaic Specialist Conference, (1993) p521.
[59] L.C. Yang, G. Berry, L.J. Chou, G. Kenshole, A. Rockett, C.S. Mullan, C.J. Kiely, 23rd IEEE Photovoltaic Specialist Conference, (1993) p505.
[60] L.C. Yang, L.J. Chou, A. Agarwal, A. Rocket, 22nd IEEE Photovoltaic Specialist Conference, (1991), p. 1185.
[61] R.A. Mickelsen, B.J. Stanbery, J.E. Avery, W.S. Chen, 19th IEEE Photovoltaic Specialist Conference, (1987) 744.

[62] R.H. Mauch, J. Hedstrom, D. Lincot, M. Ruckh, J. Kessler, R. Klinger, L. Stolt, J. Vedel, H.W. Schock, 22nd IEEE Photovoltaic Specialists Conference, (1991), p. 898.
[63] R.A. Mickelsen, W.S. Chen, Y.R. Hsiao, V.E. Lowe, IEEE Trans. Electron. Devices 31 (1984) 542.
[64] R.A. Mickelsen, W.S. Chen, 16th IEEE Photovoltaic Specialist Conference, (1982) 781.
[65] B.M. Basol, V.K. Kapur, Appl. Phys. Lett. 54 (1989) 1918.
[66] A.J. Nelson, C.R. Schwerdtfeger, S.-W. Wei, A. Zunger, D. Rioux, R. Patel, H. Hochst, Appl. Phys. Lett. 62 (1993) 2557.
[67] A. Nouhi, R.J. Stirn, A. Hermann, 19th IEEE Photovoltaic Specialist Conference, (1987) 1461.
[68] L.C. Olsen, F.W. Addis, D.A. Huber, 23rd IEEE Photovoltaic Specialist Conference, (1993), p. 603.
[69] G. Gordillo, C. Calderon, Solar Energy Mater. Solar Cells 77 (2003) 163.
[70] G. Gordillo, C. Calderon, 17th European Photovoltaic Solar Energy Conference, (2001) Munich, p. 1236.
[71] K.O. Velthaus, J. Kessler, M. Ruck, D. Hariskos, D. Schmid, H.W. Schock, 11th E. C. Photovoltaic Solar Energy Conference, (1992) Montreux, p. 842.
[72] T. Nakada, T. Kume, A. Kunioka, Solar Energy Mater. Solar Cells 50 (1998) 97.
[73] D. Lincot, J.-F. Guillemoles, P. Cowache, S. Massaccesi, 1st World Conference on Photovoltaic Energy Conversion, Hawaii, (1994), p. 136.
[74] S.N. Qiu, W.W. Lam, C.X. Qiu, I. Shih, Appl. Surface Sci. 113/114 (1997) 764.
[75] D. Guimard, P.P. Grand, N. Bodereau, P. Cowache, J.-F. Guillemoles, D. Lincot, S. Taunier, M.B. Farah, P. Mogensen, 29th IEEE Photovoltaic Specialist Conference, (2002), p. 692.
[76] I. Shih, C.X. Qiu, 19th IEEE Photovoltaic Specialist Conference, (1987) 1291.
[77] S.N. Qiu, L. Li, C.X. Qiu, I. Shih, C.H. Champness, Solar Energy Mater. Solar Cells 37 (1995) 389.
[78] A.M. Hermann, R. Westfall, R. Wind, Solar Energy Mater. Solar Cells 51 (1998) 355.
[79] R.O. Borges, D. Lincot, J. Veddel, 11th E. C. Photovoltaic Solar Energy Conference Montreux, (1992) p862.
[80] I.M. Dharmadasa, N.B. Chaure, A.P. Samantilleka, C. Furlong, P.H. Gardiner, 19th European Photovoltaic Solar Energy Conference, (2004) Paris, p. 1745.
[81] P.K.V. Pillai, K.P. Vijayakumar, Solar Energy Mater. Solar Cells 51 (1998) 47.
[82] R.W. Birkmire, B.E. McCandless, Appl. Phys. Lett. 53 (1988) 140.
[83] R.R. Arya, T. Lommasson, B. Fieselmann, L. Russell, L. Carr, A. Catalano, 22nd IEEE Photovoltaic Specialists Conference, (1991), p. 903.
[84] L. Russell, B. Fieselmann, R.R. Arya, 23rd IEEE Photovoltaic Specialist Conference, (1993), p. 581.
[85] F. Karg, V. Probst, H. Harms, J. Rimmasch, W. Riedl, J. Kotschy, J. Holz, R. Treichler, O. Eibl, A. Mitwalsky, A. Kiendl, 23rd IEEE Photovoltaic Specialist Conference, (1993), p. 441.
[86] T. Nakada, N. Murakami, A. Kunioka, Mater. Res. Soc. Symp. Proc. 426 (1996) 411.
[87] J. Kessler, M. Ruckh, D. Hariskos, U. Ruhle, R. Menner, H.W. Schock, 23rd IEEE Photovoltaic Specialist Conference, (1993), p. 447.
[88] K. Ramanathan, H. Wiesner, S. Asher, D. Niles, R.N. Bhattacharya, J. Keane, M.A. Contreras, R. Noufi, 2nd World Conference and Exhibition on Photovoltaic Solar Energy Conversion, (1998), p. 477.
[89] L.C. Olsen, F.W. Addis, W. Lei, J. Li, AIP Conf. Proc. 353 (1996) 436.

[90] T. Nii, I. Sugiyama, T. Kase, M. Sato, Y. Kariyagawa, K. Kushiya, H. Takeshita, 1st World Conference on Photovoltaic Energy Conversion, Hawaii, (1994), p. 254.
[91] M. Nishitani, M. Ikeda, T. Negami, S. Kohoki, N. Kohara, M. Terauchi, H. Wada, T. Wada, Solar Energy Mater. Solar Cells 35 (1994) 203.
[92] T. Negami, M. Nishitani, M. Ikeda, T. Wada, Solar Energy Mater. Solar Cells 35 (1994) 215.
[93] T. Nakada, N. Okano, Y. Tanaka, H. Fukuda, A. Kunioka, 1st World Conference on Photovoltaic Energy Conversion, (1994), p. 95.
[94] T. Negami, M. Nishitani, T. Wada, T. Hirao, 11th E. C. Photovoltaic Solar Energy Conference Montreux, (1992) p783.
[95] H. Sano, S. Nakamura, K. Kondo, K. Sato, 1st World Conference on Photovoltaic Energy Conversion, Hawaii, (1994), p. 179.
[96] B.M. Basol, V.K. Kapur, A. Halani, A. Minnick, C. Leidholm, 23rd IEEE Photovoltaic Specialists Conference, (1993), p. 426.
[97] R. Gay, J. Ermer, C. Fredric, K. Knapp, D. Pier, C. Jensen, D. Willett, 22nd IEEE Photovoltaic Specialists Conference, (1991) p848.
[98] S. Wiedeman, J. Kessler, T. Lommasson, L. Russell, J. Fogleboch, S. Skibo, R. Arya, AIP Conference Proceedings 353 (1996) 12.
[99] R.R. Arya, T.C. Lommasson, S. Wiedeman, 23rd IEEE Photovoltaic Specialist Conference, (1993), p. 516.
[100] W. Chesarek, A. Mason, K. Mitchell, L. Fabick, 19th IEEE Photovoltaic Specialist Conference, (1987), p. 791.
[101] T.J. Gillespie, B.R. Lanning, C.H. Marshall, 26th IEEE PVSC, (1997), p. 403.
[102] M. Rusu, S. Doka, C.A. Kaufmann, N. Grigorieva, T.S. Niedrig, M.Ch. Lux-Steiner, Thin Solid Films 480–481 (2005) 341.
[103] J.H. Schon, M. Klenk, O. Schenker, E. Bucher, Appl. Phys. Lett. 77 (2000) 3657.
[104] M. Saad, H. Riazi- Nejad, E. Bucher, M.Ch. Lux-Steiner, 1st World Conference Photovoltaic Energy Conversion, (1994), p. 214.
[105] M. Saad, H. Riazi, E. Bucher, M.Ch. Lux-Steiner, Appl. Phys. A 62 (1996) 181.
[106] D.F. Marron, A. Meeder, I.G. Perez, A. Rumberg, A.J. Waldau, M.Ch. Lux-Steiner, 17th European Photovoltaic Solar Energy Conference, Munich, (2001), p. 1159.
[107] D.L. Young, J. Keane, A. Duda, J.A.M. AbuShama, C.L. Perkins, M. Romero, R. Noufi, Prog. Photovolt: Res. Appl. 11 (2003) 535.
[108] M.A. Contreras, M. Romero, D. Young, 3rd World Conference on Photovoltaic Energy Conversion, (2003), p. 2864.
[109] R. Noufi, R. Powell, C. Herrington, T. Coutts, Solar Cells 17 (1986) 303.
[110] M. Klenk, O. Schenker, V. Alberts, E. Bucher, Thin Solid Films 387 (2001) 47.
[111] S. Siebentritt, A. Bauknecht, A. Gerhard, U. Fiedeler, T. Kampschulte, S. Schuler, W. Harneit, S. Brehme, J. Albert, M.Ch. Lux-Steiner, Solar Energy Mater. Solar Cells 67 (2001) 129.
[112] D. Fischer, N. Meyer, M. Kuczmik, M. Beck, A.-J. Waldau, M.Ch.-L. Steiner, Solar Energy Mater. Solar Cells 67 (2001) 105.
[113] R. Klenk, R. Mauch, R. Schaffler, D. Schmid, H.W. Schock, 22nd IEEE Photovoltaic Specialist Conference, (1991), p. 1071.
[114] M.A. Contreras, K. Ramanathan, J. AbuShama, F. Hasoon, D.L. Young, B. Egaas, R. Noufi, Prog. Photovolt: Res. Appl. 13 (2005) 209.
[115] P. Jackson, R. Wurz, U. Rau, J. Mattheis, M. Kurtha, T. Schlotzer, G. Bilger, J.H. Werner, Prog. Photovolt. Res. Appl. 15 (2007) 507.
[116] A. N. Tiwari, Swiss Federal Institute of Technology, Zurich, Private communication.

[117] C.A. Kaufmann, A. Neisser, R. Klenk, R. Scheer, H.W. Schock, Mater. Res. Soc. Symp. Proc. 865 (2005) F7.5.1K. Orgassa, H.W. Schock, J.H. Werner, Thin Solid Films 431–432 (2003) 387.
[118] S. Chaisitsak, Y. Tokita, H. Mikami, A. Yamada, M. Konagai, 17th European Photovoltaic Solar Energy Conference, (2001), p. 1011.
[119] K. Ramanathan, G. Teeter, J.C. Keane, R. Noufi, Thin Solid Films 480–481 (2005) 499.
[120] K. Ramanathan, M.A. Contreras, C.L. Perkins, S. Asher, F.S. Hasoon, J. Keans, D. Young, M. Romero, W. Metzger, R. Noufi, J. Ward, A. Duda, Prog. Photovolt: Res. Appl. 11 (2003) 225.
[121] T. Negami, Y. Hashimoto, S. Nishiwaki, Solar Energy Mater. Solar Cells 67 (2001) 331.
[122] M.A. Contreras, B. Egaas, K. Ramanathan, J. Hiltner, A. Swartzlander, F. Hasoon, R. Noufi, Prog. Photovolt: Res. Appl. 7 (1999) 311.
[123] J.A. Abushama, J. Wax, T. Berens, J. Tuttle, 4th World Conference on Photovoltaic Energy Conversion, (2006), p. 487.
[124] M.A. Contreras, J. Tuttle, A. Gabor, A. Tennant, K. Ramanathan, S. Asher, A. Franz, J. Keane, L. Wang, J. Scofield, R. Noufi, First World Conference on Photovoltaic Energy Conversion, Hawaii, (1994), p. 68.
[125] M.A. Contreras, J. Tuttle, A. Gabor, A. Tennant, K. Ramanathan, S. Asher, A. Franz, J. Keane, L. Wang, R. Noufi, Solar Energy Mater. Solar Cells 41–42 (1996) 231.
[126] N. Romeo, A. Bosio, V. Canevari, R. Tedeschi, S. Sivelli, A. Romeo, F.V. Kurdesau, 19th European Photovoltaic Solar Energy Conference, Paris, (2004), p. 1796.
[127] I.L. Repins, B.J. Stanbery, D.L. Young, S.S. Li, W.K. Metzger, C.L. Perkins, W. N. Shafarman, M.E. Beck, L. Chen, V.K. Kapur, D. Tarrant, M.D. Gonzalez, D. G. Jensen, T.J. Anderson, X. Wang, L.L. Kerr, B. Keyes, S. Asher, A. Delahoy, B. Von Roedern, Prog. Photovolt: Res. Appl. 14 (2006) 25.
[128] V. Kapur, R. Kemmerle, A. Bansal, J. Haber, J. Schmitzberger, P. Le, D. Guevarra, V. Kapur, T. Stempien, 33rd IEEE Photovoltaic Specialist Conference, (2008) 627–080519130516.
[129] T. Yamaguchi, T. Kobata, S. Niiyama, T. Nakamura, A. Yoshida, Solar Energy Mater. Solar Cells 75 (2003) 87.
[130] A.E. Delahoy, L. Chen, M. Akhtar, B. Sang, S. Guo, Solar Energy 77 (2004) 785.
[131] T. Unold, T. Enzenhofer, C.A. Kaufmann, R. Klenk, A. Neisser, K. Sakurai, H. W. Schock, 4th World Conference on Photovoltaic Energry Conversion, (2006), p. 356.
[132] M. Edoff, S. Woldegiorgis, P. Neretnieks, M. Ruth, J. Kessler, L. Stolt, 19th European Photovoltaic Solar Energy Conference, Paris, (2004), p. 1690.
[133] C.L. Jensen, D.E. Tarrant, 23rd IEEE Photovoltaic Specialist Conference, (1993), p. 577.
[134] K. Kushiya, A. Shimizu, A. Yamada, M. Konagai, Jpn. J. Phys. 34 (1995) 54.
[135] W.S. Chen, J.M. Stewart, W.E. Devaney, R.A. Mickelsen, B.J. Stanbery, 23rd IEEE Photovoltaic Specialists Conference, (1993), p. 422.
[136] S. Ishizuka, K. Sakurai, A. Yamada, K. Matsubara, P. Fons, K. Iwata, S. Nakamura, 19th European Photovoltaic Solar Energy Conference, Paris, (2004), p. 1729.
[137] J.E. Phillips, J. Titus, D. Hofmann, 26th IEEE PVSC, Anaheim, (1997) p463.
[138] K. Kushiya, A. Shimizu, K. Saito, A. Yamada, M. Konagai, 1st World Conference on Photovoltaic Energy Conversion, Hawaii, (1994), p. 87.
[139] J. Malmstrom, J. Wennerberg, M. Bodegard, L. Stolt, 17th European Photovoltaic Solar Energy Conference, Munich, (2001), p. 1265.
[140] R. Caballero, C. Guillen, M.T. Gutierrez, C.A. Kaufmann, Prog. Photovolt.:Res. Appl. 14 (2006) 145.

[141] K. Kushiya, Y. Ohtake, A. Yamada, M. Konagai, Jpn. J. Appl. Phys. 33 (1994) 6599.
[142] B.M. Basol, V.K. Kapur, C. Leidholm, A. Halani, A. Minnick, AIP Conf. Proc. 353 (1996) 26.
[143] R.W. Birkmire, H. Hichri, R. Klenk, M. Marudachalam, B.E. Mc Candless, J. E. Phillips, J.M. Schultz, W.N. Shafarman, AIP Conf. Proc. 353 (1996) 420.
[144] M. Nisitani, T. Negami, M. Ikeda, N. Kohara, M. Terauchi, T. Wada, 1st World Conference on Photovoltaic Energy Conversion, Hawaii, (1994), p. 222.
[145] R.N. Bhattacharya, W. Batchelor, H. Wiesner, F. Hasoon, J.E. Granata, K. Ramanathan, J. Alleman, J. Keane, A. Mason, R.J. Matson, R. Noufi, J. Electrochem. Soc. 145 (1998) 3435.
[146] R.N. Bhattacharya, W. Batchelor, J.E. Granata, F. Hasoon, H. Wiesner, K. Ramanathan, J. Keane, R.N. Noufi, Solar Energy Mater. Solar Cells 55 (1998) 83.
[147] N.G. Dhere, AIP Conf. Proc. 353 (1996) 428.
[148] N.G. Dhere, S. Kuttath, K.W. Lynn, 1st World Conference on Photovoltaic Conversion, Hawaii, (1994), p. 190.
[149] N.G. Dhere, K.W. Lynn, Solar Energy Mater. Solar Cells 41 (1996) 271.
[150] T. Satoh, S. Hayashi, S. Nishiwaki, S.-I. Shimakawa, Y. Hashimoto, T. Negami, T. Uenoyama, Solar Energy Mater. Solar Cells 67 (2001) 203.
[151] A. Virtuani, E. Lotter, M. Powalla, U. Rau, J.H. Werner, Thin Solid Films 451 (2004) 160.
[152] A. Virtuani, E. Lotter, M. Powalla, 19th European Photovoltaic Solar Energy Conference, Paris, (2004), p. 1898.
[153] T.J. Gillespie, W.A. Miles, 26th IEEE PVSC, Anaheim, (1997), p. 487.
[154] W.N. Shafarman, R.W. Birkmire, S. Marsillac, M. Marudachalam, N. Orbey, T.W. F. Russell, 26th IEEE PVSC, Anaheim, (1997), p. 331.
[155] D.Y. Lee, B.T. Ahn, K.H. Yoon, J.S. Song, Solar Energy Mater. Solar Cells 75 (2003) 73.
[156] S. Nishiwaki, T. Satoh, S. Hayashi, Y. Hashimoto, S. Shimakawa, T. Negami, T. Wada, Solar Energy Mater. Solar Cells 67 (2001) 217.
[157] J. Britt, A. Delahoy, G. Butler, F. Faras, A. Muttaiah, 1st World Conference on Photovoltaic Energy Conversion, (1994), p. 140.
[158] A.E. Delahoy, J.S. Britt, A.M. Gabor, Z.J. Kiss, AIP Conference Proc. 353 (1996) 3.
[159] K. Ramanathan, J.C. Keane, B. To, R.G. Dhere, R. Noufi, 20th European Photovoltaic Solar Energy Conference, Barcelona, (2005), p. 1695.
[160] O. Lundberg, M. Bodegard, J. Malmstrom, L. Stolt, Prog. Photovolt.: Res. Appl. 11 (2003) 77.
[161] D. Ohashi, T. Nakada, A. Kunioka, Solar Energy Mater. Solar Cells 67 (2001) 261.
[162] V. Probst, W. Stetter, J. Plm, S. Zweigart, M. Wendl, H. Vogt, K.-D. Ufer, H. Calwer, B. Freienstein, F.H. Karg, 17th European Photovoltaic Solar Energy Conference, Munich, (2001), p. 1005.
[163] D. Tarrant, J. Ermer, 23rd IEEE Photovoltaic Specialist Conference, (1993), p. 372.
[164] M. Gossla, W.N. Shafarman, Thin Solid Films 480–481 (2005) 33.
[165] M. Bar, M. Rusu, J. Reib, Th. Glatzel, S. Sadewasser, W. Bohne, E. Strub, H.-J. Muffler, S. Lindner, J. Rohrich, T.P. Niesen, F. Karg, M.Ch. Lux-Steiner, Ch.-H. Fischer, 3rd World Conference on Photovoltaic Energy Conversion, (2003), p. 335.
[166] M. Turcu, O. Pakma, U. Rau, Appl. Phys. Lett. 80 (2002) 2598.
[167] U. Rau, H.W. Schock, Appl. Phys. A69 (1999) 131.
[168] G.T. Koishiyev, J.R. Sites, S.S. Kulkarni, N.G. Dhere, 33rd IEEE Photovoltaic Specialist Conference, (2008) 45-080506180551.
[169] S.S. Kulkarni, G.T. Koishiyev, H. Moutinho, N.G. Dhere, Thin Solid Films 517 (2009) 2121.

[170] A.O. Pudov, J.R. Sites, M.A. Contreras, T. Nakada, H.-W. Schock, Thin Solid Films 480–481 (2005) 273.
[171] D.B. Mitzi, M. Yuan, W. Liu, A.J. Kellock, S.J. Chey, L. Gignac, A.G. Schrott, Thin Solid Films 517 (2009) 2158.
[172] D. Rudmann, D. Bremaud, H. Zogg, A.N. Tiwari, 19th European Photovoltaic Solar Energy Conference, Paris, (2004) p1710.
[173] M. Lammer, U. Klemm, M. Powalla, Thin Solid Films 387 (2001) 33.
[174] B.M. Keyes, F. Hasoon, P. Dipp, A. Balcioglu, F. Abulfotuh, 26th IEEE PVSC, Anaheim, (1997), p. 479.
[175] V. Probst, J. Rimmasch, W. Riedle, W. Stetter, J. Hoz, H. Harms, F. Karg, H.W. Schock, 1st World Conference on Photovoltaic Energy Conversion, Hawaii, (1994), p. 144.
[176] S. Marsillac, S. Don, R. Rocheleau, E. Miller, Solar Energy Mater. Solar Cells 82 (2004) 45.
[177] M. Lammer, A. Eicke, M. Powalla, 29th IEEE Photovoltaic Specialist Conference, (2002), p. 696.
[178] Y. Hashimoto, T. Satoh, S. Shimakawa, T. Negami, 3rd World Conference on Photovoltaic Energy Conversion, (2003) p574.
[179] M.A. Contreras, B. Egaas, P. Dippo, J. Webb, J. Granata, K. Ramanathan, S. Asher, A. Swartzlander, R. Noufi, 26 th PVSC, (1997), p. 359.
[180] T. Nakada, D. Iga, H. Ohbo, A. Kunioka, Jpn. J. Appl. Phys. 36 (1997) 732.
[181] M. Hartmann, M. Schmidt, A. Jasenek, H.W. Schock, F. Kessler, K. Hertz, M. Powalla, 28th IEEE Photovoltaic Specialists Conference, (2000), p. 638.
[182] J.H. Yun, K.H. Kim, M.S. Kim, B.T. Ahn, S.J. Ahn, J.C. Lee, K.H. Yoon, Thin Solid Films 515 (2007) 5876.
[183] R. Caballero, C.A. Kaufmann, T. Eisenbarth, M. Cancela, R. Hesse, T. Unold, A. Eicke, R. Klenk, H.W. Schock, Thin Solid Films 517 (2009) 2187.
[184] D. Abou-Ras, G. Kostorz, A. Romeo, D. Rudmann, A.N. Tiwari, Thin Solid Films 480–481 (2005) 118.
[185] Y. Hashimoto, Y. Satoh, T. Minemoto, S. Shimakawa, T. Negami, 17th European Photovoltaic Solar Energy Conference, Munich, (2001), p. 1225.
[186] C.H. Huang, S.S. Li, L. Rieth, A. Halani, M.L. Fisher, J. Song, T.J. Anderson, P.H. Holloway, 28th IEEE Photovoltaic Specialists Conference, (2000), p. 696.
[187] M. Bar, U. Bloeck, H.-J. Muffler, M.Ch. Lux-Steiner, Ch.-H. Fischer, M. Giersig, T.P. Niesen, F. Karg, J. Appl. Phys. 97 (2005) 014905.
[188] K. Urabe, T. Hama, M. Roy, H. Sato, H. Fujisawa, M. Ohsawa, Y. Ichikawa, H. Sakai, 22nd IEEE Photovoltaic Specialist Conference, (1991), p. 1082.
[189] W.E. Devaney, W.S. Chen, J.M. Stewart, R.A. Mickelsen, IEEE Trans. Electron Dev. 37 (1990) 428.
[190] J. Song, S.S. Li, L. Chen, R. Noufi, T.J. Anderson, O.D. Crisalle, 4th World Conference on Photovoltaic Energy Conversion, (2006), p. 534.
[191] B. Dmmler, H. Dittrich, R. Menner, H.W. Schock, 19th IEEE Photovoltaic Specialist Conference, (1987) 1454.
[192] W.S. Chen, J.M. Stewart, B.J. Stanbery, W.E. Devaney, R.A. Mickelsen, 19th IEEE Photovoltaic Specialist Conference, (1987), p. 1446.
[193] M.A. Contreras, T. Nakada, M. Hongo, A.O. Pudov, J.R. Sites, 3rd World Conference on Photovoltaic Energy Conversion, (2003) 570.
[194] T. Nakada, M. Mizutani, Y. Hagiwara, A. Kunioka, Solar Energy Mater. Solar Cells 67 (2001) 255.

[195] T. Nakada, K. Furumi, A. Kunioka, IEEE Transactions on Electron Devices 46 (1999) 2093, T. Nakada, M. Mizutani, 28th IEEE Photovoltaic Specialist Conference 2000, Anchorage, AK, (2000), p. 529 T. Nakada, M. Mizutani, Jpn. J. Appl. Phys. 41 (2002) L165.

[196] A. Pudov, J. Sites, T. Nakada, Jpn. J. Appl. Phys. 41 (2002) L672.

[197] A. Ichiboshi, M. Hongo, T. Akamine, T. Dobashi, T. Nakada, Solar Energy Mater. Solar Cells 90 (2006) 3130.

[198] S. Kundu, L.C. Olsen, 29th IEEE Photovoltaic Specialist Conference, (2002), p. 648.

[199] K. Kushiya, T. Nii, I. Sugiyama, Y. Sato, Y. Inamori, H. Takeshita, Jpn. J. Appl. Phys. 35 (1996) 4383.

[200] B. Sang, W.N. Shafarman, R.W. Birkmire, 29th IEEE Photovoltaic Specialist Conference, (2002), p. 632.

[201] S. Nishiwaki, T. Satoh, Y. Hashimoto, S.-I. Shimakawa, S. Hayashi, T. Negami, T. Wada, Solar Energy Mater. Solar Cells 77 (2003) 359.

[202] R.N. Bhattacharya, K. Ramanathan, Solar Energy 77 (2004) 679.

[203] C.P. Bjorkman, T. Torndahl, D. Abou-Ras, J. Malmstrom, J. Kessler, L. Stolt, J. Appl. Phys. 100 (2006) 044506.

[204] R.N. Bhattacharya, M.A. Contreras, G. Teeter, Jpn. J. Appl. Phys. 43 (2004) L1475.

[205] A. Ennaoui, M. Weber, M. Saad, W. Harneit, M.Ch. Lux-Steiner, F. Karg, Thin Solid Films 361–362 (2000) 450.

[206] A. Ennaoui, U. Blieske, M.Ch. Lux-Steiner, Prog. Photovolt.: Res. Appl. 6 (1998) 447.

[207] W. Eisele, A. Ennaoui, P.S. Bischoff, M. Giersig, C. Pettenkofer, J. Krauser, M. Ch. Lux-Steiner, S. Zweigart, F. Karg, Solar Energy Mater. Solar Cells 75 (2003) 17.

[208] A. Ennaoui, S. Siebentritt, M.Ch. Lux-Steiner, W. Riedl, F. Karg, Solar Energy Mater. Solar Cells 67 (2001) 31.

[209] W. Eisele, A. Ennaoui, P.S. Bischoff, M. Giersig, C. Pettenkofer, J. Krauser, M.Ch Lux-Steiner, T. Riedle, N. Esser, S. Zweigart, F. Karg, 28th IEEE Photovoltaic Specialists Conference, (2000), p. 692.

[210] S. Siebentritt, T. Kampschult, A. Bauknecht, U. Blieske, W. Harneit, U. Fiederler, M.Ch Lux-Steiner, Solar Energy Mater. Solar Cells 70 (2002) 447.

[211] S. Siebentritt, P. Walk, U. Fiedeler, I. Lauermann, K. Rahne, M.Ch. Lux-Steiner, T.P. Niesen, F. Karg, Prog. Photovolt.: Res. Appl. 12 (2004) 333.

[212] Y. Ohtake, K. Kushiya, M. Ichikawa, A. Yamada, M. Konagai, Jap. J. Appl. Phys. 34 (1995) 5949.

[213] Y. Ohtake, K. Kushiya, A. Yamada, M. Konagai, 1st World Conference on Photovoltaic Energy Conversion, (1994), p. 218.

[214] M. Konagai, Y. Ohtake, T. Okamoto, Mater. Res. Soc. Symp. Proc. 426 (1996) 153.

[215] Y. Ohtake, T. Okamoto, A. Yamada, M. Konagai, K. Saito, Solar Energy Mater. Solar Cells 49 (1997) 269.

[216] Y. Ohtake, S. Chaisitsak, A. Yamada, M. Konagai, Jpn. J. Appl. Phys. 37 (1998) 3220.

[217] A. Yamada, T. Sugiyama, S. Chaisitsak and M. Konagai, 28th IEEE Photovoltaic Specialists Conference 2000, p462.

[218] A. Delahoy, J. Bruns, L. Chen, M. Akhtar, Z. Kiss, 28th IEEE Photovoltaic Specialist Conference, (2000) 1437.

[219] S. Chaisitak, A. Yamada, M. Konagai, K. Saito, Jpn. J. Appl. Phys. 39 (2000) 1660.

[220] J. Sterner, J. Malmstrom, L. Stolt, Prog. Photovolt.: Res. Appl. 13 (2005) 179.

[221] A. Strohm, L. Eisenmann, R.K. Gebhardt, A. Harding, T. Schlotzer, D. Abou-Ras, H.W. Schock, Thin Solid Films 480–481 (2005) 162.

[222] Q. Nguyen, K. Orgassa, I. Koetschau, U. Rau, H.W. Schock, Thin Solid Films 431–432 (2003) 330.
[223] D. Hariskos, R. Menner, E. Lotter, S. Spiering, M. Powalla, 20th European Photovoltaic Solar Energy Conference, Barcelona, (2005), p. 1713.
[224] D. Hariskos, R. Menner, S. Spiering, A. Eicke, M. Powalla, K. Ellmer, M. Oertel, B. Dimmler, 19th European Photovoltaic Solar Energy Conference, Paris, (2004), p. 1894.
[225] N. Naghavi, J.-F. Guillemoles, D. Lincot, B. Canava, A. Etcheberry, S. Taunier, S. Spiering, M. Powalla, M. Lamirand, L. Legras, 19th European Photovoltaic Solar Energy Conference, Paris, (2004), p. 1733.
[226] N. Barreau, J.C. Bernede, S. Marsillac, C. Amory, W.N. Shafarmann, Thin Solid Films 431–432 (2003) 326.
[227] S. Buecheler, D. Corica, D. Guettler, A. Chirila, R. Verma, U. Muller, T.P. Niesen, J. Palm, A.N. Tiwari, Thin Solid Films 517 (2009) 2312.
[228] N.A. Allsop, A. Schonmann, H.-J. Muffler, M. Bar, M.Ch. Lux-Steiner, Ch.-H. Fischer, Prog. Photovolt.: Res. Appl. 13 (2005) 607.
[229] S. Gall, N. Barreau, S. Harel, J.C. Bernede, J. Kessler, Thin Solid Films 480–481 (2005) 138.
[230] L. Larina, K.H. Kim, K.H. Yoon, M. Konagai, B.T. Ahn, 3rd world Conference on Photovoltaic Conversion Energy, (2003), p. 531.
[231] D. Hariskos, M. Ruck, U. Ruhle, T. Walter, H.W. Schock, 1st World Conference on Photovoltaic Energy Conversion, Hawaii, (1994), p. 91.
[232] D. Hariskos, M. Ruck, U. Ruhle, T. Walter, H.W. Schock, J. Hedstrom, L. Stolt, Solar Energy Mater. Solar Cells 41–42 (1996) 345.
[233] C.H. Huang, S.S. Li, W.N. Shafarman, C.-H. Chang, E.S. Lambers, L. Rieth, J.W. Johnson, S. Kim, B.J. Stanbery, T.J. Anderson, P.H. Holloway, Solar Energy Mater. Solar cells 69 (2001) 131.
[234] Y. Tokita, S. Chaisitsak, A. Yamada, M. Konagai, Solar Energy Mater. Solar Cells 75 (2003) 9.
[235] N. Naghavi, S. Spiering, M. Powalla, B. Cavana, D. Lincot, Prog. Photovolt.: Res. Appl. 11 (2003) 437.
[236] J.C. Lee, K.H. Kang, S.K. Kim, K.H. Yoon, I.J. Park, J. Song, Solar Energy Mater. Solar Cells 64 (2000) 185.
[237] S. Ishizuka, K. Sakurai, A. Yamada, K. Matsubara, P. Fons, K. Iwata, S. Nakamura, Y. Kimura, T. Baba, H. Nakanishi, T. Kojima, S. Niki, Solar Energy Mater. Solar Cells 87 (2005) 541.
[238] S. Chaisitsak, T. Sugiyama, A. Yamada, M. Konagai, Jpn. J. Appl. Phys. 38 (1999) 4989.
[239] J.N. Duenow, T.A. Gessert, D.M. Wood, B. Egaas, R. Noufi, T.J. Coutts, 33rd IEEE Photovoltaic Specailist Conference, (2008) p153–080502160555.
[240] T. Hagiwara, T. Nakada, A. Kunioko, Solar Energy Mater. Solar Cells 67 (2001) 267.
[241] L. Stolt, J. Hedstrom, J. Skarp, 1st World Conference on Photovoltaic Energy Conversion, Hawaii, (1994), p. 250.
[242] A. Yamada, H. Miyazaki, Y. Chiba, M. Konagai, Thin Solid Films 480–481 (2005) 503.
[243] H. Miyazaki, R. Mikami, A. Yamada, M. Konagai, Jpn. J. Appl. Phys. 45 (2006) 2618.
[244] A.E. Delahoy, L. Chen, B. Sang, S.Y. Guo, J. Cambridge, F. Ziobro, R. Govindarajan, S. Kleindienst, M. Akhtar, 19th European Photovoltaic Solar Energy Conference, Paris, (2004), p. 1686.

[245] N.F. Cooray, K. Kushiya, A. Fujimaki, I. Sugiyama, T. Miura, D. Okumura, M. Sato, M. Ooshita, O. Yamase, Solar Energy Mater. Solar Cells 49 (1997) 291.
[246] L.C. Olsen, P. Eschbach, S. Kundu, 29th IEEE Photovoltaic Specialist Conference, (2002), p. 652.
[247] T. Negami, T. Minemoto, Y. Hashimoto, T. Satoh, 28th IEEE Photovoltaic Specialists Conference, (2000), p. 634.
[248] Y. Chiba, H. Miyazaki, A. Yamada, M. Konagai, 19th European Photovoltaic Solar Energy Conference, (2004), p. 1737.
[249] T. Minemoto, Y. Hashimoto, W.-S. Kolahi, T. Satoh, T. Negami, H. Takakura, Y. Hamakawa, Solar Energy Mater. Solar Cells 75 (2003) 121.
[250] T. Glatzel, S.V. Roon, S. Sadewasser, R. Klenk, A.J. Waldau, M.Ch. Lux-Steiner, 17th European Photovoltaic Solar Energy Conference, Munich, (2001), p. 1151.
[251] A. Strohm, T. Schlotzer, Q. Nguyen, K. Orgassa, H. Wiesner, H.W. Schock, 19th European Photovoltaic Solar Energy Conference, Paris, (2004), p. 1741.
[252] C. Platzer-Bjorkman, T. Torndahl, A. Hultqvist, J. Kessler, M. Edoff, Thin Solid Films 515 (2007) 6024.
[253] T. Negami, T. Aoyagi, T. Satoh, S. Shimakawa, S. Hayashi, Y. Hashimoto, 29th IEEE Photovoltaic Specialist Conference, (2002), p. 656.
[254] K. Ramanathan, F.S. Hasoon, S. Smith, A. Mascarenhas, H. Al-Thani, J. Alleman, H.S. Ullal, J. Keane, 29th IEEE Photovoltaic Specialist Conference, (2002), p. 523.
[255] P.K. Johnson, A.O. Pudov, J.R. Sites, K. Ramanathan, F.S. Hasoon, D.E. Tarrant, 17th European Photovoltaic Solar Energy Conference, Munich, (2001), p. 1035.
[256] R.N. Bhattacharya, W. Batchelor, K. Ramanathan, M.A. Contreras, T. Moriarty, Solar Energy Mater. Solar Cells 63 (2000) 367.
[257] T. Wada, Y. Hashimoto, S. Nishiwaki, T. Satoh, S. Hayashi, T. Negami, H. Miyake, Solar Energy Mater. Solar Cells 67 (2001) 305.
[258] T. Nakada, K. Matsumoto, M. Okumura, 29th IEEE Photovoltaic Specialist Conference, (2002) p527.
[259] B. Canava, O. Roussel, J.F. Guillemoles, D. Lincot, A. Etcheberry, Phys. Stat. Sol. (C) 3 (2006) 2551.
[260] M. Bar, H.-J. Muffler, Ch.-H. Fischer, S. Zweigart, F. Karg, M.Ch. Lux-Steiner, Prog. Photovolt.: Res. Appl. 10 (2002) 173.
[261] B.E. McCandless, W.N. Shafarman, 3rd World Conference on Photovoltaic Energy Conversion, (2003), p. 562.
[262] X. Wang, S.S. Li, W.K. Kim, S. Yoon, V. Craciun, J.M. Howard, S. Easwaran, O. Manasreh, O.D. Crisalle, T.J. Anderson, Solar Energy Mater. Solar Cells 90 (2006) 2855.
[263] X. Wang, S.S. Li, C.H. Huang, S. Rawal, J.M. Howard, V. Craciun, T.J. Anderson, O.D. Crisalle, Solar Energy Mater. Solar Cells 88 (2005) 65.
[264] A.M. Hermann, C. Gonzalez, P.A. Ramakrishnan, D. Balzar, C.H. Marshall, J.N. Hilfiker, T. Tiwald, Thin Solid Films 387 (2001) 54.
[265] E. Moons, D. Gal, J. Beier, G. Hodes, D. Cahen, L. Kronik, L. Burstein, B. Mishori, Y. Shapira, D. Hariskos, H.-W. Schock, Solar Energy Mater. Solar Cells 43 (1996) 73.
[266] T. Nakada, T. Kume, T. Mise, A. Kunioka, Jpn. J. Phys. 37 (1998) L499.
[267] T. Nakada, T. Mise, 17th European Photovoltaic Solar Energy Conference, (2001) p1027.
[268] F.-J. Haug, H. Zogg, A.N. Tiwari, 29th IEEE Photovoltaic Specialist Conference, (2002), p. 728.
[269] F.-J. Haug, D. Rudmann, H. Zogg, A.N. Tiwari, Thin Solid Films 431–432 (2003) 431.

[270] D.L. Young, J. Abushama, R. Noufi, X. Li, J. Keane, T.A. Essert, J.C. Ward, M. Contreras, M. Symko-Davies, T.J. Coutts, 29th IEEE Photovoltaic Specialist Conference, (2002) p608.
[271] T. Tokado and T. Nakada, 3rd world Conference on Photovoltaic Conversion Energy, 2003, p 539.
[272] T. Nakada, Thin Solid Films 480–481 (2005) 419.
[273] T. Nakada, Y. Kanda, S. Kijima, Y. Komiya, D. Ohmori, H. Ishizaki, N. Yamada, 20th European Photovoltaic Solar Energy Conference, Barcelona, (2005), p. 1736.
[274] T. Nakada, Y. Hirabayashi, T. Tokado, D. Ohmori, 3rd world Conference on Photovoltaic Conversion Energy, (2003) S4OC124-2880.
[275] R.J. Matson, M.A. Contreras, J.R. Tuttle, A.B. Swartzlander, P.A. Parilla, R. Noufi, Mat. Res. Soc. Symp. Proc. 426 (1996) 183.
[276] M.A. Contreras, H. Wiesner, D. Niles, K. Ramanathan, R. Matson, J. Tuttle, J. Keane, R. Noufi, 25th IEEE Photovoltaic Specialist Conference May13–17, Washington DC, (1996), p. 809.
[277] S. Ishizuka, K. Sakurai, K. Matsubara, A. Yamada, M. Yonemura, S. Kuwamori, S. Nakamura, Y. Kimura, H. Naknishi, S. Niki, Mater. Res. Soc. Symp. Proc. 865 (2005) F8.3.1.
[278] T. Satoh, Y. Hashimoto, S.-I. Shimakawa, S. Hayashi, T. Negami, 28th IEEE Photovoltaic Specialists Conference, (2000), p. 567.
[279] T. Satoh, Y. Hashimoto, S.-I. Shimakawa, S. Hayashi, T. Negami, Solar Energy Mater. Solar Cells 75 (2003) 65.
[280] A. Kampmann, J. Rechid, S. Wulf, M. Mihhailova, R. Thyen and A. Lossin, 19th European Photovoltaic Solar Energy Conference, Paris, 2004, p1806.
[281] J. Rechid, R. Thyen, A. Raitzig, S. Wulff, M. Mihhailova, K. Kalberlah, A. Kampmann, 3rd World Conference on Photovoltaic Energy Conversion, (2003), p. 559.
[282] C.Y. Shi, Y. Sun, Q. He, F.Y. Li, J.C. Zhao, Solar Energy Mater. Solar Cells 93 (2009) 654.
[283] M. Powalla, 17th European Photovoltaic Solar Energy Conference, Munich, (2001), p. 983.
[284] C.A. Kaufmann, R. Klenk, M.Ch. Lux-Steiner, A. Neisser, P. Korber, R. Scheer, H.W. Schock, 20th European Photovoltaic Solar Energy Conference, Barcelona, (2005), p. 1729.
[285] T.P. Niesen, A. Lerchenberger, M. Wendl, F. Karg, V. Probst, 19th European Photovoltaic Solar Energy Conference, Paris, (2004), p. 1706.
[286] B. Bremaud, D. Rudmann, M. Kaelin, K. Ernits, G. Bilger, M. Dobeli, H. Zogg, A.N. Tiwari, Thin Solid Films 515 (2007) 5857.
[287] A.N. Tiwari, M. Krejci, F.-J. Haug, H. Zogg, Prog. Photovolt.: Res. Appl. 7 (1999) 393.
[288] M. Powalla, B. Dimmler, K.-H. Grob, 20th European Photovoltaic Solar Energy Conference, Barcelona, (2005) 1689.
[289] M. Powalla, B. Dimmler, R. Schaeffler, G. Voorwinden, U. Stein, H.-D. Mohring, F. Kessler, D. Hariskos, 19th European Photovoltaic Solar Energy Conference, Paris, (2004), p. 1663.
[290] F. Kessler, D. Hermann, M. Lammer, M. Powalla, 19th European Photovoltaic Solar Energy Conference, Paris, (2004), p. 1702.
[291] G.M. Hanket, U.P. Singh, E. Eser, W.N. Shafarman, R.W. Birkmire, 29th IEEE Photovoltaic Specialist Conference, (2002), p. 567.
[292] S. Wiedeman, M.E. Beck, R. Butcher, I. Repins, N. Gomez, B. Joshi, R.G. Wendt, J.S. Britt, 29th IEEE Photovoltaic Specialist Conference, (2002), p. 575.

[293] F. Kessler, D. Herrmann, U. Klemm, M. Powalla, 20th European Photovoltaic Solar Energy Conference, Barcelona, (2005), p. 1732.
[294] B.M. Basol, V.K. Kapur, C.R. Leidholm, A. Minnick, A. Halani, 1st World Conference on Photovoltaic Energy Conversion, Hawaii, (1994), p. 148.
[295] N. Kohara, S. Nishiwaki, Y. Hasimoto, T. Negami, T. Wada, Solar Energy Mater. Solar Cells 67 (2001) 209.
[296] H.A. Al-Thani, F.S. Hasoon, M. Young, S. Asher, J.L. Alleman, 29th IEEE Photovoltaic Specialist Conference, (2002), p. 720.
[297] T. Wada, N. Kohara, T. Negami, M. Nishitani, Jpn. J. Appl. Phys. 35 (1996) L1253.
[298] P.J. Roston, J. Mattheis, G. Bilger, U. Rau, J.H. Werner, Thin Solid Films 480–481 (2005) 67.
[299] D. Abou-Ras, G. Kostorz, D. Bremaud, M. Kalin, F.V. Kurdesau, A.N. Tiwari, M. Dobeli, Thin Solid Films 480–481 (2005) 433.
[300] P.J. Sebastian, M.E. Calixto, R.N. Bhattacharya, R. Noufi, J. Electrochem. Soc. 145 (1998) 3613.
[301] M.E. Calixto, R.N. Bhattacharya, P.J. Sebastian, A.M. Fernandez, S.A. Gamboa, R.N. Noufi, Solar Energy Mater. Solar Cells 55 (1998) 23.
[302] P.J. Sebastian, M.E. Calixto, R.N. Bhattacharya, R. Noufi, Solar Energy Mater. Solar Cells 59 (1999) 125.
[303] M. Ganchev, J. Kois, M. Kaelin, S. Bereznev, E. Tzvetkova, O. Volobujeva, N. Stratieva, A. Tiwari, Thin Solid Films 511–512 (2006) 325.
[304] J.S. Ward, K. Ramanathan, F.S. Hasoon, T.J. Couts, J. Keane, M.A. Contreras, T. Moriarty, R. Noufi, Prog. Photovolt.: Res. Appl. 10 (2002) 41.
[305] T. Negami, T. Satoh, Y. Hashimoto, S. Nishiwaki, S.-I. Shimakawa, S. Hayashi, Solar Energy Mater. Solar Cells 67 (2001) 1.
[306] R. Menner, M. Oertel, G. Voorwinden, M. Powalla, 17th European Photovoltaic Solar Energy Conference, Munich, (2001), p. 1207.
[307] B. Dimmler, M. Powalla, H.W. Schock, Prog. Photovolt.: Res. Appl. 10 (2002) 149.
[308] M. Powalla, B. Dimmler, Solar Energy Mater. Solar Cells 75 (2003) 27.
[309] J. Palm, V. Probst, F.H. Karg, Solar Energy 77 (2004) 757.
[310] R.W. Birkmire, E. Eser, Annu. Rev. Mater. Sci. 27 (1997) 625.
[311] C. Fredric, R. Gay, D. Tarrant, D. Willet, 23rd IEEE Photovoltaic Specialist Conference, (1993), p. 437.
[312] J.R. Tuttle, A. Szalaj, J. Keane, 28th IEEE Photovoltaic Specialists Conference, (2000), p. 1042.
[313] C. Deibel, V. Dyakonov, J. Parisi, J. Palm, S. Zweigart, F. Karg, Thin Solid Films 403–404 (2002) 325.
[314] K. Kushiya, S. Kuriyagama, Y. Tanaka, Y. Nagoya, M. Tachiyuki, M. Akema, 19th European Photovoltaic Solar Energy Conference, Paris, (2004), p. 1672.
[315] C. Radue, E.E.V. Dyk, E.Q. Macabebe, Thin Solid Films 517 (2009) 2383.
[316] K. Kushiya, M. Tachiyuki, T. Kase, I. Sugiyama, Y. Nagoya, D. Okumura, M. Sato, O. Yamase, H. Takeshita, Solar Energy Mater. Solar Cells 49 (1997) 277.
[317] B. Sang, Y. Nagoya, K. Kushiya, O. Yamase, Solar Energy Mater. Solar Cells 75 (2003) 179.
[318] K. Kushiya, S. Kuriyagawa, I. Hara, Y. Nagoya, M. Tachiyuki, Y. Fujiwara, 29th IEEE Photovoltaic Specialist Conference, (2002), p. 579.
[319] B. Sang, K. Kushiya, D. Okumura, O. Yamase, Solar Energy Mater. Solar Cells 67 (2001) 237.

[320] K. Kushiya, M. Ohshita, I. Hara, Y. Tanaka, B. Sang, Y. Nagoya, M. Tachiyuki, O. Yamase, Solar Energy Mater. Solar Cells 75 (2003) 171.
[321] V. Probst, J. Palm, S. Visbeck, T. Niesen, R. Tolle, A. Lerchenberger, M. Wendl, H. Vogt, H. Calwer, W. Stetter, F. Karg, Soar Energy Mater. Solar Cells 90 (2006) 3115.
[322] K. Kushiya, 3rd world Conference on Photovoltaic Conversion Energy, (2003), p. 319.
[323] K. Kushiya, Solar Energy 77 (2004) 717.
[324] K. Kushiya, Y. Tanaka, H. Hakuma, Y. Goushi, S. Kijima, T. Aramoto, Y. Fujiwara, Thin Solid Films 517 (2009) 2108.
[325] V. Probst, W. Stetter, J. Palm, R. Toelle, S. Visbeck, H. Calwer, T. Niesen, H. Vogt, O. Hernandez, M. Wendl, F.H. Karg, 3rd world Conference on Photovoltaic Conversion Energy, (2003) 329.
[326] J. Palm, W. Stetter, S. Visbeck, T. Niesen, M. Furfanger, H. Vogt, J. Baumbach, H. Calwer, V. Probst, F. Karg, 20th European Photovoltaic Solar Energy Conference, Barcelona, (2005), p. 1699.
[327] M. Powalla, G. Voorwinden, D. Hariskos, P. Jackson, R. Kniese, Thin Solid Films 517 (2009) 2111.
[328] R. Saez, D. Abou-Ras, T.P. Niesen, A. Neisser, K. Wilchelmi, M.Ch. Lux-Steiner, A. Ennaoui, Thin Solid Films 517 (2009) 2300.
[329] W. Stetter, V. Probst, J. Palm, S. Visbeck, T. Niesen, R. Tolle, M. Wendl, H. Vogt, H. Calwer, 19th European Photovoltaic Solar Energy Conference, Paris, (2004), p. 1682.
[330] M. Powalla, E. Lotter, R. Waechter, S. Spiering, M. Oertel, 29th IEEE Photovoltaic Specialist Conference, (2002), p. 571.
[331] S. Spiering, A. Eicke, D. Hariskos, M. Powalla, N. Naghavi, D. Lincot, Thin Solid Films 451 (2004) 562.
[332] N. Naghavi, S. Spiering, M. Powalla, D. Lincot, 3rd world Conference on Photovoltaic Conversion Energy, (2003), p. 340.
[333] S. Spiering, D. Hariskos, M. Powalla, N. Naghavi, D. Lincot, Thin Solid Films 431–432 (2003) 359.
[334] S. Spiering, D. Hariskos, S. Schroder, M. Powalla, Thin Solid Films 480–481 (2005) 195.
[335] R. Scheer, R. Klenk, J. Klaer, I. Luck, Solar Energy 77 (2004) 777.
[336] R. Scheer, T. Walter, H.W. Schock, M.L. Fearheiley, H.J. Lewerenz, Appl. Phys. Lett. 63 (1993) 3294.
[337] J. Klaer, J. Bruns, R. Henninger, K. Siemer, R. Klenk, K. Ellmer, D. Braunig, Semicond. Sci. Technol. 13 (1998) 1456.
[338] M.H. Jin, K.K. Banger, C.V. Kelly, J.H. Scofield, J.S. McNatt, J.E. Dickman, A.F. Hepp, 19th European Photovoltaic Solar Energy Conference, Paris, (2004), p. 1943.
[339] Y. Ogawa, A. Jager-waldau, T.H. Hua, Y. Hashimoto, K. Ito, Appl. Surface Sci. 92 (1996) 232.
[340] I. Lauermann, I. Luck, K. Wojczykowski, 17th European Photovoltaic Solar Energy Conference, Munich, (2001), p. 1163.
[341] Y. Ogawa, A. Jager-Waldau, Y. Hashimoto, K. Ito, Jpn. J. Appl. Phys. 33 (1994) L1775.
[342] T. Nakabayahi, T. Miyazawa, Y. Hashimoto, K. Ito, Solar Energy Mater. Solar Cells 49 (1997) 375.
[343] R. Scheer, I. Luck, S. Hessler, H. Sehnert, H.J. Lewerenz, 1st World Conference on Photovoltaic Energy Conversion, Hawaii, (1994), p. 160.
[344] T. Watanabe, M. Matsui, K. Mori, Solar Energy Mater. Solar Cells 35 (1994) 239.

[345] T. Negami, Y. Hashimoto, M. Nishitani, T. Wada, Solar Energy Mater. Solar Cells 49 (1997) 343.
[346] T. Enzenhofer, T. Unold, R. Scheer, H.-W. Schock, 20th European Photovoltaic Solar Energy Conference, Barcelona, (2005), p. 1751.
[347] S. Nakamura, A. Yamamoto, Solar Energy Mater. Solar Cells 75 (2003) 81.
[348] J.D. Harris, K.K. Banger, D.A. Scheiman, M.A. Smith, M.H.-C. Jin, A.F. Hepp, Mater. Sci. Engg. B98 (2003) 150.
[349] X. Hou, K.-L. Choy, Thin Solid Films 480–481 (2005) 13.
[350] A. Mere, O. Kijatkina, H. Rebane, J. Krustok, M. Krunks, J. Phys. Chem. Solids 64 (2003) 2025.
[351] T. Watanabe, M. Matsui, Jpn. J. Appl. Phys. 38 (1999) L1379.
[352] T. Watanabe, H. Nakazawa, M. Matsui, H. Ohbo, T. Nakada, Solar Energy Mater. Solar Cells 49 (1997) 357.
[353] G.-C. Park, H.-D. Chung, C.-D. Kim, H.-R. Park, W.-J. Jeong, J.-U. Kim, H.-B. Gu, K.-S. Lee, Solar Energy Mater. Solar Cells 49 (1997) 365.
[354] N. Meyer, I. Luck, U. Ruhle, J. Klaer, R. Klenk, M.Ch Lux-Steiner, R. Scheer, 19th European Photovoltaic Solar Energy Conference, Scheer, (2004), p. 1698.
[355] J. Klaer, I. Luck, K. Siemer, R. Klenk, D. Braunig, 28th IEEE Photovoltaic specialist conference, Anchorage AK, (2000), p. 559.
[356] J. Klaer, I. Luck, A. Boden, R. Klenk, I.G. Perez, R. Scheer, Thin Solid Films 431–432 (2003) 534.
[357] K. Siemer, J. Klaer, I. Luck, J. Bruns, R. Klenk, D. Braunig, Solar Energy Mater. Solar Cells 67 (2001) 159.
[358] J. Klaer, K. Siemer, I. Luck, D. Braunig, Thin Solid Films 387 (2001) 169.
[359] A. Ennaoui, M. Bar, J. Klaer, T. Kropp, R.S. Araoz, M.Ch. Lux-Steiner, 20th European Photovoltaic Solar Energy Conference, Barcelona, (2005), p. 1882.
[360] B. Asenjo, A.M. Chaparro, M.T. Gutierrez, J. Herrero, J. Klaer, Solar Energy Mater. Solar Cells 92 (2008) 302.
[361] O. Papathanasiou, S. Siebentritt, W. Bohne, J. Lauermann, K. Rahne, J. Rohrich, M. Rusu, E. Strub, M.Ch. Lux-Steiner, 19th European Photovoltaic Solar Energy Conference, Paris, (2004), p. 1951.
[362] A. Ennaoui, M. Weber, R. Scheer, H.J. Lewerenz, Solar Energy Mater. Solar Cells 54 (1998) 277.
[363] A. Mere, A. Katerski, O. Kijatkina, M. Krunks, 19th European Photovoltaic Solar Energy Conference, Paris, (2004), p. 1973.
[364] C. Kaufmann, P.J. Dobson, S. Neve, W. Bohne, J. Klenk, C. Pettenkofer, J. Rohrich, R. Scheer, U. Storkel, 28th IEEE Photovoltaic Specialists Conference, (2000), p. 688.
[365] D. Braunger, D. Hariskos, T. Walter, H.W. Schock, Solar Energy Mater. Solar Cells 40 (1996) 97.
[366] B.R. Sankapal, A. Ennaoui, T. Guminskaya, Th. Dittrich, W. Bohne, J. Rohrich, E. Strub, M.Ch. Lux-Steiner, Thin Solid Films 480–481 (2005) 142.
[367] J. Verschraegen, M. Burgelman, J. Penndorf, 20th European Photovoltaic Solar Energy Conference, (2005), p. 1835.
[368] O. Tober, J. Wienke, M. Winkler, J. Penndorf, J. Griesche, Mat. Res. Soc. Symp. Proc. 763 (2003) B8.16.1.
[369] M. Winkler, J. Griesche, I. Konovalvo, I. Penndorf, I. Wienke, O. Tober, Solar Energy 77 (2004) 705.
[370] W.W. Hou, B. Bob, S.-H. Li, Y. Yang, Thin Solid Films 517 (2009) 6853.

[371] T. Ohashi, K. Inakoshi, Y. Hashimoto, K. Ito, Solar Energy Mater. Solar Cells 50 (1998) 37.
[372] T. Walter, A. Content, K.O. Velthaus, H.W. Schock, Solar Energy Mater. Solar Cells 26 (1992) 357.
[373] J. Kessler, J. Sicx-Kurdi, N. Naghavi, J.-F. Guillemoles, D. Lincot, O. Kerrec, M. Lamirand, L. Legras, P. Mogensen, 20th European Photovoltaic Solar Energy Conference, Barcelona, (2005), p. 1704.
[374] S. Taunier, J.S. Kurdi, P.P. Grand, A. Chomont, O. Ramdani, L. Parissi, P. Panheleux, N. Naghavi, C. Hubert, M.B. Farah, J.P. Fauvarque, J. Connolly, O. Roussel, P. Mogensen, E. Mahe, J.F. Guillemoles, D. Lincot, O. Kerrec, Thin Solid Films 480–481 (2005) 526.
[375] S.R. Kodigala, V.S. Raja, Solar Energy Mater. Solar Cells 32 (1994) 1.

Subject Index

A
Absorber 15
Absorption 196, 490
Absorption coefficient 16, 203
Acoustic scattering 336
Air annealing 529
Allowed transition 198
Amorphous 524
AR coating 252
Asymmetry 90, 109
Atomic force microscopy 55
Atomic vibrations 277
Auger electron spectroscopy (AES) 84

B
Back scattering 87
Backwall configuration 8–114
Band diagram 505
Band gap 13, 199
Band structure 195
Band to band transition 236
Bifacial 622
Bilayer 29
Binding energy 90
Biomass 1
Bohr radius 337
Boltzmann's constant 10
Bose-Einstein 458
Bragg angle 78
Bridgman technique 130
Bright field image 119
Brooks Herring 337
Buffer layer 653
Burstein-Moss shift 236, 412, 415, 446

C
Cantilever 55
Capacitance transient 372
Carrier concentration 488
Cathodeluminescence 221
Cathodic potential 163

Chalcogenides 43
Chalcopyrite 115, 134, 174
Chemical bath deposition 393
Cliff 507
Closed space chemical vapor 80
CO_2 1
Coalescence 57
Columnar structure 69
Composition 30
Conduction band 218
Conduction band offset 506, 507
Conventional evaporation 65
Core electron 92
Corrosion 630
Coulombic interaction 212
Cross over 518
Crystal field splitting 195, 243
Crystallographic 514
CuAu CA 115
Cubic 15
Cubic phase 410
Current density 509
Cu_xS/CdS 2
C-V measurements 529
Czochralski technique 9

D
Damp-heat test 591
DC magnetron sputtering 28, 433
Deep level transient spectroscopy 372
Deformation potentials 337
Degenerate 195
Depth profile 82
Dielectric constant 212
Differential thermal analysis 117
Diffraction pattern 118
Diffusion 83
Diode characteristics 536
Diode quality factor 509
Doctor blade 41, 269
Donor acceptor pair (DAP) 212

Donor like states 511
Donor to free transition 245
Doublets 102
Dye sensitized solar cells 8

E
EBIC 545
Effective mass 336
Efficiency 4, 12
Effusion 173
Electrodeposition 27
Electrolyte 613
Electron affinity 13
Electron beam 78
Electron hole pair 7, 524
Electron probe microanalysis 78
Electron trap level 381
Electron-phonon interaction 459
Electroreflectance 195
Emission 1
Endothermic 42
Excited state 213
Excitonic emission 224
Exothermic 143

F
Fermi level 364
Fill factor 12
Flash evaporation 22
Flat band potential 632
Flexible substrates 626
Forbidden transition 198
Fossil fuels 1
Free exciton 219
Free to bound transition 213

G
GaAs 2
Gaussian shape 250
Global energy 1
Grading 81
Grazing incidence 160
Green band 420
Ground state 213

H
Hall mobility 340
Haynes formula 451
He-Ne laser 214

Heterogeneous 394
Heterojunction 7, 374
Heterostructure 404
Hexagonal 15, 399
Homo 11
Homogeneous 394
Homojunction 378
Hopping conduction 362
Hybrid process 522
Hydrazine 40, 326
Hydrogen implantation 270
Hydrothermal method 96

I
Illumination 518
In-line process 633
Interaction force 55
Interface 10
Ionized impurity scattering 337, 427

J
Junction 515

K
KCN etching 97
Keating model 280
Kinetic energy 92

L
Lamination 636
LO phonon 219, 233, 423, 453
Local vibrational modes 472
Localized states 364
Lumo 11

M
Magnetic field 424
MBE 14, 28
Miller indices 115
Minmodules 653
Minority carrier injection 377
MOCVD 14, 252
Modules 505
Molecularity 321
Monochromator 276
Monolithic 537, 632
Mott parameters 362
Multiphonons 422
Multiple junction 5

N

Na doping 577
Nanocrytal 463
Nanoparticles 40
Nanorods 453
Near band emission 225
Neutral acceptor to bound exciton 418
Neutral donor to bound exciton 418, 457
Non-degenerate 195
Non-polar optical scattering 336
Nonvacuum process 39
Nozzle 43

O

Open circuit voltage 11
Optical branches 422
Optical method 411
Optical pulse 388
Order vacancy compound 115
Orientation 31
Orthorhombic 15
OVC 132
Oxidation 100
Ozone 1

P

Parabolic bands 198
Pentenary 15
Periodicity 120
pH 187
Phonon characteristic temperature 337
Phonon frequency 280
Photoacoustic 202
Photoconductivity 243
Photocurrent 9
Photocurrent-capacitance 513
Photoelectron 92
Photolithographic 629
Photoluminescence 212, 450
Photomultiplier tube 78
Photothermal 1
Photovoltage 9
Photovoltaic 1
Physica vapor deposition 29
Piezoelectric scattering 427
Plasma 65
Plastic solar cells 6
Point defects 136
Polarizer 276

Polymer semiconductor 8
Polymorphism 414
Precursor 81
Preferred orientation 129
Proton irradiation 335
Psuedobinary 119
Pulsed laser deposition 26
Pyramidal 62

Q

Quantum efficiency 33
Quarternary 38
Quasi-neutral region 510
Quatum size 415

R

Raman spectroscopy 62, 195, 276, 470
Rapid thermal process 41
Reflectance 196, 415, 490
Refractive indices 198, 205
Renewable 1
Repulsive force 55
Resistivity 325
Reverse saturation current 567
RF sputtering 433
Rheology 41
Roll to roll 658
Roquesite 176

S

Scanning electron microscope 63
Scherrer formula 406
Schottky 1, 169, 380
Scribing 565
Secondary ion mass spectroscopy 81
Secondary phase 84, 415
Selected area electron diffraction (SAED) 119
Selenization 138
Semiconductor 212
Semimetallic 133, 323
Sensitive factor 85, 92
Series resistance 509
Shockley 10
Shockley-Read-Hall recombination 509
Short circuit current 12
Shunt resistance 509
Sintering 647
Soft x-ray photon 92

Sol gel 432
Solar energy 1
Solar radiation 3
Solar spectrum 4
Solvothermal technique 28
Space charge region 509
Spectral distribution 3
Spectral response 517
Sphalerite 115, 129
Spike 507
Spin coating 40
Spin orbit splitting 195, 243
Splitting 153
Spotty ring pattern 400
Spray pyrolysis 26, 103, 172
Spring constant 55
Sputtering 23, 82, 259
Stoichiometry 21, 136
Sulfurizaiton 45, 167
Superstrate 619
Surface analysis 55, 395
S-vacancy 421

T
Tetragonal 15
Tetrahedral 431
Thermal activation energy 212, 328
Thermal expansion co-efficient 630
Thermocouple 44
Thermoelectric power 361, 369
Three stage process 29
TiO_2 7
Topography 55
Transition levels 221
Transmission 196, 490
Transmission electron microscopy (TEM) 117
Traps 372
Triangle pits 76
Tunneling process 509
Two fold degenerate 441

V
Vacuum evaporation 21, 45
Valence band 218
Valence band offset 507
Varshni's fitting 203
Vegard's law 151, 211

W
Wavelength 196
Window 15
Window layer 393
Wurtzite 15

X
X-ray emission 78
X-ray photoelectron spectroscopy (XPS) 90
XRD 117
XRF 79

Y
Yellow band 420

Z
ZAF corrections 78
Zone boundary 471

LaVergne, TN USA
02 December 2010

207019LV00007B/2/P